Lecture Notes in Computer Science 4505

Commenced Publication in 1973
Founding and Former Series Editors:
Gerhard Goos, Juris Hartmanis, and Jan van Leeuwen

Guozhu Dong Xuemin Lin Wei Wang
Yun Yang Jeffrey Xu Yu (Eds.)

Advances in Data and Web Management

Joint 9th Asia-Pacific Web Conference, APWeb 2007
and 8th International Conference
on Web-Age Information Management, WAIM 2007
Huang Shan, China, June 16-18, 2007
Proceedings

Springer

Volume Editors

Guozhu Dong
Wright State University
Department of Computer Science and Engineering, USA
E-mail: gdong@cse.wright.edu

Xuemin Lin
University of New South Wales & NICTA, Australia
E-mail: lxue@cse.unsw.edu.au

Wei Wang
University of New South Wales
School of Computer Science and Engineering, Australia
E-mail: weiw@cse.unsw.edu.au

Yun Yang
Swinburne University of Technology, Melbourne, Australia
E-mail: yyang@ict.swin.edu.au

Jeffrey Xu Yu
The Chinese University of Hong Kong
Department of Systems Engineering and Engineering Management, China
E-mail: yu@se.cuhk.edu.hk

Library of Congress Control Number: 2007927715

CR Subject Classification (1998): H.2-5, C.2, I.2, K.4, J.1

LNCS Sublibrary: SL 3 – Information Systems and Application, incl. Internet/Web and HCI

ISSN 0302-9743
ISBN-10 3-540-72483-4 Springer Berlin Heidelberg New York
ISBN-13 978-3-540-72483-4 Springer Berlin Heidelberg New York

Springer is a part of Springer Science+Business Media

springer.com

Printed in Germany

Typesetting: Camera-ready by author, data conversion by Scientific Publishing Services, Chennai, India
Printed on acid-free paper SPIN: 12063062 06/3180 5 4 3 2 1 0

Preface

The rapid prevalence of Web applications requires new technologies for the design, implementation and management of Web-based information systems, and for the management and analysis of information on the Web. The joint APWeb/WAIM 2007 conference, combining the traditions of APWeb and WAIM conferences, was an international forum for researchers, practitioners, developers and users to share and exchange cutting-edge ideas, results, experience, techniques and tools in connection with all aspects of Web data management. The conference drew together original research and industrial papers on the theory, design and implementation of Web-based information systems and on the management and analysis of information on the Web. The conference was held in the beautiful mountain area of Huang Shan (Yellow Mountains) — the only dual World Heritage listed area in China for its astonishing natural beauty and rich and well-preserved culture. These proceedings collected the technical papers selected for presentation at the conference, held at Huang Shan, June 16–18, 2007.

In response to the call for papers, the Program Committee received 554 full-paper submissions from North America, South America, Europe, Asia, and Oceania. Each submitted paper underwent a rigorous review by three independent referees, with detailed review reports. Finally, 47 full research papers and 36 short research papers were accepted, from Austria, Australia, Canada, China, Cyprus, Greece, Hong Kong, Japan, Korea, Singapore, Taiwan, and USA, representing a competitive acceptance rate of 15%. The contributed papers address a broad spectrum on Web-based information systems, including data mining and knowledge discovery, information retrieval, P2P systems, sensor networks and grids, spatial and temporal databases, Web mining, XML and semi-structured data, query processing and optimization, data integration, e-learning, privacy and security, and streaming data. The proceedings also include abstracts of keynote speeches from four well-known researchers and four invited papers.

We were extremely excited with our Program Committee, comprising outstanding researchers in the APWeb/WAIM research areas. We would like to extend our sincere gratitude to the Program Committee members and external reviewers.

Last but not least, we would like to thank the sponsors, for their support of this conference, making it a big success. Special thanks go to Anhui University, The Chinese University of Hong Kong, The University of New South Wales, Anhui Association of Science and Technology, Anhui Computer

Federation, Hohai University, Huangshan University, National Science Foundation of China, and Oracle.

June 2007

Guozhu Dong
Xuemin Lin
Wei Wang
Yun Yang
Jeffrey Xu Yu

Organization

APWeb/WAIM 2007 was jointly organized by Anhui University, University of Science and Technology of China, The Chinese University of Hong Kong, and The University of New South Wales.

Organizing Committee

Conference Co-chairs

Guo Liang Chen, University of Science and Technology of China, China Ramamohanarao Kotagiri, University of Melbourne, Australia

Program Committee Co-chairs

Guozhu Dong, Wright State University, USA
Xuemin Lin, University of New South Wales, Australia
Yun Yang, Swinburne University of Technology, Australia
Jeffrey Xu Yu, Chinese University of Hong Kong, China

Workshop Chair

Kevin Chen-Chuan Chang, University of Illinois at Urbana-Champaign, USA

Panel Chair

Flip Korn, AT&T, USA

Tutorial Chair

Jian Pei, Simon Fraser University, Canada

Industrial Chair

Mukesh K. Mohania, IBM India Research Laboratory, India

Demo Co-chairs

Toshiyuki Amagasa, University of Tsukuba, Japan
Bin Luo, Anhui University, China

Publication Chair

Wei Wang, University of New South Wales, Australia

Publicity Co-chairs

Chengfei Liu, Swinburne University of Technology, Australia
Guoren Wang, Northeastern University, China
Haixun Wang, IBM T.J. Watson Research Center, USA

Local Arrangement Co-chairs

Jiaxing Cheng, Anhui University, China Nian Wang, Anhui University, China

ACM SIGMOD Liaison

Jianwen Su, UC Santa Barbara, USA

China Computer Federation Database Society Liaison

Dongqing Yang, Peking University, China

APWeb Steering Committee Liaison

Xiaofang Zhou, University of Queensland, Australia

WAIM Steering Committee Liaison

X. Sean Wang, University of Vermont, USA

WISE Society Liaison

Yanchun Zhang, Victoria University, Australia

Program Committee

Mikio Aoyama	Nanzan University, Japan
Vijay Atluri	Rutgers University, USA
James Bailey	University of Melbourne, Australia
Sourav S. Bhowmick	Nanyang Technological University, Singapore
Haiyun Bian	Wright State University, USA
Athman Bouguettaya	Virginia Tech, USA
Stephane Bressan	National University of Singapore, Singapore
Chee Yong Chan	National University of Singapore, Singapore
Kevin Chen-Chuan Chang	University of Illinois at Urbana-Champaign, USA

Akmal B. Chaudhri	IBM DeveloperWorks, USA
Sanjay Chawla	University of Sydney, Australia
Lei Chen	Hong Kong University of Science and Technology, China
Yixin Chen	Washington University at St. Louis, USA
Reynold Cheng	Hong Kong Polytechnic University, China
Byron Choi	Nanyang Technological University, Singapore
Gao Cong	Microsoft Research Asia, China
Bin Cui	Peking University, China
Alfredo Cuzzocrea	University of Calabria, Italy
Stijn Dekeyser	University of Southern Queensland, Australia
Amol Deshpande	University of Maryland, USA
Gill Dobbie	University of Auckland, New Zealand
Xiaoyong Du	Renmin University of China, China
Gabriel Fung	Chinese University of Hong Kong, China
Hong Gao	Harbin University of Technology, China
Guido Governatori	University of Queensland, Australia
Stephane Grumbach	The Sino-French IT Lab, China
Giovanna Guerrini	Universita di Genova, Italy
Michael Houle	National Institute for Informatics, Japan
Joshua Huang	Hong Kong University, China
Ela Hunt	ETH Zurich, Switzerland
Yoshiharu Ishikawa	Nagoya University, Japan
Panagiotis Kalnis	National University of Singapore, Singapore
Raghav Kaushik	Microsoft Research, USA
Hiroyuki Kitagawa	University of Tsukuba, Japan
Yasushi Kiyoki	Keio University, Japan
Flip Korn	AT&T, USA
Manolis Koubarakis	Technical University of Crete, Greece
Chiang Lee	National Cheng-Kung University, Taiwan
Yoon-Joon Lee	Korea Advanced Institute of Science and Technology (KAIST), Korea
Chen Li	University of California (Irvine), USA
Jianzhong Li	Harbin University of Technology, China
Jinyan Li	Institute for Information Research, Singapore
Qing Li	City University of Hong Kong, China
Ee Peng Lim	Nanyang Technological University, Singapore
Chengfei Liu	Swinburne University of Technology, Australia
Tieyan Liu	Microsoft Research Asia, China
Qiong Luo	Hong Kong University of Science and Technology, China
Hongyan Liu	Tsinghua University, China
Qing Liu	University of Queensland, Australia
Frank Maurer	University of Calgary, Canada
Emilia Mendes	Auckland University, New Zealand
Weiyi Meng	Binghamton University, USA

Xiaofeng Meng	Renmin University of China, China
Mukesh K. Mohania	IBM India Research Laboratory, India
Miyuki Nakano	University of Tokyo, Japan
Wolfgang Nejdl	University of Hannover, Germany
Jan Newmarch	Monash University, Australia
Zaiqing Nie	Microsoft Research Asia, China
John Noll	Santa Clara University, USA
Chaoyi Pang	CSIRO, Australia
Zhiyong Peng	Wuhan University, China
Evaggelia Pitoura	University of Ioannina, Greece
Sunil Prabhakar	Purdue University, USA
Weining Qian	Fudan University, China
Tore Risch	Uppsala University of Sweden, Sweden
Uwe Roehm	Sydney University, Australia
Prasan Roy	IBM India Research Laboratory, India
Keun Ho Ryu	Chungbuk National University, Korea
Monica Scannapieco	University of Rome "La Sapienza," Italy
Klaus-Dieter Schewe	Massey University, New Zealand
Albrecht Schmidt	Aalborg University, Denmark
Markus Schneider	University of Florida, USA
Heng Tao Shen	University of Queensland, Australia
Jialie Shen	University of Glasgow, UK
Timothy K. Shih	Tamkang University, Taiwan
Kian-Lee Tan	National University of Singapore, Singapore
David Taniar	Monash University, Australia
Changjie Tang	Sichuan University, China
Yufei Tao	Chinese University of Hong Kong, China
Minh Hong Tran	Swinburne University of Technology, Australia
Anthony Tung	National University of Singapore, Singapore
Andrew Turpin	RMIT University, Australia
Guoren Wang	Northeastern University, China
Haixun Wang	IBM T. J. Watson Research Center, USA
Jianyong Wang	Tsinghua University, China
Min Wang	IBM T. J. Watson Research Center, USA
Qing Wang	Institute of Software CAS, China
Shaojun Wang	Wright State University, USA
Wei Wang	Fudan University, China
Wei Wang	University of New South Wales, Australia
X. Sean Wang	University of Vermont, USA
Gerald Weber	Auckland University, New Zealand
Sui Wei	Anhui University, China
Jirong Wen	Microsoft Research Asia, China
Raymond Wong	University of New South Wales, Australia
Jun Yan	Wollongong University, Australia
Dongqing Yang	Peking University, China
Jian Yang	Macquarie University, Australia

Jun Yang	Duke University, USA
Cui Yu	Monmouth University, USA
Ge Yu	Northeastern University, China
Jenny Xiuzhen Zhang	RMIT University, Australia
Jianliang Xu	Hong Kong Baptist University, China
Qing Zhang	CSIRO, Australia
Rui Zhang	University of Melbourne, Australia
Yanchun Zhang	Victoria University, Australia
Aoying Zhou	Fudan University, China
Xiaofang Zhou	University of Queensland, Australia

APWeb Steering Committee

Xuemin Lin	University of New South Wales, Australia
Hongjun Lu	Hong Kong University of Science and Technology, China
Jeffrey Xu Yu	Chinese University of Hong Kong, China
Yanchun Zhang	Victoria University, Australia
Xiaofang Zhou (Chair)	University of Queensland, Australia

WAIM Steering Committee

Guozhu Dong	Wright State University, USA
Masaru Kitsuregawa	University of Tokyo, Japan
Jianzhong Li	Harbin Institute of Technology, China
Qing Li	City University of Hong Kong, China
Xiaofeng Meng	Renmin University, China
Changjie Tang	Sichuan Universty, China
Shan Wang	Renmin University, China
X. Sean Wang (Chair)	University of Vermont, USA
Ge Yu	Northeastern University, China
Aoying Zhou	Fudan University, China

External Reviewers

Carola Aiello
Halil Ali
Toshiyuki Amagasa
Saeid Asadi
Eric Bae
Manish Bhide
Niranjan Bidargaddi
Leigh Brookshaw
Penny De Byl
Badrish Chandramouli
Sanjay Chawla
Chih-Wei Chen
Chun-Wu Chen
Ding Chen
Jinchuan Chen
Jing Chen
Lijun Chen
Wei Chen

Ryan Choi
Kin Wah Chow
Soon Ae Chun
Yu-Chi Chung
Valentina Cordl
Ken Deng
Bolin Ding
Zhicheng Dou
Jing Du
Flavio Ferrarotti
Sergio Flesca
Francesco Folino
Gianluigi Folino
Mohamed Medhat Gaber
Bin Gao
Jun Gao
Xiubo Geng
Gabriel Ghinita
Kazuo Goda
Nizar Grira
Qi Guo
Himanshu Gupta
Rajeev Gupta
Yanan Hao
Sven Hartmann
Hao He
Jun He
Reza T. Hemayati
Bo Hu
Yuan-Ke Huang
Ingrid Jakobsen
Nobuhiro Kaji
Odej Kao
Hima Karanam
Roland Kaschek
Hiroko Kinutani
Markus Kirchberg
Henning Koehler
Yanyan Lan
Michael Lawley
Massimiliano De Leoni
Jianxin Li
Xue Li
Xuhui Li
Xiang Lian
Chien-Han Liao
Shi-Jei Liao
Lipyeow Lim
Xide Lin
Yi-Ying Lin
Yuan Lin
Sebastian Link
Hai Liu
Tao Liu
Xumin Liu
Yuanjie Liu
Yuting Liu
Elso Loekito
Jiaheng Lu
Ruopeng Lu
Yiming Lu
Yiyao Lu
Xijun Luo
Yi Luo
Zhong-Bin Luo
Hui Ma
Jiangang Ma
Yunxiao Ma
Zaki Malik
Da-Chung Mao
Carlo Mastroianni
Marco Mesiti
Diego Milano
Zoran Milosevic
Anirban Mondal
Yuan Ni
Shingo Otsuka
Khaleel Petrus
Giuseppe Pirrr
Kriti Puniyani
Tieyun Qian
Tao Qin
Lu Qing
Michael De Raadt
Simon Raboczi
Wenny Rahayu
Cartic Ramakrishnan
Weixiong Rao
Andrew Rau-Chaplin
Faizal Riaz-Ud-Din

Sourashis Roy
Ruggero Russo
Attila Sali
Falk Scholer
Basit Shafiq
Derong Shen
Heechang Shin
Takahiko Shintani
Houtan Shirani-Mehr
Yanfeng Shu
Adam Silberstein
Guojie Song
Shaoxu Song
Alexandre De Spindler
Bela Stantic
I-Fang Su
Sai Sun
Ying Sun
Takayuki Tamura
Nan Tang
Bernhard Thalheim
Wee Hyong Tok
Rodney Topor
Alexei Tretiakov
Paolo Trunfio
Yung-Chiao Tseng
Bin Wang
Botao Wang
John Wang
Rong Wang
Xiaoyu Wang
Xin Wang
Yitong Wang
Youtian Wang
Janice Warner
Yousuke Watanabe
Richard Watson
Wei Wei
Derry Tanti Wijaya
Di Wu
Mingfang Wu
Junyi Xie
Xiaocao Xiong
Guangdong Xu
Zhihua Xu
Xifeng Yan
Chi Yang
Hui Yang
Liu Yang
Xu Yang
Zhenglu Yang
Dai Yao
Tsung-Che Yeh
Qi Yu
Yainx Yu
Carmen Zannier
Wenjie Zhang
Ying Zhang
Jane Zhao
Peixiang Zhao
Xiaohui Zhao
George Zheng
Yong Zheng
Shijie Zhou
Jun Zhu
Liang Zhu
Qing Zhu
Cammy Yongzhen Zhuang

Sponsoring Institutions

Anhui University
The Chinese University of Hong Kong
The University of New South Wales
Anhui Association of Science and Technology
Anhui Computer Federation
Hohai University Huangshan University
National Science Foundation of China
Oracle

Table of Contents

Keynote

Invited Paper

Data Mining and Knowledge Discovery I

Information Retrieval I

P2P Systems

Sensor Networks

Spatial and Temporal Databases I

Web Mining

XML and Semi-structured Data I

Data Mining and Knowledge Discovery II

Sensor Networks and Grids

Query Processing and Optimization

Information Retrieval II

Data Stream

Spatial and Temporal Databases II

Data Integration and Collaborative Systems

Data Mining and E-Learning

XML and Semi-structured Data II

Data Mining, Privacy, and Security

Potpourri

Data Mining and Data Streams

Data Mining Using Fractals and Power Laws

Christos Faloutsos

Carnegie Mellon University
christos@cs.cmu.edu

Abstract. What patterns can we find in a bursty web traffic? On the web or on the internet graph itself? How about the distributions of galaxies in the sky, or the distribution of a company's customers in geographical space? How long should we expect a nearest-neighbor search to take, when there are 100 attributes per patient or customer record? The traditional assumptions (uniformity, independence, Poisson arrivals, Gaussian distributions), often fail miserably. Should we give up trying to find patterns in such settings?

Self-similarity, fractals and power laws are extremely successful in describing real datasets (coast-lines, rivers basins, stock-prices, brain-surfaces, communication-line noise, to name a few). We show some old and new successes, involving modeling of graph topologies (internet, web and social networks); modeling galaxy and video data; dimensionality reduction; and more.

About the Speaker

Christos Faloutsos is a Professor at Carnegie Mellon University. He has received the Presidential Young Investigator Award by the National Science Foundation (1989), the Research Contributions Award in ICDM 2006, nine "best paper" awards, and several teaching awards. He has served as a member of the executive committee of SIGKDD; he has published over 160 refereed articles, 11 book chapters and one monograph. He holds five patents and he has given over 20 tutorials and 10 invited distinguished lectures. His research interests include data mining for streams and networks, fractals, indexing for multimedia and bio-informatics data, and database performance.

G. Dong et al. (Eds.): APWeb/WAIM 2007, LNCS 4505, p. 1, 2007.

Exploring the Power of Links in Data Mining

Jiawei Han*

Department of Computer Science
University of Illinois at Urbana-Champaign
hanj@cs.uiuc.edu

Abstract. Algorithms like PageRank and HITS have been developed in late 1990s to explore links among Web pages to discover authoritative pages and hubs. Links have also been popularly used in citation analysis and social network analysis. We show that the power of links can be explored thoroughly in data mining, such as classification, clustering, information integration, and object distinction. Some recent results of our research that explore the crucial information hidden inside links will be introduced, including (1) multi-relational classification, (2) user-guided clustering, (3) link-based clustering, and (4) object distinction analysis. The power of links in other analysis tasks will also be discussed in the talk.

About the Speaker

Jiawei Han is a Professor in the Department of Computer Science at the University of Illinois at Urbana-Champaign. He has been working on research into data mining, data warehousing, database systems, mining spatiotemporal data, multimedia data, stream and RFID data, Web data, social network data, and biological data, and software bug mining, with over 300 conference and journal publications. He has chaired or served on over 100 program committees of international conferences and workshops, including PC co-chair of 2005 (IEEE) International Conference on Data Mining (ICDM), Americas Coordinator of 2006 International Conference on Very Large Data Bases (VLDB). He is also serving as the founding Editor-In-Chief of ACM Transactions on Knowledge Discovery from Data. He is an ACM Fellow and has received 2004 ACM SIGKDD Innovations Award and 2005 IEEE Computer Society Technical Achievement Award. His book "Data Mining: Concepts and Techniques" (2nd ed., Morgan Kaufmann, 2006) has been popularly used as a textbook worldwide.

* The work was supported in part by the U.S. National Science Foundation NSF ITR/CCR-0325603, IIS-05-13678/06-42771, and NSF BDI-05-15813. Any opinions, findings, and conclusions or recommendations expressed here are those of the authors and do not necessarily reflect the views of the funding agencies.

G. Dong et al. (Eds.): APWeb/WAIM 2007, LNCS 4505, p. 2, 2007.

Community Systems: The World Online

Raghu Ramakrishnan

Yahoo! Research, Santa Clara, CA, USA
ramakris@yahoo-inc.com

Abstract. The Web is about you and me. Until now, for the most part, it has denoted a corpus of information that we put online sometime in the past, and the most celebrated Web application is keyword search over this corpus. Sites such as del.icio.us, flickr, MySpace, Slashdot, Wikipedia, Yahoo! Answers, and YouTube, which are driven by user-generated content, are forcing us to rethink the Web — it is no longer just a static repository of content; it is a medium that connects us to each other. What are the ramifications of this fundamental shift? What are the new challenges in supporting and amplifying this shift?

About the Speaker

Raghu Ramakrishnan is VP and Research Fellow at Yahoo! Research, where he heads the Community Systems group. He is on leave from the University of Wisconsin-Madison, where he is Professor of Computer Sciences, and was founder and CTO of QUIQ, a company that pioneered question-answering communities such as Yahoo! Answers, and provided collaborative customer support for several companies, including Compaq and Sun. His research is in the area of database systems, with a focus on data retrieval, analysis, and mining. He has developed scalable algorithms for clustering, decision-tree construction, and itemset counting, and was among the first to investigate mining of continuously evolving, stream data. His work on query optimization and deductive databases has found its way into several commercial database systems, and his work on extending SQL to deal with queries over sequences has influenced the design of window functions in SQL:1999. His paper on the Birch clustering algorithm received the SIGMOD 10-Year Test-of-Time award, and he has written the widely-used text "Database Management Systems" (WCB/McGraw-Hill, with J. Gehrke), now in its third edition.

He is Chair of ACM SIGMOD, on the Board of Directors of ACM SIGKDD and the Board of Trustees of the VLDB Endowment, and has served as editor-in-chief of the Journal of Data Mining and Knowledge Discovery, associate editor of ACM Transactions on Database Systems, and the Database area editor of the Journal of Logic Programming. Dr. Ramakrishnan is a Fellow of the Association for Computing Machinery (ACM), and has received several awards, including a Distinguished Alumnus Award from IIT Madras, a Packard Foundation Fellowship, an NSF Presidential Young Investigator Award, and an ACM SIGMOD Contributions Award.

G. Dong et al. (Eds.): APWeb/WAIM 2007, LNCS 4505, p. 3, 2007.

A New DBMS Architecture for DB-IR Integration

Kyu-Young Whang

Computer Science Department and
Advanced Information Technology Research Center(AITrc)
KAIST, Korea
kywhang@cs.kaist.ac.kr

Abstract. Nowadays, as there is an increasing need to integrate the DBMS (for structured data) with Information Retrieval (IR) features (for unstructured data), DB-IR integration becomes one of major challenges in the database area[1,2]. Extensible architectures provided by commercial ORDBMS vendors can be used for DB-IR integration. Here, extensions are implemented using a high-level (typically, SQL-level) interface. We call this architecture *loose-coupling*. The advantage of loose-coupling is that it is easy to implement. But, it is not preferable for implementing new data types and operations in large databases when high performance is required. In this talk, we present a new DBMS architecture applicable to DB-IR integration, which we call *tight-coupling*. In tight-coupling, new data types and operations are integrated into the core of the DBMS engine in the *extensible type layer*. Thus, they are incorporated as the "first-class citizens"[1] within the DBMS architecture and are supported in a consistent manner with high performance. This tight-coupling architecture is being used to incorporate IR features and spatial database features into the Odysseus ORDBMS that has been under development at KAIST/AITrc for over 16 years[3]. In this talk, we introduce Odysseus and explain its tightly-coupled IR features (U.S. patented in 2002[2]). Then, we demonstrate excellence of tight-coupling by showing benchmark results. We have built a web search engine that is capable of managing 20∼100 million web pages in a non-parallel configuration using Odysseus. This engine has been successfully tested in many commercial environments. In a parallel configuration, it is capable of managing billons of web pages. This work won the Best Demonstration Award from the IEEE ICDE conference held in Tokyo, Japan in April 2005[3].

About the Speaker

Kyu-Young Whang is Professor of Computer Science and Director of Advanced Information Technology Research Center (AITrc) at KAIST. Previously, he was with IBM T.J.Watson Research Center from 1983 to 1990. Since joining KAIST in 1990, he has been leading the Odysseus DBMS project featuring tight-coupling of DBMS with information retrieval (IR) and spatial functions. Dr. Whang is one of the pioneers of probabilistic counting, which nowadays is being widely used in

G. Dong et al. (Eds.): APWeb/WAIM 2007, LNCS 4505, pp. 4–5, 2007.

approximate query answering, sampling, and data streaming. One of the algorithms he co-developed at IBM Almaden (then San Jose) Research Lab in 1981 has been made part of DB2. Dr. Whang is the author of the first main-memory relational query optimization model developed in 1985 and reported in 1990 in ACM TODS in the context of Office-by-Example (OBE). This model influenced subsequent optimization models of commercial main-memory DBMSs. His research has covered a wide range of database issues including physical database design, query optimization, DBMS engine technologies, and more recently, IR, spatial databases, data mining, and XML. Dr. Whang is a Co-Editor-in-Chief of the VLDB Journal, having served the journal for 17 years from its inception as its founding editorial board member. He is a Trustee Emeritus of the VLDB Endowment and served the international academic community as the General Chair of VLDB2006, DASFAA2004, and PAKDD2003, as a PC Co-Chair of VLDB2000, CoopIS1998, and ICDE2006, and as an editorial board member of journals such as IEEE TKDE and IEEE Data Engineering Bulletin. He served as an IEEE Distinguished Visitor from 1989 to 1990. He earned his Ph.D. from Stanford University in 1984. Dr. Whang is an IEEE Fellow, a member of the ACM and IFIP WG 2.6.

References

1. Abiteboul, S. et al., "The Lowell Database Research Self-Assessment," *Communications of the ACM*, Vol.48, No.5, pp. 111-118, May 2005.
2. Whang, K., Park, B., Han, W., and Lee, Y., "An Inverted Index Storage Structure Using Subindexes and Large Objects for Tight Coupling of Information Retrieval with Database Management Systems," U.S. Patent No. 6,349,308, Feb. 19, 2002 (Appl. No. 09/250,487, Feb. 15, 1999).
3. Whang, K., Lee, M., Lee, J., Kim, M., and Han, W., "Odysseus: a High-Performance ORDBMS Tightly-Coupled with IR Features," In *Proc. IEEE 21st Int'l Conf. on Data Engineering (ICDE)*, Tokyo, Japan, Apr. 5-8, 2005. This paper received the Best Demonstration Award.

Study on Efficiency and Effectiveness of KSORD

Shan Wang[1,2], Jun Zhang[1,2], Zhaohui Peng[1,2], Jiang Zhan[1,2], and Qiuyue Wang[1,2]

[1] School of Information, Renmin University of China, Beijing 100872, P.R. China
{swang,zhangjun11,pengch,zhanjiang,qiuyuew}@ruc.edu.cn
[2] Key Laboratory of Data Engineering and Knowledge Engineering (Renmin University of China), MOE, Beijing 100872, P.R. China

Abstract. KSORD(Keyword Search Over Relational Databases) is an easy and effective way for casual users or Web users to access relational databases. In recent years, much research on KSORD has been done, and many prototypes of KSORD have been developed. However, there are still critical problems on the efficiency and effectiveness of KSORD systems. In this paper, we describe the overview of KSORD research and development, analyze the efficiency and effectiveness problems in existing KSORD systems, and introduce our study on KSORD in terms of efficiency and effectiveness. In the end, we point out the emerging topics worthy of further research in this area.

1 Introduction

KSORD(Keyword Search Over Relational Databases) is an easy and effective way for casual users or Web users to access relational databases [1]. In recent years, much research on KSORD has been done, and many prototypes of KSORD have been developed. According to the query processing strategy KSORD systems adopted, they can be categorized into two types: offline systems and online systems. Offline systems retrieve results for a keyword query from an intermediate representation generated by "crawling" the database in advance, such as EKSO [2], or from some indexes created beforehand, such as ObjectRank [3] and ITREKS [4]. Online systems convert a keyword query into many SQL queries to retrieve results from the database itself. Furthermore, online KSORD systems can be classified into two categories [1] according to the data model they adopted, Schema-graph-based Online KSORD(S0-KSORD) systems like SEEKER [5], DBXplorer [6], DISCOVER [7] and IR-Style [8], and Data-graph-based Online KSORD(DO-KSORD) systems like BANKS [9], BANKS II [10] and DETECTOR [11,12].

Offline KSORD Systems usually execute queries efficiently, but they can not query the up-to-date data in time, and also require a long preprocessing time and large storage space to generate and store the intermediate representation. On the contrary, online KSORD systems can retrieve the latest data from the database, but the execution is usually inefficient because the converted SQL queries often contain many join operators as for SO-KSORD systems and the

G. Dong et al. (Eds.): APWeb/WAIM 2007, LNCS 4505, pp. 6–17, 2007.

data graph search algorithms can not scale to the number of query keywords and the size of data graph as for DO-KSORD systems.

Many prototype systems of KSORD have been developed. However, a KSORD system would not be put into practice if its efficiency or effectiveness is poor [1]. Fortunately, the effectiveness and efficiency of KSORD have attracted more and more attention recently. IR-Style [8] improved DISCOVER [7], BANKS II [10] improved BANKS [9], and Fang Liu et al. studied the effectiveness of KSORD. In this paper, we focus on the efficiency and effectiveness of online KSORD systems. We analyze the cause of the efficiency and effectiveness problems in online KSORD systems, and introduce many optimization methods we have proposed in this area. Finally, we discuss some further research topics.

The rest of this paper is organized as follows. Section 2 analyzes the performance problems in KSORD in terms of efficiency and effectiveness. Section 3 introduces the methods on improving the efficiency of KSORD in detail, while the methods for effectiveness improvement are described in Section 4. Section 5 points out further research topics, and we conclude in Section 6.

2 Efficiency and Effectiveness Problems in KSORD

In this section, we present the architectures of online KSORD systems, and analyze the bottlenecks of their efficiency and effectiveness.

2.1 Efficiency of SO-KSORD

The architecture of SO-KSORD systems is shown in Fig. 1. Once a SO-KSORD system starts up, the database schema graph can be rapidly created. When a user query Q comes, tuple set creator(TS Creator) creates tuple sets for each relation that has text attributes along with full-text indexes, and only those non-empty tuple sets are left. Then, Candidate Network Generator(CN Generator) outputs a complete and non-redundant set of Candidate Networks(CNs) [7, 8] whose sizes are not greater than Maximum allowed CN size(MaxCNsize, the maximum allowed number of nodes in a CN) through a breadth-first traversal of tuple set graph G_{ts}. CNs are join trees of tuple sets which will be used to produce potential answers to Q. G_{ts} is generated by extending database schema graph G_s. Finally, CN executor employs a certain strategy to run CNs to get results for Q.

The user query response time(T_{res}) of SO-KSORD systems is mainly composed of three parts. The first is T_{ts} which denotes the time to create tuple sets(TSs). The second is T_{cn} which denotes the time to generate CNs. The third is T_{sql} which denotes the time to execute converted SQL queries from CNs. T_{ts} is usually small and determined by the Information Retrieval(IR) engine of RDBMS. T_{sql} is the most significant part affecting the execution efficiency of SO-KSORD systems. Still, T_{cn} also has an important effect on the efficiency of SO-KSORD systems.

In order to improve the efficiency of DISCOVER [7], IR-Style [8] only creates a single tuple set for each relation with text attributes so that G_{ts} is much smaller

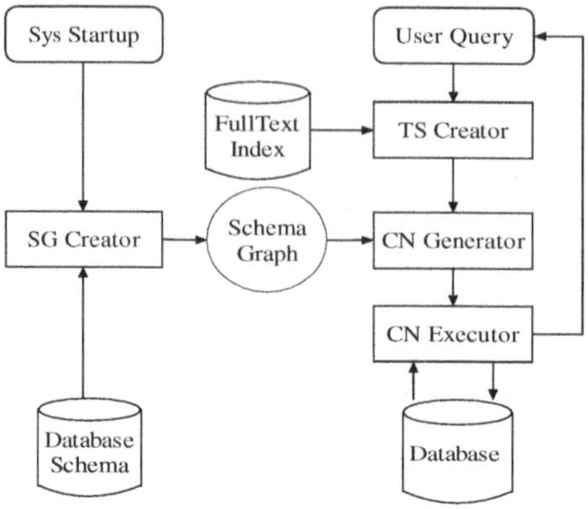

Fig. 1. Architecture of SO-KSORD

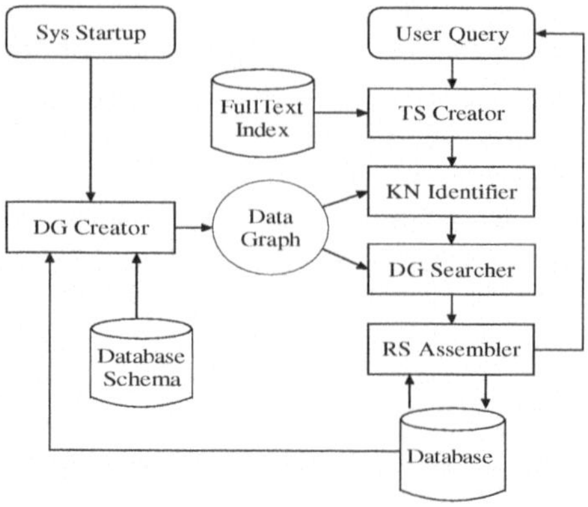

Fig. 2. Architecture of DO-KSORD

than that in DISCOVER. Thus, CN generation is much faster in IR-Style. Unlike that DISCOVER executes all CNs and returns all results for a query Q, IR-Style runs a top-k algorithm, such as SParse algorithm(SP) or Global pipelined Algorithm(GA) [8], to execute CNs to get top-k results for Q. Then T_{sql} can be greatly reduced. Therefore, IR-Style can perform much faster than DISCOVER. However, IR-Style still performs very poorly in some cases. In fact, the efficiency bottlenecks of SO-KSORD systems are as follows.

Inefficiency of CN generation. As the number of query keywords or MaxCNsize increases, or the database schema becomes complicated, it will take much more time to generate CNs for a query Q. Currently, existing SO-KSORD systems generate CNs temporarily through a breadth-first traversal of G_{ts} for any user query. There can be two ways to reduce the time for the generation of CNs. One is to develop a more efficient CN generation algorithm, the other is to develop preprocessing techniques to generate CNs in advance.

Inefficiency of CN Execution. On one hand, tens or hundreds of CNs may be generated for a query Q, while existing top-k algorithms are inefficient to execute so many CNs. On the other hand, the SQL queries converted from CNs may contain many JOIN operators, but, as we know, JOIN operation is an expensive operation in RDBMSs.

2.2 Efficiency of DO-KSORD

The architecture of DO-KSORD systems is shown in Fig. 2. This figure looks like the architecture of SO-KSORD systems. However, they are essentially different. When a DO-KSORD system starts up, it creates data graph instead of schema graph. Similarly, when a user query Q comes, tuple set creator(TS Creator) creates tuple sets for each relation that has text attributes along with full-text indexes, and only those non-empty tuple sets are left. Then, Keyword Node Identifier(KN identifier) identifies those data graph nodes which contain query keywords according to the tuple sets. And the key step is that Data Graph Searcher(DG Searcher) searches the data graph with a certain strategy to get Join Trees of Tuples(JTTs) as top-k results for Q. Finally, Result Assembler(RS Assembler) can retrieve the real information of each tuple in a JTT from the relational database to assemble complete results for end users.

Obviously, DG Searcher determines the efficiency of DO-KSORD systems. In other words, the efficiency bottleneck of DO-KSORD systems lies in the efficiency of data graph search algorithm employed by DG Searcher. BANKS [9] employs a heuristic backward expanding search algorithm to produce JTTs as results for a query Q. But it may perform poorly if some keywords match many nodes, or some nodes have very large degrees [10]. In order to improve BANKS, bidirectional search algorithm was proposed in BANKS II [10]. Bidirectional search improves backward expanding search by allowing forward search from potential roots towards leaves, and a novel search frontier prioritization technique based on spreading activation was devised to exploit this flexibility [10]. However, this data graph search algorithm still is not efficient enough. As the number of query keywords increases, or the data graph becomes larger, BANKS II [10] performs more poorly in terms of time complexity and space complexity.

In addition, all of existing DO-KSORD prototypes assume that data graph fits in memory. However, data graph for a large database can be too huge to be accommodated in limited main memory. To a certain extent, data graph is similar to Web graph [13], however, data graph, containing database schema information, is quite different from Web graph. So, it is possible to develop special techniques to compress data graph by exploiting the characteristics of data graph so that larger data graph can be loaded into memory.

Therefore, there are two ways that can be explored for improving the performance of DO-KSORD systems. One is to develop more efficient data graph search algorithms which can scale to the number of query keywords and the size of data graph. The other is to develop data graph compression techniques.

2.3 Effectiveness of KSORD

Fang Liu et al. are the first ones to study the effectiveness of KSORD in detail [14]. They found out that KSORD is different from keyword search over text databases in the following ways: (1) Answers for a query are JTTs. (2) A single score for each JTT is needed to estimate its relevance to a given query. (3) Relational databases have much richer structures than text databases. As a result, existing IR strategies are inadequate in ranking relational outputs. So, they proposed a novel IR ranking strategy for effective KSORD. Their main ideas are as follows. Firstly, four new normalization factors(tuple tree size, document length, document frequency and inter-document weight) are identified and used. Secondly, schema terms are identified and are processed differently from value terms. Finally, phrase-based and concept-based models are used to further improve search effectiveness.

However, Fang Liu et al. also pointed out that the link structures (primary key to foreign key relationships as well as some hidden join conditions), and some non-text columns can be further utilized to improve the effectiveness of KSORD [14].

The result presentation also affects the effectiveness of KSORD to a great extent [1, 11]. Firstly, the results need to be semantically meaningful to users. However, a result which is a tuple or a tuple connection tree is not so easy to understand for end users. Secondly, it is important to avoid overwhelming users with a huge number of trivial results. However, lots of similar results are often produced, which makes users tired or confused. In KSORD research, many ways are used to present query results. BANKS [9] shows the query results in a nested table and improves the answer format by addressing readability. DbSurfer [15] uses tree-like structures to display all trails, while DataSpot [16] uses a distinguished answer node to represent a result. However, their work does not solve the problem of lots of similar results.

Currently, KSORD systems are based on full-text indexes created by IR engine of RDBMS. In general, keyword search has inherent limitations. Keyword search is only based on keyword matching and does not exploit the semantic relationships between keywords such as hyponymy, meronymy, or antonymy, so the effectiveness is often dissatisfactory in terms of recall rate and precision rate. With the increasing research interest on ontology and semantic web, ontology-based semantic search over relational databases has become a 'hot' research topic in database community [17, 18]. So, exploiting ontology to improve the effectiveness of KSORD receives increasing attention.

3 Efficiency Improvements on KSORD

In recent years, we have developed many techniques to optimize the efficiency of online KSORD systems. Aiming at improving SO-KSORD systems, we proposed a new preprocessing approach PreCN [21] to improve the generation efficiency of CN, and CLASCN [24] and QuickCN [26] methods to improve the execution efficiency of CN. As for DO-KSORD systems, a novel and efficient data graph search algorithm called DPBF [11, 12] was developed to improve the efficiency of DO-KSORD. CodCor [27] method was proposed to compress data graph by exploiting connection relations in relation databases. CodCor not only makes a large data graph fit in memory, but also improves the efficiency of existing data graph search algorithms, such as that in BANKS [9] and BANKS II [10]. Of course, CodCor is also helpful to improve the efficiency of QuickCN [26] and DPBF [12]. Based on the above methods, we implemented a new efficient and effective online KSORD system called $QuicK^2SORD$. Fig. 3 shows the architecture of $QuicK^2SORD$. We will discuss those methods in the following subsections.

3.1 Improving the Generation Efficiency of CN

Since SO-KOSRD systems generate CNs for a query temporarily, we exploit preprocessing techniques to generate CNs in advance [22, 21]. Offline systems(such

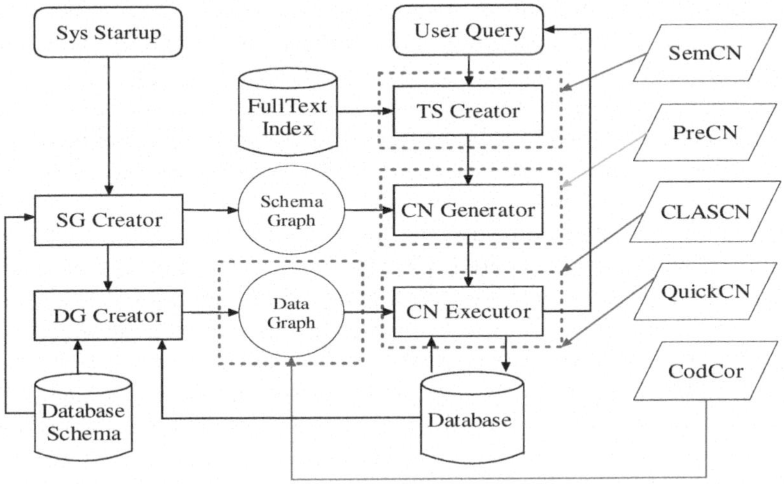

Fig. 3. Architecture of $QuicK^2SORD$

as EKSO [2]) usually preprocess the data and generate an intermediate representation for the database. However, what is preprocessed in SO-KSORD systems is the database schema information, not the data. As we know, the schema information is usually stable, but the data is changing dynamically.

As for boolean-AND semantic keyword queries in DISCOVER [7], a new preprocessing technology [22] was proposed to generate CN patterns in advance through a breadth-first traversal of G_s and to store them in the database. When a user issues a keyword query, proper CN patterns are retrieved from the database and evaluated according to the specific tuple sets created temporarily for the query, thus CNs are generated for the query.

As for boolean-OR semantic keyword queries in IR-Style [8] and SEEKER [5], we find out that G_{ts} patterns can be viewed as user keyword query patterns for a given G_s. All CNs under the limitation of MaxCNsize and Maximum allowed Keyword Number(MaxKeywNum) can be generated in advance through a breadth-first traversal of the maximum G_{ts}, and then stored in the database. When a user query arrives, its proper CNs are directly retrieved from the database. This method is called PreCN [21]. PeCN is simpler and more efficient than that in [22].

PreCN requires less physical storage space to store the pre-generated CNs, and can also search the up-to-date data in the database.

3.2 Improving the Execution Efficiency of CN

We proposed two methods to improve the execution efficiency of CN, one is CLASCN [24], the other is QuickCN [26].

CLASCN: Select the most Promising CN to be executed. Although tens or hundreds of CNs can be generated for a keyword query in SO-KSORD systems, the top-k results only distribute in a few CNs. So, a novel approach CLASCN (Classification, Learning And Selection of Candidate Network) was proposed to improve the efficiency of SO-KSORD systems [24]. The main ideas of CLASCN are as follows. Each CN can be viewed as a database, and CN Language Models(CNLMs) can be constructed by performing trained keyword queries in advance and user queries dynamically. When a user query arrives, the similarities between the query and its CNs are computed by employing Vector Space Model (VSM) [25], and only the most promising CNs to produce top-k results are picked out and executed. CLASCN can be combined with any exact top-k algorithm to support efficient top-k keyword queries, and at the same time acceptable recall and precision rates of top-k results can be achieved.

Currently, because CNLMs are constructed using query keywords, CLASCN is only applicable for previously executed queries and New All-keyword-Used queries(never executed before but all the keywords occurred in previous queries), which are frequently submitted. Our extensive experiments showed that the CLASCN approach was efficient and effective.

QuickCN: Exploiting Data Graph to Execute CN. A novel method QuickCN was proposed to quickly execute CNs on data graph [26]. The basic ideas are as follows. CNs are considered as join expressions , and also can be viewed as query patterns and result patterns. At the same time, the database can be modeled as a data graph [9,10] which is actually a huge tuple-joined network generated in advance. So, the data graph can be searched with CN patterns to quickly produce JTTs as the final results. This is different from BANKS [9,10]. BANKS searches the data graph without knowing of result patterns. As a result, lots of intermediate results will be produced during the search process. In QuickCN, the data graph search process is schema-driven due to the search result patterns known in advance. The adjacent nodes of each node in the data graph can be classified by their relation names and Primary-Key-to-Foreign-Key relationship types, and the foreign-key nodes have n-to-1 map to their primary-key adjacent nodes. These properties can be exploited to reduce the search space in data graph. Our experiments showed that QuickCN was efficient.

3.3 Exploiting Connection Relation to Compress Data Graph

For DO-KSORD systems, we proposed an approach CodCor(abbr. for Compressing Data graph by Connection Relations) to compress data graph by exploiting connection relations [27]. A connection relation R is a relation in the database that satisfies the following conditions: (i) there are exactly two foreign keys in relation R; (ii) R's primary key consists of its foreign keys. (iii) R is not referenced by any relation. Many databases, such as DBLP or Northwind, have connection relations. The main ideas of CodCor are as follows. The nodes coming from connection relations are removed and the edges linked with each removed node are connected into one edge. In theory, CodCor can compress data graph by more than half of its storage if there are enough tuples in connection relations.

CodCor can not only make a large data graph fit in memory, but also improve the efficiency of existing in-memory data graph search algorithms, such as BANKS [9] and DPBF [12]. Of course, CodCor can also improve the efficiency of QuickCN. For the future work, we consider the relaxation of connection relation's definition so that more kinds of databases can benefit from CodCor, and also try to combine CodCor with some compression techniques used for Web Graph [13] to further compress the data graph.

3.4 Developing Efficient Data Graph Search Algorithm

An efficient data graph search algorithm named DPBF was proposed [11,12]. Like in *BANKS* [9,10], we model a relational database as a weighted graph, $G(V, E)$. Here V is a set of nodes representing tuples and E is a set of edges representing foreign-key references among tuples. An edge, $(u, v) \in E$, represents a foreign key reference between two tuples, u and v, if u has a foreign key matching the primary key attributes of v, or v has a foreign key referring to the primary key of u. The weights on nodes and edges are predetermined [9,10].

Given a l keyword query, $p_1, p_2, \cdots, p_l$, against a relational database or equivalently the corresponding graph $G(V, E)$. Let $V_i \subseteq V$ be a set of nodes that contain the keyword p_i. An answer to such a query is a weighted and connected tree containing at least one node from each V_i. The problem we are targeting is how to find top-k minimum cost tuple connection trees.

A dynamic programming approach was proposed to find the optimal top-1 with the time complexity of $O(3^L \cdot N + 2^L((L + \log N) \cdot N + M))$, where N and M are the numbers of nodes and edges in the graph G respectively. Because the number of keywords, L, is small in keyword queries, this solution can handle graphs with a large number of nodes efficiently. It is important to note that our solution can be easily extended to support top-k. That is, we compute top-k minimum cost tuple connection trees one-by-one incrementally, and do not need to compute or sort all results in order to find the top-k results.

4 Effectiveness Improvements on KSORD

We have developed techniques to improve the effectiveness of online KSORD systems. For example, SemCN [19] was proposed to improve the effectiveness of SO-KSORD systems, and a novel clustering method named TreeCluster [23,11] was proposed to improve the effectiveness of DO-KSORD systems. We discuss them in detail as follows.

4.1 Improving the Effectiveness of KSORD Based on Ontology

We proposed a novel approach SemCN(semantic CN) in SO-KSORD systems to implement semantic search over relational databases, and developed a prototype Si-SEEKER [19] which extends SEEKER [5]. In Si-SEEKER, the data are annotated with the concepts in the ontology. Semantic indexes are created before query processing. A user keyword query is transformed into a concept query in the same concept space of the ontology, and the hierarchical structure of domain-specific ontology and generalized vector space model(GVSM) [20] are employed to compute semantic similarity between the concept query and annotated data. As a result, semantic results will be returned in higher recall and precision rates than that in KSORD systems. We also combine semantic search with keyword search to tolerate the incompleteness of ontology and annotations of data. Our experiments show that the framework is effective.

4.2 Clustering and Presenting Search Results

Organizing search results into clusters can facilitate users' quick browsing through search results. Hunter [22] proposed a result classification method. In preprocessing, the system produces various patterns, and in querying, users select a particular pattern and the system searches the results matching the selected pattern.

We proposed a novel approach for clustering results named TreeCluster [23,11]. Our approach can be implemented outside the system and applicable to various KSORD systems. Clustering results has been widely applied in related research areas, such as in presenting Web search results, but those clustering methods are not applicable to KSORD results. Take into account the characteristics of KSORD search results, TreeCluster combines the structure and content information together and includes two steps of pattern clustering and keyword clustering. In the first step, we use labels to represent schema information of each result tree and cluster the trees into groups. The trees in each group are isomorphic. In the second step, we rank user keywords according to their frequencies in the database, and further partition the large groups based on the content of the keyword nodes. Furthermore, we give each cluster a meaningful description, and present the description and each result tree graphically to help users understand the results more easily. Experimental results verify our methods' effectiveness and efficiency.This is the first proposal for clustering search results of KSORD.

5 Future Research Topics and Challenges

Current study on KSORD generally focuses on a single database, however, many practice settings require keyword search over multi-databases [28]. Exploiting ontology to do semantic search over relational databases has attracted more and more attention in database community. Above scenarios give birth to more challenges to improve the efficiency and effectiveness of KSORD. In addition, up to now, there is not any standard testbed for KSORD yet. We discuss some topics worthy of further research as follows in this section.

5.1 Keyword Search over Multi-databases

M.Sayyadian et al. addressed the problem of keyword search over heterogeneous relational databases, and proposed Kite algorithm which combines schema matching and structure discovery techniques to find approximate foreign-key joins across heterogeneous databases [28]. There will be more challenges to improve the performance of KSORD in multi-database setting than in single-database setting [28], such as exponential search space with the growing number of databases and their associated foreign-key joins, expensive foreign-key joins due to communication and data-transfer costs, and the difficulty to estimate accurate statistics(e.g., the estimated result size of a SQL query). Kite implemented keyword search over multi-databases, but the query efficiency can be further improved by takeing into account more factors, such as communication and data-transfer costs.

5.2 Ontology-Based KSORD

Ontology has been widely studied in semantic Web and IR community. Due to great differences between relational data and the documents residing in semantic Web and text databases [14], however, ontology-based semantic search over databases has many new challenges, such as efficient semantic indexes, semantic similarity computation, and so on. Ontology can be used to improve the effectiveness of KSORD, whereas it may impair the efficiency of KSORD. Efficient semantic indexes ought to be developed to improve the efficiency of ontology-based KSORD systems. In addition, ontology-based data graph search algorithm is also an interesting topic to study on.

5.3 Result Presentation

Although some research has been done to improve result presentation of KSORD system, much work is still to be done. We only enumerate two cases here.

For one example, to improve the clustering effectiveness, we could try to employ more schema information in the search process of KSORD to facilitate clustering or classifying the results.

For another example, relevance feedback could be used to improve result presentation [25]. In a relevance feedback cycle, the user is presented with a list of the retrieved results and, after examining them, marks those that are relevant. The main idea is to detect important characteristics of the results that have been identified as relevant by the user, and then enhance the importance of these characteristics in a new query formulation. The expected effect is that the new query will be moved towards the relevant results and away from the non-relevant ones. Early experiments in traditional IR have shown good improvements in precision for small test collections when relevance feedback is used. However, those feedback methods used in IR may not be applicable to KSORD because of the different characteristics of search results. Furthermore, in KSORD, there is also database schema information which could be used to assist the feedback.

5.4 Benchmark of KSORD

Benchmark-based experimental results for all current KSORD systems are needed. There are many reference collections used to evaluate information retrieval systems, such as the TREC collection [29]. It is necessary to build a reference database collection used for evaluating the performance of KSORD systems in terms of efficiency and effectiveness. It is a significant fundamental work and requires many efforts.

6 Conclusion

Many approaches have been proposed to implement KSORD, however, the efficiency and effectiveness of KSORD remains critical issues. We analyze the efficiency and effectiveness problems in KSORD systems, and present several approaches that we proposed to improve the efficiency and effectiveness of online KSORD systems. In the end, some topics worthy of further research are discussed in this area.

Acknowledgements

This work is supported by the National Natural Science Foundation of China (No.60473069 and 60496325), and China Grid(No.CNGI-04-15-7A).

References

1. S. Wang, K. Zhang. Searching Databases with Keywords. Journal of Computer Science and Technology, Vol.20(1). 2005:55-62.
2. S. Qi, W. Jennifer. Indexing Relational Database Content Offline for Efficient Keyword-Based Search. IDEAS, 2005:297-306.
3. A. Balmin, V. Hristidis, Y. Papakonstantinou. ObjectRank: Authority-Based Keyword Search in Databases. VLDB, 2004:564-575
4. J. Zhan, S. Wang. ITREKS: Keyword Search over Relational Database by Indexing Tuple Relationship. 12th International Conference on Database Systems For Advance Applications(DASFAA), 2007.
5. J. Wen, S. Wang. SEEKER: Keyword-based Information Retrieval Over Relational Data-bases. Journal of Software, Vol.16(4). 2005:540-552(in Chinese).
6. S. Agrawal, S. Chaudhuri, and G. Das. DBXplorer:A System for keyword Search over Relational Databases. ICDE, 2002:5-16.
7. V. Hristidis, Y. Papakonstantinou: DISCOVER: Keyword Search in Relational Databases. VLDB, 2002:670-681.
8. V. Hristidis, L. Gravano, Y. Papakonstantinou. Efficient IR-Style Keyword Search over Relational Databases. VLDB, 2003:850-861.
9. G. Bhalotia, A. Hulgeri, C. Nakhe et al.. Keyword Searching and Browsing in Databases using BANKS. ICDE, 2002:431-440.
10. V. Kacholia, S. Pandit, S. Chakrabarti et al. Bidirectional Expansion For Keyword Search on Graph Databases. VLDB, 2005:505-516.
11. S. Wang, Z. Peng, J. Zhang et al. NUITS: A Novel User Interface for Efficient Keyword Search over Databases. VLDB, 2006:1143-1146.
12. B. Ding, J. Yu, S. Wang et al. Finding Top-k Min-Cost Connected Trees in Databases. ICDE, 2007.
13. K.H. Randall, R. Stata, R. Wickremesinghe, J.L. Wiener. The link database: Fast access to graphs of the web. The Data Compression Conference, 2002:122-131.

14. F. Liu, C. Yu, W. Meng, A. Chowdhury. Effective Keyword Search in Relational Databases. SIGMOD, 2006:563-574.
15. R. Wheeldon, M. Levene, and K. Keenoy. DbSurfer: A Search and Navigation Took for Relational Databases. 21st Annual British National Conference on Databases, 2004:144-149.
16. S. Dar et al. DTL's DataSpot:Database Exploration Using Plain Language. VLDB, 1998.
17. S. Das, E.I. Chong, G. Eadon, J. Srinivasan. Supporting Ontology-Based Semantic matching in RDBMS. VLDB, 2004:1054-1065
18. A. Ranganathan, Z. Liu. Information Retrieval from Relational Databases using Semantic Queries. CIKM, 2006:820-821.
19. J. Zhang, Z. Peng, S. Wang, H. Nie. Si-SEEKER: Ontology-based semantic search over databases. 1st International Conference on Knowledge Science, Engineering and Management(KSEM), 2006:599-611.
20. P. Ganesan, H. Garcia-Molina, and J. Widom. Exploiting Hierarchical Domain Structure to Compute Similarity. ACM Trans. Inf. Syst. 21(1). 2003:64-93
21. J. Zhang, Z. Peng, S. Wang, H. Nie. PreCN: Preprocessing Candidate Networks for Efficient Keyword Search over Databases. 7th International Conference on Web Information Systems Engineering (WISE), 2006:28-39.
22. K. Zhang. Research on New Preprocess-ing Technology for Keyword Search in Databases. PhD thesis of Renmin University of China. 2005(in Chinese).
23. Z. Peng, J. Zhang, S. Wang, L. Qin. TreeCluster: Clustering Results of Keyword Search over Databases. 7th International Conference on Web-Age Information Management(WAIM), 2006:385-396.
24. J. Zhang, Z. Peng, S. Wang, H. Nie. CLASCN: Candidate Network Selection for Efficient Top-k Keyword Queries over Databases. Journal of Computer Science and Technology, Vol.22(2). 2007:197-207.
25. Baeza-Yates R,Ribeiro-Neto B et al. Modern Information Retrieval. ACM Press,1999,pp.27–30.
26. J. Zhang, Z. Peng, S. Wang. QuickCN: A Combined Approach for Efficient Keyword Search over Databases. 12th International Conference on Database Systems For Advance Applications(DASFAA), 2007.
27. J. Zhang, Z. Peng, S. Wang, J. Zhan. Exploiting Connection Relation to Compress Data Graph. APWeb/WAIM 2007 Workshop on DataBase Management and Application over Networks(DBMAN), 2007.
28. M. Sayyadian, H. LeKhac, A. Doan, L. Gravano. Efficient Keyword Search Across Heterogeneous Relational Databases. ICDE, 2007.
29. E.M. Voorhees, D.K. Harman. Overview of the 6th Text REtrieval Conference (TREC-6). In Proceedings of the 6th Text REtrieval Conference, 1997.

Discovering Web Services Based on Probabilistic Latent Factor Model

Yanchun Zhang and Jiangang Ma

School of Computer Science & Mathematics,
Victoria University, Australia
{yzhang,ma}@csm.vu.edu.au

Abstract. Recently, web services have been increasingly used to integrate and build business applications on the Internet. Once a web service is published and deployed, clients and other applications can discover and invoke it. With the incredibly increasing number of Web services on the Internet, it is critical for service users to discover desired services that match their requirements. In this paper, we present a novel approach for discovering web services. Based on the current dominating mechanisms of the discovering and describing web services with UDDI and WSDL, the proposed method utilizes Probabilistic Latent Semantic Analysis (PLSA) to capture semantic concepts hidden behind words in a query and the advertisements in services so that services matching is expected to be carried out at concept level. We also present related algorithms and preliminary experiments to evaluate the effectiveness of our approach.

Keywords: Web services, web services discovering.

1 Introduction

Web services have emerged as one of distributed computing technologies and sparked a new round of researches. Web services are actually self-contained, self-describing and modular applications. Because web services adopt open standard interfaces and protocols, they are increasingly used to integrate and build business applications on the Internet. With web services, business organizations can build their applications by outsourcing some other services published on the Internet. As an ever-increasing number of web services published and deployed on the Internet, it is critical for service users to discover desired services that match their requirements.

The main processes of discovering and matching services involve several activities which are performed in the collaboration between clients and web services databases. When a service user intends to utilize a service, he will first communicates with a service registry like Universal Description, Discovery and Integration (UDDI) [17] to locate the services that are closely complied with the search criteria. Then the user would create a request/ response invocation on the matched services described by the Web Services Description Language (WSDL).

At present, one of the dominating industrial techniques for web services discovery is to use UDDI registry. The UDDI is an online electronic registry, where web services are registered and described as core type of information: white pages with contacting details,

G. Dong et al. (Eds.): APWeb/WAIM 2007, LNCS 4505, pp. 18–29, 2007.

yellow pages containing classification information based on standard taxonomies and green pages providing the specification of interface for web services. The UDDI also allows syntactically search and category-based match web services. In addition, a service requester can use the Inquiry API provided in UDDI for retrieving services via submitting instructions like *find_service ()*.

However, the keywords-based mechanism supported by UDDI and most of the existing service discovering and matching approaches [10, 21, 22] show some disadvantages. First, it is difficult for users to get desired services. For instance, if a user types inaccurate keywords for searching services, he either receives numerous responses that may be totally irrelevant to his needs or get no answers at all. Second, users need to have full knowledge on the categories in a registry and manually go through the detailed database. In the majority of situations, it is impractical, ineffective and time-consuming. Another drawback of the most existing approaches is that they take into account only the keywords in users' requirements and textual advertisements in web services, instead of considering the semantic concepts hidden behind the descriptions of web services.

To address these issues, we need an effective mechanism by which the real intention of service users can be associated to the advertisements of web services and an effective approach by which a user's query is conceptually matched to the contents advertised in web services available.

In this paper, we present a novel approach for discovering and matching web services. Based on the current dominating mechanisms of discovering and describing web services with UDDI and WSDL, the proposed method utilizes Probabilistic Latent Semantic Analysis to capture the semantic concepts hidden behind the words in queries and the advertisements in services so that services matching is expected to be carried out at concept level.

The organization of this paper is as follows: We first introduce basic discovering problems and related research work. In Section 3, we briefly discuss our probabilistic semantic discovering approach. The detailed matching principle, probabilistic model and matching algorithm are introduced in section 4. The preliminary experiment evaluation is presented in section 5. Finally, the conclusion and future work can be found in Section 6.

2 Related Work

Service discovery and matching is one of the challenging issues in Service-Oriented computing [3]. Finding desired services is just similar to looking for a needle in a haystack [3]. This discovery process mainly involves locating the relevant services either published in a registry, like UDDI, or scattered in P2P systems, matching the requirements of users with a set of services and recommending desired services to the consumers.

To effectively discover and match web services, it is a necessary to establish some kinds of the correlations between a user and potential services available, which can be achieved through two steps. On the client side, a service user can express his requirements described in the form of nature language and then use a service search engine to interact with a set of potential web services. On the side of services providers, on the other hand, they advertise services' capabilities through some descriptions such as the web services' names, the operations' descriptions and the operations' names described by WSDL. In this situation, they also assume that clients would agree on the words used to describe the web services. However, the problem of how to deal with the agreement and how to associate the users' requirements to the advertisements of web services would have a critical impact on discovering web services. Therefore, locating desired services might be difficult.

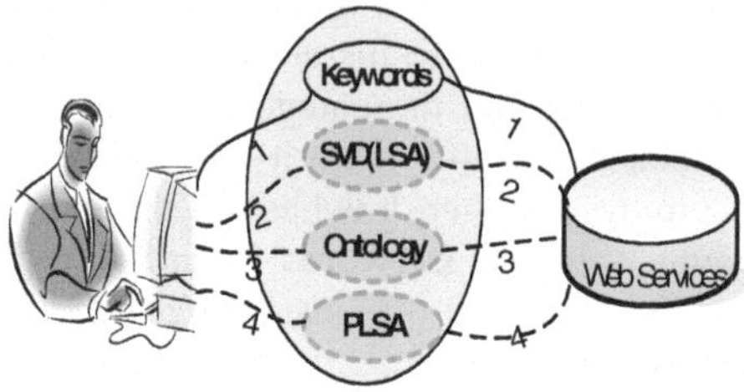

Fig. 1. The approaches of discovering and matching web services

A commonly used approach on discovering and matching web services is to *directly* associate a user's requirements to the advertisements of web services (solid line 1 shown in Fig. 1.). For example, a user types keywords in a web service search engine [24] to look for the desired web services. If the typed words are included in or identical to the descriptions of some services, the return services might be relevant to his need. Nevertheless, this approach based on the term frequency analysis is insufficient in the majority of cases. For one thing, syntactical different words may have similar semantics (synonyms), which results in low recall. For another, semantically different concepts could possess the identical representation (homonyms), thus leading to low precision. In short, this discovery mechanism fails to contemplate the semantic concepts hidden behind the words in a query and the descriptions in web services.

An alternative to the keywords-based approach is to *indirectly* associate a user's requirements to the advertisements of web services (dashed lines 2, 3, 4 shown in Fig. 1.). This relies on finding common semantic concepts between the terms in a query and services' advertisements. Then the similarity between a query and services can be compared at concept level. An interesting approach to capture the semantic concepts in the descriptions of web services has been implemented in [16]. The method starts with constructing a vector for each service description, wherein each element in a vector is assigned a TF-IDF weight. With this method, returned m services can form an m *by* n service matrix. Based on the service matrix, singular value decomposition (SVD) of the matrix is employed to discover the associated patterns between the words and their corresponding concepts. Thus, a commonly used cosine measure of the similarity between two vectors can conceptually represent how close they are in a semantic space even if a service doesn't contain terms in a query. Although this approach shows some advantages compared to the keywords-based one, its lacking of completely reasonable probabilistic interpretation [5] might limit its further applications.

More recently, ontology-based approaches [13, 15] have been seeking to use ontology to annotate the elements in web services. Such techniques, based on a theory on existence, organize domain knowledge into the categories by virtue of objects' components and their semantic connectives so that the suggested approaches aim to not only capture the information on the structure and semantics of a domain, but facilitate software agents to make inference at concept level. However, creating and maintaining ontology may involve huge amount of human effort [9]. We leave this interesting issue to be addressed in the near future.

Some other recent work can also be fond in [11, 4].

We propose to extend the SVD of the matrix approach of matching web services [16] with a different methodology called Probabilistic Latent Semantics Analysis (PLSA) [5, 6, 7], which turns out to have sound probabilistic interpretation [6] and better performance.

3 Probabilistic Latent Factor Discovering Approach

In this section, we first investigate main specifications of WSDL and then briefly introduce our probabilistic latent factor discovering approach.

3.1 Services Description and Specification

Since the WSDL and UDDI are currently the dominating mechanism for web services description and discovery, we focus on discovering and matching web service in this context, rather than using ontology to annotate elements in web services.

Normally, a web service can be described by WSDL as a collection of network endpoints. The description consists of two main parts: the abstract definition of interfaces and the concrete implementations of network. In the abstract definition, interfaces and a set of operations are defined by *portType* element and *operation* element respectively. Besides, each operation may contain input/output messages that are defined by *message* element. On the other hand, the concrete implementations specify how the abstract interfaces are mapped to the specific bindings, which may include particular binding protocols like SOAP and network address. Similarly, a set of elements such as *service, port* and *binding* are used to define these deployment details.

The key advantages of this mechanism adopted in web services lie in the separating the interface definition from the network implementation and making it possible to multiple deployments on the identical interface. Moreover, it would facilitate the reuse of the software in the web service community. Figure 2 shows the specification of web services and an example of WSDL file for a CargoShipping service is shown in Figure 3.

3.2 Overview of Our Probabilistic Approach

Our approach is based on our observation of uncertainty on the usage of web services in the Web environment. This uncertainty is reflected in two aspects. On the client side, a service user may not have a specific goal in his mind while he browses web service categories on the Web. For this reason, the query the user selects may not

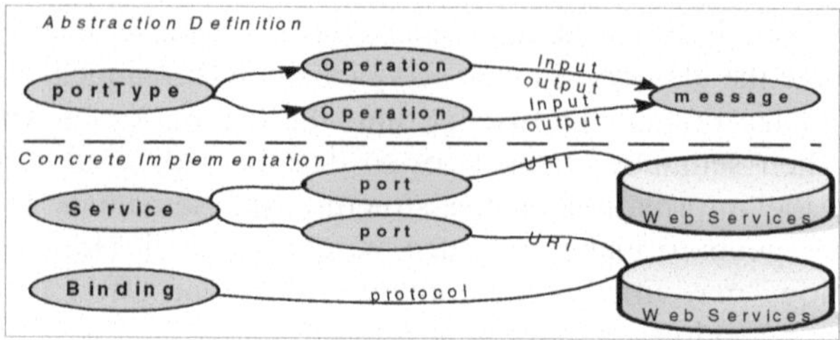

Fig. 2. The specification of web services

```
<message    name="ListOfDeliveryDestination">
        <part name="destinations" type=
              " tns:Vector"/>
</message>
<message name="DeliveryDestinaiton">
        <part name="destinaiton" type=
              " xsd:String"/>
</message>
<message name="DeliveryPrice">
        <part name="price" type="tns:String"/>
</message>
<portType name="CargoShippingService_port">
        <operation name="getDestinationList">
            <output message="
                  ListOfDeliveryDestination"/>
        </operation>
        <operation name="getDeliveryPrice">
            <input message="
                  DeliveryDestinaiton"/>
            <output message="DeliveryPrice"/>
        </operation>
</portType>
```

Fig. 3. An example of WSDL file for cargoshipping web service

fully represent his real intention. Second, it is difficult for users to choose appropriate words to indicate semantic concepts because of the dictionary problem [2]. Furthermore, as mentioned, homonyms and synonym also have negative impact on effectively discovering and matching web services. On the other hand, different service providers may choose different phrases to describe their services and in the Web context, web services are priori unknown [12], which makes the discovery of services more challenging.

Based on this observation, the key idea of our approach is to indirectly associate the intention of a user to the advertisements in web services by applying Probabilistic Latent Semantics Analysis, which is expected to capture the semantic concepts hidden in the descriptions in web services. As a result, web services can be matched against a query at concept level.

Figure 4 illustrates the outline of the proposed probabilistic latent semantic approach. To begin with, the approach will filter out those web services whose types are not compatible to a user's query, which will lead to a smaller size of services available. Then the Probabilistic Latent Semantic Analysis is used to the match semantic similarity between a query and web services. Finally, the Quality of Services

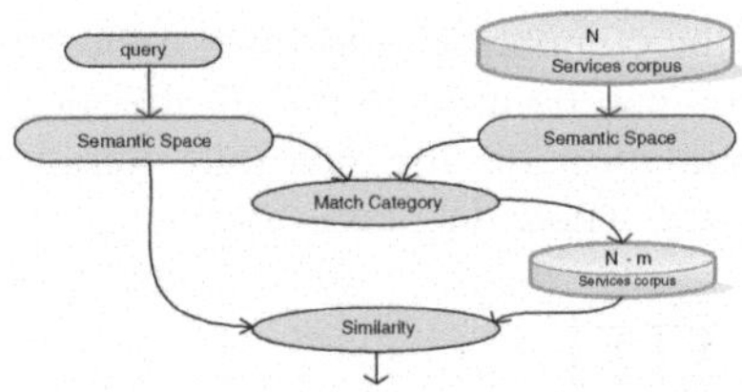

Fig. 4. The outline of matching approach

(QoS) measure will be combined with the proposed semantic measure to produce a final score that reflects how semantically close the query is to available services.

4 Web Services Discovering Based on PLSA

4.1 PLSA Introduction

Our probabilistic approach is based on the PLSA model that is called *aspect model* [5]. PLSA utilizes the Bayesian Network to model an observed event of two random objects with a set of probabilistic distributions. In the text context, an observed event corresponds to occurrence of a word *w* occurring a document *d*. The model indirectly associates keywords to their corresponding documents through introducing an intermediate layer called the hidden factor variable $Z=\{z_1, z_2, ..., z_k\}$ with each observation of a word *w* in a document *d*. In our context, each observation corresponds to a service user accessing to services by submitting a query for locating desired web services. Thus, the generative probabilistic model is expected to infer the common semantic concepts between a query and services. PLSA model works like this:

- Select a document d_i from a corpus of documents with probability $P(d_i)$
- Select a latent factor z_f with probability $P(z_f | d_i)$
- Generate a word w_j distribution with probability $P(w_j | z_f)$

Based on the assumption that a document and a word are conditionally independent when the latent concept is given, the joint probability of an observed pair (d_i, w_j) obtained from the probabilistic model is shown as following:

$$P(d_i, w_j) = P(d_i)P(w_j \mid d_i), \tag{1}$$

Where

$$P(w_j \mid d_i) = \sum_{f=1}^{K} P(z_f \mid d_i)P(w_j \mid z_f), \tag{2}$$

Now we face the task on fitting the model from a set of training data. The basic principle is to maximize probability $P(training-data \mid parameters)$ by finding a set of parameters. To simplify the computing procedure, an iterating approach is adopted. First of all, initial estimation can be used to update the model, and then the updated model presents a new estimation on the previous iteration. In our context, an objective function based on the whole data collection is:

$$\prod_{j=1}^{N}\prod_{i=1}^{M} P(w_i \mid d_j)^{m(w_i,d_j)} \tag{3}$$

Thus, a log likelihood function of an observation is defined as following:

$$\ell = \sum_{j=1}^{N}\sum_{i=1}^{M} m(w_i, d_j) \bullet \log P(w_i \mid d_j) \tag{4}$$

Where $m(w_i, d_j)$ denotes the frequency of a word w_i occurring in a document d_j.

PLSA uses the Expectation-Maximization (EM) [5] algorithm to learn the model. The whole learning process starts with randomly assigning initial values to the parameters $P(z_f)$, $P(d_i \mid z_f)$ and $P(w_j \mid z_f)$, then is followed by alternative two steps: E-step and M-step. In E-step, based on the current estimation of the parameters, the posterior probabilities for latent variables are computed for all observed pairs (d_i, w_j). In M-step, the parameters are updated based on the probabilities computed in the previous E-step.

This learning process can be summarized as following:

- Parameter $P(z_f)$, $P(d_i \mid z_f)$ and $P(w_j \mid z_f)$ are randomly assigned an initial value
- In E-step, the posterior probability over the latent variable conditioned on the occurrence of w_j occurring in d_i is computed as:

$$P(z \mid d, w) = \frac{P(z)P(d \mid z)P(w \mid z)}{\sum_{z_i \in Z} P(z_i)P(d \mid z_i)P(w \mid z_i)} \tag{5}$$

- And in M-step, according to the previous values, the parameters are updated for the conditional likelihoods of the observation:

$$P(d \mid z) = \frac{\sum_{w_i \in W} P(z \mid d, w_i)m(d, w_i)}{\sum_{w_i \in W}\sum_{d_j \in D} P(z \mid d_j, w_i)m(d_j, w_i)} \tag{6}$$

$$P(w \mid z) = \frac{\sum_{d_j \in D} P(z \mid d_j, w)m(d_j, w)}{\sum_{w_i \in W}\sum_{d_j \in D} P(z \mid d_j, w_i)m(d_j, w_i)} \tag{7}$$

$$P(z) = \frac{\sum_{d_j \in D} \sum_{w_i \in W} P(z \mid d_j, w_i) m(d_j, w_i)}{\sum_{d_j \in D} \sum_{w_i \in W} m(d_j, w_i)} \tag{8}$$

PLSA was originally used in text context for information retrieval and now has been used in web data mining [19]. In this paper, we utilize PLSA for discovering and matching web services.

4.2 Web Services Information Processing

The overall process of discovering web services includes information collecting, data processing, data representation and similarity matching (see Section 4.3).

The information collecting: The main consideration on the information source in our discovering approach is based on the current specification of WSDL description and UDDI discovery mechanism. As each web service has its associated WSDL file describing its functions, we firstly extract the overall service interface information such as name and textual description in the WSDL file. This kind of information will be used to decide whether a web service's category is relevant to a user's query.

The data processing stage consists of transforming raw web service information into appropriate the format of data suitable for the model learning. For this purpose, the commonly used approaches for the words processing are applied. As descriptions and names are likely concatenated by a sequence of strings where individual word starts by uppercase letter, for instance, getCityWeather, the names and descriptions are separated so that each token conveys some meaning. Other methods of data processing include the word stemming and the stopwords removing. The former intends to remove common term suffixes while the latter eliminates very frequently used words.

The data representation: In the PLSA model, the pre-processing information through the previous phase will be represented by the bag of words, in other words, the frequency of the words in each web service is considered while the positional relationship between them is ignored. According to this, each web service can be represented as a vector.

Definition 1. *(Service Document) A service document (SD) is defined as a vector* $v = \{v_1, v_2, \ldots, v_m\}$ *, where* v_i *is the number of times the word* w_i *appearing in the document and m denotes the size of a vocabulary. The words in SD are extracted from service name, service description, operation name and input/output names in the WSDL file.* ◊

Suppose we have a corpus of N web services with $ws_i \in WS = \{ws_1, ws_2, \ldots, ws_n\}$ and a collection containing M different words (vocabulary), with $w_j \in W = \{w_1, w_2, \ldots, w_m\}$. Based on this, we can define a service matrix as following:

Definition 2. *(Service Matrix) A service corpus is defined as an M by N Service Matrix(SM), where* $n(w_i, d_j)$ *denotes the number of occurrences of a word* w_i *appearing in the document* d_j. ◊

4.3 PSMA – Probabilistic Semantic Matching Algorithm

Our discovering approach is first to cluster services to a group of learnt latent variables, then the similarity of a query in respect to the services in its relevant group can be computed in a smaller size of collection of web services.

The learnt latent variables can be used to characterize web services. From formula 2 introduced in the previous section, one can obtain some interesting interpretations. First, the right-hand side of the formula indicates a matrix decomposition, that is, the aspect model expresses dimensionality reduction by mapping a high dimensional term document matrix into a lower dimensional one(k dimension) in a latent semantic space. Second, $P(z_f \mid d_i)$ implies the association of the service d_i and its hidden factors z_f. For example, for a latent factor such as z_x, if the probability $P(z_x \mid d_i)$ is very high, the concept implied in the hidden factor z_x is regarded as to be highly correlated to the service d_i. So, based on a group of hidden variables, we can compute $P(z_f \mid d_i)$ over each hidden factor z_f, $f \in \{1,2,...,k\}$. With the k computing values, we can find a maximum value $P_{\max}(Z_f \mid d_i)$, which can be used as the class label for this service. What is more, in a dimension-reduced semantic space, each dimension represents a semantic concept and the services with similar semantic concepts are projected to be close each other. Based on the discussion, we employ the following formula to infer the relationship between a web service and hidden factors.

$$P(z \mid d_{new}) = \frac{P(d_{new} \mid z)P(z)}{P(d_{new})} = \frac{P(d_{new} \mid z)P(z)}{\sum_{z_f \in Z} P(d_{new} \mid z_f)} \tag{9}$$

An algorithm of the category matching is shown as following:

```
CategorizingServices(SM, K)
 begin:
    Input: service matrix SM = { sm_1, sm_2,..., sm_n }
    k: the number of latent factors
    μ : threshold
    output: k service communities: SC = { sc_1, sc_2,..., sc_k }
    sc_f ← φ , sc_f ∈ SC
    Repeat
       for each service sm_i in SM {
          for each hidden variable hv_j , j=1,…,k
             hs_ji[j] = calculate_P( hv_j | sm_i )
```

```
            f ← find_max( probability_hs_ji[j] )
            sc_f  =  sc_f .append( sm_i )
        }
    Return SC
end.
```

After clustering the services into their corresponding concept groups, we will match a user's requirements against the relevant services. This can be first achieved by finding the query's correlated concept group. As a new query may be outside the model, in this case, it needs to be added to or folded in the model through the iterative EM steps, where the probabilistic distribution $P(w \mid z)$ over the words conditioned on the latent variables is fixed, but mixing probabilistic distribution $P(z_f \mid query)$ are computed. With this way, the web services whose types are not compatible to a user's query will be filtered out so that we can get a much smaller size set of services. On the next step, we can utilize commonly used cosine measure to decide how semantically similar a query is to each of services in the group.

$$sim_{PLSA}(d_i, q) = \frac{|d_i \bullet q|}{\|d_i\|_2 \|q\|_2} \tag{10}$$

Finally, we believe that service selection should comply with the standard of Quality of Services (QoS). Therefore, in our application, the process of selecting services involves two steps:

- Step 1: calculate service's QoS.
- Step 2: combine QoS with semantic matching to produce a final score.

As a result, quality of services can be combined with the PLSA similarity score to produce a final ranking for the specific web service:

$$Sim(d_i, q) = \lambda \bullet S_{QoS} + (1 - \lambda) \bullet Sim_{PLSA}(d_i, q), \tag{11}$$

5 Preliminary Evaluation

Preliminary experiments were carried out on the corpus of web services whose WSDL files can be accessed via the service collections published in XMethods [20] and other service registries. In our case, we identify the services of four categories: General Information, Location Finder, Translation Services and Business Services. We extract related information such as names and textual descriptions in the WSDL files and all extracted information is processed with the approaches introduced in the previous section. Thus, we obtain a corpus of services consisting of 77 services which are divided into two data sets: training data and testing data.

The extracted training data are used to fit the PLSA model. We train the model with 6 aspects, which is slightly greater than the number of selected four service categories.

In order to evaluate the outcome, we compute precision and recall, and apply traditional vector-based similarity baseline used in information retrieval approach to compare to the proposed approach in this paper. As it turns out, our probabilistic semantic discovery method increases the overall recall because the approach considers semantic concepts hidden behind the words in a query and the advertisements in services.

6 Conclusion and Future Work

It is a challenging work to effectively discover the web services that conceptually match users' requirements. In this paper, we studied the current issues of the existing methods for discovering web services and proposed a novel approach to deal with the issue. Based on the current industrial standards, our approach is to indirectly associate the intention of users to the advertisements in web services by applying Probabilistic Latent Semantics Analysis, which is expected to capture the semantic concepts hidden in the descriptions in web services. Consequently, web services can be matched against a query at concept level in a dimension-reduced semantic space.

We also showed a matching algorithm based on the probabilistic model and preliminary evaluation indicates that the proposed approach improve the recall of web service discovering.

Finally, our probabilistic semantic approach is the first step towards effectively discovering web services. The ongoing project is to investigate the unification of the proposed approach with ontology towards effectively discovering web services.

References

1. S. Deerwester, S.T. Dumais. Indexing by Latent Semantic Analysis. In Journal American Society for Information Retrieval, pages: 391-407, 1990.
2. G.W. Furnas, T.K. Landauer, L.M. Gomez and S.T. Dumais. The Vocabulary Problem in Human-System Communication. In Communication of ACM, 30(11), pages: 964-971, 1987.
3. J. Garofalakis, Y. Panagis, E. Sakkopoulo and A. Tsakalidis. Web Service Discovery Mechanisms: Looking for a Needle in a Haystack? In International Workshop on Web Engineering, August 10, 2004.
4. Y. Hao and Y. Zhang. Web Services Discovery based on Schema Matching. In the Proceedings of the 30th Australiasian Computer Science Conference(ACSC 2007), Feb, Australia, 2007
5. T. Hofmann. Probabilistic Latent Semantic Analysis. In Proceedings of the 22nd Annual ACM Conference on Research and Development in Information Retrieval. Berkeley, California, pages: 50-57, ACM Press, August, 1999.
6. T. Hofmann. Probabilistic Latent Semantic Indexing. In Proceedings of the 22nd Annual International SIGIR Conference on Research and Development in Information Retrieval. 1999.
7. T. Hofmann. Unsupervised Learning by Probabilistic Latent Semantic Analysis. In Machine Learning. Volume 42, Number 1-2/ January, pages: 177-196 , 2001.

8. R. Hull, M. Benedikt, V. Christophides and J. Su. E-services: A look behind the curtain. In Proceedings of the Twenty-Second ACM SIGMOD-SIGACT-SIGART Symposium on Principles of Database Systems, June 2003.
9. M. Klein and A. Bernstein. Toward High-Precision Service Retrieval. In IEEE Internet Computing, Volume: 8, No. 1, Jan. – Feb. pages: 30 – 36, 2004
10. L.S. Larkey. Automatic essay grading using text classification techniques. In Proceedings of ACM SIGIR, 1998.
11. J. Ma, J. Cao and Y. Zhang. A Probabilistic Semantic Approach for Discovering Web Services. To appear In the 16th International World Wide Web Conference(WWW2007). Banff, Alberta, Canada, May 8 -12, 2007.
12. M. Oussani and A. Bouguettaya. Efficient Access to Web Services. In IEEE Internet Computing, Volume 8, Issue 2, pages: 34 – 44, March-April, 2004.
13. M. Paolucci, T. Kawamura, T. Payne and K. Sycara. Semantic Matching of Web Services Capabilities. In Proceddings of the 1st International Semantic Web Conference (ISWC2002). 2002.
14. K. Sivashanmugam, K. Verma, A.P and J.A. Miller. Adding Semantics to Web Services Standards. In Proceedings of the International Conference on Web Services ICWS'03, pages: 395-401, 2003.
15. S. Staab, W. Van der Aalst, V.R. Benjamins, A. Sheth, J.A. Miller, C. Bussler, A. Maedche, D. Fensel and D, Gannon. Web services: been there, done that? In IEEE, Intelligent Systems, Volume: 18, Issue: 1, Jan. – Feb. pages: 72 – 85, 2003.
16. A. Sajjanhar, J. Hou and Y. Zhang. Algorithm for Web Services Matching. In Proceedings of the 6th Asia-Pacific Web Conference, APWeb 2004, Hangzhou, China, April 14-17, 2004.
17. UDDI Version 2.03 Data Structure Reference UDDI Committee Specification, 19 July 2002, http://uddi.org/pubs/DataStructure-V2.03-Published-20020719.htm
18. Y. Wang and E. Stroulia. Semantic Structure Matching for Assessing Web Service Similarity. In the First International Conference on Service Oriented Computing, Trento, Italy, December 15-18, 2003.
19. G. Xu, Y. Zhang, J. Ma and X. Zhou. Discovering User Access Pattern Based on Probabilistic Latent Factor Model. In Proceedings of the 16th Australasian Database Conference – Volume: 39 pages: 27 – 35, Newcastle, Australia, 2005.
20. XMethods. http://www.xmethods.com/
21. Y. Yang and J. Pedersen. A Comparative Study on Feature Selection in Text Categorization. In International Conference on Machine Learning, 1997.
22. A.M. Zaremski and J.M. Wing. Signature Matching: a Tool for Using Software Libraries. In ACM Transactions on Software Engineering and Methodology, Volume 4, Number 2, pages: 146-170, April, 1995.
23. http://www.census.gov/epcd/www/naics.html.
24. http://www.webservicelist.com

SCORE: Symbiotic Context Oriented Information Retrieval

Prasan Roy and Mukesh Mohania

IBM India Research Lab
{prasanr,mkmukesh}@in.ibm.com

1 Introduction

Much of the data in an enterprise is strictly-typed and thus can be meaningfully decomposed at a fine granularity and stored in a relational database. Such data is mostly operational business data (e.g. sales, accounting, payroll, inventory), and has been the mainstay of the RDBMS products like DB2.

However, this "structured" data constitutes only a fraction of the entire information content within an enterprise, which also includes "unstructured" content like analytical reports, email, meeting minutes, web-pages etc. Due to its free-flow, untyped structure, this unstructured content is not as amenable to structured storage and retrieval in the RDBMS as the strictly-typed operational data. Such data is stored in a content manager, like the IBM Content Manager [4], which associates the unstructured data, say a document, with structured metadata (such as relevant keywords) that describes the document. The unstructured content is then retrieved by querying the metadata.

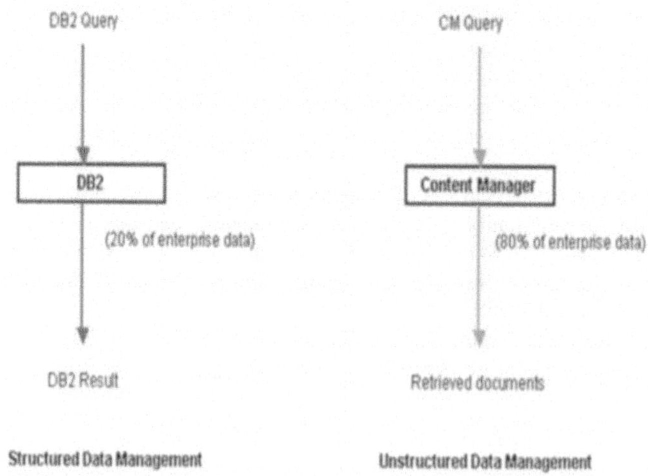

Fig. 1. Isolated Management of Structured and Unstructured Data

In any enterprise today, the structured data is managed by the database system (say, IBM DB2), the unstructured data is managed by the content manager (say, IBM Content Manager [4]) and these two exist as silos (see Figure 1). This is unfortunate

G. Dong et al. (Eds.): APWeb/WAIM 2007, LNCS 4505, pp. 30–38, 2007.

since the two kinds of data are complementary in terms of information content. Due to the separation between the two kinds of data today, an application would need to straddle across the two disparate data sources, querying one and then the other on similar context.

For example, consider a stock-market information system. Such a system not only maintains the market statistics (structured data) but also the analyst advisories, risk-assessment reports, articles, related news, etc. (unstructured data). It would be nice if the stock trader, while querying the market statistics on, say, the fastest moving stock within a given sector at the moment, would also get the related advisories and reports. If she wants to trade on the stock, then depending upon the size of the trade, she gets the appropriate risk-assessment report. Note that these reports are available without her making an effort to hunt for them in the content repository, or on the web – *saving valuable time and effort*. Similarly, while browsing through an analyst report on a sector, it would be nice if the operator has access to the current statistics on the mentioned stocks without having to access them explicitly. Similar scenarios can be thought of in other domains also, e.g.:

- *Health:* Patient specific report and medical articles,
- *Manufacturing:* Defect statistics and engineering specifications,
- *Marketing:* Customer transaction history and marketing documents,
- *Travel:* Traveler itinerary and promotional flyers, travel advisories,
- *Management:* Employee records and status reports (details in Section 2).

The goal of this paper is to present a novel, context-oriented, loosely coupled integration of structured (DB2) and unstructured data (CM) through *symbiotic* consolidation of related information[1] in an enterprise. Specifically:

- To enhance structured data retrieval by associating additional documents relevant to the user context with the query result [9], and
- To enhance document contents by associating additional information derived from structured data [10].

This integrated information can be used for deriving the business insights, targeted marketing, CRM applications, etc [10].

At a high level, this is achieved through a broker that interfaces the information present in DB2 with that present in CM and vice versa, as illustrated in Figure 2. This broker is loosely coupled with DB2 on one side and CM on the other, and resides as a separate entity.

It is important to note that data integration products like the *IBM Enterprise Information Integrator* [7, 8, 11] solve only part of the problem. They consolidate multiple diverse, structured and unstructured information sources into a single point of access, enabling the user to write a single query that spans these sources. However, the onus of specifying appropriate context remains with the user, which is a limitation since, as shown in Section 2, the user might not be aware of the overall context at the point of submitting the query.

---

[1] It should perhaps be emphasized that such a feature is purely optional, and the consolidation can be controlled using a configuration parameter.

In the rest of this paper we describe the project in greater detail. A motivating example is presented in Section 2. Preliminary ideas and alternatives on the design of the broker are presented in Section 3, and research issues that need to be addressed going forward are discussed in Section 4. Related work is discussed in Section 5, with emphasis on how this work differentiates from the same. Finally, the conclusions appear in Section 6.

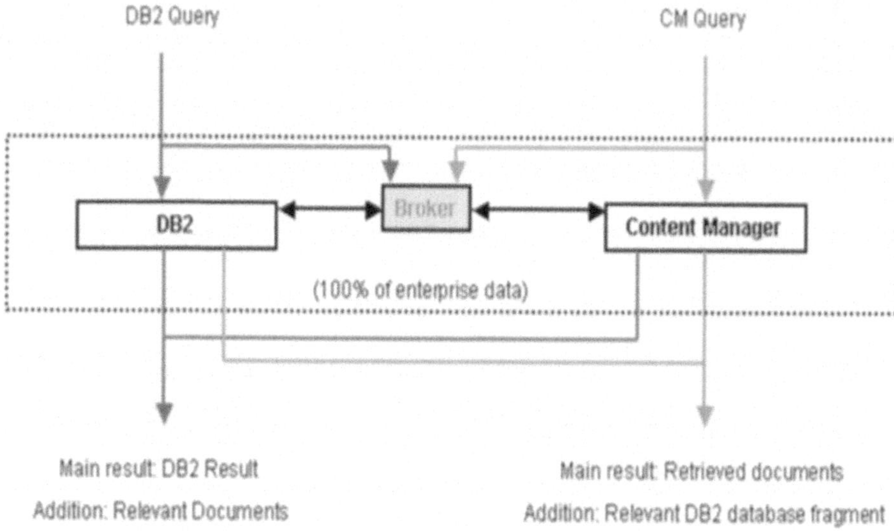

Fig. 2. Consolidated Management of Structured and Unstructured Data

2 Motivating Example

Consider a DB2 database containing information about employees in IBM. It details what projects an individual is working on, which group he/she is a part of and details of the organization. Additionally, the content manager contains documents on several topics related to the status of ongoing projects, reviews, procedures, policies, etc. in the organization.

The user queries DB2 for information on a project named CORTEX, and gets back not only the query result, but also a set of relevant documents. In this case, it is a single document *Report.doc* that contains a report on database research at IRL, of which CORTEX is a part. This scenario is illustrated in Figure 3. Notice that the document has no explicit mention of CORTEX, but contains the relevant *context*. Essentially, the system has taken cue from what the user has asked for (the shown tuple in the PROJECTS relation) and looked around in the neighborhood to determine the relevant context (in this example, the context is characterized by a set of keywords, but there could be more sophisticated characterizations, as will be discussed in a later section).

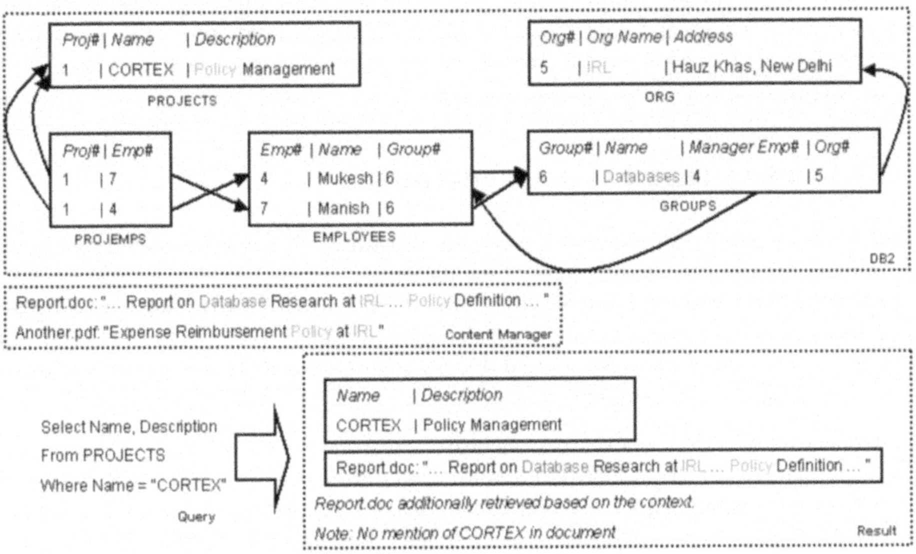

Fig. 3. Enhancing Structured Data Retrieval

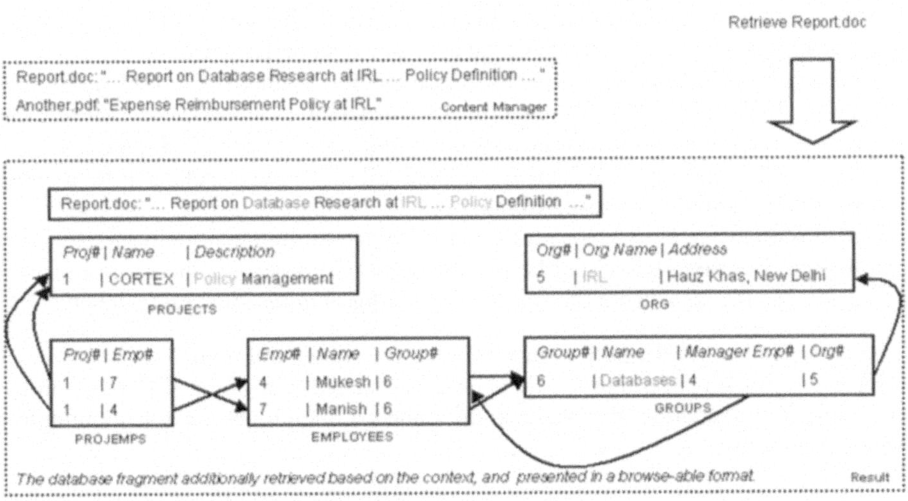

Fig. 4. Enhancing Unstructured Data Retrieval

Next, consider another user who retrieves the document Report.doc, and gets back not only the document asked for, but also (a handle of) the fragment of DB2 relational database relevant to the document. This is shown in Figure 4. Essentially, the system extracts the context of the document (again, a set of keywords for the purpose of this example) and uses this context as an anchor into the DB2 database. The relational data can be presented in to the user in a browse-able manner.

Clearly, such functionality cannot be achieved by querying DB2 and CM independently. In the first case CM uses the context retrieved from DB2 to identify relevant documents, while in the second case DB2 uses the context retrieved from CM to identify the relevant database fragment.

3 Preliminary High Level Architecture

This section expands on the high level idea illustrated in Figure 2, and gives some preliminary ideas on the architecture of the broker.

Recall from the discussion in Section 1 that the purpose of the broker is to correlate the information content of the relational data in the DB2 database with the information content of the unstructured documents in the CM. In this initial design we keep things simple, and assume that:

- Context is modeled as a set of keywords.
- A domain expert has identified a set of categories in which the contexts can be classified. Each category is characterized using a representative set of keywords.
- There is an efficient algorithm to find the categories most relevant to a given context. With the assumptions above, this amounts to simply finding the category whose representative keyword set has the maximum overlap with the context.
- There is an efficient algorithm to determine the context of a DB2 query. As in the motivating example (Section 2), this can be done, for instance, by constrained navigation of the neighborhood of the accessed data in the database.
- There is an efficient algorithm to determine the context of a CM query. We assume that each retrieved document's metadata contains a set of relevant keywords; a union of these keywords forms the context of the CM query.
- There is an efficient mechanism to find all documents relevant to a given category; essentially, an inverted index.
- There is an efficient mechanism to find the database fragment relevant to a given category. For now, this could be just an index. However, for increased precision we can use the ideas from the *keyword browsing in databases* research here.

Accordingly, we get the preliminary architecture shown in Figure 5.

Here, the broker in Figure 2 has been expanded into four different entities responsible for executing the various mechanisms mentioned in the assumptions above. Specifically:

- *DB2 Context Analyzer* analyses the input DB2 query and the accessed database fragment and generates the context for the DB2 query (as mentioned in the assumptions, this is essentially as set of keywords obtained by navigating a constrained neighborhood of the accessed database fragment).
- *CM Context Analyzer* analyses the input CM query and the retrieved documents, and generates the context for the CM query.
- *CM Context Index* determines the set of documents in CM most relevant to the context given as the input. Essentially, the input context is first mapped to a (small) set of relevant categories. Next, handles of the documents relevant to

these categories are retrieved from a precomputed index, optionally pruned, and output.

- *DB2 Context Index* determines the database fragments in the DB2 database most relevant to the context given as the input. The naïve implementation is similar to the CM Context Index mentioned above; the input context is first mapped to a (small) set of relevant categories. Next, handles of the database fragments relevant to these categories are retrieved from a precomputed index, optionally pruned, and output. As mentioned earlier, a less naïve implementation would use the ideas in the *keyword searching in databases* research [1, 2].

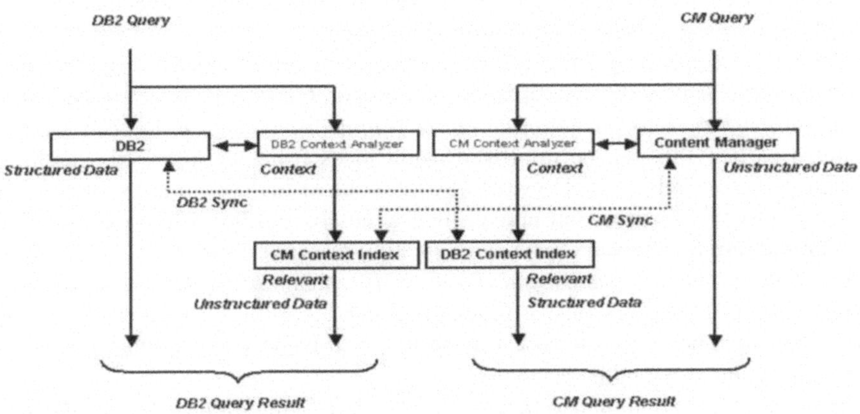

Fig. 5. Preliminary Architecture

4 Research Issue

While being an acceptable proof of concept, the preliminary architecture presented in Section 3 has some limitations. In this section, we enumerate some of these limitations and identify the research issues that need to be explored in order to overcome them.

- Limitation: *Context is a set of keywords.* This is not very expressive. Can we do better than that? Including semantic information in the context appears to be an interesting issue for further exploration.
- Limitation: *The context of a DB2 query is determined by scanning a constrained neighborhood of the accessed database fragment for keywords.* This needs further study. Moreover, there exist other avenues that could be helpful in ascertaining the query context; such as the results retrieved and the query workload. If the user has provided a profile, that can be helpful too. Determining the query context from each of these dimensions, and consolidating the same appears to be an interesting research issue.
- Limitation: *The context of a CM query is determined as the union of the keywords associated with the retrieved documents.* The idea is to extract the context of each retrieved document and merge the same. In case the context includes more

semantics than merely keywords, merging this semantic information could be an interesting research issue. As with the DB2 query, context could also be determined using the query workload and user profile, if provided. As earlier, determining the query context from each of these dimensions, and consolidating the same appears to be an interesting research issue.

- Limitation: *Context of a DB2 query is mapped to one or more categories; the set of documents associated with each such category is retrieved, and the union is returned as the set of documents relevant to the DB2 query.* This strategy clearly suffers from a loss of precision. To improve precision, the final set of documents after the union can be further pruned based on the context, but it is not clear if it would be of significant help. More precise context-based document retrieval techniques need to be studied and/or developed.
- Limitation: *Context of a CM query is mapped to one or more categories; the set of database fragments associated with each such category is retrieved, and the union is returned as the set of database fragments relevant to the CM query.* As stated earlier, this can be improved using ideas from the keyword searching in databases research[1, 2]. However, as the context includes more semantics than just keywords, as proposed above, then generalizing the ideas to context-based database retrieval appears to be an interesting research issue.
- Limitation: *The documents and database fragments are returned as unordered sets.* For usability reasons, efficient ranking algorithms to order the returned results (documents or database fragments) with respect to the input query (DB2 or CM) context would be needed.
- Limitation: *Entire database fragments returned in addition to the documents on response to a CM query.* The database fragments need to be presented in a browse-able manner, or may even be presented as smart tags dynamically attached to the documents. This appears to be an interesting user interface research issue.

5 Related Work

This work consolidates, and extends several prior efforts. In this section, we review these efforts in perspective of this project, and emphasize how this project differentiates from them.

The Unstructured Information Management Architecture (UIMA) [6] is a framework for classifying, describing, developing and combining natural language components in applications for processing unstructured information. As the name implies, UIMA is necessarily focused on unstructured data analysis. Although UIMA proposes an interaction with structured information, it seems limited to (a) extracting structured information from unstructured data, and (b) using this extracted structured information to aid in further analysis of the unstructured data. As such, the UIMA framework merely works as an application using the relational DBMS as a repository of its private information, and the semantics of this stored information is totally understood by the UIMA framework. It is not clear if the framework addresses the problem of correlating existing data in the relational DBMS with the unstructured information being analyzed, and vice versa – the focus of this proposal.

Recently, there has been noticeable effort on keyword search in relational databases [1, 2]. The idea is to consider each tuple in each relation in the database as a document, and consider the entire database as a graph with these tuples as nodes and edges joining the tuples connected through foreign-key relationships. The result of a query with multiple keywords is a set of trees with leaves as tuples containing one of more keywords. These trees are ranked on relevance to the query. As mentioned in Section 3, ideas from this work can be applied to develop a more sophisticated DB2 Context Index.

Sapient [5] integrates data and text through an OLAP-style interaction model. The authors propose a framework for automatically extracting structured information from documents to form a "document warehouse" that can complement the data warehouse in business analysis. Our focus is not as much on information extraction, but on providing context-based correlation of the structured and unstructured content.

We reiterate that this effort is different from information integration [8], which primarily brings together multiple disparate databases as one virtual database. This enables the user, for instance, to access structured data from a DB2 database and (the metadata of) unstructured data in a different CM database in a single query. Clearly, this only solves part of the problem, since it remains the onus of the user to specify the context of all the information needed.

The *IBM Enterprise Information Integrator (EII) for Content* [7, 11] provides support for information mining, including categorization and clustering of the unstructured content stored in content managers federated into the system. However, it does not provide any support for context-based consolidation of the structured and unstructured information, which is the focus of this work. Nevertheless, this work can build on IBM EII for Content, using it as a platform for developing the CM Context Index (Section 3).

6 Conclusion

This paper presented a framework for consolidating structured and unstructured data retrieval in a novel, symbiotic manner. The problem is well-motivated and a comparison with prior and ongoing related efforts shows that this problem has, to the best of our understanding, not been addressed earlier. The paper also discussed a preliminary architecture for the framework, its current limitations and the issues that need to be addressed in order to overcome these limitations. We believe that this new way of information integration has several interesting research problems as well.

Acknowledgment. We would like to thank Prof. Xuemin Lin and Prof. Jeffrey Xu Yu for inviting us to deliver a talk in this conference. We are very thankful to Dr. Wei Wang and Mr. Di Wu for their great help in revising and formatting this paper.

References

1. Agrawal, S. et al., DBXplorer: A System for Keyword-based Search over Relational Databases, *ICDE 2002*
2. Bhalotia, B. et al., Keyword Searching and Browsing using BANKS, *ICDE 2002*

3. Bhide, M et al., Linking and Processing Tool for Unstructured and Structured Data, *To appear in SIGMOD 2007*
4. Chen, K. et al., IBM DB2 Content Manager V8 Implementation on DB2 Universal Database: A Primer, *IBM Technical White Paper, May 2003*
5. Cody, W. F. et al, The Integration of Business Intelligence and Knowledge Management, *IBM Systems Journal, Vol. 41, No. 4, 2002*
6. Ferruci, David et al., UIMA: An Architectural Approach to Unstructured Information Processing in the Corporate Research Environment, In *Journal of Natural Language Engineering, 2003*
7. Greenstreet, Carol S., Enterprise data access with IBM DB2 Information Integrator for Content, *IBM Technical White Paper, March 2003*
8. Jhingran, A. et al, Information integration: A research agenda, *IBM Systems Journal, Vol. 41, No. 4, 2002*
9. Roy, P., et al., Efficiently Linking Text Documents With Relevant Structured Information. *Very Large Data Bases,* Seoul, Korea, September 2006
10. Roy, P. el al., Associating Relevant Unstructured Contents with Structured Database Query Results. *CIKM, Germany 2005.*
11. Somani et al., Bringing together content and data management systems: Challenges and opportunities, *IBM Systems Journal, Vol. 41, No. 4, 2002*

Process Aware Information Systems: A Human Centered Perspective

Clarence A. Ellis[1] and Kwanghoon Kim[2]

[1] Collaboration Technology Research Group
Department of Computer Science
University of Colorado at Boulder
Campus Box 430, Boulder, Colorado, 80309-0430, USA
skip@cs.colorado.edu
[2] Collaboration Technology Research Lab.
Department of Computer Science
Kyonggi University
San 94-6 Yiui-dong Youngtong-ku Suwon-si Kyonggi-do, 442-760, South Korea
kwang@kyonggi.ac.kr
http://ctrl.kyonggi.ac.kr

Abstract. Process Aware Information Systems (PAISs) are a useful form of software system; they are being found to be useful by an increasingly large population of diverse humans. The strength and the challenge of PAISs are within the collective endeavors domain where PAISs are used by groups of people to support communication, coordination, and collaboration. Even today, after years of research and development, collective endeavors PAISs are plagued with problems and pitfalls intertwined with their benefits. These problems are frequently elusive and complex due to the fact that "PAISs are first and foremost people systems." This paper addresses some of the people issues and proposes a framework for research in this domain[1].

Keywords: Process Aware Information System, Workflow, Business Process, People System.

1 Introduction

A Process Aware Information System (PAIS) is a software system that manages and executes operational processes involving people, applications, and/or information sources on the basis of an explicit imbedded process model. The model is typically instantiated many times, and every instance is typically handled in a predefined way. Thus this definition shows that a typical text editor is not process aware, and likewise a typical e-mail client is not process aware. In both of these examples, the software is performing a task, but is not aware that the

[1] The authors, as the co-organizers of the PAIS2007 workshop, specially research into this topic for the workshop. Also the research was partly supported by the BEIT special fund of Kyonggi University.

G. Dong et al. (Eds.): APWeb/WAIM 2007, LNCS 4505, pp. 39–49, 2007.

task is part of a process. Note that the explicit representation of process allows automated enactment, automated verification, and automated redesign; all of which can lead to increased organizational efficiency. Other potential benefits of PAIS are elaborated in [1].

An increasingly large population of diverse humans is interacting with technology these days; it is therefore increasingly important for computerized systems to address issues of human interaction / collaboration. The incorporation of process into information systems is particularly challenging because most PAIS involve processes performed by people. People processes are complex, semi-structured, and dynamically changing. Within PAIS design, it is thus necessary to take into account factors that impinge upon the organizational structures, the social context, the cultural setting, and other dimensions.

This paper first introduces examples and concepts of collective endeavors noting characteristics that can be quite challenging to capture within process descriptions. In the following sections, a framework for analysis, enactment, and mining is described; then a multidimensional meta-model is introduced. The paper ends with summary and conclusions.

2 Collective PAIS Environments

This section discusses issues related to collective endeavors in a broad sense. It introduces the complexity and variability of collective group interactions through examples, and discusses how structure can be identified in these interactions. It remains true that "how people work is one of the best kept secrets" (David Wellman; cited by Suchman [18]). Collective work is characterized by its fluidity and complex weaving of organizational, social, political, cultural and emotional aspects. Interaction at work takes a wide variety of forms. Consider for instance the following examples:

Extreme Collaboration. Mark[12] describes a "war room" environment employed by the NASA's Jet Propulsion Laboratory (JPL) to develop complex space mission designs in a very short time—nine hours over a single week for a complete and detailed mission plan. During these interaction sessions, sixteen specialists are physically co-located in a room that contains a network of workstations and public displays.

Collaboration is prompted by a complex combination of physical awareness, by monitoring of the parallel conversations in the noisy environment and in response to data that is published through customized networked spreadsheets that allow team members to publish data they produce and subscribe to data published by others. Team members move around the room to consult other specialists, or flock to the public display to discuss problems of their interest. Their movements impart important awareness information to others in the room, which may choose to join a group based on the perceived dependencies of ones own work on what a specific set of specialists is discussing. While working, team members are peripherally aware of multiple simultaneous conversations and react to keywords that concern their part of the job by giving short answers from their workplace or moving to join a group. Finally, data that is made available through the computerized spreadsheet system may also trigger collaboration.

Congressional Sessions. These are highly formalized interactions based on the Robert's Rules of Order [15]. Participants have very specific roles and duties, e.g. the Speaker of the House is the presiding officer, responsible for maintaining proper order of events; the Chief Clerk is responsible for day-to-day operation of the house. The structure of each session is predefined—the allowable items of discussion are known to all in advance. There are strict rules that determine how issues may be debated, including the order of speakers (pro, against), the time they are allowed to speak, and to some degree, the content of their addresses. Deliberation is based on voting, which is regulated as well, e.g. by rules that specify when a vote can be called, or waived, and the proportion of voters needed for approval in many different situations.

Policy making and design. Horst Rittel [14] discusses the inherently intractable nature of design and planning problems, that he names "wicked problems". This class of problems is characterized by their ill-defined nature - in many cases it is not possible to separate the understanding of the problem from the solution, as the formulation of the problem is equated to statements of potential solutions [13]. Multiple solutions are in general possible, and it might be even hard to determine which solutions are superior to others. Sometimes these problems emerge as a result of conflicting interests, e.g. when deciding how to allocate a limited number of rooms to different individuals that might have coinciding preferences. Possible solutions for these problems involve compromise—ideal solutions are replaced by acceptable ones.

Rittel [13] proposes tackling wicked problems through an argumentative method in which questions are continually raised, and advantages and disadvantages of multiple possible responses are discussed. The method, called Issue Based Information Systems (IBIS) is based on documenting and relating issues, positions about issues, and arguments that either support a position or object to it. Each separate issue is the root of a (possibly empty) tree, with the children of the issue being positions and the children of the positions being arguments. Links among these three basic elements are labeled, e.g. issues and positions are connected by responds-to links; arguments are connected to positions either by supports or objects-to links.

An IBIS discussion starts with the elicitation of one or more (abstract) issues, to which participants respond with positions, arguments and refine into more concrete sub-issues. Contradictory positions are resolved by consensus or voting. The end result is a forest of linked elements that represent the evolution of the discussion, alternatives that were considered and the rationale for decisions.

In all the above examples, the actual interactions represent a small fraction of a much larger interaction over time and space. The members of the extreme collaboration team at NASA have been working together for many years, and have detailed knowledge of each others peculiarities and expertise. They also share common engineering knowledge of their field, as well as the common approaches and problem solving strategies that are part of the cultural heritage of their field. As for the congressional sessions, these represent just the visible tip of the political iceberg; complex backstage negotiations shape the performance at the session and result from economic and political pressures of a multitude of stake-holders. Finally, the policy making and design based on IBIS is guided by a deep understanding of the issues in discussion that the participants bring to the interaction based on a lifetime of experiences, shared or not, and by expert opinion and supporting documentation that is sought as part of the process. All these interaction contexts are in turn embedded in larger societal settings, as parts of organizations, the government, as part of a nation, and so on.

Although these examples are all extracted from work life, clearly the complexity and variability exhibited here also extends to human interaction beyond work environments.

3 In Search of Structure: Observations and Concepts

Although complete details of interactions and the intricate factors that govern them are usually beyond what can be understood, constituting implicit processes that are mostly inaccessible, certain emergent regularities and patterns of group behavior can be observed. Rather than being unconstrained, interactions usually follow a structure that is repetitively reproduced by participants at each new instance [10,11]. This structure is a result of shared belief and value systems, and is frequently learned from previous experiences of participants in similar situations. This structure reinforces the enacted behaviors, helping to shape future interactions. More than repeating patterns, participants make implicit or explicit statements about/through their actions, as they go about their activities. Participants exert "reflective self-monitoring" [10] so as to act accountably, i.e. in a manner that is "observable-and-reportable" [9]. Acting accountably means acting explicitly (even if unconsciously) according to values and rules shared by a social group, that get at the same time instantiated and reinforced by actions of individuals [16]. Participation in interactions may be constrained by organizational rules, goals, and norms. Participants are able to make sense of each other's actions (and reorient their own accordingly) because individual actions are recognizable by the group as being one of the meaningful actions that are sensible within a context. Bittner [4] suggests that "a good deal of the sense we make of things happening in our presence depends on our ability to assign them to the phenomenal sphere of influence of some rule." Participation in interactions is further constrained to specific sets of behaviors that are associated to the roles played by participants (e.g. teacher, student; meeting chair, meeting participant). Roles to some extent determine the behaviors of any person occupying a certain position within a context, independently of personal characteristics [2,3]. Some of these roles may be non-institutionalized and sometimes even pathological, e.g. the devil's advocate, and the scapegoat respectively.

The linguistic interchanges among participants of an interaction can be seen as forming an elaborate game as well, where each speech act [17] constrains and directs subsequent acts. Intuitively, the act of asking a question is bound to elicit some response related to the nature of this question, even if indirectly. Searle [17] and others associated to the language/action perspective (e.g. [5,7,19]) identify a set of illocutionary points that would constitute the essential components of conversations for actions. Individual acts are inter-related into acceptable "move sequences", so that e.g. a request by a participant can be accepted, declined, or counter-offered. Each of these in turn has its possible continuations, e.g. a counter-offer can be accepted, the original request might be canceled, or a new counter-offer might be generated [17]. Collective discourses thus display structure and can be equated to an evolving process. In practical terms that means that interactions, even seemingly unstructured ones, are regulated by linguistic, social and cultural norms that dictate to a large extent the way interactions are "played out". In other words, interactions constitute social processes. Such processes take place at many distinct levels, embedded within each other in a recursive structure. Debate and voting, for instance, can be considered sub processes within a meeting in which they occur; meetings in turn are part of larger processes within organizations, which are embedded within yet larger organizational and societal settings.

This paper next focuses upon a framework and meta-model for multi-dimensional PAIS analyses and enactment. These PAISs tend to have a high degree of complex human involvement, which poses special challenges to technological augmentation.

4 A PAIS Framework

In general, a framework is a way of thinking; it is a way of conceptualizing an area of endeavor. It sometimes allows one to categorize, and to see the bigger picture within which studies, research, and development is being performed. A useful effect of a framework can be introduction of a means of communication, comparison, evaluation, and synthesis. PAIS mining is concerned with enterprize analysis via automated inspection of organizational work logs. In the fledgling area of PAIS mining, different developers have different terminology and different methodologies. There is a need at this time for synthesis. The stage is set for productive communication, comparison, and combination. We hope this framework helps.

Most PAIS mining research is narrowly aimed at the rediscovery of explicit control flow models. We believe that this approach limits the scope and utility of PAIS mining. It is indeed true that PAIS technology is highly concerned with process execution, analysis, and improvement; but to address these process concerns adequately, it is frequently necessary to take into account the larger picture of social and organizational structures, goals, and resources. Thus, PAIS mining needs to be concerned with gathering and discovering useful information about the organizational processes, and also the social structures that support these processes. We would like to extend the mining domain to give serious effort to mining of data-flow information, of organizational information, of human and social information, and of other perspectives. In this section, we introduce a framework that enables and facilitates PAIS mining in this broader sense.

In general, we consider PAIS mining to be a sub-area of Knowledge Discovery in Data (KDD). KDD is concerned with extracting knowledge from stored data. The KDD process consists of (1) understanding the domain, (2) data selection, (3) data cleaning, (4) data transformation, (5) data mining, and (6) result interpretation/evaluation [23]. Conceiving PAIS mining in these broader terms opens up new vistas of possibility. A particularly important phase forPAIS is the data mining step in KDD. Data mining is the process of fitting models to, or discovering patterns from, stored data [23]. These models can be either statistical or logical. Statistical models are inherently nondeterministic, while logical models are purely deterministic. The selection of a model for data mining primarily depends on the knowledge discovery goal. In general, knowledge discovery goals are either descriptive or predictive. If the knowledge discovery goal is descriptive, the data mining step aims at finding a model that can describe the stored data in a human interpretable form. If the knowledge discovery goal is predictive, data mining aims at discovering a model that is used to predict some future behavior. Some useful techniques for discovering descriptive and predictive models are:

1. Classification: A function that maps data elements into predefined classes.
2. Clustering: A function that maps data elements into their natural classes.
3. Summarization: A function that summarizes the data elements (i.e. mean)
4. Dependency Modeling: A technique that attempts to define structural relationships between data elements.
5. Anomaly Detection: A technique that focuses on detecting patterns that deviate from normative behavior.

Data mining, in the context of PAIS, is concerned with using the entire set of discovery techniques mentioned above to gain useful knowledge about an organizational process from an event log. However, in this paper we will focus on the discovery of a logical descriptive model via the dependency modeling technique.

In order to put the PAIS mining step in KDD in its proper context, we must understand its interfaces. The interfaces to the PAIS mining step are the output of the data transformation step, an event log, and the input of the result interpretation/evaluation step, a PAIS model. When given an event log, `WL`, generated by a set of process instances, `I`, and the knowledge discovery goal of finding a logical descriptive model of `WL`, a PAIS mining algorithm attempts to discover a complete and consistent model, `M`, with respect to `WL`. Completeness of a PAIS model means that the discovered model can describe all of the event sequences and salient relationships in `WL` without simply enumerating them. Consistency means that the discovered workflow model only describes the event sequences and relationships in `WL` (or ones that are "consistent" with `WL`) and does not introduce superfluous or spurious event sequences. Stated more plainly, a PAIS mining algorithm accepts as input an event log and produces as output a complete and consistent PAIS model.

The details of the log are dependent upon the details of the particular family of models. Thus, in the next section, we describe our PAIS meta-model.

5 A PAIS Meta-model

Different organizations have different goals, different resources, and different needs for automated assistance within their business processes. Therefore, different organizations typically need different workflow products, different CSCW tools, different mining techniques, and different models to express different business perspectives. The concept of a meta-model provides a coherent, uniform notation and a set of conceptual tools for creating various models appropriate to various organizations.

A comprehensive PAIS representation language is defined in [24] as a representation language that can be used to express the major organizational perspectives from which to examine a process. The Information Control Net (ICN) is an open-ended, graphical formalism conceived over 25 years ago (by one of the co-authors of this paper) as a family of models for organizational process description, analyses, and implementation[21,22]. In the Collaboration Technology Research Group (CTRG) at the University of Colorado, and the Collaboration Technology Research Lab (CTRL) at Kyonggi University in Korea, there has been ongoing research concerned with ICNs. In this section, we combine and extend ideas of comprehensive PAIS representation languages with ICN concepts to present a meta-model suitable for *multidimensional PAIS mining*.

The heart of the ICN meta-model is the observation that understanding of organizational processes begins with understanding of organizational goals, structures, and resources. Thus, in order to create a specific model of a specific enterprize, a modeler chooses certain objects of interest and structures from an organizational framework, from an organizational schema, and from an organizational net. The organizational framework is used to specify various classes of organizational objects (e.g. goals, constraints, resources, activities). The organizational schema is used to specify the set of mappings over the classes of organizational objects (e.g. who does what, which activities precede which). The organizational net is used to specify the dynamic behavior of an organization.

Within the ICN modeling methodology, basic workflow areas are organized as object sets called dimensions which are then organized into perspectives. Dimensions of interest might include the activities dimension (e.g. tasks done within an organization), the data dimension (e.g. descriptions of what information is used within the organization), the participants dimension (e.g. who are the human employees), and the roles dimension (e.g. job descriptions such as secretary and manager). In our related research document [26], we argue that our models and our research must incorporate more than the above standard dimensions. Our meta-model builds in extensibility to choose models incorporating various perspectives. The following is a partial list of important perspectives for incorporation:

- Functional perspective,
- Structural perspective,
- Dataflow perspective,
- Social perspective,
- Organizational perspective,
- Role/Actor perspective,
- Reputational perspective,
- Cultural perspective,
- Security perspective,
- Political perspective,
- Inter-organizational enterprize perspective.

More formally, a *dimension* is defined as a set of homogeneous objects (e.g. employees), a set of attributes associated with objects (e.g. employees' ages), a set of zero or more automorphisms (relationships, such Abe reports to Bob) on the object set, and a set of constraints (rules that all employees must obey) associated with the automorphisms. When we inter-relate the organizational objects of one or more dimensions, we form an organizational *perspective*. Organizational models are constructed by selecting dimensions of interest, and relating them via mappings (multi-valued relationships) and constraints to form perspectives. A *multidimensional PAIS model* for an enterprize is defined as an interrelated family of models, each depicting a perspective on the enterprize.

For example, in studying PAIS for collective endeavors, it is possible to relate the organizational objects of the activities dimension, activities, to the organizational objects of other dimensions. To the extent that an activity is actually captured in its processes, these types of relationships give insight into what an organization does. As another example, a *data flow perspective* is formed when we impose relationships between three dimensions: activities, data items, and repositories. Data dependence is one useful mapping in this perspective; it reveals the data dependencies of activities. Another example of interest is the *activity assignment perspective* formed by defining a set of relationships between three dimensions: employees, roles, and activities. Depending on the size and nature of an organization, the dimensions involved in the definition of this perspective can vary. For a small organization, with two people and a relatively simple process, it is probably quite adequate and convenient to relate participants directly to the activities they perform. However, in organizations with significant complex interactions, this type of relationship is very impractical [25]; it is more appropriate to relate activities to roles, then relate those roles to participants. Therefore, through one level of indirection, activities are related to participants. Mining this perspective in a small organization is typically easy and intuitive, but in a large complex organization, it may be quite nontrivial. Also, we have found within our research group that formulation of multiple related models of the cultural perspective [26] are challenging, but quite promising.

6 Summary and Conclusion

In summary, people oriented PAISs are important because they focus upon socio-technical issues within information systems. An increasingly large population of diverse humans is interacting with technology these days; it is therefore increasingly important for systems to address issues of person-to-person collaboration. The incorporation of people perspectives into PAIS models is particularly challenging because people processes are complex, semi-structured, and dynamically changing. It is thus necessary to take into account factors that impinge upon the organizational structures, the social context, and the cultural setting.

This leads to the question of "when should an organization move to process aware technology?" There is an issue of goodness of fit of technology to process. Examples have been observed of significant gains in efficiency and effectiveness by the incorporation of process aware systems [6]. For example, in some of these cases, the ability of people to view the current state of the entire process has been extremely valuable. Examples have also been observed of cumbersome inhibition of people work by overly inflexible information systems imposing strict process orderings. In some of these cases, the ability of people to get their work done in a timely fashion has been seriously impeded by unnecessary formality and complexity introduced by the system [8]. In general, process aware technology is most useful in situations of non-trivial multi-person collaborative processes that are regularly followed within a structured stable environment.

In this paper, we discussed a framework and a multidimensional organizational meta-model as a means of analysis and enactment of PAIS for collective endeavors. We hope that in the near future, there is enhanced research on further dimensions and perspectives.

We envision many future benefits of well designed people oriented PAIS technology. Some interesting fledgling examples of this technology are being investigated in research laboratories today. For example, role based information systems can help to partition complexity: affective computing techniques enable virtual agents to actively participate in group communications in a fashion that is familiar and natural to humans, rather than requiring people to learn details of the technology's interface. Also, multimedia and multimodal systems are becoming feasible, available, and useful. As organizations become more distributed and intertwined, we see an increasing need for intelligent process technology. We see exciting research progress, and significantly enhanced technology in the future. A thoughtful marriage of information technology and social science is necessary to produce information systems that are functionally aware, socially aware, culturally aware, and truly process aware.

Acknowledgements. We would like to extend our appreciation to our fellow researchers in the Collaboration Technology Research Group (CTRG) at the University of Colorado, and the Collaboration Technology Research Lab (CTRL) at Kyonggi University in Korea. Also thanks to international colleagues, especially to the program committee members of the PAIS2007 workshop, who have contributed greatly to our insights and research.

References

1. M. Dumas, W.Aalst, and A. Hofstede (eds.). Process-Aware Information Systems. Wiley, 2005.
2. D. Berlo. The Process of Communication. Holt, Rinehart and Winston Inc, 1960.
3. B. Biddle and E. Thomas, editors. Role Theory: Concepts and Research. John Wiley & Sons, Inc., New York, 1966
4. E. Bittner. The concept of organisation. Social Research, 32, 1965. (reproduced in Turner, ed, Ethnomethodology. Harmondsworth: Penguin).
5. G. De Michelis and M.A. Grasso. Situating conversations within the language/ action perspective: The milan conversation model. In Proceedings of the Conference on Computer Supported Cooperative Work - CSCW, pages 89-100, 1994.
6. S. Dustdar. Caramba - A Process-Aware Collaboration System Supporting Ad Hoc and Collaborative Processes in Virtual Teams, In Distributed and Parallel Databases, 15:1, 45-66, Kluwer Academic Publishers, Special Issue on Teamware Technologies, January 2004.
7. F. Flores and J.J. Ludlow. Doing and speaking in the office. In G. Fick and H. Spraque Jr., editors, Decision Support Systems: Issues and Challenges, pages 95-118. Pergamon Press, New York, 1980.
8. C. Ellis and G. Nutt "Multi-Dimensional Workflow" In Proceedings of the Second World Conference on International Design and Process Technology, Society for Design and Process Science, Austin, Texas, December, 1996.
9. H. Garfinkel. Studies in Ethnomethodology. Prentice Hall, New Jersey, 1967.
10. A. Giddens. The constitution of society: outline of the theory of structuration. Polity Press, 1984.
11. M.A.K Halliday. Language as Social Semiotic: The Social Interpretation of Language and Meaning. University Park Press, Baltimore, MD, 1978.
12. G. Mark. Extreme collaboration. Communications of the ACM, 45(6):89-93, June 2002.
13. H. Rittel and W. Kunz. Issues as elements of information systems. Working paper 131, Institut fur Grundlagen der Planung, University of Stuttgart, 1979.
14. H. Rittel and M. Webber. Dilemmas in a general theory of planning. Policy Sciences, 4:155-169, 1973.
15. H. M. Robert. Robert's Rules of Order Revised for Deliberative Assemblies. Scott, Foresman, 1915. Online edition at http://www.bartleby.com/176/.
16. J. Rose and R. H. Hackney. Towards a structurational theory of information systems: a substantive case analysis. In Proceedings of the Hawaii International Conference on Systems Science, Hawaii, 2003.
17. John Searle. Speech acts: An essay in the philosophy of language. Cambridge University, Cambridge, England, 1969.
18. L. Suchman. Plans and Situated Actions: The Problem of Human Machine Communication. Cambridge University Press, Cambridge (UK), 1987.
19. T. Winograd and F. Flores. Understanding Computers and Cognition: A New Foundation for Design. Ablex, Norwood, 1986.
20. Terry Winograd. A language/action perspective on the design of cooperative work. In Proceedings of the 1986 ACM conference on Computer-supported cooperative work, pages 203-220. ACM Press, 1986.
21. C. Ellis. Information Control Nets: A Mathematical Model of Information Flow. Conference on Simulation, ACM Proc. Conf. Simulation, Modeling and Measurement of Computer Systems, pages 225-240. ACM, 1979.

22. C. Ellis. Formal and Informal Models of Office Activity. Information Processing 83, pages 11-22, 1983.
23. U. Fayyad, G. Piatetsky-Shapiro, and P. Smyth. From Data Mining to Knowledge Discovery in Databases. AAAI 97, 1997.
24. S. Jablonski and C. Bussler. Workflow Management: Modeling Concepts, Architecture and Implementation. Thomson Computer Press, 1996.
25. K. Kim and H. Ahn. An EJB-Based Very Large Scale Workflow System and Its Performance Measurement. In Advances in Web-Age Information Management, 2005.
26. C. Ellis, K. Kim, A. Rembert, and J. Wainer. A Cultural Perspect on PAIS. Internal University of Colorado, Department of Computer Science, ICSa Report, 2007.

IMCS: Incremental Mining of Closed Sequential Patterns*

Lei Chang, Dongqing Yang, Tengjiao Wang, and Shiwei Tang

Department of Computer Science & Technology, Peking University, Beijing, China
{changlei,dqyang,tjwang,tsw}@pku.edu.cn

Abstract. Recently, mining compact frequent patterns (for example closed patterns and compressed patterns) has received much attention from data mining researchers. These studies try to address the interpretability and efficiency problems encountered by traditional frequent pattern mining methods. However, to the best of our knowledge, how to efficiently mine compact sequential patterns in a dynamic sequence database environment has not been explored yet.

In this paper, we examine the problem how to mine closed sequential patterns incrementally. A compact structure CSTree is designed to keep the closed sequential patterns, and an efficient algorithm IMCS is developed to maintain the CSTree when the sequence database is updated incrementally. A thorough experimental study shows that IMCS outperforms the state-of-the-art algorithms – PrefixSpan, CloSpan, BIDE and a recently proposed incremental mining algorithm IncSpan by about a factor of 4 to more than an order of magnitude.

1 Introduction

Sequential pattern mining is an important data mining task with broad applications, for example, market and customer analysis, biological sequence analysis, stock analysis and discovering access patterns in web logs. There have been a lot of efficient algorithms proposed to mine frequent sequential patterns [9,2,3,15,1].

Recently, mining compact frequent patterns has become an active research topic in data mining community [12,14,11,13,5]. In these studies, researchers try to solve the interpretability and efficiency problems of traditional frequent pattern mining methods, i.e. the number of all frequent patterns can grow exponentially at low support threshold, and even at relatively high support threshold in dense databases [14,5], due to the well-known downward closure property of frequent patterns.

An effective solution is to mine only closed sequential patterns [12,14], i.e. those containing no supersequence with the same occurrence frequency. Closed

* This work is supported by the NSFC Grants 60473051, 60642004 and 60473072, and the National High Technology Research and Development Program of China Grant 2006AA01Z230.

G. Dong et al. (Eds.): APWeb/WAIM 2007, LNCS 4505, pp. 50–61, 2007.

pattern mining is a lossless compression method and closed pattern mining algorithms [12,14], which make full use of search space pruning techniques, often outperform the algorithms mining the complete set of frequent patterns [9,15].

However, in some domains such as customer behavior analysis, collaborative surgery, stock analysis etc., databases are updated incrementally [4], and sometimes nearly real-time constraints are imposed on the mining process. Existing methods either mine frequent (closed) sequential patterns from scratch [9,12,14], or mine the complete set of frequent sequential patterns incrementally [4,10,8,7,6]. It is obvious that mining frequent (closed) sequential patterns from scratch is not a feasible solution, because the mining task is extremely time-consuming. On the other hand, mining the complete set of frequent sequential patterns incrementally leads to the interpretability problem [11,13,5].

Thus, we explore how to mine closed sequential patterns incrementally in this study, with the goal of alleviating the above problems. Incremental mining of closed sequential patterns is a challenging task. The closed patterns in the original sequence database can become non-closed in the new updated database, and the non-closed patterns can become closed too. How to determine the state switching efficiently is not an easy work. In addition, newly added sequences can make previous non-frequent sequences become frequent; and it raises naturally the following questions: do we need to store all the frequent patterns and potential frequent patterns in memory and is it memory efficient?

To answer these questions, a compact structure CSTree (Closed Sequence Tree) is introduced to keep the closed sequential patterns and other auxiliary information, and an efficient algorithm IMCS (Incremental Mining of Closed Sequential Patterns) is designed to maintain the CSTree when the database is updated incrementally. Extensive experimental results show that IMCS outperforms the frequent (closed) sequential pattern mining algorithms – PrefixSpan, CloSpan, BIDE and a recently proposed incremental mining algorithm IncSpan by about a factor of 4 to more than an order of magnitude.

The remainder of this paper is organized as follows. Section 2 gives some preliminary concepts and formulates the incremental closed sequential pattern mining problem. We introduce the CSTree structure, study its properties and present an enumeration algorithm to construct it in Section 3. In Section 4, the IMCS algorithm is proposed and experimental results are given in Section 5. Related work is introduced in Section 6. Finally, we conclude this paper and give some future work.

2 Preliminary Concepts and Problem Statement

Let $I = \{i_1, i_2, \ldots, i_n\}$ be a set of all **items**. A **sequence** is an ordered list of items[1]. A sequence s is denoted by $\langle s_1 s_2 \ldots s_l \rangle$, where s_i is an item, i.e. $s_i \in I$ for $1 \leq i \leq l$. The number of instances of items in a sequence is called the **length** of the sequence. A sequence $\alpha = \langle \alpha_1 \alpha_2 \ldots \alpha_m \rangle$ is called the

[1] For ease of discussion, in this paper, we only discuss sequences which are list of items (not itemsets). The proposed methods can be easily extended to the itemset case.

ID	Original DB	Inc DB
1	A B D	
2	A C B	D
3	C	A

(a) A sequence database

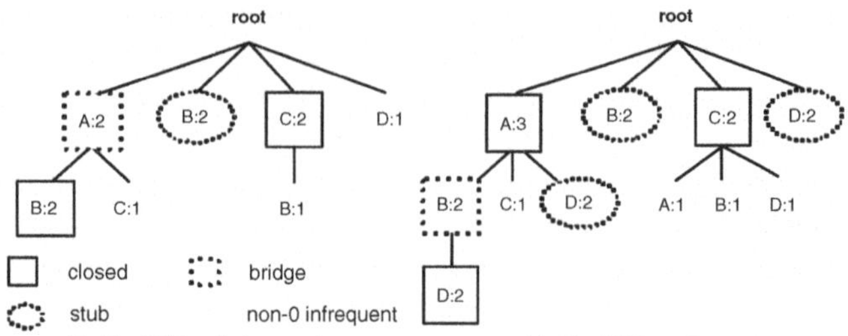

(b) The CSTree before update (b) The CSTree after update

Fig. 1. The sequence database and the corresponding CSTrees

subsequence of another sequence $\beta = \langle \beta_1 \beta_2 \ldots \beta_n \rangle$, if there exist integers $1 \leq i_1 < i_2 < \ldots < i_m \leq n$, such that $\alpha_1 = \beta_{i_1} \wedge \alpha_2 = \beta_{i_2} \wedge \ldots \wedge \alpha_m = \beta_{i_m}$, denoted as $\alpha \sqsubseteq \beta$; and we call β the **supersequence** of α, denoted as $\beta \sqsupseteq \alpha$. If $\alpha \sqsubseteq (\sqsupseteq) \beta$ and $\alpha \neq \beta$, we say that α is the **proper subsequence(supersequence)** of β, denoted as $\alpha \sqsubset \beta$ $(\alpha \sqsupset \beta)$.

A **sequence database** D is a set of sequences, each of which has a unique identifier. The **frequency count** of α in D is the number of sequences (in D) containing α, denoted as $freq_D(\alpha)$. The **support** of α in D is defined as $\alpha's$ frequency count divided by the number of sequences in D, (i.e. $\frac{freq_D(\alpha)}{|D|}$), denoted as $sup_D(\alpha)$. A sequence α is called a **frequent sequential pattern** in D, if $sup_D(\alpha) \geq \sigma$, where σ is a user-specified min_sup value, $0 < \sigma \leq 1$. A frequent sequential pattern α is **closed**, if there does not exist a frequent sequential pattern β, such that $sup_D(\beta) = sup_D(\alpha)$ and $\alpha \sqsubset \beta$.

Given two sequences $\alpha = \langle \alpha_1 \alpha_2 \ldots \alpha_m \rangle$ and $\beta = \langle \beta_1 \beta_2 \ldots \beta_n \rangle$, $s = \alpha \diamond \beta$ means α concatenates with β. α is a **prefix** of s, and β is a **suffix** of s. For example, $\langle AB \rangle$ is a prefix of $\langle ABAD \rangle$ and $\langle AD \rangle$ is its suffix. A **s-projected database** in D [9] is defined as $D_s = \{p \mid s' \in D, s' = r \diamond p$ such that r is the minimum prefix (of s') containing $s\}$. In the above definition, p can be empty.

For a sequence s which contains a sequence q, the **first instance** of the sequence q in s is defined as the minimum prefix p of s such that $q \sqsubseteq p$. For example, the first instance of $\langle AB \rangle$ in $\langle DBAABC \rangle$ is $\langle DBAAB \rangle$. Let $q = \langle q_1 q_2 \ldots q_n \rangle$ and s be a sequence containing q. The ***ith* semi-maximum period of *q* in *s*** [14] is defined as: (1) if $1 < i \leq n$, it is the piece of sequence between the end of the first instance of $\langle q_1 q_2 \ldots q_{i-1} \rangle$ in s and LF_i of q in s (LF_i *of* q *in* s is the last appearance of q_i in the first instance of q in s, and LF_i must appear before LF_{i+1}); (2) if $i = 1$, it is the piece of sequence in s locating before LF_1 of q in s; For example, if $s = \langle ABCB \rangle$, and $q = \langle AC \rangle$, the $2nd$ semi-maximum period of q in s is $\langle B \rangle$.

A sequence database can be updated in two ways [4]. One is inserting new sequences (referred as INSERT), and the other is appending items to the existing sequences (referred as APPEND). INSERT is a special case of APPEND, since INSERT can be regarded as appending items to zero-length sequences. Thus, we only need to consider the APPEND case.

Figure 1(a) shows an example sequence database D (the second column), and an incremental database Δ (the third column). Two sequences in D are appended with new items. The database after update is called an **appended sequence database**, denoted as D'. The **incremental changed database** is defined as $IDB = \{s' \mid s' \in D', \exists s \in D, \delta \in \Delta, s' = s \diamond \delta\}$. For example, in Figure 1(a), $IDB = \{\langle ACBD \rangle, \langle CA \rangle\}$.

Definition 1 (Incremental Closed Sequential Pattern Mining Problem). *Given a sequence database D, an incremental sequence database Δ, and a min_sup value σ, the* ***incremental closed sequential pattern mining problem*** *is to mine the complete set of closed sequential patterns in the appended sequence database D' using the closed sequential pattern information of D.*

3 Closed Sequence Tree

In this section, we first introduce the structure of CSTree, which is designed by us to keep the closed sequential patterns and other auxiliary information. Then, we study some nice properties of CSTree, which will be used in the design of IMCS algorithm. Finally, the CSTree enumeration algorithm is presented.

Each node n in the tree corresponds to a sequence, denoted by s_n, which starts from the root to the node n. The root is a *null* sequence. Figure 1(b) shows the CSTree of the original database D with $min\_sup = 0.5$. Except for the root node, the nodes in a CSTree can be classified into four types, defined as follows. Let the sequence corresponding to a node n be $s_n = \langle \alpha_1 \alpha_2 \ldots \alpha_l \rangle$.

Closed node: If s_n is closed, n is a closed node. For example, in Figure 1(b), the node C at depth 1 and the node B at depth 2 (under A) are two closed nodes.

Stub node: n is a stub node if there exist an integer i ($1 \le i \le l$) and an item e which appears in each of the ith semi-maximum periods of s_n in all the sequences in D which contain s_n. In addition, we restrict that a stub node must be a leaf node in a CSTree. This means we do not extend a stub node further when a CSTree is constructed.

In Figure 1(b), the node B at depth 1 is a stub node, since item A appears in each of the $1st$ semi-maximum periods of $\langle B \rangle$ in the set of sequences $\{\langle ABD \rangle, \langle ACB \rangle\}$. If a sequence α has $\langle B \rangle$ as a prefix, it can not be closed, since $\langle AB \rangle \diamond \alpha$ has the same support as α. Thus, we do not need to extend the node B further.

Bridge node: If s_n is frequent, and n is neither a closed node nor a stub node, n is a bridge node. In Figure 1(b), the node A at depth 1 is a bridge node.

Non-0 infrequent node: n is a non-0 infrequent node if either of the following conditions is satisfied: i) $l = 1 \wedge 0 < sup_D(s_n) < \sigma$; ii) $l \ge 2 \wedge 0 < sup_D(s_n) < \sigma \wedge sup_D(\langle \alpha_1 \ldots \alpha_{l-1} \rangle) \ge \sigma \wedge sup_D(\langle \alpha_1 \ldots \alpha_{l-2} \alpha_l \rangle) \ge \sigma$. In other words, n is a non-0 infrequent node if $n's$ parent node p and the sibling node of p (including p) which has the same item as node n are frequent, and s_n's support is greater than 0 and less than σ. In Figure 1(b), the node D at depth 1, the node C and the node B (under C) at depth 2 are three non-0 infrequent nodes.

Our extensive experiments on various kinds of datasets show that if zero-support infrequent nodes at the tree boundary are kept in the tree, the number of these nodes can range from about 20 to 90 percent of the total number of nodes in the tree. Since these zero-support nodes can be obtained on the fly in IMCS algorithm, we do not keep them in CSTree. Keeping only non-0 infrequent nodes contributes partly to the memory efficiency of our CSTree structure which will be explored further in the experiment section.

CSTree has several nice properties stated below. The detailed proofs are omitted here due to the space limitation.

Property 1 (***closed node state switching***). After appending items to existing sequences in D, an original closed node can change to be a bridge node or stay to be a closed node. It never becomes a stub node.

Property 2 (***stub node state switching***). After appending items to existing sequences in D, if the support of an original stub node does not change, it keeps to be a stub node; and if its support increases, it can stay to be a stub node or change to be a closed node or a bridge node.

Property 3 (***bridge node state switching***). After appending items to existing sequences in D, if the support of an original bridge node does not change, it keeps to be a bridge node; and if the support increases, it can keep to be a bridge node or change to be a closed node. It never becomes a stub node.

Algorithm 1. ConstructCSTree(n, s, D_s, σ)

if $n = root$ **then**
    for each item $i \in I$ **do**
        construct $\langle i \rangle$-projected database $(D_s)_{\langle i \rangle}$;
        if $(sup_{D_s}(\langle i \rangle) > 0)$ **then**
            create a node t, and $t.item \Leftarrow i$, $t.sup \Leftarrow sup_{D_s}(\langle i \rangle)$;
            add t as a child of n;
            call ConstructCSTree($t, \langle i \rangle, (D_s)_{\langle i \rangle}, \sigma$);
else
    if n is a stub node or $0 < n.sup < \sigma$ **then**
        $n.state \Leftarrow$ STUB or NON-0-INFREQUENT respectively;
    else
        if n is a closed node **then**
            $n.state \Leftarrow$ CLOSED;
        else
            $n.state \Leftarrow$ BRIDGE;
        for each child x of $n's$ parent s.t. $x.sup \geq \sigma$ **do** {here let $i = x.item$}
            construct $\langle i \rangle$-projected database $(D_s)_{\langle i \rangle}$
            if $(sup_{D_s}(\langle i \rangle) > 0)$ **then**
                create a node t, and $t.item \Leftarrow i$, $t.sup \Leftarrow sup_{D_s}(\langle i \rangle)$;
                add t as a child of n;
                call ConstructCSTree($t, \langle i \rangle, (D_s)_{\langle i \rangle}, \sigma$);

These properties can be used in IMCS algorithm to accelerate the CSTree extension and state update operations (in Section 4).

Algorithm 1 outlines the pseudo code for constructing a CSTree. We can call ConstructCSTree($root, \phi, D, min\_sup$) to construct a CSTree for a sequence database D, where *root* is the root node (initially no child) of the CSTree.

Line 1-7 construct initial single item projected databases, and call the function recursively. Line 9-10 check if the node is a stub node or a non-0 infrequent node, and update the node's state correspondingly. Line 12-15 check the closedness of the node. Here, we use the BI-Directional Extension closure checking technique introduced in [14]. Line 16-21 generate the candidate frequent sequences and the local projected databases, based on the property that if a sequence $\alpha = \langle \alpha_1, \alpha_2, \ldots, \alpha_l \rangle$ is frequent, $\langle \alpha_1 \alpha_2, \ldots, \alpha_{l-1} \rangle$ and $\langle \alpha_1 \alpha_2, \ldots, \alpha_{l-2} \alpha_l \rangle$ must be frequent. Then the algorithm calls itself recursively. Figure 1(b) and (c) are two CSTrees for the original sequence database and the appended sequence database respectively with $min\_sup = 0.5$.

4 The IMCS Algorithm

In this section, we introduce IMCS algorithm for incremental mining of closed sequential patterns. Algorithm 2 gives the framework of IMCS. IMCS first calls the subroutine UpdateCSTree() to update the supports of the CSTree nodes of the original database D, and at the same time extends the new frequent nodes and the stub nodes which have changed their states. Then, it calls ChangeCSTreeNodeState() to update the node states of the CSTree using a hash table, since some nodes have changed their states (from closed to non-closed or non-closed to closed).

Each of the two subroutines scans the CSTree once. The reason we do not combine the two scans into one is that the states of the new nodes generated in UpdateCSTree() are up-to-date already, and do not need to be updated. Thus, we simply insert them into the hash table H described below and the expense of closedness checking for these nodes is saved.

In UpdateCSTree(), Line 1-4 determine the child nodes of n whose supports can be updated using only IDB (the incremental changed database). If x is a new frequent node, the support of $s_n \diamond \langle x.item \rangle$ in D is not present in the tree. Consequently, we do not include its item in the set $Items$.

Line 6-9 construct corresponding projected databases in IDB, update corresponding nodes' supports, and insert new nodes (frequent or non-0 infrequent) into the tree. To calculate the support change of p_i, we only need to check its projected database in IDB. For a given sequence $\alpha = \alpha_1 \diamond \alpha_2$ in IDB, where α_1 is an old sequence in D, and α_2 is the appended sequence in Δ. If the first instance of s_{p_i} occurs before α_2, then α does not contribute to the support increase of node t; otherwise, it contributes 1 to the support increase of s_{p_i}.

If p_i is a new frequent node or it is a stub node and its support is increased (*Property 2*), IMCS extends it by calling ConstructCSTree() (Line 16).

Before calling ConstructCSTree, we need to construct $(s_{p_i})'s$ projected database in D', called a **full projection** of s_{p_i} (Line 10-13). It is a time-consuming operation. Here, we propose a new full projection computation technique. At the start of IMCS, the full projections of length-1 frequent sequences are precomputed and linked to the corresponding tree nodes. When we need to compute the full projection for node p_i, we first check if there exists a full projection for its parent p. If there does, we calculate the full projection from $p's$ full projection; otherwise, we recursively construct $p's$ full projection, then construct $(p_i)'s$ full projection from $p's$ full projection. The full projections constructed for p_i and its ancestors can be reused (shared) by other new frequent nodes or changed stub nodes in the subtrees rooted at p_i and its ancestors. Line 15-16 call

Algorithm 2. IMCS(n, D, Δ, σ)

1: call UpdateCSTree(n, ϕ, IDB, D', σ);
2: call ChangeCSTreeNodeState(n);

Procedure UpdateCSTree(n, s, IDB_s, D', σ)

1: **if** $n = root$ **then**
2: let $Items$ be the set of all items that appear in Δ;
3: **else**
4: let $Items$ be the set $\{x.item|\ (x = n \vee x$ is a frequent sibling node of $n) \wedge (x$ is not a new frequent node$)\}$;
5: let p_i be the child node of n with item i (if it has one, otherwise $p_i = \phi$ and $s_{p_i} = s_n \diamond \langle i \rangle$);
6: **for** each item $i \in Items$ **do**
7: construct $\langle i \rangle$-projected database $(IDB_s)_{\langle i \rangle}$, and update $sup_{D'}(s_{p_i})$;
8: **if** $sup_{D'}(s_{p_i}) > 0 \wedge p_i = \phi$ **then**
9: create a new node t, $t.item \Leftarrow i$, $t.sup \Leftarrow sup_{D'}(s_{p_i})$, and add t as a child of n;
10: **for** each item $i \in IExt = \{x.item|x$ is a new frequent sibling node of $n\} \cup \{i \in Items|(p_i.state = STUB \wedge sup_D(s_{p_i}) < sup_{D'}(s_{p_i})) \vee (sup_{D'}(s_{p_i}) \geq \sigma \wedge sup_D(s_{p_i}) < \sigma)\}$ **do**
11: construct s_{p_i}-projected database $D'_{s_{p_i}}$;
12: **if** $sup_{D'}(s_{p_i}) > 0 \wedge p_i = \phi$ **then**
13: create a new node t, $t.item \Leftarrow i$, $t.sup \Leftarrow sup_{D'}(s_{p_i})$, and add t as a child of n;
14: **for** each item $i \in IExt \cup Items$ **do**
15: **if** $(sup_{D'}(s_{p_i}) \geq \sigma \wedge sup_D(s_{p_i}) < \sigma) \vee (p_i.state = STUB \wedge sup_D(s_{p_i}) < sup_{D'}(s_{p_i}))$ **then**
16: call ConstructCSTree($p_i, s_{p_i}, D'_{s_{p_i}}, \sigma$), and insert new closed patterns into H;
17: **else if** $sup_{D'}(s_{p_i}) \geq \sigma \wedge p_i.state \neq STUB$ **then**
18: call UpdateCSTree($p_i, s_{p_i}, (IDB_s)_{\langle i \rangle}, D', \sigma$);

Procedure ChangeCSTreeNodeState(n)

1: **if** n is a new frequent node or $n.state = STUB$ or n is a stub node in D but not in D' **then** return;
2: **if** $n \neq root$ **then**
3: **if** $(n.state = CLOSED) \vee (n.state = BRIDGE \wedge sup_D(s_n) < sup_{D'}(s_n))$ **then**
4: insert n to H, update n and the nodes having the same ID-sum as n;
5: **for** each frequent child node t of n **do**
6: call ChangeCSTreeNodeState(t);

ConstructCSTree() to further extend the node p_i, and, at the same time, it inserts the newly found closed sequences into a global hash table H which is empty at the start of IMCS. Here, we use the ID-sum [12] of a sequence s as the hash key, i.e. the sum of the IDs of the sequences in which s appears, and only store in the hash table a pointer pointing to the corresponding node in the CSTree.

In the subroutine ChangeCSTreeNodeState(), Line 1 simply returns if the node is a new frequent node, a stub node or a changed stub node, since the node's state must have been determined in UpdateCSTree(), and the states of the tree nodes under the node have been determined by ConstructCSTree(). In addition, the closed nodes under the node have been inserted into the hash table. Line 3 checks if n's state needs to be updated (*Property 1,3*). If $n.state$=BRIDGE and s_n's support is not increased, its state does not change. Line 4 inserts n into the hash table. For each node t which has the same ID-sum value as n, if $s_n \sqsubset s_t \wedge supp_{D'}(s_n) = supp_{D'}(s_t)$, then $n.state \Leftarrow BRIDGE$, and if $s_n \sqsupset s_t \wedge supp_{D'}(s_n) = supp_{D'}(s_t)$ then $t.state \Leftarrow BRIDGE$.

5 Experimental Results

In this section, we perform a thorough evaluation of the IMCS algorithm on various kinds of datasets, compared with one frequent sequence mining algorithm PrefixSpan, two closed sequence mining algorithms CloSpan and BIDE, and a recently proposed incremental mining algorithm IncSpan.

PrefixSpan and IncSpan were provided as binaries and CloSpan was provided as source code. We implemented BIDE and IMCS in C++. All experiments were done on a 2.8GHz Intel Pentium-4 PC with 512MB memory, running Windows Server 2003. The datasets were produced by using the well-known IBM synthetic dataset generator [2][2]. Please see [2] for more details. In order to test incremental algorithms, from a dataset D', we obtain another dataset D (the original dataset) by cutting v percent of items from the tail of h percent of sequences in D'. v and h are called **vertical ratio** and **horizontal ratio** respectively.

Figure 2 shows the running time of the five algorithms when min_sup is varied from 0.02% to 0.1% on the dataset D10C10T2.5N10, with $v = 10\%$ and $h = 0.5\%$. IMCS runs 4 or more times faster than IncSpan, BIDE and CloSpan, and 11 or more times faster than PrefixSpan. When the min_sup is low, the gap between IMCS and non-closed pattern mining algorithms is much more obvious. For example, with min_sup=0.02%, IMCS completes in 5.98s, while IncSpan and PrefixSpan complete in 100.75s and 319.19s respectively. It is because at extremely low support, there are too many non-closed patterns generated, IMCS can successfully prune the non-closed search space. In comparison with the closed sequence mining algorithms BIDE and CloSpan, IMCS is about 4 to 10 times faster. It is because IMCS does not start its work from scratch, and it simply

[2] We slightly modified the output of the generator (http://www.almaden.ibm.com/cs/quest) such that each item in a sequence is regarded as a single transaction, since we only implemented IMCS for item sequence databases.

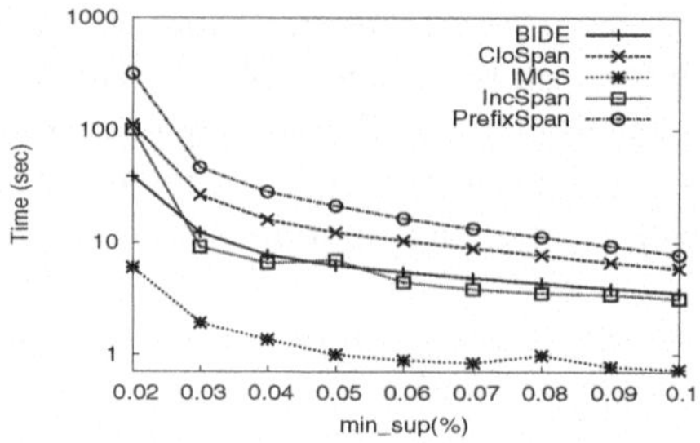

Fig. 2. D10C10T2.5N10

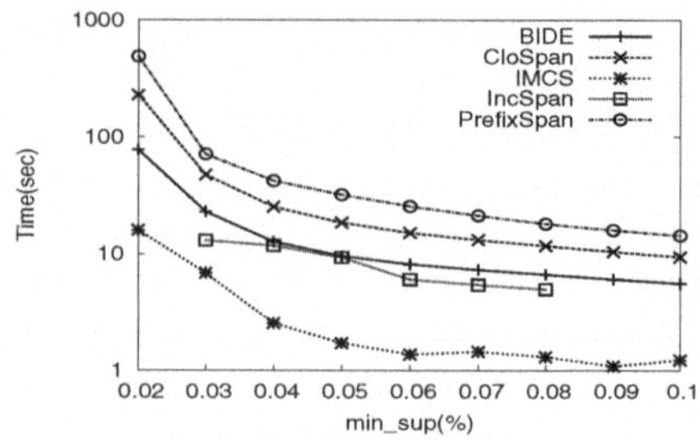

Fig. 3. D10C12T2.5N10

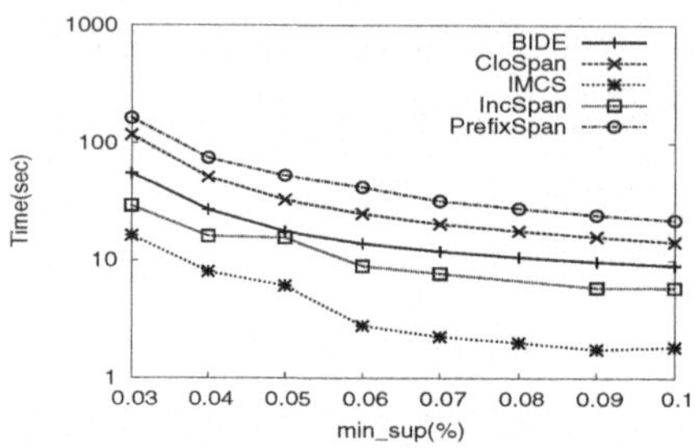

Fig. 4. D10C15T2.5N10

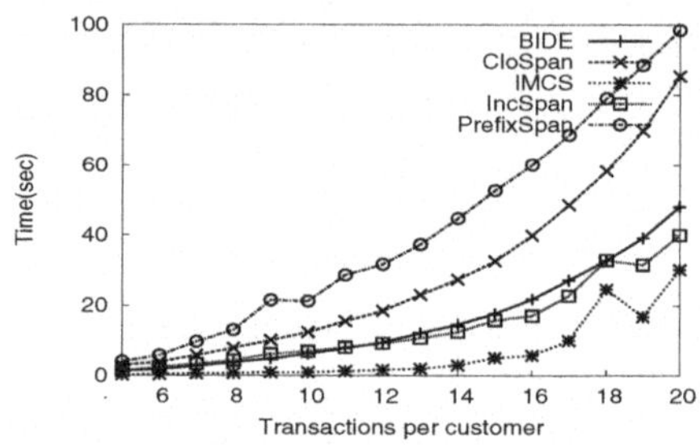

Fig. 5. D10C5-20T2.5N10

checks the incremental changed database, uses the nice properties of CSTree to extend only a few nodes when necessary and to change the states of only a small part of the nodes in the CSTree using a fast hashing technique.

Figure 3 and Figure 4 show the running time of the algorithms on D10C12T2.5-N10 and D10C15T2.5N10 respectively.[3] These figures show the same trend as Figure 2. On D10C12T2.5N10, when $min\_sup = 0.02\%$, IncSpan can not complete because it ran out of memory. With min_sup fixed at 0.05%, the running time of the algorithms is illustrated in Figure 5 when the number of transactions per customer is increased from 5 to 20 (D10C5-20T2.5N10). Figure 6 shows the running time when we varied the number of distinct items (D10C10T2.5N5-15). We can observe that IMCS is always the clear winner over other algorithms.

Figure 7 illustrates how the five algorithms are affected by horizontal ratio on D10C10T2.5N10 with $v = 10\%$. When h exceeds 10%, BIDE outperforms IMCS. It is better to mine the dataset from scratch, because, when the incremental database is too large, the support update of CSTree uses too much time, and too many nodes are needed to be extended. Consequently, the expense IMCS can save does not compensate the extra overhead it brings. In Figure 8, vertical ratio is varied from 0.04 to 0.8 when h is fixed at 2%. All the algorithms show very little variation.

Figure 9 shows the running time of the subroutine ChangeCSTreeNodeState(), compared with the non-closed sequence elimination time of CloSpan. CloSpan

[3] Except explicitly mentioned, the reason for the missing points of IncSpan in the figures of this section is that it terminated abnormally on some datasets.

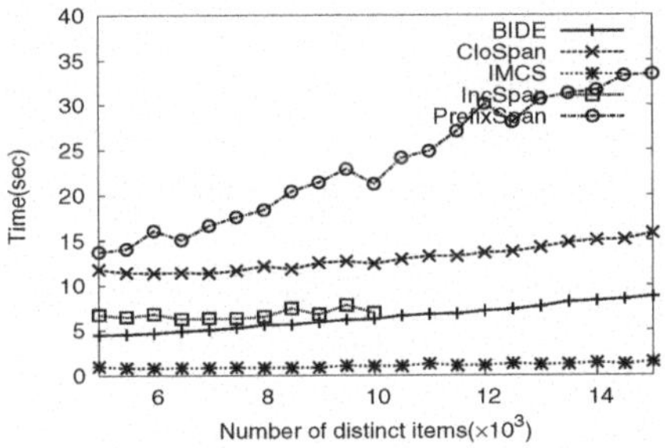

Fig. 6. Varying the number of items

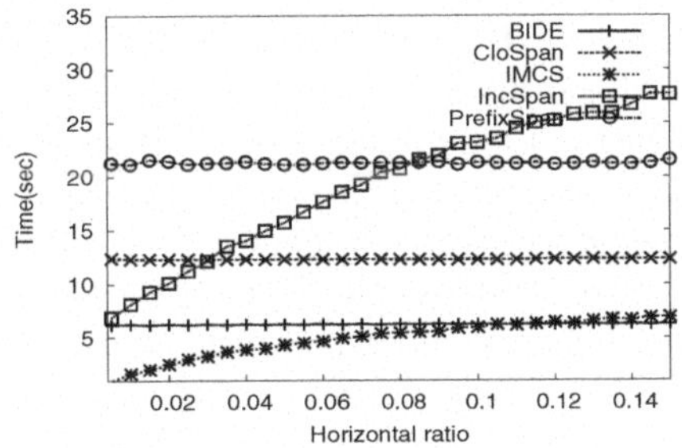

Fig. 7. Varying horizontal ratio

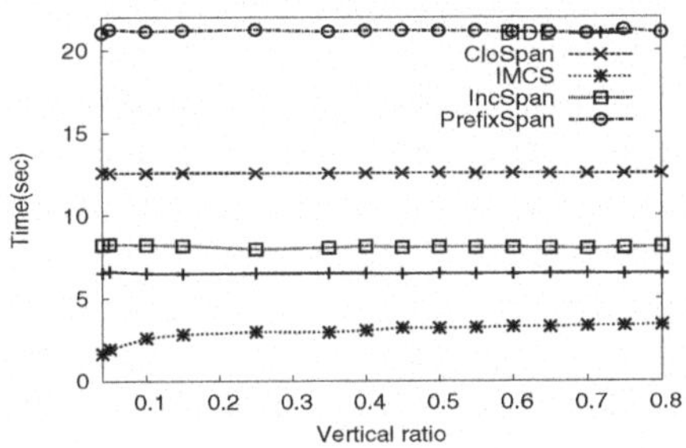

Fig. 8. Varying vertical ratio

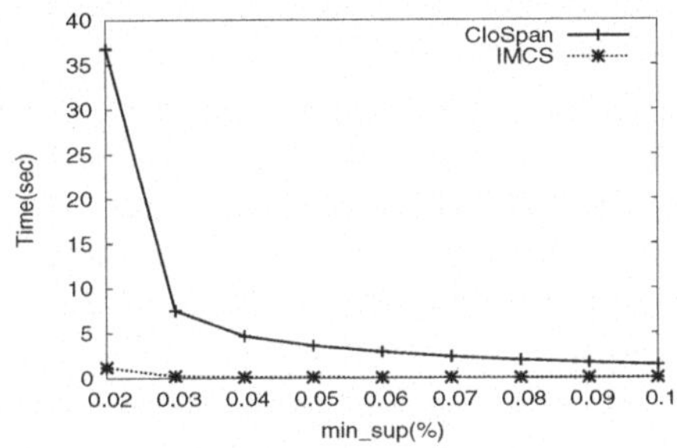

Fig. 9. ChangeCSTreeNodeState() time

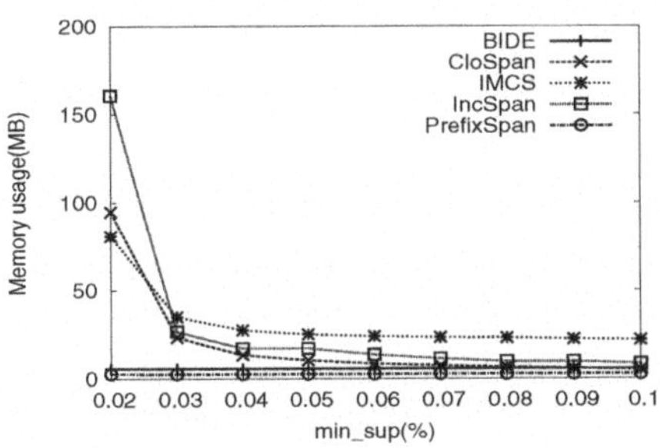

Fig. 10. Memory usage

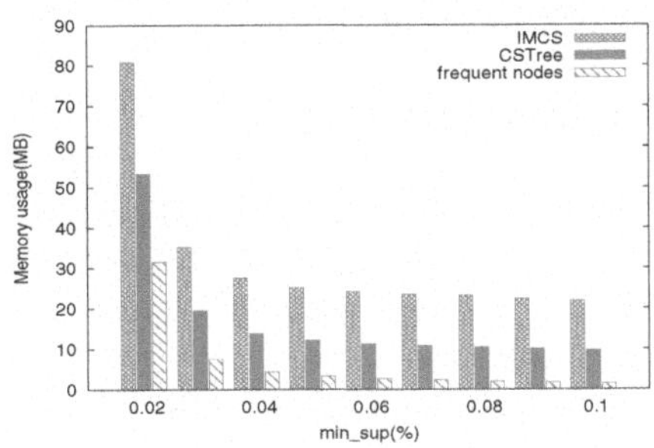

Fig. 11. Memory usage of CSTree

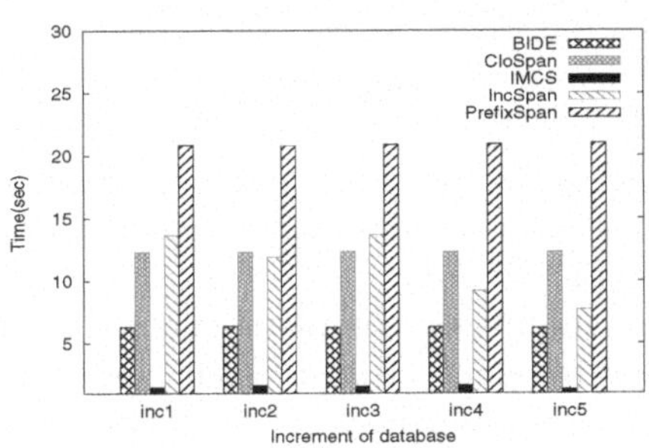

Fig. 12. Multiple increments

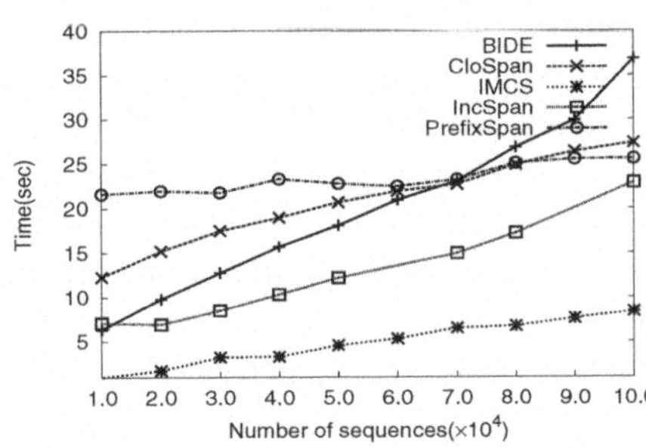

Fig. 13. Varying the number of sequences

also uses a hash structure in its postprocessing phase. Although they are not directly comparable, this figure does show the effectiveness of the nice properties of CSTree in some aspects. A great part of the nodes of CSTree can be skipped and do not need to be inserted into the hash table. This leads to the efficiency of the subroutine ChangeCSTreeNodeState().

Figure 10 shows the memory usage of the five algorithms. Overall, incremental algorithms use more memory, since they need to store the previously discovered patterns. When the *min_sup* value is high, IMCS is not as memory efficient as IncSpan. It is because IMCS needs to maintain more node information than IncSpan. However, when the *min_sup* value is low, IMCS is much better than IncSpan. For example, with *min_sup* = 0.02%, IncSpan uses 168.5MB, while IMCS uses only 84.8MB. It is because there are 3,238,315 frequent (closed and non-closed) patterns IncSpan needs to handle, of which only 310,898 patterns are closed. The memory usage of the CSTree structure is further analyzed in Figure 11. We can observe that as *min_sup* decreases, the percent of memory used by frequent nodes increases, and the percent of memory used by non-0 infrequent nodes decreases. This shows the effectiveness of our strategy of keeping only non-0 infrequent nodes at the tree boundary.

Figure 12 illustrates the performance of the five algorithms to deal with multiple database increments. As the increments accumulate, the algorithms also show a little variation. Overall, they are not affected significantly. We also tested the scalability of the five algorithms (Figure 13). The number of sequences is varied from 10,000 to 100,000 with *min_sup*=0.05%. We can see from the figure that all algorithms scale linearly.

6 Related Work

For non-incremental sequential pattern mining, efficient algorithms, such as GSP [3], PrefixSpan [9], SPADE [15] and SPAM [1] were developed. CloSpan [12] and BIDE [14] are two scalable algorithms for mining closed sequential patterns in static databases.

ISE [8], MFS+ [6] and IncSP [7] are three incremental sequential pattern mining algorithms. They are all based on the candidate-generate-and-test paradigm introduced in [3]. This kind of method suffers from the huge number of candidate patterns and the inefficient support counting operation. Especially for long sequential patterns, the multiple scans of the database can be very costly.

ISM [10] is an interactive and incremental algorithm using vertical format data representation. Based on SPADE [15], ISM maintains in memory a sequence lattice which includes both the frequent sequential patterns and the negative border. In addition, ISM also needs to manage the ID-lists of items/sequences. This leads to the huge memory consumption of ISM [4].

IncSpan [4] is another incremental algorithm mining the complete set of frequent sequential patterns. Based on the intuition that frequent patterns often come from "almost frequent" sequences, IncSpan buffers semi-frequent patterns. However, it is really an expensive operation for IncSpan to decide if a sequence has changed its state from infrequent to semi-infrequent. Furthermore, like other incremental algorithms mining the complete set of sequential patterns, it suffers from the huge memory usage when the *min_sup* threshold is low, or the datasets are dense.

7 Conclusions and Future Work

In this paper, we examined the problem of incremental mining closed sequential patterns in a dynamic environment. A new structure CSTree was employed to keep the closed sequential patterns compactly. Several nice properties of CSTree were studied and used to facilitate the design of the IMCS algorithm. A thorough experimental study on various kinds of datasets has been conducted to show the efficiency of IMCS compared with four other sequence mining algorithms. In the future, we will examine how to extend our algorithm to mine closed sequential patterns in data streams with time, memory and other types of constraints.

References

1. Ayres J., Gehrke J., Yiu T., Flannick J.: Sequential PAttern Mining using A Bitmap Representation. Int. Conf. on Knowledge Discovery and Data Mining. (2002)
2. Agrawal R., Srikant R.: Mining Sequential Patterns. Int. Conf. on Data Engineering. (1995)
3. Agrawal R., Srikant R.: Mining Sequential Patterns: Generalizations and Performance Improvements. Int. Conf. on Extending Database Technology. (1996)
4. Cheng H., Yan X., Han J.: IncSpan: Incremental Mining of Sequential Patterns in Large Database. Int. Conf. on Knowledge Discovery and Data Mining. (2004)
5. Chang L., Yang D., Tang S., Wang T.: Mining Compressed Sequential Patterns. Int. Conf. on Advanced Data Mining and Applications. (2006)
6. Kao B., Zhang M., Yip C., Cheung D. W.: Efficient Algorithms for Mining and Incremental Update of Maximal Frequent Sequences. Data Mining and Knowledge Discovery.(2005)
7. Lin M., Lee S.: Incremental Update on Sequential Patterns in Large Databases by Implicit Merging and Efficient Counting. Information System. (2004)
8. Masseglia F., Poncelet P., Teisseire M.: Incremental Mining of Sequential Patterns in Large Databases. Data & Knowledge Engineering. (2003)
9. Pei J., Han J., Mortazavi-Asl B., Pinto H., Chen Q., Dayal U., Hsu M.: PrefixSpan: Mining Sequential Patterns Efficiently by Prefix-Projected Pattern Growth. Int. Conf. on Data Engineering. (2001)
10. Parthasarathy S., Zaki M. J.,Ogihara M., Dwarkadas S.: Incremental and Interactive Sequence Mining. Int. Conf. on Information and Knowledge Management. (1999)
11. Xin D., Han J., Yan X., Cheng H.: Mining Compressed Frequent-Pattern Sets. Int. Conf. on Very Large Data Bases. (2005)
12. Yan X., Han J., Afshar R.: CloSpan: Mining Closed Sequential Patterns in Large Datasets. SIAM Int. Conf. on Data Mining. (2003)
13. Yan X., Cheng H., Han J., Xin D.: Summarizing Itemset Patterns: A Profile-Based Approach. Int. Conf. on Knowledge Discovery and Data Mining.(2005)
14. Wang J., Han J.: BIDE: Efficient Mining of Frequent Closed Sequences. Int. Conf. on Data Engineering. (2004)
15. Zaki M. J.: SPADE: An Efficient Algorithm for Mining Frequent Sequences. Machine Learning. (2001)

Mining Time-Shifting Co-regulation Patterns from Gene Expression Data*

Ying Yin, Yuhai Zhao, Bin Zhang, and Guoren Wang

Northeastern University, Shengyang, China 110004
yy_00000000@163.com

Abstract. Previous work for finding patterns only focuses on grouping objects under the same subset of dimensions. Thus, an important bio-interesting pattern, i.e. time-shifting, will be ignored during the analysis of time series gene expression data. In this paper, we propose a new definition of coherent cluster for time series gene expression data called ts-cluster. The proposed model allows (1) the expression profiles of genes in a cluster to be coherent on different subsets of dimensions, i.e. these genes follow a certain time-shifting relationship, and (2) relative expression magnitude is taken into consideration instead of absolute one, which can tolerate the negative impact induced by "noise". This work is missed by previous research, which facilitates the study of regulatory relationships between genes. A novel algorithm is also presented and implemented to mine all the significant ts-clusters. Results experimented on both synthetic and real datasets show the ts-cluster algorithm is able to efficiently detect a significant amount of clusters missed by previous model, and these clusters are potentially of high biological significance.

1 Introduction

With the rapid development of microarray technologies, the amounts of high-dimensional microarray data are being generated, which in turn pose great challenges for existing analysis technique. Clustering is one of the most important methods as similar expression profiles imply a related function and indicate the same cellular pathway [1].

Table 1(a) shows an example of microarray dataset, D, consisting of a set of rows and a set of columns, where the rows denote genes, $G = \{g_1, g_2..., g_m\}$, and the columns denote time points with uniform time intervals, $T = \{t_1, t_2..., t_n\}$. Note that the expression value of a gene, g_i, on a certain time point, t_j, is denoted by $d_{i,j}$. For simplicity, certain cells have been left blank in the table. We assume that those blank cells are filled by some random expression values. Table 1(b) is a transposed version of the running example in Table 1(a) after some row permutations, where two different regulation groups emerge. The first one, shadowed and enveloped by a solid polygon, is plotted in Figure 1(b) against every gene's expression profile of it. Similarly, Figure 1(a) corresponds to the second one not shadowed but enveloped by a dashed rectangle. Note: any pair of genes within a regulation group show either coherent patterns or time-shifting coherent patterns.

* Supported by the National Grand Fundamental Research 973 Program of China under Grant No. 2004BA721A05.

G. Dong et al. (Eds.): APWeb/WAIM 2007, LNCS 4505, pp. 62–73, 2007.

Table 1. A Matrix for a Simple Microarray Dataset

(a) Example Microarray Dataset

	t_1	t_2	t_3	t_4	t_5	t_6	t_7
g_1	1.6		3.0	2.2			
g_2	1.0		2.4		1.0	1.5	2.0
g_3	0.6		1.2		0.8	0.9	1.2
g_4		1.5		3.0	2.2		
g_5		0.6		2.0	1.4		
g_6				1.2		2.4	1.7
g_7			0.4		1.0	0.5	
g_8	0.3		0.6		0.3	0.45	0.6
g_9	0.2		0.4		0.2	0.3	0.4

(b) Some Clusters

	t_1	t_2	t_3	t_4	t_5	t_6	t_7
g_1	1.6		3.0	2.2			
g_4		1.5		3.0	2.2		
g_5		0.6		2.2	1.4		
g_7			0.4		1.0	0.5	
g_6				1.2		2.4	1.7
g_2	1.0		2.4		1.0	1.5	2.0
g_3	0.6		1.2		0.8	0.9	1.2
g_8	0.3		0.6		0.3	0.45	0.6
g_9	0.2		0.4		0.2	0.3	0.4

Traditional clustering algorithms work in the full dimensional space, which consider the value of each point in all the dimensions and try to group the similar points together. Unfortunately, most of these conventional clustering algorithms [2,3] do not scale well to cluster high dimensional data sets in terms of effectiveness and efficiency, because of the inherent sparsity of high dimensional data. Biclustering [4, 5], however, does not have such a strict requirement. If some points are similar in several dimensions (a subspace), they will be clustered together in that subspace. This is very useful, especially for clustering in a high dimensional space where often only some dimensions are meaningful for some subset of points.

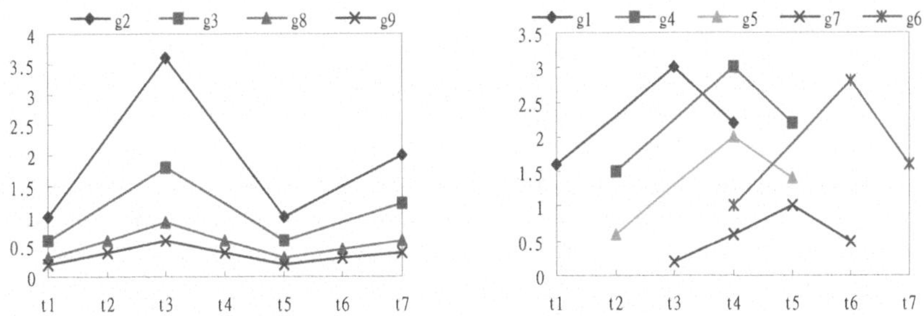

(a) The first regulation group. (b) The second regulation group.

Fig. 1. Two regulation groups

As a step forward, pattern-based biclustering [6] algorithms take into consideration the fact that genes with strong correlation do not have to be spatially close in correlated subspace. However, the existing pattern-based biclustering algorithms are only limited to address pure shifting [7, 8] or pure scaling [9] patterns under the same conditions. For example, the second regulation group of Table 1(b) is visualized in Fig. 1(a). It is a typical pure scaling regulation group since the three patterns are of the relationship: $g_8 = g_3 * 0.5 = g_9 * 3/2$.

Note that since the shifting cluster and the scaling cluster can be transformed mutually [9], we only give the example for scaling pattern here.

Although the previous methods mentioned above prove successful somewhat, they ignore an important bio-interesting pattern implicit in time series microarray data, i.e. *time-shifting pattern*. For example, the expression profiles of genes in the first regulation group of Table 1(b) are illustrated by Figure 1(b), where the five expression profiles of genes g_1, g_4, g_5, g_7 and g_6 show the similar rising and falling patterns but with a successive time-lag. Biologically, the relationship implicit in Figure 1(b) is very interesting since from time series gene expression data it is apparent that most genes do not regulate each other simultaneously but after a certain time-lag [10]. That is, a gene may control or activate another gene downstream in a pathway [11]. In this case, their expression profiles may be staggered, indicating a time-lagged response in the transcription of the second gene [11]. Accordingly, we call the relationship among genes in Figure 1(b) *time-shifting*, which facilitates the study of genetic regulatory networks.

In this paper, we are interested in mining this kind of time-shifting patterns, which is really bio-interesting but have received little attention so far. Current pattern-based models only validate the case when the time-lag is 0, which is just the special case of our time-shifting pattern.

The main contributions of this work are: (1) We propose a new clustering model, namely ts-cluster, to capture not only the coherent patterns but also the time-shifting coherent patterns. It is a generalization of existing time series pattern-based methods; (2) We propose a tree-based clustering algorithm, i.e. FBLD, which discovers all qualified significant ts-clusters in a "**f**irst **b**readth-first and **l**ast **d**epth-first" manner. Further, several novel pruning rules are also designed; (3) We consider the relative expression magnitude instead of the absolute one, which makes the proposed model more flexible and robust; (4) We conducted extensive experimental studies on both real data sets and synthetic data sets to confirm the effectiveness and efficiency of our algorithm.

The remainder of this paper is organized as follows: Section 2 gives some preliminary conceptions and the problem statement. In section 3, we present the ts-cluster model and propose the FBLD algorithm for mining this kind of time-shifting coherent clusters. We also design several advanced pruning rules to improve the performance of the algorithm. Experimental results and analysis are described in Section 4. Finally, we summarize our research in Section 5.

2 The TS-Cluster Model

In this section, we define the ts-Cluster model for mining time-shifting co-regulation patterns.

2.1 The Preliminary and Problem Statement

Let $G=\{g_1, g_2, ..., g_m\}$ be a set of m genes, and $T=\{t_1, t_2, ..., t_n\}$ be a set of n experimental time points with uniform time intervals. A two dimensional microarray time series dataset is a real-valued $m \times n$ matrix $D=G\times T=\{d_{ij}\}$, with $i \in [1, m]$, $j \in [1, n]$, whose two dimensions correspond to genes and times respectively. Each entry d_{ij} records the expression value of gene g_i at time t_j.

Definition 1. *Time Sequence. Let* Y=<$t_{i_1}, t_{i_2}, \cdots, t_{i_l}$> *be a subset of* T, *where* $1 \leq i_1, i_2, \cdots, i_l \leq n$. *We call* Y *a time sequence iff* $i_1 < i_2 <, \cdots, < i_l$ *(i.e.,* $t_{i1} < t_{i2} <, \cdots, < t_{il}$*). The length of* Y *is denoted as* $|Y|=l$.

Definition 2. *L-segment. Let* T= <$t_1, t_2, ..., t_n$> *and a subsequence* Y *be the subsequence of* T. *We call* Y *an L-segment of* T *if its length (the number of elements), denoted* $|Y|$, *is* L+1.

Let Y_p = <$t_{i_1}, t_{i_2}, ..., t_{i_l}$>, and Y_q = <$t_{j_1}, t_{j_2}, ..., t_{j_l}$> be two L-segment time sequences, and $t_{i_1} \leq t_{j_1}$. We say that there is a *time-shifting* relationship between Y_p and Y_q iff we have $j_k = i_k + t'$ for all $k \in [1, l]$, where t' is a constant. That is, Y_q is lagged t' time intervals behind Y_p. Obviously, Y_q is the same as Y_p when $t' = 0$. For unity, we also regard Y_p and Y_q as two time sequences with time-lagged relationship in this case.

For example, given $T=\{t_1, t_2, t_3, t_4, t_5, t_6\}$, then Y_1= <t_1, t_2, t_4> is a 2-segment of T, Y_2= <t_3, t_4, t_6> is also a 2-segment of T and $Y_2 = Y_1 + 2$. So we say Y_2 is lagged 2 time intervals behind Y_1.

Definition 3. $\mathcal{O}$ *Operation. Given a gene* g_x, *a* 1*-segment* <t_i, t_j> *specifies the following three types of regulations, where* d_{xi} *and* d_{xj} *denote the expression values of gene* x *on the time point* t_i *and* t_j *respectively, and* $\delta(>0)$ *is a user-specified regulation threshold:*

(1) up-regulation, *denoted* $\mathcal{O}_x(t_i, t_j) = \nearrow$, *if* $d_{xj} - d_{xi} > \delta$
(2) non-regulation, *denoted* $\mathcal{O}_x(t_i, t_j) = \rightarrow$, *if* $|d_{xj} - d_{xi}| \leqslant \delta$
(3) down-regulation, *denoted* $\mathcal{O}_x(t_i, t_j) = \searrow$, *if* $d_{xj} - d_{xi} < -\delta$

Note: we only consider the genes related to regulation, so those occurring non-regulation are overlooked in this paper.

Definition 4. General Similarity. *Let* g_x *and* g_y *be any two genes in the data set,* Y=<$t_{i_1}, t_{i_2}, ..., t_{i_n}$> *a (n-1)-segment. We say genes* x *and* y *to be general similar if the condition* $\mathcal{O}_x(t_j, t_k) = \mathcal{O}_y(t_{(j+\Delta t)}, t_{(k+\Delta t)})$ *holds, where* $j, k \in \{i_1, i_2, ..., i_n\}$ *and* Δt *is the time-lag.*

Definition 5. *TS-Cluster. Let* $C = \bigcup_{i=1}^{r} X_i Y_i = \{c_{xy}\}$, *where* X_i *is a subset of genes* ($X_i \subseteq G$), *and* Y_i *is a subsequence of time points* ($Y_i \subseteq T$), *then* $X_i \times Y_i$ *specifies a submatrix of* $D = G \times T$. C *is a ts-cluster if and only if: (1)* $\forall Y_i, Y_j$, $1 \leq i, j \leq r$, $|Y_i| = |Y_j|$, *(2)* $\forall Y_i, Y_j, 1 \leq i, j \leq r$, *there is a time-shifting relationship between* Y_i *and* Y_j, *and (3)* $\forall g_x \in X_i$, $\forall g_y \in X_j$, $1 \leq i, j \leq r$, *suppose* Y_j *is lagged* Δt *time intervals behind* Y_i, $\forall t_i, t_j \in Y_i$, *the condition* $\mathcal{O}_x(t_i, t_j) = \mathcal{O}_y(t_{(i+\Delta t)}, t_{(j+\Delta t)})$ *holds.*

For example, Fig 1(b) shows a ts-cluster $C_1 = \{g_1\} \times \{t_1, t_3, t_4\} \cup \{g_4, g_5\} \times \{t_2, t_4, t_5\} \cup \{g_7\} \times \{t_3, t_5, t_6\} \cup \{g_6\} \times \{t_4, t_6, t_7\}$ embedded in Table 1(b). Apparently, their similarity can not be revealed by previous models. In the ts-cluster model, any two genes have a time-shifting co-regulation relationship on their corresponding time sequences, and moreover the previous pattern-based mothod is just a special case of the ts-cluster model when Δt is equal to 0. Although a ts-cluster in Definition 5 represents a time-shifting cluster, our definition can be easily generalized to cover other types of time-lag patterns, such as time shifting-and-inverting pattern etc, just by modifying the third condition of the ts-cluster definition, which determines the types of ts-cluster.

Let $\mathcal{B}$ be the set of all ts-clusters that satisfy the given homogeneity conditions, then $\mathcal{C} \in \mathcal{B}$ is called a *maximal ts-cluster* iff there doesn't exist another cluster $\mathcal{C}' \in \mathcal{B}$ such that $\mathcal{C}$ is contained by $\mathcal{C}'$.

Problem Definition. Given: (1) D, a microarray data matrix, (2) δ, a user-specified maximum regulation threshold, (3) min_t, a minimal number of time points, and (4) min_g, a minimal number of genes, the task of mining is to find all maximal ts-clusters, which satisfy Definition 5 and all the given thresholds.

3 Algorithm for Mining TS-Clusters

The ts-Cluster algorithm has two main steps: (1) Construct initial TS-tree. The co-regulation information and preliminary ts-clusters of all 1-segments are preserved in this step; (2) Develop initial TS-tree recursively to find all maximal ts-Clusters. Unlike the previous algorithms, we take a "first breadth-first and last depth-first" searching strategy, which combines the pruning rules special for pure breadth-first or pure depth-first, to make the algorithm more efficient.

Algorithm 1. FBLD algorithm

Input: a microarray expression matrix D, δ, min_t, min_g

Output: The complete set, M, of maximal TS-Clusters

1: M ← ∅; l=1;
2: Create initial TS-tree for 1-segment with height 2, T_2
3: Applying Pruning 3,4
4: **if** $min_t = 2$ **then**
5:     Insert those in T_2 into M as maximal TS-clusters if they satisfy the conditions
6: **end if**
7: **while** $(l < min_t - 1)$ **do**
8:     Jump to a TS-tree, T_l', with height $l' = min(min_t - 1, 2 * l)$
9:     $l \leftarrow l'$;
10: Construct TS-tree T_l in a breath-first method
11: Applying Pruning 1, pruning 4
12: **end while**
13: **Call DFS**(T_l, l_f)
14: Insert those in T_l into M as maximal TS-clusters if they satisfy the conditions;
15: Return M;

Procedure: DFS(T_l, l_f)
1: **for** the leftmost segment l_f **do**
2:     branch l_f to l_f' as decribed in subsection 3.2
3:     Applying Pruning 2
4:     **if** the result on l_f' is maximal **then**
5:         output it to M
6:     **end if**
7:     DFS(T_l, l_f')
8: **end for**

We give the outline of the algorithm followed by the discussions on every main step. First, we construct the initial TS-tree for 1-segments (line 1-2), which will be discussed in Section 3.1, where pruning rules 3 and 4 are used for trimming nonsignificant ts-clusters and unpromising developments (line 3). Note: Algorithm 1 will find maximal ts-clusters for 1-segments if min_t= 2 (line 4-6). Second, we develop the initial TS-tree recursively (line 7-13), which will be discussed in Section 3.2. There are two steps in the tree-based clustering. In step 1 (line 7-12), we construct a tree in a breadth-first method. Unlike the previous work, it construct a TS-tree containing the minimal required number of conditions as soon as possible via the proposed min_t-based jumping techniques. In step 2 (line 3), we switch to the development of TS-tree with depth-first. In this step, the pruning rule 2, which is designed for filtering those non-maximal TS-clusters, can be used. line 1-8 in Procedure DFS(T_l,l_f) detail the development of TS-tree when the height of tree, l, satisfies the limitation of min_t. Finally, we obtain the complete set, M, of maximal TS-clusters and return it (line 14-15).

3.1 Construct Initial TS-Tree

For clarity, we first look at the appearance of the initial TS-tree, and then describe how it is developed recursively. Figure 2 shows the initial TS-tree constructed from Table 1, which contains the ts-clusters according to Definition 5 for all pairs of time points. There are two branches under each leaf node. One, with '$\nearrow$', represents all genes under it are up-regulated, and the other, with '$\searrow$', represents all genes under it are down-regulated. Each ts-cluster C = $\bigcup_{i=1}^{r} X_i \times Y_i$ is composed of a set of numbered buckets. And we call the buckets with number '0' *baseline buckets* since the time sequence of a baseline bucket (Y_1) is composed of the time points in the path from the root to the node that the ts-cluster C linked to. The number within each bucket denotes the time intervals that Y_i is lagged behind Y_1. For example, in Figure 2, the leftmost ts-cluster under t_1t_3 is composed of five buckets. The time sequence of the baseline bucket (Y_1) is $<t_1, t_3>$, and the time sequence of the second bucket is $<t_2, t_4>$ since the bucket's number is 1. Analogically, the time sequence of the third bucket is $<t_3, t_5>$, and so on.

The process of constructing the initial TS-tree is described as follows:

Step 1. We begin with the time sequence $<t_1, t_2>$, which corresponds to path t_1t_2 in Figure 2, and find the two baseline buckets under it: one with '$\nearrow$', which contains all genes up-regulated on time sequence $<t_1, t_2>$, and the other with '$\searrow$', which contains all genes down-regulated on time sequence $<t_1, t_2>$. Next, it is the two baseline buckets to be extended.

Step 2. Generate baseline buckets for all time sequences $<t_{1+\Delta t}, t_{2+\Delta t}>$ one by one as step 1 does, where $\Delta t \in$ [1,m-2]. After generating the two baseline buckets for each $\{t_{(1+\Delta t)}, t_{(2+\Delta t)}\}$, we need to use these buckets to backwards generate the bucket (1 + Δt -i) for all time sequences $\{t_i, t_{(1+i)}\}$ before $\{t_{(1+\Delta t)}, t_{(2+\Delta t)}\}$, where i $\in$ [1, Δt]. For example, the baseline bucket of t_4t_5, i.e. $\{g_4, g_5\}$, can be used as a bucket with number 3 to extend the ts-cluster with '$\searrow$' under t_1t_2, as a bucket with number 2 to extend the ts-cluster with '$\searrow$' under t_2t_3, and as a bucket with number 1 to extend the ts-cluster with '$\searrow$' under t_3t_4. Note: we need not to generate new buckets. Instead, we only need to keep pointers to the corresponding baseline buckets. Figure 2 shows the logical structure of the initial TS-tree and only baseline buckets exist in main memory.

Step 3. For all time sequences $<t_1, t_i>$ (i=2,3,...,m), repeat the process in step 1 and 2 just as $<t_1, t_2>$ does. After finishing all the above steps, the initial TS-tree is constructed, as Figure 2 shown.

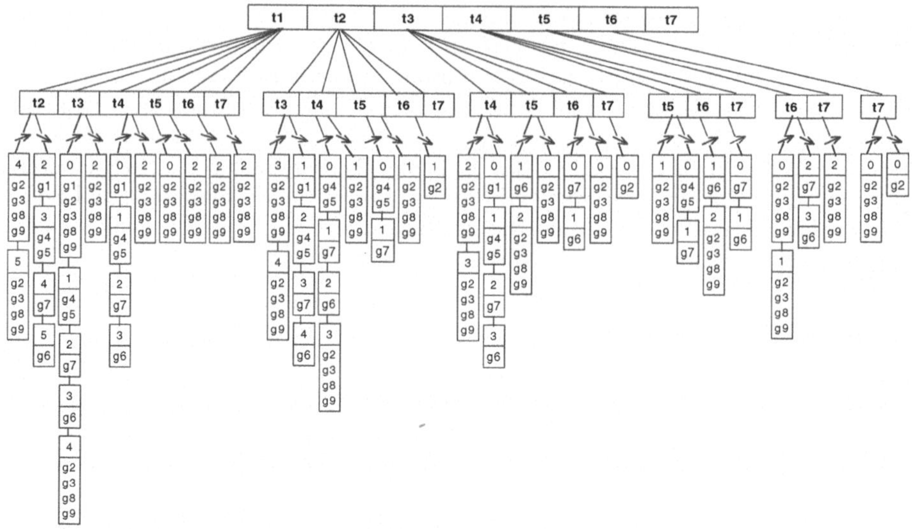

Fig. 2. Initial TS-tree

3.2 Develop TS-Tree Recursively

In this subsection, we introduce how to develop the initial TS-tree recursively to generate all maximal ts-clusters. A TS-tree with height l, represents all l-segments. Here, an l-segment is represented by a corresponding path in TS-tree from an element in the root node to an element in the leaf node.

Unlike the previous algorithms, we propose a "first breadth-first and last depth-first" searching strategy to make the ts-Cluster algorithm more efficient. As mentioned above, the development consists of two phases, i.e. the first phase, BFD(an acronym for "breadth-first development"), and the second phase, DFD (an acronym for "depth-first development").

In BFD phase, different from previous work [9], there is no need to grow TS-tree level by level until the TS-tree is of height min_t. There is a trick during the development that we can skip several levels of TS-tree based on the following min_t-based jumping pruning rule.

Pruning Rule 1. min$_t$-based jumping. *Given a k-segment* $<t_{i_1}, t_{i_2}, ..., t_{i_{k+1}}>$ *and an l-segment* $<t_{j_1}, t_{j_2}, ..., t_{j_{l+1}}>$*, we can directly obtain a* $MIN(min_t, (k+l))$*-segment, jumping over* $(k+1)$*-segment*$\sim MIN(min_t, (k+l-1))$*-segment, if and only if* $t_{i_{k+1}} = t_{j_{l+1}}$.

With Pruning 1, we can quickly jump to the $(min_t - 1)$ level, and meanwhile we can skip over $(k+1)$-segment$\sim MIN(min_t, (k+l-1))$-segment. As shown in Figure 3, we develop the TS-tree in a breadth-first method via min_t-based jumping strategy if min_t=3, as shown above the broken line of Figure 3.

Generally, the proposed technique is efficiency since min_t is usually suitable large for significant co-regulated gene clusters [7]. For example, if min_t=8, we only need to create TS-tree, T_1, T_2, T_4, T_7, with height 1, 2, 4, 7, respectively and jump over the TS-tree, T_3, T_5, T_6, with height 3, 5, 6 levels ,which is efficiently. Note: *the height of initial TS-tree is 1 since it represents all 1-segments.* Once the height of TS-tree grows up to $min_t - 1$, the development switches to the next phase, i.e. DFD.

In DFD, we always develop the leftmost path. Let $T = \{t_{i1}, t_{i2}, \cdots, t_{il}\}$ be the current time sequence to be extended. Let $T_j = \{t_{il}, t_j\}$ be any time sequence that can extend T. Then we create a new node which contains the set of t_j and link the new node to the tail of T (cell t_{il}). Next we generate the ts-clusters for each extended time sequence $T'=\{t_{i1}, t_{i2}, \cdots, t_{il}, t_j\}$.

We continue the process by depth-first method until the rightmost path is developed. DFD allows the following pruning rule, which avoids the non-maximal TS-clusters generated.

Pruning Rule 2. *Given two pathes X and Y of a TS-tree, where X corresponds the segment $<t_{i_1}, t_{i_2}, t_{i_m}>$ and Y corresponds the segment $<t_{j_1}, t_{j_2}, t_{j_n}>$. If $X \subseteq Y$ and the ts-clusters under X is the same as those under Y, then the ts-clusters on $<t_{i_1}, t_{i_2}, t_{i_m}>$ are not maximal and all searches down the path $t_{i_1}, t_{i_2}, t_{i_m}$ can be pruned because they are guaranteed not to contain any maximal ts-cluster.*

FBLD is a hybrid of BFD and DFD. It first develops TS-tree in a breadth-first way. Once the height of TS-tree grows up to $min_t - 1$, the development switches to the next phase, DFD. Pruning rule 1 and pruning rule 2 can be used in FBLD successively, so it outperforms single BFD or single DFD in performance.

3.3 More Pruning Strategies

Besides the pruning rules 1 and 2 mentioned above, the following pruning rules can also be used in FBLD, which are essential for the efficiency.

Pruning Rule 3. *min_g-based pruning. For a ts-cluster linked to a cell (say v), we can prune it if it contains less than min_g genes, as further extension of the corresponding time sequence will only reduce the number of genes in the ts-cluster. Furthermore, we prune the search after v if all the ts-clusters linked to it are pruned.*

Pruning Rule 4. *Pruning short Sequence*
(a) For a time sequence $<t_i, t_j>$ of length 2, let T' be any arbitrary extended time sequence from $<t_i, t_j>$, then according to our algorithm, the expression of T' must be $<t_{il}, ..., t_i, t_j, ..., t_{jl}>$. Thus, the longest extended time sequence from $<t_i, t_j>$ is $T_{max} = <t_1, ..., t_{(i-1)}, t_i, t_j, t_{(j+1)}, ..., t_m>$. If the length of T_{max} is less than min_t (i.e., $i + (m\text{-}j + 1) < min_t$), then $<t_i, t_j>$ cannot lead to any coherent gene cluster having min_t or more times points, and thus all the ts-Clusters on it can be pruned and the search after it can also be pruned.
(b) When constructing the initial TS-tree, we only need to generate the ts-Clusters on $<t_i, t_j>$ when the condition $i + (m\text{-}j + 1) < min_t$ is met, applying pruning rule 4(a). And we can further prune the buckets in the ts-Clusters on these time sequences. Let c be a ts-Cluster on $<t_i, t_j>$ and let b be a bucket in c. Assume the bucket number of b is d,

then the time sequence of b is $<t_{(i+d)}, t_{(j+d)}>$. *Since the longest extended time sequence from* $<t_i, t_j>$ *is* $T_{max}=<t_1, ..., t_{(i-1)}, t_i, t_j, t_{(j+1)}, ..., t_m>$, *the longest time sequence from* $<t_{(i+d)}, t_{(j+d)}>$ *extended along with* $<t_i, t_j>$ *is* $T_{max}=<t_{(1+d)}, ..., t_{(i-1+d)}, t_{(i+d)}, t_{(j+d)}, t_{(j+d+1)}, ..., t_m>$. *If the length of* T_{max} *is less than* min_t *(i.e.,* $i+m-(j+d)+1 < min_t$*), then* $<t_{(i+d)}, t_{(j+d)}>$ *can not lead to any coherent gene cluster having* min_t *or more times, and thus b can be pruned if its bucket number* $d > m - min_t + 1 - (j - i)$.

For example, in Figure 3, the time sequence of the bucket 3 in the leftmost ts-Cluster on $<t_1, t_2>$ is $<t_4, t_5>$, and thus the bucket can be pruned in the case of $min_t = 5$, as the longest time sequence from $<t_4, t_5>$ extended along with $<t_1, t_2>$ is $<t_4, t_5, t_6, t_7>$.

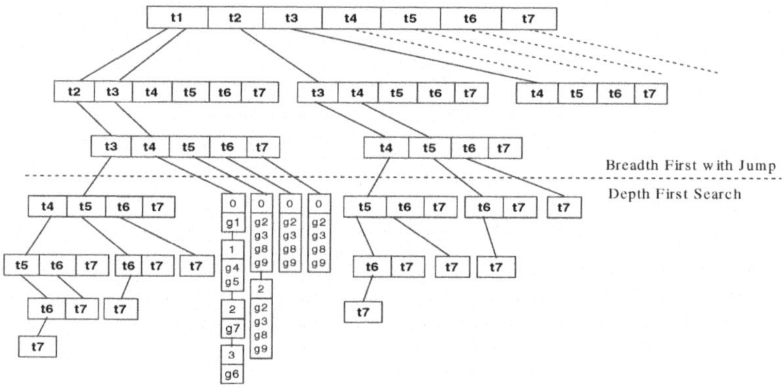

Fig. 3. TS-tree during development

4 Experiments

We implemented and tested our approaches on both real and synthetic microarry data sets in Java. For convenience, the basic breadth-first approach is called BBFS, the basic depth-first approach is called BDFS, and the first breadth-first and last depth-first approach is called FBLD. The tests are conducted on a 2.4-GHz DELL PC with 512 MB main memory running Windows XP.

For the **real dataset** we used a yeast gene expression data from [12], Yeast Datasets contains expression levels of 2884 genes under 17 conditions, which is a subset of [12]. The **synthetic datasets**, which are obtained with a data generator algorithm with three input parameters: k, the number of embedded clusters (#clusters) in the data set; max_g and min_g, the maximal and minimal numbers of genes in a ts-Cluster, respectively. We generate the data sets by setting $min_g = 10$ and $min_t = 5$. max_t is set to the value of 20, and max_g is set to 1000. Clusters are generated by restricting the value of relevant dimensions for each instance in a cluster. Values for irrelevant dimensions are chosen randomly.

4.1 Efficiency

We first evaluate the performance of the three approaches, i.e. BBFS, BDFS and FBLD, on synthetic data sets as we increase the number of genes and the number of time points in the data sets. The average run times of the three algorithms are illustrated in Figure 4 respectively, where we vary the parameters invoked with min_g=30, min_t=6, and δ=0.01.

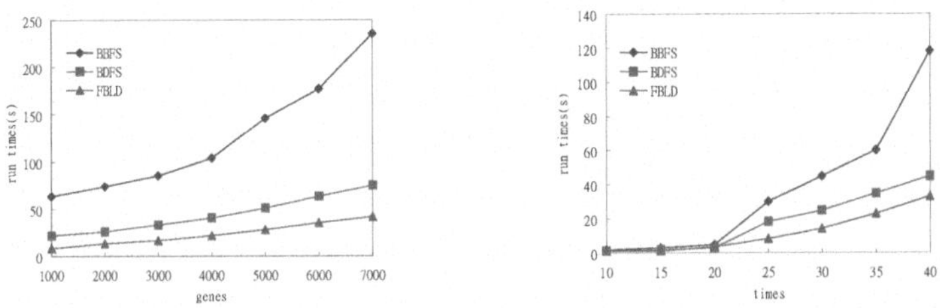

(a) Scalability # genes (b) Scalability # times

Fig. 4. Evaluation of efficiency on synthetic datasets

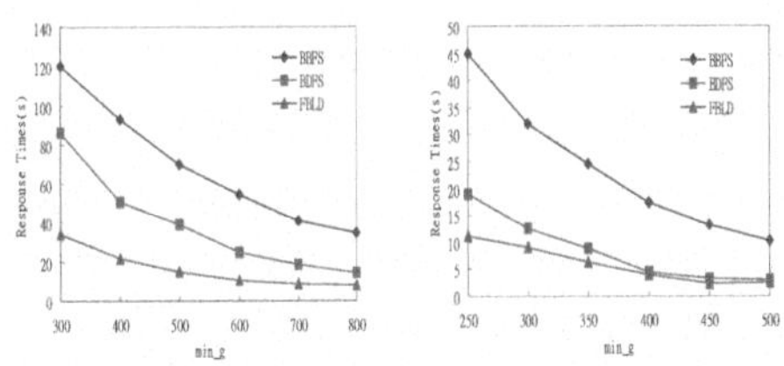

(a) Yeast dataset (b) Synthetic data

Fig. 5. Response time vs. min_g

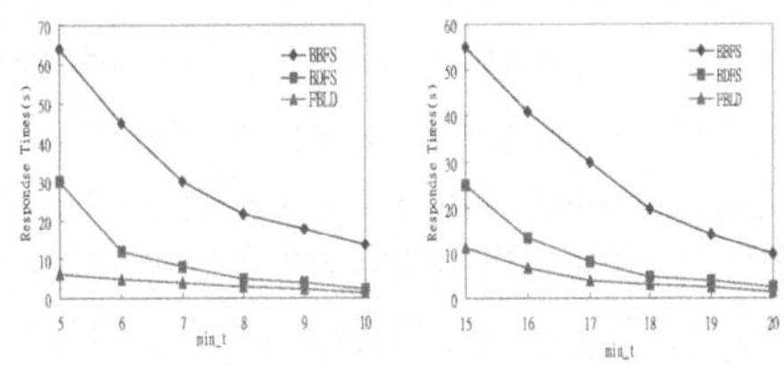

(a) Yeast dataset (b) Synthetic data

Fig. 6. Response time vs. min_t

Figure 4(a) shows the scalability for three approaches under different number of genes, when the number of time points is fixed to 6. Figure 4(b) shows the scalability for three approaches under different number of time points, when the number of genes is fixed to 30. The response time of the mining algorithms is mostly determined by the size of TS-tree. As the number of genes and the number of time points increase, the size of developed TS-tree will get deeper and broader. Hence, the response time will unavoidably become longer. FBLD cut down the search space significantly, so it spends the least response time. BBFS need to decide which buckets(ts-Clusters) can be joined with a given bucket during the development of TS-tree, however, BDFS need not. So BBFS will spend more time than BDFS.

Next, we study the impact of the parameters(min_g and min_t) towards the response time on the real datasets and one synthetic dataset. The results are shown in Figure 5 and Figure 6. As min_g and min_t increase, the response time shortened. According to pruning rule 3, more number of unqualified ts-clusters are eliminated when min_g is larger. As a result, a smaller TS-tree is constructed, which spends less time. A similar effect can be observed w.r.t min_t in Figure 6.

4.2 Effects of the Parameters

The mined ts-cluster is validated w.r.t three parameters, i.e., min_g, min_t and δ. We set their default values as min_g=30, min_t=5 and δ=0.01. Then we test the effect of the parameters on the real GT data set by varying only one parameter while keeping the other one as default. Figure 7 show the effect of each parameter on the number of ts-clusters.

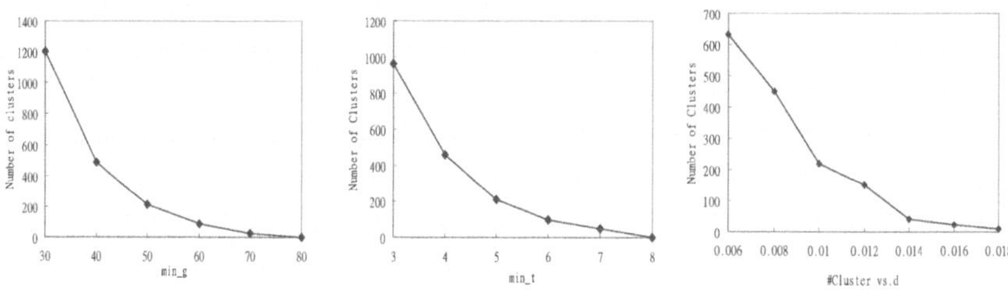

(a) clusters vs. min_g (b) clusters vs. min_t (c) clusters vs. δ

Fig. 7. Effects of parameters on the number of clusters

Interestingly, the two curves in Figures 7(a) and 7(b) share similar trends. That is, when the value of the parameter (represented by the X axis) increases, the number of ts-clusters (represented by the Y axis) goes down. The curve drops sharply until a "knot" is met, then the curve goes stably to the right. For example, we can see the "knots" of min_g=50 in Figure 7(a), min_t=5 in Figure 7(b) and δ=0.01 in Figure7(c). These "knots" indicate that there exist stable and significant ts-clusters in the real data set. They are highly correlated, involving a statistically significant number of genes and time points. The "knots" also suggest the best settings of the parameters to avoid the coherent gene clusters formed just by chance.

5 Conclusions and Future Research Directions

In this work, we proposed a ts-Cluster model for identifying arbitrary time-lagged shifting patterns from time series gene expression data. However, our definition can be generalized to cover time-lagged inverting or other types of time-lagged patterns just by modifying the third condition of the ts-Cluster definition. We have overcome the problem of previous pattern-based biclustering algorithms which can only find pure shifting, scaling or inverting patterns that are just special cases of the ts-Cluster model. And, based on a TS-tree structure, we have developed a "first breadth-first and last depth-first" searching strategy with effective pruning rules to make the maximal ts-clusters mining more efficient. Experimental results prove that our algorithm is able to discover a significantly number of biologically meaningful ts-Clusters missed by previous work.

References

[1] T. R. Hughes, M.J.M.: Functional discovery via a compendium of expression profiles. In: Cell. (2000)
[2] V. Filkov, S.S., Zhi, J.: Analysis techniques for microarray time-series data. In: 5th Annual International Conference on Computational Biology. (2001)
[3] S. Erdal, O.O., Ray, W.: A time series analysis of microarray data. In: 4th IEEE International Symposium on Bioinformatics and Bioengineering, May. (2004)
[4] R. Agrawal, J. Gehrke, D.G., Raghavan, P.: Authomatic subspace clustering of high dimensional data for data mining applications. In: SIGMOD. (1998)
[5] L. Parsons, E.H., Liu, H.: Subspace clustering for high dimensional data: a review. In: SIGKDD. (2004)
[6] Cheng, Y., Church, G.M.: Biclustering of expression data. In: 8th International Conference on Intelligent Systems for Molecular Biology. (2000)
[7] J. Pei, X. Zhang, M.C.H.W., Yu, P.S.: Maple: A fast algorithm for maximal pattern-based clustering. In: ICDM 2003 Conf., Florida,. (2003) 259–266
[8] H. Wang, W. Wang, J.Y., Yu, P.S.: Clustering by pattern similarity in large data sets. In: In SIGMOD. (2002)
[9] Zhao, L., Zaki, M.J.: Tricluster: An effective algorithm for mining coherent clusters in 3d microarray data. In: ACM SIGMOD Conference. (2005)
[10] H. Yu, N. Luscombe, J.Q., Gerstein, M.: Genomic analysis of gene expression relationships in transcriptional regulatory networks. In: Trends Genet. (2003) 19 (8): 422–427
[11] J. Qian, M.D.F.: Beyond synexpression relationships local clustering of time-shifted and inverted gene expression profiles identifies new, biologically relevant interactions. In: Journal of Molecular Biology. (2001)
[12] P. Spellman, G. Sherlock, M.Z., et al: Comprehensive identification of cell cycle-regulated genes of the yeast sacccharomyces cerevisiae by microarray hybridization. In: Molecular Biology of the Cell. (1998) 3273–3297

Tight Correlated Item Sets and Their Efficient Discovery*

Lizheng Jiang[1], Dongqing Yang[1], Shiwei Tang[1,2], Xiuli Ma[2], and Dehui Zhang[2]

[1] School of Electronics Engineering and Computer Science, Peking University
[2] National Laboratory on Machine Perception, Peking University
Beijing 100871, China
{dqyang,tsw}@pku.edu.cn, {jianglz,maxl,dhzhang}@cis.pku.edu.cn

Abstract. We study the problem of mining correlated patterns. Correlated patterns have advantages over associations that they cover not only frequent items, but also rare items. Tight correlated item sets is a concise representation of correlated patterns, where items are correlated each other. Although finding such tight correlated item sets is helpful for applications, the algorithm's efficiency is critical, especially for high dimensional database. Thus, we first prove Lemma 1 and Lemma 2 in theory. Utilizing Lemma 1 and Lemma 2, we design an optimized **RSC** (Regional-Searching-Correlations) algorithm. Furthermore, we estimate the amount of pruned search space for data with various support distributions based on a probabilistic model. Experiment results demonstrate that **RSC** algorithm is much faster than other similar algorithms.

1 Introduction

Association mining and correlation mining has been studied widely and intensively. Frequent item sets and association rule mining will find co-occurrence relationship of frequent items, while correlation mining will find correlated patterns not only restricted to frequent items, but also about those infrequent or rare items. Now, they have been applied to a wide range of applications, such as product analysis, market and customer segmentation, climate studies, gene expression analysis, and so on. However, most existing algorithms are too expensive in run time cost to be practical in some cases, especially for a great number of items. On the other hand, existing methods will produce so many association rules or correlated patterns that it is difficult for user to capture the relationship of these rules or patterns.

In order to overcome the shortcomings of classical association and correlation mining, our research is motivated by the question of devising a concise representation for correlated patterns that will provide user a general view, and designing efficient algorithm that will be practical for applications. For example, in market-basket data set, standard association rule mining algorithm is only useful for the high-support items, such as "*beer* and *diapers*"; correlation mining isn't limited to high-support items, it can also find correlations among infrequent items, such as "*necklace* and *earrings*". Our

* This work was supported by the National Natural Science Foundation of China under Grant No. 60473072, Grant No. 60473051, and the National High Technology Research and Development Program of China (863 Program) 2006AA01Z230.

G. Dong et al. (Eds.): APWeb/WAIM 2007, LNCS 4505, pp. 74–82, 2007.

research tries to go ahead. For example, if we find the correlations: 1) *Bread* is correlated with *cheese*; 2) *Cheese* is correlated with *milk*; 3) *Bread* is correlated with *milk*; we further point out {*bread, cheese, milk*} is a tight correlated item set. In a tight correlated item set, items are correlated each other. Thus, each tight correlated item set includes several correlated items that have common correlated patterns.

Our research is first related to frequent item sets and association rule mining [1]. The classical association rule framework use a minimum *support* threshold ε to control the minimum number of data cases that a rule must cover, and a minimum *confidence* threshold δ to control the predictive strength of the rule. Item sets with *support* above ε are called frequent item sets and will be kept in the mining process, while others are called infrequent item sets and will be discarded. Association rules are derived from those frequent item sets. The first problem for theses settings is the rare item dilemma [2] that is a low minimum *support* will typically generate many rules of no interest, but a high minimum *support* will eliminate the rules with rare items. Another problem is that some association rules with *confidence* above δ are actually misleading [3]. For an example, rule $A \rightarrow B$ is a valid association rule, if its confidence $SUP(AB)/SUP(A) > \delta$. Noticed that $SUP(AB)/SUP(A)$ is probability of A in condition B, $SUP(B)$ is probability of B, if $SUP(AB)/SUP(A) < SUP(B)$, rule $A \rightarrow B$ is wrong.

Recently, efforts have been devoted to address these problems. Correlation mining methods are developed to overcome difficulties of association mining. Instead of discovering frequent patterns in data, correlation mining will find the underlying dependency between items. Mining correlations does not rely on the support measure to perform pruning; thus, correlated patterns are not restricted to frequent patterns, and those infrequent but significant patterns can also be discovered. Brin et al. [3] introduce the measure *lift* (called *interest* there) and χ^2 for mining correlations that is significant in classical statistics. Omiecinski et al. [4] use two interesting measures *all-confidence* and *bond* to find highly correlated but rarely occurring item sets from database. Later, paper [5] show that mining *all-confidence* and coherent patterns generates highly correlated patterns and are better measures than association-based *confidence* measure and two other claimed correlation measures: *lift* and χ^2. Xiong et al. [6] introduce *Pearson's coefficient* as the measure of correlated item pairs and exploit an upper bound for efficient computation. Paper [7] presents 3 new support measures $SUPP_\tau$, $SUPP_\rho$, and $SUPP_F$ to mine rank correlations based on statistical measures. Yiping et al. [9] define *Normalized Mutual Information* (*NMI*) to mine correlations based on information theory.

In general, our research focuses on developing techniques of correlation mining. We propose tight correlated item set to represent items that share common correlated patterns and design an optimized algorithm to find such tight correlated item sets efficiently from large data sets.

In details, our technique involves three aspects of the problem. The first one is about the choice of correlation measure. Suited measures vary from application to application. In this paper, we utilize *Cosine* as the measure of items correlations. In the next sections, we will explain the nature of *Cosine* in terms of conditional probability, and explore *Cosine*'s other properties.

The second aspect is about algorithm's efficiency. Suppose we have a database with m items (columns) and n transactions (rows), there are $C_m^2 = \frac{m*(m-1)}{2}$ item pairs need to be calculated. Obviously, as the number of items increases, the cost for computing correlations becomes expensive. So we explore the relationship between *Cosine* and *support* of items, and prove a *support* boundary for correlated item pairs. According to the *support* boundary, the search space is confined to limited *support* ranges. Thus, we develop a support regional query technique to optimize the algorithm of computing correlation.

The third aspect is about tight correlated item sets. If we get these correlated item pairs, it's easy to know the correlated neighbors for each individual item. But can we get the tight correlated item sets from correlated item pairs easily? Unfortunately, the complexity of algorithm to get the whole tight correlated item sets is NP-hard. However, for many applications, we don't need to get the whole tight correlated item sets. In this paper, we introduce a skill to get tight correlated item sets directly from correlated pairs by increasing correlation threshold θ. The algorithm is fast enough to be applicable to applications with a great number of items. Its price is that the results include only part of the whole tight correlated item sets. As the complement, we explain that any item A and its correlated members form an upper bound of the tight correlated item sets containing item A.

In summary, our work has the following contributions.

1. This paper proposes a method to mine tight correlated item sets from binary database. The tight correlated item set is a concise representation of correlated patterns. Items in a tight correlated item set are tightly correlated each other. Mining tight correlated item sets is useful for a lot of applications.

2. In order to improve algorithm's efficiency, we prove Lemma 1 and Lemma 2 in theory. Lemma 1 is about *support* boundary for correlated item pairs, which can be used to prune unrelated item pairs in correlation computation. Lemma 2 is about transitivity of correlations, which can be used to get tight correlated item sets directly and provide an upper bound of the complete results.

3. Utilizing Lemma 1 and Lemma 2, we design **RSC** algorithm. Analysis and experiments are also performed to examine the efficiency and effects of **RSC** algorithm.

The rest of the paper is organized as the following. Section 2 briefly introduces some basic terminologies and concepts, and explains the reason that we choose *Cosine* as the correlation measure. Section 3 deeply examines the properties of *Cosine* and explains the **RSC** algorithm. Section 5 is the experiment results. In the last section, we summarize our work.

2 Preliminaries

In this section, we present formal definitions and terminology about correlation measure *Cosine* and tight correlated item sets, and describe the concepts briefly.

Let $\mathbf{I}=\{i_1, i_2, \ldots, i_m\}$ be a set of items, and $\mathbf{DB}=<t_1, t_2, \ldots, t_n>$ be a database that consists a set of transactions. Each transaction t_i consists a set of items such that $t_i \subset \mathbf{I}$. We can also consider DB as a $n*m$ matrix with binary values, whose row index is $t_1, t_2, \ldots, t_n$ and column index is $i_1, i_2, \ldots, i_m$. If transaction t_q contains item i_p, then the cell c_{qp} is 1, otherwise the cell c_{qp} is 0. In this view, an item i_s can be represented by a vector $(c_{1s}, c_{2s}, \ldots, c_{ns})$ with n dimensions, as illustrated in Fig. 1.(b). The value in each dimension

Trans	Items	i_1	i_2	i_3	i_4	I_5	i_6
T_1	1, 2, 3	1	1	1	0	0	0
T_2	1, 2, 4, 5	1	1	0	1	1	0
T_3	1, 2	1	1	0	0	0	0
T_4	1, 3	1	0	1	0	0	0
T_5	1, 2, 3, 4	1	1	1	1	0	0
T_6	2, 3, 4, 5	0	1	1	1	1	0
T_7	1,2, 3	1	1	1	0	0	0
T_8	1, 6	1	0	0	0	0	1

$$\begin{array}{c} \\ t_1 \\ t_2 \\ \cdots \\ t_n \end{array} \begin{array}{c} \begin{array}{ccccccc} i_1 & i_2 & \cdots & i_s & \cdots & i_m \end{array} \\ \begin{pmatrix} c_{11} & c_{12} & \cdots & c_{1s} & \cdots & c_{1m} \\ c_{21} & c_{22} & \cdots & c_{2s} & \cdots & c_{2m} \\ \cdots & \cdots & \cdots & \cdots & \cdots & \cdots \\ c_{n1} & c_{n2} & \cdots & c_{ns} & \cdots & c_{nm} \end{pmatrix} \end{array}$$

Fig. 1. (a) Transaction data and matrix. **(b)** Item vectors.

refers to the item's occurrence in the corresponding transaction. The example transaction data and its corresponding matrix are illustrated as the follow (Fig. 1.(a)).

Without leading to misunderstanding, we also use A, B, C, … to distinguish different items, and their vectors are $(a_1, a_2, \ldots, a_n)$, $(b_1, b_2, \ldots, b_n)$, … .

Definition 1. *Cosine* coefficient between items

Given items $A=(a_1, a_2, \ldots, a_n)$ and item $B=(b_1, b_2, \ldots, b_n)$, their *Cosine* coefficient is defined as:

$$Cos(A, B)=\frac{\sum_{i=1}^{n}(a_i * b_i)}{\sqrt{\sum_{i=1}^{n} a_i^2}\sqrt{\sum_{i=1}^{n} b_i^2}} \quad (1)$$

This definition defines the original form of *cos(A,B)*. We can transform it for binary variables.

$$Cos(A, B)=\frac{SUP(AB)}{\sqrt{SUP(A)}\sqrt{SUP(B)}}=\frac{P(AB)}{\sqrt{P(A)}\sqrt{P(B)}}=\sqrt{P(A \mid B)}\sqrt{P(B \mid A)} \quad (2)$$

SUP(A) is the support of item *A*, *SUP(B)* is the support of item *B*, and *SUP(AB)* is the support of $(A \cup B)$ (*A* and *B* happen together). *P(A)* is the probability that item *A* occurs in the database. *P(B)* is the probability of item *B*. *P(AB)* is the probability of $(A \cup B)$. $P(A|B)$ is the probability of *A* under condition *B*. $P(B|A)$ is the probability of *B* under condition *A*.

$0 \leq Cos(A, B) \leq 1$. When *Cos(A, B)* is small, items *A* and *B* have little chance to occur together, and their correlation is weak. When *Cos(A, B)* is big, items *A* and *B* have more chance to occur together, and their correlation is strong. Thus, we chose *Cosine* as measure of correlation between items.

Next, we define correlated item pairs and tight correlated item sets for a user specified threshold θ.

Definition 2. θ-Correlated-Items for item J ($CI_\theta(J)$)

Given item $J \in \mathbf{I}$, $CI_\theta(i_s)=\{X \mid Cos(X, J) \geq \theta, X \in \mathbf{I}\}$.

Definition 3. θ-Correlated-Item-Pair (CIP_θ)

For items $A \in \mathbf{I}$ and $B \in \mathbf{I}$, if $Cos(A, B) \geq \theta$, $\{A, B\}$ is a θ-Correlated-Item-Pair (CIP_θ).

Definition 4. θ-Tight-Correlated-Item-Set ($TCIS_\theta$)

For item set $\mathbf{K}=\{A, B, \ldots\}$ $K \subseteq I$, if $\forall X \in K, Y \in K$, $Cos(X, Y) \geq \theta$, item set K is a θ-Tight-Correlated-Item-Set ($TCIS_\theta$).

From definition 2, 3, 4, we can directly get properties as the follow.

(1) For any item $X \in CI_\theta(J)$, item pair (X, J) is a CIP_θ.

(2) A CIP_θ is a 2-items $TCIS_\theta$, and any item pairs from a $TCIS_\theta$ is a $TCIS_\theta$.

Our target is to mine tight correlated item sets $TCIS_\theta$ from database, such that items in a $TCIS_\theta$ are correlated each other for threshold θ.

3 Techniques of Optimization and RSC Algorithm

In this section, we first introduce the correlation measure *Cosine* and examine its two interesting properties. Using these properties, we design an optimized algorithm to discover tight correlated item sets.

3.1 Correlation Properties

The *Cosine* function is one of the basic functions encountered in trigonometry. It is the ratio of the base to the hypotenuse. In data mining area, *Cosine* often acts as the measure of similarity between vectors. According to the *Cosine* rule, big *Cosine* means small distance. The more *Cosine* between the vectors is, the more similarity they have. In order to apply *Cosine* to binary variables for correlation mining, we first explore the relationship between *Cosine* and *support*.

Lemma 1. Correlation's *support* boundary

Given an item pair $\{A, B\}$, the *support* value $SUP(A)$ for item A, and the *support* value $SUP(B)$ for item B, we have

$$Cos(A, B) \leq \frac{\sqrt{\min(SUP(A), SUP(B))}}{\sqrt{\max(SUP(A), SUP(B))}} \tag{3}$$

Proof: (omitted)

We can utilize Lemma 1 to reduce the search space when computing correlations for item pairs. Without loss of generality, suppose $SUP(A) \geq SUP(B)$, Lemma 1 can be expressed as $Cos(A, B) \leq \frac{\sqrt{SUP(B)}}{\sqrt{SUP(A)}}$. Given a user specified threshold θ, for item A, if $\frac{\sqrt{SUP(B)}}{\sqrt{SUP(A)}} < \theta$, that is $SUP(B) < \theta^2 * SUP(A)$, then $Cos(A, B) < \theta$. That means the items with support below $\theta^2 * SUP(A)$ are unrelated with item A. When computing correlated items for item A, we need only to test those items whose *support* measures are in range of [$SUP(A)$, $\theta^2*SUP(A)$]. In a naive algorithm, the searching range is [$SUP(A)$, 0].

Lemma 2. Correlation's attenuated transitivity

Given items A, B and C, threshold θ, if $Cos(A, B)\geq\theta$, $Cos(A, C)\geq\theta$, We have $Cos(B, C)\geq 2\theta^2-1$ (4)

Proof: (omitted)

Corollary 1. For item A and its correlated item set $CI_\theta(A)$, item set $\mathbf{H}=\{A\}\cup CI_\theta(A)$ is a tight correlated item set in threshold $2\theta^2-1$.

Proof: (omitted)

Corollary 2. For any θ-Tight-Correlated-Item-Set ($TCIS_\theta$) **T**, if item $A\in\mathbf{T}$, we can get $\mathbf{H}=\{A\}\cup CI_\theta(A)$, $\mathbf{T}\subset\mathbf{H}$. Thus, $\{\{X\}\cup CI_\theta(X) | X\in\mathbf{I}\}$ is the upper bound of all $TCIS_\theta$ in item set **I**.

Proof: (omitted)

Using these properties, we design the Regional-Searching-Correlations (**RSC**) algorithm.

3.2 RSC Algorithm

RSC algorithm first scans database to compute support for every item, and then sorts the items by their support in a descending order ($i_1, i_2, \ldots, i_m$).

In the next steps, **RSC** algorithm computes correlated item pairs for items from i_1 to i_{m-1}. According to Lemma 1, correlated item pair has a support boundary. For item i_1 to i_{m-1}, their correlation support boundary are [$SUP(i_1)$, $\theta^2*SUP(i_1)$], [$SUP(i_2)$, $\theta^2*SUP(i_2)$], …, and [$SUP(i_{m-1})$, $\theta^2*SUP(i_{m-1})$].

For an example, consider the transaction data in Fig.1.(a), we first compute each item's support and sort items by their support in a descending order. The sorted items are {1,2,3,4,5,6} (data are selected to simplify the discussion). Given the minimum correlation threshold $\theta=2^{0.5}/2$, $\theta^2=1/2$, for item 1, its *support* is $7/8$; correlation boundary is [$7/8$, $7/16$]; the last item to be examined is item 3. Thus, we only compute *Cosine* for item pairs {1,2} and {1,3}, while pairs {1,4}, {1,5}, {1,6} will be discarded. Similarly, we will process item 2, item 3, and so on. Some intermediate results and the pruned search space are illustrated in Fig. 2.

Item	*Support*
1	7/8
2	6/8
3	5/8
4	3/8
5	2/8
6	1/8

1	2	3	4	5	6
Cosine	$5/42^{0.5}$	$4/35^{0.5}$	NA	NA	NA
		$4/30^{0.5}$	$3/18^{0.5}$	NA	NA
			$2/15^{0.5}$	NA	NA
				$2/6^{0.5}$	NA
					0

Fig. 2. Item's support and correlation, the degree of correlation is represented by color of cell. **Grey**-Pruned pairs; **Powder blue**-Unrelated pairs, **Green**-Correlated pairs.

Recall that the target is to mine tight correlated item sets $TCIS_{\theta}$ for a correlation threshold θ, item i_s and $CI_{\theta}(i_s)$ can't form a $TCIS_{\theta}$ directly. According to corollary 2, our skill is increasing threshold from θ to $\delta = \sqrt{\frac{\theta+1}{2}}$. Running **RSC** algorithm with a new threshold δ, we can get $CI_{\delta}(i_s)$ for every item $i_s \in$ I, thus we can get $TCIS_{\theta}$ from $\{i_s\} \cup CI_{\delta}(i_s)$. On another hand, according to corollary 2, $\{\{i_s\} \cup CI_{\theta}(i_s) | \ i_s \in \mathbf{I}\}$, is the upper bound of $TCIS_{\theta}$. Running **RSC** algorithm with original threshold θ, we can get the upper bound of $TCIS_{\theta}$.

4 Experiment Results

Experiments are performed to evaluate the methods proposed in this paper. Specifically, we will 1) compare the performance of **RSC** algorithm, TAPER algorithm [6] and a naïve approach; 2) demonstrate the pruning efficiency of the regional searching technique; 3) test the scalability of **RSC** algorithm.

In our experiments, we use synthetic data sets and real data sets. Synthetic data set has *m* items and *n* transactions, and it is generated such that the *support* of items is distributed uniformly. Real data sets are *pumsb* and *pumsb** that are often used as benchmark for evaluating the performance of association rule algorithms on dense data sets. *Pumsb* data set has 2113 items and 49046 records. *Pumsb** data set does not contain items with support greater than 80%, and it has 2089 items and 49046 records.

We implement **RSC** algorithm using Microsoft visual c++ 6.0, and run the program on the windows XP platform with a 1.7 GHz CPU and 512 MB main memory.

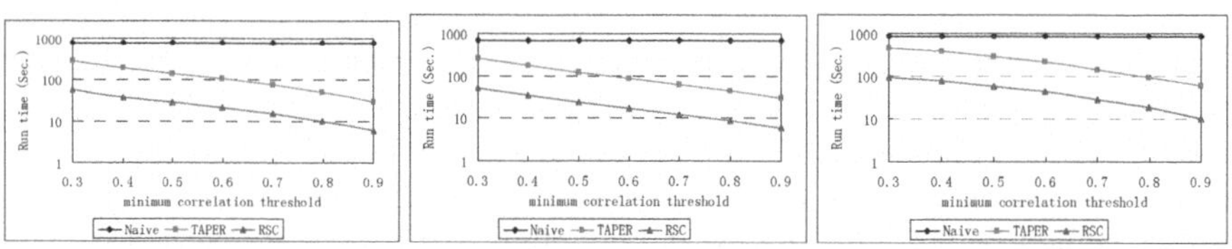

(a) (b) (c)

Fig. 3. **(a)** Pumsb. **(b)** Pumsb*. **(c)** Synthetic.

First, we compare the relative computation performance of **RSC** algorithm with TAPER [6] and a naïve approach on the *pumsb*, *pumsb** and the synthetic data sets (m=5000, n=50000). As can be seen in Fig. 3, the naïve algorithm's run time doesn't change much in different minimum correlation threshold settings for these data sets. The run time cost of **RSC** and TAPER decreases as we increase the threshold θ. Both of **RSC** and TAPER are faster than the naïve approach, and the performance of **RSC** is much better that of TAPER.

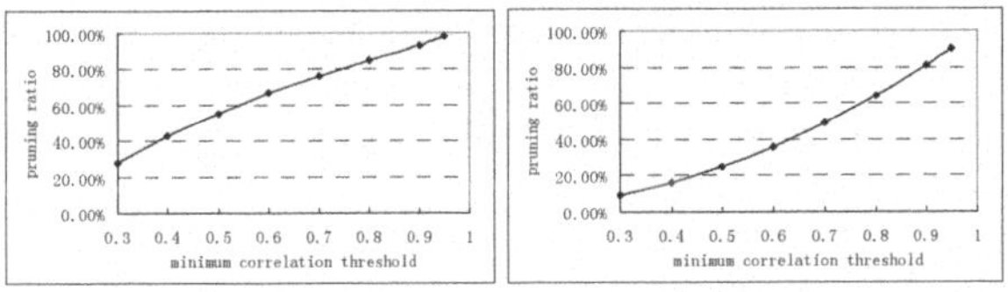

(a) (b)

Fig. 4. (**a**) Pumsb. (**b**) Synthetic.

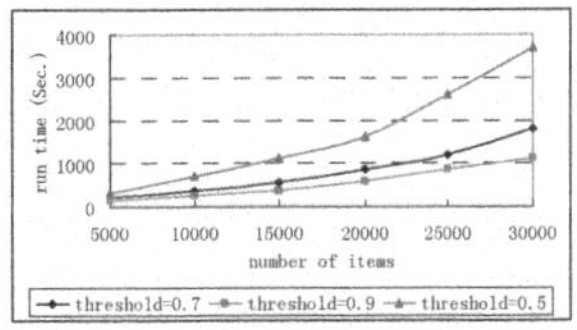

Fig. 5. Run time via number of items

We illustrate the pruning ratio of the regional searching technique in Fig. 4. Fig. 4(a) is on *Pumsb* data set. Fig. 4(b) is on the synthetic data set (m=5000, n=50000). The pruning ration is defined to be: $\frac{number\ of\ pruned\ item\ pairs}{C_m^2}$. We can see that more than 80% pairs are pruned when threshold θ=0.9. This will help us to understand the reason that **RSC** algorithm has high performance.

Fig. 5 is the result of experiment to test the scalability of **RSC** algorithm with respect to database dimensions. It is on the synthetic data set, where we fix the number of transactions n=50000, and change the number of items m from 5000 to 30000. It shows that the run time cost increases nearly in a linear scale with the number of items at a high minimum correlation threshold. **RSC** algorithm has a good scalability for high dimensional database.

5 Conclusion

In this paper, we propose an approach to mine tight correlated item sets that share common correlated patterns. However, mining tight correlated item sets is a challenging problem, especially for a great number of items. So, we emphasize the efficiency of algorithm.

Correlation's support boundary and *Correlation's attenuated transitivity* are two interesting properties of the selected *Cosine* function, which can be utilized to design the algorithm. Using *Correlation's support boundary*, we can determine the unrelated item pairs according to item's support, and then exclude them from computation. Using *Correlation's attenuated transitivity*, we provide user an upper bound of the complete result.

Generally, mining tight correlated item sets will find correlated patterns where items in a tight correlated item set have strong correlations each other, which is helpful for a wide range of applications. This paper provides a method to be practical for such applications.

References

1. R. Agrawal, T. Imielinski, and A. Swami. Mining Association Rules between Sets of Items in Large Databases. In Proc. of 1993 Int. Conf. on Management of Data (SIGMOD'93), pp. 207-216, 1993.

2. B. Liu, W. Hsu, and Y. Ma, "Mining Association Rules with Multiple Minimum Supports," Proc. Knowledge Discovery and Data Mining Conf., pp. 337-341, Aug. 1999.
3. S. Brin, R. Motwani, and C. Silverstein. Beyond market baskets: generalizing association rules to correlations. In SIGMOD, pages 265–276, 1997.
4. E. Omiecinski. Alternative interest measures for mining associations. IEEE Trans. Knowledge and Data Engineering, 15:57-69, 2003.
5. Y.-K. Lee, W.-Y. Kim, Y. D. Cai, and J. Han. Comine: Efficient mining of correlated patterns. In ICDM, page 581, 2003.
6. H. Xiong, S. Shekhar, P. N. Tan, and V. Kumar. Exploiting a Support-based Upper Bound of Pearson's Correlation Coefficient for Efficiently Identifying Strongly Correlated Pairs. In Proc. of 2004 Int. Conf. Knowledge Discovery and Data Mining (KDD'04), pp. 334-343, 2004.
7. Toon Calders, Bart Goethals, Szymon Jaroszewicz: Mining rank-correlated sets of numerical attributes. KDD 2006: 96-105.
8. M. Kendall: Rank Correlation Methods. Oxford University Press, 1990.
9. Yiping Ke, James Cheng, Wilfred Ng: Mining quantitative correlated patterns using an information-theoretic approach. KDD 2006: 227-236.

Improved Prediction of Protein Secondary Structures Using Adaptively Weighted Profiles

Gouchol Pok[1], Keun Ho Ryu[2], and Yong J. Chung[3]

[1] Yanbian University of Science and Technology, Department of Computer Science, Yanji, Jilin Province, China
pok@ybust.edu.cn
[2] Chungbuk National University, Department of Computer Science Cheongju, Chungbuk, Korea
khryu@dblab.chungbuk.ac.kr
[3] Chungbuk National University, Department of Biochemistry Cheongju, Chungbuk, Korea
chungyj@chungbuk.ac.kr

Abstract. Prediction of protein secondary structures from amino acid sequences is a useful intermediate step for further elucidation of native, three-dimensional conformation of proteins. Currently, most predictors are based on machine learning approaches with a short fixed-size input window scanning over the amino acid sequence. The center of the window corresponds to the prediction site where the prediction is performed by utilizing the properties of neighboring amino acid residues. By nature, most machine learning approaches consider feature vectors as position-independent in terms of feature components. As such, for the secondary structure prediction problem, most existing approaches do not take into account the distance of amino acid residues from the center residue. We have studied on how the prediction performance can be affected by imposing different weights on the features according to the distance of residues from the center residue, and in this work, we propose an adaptive weighting scheme to improve prediction accuracy.

1 Introduction

Structural characterization of a protein through prediction is one of the most challenging problems in computational molecular biology. The prediction process itself involves many lesser but still challenging intermediate problems. Among others, one important problem worth of special attention is the prediction of secondary structures from a given amino acid residue sequence. If one knows how a protein is composed of the secondary structural elements, their packing ways give an insight into the possible tertiary structures. The secondary structures refer to the regularly repeated folding patterns sustained by hydrogen bonds and conventionally grouped into three classes: α-helices, β-sheets and coils representing all the other structures without regularities. The arrangement of twenty amino acids in a primary sequence is not randomly composed, but rather shows some kinds of

G. Dong et al. (Eds.): APWeb/WAIM 2007, LNCS 4505, pp. 83–94, 2007.

preferences and correlations between the residues in the sequence[2]. The rule or principle governing these patterns behind the structure formation has been one of the most classical and frequent research issues in computational molecular biology. This is also the area where machine learning approaches have been successfully applied to [3]. The main reason why these machine learning approaches have been widely used is partly because the prediction process is independent of any direct information relating to protein conformation [21]. Following the early works based on the statistical analysis ([6]) and information theory([10]), which are categorized as the first generation methods, the advent of neural networks provided a useful blackbox model to estimate a unknown function mapping the input sequences to secondary structures. [23] [18]. The neural network architecture proposed by Qian and Sejnowski was the fully connected feed-forward network with one hidden layer and a local window of length 13 amino acids. Prediction performance of the approaches based on single sequences and local windows are limited to about 65% accuracy. Qian and Sejnowski also remarked that prediction performance between a single layer network and a network with a hidden layer had no considerable difference. This indicates that the input data do not contain the second order correlations among amino acid residues in the window. The authors experimented further with artificial structures of second order and drew a conclusion that no second order information was present. In addition to this discouraging observation, other main problems with the second generation approaches include that prediction results were not good enough for practical use [29]. In particular, the prediction accuracy for β-sheets was slightly better than a naive guess. When a protein folds into a natural conformation, amino acids that are far away on the primary chain may come into close contact in the tertiary space. This distal relationship can not be captured by a short window, which seems to be a main reason for poor performance in predicting β-sheets. This problem can not be addressed by merely considering longer windows of residues because of the over-fitting problem in training neural networks. Therefore, one needs to devise a different way other than merely increasing the window size when incorporating long range interactions into prediction process. The first method that achieved a prediction accuracy of over 70% is the PHD prediction system by Rost and Sander [26], which later reached up to a level of 74% [27]. While they introduced a number of techniques including early stopping and ensemble of modular networks, the breakthrough in the performance improvement resulted mainly from the use of profiles of the input amino acid sequence derived from the multiple sequence alignment of homologous proteins [2]. As the amino acid sequence profiles contain information pertaining to inter-sequence relationships, one can naturally expect that profile-based approaches outperform the single sequence-based ones. More importantly, as evidenced by the observation that the space of possible amino acid sequences is highly constrained by protein structures, profiles contain a certain kind of long-range information in themselves. However, it should be noted that profiles also discard the information pertaining to intra-sequence correlations among amino acid residues [2]. Once the use of profiles made a breakthrough of 70% level for the prediction accuracy, virtually all new prediction methods since then adopted the input profiles. On the other hand, there have been considerable efforts to further refine the predictor architectures. Riis and Krogh [24] employed the weight-sharing technique to encode the inputs and reduce the number of free parameters. Separate networks are developed for each secondary structure class to take into account a priori knowledge of amino acids within each class and the periodicity of α-helices. This scheme also employed output filtering and ensemble of single structure networks, the primary goal of which was to avoid the over-fitting problem by a careful design of the network topology. Motivated by the observation that adaptive dynamics and non-causal relations are required to overcome the disadvantages of the local fixed-size window, Baldi *et al.* ([2]) proposed a bidirectional recurrent neural network to predict secondary structures. The PSIPRED method designed by Jones ([14]) used an ensemble of feedforward neural networks and position-specific scoring matrices (PSSM) derived from PSI-BLAST profiles. In spite of considerable volume of efforts in architectural design since the seminal PHD method, the prediction performance scored 74% to 77% accuracy in Q_3 measure, which is practically the same level as the PHD ([2]). Other methods that achieved similar performance include NNSP ([30]), PREDATOR ([9]), JPred ([7], [8]), PROF([28]), and SSPro ([22]). Recently, a number of novel architectures using the support vector machine (SVM) have been proposed ([13] [20] [35] [34] [12]). Overall, secondary structure prediction systems have been developed along two kinds of elaboration: one is the refinement of architectural design and the other is the introduction of expressive coding schemes to amino acids.

To the best of our knowledge, all exiting methods implicitly assume the effect of neighboring residues to be uniform regardless of their distance from the center residue. Our proposal to improve the prediction performance is motivated by the hypothesis that performance gain can be achieved by imposing different weights on profiles according to the distance from the center residue. We have also observed that prediction performance is poor at the locations where a second structure class changes to another class. When a Gaussian weight function is applied to the input profiles, the possible negative effects from the boundary area could be alleviated while the local effects from the residues around the center residue is strengthened. We have tested the proposed method using the test data consisting of 513 protein sequences and observed 2 to 5% increase in prediction accuracy.

2 Methods and Algorithms

2.1 Classes of Secondary Structure

One of the most widely used methods for secondary structure assignment is the DSSP algorithm developed by Kabsch and Sander [16]. DSSP uses as input the experimentally determined 3D coordinates of the backbone atoms, and assigns secondary structure based on the hydrogen bonding patterns between carboxyl and amino group. In the DSSP classification, second structures are grouped into eight different classes as shown in Table 1.

Table 1. Eight classes of secondary structure and three simplified classes for prediction

DSSP 8-class	Notation	3-class
α-helix	H	H
3_{10}-helix	G	H
π-helix	I	H
Extended β-strand	E	E
Isolated β bridge	B	C
Turn	T	C
Bend	S	C
Coil	-	C

Conventional way of performance measure is based on the three classes mapped from the eight classes as shown in the right-most column in Table 1.

2.2 Data Set

Preparing a test data should be carefully done to include a representative set of data appropriate for the applications. A data set should reflect the underlying distribution of examples in the sample space and exclude any bias or artificial correlations. In the problem of secondary structure prediction, according to Rost and Sander [25], this means that a typical test suite is prepared so that no two amino acid sequences in the suite have pairwise identity over 25%. Many standardized test suite containing representative samples have been proposed in the literature in order to facilitate a valid comparison of different methods [2]. The test data used in this work are the set of 126 proteins (23,348 amino acids) of Rost and Sander (referred to as RS126) [25] and the complementary set of 396 proteins (62,189 amino acids)of Cuff and Barton (CB396) [8]. The set RS126 was selected based on the percentage identity as above. However, percent identity has been considered as a poor measure in computing sequence similarity, in particular for the sequences with identity of below 30% [8]. To address this drawback, instead of percent identity, the SD score can be used, which is given by $(V - \mu)/\sigma$. Here, V is the alignment score of two sequences computed by a standard dynamic programming [19], μ and σ are the mean and standard deviation of alignment score computed for a set of typically 100 or more randomly re-ordered sequences from each of the two sequences. Accordingly, unlike the percentage identity, SD score reduces the possible effects of bias due to sequence length and amino acid compositions. Cuff and Barton selected a non-redundant set (CB396) based on the SD score to remove redundancy of sequences, and constructed a new test set CB513 by adding two sets CB396 and RS126. After removing 16 correlated proteins (details of which are referred to [8]) from CB513, we have prepared a test suite of 497 protein chains. This data set is randomly partitioned into seven subsets in order to evaluate the prediction performance based on cross-validation.

2.3 Proposed Method: Profiles and Adaptive Weights

Using the set of 497 protein chains described above, we generated profiles following the steps specified by Cuff and Barton [8]. First, a search using the BLAST program ([1]) is performed over the NR (non-redundant) database ([4]) with standard default parameters (E = 10.0, BLOSUM62 matrix), generating a set of similar sequences. The Smith Waterman dynamic programming algorithm ([31]) is applied to the sequences and those whose score is lower than 10^{-4} are discarded. Short sequences are also discarded if their length is less than two-thirds of the target sequence. Sequences longer than 1.5 times the length of the target are truncated by removing the end residues. After the cut-off process, the sequences are then aligned to the target sequence using the CLUSTALW program [32]. In the multiple sequence alignment process, the target sequence is not allowed to contain gaps to preserve its length. A profile is then generated from the aligned sequences by calculating the frequency of occurrences of each residue type at each location along the chain.

When inputs are presented to the predictor program, each amino acid residue i in the input window of length l is encoded as the corresponding profile entry, which is a 21-dimensional vector,

$$\mathbf{f^i} = (f_1^i, f_2^i, \cdots, f_{21}^i), \quad i = 1, 2, \cdots, l \tag{1}$$

Therefore, the input feature vector $\mathbf{f}$ has the dimension of $21 \cdot l$, and represented as,

$$\mathbf{f} = (\mathbf{f^1}, \mathbf{f^2}, \cdots, \mathbf{f^{21}}) = (f_1^1, f_2^1, \cdots, f_{21}^1, \cdots, f_1^{21}, f_2^{21}, \cdots, f_{21}^{21}) \tag{2}$$

All existing prediction methods incorporating the profile as input do not differentiate between short-range interaction and long-distance interaction within the input window. This implies that each $\mathbf{f^i}(i = 1, 2, \cdots, l)$ contributes equal portion to the decision process. Therefore, the vector $\mathbf{f} = (\mathbf{f^1}, \mathbf{f^2}, \cdots, \mathbf{f^{21}})$ is identical to its arbitrary permutation $\mathbf{f}' = (\mathbf{f^{(1)}}, \mathbf{f^{(2)}}, \cdots, \mathbf{f^{(21)}})$ in terms of prediction performance. Instead of this flat model, we propose an adaptive weighting scheme in which the features of the residues proportionally weighted according to their distance from the center residue. The weighted feature vector $\mathbf{g}$ is obtained by imposing weights on feature vectors,

$$\mathbf{g} = (w^1\mathbf{f^1}, w^2\mathbf{f^2}, \cdots, w^{21}\mathbf{f^{21}}) \tag{3}$$

where w^i is a function of distance d from the center residue. The weight function could be any bell-shaped one such as the 1-st order spline, which was defined in this work as the Gaussian function,

$$w^i \equiv \mathbf{w}(d = ||m - i||) = \frac{1}{\sqrt{2\pi}} \exp^{-d^2/2} \tag{4}$$

where $m = \lceil \frac{21 \cdot l}{2} \rceil$, namely the location of the center residue.

2.4 Prediction Scheme: Support Vector Machine

Support vector machine (SVM) is a kernel-based machine learning technique that is capable of classification and functional mapping like traditional feedforward networks [33]. As the SVM is strongly founded on mathematical theory and has many interesting properties such as effective avoidance of overfitting and the ability to deal with high-dimensional features[13], it has been superior in a wide range of practical pattern recognition problems. SVM searches for a model which guarantees the lowest error through two stages as shown in Fig 1. At the first step, the input space is transformed by a non-linear mapping so that it maps the data into linear-separable regions. At the second step, the optimal hyperplanes separating classes are calculated, in such a way as to maximize the margin between two closest training samples.

The non-linear mapping is specified in terms of a kernel function $K(\mathbf{x}, \mathbf{y})$, which is designated in this work by a radial basis function,

$$K(\mathbf{x}, \mathbf{y}) = \exp(-\gamma||\mathbf{x} - \mathbf{y}||^2) \tag{5}$$

where γ is a user-specified parameter. Regularization parameter C controls the trade-offs between margin (model complexity) and classification error, which is specified by the following optimization problem,

$$\min_{w,\xi} \quad \frac{1}{2}\mathbf{w^t w} + C\sum_{i=1}^{n} \xi_i \tag{6}$$

$$\text{s.t.} \qquad y_i[w^t x_i + b] \geqq 1 - \xi_i, \quad \xi_i \geqq 0, \quad i = 1, ..., n \tag{7}$$

where x_i is an input vector, $y_i = \pm 1$ depending on the binary classification of x_i, n is the number of training samples. Based on these values, the SVM method determines a separating hyperplane that optimally divides two classes. Because the SVM is inherently a binary classifier, multi-class classification is performed by combining SVMs in the cascading way. Here, choosing the optimal values of γ and C is essential for each binary SVM classifier [17].

2.5 Performance Measures

After secondary structure prediction is performed for the target sequence, we assess the performance using the following statistics. Let a_{ij} denote the number of structure elements of class i which was classified to be of class j. Let c_i denote the number of structure elements in class i , and p_j the number of structure elements predicted to be class j,

$$c_i = \sum_j a_{ij}, \tag{8}$$

$$p_j = \sum_i a_{ij} \tag{9}$$

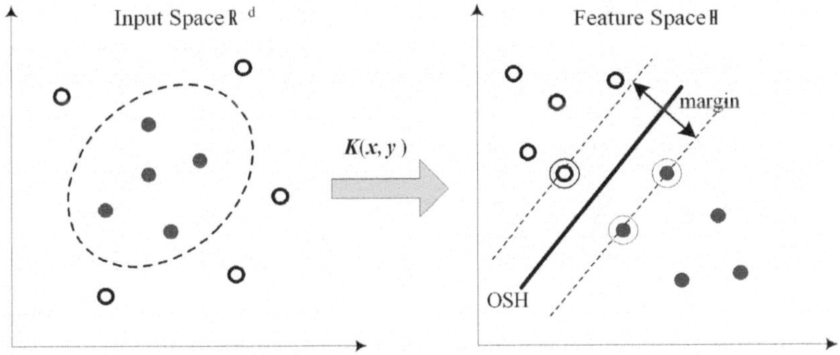

Fig. 1. Non-linear mapping $K(\mathbf{x}, \mathbf{y})$ of the input space $\mathbf{R}^d$ to a feature space H and a separating hyperplane. SVM constructs an optimal separating hyperplane (OSH) which separates two classes (black and white dots) by a maximal margin.Support vectors are represented by bull's eyes.

Then, the total number of residues, N, is defined as,

$$N = \sum_i c_i = \sum_j p_j = \sum_{i,j} a_{ij} \tag{10}$$

Based on these statistics, the performance measures are defined. The most intuitive measure is Q_3, which is the percentage of correctly classified residues regardless of their types,

$$Q_3 = 100 \times \frac{1}{N} \sum_{i=1}^{n} a_{ii} \tag{11}$$

Although Q_3 is intuitively clear, merely relying on this measure can be easily misleading, in particular for the cases where class distribution is heavily biased. Class accuracy, Q_i is a measure to go around the unbalanced distribution of classes,

$$Q_i = 100 \times \frac{a_{ii}}{c_i} \tag{12}$$

where class i can have state H, E, or C. Another widely used measure is Matthew's correlation coefficient,

$$C_i = \frac{TP_i \times TN_i - FP_i \times FN_i}{\sqrt{(TP_i + FN_i)(TP_i + FP_i)(TN_i + FP_i)(TN_i + FN_i)}} \tag{13}$$

where TP_i, FP_i, FN_i, and TN_i are the number of true positives, false positives, false negatives, and true negatives, respectively, in class i. The Matthew's correlation coefficient is calculated separately for each class. The values range from -1 to 1, where 1 refers to complete agreement of prediction with ground truth, -1 complete disagreement, and 0 un-correlation. This measure is useful for comparing with the random baseline, because random prediction according to class distribution would give its value close to zero. In secondary structure prediction, recognition of secondary structure segment is usually more useful than exact prediction of individual residues. For this purpose, Rost and Sander ([27]) proposed the segment overlap measure (SOV) as follows,

$$SOV = \frac{1}{N} \sum_{i \in \{H,E,C\}} \sum_{S(i)} \frac{minov(s_1, s_2) + \delta}{maxov(s_1, s_2)} \times len(s_1) \times 100 \qquad (14)$$

where $S(i)$ is the set of all overlapping pairs of segments (s_1, s_2) where both segments are in state i, $len(s_1)$ is the number of residues in segment s_1, $minov(s_1, s_2)$ is the length where s_1 and s_2 are both in state i, and $maxov(s_1, s_2)$ is the total length spanned by s_1 and s_2.

3 Experiments and Results

3.1 Optimal Parameter Selection

As aforementioned, the performance of binary classifiers and eventually that of the overall classification system depends on the values of parameters γ and C in Eq. (5) and (6). The optimal values for these parameters should be determined according to the performance measure of the final results. Most commonly used performance measures are Q_3 [17] and Matthew's correlation coefficient [5]. The latter has the advantage to give a better indication of the prediction quality because Q_3 tends to favor the large class. The optimal values of γ range from 0.185 to 0.410, and those of C from 0.86 to 1.62 depending on the window size when computed using the CB513 set for training the SVM networks [5]. However, as small variation of these values does not give considerable influence on the prediction performance, we used fixed values for training all SVM classifiers, $\gamma = 0.18$ and $C = 1.43$.

3.2 Classifier Design and Testing

Once the kernel function and optimal values of parameters are determined, six SVM binary classifiers are constructed based on the SVMlight [15]. When designing a classifier with SVM, the number of support vectors indicates how complex a problems is [5], i.e., more number of support vectors means more complex the problem is. Table 2 shows the number of support vectors required for each binary classifier.

All experiments were carried out using the three-fold cross validation method using the CB513 set. The main objective of this work is to investigate the prediction improvement when weighted profiles are used instead of normal profiles. Therefore, we have focused on comparing those two schemes under various conditions rather than on fine tuning the classifiers to obtain higher accuracies in prediction. Table 3 shows the prediction results for both weighted profiles (boldfaced numbers on upper rows) and normal profiles (on lower rows). One can

Table 2. Number of required support vectors for binary classifiers. The numbers are percentage (%) of training examples.

Classifier	H/∼H	E/∼E	C/∼C	E/C	H/E	C/H
Percentage	72.5	69.3	73.1	64.1	63.8	72.9

Table 3. Prediction results of binary classifiers on the CB513 set for different window sizes (l). The numbers in bold face are the results from the weighted profiles scheme compared to those from normal profiles scheme. *ol* denotes the optimal size of a window.

Classifier	$l = 7$	$l = 9$	$l = 11$	$l = 13$	$l = 15$	*ol*
H/∼H	**73.88**	**75.63**	**76.01**	**76.43**	**75.41**	**13**
	71.24	73.56	73.87	73.92	73.11	13
E/∼E	**76.53**	**78.72**	**79.71**	**78.86**	**78.55**	**11**
	73.89	74.21	74.27	73.30	73.54	11
C/∼C	**61.93**	**62.19**	**61.50**	**60.85**	**62.39**	**9**
	62.47	62.64	61.39	61.69	62.12	9
E/C	**71.78**	**71.81**	**71.03**	**70.20**	**70.11**	**9**
	69.37	69.61	69.35	68.94	68.74	9
H/E	**76.67**	**77.92**	**78.46**	**78.27**	**78.25**	**11**
	74.35	74.94	75.89	75.23	75.14	11
C/H	**70.08**	**71.42**	**71.49**	**72.51**	**71.58**	**13**
	69.05	70.34	70.79	71.31	70.96	13

Table 4. Prediction results of tertiary classifiers on the CB513 set. The numbers in bold face are the results from the weighted profiles scheme.

Classifier	Q_3(%)	Q_H(%)	Q_E(%)	Q_C(%)	SOV
SVM_MAX_D	**68.94**	**70.85**	**55.38**	**72.73**	**71.45**
	66.32	67.13	53.97	71.47	70.23
SVM_TREE1	**65.26**	**69.48**	**52.29**	**69.81**	**68.27**
	61.49	64.80	50.35	69.76	66.34
SVM_TREE2	**65.04**	**69.31**	**58.07**	**67.52**	**68.13**
	62.25	65.41	54.58	67.04	65.72
SVM_TREE3	**64.58**	**66.93**	**50.47**	**68.37**	**67.15**
	60.28	63.44	47.81	67.89	**64.56**

see that, when the weighted profiles are used, consistent increase in accuracy is achieved for all the SVM classifiers except the one for C/∼C. This could be explained by the fact that α-helices and β-sheets are formed as regular structure whereas the coiled regions have no specific structure. Due to this lack of meaningful structure, the adaptively weighted profiles, namely emphasizing short-range interactions while slightly suppressing long-distance interactions have no effects on prediction for coiled regions. Consequently, performance gain of the weighted profiles scheme is not consistent, but as shown on the row of C/∼C, the prediction accuracy degrades slightly for the cases of $l = 7, 9$, and 15.

The final prediction is performed using the tertiary classifier which is constructed based on the binary classifiers. Hua and Sun ([13]) proposed a number of ways to construct tertiary classifiers as a cascade of two binary classifiers. For example, in the SVM_TREE1 scheme, patterns first are classified into two classes of α-heix and others, then those in the non-α-helix class are further classified into β-sheet and coils. In the same way, SVM_TREE2 combines E/$\sim$E and C/H classifier, and SVM_TREE3 is made up of C/$\sim$C and H/E classifier. The SVM_MAX_D scheme operates in such a way that a pattern is assigned to the class corresponding to the largest distance from the optimal separating hyperplane (OSH).Table 4 shows the final prediction results for overall prediction (Q_3), state-wise prediction (Q_H, Q_E, and Q_C), and the results measured using SOV. Here again, the weighted profiles schemes perform better for all classifiers and measures except Q_C.

4 Conclusion

We have presented a new method for encoding profiles based on the adaptive weights. The input profiles for amino acid residues are differently weighted according to their distance from the center residue. Our contribution is that we formulated a non-uniform model in which short-range interactions and long-distance interactions are differentiated. Effectiveness of the proposed scheme has been proved by improvement of prediction accuracies up to 5% in Q_3 on the widely used test data set. Moreover, we have observed that regular structure elements such as α-helices and β-sheets are considerably sensitive with the weighted profiles, whereas the coiled region do not show any difference between weighted and normal profiles. For future research, we are going to investigate how the proposed scheme is related to analyzing the boundary of two different structure elements. The shape of the optimal weight function can be another direction of enhancing the current work.

Acknowledgment

This work was supported by the Korea Research Foundation Grant funded by the Korean Government(MOEHRD) under the Regional Research Universities Program/Chungbuk BIT Research-Oriented University Consortium.

References

1. S. F. Altschul, W. Gish, W. Miller, E. W. Myers, D. J. Lipman, "Basic local alignment search tool," *J. olecular Biology* vol. 215, pp. 403-410, 1990.
2. P. Baldi, S. Brunak, P. Frasconi, G. Soda, and G. Pollastri, "Exploiting the Past and the Future in Protein Secondary Structure Prediction," *Bioinformatics*, vol. 15, no. 11, pp.937-946, 1999.
3. P. Baldi and S. Brunak, *Bioinformatics: The Machine Learning Approach*, MIT Press, Cambridge, MA, 1998.

4. H. M. Berman, J. Westbrook, Z. Feng, G. Gilliland, T. N. Bhat, H. Weissig, I. N. Shindyalov, and P. E. Bourne, "The protein data bank," *Nucl. Acid. Res.* vol. 28, pp. 235-242, 2000.
5. J. Casbon, *Protein Secondary Structure Prediction with Support Vector Machines,* MSc Thesis, University of Sussex, 2002.
6. P. Y. Chou and G. D. Fasman, "Prediction of Protein Conformation," *Biochemistry,* vol. 13, pp.222-245, 1974.
7. J. A. Cuff, M. E. Clamp, A. S. Siddiqui, M. Finlay, and G. J. Barton, "JPred: a consensus secondary structure prediction server," *Bioinformatics,* vol. 14, pp. 892-893, 1998.
8. J. A. Cuff and G. J. Barton, "Evaluation and improvement of multiple sequence methods for protein secondary structure prediction," *Proteins,* vol. 34, pp. 508-519, 1999.
9. D. Frishman and P. Argos, "Seventy-five percent accuracy in protein secondary structure prediction," *Proteins* vol. 27, pp. 329-335, 1997.
10. J. Garnier, D. J. Osguthorpe, and B. Robson, "Analysis of the accuracy and Implications of simple methods for predicting the secondary structure of globular proteins," *J. Molecular Biology,* vol. 120, pp.97-120, 1978.
11. M. Gromiha and S. Selvaraj, "Protein Secondary Structure Prediction in Different Structural Classes," *Protein Engineering,* vol. 11, no. 4, pp.249-251, 1998.
12. J. Guo, H. Chen, Z. Sun, and Y. Lin, "A Novel Method for Protein Secondary Structure Prediction Using Dual-Layer SVM and Profiles," *Poteins: Structure, Function, and Bioinformatics,* vol. 54, pp. 738-743, 2004.
13. S. Hua and Z. Sun, "A novel method of protein secondary structure prediction with high segment overlap measure: support vector machine approach," *J. Molecular Biology,* vol. 308, pp. 397-407, 2001.
14. D. T. Jones, "Protein Secondary Structure Prediction Based on Position-specific Scoring Matrices," *J. Molecular Biology* vol. 292, pp. 195-202, 1999.
15. T. Joachims, "SVMlight: Support Vector Machine," **http://svmlight.joachims.org/**.
16. W. Kabsch and C. Sander, "Dictionary of protein secondary structure: pattern recognition of hydrogen-bonded and geometrical features," *Biopolymers,* vol. 22, no. 12, pp. 2577-2637, 1983.
17. H. Kim and H. Park, "Protein secondary structure prediction based on an improved support vector machines approach," *Protein Engineering,* vol. 16, no. 8, pp. 553-560, 2003.
18. D.G. Kneller, F.E Cohen, and R. Langridge"Improvements in Protein Secondary Structure Prediction by an Enhanced Neural Network," *J. Molecular Biology,* vol. 214, pp.171-182, 1990.
19. S. B. Needleman and C. D. Wunsch, "A General Method Applicable tothe Search for Similarities in the Amino Acid Sequence of Two Proteins," *J olecular Biology* vol. 48, pp. 443-453, 1970.
20. M. H. Nguyen and J. C. Rajapakse, "Multi-Class Support Vector Machines for Protein Secondary Structure Prediction," *Genome Informatics,* vol. 14, pp. 218-227, 2003.
21. M. Nordin and M. Sundstrom, "Structural Proteomics: Developments in Structure-to-Function Predictions," *TRENDS in Biochemistry,* vol. 20, no. 2, pp.79-84, 2002.
22. G. Pollastri, D. Przybylski, B. Rost, and P. Baldi, "Improving the Prediction of Protein Secondary Structure in Three and Eight Classes Using Recurrent Neural Networks and Profiles," *Proteins,* vol. 47, pp. 228-235, 2002.

23. N. Qian and T.J. Sejnowski, "Predicting the secondary structure of globular proteins using neural network models," *J. Molecular Biology*, vol. 202, pp865-884, 1988.
24. S. K. Riis and A. Krogh, "Improving prediction of protein secondary structure using structured neual networks and multiple sequence alignment," *J. Comput. Biol.* vol. 3, pp.163-183, 1996.
25. B. Rost and C. Sander, "Prediction of protein secondary structure at better than 70% accuracy," *J. Molecular Biology*, vol. 232, pp. 584-599, 1993.
26. B. Rost and C. Sander, "Improved prediction of protein secondary structure by use of sequence profiles and neural networks," *Proc. Natl. Acad. Sci USA*, vol. 90, pp.7558-7562, 1993.
27. B. Rost and C. Sander, "Combining evolutionary information and neural networks to predict protein secondary structure ," *Proteins*, vol. 19, pp.55-72, 1994.
28. B. Rost, "Better secondary structure prediction through more data," Columbia University, **http://cubic.bioc.columbia.edu/predictprotein**.
29. B. Rost, "Rising accuracy of protein secondary structure prediction," in *Protein structure determination, analysis, and modeling for drug discovery* (ed. D Chasman), New York: Dekker, pp. 207-249, 2003.
30. A. A. Salamov and V. V. Solovyev, "Prediction of protein secondary structure by combining nearest-neighbor algorithms and multiple sequence alignments," *J. Molecular Biology* vol. 247, pp. 11-15, 1995.
31. T. Smith and M. Waterman, "Identification of common molecular subsequences," *J. Molecular Biology*, vol. 147, pp. 195-197, 1981.
32. J. Thompson, D. Higgins, and T. Gibson, "Clustal w: Improving the sensitivity of progressive multiple sequence alignments through sequence weighting, position specific gap penalties and weight matrix choice," *Nucleic Acids Res.*, vol. 22, pp. 4673-4680, 1994.
33. V. Vapnik, *Statistical learning theory,* John Wiley & Sons, New York, 1998.
34. L.-H. Wang and J. Liu, "Predicting Protein Secondary Structure by a Support Vector Machine Based on a New Coding Scheme," *Genome Informatics*, vol. 15, no. 2, pp. 181-190, 2004.
35. J. J. Ward, L. J. McGuffin, B. F. Buxton and D. T. Jones, "Secondary structure prediction with support vector machines," *Bioinformatics*, vol. 19, no. 13, pp. 1650-1655, 2003.

Framework for Building a High-Quality Web Page Collection Considering Page Group Structure

Yuxin Wang and Keizo Oyama

National Institute of Informatics, 2-1-2 Hitotsubashi, Chiyoda-ku, Tokyo, 101-8430 Japan
{mini_wang,oyama}@nii.ac.jp

Abstract. We propose a framework for building a high-quality web page collection considering page group structure in a two-step process: rough filtering and accurate classification. In both processes, we apply the idea of local page group structure. The rough filtering comprehensively gathers all potential homepages from the web with as few noise pages as possible. It uses property-based keyword lists according to four page group models that are based on the page group structure. The accurate classification uses a textual feature set for the support vector machine, which is composed by independently concatenating the feature subsets on the surrounding pages grouped according to the page group structure. Using a combination of a recall-assured classifier and a precision-assured classifier, we build a three-way classifier to accurately select the pages that need manual assessment to assure the required quality. The effectiveness of proposed method is shown by the experimental results.

Keywords: web page collection, page group model, three-way classifier, quality assurance, precision and recall.

1 Introduction

The web is becoming more and more important as a potential information source to add value to high-quality scholarly information services. What is required is a web page collection with guaranteed quality (i.e., high recall and high precision). However, even with automation, this task requires a large amount of manual labor because of the diverse styles, granularities and structures of web pages, the vastness of the web data, and the sparseness of relevant pages.

Many researchers have studied the topic of search and classification of web pages; however, most of these studies are of a best-effort type and pay no attention to quality assurance. Instead, we sought an efficient method to comprehensively build a homepage collection that would assure both high recall and high precision. We mainly focused on researchers' homepages in the present study.

Some studies show that it is generally effective to collect homepages by using various web-related features. Taking into account that a homepage is often represented by a logical page group, the information in the surrounding pages of a page group structure (local link structure) must be considered in addition to the contents of the entry pages. Therefore, we propose a method to utilize features considering page group

G. Dong et al. (Eds.): APWeb/WAIM 2007, LNCS 4505, pp. 95–107, 2007.

structures as a means of building a high-quality homepage collection with the support vector machine (SVM).

Since there are a huge number of web pages, one foreseeable problem with our method is the high computational cost of feature extraction. Therefore, we split the process into two steps: rough filtering for efficiently narrowing down the candidate pages with a very high recall, followed by accurate classification of the candidate target pages output from the rough filtering with both high recall and high precision. Both processes consider the page group structures. Our method achieves high classification performance, especially in terms of recall, and would meet the quality requirements for the collection.

2 Related Work

The method proposed in this paper belongs to the web page classification domain, and closely related to the web page search and clustering domains. In these domains, what information sources to use is the most important consideration and how to use them is the second most important.

The prior works tried to exploit, besides textual contents, various web-related information sources, such as html tags [2, 3], URLs [4, 5], subgraphs of web pages [6, 7], directory structure [7], anchor texts [2, 3, 4], contents of globally link-related pages [8], and contents of local surrounding pages [6, 7]. All of these information sources except the last one are used to capture features that are characteristic to the target pages. The last one, contrarily, is used to collect information dispersed over a logical page group, and while it is used to gather potential pages comprehensively, it tends to increase noise. Since comprehensiveness is a key factor for quality assurance of a web page collection, we mainly exploited the last source.

Some studies on exploiting surrounding pages first classify a page based only on its content and then further use results derived from other information sources such as the link structure and directory structure [7]. However, such an approach does not work when an entry page contains no textual information except hyperlinks. Other studies first cluster web pages based on the local link structure, etc., and then merge

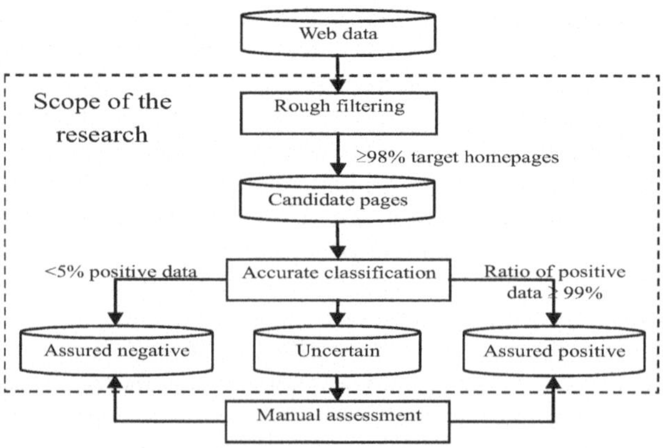

Fig. 1. Framework of the method

the score (or weight) of each word to generate a document vector [6]. However, the effectiveness of this approach is limited, probably because it also merges many irrelevant words from the surrounding pages.

We also exploit the content of surrounding pages by considering the local link structures, but with a different approach. In the rough filtering, to collect as many target homepages as possible, page group models are applied for considering the local link structure, so that the homepages presented on a single page or on a set of pages that constitutes a logical page group can be gathered. In the accurate classification, the features on surrounding pages belonging to different page groups are concatenated, so that the contexts corresponding to the relative location are represented.

Almost no prior study has tried to assure a quality level that would be high enough for practical applications. Our approach to assuring quality is by building a three-way classifier by combining a recall-assured classifier and a precision-assured classifier.

3 Framework

The framework of our method has two steps (Fig. 1): rough filtering and accurate classification.

The rough filtering efficiently gathers the candidate pages with very high recall with limited noise from the web. The input is all the web pages and the output is the candidate pages satisfying the required recall level. We set the rough filtering performance to, for example, at least 98% and desirably 99%. Precision does not matter so much, but fewer output pages is desirable given the condition on recall.

The accurate classification is for accurately classifying the candidate pages output from the rough filtering into three classes: assured positive, assured negative, and uncertain. For example, the recall should be at least 95% and the precision should be at least 99%. Even with the state-of-the-art classification technology, it is impossible to make target data collections of the required quality solely by means of automatic processing. Therefore, human involvement is indispensable for overcoming the gap between the requirement and the technology level. To assure high performance, a relatively high computer processing cost is allowed for the accurate classification compared with the rough filtering, whereas the number of pages that need manual assessment should be reduced as much as possible.

4 Rough Filtering

4.1 Overview of the Rough Filtering

The rough filtering uses property-based keyword lists and several page group models. Fig. 2 illustrates the conceptual overview of the rough filtering.

Each web page is first mapped to a document vector consisting of binary values, each of which corresponds to a keyword list and indicates if any of the keywords in the keyword list are present in the web page. Next, for each of the page group models, the document vectors are merged by making a logical sum of each vector element. In

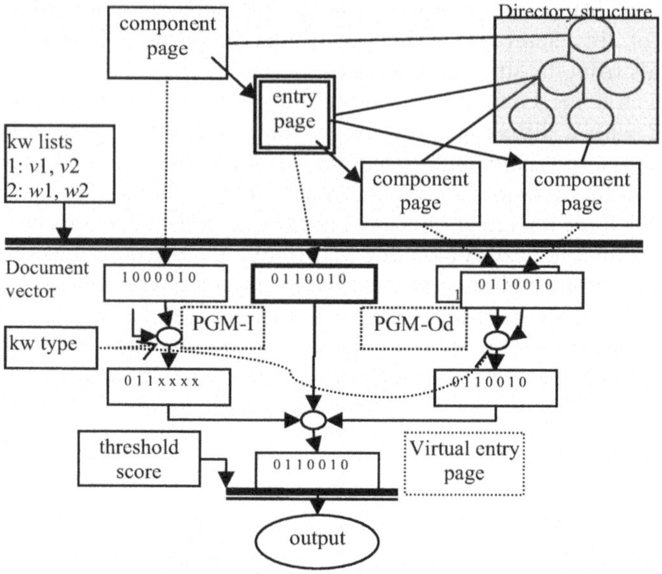

Fig. 2. Overview of the rough filtering

this process, only the elements corresponding to the keyword lists of suitable types for each page group model are considered (in the figure, ignored elements are indicated with `x' at the output from PGM-I). They are further merged to the entry page's document vector to compose a final document vector. Here, a conceptual document represented by the final document vector is called a virtual entry page, and the process to merge the document vectors is called keyword propagation. Finally, scores of virtual entry pages are obtained by counting the number of 1's in the document vector, and those that scored more than or equal to a threshold score are output. The threshold score is selected in regard to the evaluation results so that the recall satisfies the requirement, e.g., 99%, and the output amount is reasonable.

4.2 Property-Based Keyword Lists

In order to grasp the basic information elements common to homepages in the same category, we introduce property-based keyword lists, expecting that a certain number (not necessarily all) of them will be included in each target page or in its surrounding pages. Since no method is available for automatically extracting property-based keyword lists where each of them contains keywords grouped according to the same property, we used an ad hoc method to create keyword lists for the present work and obtained 12 keyword lists, such as research topic, history, major, achievement, etc., containing 86 keywords for the researchers' homepage category. Each of the keyword lists is then assigned a type, either organization-related or non-organization-related based on whether the keyword list corresponding to the properties common to the members in the same organization. Note that the actual keywords are in Japanese.

4.3 Page Group Models

Taking into account logical page group structure in the same site, we propose four simple page group models (PGMs). Table 1 lists the definitions of surrounding pages and Table 2 those of PGMs.

Table 1. Definitions of surrounding pages

Notations	Definition
r	current page
$P_{out}(r)$	set of pages linked from r in the same site (r's out-linked pages)
$P_{in}(r)$	set of pages linking to r in the same site (r's in-linked pages)
$P_{ent}(r[,s,l])$	set of directory entry pages in r's directory path from s to l level
$P_{same}(r)$	set of pages in the same directory as r
$P_{low}(r[,s,l])$	set of pages in the lower directory subtree of r from s to l level
$P_{upper}(r[,s,l])$	set of pages in the upper directory path of r from s to l level
$N_{lod}(r)$	number of links from page r to the pages in the same and lower directories of r

Note: The level of the same directory is defined as 0. s and l are options for specifying the ranges of the directory levels to propagate the keywords from. Default options mean the pages in all levels of the specified directories.

Table 2. Definitions for PGMs with parameters

Models		Description	Propagated pages	Parameters
SPM (baseline)		Single page model	Ø	
SSM (baseline)		Single site model based on out-linked pages in the same site	$P_{out}(r)$	
Simple PGM	Od(s,l)	A PGM based on out-links downward	$P_{out}(r) \cap P_{low}(r, s, l)$	$s = 0, 1$; $l = s .. 2$
	Ou(s,l)	A PGM based on out-links upward	$P_{out}(r) \cap P_{upper}(r, s, l)$	$s = 0, 1$; $l = s .. 4$
	I(s,l)	A PGM based on in-links upward	$P_{in}(r) \cap P_{upper}(r, s, l)$	$s = 0, 1$; $l = s .. 3$
	U(s,l)	A PGM based on directory entry pages	$P_{ent}(r, s, l)$	$s = 0, 1$; $l = s .. 8$
Modified PGM	Od@θ	Od with additional conditions on the number of out-links in the URL directory subtree	If $N_{Lod}(r) \leqq \theta$, same as Od; otherwise, same as SPM.	θ=5,10,20 ;
	Ou#, I#,U#	Ou, I, and U, each propagating organization-related keywords only	Same as Ou, I, and U for organization-related keywords; for others, same as SPM.	

Single page model (SPM) and single site model (SSM) were used as two baselines. PGM-Od is intended to exploit all kinds of keywords in out-linked component pages in the lower levels of the directory subtree. PGM-Ou, PGM-I, and PGM-U are intended to exploit all kinds of keywords in component pages in the upper levels of the directory path: PGM-Ou for out-linked pages, PGM-I for in-linked pages, and PGM-U for directory entry pages, respectively.

Since simple PGMs usually introduce many irrelevant pages, we propose modified PGMs to limit such noise. PGM-Od@θ propagates keywords only when the out-link page number is less than threshold θ, based on the observations that some noise sources are large groups of pages mutually linked within a directory, and that an entry page having many out-links always contains sufficient keywords in itself. PGM-Ou#, PGM-I#, and PGM-U# propagate only organization-related keywords, based on the observation that non-organization-related keywords are not included in the upper directory hierarchies. Since any single PGM can utilize only a part of the available information in component pages and collect insufficient information, we combine all of the modified PGMs to utilize as much as useful information.

4.4 Experiments

For the experiments, we used a 100GB web corpus (NW100G-01, gathered from .jp domain and used for WEB Tasks of NTCIR Workshops). We randomly collected 11,338 pages containing some typical Japanese family names (Jname data) and manually assessed them, obtained 426 positive samples and 10,912 negative samples. For further evaluating, we used another corpus (NW1000G-04) addtionally.

First, we experimented on individual simple PGMs. As what we expected, the results show that in terms of page amount all simple PGMs are inferior to SPM but superior to SSM since a lot of noise is introduced by the keyword propagation.

Next, we experimented on modified PGMs and compared each of them with the corresponding simple PGM with typical parameters. We selected s=0 for PGM-Od since almost all non-organization-related keywords are collected from within the same directory. Around the 99% recall region, the page amount with simple PGM-Od increases by 80% over SPM, whereas modified PGM-Od can limit the increase to 50%. For PGM-Ou, PGM-I, and PGM-U around the 99% recall region, although the page amount for each simple PGM increases by 40% to 120% over that of SPM, the modified PGMs can limit the increase to almost the same level as SPM.

Finally, we experimented on combinations of PGMs with several promising parameter sets. Fig. 3 shows the run results of the top three best-performing combinations of PGMs. We will refer to each of them hereinafter as follows:

PGM-C1: PGM-Od@5(0,2),Ou#(1,3),I#(0,3),U#(0,3)
PGM-C2: PGM-Od@10(0,2),Ou#(1,3),I#(0,3),U#(0,3)
PGM-C3: PGM-Od@20(0,2),Ou#(1,3),I#(0,3),U#(0,3)

Each of them uses all four modified PGMs with the same parameters except for θ of PGM-Od. As s= 0 is used for PGM-Od, s= 1 is selected for PGM-Ou. The x-axis is the page amount $n_c(i)$ ($1 \leq i \leq 12$), namely, the number of pages in the corpus that scored at least i. The y-axis is the recall defined by $n_p(i)$ / N_p, where N_p is the total number of positive sample data, $n_p(i)$ is the number of positive sample data that scored at least i. For each curve, the upper right most plot corresponds to a threshold score of 1 and every subsequent one corresponds to a threshold score incremented by 1. In general, a higher recall and a lower page amount indicate better performance; however, we put a priority on recall. The results show that even the best performing run PGM-C1 is inferior to SPM in all the recall ranges except for 100%. However, the proposed method reduced the page amount to a certain degree despite its use of PGMs.

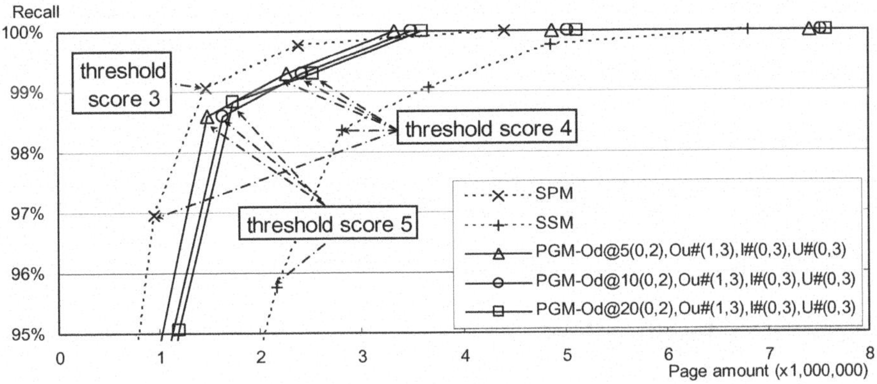

Fig. 3. Performance of typical PGM combinations

4.5 Considerations

4.5.1 Ability to Find Potential Homepages

We assessed the pages that were contained in Jname data and scored less than 3 with SPM, but scored at least 4 with PGM-C3 and found *13* new positive data. The results show that the proposed method can collect positive pages that cannot be gathered by SPM even if the threshold score is set to 3 to guarantee 99% recall for the manually assessed positive samples. Taking into account the new 13 positive data, the recalls of SPM at threshold scores of 2 and 3 should be corrected from 99.8% (425/426) to 97.0% (426/439) and from 99.1% (422/426) to 96.1% (422/439). By comparing these values with the recalls of the proposed method at threshold scores of 4 and 5 respectively, it is obvious that the proposed method outperforms SPM with 5% significance.

Furthermore, four of the positive pages could not be gathered with SPM even when the threshold score was set to 1. This implies SPM can hardly achieve the recall goal for any feasible page amount. A failure analysis on all the three pages that scored only 3 with PGM-C1 through PGM-C3 revealed that they have similar patterns and scored only 2 with SPM. Although they have hyperlinks to the researchers' personal homepages, our method cannot exploit them because they are on separate sites.

To guarantee that the overall recall to be more than 98% regarding the confidence interval based on the positive sample data number, we set the threshold score to 4. Consequently we selected PGM-C2 as the most appropriate one for the current goal.

4.5.2 Applicability to a Larger Data Set

We applied the rough filtering to the larger data set (NW1000G-04) with a procedure similar to that for NW100G-01. The approximate computational complexities of the overall processing cost for the rough filtering is O(N log N) where N is the number of web pages in the corpus.

The same parameters of PGMs for NW100G-01 were applied to NW1000G-04. The threshold number θ of out-link pages for PGM-Od was set to 20. The candidate pages were gathered with the threshold score 4. Table 3 compares the experimental results and it shows the rough filtering is not only applicable to but also more efficient for the larger data set. However, as we have not assessed the correctness of the output, the stability of the accuracy remains to be investigated.

Table 3. Comparison of pages output from the rough filtering for two data sets

Data set	#Total pages	#Output pages	Output proportion
NW100G-01	11,038,720	2,530,850	22.9%
NW1000G-04	95,870,352	14,128,826	14.7%

5 Accurate Classification

5.1 Composition of the Accurate Classification

Fig. 4 shows the composition of the accurate classification (95% recall and 99% precision are example quality requirements). We use two component classifiers to construct a three-way classifier. The recall-assured (precision-assured) classifier assures the target recall (precision) with the highest possible precision (recall).

The pages output from the rough filtering are first input to the recall-assured classifier, and its negative predictions are classified as "assured negative". The rest are then input to the precision-assured classifier and its positive predictions are classified as "assured positive". The remaining pages are classified as "uncertain", which require manual assessment.

The current work uses the SVM^{light} package with a linear kernel, tuning with its options *c* and *j*. The experiment performance was evaluated by precision (#correct positive predictions / #positive predictions), recall (#correct positive predictions / #positive samples) and F-measure (2*precision*recall / (precision+recall)).

5.2 Surrounding Page Group and Feature Set

When a page (current page) is given, its surrounding pages are categorized into groups $G_{c,l}$ based on connection type *c* (in-link (*in*), out-link (*out*), and directory entry

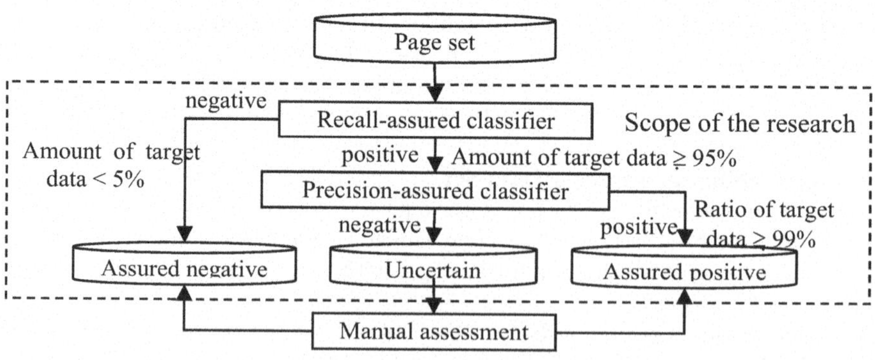

Fig. 4. Composition of the accurate classification

(*ent*)) and URL hierarchy level l (same, upper, and lower) relative to it. The current page constitutes an independent group G_{cur}, and all defined surrounding page groups are shown in Table 4. In a logical page group, for example, $G_{in,low}$ consists of in-link pages in lower directories which might represent component pages having back links to the entry page, and $G_{ent,upper}$ consists of directory entry pages in upper directories which might represent entry pages of the organization the researcher belongs to.

We use textual features $f_{t,v}(g,w_t)$, where t indicates a text type *plain* (plain-text-based) or *tagged* (tagged-text-based), v indicates a value type of *binary* or *real*, g denotes a surrounding page group, and $w_t \in W_t$ denotes a feature word. A feature set is composed by concatenating one or more feature subsets $F_{t,v}(g) = \{ f_{t,v}(g,w_t) \mid w_t \in W_t \}$. For instance, the feature set "u-1" shown in Subsection 5.4 is composed by concatenating feature subsets on pages in G_{cur} and $G_{*,upper}$, and the feature set "o-i-e-1" is composed by concatenating feature subsets on pages in G_{cur} and $G_{*,*}$.

Table 4. Groups of surrounding pages

	r	$P_{\mathrm{ent}}(r)$	$P_{\mathrm{in}}(r)$	$P_{\mathrm{ent}}(r)$
$P_{\mathrm{same}}(r)$	G_{cur}	$G_{out,same}$	$G_{in,same}$	$G_{ent,same}$
$P_{\mathrm{low}}(r)$		$G_{out,low}$	$G_{in,low}$	
$P_{\mathrm{upper}}(r)$		$G_{out,upper}$	$G_{in,upper}$	$G_{ent,upper}$

5.3 Feature Word, Text Type, and Value Type

We use two kinds of textual feature and use *Chasen*[1] for Japanese and *Rainbow*[2] for English to tokenize the feature words from the corresponding texts.

Plain-text-based features $F_{plain,*}(*)$ are extracted from textual content excluding tags, scripts, comments, etc. We use top 2,000 feature words W_{plain} selected based on mutual information. Tagged-text-based features $F_{tagged,*}(*)$ are extracted from "*text*" segments that "*text*" matches either ">*text*<" or "<img…alt= "*text*"...>" and that are not more than 16 bytes long, omitting spaces in the Japanese case and not more than 4 words in the English case. The obtained words with less than 1% file frequency in Japanese and all the obtained words in English are used as feature words W_{tagged} since all of them are considered to be property words. In the experiments, a feature set was composed by using feature words either *plain* alone or *plain* and *tagged* together. The latter case is indicated by the suffix "_tag" of the run name.

A *binary* value $f_{*,binary}(g,w_t)$ represents the presence of w_t in g. A *real* value $f_{*,real}(g,w_t)$ represents the proportion of pages containing w_t within g. The *real* value is tested to see if the feature word distribution within surrounding page groups is informative. The two value types are exclusively used for composing a feature set. Use of the *real* value type is indicated by the suffix "_real" in the run name.

---

[1] http://chasen.naist.jp/hiki/ChaSen/

[2] http://www.cs.cmu.edu/afs/cs/project/theo-11/www/naive-bayes/gentle_intro.html

5.4 Experiments

5.4.1 Experiments Using WebKB Data Set

We used the WebKB data set [9] (in English) for testing the effectiveness of the proposed features. All the 8,282 pages are classified into 7 categories, and 4 categories, student, faculty, course, and project, were used as positive samples respectively. As recommended by WebKB project, we used the *leave-one-university-out* cross-validation method and the pages of miscellaneous universities were always used as training data. A feature set composed by $F_{plain,binary}(G_{cur})$ only was used as the baseline. The experimental results of well performing feature sets compared with prior works are shown in Table 5. The overall experimental results show that tag-based features are consistently effective and the differences caused by the feature values, *binary* or *real*, are negligible. o-i-e-1_tag and o-i-e-1_tag_real performed the best and u-1_tag also performed rather well considering its relatively simple feature composition. The results show that our method out-performed almost all of the prior works.

Table 5. Performance comparison with prior works (percentage of F-measure)

method	course	faculty	project	student	Macro(4)	Macro(3)
baseline	68.41	76.01	39.62	74.95	64.75	73.12
o-i-e-1	76.97	78.27	53.30	72.53	70.27	75.92
u-1_tag	75.13	77.37	41.49	71.51	66.38	74.67
o-i-e-1_tag	77.8	79.6	59.5	74.5	*72.9*	*77.3*
o-i-e-1_tag_real	77.1	79.4	57.2	75.0	72.2	77.2
SVM(TA)[2]	68.2	65.9	32.5	73.0	59.9	69.0
SVM-FST(XATU)[3]	60.9	40.9	66.5	25.3	48.4	42.4
ME(TU)[4]					62.7	
SVM-iWUM(α=1)[7]	54.7	87.6	17.1	95.8	63.8	*79.4*

5.4.2 Experiments Using NW100GB-01 Sample Data Set

We manually prepared 906 positive samples and 20,366 negative samples from NW100G-01 (in Japanese). *Five-fold* cross validation was used. A feature set only composed of $F_{plain,binary}(G_{cur})$ was used as the baseline too. Fig. 5 shows the overall performances of the two best performing classifiers. Their performances compared with the baseline are listed in Table 6. The results show that, in the high precision area, both o-i-e-1_tag_real and u-1_tag slightly outperform the baseline; while in the high recall region, o-i-e-1_tag_real evidently outperforms the baseline and u-1_tag also performs rather well.

Table 6. Performance of well performing feature sets (in percentage of F-measure)

Feature set	Best F-measure	Recall at 99% precision	Precision at 95% recall
o-i-e-1_tag_real	88.65	20.86	41.33
u-1_tag	88.58	23.18	33.22
baseline	83.26	17.98	26.49

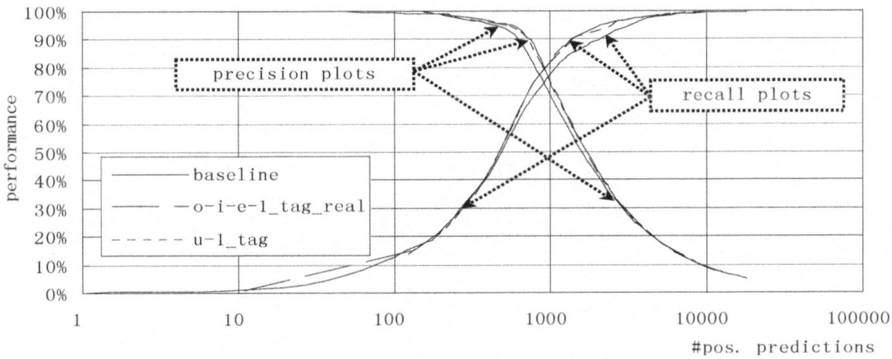

Fig. 5. Overall performances on NW100G-01 sample data set

5.5 Considerations

5.5.1 Effectiveness of the Proposed Features

The experiment results on both data sets indicate that the proposed features are effective for general performance (F-measure) and that the proposed method is applicable to research-related homepage categories in English or Japanese. Furthermore, the proposed features improve the performances of the precision/recall-assured classifiers. Features on the surrounding page groups are generally effective, but their contributions vary. For precision-assured classifiers, the upper hierarchy page groups ($G_{*,upper}$) contribute the most to the recall. This probably indicates that such pages provide contextual information, e.g., organization names and research fields, which are lacking from the current page itself but are very important for classifying the page with very high confidence. For recall-assured classifiers, all surrounding page groups ($G_{*,*}$) notably contribute to precision, despite their noisy nature.

Other experiment results not presented here have shown the following. (1) Adding tagged-text-based features consistently gained performance. This result can be interpreted that noisy information from the surrounding pages is suppressed by the tagged-text-based features and consequently the useful information in the surrounding pages can be exploited. (2) The effect of value types varies depending on the feature sets, and the performance gain is marginal.

5.5.2 Reduction of Manual Assessment

To know by how much the proposed method reduced the amount of pages requiring manual assessment (i.e., the pages classified as uncertain), we compared two compositions of three-way classifiers.

Table 7 shows estimated page numbers of classification output from NW100G-01 at three different quality requirements for two three-way classifiers, one using the baseline and the other using the best performing feature set, i.e. o-i-e-1_tag_real, as both recall/precision-assured classifiers.

Table 7. Estimated page numbers of classification output from the corpus

Required quality	baseline			o-i-e-1_tag_real			Reduction ratio
precision./ recall	assured positive	uncertain (Nb)	assured negative	assured positive	uncertain (No)	assured negative	No/Nb
99.5% / 98%	3800	461832	1618988	9206	358207	1717187	77.6%
99% / 95%	6163	274524	1803913	11251	156782	1916567	57.1%
98% / 90%	11116	155418	1918066	15503	81157	1987940	52.2%

Comparing the "uncertain" class sizes, o-i-e-1_tag_real significantly reduces the amount of pages requiring manual assessment, especially when the required quality is relaxed.

6 Conclusion

We proposed a realistic framework with a two-step process of rough filtering and accurate classification to assure the quality of a web page collection. In both processes, we introduced an idea of the local page group structure and demonstrated its uses for filtering and classifying web pages, where researchers' homepages were used as an example. Although we have just applied our method to research-related categories, we expect it will be effective in categories such as shopping, product catalogs, and so on. In terms of an information service, we believe that high-quality collections built with our method would give a guarantee of high quality for the results of various domain-specific search engines.

Even though our method fulfills the objectives of our research, for the accurate classification, we will further investigate a way to estimate the likelihood of the component pages and incorporate it in the current method.

In conclusion, building a high-quality homepage collection by exploiting rich web-based features under the restriction of a reasonable processing cost is a challenging problem when we look at the diversity of web data. The method presented in this paper gives a general framework for solving such problems, and we hope that it will contribute to research on web information utilization.

Acknowledgements

This study was partially supported by a Grant-in-Aid for Scientific Research B (No. 18300037) from the Japan Society for the Promotion of Science (JSPS). We used the NW100G-01 and NW1000G-04 document data sets with permission from the National Institute of Informatics (NII). We would like to thank Professors Akiko Aizawa and Atsuhiro Takasu of NII for their precious advice.

References

1. S. Chakrabarti. Data mining for hypertext: a tutorial survey. ACM SIGKDD Explorations, Vol. 1, No. 2, (2000) 1-11.
2. A. Sun, E.-P. Lim, and W.-K. Ng. Web classification using support vector machine. In Proc. of the fourth international workshop on web information and data management, ACM Press, (2002) 96-99, McLean, Virginia, USA.

3. M.-Y. Kan. Web Page Categorization without the Web Page. In Proc. of 13th World Wide Web Conference (WWW2004), New York, NY, USA, May 17-22, (2004).
4. M.-Y. Kan and H.O.N. Thi. Fast webpage classification using URL features. CIKM'05, (2005) 325-326, Bremen, Germany.
5. L. K. Shih and D. R. Karger. Using URLs and table layout for web classification tasks. WWW2004, (2004) 193-202, New York, NY, USA.
6. T. Masada, A. Takasu, and J. Adachi. Improving web search performance with hyperlink information. IPSJ Transactions on Databases, Vol.46, No.8, (2005) 48-59.
7. A. Sun and E.-P. Lim. Web unit mining: finding and classifying subgraphs of web pages. In Proc. of International Conference on Information and Knowledge Management (CIKM2003), (2003) 108-115, New Orleans, Louisiana, USA.
8. M. Chau. Applying web analysis in web page filtering. In Proc. of the ACM/IEEE-CS Joint Conference on Digital Libraries (JCDL'04), (2004) 376, Tucson, Arizona, USA.
9. M. Craven, D. DiPasquo, D. Freitag, A. McCallum, T. Mitchell, K. Nigam, and S. Slattery. Learning to Extract Symbolic Knowledge from the World Wide Web. Proceedings of AAAI-98, (1998) 509-516, Madison, Wisconsin.

Multi-document Summarization Using Weighted Similarity Between Topic and Clustering-Based Non-negative Semantic Feature

Sun Park[1], Ju-Hong Lee[2,*], Deok-Hwan Kim[3], and Chan-Min Ahn[1]

[1,2] Dept. of Computer Science & Information Engineering, Inha University, Incheon, Korea
[3] Dept. of Electronics Engineering, Inha University
{sunpark,ahnch1}@datamining.inha.ac.kr,
{juhong,deokhwan}@inha.ac.kr

Abstract. This paper presents a new multi-document summarization method using weighted similarity between topic and non-negative semantic features to extract meaningful sentences relevant to a given topic. The proposed method decomposes a sentence into the linear combination of sparse non-negative semantic features so that it can represent a sentence as the sum of a few semantic features that are comprehensible intuitively. It can avoid extracting the sentences whose similarities with topic are high but are meaningless by using the weighted similarity measure between the topic and the semantic features. Clustering sentences remove noises so that it can avoid the biased semantics of the documents to be reflected in summaries. Besides, it can enhance the coherence of document summaries by arranging extracted sentences in the order of their rank. The experimental results using DUC data show that the proposed method achieves better performance than the other methods.

Keywords: multi-document summarization, non-negative matrix factorization, clustering, topic-based summarization, weighted similarity measure.

1 Introduction

Radev et al. [12] mentioned three points to be considered for the multi-document summarization: (i) recognizing and removing redundancy, (ii) identifying important difference between documents, (iii) ensuring the coherence of summaries. The redundancy represents how many terms or concepts are repeated in a document while the diversity or the difference represents how many terms or concepts are different from each other [2]. The coherence represents how many summaries are readable and relevant to the user [8].

The recent studies for document summarization are as follows: Gong and Liu proposed the method using the Latent Semantic Analysis (LSA) technique to semantically identify important sentences for summary creation [3]. However, it may extract less meaningful sentences since the term weights of the semantic feature vector in latent semantic space may be negative [16]. Sassion proposed a multi-document

---

* Corresponding author.

G. Dong et al. (Eds.): APWeb/WAIM 2007, LNCS 4505, pp. 108–115, 2007.

summarization method based on topics. His method summarizes documents by removing irrelevant sentences one by one from set of candidate sentences until a user's specified compression ratio [14]. This method has a weakness that it cannot select most meaningful sentences because it uses simply the cosine similarity between candidate sentences which are extracted by using n-gram. Park et al. proposed a generic document summarization method based on NMF [9]. Park et al. proposed a query based summarization method by using NMF [10]. This method extracts sentences using the cosine similarity between a query and semantic features, whereas it does not consider the noises with respect to inherent structures. Park et al. proposed a multi-document summarization method based on clustering using NMF [11]. This method clusters the sentences and extracts sentences using the cosine similarity measure between a topic and semantic features. This method improves the quality of summaries and avoids the topic to be deflected in the sentence structure by clustering sentences and removing noise, whereas it may extract more or less similar but meaningless sentences from documents and does not consider the coherence with respect to extracted sentences.

In this paper, we propose a new topic based multi-document summarization method using weighted similarity between topic and non-negative semantic features which are obtained by means of clustering and non-negative matrix factorization (NMF). The NMF can represent individual object as the non-negative linear combination of part information extracted from a large volume of objects. It is observed by human's cognition that human use only the addition of non-negative data when they recognize an object as the combination of part information. This method can deal with a large volume of information efficiently since original non-negative matrix is decomposed into sparsely distributed representation of two non-negative matrices [6, 7, 15, 16].

The proposed method has the following advantages. First, the semantic feature vectors and the semantic variable vectors are sparse with non-negative values. Sentences can be decomposed into intuitively comprehensible semantic features having a few terms. The inherent structure of documents can be analyzed into a linear combination of semantic features. Therefore, it can select more meaningful sentences closest to the topic. Second, it can avoid the biased inherent structure of documents to be reflected in summaries since it deals with redundant information and removes noise by clustering sentences. Therefore, it can improve the accuracy of the document summarization. Third, it can enhance the coherence by sorting extracted sentences in the order of their rank.

The rest of the paper is organized as follows: Section 2 describes the non-negative matrix factorization. In Section 3, a new multi-document summarization method is introduced. Section 4 shows the evaluation and experimental results. Finally, we conclude in Section 5.

2 Non-negative Matrix Factorization

In this paper, we define the matrix notation as follows: Let X_{*j} be j'th column vector of matrix X, X_{i*} be i'th row vector, and X_{ij} be the element of i'th row and j'th column.

Non-negative matrix factorization is to decompose a given $m \times n$ non-negative matrix A into multiplication of a $m \times r$ non-negative semantic feature matrix (NSFM), W, and a $r \times n$ non-negative semantic variable matrix (NSVM), H, as shown in Equation (1).

$$A \approx WH \tag{1}$$

where r is usually chosen to be smaller than m or n so that the total sizes of W and H are smaller than that of the matrix A.

We use the Frobenius norm as an objective function to satisfy the approximation condition $\tilde{A} = WH$. The Frobenius norm is shown in Equation (2) [6, 7, 15]:

$$\Theta_E(W,H) \equiv \|A - WH\|_F^2 \equiv \sum_{j=1}^{m}\sum_{i=1}^{n}(X_{ji} - \sum_{l=1}^{r} W_{jl}H_{li})^2 \tag{2}$$

We keep updating W and H until $\Theta_E(W,H)$ converges under the predefined threshold or exceeds the number of repetition. The update rules are as follows:

$$H_{\alpha\mu} \leftarrow H_{\alpha\mu} \frac{(W^T A)_{\alpha\mu}}{(W^T WH)_{\alpha\mu}}, \quad W_{i\alpha} \leftarrow W_{i\alpha} \frac{(AH^T)_{i\alpha}}{(WHH^T)_{i\alpha}} \tag{3}$$

3 Topic-Based Multi-document Summarization Using the NMF

In this section, we propose a new topic-based multi-document summarization method. The proposed method consists of the preprocessing phase, the clustering phase, and the sentence extraction phase. We explain three phases in full, respectively.

Table 1. Symbols and their meanings

symbol	Meaning
A	$m \times n$ weighted matrix
A_{ji}	the term frequency of term j in the sentence i
m	the number of terms
n	the number of sentences
r	the number of semantic features
W	$m \times r$ semantic feature matrix
H	$r \times n$ semantic variable matrix
C^i	the weighted matrix of i'th the cluster of sentences
s_i	the number of columns of C^i
k	the number of clusters
k_{noise}	the number of noise clusters
k'	$k' = k - k_{noise}$
r_i	the number of semantic features in cluster C^i
f	the number of sentences in the summary
T	topic vector $T = (T_1, \ldots, T_m)$

In the preprocessing phase, after given documents are decomposed into a set of sentences, we remove stop-words, and perform words stemming and weight calculation. Then we construct the term-frequency matrix for documents [13].

We use *K*-means to cluster sentences. *K*-means clustering is a partition algorithm that splits given a set of *n* object into *K* clusters [4]. We perform *K*-means clustering using the cosine similarity measure with respect to the matrix *A* as shown in Equation (4).

$$d(A_{*a}, A_{*b}) = 1 - sim(A_{*a}, A_{*b}) \tag{4}$$

$$sim(A_{*a}, A_{*b}) = \frac{A_{*a} \cdot A_{*b}}{|A_{*a}| \times |A_{*b}|} = \frac{\sum_{j=1}^{m} A_{ja} \times A_{jb}}{\sqrt{\sum_{j=1}^{m} A_{ja}^2} \times \sqrt{\sum_{j=1}^{m} A_{jb}^2}} \tag{5}$$

where A_{*a} and A_{*b} denote weights for *a'th* sentence and *b*'th sentence, respectively. Besides, since these are non-negative values, it follows that $0 \le sim(\,) \le 1$. Hence, $0 \le d(\,) \le 1$.

The weighted matrix of *i'th* cluster of sentences, C^i, is a subset of column vectors of matrix *A*. C^i and C^j are disjoint and satisfy the following property:

$$\{A_{*j} \mid j = 1,\ldots,n\} = \bigcup_{i=1}^{k} \{C_{*l}^i \mid l = 1,\ldots,s_i\}, \; C^i \cap C^j \neq \phi, i \neq j \tag{6}$$

We construct matrices W^i and H^i by applying NMF algorithm with respect to sentence matrices of k' clusters after removing noise, respectively, as shown in Equation (7).

$$C^i \approx W^i H^i, \; i = 1, 2, \ldots, k', \quad k' = k - k_{noise} \tag{7}$$

The *j*'th column vector C_{*j}^i of matrix C^i is a weighted vector for *j*'th sentence, and is represented as the linear combination of the semantic feature vector W_{*l}^i and the semantic variable H_{lj}^i shown in Equation (8). That is, the weight of *l*'th semantic feature vector W_{*l}^i in sentence C_{*j}^i is H_{lj}^i.

$$C_{*j}^i \approx \sum_{l=1}^{r} H_{lj}^i W_{*l}^i \tag{8}$$

The semantic analysis of non-negative matrix *W* and *H* is as follows: semantic feature matrix *W* is composed of groups of semantically related terms. This characteristic can discriminate the same term to be used in a different meaning in the context. Besides, semantic variable matrix *H* represents weights of semantic features denoting each sentence. That is, the importance of a semantic feature in a sentence is proportional to the value of the semantic variable for the semantic feature. We can quantitatively differentiate the importance of sentences by means of values of the semantic variables.

Equation (9) is a formula that denotes the weighted similarity measure between topic *T* and semantic feature vector W_{*p}^i. The simple cosine similarity measure,

$sim(T, W^i_{*p})$, shown in Equation (5) does not reflect well the importance of the semantic feature in sentences. Hence, we multiply the ratio of summation of weights of the semantic feature to reflect the importance of the semantic feature in sentences. We can avoid the error selecting semantic features whose cosine similarity with topic is high but unimportant. The sentence, in which the weight of the semantic feature selected as the most important one is the largest, is extracted as one of the summary.

$$WeightedSim(T, W^i_{*p}) = sim(T, W^i_{*p}) \times \frac{\sum_{v=1}^{s_i} H^i_{pv}}{\sum_{u=1}^{r_i}\sum_{v=1}^{s_i} H^i_{uv}} \tag{9}$$

To estimate the number of sentences extracted from each cluster, we define the topic-cluster similarity measure between the topic and the cluster.

$$TCsim(T, C^i) = \sum_{p=1}^{r_i} WeightedSim(T, W^i_{*p}) \tag{10}$$

The number of sentences extracted from cluster C^i, e_i, is defined as follows. The larger the topic-cluster similarity value for a cluster is, the more sentences are extracted from the cluster. Besides, the more sentences in a cluster are there, the more sentences are extracted from the cluster.

$$e_i = \left\lceil f \times \frac{s_i \times TCsim(T, C^i)}{\sum_{j=1}^{k'} \left(s_j \times TCsim(T, C^j)\right)} \right\rceil \tag{11}$$

The proposed multi-document summarization algorithm is as follows:

1. Decompose a document into a set of sentences, and let f be the number of sentences to be summarized.
2. Construct the weighted matrix A by preprocessing the set of sentences.
3. Construct k clusters C^i, i=1, …, k, from matrix A, remove noise clusters, and then compute the number of sentences extracted from each cluster C^i, e_i, as shown in Equation (11), i=1, …, k', and $k'=k-k_{noise}$.
4. Obtain the non-negative matrix W^i and H^i from the matrix C^i as shown in Equation (7)
5. Perform the following steps for each cluster C^i.
   (a) Select $p = \arg\max_{1 \le j \le r_i} \{WeightedSim(T, W^i_{*j})\}$.
   (b) Select $q = \arg\max_{1 \le j \le s_i} \{H^i_{pj}\}$.
   (c) Put the sentences corresponding to C^i_{*q} into a set of summarized sentences.

(d) Repeat from step (a) to step (c) until the number of sentences e_i is reached. Here, we choose the largest value excluding previously selected ones when step (a) through step (b) are repeated.

6. Arrange extracted sentences in the order of similarity values between the topic and the extracted sentences, and return final summaries.

In step 5.a, we select the semantic feature vector W^i_{*p} most similar to topic T by using Equation (9). In step 5.b, we select the q'th column having the largest value H^i_{pq} among p'th row of semantic variable vector H^i_{p*} to select sentence which reflect chiefly p'th semantic feature vector.

4 Experiment and Results

We use the ROUGE (recall-oriented understudy for gisting evaluation) to evaluate the proposed method. The ROUGE has been applied by Document Understanding Conference (DUC) for performance evaluation [1]. The DUC is the international conference for performance evaluation of the proposed system by comparing manual summaries with summaries of the proposed system. As experimental data, we use the testing data of DUC 2005 since it is one of the standard dataset for testing the performance of the document summarization. The testing data of DUC 2005 consist of 50 topics and 25 to 50 documents related with each topic [5]

We conducted the performance evaluation of the topic-based multi-document summarization using given 50 topics of DUC 2005. ROUGE includes five automatic evaluation methods such as ROUGE-N, ROUGE-L, ROUGE-W, ROUGE-S, and ROUGE-SU [1].

We evaluated four different summarization methods such as the *LSA*, the *Kmeans*, the *Clustering+NMF*, and the proposed method. In Table 2, the *LSA* denotes Gong's method [3]. The *Kmeans* denotes the multi-document summarization method using K-means clustering[4], which clusters the sentences and extracts from each cluster the sentence whose similarity value with respect to a given topic is the largest. The *Clustering+NMF* denotes the multi-document summarization method using NMF and K-means clustering [11].

Table 2 shows the f-measure values of four methods using ROUGE evaluation. The proposed method shows the best performance whereas the *LSA* shows the lowest performance. The *Clustering+NMF* shows better performance than the *Kmeans*. Because the *Clustering+NMF* generates more meaningful summary by removing noise and redundancy and reflecting the inherent semantics of documents whereas the *Kmeans* may produce poor summary since it remove only simple noise.

The proposed method shows better performance than other methods. It generates more meaningful summary by reflecting the inherent semantics of documents without noise and redundancy, and it prevents from extracting the less meaningful sentences since the semantic feature whose cosine similarity with respect to a topic is high but meaningless is not selected by using the weighted similarity measure.

Table 2. *f*-measure values of four methods using ROUGE evaluation

ROUGE	the *LSA*	the *K*means	the *Clustering+NMF*	the proposed method
1-gram	0.2829	0.3033	0.3339	**0.3376**
L	0.2681	0.2889	0.3170	**0.3196**
W	0.1079	0.1037	0.1243	**0.1268**
SU	0.0873	0.0990	0.1113	**0.1116**

5 Conclusion

This paper presents a new topic based multi-document summarization method. In the method, it is important to extract sentences which are common to all documents and relevant to a given topic, and to remove noise and redundant information for improving the quality of summaries. The advantages of the proposed method are as follows. First, it can represent documents by means of intuitively comprehensible form since it uses very sparse semantic features and semantic variables with non-negative values. Therefore, it can extract the sentences that are semantically closer to the topic and prevent from extracting more or less similar but meaningless sentences. Second, it removes the redundancy of sentences within a cluster and identifies the important difference of sentences between clusters. So, it can avoid the biased inherent structure of documents to be reflected in summaries.

In the future, we will study the term re-weighting method of semantic features and semantic variables to summarize documents. We anticipate that re-weighting terms will enable us to extract sentences more relevant to the topic and improve the accuracy of document summarization.

Acknowledgements. This research was supported by the MIC(Ministry of Information and Communication), Korea, under the ITRC(Information Technology Research Center) support program supervised by the IITA(Institute of Information Technology Assessment).

References

1. Chin-Yew, L.: ROUGE: A Package for Automatic Evaluation of Summaries. In Proceedings of Workshop on Text Summarization Branches Out, Post-Conference Workshop of ACL (2004)
2. Goldstein. J., Mittal. V., Carbonell. J., Kantrowitz. M: Multi-Document Summarization By Sentence Extraction. The Proceeding of the ANLP/NAACL Workshop (2000)
3. Gong, Y., Liu, X.,: Generic Text Summarization Using Relevance Measure and Latent Semantic Analysis. In proceeding of ACM SIGIR (2001) 19-25
4. Han. J., Kamber., M.: Data Mining Concepts and Techniques. Morgan Kaufmann (2001)
5. Hoa., H., D.: Overview of DUC 2005. In Proceedings of the DUC (2005)
6. Lee, D. D., Seung, H. S.: Learning the parts of objects by non-negative matrix factorization. Nature 401 (1999) 788-791

7. Lee, D. D., Seung, H. S.: Algorithms for non-negative matrix factorization. In Advances in Neural Information Processing Systems, volume 13: (2000) 556-562
8. Nomoto, T., Matsumoto, Y.: A New Approach to Unsupervised Text Summarization. In proceeding of ACM SIGIR (2001) 26-34
9. Lee, J. H., Part, S., Ahn, C. M.: Automatic Generic Document Summarization Based on Non-negative Matrix Factorization. In Proceeding of BIS (2007)
10. Park, S., Lee, J. H., Ahn, C. M., Hong, J. S., Chun, S. J.: Query Based Summarization using Non-negative Matrix Factorization. In proceeding of KES (2006) 84-89
11. Park, S., Lee, J. H., Kim, D. H., Ahn, C. M.: Multi-document Summarization Based on Cluster Using Non-negative Matrix Factorization. In Proceeding of SOFSEM (2007) 761-770
12. Radev, D. R., Hovy, E. and Mckeown, K.: Introduction to the Special Issue on Summarization. Computational Linguistics, volume 28 (2002) 399-408
13. Ricardo, B. Y., Berthier, R. N.: Moden Information Retrieval. ACM Press (1999)
14. Sassion, H.: Topic-based Summarization at DUC 2005. In Proceedings of DUC (2005)
15. Wild, S., Curry, J., Dougherty, A.: Motivating Non-Negative Matrix Factorizations. In proceeding of SIAM ALA (2003)
16. Xu, W., Liu X., Gong, Y.: Document Clustering Based On Non-negative Matrix Factorization. In proceeding of ACM SIGIR (2003) 267-273

A Fair Load Balancing Algorithm for Hypercube-Based DHT Networks

Guowei Huang, Gongyi Wu, and Zhi Chen

Department of Computer Science, Nankai University, 300071 Tianjin, China
linger@mail.nankai.edu.cn, wgy@nankai.edu.cn,
mufchen@mail.nankai.edu.cn

Abstract. Load balance is an important problem in the DHT-based P2P networks. In recent years, many solutions have been proposed to address this problem. However, these solutions have some limitations in our opinion. They either make some unrealistic assumptions about the network, or have high communication or maintenance overhead. In this paper, we present a distributed load balancing algorithm for the hypercube-based DHT networks. Our algorithm is based on the concept of fairness and uses the fairness index as the fairness metric. The purpose of our algorithm is to distribute the query load fairly to nodes. Each node periodically monitors the fairness index of current load distribution by using only local computation and it tries to achieve a fairer load distribution by dynamically adjusting its indegree according to its experienced load and the fairness index. The results of our experiments show that our algorithm has low overhead and it can achieve good load balance without unrealistic assumptions about the network.

1 Introduction

In recent years, a number of DHT-based P2P systems [6], [8], [12], [13] have been proposed to support a large variety of applications including file sharing, object storage, and name service. One important issue in the DHT design is how to balance the load across the nodes in the system. To achieve load balance, existing DHT-based P2P systems simply resort to the hash function to map objects to nodes randomly. However, this could result in an $O(\log N)$ imbalance factor in the number of objects stored at a node [15], where N is the number of nodes in the system. In addition, these systems assume that nodes are homogeneous and the workload is uniform. Recent measurement studies [10], [11] have shown that the capabilities of nodes can differ by multiple orders of magnitude in typical P2P systems and the query load is skewed. These facts further aggravate the problem of load imbalance in the DHT-based P2P systems.

Many algorithms [1], [2], [5], [7], [9], [14] have been proposed to address the load balancing problem in the DHT systems. However, these algorithms have some limitations in our opinion. They either make some unrealistic assumptions about the systems or have costly maintenance or communication overhead. In this paper, we propose a fairly layered load balancing algorithm (FLLB) to cope with load balancing

G. Dong et al. (Eds.): APWeb/WAIM 2007, LNCS 4505, pp. 116–126, 2007.

problem in the hypercube-based DHT systems. FLLB is designed based on the concept of fairness and uses the fairness index [4], [16] to measure how equally the load is distributed to the nodes. It splits the total load of a node into several sublayers. Each node measures the fairness index of each layer periodically, then dynamically adjusts its indegree according to the fairness index and directs the query flow to the lightly loaded nodes.

In this paper we make the following contributions:

1. We design a distributed load balancing algorithm based on the fairness. Relying on the measurement of fairness index and dynamic indegree adaptation, it achieves fair load distribution, that is, has higher capacity nodes carry more loads.
2. We propose a load splitting mechanism. It splits the load of a node into sublayers and the load balancing algorithm is scheduled at each layer. Relying on this mechanism, our load balancing algorithm is fine-grained.
3. We study the proposed algorithm by using extensive experiments. Experiment results show that the algorithm achieves fairer load distribution and has lower overhead in comparison with some existing load balancing algorithms.

This paper is organized as follows. Section 2 provides a review of related work. In Section 3, we present the FLLB algorithm in detail. Experiment results are presented in Section 4. Section 5 concludes the paper.

2 Related Work

2.1 Hypercube-Based DHT

The routing algorithm of the hypercube-based DHT system is based on the hypercube routing. Given a query, the hypercube-based DHT system routes the query to its destination ID digit by digit in the left-to-right order (e.g., $1xxx \Rightarrow 12xx \Rightarrow 123x \Rightarrow 1234$, where x represents a wildcard). To do this, each node in the system maintains a routing table which is organized into $\lceil \log_{2^b} N \rceil$ rows, where N is the number of nodes in the system, and the entries at row n refer to nodes whose IDs share the first n digits with the present node's ID, but whose $(n+1)^{th}$ digits are different from the $(n+1)^{th}$ digit in the present node's ID. Every node satisfying this condition could be added into the entries at row n of present node's routing table. This provides the routing redundancy. Representative hypercube-based DHT systems include Pastry [12] and Tapestry [6].

2.2 Representative Load Balancing Algorithms

To achieve load balance, existing DHT-based P2P systems resort to the hash function to generate object IDs uniformly. Such a random choice of object IDs, however, results in $O(\log N)$ load imbalance. Moreover, these systems ignore the heterogeneous nature of the P2P networks.

Many load balancing algorithms have been proposed to cope with the load imbalance problem in the DHT-based P2P systems. Chord [8] was the first to propose the concept of virtual server to achieve load balance by having each physical node maintain $O(\log N)$ virtual servers. Theoretically, it reduces the load imbalance to a constant factor. However, it assumes that nodes are homogeneous and query load is uniform.

CFS [14] accounts for node heterogeneity by allocating to each node, some virtual servers proportional to the node's capacity. The underloaded node creates new virtual servers to gain more query load and the overloaded node destroys some of its virtual servers to shed the load. It focuses on heterogeneity primarily, not load skew.

The load balancing algorithms proposed by Shen et al. [2], Zhu et al. [5], Godfrey et al. [7] and Rao et al. [9] are based on dynamic load reassignment. The basic idea of these algorithms is to move load from heavily loaded nodes to lightly loaded nodes by virtual server migration. In these algorithms, each node periodically reports the load information of its virtual servers to some nodes that are in charge of aggregating load information. The reassignments of virtual servers are scheduled in the latter nodes according to the load information. These algorithms achieve good load balance. However, periodic load information aggregation and virtual server migration bring heavy burden to the system.

Shen et al. [1] proposed an novel load balancing algorithm based on irregular routing table to cope with node heterogeneity, skew workload and churn in DHT network. Its basic idea of dynamically adjusting node's indegree to achieve load balance is similar to our FLLB algorithm. Their algorithm adjusts the node's indegree according to its experienced load. In contrast, FLLB algorithm adjusts the indegree of a node not only considering its experienced load, but also considering how fair the current load distribution is.

3 The FLLB Algorithm

In this section, we introduce our load balancing algorithm FLLB. FLLB is based on the concept of fairness and its purpose is to distribute the load fairly to the nodes in the system. FLLB uses the fairness index to measure how equally the load is distributed. If the fairness index is low, this indicates that there exists load imbalance in the current load distribution, that is, the load is not distributed among the nodes proportional to their capacities. In this case, to achieve a fairer load distribution, each node dynamically adjusts its indegree.

3.1 Layered Load, Fairness Index and Indegree of Node

We consider a hypercube-based DHT network with n nodes. Each node i ($1 \le i \le n$) has a capacity c_i which corresponds to the maximum number of the queries that it can process in a given time interval T. The load of node i, l_i, is defined as the number of the queries that it receives from other nodes over T. For a node i, overload occurs when $l_i > c_i$. An overloaded node is not able to process the queries that impose work beyond its capacity. The total utilization of node i, u_i, is defined as the ratio of l_i to c_i. We define $\mathrm{N}(n_i,k)$ as the set which contains all of the nodes whose IDs share the first k digits with node i's ID and define $M(n_i,k)$ as the set which contains all of the nodes whose IDs share the first k digits with node i's ID, but whose $(k+1)^{th}$ digits are different from the $(k+1)^{th}$ digit in node i's ID.

According to the routing algorithm of the hypercube-based DHT network, node i with ID $(a_0a_1a_2\ldots a_{m-1}a_m\ldots a_{d-1})$ may be taken by some nodes in $M(n_i,m)$ as their routing table neighbor and be added into the entries at row m of their routing tables. As a result, the keys of the queries that these nodes forward to it have a common prefix $(a_0a_1a_2\ldots a_{m-1}a_m xx\ldots x)$ (where x represents a wildcard). Therefore, for each node i, its load l_i can be split into several sublayers: $l_i=\sum_{k=1}^{d-1} l_i^k$, where l_i^k is defined as the query load that node i receives from the nodes in $M(n_i,k-1)$. We refer to l_i^k as the local layer-k load of node i. We define the global layer-k query load for all nodes in $\mathrm{N}(n_i,k)$, $l_{\mathrm{N}(n_i,k)}$, as the total query load that the nodes in $\mathrm{N}(n_i,k)$ receive from the nodes in $M(n_i,k-1)$. Thus, $l_{\mathrm{N}(n_i,k)}=\sum_{p\in \mathrm{N}(n_i,k)} l_p^k$ and we say the nodes in $\mathrm{N}(n_i,k)$ share the global layer-k load $l_{\mathrm{N}(n_i,k)}$.

The purpose of FLLB algorithm is to distribute the query load $l_{\mathrm{N}(n_i,k)}$ fairly among the nodes in $\mathrm{N}(n_i,k)$. In order to measure how fair the current distribution of $l_{\mathrm{N}(n_i,k)}$ is, we use the fairness index [5], [16] as the fairness metric:

$$F\left(l_{\mathrm{N}(n_i,k)}\right)=\frac{\left(\sum_{p\in \mathrm{N}(n_i,k)} u_p^k\right)^2}{\left|\mathrm{N}(n_i,k)\right|\cdot\sum_{p\in \mathrm{N}(n_i,k)}\left(u_p^k\right)^2}. \tag{1}$$

where u_p^k is the layer-k utilization of node p, defined as the ratio of l_p^k to c_p. The value of the fairness index ranges between 0 and 1. A totally fair load distribution has a fairness index of 1 and the fairness index of a totally unfair load distribution is 0. Given the above definition, one can verify that if the nodes in $\mathrm{N}(n_i,k)$ have the same layer-k utilization, i.e., $l_{\mathrm{N}(n_i,k)}$ is distributed to nodes proportional to their capacities, this distribution of $l_{\mathrm{N}(n_i,k)}$ is totally fair.

In [1], Shen et al. showed that the node with a number of inlinks would likely experience high query load. We define the indegree of node i, d_i, as the number of its inlinks. Thus d_i is related to l_i: node i could adjust its indegree to increase or decrease its load. In practice, the P2P network is a dynamic system: the query load is time-varying, nodes join and depart at high rate. For this reason, a fair load distribution at present may become unfair later. To react to changes quickly, every node should actively monitor the fairness index of the load distribution. If a node finds the fairness index is low, it should adjust its indegree to achieve a fairer load distribution.

3.2 Computing the Fairness Index

According to the above definition, to compute the fairness index of current distribution of $l_{\mathrm{N}(n_i,k)}$ ($1 \le k \le (d-1)$), node i should measure its layer-k utilization u_i^k and be aware of the layer-k utilization of other nodes in $\mathrm{N}(n_i,k)$.

In order to know the nodes which forward queries to it, node i with ID $(a_0 a_1 a_2 \ldots a_{k-2} a_{k-1} \ldots a_{d-1})$ maintains a backward link for each of its inlink and the nodes which these backward links connect to are classified into several sets according to their IDs: nodes with IDs $(a_0 a_1 a_2 \ldots a_{k-2} a_{k-1} xx \ldots x)$ are inserted into the set $B(n_i,k)$ of node i. Thus, the local layer-k load of node i comes from the nodes in $B(n_i,k-1)$. Node i records its local layer-k query load ($1 \le k \le (d-1)$) over T periodically and then generates messages to report its layer-k utilization to other nodes in $\mathrm{N}(n_i,k)$. These messages contain not only its layer-k utilization u_i^k, but also the address information (e.g. IP address) of some nodes chosen randomly from $B(n_i,k-1)$. The address information is later used by the indegree adaptation algorithm. Sending these messages to all nodes in $\mathrm{N}(n_i,k)$ is a costly process. Instead, node i only sends messages to the nodes in its neighborhood, that is, the nodes referred by the entries at the row k of its routing table and the nodes in its set $B(n_i,k)$. One can verify that these nodes are also in the set $\mathrm{N}(n_i,k)$.

At the same time, node i also receives messages from the nodes in its neighborhood. It estimates the fairness index of current distribution of $l_{\mathrm{N}(n_i,k)}$ and the average layer-k utilization of all nodes in $\mathrm{N}(n_i,k)$ using the information contained in these messages:

$$\overline{F}\left(l_{\mathrm{N}(n_i,k)}\right) = \frac{\left(\sum_{q \in R(i,k)} u_q^k\right)^2}{|R(i,k)| \cdot \sum_{q \in R(i,k)} \left(u_q^k\right)^2} \qquad u\left(l_{\mathrm{N}(n_i,k)}\right) = \frac{\sum_{q \in R(i,k)} u_q^k}{|R(i,k)|}. \tag{2}$$

where $R(i,k)$ is the set of nodes consisting of node i as well as the nodes that report their layer-k utilization to node i in T.

3.3 Indegree Adaptation and Query Forwarding

After computing the fairness index and the average utilization for each layer, each node i begins to adjust its indegree according to these two metrics.

If node i is underloaded, then it checks the fairness index of each layer. If the fairness index of a layer, say layer k, is lower than a threshold $\tau_{fairness}$ and u_i^k is lower than $u\left(l_{\mathrm{N}(n_i,k)}\right)$, node i considers that it does not have the fair load share of $l_{\mathrm{N}(n_i,k)}$ and it should increase its indegree to receive more layer-k query load. Therefore, it sets the increment of its layer-k utilization, Δu_i^k, to $u\left(l_{\mathrm{N}(n_i,k)}\right) - u_i^k$. After checking each layer, if the sum of the increments does not exceed $1-u_i$, then node i can increase its utilization at the relevant layers according to the increments that are set previously without being overloaded. Otherwise, to avoid overload, each increment is set to $(1-u_i)/n_{incr}$, where n_{incr} is the number of the layers at which node i needs to increase its utilization.

If node i needs to increase its layer-k utilization, to achieve fair load distribution, node i should share the surplus local layer-k load of the heavier nodes in $R(i,k)$, i.e., the nodes whose layer-k utilization is higher than $u\left(l_{N(n_i,k)}\right)$. To achieve this goal, node i creates a set which contains the nodes whose address information is included in the messages that node i receives from the heavier nodes. Then it randomly chooses $\lambda\Delta u_i^k c_i / P_i^k$ nodes from this set and sends requests to these nodes, where P_i^k is the average load of the inlinks which direct l_i^k to node i and λ is the indegree adaptation coefficient. On receiving such requests, these nodes add node i into the entries at the row $k-1$ of their routing tables and send back replies to node i. Once node i receives a reply from a node, it builds a backward link to this node and inserts this node into its backward set $B(n_i, k-1)$.

If node i is overloaded, this indicates its indegree may be too high. In this case, node i decreases its indegree by requesting $\lambda(l_i - c_i)/P_{avg}$ nodes in its backward sets to delete it from their routing tables, where P_{avg} is the average load of all of the inlinks of node i. To choose nodes for deletion, node i first chooses the nodes in its set $B(n_i,0)$. If there are not enough nodes in $B(n_i,0)$ for deletion, then it chooses the nodes in $B(n_i,1)$, and so on.

As a result of dynamic indegree adaptation, each node's routing table is irregular and each entry of the routing table may refer to several nodes with high probability. According to the routing algorithm of the hypercube-base DHT network, once node i receives a query, it has to lookup one of its routing table entries to find the next hop for this query. If this entry refers to several nodes, node i has to determine which node it should forward the query to. Obviously, if node i always forwards the query to the least lightly loaded node, the query load is fairly distributed among these nodes. However, this requires that node i has to communicate with all candidates to find the least lightly loaded node. This process causes high communication overhead. Instead, we propose a query forwarding mechanism based on the randomized probing mechanism adopted by Shen et al. [3]. Shen et al. suggested that node i should only probe two nodes chosen randomly from the candidates, and the query should be forward to the one with lower utilization. They found that this simple query forwarding mechanism could achieve good load distribution. We modified their mechanism to take into consideration the layer utilization of the node. When receiving a query, node i only communicates with two nodes chosen randomly from the candidates and learns about their total utilization and the corresponding layer utilization, say layer-k utilization. If these two nodes are both underloaded, the query is forward to the one with lower layer-k utilization. Otherwise, the query is forward to the one with lower total utilization.

4 Performance Evaluation

4.1 Experiment Setup

In this section, we evaluate the performance of FLLB algorithm on top of a Pastry simulator (64-bit identifier space). We assumed a bounded Pareto distribution for the capacities of nodes. This distribution reflects the practical situations where nodes' capacities vary by different orders of magnitude. Queries of a node are generated based on Poisson process and their destinations are chosen from either uniform or Zipf distribution, depending on the experiment. Each node uses the appropriate routing table entry to route queries toward their destination and each use of routing table adds one unit load to the node. If the next hop of the query is to an overloaded node, the query fails. Queries succeed when they reach their destinations finally. We list the parameters of the experiment in Table 1.

Table 1. Experiment parameters

Experiment Parameters	Default values
Pastry ID's base	16
Number of nodes	2048
Time interval T	1 second
Capacity of node	Bounded Pareto: shape 2 Mean: 100 Upper bound: 1000
Exponent of Zipf distribution	1
Indegree adaptation coefficient λ	0.5
Fairness threshold $\tau_{threshold}$	0.9

For comparison, we implemented a virtual server load balancing algorithm [14] (VS) and the ERT-based load balancing algorithm [1] (ERT) on top of the Pastry simulator. We also evaluated the performance of the base Pastry network without load balancing (No LB) in our experiments to show the improvement that these load balancing algorithms can make. In order to examine how these algorithms respond to different degrees of query load, we measured their performance as functions of the average number of queries that node initiates in T. We varied the number of queries per node from 1 to 50 to increase the total query load in the network.

4.2 Experiment Results

4.2.1 Success Rate

The success rate of the query is determined by the probability of the query encountering no overloaded nodes on its routing path. We recorded the percentage of queries that reached their destination successfully and present the results in Fig. 1.

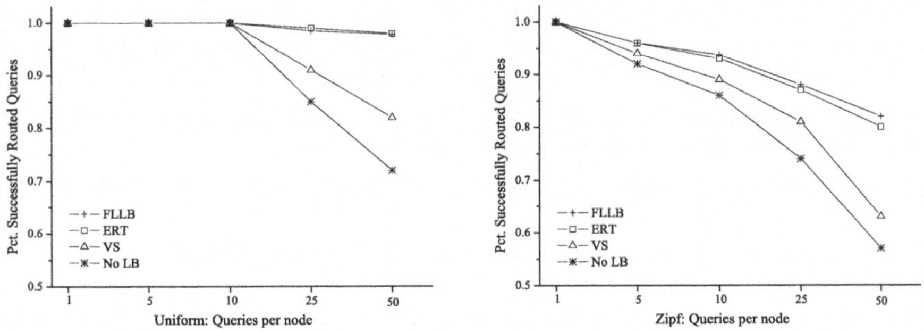

Fig. 1. Percentage of successfully routed queries for varying query loads

When queries are uniform and the load is light, all of the load balancing algorithms can sustain high success rates. However, when the load is heavy, the success rate of VS decreases drastically. This is mainly because the overloaded node in VS deletes some of its virtual servers to shed the surplus load and this may result in cascades of overload. In contrast, FLLB and ERT enable each node match its indegree to its capacity and adjusts its indegree in response to the changes of its experienced load. This helps distribute load among nodes proportional to their capacities and avoid overloaded nodes in query routings.

When queries are skewed, the query load concentrates on some ID space intervals. Fig. 1 shows that FLLB and ERT sustain much higher success rates than VS in this case. Due to the concentration of query load, the deletion of virtual servers in these intervals may cause the cascades of overload with high probability in VS.

4.2.2 The Fairness Index

As the number of nodes in our experiments is 2048, the expected number of forwarding hops is about 2.75. Therefore, the load of a node, say node i, mainly comes from the nodes in its sets $B(n_i,0)$ and $B(n_i,1)$. We measured how fairly the layer-1 load and layer-2 load are distributed among the nodes in these algorithms. Fig. 2 and Fig. 3 plot the results.

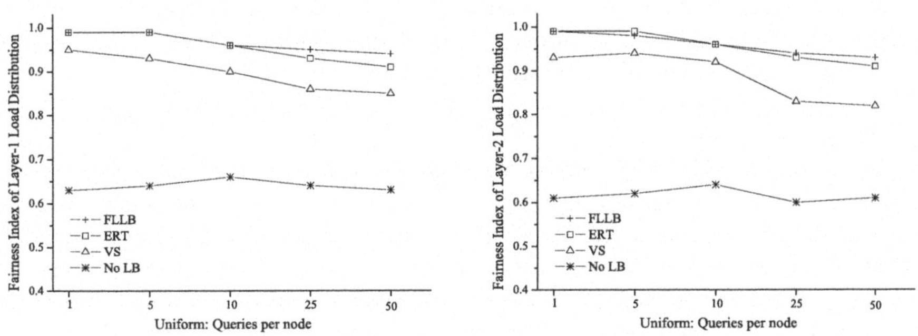

Fig. 2. The fairness index of load distribution for uniform query load

When queries are uniform, FLLB and ERT can achieve high fairness index and this indicates the load is distributed fairly among nodes. The fairness index of VS is not as high as those of FLLB and ERT. The reason is that each node tries to achieve load balance by creating or deleting virtual servers in VS. This is less fine-grained than the dynamic indegree adaptation mechanism of FLLB and ERT.

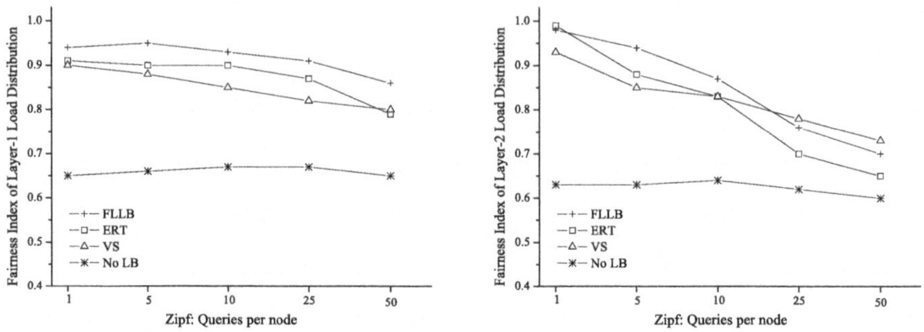

Fig. 3. The fairness index of the load distribution for Zipf query load

When queries are skewed, all of these algorithms can achieve high fairness index of the layer-1 load distribution. However, the fairness indexes of layer-2 load distribution of FLLB and ERT degrade when the query load is heavy. As mentioned above, the skewed load make the nodes in some ID space intervals are easily overloaded. If the load is too heavy, most of nodes in these intervals are overloaded at all time. In this case, the only thing that these nodes can do is to decrease their indegree regardless of the fairness index. To our surprise, VS performs better than FLLB and ERT when the query load is heavy. The reason is that the underloaded nodes in VS may create new virtual servers in these heavily loaded intervals and share the load.

We found that FLLB always performed better than ERT in either case. Each node in ERT dynamically adjusts its indegree only based on its experienced load. They have no knowledge about other nodes in the network. In contrast, the node in FLLB is aware of other nodes in the network and it adjusts its indegree not only based on its load, but also based on the load information of other nodes. Therefore, nodes in FLLB have more accurate knowledge about the load distribution. Furthermore, the load splitting mechanism makes FLLB more fine-grained than ERT.

4.2.3 Maintenance Overhead

In the hypercube-based DHT network, to build the routing table, each node needs to maintain a number of neighboring relationships. In this experiment, we measure the average indegree and outdegree of nodes for the evaluation of the maintenance overhead of all algorithms. As the average indegree is equal to the average outdegree, Fig. 4 only shows the results of indegree. The indegree of VS is much higher than others. The reason is that a node in VS may have many virtual servers and it needs to maintain a routing table for each virtual server. The indegree of ERT is also high. The reason is that the purpose of ERT is to make full utilization of node's capacity. The underloaded nodes always increase their indegree to receive more query load. In contrast, nodes in FLLB increase their indegree only when the fairness index of the load distribution is low. Thus, to achieve load balance, FLLB needs much lower overhead for maintenance.

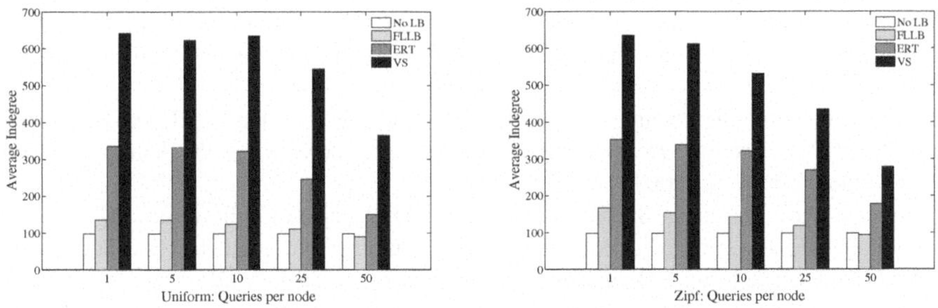

Fig. 4. Average Indegree of nodes in different load balancing algorithms

5 Conclusion

In this paper, we present a fully distributed load balancing algorithm for the hypercube-based DHT network. It is based on the fairness of the load distribution and its purpose is to distribute the load fairly among the nodes. To achieve this goal, each node periodically monitors the fairness index of current load distribution and adjusts its indegree if necessary. Experiment results show the superiority of our algorithm compared with other algorithms under conditions of node heterogeneity, skewed query. It achieves fairer load distribution with much lower maintenance overhead. Furthermore, it sustains high success rate of the query.

References

1. Shen, H., Xu, C.: Elastic Routing Table with Probable Performance for Congestion Control in DHT Networks. In: Proceedings of the 26th International Conference on Distributed Computing Systems (ICDCS). (2006)
2. Shen, H., Xu, C.: Hash-based Proximity Clustering for Load Balancing in Heterogeneous DHT Networks. In: Proceedings of the 20th IEEE International Parallel and Distributed Processing Symposium (IPDPS). (2006)
3. Shen, H., Xu, C.: Locality-aware Randomized Load Balancing Algorithms for Structured P2P Networks. In: Proceedings of International Conference on Parallel Processing (ICPP). (2005)
4. Drougas, Y., Kalogeraki, V.: A Fair Resource Allocation Algorithm for Peer-to-Peer Overlays. In: Proceedings of IEEE Conference on Computer Communications (INFOCOM). (2005)
5. Zhu, Y., Hu, Y.: Efficient, Proximity-aware Load Balancing for DHT-based P2P Systems. IEEE Transactionson Parallel and Distributed Systems **16** (2005) 349–361

6. Zhao, B.Y., Huang, L., Stribling, J., Rhea, S.C., Joseph, A.D., Kubiatowicz, J.: Tapestry: An Infrastructure for Fault Tolerant Wide-area Location and Routing. IEEE Journal on Selected Areas in Communications **12** (2004) 41–53
7. Godfrey, B., Lakshminarayanan, K., Surana, S., Karp, R., Stoica, I.: Load balancing in Dynamic Structured P2P Systems. In: Proceedings of IEEE Conference on Computer Communications (INFOCOM). (2004)
8. Stoica, I., Morris, R., Liben-Nowell, D., Karger, D.R., Kaashoek, M.F., Dabek, F., Balakrishnan, H.: Chord: A Scalable Peer-to-Peer Lookup Protocol for Internet Applications. IEEE/ACM Transactions on Networking **1** (2003) 17–32
9. Rao, A., Lakshminarayanan, K., Surana, S., Karp, R., Stoica, I.: Load balancing in Structured P2P Systems. In: Proceedings of the 2nd International Workshop on Peer-to-Peer Systems (IPTPS). (2003)
10. Gummadi, P.K., Dunn, R., Saroiu, S., Gribble, S.D., Levy, H., Zahorjan, J.: Measurement, Modeling, and Analysis of a Peer-to-Peer File-Sharing Workload. In Proceedings of the 19th ACM Symposium on Operation Systems Principles (SOSP). (2003)
11. Saroiu, S., Gummadi, P.K., Gribble, S.D.: A Measurement Study of Peer-to-Peer File Sharing Systems. In: Proceedings of Multimedia Computing and Networking (MMCN). (2002)
12. Rowstron, A., Druschel, P.: Pastry: Scalable, Decentralized Object Location and Routing for Large-scale Peer-to-Peer Systems. In: Proceedings of the 18th IFIP/ACM International Conference on Distributed Systems Platforms (Middleware). (2001)
13. Ratnasamy, S., Francis, P., Handley, M., Karp, R., Shenker, S.: A Salable Content-addressable Network. In: Proceedings of the ACM SIGCOMM. (2001)
14. Dabek, F., Kaashoek, M.F., Kaerger, D., Morris, R., Stocia, I.: Wide Area Cooperative Storage with CFS. In: Proceedings of the 18 th ACM Symposium on Operation Systems Principles (SOSP). (2001)
15. Karger, D., Lehman, E., Leighton, T., Levine, M., Lewin, D., Panigrahy, R.: Consistent Hashing and Random Trees: Distributed Caching Protocols for Relieving Hot Spots on the World Wide Web. In: ACM Symposium on Theory of Computing. (1997)
16. Jain, R., Chiu, D., Hawe, W.: A Quantitative Measure of Fairness and Discrimination for Resource Allocation in Shared Computer Systems. Technical Report, Digital Institution Corporation, Hudson (1984) http://www.cse.wustl.edu/~jain/papers/ftp/fairness.pdf.

LINP: Supporting Similarity Search in Unstructured Peer-to-Peer Networks

Bin Cui[1,⋆], Weining Qian[2,4], Linhao Xu[3], and Aoying Zhou[2,4]

[1] Peking University, China
bin.cui@pku.edu.cn
[2] East China Normal University, China
wnqian@fudan.edu.cn
[3] National University of Singapore
xulh@comp.nus.edu.sg
[4] Fudan University, China
ayzhou@fudan.edu.cn

Abstract. An important problem that confronts peer-to-peer (P2P) systems is efficient support for content-based search. In this paper, we look at how similarity query in high-dimensional spaces can be supported in unstructured P2P systems. We design an efficient index mechanism, named *Linking Identical Neighborly Partitions* (LINP), which takes advantage of both space partitioning and routing indices techniques. We evaluate our proposed scheme over various data sets, and experimental results show the efficacy of our approach.

1 Introduction

P2P computing technique has been thought of as a powerful paradigm for data sharing, and peer-based data management has attracted great interests from database community. A large number of P2P systems have been proposed in recent years, which can be roughly classified into two categories: *structured* (e.g., DHTs [12] and BATON [7]) and *unstructured* (e.g., Gnutella [4]). Structured P2P systems simplify object lookup as they tightly control both data placement and overlay topology. However, such systems destroy data locality and require high maintenance cost in a dynamic environment. In reality, most of P2P systems are unstructured overlay [4], and therefore we focus the search problem on unstructured P2P systems only in this work.

Most of recent researches dedicate to implementing content-based query processing, which is especially important to users for making decisions, mining data or searching similar multimedia data over the Internet. Since unstructured P2P system is a fully decentralized network, each user has no prior knowledge of which nodes contain the desired answers. Hence, the primary problem in this case is not only the processing of similarity queries in high-dimensional data spaces at each node, but also the way to form a partial knowledge of the data distribution of the network for routing queries to promising nodes, instead of selecting neighbors blindly. In this paper, we focus on how similarity query in high-dimensional spaces can be supported in unstructured P2P

⋆ This work is supported by the NSFC under grant No. 60603045.

G. Dong et al. (Eds.): APWeb/WAIM 2007, LNCS 4505, pp. 127–135, 2007.

systems. To the best of our knowledge, the high-dimensional search over unstructured P2P systems has received relatively little attention compared with structured ones, and the main contributions of this paper are as follows:

- We design an efficient index mechanism, named *Linking Identical Neighborly Partitions* (LINP), which makes use of both space partitioning and routing indices techniques. Specifically, the VA-file [13] is adopted as the partition scheme, which is effective in a high-dimensional space, and the routing index [6] is employed to index neighbors with identical partitions to form a knowledge of the data distribution of the surrounding peers. To construct the LINPs, all peers need to exchange *exact* data summaries with their neighboring peers. We also present the detailed algorithms for LINP construction and maintenance, and similarity search.
- We implement our proposed scheme and conduct experiments over a set of synthetic high-dimensional data sets. The experimental study shows that the proposed method yields superior performance.

The rest of this paper is organized as follows: Section 2 reviews the related work. Section 3 presents the structure and algorithms of LINP. In Section 4, we present the experimental study. Section 5 summarizes this work.

2 Related Work

To avoid blindly flooding queries with the whole network, routing schemes are designed to guide a search to only a fraction of node population. In [10], documents are assigned to each node whose identifier is equal to the hash value of terms in the document, and hence nodes can intersect the inverted lists of the query terms to find desired documents. Instead of using Bloom Filter, Crespo et al. [6] introduced the notion of Routing Indices that gives a promising "direction" towards relevant documents. Obviously, the cost of the above approaches grows proportionally with corpus size and node number.

Recently, some researches aim at merging the merits of both unstructured and structured P2P systems. Structella [2] employed Pastry-based broadcast mechanism [11] to flood user queries in Gnutella. Hence, the maintenance cost is low since no network constraint is obeyed and lots of duplicate query messages are avoided since nodes are guaranteed to be visited only once. In PIERSearch [8], a Gnutella-liked system is responsible for finding highly replicated items, while DHTs for locating rare items. However, these P2P systems are still for matching-based search instead of content-based search. On the other side, by drawing the inspiration from the indexing technique in the database research, many indexing approaches were invented. P-tree [5] indexes the data range of peers to process one-dimensional range queries on Chord [12], while BATON [7] organizes all nodes into a B+-tree and therefore the range search can be naturally supported.

The proposed LINP approach distinguishes itself at three aspects: First, the LINP is more suitable for real-life P2P applications, since it does not depend on any specific overlay structure or global indices; Second, the LINP enables peers to efficiently handle similarity queries in high-dimensional data spaces by conquering the problem of dimensional curse; Finally, the LINP is computationally efficient for maintenance issue, which keeps the efficiency of query processing and the P2P network scalable against the peer population and dynamism.

3 LINP: Enabling Similarity Search over Peers

3.1 Space Partitioning

Since the unstructured P2P system is decentralized, the space partitioning mechanism must be autonomic and also computationally efficient as the network is dynamic. Hence we employ the *Vector Approximation file* (VA-file) approach [13] to split space, which independently divides a high-dimensional space into 2^b partitions where b is the total number of bits. A number of bits b_i is assigned to each dimension, which is split into 2^{b_i} equal segments. Each partition, termed *vector approximation* (VA), has a bit representation of length b that approximates all data belonging to it. Thus a high-dimensional data set is represented by an array of VAs. Figure 1(a) shows that a 2-dimensional space is divided into 16 partitions and data d_5 and d_7 are represented by the same VA "0111".

There are two reasons for us to adopt the VAs as the data summary at peers: first, it will not suffer from dimensionality curse and outperform sequential scan in high-dimensional spaces while other indexing techniques fail, when dimensionality exceeds a certain threshold [13]; and second, it is of extremely computational efficiency in terms of maintenance issue as peers join or quit the P2P network. In other words, when a new partition appears, it will be simply appended to the partition array without any other computational cost.

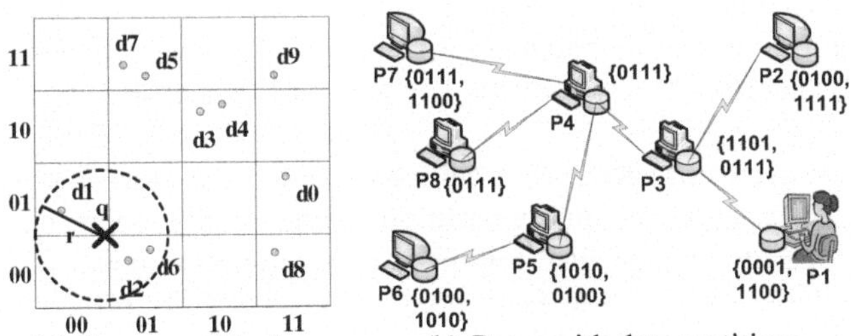

(a) Space partitioning (b) Peers with data partitions

Fig. 1. Peers with data partitions

Naturally we can utilize the space partitioning scheme to solve our first problem. However, with partitions, we can only query each individual peer, but cannot query the P2P network since peers have no complete knowledge of the data distribution of the whole network. The next section will present how LINP can solve the second problem.

3.2 Linking Identical Neighborly Partitions

The basic idea of the *Linking Identical Neighborly Partitions* (LINP) is to form a knowledge of the data distribution at the surrounding peers by linking the identical partitions of neighbor peers together. The LINP scheme uses *partitions* as indexing item and is organized as an inverted index. Each entry is a binary tuple <*id*, *list*>, where *id* denotes the bit representation of the partition and *list* is a linking list that contains routing information. Each item of *list* is also a binary tuple <*peer_id*, *soi*>, where *peer_id* refers to the identifier of the current peer who is exchanging the partitions with other peers, *soi* is an integer specifying the number of hops from the owner peer of the partition *id* to the current location (i.e., *soi* is used for the maintenance of the LINP by specifying the maximum propagation horizon soi_{max} of a partition).

In Figure 1(b), consider that the joining sequence of peers is P_1, P_3 and P_2. When P_3 joins, it will exchange its LINP with P_1. Then P_3 has the knowledge of two partitions "0001" and "1100" and their *soi* values are 1 for they are propagated by one hop from P_1 to P_3. When P_2 joins, it will obtain the knowledge of four partitions "1101", "0111", "1100" and "0001" from P_3, but the *soi* values of the first two cells are 1 and the last two are 2. This is because, for each hop, the *soi* value of each partition is increased by 1. Thus each peer controls the scope of propagating the knowledge of any partition, by comparing its current *soi* value with the threshold soi_{max}.

The LINP has two important features. First, each peer only consider the data distribution of its neighbors, instead of indexing the actual owner of any partition. For example, in Figure 2 ($soi_{max} = 2$), P_5 knows of P_4 having the knowledge of the partition "1101", instead of the actual owner P_3. The rationale is that when any peer receives a query, it will determine to which neighbor(s) the desired partitions can be found. Second, using partitions as the content summarization can reduce the size of the LINP since different data belonging to the same partition can be absorbed. For example, P_5's LINP contains 5 entries that actually include 10 data objects under the constraint of $soi_{max} = 2$. Therefore the size of the LINP is half that of directly indexing each individual data.

Based on the above observations, we can observe that the most important property of the LINP is that through exchanging partitions with neighbor peers, each peer forms a partial knowledge of the data distribution at its surrounding peers. Thus peers can route similarity queries to promising neighbor peers purposefully, rather than by blindly propagating queries.

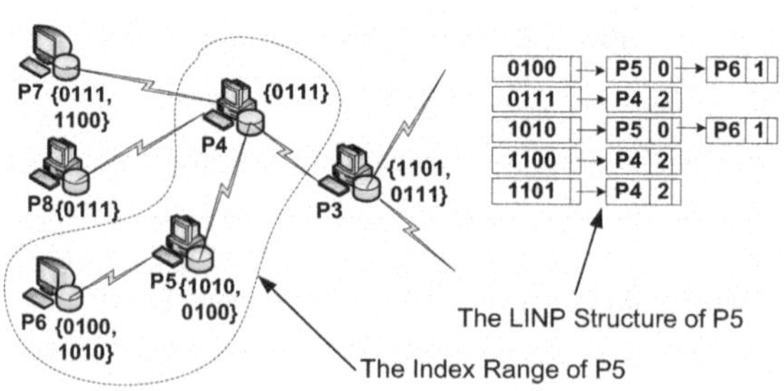

Fig. 2. An example of the LINP structure

3.3 Constructing LINP

The construction of the LINP takes place at the time when a peer enters the P2P network. The information exchanged between peers is an array of triple <*id, peer_id, soi*>. Suppose that a peer P joins the network, and its neighbors are P_1, ..., P_n, the process of building the LINP is as follows:

1. P first initializes its LINP. Then it requests neighbor peers P_1, ..., P_n for their LINPs, and waits for their replies.
2. Upon receiving the <*id, peer_id, soi*> array from a neighbor P_i, P first checks if the *id* appears in P's LINP. If true, a new item $< P_i.peer\_id, soi + 1 >$ is linked to the *list* of the corresponding partition link; otherwise, a new partition link $< id, list >$ is created and inserted into the LINP, then the new item $< P_i.peer\_id, soi + 1 >$ is linked to the $list$.
3. P checks the *soi* value of each item of each partition link to determine which partition should be disseminated to neighbors: for a partition link l, if the *soi* value of all items equals to soi_{max}, then l will not be propagated to neighbors; for a neighbor P_i, if its $peer\_id$ appears in a certain item, then l will not be sent to P_i again for P_i has transferred l to P before; otherwise P sends an array of all valid tuples <*id, peer_id, soi*> to P_1, ..., P_n.

When a peer joins the network, it first requests its neighbors' LINPs because a new path may occur between some peers bridged by it and the LINPs of all neighbors need to be updated through the newcomer. Further, through checking and updating the *soi* value of each partition at each hop, the infinite propagation of a partition's information in a loop of peers can be avoided.

3.4 Updating LINP

The updating operation of the LINP is triggered by peer joining, departure, failure or data change. For peer joining we have discussed above.

Peer Leave/Failure. Peer leave and failure follow the similar maintenance procedure except for the way of detecting a peer's leave. We first describe the process of updating LINP in terms of peer departure. Suppose that a peer P departs from the network and a typical updating procedure is as follows:

1. P notifies all neighbors P_1, ..., P_n about its leave and quits the network.
2. If a neighbor P_i receives the notification from P, it will remove all index items whose $peer\_id$ equals to P's identifier from its LINP.
3. After removing some index items, if the *list* of a partition link l becomes empty, then P_i removes l from its LINP (because P_i does not have the knowledge of l) and propagates the tuple $< id, soi + 1 >$ to its neighbors recursively until soi_{max} is reached. All peers receiving the message directly remove the entry whose identifier equals to id from their LINPs.
4. If the *list* of any partition link is not empty, then stop updating.

For peer failure, the steps 1 and 2 should be replaced by: P periodically polls neighbors P_1, ..., P_n to detect whether a neighbor quits the network. If a neighbor P_i does not respond for a period, then P regards P_i as failure and execute steps 3 and 4, to update the LINPs of its own and neighbors.

Data change. Another possible event that triggers the updating operation of the LINP is data change at peers. There are two cases for a peer P changing its content: partition(s) emerging or disappearing. In the first case, when a neighbor P_i receives the tuples $< id, P.peer\_id, soi >$ from P, it executes the step 2 of the algorithm of constructing LINP to update its LINP. Then P_i propagates all tuples $< id, P_i.peer\_id, soi >$ that do not appear in its LINP and whose soi value is less than soi_{max} to its neighbors recursively, till soi_{max} is reached. In the second case, P_i runs the steps 3 and 4 of the above algorithm of peer departure to update the LINPs of its own and its neighbor peers.

3.5 Similarity Search

Without loss of generality, for convenience, we assume the data space is a d-dimensional unit hypercube, and the Euclidean distance function is used for measuring the similarity of pairs of data points, although other metric distance functions can be used. We only present the range search algorithm as KNN query algorithm is similar.

Range Search. Given a range query q and its search radius r, suppose node P is now processing q. The LINP range search performs the following steps:

1. P sequentially scans each partition link l in the LINP and calculates the lower bound $lb_{q,l.id}$ of q and $l.id$ [13]. If the $lb_{q,l.id}$ is less than r, then $l.list$ is scanned. As each item e of $l.list$ is scanned, if $e.peer\_id$ equals to its own identifier then $l.id$ is added into a candidate partition set $partition_q$; otherwise, $l.id$ is filtered because P does not have data belonging to the partition $l.id$. If $e.peer\_id$ is not equal to P's identifier, then $e.peer\_id$ is added into a candidate neighbor set $neighbor_q$.
2. P scans all data objects belonging to each candidate partition in $partition_q$ and computes the distance $d_{q,d}$ between each data object d and q. If the $d_{q,d}$ is less than r, insert d into a result set $result_q$; otherwise, d is filtered.
3. The TTL (Time-to-Live) value of q is decreased by 1. If the $neighbor_q$ is not empty and the TTL is greater than 0, P then sends q to all candidate neighbors in the $neighbor_q$; otherwise, P drops q.
4. If the $result_q$ is not empty, directly return the $result_q$ to the query peer.

4 A Performance Study

We build a simulator to evaluate the performance of similarity search by using the proposed LINP technique over a large-scale network with 1024 nodes and the network topology is power-law that is generated according to the PLOD algorithm [9] with average outdegree of 4.07. The simulator is written by JAVA SDK 1.4, and all experiments are performed on a Linux Server that has an Intel Xeon 2.8GHz Processor and 1.5GB main memory. In the simulation, we use Gnutella-based search manner as the baseline to compare with our proposed scheme. The reason is that there is no similar work on supporting similarity search in unstructured P2P networks. For evaluation metrics, we are interested in recall (the percentage of answers returned), ratio of distance error, query time and coverage (the percentage of peers probed).

We conduct experiments on both synthetic and real life datasets, e.g. 32 dimensional color histograms [1]. However, due to space constraint, we will only report results on synthetic datasets here. Results from real life datasets mirror the result of the synthetic datasets closely. To examine the effects of various factors on the performance of LINP, we generate a typical 20-dimensional dataset using the data generator of [3] which has 1M objects distributed in 10 clusters in subspaces of different orientations and dimensions. Then we randomly choose 1,000 data objects for each peer. All similarity queries are generated in terms of the data distribution. For range queries, the search radii are varied from 0.05 to 0.1; and for kNN queries, the values of k are varied from 5 to 50. Each query is randomly initiated from any node in the network. For each measurement, we report the average results over 500 similarity queries.

There is an intrinsic parameter of the LINP scheme to generate the index, which is the number of partitions, i.e. the number of bits for VA-file. When the number of bits increases, the recall rate reduces. Because the LINP only store the information of neighborly peers with identical partitions, if there are more partitions, some partitions may fall in the query range but not exist in the LINP. In extreme case, if we treat the whole space as a single partition, the LINP works exactly same as Gnutella as it has links to all neighbors. However we should not ignore another goal that is to minimize the number of the peers involved into the query processing. This is very important to unstructured P2P systems for its efficiency and scalability against the peer population and dynamism. The optimal bit number is a tradeoff between recall and coverage. Due to the space constraint, we omit the details of experimental results here. According to the results, we set the number of bits for each dimension 5 as a default value for the clarity of presentation.

We evaluate the effectiveness of the proposed LINP scheme with Gnutella on the performance of range search. Notice that, although LINP and Gnutella are running on the same network infrastructure, queries are conducted on different paths for two methods since the LINP has an additional routing index. The experimental results are shown in Figure 3.

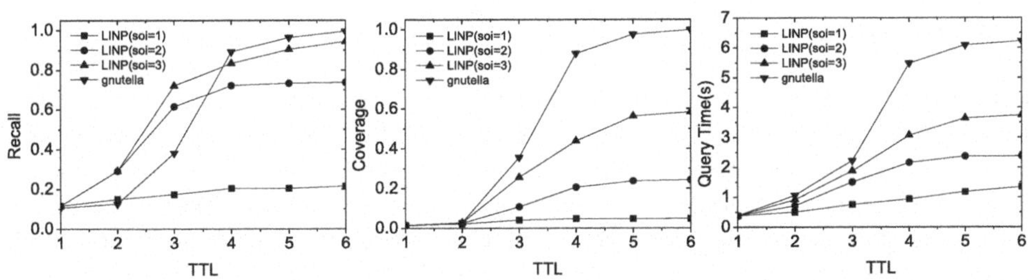

(a) Recall (b) Coverage (c) Query time

Fig. 3. Performance on range search

Figure 3(a) and Figure 3(b) show that when the TTL value is less than 4, the recall of the LINP is better than that of Gnutella, although Gnutella accesses more peers than the LINP. As the number of TTL increases, the Gnutella floods the query to more and more peers in the network, and finally almost covers the whole network when the TTL is 6. Although the Gnutella retrieves more answers than the LINP approaches, it has to pay higher query cost. Clearly, when the soi value is set as 3, the LINP gets 95% recall rate by only access 60% of peers, which is much less than those of Gnutella. On the other side, for the LINP, both recall and coverage increase greatly as the soi value increases. The reason is that, in the LINP mechanism, the number of links to the neighboring peers is determined by the soi, e.g. when soi is equal to 1, the LINP only stores the link to directly connected peers which have the identical partitions. With the large value of soi, each peer stores more information about the neighboring peers which have the potential query answers. The LINP can easily access such peers with the links in the local index. Thus a good soi value can improve the system performance greatly. In a real-life P2P system, query time is typically the most important issue that users concern about. We next evaluate the query time that the system requests to answer the similarity query. In our simulation we assume that each peer has enough bandwidth to relay similarity queries and no network congestion occurs. Figure 3(c) shows the results of the query time for Gnutella and LINP. When the TTL value increases, the query time of all approaches increases, as they have to access more peers and conduct the query on the certain nodes. Gnutella yields worst performance in all cases, as it floods the query to all neighbors at each step. Though Gnutella can retrieve more answers for a large TTL, the tradeoff is higher query processing cost.

Finally, from the experimental results, we can observe that the system performance is more preferable under the constraints of $soi_{max} = 3$ than those of $soi_{max} = 1$ or $soi_{max} = 2$, i.e. the recall rate is closer to that of Gnutella, but much lower cost.

We also test the performance of kNN search. Since both Gnutella and LINP may not access all the peers due to the restriction of TTL, we can only get approximate kNN in this case. As expected, though the result quality of Gnutella is better than that of the LINP, the query cost of Gnutella is much higher than that of the LINP, e.g. the LINP achieves similar query quality while its cost is 50% better. Due to the space constraint, we omitted the details in this paper.

5 Conclusion

In this paper, we have addressed the high-dimensional similarity query problem in unstructured P2P systems. To this end, we proposed the LINP index scheme, which not only enables peers to efficiently handle similarity queries in high-dimensional spaces, but also efficiently routes to desired neighbors. Additionally, the LINP structure can be easily generated and maintained locally, thus avoids a large amount of computation and communication costs, which makes the P2P network scalable against the peer population and dynamism.

References

1. *Corel Image Features*. available from http://kdd.ics.uci.edu.
2. M. Castro, M. Costa, and A. Rowstron. Should we build gnutella on a structured overlay? In *Proc. of HotNets-II*, 2003.

3. K. Chakrabarti and S. Mehrotra. Local dimensionality reduction: A new approach to indexing high dimensional spaces. In *Proc. of VLDB*, pages 89–100, 2000.
4. Y. Chawathe, S. Ratnasamy, L. Breslau, N. Lanham, and S. Shenker. Making gnutella-like p2p systems scalable. In *Proc. of SIGCOMM*, pages 407–418, 2003.
5. A. Crainiceanu, P. Linga, J. Gehrke, and J. Shanmugasundaram. P-tree: A p2p index for resource discovery applications. In *Proc. of WWW*, pages 390–391, 2004.
6. A. Crespo and H. Garcia-Molina. Routing indices for peer-to-peer systems. In *Proc. of ICDCS*, pages 23–34, 2002.
7. H. V. Jagadish, B. C. Ooi, and Q. H. Vu. Baton: A balanced tree structure for peer-to-peer networks. In *Proc. of VLDB*, 2005.
8. B. T. Loo, J. M. Hellerstein, R. Huebsch, S. Shenker, and I. Stoica. Enhancing p2p file-sharing with an internet-scale query processor. In *Proc. of VLDB*, 2004.
9. C. R. Palmer and J. G. Steffan. Generating network topologies that obey power law. In *Proc. of IEEE GLOBECOM*, 2000.
10. P. Reynolds and A. Vahdat. Efficient peer-to-peer keyword searching. In *Proc. of ACM Middleware*, 2003.
11. A. Rowstron and P. Druschel. Pastry: Scalable, distributed object location and routing for large-scale peer-to-peer systems. In *Proc. of Middleware*, 2001.
12. I. Stoica, R. Morris, D. Karger, F. Kaashoek, and H. Balakrishnan. Chord: A scalable peer-to-peer lookup service for internet applications. In *Proc. of SIGCOMM*, 2001.
13. R. Weber, H.-J. Schek, and S. Blott. A quantitative analysis and performance study for similarity-search methods in high dimensional spaces. In *Proc. of VLDB*, 1998.

Generation and Matching of Ontology Data for the Semantic Web in a Peer-to-Peer Framework

Chao Wang, Jie Lu, and Guangquan Zhang

Faculty of Information Technology, University of Technology, Sydney
PO Box 123, Broadway, NSW 2007, Australia
{cwang,jielu,zhangg}@it.uts.edu.au

Abstract. The abundance of ontology data is very crucial to the emerging semantic web. This paper proposes a framework that supports the generation of ontology data in a peer-to-peer environment. It not only enables users to convert existing structured data to ontology data aligned with given ontology schemas, but also allows them to publish new ontology data with ease. Besides ontology data generation, the common issue of data overlapping over the peers is addressed by the process of ontology data matching in the framework. This process helps turn the implicitly related data among the peers caused by overlapping into explicitly interlinked ontology data, which increases the overall quality of the ontology data. To improve matching accuracy, we explore ontology related features in the matching process. Experiments show that adding these features achieves better overall performance than using traditional features only.

1 Introduction

Ontology has been realized to be an essential layer of the emerging semantic web [1]. According to description logics (DL) [2], an ontology as a knowledge base normally consists of a "TBox" and a "ABox". The TBox consists of concepts and their relations while the ABox consists of instances of concepts or individuals. For an ontology expressed in the web ontology language (OWL) [3], we use the term *ontology schema* and *ontology data* for the part corresponding to the notion of TBox and ABox respectively. Through the semantic web search engine Swoogle (http://swoogle.umbc.edu), we've found that there are plenty of ontology schemas available over the web. But in contrast, ontology data does not seem to be very abundant.

The abundance of information and data in the current web has made the web an important place for people to seek information. This leads us to believe that the abundance of ontology data in the semantic web is very crucial. Therefore, besides the development of ontology schemas, the generation of ontology data also plays an important role for the semantic web.

Generation of ontology data can be achieved through different ways. This paper proposes a framework that supports the generation of ontology data through conversion and authoring. It is designed with the peer-to-peer architecture, which

G. Dong et al. (Eds.): APWeb/WAIM 2007, LNCS 4505, pp. 136–143, 2007.

is more flexible and scalable in terms of ontology data management and distribution. Although several (ontology) data management or sharing frameworks based on peer-to-peer architecture have been proposed (e.g. Edutella [4], Piazza [5],and etc), our framework has an advantage that differentiate it from them in that it deals with the issue of ontology data matching across peers. Normally, the ontology data for a certain domain contributed by one peer may not be complete but may be implicitly related to those contributed by other peers as data overlapping often occurs. Our framework, designed with a matching process to discover the implicit relations, is able to help reduce the redundancy and increase the amount of richly inter-linked ontology data.

The rest of the paper is organized as follows. Section 2 discusses related work. An overview of the framework is given in Section 3. Section 4 describes how the framework supports the generation of ontology data. Section 5 presents the function of ontology data matching. Experimental results about ontology data matching are shown in Section 6. Section 7 concludes the paper.

2 Related Work

We discuss relate work from two aspects: ontology data generation and data matching.

There are a few ways to generate ontology data. Some ontology development tools such as Pretege (http://protege.stanford.edu), SWOOP [6] can be used to create new ontology data as well as to develop ontology schemas. However, as these tools are designed with an emphasis on knowledge representation, they are often used by experienced domain experts familiar with ontology related techniques. Ordinary users may have difficulty using it, which hinders the generation of ontology data in a large scale. Annotation of existing data with ontologies is another way to generate ontology data (e.g., CREAM [7]). Ontology data can also be generated by automatically annotating web pages (e.g. [8]). Although a large amount of ontology data can be generated by this type of methods, some applications may not be able to use them due to its relatively poor quality.

On the other hand, data matching is mostly a research topic in the traditional database community. It involves creating semantic matching between objects, instances, or records from different data sources (mainly different databases) [9]. Different techniques have been employed to perform data matching in different applications (e.g., [10,11]). Although we can directly use these methods to perform ontology data matching by treating instances in ontology data as data records in databases, this ignores the particular features that ontology data inherently has, which may affect the performance of ontology data matching.

3 Framework Overview

We first give a brief overview of our framework that supports the generation and matching of ontology data. The framework uses the super peer topology [4] for the peer-to-peer architecture. Different from traditional client/server architecture, it shifts part of the tasks of the server (super peer) to the clients (ordinary peers). For example, when processing a query, a super peer can simply tell the query issuer from which peers they can get potential query results instead of returning the complete query results directly.

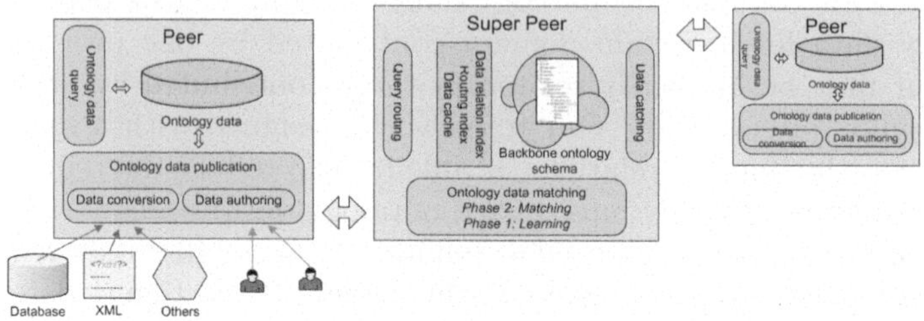

Fig. 1. The framework that supports ontology generation and matching with a design of a peer-to-peer architecture

Accordingly, as shown in Fig. 1, our framework introduces two types of peers: super peer and ordinary peer (or just denoted as peer). The super peer acts as a coordinator for peers connected to it. It hosts the backbone ontology schema used for the generation of ontology data and provides related functions, one of which is *ontology data matching* (Section 5). On the other hand, peers offer functions that support ontology data generation (*ontology data publication*) and query (*ontology data query*).

4 Generation of Ontology Data from Peers

4.1 Data Conversion

As many existing data are stored in databases or formatted in XML with no ontologies to interpret them, it is desirable to convert them into formats that can be explained by given ontology schemas and ready to be integrated. Mostly we use OWL to encode these converted data according to the backbone ontology. Here we only discuss the case of XML data due to limited space.

XQuery [12] is used to convert ordinary XML data into OWL format.Executed by a XQuery-compatible engine, a query written in XQuery takes XML files as input and generates results in an XML format defined by the query itself. Since OWL is also based on XML syntax, the problem is then transformed into how to design the query so that the output results are actually the desired OWL format. The advantage of using XQuery instead of developing our own programs for conversion is obvious. We don't have to design any custom conversion rules and to maintain the program, which might be time-consuming and error-prone. As a W3C candidate recommendation, XQuery is versatile enough to satisfy our needs and has several implemented query engines for us to choose. Therefore, to convert XML data to desired OWL format, we only focus on the design of the queries and use *Nux* (http://dsd.lbl.gov/nux/), a java toolkit capable of XQuery processing, to process them.

4.2 Data Authoring

While existing data is very useful for the ontology data generation, it is equally important to allow new data to be created and published in the framework.

Data authoring is the process during which peers create their own ontology data and publish it into the framework. Like the current Web, where data is directly contributed and published by various individuals and organizations, our framework should also enable different individuals or organizations to publish their new data via their corresponding peers. Therefore, the data in the framework will be very dynamic, often reflecting its very recent status while still retaining reasonable semantics thanks to the backbone ontology schemas.

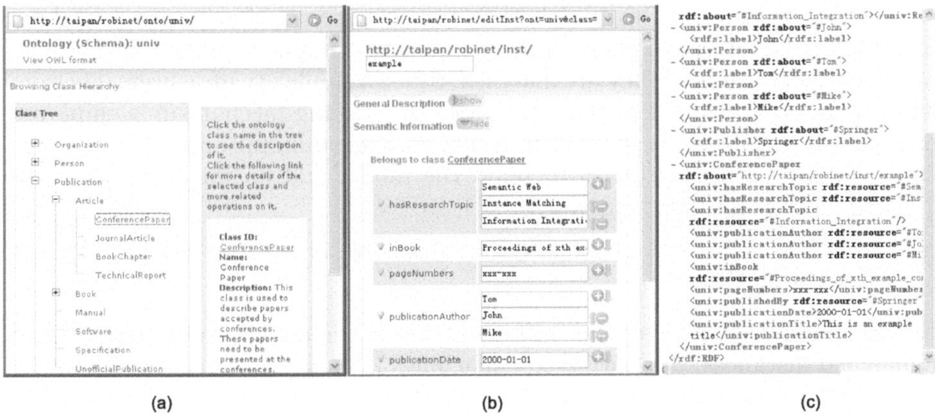

Fig. 2. The user interface for data authoring and the generated data

Users who want to author their concerned data through the peers should be familiar with the backbone ontology schemas. We develop web-based programs at each peer server so that users can get familiar with them easily and quickly. For example, Fig. 2 (a) shows the interface that allows users to learn a backbone ontology schema describing the university settings by browsing intuitively. Therefore, the users don't have to study the original ontology schema encoded in OWL with more efforts. A professor, if he/she wants to publish some data about his/her recent publications, can choose the class that is most appropriate for the data. He/She then can use the selected class to create the ontology data, through a friendly interface as shown in Fig. 2 (b). With the data supplied by the user, the program at the peer server generates the corresponding ontology data in OWL format as shown in Fig. 2 (c). In summary, supported with these functionalities, users can create and publish ontology data with ease.

5 Ontology Data Matching

Data matching process is designed to deal with the problem of implicit relations among the data from different peers. The necessity of it can be illustrated by the following example. Suppose a professor has a peer contributing data about his/her own details including contact information, research interests, supervised students, research groups, selected publications (without details such as abstracts and full texts), and etc. Meanwhile a publisher's peer contributes information of a detailed publication list, which includes some publications (with full texts) of that professor. The publisher's peer lets us know details of publications by that professor. All the data are published as instances of concepts of the given ontology. Because the instance describing the professor from the publisher's peer is not explicitly related to that from the professor's peer due to the decentralized environment, we only get a partial view of the professor's information from these separate peers. It is impossible to issue an enquiry like getting some publication details of a professor whose research interests are of a given area.

Therefore, the task is to match the instances from different peers, making their implicit relations explicit. It is common to compute similarities between instances to determine if they are matched. Several similarity measurements from different aspects are used in the framework. A learning mechanism is employed to implicitly combine these measurements in a meaningful and adaptive way. This involves a learning phase to build the model and an matching phase to apply the model for data matching.

5.1 Learning Phase

The learning phase involves the training of a binary classifier from matched and unmatched instances. Support vector machines [13] are chosen as the classification model in the proposed framework. First, a certain amount of initial data are gathered by super peer from different peers. This data set should contain a portion of matched instances. These matched instances are not discovered and specified initially, while training an SVM classifier requires both specified matched instances and unmatched instances as positive and negative samples. Therefore, these initial data should be checked and tagged manually for the training. An initial similarity checking and sorting process based on selected instances properties is performed to make the manual tagging easier. After all these initial data are tagged and the matched and unmatched instance pairs are created, it is ready to train the classifier.

Several similarity measurements are used to compute similarity/distance scores for instance pairs as different feature scores. string edit distance [14] (denoted as SED) and cosine similarity based on TF-IDF [15] (denoted as SIM) are used to measure the string similarity of instance properties at character level and at word level respectively. In addition to string-based similarities, ontology-based similarities are also used. We define the term of "concept distance" (denoted as CD), which can measure the distance of concepts of two given instances according to the ontology. Instances belonging to the same concepts have the closest distances while those belonging to disjoint concepts have the largest distances. As ontology technology allows instances to be related by object properties [3], it is useful to check the "context similarity" (denoted as CS) for object properties of two instances. If the instances that are related to two target instances via the same object property are similar according to string-based similarities, the two target instances will have high context similarity. Details of these similarity measurements are presented in another paper due to limited space.

Given the above different similarity measurements, we create feature vector for each pair of instances from the initial tagged data set. For a pair of instance a and instance b, its feature vector is composed as follows:

$$\begin{aligned}\mathbf{p}(a,b) = [&SIM_{d_1}(a,b), \ldots, SIM_{d_m}(a,b),\\ &SED_{d_1}(a,b), \ldots, SED_{d_m}(a,b),\\ &CS_{o_1}(a,b), \ldots, CS_{o_n}(a,b),\\ &CD(a,b)]. \end{aligned} \quad (1)$$

where $d_1, \cdots, d_m$ are data type properties [3] of the instances; and $o_1, \cdots, o_n$ are object properties. These feature vectors together with the tagged information (matched or unmatched) enable us to build the classification model for the matching.

5.2 Matching Phase

During the Matching phase, the super peer matches instances from the peers when they contribute their own data. When a peer has some data published, the super peer will be notified. Instances in those data will be sent to the super peer for an initial check upon its request. The initial check searches potentially matched instances that are previously indexed for the new instances through an inverted index. If no instances are found for the new instances or the found instances have very low hits, these new instances will be ignored. This initial check screens off a number of instances. For those instances with potentially matched instances found, instance pairs are created as the input of the classifier.

The trained SVM classifier is used to determine if the instance pairs are matched pairs with the following classification function:

$$f(\mathbf{q}) = \sum_{i=1}^{l} \alpha_i y_i K(\mathbf{p_i}, \mathbf{q}) + b \quad (2)$$

where $K(\mathbf{p}, \mathbf{q})$ is a kernel function used for mapping features into different spaces, α_i is the Lagrangian coefficient of the i-th training instance pair, $y_i \in \{-1, +1\}$ is the label of the training instance pair. In this function, α_i, b are obtained during the learning phase and $f(\mathbf{q})$ indicates the distance of $\mathbf{q}$ from the optimal hyperplane. So we can use this value to evaluate the confidence level of the pair being matched [11]. That is, if $f(\mathbf{r}) > f(\mathbf{q})$, then $\mathbf{r}$ is more likely to be a matched pair than $\mathbf{q}$. For a potentially matched instance pair $\mathbf{q}$, we regard $\mathbf{q}$ as a matched pair if $f(\mathbf{q}) > \delta$, where δ is obtained from experiments. This δ allows the classifier to achieve the maximum F measure [16] in the cross validation.

After matched instance pairs are found through the above process, an index storing data relation information among peers is updated by adding information about these pairs. This index reveals the semantic relations of the data across the peers. It is therefore possible to query related information from more than one peer.

6 Experiments

Experiments are conducted to test the effectiveness of ontology data matching. Data related to the university setting (Professors, Publications, and etc) are collected from five different sources over the Web. As these existing data are in various formats, data conversion has been performed to make them aligned to a backbone ontology schema that describes the university setting. Totally there are 453 instances in the data set. After manually checking these instances, 136 instances are found to match with each other. Instance pairs including matched and unmatched pairs are then generated from these tagged instances. The SVM^{light} package [17] is used in our experiments. 20 random experimental trials are conducted. For one trial we split the pair set into two folds randomly, one for training, the other for testing and then reverse. Traditional measurements such as *precision*, *recall* and *F measure* [16] are used for evaluation. We record the maximum F measure achieved in each trial and its corresponding recall and precision. The overall results, shown in table 1, are obtained by averaging all these trials. The first row indicates the different methods of similarity measurements (or their combinations) used in creating the feature vectors. "ONTO" indicates the features related to ontology (CD, CS) are used. Overall, the method that explores the ontology features yields the best result.

Table 1. Overall results of different methods of similarity measurements used for ontology data matching

Similarity	SIM	SED	SIM+SED	SIM+SED+ONTO
Precision	0.909	0.945	0.919	**0.929**
Recall	0.910	0.752	0.933	**0.946**
F measure	0.909	0.837	0.926	**0.937**

7 Conclusions and Future Work

This paper proposes a framework that supports the generation and matching of ontology data in a peer-to-peer environment. It helps users generate ontology data in two ways. Besides data generation, the issue of ontology data matching in the peer-to-peer environment is also addressed. Experiments show that the proposed matching method which explores the ontology features outperforms other traditional methods. With a matching process that employs this method in the framework, ontology data across peers can be interrelated to offer better information services.

Future work includes the refinement of the data matching process and the design of particular query services based on interrelated ontology data after matching. Since the ontology data matching method can not completely guarantee correct decisions, it is desirable to incorporate peer interaction to correct them. Given the interrelated data across different peers, particular query or reasoning services will be designed to take advantage of them.

References

1. Berners-Lee, T., Hendler, J., Lassila, O.: The semantic web. Scientific American **284**(5) (2001) 34–43
2. Baader, F., Calvanese, D., McGuinness, D., Nardi, D., Patel-Schneider, P., eds.: The description logic handbook : theory, implementation, and applications. Cambridge University Press, New York (2002)
3. McGuinness, D.L., Harmelen, F.v.: Owl web ontology language overview. w3c recommendation. `http://www.w3.org/TR/owl-features/` (2004)
4. Nejdl, W., Wolf, B., Qu, C., Decker, S., Sintek, M., Naeve, A., Nilsson, M., Palmer, M., Risch, T.: Edutella: a p2p networking infrastructure based on rdf. In: WWW 2002. ACM Press, Honolulu, Hawaii, USA (2002) 604–615
5. Halevy, A.Y., Ives, Z.G., Mork, P., Tatarinov, I.: Piazza: data management infrastructure for semantic web applications. In: WWW2003. (2003) 556–567
6. Kalyanpur, A., Parsia, B., Sirin, E., Cuenca-Grau, B., Hendler, J.: Swoop: A 'web' ontology editing browser. Journal of Web Semantics **4**(2) (2006) 144–153
7. Handschuh, S., Staab, S., Maedche, A.: Cream: creating relational metadata with a component-based, ontology-driven annotation framework. In: Proceedings of the international conference on Knowledge capture. ACM Press (2001) 76–83
8. Cimiano, P., Handschuh, S., Staab, S.: Towards the self-annotating web. In: Proceedings of the 13th international conference on World Wide Web. (2004) 462–471
9. Doan, A., Halevy, A.Y.: Semantic-integration research in the database community. AI Mag. **26**(1) (2005) 83–94
10. Tejada, S., Knoblock, C.A., Minton, S.: Learning domain-independent string transformation weights for high accuracy object identification. In: KDD '02, New York, NY, USA, ACM Press (2002) 350–359
11. Bilenko, M., Mooney, R.J.: Adaptive duplicate detection using learnable string similarity measures. In: KDD '03, New York, NY, USA, ACM Press (2003) 39–48
12. Boag, S., Chamberlin, D., Fernndez, M.F., Florescu, D., Robie, J., Simeon, J.: Xquery 1.0: An xml query language. `http://www.w3.org/TR/xquery` (2006)
13. Vapnik, V.N.: The nature of statistical learning theory. 2nd edn. Statistics for engineering and information science. Springer, New York (1999)
14. Gusfield, D.: Algorithms on strings, trees, and sequences : computer science and computational biology. Cambridge University Press (1997)
15. Salton, G., Buckley, C.: Term-weighting approaches in automatic text retrieval. Inf. Process. Manage. **24**(5) (1988) 513–523
16. Baeza-Yates, R., Ribeiro, B.d.A.N.: Modern information retrieval. Addison-Wesley Longman, Reading, Mass. (1999)
17. Joachims, T.: Text categorization with suport vector machines: Learning with many relevant features. In: ECML '98: Proceedings of the 10th European Conference on Machine Learning, London, UK, Springer-Verlag (1998) 137–142

Energy-Efficient Skyline Queries over Sensor Network Using Mapped Skyline Filters

Junchang Xin, Guoren Wang, and Xiaoyi Zhang

Institute of Computer System, Northeastern University, Shenyang, China
wanggr@mail.neu.edu.cn

Abstract. In recent years, wireless sensor network has been widely used in military and civil applications. For many wireless sensor applications, the skyline query is a very important operator for retrieving data according to multiple criteria. In traditional database system skyline queries have been well studied, but in sensor environment the existing solutions are not suitable, because of the essential characteristics of wireless sensor network, such as wireless, multi-hop communication, resource-constrained and distributed environment. An Energy-Efficient Sliding Window Skyline Maintaining Algorithm (EES), which continuously maintains sliding window skylines over a wireless sensor network, is proposed in this paper. In particular, we propose a mapped skyline filter (MSF) in EES. MSF resides in each sensor node and filters the tuples having no contribution to the final result, therefore energy consumption is saved significantly. Our extensive performance studies show that EES can effectively reduce communication cost and save the energy on maintaining sliding window skylines over wireless sensor network.

1 Introduction

A wireless sensor network (WSN), a network consists of several base stations and a large number of wireless sensors, has been widely used in many fields, such as military applications, environmental applications, health applications, traffic surveillance, etc [15, 21]. Each sensor node plays multiple roles as data originator, data router and data processor, which consume a lot of energy. The lifetime of batteries which supplies power for sensor nodes is limited, and it mainly determines the lifetime of sensor network, however battery replacement is impossible or at least very difficult in some circumstances. Thus, a method is needed to manage the tremendous data generated by sensors as well as minimize power consumption to prolong the lifetime of the sensor network.

Skyline query is proposed to retrieve data according to multiple criteria. It can be especially useful in the context of sensor network where multiple criteria query is essential, for example, the drier and hotter the forest is, the more possible to catch fire. In sensor networks, sensor nodes are generally cheap, wireless, multi-functional, resource-constrained and distributed. Due to these characteristics, existing solutions which are mainly focus on traditional database are inapplicable to the sensor network environment directly.

G. Dong et al. (Eds.): APWeb/WAIM 2007, LNCS 4505, pp. 144–156, 2007.

In wireless sensor network, data are collected by sensing devices periodically, a wireless sensor network is more like a distributed, multiple data stream system than a traditional database. While in a data stream system, sliding window computation is often considered. Thus, in this paper, we explore *sliding window skylines*, which seek the skylines over the latest data that are constrained by a sliding window. As mentioned above, energy is the most precious resource in sensor networks, and it is mainly consumed by wireless communication. Therefore, the main challenge in maintaining sliding window skyline is how to minimize the communication cost in the sensor network.

In this paper, we propose an energy-efficient algorithm (EES) to continuously maintain sliding window skylines over a sensor network. EES uses a mapping function to map the data to a smaller range of integer, and carries out the skyline of the mapped set as the mapped skyline filter (MSF). MSF within each node can filter the data that is strictly dominated by the elements in it. The benefit brought by MSF is much more than the cost, because only several bits are needed for small range of data. Consequently the amount of data transferred is reduced greatly, and as a consequence the energy consumption is saved. The contributions of this paper are summarized as follows:

- We prove theoretically that the mapped skyline can be used as a filter to reduce the data communication and can be maintained dynamically.
- We propose an energy-efficient approach which uses the mapped skyline as the filter to compute initial sliding window skyline over wireless sensor network and a strategy to maintain the MSF so as to maintain the sliding window skyline over wireless sensor network.
- Last but not the least, our extensive simulation studies show that EES performs effectively on reducing communication cost and saving the energy on maintaining sliding window skylines.

The rest of the paper is arranged as follows. Section 2 briefly reviews the previous related work. The basic and our energy-efficient algorithm to maintain sliding window skyline is introduced in Section 3. The extensive simulation results to show the effectiveness of the proposed algorithm are reported in Section 4. Finally we conclude in Section 5.

2 Related Work

Borzonyi et al. [3] first investigate the skyline query and present several methods, including SQL implement, divide-and-conquer (DC) and block-nested-loop (BNL), to compute skylines. A pre-sort method is presented in [5], which sorts the dataset according to a monotone preference function and then computes the skyline in another pass over the sorted list. Tan et al. [18] present two progressive methods, Bitmap and Index. Since the nearest neighbor (NN) is sure to belong to skyline, a progressive on-line method based on NN is presented in [9], which allows user to interact with the process. The performance of algorithm presented in [9] is further improved by the algorithm using R-tree in [17].

The methods mentioned above are just suitable for centralized scenarios. So far, we do not find any proposed approach addressing skyline queries over a sensor network. The most related works to ours are some studies about skylines in a distributed scenario. In [4], the skyline problem is extended to the world wide web in which the attributes of an object are distributed in different web-accessible servers, and a basic distributed skyline algorithm (BDS) and an improved distributed skyline algorithm (IDS) are presented, which compute the skyline in such a distributed environment. In BDS, a simple method is used to identify a subset of the objects that include the skyline, and then all the non-skyline objects in that subset are filtered away. IDS uses a heuristic approach and finds the subset more quickly than BDS. Later, in [10], a progressive distributed skyline algorithm (PDS) based on progressiveness and rank estimation is proposed to improve the performance of BDS and PDS. In [8], a hybrid storage model is proposed to reduce the execution time on each single mobile device, and a filtration policy is proposed to reduce communication cost among mobile devices, which is similar to our proposal. However, their approach mainly focuses on answering skyline queries on one timestamp, i.e. snapshot skylines, our approach focuses on continuously maintaining sliding window skyline queries.

There are some works proposed to answer sliding window skyline queries with the focus on handling the characteristics of stream data. A framework is proposed to continuously monitor skyline changes over stream data in [19]. Lin et al. [11] present a pruning technique to reduce the amount of data, an encoding scheme to reduce memory space, and a new trigger based technique to continuously process an n-of-N skyline query, which compute skyline against any most recent n elements in the set of the most recent N elements.

In wireless sensor network literature, the aggregate functions, such as MAX, MIN, AVERAGE, SUM, and COUNT, have been widely studied in the past few years [12, 13, 14, 22], they often use in-network computation to reduce the communication cost. There are also works on implementing join operator in sensor network. For example, Bonfils et al. [2] present a dynamic adjustment algorithm for join operator in in-network query processing, and REED [1] studies efficiently evaluating join queries over static data tables. Moreover, the in-network implementation of general join and range join in sensor network are conducted in [7] and [16], respectively.

3 Sliding Window Skyline Maintaining

In this section, we first describe the sliding window skyline. Then, the Basic Sliding Window Skyline Monitoring Algorithm (BS) is presented in Section 3.2, and our Energy-Efficient Sliding Window Skyline Maintaining Algorithm (EES) is provided in Section 3.3.

3.1 Sliding Window Skyline over Wireless Sensor Network

Skyline query plays an important role in many sensing applications that require retrieval with respect to user preferences. It has been well studied in the traditional database literature with the assumption that data are located in one central site. Skyline query is defined as following:

Definition 1. *Assume that we have a relational database, given a set of tuples T, a skyline query retrieves tuples in T that are not dominated by any other tuple. For two tuples t_i and t_j in T, tuple t_i dominates tuple t_j if it is no worse than t_j in all dimensions and better than t_j in at least one.*

In wireless sensor network, data are collected by sensing devices periodically, and each tuple that has been collected has a timestamp $t.arr$ indicating its arrival time. Since energy is the precious resource in the sensor network and wireless communication is the main consumer, the data will not be transmitted unless necessary. Sensed data are dispersedly stored in each sensor node, therefore, strictly speaking, a wireless sensor network is more like a distributed, multiple data streams system than a traditional database.

It is impossible to carry out skyline operation after all data have been collected, because the sensor stream is infinite, and the volume of the complete stream is theoretically boundless. Thus, *sliding window skyline*, which aims to provide the most recent on-line information, is considered. If the size of sliding window is set to W, and the current time is $t.curr$, Sliding window skyline only considers the data which satisfy $t.arr + W > t.curr$.

3.2 Basic Sliding Window Skyline Maintaining Algorithm

The naive approach to maintain sliding window skyline is to transmit all sensed data to the base station, and then compute and maintain the sliding window skyline there. Since the skyline is only a little part of the entire tuple set, many tuples which have no contribution to the final result will be transmitted to the base station. Thus the number of messages and the power consumption are large. Therefore, this method is unpractical for wireless sensor network.

A better approach is to carry out the computation within each sensor node and then merge the result on the intermediate nodes. But this approach requires the operation to be decomposable. Fortunately, the sliding window skyline query over wireless sensor network has this attractive property.

Denotes the tuple set in the entire sensor network as T, the tuple set for each node as T_i. Using $skyline()$ to stand for skyline operator, we can easily get $skyline(T) = skyline(\bigcup skyline(T_i))$. It satisfies the formula $f(v_1, v_2, \ldots, v_n) = g(f(v_1, v_2; \ldots, v_k), f(v_{k+1}, \ldots, v_n))$ given in [6]. Thus, using in-network computation like TAG [13] is feasible.

Like TAG [13], a tree-based structure rooted at the base station is first established as the rooting tree. The base station broadcasts a message with its own id and level (in general case, zero) to construct the routing tree. Any node that hears this message will assign its own level to be the level in the message plus one, and choose the sender as its parent, then replace the id and level in message with its own id and level, finally rebroadcast the routing message to its neighbors. The routing tree is constructed step by step until all nodes have been assigned a level and a parent. The process above will be initiated periodically by the base station, thus the network topology will be constructed periodically. Therefore, this structure can easily adapt to the moving, entering or removing of the sensor nodes.

Once the construction of rooting tree is finished, the skyline will be computed in-network whenever possible. Each leaf node computes its own skyline and forwards the skyline to its parent. The intermediate nodes receive the skyline of their children and combine these results with their own using the merging function. Then, they submit the new partial results to parents of their own. Most transmission of the tuples that belong to local skyline but not global skyline is terminated on the intermediate nodes.

After the process of sliding window skyline computation, new tuples are collected by sensor nodes, while the old ones expire. A simple approach to maintain the global skyline is to recompute the skyline periodically using the method presented above. Obviously it is unpractical, because there is a great intersection between the old skyline and the new one, same as the old window and the new one. The redundant data need not to be transmitted again, so an effective way should be "*update-only*", which means only the tuples that have not been transmitted are transmitted in maintenance. Therefore, the communication cost is further reduced.

3.3 Energy-Efficient Sliding Window Skyline Maintaining Algorithm

The in-network computation can reduce the amount of data transferred among sensor nodes, however, there are still a great number of tuples that do not belong to the final skyline having been transmitted.

Generally speaking, data collected by sensor nodes is float. If a good mapping function with careful design is used to map the float data to a range of integer, we can carry out the skyline of mapped set and use it to filter data that do not belong to skyline. Since only several bits are needed to present an arbitrary integer in this range, the cost of computation and broadcasting the filter is very low. The benefit brought by this process is much more than the cost, thus the transmission cost is reduced greatly.

With the following mapping function, $x \in [l, u]$ can be mapped to the integer range of $[0, m]$.

$$f(x) = \left\lceil \frac{x-l}{u-l} \times m \right\rceil$$

Note in the above function that it has a serial of properties, which can be used to reduce the amount of data transferred in sensor network, therefore the performance is greatly improved.

Lemma 1. *If $f(x_i) < f(x_j)$, then $x_i < x_j$.*

Proof: According to properties of the function, we have

$$0 < \left\lceil \frac{x_j-l}{u-l} \times m \right\rceil - \frac{x_j-l}{u-l} \times m < 1$$

$$f(x_i) < f(x_j) \Leftrightarrow \left\lceil \frac{x_i - l}{u - l} \times m \right\rceil < \left\lceil \frac{x_j - l}{u - l} \times m \right\rceil$$

$$\left\lceil \frac{x_j - l}{u - l} \times m \right\rceil - \left\lceil \frac{x_i - l}{u - l} \times m \right\rceil \geq 1$$

$$\frac{x_j - l}{u - l} \times m > \left\lceil \frac{x_i - l}{u - l} \times m \right\rceil$$

$$\left\lceil \frac{x_i - l}{u - l} \times m \right\rceil \geq \frac{x_i - l}{u - l} \times m$$

$$\frac{x_j - l}{u - l} \times m > \frac{x_i - l}{u - l} \times m$$

Therefore, we can conclude that $x_i < x_j$. □

We use $\succeq$ to stand for the dominance relationship, t for a tuple in T, and $t.x_d$ for the dth attribute of tuple t. Assume that the dimensionality of tuple set is D. We denote

$$f(t) = (f_1(t.x_1), f_2(t.x_2), \cdots, f_D(t.x_D))$$

$$f_k(t.x_k) = \left\lceil \frac{t.x_k - l_k}{u_k - l_k} \times m_k \right\rceil \qquad k = 1, 2, \cdots, D$$

where $[l_k, u_k]$ is range of the kth dimension, the total bits needed to present the element that a tuple mapped to is $\sum_{k=1}^{D} \lceil \log_2 m_k \rceil$. In general, the value of l_k and u_k are set according to the history data. Usually, u_k is set to twice of the maximum value, and l_k is set to half of the minimum value [20].

Definition 2. *For two tuples t_i and t_j in T, tuple t_i strictly dominates tuple t_j if it is better than t_j in all dimensions. Using $\succ$ to denote the strictly dominance relationship.*

We can easily conclude,

Lemma 2. *If $t_i \succ t_j$, then $t_i \succeq t_j$.*

Proof: Immediate deduct from the definition 1 and 2. □

Lemma 3. *If tuple t_i and t_j are two tuples in T, then $f(t_i) \succ f(t_j) \Rightarrow t_i \succeq t_j$.*

Proof: Since $f(t_i) \succ f(t_j)$, we have

$$(f_1(t_i.x_1), f_2(t_i.x_2), \cdots, f_{|D|}(t_i.x_{|D|})) \succ (f_1(t_j.x_1), f_2(t_j.x_2), \cdots, f_{|D|}(t_j.x_{|D|}))$$

That is $\forall k \in \{1, 2, \cdots, D\}, f_k(t_i.x_k) < f_k(t_j.x_k)$,
From lemma 2, $\forall k \in \{1, 2, \cdots, D\}, t_i.x_k < t_j.x_k \Rightarrow t_i \succ t_j$,
Therefore, we can conclude that $t_i \succeq t_j$. □

We denote T' as tuple set that is obtained through mapping, and denote *skyline* (T') as the skyline of T'.

Theorem 1. *For a tuple t_j in T, if there exists one tuple t_i such that $f(t_i) \in$ skyline(T') and $f(t_i) \succ f(t_j)$, then $t_j \notin$ skyline(T)*

Proof: Immediate deduct from lemma 3. □

According to theorem 1, it will reduce the amount of data transferred in wireless sensor network if $skyline(T')$ is used as a filter.

Definition 3. *We define skyline(T') as mapped skyline, and define mapped skyline used in filter as mapped skyline filter (MSF).*

The process of sliding window skyline computation which integrates MSF as the filter is presented as follows:

1. Determine the parameter l, u and m, that is to determine the range $[l_k, u_k]$ and the corresponding integer range $[0, m_k]$ to be mapped on each dimension.
2. Each sensor node maps its own T_i to T_i' using $f(t)$, and carrys out $skyline(T_i')$.
3. Get $skyline(T')$ using the method of in-network computation, and set it as MSF.
4. Broadcast MSF to the entire network.
5. Remove tuples that are filtered by the MSF in sensor nodes.
6. Use in-network computation to carry out skyline query.

How to maintain MSF incrementally becomes a critical problem in the process of maintaining sliding window skyline in sensor network. The following lemma and theorem help us to maintain MSF dynamically in sliding window skyline maintaining process.

Lemma 4. *If $t_i \succeq t_j$, then $f(t_i) \succeq f(t_j)$.*

Proof: Immediate deduct from the mapping function. □

Theorem 2. *Let $S = skyline(T)$, and S' denotes the set mapped from set S, Then $skyline(T') = skyline(S')$.*

Disproof: Assume there exists a tuple t', $t' = f(t) \in T' - S'$, and $t' \in skyline(T')$.

$$\because t \in S \Rightarrow t' = f(t) \in S' \Rightarrow f(t) \notin T' - S'$$
$$\therefore t' \in T' - S' \Rightarrow t \notin S \Leftrightarrow t \in T - S$$

According to the definition of skyline,

$$\exists s \in S, s \succeq t$$

From lemma 4, we have $f(s) \succeq f(t)$, So $t' \notin skyline(T')$, conflicting with the assumption $t' \in skyline(T')$. So, $skyline(T') = skyline(S')$. □

According to theorem 2, we know that the base station can get the corresponding MSF using the skyline result at each timestamp. Comparing the new MSF with the old one, when base station finds invalid or new elements in MSF, it broadcasts them to the entire sensor network. Sensor node updates its own MSF to guarantee the correctness of filtering constantly according to certain strategy. In the broadcast package, only the invalid or new elements need to be transmitted, and the operation of addition or deletion will not be identified. Let MSF^+ denote the set of elements which need to be transmitted, then we have

$$MSF^+ = (MSF_{old} - MSF_{new}) \cup (MSF_{new} - MSF_{old})$$

The revised approach is as follow.

$$MSF_{new} = (MSF_{old} - MSF^+) \cup (MSF^+ - MSF_{old})$$

Therefore, the MSF are correctly maintained dynamically in the process of sliding window skyline maintenance.

In most cases, the data generated by sensor nodes is in $[l_k, u_k]$, however some special cases may take place. Once data is out of this range, original mapping function is out of use, because the data that is mapped to is also out of range. So we need to renew the parameters and broadcast them as well as the new MSF carried out according to the new parameters to entire wireless sensor network. Then each node recalculates mapping according to the new parameters and uses MSF to filter data. In this way, sliding window skyline can be maintained very well in all cases.

4 Performance Evaluation

In this section, we present our simulation results evaluating the performance of energy-efficient sliding window skyline maintaining algorithm (EES) against the basic sliding window skyline maintaining algorithm (BS) under two data distributions, independent and anti-correlated, which are the common benchmarks for skyline query [3, 19].

We simulate sensor network by randomly placing n sensors nodes in an area of $\sqrt{n} \times \sqrt{n}$ units, and the communication radius of each sensor node is set to $2\sqrt{2}$. The experimental data evenly distribute on n sensor nodes. The number of sensor nodes n is in the range from 600 to 1000, the dimension of sensory data d ranges from 2 to 4, and the size of sliding window c varies from 100 to 500. Each sensor node generates a new tuple on each timestamp, thus there will be n new tuples generated in the whole sensor network.

All experiments are run on a PC with 2.8GHz CPU, 512M of memory and 80G harddisk. The default setting of experiment is $n = 1000$, $c = 300$ and $d = 3$.

First,we study the effect of the selection of integer range m in the process of skyline computation under independent and anti-correlated data distribution respectively. For simplicity, we let $m = 2^x - 1$. Figure 1 shows the total communication cost (the number of bytes of the messages transferred among sensor nodes) under different m. The best m for independent distribution is 1023 and the best m for anti-correlated distribution is 63. This is because the larger m means the larger broadcast cost of filter, while the smaller m means the lower filter ability. On the best choice of m, the cost and the benefit of the filter balance well. Therefore, in the experiment of skyline computation, m is 1023 for independent data, and set to 63 for anti-correlated data.

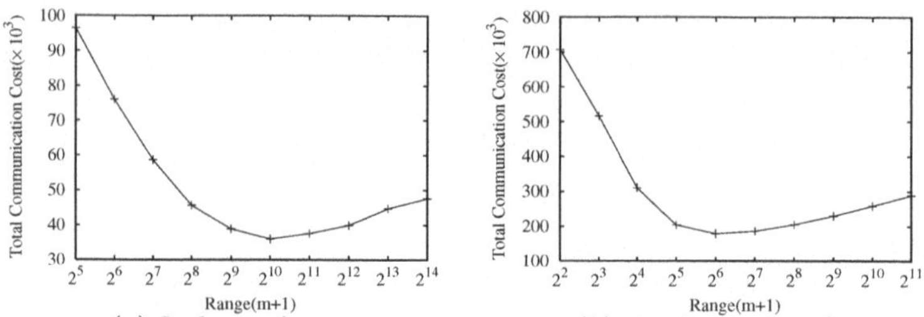

(a) Independent (b) Anti-correlated

Fig. 1. Effect of Range (m) in skyline computation

Figure 2, 3 and 4 present the influence on performance by dimension, cardinality and the number of nodes under independent and anti-correlated data distribution, respectively. They show that EES always performs better than BS under all circumstances. Communication cost increases with the increase of dimensions, since the skyline result will increase with a high dimension which leads to the increment of communication cost. Change of cardinality and the number of nodes also affect the cost, and EES increases more slowly than BS with the increase of c and n respectively.

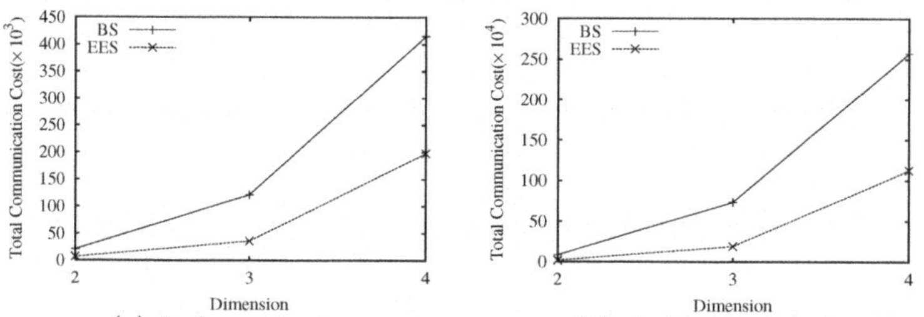

(a) Independent (b) Anti-correlated

Fig. 2. Communication Cost Vs. Dimension in skyline computation

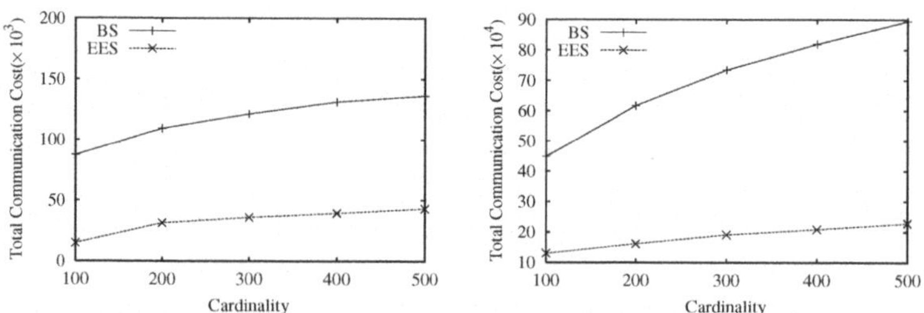

(a) Independent (b) Anti-correlated

Fig. 3. Communication Cost Vs. Cardinality in skyline computation

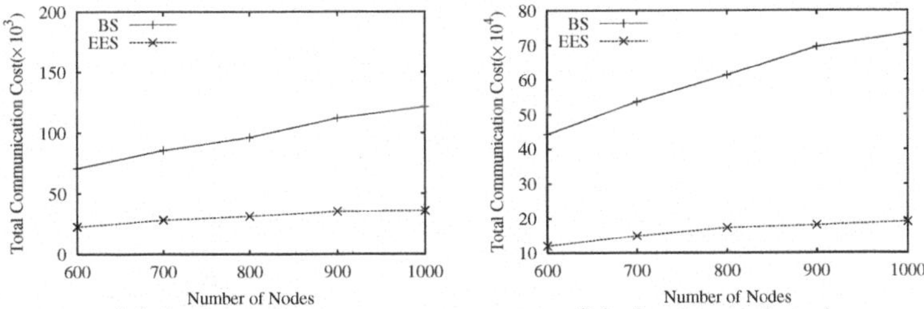

(a) Independent (b) Anti-correlated

Fig. 4. Communication Cost Vs. Number of sensor nodes in skyline computation

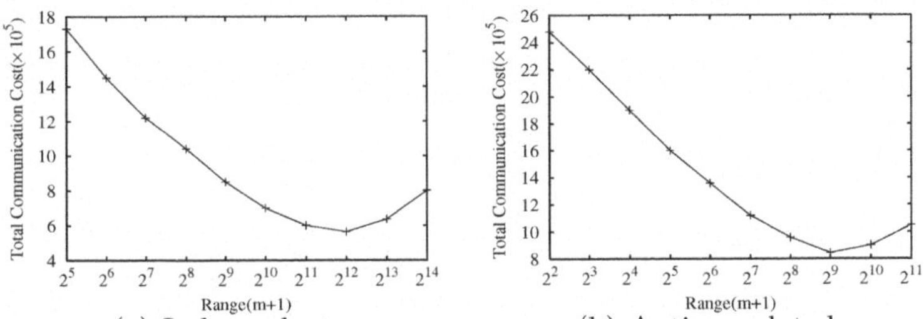

(a) Independent (b) Anti-correlated

Fig. 5. Effect of Range (m) in skyline maintenance

Next, we study the performance of EES and BS in the process of skyline maintenance. Before comparing the performance of BS and EES, the effect of the selection of integer range m in the process of skyline maintenance under independent and anti-correlated data distribution is studied respectively. The sliding length is 500 time-stamps. Figure 5 shows that the best m for independent distribution is 4095, and it is 511 for anti-correlated distribution. Both are larger than the ones in skyline computation, because each element in MSF has a long aging in the process of skyline maintenance, and thus the broadcast cost will be shared by each timestamp, as a result the optimum m changes. The same, m is set to 4095 for independent data, and 511 for anti-correlated data in process of skyline maintenance.

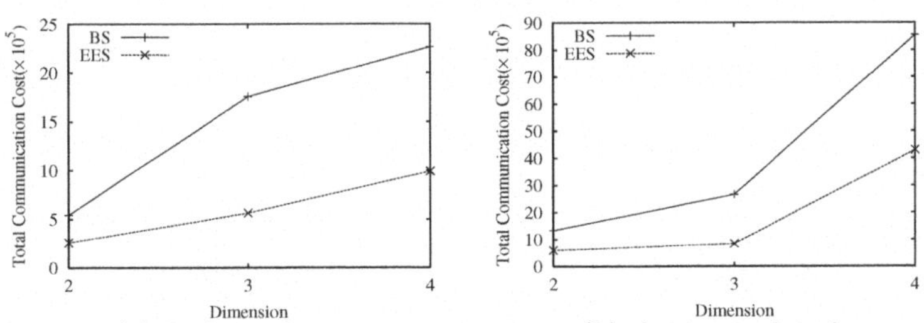

(a) Independent (b) Anti-correlated

Fig. 6. Communication Cost Vs. Dimension in skyline maintenance

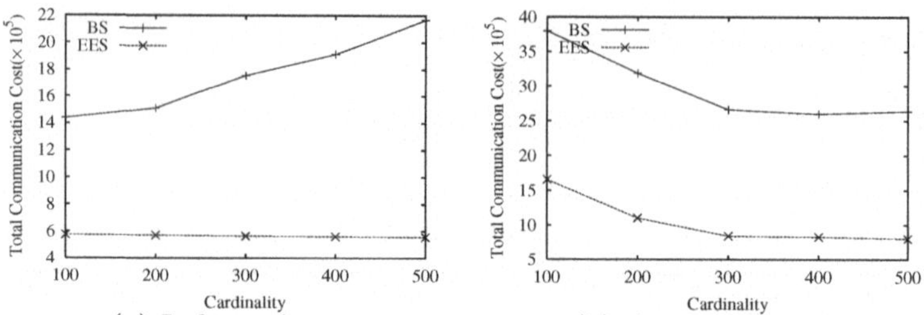

(a) Independent (b) Anti-correlated

Fig. 7. Communication Cost Vs. Cardinality in skyline maintenance

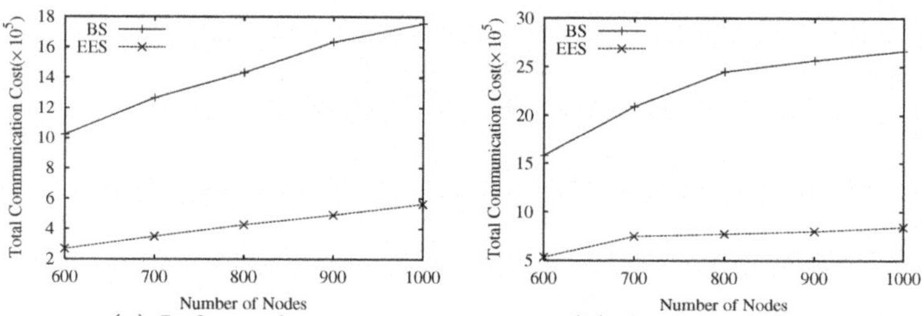

(a) Independent (b) Anti-correlated

Fig. 8. Communication Cost Vs. Number of sensor nodes in skyline maintenance

Figure 6, 7 and 8 demonstrate the influence on performance by dimension, cardinality and the number of nodes under independent and anti-correlated data distribution, respectively. they show the similar result as the process of skyline computation except cardinality. The communication cost slightly decreases with the increase of cardinality in some cases, because the tuple has a longer lifespan, and the chance of a new tuple to join the skyline result may be smaller.

Finally, we study the time-varying regularity of communication cost in the process of skyline maintenance. Figure 9 shows that the communication cost increases smoothly with time. EES always performs better than BS.

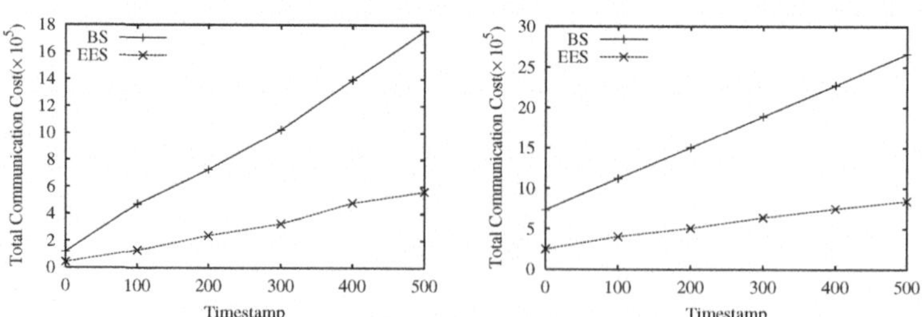

(a) Independent (b) Anti-correlated

Fig. 9. Communication Cost Vs. Time

5 Conclusion

In this paper, we focus on continuously maintaining sliding window skyline over sensor network. In particular, we propose a mapped skyline filter (MSF) to reduce the communication cost in process of skyline computation. Moreover, the method to maintain MSF so as to maintain the sliding window skyline is discussed. Our experiment result proves that EES is an energy-efficient approach for computing and maintaining sliding window skyline on sensor streams.

Acknowledgement. This work is partially supported by National Natural Science Foundation of China under grant No. 60573089 and 60473074 and supported by Natural Science Foundation of Liaoning Province under grant no. 20052031.

References

1. D. J. Abadi, S. Madden, and W. Lindner: REED: Robust, Efficient Filtering and Event Detection in Sensor Networks. In *Proc. of VLDB*, pages 769-780, 2005.
2. Boris Jan Bonfils and Philippe Bonnet: Adaptive and Decentralized Operator Placement for In-Network Query Processing. In *Proc. of IPSN*, pages 47-62, 2003.
3. S. Borzonyi, D. Kossmann, and K. Stocker: The skyline operator. In *Proc. of ICDE*, pages 421-430, 2001.
4. W.-T. Balke, U. Guntzer, J. X. Zheng: Efficient distributed skylining for web information systems. *EDBT*, pages 256-273, 2004.
5. J. Chomicki, P. Godfrey, J. Gryz, and D. Liang: Skyline with presorting. In *Proc. of ICDE*, pages 717-719, 2003.
6. J. Considine, F. Li, G. Kollios, and J. Byers: Approximate aggregation techniques for sensor databases. In *Proc. of ICDE*, pages 449-460, 2004.
7. Vishal Chowdhary, Himanshu Gupta: Communication-Efficient Implementation of Join in Sensor Networks. In *Proc. of DASFAA*, pages 447-460, 2005.
8. Zhiyong Huang, Christian S. Jensen, Hua Lu, Beng Chin Ooi1: Skyline Queries Against Mobile Lightweight Devices in MANETs. In *Proc. of ICDE*, pages 66, 2006.
9. D. Kossmann, F. Ramsak, S. Rost: Shooting Stars in the Sky: An Online Algorithm for Skyline Queries. In *Proc. of VLDB*, pages 275-286, 2002.
10. Eric Lo, Kevin Ip, King-Ip Lin, David Cheung: Progressive Skylining over Web-Accessible Database. *DKE*, 57(2): 122-147, 2006.
11. Xuemin Lin, Yidong Yuan, Wei Wang, Hongjun Lu: Stabbing the Sky: Efficient Skyline Computation over Sliding Windows. In *Proc. of ICDE*, pages 502-513, 2005.
12. S. Madden, M. Franklin, J. Hellerstein, and W. Hong: The design of an acquisitional query processor for sensor networks. In *Proc. of SIGMOD*, pages 491-502, 2003.
13. S. Madden, M. J. Franklin, J. M. Hellerstein, and W. Hong. TAG: A Tiny AGgregation Service for Ad-Hoc Sensor Networks. In *Proc. of OSDI*, 2002.
14. S. Madden et al.: Supporting aggregate queries over ad-hoc wireless sensor networks. In *Proc. of WMCSA*, pages 49-58, 2002.
15. R. Oliver, K. Smettem, M. Kranz, K. Mayer: A Reactive Soil Moisture Sensor Network: Design and Field Evaluation. *JDSN*, 1: 149-162, 2005.
16. Aditi Pandit, Himanshu Gupta: Communication-Efficient Implementation of Range-Joins in Sensor Networks. In *Proc. of DASFAA*, pages 859-869, 2006.

17. D. Papadias, Y. Tao, G. Fu, et.al.: An Optimal and Progressive Algorithm for Skyline Querie. In *Proc. of SIGMOD*, pages 467-478, 2003.
18. K.-L. Tan, P.-K. Eng, and B. C. Ooi: Efficient progressive skyline computation. In *Proc. Of VLDB*, pages 301-310, 2001.
19. Yufei Tao, Dimitris Papadias: Maintaining Sliding Window Skylines on Data Streams. *TKDE*, 18(3): 377-391, 2006.
20. M. Wu, J. Xu, X. Tang and Wang-Chien Lee: Monitoring Top-k query in wireless sensor network. In *Proc. of ICDE*, pages 143, 2006.
21. W. Xue, Q. Luo, L. Chen, and Y. Liu: Contour Map Matching For Event Detection in Sensor Networks. In *Proc. of SIGMOD*, pages 145-156, 2006.
22. Y. Yao and Johannes Gehrke: Query processing in sensor networks. In *Proc. of CIDR*, 2003.

An Adaptive Dynamic Cluster-Based Protocol for Target Tracking in Wireless Sensor Networks

WenCheng Yang, Zhen Fu, JungHwan Kim, and Myong-Soon Park*

Dept. of Computer Science and Engineering, Korea University
Seoul 136-701, Korea
{wencheng,fuzhen,glorifiedjx,myongsp}@ilab.korea.ac.kr

Abstract. The rapid progress of wireless communication and embedded micro sensing technologies has made wireless sensor networks possible. Target tracking is an important application of wireless sensor networks. Good tracking quality and energy efficiency are the key requirements for any protocol designed for target tracking sensor networks. In this paper, we present a novel protocol, Adaptive Dynamic Cluster-based Tracking (ADCT), for tracking a mobile target. This protocol uses the optimal choice mechanism and dynamic cluster-based approach to achieve a good tracking quality and energy efficiency by optimally choosing the nodes that participate in tracking and minimizing the communication overhead, thus prolongs the lifetime of the whole sensor network. Simulation results show that our protocol can accurately track a target with random moving speeds and cost much less energy than other protocols for target tracking.

1 Introduction

With the advances in the fabrication technologies that integrate the sensing and the wireless communication technologies, tiny sensor nodes can be densely deployed in the battle fields or the urban areas to form a large-scale wireless sensor network. Hundreds of thousands of sensor nodes may be deployed in a surveillance region. So the density of the sensor networks may be very large, for example in a ten square meters region more than five or six sensor nodes may be deployed. This feature of sensor networks makes tracking with wireless sensor networks having several advantages: (1) the quality of the sensed data will be more reliable, because the senor nodes can be deployed much closer to the target (2) With a dense deployment of sensor nodes, the information about the target is simultaneously generated by multiple sensors and thus contains redundancy, which can be used to increase the accuracy of tracking. However challenges and difficulties also exist in target tracking sensor networks: (1) the sensor nodes have limited power, processing and sensing ability. Because the sensor node is usually sustained by battery which cannot be changed during its lifetime so the limit of power is especially intense. In order to save energy, every node cannot be always in active mode (2) the sensor nodes are prone to failure because of lack of power, physical damage or environment interference. So the topology of sensor networks will be

* Corresponding author.

G. Dong et al. (Eds.): APWeb/WAIM 2007, LNCS 4505, pp. 157–167, 2007.

very easy to change (3) the information generated by a single node is usually incomplete or inaccurate. Thus tracking needs collaborative communication and computation among multiple sensors.

Thus, the target tracking sensor networks need protocols which can efficiently organize the sensor nodes to track the target with the less energy dissipation, so that prolong the lifetime of the sensor networks, and also the protocols should be fault tolerant that one or some nodes' death cannot influence the overall task of sensor networks. Finally the protocols should make sure high tracking quality.

In this paper, we propose a distributed and scalable cluster-based protocol (ADCT) to accurately and efficiently track a mobile target using wireless sensor networks. Energy conservation and tracking quality are the two key guidelines of our protocol. In the initial phase, all the nodes in power save model. Given a target to track into the sensor networks, the protocol provides an optimal choice mechanism for locally determining an optimal set of sensors suitable to incorporate in the tracking collaboration. Only these nodes are then active thus minimizing the energy spent on tracking. Additionally, the protocol uses predictive-based and low-delay mechanisms to select new cluster head, and then the new cluster head form a new cluster around the target and reuses the optimal choice mechanism to choose appropriate number of nodes to join in the tracking collaboration. Thus the tracking maintenance can be kept.

The rest of the paper is organized as follow: we give an overview of the existing protocols for target tracking in section 2. We introduce our protocol in detail in section 3, and present the simulation results in section 4. Finally, we conclude the paper and give the future work in section 5.

2 Related Work

In the current body of research done in the area of target tracking in wireless sensor networks, we see that the researches are mainly dedicated to the design of some energy efficient schemes for target tracking which try to explore good trade-off between energy conservation and tracking quality. According to the survey by Chuang [1], the researches about the target tracking can be divided into three categories (1) tree-based scheme (2) cluster-based scheme (3) prediction-based scheme. The examples of each scheme are introduced as following:

Tree-based scheme: In [2] a dynamic convey tree-based collaboration (DCTC) framework has been proposed. The convoy tree includes sensor nodes around the detected target, and the tree progressively adapts itself to add more nodes and prune some nodes as the target moves. Relying on the convoy tree, the information about the target generated from all the on-tree nodes will be gathered to the root node, which then sends the gathered information to the base station. DCTC, however, has some limitations, first the tree in the DCTC is a logic tree and does reflect the physical structure of the sensor network, second as the target moves many nodes in the tree may become far from the root of the tree and hence a lot of energy would be wasted for them to send their sensed date to the root.

Cluster-based scheme: In [3] a dynamic cluster-based algorithm is proposed, it assumes that the sensor network is composed of sparsely placed high-capability sensors (CH) and normal nodes. When a CH is triggered by certain signal events, the CH becomes active and then broadcasts an information solicitation packet, asking sensors in its vicinity to join the cluster and provide their sending information. The CH then compresses data and sends the aggregated data to the base station. However, only the CH can be the cluster head, so that the CH can easily overly-utilized. Also in the real environment in some places may have no such CH, thus the tracking of target will be lost.

Prediction-based scheme: In [4] [5] [6] a prediction-based method has been used to predict the next position of the target basing on the following assumptions: (1) the current moving speed and direction of the target don't change for the next few seconds (2) the speed and the direction for the next few seconds can be computed by the current sensed data. In some papers, the wake-up mechanism has been taken to work with prediction-based method, according to the prediction result of current sensor nodes to wake up the sensor nodes lying on the predicted moving path, before the target leaves the current detection range and enters the adjacent range. Different wake-up mechanism may be chosen according to different requirements of the application. Three main methods are listed below: (1) awakes only one sensor nearest to the predicted destination (2) awakes all the nodes on the route of the moving target from current location to the destination (3) wakes all the nodes on and around the route of the moving target from current location to the destination [7] [8]. All these three methods may have achieved some energy efficiency in a certain extent. However, for the first method, because one time there is just only one node to monitor the target so the tracking quality cannot be guaranteed. For the second method, the missing rate of the tracking will increase when the target's moving speed and direction beyond the prediction because in the real environment the target's moving speed and direction maybe change very fast. For the third method, although it considers the tracking quality very well, it ignores that it will waste much energy to wake up so many nodes which cannot join in the target tracking in the nearest future.

3 Adaptive Dynamic Cluster-Based Tracking Protocol (ADCT)

In this section we describe our proposed protocol, ADCT which is aimed at addressing the various challenges that we proposed in the introduction for wireless sensor networks while especially for efficiently and accurately tracking the moving target. The ADCT protocol includes the following 4 phases.

3.1 Initial Cluster Forming

We assume that each node knows the location information of its one-hop neighboring nodes and can estimate the cost of communicating to its one-hop neighboring nodes. At the initial time, all the nodes in the sensor networks are in the power save mode. They just periodically wake up and do the sensing for a short time. If nothing happens, then they will fall asleep for another period time. Once some nodes detect the target they will form a cluster and enter the target tracking state. Firstly, a cluster head should be elected among the initial nodes. Since the communication cost of deterministic leader election is very high, we propose to use a simple heuristic 2 phase-based mechanism to simply determine the cluster head. In the first phase, the sensor nodes that detect the target will be required to broadcast an election message (di, idi) with its distance to the target (di) and its own ID (idi) to its neighboring nodes. If a node does not receive any election message with (dj, idj) that is smaller than (di, idi), it becomes a cluster head candidate. Otherwise, it gives up and selects the neighbor with the smallest (dj, idj) to be its head. However, multiple head candidates are possible to appear at the same time. Thus, the second phase is needed by letting the candidate i flood a winner (di, idi) message to other nodes. A head candidate i will give up the candidacy when it receives a winner (dj, idj) message with smaller (di, idi) values. Then, a candidate node with the smallest (di, idi) will be selected to be the cluster head. Finally a join-request message will be sent by the selected cluster head to ask its one-hop neighbor to join the cluster. After receiving the join-request message, all its one-hop neighboring nodes join the cluster.

3.2 Optimal Sensor Selection

The tracking of target requires cluster head to aggregate data among the cluster member nodes. However, not all cluster members that detect the target have useful information.

An informed selection of optimal nodes that have the best data and cost less energy for collaboration will save both power and bandwidth cost. Thus not only the tracking quality can be guaranteed but also prolong the lifetime of the sensor network. In our ADCT protocol, we take an optimal choice mechanism to choose the appropriate cluster members which own the best data and cost less energy to join the tracking collaboration.

After the cluster is formed, the cluster head sends a message which contains the estimate of the target and an optimal node selection command to its cluster members. While receiving the message, member nodes combine their own measures of target with the cluster head's estimate to compute a value using an optimal selection function and then respond the cluster head by a bid. The cluster head evaluates the received bids and ranks them according to the value of bids. We use a similar method as [9] to define the optimal selection function as a mixture of both data usefulness and energy cost:

$$Q(\mu(x|m_i, ch)) = \alpha * \lambda_{use}(\mu(x|m_i, ch)) - (1-\alpha) * \eta_{cost}(m_i) \quad (1)$$

where $\mu(x|m_i, ch)$, $m_i \in members(ch)$, is the estimate of the target formed by each cluster member combines its own estimate $\mu(x|m_i)$ with the estimate $\mu(x|ch)$ from cluster head, using the Bayesian filter: $\mu(x|m_i, ch) = \mu(x|m_i) \oplus \mu(x|ch)$. $\lambda_{use}(\mu(x|m_i, ch))$ is the data usefulness measure function, $\eta_{cost}(m_i)$ is the cost of communication between the cluster head ch and member node m_i. α is the relative weighting of the usefulness and cost.

In the selection function $Q(\mu(x|m_i,ch))$, it includes two terms. The first term $\lambda_{use}(\mu(x|m_i,ch))$ characterizes the usefulness of the data provided by the member node m_i . According to the Mahalanobis distance measure, the usefulness of the sensor data can be measured by how close the member node m_i is to the target position estimated by m_i and ch . The measure function is below:

$$\lambda_{use}(\mu(x|m_i,ch)) = \lambda_{use}(x_i,\bar{x}) = -(x_i-\bar{x})^T \sum^{-1}(x_i-\bar{x}) \quad (2)$$

where x_i is the position of member node m_i , and $\bar{x}$ is the target position estimated by m_i and ch . $\sum$ is the covariance of $\bar{x}$. The second term $\eta_{cost}(m_i)$ measures the energy cost of communication between the cluster head and member node. We use the Euclidean distance as a crude measure of the amount of energy which is required to communicate between the cluster head and member node. $\eta_{cost}(m_i)$ can be denoted as a function related with the distance between the member node and cluster head. The function is below:

$$\eta_{cost}(x_0,x_i) = (x_i-x_0)^T(x_i-x_0) \quad (3)$$

where x_i is the position of the member node, x_0 is the position of the cluster head, (x_i - x_0) is the distance between the member node and cluster head. Combining the function (1) with the functions (2) and (3), the selection function (1) is reduced as below:

$$Q(x_0,x_i,\bar{x}) = -\alpha*(x_i-\bar{x})^T \sum^{-1}(x_i-\bar{x}) - (1-\alpha)*(x_i-x_0)^T(x_i-x_0) \quad (4)$$

This function only relates with cluster head, cluster member and target's positions: x_0 , x_i and $\bar{x}$.

After the rank of contributions of nodes in estimating the state of target is created by using the above optimal selection function (4). Appropriate number of nodes can be chosen to incorporate in the tracking collaboration. According to different specific application and assumption of the sensor networks, the number of the sensor nodes that are sufficient to achieve the tracking task is different. For example, in [10] the author assumes that the sensor knows the location of each sensor and can identify whether a target is moving away from or towards it. With this assumption, a secondary senor can be used in conjunction with the first sensor to discover the precise location of the target. In [11] three sensor nodes are sufficient to determine the location of the target with the assumption that each node can know the distance between itself and the target and all the sensor nodes in the sensor network have synchronous time.

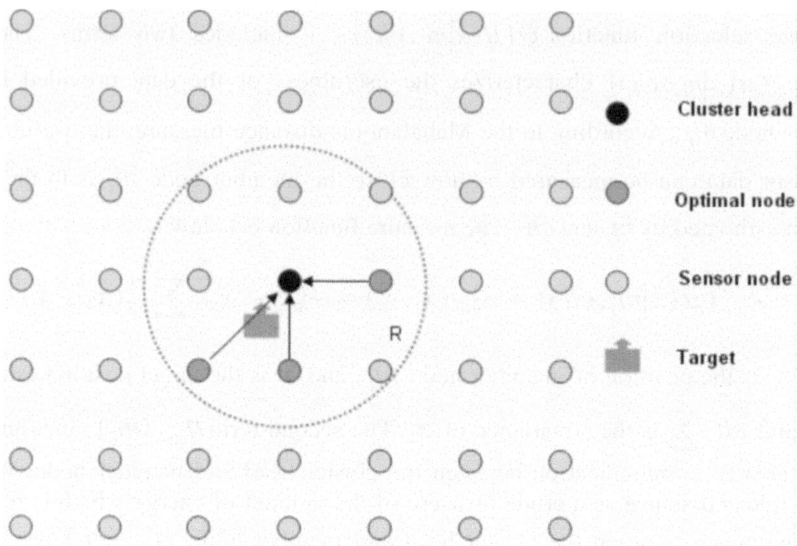

Fig. 1. The formation of a cluster

In order to make our protocol more suitable to different cases, we do not make such specific assumptions. We only define a parameter: ThresholdNumber (TN). This is a threshold value that the number of optimal nodes chosen should be no more than the stated TN. As in figure 1, we set the TN = 3. According to different tracking tasks the value of TN can be changed through the base station broadcasts a TN message to the sensor networks. If the number of bidders is less than the stated TN, all the bidders are chosen. Once the optimal nodes are determined, they will send their sensed data to the cluster head, and then the cluster head compresses the multiple data and generates a more precise estimate of the target state.

3.3 Cluster Reconfiguration

In this paper we focus on the target that moves with varying speed. As shown in figure 2, with the movement of the target, some nodes in the current cluster will drift farther away for the target. In order to keep the track maintenance the nodes lying on the predicted moving path will soon need to form a cluster to join the collaborative tracking. However the election of the cluster head is very important, it not only impacts the accuracy of the tracking but also the energy efficiency of the cluster, thus impact the lifetime of the whole network.

In most of the existing researches related to the cluster head election in target tracking, sensor networks just simply select a node with the highest residual energy or strongest sense ability to be the new cluster head. Also some researches select the node which is nearest to the predicted target position to be the cluster head. However, in the real environment, the prediction may be inaccurate because a target may travel at a varying speed all the time. So the existing methods may not be suitable for target tracking application in real sensor network.

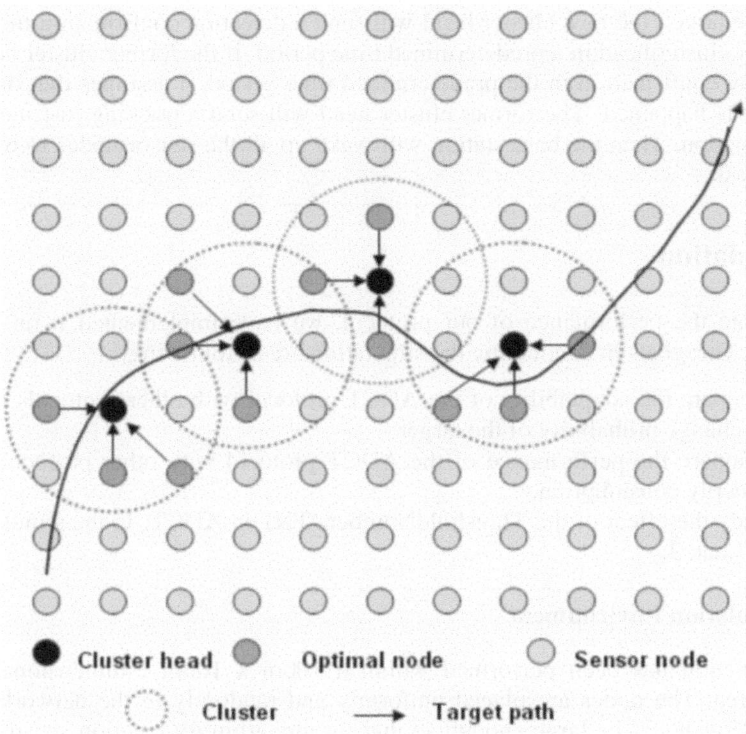

Fig. 2. The maintenance of target tracking

In our ADCT protocol, we take the prediction-based and low-delay algorithm to select a new cluster head. When the predicted position of the target is on the boundary of the current cluster, the current cluster head would send a command message which contains the new cluster selection information to its neighboring node which is nearest to the prediction location. The node which has received the command message would send a new cluster head solicitation packet to its neighboring nodes and select the node which firstly replies the message and is not the neighbor of current cluster head to be the new cluster head. The new cluster head then combine with its one-hop neighbor to form a cluster and reuses the optimal choice mechanism to select optimal member nodes to incorporate in the tracking collaboration. Thus the process of target tracking continues and periodically the current cluster head will send the state information of the target back to the base station.

3.4 Tracking Lost Detection and Recovery

Our ADCT protocol uses a prediction-based and low-delay algorithm to select a new cluster head. However when the target changes its direction or/and velocity so abruptly that it moves significantly away from the predicted location and out of the detectable region of the sensors selected for the sensing task. So a mechanism which can detect and recover the lost of tracking is used in this paper. When the new cluster detects the target, the new cluster head will send a detection confirmation message to the former cluster head in a predetermined time period. If the former cluster head does not get any confirmation in the predetermined time period, it assumes that the lost of tracking has happened. The former cluster head will send a tracking lost message to the base station. Then the base station will wake up all the sensor nodes to restart the tracking task.

4 Simulation

To evaluate the performance of our protocol, we have implemented it on the ns-2 simulator. Our goals in conducting the simulation are as following:

(1) Compare the adaptability of the ADCT protocol with other protocols to different speed change probability of the target

(2) Compare the performance of the ADCT protocol with other protocols on the basis of energy consumption.

(3) Study the effect of the ThresholdNumber (TN) on ADCT. In the simulation we set TN = 1 and 3.

4.1 Simulation Environment

The simulation has been performed within a 100m x 100m 2-dimensional square sensing area. The nodes are placed uniformly and randomly in the network but the node density should be large enough so that for any arbitrary location sensing region there are at least 3 sensors which can monitor it. At the beginning of the simulation, the target shows up at a random position of the sensing area with an initial moving direction and velocity V. At very ts, the target will change its moving direction or/and velocity with a probability of P. If the change happens its velocity will be selected between 0 and Vmax and its direction will change x°. The monitoring radius of sensor node is 10m, and the distance between each sensor is 5m.

4.2 Simulation Results

In the simulation, we compare our ADCT protocol with other existing tree-based method, prediction-based method and dynamic cluster-based methods. For the prediction-based method we choose the wake-up mechanism that wakes all the nodes on the route of the moving target from current location to the destination. Below are the simulation results:

1) Miss probability with different speed change probability of the target: From figure 3 we can see that when the speed change probability is low, both the dynamic cluster-based and prediction-based method perform better than the proposed (TN=3) method and the proposed (TN=1) method performs the worst. However as the speed change probability increasing, the proposed (TN=3) method performs better than other methods and the proposed (TN=1) performs better than the prediction-based and tree-based methods.

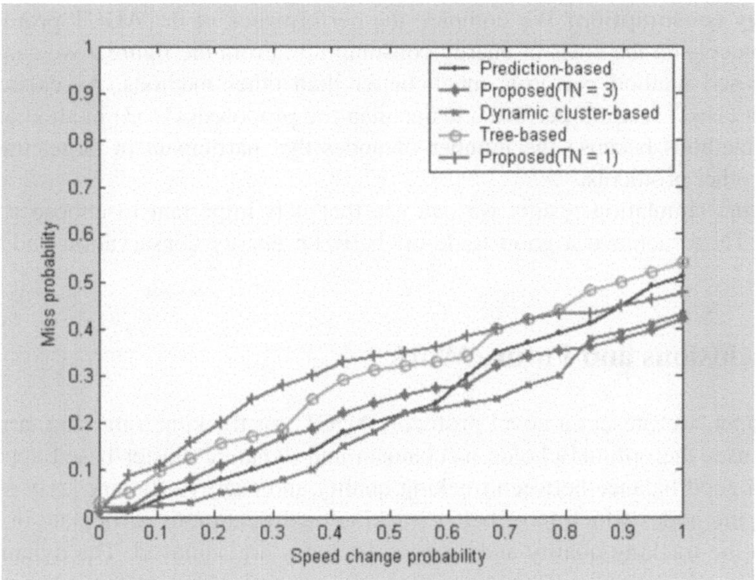

Fig. 3. The relationship between miss probability and speed change probability

The result of the simulation indicates that our proposed method has a better adaptation to the fast varying speed target tracking compared to other existing methods and we can change the value of the TN to adjust tracking quality.

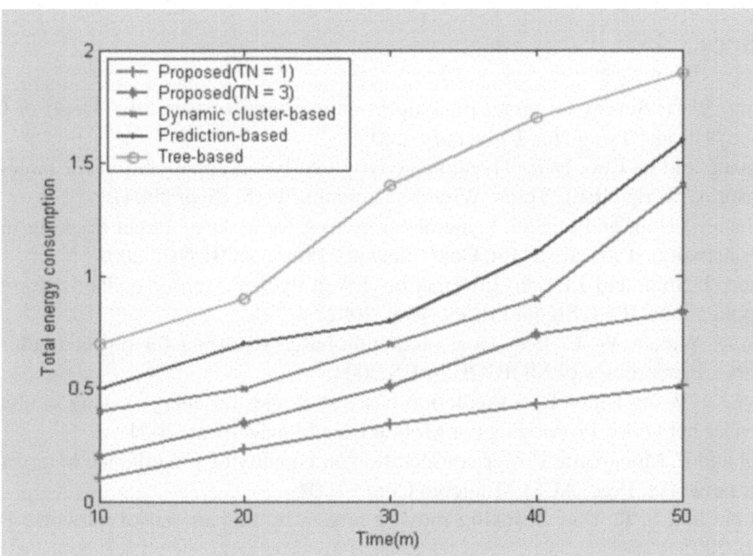

Fig. 4. The total energy spent by the tracking sensor network

2) Energy consumption: We compare the performance of the ADCT protocol with other protocols on the basis of energy consumption. From the figure 4 we can see that our proposed method performs much better than other methods. As expected, the proposed(TN=1) method performs better than the proposed(TN=3) method and other existing methods because the number of nodes that participant in target tracking is less than other protocols.

From the simulation results we can see that it is important to choose a suitable value of TN to achieve a good trade-off between energy conservation and tracking quality.

5 Conclusions and Future Work

In this paper, we present a novel protocol, ADCT, for tracking a moving target. This protocol uses the optimal choice mechanism and dynamic cluster-based approach to achieve a good balance between tracking quality and energy efficiency. By optimally choosing the nodes which have better tracking qualification to participate in tracking task both the tracking quality and energy efficiency are achieved. The dynamic cluster-based mechanism significantly reduces the amount of message exchange during the new cluster head election and cluster reconfiguration hence further achieves the energy efficiency.

In the future, we are going to work on a new cluster head selection algorithm based on other specific sensor network applications. Another possible direction of our work is to accommodate mobile sensors. With mobile sensors the more complicate and wiser methods may be necessary for the solution.

References

1. Chuang, S. C.: Survey on target tracking in wireless sensor networks, Dept. of Computer Science National Tsing Hua University. 2005.
2. W. Zhang and G. Cao, Dctc: Dynamic convoy tree-based collaboration for target tracking in sensor networks, IEEE Trans. Wireless Commun. 11(5) (Sept.2004)
3. W. Zhang, J.Hou, and L. Sha: Dynamic clustering for acoustic target tracking in wireless sensor networks, Proc. IEEE Int. Conf. Network Protocols (ICNP), 2003
4. F. Zhao, J. Shin, and J.Reich: Information-driven dynamic sensor collaboration for tracking applications, IEEE Signal Proces. Mag. 2002
5. Y. Xu, J. Winter, W.-C. Lee: Dual prediction-based reporting for object tracking sensor networks. Proceedings of MOBIQUTOUS 2004.
6. Yingqi Xu, Wang-Chien Lee: Prediction-based strategies for energy saving in object tracking sensor networks. Proceedings of Mobile Data Management, 2004
7. C. Gui and P. Mohapatra: Power conservation and quality of surveillance in target tracking sensor networks, Proc. ACM Mobicom Conf., 2004
8. R. Gupta and S. R. Das: Tracking moving targets in a smart sensor network, Proc VTC Symp., 2003

9. Maurice Chu, Horst Haussecker, and Feng Zhao: Scalable Information-Driven Sensor Querying and Routing for ad hoc Heterogeneous Sensor Networks, 2001
10. J. Aslam, Z. Buter, V. Crespi, G.CYBENKO, Aand D.Rus: Tracking a moving object with a binary sensor network, proc. ACM Int. Conf. Embedded Networked Sensor Systems (SenSys), 2003
11. Yu-Chee Tseng, Sheng-Po Kuo, Hung-Wei Lee, Chi-Fu Huang: Location tracking in a wireless sensor network by mobile agents and its data fusion strategies, International Workshop on Information Processing in Sensor Networks, 2003

Distributed, Hierarchical Clustering and Summarization in Sensor Networks*

Xiuli Ma[1], Shuangfeng Li[1], Qiong Luo[2], Dongqing Yang[1], and Shiwei Tang[1]

[1] School of Electronics Engineering and Computer Science, State Key Laboratory on Machine Perception, Peking University, Beijing, China, 100871
[2] Department of Computer Science, The Hong Kong University of Science and Technology, Clear Water Bay, Kowloon, Hong Kong
maxl@cis.pku.edu.cn, lishuangfeng@gmail.com, luo@cse.ust.hk, {dqyang,tsw}@pku.edu.cn

Abstract. We propose DHCS, a method of distributed, hierarchical clustering and summarization for online data analysis and mining in sensor networks. Different from the acquisition and aggregation of raw sensory data, our method clusters sensor nodes based on their current data values as well as their geographical proximity, and computes a summary for each cluster. Furthermore, these clusters, together with their summaries, are produced in a distributed, bottom-up manner. The resulting hierarchy of clusters and their summaries facilitates interactive data exploration at multiple resolutions. It can also be used to improve the efficiency of data-centric routing and query processing in sensor networks. Our simulation results on real world data sets as well as synthetic data sets show the effectiveness and efficiency of our approach.

Keywords: Sensor networks, clustering, summarization.

1 Introduction

Many data-centric sensor network applications are not only interested in raw sensory readings of individual nodes, but are also interested in the patterns, outliers, and summaries of network-wide sensory data. For example, on the left of Fig.1 shows part of the deployment of a sensor network at the Intel Berkeley Lab together with a snapshot of the temperature sensor readings of individual nodes [3]. If we cluster the sensor nodes by their temperature readings and report the data range and average of each cluster, it gives a clear overview of the sensory data distribution as shown on the right of Fig.1. In addition to facilitating interactive data analysis, this kind of clustering and summary information is useful for data-centric routing and in-network query processing. Therefore, we study the problem of online clustering and summarization in sensor networks.

Since both clustering and summarization are computation-intensive tasks over a large amount of data, a natural solution is to conduct these tasks at a PC-grade base

* This work is supported by the National Natural Science Foundation of China under Grant No.60473072, 60473051, and the National High Technology Development 863 Program of China under Grant No. 2006AA01Z230.

G. Dong et al. (Eds.): APWeb/WAIM 2007, LNCS 4505, pp. 168–175, 2007.

station after the sensory readings are collected there. However, this centralized approach has two major drawbacks: one is timeliness and the other network power efficiency. In consideration of these drawbacks, we take a distributed approach to clustering and summarization in sensor networks.

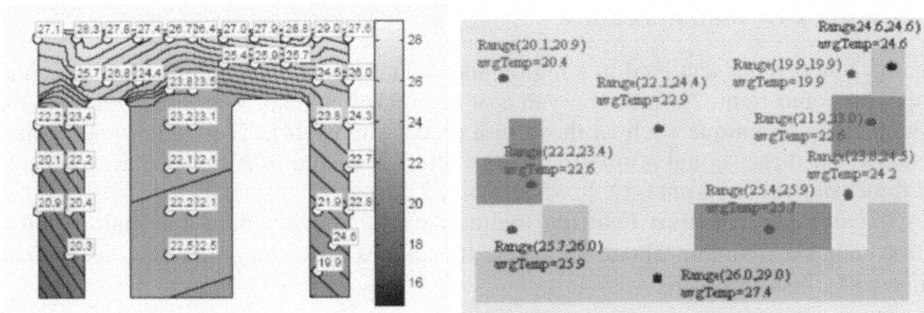

Fig. 1. Example of Clustering and Summarization

When considering distributing the clustering and summarization task to individual nodes in the network, we need to take into account the limited computing resources on each node and the multi-hop communication scheme of sensor networks. As a result, we first treat each sensor node as an initial cluster and then let geographically adjacent clusters gradually merge based on their readings. This bottom-up process is efficient because spatial correlations exist in real world sensory data so that computation and communication happens among proximate clusters mostly. In addition, summary information are also computed and maintained for the resulting hierarchy of clusters. As such, we call our method Distributed, Hierarchical Clustering and Summarization (DHCS).

Specifically, we propose a summary structure called Cluster Distribution Feature, or CDF, for each cluster. This feature includes both the data range and the statistical features of a cluster, and can be incrementally maintained along the hierarchy. Subsequently, we design a dissimilarity metric called *expansion*, to compute the dissimilarity between two clusters based on CDF. Furthermore, the computation complexity of both CDF and *expansion* is low and can be done efficiently in the network in a distributed manner. With these and other information, the DHCS algorithm clusters sensor nodes and computes summaries for clusters in a distributed, hierarchical manner.

Research on scalable clustering algorithms has been fruitful [2, 10, 11]. Unfortunately, these traditional clustering methods are infeasible in sensor networks, because they are mostly centralized. DHCS is distributed, thus can fully utilize nodes' computation ability. More, DHCS considers both data similarity and spatial proximity, whereas previous work has only considered data similarity. In recent years, there has been some work about clustering in sensor networks [1, 9]. However, they focus on network topology information. In comparison, DHCS is more data-centric and brings more opportunities for data reduction. Kotodis introduces the idea of snapshot queries towards data-centric sensor networks [7]. However, those representative nodes can only represent their one-hop neighbors, which limits the reduction ratio.

2 Preliminaries

In this section, we first introduce the concept of Cluster Distribution Feature (CDF), then the dissimilarity metric *expansion* based on CDF, and finally the compact cluster.

2.1 Cluster Distribution Feature

Assume N nodes are scattered in a region and each node can sense d attributes, such as temperature and light. Assume the value of each attribute can be normalized by some normalization technique such as those in a recent survey [4]. Then, each node corresponds to a d-dimensional normalized data vector. A cluster of N nodes corresponds to N d-dimensional data points, $\{X_i\}$, where $i = 1, 2, \ldots, N$.

A **Cluster Distribution Feature** summarizes the sensory data distribution information that we maintain about a cluster. It includes two components: Cluster Data Range and Cluster Feature.

Definition 1. Assume that there are N nodes in a cluster, each of which can sense d attributes. Let $\{X_i\}$, where $i = 1, 2, \ldots, N$, be the corresponding set of N d-dimensional data points. The **Cluster Data Range** (**CDR**) of the cluster is the smallest closed region in the data space into which all X_i fall. The **Spherical Cluster Data Range** (**SCDR**) of the cluster is a tuple (*Center*, *R*), where *Center* is the center and R is the radius of the smallest sphere into which all X_i fall.

Intuitively, CDR provides a tight boundary for the data of the nodes in a cluster. For example, a circle or a rectangle in a 2D data space, a sphere or a cube in a 3D data space can be a CDR. In this paper we choose sphere for simplicity and intuitiveness. In the following of this paper, we use CDR and SCDR interchangeably.

We adopt **Cluster Feature (CF)** from Birch [10] to describe the statistical features of a cluster. Given the corresponding N d-dimensional data points of a cluster, $\{X_i\}$, where $i = 1, 2, \ldots, N$, the CF of the Cluster is a triple CF = (N, LS, ss), where N is the number of data points in the cluster, LS is the linear sum of the N data points, and ss is the square sum of the N data points. CF facilitates the computation of the mean, deviation and other statistical features. More, it can be incrementally maintained [10].

Having the definitions of CDR and CF, we define the Cluster Distribution Feature.

Definition 2. Given N nodes in a cluster, the **Cluster Distribution Feature** (**CDF**) of the cluster is a tuple CDF = <CDR, CF>.

Proposition 1. (CDF Additivity) Assume that the CDF of cluster A is CDF_A $=<(Center_A, R_A), (N_A, LS_A, ss_A)>$, The CDF of cluster B is $CDF_B = <(Center_B, R_B), (N_B, LS_B, ss_B)>$, *dist* is the distance between $Center_A$ and $Center_B$. Then the CDF of cluster C that is formed by merging A and B, is the smallest sphere in the data space that can enclose the CDR of A and the CDR of B. That is,

- when $(dist + R_B) \leq R_A$, $Center_C = Center_A$, $R_C = R_A$;
- when $(R_A - R_B) < dist < (R_A + R_B)$ or $dist \geq (R_A + R_B)$, $Center_C$ is the middle point between $Center_A$ and $Center_B$, $R_C = (R_A + R_B + dist)/2$.

- Addition of CF: the following additive law [10] holds:

$(N_C, LS_C, ss_C) = (N_A, LS_A, ss_A) + (N_B, LS_B, ss_B) = (N_A + N_B, LS_A + LS_B, ss_A + ss_B)$

From the CDF definition and the additivity proposition, we know that the CDF vectors of clusters can be stored and calculated incrementally as clusters are merged. These CDF vectors as summaries are not only efficient but also accurate for calculating the dissimilarity metric that we need for making clustering decisions in DHCS.

Next we define the dissimilarity metric between two clusters, *expansion*.

2.2 Expansion

Definition 3. Assume that the CDF of cluster A is $\text{CDF}_A = <(Center_A, R_A), (N_A, LS_A, ss_A)>$. The CDF of cluster B is $\text{CDF}_B = <(Center_B, R_B), (N_B, LS_B, ss_B)>$. The CDF of cluster C, which is the cluster formed by merging A and B, is $\text{CDF}_C = <(Center_C, R_C), (N_C, LS_C, ss_C)>$. Then ***expansion*** is the difference between R_C and the larger one of R_A and R_B. That is, $expansion = R_C - \max(R_A, R_B)$.

Essentially, *expansion* describes how much the CDR will expand after merging of two clusters. The smaller the *expansion*, the more similar the two clusters.

2.3 Compact Cluster

Assume that the normalized vector $\vec{et} = (\Delta e_1, \Delta e_2, ..., \Delta e_d)$ is a predefined *difference threshold*, where Δe_i is the maximum tolerable difference in the i-th attribute between any two nodes within a cluster. Assume that *hopcount threshold* is the maximum tolerable hop count between any two nodes within a cluster. The compact cluster is defined as follows:

Definition 4. Let D be a cluster of N nodes, each of which can sense d attributes. Given the *difference threshold* $\vec{et} = (\Delta e_1, \Delta e_2, ..., \Delta e_d)$ and *hopcount threshold*, a **compact cluster** C is a non-empty subset of D satisfying the following conditions:

- Similar sensory values: $\forall i$, $1 \le i \le d$, $(R \times 2 \le \Delta e_i)$, where R is the radius of CDR of C;
- Geographical proximity. Two conditions must hold. First, any two nodes within C can communicate with each other, possibly through intermediate nodes; if intermediate nodes are needed, they must be also in C. Second, the hop count between any two nodes in C should be no greater than *hopcount*.

Different from traditional clustering methods, DHCS clusters nodes based on their current data values as well as their geographical proximity. Adjacency is defined as:

Definition 5. Assume n_i (n_j) is a sensor node and cluster C_i (C_j) is a set of sensor nodes.

- **Adjacent nodes:** n_i and n_j are *adjacent,* or n_i is a neighbor of n_j, if n_i and n_j can communicate with each other directly (within one hop).
- **Adjacent clusters:** C_i and C_j are *adjacent,* or C_i is a neighbor of C_j, if there exist node n_p in C_i and node n_q in C_j, n_p and n_q are *adjacent*.

DHCS produces compact clusters and their summaries, CDF vectors, in a bottom-up manner. Each compact cluster covers a local continuous region with similar sensory data. By partitioning sensors into several compact clusters and giving a summary for each cluster, DHCS divides the entire region into several sub-regions and keeps multi-resolution summaries for each sub-region. These summaries are organized in trees.

Definition 6. A **summary tree** of a compact cluster C is a tree structure of the sensor nodes in C satisfying the following condition: the nodes in any sub-tree form a compact cluster C_i with the root of the sub-tree being the cluster head for C_i, and storing the CDF of C_i.

3 DHCS

Given the *difference* and *hopcount* thresholds, DHCS produces compact clusters and their summaries in a distributed, bottom-up manner. Initially, each node treats itself as an active cluster. Then, similar adjacent clusters are merged into larger clusters round by round. In each round, each cluster will try to combine with its most similar adjacent cluster simultaneously. Two clusters can be merged only if both consider each other as the most similar neighbor. A compact cluster produced through merging must satisfy the thresholds. DHCS terminates when no merging happens any more. The final clusters, which cannot be merged any more, are called steady clusters.

In each round, each *CH* (short for cluster head) represents its cluster to coordinate with other clusters. In order for a *CH* to route to other *CH*s in its adjacent clusters efficiently, we maintain the adjacency information of a cluster in its *CH* and adapt *DSR* (Dynamic Source Routing) [6] for DHCS. Thus, a *CH* keeps the CDF and adjacency information of its cluster. In DHCS, there are three kinds of nodes by their states:

- *ACTIVE* nodes: the *CHs* of the active clusters;
- *PASSIVE* nodes: the nodes that are not *CH* of any cluster;
- *STEADY* nodes: the *CHs* of the steady clusters.

Initially each node will be *ACTIVE*. *ACTIVE* nodes become *PASSIVE* or *STEADY* along the merging of clusters. DHCS terminates when there is no *ACTIVE* node. The *STEADY* nodes represent the final compact clusters.

Each round has four stages: advertising, inviting, accepting and merging. In the advertising stage, clusters exchange CDFs with neighbors. Then adjacent clusters may try to reach an agreement about merging by shaking hands in inviting and accepting stages. In the merging stage, new clusters are generated.

Next we describe the detailed operations in a round. Note that, each node has a globally unique hardware *ID* [8], which we use as the node *ID*. Cluster *ID* is defined as the *ID* of its *CH*.

(1) Advertising CDF: Each *ACTIVE* node n_i advertises the CDF of its cluster to the *ACTIVE* nodes of all its adjacent clusters simultaneously. After exchanging CDFs, each *ACTIVE* cluster determines the most similar neighbor by computing *expansion* based on CDFs. If more than one neighbor have the same *expansion*, we choose the one whose *ID* is the largest. If a cluster cannot be merged any more given the thresholds, or if a cluster receives no messages, the state of its *CH* turns into *STEADY*.

(2) Sending invitation: For the purpose of coordination and avoiding redundant invitations, we take the following policy when sending invitations. Assume that n_i considers n_j as the most similar. n_i sends an inviting message to n_j only if (a) the cluster of n_i

has more nodes than n_j; or (b) the two clusters have the same number of nodes, and n_i has the larger *ID*. Otherwise n_i waits for invitations. After this stage, some *ACTIVE* nodes will receive several invitations. Note that n_i will become the new CH if this merging succeed.

(3) Accepting an invitation: Suppose n_j receives several invitations. Assume that n_j considers n_i as the most similar. If there is no invitation from n_i, n_j will not accept any invitation; otherwise it will reply an accepting message to n_i. If n_j accepts n_i's invitation, adjacency information of n_j is piggybacked in the accepting message, and then n_j sets its state as *PASSIVE*. By shaking hands, some pairs of adjacent clusters agree to merge.

(4) Merging clusters: If n_j accepts n_i's invitation, a new cluster is generated by merging the nodes from the clusters of n_i and n_j. n_i becomes the new *CH*. The *ID* of the new cluster is the *ID* of node n_i. n_j set its parent as n_i, while n_i appends n_j to its children. The *CHs of* these newly generated clusters finish the cluster merging by the following operations: (1) Compute the new CDF by the addition of the original two CDFs; and (2) Maintain the adjacency information of the new cluster.

After DHCS terminates, nodes' information about their parents, children and CDF will form several summary trees. Summary trees organize nodes considering data correlation as well as geographical proximity. They keep multi-resolution summaries to facilitate interactive data exploration at multiple resolutions. We leave the maintenance mechanism of summary trees for future work.

4 Experiments

We build a preliminary simulation environment to evaluate the performance of DHCS. First comes the effectiveness, then the efficiency. The following two datasets are used:

- The real geographical data set downloaded from Climatic Research Unit [12]. This website collects the climate data of 100 years of the entire world. We use a 30×30 grid taken from China's map covering the region with 24.5 N, 101.5E as the lower left corner located in Yunnan, and 39.5 N, 116.5 E the upper right corner located in Beijing. Each grid cell corresponds to a half geographical degree and contains a point, which makes a total of 900 points. We extract the data of the average temperature in January 2002.
- Synthetic spatially-correlated data: In 4.2, we use the tool in [5] to generate larger synthetic datasets from a small 10*10 sample of the real data set, to keep the same spatial correlation.

Let w denote the network width. Nodes are arranged on a w×w grid, totally w*w nodes. d is the transmission range. A large d allows a large number of adjacent nodes for a node. dt is the *difference* threshold. Assume that the communication is reliable.

We define two metrics to evaluate the quality of clusters: The reduction ratio is defined as N/N_C, where N is the number of nodes and N_C is the number of clusters. The average absolute deviation is the average absolute error when using the value of a cluster head to estimate those of any other nodes in the cluster. We consider only the impact of dt by setting the *hopcount* threshold to be sufficiently large.

(1) Effectiveness of DHCS
We vary *dt* from 1 to 5 and *d* from 1 to 3. We can see in Fig.2, DHCS achieves the reduction ratios of 10 to 50 for various parameters. The reduction ratio will increase with the increase of *dt* and *d*, due to the increase of cluster size. Fig. 3 shows that the average absolute deviation is significantly smaller than *dt* used, about 1/6 of *dt*. Additionally, the deviation has little correlation with *d*. Both the reduction ratio and the average absolute deviation are mainly influenced by data correlation and *dt*.

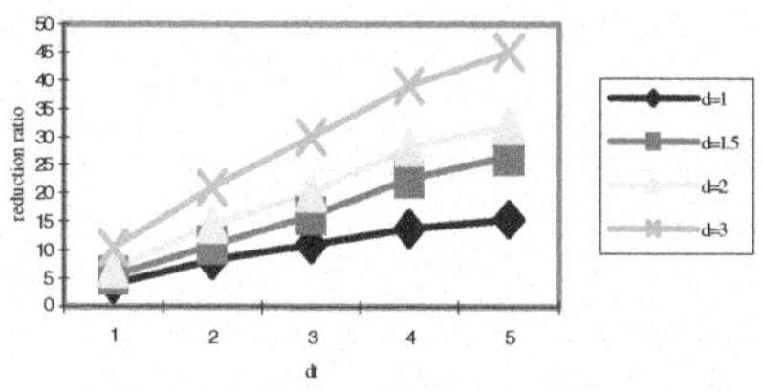

Fig. 2. Reduction ratio

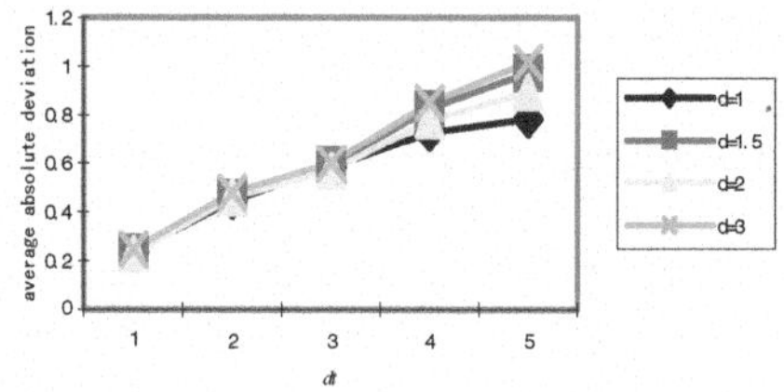

Fig. 3. Average absolute deviation

(2) Efficiency of DHCS
The most important metric for sensor networks is power efficiency. Therefore, we evaluate the efficiency of DHCS by the number of messages transmitted. We generate *w*w* datasets with *w* varied from 40 to 160, with step 20. For the centralized clustering, the main cost is in collecting data.

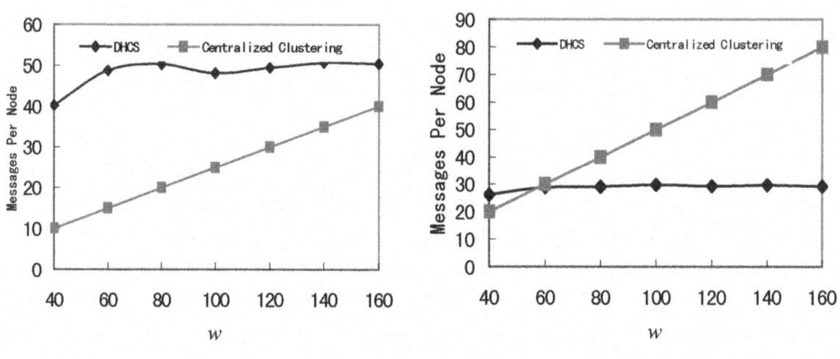

Fig. 4. DHCS vs. Centralized Clustering (*dt* = 3)

Suppose the sink node resides at the center of the upper side of a *w*w* square, then the depth of the routing tree is about *w/d* and on average *w/(2d)* messages per node for collecting all data to the sink. We use *w/(2d)* to represent the cost of centralized clustering.

When *d* is 2 in Fig. 4(a), DHCS is worse than centralized clustering. The main reason is that only two clusters are merged at a time in DHCS, which slows down the convergence of clustering. Fortunately, this downside is compensated by the scalability of DHCS. As shown in the figures, DHCS will eventually outperform centralized clustering given a sufficiently large network, although we cannot finish the experiment in larger network limited by our simulation platform. Fig. 4(b) shows that, when d is 1, the cost of DHCS is rather stable, about 30 messages per node. In contrast, the cost of the centralized clustering increases linearly with w, about 80 messages when w is 160. The larger the network is, the better DHCS performs.

5 Conclusion

We propose DHCS, a method of distributed, hierarchical clustering and summarization for sensor networks. DHCS clusters nodes based on their current data values as well as their geographical proximity in a distributed, bottom-up manner. The resulting hierarchy of clusters and their summaries can provide quick overviews about the network and facilitate interactive data exploration at multiple resolutions.

Future work includes extending DHCS to merging more than two clusters at a time, designing maintenance mechanisms for the clustering and summary information, and evaluating DHCS in real sensor networks.

References

1. S. Bandyopadhyay and E. J. Coyle. An Energy Efficient Hierarchical Clustering Algorithm for Wireless Sensor Networks. *INFOCOM* 2003.
2. M. M. Breunig, H. Kriegel, P. Kroger, and J. Sander. Data Bubbles: Quality Preserving Performance Boosting for Hierarchical Clustering. *SIGMOD* 2001.
3. C. Guestrin, P. Bodik, R. Thibaux, M. Paskin, and S. Madden. Distributed Regression: An Efficient Framework for Modeling Sensor Network Data. *IPSN* 2004.
4. J. Han and M. Kamber. Data Mining: Concepts and Techniques. China Machine Press, 2001.
5. A. Jindal and K. Psounis. Modeling Spatially-Correlated Sensor Network data. *SECON* 2004.
6. D. B. Johnson and D. A. Maltz. Dynamic Source Routing in Ad-hoc Wireless Networks. *Mobile Computing*, Kluwer Academic Publishers, pp. 153-181, 1996.
7. Y. Kotidis. Snapshot Queries: Towards Data-Centric Sensor Networks. *ICDE*, 2005.
8. S. Madden, M. J. Franklin, J. M. Hellerstein, and W. Hong. Tag: A Tiny Aggregation Service for ad hoc Sensor Networks. *OSDI* 2002.
9. O. Younis and S. Fahmy. Distributed Clustering in Ad-hoc Sensor Networks: A Hybrid, Energy-efficient Approach. *INFOCOM* 2004.
10. T. Zhang, R. Ramakrishnan, and M. Livny. BIRCH: An efficient data clustering method for very large databases. *SIGMOD* 1996.
11. T. Zhou, R. Ramakrishnan, and M. Livny. Data Bubbles for Non-Vector Data: Speeding-up Hierarchical Clustering in Arbitrary Metric Spaces. VLDB 2003.
12. http://www.cru.uea.ac.uk/

A New Similarity Measure for Near Duplicate Video Clip Detection

Xiangmin Zhou, Xiaofang Zhou, and Heng Tao Shen

School of Information Technology & Electrical Engineering
University of Queensland, Australia
{Emily,zxf,shenht}@itee.uq.edu.au

Abstract. Near-duplicate video clip(NDVC) detection is a special issue of content-based video search. Identifying the videos derived from the same original source is the primary task of this research. In NDVC detection, an important step is to define an effective similarity measure that captures both frame and sequence information inherent to the video clips. To address this, in this paper, we propose a new similarity measure, named as *Video Edit Distance*(VED), that adopts a complementary information compensation scheme based on the visual features and sequence context of videos. Visual features contain the discriminative information of each video, and sequence context captures the feature variation of it. To reduce the computation cost of inter-video comparison by VED, we extract key frames from video sequences and map each key frame into one single symbol. Various techniques are proposed to compensate the information loss in the measurement. Experimental results demonstrate that the proposed measure is highly effective.

Keywords: Near-duplicate Detection, Context Information Compensation.

1 Introduction

Searching for the near-duplicates of a given video clip is an important research issue in content-based video search. Consider an application of NDVC detection in TV broadcast monitoring. A company that contracts TV stations for certain commercials would like to contract a market survey company to monitor whether their commercials are actually broadcasted as contracted, and how much their commercials has been edited. While the applications of NDVC detection have become widespread, effective NDVC detection approaches are high demanded to handle this task. Defining a suitable similarity measure for detecting similar videos is the first step towards effective NDVC detection.

A video is defined as a sequence of frames which represent high dimensional feature vectors of specific images over a particular time. NDVCs are those from the same original video source but possibly compressed at different qualities, reformatted to different sizes and frame-rates, or undergone different editing in either spatial or temporal domain. In NDVC detection, two factors, *information*

G. Dong et al. (Eds.): APWeb/WAIM 2007, LNCS 4505, pp. 176–187, 2007.

loss and *computation cost*, are required to be considered, i.e., the information loss is minimized in the similarity measure, and the search is fast enough. Due to the complexity of video data, using original data to compare clips is not appreciable for large data set. To address this problem, an appropriate way is to represent each video in compact summaries, on which the inter-video similarity is measured. Although many approaches have been proposed to further the response of video matching, they suffer from the drawback of information loss, thus producing undesirable search results.

A typical approach for video matching is measuring the similarity by the number of similar frames [4,9]. Meanwhile, many compact representations are used to further the similarity search processing. A common approach is to extract several key frames from the segmented video shots, and perform the similarity match between two videos by comparing the key frame sets of them [11]. However, matching videos over the selected key frames incurs a heavy information loss, and also neglects the sequence information of videos, thus producing poor query results. In [5], each video is represented by a sequence of single values, each of which describes the change in color from one image to the next. Video matching proceeds by using local alignment to find sequence of similar values in video clips. This approach is robust to color variance. However, it can not discriminate the similarity between near-duplicates with a desirable ranking.

In this paper, we believe NDVCs have similar sequence context, and also those with both similar sequence context and similar visual features are more similar. Based on this, we propose a new similarity measures that takes the temporal order, inter-frame similarity and sequence context into consideration. To improve the efficiency of the matching, the effective estimation of it based on the symbolization and probability is utilized. The main innovation of the proposed measure can be presented in three aspects. (1) We derive the traditional edit distance into video matching; (2) The sequence context is embedded into the measure to not only solve the problem of feature variation, but also effectively compensate the information loss caused by compact representation; (3) With this complementary information compensation scheme, key symbols approach is employed. Experiments on real video show the promising results of our approach.

2 Related Work

Several related video matching approaches [7,9,3] are proposed in recent years for effective video retrieval. However, they only consider part of properties of video data that can not capture the flexible similarity of NDVCs in a manner suitable to human perception. In [9,4], the inter-video similarity is measured by the number of similar frames shared by two videos. The distance between two videos is defined as the ratio of the number of similar frames shared by them to the total number of their frames. The main disadvantage of this approach is that, since each video is considered as a set of frames by this approach, it does not capture the information of sequence.

Taking temporal order into consideration, there are a wealth of papers [7,3,2,8] working on sequence matching. Edit distance variants [3,2,8] are the most robust approaches, since they preserve temporal order in a flexible manner. ERP is the first one proposed to combine the edit distance with L1 norm distance together. In [8], a symbolization representation has been proposed based on dimension-wise quantization, called *vString*. The real-valued feature values are mapped into some discrete classes. Each dimension of the feature is transformed to a symbol that represents a class. Accordingly, each video is represented by a multidimensional video string. Then, vstring edit distance is utilized. This work also does not reduce the dimensionality of video features, thus the representation is not compact. Therefore, the existing edit distance variants incur the high computation cost. Moreover, in the real applications, visual features may be variant largely among the NDVCs due to the various forms of quality degradation or video editing, thus these features that use visual features directly are not workable.

To solve the problem of color shift, a *Signature Alignment* (*SA*) based video matching approach is proposed in [5]. *Signature Alignment* first transforms each frame into a single value sequence by computing the similarity from one image to the next. This approach uses the local context of a video and is robust to feature variation. However, in real applications, except for the cases of shot transition, neighboring frames are always very similar with each other. Therefore, the single video matching by *Signature Alignment* is not discriminative.

3 A New Video Matching Approach Based on Edit Distance and Video Context

In this section, we present a novel complementary information compensation scheme based video matching method, VED, that not only considers the temporal and spatial information of sequences, but also the relationship of neighboring frames in videos, for NDVC detection. The visual feature based video similarity, ED, is first defined on original sequences. Then, the complexity reduction strategy based on video frame summarization and estimation is described in detail, including how a video sequence is compactly represented, and how the inter-video similarity is estimated on the summaries. Finally, VED is proposed by embedding video context information into ED measure.

3.1 Video Similarity Based on Spatial and Temporal Information

Edit distance is widely used in string matching and pattern recognition. We extend it to the inter-video similarity measure by redefining the match or mismatch of different frames. Given two video sequences, A and B, the edit distance between them, ED(A,B), is the number of insertion, deletion or substitution operations that are required to transform A into B. To formally define ED(A,B), it is crucial to decide whether two compared frames are matched or not by measuring the similarity of them. Thus, two suitable distance functions, for measuring the inter-frame similarity and the inter-sequence similarity respectively, are essential.

Generally, the distance between frames is defined by the Lp-norm in a d-dimensional space. Given a matching threshold, ϵ, whether two elements are matched or not is judged by the Lp-norm distance between them. If the Lp-norm distance between two frames is no more than ϵ, they are matched; otherwise, they are mismatched. Given two videos $\overrightarrow{S_1} = \{s_1, s_2, ...s_m\}$ and $\overrightarrow{S_2} = \{f_1, f_2, ...f_n\}$ ($m \geq n$), the edit distance between them, $ED(S_1,S_2)$, is defined as follows:

$$ED(S_1, S_2) = \begin{cases} m & n = 0 \\ \min\{ED(S_1^{m-1}, S_2^{n-1}) + \alpha, & \\ \quad ED(S_1^{m}, S_2^{n-1} + 1), & \\ \quad ED(S_1^{m-1}, S_2^{n} + 1)\} & otherwise \end{cases} \quad (1)$$

Here, α is the cost for substitution operation. If s_m and f_n are matched, α=0; otherwise α=1. The definition of ED is very similar to that of EDR, but more suitable to the similarity match between videos, since the comparison between high dimensional data is based on the distance over all dimensions rather than the difference over each single dimension.

3.2 Video Similarity with Sequence Information

When the edit distance variants like ERP, EDR, and our proposed ED are used for sequence matching, practically, two main problems are required to be solved. Since visual features are subjective to various forms of quality degradation, the visual features do not capture enough information. Also, frame based measure with these distance functions suffers from the high complexity of high dimensional distance computations. Although key frames and other video representations are very effective for reducing the cost of measures, they incurs considerable information loss that reflects human perception for effective NDVC detection. A desirable approach for solving these two problems is to compensate the information loss, which are from the video representation and the video recording as well, by using the context information of sequences. With this consideration, the distance between two videos is re-defined by embedding the sequence context difference between them.

Given a video X, let $< f_1,...f_i,...f_n >$ be the sequence that consists of its key frames, the sequence context information is formed by the similarity between the neighboring key frames that can be represented as $X_c =< x_1,..x_i...x_n >$, where x_i denotes the distance between the ith and $(i\text{-}1)^{th}$ key frame, and X_c is called as the context vector of X. Suppose that X_c and Y_c are the context vectors of video X and Y respectively, we define the context difference between them as $d(X_c, Y_c)$. Given two videos X and Y, the distance between them, VED, is defined as:

$$VED(X,Y) = w_1 ED(X,Y) + w_2 d(X_c, Y_c)$$

where w_1 and w_2 are the weights of the visual feature difference and context difference in the distance function. VED considers the difference of visual features between two videos and that of their sequence relationship. This measure scheme compensates the information loss from video representations, thus key frames can be effectively utilized in this measure to reduce the complexity of the measure significantly. To further save the cost of video matching, the key frames can be summarized and symbolized. Accordingly, VED is computed over the summarized symbol space. This part will be described in 3.3.

3.3 Complexity Reduction

A major step in computing the VED between two videos is to decide whether each frame pair is matched or not by the distance between them. The complexity of the frame distance computation is proportional to the dimension of the feature vector, which is usually quite high in video applications. For the video sequence matching, the number of frame distance computations is exponential to the length of them, usually several hundreds to thousands. Clearly, for the real video applications, reducing the complexity of video matching is a crucial task.

We propose two schemes, frame symbolization and key symbol representation, to reduce the complexity of video matching. The basic idea of frame symbolization is to transform each video frame into a symbol, which is a process of aggressive dimensionality reduction. Then, key symbols are selected to represent the video clip, thus reducing the length of compared sequences. Turning a longer video sequence into a shorter symbol string, one may think that the matching can not work because of the severe information loss. However, this is not necessarily true, since the information can be fully compensated by the complementary scheme which has been introduced in 3.2 and some effective strategies such as multi-symbol representation and optimal probability selection(T=0.5), which has been described in [10].

Video representation. We have introduced the technique of frame symbolization in [10]. As a video symbol sequence usually contains same symbols which occurs sequentially, in this paper, we represent the symbol sequence by selecting the key symbols at equal intervals. Because we only use the selected key frames of video sequences in the sequence matching, the process of summarization and symbolization is not performed over the whole frame dataset, but only the selected key frames of each video sequences are utilized. Given a video dataset and a valve, $\epsilon \in (0, \sqrt{D})$ (D is the dimensionality of video frame), the key frames of each videos are selected first, and the video symbolization is performed by first clustering over this key frame dataset and then mapping each key frame to its corresponding cluster id. To ensure the high similarity of frames in the same cluster, the maximal cluster radius that is equal to $\frac{\epsilon}{2}$ is usually set very small.

With the traditional key frame selection methods, the similarity between two videos may vary due to the change of sequence lengthes. The increased difference of the sequence lengthes increases the dissimilarity between them, accordingly, producing the inaccurate measure results. Suppose that we have two videos of equal lengthes, A and B. After key frame selection and symbolization, they are transformed into (aba) and (a) respectively. Obviously, the dissimilarity between them is increased because they are transformed into symbol sequences of different lengthes. To simplify the issue of key frame selection and eliminate the effect of length changing, we select the key frames by simply sampling video frames with equal intervals. As such, the ratio of the sequence lengthes is maintained.

Many clustering methods have been reviewed in [6]. We adopt a recursive 2-mean algorithm that recursively performs the binary clustering algorithm until the radius of the cluster is no more than ϵ. As such, a set of clusters, each containing similar frames, can be produced. We represent each cluster by a four-tuplet $\{id, O, r, n\}$, where id is the cluster identifier; O is the cluster centre which indicates the position of the cluster in the original high dimensional space; r is the radius of boundary hypersphere of the cluster. n is the number of frames in the cluster. The information of the clusters kept in the cluster data set is used as a *video dictionary* to map a key frame into an id during the preprocessing procedure of the similarity query.

By looking up the video dictionary and representing each video key frame with its cluster id, each video is symbolized as a digital string which consists of the cluster ids of its key frames. With this approach, the similarity comparison between query and video data is simplified as the issue of string matching, and the comparison between each frame pair is transformed into that between clusters. Although this key symbol representation may leak certain important information, this information loss can be effectively compensated by utilizing the context between neighboring key symbols in the video matching.

Probability Measure on Clusters. As each summary obtained from video sequence symbolization is not only a symbol, but corresponds to a cluster having a set of frames, traditional string matching is not suitable for measuring the similarity of transformed sequences that are series of cluster ids. A cluster id has two features: (1) one id represents a set of frames within a high dimensional space; (2) subspaces to different ids may have certain overlapping. Based on this, we proposed probability based approach for the inter-cluster comparison in [10], i.e., comparing two frames by the probability that neither of them falls into the intersection of the clusters to their ids. For two symbols, the similarity between them is constructed by a probability value, P∈ [0,1]. The value of P can be obtained by the probability function, which is defined as follows:

$$P(i,j) = \frac{|O_i - O_j| * |O_j - O_i|}{|O_i| * |O_j|} \tag{2}$$

Here, $|O_i - O_j|$ is the number of frames in cluster i but out of cluster j. $|O_i|$ refers to the number of frames in cluster i. Figure 1 shows the comparison between different clusters. The data distributed in the small clusters is uniform,

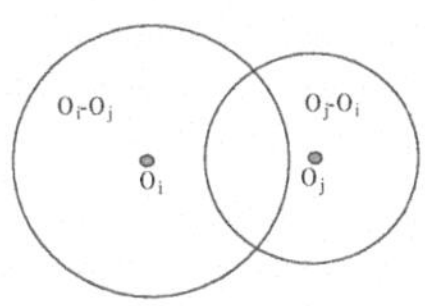

Fig. 1. p value between two clusters/symbols

thus the $|O_i - O_j|$ can be estimated by the ratio of the volume of the part of the cluster O_i outside of the intersection to that of the whole cluster, which is shown as follows:

$$|O_i - O_j| = \frac{V(O_i - O_j) * |O_i|}{V(O_i)} \tag{3}$$

$V(O_i - O_j)$ is the volume of a hyper concavo-convex. When performing a similarity match on the symbols, for the purpose of maintaining more information of frames in the original high dimensional space, the inter-symbol distance is not determined by whether they are same or not, but by their probability distance. This probability value shows the extent of overlapping between two clusters and can be obtained before the sequence comparison.

4 Evaluation

4.1 Experimental Set-Up

Our experiments are conducted on a real video collection which consists of 1083 real-world commercial videos captured from TV stations and recorded using the Virtual Dub at the PAL frame rate of 25fps [9]. Each video frame is compressed using PLCVideo Mjpegs and the resolution is 192×144 pixels. A video in the dataset is a 60s-clip, which is represented as a 32-dimensional feature vector in the RGB color space.

Six video clips are selected as queries. The selection criteria is that the selected clips are not near-duplicates with each other and, for each query, at least one near-duplicate can be found in the video collection. The major parameters and their default values used in the experiments are summarized in Table 1. We run experiments with different ϵ, T, w_1 and w_2, and the default values of them in Table 1 are chosen according to the best performance of them.

4.2 Evaluation Criteria

The standard evaluation method in IR that has been used for evaluating the VideoQ system [1] is applied to measure the effectiveness of the proposed video matching approach. The evaluation is based on two factors: *Precision* and

Table 1. Parameters used in the experiments

Para	Description	Default value
w_1	Weight of ED in VED	1
w_2	Weight of context difference in VED	1
ϵ	Inter-frame similarity threshold	0.2
T	Probability threshold	0.5
I	Frame interval	20
K	Number of most similar sequences	30
N	Number of sequences	1083 (60s-clips)

Recall. Given a query Q, let *rel* be the set of the relevant video clips to the query and $|rel|$ be the size of the set; let *ret* be the set of top 30 results returned by the system. *Precision* and *Recall* are defined as below:

$$Precision = \frac{|rel \bigcap ret|}{ret} \qquad Recall = \frac{|rel \bigcap ret|}{rel} \tag{4}$$

For each query, we find its top 30 ranked nearest neighbors. The precision is calculated after each relevant clip is retrieved. If a relevant clip is not retrieved, its precision is 0.0. A precision-recall curve is then produced by measuring precisions at 11 evenly spaced recall points (0,...1.0). All precision values are then averaged together to get a single number for the performance of a query. The values are then averaged over all queries, leading to the average precision of a search system.

Three sets of experiments are conducted to evaluate the effectiveness of proposed approach. Our objectives of this evaluation are: (1) to study the effect of the sequence context information compensation; (2) to study the effect of the number of key symbols selected; (3) to study the superiority of VED with the existing video matching approaches.

4.3 Effect of Information Compensation

We performed experiments to evaluate the impact of context information compensation in terms of *Precision* and *Recall* during the similarity search by comparing VED against ED. In this experiment, all parameters described in Table 1 are fixed as default values. Figure 2 shows the precision-recall curve, and Table 2 reports the average precision of these two measures.

From Figure 2 and Table 2, we found that VED approach achieves the much better average precision (0.9037) as well as the higher precision values at all recall

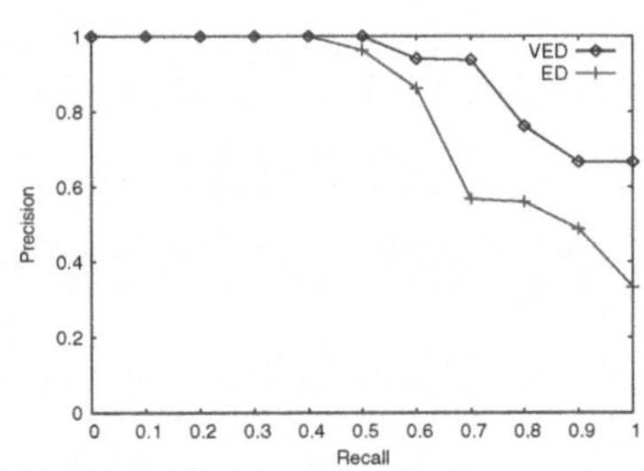

Fig. 2. Effect of Sequence Context Information Compensation

Table 2. Average Precision of ED and VED

	Q_1	Q_2	Q_3	Q_4	Q_5	Q_6	Average Precision
ED	0.5899	0.6479	1.0000	0.6667	0.8925	0.7813	0.7446
VED	0.7536	0.7226	1.0000	1.0000	1.0000	1.0000	0.9037

levels, since sequence context are utilized to effectively compensate the information loss, thus enhancing the quality of NDVC detection. Taking the average precision of each individual query into consideration, VED always outperforms ED, leading to much better average precision of the system.

4.4 Effect of Key Symbol Interval

Then, we examine effectiveness of the VED by varying key symbol intervals from 10 to 50, with other parameters in Table 1 to default values. Figure 3 shows the average precision of VED at different key symbol interval levels. From this figure, it can be seen that, from 10 to 20, the average precision of VED keeps steady due to the information compensation of sequence context. Consequently, the information loss from key symbol representation affects the search results very slightly. When the key symbol interval reaches to 30, with the increasing of key symbol interval, the average precision degrades at an obvious rate for the sequence representation containing only few key symbols incurs too heavy information loss that can not be well compensated.

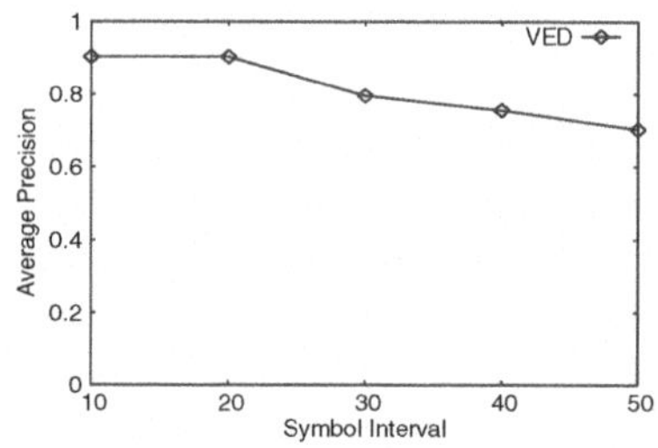

Fig. 3. Average Precision vs. Key Symbol Intervals

4.5 Comparison of VED and Existing Measures

Having shown that the information loss originated from the video representation can be compensated effectively, and key symbols can be used to represent video sequences more compactly in VED under the limited symbol intervals, we will

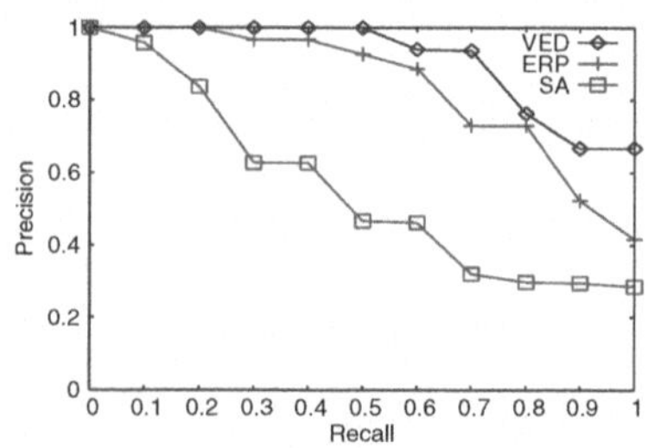

Fig. 4. VED, ERP and Signature Alignment

Table 3. Average Precision of VED, ERP and Signature Alignment

Recall	Average Precision		
	VED	ERP	Signature Alignment(SA)
Q_1	0.7536	0.7486	0.2096
Q_2	0.7226	0.7805	0.8228
Q_3	1.0000	1.0000	0.2222
Q_4	1.0000	0.8333	1.0000
Q_5	1.0000	0.5379	0.2776
Q_6	1.0000	1.0000	0.7193
Average Precision	0.9037	0.7999	0.4439

Fig. 5. Query Results of VED

also demonstrate that VED is more effective by comparing with the existing competitors, including ERP and *Signature Alignment* (SA). For each of these three approaches, the precision at each recall level and the average precision over each individual query and all queries are reported in Figure 4 and Table 3.

From Table 3 and Figure 4, we can see that VED has the best average precision and the best precision at each recall level, with the ERP following it, and the Signature Alignment performs much worse than the other two. This is caused by

Fig. 6. Query Results of ERP

the effective complementary information compensation scheme in VED measure. Since ERP only captures the information of frame similarity and the alignment of different video sequences, the NDVCs having much visual feature variation can not be retrieved with this measure. While Signature Alignment with the relationship of neighboring frames considered neglects the visual features of each clip, the spatial information of each clip can not be captured, thus leading to worse matching results. VED overcomes the weakness of ERP and Signature Alignment by introducing a complementary information compensation scheme into the measure, which produces great improvement of effectiveness.

To visualize the superiority of VED, we give the results of a query sample in Figure 5 and 6. To save the space, only the top 10 results, produced by VED and ERP, of the first query Q_1 that is the No 1 clip in the results are shown in the figures. Clearly, 10 correct results are obtained by VED, while ERP only finds 8 relevant clips with non-relevant clips occurring at position 7 and 10 respectively. Clearly, VED is more robust for NDVC detection.

5 Conclusion

In this paper, we proposed a new video similarity measure, VED, which is based on a complementary information compensation scheme for NDVC detection.

VED takes in consider not only the visual information of a video clip, but also the relationship of neighboring frames in video matching based on similarity. With this measure, compact video representation using frame symbolization and key symbols can be deployed effectively, and thus efficient similarity match is performed over the summaries. The extensive experimental results have shown that the proposed measure is high effective for NDVC detection.

References

1. S.-F. Chang, W. Chen, H. J. Meng, H. Sundaram, and D. Zhong. Videoq: An automated content based video search system using visual cues. In *MM*, pages 313–324, 1997.
2. L. Chen and R. Ng. On the marriage of lp-norm and edit distance. In *VLDB*, pages 792–803, 2004.
3. L. Chen, M. T. Özsu, and V. Oria. Robust and fast similarity search for moving object trajectories. In *SIGMOD*, 2005.
4. S. Cheung and A. Zakhor. Efficient video similarity measurement with video signature. *IEEE Trans. Circuits Syst. Video Techn.*, 13(1):59–74, 2003.
5. T. C. Hoad and J. Zobel. Fast video matching with signature alignment. In *MIR*, pages 262–269, 2003.
6. A. K. Jain, M. N. Murty, and P. J. Flynn. Data clustering: a review. *ACM Computing Surveys*, 31(3):264–323, 1999.
7. S.-L. Lee, S.-J. Chun, D.-H. Kim, J.-H. Lee, and C.-W. Chung. Similarity search for multidimensional data sequences. In *ICDE*, pages 599–608, 2000.
8. W. Ren and S. Singh. Video sequence matching with spatio-temporal constraints. In *ICPR*, pages 834–837, 2004.
9. H. T. Shen, B. C. Ooi, X. Zhou, and Z. Huang. Towards effective indexing for very large video sequence database. In *SIGMOD*, pages 730–741, 2005.
10. X. Zhou, X. Zhou, and H. T. Shen. Efficient similarity search by summarization in large video database. In *ADC*, pages 161–167, 2007.
11. X. Zhu, X. Wu, J. Fan, A. K. Elmagarmid, and W. G. Aref. Exploring video content structure for hierarchical summarization. *Multimedia Syst.*, 10(2):98–115, 2004.

Efficient Algorithms for Historical Continuous *k*NN Query Processing over Moving Object Trajectories

Yunjun Gao[1], Chun Li[1], Gencai Chen[1], Qing Li[2], and Chun Chen[1]

[1] College of Computer Science, Zhejiang University, Hangzhou 310027, P.R. China
{gaoyj,lichun,chengc,chenc}@cs.zju.edu.cn
[2] Department of Computer Science, City University of Hong Kong, Hong Kong, P.R. China
itqli@cityu.edu.hk

Abstract. In this paper, we investigate the problem of efficiently processing historical continuous *k*-Nearest Neighbor (HC*k*NN) queries on R-tree-like structures storing historical information about moving object trajectories. The existing approaches for HC*k*NN queries need high I/O (i.e., number of node accesses) and CPU costs since they follow *depth-first* fashion. Motivated by this observation, we present two algorithms, called HC_P-*k*NN and HC_T-*k*NN, which deal with the HC*k*NN retrieval with respect to the stationary query point and the moving query trajectory, respectively. The core of our solution employs *best-first* traversal paradigm and enables effective update strategies to maintain the *nearest lists*. Extensive performance studies with real and synthetic datasets show that the proposed algorithms outperform their competitors significantly in both efficiency and scalability.

1 Introduction

Advances in wireless communication, mobile computing, and positioning technologies have made it possible to manipulate (e.g., model, index, query, etc.) moving object trajectories in practice. A number of interesting applications are being developed based on the analysis of trajectories. For instance, zoologists can figure out the living habits and the migration patterns of wild animals by analyzing their motion trajectories. An important type of query thereinto is the so-called *k*-Nearest Neighbor (*k*NN) search, which retrieves from a dataset within a predefined time interval, the k ($k \geq 1$) objects that are closest to a given query object. Assume that the trajectories of animals are known in advance, the zoologists may pose the following query: *find which k animal's trajectories are nearest either to a given stationary point (e.g., food source, lab, etc.) or to a predefined animal's one at any time instance of the time period* $[t_i, t_j]$. This motivating example fosters the need of a new type of query, i.e., historical continuous *k*NN (HC*k*NN).

Given a set S of trajectories, a query object (point or trajectory) q, and a temporal extent T, a HC*k*NN query over trajectories retrieves from S within T, the k nearest neighbors (NNs) of q at any time instance of T. The result lists contain a set of tuples in the form of $\langle Tr, [t_i, t_j)\rangle$, where $Tr \in S$ and $[t_i, t_j)$ is the time interval in which Tr is the NN of q. As an example, Figure 1 shows two HC2NN queries, labeled as Q_1 and Q_2, on $S = \{Tr_1, Tr_2, \ldots, Tr_5\}$ in 3-dimensional space (two dimensions for spatial

G. Dong et al. (Eds.): APWeb/WAIM 2007, LNCS 4505, pp. 188 – 199, 2007.

positions, and one for time). Then, the 1-NN list for Q_1 (that takes a point f and a time extent $[t_1, t_3]$ as input) includes $\{\langle Tr_1, [t_1, t_2)\rangle, \langle Tr_2, [t_2, t_3]\rangle\}$ (highlighted in black thick line), and the 2-NN list for Q_1 includes $\{\langle Tr_2, [t_1, t_2)\rangle, \langle Tr_1, [t_2, t_3]\rangle\}$ (denoted in grayed thick line). Similarly, for Q_2 (which takes as input a trajectory segment *ts* belonging to Tr_3 and a time extent $[t_2, t_4]$), the 1-NN list contains $\{\langle Tr_5, [t_2, t_3)\rangle, \langle Tr_4, [t_3, t_4]\rangle\}$, and the 2-NN list contains $\{\langle Tr_4, [t_2, t_3)\rangle, \langle Tr_5, [t_3, t_4]\rangle\}$.

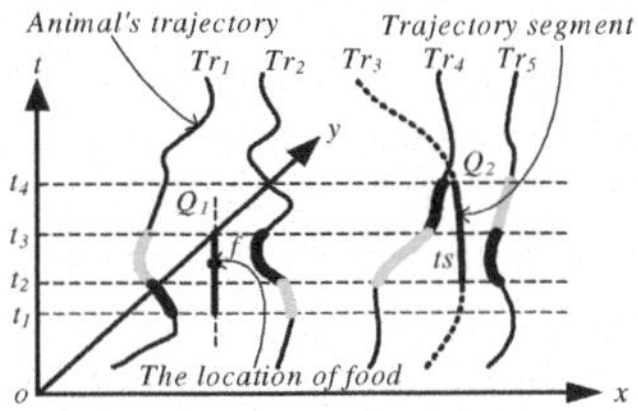

Fig. 1. Example of HC2NN search over moving object trajectories

Even though much work on continuous *k*NN (C*k*NN) search for spatial and spatio-temporal objects has been done over the last decade [2, 6, 8, 9, 12, 13, 14, 17, 18], work on HC*k*NN queries for moving object trajectories has been left largely untouched to the best of our knowledge. Recently, Frentzos *et al.* [4] studied the HC*k*NN retrieval for historical trajectories of moving objects. However, they follow the *depth-first* (DF) manner [11] to handle such a query. As it is well known, DF traversal induces a backtracking operation, resulting in reaccessing some nodes that were visited before. Thus, the I/O cost (i.e., number of node accesses) and CPU time incurred in the algorithms are rather high.

In our earlier work [19], we have studied *k*NN search on static or moving object trajectories with respect to non-continuous history, and developed two algorithms called BFP*k*NN and BFT*k*NN which are shown to be superior to the algorithms of PointkNNSearch and TrajectorykNNSearch proposed by Frentzos *et al.* [4]. In this paper, we move on to study, with respect to continuous history, *k*NN search on static or moving object trajectories through R-tree-like structures [7] storing historical information about trajectories. Specifically, we present two algorithms, called HC_P-*k*NN and HC_T-*k*NN, which deal with the HC*k*NN retrieval with respect to the stationary query point and the moving query trajectory, respectively. The core of our solution employs the *best-first* traversal paradigm [5] and enables effective update strategies to maintain the *nearest lists* (to be discussed in Section 3). Our goal is to reduce the number of node accesses and accelerate the execution of the algorithms (i.e., lead to less running time). Finally, we conduct extensive experiments by using real and synthetic datasets, the results of which confirm that our proposed algorithms outperform their competitors (including ContPointNNSearch and ContTrajectoryNNSearch algorithms [5]) significantly in terms of efficiency and scalability.

The rest of the paper is organized as follows. Section 2 surveys the previous work related to ours. Section 3 discusses the update of k nearest lists in detail. Sections 4 and 5 describe the HC_P-*k*NN and HC_T-*k*NN algorithms, respectively. Section 6 experimentally evaluates the performance of the algorithms under various settings. Section 7 concludes the paper with a few directions for future work.

2 Related Work

One area of related work concerns indexing of moving object trajectories. The trajectory *Tr* of a moving object can be represented as a sequence of the form $[(t_1, Tr_1), (t_2, Tr_2), \ldots, (t_n, Tr_n)]$ where *n*, the number of sample timestamp in *Tr*, is defined as the length of *Tr*, with Tr_i being a position vector sampled at timestamp t_i. Therefore, trajectories can describe the motion of objects in 2- or 3-dimensional space, in addition to be considered as 2- or 3-dimensional time series data. Although our proposed algorithms can be suitable for any above indexing structure, we focus on R-tree-like structures [7] storing historical information about trajectories such as 3DR-tree [16], STR-tree and TB-tree [10]. In particular, we assume that the dataset is indexed by a TB-tree due to its high efficiency in *trajectory-based queries,* and that the TB-tree aims at strict *trajectory preservation.*

Another area of related work is C*k*NN queries in spatial and spatio-temporal databases. Song *et al.* [12] utilized a periodical sampling technique to process C*k*NN search. Tao *et al.* [13] considered C*k*NN query algorithms using R-trees [1] as the underlying data structure. Benetis *et al.* [3] developed an algorithm to answer NN retrieval for continuously moving points. Tao *et al.* [14] presented a technique, termed as *time-parameterized queries*, which can be applied with mobile queries, mobile objects or both, given an appropriate indexing structure. Iwerks *et al.* [6] investigated the problem of C*k*NN queries for continuously moving points, assuming that the updates that change the functions describing the motions of the points are allowed. Currently, the problem of *CkNN monitoring* has been studied in either a centralized [8, 17, 18] or distributed [9] environment. All work mentioned above, nevertheless, does not cover the C*k*NN query on moving object trajectories. Recently, Frentzos *et al.* [4] explored the issue of HC*k*NN query processing over historical trajectories of moving objects, by proposing two algorithms called ContPointNNSearch and ContTrajectoryNNSearch which can handle, respectively, such a query w.r.t. a static query point and a moving query trajectory. Unfortunately, the I/O cost and CPU cost of their algorithms are expensive because they use the DF traversal paradigm.

3 Updating *k* Nearest Lists

In this section, we discuss how to maintain as the final result of a HC*k*NN query the *k* nearest lists (denoted by *kNearestLists*). Figure 2 shows the procedure of our **Update*k*Nearests** algorithm, in which arguments *M* and *kNearestLists* are taken as the input. Note that the structure *M* retains the parameters of the distance function including *a*, *b*, and *c* (computed using the method described in [4]), the associated minimum *Dmin* and maximum *Dmax* of the function during the lifetime, a time period, and the actual entry in order to report it as the answer instantly. To avoid adding unnecessary elements to *kNearestLists*, we maintain a set *PruneDist* of thresholds, to keep track of the maximum among each nearest list. In fact, the set *PruneDist* is an array which is initialized in line 1. Moreover, let *PruneDist*(*i*) be the maximum in the *i*-th nearest list, then we can easily derive the following relationship: *PruneDist*(1) < *PruneDist*(2)

< … < *PruneDist*(*k*), because all the nearest lists are sorted in ascending order by the distances. Initially, **Update*k*Nearests** compares *M.Dmin* with *PruneDist*(*k*). If *M.Dmin* ≥ *PruneDist*(*k*) holds, then the entry contained in *M* is an unnecessary one since its distance is larger than all the ones enclosed in *kNearestLists*, and the algorithm terminates. Otherwise, *M* is added to the structure *UpdateList* which stores the checked elements (line 3). Next, **Update*k*Nearests** recursively inserts every element in *UpdateList* into appropriate nearest list (lines 5-17). Note that in line 12, a sub-procedure **UpdateNearest** is invoked to update a single nearest list, denoted by *NearestList*.

```
Algorithm UpdatekNearests (Structure M, List kNearestLists)
1. Initialize structure UpdateList and list PruneDist
2. If M.Dmin < PruneDist (k) then
3.    Add M to UpdateList
4.    i = 0
5.    Do Until i = k OR UpdateList.count = 0
6.       i = i + 1
7.       Initialize lists AuxiliaryList and TempList
8.       For j = 1 to UpdateList.count
9.          M' = UpdateList (j)   // Fetch the j-th element
10.         If M'.Dmin ≥ PruneDist (i) Then Continue
11.         Else
12.            TempList = UpdateNearest (M', kNearestLists (i))
13.         Endif
14.         Transfer all elements from TempList to AuxiliaryList
15.      Next
16.      Transfer all elements from AuxiliaryList to UpdateList
17.   Loop
18. Endif
End UpdatekNearests
```

Fig. 2. Update*k*Nearests algorithm

Figure 3 shows the pseudo-code of the **UpdateNearest** algorithm, which outputs the list *NextUpdateList* storing the elements that need to be checked later. Initially, **UpdateNearest** examines whether *M* overlaps with some elements in *NearestList* w.r.t. time extent. If not, it adds *M* to *NearestList* directly, returns *NextUpdateList*, and terminates (lines 3). Otherwise, it takes various cases into consideration in updating *NearestList* (lines 5-19). Specifically, **UpdateNearest** first determines whether the time interval of *T* (storing the element already in *NearestList*), denoted by [$T.T_s$, $T.T_e$], intersects with that of *M* (i.e., [$M.T_s$, $M.T_e$]). If so, it calculates the temporal overlapping (denoted by [T_s, T_e]) between *T* and *M* (line 10). Subsequently, **UpdateNearest** updates *NearestList* by analyzing all the relationships between the distance functions of *T* and *M*, as illustrated in Figure 4. Specifically, Figures 4c and 4d correspond to line 11 of the **UpdateNearest** algorithm, where the starting time of *M* (i.e., $M.T_s$) is smaller than T_s, leading to the addition of the partial *M* having the time interval [$M.T_s$, T_s), denoted as Part (*M*, [$M.T_s$, T_s)). Similarly, Figures 4a and 4b, Figures 4b and 4d, as well as

```
Algorithm UpdateNearest (Structure M, List NearestList)
1. Initialize list NextUpdateList
2. If M does not overlap NearestList w.r.t. time extent Then
3.    Add M to NearestList and Return NextUpdateList
4. Else   // M overlaps NearestList w.r.t. time extent
5.    i = 0
6.    Do Until i = NearestList.count OR M.T_s = M.T_e
7.       i = i + 1 : T = NearestList (i)
8.       If M does not overlap T w.r.t. time extent Then Continue
9.       Else
10.         T_s = Max (M.T_s, T.T_s) : T_e = Min (M.T_e, T.T_e)
11.         If M.T_s < T_s Then Add Part (M, [M.T_s, T_s)) to NearestList
12.         If T.T_s < T_s Then Add Part (T, [T.T_s, T_s)) to NearestList
13.         nM = Interpolate (M, T_s, T_e) : nT = Interpolate (T, T_s, T_e)
14.         Consider all the relationships between nM and nT in order to
            determine if nT is to be replaced by nM or not   // See Fig. 5
15.         If T.T_e > T_e Then Add Part (T, [T_e, T.T_e)) to NearestList
16.      Endif
17.      M.T_s = T_e   // Update M such that its time period is [T_e, M.T_e]
18.   Loop
19.   If M.T_e > M.T_s Then Add Part (M, [M.T_s, M.T_e)) to NearestList
20. Endif
21. Return NextUpdateList
End UpdateNearest
```

Fig. 3. UpdateNearest algorithm

Figures 4a and 4c graphically represent the cases of lines 12, 15, and 19 in Figure 3, respectively. In lines 13-14, the algorithm first applies linear interpolation in both *T* and *M*, producing *nT* and *nM* having the same time interval $[T_s, T_e]$; and then it considers all the relationships between *nT* and *nM* in order to determine whether *nT* is to be replaced by *nM* fully (partially). For the sake of easy comprehension, Figure 5 illustrates the possible relationships between *nT* and *nM*. The details are elaborated immediately below.

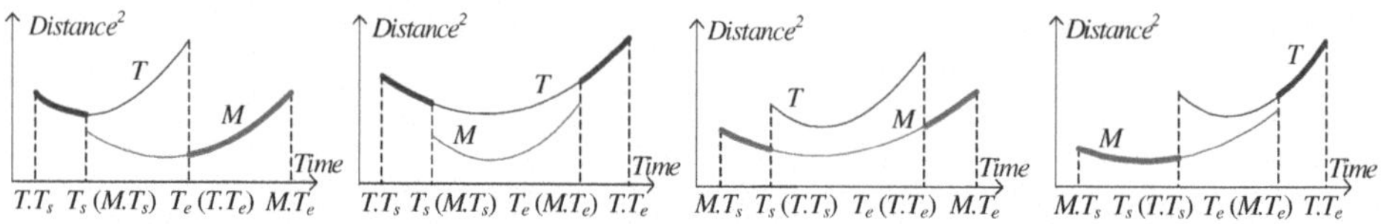

Fig. 4. Illustration of the relationships between *T* and *M*

Let $at^2 + bt + c$ and $a't^2 + b't + c'$ be the distance functions of *nT* and *nM*, respectively. As demonstrated in [4], if a (a') equals 0, then b (b') must be 0. Thus, *nT* (*nM*) may be a parabola when a (a') does not equal 0, or a line that is in parallel with the time axis when both a (a') and b (b') are equal to 0. For simplicity, we assume that both *nT* and *nM* are parabolas in the following discussion. However, similar conclusions also hold if they are lines.

To speed up the update between *nT* and *nM*, we first consider the case that the maximum of *nT* (i.e., *nT.Dmax*) is completely smaller than the minimum of *nM* (i.e., *nM.Dmin*). Then, *nT* is still stored in *NearestList*, but *nM* is added to *NextUpdateList*. Similarly, *nT* is replaced by *nM* if *nT.Dmin* > *nM.Dmax* holds. Next, we distinguish four cases to process the update (cf. Figure 5).

Case 1. This case (Figure 5a) occurs when $a = a'$, $b = b'$, $c = c'$, i.e., *nT* and *nM* are identical. Then, *nT* is stored in *NearestList* and *nM* is added to *NextUpdateList*.

Case 2. This case (Figure 5b) occurs when $a = a'$, $b = b'$, $c \neq c'$. This means *nT* and *nM* only have the different offset in the distance axis. In this case, we need to check their maximum in order to determine whether *nT* is replaced by *nM*.

Case 3. This case (Figure 5c) occurs when $a = a'$, $b \neq b'$, meaning that *nT* and *nM* have the different offsets both in the distance and time axes. After computing the *Root* $(= (c - c') / (b' - b))$ of the difference between the distance functions of *nT* and *nM*, we distinguish several sub-cases to handle the update. Assume that $[T_s, T_e] = [T_1, T_5]$, for example, we must split the parabolas into different parts and determine the part of *nT* to be replaced by that of *nM* because the timestamp of the *Root* (i.e., T_3), denoted by *Root.time*, falls into the interval $[T_1, T_5]$. Hence, Part (*nT*, $[T_3, T_5)$) is replaced by Part (*nM*, $[T_3, T_5)$). The other sub-cases are similarly handled and are omitted due to the space limitation.

Case 4. This case (Figures 5d-5f) happens when $a \neq a'$. This implies that *nT* and *nM* not only have the different offsets in both the distance and time axes, but also are of the different radians of the parabolas. In this case, we first compute the discriminant D of the difference between the distance functions of *nT* and *nM*, and then distinguish among the following sub-cases: (i) $D < 0$ (Figure 5d), meaning that the distance functions of *nT*

node entry containing a leaf node one (lines 12-25) and a non-leaf node one (lines 27-31). More specifically, (i) if *E* is a trajectory segment entry, then HC_T-*k*NN calls **Update*k*Nearests** algorithm to add *E* to *kNearestLists* and update *kNearestLists* (if necessary); (ii) if *E* is a leaf node, then HC_T-*k*NN inserts for every entry *e* in *E* into *hp* if *e* has the time period across *T*, *e*'s time interval overlaps with that of each entry *qe* in *QT*, and its distance from *qe* is smaller than *PruneDist*(*k*); similarly, (iii) HC_T-*k*NN adds all the necessary entries in *E* to *hp* when *E* is an intermediate node. It is important to note that the operation concerned in line 15 is necessary because the temporal extent of some *qe* in *QT* may not intersect with that of *e* in *E* (therefore it needs not be visited). Also note that, the computation of the *Mindist_Trajectory_Rectangle* metric included in line 29 uses the method proposed in [4].

```
Algorithm HC_T-kNN (TB-tree R, query trajectory Q, time period T, kNNcount k)
1.  Initialize heap hp, lists kNearestLists and PruneDist
2.  Get the set QT of query trajectory segments having the time intervals across T
3.  Insert all the entries of the root in R into hp
4.  Do While hp.count > 0
5.    De-heap the top entry E in hp
6.    If E.Dmin ⩾ PruneDist (k) then
7.      Return kNearestLists   // Report the final k nearest lists
8.    Endif
9.    If E is an actual trajectory segment entry then
10.     UpdatekNearests (E, kNearestLists)   // see Fig. 2
11.   ElseIf E is a leaf node
12.     For each entry e in E
13.       If (e.t_s, e.t_e) overlaps T then   // e crosses partially (or fully) T
14.         For each entry qe in QT
15.           If (qe.t_s, qe.t_e) overlaps (e.t_s, e.t_e) then
16.             ne = Interpolate (e, Max (qe.t_s, e.t_s), Min (qe.t_e, e.t_e))
17.             nqe = Interpolate (qe, Max (qe.t_s, e.t_s), Min (qe.t_e, e.t_e))
18.             MDist = ConstructMovingDistance (nqe, ne)
19.             If MDist.Dmin < PruneDist (k) then
20.               Insert ne into hp together with its MDist
21.             Endif
22.           Endif
23.         Next
24.       Endif
25.     Next
26.   Else   // E is an intermediate (i.e., a non-leaf) node
27.     For each entry e in E
28.       If (e.t_s, e.t_e) overlaps T then
29.         Add e whose time interval overlaps that of each entry qe in QT and
            and distance from qe, denoted by Mindist_Trajectory_Rectangle,
            is smaller than PruneDist (k) to hp
30.       Endif
31.     Next
32.   Endif
33. Loop
End HC_T-kNN
```

Fig. 7. HC_T-*k*NN algorithm

6 Experimental Evaluation

In this section, we experimentally evaluate the efficiency and scalability of our proposed algorithms both in terms of the I/O cost (i.e., number of node access) and CPU cost, using real and synthetic datasets. Since the work of [4] is most related to ours, we evaluate the performance of our algorithms by comparing the results against those of the algorithms proposed in [4]. All algorithms used in the experiments were implemented in Visual Basic, running on a PC with 3.0 GHz Pentium 4 processor and 1 GB memory.

6.1 Experimental Settings

We use two real datasets[1] that consist of a fleet of trucks containing 276 trajectories and a fleet of school buses containing 145 trajectories. We also deploy several synthetic datasets generated by a GSTD data generator [15] to examine the scalability of the algorithms. Specifically, the synthetic data sets correspond to 100, 200, 400, 800, and 1600 moving objects, with the position of each object being sampled approximately 1500 times. Furthermore, the initial distribution of moving objects is *Gaussian* while their movement is ruled by a random distribution. Table 1 summarizes the statistics of both real and synthetic datasets.

Table 1. Statistics of real and synthetic datasets

Datasets	# Trajectories	# Entries	# Pages
Trucks	276	112203	835
School buses	145	66096	466
GSTD 100	100	150052	1008
GSTD 200	200	300101	2015
GSTD 400	400	600203	4029
GSTD 800	800	1200430	8057
GSTD 1600	1600	2400889	16112

Table 2. Parameters in experiments

Parameters	Description	Values
k	number of NNs	1, 2, **4**, 8, 16
TE	temporal extent	1%, 2%, **3%**, 4%, 5%
#MO	number of moving objects	100, 200, **400**, 800, 1600

Each dataset is indexed by a TB-tree [10], using a page size of 4 KB and a (variable size) buffer fitting 10% of the tree size with the maximal capacity of 1000 pages. The experiments study three factors, involving *k*, temporal extent (TE), and the number of moving objects (#MO), which can affect the performance of the algorithms. The parameters used in the experiments are described in Table 2, in which the values in bold denote the default ones used. In each experiment, only one parameter varies while the others are fixed to their default values. Performance is measured by executing workloads, each comprising of 100 HC*k*NN queries. For each experimental instance, the reported results are the mean cost per query for a workload with the same settings. In addition, the query points used in the HC_P-*k*NN algorithm utilize random ones in 2-dimensional space. For the HC_T-*k*NN algorithm on trucks dataset, we take random parts of random trajectories belonging to the school bus dataset as the query trajectory collection; while in the case of GSTD datasets, the query sets of trajectories are created by the GSTD data generator.

6.2 Experimental Results on HC_P-*k*NN Algorithm

The first set of experiments investigates the effect of *k*. Figure 8 shows the number of node accesses and CPU time (in seconds) of the algorithms as a function of *k*.

[1] The real datasets are downloaded from the R-tree portal (http://www.rtreeportal.org).

Obviously, HC$_P$-*k*NN outperforms its competitor (i.e., ContPointNNSearch of [4]) significantly, and the difference increases with *k*. As expected, the query overhead of each algorithm grows with *k*, due to the increase in the update cost of *k* nearest lists.

Next, Figure 9 compares the performance of the two algorithms with respect to TE. Also, HC$_P$-*k*NN is evidently superior to ContPointNNSearch in all cases. Overall, the CPU time of each algorithm increases linearly as TE grows, which is caused by the growth of temporal overlapping.

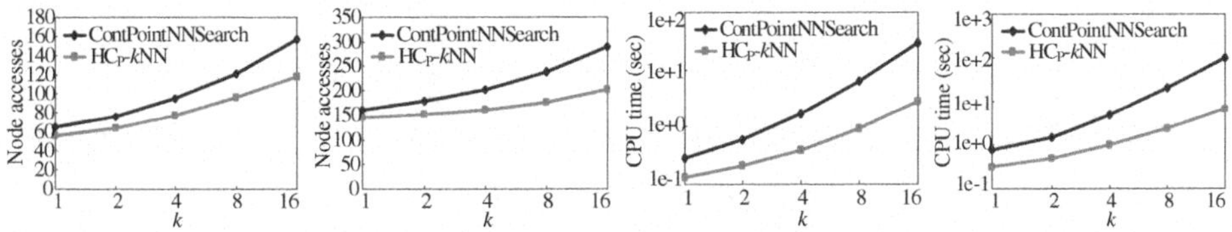

(a) Trucks (b) GSTD 400 (c) Trucks (d) GSTD 400

Fig. 8. Effect of *k* (TE = 3%)

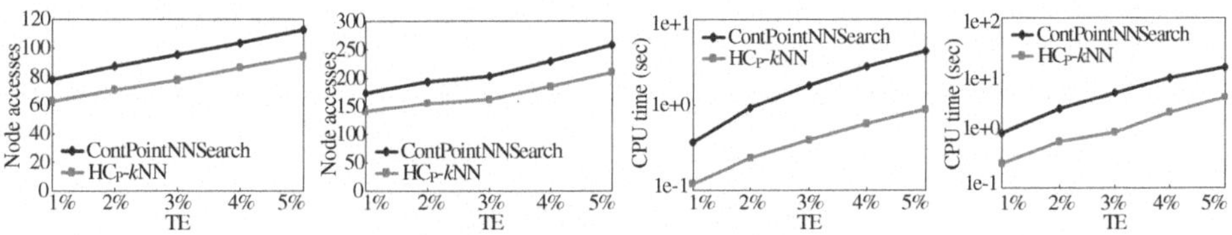

(a) Trucks (b) GSTD 400 (c) Trucks (d) GSTD 400

Fig. 9. Effect of TE (*k* = 4)

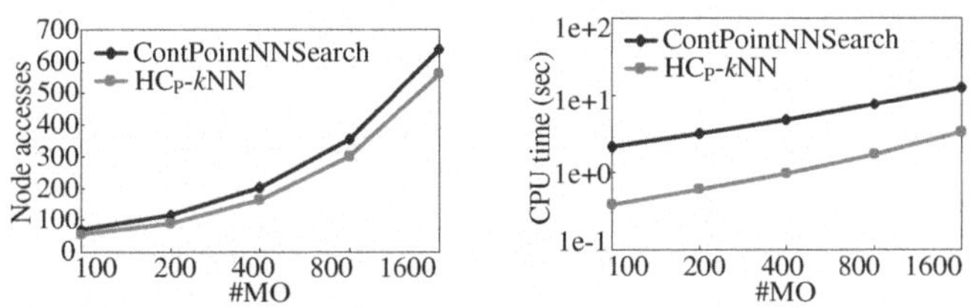

Fig. 10. Effect of # MO (*k* = 4, TE = 3%, GSTD)

Finally, Figure 10 plots the performance of the two algorithms with respect to #MO using the synthetic dataset. HC$_P$-*k*NN again wins, and is several orders of magnitude faster than ContPointNNSearch in terms of CPU time.

6.3 Experimental Results on HC$_T$-*k*NN Algorithm

Having confirmed the superiority of HC$_P$-*k*NN for the HC*k*NN retrieval w.r.t. the static query point, we proceed to evaluate the performance of HC$_T$-*k*NN for the HC*k*NN query w.r.t. the moving query trajectory. Figure 11 shows the efficiency of our algorithm as a function of *k* for trucks and GSTD 400 datasets. Similar to the diagrams in Figure 8, HC$_T$-*k*NN is clearly better than its competitor (i.e., ContTrajectoryNNSearch of [4]) significantly, and the difference increases with *k*.

Subsequently, Figure 12 compares the cost of the two algorithms by varying TE. The diagrams and their explanations are similar to those of Figure 9. As with the settings of Figure 10, the last set of experiments (Figure 13) shows the performance of the two algorithms versus #MO, which exhibits a similar pattern as that of Figure 10.

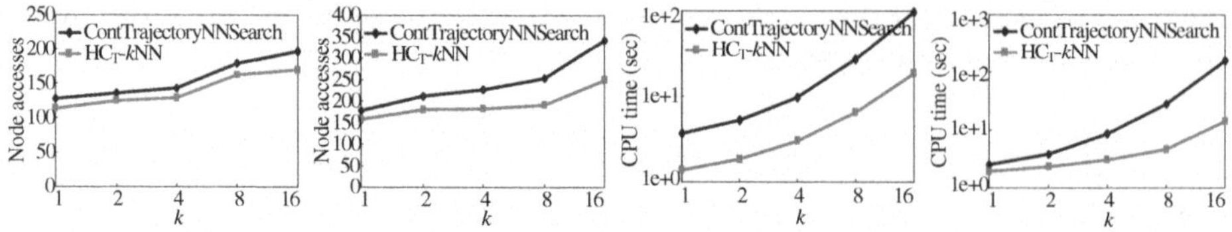

(a) Trucks (b) GSTD 400 (c) Trucks (d) GSTD 400

Fig. 11. Effect of k (TE = 3%)

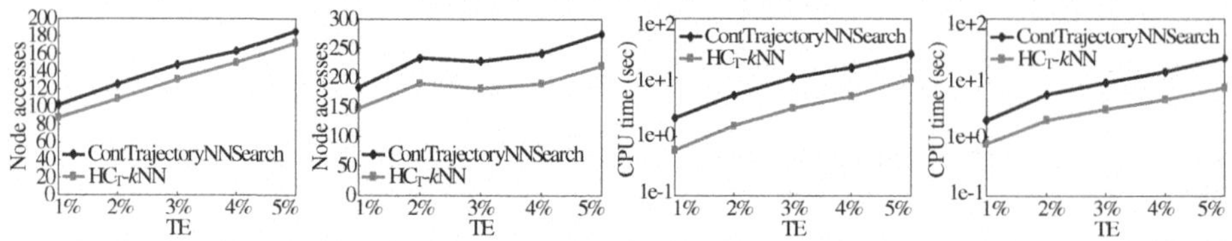

(a) Trucks (b) GSTD 400 (c) Trucks (d) GSTD 400

Fig. 12. Effect of TE ($k = 4$)

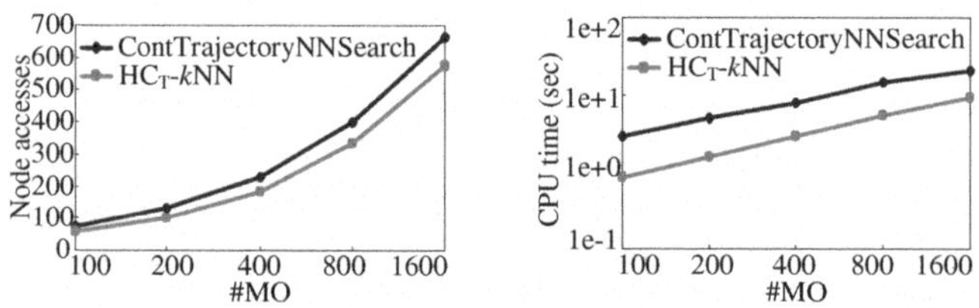

Fig. 13. Effect of # MO ($k = 4$, TE = 3%, GSTD)

7 Conclusions

Although CkNN queries for spatial and spatiotemporal objects have been well-studied in the last decade, there is little prior work on HCkNN retrieval over moving object trajectories. In this paper, we have developed two efficient algorithms to process HCkNN search on R-tree-like structures storing historical information about trajectories. In contrast to the existing HCkNN query algorithms [4] which adopted the *depth-first* traversal paradigm hence incurs expensive I/O and CPU cost, our solution uses the *best-first* traversal paradigm, and enables effective update strategies to maintain the *nearest lists*. Extensive experiments with real and synthetic datasets show that the proposed algorithms outperform their competitors significantly in both efficiency and scalability. An interesting direction for future work is to explore other query algorithms (e.g., k-closest pair queries [3]) for moving object trajectories. For instance, *"find the k pairs of trajectories that have the k smallest distances among all possible pairs during a pre-defined time period"*. Another challenging issue is to develop a cost model to estimate the execution time of the *k*NN retrieval over trajectories, so as to facilitate query optimization and reveal new problem characteristics that could lead to even faster algorithms.

Acknowledgment. We would like to thank the authors for sharing the implementation of their proposed algorithms in [4].

References

1. Beckmann, N., Kriegel, H-P, Schneider, R., Seeger, B.: The R*-tree: An Efficient and Robust Access Method for Points and Rectangles. In: SIGMOD (1990) 322-331
2. Benetis, R., Jensen, C.S., Karciauskas, G., Saltenis, S.: Nearest Neighbor and Reverse Nearest Neighbor Queries for Moving Objects. In: IDEAS. (2002) 44-53
3. Corral, A., Manolopoulos, Y., Theodoridis, Y., Vassilakopoulos, M.: Closest pair queries in spatial databases. In: SIGMOD. (2000) 189-200
4. Frentzos, E., Gratsias, K., Pelekis, N., Theodoridis, Y.: Nearest Neighbor Search on Moving Object Trajectories. In: SSTD. (2005) 328-345
5. Hjaltason, G.R., Samet, H.: Distance Browsing in Spatial Databases. ACM TODS **24** (1999) 265-318
6. Iwerks, G.S., Samet, H., Smith, K.: Continuous k-Nearest Neighbor Queries for Continuously Moving Points with Updates. In: VLDB. (2003) 512-523
7. Manolopoulos, Y., Nanopoulos, A., Papadopoulos, A.N., Theodoridis, Y.: R-trees: theory and applications. Springer. (2005)
8. Mouratidis, K., Hadjieleftheriou, M., Papadias, D.: Conceptual Partitioning: An Efficient Method for Continuous Nearest Neighbor Monitoring. In: SIGMOD. (2005) 634-645
9. Mouratidis, K., Papadias, D., Bakiras, S., Tao, Y.: A Threshold-based Algorithm for Continuous Monitoring of k Nearest Neighbors. TKDE **17** (2005) 1451-1464
10. Pfoser, D., Jensen, C.S., Theodoridis, Y.: Novel Approaches in Query Processing for Moving Object Trajectories. In: VLDB. (2000) 395-406
11. Roussopoulos, N., Kelley, S., Vincent, F.: Nearest neighbor queries. In: SIGMOD. (1995) 71-79
12. Song, Z., Roussopoulos, N.: K-Nearest Neighbor Search for Moving Query Point. In: SSTD. (2001) 79-96
13. Tao, Y., Papadias, D., Shen, Q.: Continuous Nearest Neighbor Search. In: VLDB. (2002) 287-298
14. Tao, Y., Papadias, D.: Time Parameterized Queries in Spatio-Temporal Databases. In: SIGMOD. (2002) 334-345
15. Theodoridis, Y., Silva, J.R.O., Nascimento, M.A.: On the Generation of Spatiotemporal Datasets. In: SSD. (1999) 147-164
16. Theodoridis, Y., Vazirgiannis, M., Sellis, T.K.: Spatio-Temporal Indexing for Large Multimedia Applications. In: ICMCS. (1996) 441-448
17. Xiong, X., Mokbel, M., Aref, W.: SEA-CNN: Scalable Processing of Continuous K-Nearest Neighbor Queries in Spatio-temporal Databases. In: ICDE. (2005) 643-654
18. Yu, X., Pu, K., Koudas, N.: Monitoring k-Nearest Neighbor Queries Over Moving Objects. In: ICDE. (2005) 631-642
19. Gao, Y., Li, C., Chen, G., Chen, L., Jiang, X., Chen C.: Efficient k-Nearest-Neighbor Search Algorithms for Historical Moving Object Trajectories. JCST **22** (2007) 232-244

Effective Density Queries for Moving Objects in Road Networks*

Caifeng Lai[1,2], Ling Wang[1,2], Jidong Chen[1,2], Xiaofeng Meng[1,2], and Karine Zeitouni[3]

[1] School of Information, Renmin University of China
[2] Key Laboratory of Data Engineering and Knowledge Engineering, MOE
{laicf,jingyiwang,chenjd,xfmeng}@ruc.edu.cn
[3] PRISM, Versailles Saint-Quentin University, France
karine.zeitouni@prism.uvsq.fr

Abstract. Recent research has focused on density queries for moving objects in highly dynamic scenarios. An area is dense if the number of moving objects it contains is above some threshold. Monitoring dense areas has applications in traffic control systems, bandwidth management, collision probability evaluation, etc. All existing methods, however, assume the objects moving in the Euclidean space. In this paper, we study the density queries in road networks, where density computation is determined by the length of the road segment and the number of objects on it. We define an effective road-network density query guaranteeing to obtain useful answers. We then propose the cluster-based algorithm for the efficient computation of density queries for objects moving in road networks. Extensive experimental results show that our methods achieve high efficiency and accuracy for finding the dense areas in road networks.

1 Introduction

The advances in mobile communication and database technology have enabled innovative mobile applications monitoring moving objects. In some applications, the object movement is constrained by an underlying spatial network, e.g., vehicles move on road networks and trains on railway networks. In this scenario, objects can not move freely in space, and their positions must satisfy the network constrains. A network is usually modeled by a graph representation, comprising a set of nodes (intersections) and a set of edges (segments). Depending on the application, the graph may be directed, i.e. each edge has an orientation. Additionally, moving objects are assumed to move in a piecewise linear manner [6], i.e., each object moves at a stable velocity at each edge. The distance between two arbitrary objects is defined as the network distance between them on the network. Several types of queries have been studied in the road network, such

* This research was partially supported by the grants from the Natural Science Foundation of China under grant number 60573091; Program for New Century Excellent Talents in University (NCET).

G. Dong et al. (Eds.): APWeb/WAIM 2007, LNCS 4505, pp. 200–211, 2007.

as kNN queries [7], range queries [9], aggregate nearest neighbor queries [11], reverse nearest neighbor queries [12].

In this paper, we focus on the problem of dynamic density queries for moving objects in road networks. The objective is to find dense areas with high concentration of moving objects in a road network efficiently. The density query can be used in the traffic management systems to identify and predict the congested areas or traffic jams. For example, the transportation bureau needs to monitor the dense regions periodically in order to discover the traffic jams.

Existing research works on the density query [2,4] assume the objects moving in a free style and define the density query in the Euclidean space. In this setting, the general density-based queries are difficult to be answered efficiently and their focus is hence turned to simplified queries [2] or specialized density queries without answer loss [4]. These methods use the grid to partition the data space into disjoint cells and report the rectangle area with the fixed size. However, the real dense areas may be larger or smaller than the fixed-size rectangle and appear in different shapes. Simplifying the dense query to return the area with fixed size and shape can not reflect the natural congested area in real-life application. We focus on the density query in the road-network setting, where the dense area consists of road segments containing large number of moving objects and may be formed in any size and shape. The real congested areas can therefore be obtained by finding the dense segments. In addition, for querying objects moving in a road network, the existing methods based on a regular spatio-temporal grid ignore the network constraint and therefore result in inaccurate query results.

Considering the real-life application, finding dense regions for a point in time is more useful than finding the dense regions for a period of time [4]. In this paper, we study the querying for dense regions consisting of dense segments for a point in time. For monitoring the dense areas of moving objects in the road network, the dense query requests need to be issued periodically in order to find the changes of dense areas. If we use the existing methods, the total cost is quite high since each query request requires accessing all objects in the road network. Since clustering can represent the dense areas naturally, we propose a cluster-based method to process density queries in a road network. The moving objects are first grouped into cluster units on each road segment according to their locations and moving patterns. Then the cluster units are maintained continuously. The process can be treated as a separate pre-processing for the periodical density queries. For density query processing, we use a two-phase algorithm to identify the dense areas based on the summary information of the cluster units. Maintaining cluster units comes with a cost, but our experimental evaluations demonstrate it is much cheaper than keeping the complete information about individual locations of objects to process the dynamic density queries.

Our contributions are summarized as follows:

- We define the density query for moving objects in road networks that is amenable to obtain the effective answers.
- We propose a cluster-based algorithm to efficiently monitor the dense areas in a road network.

– We show, through extensive experiments, that our query algorithms achieve high efficiency and accuracy.

The rest of the paper is organized as follows. Section 2 reviews related work on density query processing and clustering moving objects. Section 3 gives the problem definition. Section 4 details our density query method including dynamic cluster maintenance and two-phase query algorithm. Experimental results are shown in Section 5. We conclude this paper in Section 6.

2 Related Work

Density query for moving objects is first proposed in [2]. The objective is to find regions in space and time that with the density higher than a given threshold. They find the general density-based queries difficult to be answered efficiently and hence turn to simplified queries. Specifically, they partition the data space into disjoint cells, and the simplified density query reports cells, instead of arbitrary regions that satisfy the query conditions. This scheme may result in answer loss. To solve this problem, Jensen et al. [4] define an effective density query to guarantee that there is no answer loss. The two works both assume the objects moving in a free style and define the density query in Euclidean space. However, efficient dynamic density query in spatial networks is crucial for many applications. As an example of a real world, considering that the queries correspond to vehicles distribution in the road network, the users would like to know real-time traffic density distribution. Clearly, in this case the Euclidean density query methods are inapplicable, since the path between two cars is restricted by the underlying road network. Additionally, these existing query methods can not reflect the natural dense area in a road network since they simplify the dense query to return the area with fixed size and shape. The grid-based algorithms also ignore the network constraint and result in inaccurate query results. It is natural to represent the dense area in a road network as road segments containing large number of moving objects. We exploit the network property and define *effective road-network density query*(*e-RNDQ*) to return the natural density areas with arbitrary size and shape in the road network.

Existing network based clustering algorithms are also related to our work. Jain et al. [3] use the agglomerative hierarchical approach to cluster nodes of a graph. CHAMELEON [5] is a general-purpose algorithm, which transforms the problem space into a weighted *kNN* graph, where each object is connected with its k nearest neighbors. Yiu and Mamoulis [10] define the problem of clustering objects based on the network distance and propose algorithms for three different clustering paradigms, i.e., *k-medoids* for K-partitioning, ϵ*-link* for density-based, and *single-link* for hierarchical clustering. The ϵ*-link* method is most efficient to find dense segments in road network. However, all these solutions assumed a static dataset. Li *et al.* [6] propose the micro moving cluster (MMC) to clustering moving objects in Euclidean spaces. Our clustering algorithm focuses on moving objects in the road network which exploits the road-network features and provides the summary information of moving objects to density query processing.

There are some other related works on query processing in spatial network databases [1,7,11]. Their focus is to evaluate various types of queries based on the network distance by minimizing the cost of the expensive shortest path computation. To the best of our knowledge, this is the first work which specifies on the cluster-based method for dynamic density queries in spatial networks.

3 Problem Definition

As the result of density queries in the road network are the set of dense segments, we first introduce the concepts of *density* and *dense segment.*

Definition 1. *Density. The density of a road segment s is represented as density* $(s) = N/len(s)$, *where N is the number of objects on s and* $len(s)$ *is the length of s.*

Definition 2. *Dense Segment (DS). The road segment s is a dense segment (DS) if and only if density(s)* $\geq \rho$, *where* ρ *is a user-defined density parameter.*

A straightforward method to process the query is to traverse all objects moving on a road network to compute dense regions by their number, the length of the segment and a given density threshold. Figure 1 shows a density query in a road network. Obviously, the cost is very high and it is difficult to find effective results. Specifically, the query results may have three problems as follows:

1) The different *DS* may be overlapped, such as Case 1 in Figure 1.
2) The distribution of moving objects may be very skewed in some *DS*, namely, the distribution of objects is dense in one part of a *DS*, but it is sparse in another part, such as Case 2 in Figure 1.
3) Some *DS* may contain very few objects, such as Case 3 in Figure 1.

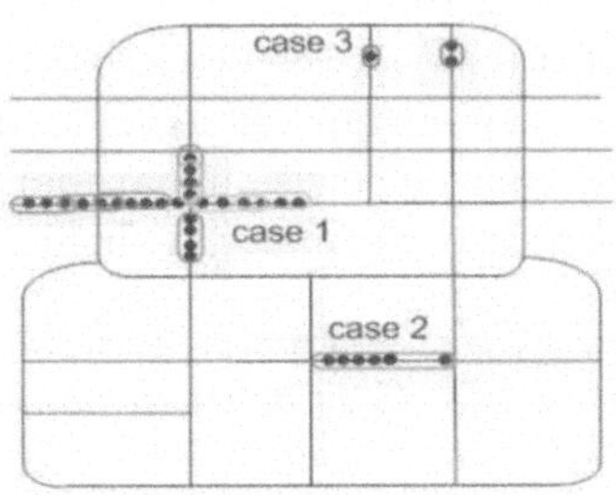

Fig. 1. An example of density query

Such query results are less useful. Thus, we define an *effective density query* in a road network to find the useful dense regions with a high concentration of objects and symmetrical distribution of objects as well as no overlapping.

Definition 3. *Effective Road-Network Density Query (e-RNDQ): Given density parameter ρ, find all dense segments that satisfy the following conditions:*

1. Any dense segment set can not be intersecting (namely no overlaps).

2. In each dense segment set, the distance between any neighboring object is not more than a given distance threshold δ.

3. The length of dense segments is not less than a given length threshold L.

4. Any dense segment containing moving objects is in the query result set.

The first condition ensures that the result is not redundant. It avoids the case 1 in Figure 1. The second condition guarantees that objects are symmetrically distributed in a dense segment set. The third condition provides restriction that there is no small segments that only contain few objects in the result. The fourth condition ensures that query results do not suffer from answer loss.

4 Density Query Processing in Road Networks

4.1 Overview

Considering the feature of road networks, we propose a cluster-based density querying algorithm, which regards clustering operation as a pre-processing to provide the summary information of moving objects. In the query processing, we develop a two-phase filter-and-refinement algorithm to find dense areas.

4.2 Cluster-Based Query Preprocessing

To reduce the cost of clustering maintenance, we introduce the definition of Cluster Unit. A cluster unit is a group of moving objects close to each other at present and near future time. It will be incrementally maintained with moving of objects in it. Specifically, we constrain the objects in a cluster unit moving in the same direction and on the same segment. For keeping the objects in a cluster unit dense enough, the network distance between each pair of neighboring objects in a cluster unit does not exceed a system threshold ϵ. As mentioned in Introduction, we assume that objects move in a piecewise linear manner and the next segment to move along is known in advance. Formally, a cluster unit is defined as follows:

Definition 4. *Cluster Unit (CU). A cluster unit is represented by CU= (O, n_a, n_b, head, tail, ObjNum), where O is a list of objects $\{o_1, o_2, \cdots, o_i, \cdots, o_n\}$, o_i=(oid_i, n_a, n_b, pos_i, $speed_i$, $next\_node_i$), where pos_i is the relative location to n_a, $speed_i$ is the moving speed and $(n_b, next\_node)$ is the next segment to move along. Without loss of generality, assuming $pos_1 \leq pos_2 \leq \cdots \leq pos_n$, it must satisfy $|pos_{i+1} - pos_i| \leq \epsilon$ $(1 \leq i \leq n-1)$. Since all objects are on the same segment (n_a, n_b), the position of the CU is determined by an interval $(head, tail)$ in terms of the network distance from n_a. Thus, the length of the CU is $|tail - head|$. $ObjNum$ is the number of objects in the CU.*

Initially, based on the definition, a set of CU are created by traversing all segments in the network and their associated objects. The CUs are incrementally maintained after their creation. As time elapsed, the distance between adjacent objects in a CU may exceed ϵ. Thus, we need to split the CU. A CU may also merge with its adjacent CUs when they are within the distance of ϵ. Hence, for each CU, we predict the time when they may split or merge. The predicted split and merge events are then inserted into an event queue. Afterwards, when the first event in the queue takes place, we process it and update the affected CUs. This process is continuously repeated. The key problems are: 1) how to predict split/merge time of a CU, and 2) how to process a split/merge event of a CU.

The split of a CU may occur in two cases. The first one is when CU arriving at the end of the segment (i.e., an intersection node of the road network). When the moving objects in a CU reach an intersection node, the CU has to be split since they may head in different directions. Obviously, a split time is the time when the first object in the CU arrives at the node. In the second case, the split of a CU is when the distance between some neighboring objects moving on the segment exceed ϵ. However, it is not easy to predict the split time since the neighborhood of objects changes over time. Therefore, the main task is to dynamically maintain the order of objects on the segment. We compute the earliest time instance when two adjacent objects in the CU meet as t_m. We then compare the maximum distance between each pair of adjacent objects with ϵ until t_m. if this distance exceeds ϵ at some time, the process stops and the earliest time exceeding ϵ is recorded as the split time of CU. Otherwise, we update the order of objects starting from t_m and repeat the same process until some distance exceed ϵ or one of the objects arrives at the end of the segment. When the velocity of an object changes over the segment, we need to re-predict the split and merge time of the CU.

To reduce the processing cost of splitting at the end of segment, we propose group split scheme. When the first object leaves the segment, we split the original CU into several new CUs according to objects' directions (which can be implied by *next_node*). On one hand, we compute a to-be-expired time (i.e., the time until the departure from the segment) for each object in the original CU and retain the CU until the last object leaves the segment. On the other hand, we attach a to-be-valid time (with the same value as *to-be-expired time*) for each object in the new CUs. Only valid objects will be counted in constructing CUs.

The merge of CUs may occur when adjacent CUs in a segment are moving together (i.e., their network distance $\leq \epsilon$). To predict the initial merge time of CUs, we dynamically maintain the boundary objects of each CU and their validity time (the period when they are treated as boundary of the CU), and compare the minimum distance between the boundary objects of two CUs with the threshold ϵ at their validity time. The boundary objects of CUs can be obtained by maintaining the order of objects during computing the split time.

The processing of the merge event is similar to the split event on the segment. We get the merge event and time from the event queue to merge the CUs into one CU and compute the split time and merge time of the merged CU. Finally, the corresponding affected CUs in the event queue are updated.

Besides the split and merge of CUs, new objects may come into the network or existing objects may leave. For a new object, we locate all CUs of the same segment that the object enters and see if the new object can join any CU according to the CU definition. If the object can join some CU, its split and merge events are updated. If no such CUs are found, a new CU for the object is created and the merge event is computed. For a leaving object, we update the split and merge events of its original CU if necessary. Due to the limitation of the space, we omit the algorithm pseudo of maintaining CUs.

4.3 Density Query Processing

Based on the dynamic CUs, density query at any time point can be processed efficiently to return dense areas in the road networks. And then the dense segment (DS) we defined in Section 3 is represented as (*CU*, n_a, n_b, *startpos*, *endpos*, *len*, *N*), where *CU* is the set of cluster units on segment (n_a, n_b), *startpos* is the start position of the DS, and *endpos* is the end position of the DS, *len* is the length of DS, *N* is the number of objects. To obtain the effective dense areas restricted in the e-RNDQ, we introduce the parameter δ to DS.

Definition 5. *δ-Dense Segment (δ-DS). A DS is δ-DS if and only if the distance between any adjacent CUs is not more than δ (i.e. guarantee that the distance between any adjacent object satisfies Distance(o_i, o_{i+1})$\leq\delta$), and density is not less than ρ. (For convenience, we abbreviate the term δ-DS to DS in the sequel)*

In fact, δ is a user-defined parameter of the density query and ϵ is a system parameter to maintain the CUs. Since the distance of adjacent objects is not more than ϵ in a CU, in order to retrieve dense areas based on CUs, we require $\epsilon \leq \max\{\delta, \frac{1}{\rho}\}$. In the road network, a dense area is represented as a dense segment set, which may contain several DSs in different segments. Therefore, we exploit network nodes to optimize the combination of these DSs.

Definition 6. *δ-ClusterNode (δ-CN). In each DS, n_a is δ-CN of the DS, if and only if $|startpos\text{-}n_a| \leq \delta$; n_b is δ-CN of the DS, if and only if $|endpos\text{-}n_b| \leq \delta$.*

Definition 7. *Dense Segment Set (DSS). A DSS consists of different DSs where the distance between adjacent DSs is not more than δ and the total length of DSs in the DSS is not less than L, the density in the DSS is not less than ρ.*

Actually, DSS may contain DSs located in different segments where DSs are joined by δ-CN. DSS constitutes the road-network density query results. Suppose the density query parameter is given as (ρ, δ, L, t_q), where t_q is the query time. For query processing based on CUs, our algorithm includes two steps:

(1) **The filtering step:** Merge CUs into DSs by checking the parameter of ρ and δ, which can prune some unnecessary segments. In this step, we can obtain a series of dense segments, specifically, a list of DSs and δ-CNs.

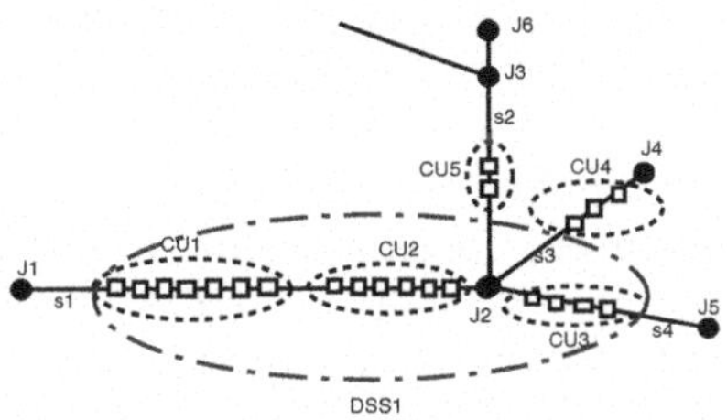

Fig. 2. An example to construct DS and DSS

Algorithm 1. $Filter(\rho, \delta, t_q)$

Input: density threshold ρ, query time t_q

begin
    foreach $e(n_x, n_y)$ *of edgeList* **do**
        if $e.cuList \neq null$ **then**
            create a new DS: *ds*
            $cu \leftarrow getFirstCU(e)$
            $ds.addCU(cu)$; $ds.startpos = cu.pos$
            if $ds.startpos < \delta$ **then**
                $ds.putCN(n_x)$; $\delta\text{-}CN[n_x].putDS(ds)$
            while $getNextCU(e) \neq null$ **do**
                $nextcu \leftarrow getNextCU(e)$
                if $Dd(ds, nextcu) > \delta$ *or*
                $Dens(ds, nextcu) < \rho$ **then**
                    $ds.endpos = cu.pos + cu.len$; $e.addDS(ds)$
                    create a new DS: *ds*
                    $ds.startpos = nextcu.pos$
                $ds.addCU(nextcu)$; $cu = nextcu$
            $ds.endpos = cu.pos + cu.len$
            if $1 - ds.endpos < \delta$ **then**
                $ds.putCN(n_y)$; $\delta\text{-}CN[n_y].putDS(ds)$
            $e.addDS(ds)$
end

(2) **The refinement step:** Merge the adjacent DSs around δ-CNs to construct DSS by checking the parameter of ρ, δ, L and finally find out the effective density query result consisting of dense segment sets.

We explain the two steps of density query processing in detail. Firstly, according to network expansion approach [8], we traverse each segment to retrieve CUs sequentially, then compute the distance between adjacent CUs and the density of them. If the distance is not more than δ and the density is not less than ρ, the CUs are merged to be a DS. Figure 2 shows an example. Given ρ=1.5 and δ=2, we compute DS at query time t_q. The road segment s_1 (represented as $< J_1, J_2 >$) includes two CUs named cu_1 and cu_2. Assume that the distance between

cu_1 and cu_2 is 1.2 at t_q which is less than δ, and the density is 1.8 after merging cu_1 with cu_2 which is more than ρ, cu_1 and cu_2 can construct a DS (we call it DS_1). The start position of DS_1 is the head of cu_1 and the end position of DS_1 is the tail of cu_2. The number of objects in DS_1 is the sum of the number of objects in cu_1 and that in cu_2. Assume that the distance between DS_1 and node J_2 is 1.0 which is less than δ, J_2 is the δ-CN of DS_1 (we call it δ-CN_1). We insert DS_1 into the DS list of δ-CN_1. In this way, we can obtain DS_2 on s_3 including cu_4 and DS_3 on s_4 including cu_3 respectively. The δ-CN of DS_2 (δ-CN_2) is J_4 and that of DS_3 is J_2. So the DS list of δ-CN_1 includes DS_1 and DS_3, while the DS list of δ-CN_2 includes DS_2. Algorithm 1 shows the pseudo.

Algorithm 2. *Refinement*(ρ, δ, L, t_q)

```
Input: density threshold ρ, length threshold of DSS L
Output: Result: The set of DSSs
begin
    foreach δ-CN_i of δ-CNList do
        if (δ-CN_i.dsList ≠ null) and (not δ-CN_i.accessed) then
            /*Q is a priority queue to store all DSs around δ-CN_i*/
            /*δ-Q is a priority queue to store all unaccessed δ-CNs*/
            Q ← null; δ-Q.put(δ-CN_i)
            while δ-Q ≠ null do
                cn = δ-Q.pop(); cn.accessed = true
                Q.addDSs(cn); /*add all DSs around cn and sorted*/
                create a new DSS: dss
                ds = Q.pop(); dss.addDS(ds)
                δ-Q.putdscn(ds); /*add all unaccessed δ-CN around ds*/
                while Q ≠ null do
                    nextDS = Q.pop()
                    if Dist(dss, nextDS) ≤ δ and Dens(dss, nextDS) ≥ ρ then
                        dss.addDS(nextDS)
                        δ-Q.putdscn(nextDS)
            if dss.len > L then
                Result.insert(dss)
    return Result
end
```

In the refinement step, we compute dense segment sets so that the effective dense areas can be obtained. We traverse the list of each δ-CN and evaluate whether those DSs around the δ-CN can construct DSS based on the definition 8. Given L=100 in Figure 2. As the Distance(DS_1, δ-CN_1)=1.0 and Distance(DS_3,δ-CN_1)=0.7, the distance between DS_1 and DS_3 is 1.7, which is less than δ. In addition, if DS_1 is merged with DS_3, the density is more than ρ. Therefore, DS_1 and DS_3 can be merged to be a DSS named DSS_1. In the same way, we check if there are other dense segments can be merged with DSS_1 by utilizing its δ-CN and insert it into DSS_1. Finally, we check if the total length of DSS_1 is more than L. If so, DSS_1 is one of the answers of the density query. Repeat the process until all δ-CNs containing dense segments are accessed. Then we can obtain all dense areas which are represented as dense segment sets at t_q. Note that a DS may be involved in the lists of two δ-CNs. To avoid scanning the same nodes repeatedly, we mark the scanned δ-CN as accessed node. Algorithm 2 shows the pseudo of the refinement step.

5 Experimental Evaluation

In this section, we compare the cluster-based density query processing with the existing density-based road-network clustering algorithm, ϵ-link proposed by Yiu et al. [10] in terms of query performance and accuracy since ϵ-link also returns the dense areas which consist of the density-based clusters of objects. We monitor the query results by running the ϵ-link algorithm periodically and by maintaining CUs and finding the dense segments based on CUs.

Experimental Settings. We implement the algorithms in C++ and carry out experiments on a Pentium 4, 2.4 GHz PC with 512MB RAM running Windows XP. For monitoring the dense areas in a road network, we design a moving object generator to produce synthetic datasets. The generator takes a map of a road network as input, and our experiment is based on the map of Beijing city. We set K hot spots (destination nodes) in the map. Initially, the generator places 80 percent objects around the hot spots and 20 percent objects at random positions, and updates their locations at each time unit. The query workload is 100 and each query has three parameters: (i) the density threshold ρ; (ii) the threshold for dense segment length L; (iii) the threshold for the distance of adjacent objects δ. The query cost is measured in terms of CPU time. We also measure the accuracy of query answers.

Comparison with the ϵ-link Algorithm. To evaluate the performance, we first measure the total workload time and average query response time of two algorithms when varying the number of moving objects from 100K to 1M. We execute the CU maintenance and query processing in comparison with the static ϵ-link on all objects at each time unit during 0 to 20 time units. For total workload time (shown in Figure 3), we measure the total CPU time including CU maintenance and query processing based on CUs. Figure 4 shows the average query response time for periodic query processing. In fact, considering the feature of road network, a CU represents the summary information of its objects and is incrementally updated over time with low cost, which can help speeding up the query processing. Therefore, our method is substantially better than the static one in terms of average query response time and still better in terms of total workload time.

Accuracy Density Query. We evaluate the accuracy of density queries by computing average correct ratio of the number of objects in query result to that in the dataset around hot spots. Let $avgCorrectRate$ represent the average correct ratio of query result, $Query\_objNum$ be the number of objects of the

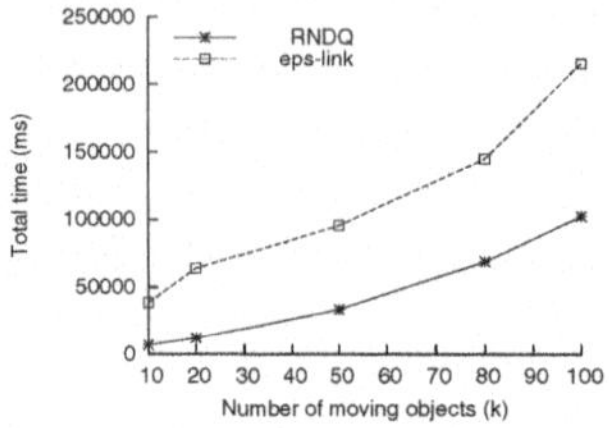

Fig. 3. Total time varies in data size

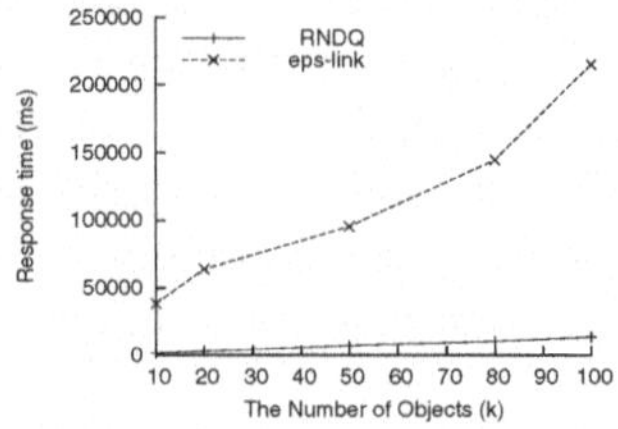

Fig. 4. Response time varies in data size

result, $Real\_objNum$ be the average number of objects around each hot spot in the dataset, $avgCorrectRate$ can be calculated by the following equation:

$$avgCorrectRate = \frac{1}{M}\sum_{i=1}^{M}(1 - \frac{|Query\_objNum - Real\_objNum|}{Real\_objNum}) \quad (1)$$

where M denotes the number of dense areas (i.e., DSS) in the query result. Figure 5 shows the comparison of the two methods in the query accuracy. We can see that the accuracy of our query algorithm is higher and stable with the different data distributions.

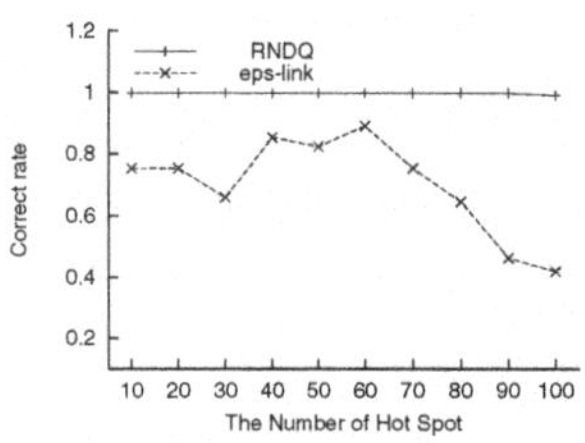

Fig. 5. Accuracy comparison

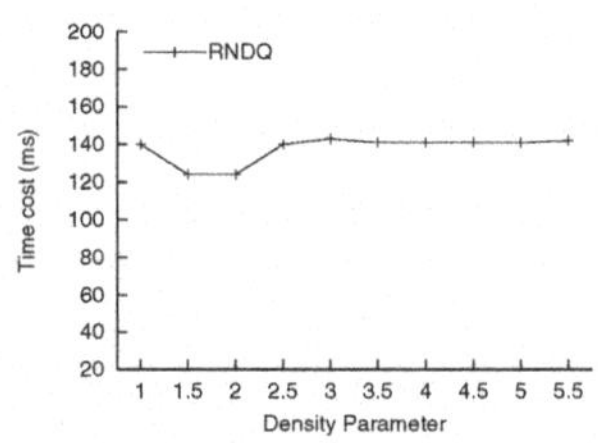

Fig. 6. Query Performance with ρ

Effect of Parameters. Finally, we study the effect of parameters (ρ, L, δ and ϵ) of our methods on the query performance. Given ϵ value at 2.5, δ value at 4.5, and L value at 100, we change density threshold ρ from 1 to 5.5 to evaluate time cost of query processing. Figure 6 shows the experimental result. In addition, we also evaluate time cost by adjusting the parameter L from 100 to 1000, and the result is similar to Figure 6. Next, when fixing the ϵ value at 2.5, we vary δ from 2 to 6.5 to study its effect on the query processing. Finally, as the number of CUs depends on the system parameter ϵ, we change the value of ϵ from 0.5 to 3 to measure the maintenance cost of CUs. Figure 7 and Figure 8 show the effect of the two parameters. We observe that when δ and ϵ are set to 4.5 and 2.5, the method achieves the highest efficiency in our experimental settings.

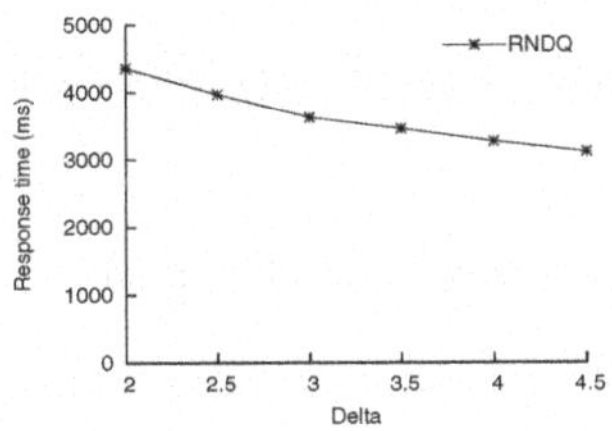

Fig. 7. Clustering performance with δ

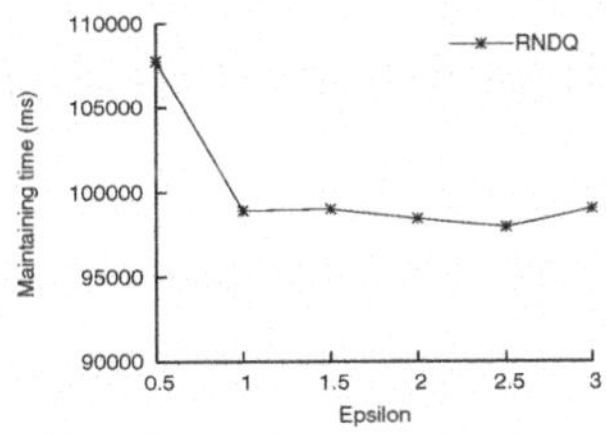

Fig. 8. Query performance with ϵ

6 Conclusion

In this paper, we introduce the definition of the dense segment and propose the problem of the effective road-network density query. Under our definition, we are able to answer queries for dense segments and find out dense areas in road network with arbitrary shape and arbitrary size. We present an cluster-based algorithm to response dynamic density queries and analyze the cost of cluster maintenance based on the object's movement feature in the road network. The cluster-based pre-processing can efficiently support density queries in road network. The experimental results show the efficiency and accuracy of our methods.

References

1. Hyung-Ju Cho, Chin-Wan Chung: An Efficient and Scalable Approach to CNN Queries in a Road Network. VLDB 2005: 865-876.
2. Marios Hadjieleftheriou, George Kollios, Dimitrios Gunopulos, Vassilis J. Tsotras: On-Line Discovery of Dense Areas in Spatio-temporal Databases. SSTD 2003: 306-324
3. Anil K. Jain, Richard C. Dubes: Algorithms for Clustering Data. Prentice Hall, 1988
4. Christian S. Jensen, Dan Lin, Beng Chin Ooi, Rui Zhang: Effective Density Queries on Continuously Moving Objects. ICDE 2006: 71
5. George Karypis, Eui-Hong Han, Vipin Kumar: Chameleon: Hierarchical clustering using dynamic modeling. IEEE Computer, 1999, 32(8):68-75.
6. Yifan Li, Jiawei Han, Jiong Yang: Clustering moving objects. KDD 2004: 617-622
7. Kyriakos Mouratidis, Man Lung Yiu, Dimitris Papadias, Nikos Mamoulis: Continuous Nearest Neighbor Monitoring in Road Networks. VLDB 2006: 43-54
8. Dimitris Papadias, Jun Zhang, Nikos Mamoulis, Yufei Tao: Query Processing in Spatial Network Databases. VLDB 2003: 802-813
9. Dragan Stojanovic, Slobodanka Djordjevic-Kajan, Apostolos N. Papadopoulos, Alexandros Nanopoulos: Continuous Range Query Processing for Network Constrained Mobile Objects. ICEIS (1) 2006: 63-70
10. Man Lung Yiu, Nikos Mamoulis: Clustering Objects on a Spatial Network. SIGMOD 2004: 443-454.
11. Man Lung Yiu, Nikos Mamoulis, Dimitris Papadias: Aggregate Nearest Neighbor Queries in Road Networks. IEEE Trans. Knowl. Data Eng. 17(6): 820-833 (2005)
12. Man Lung Yiu, Dimitris Papadias, Nikos Mamoulis, Yufei Tao: Reverse Nearest Neighbors in Large Graphs. IEEE Trans. Knowl. Data Eng. 18(4): 540-553 (2006)

An Efficient Spatial Search Method Based on SG-Tree*

Yintian Liu, Changjie Tang, Lei Duan, Tao Zeng, and Chuan Li

College of Computer Science, Sichuan University, Chengdu, 610065, China
{liuyintian,tangchangjie}@cs.scu.edu.cn

Abstract. To solve the overlapping search of multidimensional spatial database containing large quantity of objects with dynamic spatial extent, this paper proposes an index structure named Space Grid Tree (SG-Tree) based on Peano Space-Filling Curve (SFC) to index the region of object. By appropriate linearization strategy, the bounding box of a spatial object can be presented by a union of mutually disjoint hypercube grids with different granularity, and only the object's *oid*, i.e. the identifier which is used to refer to this object in the database, is registered on the nodes relative to these grids. The overlapping queries of spatial objects can be operated real-time directly on SG-Tree. Experiments show that SG-Tree is feasible and efficient to solve the overlapping search of multidimensional spatial objects.

1 Introduction

For objects in a spatial database, existing Access Methods impliedly assume that their spatial sizes are approximately similar. In real applications the region size of objects are often much different, and the size and position of regions are dynamic changeable, which make the efficiency of access methods based on MBR (such as the most popular access method R-tree and its extension) or hyperplane (such as skd-tree and its extension) decrease largely, for the high cost of node splitting and entity rectangle modifying caused by the insert, delete, and update operation.

To solve above problem, this paper proposes a novel index structure called Space Grid Tree (SG-Tree). The main ideas are:

(a) Orderly storing the spatial objects in database file according to the z-value of object's centroid to realize the spatial clustering storage of objects;
(b) Constructing the index structure (SG-Tree) of the multidimensional space which can reflect the overcast region of objects within the space. The structure of SG-Tree can avoid the operation of node splitting;
(c) Executing overlapping search operation on SG-Tree to retrieve the objects satisfying query condition. The result is the union of the *oids* in nodes bucket, which can avoid additional region compare with the bounding box of object.

The rest of the paper is organized as follows: Section 2 presents related work. Section 3 introduces the concept of spatial hypercube grid and proposes the structure of Spatial Grid Tree (SG-Tree). Section 4 gives the linearization method of spatial

* Supported by the National Natural Science Foundation of China under Grant No. 60073046.

G. Dong et al. (Eds.): APWeb/WAIM 2007, LNCS 4505, pp. 212–219, 2007.

objects. Section 5 introduces the search operations of SG-Tree. Section 6 presents the performance study. Section 7 gives the conclusion and future work.

2 Related Work

Multidimensional index structures can be classified into point access methods (PAMs) and spatial access methods (SAMs). To the SAMs, many classical index structures and their extension have been proposed. V.Gaede and O.Gunther give an overview of multidimensional access methods [1]. The most popular SAMs include R-Tree and its extension [2, 3, 4], SR-Tree [5], Pyramid-Tree [6], A-Tree [7], and VA-File [8]. There are two characters influence the efficient of the index structure. The first is that the region of each entity in node must be included for the use of node splitting or object filtering, which causes the additional storage consumption. The second is the node splitting algorithm and the region match of rectangles consume CPU resource. We want to find index structure satisfying the two conditions: the node splitting cost is negligible and the range of entities can be ignored.

3 Space Grid Tree

3.1 Space Hypercube Grid

A spatial database consists of a collection of spatial objects. Each object has a unique identifier *oid* which is used to retrieve object entity from the database file. The overcast region or bounding box of an object in *n*-dimensional space can be described as $Rect = ((x_{1lower}, x_{1upper}), (x_{2lower}, x_{2upper}) \dots (x_{nlower}, x_{nupper}))$.

Space-filling curves (SFCs) [9] such as Hilbert curve (H-curve), Peano curve (Z-curve) are a type of curve that can pass through every grid of a multidimensional space recursively. A SFC with order *k* passes through n-dimensional space and maps an integer set $[0, 2^{kn}-1]$ into an *n*-dimensional integer space $[0, 2^k-1]^n$. each dimension is divided into 2^k intervals and each interval is mapped into an integer within $[0, 2^k-1]$ described with a *k*-bitstring. A point with coordinate V on a dimension can be mapped into an interval and the normalization value T can be calculated by the following mapping formula:

$$T = \left\lceil \frac{(V - D_{min}) \times 2^k}{(D_{max} - D_{min})} \right\rceil \quad (D_{max} \text{ and } D_{min} \text{ are the upper and lower limit of a dimension respectively})$$

The whole space is then divided into 2^{kn} *n*-dimensional grids, the derived key of each grid is specified with a $(k \times n)$-bitstring according to relative mapping function. If the SFC is Peano curve, the derived key of the grid can be gained by simply shuffling the *k*-bitstring of its n edges.

Given two *k*-bitstring data *A* and *B*, the shuffle operation on them results into a *2k*-bitstring data $w = a_1b_1 \dots a_kb_k$, where $a_1 \dots a_k \in A$ and $b_1 \dots b_k \in B$.

Definition 1. Peano SFC with order *k* passing through an *n*-dimensional space partitions the whole space into 2^{kn} mutual disjoint *n*-dimensional hyperrectangle-shaped grids. Each of these grids is called a **Hypercube Grid Unit** with granularity *k* and its value can be calculated by orderly shuffling the k-bitstring value of n intervals constructing this hypercube unit.

Figure 1 illustrates the value calculation of three hypercube unit with granularity 1, 2, 3 respectively. We can find that the hypercube unit passed by Peano SFC has the following characters:

(a) A hypercube unit with granularity k is divided into 2^n mutual disjoint hypercube units with granularity $k+1$.
(b) The z-values of these 2^n child hypercube units share the prefix which is the z-value of their parent hypercube unit.

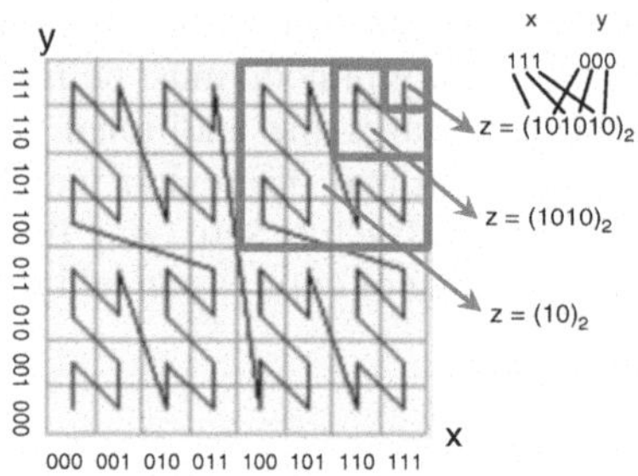

Fig. 1. The derived key (z-value) of hypercube grid

3.2 Structure of Space Grid Tree

Section 3.1 clarifies that a hypercube grid in an n-dimensional space passed by a Peano curve can be partitioned into 2^n mutual disjoint child hypercube grids and all these 2^n child hypercube grids' z-value share the same prefix corresponding to the z-value of their parent hypercube grid. According to these we give the definition of Space Grid Tree.

Definition 2. The **Space Grid Tree (SG-Tree)** of n-dimensional space partitioned by Peano SFC with order k is a tree structure where:

(1) The level of tree is k, a node of the form (*node_mark*, *oid_bucket*) on level i maps a hypercube grid unit with granularity i.
(2) There are at most 2^n child nodes for each node, the *node_mark* of a node is the order number its parent node being passed by SFC.
(3) Each node has an *oid_bucket* recording objects overcastting relative grid.

On each level of a full SG-Tree, the nodes mutually disjoint and the union of their volume consist the whole space. For a full SG-Tree with order k the node number should be $N = \frac{2^n(2^{ni}-1)}{2^n-1}$. In fact the spatial objects distribute in the whole space unevenly and there are many subspaces without any object overcastting them. These empty subspaces can be ignored for the construction of SG-Tree.

To find a hypercube grid unit from SG-Tree, we need to travel the relative nodes from a first level node to the target node according to the z-value of given hypercube grid. Algorithm 1 gives the grid search operation of SHG-Tree in n-dimensional space. The search of a hypercube grid unit ***g*** forms a path ***Path(g)*** from the first level node to the target node. The target node is enclosed by all the nodes on ***Path(g)***. The search operation also forms a subtree ***SubT(g)*** with target node as root node. The target node encloses all the nodes on ***SubT(g)***.

Algorithm 1. Grid_find(z-value)

```
Begin:
  i ← 1, pointer ← Null, order ← substring(z-value, 0, n)
  pointer ← the node with node_mark equalling to order on level 1
  while i < length(z-value)/n do
     order ← substring(z-value, i, n)
     finding child node x of pointer with node_mark equaling to order
     if x existing then
        pointer ← x
        i++
     else
        return Null
  return pointer
End
```

4 Space Object

4.1 Linearization of Spatial Objects

To index the overcast region of spatial objects with SG-Tree, the bounding box of an object need to be transformed into a union of hypercube grid with variant granularity. There are three different linearization strategies according to how accurately a linearization method presents the overcast region of object:

(a) **Full Linearization Strategy.** It presents the region accurately with a union of variant granularity hypercube grids. This strategy accurately reverts the spatial extent of object if the order of SFC is high enough.

(b) **Core Linearization Strategy.** It presents the core part region with several hypercube grids with the same granularity. Usually these finite grids can overcast the large part region of object.

(c) **Simple Linearization Strategy.** It presents the whole region with just a Minimum Bounding hypercube Grid unit (**MBG**) that can enclose the whole region. The result grid of this strategy includes not only the whole region of object but also additional space the object not overcastting. The quality of MBG approximation varies considerably according to the position and region of objects. For example, the volume of the object given in Figure 2 is only about 1/3 as large as its MBG.

Figure 2 illustrates the three linearization strategies to present the bounding box of a spatial object in 2-dimensional space with granularity 5. The overcast region of object is the shady rectangle, the rectangle of r_1 is the result of exact linearization strategy, the rectangle of r_2 is the result of core linearization strategy, and the square of r_3 is the result of simple linearization strategy.

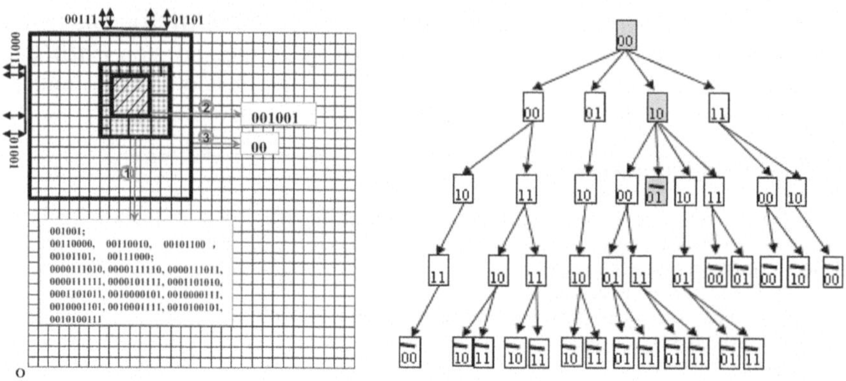

Fig. 2. Object linearization with different linearization strategy and the relative SG-Tree

According to Figure 2, we can find the characters of each strategy. Exact linearization strategy can correctly reflect the overcast region while the number of grids is too much. Core linearization strategy can reflect the nuclear part with just a few grids although some of its overcast region will lose which will cause the occurrence of false-negative. Simple linearization strategy conquers the false-negative shortage of core linearization strategy while it magnifies the overcast region which will cause the occurrence of false-positive.

4.2 Index of Spatial Objects

The bounding box of a spatial object can be presented as a set of hypercube grids with variant granularity. The overcast region of object then can be mapped into the relative nodes of SG-Tree.

We needn't construct the whole SG-Tree of a space at first. On the contrary, we create the nodes only when there existing objects overcast the relative grids. Algorithm 2 illustrates the insert operation of spatial object.

Algorithm 2. Insert(*o*)

```
Begin
 linerize object o into a grids union G according to relative linearization strategy
 for each grid g in G do
    i ← 0, L ← length(g), tempnode ← root
    while I < L do
       temporder ← substr(g, i, n)
       find the child node n' of tempnode with node_mark value temporder
       if n' exist then
          tempnode ← n', i++
       else
          create new node n' and make tempnode as it parent
          tempnode ← n', i++
    add object's oid into the oid_bucket of tempnode
 return
End
```

Figure 2 illustrates the SG-Tree after the insert operation of the object shown on the left. We can find that an object need register many times on different level nodes with exact linearization strategy but only a time on a 3rd-order approximation node with core linearization strategy (the shady path).

5 Query of SG-Tree

To find all spatial objects having at least one point in common with a given object, i.e. Intersection Query (IQ), the SG-Tree of spatial objects should be constructed with exact linearization strategy. Exact linearization strategy can perfectly reflect the overcast region of spatial objects if the order of approximation is proper. We also need to partition the given object into a union of hypercube grids with exact linearization strategy. For each hypercube grid we execute search on SG-Tree. Algorithm 3 gives Intersection Query operation.

Algorithm 3. IQ(*o'*)

```
Begin
  1. Result ← ∅
  2. G(A) ← ObjectLine_Exact(o)  // Exact linearize strategy
  3. for each grid g in G(A) do
  4.     Path(g), SubT(g) ← Grid_find(g)
  5.     for each node on Path(g) and SubT(g), add data in oid_bucket into Result
  6. return Result
End
```

The other common search operations includes Exact Match Query (EMQ), Point Query (PQ), Range Query (RQ), Enclosure Query (EQ), Containment Query (CQ), Adjacency Query (AQ), k-NN Query (NQ), Top-k Query (TQ), and Spatial Join. The SG-Tree structure can reflect the region overlapping relation between the nodes locating on the same or different level, which make these common search operations can be fulfilled on SG-Tree efficiently. The methods of these search operations are similar to Intersection Query. An important thing for these operations is to select a proper linearization strategy. For example, to find out *k* nearest neighbors of a given spatial object (**k-NN Query**), we should use core linearization strategy to linearize objects.

6 Experimental Study

The test data set is generated according to following rules: the domain of each dimension is [0, 100000], the domain of object's edge is [0, 600], thus the base granularity of SG-Tree should be at least $f = \lfloor \log_2 \frac{100000}{600} \rfloor = 7$. The scale of data set is 1 million and all the spatial objects obey the uniform distribution in the whole multidimensional space. We compare our method with directly matching method i.e. the overlapping judgment is implemented by comparing with each object's bounding box storing in main memory.

Figure 3 illustrates the time consumption of Intersection Query over SG-Tree in 3-dimensional space under different granularity. Each time we select 100 objects randomly and for each object we find out all the objects it intersecting. We count the total runtime of these 100 times query as the runtime of IQ operation. The result indicates that Exact Linearization Strategy and Core Linearization Strategy are real-time and are more efficient than memory match method. Simple Linearization Strategy is relatively inefficient but usually better than memory match method.

Figure 4 shows the time consumption of top-*k* query with Simple Linearization Strategy and *k*-NN query with Core Linearization Strategy. The result shows that the time of *k*-NN query is very small and the influence of dimension number and granularity is small. Simple Linearization Strategy is more inefficient than Core Linearization Strategy for this strategy need a second match for the query result of SG-Tree to avoid **False-positive**.

For a spatial database, an object intersects with many objects. The best method for top-*k* and *k*-NN query is to construct SG-Tree with Core linearization Strategy for we can always find out *k* objects even though the marginal region is ignored.

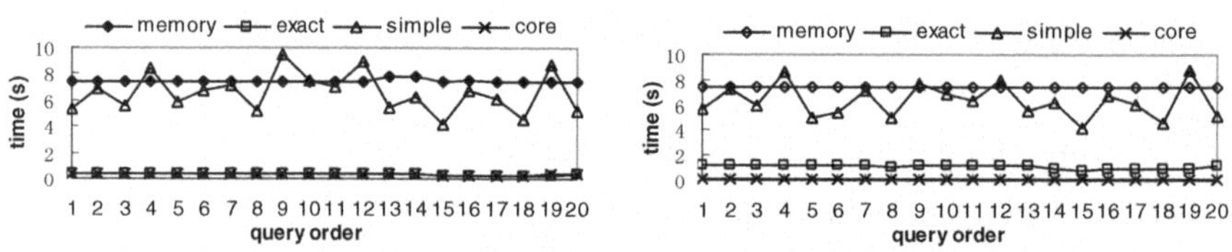

(a) Granularity 8 (b) Granularity 10

Fig. 3. Runtime for intersection query

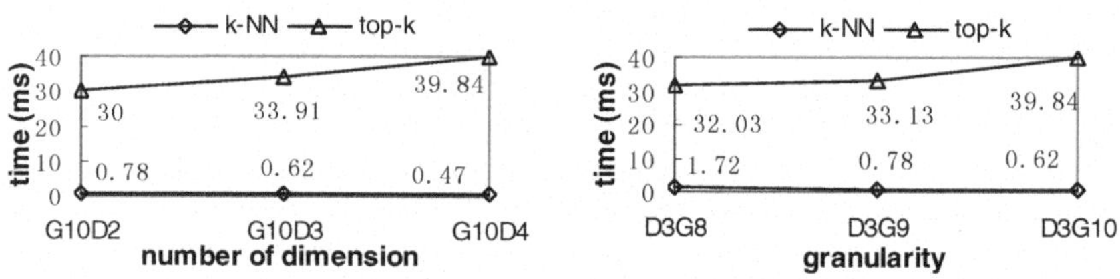

(a) Influence of dimension (b) Influence of granularity

Fig. 4. Runtime for top-*20* and *20*-NN query

7 Conclusion and Future Work

This paper proposes a spatial index structure Space Grid Tree (SG-Tree) and its search operations. This structure can efficiently implement the common spatial database operations, including Dynamic Insert/Update/Delete of spatial object, EMQ, PQ, RQ, IQ, EQ, CQ, AQ, NQ, TQ, and Spatial Join. On the other hand, in a SG-Tree, we suppose that each node contains just an entry denoting a grid, i.e. the fanout of SG-Tree is 1. We can find that the lower level a node locates, the fewer objects falls into its *oid_bucket*. These characters lead SG-Tree to a main memory index structure. While taking paging of secondary memory into account, SG-Tree would lead to low storage efficiency, which also degrades query performance. This disadvantage is the future work we should solve.

References

[1] V. Gaede and O. Gunther. Multidimensional Access Methods. *ACM Computing Surveys*, Vol. 30, No. 2:170-231, June 1998.

[2] Guttman. R-trees: A Dynamic Index Structure for Spatial Searching. *ACM SIGMOD International Conference on Management of Data*, 47-57, Aug. 1984.

[3] Timos K. Sellis, Nick Roussopoulos, and Christos Faloutsos. The R^+-Tree: A Dynamic Index for Multi-Dimensional Objects. In *Proceeding of the 13th International Conference on VLDB 1987*: 507-518.

[4] Beckmann, N., HP Kriegel, R. Schneider, and B. Seeger. The R^*-Tree: An Efficient and Robust Access Method for Points and Rectangles. *ACM SIGMOD Conf. 1990*, 322-331.

[5] Norio Katayama and Shin'ichi Satoh. The SR-tree: An Index Structure for High-Dimensional Nearest Neighbor Queries. *ACM SIGMOD International Conference on Management of Data (May 1997)*, 369-380.

[6] S. Berchtold, C. Kriegel, and H. Hriegel. The Pyramid-tree: Breaking the Curse of Dimensionality. *ACM SIGMOD*, 142-153, 1998.

[7] Yasushi Sakurai, Masatoshi Yoshikawa, Shunsuke Uemura, and Haruhiko Kojima. The A-tree: An Index Structure for High-Dimensional Spaces Using Relative Approximation. *VLDB 2000*, 516-526.

[8] Weber R, Schek H J, Blott S. A quantitative analysis and performance study for similarity search methods in high dimensional space. *VLDB.1998*, 194-205.

[9] Bongki Moon, H.V.Jagadish, Christos Faloutsos, and Joel H.Saltz. Analysis of the Clustering Properties of Hilbert Space-Filling Curve. *IEEE Trans. on Knowledge and Data Engineering (IEEE-TKDE)*, vol. 13, no. 1, pp. 124-141, Jan./Feb. 2001.

Getting Qualified Answers for Aggregate Queries in Spatio-temporal Databases

Cheqing Jin, Weibin Guo, and Futong Zhao

Dept. of Computer Science, East China University of Science and Technololy, China
{cqjin,gweibin}@ecust.edu.cn, beily524@mail.ecust.edu.cn

Abstract. In many applications, such as road traffic supervision and location based mobile service in large cities, moving objects continue to generate large amount of spatio-temporal information in the form of data streams. How to get qualified answers for aggregate queries appears to be a big challenge due to the high dynamic nature of data streams. Previous methods (e.g., AMH[11]) mainly focus on efficient organization of spatio-temporal information and rapid response time, not the quality of the answer. Our main contribution is a novel method to process important aggregate queries (e.g. SUM and AVG) based on a new structure (named AMH*) to summarize spatio-temporal information. The analysis in theory shows that the relative error and (/or) absolute error of answers can be ensured smaller than predefined parameters. A series of extended experiments evaluate the correctness of our approach.

1 Introduction

Spatio-temporal databases play an important role in applications involving space and time, such as road traffic supervision and location based mobile services in large cities. Consider that a traffic manager may use a mouse to draw a region of the downtown area on the city map to find the amount of vehicles running in the area right now. Furthermore, he may also want to learn how this value has changed in the past 1 hour so that he (she) is capable of providing suggestions for drivers by doing some prediction. Such tasks can be performed well provided that aggregate queries (e.g., SUM, AVG) are processed efficiently.

One direct kind of solutions to processing aggregate queries is to calculate answers based on a database reserving the moving traces of all objects, such as TPR-tree[10] and R^{FFP}-tree[9]. However, such methods may consume too much storage resource and computation resource. An alternative kind of methods prefers to calculate answers based on compact structures. For example, Sun et al. process SUM query by using an AMH structure, which represents $w \times w$ cells by at most B buckets, $B \ll w^2$[11]. But Sun et al.' approach still suffers following weaknesses. First, although parameter B is critical for the quality of answers, no general value of B is mentioned to ensure the quality for all queries. Second, the reorganization of cells partition is only performed when system is free, so that the quality of the answer continues to deteriorate during two consecutive reorganization operations.

G. Dong et al. (Eds.): APWeb/WAIM 2007, LNCS 4505, pp. 220–227, 2007.

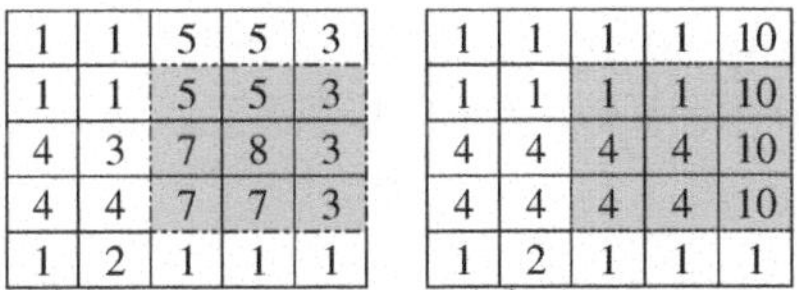

(a) At time 0 (b) At time 1

Fig. 1. An example data distribution

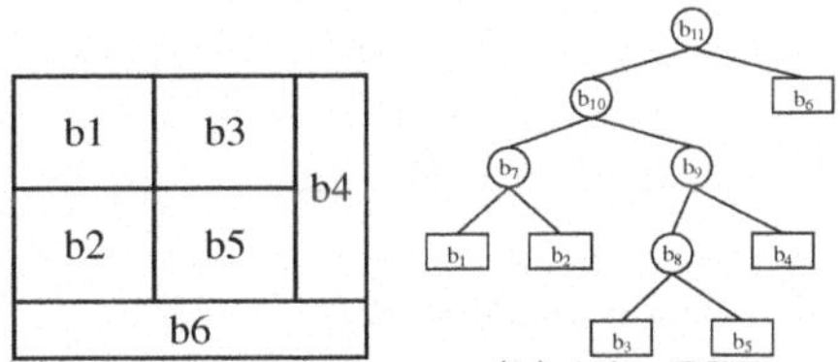

(a) Bucket extents (b) The BPT

Fig. 2. The AMH at time 0

The purpose of this paper is to calculate qualified answers for aggregate queries with small space resource and computation resource. We mainly consider two kinds of queries (e.g. SUM and AVG) and retain others as a piece of future work. The SUM query returns the number of objects within a certain area at a time point, while the AVG query returns the average number of objects per basic *cell* at a time point. Let A' and A denote the query answer and the correct answer respectively. The absolute error ε_a and relative error ε_r, are defined as $\varepsilon_a = |A - A'|$ and $\varepsilon_r = \frac{|A-A'|}{A}$. A novel structure, named AMH*, is proposed to summarize spatio-temporal information. As an improved version of AMH structure, AMH* also splits the whole area into lots of buckets, but the number of buckets can grow or shrink according to the change of the data distribution. Based on AMH*, the absolute error and (or) the relative error of a query can be restricted smaller than a predefined parameter.

The rest of the paper is organized as follows. Section 2 formally defines the problem and reviews related work. Our solution is proposed in Section 3. In Section 4, we present extensive experimental studies and report our findings. Finally, Section 5 concludes the paper with a summary.

2 Preliminaries

2.1 Query Definition

We consider an environment containing n objects and 1 central site. When an object moves, it sends its identity and new location, but not its velocity, to the central site through wireless network. Static objects do not send information to the central site. At any time point, the central site knows the locations of all objects. The central site partitions the whole data space into a 2D grid containing $w \times w$ cells, each with width $1/w$ on each axis. Each cell c is associated with a frequency F_c, representing the amount of objects in its extent currently.

The two kinds of aggregate queries are defined as follows.

SUM query $sum(q_T, q_R)$**:** q_T is the time point, q_R is the query range. It returns the number of objects within a range q_R at time q_T.

AVG query $avg(q_T, q_R)$**:** q_T is the time point, q_R is the query range. It returns the average number of objects per cell in a range q_R at time q_T.

If q_T=0, the query constitutes a present query; else if $q_T < 0$, it turns to be a historical query. Consider a small example where the data space is partitioned into 5×5 cells, and the data distributions at time points 0 and 1 are listed in Figure 1. Let q_R be the rectangle of the shadowed part. At time point 1, the query $sum(0, q_R)$ returns 48, and the query $avg(-1, q_R)$ returns 5.3.

2.2 Related Work

Processing spatio-temporal aggregate queries has been widely studied for a long period[6]. One kind of methods is based on building various indexes for all moving objects, such as TPR-tree[10] and R^{PPF}-tree[9]. However, such methods are inefficient to cope with the problem because of large storage consumption, expensive updating cost, and slow response time. An alternative kind of methods only constructs an compact structure to present the whole spatio-temporal database, such as query adaptive histograms (e.g., STGrid[1] and STHoles[3]) and other multi-dimensional approximation structures (e.g., DCT-based histogram[5], the wavelet-based histogram[7]).

The previous work mostly related to our work appears in [4,8,11,12]. The methods in [8,12] are similar to "conventional" processing framework where every query invokes disk I/Os and returns an exact answer. Contrarily, Sun et al. build AMH structure (be reviewed in Section 2.3) to compress data and return approximate answers[11]. The work in [4] also considers how to mine frequent items in spatio-temporal databases with small error.

2.3 Adaptive Multi-dimensional Histogram (AMH)[11]

An AMH contains at most B buckets. Each bucket b_k is defined as (R, n_k, f_k, g_k, v_k), where R is its rectangular extent, n_k is the number of cells in R, f_k is the average frequency of these cells (i.e., $f_k = (1/n_k) \sum_{\forall \text{ cell } c \text{ in } b_k} F_c$), g_k is the average "squared" frequency of these cells (i.e, $g_k = (1/n_k) \sum_{\forall \text{ cell } c \text{ in } b_k} F_c^2$), and v_k is their variance (i.e., $v_k = (1/n_k) \sum_{\forall \text{ cell } c \text{ in } b_k} (F_c - f_k)^2$). Clearly, v_k can be calculated through $v_k = g_k - f_k^2$. A binary partition tree (BPT) is used to index all buckets. In a BPT tree, each leaf node represents a bucket, and each intermediate node is associated with a rectangular extent R that encloses the extents of its (two) children. Buckets are reorganized when the system is free. Figure 2 shows the AMH structure and BPT tree upon the data distribution in Figure 1(a). All 25 cells are separated into 6 buckets.

3 Algorithm Description

3.1 Architecture

Figure 3 shows the architecture of the approach. The scenario contains multiple (moving) objects and a single server site. Each moving object reports its location (not the velocity) to the server site only when it changes location. The server site consists of three components, *spatio-temporal database*, *item processing engine* and *query processing engine*, which are described as follows.

Spatio-temporal Database: The *spatio-temporal database* contains two parts, AMH* and past index. As an improved version of AMH, AMH* summarizes the *current* data distribution by using multiple buckets in format All buckets are organized in a BPT-tree. The number of buckets in AMH* can grow or shrink during the running time without any restriction on the maximum amount of buckets. When a bucket becomes "old", it moves to the *past index* at once. A bucket becomes "old" because of following reasons: (1) the frequency of any cell in the bucket changes; (2) *split* and (/or) *merge* operations are executed to reorganize the bucket extents. Such "old" buckets must be saved in the past index to serve for the past timestamp queries (i.e., $q_T < 0$). Many methods have been devised to organize the buckets in the past index, such as packed B-tree[11] (used in this paper) and 3D R-tree[2].

Item Processing Engine: *Item processing engine* maintains the spatio-temporal database during the running time. When the frequency of any cell c changes, it invokes Algorithm `maintain` (Algorithm 1.1) to find the bucket b covering cell c, update fields of bucket b and invoke `isValidBucket` subroutine to check whether the bucket b is valid or not (Lines 2-4). The `isValidBucket` subroutine will be introduced in Section 3.2 in detail. If bucket b is invalid, Algorithm `split(`b`)` (Algorithm 1.2) is invoked to split b into several valid buckets. For any rectangular bucket ($col \times row$), there exist ($col + row - 2$) different partitioning ways because the bucket can be divided through x-axis or y-axis. By applying greedy algorithm, each time we choose the way with smallest value of the weighted variant sum WVS (e.g., $WVS = n_l \cdot v_l + n_r \cdot v_r$), where (n_l, v_l) and (n_r, v_r) belong to two children buckets b_l and b_r. Otherwise if the bucket b is valid, Algorithm `merge` (Algorithm 1.3) is invoked to merge some redundant buckets into one larger valid bucket (Lines 6-7).

Query Processing Engine: *Query processing engine* calculates answer for a query. Remember that the current data distribution is stored in AMH*, and the history data distributions are stored in *past index*. We can always find many buckets to construct q_R at that time. Let S denote the set of cells belonging to q_R, and $f(c) = f_b$, where cell c is covered by the bucket b, the SUM query and AVG query can be answered as follows.

$$sum(q_T, q_R) = \sum_{\forall c \in S} f(c) \qquad (1)$$

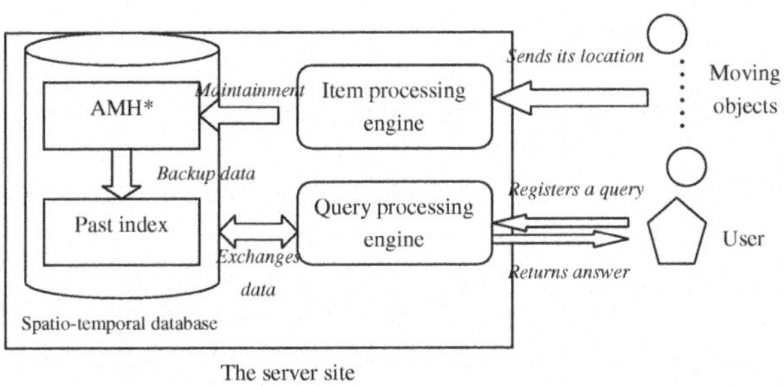

Fig. 3. The architecture of our approach

Algorithm 1. Outline of the algorithm

Algorithm 1.1: *maintain()*

if (the frequency of cell c changes from F_c to $F_c + \Delta_F$) **then**
  finds the bucket b covering c in BPT, and stores b in the *past index*;
  $f_b = (R_b \cdot w^2 \cdot f_b + \Delta_F)/(R_b \cdot w^2)$; $g_b = (R_b \cdot w^2 \cdot g_b + (F_c + \Delta_F)^2 - F_c^2)/(R_b \cdot w^2)$;
  if (`isValidBucket`(b)= **false**) **then**
    `split`(b);
  else
    `merge`(b);

Algorithm 1.2: *split(Bucket b)*

`if` (b has not moved to *past index*)
  stores b in the *past index*;
`push`(b);
while ((b' = `pop`()) $\neq NULL$)
  if (`isValidBucket`(b')= **false**) **then**
    splits b into two buckets b_l and b_r; `push`(b_l); `push`(b_r);

Algorithm 1.3: *merge(Bucket b)*

$b' = b$;
while (`isValidBucket`(`parent`(b')) = **true**);
  b' = `parent`(b');
stores all buckets whose ancestor entry is b' into *past index*;
creates a new bucket in AMH*;

$$avg(q_T, q_R) = \frac{\sum_{\forall c \in S} f(c)}{|S|} \tag{2}$$

3.2 Check the Validation of a Bucket

The goal of Algorithm `isValidBucket`(b) is to test the validation of bucket b. Here, we claim four cases (Case (1)-(4)). If one or multiple cases are employed by Algorithm `isValidBucket` to test a bucket, the bucket is valid (/invalid) when such case(s) is(are) satisfied (/unsatisfied). For example, if `isValidBucket` only tests Case 2 for all buckets, the relative error for a SUM query must be smaller than $\varepsilon_{sum,r}$.

Lemma 1. *Let X denote a random variable with an expect $E(X)$ and a deviation $\sigma(X)$. The function $\eta(\rho)$ is defined as $Pr[|X - E(X)| < \eta(\rho) \cdot \sigma(X)] = \rho$. If we use $E(X)$ to estimate the value of X, with a probability ρ, the maximum absolute error $\varepsilon_a < \eta(\rho) \cdot \sigma(X)$, the maximum relative error $\varepsilon_r < \frac{\eta(\rho) \cdot \sigma(X)}{E(X) - \eta(\rho)\sigma(X)}$.*

The correctness of the lemma comes from the definition of ε_a and ε_r.

According to Equ. (1)-(2), the answer for a query is calculated by a set of cells. Here, let X_c denote a random variable for the frequency of cell c (i.e., F_c). Then, its expect value $E(X_c)$ is equal to the average frequency of the bucket it belongs to (i.e., $f(c)$); its deviation $\sigma(X_c)$ is equal to the deviation of the bucket (i.e., $v(c)$). The answer of a query can be treated as a random variable following normal distribution (i.e., function $\eta(\rho)$ is defined as: $\rho = \frac{1}{\sqrt{2\pi}} \int_{-\eta(\rho)}^{\eta(\rho)} e^{-\frac{t^2}{2}} dt$).

Case 1. The absolute error for any SUM query is smaller than $\varepsilon_{sum,a}$ if for any bucket b, $v_b \leq (\frac{\varepsilon_{sum,a}}{\eta(\rho)\cdot n})^2$.

Case 2. The relative error for any SUM query is smaller than $\varepsilon_{sum,r}$ if for any bucket b, $v_b \leq (\frac{\varepsilon_{sum,r}}{(1+\varepsilon_{sum,r})\cdot\eta(\rho)} \cdot f_b)^2$.

We sketch the proof here. Let Y be a random variable to represent the answer of a SUM query. Then, $E(Y) = \sum_{i=1..p} E(X_i) = \sum_{i=1..p} f_i$, $\sigma(Y) = \sqrt{\sum_{i=1..p} \sigma^2(X_i)} = \sqrt{\sum_{i=1..p} v_i}$. According to Lemma 1, we have:

$$\varepsilon_a < \eta(\rho) \cdot \sigma(Y) = \eta(\rho) \cdot \sqrt{\sum_{i=1..p} v_i} \leq \eta(\rho) \cdot \sqrt{p \cdot (\frac{\varepsilon_{sum,a}}{\eta(\rho) \cdot n})^2} < \varepsilon_{sum,a} \qquad (3)$$

$$\begin{aligned} \varepsilon_r &< \frac{\eta(\rho) \cdot \sigma(Y)}{E(Y) - \eta(\rho)\sigma(Y)} \\ &= \frac{\eta(\rho) \cdot \sqrt{\sum_{i=1..p} v_i}}{\sum_{i=1..p} f_i - \eta(\rho)\sqrt{\sum_{i=1..p} v_i}} \\ &< \frac{\eta(\rho) \cdot \sqrt{\sum_{i=1..p} v_i}}{\eta(\rho) \cdot (\frac{1+\varepsilon_{sum,r}}{\varepsilon_{sum,r}} \sum_{i=1..p} \sqrt{v_i} - \sqrt{\sum_{i=1..p} v_i})} \\ &< \varepsilon_{sum,r} \end{aligned} \qquad (4)$$

Case 3. The absolute error for any AVG query is smaller than $\varepsilon_{avg,a}$ if for any bucket b, $v_b \leq (\frac{\varepsilon_{avg,a}}{\eta(\rho)})^2$.

Case 4. The relative error for any AVG query is smaller than $\varepsilon_{avg,r}$ if for any bucket b, $v_b \leq (\frac{\varepsilon_{avg,r}}{(1+\varepsilon_{avg,r})\cdot\eta(\rho)} \cdot f_b)^2$.

We sketch the proof here. Let Z denote the random variable for the answer of a AVG query. Then, $E(Z) = \frac{\sum_{i=1..p} E(X_i)}{p} = \frac{\sum_{i=1..p} f_i}{p}$, $\sigma(Z) = \sqrt{\frac{\sum_{i=1..p} \sigma^2(X_i)}{p^2}} = \frac{\sqrt{\sum_{i=1..p} v_i}}{p}$. According to Lemma 1,

$$\varepsilon_a < \eta(\rho) \cdot \sigma(Z) = \frac{\eta(\rho)}{p} \sqrt{\sum_{i=1..p} v_i} \leq \varepsilon_{avg,a} \qquad (5)$$

$$\varepsilon_r < \frac{\eta(\rho) \cdot \sigma(Z)}{E(Z) - \eta(\rho)\sigma(Z)} = \frac{\eta(\rho) \cdot \sqrt{\sum_{i=1..p} v_i}}{\sum_{i=1..p} f_i - \eta(\rho) \cdot \sqrt{\sum_{i=1..p} v_i}} < \varepsilon_{avg,r} \qquad (6)$$

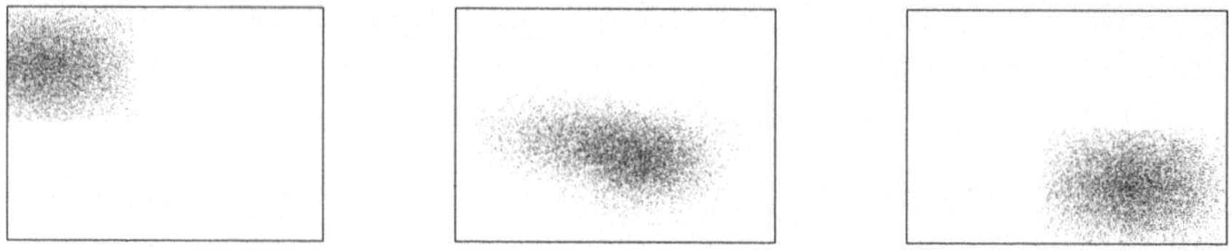

(a) Initial data (b) Median data (c) Final data

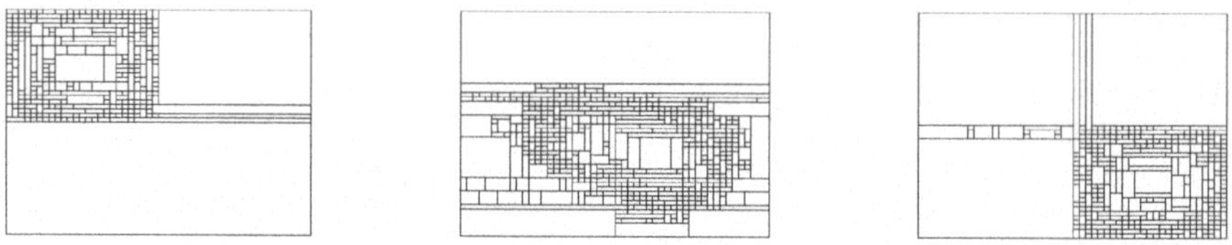

(d) Initial histogram (e) Median histogram (f) Final histogram

Fig. 4. AMH* changes during the running time

4 Experiments

In this section, we implement a series of experiments to evaluate the performance of our approach. All codes are written in C# and run in a PC with 512M memory. We use two sets of points (e.g., D_1 and D_2), each containing 10k 2D points, as shown in Figure 4(a) and (c). We then create 10k objects moving from a point in D_1 (randomly selected) to another point in D_2 (also randomly selected) in straight lines with different velocities. Here, the whole data space is separated into 50×50 cells. We set $\rho = 0.95$ (i.e., $\eta(\rho) = 2.0$) and $\varepsilon_{sum,r} = 0.5$. Figure 4 demonstrates the data distribution and bucket partitions in the initial stage, middle stage and final stage. Clearly, the histogram shape continues to change.

Figure 5 demonstrates how the number of buckets changes with the time going on. We can observe that the number of buckets changes under different distributions to reserve the precision. Figure 6 reports the qualities of two SUM queries. The sizes of q_R in query 1 and query 2 are 25×25 and 25×6 respectively. Let (x_0, y_0) denote the top left position of q_R. y_0 is fixed to 10, and x_0 is shown

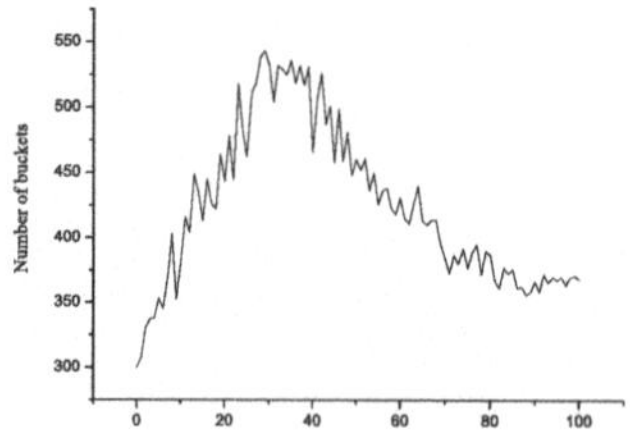

Fig. 5. Change of buckets

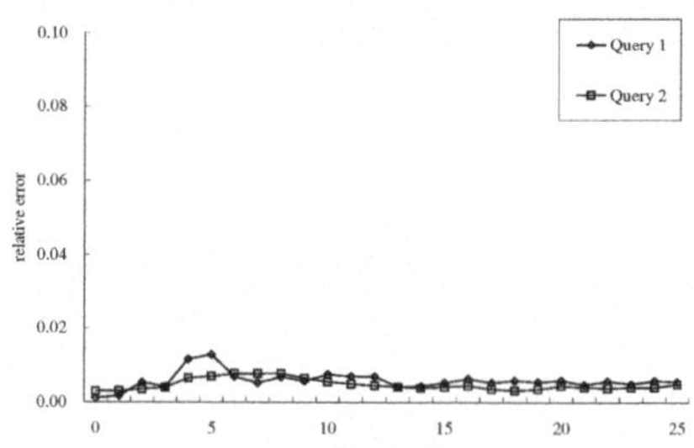

Fig. 6. The precision of queries

in the x-axis. The y-axis represents the relative errors of two queries. In all situations, the relative error is no more than 2%.

5 Conclusion

As a critical task in spatio-temporal field, getting qualified answers for aggregate queries still encounters some big challenges. Many researchers prefer to get precise answers after building indexes for all spatio-temporal logs. However, it may raise large additional costs, such as storage consumption, updating cost, and the computation consumption. Others try to get approximate answers by maintaining compact structures efficiently, but they fail to provide qualified answers.

In this paper, we devise a novel approach to get qualified answers for aggregate queries (including SUM and AVG queries) in spatio-temporal database. The core structure of the approach, named AMH*, splits the whole data space into a small number of rectangular buckets. We show that the absolute error and (/or) the relative error of a query can be kept smaller than predefined thresholds if all buckets satisfy some conditions at the same time. Besides, our approach only consumes small storage space, has low updating cost and quick response time. Experimental results evaluate the performance of our approach.

References

1. A. Aboulnaga and S. Chaudhuri. Self-tuning histograms: Building histograms without looking at data. In *Proc. of ACM SIGMOD*, 1999.
2. N. Beckmann, H. Kriegel, R. Schneider, and B. Seeger. The r*-tree: An efficient and robust access method for points and rectangles. In *Proc. of ACM SIGMOD*, 1990.
3. N. Bruno, S. Chaudhuri, and L. Gravano. Stholes: A multidimensional workload-aware histogram. In *Proc. of ACM SIGMOD*, 2001.
4. C. Jin, F. Xiong, J. Z. Huang, J. X. Yu, and A. Zhou. Mining frequent items in spatio-temporal databases. In *Proc. of WAIM*, 2004.
5. J. Lee, D. Kim, and C. Chung. Multi-dimensional selectivity estimation using compressed histogram information. In *Proc. of ACM ACM SIGMOD*, 1999.
6. I. F. V. Lopez, R. T. Snodgrass, and B. Moon. Spatiotemporal aggregate computation: A survey. *IEEE Transactions on Knowledge and Data Engineering*, 17(2), February 2005.
7. Y. Matias, J. Vitter, and M. Wang. Wavelet-based histograms for selectivity estimation. In *Proc. of ACM SIGMOD*, 1998.
8. D. Papadias, Y. Tao, P. Kalnis, and J. Zhang. Indexing spatio-temporal data warehouses. In *Proc. of ICDE*, 2002.
9. M. Pelanis, S. Šaltenis, and C. S. Jensen. Indexing the past, present, and anticipated future positions of moving objects. *ACM Transactions on Database Systems*, 31(1), March 2006.
10. S. Saltenis, C. Jensen, S. Leutenegger, and M. Lopez. Indexing the positions of continuously moving objects. In *Proc. of SIGMOD*, 2000.
11. J. Sun, D. Papadias, Y. Tao, and B. Liu. Querying about the past, the present, and the future in spatio-temporal databases. In *Proc. of ICDE*, 2004.
12. D. Zhang, V. Tsotras, and D. Gunopulos. Efficient aggregation over objects with extents. In *Proc. of PODS*, 2002.

Dynamic Adaptation Strategies for Long-Term and Short-Term User Profile to Personalize Search

Lin Li, Zhenglu Yang, Botao Wang, and Masaru Kitsuregawa

Institute of Industrial Science, The University of Tokyo,
4-6-1 Komaba, Meguro-Ku, Tokyo 153-8305, Japan
{lilin,yangzl,botaow,kitsure}@tkl.iis.u-tokyo.ac.jp

Abstract. Recent studies on personalized search have shown that user preferences could be learned implicitly. As far as we know, these studies, however, neglect that user preferences are likely to change over time. This paper introduces an adaptive scheme to learn the changes of user preferences from click-history data, and a novel rank mechanism to bias the search results of each user. We propose independent models for long-term and short-term user preferences to compose our user profile. The proposed user profile contains a taxonomic hierarchy for the long-term model and a recently visited page-history buffer for the short-term model. Dynamic adaptation strategies are devised to capture the accumulation and degradation changes of user preferences, and adjust the content and the structure of the user profile to these changes. Experimental results demonstrate that our scheme is efficient to model the up-to-date user profile, and that the rank mechanism based on this scheme can support web search systems to return the adequate results in terms of the user satisfaction, yielding about 29.14% average improvement over the compared rank mechanisms in experiments.

1 Introduction

With the advent of the era of the information explosion, never before have there been so many information sources availably indexed by search engines on the Internet. Ideally, users should be able to take advantage of the wide range of the valuable information while being able to find only those which are appealing to them. On the contrary, it becomes more difficult than ever to obtain desired results due to the ambiguity of user needs. Moreover, present search engines generally handle search queries without considering user preferences or contexts in which users submit their queries. For example, suppose that a database researcher who wants to search for information about a conference on Mobile Data Management and a banker who is interested in searching for the MDM bank, both input "MDM" on Google. Regardless of the different intentions of the two users on the same query, the results turn out to be a multimedia software company, a broadband services company, a national observatory, a conference on mobile data management, and so on. Current search engines prove unfortunately inadequate for this situation.

G. Dong et al. (Eds.): APWeb/WAIM 2007, LNCS 4505, pp. 228–240, 2007.

To address this problem, personalized search has recently become an active on-going research field. Studies [2, 14] have focused on requiring users to explicitly enter their contextual preferences including interest topics, bookmarks, etc., and these contextual preferences are used to expand user queries or re-rank search results. Forcing users to submit their contextual preferences would be a task that few users would be willing to do. Furthermore, it is very difficult for users to define their own contextual preferences accurately. Much attention has been paid in [13, 16, 18, 19] to learn user preferences transparently without any extra effort from users. These studies place emphasis on modelling user profiles or user representations to indicate user preferences automatically. Speretta et al. [18] created user profiles by classifying some information into concepts from the ODP [12] taxonomic hierarchy and then re-ranked search results based on the conceptual similarity between the web page and the user profile. The authors, however, have not taken the hierarchy structure of the ODP into account when calculating the conceptual similarity.

In this paper, we focus on studying learning user profiles and utilizing the learned user profiles to re-rank search results. Most studies on learning user profiles have deemed user profiles to be static. A related problem occurs when user preferences change over time. For instance, if a user changes her vocation from being an IT specialist to a lawyer, her interests will naturally shift with this change. It becomes important to keep the user profile up-to-date, and for a search engine to adapt accordingly. In addition, a user profile covers both short-term and long-term user preferences, which may increase or reduce respectively and co-relatedly with time. Using one model to represent two differently featured parts of the user profile will be far from perfect. Accordingly, suitable strategies are needed to capture the accumulation and degradation of changes of user preferences, and then adapt the content and the structure of a user profile to these changes. For re-ranking search results, our rank mechanism is similar to that proposed by [2] in which a semantic similarity measure is introduced with consideration to the hierarchy of the ODP structure. Meanwhile, the technique proposed in [2] suffers from the problem of requiring users to select topics which best fit their interests from the ODP, and other shortcomings we will address in Section 4.

Our contributions in this paper could be summarized as follows.

(1) We devise independent models for long-term and short-term user preferences.
(2) Dynamic adaptation strategies for modelling user profiles automatically are proposed. These strategies are based on click-history data while considering the accumulation and degradation changes of user preferences.
(3) When user preferences change, our user profiles, not only in contents, but also in structures, are modified to adapt to the changes.
(4) Finally, we propose a novel rank mechanism by measuring hierarchy semantic similarities between up-to-date user profiles and web pages. About 29.14% average improvement is gained over existing rank mechanisms.

The rest of this paper is organized as follows. In Section 2, we review the related work. In Section 3 we describe two independent models and dynamic adaptation strategies for user profiles. Rank mechanisms and evaluation metrics are addressed in Section 4. Section 5 presents the experimental results. Finally, we conclude in Section 6 with some directions for future work.

2 Related Work

2.1 Context Search

Kraft et al. [8] state that the context, in its general form, refers to any additional information associated with the query in the web search field, and also present three different algorithms to implement the contextual search instead of modelling user profiles. Generally speaking, if the context information is provided by an individual user in any form, whether automatically or manually, explicitly or implicitly, the search engine can use the context to custom-tailor search results. The process is named as a personalized search.

In this way, such a personalized search could be either server-based or client-based. The system in [4] is an available server-based search engine that unifies a hierarchical web-snippet clustering system with a web interface for the personalized search. Google and Yahoo! also supply personalized search services. With the cost of running a large search engine already very high, however, it is likely that the server-based full-scale personalization is too expensive for the major search engines at present.

On a client-based personalized search, studies [3, 16, 19] focus on capturing all the documents edited or viewed by users through computation-consuming procedures. Allowing for scalability, the client-based personalized search could learn user contexts more accurately than the server-based personalized search, while it is unavoidable that keeping track of user contexts has to be realized by a middleware in the proxy server or the client. Users, however, may feel unsafe to install such a kind of softwares even if they are guaranteed to be non-invasive, and may intend to enjoy the services provided by search engines instead. Moreover, if a user at home uses her private computer which is different from that in her office, keeping her contexts consistent becomes a problem. Therefore, our work is server-based.

In this paper, we focus on the use of suitable strategies to learn user profiles in a trade-off between scalability and accuracy for the server-based personalized search.

2.2 User Profile

There have been vast schemes of learning user profiles to figure user preferences from text documents. We notice that most of them model user profiles represented by bags of words without considering term correlations [1, 9, 17, 20]. To overcome the drawbacks of the bag of words, the taxonomic hierarchy, particularly constructed as a tree structure, has been widely accepted in [2, 11, 15]. Schickel-ZuberF et al.[15] score user preferences and concept similarity based on the structure of ontology. But their work needs users to express their preferences by rating a given number of items explicitly.

Meanwhile, these studies omit that user interests could change with time. Some topics will become more interesting to the user, while the user will completely or to varying degrees, lose interest in other topics. Studies [1, 9, 20] suggest that relevance feedback and machine learning techniques show promise in adapting to changes of user interests and reducing user involvements, while still overseeing what users dislike and their interest degradation. In [9] a two-level approach is proposed to learn user profiles for information filtering. While the lower level learns the stationary user preferences, the higher level detects changes of user preferences. In [20] a multiple three-descriptor representation is introduced to learn changes in multiple interest categories, and it also needs positive and negative relevance feedback provided by users explicitly.

Our work, particularly our dynamic adaptation strategies for user profiles, are based on the idea that sufficient contextual information is already hidden in the web log with little overhead, and all the visited pages can be considered as user preferences to various degrees because the users have accessed them. This contextual information motivates us to capture the accumulation and degradation changes of user preferences implicitly, to learn user profiles automatically.

3 User Profile and Dynamic Adaptation Strategies

As indicated in [20], for user profiles, long-term user preferences generally hold user preferences and the degree of preferences accumulated by experiences over a long time period. Hence it is fairly stable. On the other hand, short-term user preferences are unstable by nature. For instance, interests in current hot topics could change on a day-to-day basis. It is crucial to design a temporal structure for shot-term user preferences. Based on these features, we propose two novel models for long-term and short-term user preferences respectively and discuss them together with the adaptation strategies for their close correlations. Our strategies are in accord with the changes of user preferences in nature.

3.1 Long-Term Model of User Profile

The taxonomic hierarchy for our long-term model is a part of the Google Directory [5]. This part is composed of topics that have only been associated with the clicked search results, instead of the whole Google Directory. And these topics are linked as a tree structure to form our long-term model that is also called the user topic tree from now on. In other words, each node in the user topic tree means a topic in the Google Directory. We use search results and web pages interchangeably when referring to the URLs returned from the web search engine on a specific query.

In the Google Directory, each web page is classified into a topic [1]. In the "adding" operation, topics associated with the clicked pages are added into the user topic tree click by click. Moreover, each node in the user topic tree has a

[1] If necessary, all the symbolic links may be loaded into memory or the shortest distance on the graph is computed.

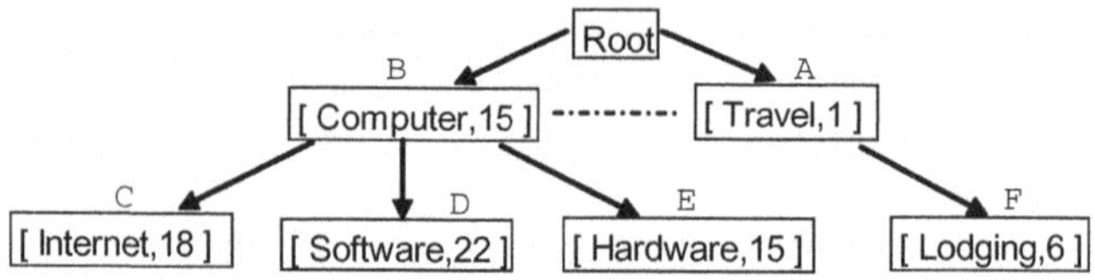

Fig. 1. Schema of Long-Term User Profile

value of the number of times the node has been visited. This value is called the "$TopicCount$", and represents the degree of preferences. The "deleting" operation is effected by the changes of the short-term model. It will be addressed in Section 3.2. Figure 1 illustrates the schema of the user topic tree. For example, node C is represented by the $[Internet, 18]$ which means one user has clicked a page associated with the topic "$Internet$" and the user has visited the "$Internet$" 18 times before this search. In our experiments node C is actually stored as the $[\backslash Root\backslash Compuetr\backslash Internet, 18]$ with a full path in the Google Directory.

3.2 Short-Term Model of User Profile

We frame the Page-History Buffer (PHB) for the short-term model. The PHB caches the most recently clicked pages with a fixed size that is determined by the ability of the search engine. We now meet the same problem as the cache in the processor, and that is how to kick off the "old" pages in time to keep up with the changes of short-term user preferences. As it is known, in the cache management, there are popular cache replacement algorithms that are all designed for the processor, the web cache and the database disk buffering. No such research could be available in the personalized search, especially in the short-term model of the user profile. Our goal, keeping track of the most recent accesses of search results in the PHB, is basically similar to that in the cache management. As a result, the LFU (Least Frequently Used), one of these replacement algorithms, is adjusted to our scheme, which is named the Least Frequently Used Page Replacement (LFUPR). The details are shown in Table 1. The LFUPR reflects the changes of the short-term model, including how to add (line 3 $\sim$ line 6) and replace (line 10 $\sim$ line 12) web pages in the PHB.

From Figure 1 and the LFUPR algorithm in Table 1, our dynamic adaptation strategies maintain user profiles such that the short-term model is updated by the LFUPR (line 1 $\sim$ line 15), while the degree of preferences in the long-term model could be degraded (line 13) when the page in the PHB is replaced, and could be accumulated when the user clicks the page ("adding" operation). On the other hand, if the user accesses the web page whose associated topic is not in the current user topic tree, the new node could be added into the tree ("adding" operation). From line 16 to line 18, if the "$TopicCount$" of one node becomes zero, the node would be deleted from the tree. This procedure is called the "deleting" operation. The "adding" and "deleting" operations dynamically

Table 1. LFUPR Algorithm

```
Input:       current short-term model, current long-term model, search results
Output:      updated user profile
Parameters:  PageCount=Vector of the number of clicked times for pages in the PHB
             TopicCount=Vector of the number of clicked times for nodes in the user topic tree
             BufferPages=Vector of pages in the PHB
             Results=Vector of pages returned by a search engine
             UserTopics=Vector of nodes in the user topic tree
For i=1 to Size(Results)
Begin Loop
   If Result[i] is the nth page IN the PHB
     PageCount[n]++;
   Else If PHB is NOT FULL
     Add the clicked page into the end of the PHB;
   Else
     Begin
       For j=1 to Size (PHB)
         BufferPages[k] ← Find one page in the BufferPages with the Minimum
         PageCount[j];
       Replace the BufferPages[k] with the Results[i];
       TopicCount[m]- -; //BufferPages[k] is the mth node IN the UserTopics
     End
End loop
For t=1 to Size (UserTopics)
If TopicCount[t] ==0
   Clear the UserTopics[t] out from the user topic tree
```

adapt the structure of the long-term model to the user click behaviors. Although we design independent models for short-term and long-term user preferences, our strategies ensure that the inherent correlations between them are not ignored, and that the changes of the short-term model have an even influence on the long-term model. Here, the meaning of "even" is that we degrade the "*TopicCount*" not on an hour-to-hour or a day-to-day basis, only after a period of time during which the user has not accessed the *topic* in the whole search process.

4 Rank Mechanisms and Evaluation for Personalized Search

4.1 Distance Metrics

The tree distance which we deal with, is the distance between each search result and the user topic tree, as described in [2]. The search result with the shorter distance, meaning the higher similarity to user preferences, should be put in the topmost position of the ranking list. For each search result, there is an associated node in the Google Directory. The user topic tree is also composed of nodes. The distance computation is actually how the distance between two nodes in the tree structure is measured.

Chirita et al. [2] point out that the main drawback of the naïve tree distance is that it overlooks the depth of the subsumer (the deepest node common to two nodes). With the help of Figure 1, let us explain the problem clearly. $sub_{i,j}$ represents the subsumer of the node i and the node j. $Edges(i, sub_{i,j})$ represents the number of edges between the node i and the node $sub_{i,j}$. The naïve distance is defined as

$$Distance(i,j) = Edges(i, sub_{i,j}) + Edges(j, sub_{i,j}) . \quad (1)$$

$Distance(A,B)$ is 2, which is the same as $Distance(C,D)$, making it difficult to re-order search results by Equation (1).

4.2 Hierarchy Semantic Similarity

Li et al. [10] takes the depth of the subsumer h and the naïve distance between two nodes l into the calculation. α and β are the parameters scaling the contribution of the naïve distance and the depth respectively. The semantic similarity is defined as

$$Sim(i,j) = e^{-\alpha \cdot l} \cdot \frac{e^{\beta \cdot h} - e^{-\beta \cdot h}}{e^{\beta \cdot h} + e^{-\beta \cdot h}}, \quad \alpha \geq 0 \, , \beta > 0 \, . \quad (2)$$

Their experiment results show that the optimized values of the two parameters are, α=0.2 and β=0.6. For example, $Sim(A,B)$ is unequal to $Sim(C,D)$ based on Equation (2). Because the subsumer of A and B, i.e., "*Root*", is in the different level from the subsumer of C and D, i.e., "*Computer*". However, Equation (2) only solves problem partially. Let us see another example. Due to the same value (i.e., 3) between $Distance(A,C)$ and $Distance(B,F)$, and the same subsumer (i.e., "Root") between the pairs (A, C) and (B,F), $Sim(A,C)$ is equal to $Sim(B,F)$.

Under this situation, Chirita et al. [2] separate l into l_1 and l_2, and then gives different weights to the two variables through the parameter δ defined as

$$Sim^*(i,j) = ((1-\delta) \cdot e^{-\alpha \cdot l_1} + \delta \cdot e^{-\alpha \cdot l_2}) \cdot \frac{e^{\beta \cdot h} - e^{-\beta \cdot h}}{e^{\beta \cdot h} + e^{-\beta \cdot h}} \, . \quad (3)$$

Equation (3) can work well for common cases. However, we find that the parameter δ is sensitive to the semantic meanings between the two topics, as illustrated in [2]. Furthermore, even if we compute the similarity by Equation (3), $Sim(C,D)$ is still equal to $Sim(E,D)$ because of the same value between l_1 and l_2. In our system, we extend Equation (2) in another way, as the "*TopicCount*" has much better effect on the overall performance than the weak parameter δ. Comparative experiments are in Section 5.

4.3 Our Rank Mechanism

When a user submits a query to the search engine, the search results are re-ranked by our semantic similarity defined as

$$CSim(i,j) = WT(i) * Sim(i,j) \, , \quad (4)$$

the degree by which the search result is similar to the user profile. i is a node in the user topic tree ($i = 1, 2, \cdots, size(UserTopics)$). j is the associated node with a search result in the Google Directory ($j = 1, 2, \cdots, size(Results)$). $WT(i)$ is defined as $TopicCount(i)/\sum_{i=1}^{size(UserTopics)} TopicCount(i)$, weighing the degree of preferences of a node in the user topic tree. The larger the WT is, the more interested the user is in one topic. For one search result, the number of the $CSim$ values is $size(UserTopics)$ in Equation (4). One user topic tree represents one user. We define the semantic similarity between one search result and the user topic tree as the maximum value among all the values ($i = 1, 2, \cdots, size(UserTopics)$) expressed as

$$CSim^*(User, j) = Max(WT(i) * Sim(i, j)) . \quad (5)$$

To keep our rank mechanism from missing the high quality pages in Google, Equation (5) is integrated with PageRank as

$$FinalRank(User, j) = (1 - \gamma)CSim^*(User, j) + \gamma * PageRank(j) . \quad (6)$$

Here γ is a parameter in [0,1] which blends the two ranking measures. The user could vary the value of γ to merge our rank mechanism and PageRank in different weights. In our experiments, γ is set to 0.5, which gives equal weight to the two measures.

4.4 Evaluation Metrics

Accuracy of User Profile. It is natural to evaluate our user profiles by computing the difference between the real user topic tree and the modelled user topic tree. Equation (2) is suitable for this task and the relative error between the two user profiles is shown as

$$Error(M) = \frac{|Sim(M, R) - Sim(R, R)|}{Sim(R, R)} , \quad (7)$$

where $Sim(M, R)$ is denoted by $\sum_{j=1}^{K} Max(Sim(j, i))$. R means the vector of topics in the real user topic tree. M means the vector of topics in the modelled user topic tree. i is a node in R ($i = 1, 2, \cdots, N \rightarrow size(R)$). j is a node in M ($j = 1, 2, \cdots, K \rightarrow size(M)$). A smaller value of $Error(M)$ means a higher accuracy of our modelled user profile.

Quality of Our Personalized Search System. Whether a personalized system is successful or not is determined by the user satisfaction. An effective rank mechanism should place relevant pages close to the top of the rank list. We ask the users to select the pages they considers relevant to their preferences for our evaluation. The quality of our system is measured as

$$AveRank(u, q) = \sum_{p \in S} (R(p))/Count(p) . \quad (8)$$

Here S denotes the set of the pages selected by user u for query q, $R(p)$is the position of page p in the ranking list, and $Count(p)$ is the number of selected pages. A smaller $AveRank$ represents better quality.

Table 2. Procedures of Evaluation Experiments

1.	Issuing the query submitted by an online user through the Google API module ;
2.	Re-ranking search results by our rank mechanism based on the current user profile and then going into the Log module;
3.	Adapting the user profile to click-history data provided by the Log module through our strategies:
4.	For the long-term model updating the structure and the degree of preferences by the "adding" operation;
5.	For the short-term model, updating web pages in the PHB by the LFUPR algorithm;
6.	If needed, degrading the long-term model according to the changes of the short-term model by the "deleting" operation.
7.	Waiting until the online user submits a new query, and then going to 1.

5 Experiments

5.1 Experimental Setup

Our rank mechanism could be combined with any search engine. In this study we choose the Google Directory Search [5] as our baseline in that Google applies its patented PageRank technology on the Google Directory to rank the sites based on their importance. It is convenient for us to combine and evaluate our rank mechanism with Google. The necessary steps are depicted in Table 2. Main modules in the experiments are listed as follows.

(1) Google API module: Given a query, we are offered titles, snippets, and page-associated Google directories beside the URLs of web pages by the Google API [6]. Here a Google directory is regarded as a topic in the user topic tree.
(2) Log module: We monitor user click behaviors, recording the query time, clicked search results, associated topics.
(3) User profile: It has been described in Section 3.

In our experiments, due to the large size of the whole Google Directory, only the top 4 levels are encoded into the user topic tree. The size of the PHB is 20 pages. Ideally if we could cache all the clicked web pages in the PHB and utilize the whole levels of the Googe Directory, it would be much easier to personalize a search.

5.2 Dataset

For each search, the Google API module got the order of the top 20 Google results due to the limited number of the Google API licenses we have. We randomized the order of the results before returning the 20 results to the user at run-time. For evaluation, 12 subjects are invited to search through our system. The 12 subjects are graduate students (5 females and 7 males) researching in several fields, i.e., computer, chemistry, food engineering, electrical engineering, art design, medical, math, architecture, and law. These subjects are divided into three types:

- Clear User, searching on queries that usually have one meaning,
- Semi-ambiguous User, searching on queries that have two or three meanings,
- Ambiguous User, searching on queries that have more than three meanings.

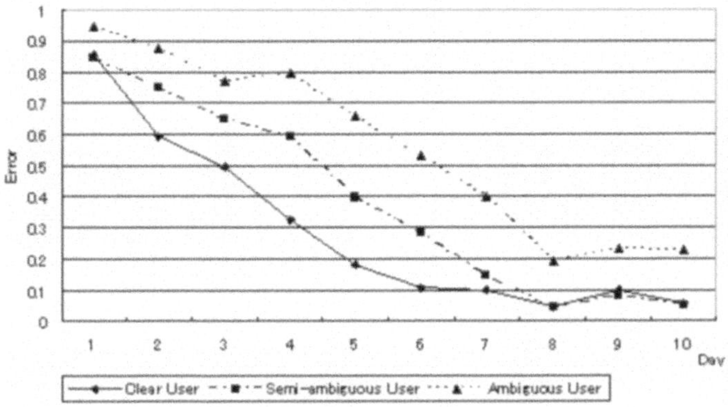

Fig. 2. Accuracy of User Profile

Our search interface was available on the Internet, and convenient for the subjects to access it at any time. They were asked to query topics closely related to their interests and majors. In the first four days, subjects input the queries on their majors, and then in the next three days the queries on their hobbies were searched. Finally, in the last three days, the subjects were required to repeat some queries done before. This repeated procedure gave a clear performance comparison between the current and earlier systems, as user profiles were updated search by search. After the data were collected over a ten-day period (From October 23nd, 2006, to November 1st, 2006), we got a log of about 300 queries averaging 25 queries per subject and about 1200 records of the pages the subjects clicked in total.

5.3 Experimental Results

Results of Accuracy of User Profile. From Figure 2, we see that as the days went on, the relative errors of our user profiles generally kept decreasing. In the last three days they even apparently stopped decreasing. The trend was expected because the subjects were asked to repeat some queries done earlier for comparison. Without a new query for a search, we are not able to learn more about the user preferences. Moreover, relative errors got even slightly larger on these days. Because the subjects might click pages different from those of the early search on the same query. This further indicates the importance of adaptation strategies to learn the changes in user preferences. Figure 2 also shows that it is easier and quicker to learn the user profile of a "Clear User" than that of a "Semi-ambiguous User" and slowest to learn the user profile of an "Ambiguous User". For example, when day=4, Error(Clear User)=0.3, Error(Semi-ambiguous User)=0.6, and Error(Ambiguous User)=0.8. Although the learning procedure of the "Ambiguous User" is slower than the other two kinds of users, as long as its user profile is converged relatively, it yields the best improvement in terms of quality among all the three kinds of users.

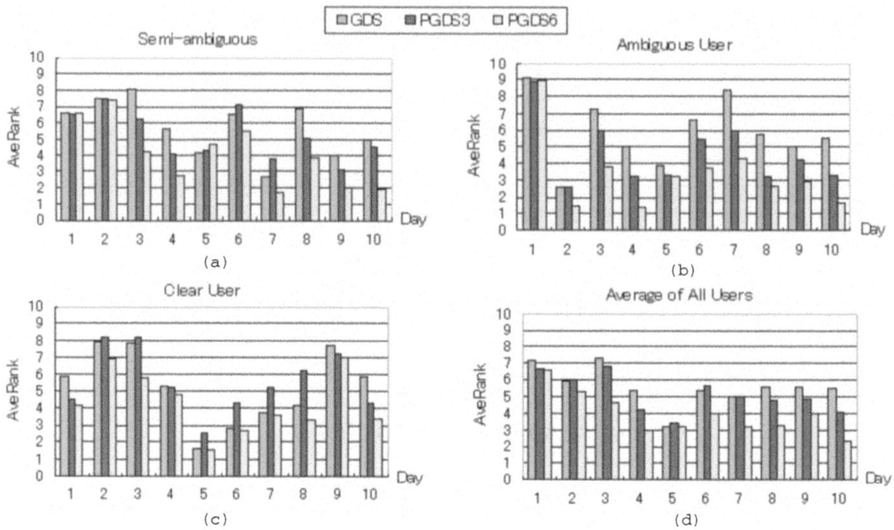

Fig. 3. Quality of Personalized Search System (Lower is better)

Results of Quality of Personalized Search System. Now, we compare the performance improvements of the following three ranking mechanisms:

- the Google Directory Search (GDS), using the Google API,
- the Personalized Google Directory Search (PGDS3), combing Equation (3) and the PageRank,
- the Personalized Google Directory Search (PGDS6), using Equation (6).

Evaluated by Equation (8), how they performed day by day is shown in Figure 3. By using the GDS as a baseline, the performance improvement of our PGDS6 in Figure 3(b) is 42.37 %, which outperforms those in both Figure 3(a) (i.e., 28.86%) and Figure 3(c)(i.e., 16.27%). The little improvement in Figure 3(c) indicates that GDS has done well with the "Clear User". However, for the "Semi-ambiguous User" and the "Ambiguous User", the significant improvements in Figure 3(a) and Figure 3(b)illustrates that GDS works worse than our strategies.

Figure 3(d) illustrates the average improvement over all users. As a result of requiring the subjects to change queries from their majors to hobbies, we see that from the fourth day to the fifth day, the values of AveRank experience a sudden increase. But after three days on learning the changes, our PGDS6 shows better results than the GDS and the PGDS3. More accurately, compared with the GDS, our PGDS6 outperforms the PGDS3 with a 60% improvement for the tenth day, while for the fifth day the improvement is only around 2%. This difference demonstrates that the changes of user preferences will lower the improvement that our strategy could achieve. Nevertheless, our rank mechanism still greatly improves over the GDS and the PGDS3 overall. The average improvements of our PGDS6 and the PGDS3 over the GDS, are 29.14% and 7.36% respectively.

6 Conclusion

In this paper we introduced how to capture the changes of user profiles from click-history data and how to use the user profiles to re-rank the search results, thus creating personalized views of the web. First, we designed independent models for short-term and long-term user preferences to consist of a user profile. Then, we adapted the user profile, including the content and the structure, to the accumulation and degradation changes of user preferences by our dynamic strategies. Finally, we proposed a novel rank mechanism to re-rank search results. Experimental results on real data demonstrate that our dynamic adaptation strategies are effective and our personalized search system performs better than the selected rank mechanisms, especially for the "Semi-ambiguous User" and the "Ambiguous User".

In the future, we plan to do some comparative experiments when the user varies the value of γ in Equation (6). In addition, when computing for the node distance in the tree, we plan to consider the edge distance, assigning a different weight for each edge, because each pair of two nodes linked by an edge has different semantic similarity. As Kelly et al. [7] summarize key papers that cover a range of approaches on implicit feedback techniques, we will study more user implicit information to construct the user profile, such as the time interval between two clicks, browsing patterns, and so on.

References

[1] D. Billsus and M. J. Pazzani. A hybrid user model for news story classification. In *Proc. of the 7th Int'l Conf. on User modeling (UM'99)*, pages 99–108, Secaucus, NJ, USA, 1999.

[2] P. A. Chirita, W. Nejdl, R. Paiu, and C. Kohlschütter. Using ODP metadata to personalize search. In *Proc. of the 28th Annual Int'l ACM SIGIR Conf. on Research and Development in Information Retrieval (SIGIR'05)*, pages 178–185, Salvador, Brazil, 2005.

[3] S. T. Dumais, E. Cutrell, J. J. Cadiz, G. Jancke, R. Sarin, and D. C. Robbins. Stuff I've seen: A system for personal information retrieval and re-use. In *Proc. of the 26th Annual Int'l ACM SIGIR Conf. on Research and Development in Information Retrieval (SIGIR'03)*, pages 72–79, Toronto, Canada, 2003.

[4] P. Ferragina and A. Gulli. A personalized search engine based on web-snippet hierarchical clustering. In *Proc. of the 14th Int'l Conf. on World Wide Web - Special interest tracks and posters (WWW'06)*, pages 801–810, Chiba, Japan, 2005.

[5] Google Directory. http://directory.google.com.

[6] Google Soap Search API(Beta). http://code.google.com/apis/soapsearch.

[7] D. Kelly and J. Teevan. Implicit feedback for inferring user preference: a bibliography. *SIGIR Forum*, 37(2):18–28, 2003.

[8] R. Kraft, C. C. Chang, F. Maghoul, and R. Kumar. Searching with context. In *Proc. of the 15th Int'l Conf. on World Wide Web (WWW'06)*, pages 477–486, Edinburgh, Scotland, UK, 2006.

[9] W. Lam, S. Mukhopadhyay, J. Mostafa, and M. J. Palakal. Detection of shifts in user interests for personalized information filtering. In *Proc. of the 19th Annual Int'l ACM SIGIR Conf. on Research and Development in Information Retrieval (SIGIR'96)*, pages 317 – 325, Zurich, Switzerland, 1996.
[10] Y. Li, Z. Bandar, and D. McLean. An approach for measuring semantic similarity between words using multiple information sources. *IEEE Trans. Knowl. Data Eng.*, 15(4):871–882, 2003.
[11] B. Markines, L. Stoilova, and F. Menczer. Bookmark hierarchies and collaborative recommendation. In *Proc. of The 21st National Conf. on Artificial Intelligence and the 8th Innovative Applications of Artificial Intelligence Conference (AAAI'06)*, Boston, Massachusetts, USA, 2006.
[12] Open Directory Project(odp). http://dmoz.org.
[13] F. Qiu and J. Cho. Automatic identification of user interest for personalized search. In *Proc. of the 15th Int'l Conf. on World Wide Web (WWW'06)*, pages 727–736, Edinburgh, Scotland, 2006.
[14] H. rae Kim and P. K. Chan. Personalized ranking of search results with learned user interest hierarchies from bookmarks. In *Proc. of the 7th WEBKDD workshop on Knowledge Discovery from the Web (WEBKDD'05)*, pages 32–43, Chicago, Illinois, USA, 2005.
[15] V. Schickel-Zuber and B. Faltings. Inferring user's preferences using ontologies. In *Proc. of The 21st National Conf. on Artificial Intelligence and the 8th Innovative Applications of Artificial Intelligence Conference (AAAI'06)*, Boston, Massachusetts, USA, 2006.
[16] X. Shen, B. Tan, and C. Zhai. Implicit user modeling for personalized search. In *Proc. of the 2005 ACM CIKM Int'l Conf. on Information and Knowledge Management (CIKM'05)*, pages 824–831, 2005.
[17] S. J. Soltysiak and I. B. Crabtree. Automatic learning of user profiles- towards the personalisation of agent services. *BT Technology Journal*, 16(3):110–117, 1998.
[18] M. Speretta and S. Gauch. Personalized search based on user search histories. In *Proc. of the IEEE / WIC / ACM Int'l Conf. on Web Intelligence (WI'05)*, pages 622–628, Compiegne, France, 2005.
[19] J. Teevan, S. T. Dumais, and E. Horvitz. Personalizing search via automated analysis of interests and activities. In *Proc. of the 28th Annual Int'l ACM SIGIR Conf. on Research and Development in Information Retrieval (SIGIR'05)*, pages 449–456, Salvador, Brazil, 2005.
[20] D. H. Widyantoro, T. R. Ioerger, and J. Yen. Learning user interest dynamics with a three-descriptor representation. *JASIST*, 52(3):212–225, 2001.

Using Structured Tokens to Identify Webpages for Data Extraction*

Ling Lin, Lizhu Zhou, Qi Guo, and Gang Li

Tsinghua University, Beijing 100084, PRC
linling03@mails.tsinghua.edu.cn
dcszlz@tsinghua.edu.cn

Abstract. As the web grows, more and more data has become available from webpages, such as the product items from the back-end databases. To provide efficient access to the data objects contained in these pages, data extraction plays an important role. However, identifying the suitable webpages to feed the data extraction is a pre-requisite and non-trivial task. As a result, there is an increasing need for methods that can automatically identify the target pages from unknown websites. In this paper, we solve the problem by exploiting the structured-token features of the webpage content, and applying decision tree based classification algorithm to induce the structure information. Furthermore, a preliminary recognition of data-object is acquired to efficiently initiate the subsequential data extraction. We experiment our approach on the real-world data, and achieve promising results.

1 Introduction

An ever-increasing number of applications on the Web target at processing the data-rich[3,6] webpages collected from the websites of target domains. By *data-rich webpages*, we mean the pages containing one or more extractable data objects[9] of certain domain. Web wrapper (or data extraction) is a generally used method to obtain these data objects and put them into structured format, such as XML files or relational databases. To automate this procedure, identifying the promising and suitable webpages to feed the data extraction is a pre-requisite and non-trivial task. In this paper, we focus on this challenging task and treat it as a two-class learning problem.

The ideal input of data extraction is a set of clean data-rich pages which are not only structured in format, but also relevant to the target domain in semantic. On the contrary, noisy input pages will reduce the efficiency of data extraction, and do harm to the wrapper maintenance as well. Moreover, the input webpages are supposed to come from as many different unknown websites as possible. Therefore, compared with data extraction, which primarily works on the structure features of a much cleaner data set, the pre-requisite task of target

* This work is supported in part by National Natural Science Foundation of China 60520130299.

G. Dong et al. (Eds.): APWeb/WAIM 2007, LNCS 4505, pp. 241–252, 2007.

webpage identification is facing a much heterogenous and noisy web environment. To collect this set of pages, the identification algorithm should judge the webpage according to both its content and structure features.

To give a more concrete vision of the challenges rising in the procedure of target pages identification, take online shopping for an example, which is one of the most popular fields where data extraction is applied. Suppose the target domain is book shopping, and we want to fetch a set of data-rich webpages providing the information about the books for sale. Fig. 1 illustrates three samples pages[1]. Fig. 1(a) and (b) shares many tokens in content, while (b) and (c) are similar in the display style and the html structure. However, only (b) is the target page, while (a) is a narrative article about this book, and (c) a page selling DVD on the same website as page (b). Therefore, to define a set of features which purely rely on either text content or structure pattern would be ineffective for this problem. That's one of the important reasons why most existing techniques, such as focused crawling or data extraction, can not be applied very successfully to this field in practice.

In this paper, we present an approach to identify domain specific target webpages for data extraction using a decision tree based learning algorithm. Our method starts with several categorized domain-specific tokens and a set of training webpages. The training webpages have only class labels, and no tedious work of the HTML structure labelling is required. For each webpage, a sequence of matched tokens are found with their offset positions within the page. A sliding window based algorithm is then applied to choose a proper set of matched tokens for the structure feature computation. Finally, a recognition of data-object within the webpage is also obtained according to the classification result and the learned decision tree, which efficiently initiates the later data extraction.

The main contributions of our work includes, (1)a combinational use of content and structure features of the webpages; (2)a novel representation of structure information, which is efficient and effective on the heterogenous web documents; (3)application of decision tree based learning to classify the webpages and help recognize the data objects. The rest of this paper is organized as follows. Section 2 describes the generation and selection method of webpage features. Section 3 describes the application of decision tree based learning algorithm and the primary data object recognition. Section 4 demonstrates the experimental results and Section 5 discuss researches related to our work. Finally, Section 6 concludes our work and plans the future work.

2 The Task of Target Webpage Identification

In this section we discuss in greater detail the feature generation and selection methods used in our webpage identification task, typically the structural information representation and feature selection. The generation of the content-and-structure combined features is accomplished in two phases: *Expanded Matching* and *C-Sliding-Window* algorithms. In order to achieve a balance between

[1] From *www.scholastic.com* and *www.amazon.com*

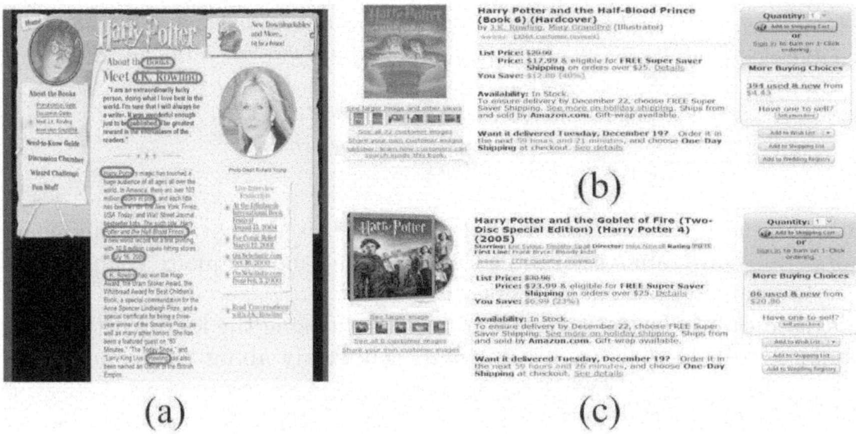

Fig. 1. Comparison of Target and Noisy Pages

precision and recall, we first perform a loose match using *Expanded Matching* algorithm to find all possible *domain-specific featured tokens* in the webpage, and then run a restrict selection using *C-Sliding-Window* to choose the closely located matched tokens. After the set of matched tokens are chosen, a relative distance is calculated to measure their closeness. Thus the set of domain-specific tokens holding their structure information (so called *structured tokens*) are exploited as the features of later decision tree learning.

2.1 Expanded Matching with Domain-Specific Featured Tokens

Domain-Specific Featured Tokens. The purpose of defining *Domian-specific Featured Tokens* is to capture the content (or semantic) feature of the target domain. As illustrated in Fig. 1(b) and Fig. 1(c), content feature is critical and indispensable to distinguish the pages of different domains, especially when noisy webpages share similar structure with relevant ones. There are two important observations of the data-rich webpages in different websites of the same domain. (1)Data objects are usually presented in similar schemas which share equivalent semantic. Following the previous example, the words such as "ISBN", "publisher", "list price", "add to cart" are good candidates for the featured tokens, which are also used as heuristics in the works such as data labelling and complex query interface discovery. (2)The same concept (or semantic) of the data object can be expressed in multiple ways. For example, the concept *publish date* of a book can also be expressed as "Pub. Date", "published:" etc. According to these observations, the domain-specific featured tokens can be defined as follows.

Definition 1. ***Domain-Specific Featured Tokens*** *denotes the words frequently used in presenting the data objects of target domain, which can be presented in a set of dual-tuples* $f_i^j \in \{< w_i, c_j > | i, j \in \mathbf{N}\}$ *complying the following constraints,*

$$\begin{cases} w_i := [Char]^+, \quad c_j \in \mathbf{N}, \quad i, j \in \mathbf{N} \\ \forall m, n \in \mathbf{N}: \quad w_m = w_n \Leftrightarrow m = n \\ \forall f_m^i, f_n^j: \quad m = n \Leftrightarrow i = j \end{cases}$$

where w_i is the words of featured tokens, and c_i is the corresponding concept category. The constraints guarantee that w_i should be unique to each other and belongs to one and only one concept category.

This definition indicates that the criteria of choosing the featured token relies on human's prior knowledge or some empirical study about the target domain. Since the same concept can be presented by various expressions, collecting a proper set of f_i^j as the content features is not an easy task. So far as our work goes, it is done half-manually. Detailed discussion of collecting f_i^j is beyond the scope of this paper, here they are taken as an input of present work. Nevertheless, there are two heuristics to collect and exploit the featured tokens, which to some extent make up for the diversity of the concept expressions.

1. Collect as many as possible, regardless of how discriminative the tokens may be. For example, simple words such as "by" for the concept of *author*, "pp" for *page number* are included in our empirical study.
2. A loose matching operation rather than a strict one should be performed to find the matched tokens from the webpages. Therefore we propose the operation called *Expanded Matching.*

Expanded Matching. The input of our matching process are the featured tokens (f_i^j) of target domain and the webpage to be visited, the output is a set of matched tokens with their offset positions within the page. Note that the matching operation processes not only the text tokens, but also non-digital attribute values of the tags. Fig. 2 illustrates a sample HTML snippet being parsed by *tag separated tokenization* and matched by *expanded matching* . The *expanded matching (EaMat)* is an operation upon the tuple $f_i^j =< w_i, c_j >$ and the Tag Separated token t, which can be defined in a formalized way as follows.

Definition 2. *Given* $\sqcup$ *denoting the single character of white space,* α *denoting the single character of white space or other non-letter character, for every* $f_i^j = f_{i-sub}[\sqcup f_{i-sub}]^* = \underbrace{f_{i0} \sqcup f_{i1} \cdots \sqcup f_{ik} \cdots \sqcup f_{in}}_{0 \le k \le n}$

$$EaMat(f_i^j, t) = \begin{cases} true & n = 0, \quad t = [\alpha]^* f_{i0} [\alpha]^* \\ true & n > 0, \quad t = [\alpha]^* \underbrace{f_{i0} [\alpha]^? \cdots [\alpha]^? f_{ik} [\alpha]^? \cdots [\alpha]^? f_{in}}_{0 \le k \le n} [\alpha]^* \\ false & otherwise \end{cases}$$

Each matched token, say $EaMat(f_i^j, t_q) = true$, is represented by a dual-tuple $< c_j, p_q >$, where p_q denotes the offset of t_q in the HTML file counted in the unit of character. As illustrated in Fig. 2, "261pp" and "hb" are matched according to above $n = 0$ condition if the featured tokens contain "pp" for concept *page number* and "hb" (hardback) for concept *format*, while "add_to_cart" is matched according to the $n > 0$ condition if the featured tokens contain "add to cart".

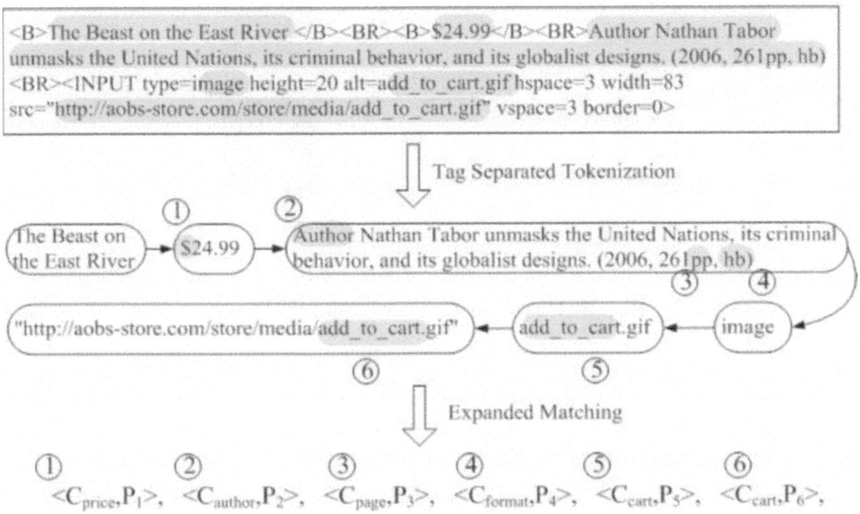

Fig. 2. Example of Expanded Matching

2.2 Feature Selection by C-Sliding-Window

C-Sliding-Window. Featured tokens without structure information are still not discriminative enough to filter out the irrelevant pages, as illustrated previously in Fig. 1(a) and Fig. 1(b). Similar feature tokens can be scattered in the context of news, forum articles, etc. Therefore, the output of *expanded matching* still requires further processing. As mentioned above, HTML structure information should be exploited along with these tokens to do the feature selection.

We start from choosing a proper data structure for representing the HTML pages. Generally speaking, tag tree and sequence are two major structure formalisms used for webpages. In this paper, the latter is adopted because not all target data objects are displayed in a tree or table-like style. Two sample pages[2] are shown in Fig. 3. Page (a) contains single book item and its information is displayed in three different sub-trees. Page (b) is an item-list page, where each item does not displayed in a table or tree like style, and no detail-link is available for the items. Therefore, tree structure is hard to be generalized among different websites. Nevertheless, one commonness shared by these pages is that the

[2] From (a)*www.half.com* and (b)*aobs − store.com*

featured tokens found on them are close to each other in location, which is a much simpler but more general feature for all heterogenous data-rich webpages. Based on this observation, we choose to present the structure information by measuring the closeness of featured tokens of different concept class on the sequence.

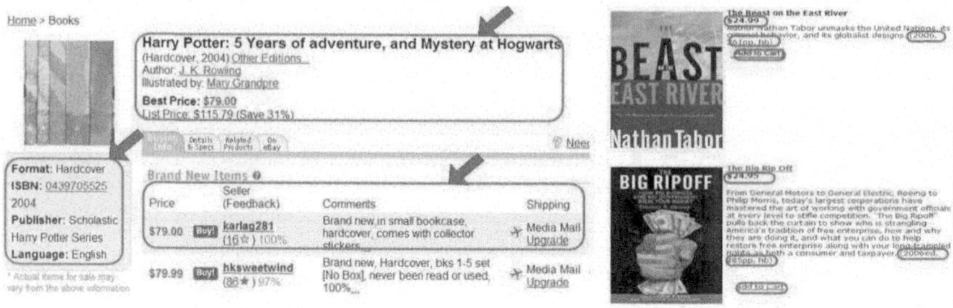

(a) (b)

Fig. 3. Example Pages of Book Selling

As mentioned above, the featured tokens are categorized into concept classes. The ith matched token is $t_i =< c_k, p_i >$, which means that at the position p_i of a page, some word *expanded matches* with a featured token of concept class k. These matches are in an ascendant sequence according to their p_i. Therefore, it is easy to find the most close set of tokens by using a sliding window W to scan the sequence. All the tokens $t_i \in W$ will be include to calculate the closeness. However, usually there will be repeated or absent matches of the same concept class, some are noisy matches, and some come from typical item-list pages. And also the nearby matches may belong to the same concept class. Therefore, the window width we use here should be counted in concept numbers, thus the algorithm is so called *C-Sliding-Window*. The *C-Sliding-Window* algorithm is described as follows. Given a set of matched tokens $t_i =< c_k, p_i >$ sorted in ascendant order of p_i, and the window width θ, the C-Sliding-Window W moves along the t_i sequence, the window has a dynamic span over the sequence to cover nearby tokens such that they belong to θ different concept classes ($\theta \leq |W|$). The closeness of the tokens in window is defined by *Window Density* $D_W = \frac{p_{s+|w|} - p_s}{|W|}$ when W moves to t_s. The minimum D_W and corresponding $t_i \in W$ are recorded, and the centroid of the window is computed as $S_W = \frac{\sum_{t_i \in W} p_i}{|W|}$. One interesting thing is how to decide the value of θ. Since in target pages, there still may be outlier matches far away from other close located items. Therefore, θ is usually set smaller than the number of concept classes which have at least one match found. According to our empirical experience, $\theta = MAX\{|C| - 2, 3\}$, in which $C = \{c_k | \exists i \in N, t_i =< c_k, p_i >\}$, will be a good choice.

Feature Representation. The output of *C-Sliding-Window* is a close set of matched tokens and the S_W. The absolute offset positions are then transformed into a relative measure, which can reflect the nature of data object displaying more faithfully. We define this kind of relative position as $r_i = \frac{p_i - S_W}{L_{html}}$, in which L_{html} is the length of HTML page file. Thus, each matched token t_i is transformed into $t_i^r =< c_k, r_i >$, a dual-tuple with a relative distance value. The set of $< c_k, r_i >$ is sent to later decision tree learning process, as the attribute values. Despite using this simple measure, our strategy is very effective, as shown later by our experiments.

3 The Application of Decision Tree Classification

The decision tree is chosen for mainly two reasons. First, it is error-robust. Since web is a immense collection of heterogenous documents, the training data will never be enough. Therefore, it is necessary to assume that the training data may contain errors. Second, the output of the algorithm contains rich information, such as the contribution of each attribute, the detailed value interval for corresponding class, which can be reused for the later processing.

As the preprocessing is done, each instance of the learning algorithm is a webpage represented by a set of dual-tuple $< c_k, r_i >$. Each attribute stands for a concept class, and the value of attribute k is r_i for the ith instance.

Selection of the Negative Examples. In the two-class learning algorithm, characterization of the negative class (e.g. "a webpage not containing extractable data object of a book ") is often problematic. Choosing representative negative examples can significantly affects the accuracy of learning algorithms, because commonly used statistical models have large estimation errors on the diverse negative class. Considering the features we define, we choose to collect the most confusing (easy to be wrongly classified)examples, so that the classifier can still performs good when new data come in. There are mainly two typical types of negative examples. (1)The unstructured text-rich webpages which contain featured tokens of the target domain in content; (2)The data-rich irrelevant webpages which may have similar schema or display style with the target domain.

Data Object Recognition Upon Identified Pages. Intuitively, the *C-Sliding-Window* we propose above shares an inherent similarity with the data record finding in data extraction. Therefore, we argue that a primary recognition of data object should be obtained to initiate the following data extraction work.

However, locating data object by the only measurement of token closeness is error prone, especially for the list page where multiple data objects resize. Take Fig. 1(b) for example, the book cover and author tokens of the second book may easily be included into the same window with the page and format tokens of the first book, because the description text for the first book is longer and make the distance larger. Fortunately, some valuable information can be inferred from the decision tree learning results.

Decision tree provides a rough statistic of the sequence of the structured tokens displaying, which can be inferred from the branching values of corresponding attributes. As the previous definition, the value of each attribute of the tree lies in the interval of [-1,1], with the magnitude indicating the distance to the window centroid, and the sign indicating the relative forward or backward displacement to the centroid. Using the above point as heuristic, we will modify the original close window to move forward or backward to get the primary recognition of the data object. A detailed description of this modifying algorithm is omitted here for lack of space. Note that for item-list pages only one data object is tended to be discovered. Although other records are ignored, we believe the discover of the one data object can still provide good initiation of the data extraction work. In most existing researches on data extraction, the wrapper induction work starts in a top-down way to learn the pattern on the tag-tree. However, with the candidate data object, data extraction only needs to do a bottom-up check to validate the pattern by comparing the sibling areas of the given candidate data object.

4 Experimental Results

The proposed algorithms are implemented in Java 1.5 platform and the experiments are performed on an AMD Sempron(TM) 1.81GHz processor with 1G RAM running Windows XP Professional Edition.

Data Preparation. The data we collected to train and test the classifier are divided as positive and negative samples. The experimental target domain is online book shopping. The sample pages and the featured tokens are collected half manually as the following steps. (1)Keywords "online book shopping" is submitted to Google and MSN, and the 76 websites are manually browsed to gather the domain specific featured tokens, as illustrated in Table 1. (2)For positive samples, 783 pages from 138 websites, which are selected from Yahoo! directory, are collected. (3)The negative samples are collected as described in Section 3. For the unstructured content relevant negative examples, the keyword "book reading club" ,"book news" and "book review" are submitted to the search engines to get the returned pages. For the structure-similar negative examples, we choose the movie, music and DVD shopping domain, which share some common attributes with book shopping, such as the publication date, author, etc. After manual checking, we obtained totally 1582 negative pages from 1143 website. Table 1 shows the collected featured tokens used by our experiments.

Target Webpage Identification. Three experiments are set up here to compare the precision of classification, including one baseline experiment using purely text as features and another two using structured tokens as features. We focus on precision here to guarantee that the data extraction can get cleaner webpages. The learning algorithm we choose is standard decision tree C4.5. For comparison, SVM based learning algorithm is also tested, because it is one of the best algorithms for traditional text classification. A publicly available implementation of the algorithms by weka3.5 is used (J48 for C4.5 and SMO for SVM). The experiment is to check how our method performs as more different webpages

Table 1. Collected Featured Tokens for Book Shopping Domain

Concept Class	Featured Tokens
(1)isbn	isbn
(2)shopping	add to cart,buy the book,basket contents,add to basket, add to shopping basket,add to trolley,add to shopping cart,cartadd,buy now
(3)page	pages,page count, number of pages, # of pages
(4)availability	availability, in stock, available in, available at
(5)price	price,list price, cover price, retail price, normal price, $, our price, club price, suggested retail price, you pay, recommended retail price, rrp, on-line price
(6)format	format, formats, other formats, hardcover, paperback, soft-cover, binding, binding:, Hard Cover, Novelty Gift, HC, PB, saddle-stitch, full-color interior ink
(7)title/cover	book name, book title, book cover
(8)author	author, written by, editor, by
(9)edition	edition, in-print editions, editions
(10)publish	publish, published, publisher, publishing, Publishing Date, Published date, Pub. Date, Date:, released:, pub date:, Imprint, Published by, Release Date, Printed:

come in. We use 10% of the layered webpages for training, and divide the rest data into layered 10 folds. The result is reported in Table 2. As the data shows, both experiments using structured tokens outperform the baseline one, and the precision is stable as the testing data grows. Surprisingly, the decision tree based learning has a similar or even better performance than SMO.

Table 2. Comparison of Classification in Precision

Algo.	0.1	0.2	0.3	0.4	0.5	0.6	0.7	0.8	0.9	1.0
J48-StructuedToken	0.977	0.991	0.987	0.990	0.992	0.990	0.990	0.990	0.991	0.989
SMO-StructuredToken	0.985	0.979	0.987	0.988	0.989	0.986	0.986	0.987	0.990	0.988
SMO-Baseline	0.910	0.910	0.911	0.911	0.912	0.912	0.912	0.914	0.910	0.909

To have a specific illustration of the advantage of the *expanded matching* and *window based structure representation*, two more experiments are performed: (1)test the sensitiveness of performance affected by different domain featured tokens, (2)use $strictMatch$ and $absolutePosition$ method to preprocess webpages for machine learning. The two experiments are run on the whole set of data using 10 fold cross validation. The result of F_1 values and the size of decision tree are reported in Table 3. For experiment(1), tokens are cut in two ways: cut the concept class, or use only one token for each concept class. $Run(0)$ uses all tokens in Table1, $Run(1)$ uses the reduced set of featured tokens from which the concept class "ISBN" is cut, for it is the root node of output decision tree in *Run*(0), then in *Run*(2), "cover" concept is cut for it is the root node of decision tree in *Run*(1), and so on. In *Run*(4) all concept classes are kept but only the first token is used for each, which is labelled as "singleToken" (or "sT" for short)in the table. For experiment(2), the method strict matching(sT_strictMatch), absolute position presentation(sT_absolutePos), and their combination(sT_s&a) are run. Given a webpage p and featured token f, strict matching finds all the tokens which are exactly equal to f from the page. Then, instead of finding a close set of matches, an absolute position information (i.e.the average offset of matched tokens in the same concept class) is used for each concept, as the input of later learning process. As the data reported, our method shows excellent stability to the change of featured tokens or the size of decision tree, and performs steadily better than the strict matching and absolute position methods.

Table 3. Test of Sensitiveness to the Domain Specific Featured Tokens

Run	(0)allToken	(1)=(0)-isbn	(2)=(1)-cover	(3)=(2)-publish
F_1	0.981	0.959	0.952	0.919
nodes	29	63	67	93
leaves	15	32	34	47
Run	(4)singleToken	sT_strictMatch	sT_absolutePos	sT_s&a
F_1	0.979	0.876	0.888	0.865
nodes	33	45	31	27
leaves	17	23	16	14

Data-Object Recognition. Based on previous experiments, a rough statistic of the sequence of concept class matches is obtained, which are used as heuristic to modify the sliding window W to get data-object recognition. An example is $< isbn_{-0.6}, [publish, edition, cover]_{-0.04}, price_{0.2} >$, which means publish, edition and cover information are usually very close to each other and become the window centroid, ISBN is usually farther ahead of them, and price is after them($-0.6 < -0.4 < 0 = S_W < 0.2$). We only checked 241 pages from 53 websites of our positive data, containing 183 single-item pages and 58 item-list pages. For websites using different schemas to represent data objects, concepts like "language", "age level", are not covered in our featured tokens and evaluation. The option items of the <select> tag is also ignored. The correctness of data object recognition is defined as if the window cover one correct object without overlapping nearby objects. There are totally 143 out of 183 for item pages and 46 out of 58 for list pages are precisely recognized. The heuristic brought a 12% raise in precision for the list pages. We did not apply it to the item pages for it does not bring much improvement on them.

5 Related Work

Recent research efforts have produced numerous works which are, directly or indirectly, related to the problem of this paper. We give a brief review of them in this section from the following aspects.

Adapting the text retrieval techniques to webpage analysis is extensively discussed. Typical works such as [1,4] exploit the text retrieval for link relevance prediction. They achieved good results on the webpages which can be described as text-rich in contrast with the data-rich ones discussed in this paper. Features other than texts are also exploited in many works for webpage content analysis. Works in [10,2] use DOM tree for webpages, tree edit distance or tree path for links are used to represent the structure information. Visual cues[13,14] are also applied to analyze the important block or object display in webpages. Our work is different in that we use a new representation for structure which can be easily obtained, and it is combined with text features.

Successful vertical search engines such as MSN shopping and Lycos have attracted much attention. Many data extraction works have been proposed [13,8,14,11,12], which motivates the work in this paper. Although many works show excellent performance in data object extraction and labelling, they are hard to be exploited in this work. An important reason may be that the pattern induction requires a cleaner data set, like a training set of multiple similar pages or item-list pages containing multiple data records. This work targets at a related but different task of identifying relevant pages prior to the data extraction.

There are also works addressing the similar problem of feeding data extraction with selected webpages. [7] proposed the hidden agents for collecting hidden webpages, which uses *navigation pattern* for locating the search forms, and learns to fill in forms using a sample repository. [5] proposed a method of structure-driven crawler which learns navigation pattern from the sample page and entry point given a prior for each known website. These works are positively complementary to our work, and this paper aims at a more general identification algorithm based on webpage content regardless of different and unknown websites.

6 Conclusion and Future Work

In this paper, we propose a novel method to identify the target pages from unknown websites accurately by exploiting the structured-token features of the web page content. We apply the decision tree based classification algorithm to induce the structure information efficiently. Furthermore, a preliminary recognition of data-object is introduced to efficiently initiate the subsequential data extraction. With the expanded matching and window-based structure information representation, our method scales well on the heterogenous web documents.

There are several interesting directions for the future work. First, an incremental learning algorithm may be introduced to update the domain featured tokens and the decision tree. Second, the primary data object recognition can be extended to do data extraction by scanning the tree in a bottom-up way.

Moreover, collaborating other information from the website will be a promising try to do the website-level resource identification.

References

1. S. Chakrabarti and B. Dom M. Berg. Focused crawling: A new approach to topic-specific web resource discovery. *Computer Networks*, 31(11-16):1623–1640, 1999.
2. V. Crescenzi, P. Merialdo, and P. Missier. Clustering web pages based on their structure. *Data Knowl. Eng.*, 54(3):279–299, 2005.
3. D. W. Embley, Y. Jiang, and Y.-K. Ng. Record-boundary discovery in web documents. In *Proceedings of SIGMOD '99*, pages 467–478, New York, USA, 1999.
4. M. Ester, H. Kriegel, and M. Schubert. Accurate and efficient crawling for relevant websites. In *VLDB*, pages 396–407, 2004.
5. Márcio L. A. Vidal et al. Structure-driven crawler generation by example. In *Proceedings of SIGIR '06*, pages 292–299, New York, USA, 2006.
6. N. Jindal. Wrapper generation for automatic data extraction from large web sites. In *DNIS*, pages 34–53, 2005.
7. J. P. Lage, A. S. Silva, P. B. Golgher, and A. H. F. Laender. Automatic generation of agents for collecting hidden web pages for data extraction. *Data Knowl. Eng.*, 49(2):177–196, 2004.
8. B. Liu, R. Grossman, and Y. Zhai. Mining data records in web pages. In *Proceedings of KDD '03*, pages 601–606, New York, USA, 2003.
9. Z. Nie, Y. Zhang, J. Wen, and W. Ma. Object-level ranking: bringing order to web objects. In *WWW*, pages 567–574, 2005.
10. D. C. Reis, P. B. Golgher, A. S. Silva, and A. F. Laender. Automatic web news extraction using tree edit distance. In *Proceedings of WWW '04*, pages 502–511, New York, USA, 2004.
11. Jiying Wang and Fred H. Lochovsky. Data extraction and label assignment for web databases. In *Proceedings of WWW '03*, pages 187–196, New York, USA, 2003.
12. Y. Zhai and B. Liu. Web data extraction based on partial tree alignment. In *WWW*, pages 76–85, 2005.
13. H. Zhao, W. Meng, Z. Wu, V. Raghavan, and C. Yu. Fully automatic wrapper generation for search engines. In *Proceedings of WWW '05*, pages 66–75, New York, USA, 2005.
14. J. Zhu, Z. Nie, and J. Wen et al. Simultaneous record detection and attribute labeling in web data extraction. In *KDD*, pages 494–503, 2006.

Honto? Search: Estimating Trustworthiness of Web Information by Search Results Aggregation and Temporal Analysis

Yusuke Yamamoto, Taro Tezuka, Adam Jatowt, and Katsumi Tanaka

Department of Social Informatics, Graduate School of Informatics, Kyoto University
Yoshida-Honmachi, Sakyo, Kyoto 606-8501, Japan
{yamamoto,tezuka,adam,tanaka}@dl.kuis.kyoto-u.ac.jp

Abstract. If the user wants to know trustworthiness of a proposition, such as whether gthe Japanese Prime Minister is Junichiro Koizumih is true or false, conventional search engines are not appropriate. We therefore propose a system that helps the user to determine trustworthiness of a statement that he or she is unconfident about. In our research, we estimate trustworthiness of a proposition by aggregating knowledge from theWeb and analyzing creation time of web pages. We propose a method to estimate popularity from temporal viewpoint by analyzing how many pages discussed the proposition in a certain period of time and how continuously it appeared on the Web.

1 Introduction

People often become unsure about a statement they encounter on the Web, for example a statement such as gthe Japanese Prime Minister is Junichiro Koizumih or gdinosaurs became extinct 65 million years agoh. In such a case, the user often types in the statement into a search engine, and examines how common it is on the Web, or tries to check if there are any contradicting answers. This is, however, a time consuming task.

We therefore propose a system that helps the user in determining trustworthiness of a topic that he or she is unconfident about. We named our system gHonto? Searchh. gHonto?h means gIs it really?h in Japanese. We focus on assisting the user to make a judgment on trustworthiness, rather than making an absolute decision.

There are various criteria for information's trustworthiness: the level of popularity, the author's reliability, or consistency of the content itself. In our research, however, we use popularity as the criterion. In other words, our system provides the user with popularity estimation of a phrase and its alternative or counter examples occurring on the Web. The system also presents changes in the frequency of these phrases in time, in order to help the user to decide if the original phrase is up-to-date, or if it has been continuously stated for a long span of time, thus ensuring its reliability.

In order to provide such functionality, the system performs the following procedure. First, it divides the phrase given by the user into parts, and constructs a

G. Dong et al. (Eds.): APWeb/WAIM 2007, LNCS 4505, pp. 253–264, 2007.

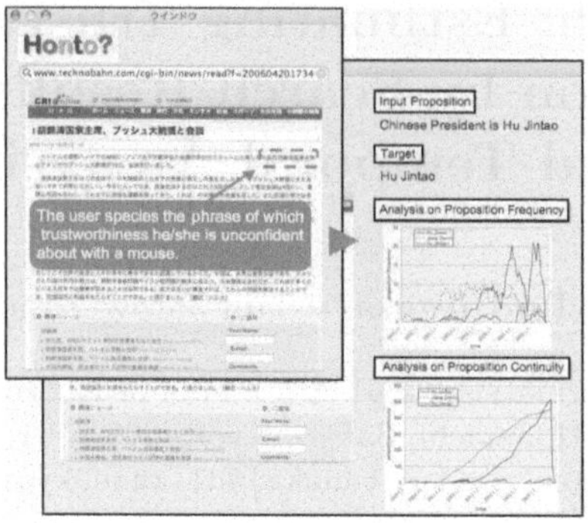

Fig. 1. Honto? Search: System Overview

query that would be sent to a web search engine. Secondly, it extracts alternative or counter examples to the original phrase out of the search results. Thirdly, the system sends the original phrase and the counter examples to the web search engine again, and obtains their present frequencies. Fourthly, it sends the phrases to a web archive and obtains the temporal change in their frequencies. The result is presented to the user as a list of phrases and a graph indicating the temporal change.

The rest of the paper is organized as follows. Section 2 describes related work. Section 3 presents the method of extracting counter examples from the Web. Section 4 presents the method of temporal analysis using a web archive. Section 5 describes the result of the experiment made to validate the effectiveness of our approach. Lastly, Section 6 concludes the paper.

2 Related Work

2.1 Web QA

Web question answering (Web QA) systems are similar to our system in that they retrieve text segments from the Web to answer the user's information requests. However, they are different from our system in terms of goals. Web QA systems use the Web to find an answer to a question posed by the user [1,11]. Most Web QA systems go no further than finding an answer to the query. For example, TRECfs question answering track provides a benchmark for evaluating efficiency in finding an answer to the user specified question[1]. It is assumed that the answer is reliable and the user is satisfied once he or she gets it. In reality, however, this is not often the case with the Web since it contains a lot of unreliable and obsolete information. In our system, the user does not type in an interrogative sentence. Instead, the user types in a phrase whose trustworthiness he or she doubts. The goal is to extract additional information from the Web to help the user judge the trustworthiness of the proposition contained in the phrase.

[1] TREC Question Answering Track, http://trec.nist.gov/data/qa.html

There is a recent trend in Web QA systems to utilize redundancy of information found on the Web [4,5,10]. Systems that do this aggregate phrases and present the most frequent one as the answer. This was made possible by the tremendous size of the Web compared to the traditional QA source data (corpora). Although these systems are similar to ours in that they utilize redundancy of information on the Web, the aim and the approach is different from our system.

2.2 Term Dynamics

Our system uses temporal changes in frequencies of phrases to filter out obsolete information. Kleinberg made a survey of recent approaches for analyzing term dynamics in text streams [9]. Kleinberg's "word burst" is a well-known method for examining changes in word frequencies over time [8]. It is a state-based approach that measures term dynamics characterized by transitions between two states: low and high frequency one. Kleinberg's work, however, was intended to model significant bursts of terms in text streams, whereas in our system we compare differences between the frequencies of phrases and their duration in time. Kizasi[2] is an online system that extracts keywords that have recently become increasingly popular recently. This system focuses only on keywords, in contrast to our method.

2.3 Web Archives

We propose utilizing web archives in order to estimate the popularity of propositions in time on the Web. Web archives preserve the history of the Web and indirectly reflect the past states and knowledge of the societies. Until now, however, Web archiving community has mostly concentrated on harvesting, storing and preserving Web pages as they pose numerous challenges. Relatively, little attention has been paid to utilizing Web archive data despite the fact that it offers a great potential for knowledge discovery purposes. Aschenbrenner and Rauber discussed possible benefits and approaches towards adopting traditional Web mining tasks in the context of Web archives [3]. Recently, Arms et al. have reported on ongoing work aiming to build a research platform for studying the evolution of the content and the structure of the Web [2]. We believe that successful completion of similar projects in the future will enable more effective knowledge discovery from the history of the Web.

2.4 What Honto? Search is Not

The following list indicates some of the functions that **are not** realized by Honto? Search.

Keyword search: In Honto? Search, the user query is a phrase. It is different from conventional web search engines where the user inputs keywords.

Page search: The search results of Honto? Search are relative frequencies of the query phrase in comparison to alternative phrases. The system presents

[2] Kizasi, http://kizasi.jp/ (in Japanese).

aggregated knowledge instead of the lists of web pages as in the case of conventional search engines.

Opinion miner: The main target of Honto? Search is factual information. It is not intended to collect people's opinions on certain topics where no definite answer exists.

3 Collecting Alternative Propositions

In this section we describe how to identify alternative propositions, or counter examples, for a phrase query given by the user. More generally, the system finds terms that fit into a specific part in the user's query. For example, for a phrase "dinosaurs became extinct 65 million years ago", the user may want to check if "65 million" is actually true. In this case, the system searches the Web to find other terms that constitute the expression, such as "60 million", "70 million", or even "10 thousand" (which is wrong). We call such terms *alternative terms*, and a phrase containing it will be called an *alternative proposition.*

In Honto? Search, the user selects a part from a phrase that he or she is unsure about. The part will be called *verification target* in this paper. Terms that would replace the verification target in the phrase are alternative terms. If the user does not specify a verification target, then the system performs the procedure to each word in the phrase.

The system performs the following procedure to identify alternative propositions. Fig. 2 explains this procedure.

1. The query is constructed by splitting the phrase into two parts by verification target T. For example, if the user inputs "dinosaurs became extinct 65 million years ago" into our system, since the target T is "65 million", we get two queries, P_1: "dinosaurs became extinct" and P_2: "years ago".
2. The system sends a query "P_1 & P_2" to a web search API. Alternative terms are extracted from the search results by using a regular expression, $/P_1$ (.*) $P_2/$. The text segment that comes in the middle is extracted as an alternative term, as long as it is contained in one sentence and is different from the original verification target.
3. Alternative terms are sorted according to their frequencies. The more they appear on different web pages, the higher they are ranked. Terms that appear below the threshold value are eliminated. A set of alternative terms may contain much noise at first, but this sorting and filtering process reduces it.

An alternative proposition is constructed by inserting an alternative term between two separated parts of the original phrase, P_1 and P_2. Each alternative proposition is again sent to the search engine, to obtain its frequency.

One of the problems with Honto? Search now is that it is still incapable to handle statements with negations. Fortunately, frequencies of such statements are relatively low compared to the positive ones, so the aggregated answers usually show good results.

In case the web search API returned too few results, searching is performed again by relaxing the query. The system extracts nouns, verbs, or adjectives from each alternative proposition and performs multiple keyword search. It constructs the query by connecting keywords with "&". In the example, the query will be "dinosaurs & became & extinct & 65 & million & years". The system then searches within the retrieved web pages to find a sentence that contains all the keyword.

This step allows the system to retrieve phrases with the same meaning but expressed in different forms. Although the method is vulnerable if many sentences contain negations, expectations, or interrogatives, we assumed that once the result is presented as a list, the user can check it by accessing the snippets which are linked from the alternative phrases.

The frequencies are then presented to the user. By comparing the frequency of the original phrase with the frequencies of alternative propositions, the user can get an idea of how much the phrase is supported on the Web.

4 Analysis of Generation Time of Propositions

Because the sentences collected by the approach proposed in Section 3 are generated at different times, it is not appropriate to use them for a majority decision without careful consideration. For example, consider the proposition, "the host city of the next Summer Olympic Games is Beijing". This proposition is correct only until the event is held in 2008. This example shows that trustworthiness of the proposition is strongly dependent on time.

In addition, if the system considers temporal information, it can estimate trustworthiness of the proposition from different aspect also. That is, the system can evaluate how continuously the proposition has been accepted over time.

We define two criteria for determination of trustworthiness. Fig. 3 explains this procedure.

4.1 Analysis of Creation Time of Web Pages

To analyze the temporal distribution of web pages relevant to a proposition, we need to determine when each page was created. The system uses Internet Archive[3]. Internet Archive is the well-known public web archive offering about 2 petabytes of data. The access to Internet Archive is provided by Wayback Machine that allows viewing snapshots of pages from the past. Using Wayback Machine it is possible to reconstruct the histories of web pages. After issuing URL address of a given page Wayback Machine returns the list of available page snapshots together with their timestamps. We can estimate the creation time of pages by considering the oldest timestamps that appear in the list. Consequently, it is possible to analyze the temporal distribution of creation time of pages that refer to a given proposition.

[3] Internet Archive, http://www.archive.org/web/web.php

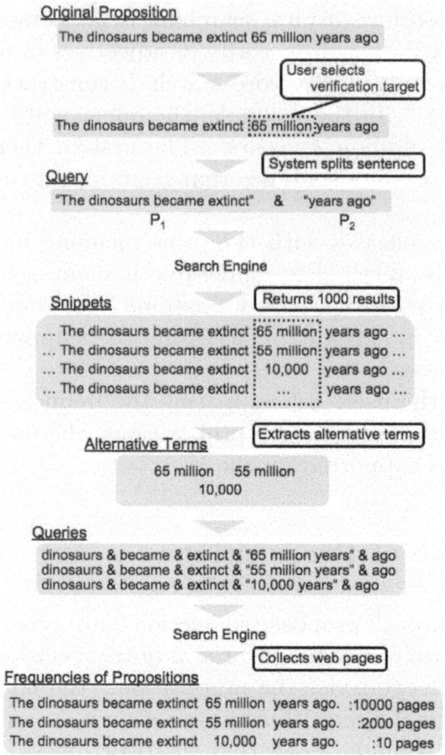

Fig. 2. Knowledge Aggregation Procedure

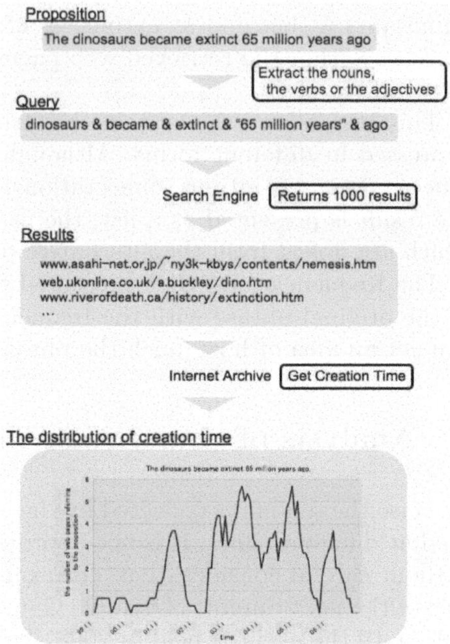

Fig. 3. Temporal Analysis Procedure

Utilizing Internet Archive for temporal analysis has, however, several limitations. First, the temporal scope of page snapshots is constrained. Crawling of Web pages has started since 1996. There is also no data provided that is younger than 6 months due to the policy of Internet Archive. Second, after a closer inspection we discovered that there are some gaps in the data due to uneven crawling pattern in the past. Third, page creation dates estimated by our method are only approximations of the actual origin dates of pages. There is always some delay between the creation of a page and its detection by Web crawlers. Lastly, it is also unsure whether propositions occurring on pages have actually appeared at the time of page creation and not later due to subsequent updates made to page content. Nevertheless, since our approach analyzes relatively large number of pages and utilizes relative frequencies of phrases, satisfactory results can still be obtained despite the above limitations.

4.2 Trustworthiness of the Proposition in a Certain Period

In this section, we propose a way of estimating trustworthiness of a proposition in a specific period by comparing the temporal distribution of a proposition with that of alternative propositions.

In the first step, we define PF_A of proposition A at time period t (PF is Proposition Frequency) .

Proposition Frequency of proposition A at time period t

$PF_A(t)$: the number of the web pages which refer to proposition A and were generated at time period t

Using PF, we can estimate which proposition is the most reliable in a certain period of alternative propositions. That is, if we want to estimate which proposition is more reliable, proposition A or proposition B in time period t, we only have to compare $PF_A(t)$ with $PF_B(t)$. If $PF_A(t)$ is greater than $PF_B(t)$, we can estimate that proposition A is truer in period t than proposition B. We calculate $PF(t)$ of the proposition, which is the proposition the user inputs into our system and $PF(t)$ of alternative propositions which our system made from the original proposition, and we identify the proposition for which $PF(t)$ has the greatest score as the most reliable proposition in period t.

Usually, the number of newly generated web pages has many up-downs in a short span, forming a zigzag line over time. Considering this phenomenon, we modify PF by using the information over a certain period. We adopted the moving average in order to solve this problem [7] .

Modification to $PF_A(t)$ with Moving Average

$$PF_A(t) = \frac{1}{2n+1} \sum_{i=0}^{n} PF'_A(t \pm i) \tag{1}$$

We modify $PF'_A(t)$ (the original value of a proposition frequency) around time period t by a window size $2n + 1$. By comparing each PF modified by moving average, the system finally estimates trustworthiness of a proposition in a certain period.

4.3 Proposition Continuity

Honto? Search employs temporal analysis not only in order to select the most recently popular proposition (as described in Subsection 4.2), but also to inform the user whether the proposition has appeared on the Web for a long enough period of time. The underlining assumption is that such information is helpful in determining trustworthiness of the proposition. For example, a proposition "aluminum is the cause of Alzheimer's disease" was once a popular theory and has been widely discussed on theWeb, yet not as much now. On the other hand, a proposition "Alzheimer's disease causes dementia" still commonly appears on the Web. Presenting the temporal change in the frequencies of the two propositions would help the user to make judgment on reliability of the two.

The difference between the two is formalized as follows: in the case of a proposition that sustains to be reliable over time, web pages referring to the proposition are constantly generated. On the other hand, in the case of a proposition which was reliable only during a certain limited period of time, the amount of new web pages containing the proposition decreases once the proposition ceases to be reliable. In order to draw a line between the two, we look to a theory in psychology.

According to Hermann Ebbinghaus, a person's memory decreases exponentially [6]. The amount of remaining memory R at time t is defined as follows, using a coefficient γ:

$$\frac{d}{dt}R(t) = -\gamma R(t)$$

Based on this theory, we built the following model. When the amount of web pages containing the proposition is over λ of the amount on the prior month, we judge that the proposition is still attracting people's attention. On the other hand, if the amount is less than λ of the prior month, we judge that the proposition has entered the receding phase; it is losing people's attention and will eventually be forgotten by the public. λ is a threshold value and may be adjusted experimentally.

We define *proposition continuity* as a measurement indicating how long a proposition has been attracting peoplefs attention. We assume that the number of new web pages containing the proposition reflects peoplefs attention to it.

$PC_A(t)$ is *proposition continuity* of a proposition A at time t. It is defined as follows:

Proposition continuity of a proposition A at time t

$$PC_A(t) = \begin{cases} PC_A(t-1) + PF_A(t) & \text{if } PF_A(t) \geq \lambda PF_A(t-1) \\ \alpha PC_A(t-1) + PF_A(t) & \text{if } PF_A(t) < \lambda PF_A(t-1) \end{cases} \quad (2)$$

λ is a threshold value for detecting the receding phase. α is a coefficient that would decrease PC exponentially when PF is dramatically decreasing. If the amount of new web pages containing the proposition A is more than λ of the prior month, $PC_{A(t)}$ increases by $PF_{A(t)}$. If the ratio is below λ, $PC_{A(t)}$ drops rapidly, since we assume that the proposition has entered the receding phase.

Honto? Search presents PC to the user as the indicator of how consistently the proposition was supported by the public.

5 Experiment

In this section, we describe the result of experiments that tested the effectiveness of our approach on estimating trustworthiness of propositions.

5.1 Discovery of Alternative Propositions and Aggregating Sentences

To get alternative propositions, we used Yahoo! Web Search APIs[4], a web service for searching Yahoo!'s index and got 1,000 results for each proposition. We

[4] Yahoo! Web Search APIs, http://developer.yahoo.com/search/web/V1/contextSearch.html

collected only web pages in Japanese. From snippets of the search results, we extracted alternative terms using the method described in Section 3. Then we counted the frequency of each alternative term and eliminated the ones which had a frequency lower than 15 % of the most frequent one. This is because we assumed that terms whose frequencies are currently low are not appropriate as alternative terms.

After creating alternative propositions containing alternative terms, we used a Japanese morphological analyzer, Mecab[5], to extract nouns, verbs and adjectives from the snippets (brief summaries of search results). Finally we collected 1,000 web pages for each alternative query and aggregated them using the procedure described in Section 3.

We performed experiments on two propositions, "there are 15 countries in the European Union" (Example 1) and "the President of China is Hu Jintao" (Example 2). Verification targets were "15" and "Hu Jintao", respectively.

Table 1. Estimation of Trustworthiness of Propositions

"There are 15 countries in the European Union."	
alternative terms	frequency
25	187
15	156
10	141

"The President of China is Hu Jintao."	
alternative terms	frequency
Hu Jintao	589
Jiang Zemin	574

Table 1 lists the frequencies of the original and alternative propositions in the web search results. For example, for the proposition gthere are 15 countries in the European Unionh, we got two alternative terms,"25" and "10". The most frequent one was "25", which is the correct answer. The alternative "15" also produced many results, since it was true until 2004. Additionally, the alternative term "10", was also frequently reported on the Web, which must have come from expressions such as "10 new countries in the European Union". The user can judge that the original proposition may not be trustworthy, since it is not the most frequent one.

For the proposition, "the President of China is Hu Jintao", we got an alternatives proposition "the President of China is Jiang Zemin", which was actually true until 2003. From the table, the user can judge that "the President of China is Hu Jintao" is reliable, since it is the most frequent one. These are simple estimations which do not consider the temporal aspect.

Furthermore, we performed a thorough experiment using a list of historical events (historical time table)[6] as a test set, to see whether the system justifies these events as "true", when compared with other mistaken information on the Web.

[5] Mecab, http://mecab.sourceforge.jp/ (in Japanese).

[6] http://www.h3.dion.ne.jp/~urutora/sekainepeji.htm (in Japanese).

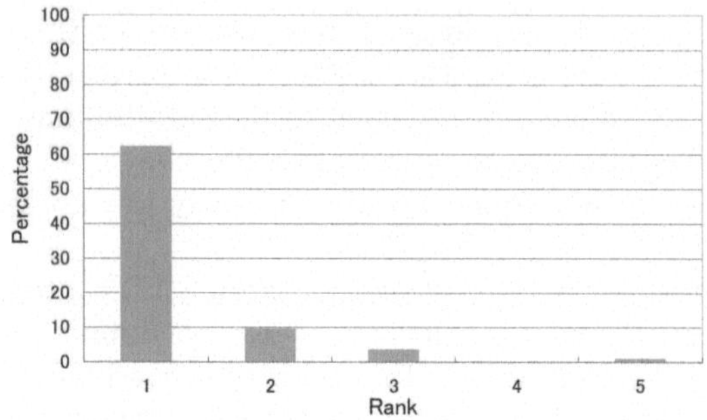

Fig. 4. The precision of our system for a test collection

In the time table, there were 360 major historical events dating from 3000 B.C. to 2003. We constructed verification targets by connecting the event name and the year when it occured, i.e. "the moon landing by Apollo 11 on 1969". A search query is constructed by replacing the year with a wild-card, i.e. "the moon landing by Apollo 11 on * ". Out of 360 queries, 116 has returned search results. The system then collected other candidates specifying different years (i.e. "the moon landing by Apollo 11 on 1970") and ranked them by their frequencies on the Web. We checked if the correct answers were ranked high when compared with other candidate phrases.

Fig. 4 illustrates the result of the experiment. 62% of the correct phrases were top ranked by the system, 9 % were ranked 2nd, and 3 % were ranked 3rd, while the rest was ranked 4th or below.

5.2 Analysis of Page Creation Times

Based on the method discussed in Section 3, we calculated PF and PC for the original and alternative propositions, in order to see the trustworthiness from the temporal point of view. We used the Internet Archive to estimate when each web page was created. We considered pages that were created between 1998 and now. In the Internet Archive, the user can not see web pages that are collected more recently than 6 months. Therefore, we could only use data until July 2006 and on. For calculating moving average, we used $n = 6$. For calculating PC, we used 0.8 for λ and 0.5 for α.

Fig. 5 shows that PF of "there are 25 countries in the European Union" overwhelms the other two propositions around 2004. The user can guess that there was possibly a change in the number of the countries in the European Union.

Fig. 6 shows that although PF of the proposition "the President of China is Jiang Zemin" is higher than that of "the President of China is Hu Jintao" at the beginning, they reversed around 2003. In fact, Jiang Zemin was the President of China until March 2003 and Hu Jintao has been the president ever since.

Fig. 7 shows that for the proposition "there are 15 countries in the European Union", PC decreases at the end, indicating that it is no longer true.

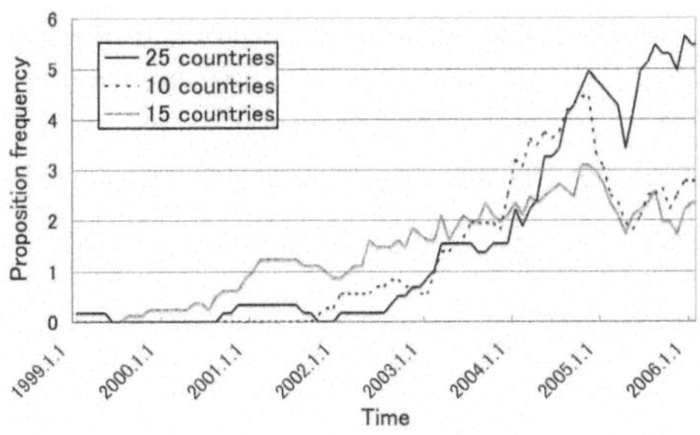

Fig. 5. Proposition Freq. for Example 1

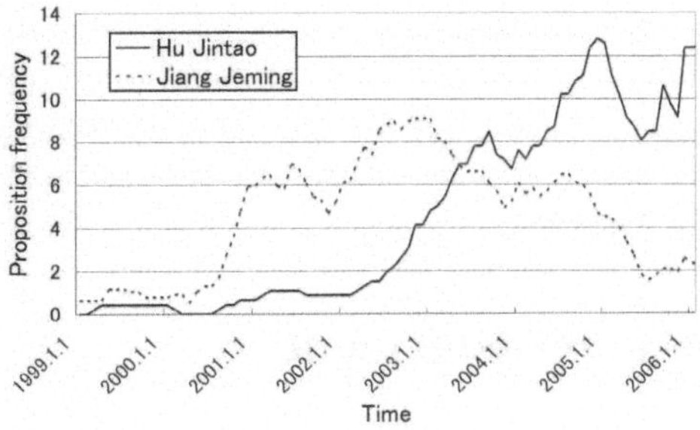

Fig. 6. Proposition Freq. for Example 2

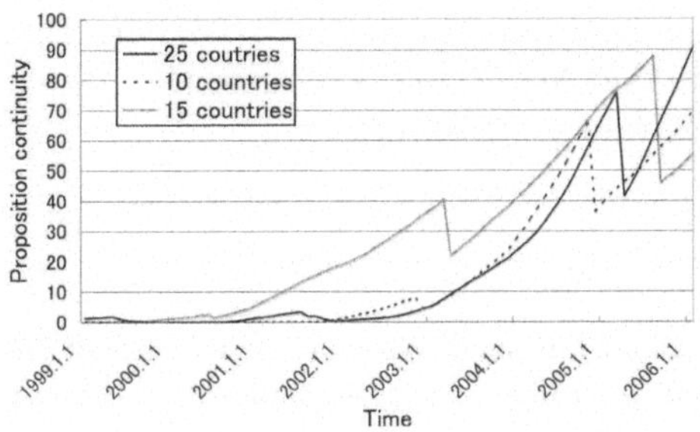

Fig. 7. Proposition Cont. for Example 1

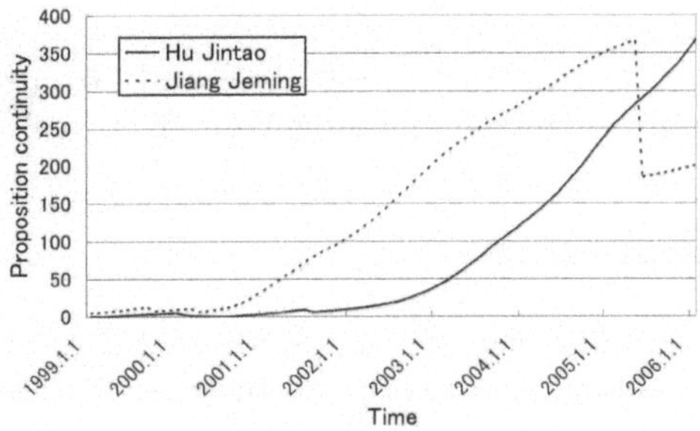

Fig. 8. Proposition Cont. for Example 2

Fig. 8 shows that *PC* for "the President of China is Hu Jintao" continues to increase, while *PC* for "the President of China is Jiang Zemin" decreases at one point, indicating the change in the presidency.

6 Conclusion

In case the user wants to know whether a proposition is true or false, there are no systems available which estimate trustworthiness of this proposition. Therefore we have proposed a method that aggregates information on the Web and estimates trustworthiness of a proposition from the viewpoint of time by aggregating web search result and using a web archive.

By analyzing when web pages were generated, we were able to determine whether a proposition is true or false during a certain period and if it has been true or false from the past until now. The problems of our approach are that we do not distinguish positive sentences from negative sentences and that temporal analysis depends onWayback Machine and so once it ceases, we can not precisely determine when pages were generated. In addition, alternative terms constructed in Section 3 can be temporally affected by Yahoo! Search. That is, if the top 1,000 results are recent data, we can not get older alternative terms.

Our method is a kind of knowledge discovery process. We think that aggregation of web knowledge can be applied not only to estimate the trustworthiness of a proposition but also to other problems. A part of our future work is to reduce the noise and to estimate trustworthiness of whole web pages rather than only their parts.

Acknowledgement

This work was supported in part by MEXT Grant for "Development of Fundamental Software Technologies for Digital Archives", Software Technologies for Search and Integration across Heterogeneous-Media Archives (Project Leader: Katsumi Tanaka), Grant-in-Aid for Young Scientists (B) "Trust Decision Assistance for Web Utilization based on Information Integration" (Leader: Taro Tezuka, Grant#: 18700086) and Grant-in-Aid for Young Scientists (B) "Information Discovery Using Web Archives" (Grant#: 18700111).

References

1. Andrenucci, A. and Sneiders, E., Automated Question Answering: Review of the Main Approaches, 3rd International Conf. on Information Technology and Applications, pp. 514-519, 2005.
2. Arms, W. Y., Aya, S., Dmitriev, P., Kot, B. J., Mitchell, R. and Walle, L., Building a research library for the history of the Web, Joint Conf. on Digital Libraries, Chapel Hill, NC, USA, pp. 95-102, 2006.
3. Aschenbrenner, A. and Rauber, A., Mining web collections. In Web archiving, Springer Verlag, Berlin Heidelberg, Germany, 2006.
4. Brill, E., Lin, J., Banko, M., Dumais, S., and Ng, A., Data-Intensive Question Answering, TREC2001, pp. 393-400, 2001.
5. Dumais, S., Banko, M., Brill, E., Lin, J. and Ng, A., Web Question Answering: Is More Always Better?, 25th Annual International ACM SIGIR Conf. on Research and Development in Information Retrieval, pp. 291-298, Tampere, Finland, 2002.
6. Ebbinghaus, H., Memory: A Contribution to Experimental Psychology, Thoemmes Press, 1913.
7. Hamilton, J. D., Time Series Analysis, Princeton University Press, 1994.
8. Kleinberg, J., Bursty and Hierarchical Structure in Streams, Data Mining and Knowledge Discovery, Vol. 7 Iss. 4, Kluwer Academic Publishers, 2003.
9. Kleinberg, J., Temporal Dynamics of on-line information streams. In Data Stream Management: Processing High-Speed Data Streams, Springer, 2005.
10. Kwok, C., Etzioni, O. and Weld, D., Scaling Question Answering to the Web, 10th International World Wide Web Conf., pp. 150-161, Hong Kong, 2001.
11. Radev, D. R., Qi, H., Zheng, Z., Blair-Goldensohn, S., Zhang, Z., Fan, W., and Prager, J., Mining the web for answers to natural language questions, Tenth International Conf. on Information and Knowledge Management, 2001.

A Probabilistic Reasoning Approach for Discovering Web Crawler Sessions

Athena Stassopoulou[1] and Marios D. Dikaiakos[2]

[1] Department of Computer Science, Intercollege, Cyprus
[2] Department of Computer Science, University of Cyprus, Cyprus
stassopoulou.a@intercollege.ac.cy, mdd@cs.ucy.ac.cy

Abstract. In this paper we introduce a probabilistic-reasoning approach to detect Web robots (crawlers) from human visitors of Web sites. Our approach employs a Naive Bayes network to classify the HTTP sessions of a Web-server access log as crawler or human induced. The Bayesian network combines various pieces of evidence that were shown to distinguish between crawler and human HTTP traffic. The parameters of the Bayesian network are determined with machine learning techniques, and the resulting classification is based on the maximum posterior probability of all classes, given the available evidence. Our method is applied on real Web logs and provides a classification accuracy of 95%. The high accuracy with which our system detects crawler sessions, proves the robustness and effectiveness of the proposed methodology.

1 Introduction and Overview

In this paper, we introduce a novel approach that addresses successfully the challenging problem of automatic crawler detection using probabilistic modeling. In particular, we construct a *Bayesian network* that classifies automatically access-log sessions as being crawler- or human-induced. To this end, we combine various pieces of evidence, which, according to earlier studies [1], were shown to distinguish the navigation patterns of crawler and human user-agents of the World-Wide Web. Our approach uses machine learning to determine the parameters of our probabilistic model. The resulting classification is based on the maximum posterior probability of each class (crawler or human), given the available evidence.

To the best of our knowledge, this is one of the few published studies that propose a crawler detection system, and the only one that uses a probabilistic approach. An alternative approach that is based on decision trees, was proposed by Tan and Kumar in [7]. The authors applied their method with success on an academic access-log collected over a period of one month in year 2001.

As it will be evident from the following sections, the application of a probabilistic approach such as Bayesian Networks, is well suited for the particular domain, due to the high degree of uncertainty inherent in the problem. The Bayesian Network does not merely output a classification label, but a probability distribution over all classes by combining prior knowledge with observed data.

G. Dong et al. (Eds.): APWeb/WAIM 2007, LNCS 4505, pp. 265–272, 2007.

This probability distribution allows decisions to be made about the final classification based on how "confident" the classification is, as demonstrated by the probability distribution. For example, one need not accept weak classifications where the resulting posterior probability is less than a pre-defined minimum.

The remaining of this paper is organized as follows. In the remaining of this section, we present an overview of our approach and describe its pre-processing steps. The proposed Bayesian network classifier is introduced in Section 2. A discussion of our experiments and experimental results is given in Section 3, and we conclude in Section 4.

Overview: The goal of this work is to classify automatically an HTTP user-agent either as a crawler or a human, according to the characteristics of that agent's visit upon a Web server of interest. These characteristics are captured in the Web-server's *access logs*, which record the HTTP interactions that take place between user agents and the server. Each access-log captures a number of sessions, where each *session* is a sequence of requests issued by a single user-agent on a particular server, i.e. the "click-stream" of one user [6]. A session ends when the user completes her navigation of the corresponding site. *Session identification* is the task of dividing an access log into sessions. This is usually performed by grouping all requests that have the same IP address and using a *timeout* method to break the click-stream of a user into separate sessions [6].

Undoubtedly, there is inherent uncertainty in this approach and in any method used to identify Web sessions based on originating IP addresses. For instance, requests posted from the same IP address during the same time period do not come necessarily from the same user-agent [6]: sometimes, different user-agents may use the same IP address to access the Web (for instance, when using the same proxy server); in those cases, their activity is registered as coming from the same IP address, even though it represents different users. Also, session identification based on the heuristic timeout method carries a certain degree of uncertainty regarding the end of a user-agent's navigation inside a Web site of interest. Uncertainty in the data and the actual detection problem itself are the reasons that we believe a probabilistic approach is an ideal application to this problem.

Our system uses training to learn the parameters of a probabilistic model (Bayesian network) that classifies the user-agent of each Web session as crawler or human. To this end, the system combines evidence extracted from each Web session. Classification is based on the maximum posterior probability given the extracted evidence. The classification process comprises three main phases: (i) Access-log analysis and session identification; (ii) Learning, and (iii) classification. An overview of the functionality of our crawler-detection system is given in Algorithm 1.

2 A Bayesian Network Classifier

Feature Selection and Labeling Training Data: We base our selection of features on the characterization study of crawler behavior reported in [1].

1. Access-log analysis and session identification.
2. Session features are selected to be used as variables (nodes) in the Bayesian network.
3. Construction of the Bayesian network structure.
4. Learning:
   (a) Labeling of the set of training examples. At this step, sessions are classified as crawler- or human-initiated sessions to form the set of examples of the two classes.
   (b) Learning the required Bayesian network parameters using the set of training examples derived from step 4a.
   (c) Quantification of the Bayesian network using the learned parameters.
5. Classification: we extract the features of each session and use them as evidence to be inserted into the Bayesian network model. A probability of each session being a crawler is thus derived.

Algorithm 1. Crawler detection system

These features (attributes) are extracted for each session and provide the distinguishable characteristics between Web robots and humans. They are as follows: (i) *Maximum sustained click rate*: This feature corresponds to the maximum number of HTML requests (*clicks*) achieved within a certain time-window inside a session. The intuition behind this is that there is an upper bound on the maximum number of clicks that a human can issue within some specific time frame t, which is dictated by human factors. To capture this feature, we first set the time-frame value of t and then use a *sliding window of time t* over a given session in order to measure the maximum sustained click rate in that session. The *sliding window* approach starts from the first HTML request of a session and keeps a record of the maximum number of clicks within each *window*, sliding the window by one HTML request until we reach the last one of the given session. The maximum of all the maximum clicks per window gives the value of this attribute/feature. (ii) *Duration of session*: This is the number of seconds that have elapsed between the first and the last request. Crawler-induced sessions tend to have a much longer duration than human sessions. Human browsing behavior is more focused and goal-oriented than a Web-robot's. Moreover, there is a certain limit to the amount of time that a human can spend navigating inside a Web site. (iii) *Percentage of image requests*: This feature denotes the percentage of requests to image files (e.g. jpg, gif). The study in [1] showed that crawler requests for image resources are negligible. In contrast, human-induced sessions contain a high percentage of image requests since the majority of these image files are embedded in the Web-pages they are trying to access.(iv) *Percentage of pdf/ps requests*: This denotes the percentage requests seeking postscript(ps) and pdf files. In contrast to image requests, some crawlers, tend to have a higher percentage of pdf/ps requests than humans [1]. (v) *Percentage of 4xx error responses*: Crawlers have a higher proportion of 4xx error codes in their requests. This can be explained by the fact that human users are able to recognize, memorize and avoid erroneous links, unavailable resources and servers [1]. (vi) *Robots.txt file request*: This feature denotes whether a request to the *robots.txt* file was made during a session. It is unlikely, that any human would check for this file, since there is no link from the Web-site to this file, nor are (most) users aware of its existence. Earlier studies showed that the majority of crawlers do not request the *robots.txt* file and so it is the *presence* of a robots.txt request in a session that will have the greater impact on it being classified as *crawler*. Therefore, a strong feature for determining the identity of a session as crawler-induced is the access to the *robots.txt*.

These features form the nodes (variables) of our Bayesian network. The Bayesian network framework enables us to combine all these pieces of evidence and derive a probability for each hypothesis (crawler vs. human) that reflects the total evidence gathered.

Our *training dataset* consists of a number of sessions, each one with its associated label (crawler or human). Since the original dataset contained thousands of sessions, it was prohibitively large to be labeled manually. Therefore, we developed a semi-automatic method for assigning labels to sessions, using heuristics. All sessions are initially assumed to be *human*. Then, we took into account a number of heuristics to label some of the sessions as crawlers: (i) IP addresses of known crawlers; (ii) The presence of HTTP requests for the Robots.txt file; (iii) Session duration values extending over a period of three hours; (iv) An HTML-to-image request ratio of more than 10 HTML files per image file.

It should be noted that we only use the first of the heuristics above to determine conclusively the label of the session as *crawler*. The other heuristics are used to give a recommended labeling of the session as *crawler*. These latter sessions are then manually inspected by a human expert to confirm or deny the suggested crawler labeling. By this semi-automatic method we aimed at minimizing the noise introduced in our training set.

Network Structure: Bayesian Networks [4] are directed acyclic graphs in which the nodes represent multi-valued variables, comprising a collection of mutually exclusive and exhaustive hypotheses. The arcs signify *direct dependencies* between the linked variables and the direction of the arcs is from *causes* to *effects*. The strengths of these dependencies are quantified by conditional probabilities. *Naive Bayes* is a special case of a Bayesian network, where a single cause (the "class") directly influences a number of effects (the "features") and the cause variable has no parents. In our proposed Bayesian network for crawler detection, each child node corresponds to one of the features presented earlier, whereas the root node represents the *class* variable. Having defined the structure of the network, we have to (i) Discretize all continuous variables; (ii) Define the conditional probability tables that quantify the arcs of the network. Subsequently, we show how we use machine learning to achieve these tasks.

Learning Network Parameters: The learning phase of the system uses the training data that have been created as described above. The training data set consists of a number of sessions, each one with its associated label (crawler or human). For each of these sessions, we obtain the values of each of the features, described above, and which are represented as nodes in the Bayesian network. We use the data for variable quantization, based on the entropy, as well as for learning the conditional probability tables, as described in the next two sections.

Variable Quantization: Since, in this implementation, the Bayesian Network is developed for discrete variables, the continuous variables need to be quantized-divided into meaningful states (meaningful in terms of our goal, i.e. to detect crawlers). One well-known measure which characterizes the purity of the class membership of different variable states is *information content* or *entropy* [3]. The number and range of classes which result in the minimum total weighted entropy were chosen to quantize the variable. This minimum entropy principle was applied on all the continuous variables (nodes), i.e. on five out of our six features: $Clicks$, $Duration$, $Images$, PDF/PS and $Code\ 4xx$.

Conditional Probabilities: Having constructed the network nodes, we need to define the conditional probabilities which quantify the arcs of the network. More specifically, we need to define the *a priori* probability for the root node, $P(Class)$ as well as the conditional probability distributions for all non-root nodes: $P(Clicks|Class)$, $P(Duration|Class)$, $P(Images|Class)$, $P(PDF/PS|Class)$, $P(Code\ 4xx|Class)$. Each of these tables gives the conditional probability of a child node to be in each of its states, given all possible parent state combinations. We derived these probabilities from statistical data. For example, the conditional probability of $Duration$ being in class (state) 1 given $Class = Crawler$, is determined from data, by counting the number of Crawler examples with a duration within class 1, and so on.

Classification: Once the network structure is defined and the network is quantified with the learned conditional probability tables, we proceed with the classification phase of our crawler detection system. For each session to be classified, we extract the set of six features that characterize the behavior of clients and that form the variables of our Bayesian Network. As described above, the network contains only discrete variables whereas the first five of the six features are continuous-valued. Each of these feature values is therefore mapped on to a discrete state according to the ranges derived by the quantization step descrbed earlier.

Following this step, each session is now characterized by six features represented as values of discrete variables corresponding to the Bayesian network. In order to classify a session, each variable in the network is instantiated by the corresponding feature value. The Bayesian network then performs inference and derives the $belief$ in the $Class$ variable, i.e. the posterior probability of the $Class$ to take on each of its values given the evidence (features) observed. In other words we derive: $P(Class = crawler|evidence)$ and $P(Class = human|evidence)$. The maximum of the two probabilities is the final classification given to the session.

3 Experimental Results

In this section we present the experiments performed in order to apply our methodology and evaluate the performance of our crawler detection system.

Training Data sets: For the purposes of evaluating the performance of our crawler detection system, we obtained access logs from two servers of two academic institutions: the University of Toronto and the University of Cyprus. The access logs were processed by our log analyzer to extract the sessions. These sessions, the majority being from the University of Toronto, were used for training. Sessions were then labeled using our approach described earlier. The learning stage proved to be challenging task. The problem encountered with this stage is one of *class imbalance* [5]. The data sets present a class imbalance when there are many more examples of one class than of the other. It is usually the case that this latter class, i.e. the unusual class, is the one that people are interested in detecting. Because the unusual class is rare among the general population, the class distributions are very skewed [5]. The study reported in [1] have concluded that crawler activity in access logs amount to less than 10 per cent of the total number of requests. To tackle the problem of imbalanced data sets we used *resampling* and adopted two resampling approaches: random oversampling and random undersampling. We performed 5 experiments, based on resampling (both oversampling and undersampling) at various ratios.

Table 1 shows the number of Crawler and Human sessions in each of the 5 training data sets created via resampling. The last column shows the prior probability distributions of variable *Class*, considering the distribution of sessions actually used for training.

We constructed five Bayesian network classifiers, one for each experiment. The networks had the same structure but differed in their parameters, i.e. prior probabilities, conditional probability tables and quantization ranges. Each time a new training data set was introduced, new network parameters were derived using training on the new set.

Testing the system: A different access log, from the ones not used during training, was randomly chosen for testing. Since the majority of the sessions used for training were extracted from the University of Toronto log, we have chosen a different institution server altogether to evaluate our detection system. This access log used for testing was obtained from the University of Cyprus and spanned a

Table 1. Data sets used for five experiments with and without resampling

Data Set No.	No. Distinct Humans	No. Distinct Crawlers	No. Humans used in training	No. Crawlers used in training	Prior Probabilities: (Human, Crawler)
1	10106	988	10106	988	(0.91, 0.09)
2	10106	988	10106	1784	(0.85, 0.15)
3	10106	988	10106	10106	(0.5, 0.5)
4	10106	988	5599	988	(0.85, 0.15)
5	10106	988	988	988	(0.5, 0.5)

Table 2. Evaluation metrics of each Bayesian network classifier

Classifier	Recall	Precision	$F_1 - measure$
C1	0.80	0.92	0.855
C2	0.81	0.93	0.866
C3	0.95	0.86	0.903
C4	0.81	0.93	0.866
C5	0.95	0.79	0.863

period of one month. A human expert did an entirely manual classification of each session, extracted by our log analyzer from this the testing set, in order to provide us with the ground truth by which we were to evaluate our classifier's performance. It should be noted that we did not do any resampling for the testing.

We tested the performance of all five Bayesian networks (one for each data set), on the same testing dataset[1]. The testing set contained 685 actual human sessions and 99 actual crawler sessions, as labeled by an independent human expert. Throughout this section we will refer to the 5 classifiers as follows: (i) Classifier $C1$: Obtained using learning of Data set 1 (no resampling); (ii) Classifier $C2$: Obtained using learning of Data set 2 (oversampling to 15%); (iii) Classifier $C3$: Obtained using learning of Data set 3 (oversampling to 50%-equally represented classes); (iv) Classifier $C4$: Obtained using learning of Data set 4 (undersampling to 85%); (v) Classifier $C5$: Obtained using learning of Data set 5 (undersampling to 50%-equally represented classes).

Two metrics that are commonly applied to imbalanced datasets to evaluate the performance of classifiers is *recall* and *precision*. These two metrics are summarized into a third metric known as the F_1-measure [8]. The values of recall, precision and F_1-measure obtained by classifiers $C1, \ldots, C5$ are given in Table 2.

As it can be seen from table 2, our crawler detection system yields promising results with both recall and precision being above 79% in all experiments performed. The lowest $F1$-measure is obtained by C1 when we train the system with the dataset without resampling. The prior probability of a session to be *Human* in that dataset was 91% and the classifier was therefore biased towards humans. It missed only 7 out of the 685 *Human* sessions but sacrificed recall, by missing 20 out of the 99 actual *Crawler* sessions. By resampling so that the *Crawler* class amounts to 85% of the sessions (either via oversampling as in C2 or by undesampling as in C4) we have slightly improved results compared to C1. Both C2 and C4 have the same precision and recall. The best results are obtained by C3, which was trained using oversampling of *Crawlers* so that they reach the number of *Human* examples in the original set. The recall, i.e. the percentage of crawlers correctly classified increases dramatically to 95%, with 94 sessions correctly classified as *Crawlers* out of 99 actual crawlers. This causes a decrease in precision, which is nevertheless not so dramatic. The same recall as C3 is achieved by C5 which was trained by undersampling *Humans* so that both classes are again, equally represented. However, this caused a significant decrease in precision to 79%, i.e. we have

[1] The networks were implemented using the $Ergo^{TM}$ tool [2].

an increase in the number of false positives, i.e. *Humans* incorrectly classified as *Crawlers*. The significant decrease in precision of C5, is not surprising since, with random undersampling there is no control over which examples are eliminated from the original set. Therefore significant information about the decision boundary between the two classes may be lost. The risk with random oversampling is to do overfitting due to placing exact duplicates of minority examples from the original set and thus making the classifier biased by "remembering" examples that were seen many times. The are other alternatives to random resampling which may reduce the risks outlined above. An investigation and a comparison of the various resampling techniques is beyond the scope of the current paper.

4 Conclusion

In this paper we have presented the use a Bayesian network, for detecting Web crawlers from access logs. This Bayesian approach is well suited for the particular domain due to the high degree of uncertainty inherent in the problem. Our system uses machine learning to determine the parameters of the Bayesian network that classifies the user-agent of each Web session as crawler or human. The system combines evidence extracted from each Web session to determine the class it belongs to. The Bayesian network does not merely output a classification label, but a probability distribution over all classes by combining prior knowledge with observed data. We have used resampling to counter the class imbalance problem and developed five classifiers by training on five different datasets. The high accuracy with which our system detects crawler sessions, proves the effectiveness of our proposed methodology.

References

1. M. D. Dikaiakos, A. Stassopoulou, and L. Papageorgiou. An Investigation of WWW Crawler behavior: Characterization and Metrics. *Computer Communications*, 28(8):880–897, May 2005.
2. Noetic Systems Incorporated. http://www.noeticsystems.com/ergo/index.shtml.
3. T. M. Mitchell. *Machine Learning*. McGraw Hill Companies Inc., 1997.
4. J. Pearl. *Probabilistic reasoning in intelligent systems: Networks of plausible inference*. Morgan Kaufmann Publishers Inc., 1988.
5. F. J. Provost and T. Fawcett. Analysis and visualization of classifier performance: Comparison under imprecise class and cost distributions. In *Proceedings of the Third International Conference on Knowledge Discovery and Data Mining*, pages 43–48, 1997.
6. J. Srivastava, R. Cooley, M. Deshpande, and P.-N. Tan. Web Usage Mining: Discovery and Applications of Usage Patterns from Web Data. *SIGKDD Explorations*, 1(2):12–23, January 2000.
7. P.-N. Tan and V. Kumar. Discovery of Web Robot Sessions Based on their Navigational Patterns. *Data Mining and Knowledge Discovery*, 6(1):9–35, January 2002.
8. P.-N. Tan, M. Steinbach, and V. Kumar. *Introduction to Data Mining*. Addison-Wesley, 2005.

An Exhaustive and Edge-Removal Algorithm to Find Cores in Implicit Communities

Nan Yang, Songxiang Lin, and Qiang Gao

The School of Information, Renmin University of China
No 59, Zhongguancun Street, Beijing, China
8601-82500902
yangnan@ruc.edu.cn

Abstract. Web community is intensely studied in web resource discovery. Many literatures use core as the signature of a community. A core is a complete bipartite graphs, denoted as Ci,j. But discovery of all possible Ci,j in the web is a challenging job. This work has been investigated by trawling [1][2]. Trawling employs repeated elimination/generation procedure until the graph is pruned to a satisfied state and then enumerate all possible Ci,j. We proposed a new method that uses exhaustive and edge removal method. Our algorithm avoids scanning dataset many times. Also, we improve crawling method by only recording potential fans to save disk space. The experiment result show that the new algorithm works properly and many new Ci,j can be found by our method.

Keywords: Web communities, Link analysis, Complete Bipartite Graph.

1 Introduction

Web is a huge information resource and increases dramatically. Although the growth of web seems chaos, but in fact web shows a great deal of self-organization [3]. Web communities are very important structure in web. There are well-known, explicitly-defined communities, for example, users interested in mercedes-benz cars or in java development. Most of them manifest themselves as newsgroups, webrings, or as resource list in directories such as Yahoo! and Infoseek [1]. Definitely, the web communities are set of pages which created by a group or people with common interest. There many literatures had got involved in web communities, for example, HITS [4][5], Companion[6][7], max flow/min cut[8] and trawling algorithm[1][2]. But there are many communities are implicit and their number overcomes that of explicit ones. Trawling algorithm mainly focused on implicit communities.

In this paper, we have analyzed some forms of structure which are not considered by trawling. Then by borrowing edge-removal idea of Newman [9][10][11], we introduced a new extraction algorithm. After a subgraph collected from web graph, we then check the possible cores in it. At each process some edges are removed from web graph. We repeat this process until the web graph is empty. Our background is same as trawling method. But our method is different from trawling in several ways. First, we improve the crawling by only recording the potential fans so that we can

G. Dong et al. (Eds.): APWeb/WAIM 2007, LNCS 4505, pp. 273–280, 2007.

save a lot of disk space. Second, we extract cores based on edge removal which reduce the times of scanning dataset. Like Kumar [2], we also use term frequency to evaluate whether or not the potential cores can organize communities.

The outline of the paper is as follow. In section 2, we introduce some basic related knowledge. In section 3, we describe the preparing procedure of dataset and link database. In section 4, we introduce new algorithm. In section 5, we describe topics of communities. In section 6, we arrange the dataset and experiment and some result examples. Conclusions and future works are shown in section 7.

2 Preliminaries

Web can be abstracted to a large directed graph G=(V, E). V is the set of nodes, E is the set of edges. A pair of nodes (u, v)∈E means that there is a hyperlink between u and v. A bipartite graph is a graph whose vertex set can be partitioned into two sets, which we denote F and C. Every directed edge in the graph is directed from a vertex u in F to a vertex v in C, depicted in Fig.1(a).

A bipartite graph is dense if many of the possible edges between F and C are present. The trawling algorithm is based on the hypothesis: the dense bipartite graphs that are signatures of web communities contain at least one core. A core is a complete bipartite graph with at least i vertices in F and at least j vertices in C. Thus, the core is a small (i, j)-sized complete bipartite graph, denoted as Ci,j. We will find a community by finding cores, and then use the cores to find the rest of the community. According to Rajagopalan [12], the data mining graph is bipartite with left hand side and right hand side, denoted as LHS and RHS respectively.

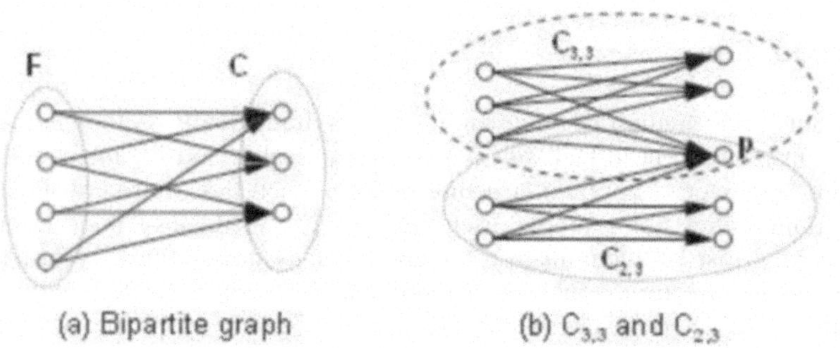

Fig. 1. (a) Birpartite Graph. (b) C2,3 and C3,3.

3 Dataset Preparation

The web pages are collected by a web crawler [13]. We only extract urls information from html text. We extract static links which begin with “http://” and are included within “<a href=” and “>”. The length of url is limited to below 100 characters and we don’t repeatedly extract the links belong to same domain. For instance, while www.edu.cn/xxx and www.edu.cn/yyy arrive, we only reserve www.edu.cn/xxx. An edge have a form of pair <source url, destination url> and set of edges are stored into edge file.

We use 128bit MD5 hash fingerprint as id to record a url. Every edge will occupy 32bytes. To get in-degree and out-degree of a page, we use two datasets to present web graph. One is the set of edges ordered by source id, denoted as DSL and other by destination id, denoted as DSR. Due to a set of edges stored contiguously in DSL(DSR), it is easy to get out-degree(in-degree). We employ BerkeleyDB to manage dataset and use its BTREE access mode with sorted key. This mode maintain key in sorted order automatically.

Because of mirror web sites and duplicated pages, there are many pages are derived from same resources [14]. Many of them are highly inter-connected and tend to form community structure. These structures are value resources but help little to find implicit communities. Hence, deletion of these pages before core extraction is necessary. There are many algorithms to delete mirror or near-duplicated pages. We chose a method as in [9]. Pages with out-links above 8 are taken into account and their common out-links is exceed to 85 percent, to say the two pages are mirror.

Many researches have showed that the distribution of in-degree obeys power-law [3]. This law was used in many community discovery algorithms to prune the dataset. The pages with very low in-links and high in-links are pruned. Too low in-links means that page is less important. And too high in-links means that page belong to popular web sites, such as www.google.com, www.yahoo.com etc.

4 Algorithm on Edge Removal

4.1 Defects of Trawling Algorithm

The complete directed bipartite graph is a metaphor of community. We call complete bipartite graph as a core, denoted as $C_{i,j}$, where i and j are nodes in fans and centers. How to find all possible cores is one of the important problems. In [1][2], Kumar employs the criterion of a core. Consider the example of a $C_{4,4}$. Let u be a node of in-degree exactly 4. Then u can belong to a $C_{4,4}$, if and only if the 4 nodes that point to it have a neighborhood intersection of size at least 4.

The trawling algorithm has three defects. First, some $C_{i,j}$s will be missing in certain subgraph. Let's look subgraph in Fig. 1(b), node p belongs to both $C_{3,3}$ and $C_{2,3}$. Unfortunately, node p has in-degree 5, according to trawling criterion, it would be pruned when find $C_{3,3}$ and $C_{2,3}$ because p does not have in-degree 3. If node p is pruned, the core $C_{3,3}$ and $C_{2,3}$ will be missing. Second, the removal of nodes will destroy the structure of other cores. For example, in Fig. 1(b), when we find $C_{3,3}$ all nodes related to $C_{3,3}$ are removed. Node p is also removed and the structure of $C_{2,3}$ is destroyed. Third, the enumeration by combining i and j is at high cost because every scan of dataset only fans with out-degree j and centers with in-degree i are taken into account. The next subsection is the new algorithm which is proposed to overcome these shortcomings.

4.2 Exhaustive Algorithm and Removal of Edges

Derived from [4][5][6], they use node removal method to find communities in a graph We propose a new algorithm to find all possible Ci,js from a subgraph. First, we employ an exhaustive idea to find all possible cores in one scan of dataset. Second, we delete edges instead of nodes. Every time scan, we will construct a bipartite subgraph from a chosen node p. Then an exhaustive algorithm will extract all Ci,js and then related edges are removed. Our method avoids scanning whole dataset whenever a set of edges are deleted. Any node will not be deleted unless all edges associated are deleted.

Before we describe our algorithm, some definitions and notations should be given first. BG is a bipartite graph. We use L(BG) to denote LHS of BG and R(BG) to denote RHS of BG. We use C(x, BG) to denote the nodes in R(BG) pointed by x and P(x, BG) to denote the nodes in L(BG) pointing to x. S(p, BG) denote the nodes in L(BG) that point to the set of nodes in R(BG) which are also pointed by node x. Let BGw to denote the web graph, BGs to denote constructed bipartite graph and BGc denote the bipartite graph of Ci,j and contain node p.

Definition BGs: For any given node p$\in$L(BGw), BGs is a bipartite subgraph constructed from p, where L(BGs)={p, S(p, BGw)} and R(BGs)=C(p, BGw).

Definition BGc: For any given Ci,j and node p, BGc is a bipartite subgraph of Ci,j, where L(BGc)=LHS of Ci,j , R(BGc)=RHS of Ci,j and p$\in$L(BGc).

The **Definition BGs** explains that procedure of constructing BGs has two steps. Step 1 is to get the children of p in BGw to construct LHS of BGs. Step 2 is to get the siblings of node p, and then use p and its siblings construct RHS of BGs.

Theorem 1: If BGs is a constructed bipartite subgraph of node p, then all possible BGc$\subseteq$BGs.

Proof: For any u $\in$L(BGc), we have C(u, BGc)=R(BGc), likewise for any v$\in$ R(BGc) we have P(v, GBc)=L(BGc). Because p$\in$(BGc), C(p, BGc)=R(BGc). BGs is constructed from BGw and the L(BGs) is all children of node p in BGw, therefore R(BGc)$\subseteq$R(BGs) . For any q(q$\neq$p, q$\in$L(BGs)), we have C(q, BGc)=R(BGc), so q is sibling of p. Because L(BGs) is p and all its siblings derived from BGw, L(BGc)$\subseteq$L(BGs).

According to **Theorem 1**, two-step construction of subgraph is complete to include all possible Ci,js. Then next job is how to extract all possible BGc from BGs. In trawling, Kumar use enumeration by combining of i and j, from intersection of potential fans with j degree or centers with i degree to find Ci,j. Here we don't use enumeration method by employ a function. We introduce a function U (x, y, t) and V(z, x, y, t), which have following form:

$$U(x,y,t)=\begin{cases} x\cap y; |x\cap y|\geq t \\ x; else \end{cases} . \tag{1}$$

$$V(z,x,y,t) = \begin{cases} z \cup y; |x \cap y| \geq t \\ z; else \end{cases} . \quad (2)$$

Here x, y and z are sets. t is an integer(t>0) to give a threshold value.

We define two sets as H and L and p1=p. Let S(p1, BGs)={p2, p3, …, pn}. For a certain t, we have:

H1=C(p1, BGs); L1={p1}
H2=U(H1, C(p2, BGs), t); L2=V(L1, H1, C(p2, BGs), t)
H3=U(H2, C(p2, BGs), t); L3=V(L2, H2, C(p3, BGs), t)
⋮
Hn=U(Hn-1, C(pn, BGs), t); Ln=V(Ln-1, Hn-1, C(pn, BGs), t)

After n iteration, if Hn $\neq \Phi$, the intersection of all nodes in Hn should be great than or equal to t, that is to say, a C(| Ln |, | Hn |) is found. Then we can check if | Ln |≥i and | Hn |≥j, we output C(Ln, Hn). We choose C(p1, BGs) as initial value of H, it is because |C(p1, BGs)|=n. So that guarantees to find max size of neighborhood intersection.

Theorem 2: The BGs is constructed bipartite graph from p, {BGc} is the set of BGc, n=|L(BGs)| and m=|R(GBs)|. If there exist {BGc}, by apply U and V function to all nodes in L(BGs) with t={m, m-1,…,j}, and output C(L, H) with |Ln|≥i, {BGc} can be found by Hn and Ln.

Proof: Hn is all possible intersections with |Hn|≥t. We iterate the calculation of Hn and Ln with t from m, m-1, …, j. Ln is the set of node selected from L(BGs) when |Hn|≥t. Therefore, if a BGc exist and L(BGc)$\subseteq$L(BGs) and R(BGc)$\subseteq$R(BGs), the BGc can be found by Hn and Ln.

Theorem 1 tell us that the constructed bipartite graph BGs from a node p is complete to include all possible {BGc} and **Theorem 2** tell us that by enumerating t from m, m-1, …, j to calculate Hn and Ln, where n=L(BGs) and m=R(BGs), we can find all possible cores. Following is the detail description of our algorithm with specified i and j.

4.3 The Implementation of Algorithm

We need two pair of dataset to store BGw and BGs respectively. Each BG is represented by a pair of dataset. The BGw is stored in dataset DSL and DSR in hard disk. BGs is stored in dataset EOS and EOD in main memory. All operation is based on both DSL and DSR. We chose a page from DSL, usually the first one in dataset DSL. Then we apply following algorithm until both DSL and DSR are empty. Thus, either a BGc is found or not, there must be a collection of pages are removed from both dataset. Our algorithm consists of following steps and we repeat all steps until web graph empty.

Step1: Get a node p from dataset DSL, if DSL is empty to the end of algorithm.
Step2: If |C(p, BGw)|≥i then construct BGs.
Step3: Prune BGs with i and j.
Step4: Extract BGc by U and V function.
Step5: Delete edges from BGw.

5 Decision Topics on Communities

After completion of community's extraction, next job is to decide topics of each community. Due to the communities generated from only linkage information, eventually the page content is used to found topics. In the researches mentioned before, this job is fulfilled by human effort. So a mechanized process for dealing with over a hundred thousand communities is necessary. Our intuition is very simple, based on terms frequency. We can count the occurrences of terms in each page and rank the frequencies of terms. From the ranking list, we choose top N terms to make a term list. So the topic of each community will be correspondent to a term list. A stop list is needed. Because many words like 'a, the' have not the meaning, they don't help to find topic. Term in different field should be assigned different weight, for example, the terms appear in the title field will have higher weight than in other field.

6 Experiment and Result

6.1 Preparation Dataset

During crawling procedure, crawler only deals with potential fans and reserve linkage information. Crawling process continue until disk full. Because in this experiment we like to verify the feasibility and effectiveness of our algorithm, so we collect part of web graph that contain about 6.7 million pages and 9.4 million edges. Owing to one source page followed by at least 6 destination pages, the resulting nodes of graph will be great than 6.7 million. We delete mirror or near-duplicated pages. Then we create two dataset DSL and DSR and prune centers by DSR. After mirror deletion and in-degree pruning, the dataset contain 6.9 million edges.

6.2 Cores Extraction

After dataset preparation, we apply new algorithm on dataset. We run 3 times with i,j=2, i,j=3 and i,j=4 respectively. Then we get 149K, 29K and 10K Cores. From the cores extracted, we can find many are not included by trawling. Fig. 2(a) depicts the distribution of cores vs. fans. Fig. 2(b) depicts cores vs. centers. From the curve of in Fig. 2, the distribution of Ci,js versus fans and centers obey power-law.

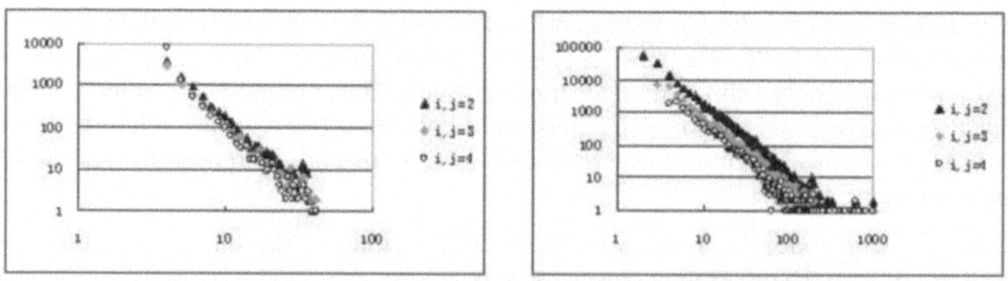

(a) Cores vs. Fans. (b)Cores vs. Centers.

Fig. 2. Cores vs. fans and centers

6.3 Core Examples

The cores extracted from SDL and SDR are numerous. To systematically arrange them into a reasonable structure is still an important job. This job is our next work. In this paper, we chose 3 cores by random. Each of them have topic on "Music", "Agriculture" and "Green building". For each topic, top 3 urls of fans and centers are shown in Table 1.

Table 1. Cores with Topic "Music", "Agriculture" and "Green building"

Music	Fans	http://www.ai88.net/wz/music.htm http://www.6cn.com/web/music_mp3.htm http://www.zpartner.com/data/24.htm
	Centers	http://www.chinamusicnet.com/ http://music.silversand.net/ http://www.mtv114.com/
Agricul-ture	Fans	http://www.3-xia.com/njz/index.asp http://www.animalsci.com/index.4.htm http://www.cqagri.gov.cn/cqagri/index.asp
	Centers	http://www.jjny.gov.cn/ http://www.wzny.gov.cn/ http://www.qjqagri.gov.cn/
Green building	Fans	http://www.asu.edu/fm/greenbuilding.htm http://www.greenbuilder.com/general/BuildingSources.html http://www.usgbc.org/Resources/links.asp
	Centers	http://www.gbapgh.org/ http://www.ci.scottsdale.az.us/greenbuilding/ http://www.builtgreen.org/

7 Conclusions and Future Works

Many communities are implicit and to discover them is a challenge job. In this paper, we discussed related researches and pointed out some defects of trawling algorithm. We proposed a new algorithm under exhaustive idea and removal of edges. A bipartite subgraph is constructed from a chosen page and our algorithm is applied to the subgraph. At each scan of dataset, we can find all possible cores and some edges are removed from web graph. In dataset collection phrase, we improve crawler by dealing with potential fans and save disk space considerably. We also use terms frequency and rank of frequencies to deduce possible topics of communities. The web pages are colleted with about 7 million pages. We have set up an experiment on i, j=2, i, j=3 and i, j=4 and find 149K, 29K and 10K cores respectively.

The web communities are very important structures in web. Finding all possible implicit communities is still a huge project. The future work could be on these aspects. First, the page's text content should be considered with linkage information. Second, the inner structure of html document is taken into account. Third, communities have the hierarchy. Forth, how to deal with the overlap is a worthwhile research.

References

1. Kumar R., Raghavan P., et al.: Trawling the web for emerging cyber-communities. Proceedings of the 8th WWW Conference, Toronto, Canada (1999) 403-415
2. Kumar R., Raghavan P., et al.: Extracting large-scale knowledge base from the web. Proceedings of 25th VLDB Conference, Edinburgh, Scotland (1999) 639-650
3. Broder A., Kumar R., et al.: Graph structure in the web. Computer Networks 33 (1-6) (2000) 309-320
4. Gibson D., Kleinberg J., et al.: Inferring Web Communities from Link Topology. Proceedings of the 9th ACM Conference on Hypertext and Hypermedia, Pittsburgh, PA, USA (1998) 225-234
5. Chakrabarti S., Dom B. E., et al.: Automatic resource compilation by analyzing hyperlink structure and associated text. Computer Networks 30 (1-7) (1998) 65-74
6. Dean J., Henzinger M. R.: Finding Related Pages in the World Wide Web. Proceedings of the 8th WWW Conference, Toronto, Canada (1999) 389-401
7. P K Reddy and Kitsuregawa M.: Inferring Web Community through relaxed-cocition and power-law. Annual Report of KITSUREGAWA Lab (2001) 27-40
8. Flake G. W., Lawrence S., et al.: Efficient Identification of Web Communities. Proceedings of the 6th ACM SIGKDD Conference on Knowledge discovery and data mining, Boston, MA, USA (2000) 150-160
9. Girvan M., Newman M. E. J.: Community structure in social and biological networks. Proc. Natl. Acad. Sci. USA 99 (2002) 7821-7826
10. Newman M. E. J.: Fast algorithm for detecting community structure in networks. Phys. Rev. E 69, 066133 (2004)
11. Newman M. E. J.: Detecting community structure in networks. Europe. Phys. J. B 38, (2004) 321-330
12. http://www.cs.cornell.edu/home/kleinber/web-graph.ps
13. http://perso.wanadoo.fr/sebastien.ailleret/index-eng.html
14. Broder A. Z., Glassman S. C., et al.: Syntactic Clustering of the Web. Computer Networks 29(8-13) (1997) 1157-1166

Active Rules Termination Analysis Through Conditional Formula Containing Updatable Variable

Zhongmin Xiong[1], Wei Wang [1], and Jian Pei[2]

[1] Department of Computing and Information Technology, Fudan University, Shanghai, 200433, China
{Zhmxiong,Weiwang1}@fudan.edu.cn
[2] School of Computing Science, Simon Fraser University, Canada
jpei@cs.sfu.ca

Abstract. While active rules have been applied in many areas including active databases, XML documentation and Semantic Web, current methods remain largely uncertain of how to terminate active behaviors. Some existing methods have been provided in the form of a logical formula for a rule set, but they suffer two problems, (i) Only those variables, which are non-updatable or finitely updatable, can be contained by a formula. (ii) They cannot conclude termination if a rule set only contains some cycles that can be executed in a finite number of times. Many active rule systems, which only contain updatable variables, can still be terminated. This paper presents an algorithm to construct a formula, which can contain updatable variable. Also, a method is proposed to detect if a cycle can only be executed in a finite number of times. Theoretical analysis shows more termination cases, which is indetctive for existing methods, can be detected by our method.

1 Introduction

Recently, active rules have been introduced into many new areas including XML [1, 2], RDF [3], Semantic Web [4], Sensor Database [5], and so on. Rules definition most commonly follows the Event-Condition-Action (ECA) paradigm.

Certain rules are triggered initially and their execution can trigger additional rules or trigger the same rules indefinitely, prevents the rules system termination. Termination is one of characteristics of an active rules set with better behavior [6]. But termination decision of a rule set is an undecidable problem [7].

Some methods propose a translation of the active rules in term of rewriting system, or deductive rules [8, 9]. However, [8] requires the definition of a well-founded term ordering in the term rewriting system and seems a rather complex task. [9] requires the equivalence of translation from ECA rules to deductive rules. This requirement cannot always be determined by a rule set. Other methods introduce some methods based on Petri net or Abstract Interpretation [10, 11]. They need use large complex data structures to examine all properties.

Most of the works about the active rules termination can be classified into two major categories. The former uses the concept of *triggering graph*(TG). [6] assumes that the absence of cycles in a triggering graph guarantees the termination. But if such

G. Dong et al. (Eds.): APWeb/WAIM 2007, LNCS 4505, pp. 281–292, 2007.

a graph has a cycle, it may be terminated. [12] augments a TG with an *activation graph* (AG) where an edge will be removed unless it is in a cycle or reachable from a cycle and it can be re-activated after a self-deactivation. [13] proposes a sophisticated technique for building Activation Graphs, which complements nicely with the techniques presented in [12]. Triggering graph is built by means of a syntactic rules analysis. All these methods [6, 12] have a common drawback: they do not concern whether all of rules can be executed at the same cyclical execution of a cycle.

The methods of the other category, which are based on a formula, concern the above problems. [14] construct a conjunct based on the trigger conditions of two rules. If the conjunct is not satisfiable, the edge between the two rules will be removed. [15, 16] construct a stronger condition for termination decision than [14]. But such a formula constructed by [15, 16], can only contain non-updatable variables or finitely-updatable variables. Unfortunately, how to determine a finitely-updatable variable is known as a NP-problem [16]. From the following examples in this paper, many active rule sets, which only contain updatable variables, can still terminate.

The rest of this paper is organized as follows: In section2, preliminaries for our work are introduced. Section3 presents two motivating examples. Section4 proposes the method to construct a formula. In section5, we show how to analyze the termination of a TG cycle by our methods, and expose our algorithm for termination analysis. Finally, section6 concludes the paper with directions to future work.

2 Preliminaries

2.1 Active Rules

The active rules are structured according to paradigm Event-Condition-Action. An ECA rule has the general syntactic structure: **on** *event* **if** *condition* **do** *actions*

Event is either data operation event (inside of the Database system) or event reported to the system by the outside. *Condition* is a request on the database; it is always expressed as database queries. *Action* is generally composed of a sequence of database updates, or of a procedure containing database updates.

Let r_i and r_j be two rules, r_i *triggers* r_j, if one of the r_i's actions has the corresponding event in r_j. This relation can be described with *Triggering Graph* (TG); TG can be constructed in means of a syntactic analysis of rules [6].

From the following examples, we can see the definitions of a rule set and its associated triggering graph.

2.2 The Rule Execution Model

Coupling modes allow to specify the evaluation moment of the Condition Part (E-C coupling), or the execution moment of the Action Part (C-A coupling). The most frequent modes are shown as follows.

- *immediate*, in which case the condition (action) is evaluated (executed) immediately after the event (condition).
- *deferred*, in which case the condition (action) is evaluated (executed) within the same transaction as the event (condition) of the rule, but not necessarily at

the earliest opportunity. Normally, further processing is left until the end of the transaction.

In this paper, E-C coupling and C-A coupling use *immediate* mode.

Cycle policy addresses the question of what happens when events are signaled by the evaluation of the condition or action of a rule. Database systems always support a *recursive* cycle policy for *immediate* rules. In this case, event signaled in the process of condition and action evaluation make them suspended. Thus, any *immediate* rules monitoring the events can be processed at the earliest opportunity.

The *scheduling* phase of rule evaluation determines what happens when multiple rules are triggered at the same time. In this paper, *scheduling* is *All Sequential.* Thus, the system fires all rule instantiations sequentially.

In this paper, the *rules process* is described as follows. When a rule is triggered, the system evaluates its condition. If the condition is satisfied, the rule is removed from the triggered rules set and its action is performed. If the condition is not satisfied, then *event-consuming* policy is adopted [18]. That is, the system eliminates the rule from the triggered rules set.

Details about the Rule Execution Model presented here can be seen in [17, 18].

3 Motivating Examples

According to Oracle8.1.7 DDL [19], all active rules in these examples are described as Oracle's triggers.

Example 1. Six triggers are defined on four tables: R (A, B, C, D), S (H, L), Q (I, F), T (E, G) and shown in Figure1. Figure 2 illustrates two TG cycles: R_1 $\{r_1, r_2, r_3\}$, R_2 $\{r_4, r_5, r_6\}$.

Under our assumption of active rules' execution semantics presented in section2, R_1 and R_2 cannot be synchronously executed at the same time.

Let us analyze the termination of R_1. *R.D* is an updatable variable of R_1 and is selected to construct a formula. For all rules of R_1, only r_3 changes *R.D* with an updating operation: *R.D* =0. So we can regard r_3 as the first triggered rule of a cyclical execution of R_1. Once r_3 is selected as the current triggered rule again, a cyclical execution of R_1 ends. It is true that the formula as (*R.D* =0) ***AND*** (*R.D* =1) is false, that is, r_3 cannot be really executed again. So R_1 must be termination. Although r_4 can update *R.D*, it cannot be synchronously executed with R_1 at the same time since it cannot be triggering reachable from R_1.

Let us analyze the termination of R_2. *T.E* is an updatable variable of R_2 and is selected to construct a formula. For all rules of R_2, only r_4 changes *T.E* with an update operation: *T.E* =0. So we can regard r_4 as the first triggered rule of a cyclical execution of R_2. Once r_4 is selected as the current triggered rule again, a cyclical execution of R_2 ends.

It is true that the formula as (*T.E* =0) ***AND*** (*T.E* =1) is false, that is, r_4 cannot be really executed again. So R_2 must be termination. Although r_2 can update*T.E*, it cannot be synchronously executed with R_2 at the same time since it cannot be triggering reachable from R_2.

Thus, this rule set always terminates. All attributes in the conditions of this rule set are updatable and no finitely-updatable attribute can be detected by [16]; No edge can be removed from TG by [12], so [12, 16] cannot deal with such a termination case.

```
create trigger r1
after update of A on R
when R.B=1
begin
   update table S
      set H=0,L=1;
   update table R
      set B=0;
end;

create trigger r2
after update of H on S
when S.H=0
begin
    update table R set C=1;
    update table S set H=1;
    update table T set E=1;
end;

create trigger r3
after update of C on R
when R.D=1
begin
  update table R
     set A=0,B=1,D=0;
 end;

create trigger r4
after update of G on T
when T.E=1
begin
  update table Q set I=0;
  update table T set E=0;
   update table R set D=1;
end;

create trigger r5
after update of I on Q
when Q.I=0
begin
   update table Q
     set I=1,F=1;
end;

create trigger r6
after update of F on Q
when Q.F=1
begin
    update table Q set F=0;
    update table T set G=0;
end;
```

Fig. 1. Example1 with Oracle's triggers

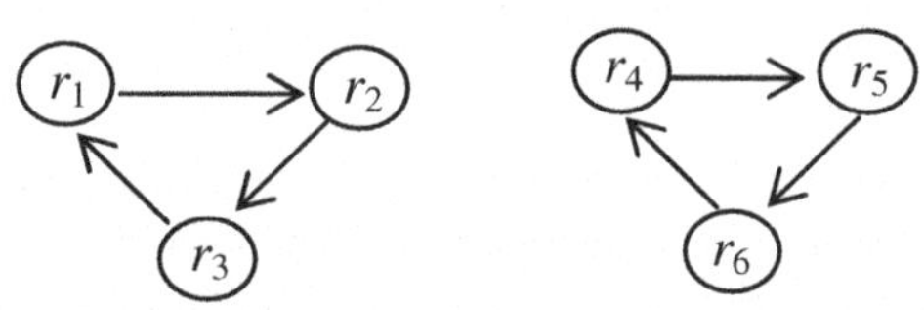

Fig. 2. Triggering Graph for Example1

Example 2. Six triggers are defined on four tables: R (A, B, C), S (H, M, L), Q (I, N), T (E, F, G) and shown in Figure3. Figure 4 illustrates two TG cycles: R_1 $\{r_1, r_2, r_3\}$, R_2 $\{r_4, r_5, r_6\}$.

Let us analyze the termination of R_1. There are three variables in the conditions of all rules in R_1: *R.B*, *S.M* and *S.L*. Firstly, *S.L* is selected to construct a formula. For all rules of R_1, r_1 assigns a value to *S.L* and r_2 updates *S.L* and *S.L* is a variable in the condition of r_3. So we can regard r_1 as the first triggered rule of a cyclical execution of R_1. Because *S.L* is updated during the cyclical execution of R_1 and it should be taken the same value in the formula containing it, the predicates contained in the formula should change their forms; and *S.L* is regarded as its original value and not updated all time. That is, if it has been updated by $+\Delta L$, the other side of a predicate containing *S.L* should be updated by $-\Delta L$. When r_3 is selected as the current triggered rule, the following formula as (*S.L* <2-0.5) ***AND*** (*S.L* =1) is true, r_3 can be executed.

Secondly, *S.M* is selected to construct a formula. Since *S.M* is not updated by this rule set and (*S.M*=0) can be regarded as true, r_2 can be really executed.

Finally, *R.B* is selected to construct a formula. For all rules of R_1, r_2 assigns a value

```
create trigger r1
after update of A on R
when R.B>2
begin
  update table S
    set H=0,L=1;
end;
```

```
create trigger r2
 after update of H on S
 when S.M=0
 begin
   update table R
     set B=2, C=1;
   update table S
     set L=L+0.5;
 end;
```

```
create trigger r3
after update of C on R
when S.L<2
begin
  update table R
    set A=0, B=B-1;
  update table T
    set G=0, F=0;
end;
```

```
create trigger r4
after update of G on T
when T.E=1
begin
  update table Q set I=0;
  update table T set F=F+2;
   update table R set D=1;
end;
```

```
create trigger r5
after update of I on Q
when Q.N>2
begin
  update table T set F=F-1;
end;
```

```
create trigger r6
after update of F on T
 when T.F<2
 begin
   update table T set G=1;
 end;
```

Fig. 3. Example2 with Oracle's triggers

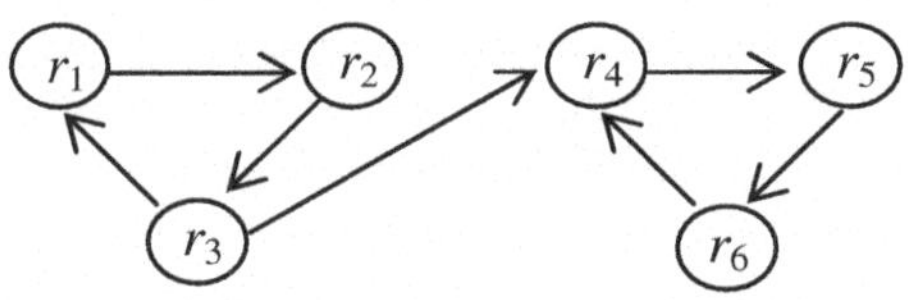

Fig. 4. Triggering Graph for Example2

to it and r_3 updates it and it is a variable in the condition of r_1. So we can regard r_2 as the first triggered rule of a cyclical execution of R_1. When r_1 is selected as the current triggered rule, the following formula as (*R.B*=2) ***AND*** (*R.B*>2+1) is false. So r_1 cannot be really executed and R_1 must be termination.

Let us analyze the termination of R_2. There are three variables in the conditions of all rules in R_2: *T.E*, *Q.N* and *T.F*. Since *T.E* and *Q.N* are not updated by any rule of R_2 or any rule which does not belong to R_2 but can be triggering reachable from R_2, so (*Q.N*>2) ***AND*** (*T.E*=1) can both be regarded as true. These two variables are non-updatable when R_2 is cyclically executed according to active rules' execution semantics presented in section2.

T.F is selected to construct a formula for R_2. No rule of R_2 assigns a value to *T.F*. r_4 and r_5 update *T.F* and *T.F* is a variable in the condition of r_6. Once r_6 is selected as the current triggered rule again, a cyclical execution of R_2 ends. There is a formula as follows: σ = (*T.F*= *T.F* +1) ***AND*** (*T.F* <2)= ((*T.F* +1)<2). The net effect of updates in a cyclical execution of R_2 makes *T.F*= *T.F* +1. Although r_3 assigns 0 to *T.F* and σ =true and R_2 can be cyclically executed once, when R_2 is cyclically executed for its third time, $T.F$ =0 and K=3 make $\sigma = ((T.F + K \times 1)<2)$ =false. So r_6 cannot be really executed and R_2 must be termination.

Thus, the rule set in example2 is always termination. Such a termination case cannot be detected by existing methods [12, 14-16] since all rules in this rule set are not self-deactivate rules [12] and are not non-updatable and finitely-updatable determined by [14-16].

4 Method to Construct a Formula

Definition 1. *A rule r_j is* **directly triggering reachable** *from a rule set R if there is a rule r_i in R such that $< r_i, r_j> \in TG$; a rule r_j is* **triggering reachable** *from a rule set R if it is directly triggering reachable from R or if there is a rule r_t, which is triggering reachable from R, such that $< r_t, r_j> \in TG$.*

Example 3. In example2, r_2 is directly triggering reachable from $\{r_1, r_2, r_3\}$, r_4 is triggering reachable from $\{r_1, r_2, r_3\}$.

Definition 2. *In TG, a path $<r_0, r_n, r_0>$ is called as a* **TG cycle**.

Example 4. In Figure4 of example2, the path $< r_1, r_2, r_3, r_1>$ is a TG cycle. To be simplified, it is always indicated by a symbol as R_1 $\{r_1, r_2, r_3\}$ in this paper.

Theorem 1. *If a rule r does not belong to a TG cycle R and cannot be triggering reachable from R, r cannot be synchronously executed with R at the same time.*

Proof. The conclusion can be proven by contradiction. Assuming that r can be executed following a rule r' which belongs to R, it must be followed that r' can trigger r according to active rules' execution semantics presented in section2. That is, r can be triggering reachable from R by the notion of triggering reachable. This is in contradiction with the precondition of theorem1. So the conclusion holds.

Through theorem1, those variables in the conditions of all rules in R, which are selected to construct a formula for R, should not be updated by r.

Definition 3. *A variable X is a* **non-updatable variable** *of a TG cycle R, if X is in the conditions of rules in R but is not in the actions of any rule in R or any rule that does not belong to R but can be triggering reachable from R.*

Based on theorem1, if a variable X is a non-updatable variable of a TG cycle R, X must not be updated when R is cyclically executed.

Definition 4. *A variable X is an* **updatable variable** *of a TG cycle R, if X is both in the conditions and in the actions of rules in R but is not in the actions of any rule that does not belong to R and can be triggering reachable byR.*

Example 5. In example2, there are three variables in the conditions of all rules in R_2: $T.E$, $Q.N$ and $T.F$. Since $T.E$ and $Q.N$ are not updated by any rule of R_2 or any rule which does not belong to R_2 but can be triggering reachable from R_2, so $Q.N$ and $T.E$ are both non-updatable variables of R_2. However, $T.F$ is an updatable variable of R_2.

An updatable variable X of a TG cycle R should be taken as the same value in a formula based on it for R. So we define the following notion.

Definition 5. *In a TG cycle R, a rule r is a* **dividing point** *of an updatable variable X if the action of r assigns an original value to X.*

Example 6. In the rule set of example2, *S.L* is an updatable variable of R_1 and r_1 is a dividing point of *S.L.*

4.1 Onstruct a Formula Based on Non-updatable Variable

Since no rule, which belongs to a TG cycle *R* or does not belong to *R* but can be triggering reachable from *R*, can update a non-updatable variable *X*, a formula for *R* based on a non-updatable variable *X* only consists of predicates containing *X* in the conditions of rules in *R*. To construct such a formula can adopt the method in [15, 16].

Example 7. In example2, *S.M*, *T.E* and *Q.N* are not updated by any rule in the rule set of example2. So, the following formula based on non-updatable variable can be constructed: (*S.M*=0) ***AND*** (*Q.N*>2) ***AND*** (*T.E*=1).

4.2 Construct a Formula Based on Updatable Variable

For an updatable variable *X* of a TG cycle *R*, the update actions of rules in *R* can assign *X* a value or make the value of *X* increase by $+\Delta X$ or decrease by $-\Delta X$.

In a formula based on *X*, *X* should be taken as the same value.1) If there is no dividing point of *X* in *R*, any rule *r* of *R* can be selected as the first triggered rule of a cyclical execution of *R* and only one formula can be constructed. In the condition of r' which following *r* in *R*, if a predicate containing *X* is selected into the formula, the other side of this predicate should be updated by $-\Delta X$ when *X* has been updated by$+\Delta X$. This is because *X* should be taken its original value when *r* is triggered. We simply call such an operation as update project on its dividing point. 2) If there is a dividing point of *X*, denoted as *r*, *r* can be selected as the first triggered rule of a cyclical execution of *R*. In the condition of r' which following *r* in *R*, if a predicate containing *X* is selected into the formula, this predicate should make an update project on *r*. So, if there are more than one dividing points of *X* in *R*, a formula should be constructed based on each dividing point of *X*. These formulae are consisted of a set of formulae for *X*.

The following algorithm is presented for constructing a formula based on an updatable variable in a TG cycle *R*.

`Algorithm1.` Formula-constructing
`Input:` A TG cycle *R* and an updatable variable *X* of *R*
`Output:` A set of formulae *SC*
`Begin`
  $SC := \phi$
  (1) **`IF`** there is no dividing point of *X* in *R*
    Let *r* be an arbitrary rule in *R* and regarded as the first triggered rule of a cyclical execution of *R*;
    Let *P* indicate a predicate containing *X* in the condition of *r*;
    Let δ_1 indicate the first formula to be constructed;
    Let Updated (*X*) indicate a set of updates has been completed for *X*;
    $\delta_1 := P$;

```
        IF the action of r updates X by +ΔX
                Updated (X)={-ΔX};
        ENDIF
        IF the action of r updates X by -ΔX
            Updated (X)={+ΔX};
        ENDIF
      ENDIF
      IF there are more than one dividing points of X in R
         Let r be an arbitrary dividing point of X and regarded as the first triggered rule of a
         cyclical execution of R;
         Let P indicate a predicate that assigns a value to X in the action of r;
         δ_1: = P; Updated (X)=ϕ;
      ENDIF
   (2) Let r' be the next triggered rule of a cyclical execution of R;
      IF there exists a predicate P in the condition of r' as the form X compare n, compare
         indicates such symbols as ">, <" and n indicates some constant;
           FOR ∀deltax∈ Updated (X)
              n: = n +deltax;
           ENDFOR
         δ_1: = δ_1∧P;
      ENDIF
      IF r' is not r and is not a dividing point of X in R
         IF the action of r' updates X by +ΔX
            Updated (X)= Updated (X)∪{-ΔX};
         ENDIF
         IF the action of r' updates X by -ΔX
            Updated (X)= Updated (X)∪{+ΔX};
         ENDIF
      Goto (2)
     ENDIF
     IF r' is not r and is a dividing point of X in R
        SC: = SC ∪{δ_1};
        Let P' indicate a predicate that assigns a value to X in the action of r';
        δ_2: = P'; Updated (X)=ϕ;
        Goto (2)
     ENDIF
     IF r' is just r /*a cyclical execution of R ends*/
        Let δ_n be the current constructing formula;
        SC: = SC ∪{δ_n};
        Return (SC);
     ENDIF
   END.
```

Example 8. In example2, $S.L$ is an updatable variable and can be selected to construct a formula through algorithm1.

1) Only r_1 assigns an original value to $S.L$. So r_1 is a dividing point of $S.L$ and regarded as the first triggered rule of a cyclical execution of R_1. By algorithm1, δ_1: = $(S.L = 1)$; Updated $(S.L)=\phi$.

2) r_1 triggers r_2. Only the action of r_2 updates $S.L$ with $S.L = S.L + 0.5$. r_2 is not a dividing point of $S.L$, By algorithm1, Updated $(S.L)=\{-0.5\}$.

3) r_2 triggers r_3. The condition of r_3 contains a predicate P= S.L<2, so P is changed through algorithm1 as follows: P= S.L<2–0.5 and $\delta_1 := (S.L = 1) \wedge (S.L < 2-0.5)$.

4) r_3 triggers r_1. The condition of r_1 does not contain any predicate involved $S.L$. As r_1 is the first triggered rule of a cyclical execution of R_1, algorithm1 ends with returning $SC=\{\delta_1\}$ and $\delta_1 = (S.L = 1) \wedge (S.L < 2-0.5)$.

5 Termination Decision of a TG Cycle

Theorem 2. *Let P be a formula constructed for a TG cycle R on the basis of a non-updatable variable X. If P is false, R must be termination.*

Proof. According to the notion of non-updatable variable of a TG cycle, X cannot be updated by any rule in R or any rule that does not belong to R but can be triggering reachable from R.

During a cyclical execution of R, there must be a rule which cannot be really executed since its condition cannot be satisfied by the result of P =false. Otherwise, for all rules in R which condition containing X, their conditions should be satisfied to guarantee that they could be really executed. That is, P should be true. This is contradictary to the precondition of theorem2. So the conclusion holds.

Non-updatable variable in [14-16] indicates a variable that cannot be updated by the actions of all rules in a rule set. Many variables, which are not determined to be non-updatable variables by [14-16], may be non-updatable variables here.

Theorem 3. *Let P be an arbitrary formula in the set of formulae constructed for a TG cycle R on the basis of an updatable variable X. If P is false, R must be termination.*

Proof. The conclusion can be proven by contradiction. Assuming that is non-termination, any rule in should be really executed. That is, the conditions of all rules in should be satisfied by the database states.

Case1: if there is no dividing point of X in R, only one formula P can be constructed through algorithm1. Let r be an arbitrary rule in R and regarded as the first triggered rule of a cyclical execution of R. P indicates the conditions which should be satisfied at the same time once r is executed and all rules in R can be really executed. P is false. It is implied that there exists at least one rule in R, whose condition must not be satisfied as a result of P =false. That is, at least one rule in R cannot be really executed. This is contradictary to the assumption that R is non-termination.

Case2: If there is a dividing point of X, denoted as r, r can be selected as the first triggered rule of a cyclical execution of R. P indicates the conditions should be satisfied at the same time once r is executed and all rules in R, which following r and its condition containing X whose value is updated on the basis of the original value assigned by the action of r, can be really executed. P is false. It is implied that there exists at least one rule in those rules, whose condition must not be satisfied as a result of P =false. This is in contradiction with the assumption that R is non-termination.

From case1 and case2, the conclusion holds.

Example 9. In example1, that two TG cycles R_1 and R_2 are both termination which can be deduced by theorem3. In example2, that R_1 is termination can be deduced by theorem 3.

If all formulae, which are constructed for a TG cycle R through algorithm1, are satisfied, R cannot be deduced to be termination by theorem2 and theorem3. Such a case can be seen from example2, so we present the following concepts and a theorem.

Definition 6. *In a TG cycle R, an updatable variable X is* **updating cumulative** *if there is no dividing point of X in R.*

Example 10. In example2, $T.F$ is an updatable variable of R_2 and is updating cumulative for R_2. Each cyclical execution of R_2 must make $T.F$ increase by 1.

If X is an updatable variable and updating cumulative in a TG cycle R, only one formula can be constructed on the basis of X through algorithm1. Let $Sum\ (\Delta X)$ indicate the net effect of all updates to X in R. For each update in a cyclical execution of R, which changes X by ΔX, let $Sum\ (\Delta X) = Sum\ (\Delta X) + \Delta X$.

Theorem 4. *Let X be an updatable variable and it is updating cumulative in a TG cycle R. Let $\delta(X)$ be a formula which is constructed for R on the basis of X through algorithm1 and $\delta(X+k\times Sum\ (\Delta X))$ be the form that should be taken by $\delta(X)$ when R has been cyclically executed for k times. If $\delta(X+k\times Sum\ (\Delta X))$= false, R must be termination.*

Proof. Assuming all formulae, which are constructed for R based on its non-updatable variables and updatable variables through algorithm1, can be satisfied, R can be cyclically executed and has been executed for k times.

At the start of the $(k+1)$ time cyclical execution of R, X should be changed as $(X+k\times Sum\ (\Delta X))$ by the net effects of times cyclical executions of R.

$\delta(X+k\times Sum\ (\Delta X))$= false, which implies that there is at least one rule in R, whose condition must not be satisfied during the $(k+1)$ time cyclical execution of R. That is, at least one rule of R cannot be really executed. So R must be termination during the $(k+1)$ time cyclical execution of R. The conclusion holds.

Based on theorem2-4, the following algorithm is proposed for termination decision of a TG cycle.

```
Algorithm2. Termination-test
Input: A TG cycle R and all updatable variables and non-updatable variables of R
Output: If R is termination then return (true); otherwise, return (false)
Begin
  (1) For each non-updatable variable X in R
        Construct a formula P based on X through methods in [9, 10];
        IF P=false
          return (true);
        ENDIF
      ENDFOR
  (2) For each updatable variable X in R
        Construct a set of formulae SC based on X through algorithm1;
        IF ∃P∈SC and P=false
          return (true);
        ENDIF
      ENDFOR
```

```
(3) IF there exists an updatable variable X and it is updating cumulative
        Let δ(X) be a formula constructed for R on the basis of X through algorithm1;
        Let Sum (ΔX) indicate the net effect of all updates to X in R;
        Let δ(X+k× Sum (ΔX)) be the form that should be taken by δ(X) when R has been
        cyclically executed for k times;
        IF δ(X+k× Sum (ΔX))=false
            return (true)
        ENDIF
    ENDIF
(4) return (false);
END.
```

Theorem 5. *Algorithm2. is correct and will terminate.*

Proof. Algorithm2 is guaranteed to be correct by theorem2-4. For non-updatable variables and updatable variables in a TG cycle *R*, the numbers of them are finite, so algorithm2 can automatically terminate.

Example 11. In example2, there are three variables in the conditions of all rules in R_2: *T.E*, *Q.N* and *T.F*. We analyze the termination of R_2 through algorithm2 as follows.

Step1, *T.E* and *Q.N* are two non-updatable variables of R_2 and their formula: (*Q.N*>2) ***AND*** (*T.E*=1) is regarded as true.

Step2, *T.F* is an updatable variable of R_2 and a formula based on *T.F* through algorithm1 can be constructed: $\delta=((T.F+1)<2)$ and *Sum* ($\Delta T.F$)=+1. *T.F* is assigned an original value by r_3 with *T.F* =0, so δ=true.

Step3, *T.F* is an only updatable variable of R_2, $((T.F+3\times1)<2)$=false when has been cyclically executed for three times. That is, R_2 must be termination as the condition of r_6 becomes false.

6 Conclusion

In this paper, we develop a new technique of termination decision that avoids two main drawbacks of existing methods. We present many new concepts such as updatable variable of a TG cycle, dividing point of an updatable variable. Then, we propose an algorithm to construct a formula that can contain updatable variables. We introduce the concept of updating cumulative for an updatable variable. Then, we present a new algorithm to detect more termination cases.

Our method only provides a sufficient condition to guarantee the termination of a TG cycle. In future, we would also like to investigate any other possible means to detect more termination cases. It is our future work to improve our method through incorporation with other factors such as composite event and time order.

Acknowledgement

This work was supported in part by the NSF of China under Grant No. 60303008 and 60673133; the National Grand Fundamental Research 973 Program of China under Grant No.2005CB321905.

References

1. A. Bonifati, S. Ceri, and S. Paraboschi. "Active rules for XML. A new paradigm for e-services", *VLDB Journal*, 10(1): 39-47, 2001.
2. J. Bailey, A. Poulovassilis, and P.T. Wood. "An event-condition-action language for XML", In *Proc.WWW2002*, pages486-495, Hawaii, 2002.
3. G. Papamarkos, A. Poulovassilis, and P.T. Wood. "RDFTL: An event-condition-action language for RDF", In *Proc. of the 3rd Web Dynamics Workshop at WWW'2004*, New York, 2004.
4. G. Papamarkos, A. Poulovassilis, and P.T. Wood. "Event-condition-action language for the Semantic Web", In *Proc.Workshop on Semantic Web and Databases*, at *VLDB'03*, Berlin, 2003.
5. M. Zoumboulakis, G. Roussos, and A. Poulovassilis. "Active rules for sensor databases", In *Proc. of the 30th VLDB Conference*, pages486-495, Toronto, 2004.
6. A. Aiken, J. Widom, J. Hellerstein. "*Behavior of database productions rules: Termination, Confluence, and Observable Determinism*", In *Proc. Int'l Conf. On Management of Data (SIGMOD)*, San Diego, California, 1992.
7. J. Bailey, Guozhu Dong et al. "On the Decidability of the Termination Problem of Active Database System", *Theor.Comput.Sci*, 311 (1-3): 389-437, 2004.
8. A.P. Karadimce, S.D. Urban. "*Conditional Term Rewriting as a Formal Basis for Analysis of Active Database Rules*", In *Proc. Int'l Workshop On Research Issues in Data Engineering (RIDE-ADS)*, Houston, Texas, 1994.
9. S. Comai, L. Tanca. "Termination and Confluence by Rule Prioritization", *IEEE Transactions on Knowledge and Data Engineering*, 15(2): 257-270, 2003.
10. Xiaoou Li, et al. "A Structural Model of ECA Rules in Active Database", In *Proc.MICAI2002*, pages486-493, Mexico, 2002.
11. J. Bailey and A. Poulovassilis. "Abstract interpretation framework for termination analysis in functional active databases", *Journal of Intelligent Information Systems*, 12(2/3): 243-273, 1999.
12. E. Baralis, S. Ceri, et al. "Compile-Time and Runtime Analysis of Active Behaviors", *IEEE Transactions on Knowledge and Data Engineering*, 10(3): 353-370, 1998.
13. E. Baralis, J. Widom. "An Algebraic Approach to Static Analysis of Active Database Rules", *ACM Transactions on Database Systems*, 25(3): 269-332, 2000.
14. A.P. Karadimce, S.D. Urban. "Refined Triggering Graph: A Logic-Based Approach to Termination Analysis in an Active Object-Oriented Database", In *Proc. Int'l Conf. On On Data Engineering (ICDE)*, New-Orlean, Louisiana, 1996.
15. S.Y. Lee, T.W. Ling. "Refined Termination Decision in Active Databases", In *Proc.Int'l Conf. On Database and Expert Systems Applications (DEXA)*, Toulouse, France, 1997.
16. S.Y. Lee, T.W. Ling. "A path Removing Technique for Detecting Trigger Termination", In *Proc. Int'l Conf. On Extended Database Technology (EDBT)*, Valencia, Spain, 1998.
17. N.W. Paton, et al. "Active Database System", *ACM Computing Surveys*, 31(1): 63-103, 1999.
18. S. Ceri, P. Fraternali, et al. "Active Rule Management in Chimera", In S. Ceri, J. Widom, "Active Database Systems", Morgan Kaufmann, 1996.
19. L. Leverenz, et al. "SQL Reference, Handbook". Oracle Corporation, 2000

Computing Repairs for Inconsistent XML Document Using Chase*

Zijing Tan, Zijun Zhang, Wei Wang, and Baile Shi

Department of Computing and Information Technology
Fudan University
Shanghai, China
{zjtan,0272415,weiwang1,bshi}@fudan.edu.cn

Abstract. An XML document is inconsistent if it violates predefined integrity constraints. In this paper, we consider how to compute *repairs* for an inconsistent XML document. Here *repair* is defined as the data consistent with the integrity constraints, and also minimally differs from the original document. Based on a repair framework by introducing a chase method, in this paper, we discuss the repairs computing problem and implement a prototype. First we discuss some key points about mends generation and repairs chasing. Next we give a cost model for this repair framework, which can be used to evaluate the cost of each repair. Finally we implement prototypes of our method, and evaluate our framework and algorithms in the experiment.

1 Introduction

Integrity constraints capture an important normative aspect of every data-based application. However, it is often the case that their satisfaction can not be guaranteed, especially when overlapping or redundant information from multiple sources is integrated. In this paper, we consider the problem of computing *repairs* for an inconsistent XML document, based on a framework introduced in our earlier work [11].

We summarize the framework here first. The concept of *repairs* for relations, originally introduced by Arenas et al.[2], is extended to XML. Intuitively, a repair for an XML document T is a document T' satisfying the constraints, which is obtained from T by applying some "minimal change". Given a constraint set Σ and an XML document T, we compute repairs for T in a two-step approach: (1)To generate *mend*, which is obtained from T by removing those conflicting information w.r.t. Σ.(2)To chase *repair* from mend, or to say, to "push" the information from Σ to mend.

In this paper, We further discuss some key points about *mends* generation and *repairs* chasing. We also give a novel cost model for our repair framework, which can be used to evaluate the cost of each repair. We implement prototypes

* This work is supported by the National Natural Science Foundation of China under Grant No. 60603043.

G. Dong et al. (Eds.): APWeb/WAIM 2007, LNCS 4505, pp. 293–304, 2007.

of our method, and evaluate our framework and algorithms in the experiment. To the best of our knowledge, our work is the first prototype implementation of data inconsistency problem for XML constraints, and with a strong formal solution.

Related Work. There are a lot of research on inconsistent databases[2,4,5,6,9,12]. The results can not be applied to XML directly because of the different data model. For example, even the basic notions of repair need to be redefined.

[3] discusses query answers in XML data exchange, to restructure XML documents that conform to a source DTD under a target DTD, and to answer queries written over the target schema. [1] presents a framework for Webhouses with incomplete information, it can represent partial information about the source document acquired by successive queries. They differ from our goal to compute repairs for inconsistent XML documents.

[10] considers the problem of querying XML documents which are not valid with respect to given DTDs. It proposes a method for measuring the invalidity of XML documents and compactly representing minimal repairing scenarios. Furthermore, [10] presents a validity-sensitive method of querying XML documents. In this paper, we consider integrity constraints for XML in the situation where a DTD is absent. Our framework can be adapted by taking DTDs into consideration, which may restrict the update operations available.

[7] considers inconsistent XML data w.r.t. a set of functional dependencies. The repairs are based on the replacing of node values, and the introduction of a function stating whether the node information is reliable. [8] investigates the existence of repairs w.r.t. a constraint set and a DTD. And for all the cases where the existence of a repair is decidable, the complexity of providing consistent answers to a query is characterized. In this paper, we consider a more general constraint model and more update operations, and provide a formal foundation for the solution in a different way.

The rest of the paper is organized as follows. Section 2 provides the basic notations, including DTD, XML path, symbol mapping and a general constraint model. The repair framework in [11] is summarized in section 3. We discuss some key points about the mends generation in section 4. We give a cost model for the repair framework in section 5. The experimental result of evaluation of our framework and algorithms is given in section 6. Finally section 7 draws a conclusion.

2 Preliminary Definitions

We adopt the usual view that an XML document is modeled as a node-labeled data tree, and assume that an element node is either followed by a sequence of element nodes and a set of attribute nodes, or is terminated with a text.

Definition 1. *Assume a finite set E of element labels, a finite set A of attribute names. An XML document(tree)is defined to be T=(V,lab,ele,att,val,v_r). 1)V is a finite set of nodes, and T is empty iff $V=\phi$. 2)Lab is a function from V to $E \cup A$,*

v is called an element node if $lab(v) \in E$, an attribute node if $lab(v) \in A$. 3)An element node v may be either followed by other nodes, or terminated with a text. If v is followed by other nodes, $ele(v)$ is a sequence of element nodes$[v_1, \ldots, v_n]$, and $att(v)$ is a set of attribute nodes$\{v'_1, \ldots, v'_m\}$. 4)Function val assigns values to attribute nodes and element nodes terminated with a text, the value may be a string constant, or a string variable. 5)v_r is a distinguished node in V and is called the root of T; without loss of generality, assume $lab(v_r)$=r.

Element nodes terminated with a text and attribute nodes are called leaf nodes. We extend the *val* function so that is can assign a string variable to node value. The reason for this is that we want to allow value modification as a repair primitive, and sometimes we are not concerned about the specific values used.

A symbol mapping from one symbol set A to another symbol set A' is a function h, such that for any symbol a in A, $h(a)$ is a symbol in A'. To map node values between different XML documents, we naturally extend symbol mapping to nodes and XML documents. Let the symbol set for XML document T be composed of all the constants and variables from node values. Given a symbol mapping h, for a node v in T, $lab(h(v)) = lab(v)$, and $val(h(v)) = h(val(v))$. Let the root of T be v_r, $h(T)$ is an XML document rooted at $h(v_r)$, and for any node v in T, if v is the kth child node of v', $h(v)$ is the kth child node of $h(v')$ in $h(T)$. In the rest of paper, we consider h that preserves constants. That is, if a is a constant, h(a)=a.

Let v and v' be two nodes in V. v and v' are value equal, denoted as $v \equiv v'$ iff (1)$lab(v) = lab(v')$, and (2)$val(v) = val(v')$.

A simple path in DTD D is a sequence of node names, with the form $P ::= \epsilon \mid e/P$. Here ϵ represents the empty path, $e \in E \cup A$, and "/" denotes concatenation of two paths. If the first element of P is r, we call P root path.

In XML document T, we say that a node v_2 is reachable from node v_1 by following the path P, iff (1)$v_1 = v_2$, and $P = \epsilon$, or (2)$P = P'/e$, there is a node v' such that v' is reachable from node v_1 by following P', and v_2 is a child of v' with label e. We write $\lfloor v\{P\} \rceil$ for the set of nodes in T that can be reached by following P from v. In particular, when there is only one node in $\lfloor v\{P\} \rceil$, we use $v\{P\}$to denote this node. If v is the *root* node, we write $\lfloor P \rceil$ for $\lfloor v\{P\} \rceil$.

Definition 2. *A General Integrity Constraint Model.*

The constraint σ is either of the form $(R_1, R_2, (Q_1, \ldots, Q_n))(X_1, \ldots, X_m \Rightarrow X_{m+1})$, or of the form $(R_1, R_2, (Q_1, \ldots, Q_n))(X_1, \ldots, X_m \Rightarrow u = w)$. Here R_1 is a root path, or $R_1 = \epsilon$. $R_1/R_2/Q_1, \ldots, R_1/R_2/Q_n$ are all root paths. $X_j (j \in [1, m+1])$ is a sequence of values $[x_{j1}, \ldots, x_{jn}]$ that may contain variables. u(and w) is either a variable or a constant.

We require all the variables in constraints are bound, that is, all the variables in X_{m+1}, u, or w should also occur in $X_1, \ldots, X_m$. Let the symbol set for constraint be composed of all the variables and constants in $X_1, \ldots, X_m, X_{m+1}$, u and w.

Definition 3. *A constraint σ is satisfied by an XML tree T, denoted as $T \models \sigma$ iff: For any symbol mapping h from σ to T, $\forall v \in \lfloor R_1 \rceil$, if (1)$\exists v_j \in \lfloor v\{R_2\} \rceil$,* *and (2)there is only one leaf node in* $\lfloor v_j\{Q_i\}\rceil$, *and (3)*$val(v_j\{Q_i\}) = h(x_{ji})$, *(*$i \in [1,n]$, $j \in [1,m]$*), and*

a. If $\sigma=(R_1, R_2, (Q_1, \ldots, Q_n))(X_1, \ldots, X_m \Rightarrow X_{m+1})$, *then* $\exists\, v_{m+1} \in \lfloor v\{R_2\}\rceil$, *with only one leaf node in* $\lfloor v_{m+1}\{Q_i\}\rceil$, $val(v_{m+1}\{Q_i\}) = h(x_{(m+1)i})$ *(*$i \in [1,n]$*);*

b. If $\sigma=(R_1, R_2, (Q_1, \ldots, Q_n))(X_1, \ldots, X_m \Rightarrow u = w)$, $h(u)=h(w)$.

For a set of constraints Σ, if $T \models \sigma$ for any $\sigma \in \Sigma$, we write $T \models \Sigma$. Below we use τ to denote the constraint of the form $(R_1, R_2, (Q_1, \ldots, Q_n))(X_1, \ldots, X_m \Rightarrow X_{m+1})$, and use γ for $(R_1, R_2, (Q_1, \ldots, Q_n))(X_1, \ldots, X_m \Rightarrow u = w)$.

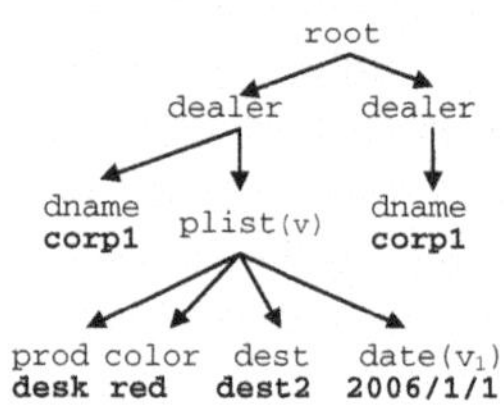

1. Dealer name determines a dealer.
(ε,root/dealer,(dname))([x],[x] ⇒ 0=1)
2. *'Red'* products will be shipped to *'dest1'* only.
(ε,root/dealer/plist,(color,dest))(['*red*',x] ⇒ x='*dest1*')
3. For a dealer, any product shipped to *'dest2'* will be shipped to *'dest1'* on the same day.
(root/dealer,plist,(prod,color,dest,date))([x,y,'*dest2*',z] ⇒ [x,y,'*dest1*',z])
4. There is no shipment on *'2006/1/1'*
(ε,root/dealer/plist,(date))(['*2006/1/1*'] ⇒ 0=1)

Fig. 1. Model of an inconsistent XML document

Example 1. Figure 1 gives an XML document about dealers. For each dealer, we give its name(dname) and each shipment of product(plist). The shipment information is composed of product name, product color, shipment destination and date. If not empty, values of element or attribute nodes are recorded under the node names in bold. For example, figure 1 says that a dealer named *'corp1'* sent out *'red' 'desk'* to *'dest2'* on *'2006/1/1'*. We also list 4 integrity constraints this document should satisfy using the notations from definition 2.

This document violates some of the integrity constraints. There are two different dealers with the same name *'corp1'*, which violate constraint 1. *'Red'* product was shipped to *'dest2'*, which violates constraint 2. *'Desk'* was shipped to *'dest2'*, but was not shipped to *'dest1'*, it violates constraint 3. And the date *'2006/1/1'* for shipment violates constraint 4.

3 From Mend to Repair

A repair R for an XML document T w.r.t a set of constraints Σ is an XML document consistent with Σ, and is also as close as possible to T. In the framework, we consider three update operations for fixing an XML document, which are node insertion, node deletion and node value modification. To describe the semantics of "as close as possible", we assume the repairing is done according to some partial order. The order $\subseteq$ below can describe node insertion and node deletion only, the order $\preceq$ is further defined based on symbol mapping, which can describe all the three update operations.

Definition 4. *Let S, T be two XML documents, if T is empty,* $T \subseteq S$*. Otherwise let the root node for T be* r_T*, the root node for S be* r_S*, the child nodes of* r_T *be* $\{v_1, \ldots, v_n\}$*, and the child nodes of* r_S *be* $\{u_1, \ldots, u_m\}$*. We use* T_{v_i} *to denote the subtree rooted at* v_i *in T, and* S_{u_j} *to denote the subtree rooted at* u_j *in S.* $T \subseteq S$ *iff*

1. $r_T \equiv r_S$;
2. $\forall T_{v_i}, \exists S_{u_j}$*, such that* $T_{v_i} \subseteq S_{u_j}$*; (*$i \in [1,n]$*,* $j \in [1,m]$*)*
3. *For* T_{v_i} *and* T_{v_k}*, there exists subtree* S_{u_j} *and* S_{u_f}*, such that* $T_{v_i} \subseteq S_{u_j}$ *and* $T_{v_k} \subseteq S_{u_f}$*. And if* $i \neq k$*,* $j \neq f$*. (*$i,k \in [1,n]$*,* $j,f \in [1,m]$*)*

Definition 5. *Let S, T be two XML documents, we write* $T \preceq S$ *iff there exists a symbol mapping h from T to S, such that* $h(T) \subseteq S$*. If* $T \preceq S$*, and* $S \preceq T$*, we write* $T \sim S$*. If* $T \preceq S$*, and* $T \not\sim S$*, we write* $T \prec S$*.*

It is clear that if $S \subseteq T$, $S \preceq T$. $\subseteq$ ignores the order of sibling nodes, and $\preceq$ allows different symbol sets for node values.

We get repairs in the following way: Starting from the original XML document T, we find document M first, and M subsatisfy Σ. Here 'subsatisfy' means that we can further find a document R based on M, and $R \models \Sigma$. To be specific, when getting M from T, we "remove" those data from T, which conflicts with Σ. This is done by nodes value modifications, to replace a constant by a variable, or by nodes deletions. The insertion of nodes will not help to "solve" conflicts, so it will not be used in this step. Please note that $M \preceq T$. When obtaining R from M, we "push" some information from Σ to M. It is done by nodes value modifications, to replace a variable in M by a constant, or to replace a variable in M by another variable in M. It can also be done by some nodes insertions. In this step, deletion of nodes will not be used, and $M \preceq R$. Further, to accommodate the "as close as possible" requirement, M should be the document meeting the requirements, and be the closest to T w.r.t $\preceq$. And R should be the document meeting the requirements, and be the closest to M w.r.t $\preceq$.

Definition 6. *Let T be an XML document and* Σ *a set of constraints, an XML document S is said to subsatisfy* Σ *iff there exists an XML document J, such that* $S \preceq J$*, and* $J \models \Sigma$*.*

A mend for T and Σ *is an XML document* $M \preceq T$*, M subsatisfy* Σ*; and there does not exist M', such that* $M \preceq M' \preceq T$*, and M' subsatisfy* Σ*.*

A repair for T and Σ *is an XML document R,* $R \models \Sigma$*; and there exists a mend M for T and* Σ*, such that* $M \preceq R$*, and for any R',* $M \preceq R' \preceq R$*,* $R' \not\models \Sigma$*.*

By definition 6, we can't judge whether a mend is valid unless we can further find a repair based on it. So a mend is just called a *candidate* before we can generate a repair from it. Next we introduce the *chase* method to justify whether a mend candidate is a mend indeed. And if the mend candidate is a mend, the *chase* will generate the corresponding repair as well. This is a novel extension of the relational chase method used for logical implication.

For those mend candidates which are not mends indeed, the chase process will fail to "chase" repairs. We introduce a special XML document, denoted $\Box$, for such situations.

Definition 7. *Let S and T be XML documents, and Σ a set of constraints. We write $T \vdash_\Sigma S$ if S can be obtained from T by a single application of one of the following rules:*

1. $\Box \vdash_\Sigma \Box$;
2. *if $T \models \Sigma$, $T \vdash_\Sigma T$; otherwise $\exists \sigma \in \Sigma$, $T \not\models \sigma$. By definition, there is a mapping h from σ to T, $\exists v \in \lfloor R_1 \rceil$, $\exists v_j \in \lfloor v\{R_2\} \rceil$, with only one leaf node in $\lfloor v_j\{Q_i\} \rceil$, and $val(v_j\{Q_i\}) = h(x_{ji})$, which violate σ:*
   (a) *σ=τ=$(R_1, R_2, (Q_1, \ldots, Q_n))(X_1, \ldots, X_m \Rightarrow X_{m+1})$. S is obtained from T by nodes insertions: Insert child node v_{m+1} for node v matching path R_2, insert child nodes for v_{m+1} matching paths $Q_1, \ldots, Q_n$, and set $val(v_{m+1}\{Q_i\})$=$h(x_{(m+1)i})$;*
   (b) *σ=γ=$(R_1, R_2, (Q_1, \ldots, Q_n))(X_1, \ldots, X_m \Rightarrow u = w)$. If $h(u)$ and $h(w)$ are two distinct constants, S=$\Box$. Otherwise assume without loss of generality that $h(u)$ is a variable, S is obtained from T by a symbol mapping g: $g(h(u))$=$h(w)$, and g preserves other variables.*

A chase of T by Σ is a maximal sequence T=T_1,T_2,…,T_n of XML documents such that for every $i \in \{1, \ldots, n\}$, $T_i \vdash_\Sigma T_{i+1}$, and $T_i \neq T_{i+1}$.

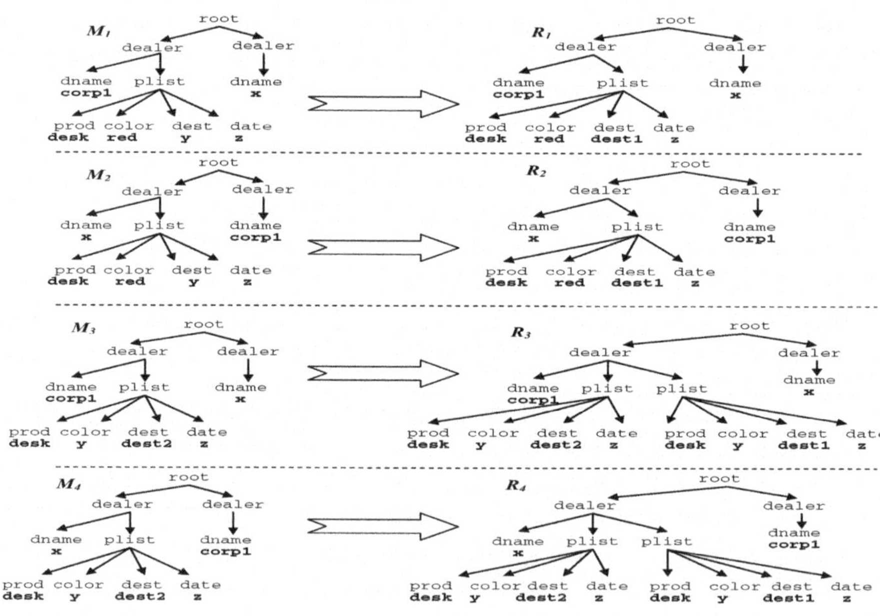

Fig. 2. Mend and Repair

Example 2. We give all the mends and corresponding repairs in figure 2, other mends are $\sim$ to them. In the mends, we replace either *'corp1'* as variable x, replace *'2006/1/1'* as variable z, and replace either *'dest2'* or *'red'* as variable y.

It can also be verified R_i is the chase result of M_i respectively in figure 2. For example, R_1 is obtained from M_1 by applying γ rule on constraint 2. And R_3 is obtained from M_3 by applying τ rule on constraint 3. The fact is that repairs will add information from Σ to mends. In R_1, we replace y as $'dest1'$, which is implied by constraint 2. And we insert new nodes describing the shipment to $'dest1'$ by constraint 3 in R_3.

According to [11], we know that the chase approach has some good properties. 1)The chase can terminate in finite steps; 2)we can 'chase' repair from T by Σ iff T subsatisfies Σ; 3)For any two possible chase results S_1 and S_2, $S_1 \sim S_2$.

4 Mends Generation

Next we discuss some key points about mends generation, which is vital to the implementation of our method. From original document T, we may use node deletions, and replace constant node values by variables, to get the candidates of mend. It is clear that the number of possible candidates are limited w.r.t. $\sim$.

To preserve more information from the document, we should use as few updates as possible. The basic intuition of mends generation is that: First we apply some update operations to the original document, and get a mend candidate M'. Next we try to find a repair based on M'. If we can find such a repair, M' is a mend indeed, and we get the corresponding repair. Or else, we further apply updates to M', get a new candidate M'', and try on it again. This routine will continue until all the possible updates are applied to the failed candidates.

Of course we are not interested in applying updates to arbitrary nodes, but to those (possibly) violated nodes only. The violated nodes can be found by validating the document against the constraint set. In the implementation, we further optimize the method of mend candidates generation as follows:

1. We disallow multiple occurrences of the same variable in a mend, to exclude some unnatural mends. For example in figure 1, to replace both $'corp1'$ and *'2006/1/1'* as variable x is unnatural, because there is no clear relationship between these two values. We can find the equivalence of two variables suggested by constraints when generating repair from mend by chase.

2. Some constraints are mutually isolated, and can be dealt with separately. For example, constraint 1 and 4 do not share any violated nodes. So the candidates for these two constraints can be generated separately, which will greatly reduce the number of possible candidates.

3. There is no need to enumerate all the candidates. By definition 6, if we know M is a mend, there is no need to check any candidate M', if $M' \preceq M$. So we can generate candidates by $\preceq$ step by step.

5 Cost Model

In this section, We propose a novel *cost* model for our repair framework. The cost model attaches a cost to each repair, which can be used in some specific applications. For example, to compute a repair with the "lowest" cost. We leave to future work the development of techniques for automatically setting some parameters in the cost model.

The cost model is based on two factors, reliability and similarity. The *reliability* of data is reflected in a conditional probability associated with each element and attribute node, with values ranging between 0 and 1. It represents the confidence placed by the user. Each probability in the source XML document is assigned conditioned on the fact that the parent element exists. That is, with a given probability p, the node's conditional probability equals p when the parent exists, and the node's probability equals 0 when the parent does not exist. Consider the chain $root \rightarrow dealer$ in figure 1. Assume the XML document contains the probabilities $prob(dealer|root)$ and $prob(root)$. To calculate $prob(dealer)$, we can use Bayes' formula to state $prob(dealer) = \frac{prob(dealer|root) \times prob(root)}{prob(root|dealer)}$. It is clear that a parent must exist with a probability of 1 if its child exists, then $prob(root|dealer) = 1$. $prob(dealer) = prob(dealer|root) \times prob(root)$. In general, the probability of a node n can be found by multiplying the conditional probabilities found in the source XML, along the path from n to the root.

The *similarity* of data is available at the value level, to accommodate update operations using node value modification. The similarity measurement for strings is itself a broad field, and our discussion does not depend on a particular approach. We assume that for two values v, v', a distance function $dis(v, v')$ is available, with lower values indicating greater similarity.

The cost of a node value modification is the *probability* of the changed node times the *distance* between the original value and the value in the repair. To be specific, given a node value modification on node n, from node value v to v', the cost is $prob(n) \times dis(v, v')$.

We next come to the cost of a node insertion. Assume a node n with empty value is inserted, with node m as its parent. Note that the possibility of a newly inserted node is always 1, then the cost is $prob(m) \times inscost$. Here $inscost > 0$ is a user-defined parameter, to reflect the preference of different update operations. For example, if the user prefer value modification in the repairs, *inscost* can be set with a larger value to increase the cost of node insertions. The insertion of a node with a specific value can be treated in two steps: 1)a node is inserted with empty values; 2)a node value modification for this node, changing the value from empty to the desired one. Then the cost is the sum of the two separate costs.

The cost of a node deletion can be calculated in a similar way. The difference is that the node being deleted will usually have a conditional probability $p \neq 1$.

6 Experimental Evaluation

6.1 Experimental Setting

We have implemented prototypes of our method in Java. All the experiments are conducted on a PC with a 3.6GHz Pentium 4 CPU and 1GB RAM, running

Windows XP(SP2) and JRE 1.5.0. The JVM memory is 512MB. We perform our experiments with artificial data sets generated by the IBM XML Generator.

XML Generator is a program that can create random instances of valid XML given a DTD. The schema used in the experiments is an extension of figure 1, and has the following characteristics: 22 schema elements with a maximum depth of 6; average and maximum number of subelements(attributes) per element at 2.2 and 7, respectively.

To assess our technique with complicated constraints setting, we include 8 constraints in the experiment, which is composed of 3 τ constraints and 5 γ constraints. Recall that we use τ to denote the constraint of the form $(R_1, R_2, (Q_1, \ldots, Q_n))(X_1, \ldots, X_m \Rightarrow X_{m+1})$, and use γ for $(R_1, R_2, (Q_1, \ldots, Q_n))(X_1, \ldots, X_m \Rightarrow u = w)$. Among the 5 γ constraints, 3 constraints are the traditional functional dependencies(keys). The constraint set is defined such that all the leaf nodes are involved in at least one constraint. Among the 8 constraints, 2, 3 and 4 share same violated nodes, and are treated as a unit in mend candidates generation and chase.

All the data sets are processed as follows: 1)*Noise* is introduced to leaf nodes involved in constraints. When noise is introduced, with probability p_n, a leaf node will have value violating constraints; 2)The data set is validated against the proposed constraints, and the violated nodes are marked; 3)Initial mend candidates are generated according to the marked nodes in step 2, then chase starts to find repairs. If chase fails, new mend candidates will be produced from the failed candidates. The chase will continue until no new candidates can be generated.

To make the implementation effective, we build a skeleton tree after step 2, which is composed of only the violated nodes and the nodes along the paths from violated nodes to the root. Each node in the skeleton contains a reference to the original node in the full document, so the skeleton is used as an index. Since all the nodes accessed in computing can be located directly using the index, it will greatly improve the performance for large document.

We generate mend candidates and chase repairs wisely as discussed in section 3 and 4. Mutually isolated constraints are dealt with separately, and their "repairs" are combined finally to get the full repairs. With this implementation, for two mutually isolated constraints, the time to compute the full repairs is the sum of the time to compute their own "repairs", while the number of full repairs is the product of the number of their "repairs". So the running time grows much more slowly compared with the number of repairs.

We note that the experimental results are very sensitive to the method of noise introduction. In the implementation, we avoid having too much violated nodes for one related constraints unit. The basic intuition is that we try to average the running time among different related constraints unit.

6.2 Experimental Results

We evaluate the time(in millisecond) to compute repairs and the number of repairs in the experiment. For timing measurements, each experiment is run three times and the average time is recorded. We consider three factors while evaluating our method.

Noise. We generate an XML document with 17295 leaf nodes(totally 36586 nodes containing element, attribute and leaf nodes), then apply different noise ratio to it. We test p_n with values ranging between 1‰ and 4‰ in figure 3. With a fixed data set, more noise will cause more violated nodes, which in turn affect both the running time and the number of repairs. Note that the running time grows much more slowly compared with the number of repairs.

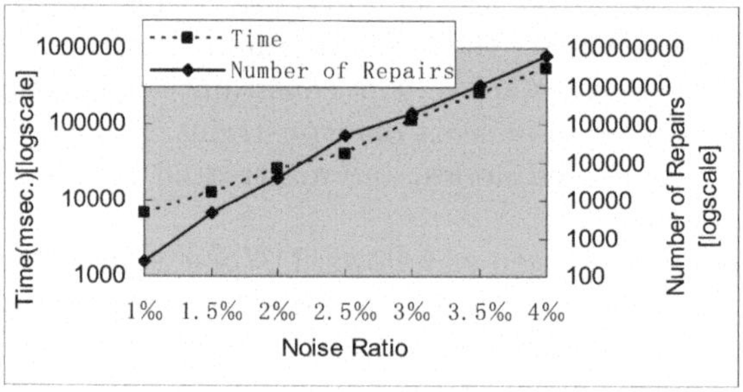

Fig. 3. Increasing noise

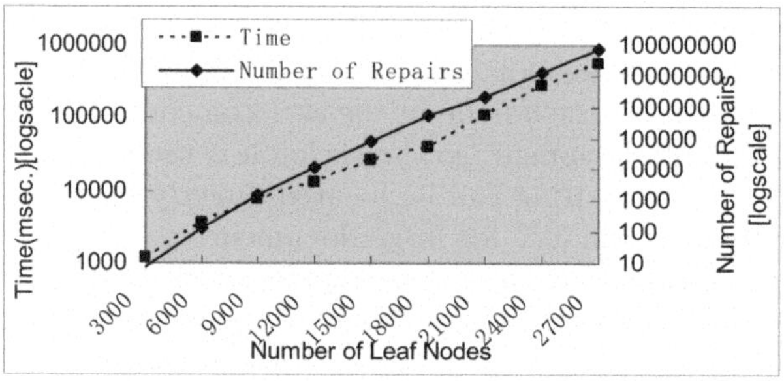

Fig. 4. Increasing data sets

Scalability. In figure 4, with a fixed noise p_n=2.5‰, we vary the size of data sets. To be more accurate, the size of each data set is measured as the number of leaf nodes it contains(the number of total nodes nearly doubles). The increase of document size will cause more violated nodes, and more time to build the index. Once the index is built, the time to locate a specific node for chase is determined by the index only. Since the index is much smaller than the whole document, in the experiment, we find that the time to compute each repair is almost not affected by the increasing document size.

Impact of Constraint Sets. The algorithm discussed in this paper gives the user flexibility in specifying the integrity constraints that need to be met. In real applications, users may find a balance between calculation efficiency and data accuracy. In figure 5, we give the running time on a data set with 17295 leaf nodes, with a fixed noise p_n=4‰, when different constraint sets are considered. From #2, each time one constraint is removed in the experiment. At #4, constraint 4 is removed. Since the only way to fix constraint 4 is to set '2006/1/1' to a variable, the total repair numbers are the same at #3 and at #4. There is a sharp decrease in the running time, because to remove a constraint from a related constraints unit will greatly reduce the number of possible mend candidates, and also simplify the chase approach for this unit. The situations at #5 and #6 are similar, where constraint 3 and 2 are removed respectively. At #6, only the three functional dependencies(keys) are considered. If we consider only functional dependencies, the chase method in definition 7 can be simplified. Since it it clear that node insertions have no effect on the correction of violations caused by functional dependencies.

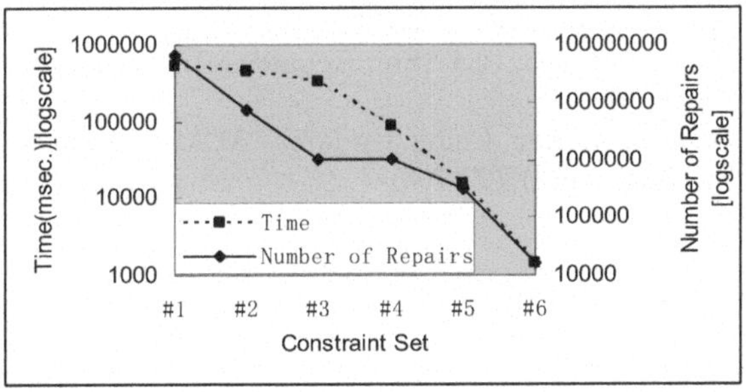

Fig. 5. Different constraint sets

7 Conclusion

In this paper, we study the data inconsistency problem for XML. Based on a repair framework, we build a prototype system to compute repairs for inconsistent XML document. We discuss some key points in the prototype implementation, and evaluate our framework and algorithms in the experiment. We also give a cost model for our repair framework, to evaluate the cost of each repair.

For future work, we will extend our framework to consider more constraints for XML, and make a further study about the computational complexity of our method.

References

1. S. Abiteboul, L. Segoufin, V. Vianu. Representing and Querying XML with Incomplete Information. In PODS 2001. 35-47.
2. M. Arenas, L. E. Bertossi, J. Chomick. Consistent Query Answers in Inconsistent Databases. In PODS 1999. 68-79.

3. M. Arenas, L. Libkin. XML Data Exchange: Consistency and Query Answering. In PODS 2005. 13-24.
4. P. Bohannon, W. F. Fan, M. Flaster, R. Rastogi. A Cost-Based Model and Effective Heuristic for Repairing Constraints by Value Modification. In SIGMOD 2005. 143-154.
5. L. Bravo, L. Bertossi. Logic programs for consistently querying data integration systems. In IJCAI 2003. 10-15.
6. J. Chomicki, J. Marcinkowski. Minimal-Change Integrity Maintenance Using Tuple Deletions. Information and Computation, 2005, 197(1-2): 90-121.
7. S. Flesca, F. Furfaro, S. Greco, E. Zumpano. Repairs and Consistent Answers for XML Data with Functional Dependencies. In XSym 2003. 238-253.
8. S. Flesca, F. Furfaro, S. Greco, E. Zumpano. Querying and Repairing Inconsistent XML Data. In WISE 2005. 175-188.
9. G. Greco, S. Greco, E. Zumpano. A logical framework for querying and repairing inconsistent databases. IEEE Transaction on Knowledge and Data Engineering, 2003, 15(6):1389-1408.
10. S. Staworko, J. Chomicki. Validity-Sensitive Querying of XML Databases. In EDBT Workshops (dataX) 2006. 164-177.
11. Z. J. Tan, W. Wang, J. J. Xu, B. L. Shi. Repairing Inconsistent XML Documents. In KSEM 2006. 379-391.
12. J.Wijsen. Database Repairing Using Updates. ACM Transactions on Databases Systems (TODS), 2005, 30(3):722-768.

An XML Publish/Subscribe Algorithm Implemented by Relational Operators*

Jiakui Zhao, Dongqing Yang, Jun Gao, and Tengjiao Wang

Database Lab, School of EECS, Peking University, Beijing 100871, China
{jkzhao,dqyang,gaojun,tjwang}@pku.edu.cn

Abstract. An XML publish/subscribe algorithm needs to store large numbers of XPath or XQuery subscriptions and match subscriptions with published XML documents. Since the number of the subscriptions may be very large, the performance and the scalability of the algorithm may be critical. The scalability of the method of constructing a large finite state automata or decision tree for all subscriptions is limited by amount of available physical memory. In this paper, we will introduce an XML publish/subscribe algorithm which is consisted of the publish algorithm, subscribe algorithm, and matching algorithm. The matching algorithm uses relational operators to match subscriptions with publications inside a relational database, so the scalability of the algorithm is no longer limited by amount of available physical memory. Experimental results show that the matching algorithm has very good performance and scalability.

1 Introduction

A publish/subscribe system receives messages from publishers and delivers to subscribers who require the messages. Earlier publish/subscribe systems are topic-based, in which subscribers subscribe topics and a published message is delivered to subscribers who subscribed the topic of the published message. Recent publish/subscribe systems are content-based, in which subscribers register "*rules*" and a published message is delivered to subscribers whose rules can be matched by the message. Content-based systems are more flexible and powerful than topic-based systems and are widely used in message oriented systems such as online auctions and financial information exchange. Almost all database and middleware vendors offer some publish/subscribe feathers in their software suits, in which users can use SQL-like language to express subscriptions and the messages may be something like relational tuples or dictionary data structures.

XML has become the standard data exchange format. There is an increasing demand for XML-based publish/subscribe systems which can support flexible document structures, and subscription rules should be expressed by powerful language such as XPath and XQuery; at the same time, the system must support high message throughput in case of millions of subscriptions. The main challenge

* Supported by NSF of China Grants 60473051 and 60503037, NSF of Beijing China Grant 4062018, and National High-Technical Project of China Grant 2006AA01Z230.

G. Dong et al. (Eds.): APWeb/WAIM 2007, LNCS 4505, pp. 305–316, 2007.

of building such a system is how to match a published XML document with millions of XPath or XQuery subscriptions. Current matching strategies can be classified into two classes. The first class of strategies see subscriptions as filters, and subscribers are notified if a message passes through the filters. The filters are implemented by decision tree or finite state automata, and common computations between filters can be shared. This class of strategies maybe quite efficient if the decision tree or finite state automata can fit in memory; however, the scalability is limited by amount of available physical memory. The second class of strategies store subscriptions into a relational database, when a message is published, matching it with stored subscriptions by queries inside the database. The matching queries may be efficiently evaluated by using some appropriate indices, and the scalability is no longer limited by amount of available memory.

In this paper, we will introduce our publish/subscribe algorithm for implementing an XML-based publish/subscribe system by relational database, in which subscriptions are expressed by XPaths. To the best of our knowledge, our work is the first that implements the algorithm by non-recursive SQL language.

The rest of the paper is structured as follows. Section 2 describes previous works on publish/subscribe systems. Section 3 presents our publish/subscribe algorithm. Experiments are presented in Section 4, followed by our conclusions.

2 Previous Works

Some content-based publish/subscribe systems treat subscriptions as data; the matching between published messages and the subscriptions is implemented by join queries, and similar queries can share common computations. NiagaraCQ [1] uses signatures to group similar subscriptions, it uses a constant table extracted from the subscriptions along with a join to implement the matching algorithm. Commercial database vendors also use relational database engines to implement publish/subscribe systems [2], [3] whose scalability is not limited by available physical memory. However, most of the systems only handle tuple-like messages and rules are expressed by SQL. Some of these systems can handle XML messages, but the systems either use a wrapper to extract tuple-like data from XML messages [2] or use a very simple language to express the subscription rules [1].

Other content-based publish/subscribe systems treat subscriptions as filters over messages. In [4], a in-memory decision tree is built for all subscriptions. The system walks down the decision tree for each message and subscriptions at the leaf node are notified. Several XML-based publish/subscribe systems employee a finite state automata for all XPath subscriptions in the main memory. A sequence of SAX parsed events of an XML message are streamed through the finite state automata to match the XPath subscriptions. [5] first introduced how to match an XML message with many XPath subscriptions using an in-memory finite state automata, a later paper [6] introduced how to share states in finite state automata construction, and [7], [8] introduced how to build an index on sub-strings of XPath expressions and share common computations between common sub-strings. [9], [10] introduced how to use in-memory finite state automata

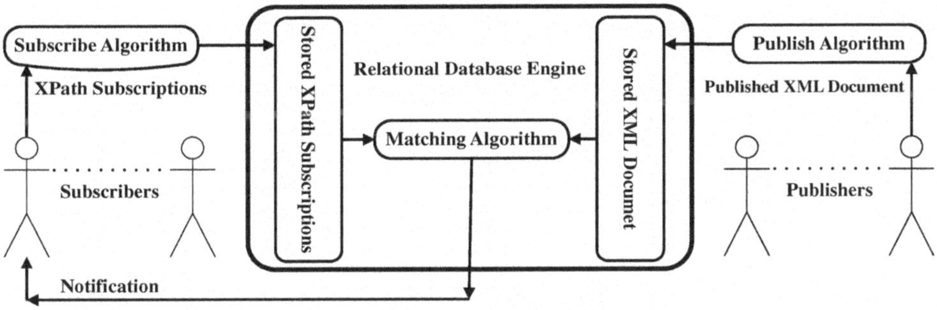

Fig. 1. Publish/subscribe framework

to evaluate XPath subscriptions over XML streams. [11] introduced a dynamic state construction approach, in which each state of a finite sate automata is constructed in memory when it is actually used; this approach is less attractive for long running systems. The main disadvantage of the in-memory approaches is the number of the subscriptions is limited by available physical memory, and insertions and deletions of subscriptions are difficult in the long-running systems.

[12] is the first that implements a long-running XML publish/subscribe system using a relational database, in which the subscriptions are expressed by an XPath subset called "*branching path expression*", and the number of the subscriptions is no longer limited by amount of available physical memory. Each XPath is rewritten as a conjunction of predicates, and each predicate contains one branch in the graph presentation of the XPath. For each published XML document, the branch path of each predicate is matched firstly, then a recursive SQL query is used to match all the predicates of each XPath. If all the predicates and branch points of an XPath are matched successfully, the owners of the XPath are notified. Our method is different from that proposed in [12] for we match XPaths with the XML document level-by-level using non-recursive SQL.

3 Publish/Subscribe Algorithm

Fig. 1 shows our publish/subscribe framework. Subscribers subscribe XML-style messages by XPaths; subscribe algorithm translates the XPath subscriptions into relational tuples and stores the tuples into a relational database. Publishers publish XML documents; publish algorithm translates an XML document into relational tuples and stores the tuples into a relational database. Matching algorithm matches stored XPath subscriptions with a stored XML document inside the relational database by relational operators to find out matched subscriptions and deliver the XML document to subscribers whose subscriptions are matched.

3.1 Publish Algorithm

We use the range-based static labeling scheme [13] to label the nodes of an XML tree. The scheme initializes a counter to one and carries out a depth-first

Table 1. The XML document shown in Fig. 2 stored in relational table *XMLFrame*

start_point	*end_point*	*depth*	*node*	*start_point*	*end_point*	*depth*	*node*
00	35	00	*-root-*	01	34	01	StudentList
02	17	02	Student	18	33	02	Student
03	05	03	@type	19	21	03	@type
06	13	03	name	22	29	03	name
14	16	03	age	30	32	03	age
07	09	04	FirstName	23	25	04	FirstName
10	12	04	LastName	26	28	04	LastName

Table 2. The XML document shown in Fig. 2 stored in relational table *XMLValue*

start_point	*node*	*value*	*start_point*	*node*	*value*
03	@type	master	19	@type	bachelor
14	age	25	30	age	20
07	FirstName	Zhao	23	FirstName	Zhong
10	LastName	Jiakui	26	LastName	Lin

traversal of the XML tree. If a node is seen for the first time, it is assigned the value of the counter as its "*start-point*", and the node is assigned the counter value as its "*end-point*" when it is encountered again. The counter is incremented by one each time its value is assigned to a node. Compared with dynamic labeling schemes such as the float number labeling scheme [14], the prefix labeling scheme [15], [16], and the prime number labeling scheme [17] which need not to re-label after insertions and deletions of nodes, the range-based labeling scheme has smaller storage requirements, so it is possible to use a fixed length representation to store the labels, and we can take advantage of the standard SQL92 data types and operations to determine the ancestor-descendant relationships

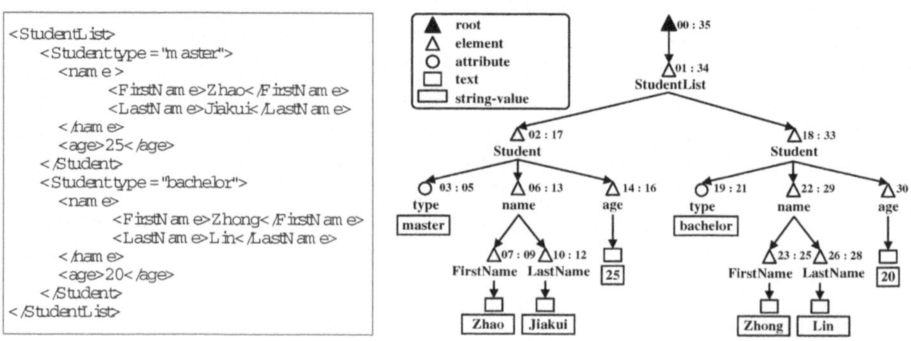

Fig. 2. An XML document and the XML tree marked by "*start-point* : *end-point*"

Table 3. The XPaths shown in Fig. 3 stored in relational table *XPathFrame*[*level*]

id	*high_num*	*low_num*	*high_node*	*low_node*	*min_diff*	*max_diff*	*final*	*branch_point*
level=0:								
01	01	01	*-root-*	StudentList	01	01	*false*	*false*
02	01	01	*-root-*	student	02	02	*false*	*false*
level=1:								
01	01	01	StudentList	FirstName	03	∞	*false*	*false*
02	01	01	student	@type	01	01	*true*	*true*
02	01	02	student	age	01	01	*false*	*true*
02	01	03	student	name	01	∞	*false*	*true*
level=2:								
02	03	01	name	FirstName	01	01	*false*	*false*

Table 4. The XPaths shown in Fig. 3 stored in relational table *XPathPred*[*level*]

level	*id*	*node_num*	*node*	*value*	*operation*
2	01	01	FirstName	Zhong	00 (=)
2	02	02	age	21	01 (>)
3	02	01	FirstName	Zhong	00 (=)

between nodes. For example, α is the ancestor of β iff $start\text{-}point_\alpha < start\text{-}point_\beta$ and $end\text{-}point_\alpha > end\text{-}point_\beta$. In addition to *start-point* and *end-point*, we record the "*depth*" of each node. α is the parent of β iff α is the ancestor of β and $depth_\beta = depth_\alpha + 1$. Each XML document is parsed by a SAX parser, and the parsed structure information and value information are stored in relational table *XMLFrame* and *XMLValue* respectively. Fig. 2 shows an XML document and the corresponding XML tree. Table 1 and Table 2 demonstrate how the XML document shown in Fig. 2 is stored in *XMLFrame* and *XMLValue* respectively.

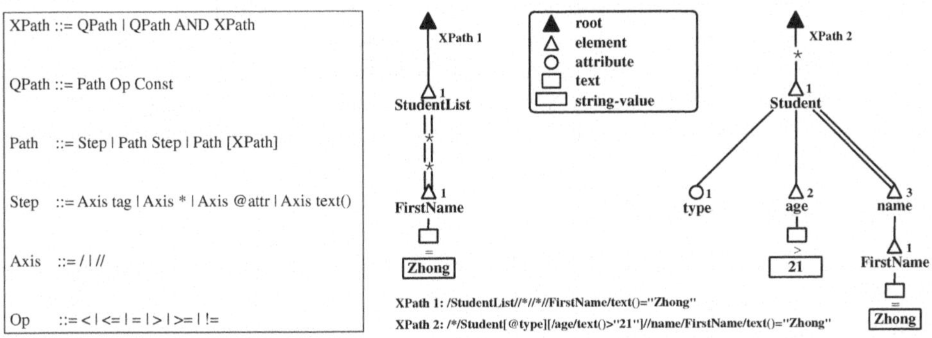

Fig. 3. The XPath subscription grammar and XPath trees of two subscription examples

3.2 Subscribe Algorithm

Fig. 3 shows our XPath subscription grammar and two XPath subscriptions with the corresponding XPath trees. The grammar is a subset of XPath called "*branching path expression*". The grammar does not include the boolean OR expression, an XPath with boolean OR expression should be rewritten to disjunctive normal form and disjunctions are submitted as separate subscriptions.

The depths of XML trees are usually very low; a statistical analysis over 200,000 XML documents had been performed [18], and discovered that 99% of the documents have less than 8 levels of nesting. For "//" may be used in XPath expressions, the depths of XPath trees should be less than the depths of XML trees; so, we match XPath subscriptions with each published XML document level-by-level; XPath subscriptions are parsed level-by-level, and the edges and the predicates in the ith level of the XPath tree are stored in relational table *XPathFrame*[i] and *XPathPred*[i] respectively. Each XPath subscription has an "*id*" which can uniquely identify the XPath. Each edge connects two nodes in the XPath tree; we use "*high-node*" and "*low-node*" to record the node that is close to the root and the node that is close to the leaf respectively. As shown in Fig. 3, the elements and the attributes that have the same depth in an XPath tree each has a unique serial number; we use "*high-num*" and "*low-num*" to record the serial number of the *high-node* and the serial number of the *low-node* respectively. The edges in the XPath tree that are connected by "⋆" are merged into a single edge; for example, the path from "*StudentList*" to "*FirstName*" in the XPath tree of XPath 1 in Fig. 3 has been merged into a single edge, and the depth difference between "*StudentList*" and "*FirstName*" in the XML tree must be not less than 3. We use "*max-diff*" and "*min-diff*" to record the maximal depth difference and the minimal depth difference between the *high-node* and the *low-node* in the XML tree respectively; in this way, "//", "/", and merged edge can be recorded in a uniform manner. In addition, we use "*branch-point*" and "*final*" to record whether the *high-node* of an edge is a branching point and whether the *low-node* of an edge is a leaf node respectively. For predicates, we use "*node*", "*node-num*", "*value*", and "*operation*" to record the tag of the node, the serial number of the node, the value, and the operation respectively; operations are recorded by integer values from 0 to 5, which mean "=", ">", "≥", "<", "≤", and "≠" respectively. Table 3 and Table 4 demonstrate how the two XPaths in Fig. 3 are stored in the *XPathFrame* tables and the *XPathPred* tables respectively. Deleting an XPath subscription on line is as easy as deleting all the tuples whose *id* equals to the *id* of the to be deleted XPath subscription.

3.3 Matching Algorithm

Fig. 4 and Fig. 5 show our algorithm for matching XPath subscriptions with an XML document inside the relational database by relational operations to find out matched subscriptions. The matching algorithm tries to reconstruct each XPath tree from the XML tree in a bottom-up manner; if an XPath tree can be reconstructed from the XML tree, the corresponding subscription is

Algorithm Matching(max_level ξ)

```
INSERT   INTO MatchedXPath(id)
SELECT   DISTINCT id
FROM     XPathFrame[0];
for i = ξ downto 0 do
begin
    FrameMatching(i);
    PredicateMatching(i);
    NeighborPruning(i);
    BranchingPointPruning(i);
    MatchedXPathCutting(i);
end
```

Algorithm FrameMatching(level i)

```
INSERT   INTO XMLMatch[i] (id, high_num, low_num, final, branch_point,
         high_start_point, low_start_point)
SELECT   id, high_num, low_num, final, branch_point,
         FH.start_point, FL.start_point
FROM     XMLFrame AS FH, XPathFrame[i], XMLFrame AS FL
WHERE    id IN (SELECT id FROM MatchedXPath)
AND      FH.node = high_node
AND      FL.node = low_node
AND      FH.start_point < FL.start_point
AND      FH.end_point > FL.end_point
AND      FL.depth − FH.depth BETWEEN min_diff AND max_diff;
```

Algorithm PredicateMatching(level i)

```
INSERT   INTO XMLMatch[i] (id, high_num, low_num, final, branch_point,
         high_start_point, low_start_point)
SELECT   id, node_num, null, true, false, start_point, null
FROM     XMLValue AS V, XPathPred[i] AS P
WHERE    id IN (SELECT id FROM MatchedXPath)
AND      P.node = V.node
AND      (CASE
           WHEN operation = 0 AND V.value = P.value    THEN 1
           WHEN operation = 1 AND V.value > P.value    THEN 1
           WHEN operation = 2 AND V.value >= P.value   THEN 1
           WHEN operation = 3 AND V.value < P.value    THEN 1
           WHEN operation = 4 AND V.value <= P.value   THEN 1
           WHEN operation = 5 AND V.value <> P.value   THEN 1
           ELSE 0
         END) = 1;
```

Fig. 4. The matching algorithm (1)

Algorithm NeighborPruning(level *i*)

```
DELETE
FROM      XMLMatch[i] AS MH
WHERE     MH.final = false
AND       NOT EXISTS (
            SELECT    *
            FROM      XMLMatch[i + 1] AS ML
            WHERE     MH.id = ML.id
            AND       MH.low_num = ML.high_num
            AND       MH.low_start_point = ML.high_start_point);
```

Algorithm BranchingPointPruning(level *i*)

```
DELETE
FROM      XMLMatch[i] AS M1
WHERE     M1.branch_point = true
AND       EXISTS ((
              SELECT    XF.low_num
              FROM      XPathFrame[i] AS XF
              WHERE     M1.id = XF.id
              AND       M1.high_num = XF.high_num)
            EXCEPT      (
              SELECT    M2.low_num
              FROM      XMLMatch[i] AS M2
              WHERE     M1.id = M2.id
              AND       M1.high_num = M2.high_num
              AND       M1.high_start_point = M2.high_start_point));
```

Algorithm MatchedXPathCutting(level *i*)

```
DELETE
FROM      MatchedXPath AS MX
WHERE     EXISTS ((
              SELECT    id
              FROM      XPathFrame[i] AS XF
              WHERE     MX.id = XF.id)
            UNION(
              SELECT    id
              FROM      XPathPred[i] AS XP
              WHERE     MX.id = XP.id))
AND       NOT EXISTS (
            SELECT    id
            FROM      XMLMatch[i] AS XM
            WHERE     MX.id = XM.id);
```

Fig. 5. The matching algorithm (2)

matched. Firstly, *id*s of all XPath subscriptions are inserted into the *MatchedXPath* table; then, all XPath subscriptions are reconstructed level-by-level in the bottom-up manner, and the reconstruction process for each level is consisted of five successive steps called frame matching, predicate matching, neighbor pruning, branching point pruning, and matched XPath cutting respectively. After the bottom-up level-by-level matching, *id*s in the *MatchedXPath* table are *id*s of the matched XPath subscriptions, and the published XML document should be delivered to all the subscribers whose subscriptions can be matched ultimately.

Step 1: Frame Matching. The frame matching step reconstructs the ith level edges of each XPath subscription from the published XML document and inserts the reconstructed edges into the *XMLMatch*[i] table. Only the ith level edges of the XPaths whose *id* is in current *MatchedXPath* table are considered to be reconstructed. The *XMLFrame* table occurs two times in the FROM clause, one for matching the *high-node* of the edge, the other for matching the *low-node* of the edge. As presented in Section 3.1, the ancestor-descendent relationship between two nodes α and β is determined by $start\text{-}point_\alpha < start\text{-}point_\beta$ and $end\text{-}point_\alpha > end\text{-}point_\beta$, and as presented in Section 3.2, the depth difference between the descendent and the ancestor is kept between *min-diff* and *max-diff*.

Step 2: Predicate Matching. The predicate matching step reconstructs the ith level predicates of each XPath subscription from the published XML document and inserts the reconstructed predicates into the *XMLMatch*[i] table. Only the ith level predicates of the XPaths whose *id* is in the *MatchedXPath* table are considered to be reconstructed. As presented in Section 3.2, integer values from 0 to 5 mean "=", ">", "≥", "<", "≤", and "≠" respectively, and a CASE-WHEN clause was used to math the XML node values with the XPath predicate values.

Step 3: Neighbor Pruning. The neighbor pruning step deletes the ith level reconstructed edges from the *XMLMatch*[i] table whose *low-node* is not a leaf node (*final*=*false*) and not exists an (i+1)th level reconstructed edge whose *high-node* is the same as the ith level reconstructed edge's *low-node* and not exists an (i+1)th level reconstructed predicate whose *node* is the same as the ith level reconstructed edge's *low-node*. Since the edges can not contribute to the reconstruction of the XPaths, they are deleted from the *XMLMatch*[i] table.

Step 4: Branching Point Pruning. The branching point pruning step deletes the ith level reconstructed edges from the *XMLMatch*[i] table whose *high-node* is a branching point and not all the branches of the branching point can be reconstructed from the published XML document. If the set difference between the set of branches in the XPath tree and the set of matched branches is not empty, we can ensure that not all the branches of the branching point can be reconstructed, and the branches should be deleted from the *XMLMatch*[i] table.

Step 5: Matched XPath Cutting. After neighbor pruning and branching point pruning, the matched XPath cutting step deletes *ids* of the XPath subscriptions from the *MatchedXPath* table which can not be reconstructed from the published XML document. If an XPath has at least one ith level edge or predicate, but not exists a reconstructed ith level edge or predicate, it is definite that the XPath can not be reconstructed from the published XML document, and the *id* of the XPath should be deleted. Without the cutting step, our algorithm still functions correctly, for we may delete *ids* of the XPath subscriptions which can not be reconstructed only in the top level. But, with the cutting step, *ids* of the XPaths which can not be reconstructed from the published XML document are deleted as early as possible; so, upper level frame matching and predicate matching need not to match the edges and the predicates of the subscriptions that had been deleted, and our algorithm can achieve better time performance.

4 Experiments

With our implementation, a wide variety of experiments can be conducted under different conditions. Due to space limitation, we only use two sets of experimental results to report the performance and the scalability of our publish/subscribe algorithm. The experiments run on a 1.4 GHz Pentium IV CPU with 2G of memory and a 160GB of SCSI hard disk driver running RedHat Enterprise Linux Advanced Server 4 and Oracle 9i Enterprise Edition; the database cache size was set to 32MB, and the database block size was set to 8KB. Only three SQL92 data types are used in the relational tables, which are *integer*, *boolean*, and *varchar* respectively. We set primary key for each relational table, and a B^+ tree unique index on the primary key will be created implicitly by the database system. {*id*}, {*start-point, end-point*}, {*start-point*}, {*id, high-num, low-num*}, {*id, high-num*}, and {*id, high-num, low-num, high-start-point, low-start-point*} were set as primary key of the *MatchedXPath* table, the *XMLFrame* table, the *XMLValue* table, *XPathFrame* tables, *XPathPred* tables, and *XMLMatch* tables respectively. In addition, two B^+ tree indices on the *XMLFrame* table are created explicitly for improving time performance of the frame matching step, one is on the *depth* attribute, the other is on the *node* attribute. The matching algorithm was implemented by a stored procedure in the Oracle database server, and the procedure has only one parameter which is the maximal level of the subscriptions.

In our experiments, XML messages and XPath subscriptions generated from Document Type Definition for the data and metadata at the Astronomical Data Center at NASA/GSFC [19] are used. Each XML message contains information about the metadata for a dataset, such as title, keywords, references, authors etc, and all of the associated tables, descriptions, and history. The XML messages are generated from the DTD by IBM's XML generator [20]; four XML messages are generated, whose size is 3KB, 15KB, 40KB and 100 KB respectively, and the maximal level of nesting is 5, 10, 11, and 12 respectively. The XPath subscriptions are generated from the DTD by XPath generator; each node in the XPath tree

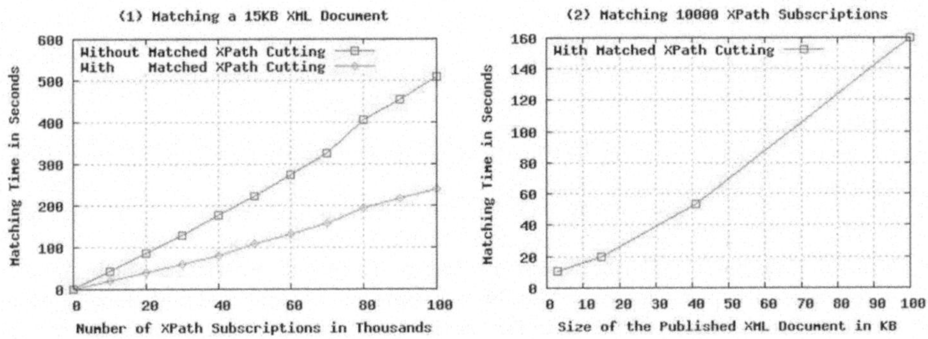

Fig. 6. Time performance of the matching algorithm. (1) The matching time in seconds for matching a 15KB XML document with XPath subscriptions whose number increases from 0 to 100000. (2) The matching time in seconds for matching 10000 XPath subscriptions with an XML document whose size increases from 3KB to 100KB.

has a 30% probability of being a branching point, has a 10% probability of being a "⋆"; each edge in the XPath tree has a 20% probability of being a "//"; the average and the maximal depth of the XPath trees are 5 and 10 respectively. In addition, due to the code of the complicated publish/subscribe system proposed in [12] belongs to IBM, and the paper only gives a brief introduction of the matching algorithm, we did not conduct comparison with the algorithm in [12].

In the first set of experiments, we match the 15KB XML message whose XML tree contains 328 inner nodes and 227 values with XPath subscriptions whose number increases from 0 to 100000. Fig 6.1 shows the time performance of the matching algorithm with and without matched XPath cutting. We can see that the matching algorithm with matched XPath cutting outperforms the algorithm without matched XPath cutting in a ratio of about 50%. In the second set of experiments, we match each of the generated 4 XML messages with 10000 XPath subscriptions. Fig 6.2 shows the time performance of the matching algorithm with matched XPath cutting. We can see that the growth rate of the matching time increases with the increase of the XML message size, the reason is that the percentage of the matched subscriptions increases at the same time.

5 Conclusions

In the paper, we introduced an XML publish/subscribe algorithm which is consisted of publish algorithm, subscribe algorithm, and matching algorithm. The matching algorithm uses relational operators to match XPath subscriptions with XML documents inside relational databases, so the scalability of the algorithm is no longer limited by amount of available physical memory. Experimental results show that the matching algorithm has good performance and scalability, so the algorithm should be a good choice for implementing publish/subscribe systems.

References

1. Jianjun Chen, David J. DeWitt, Feng Tian, and Yuan Wang. NiagaraCQ: A Scalable Continuous Query System for Internet Databases. Proceedings of SIGMOD 2000, 379-390.
2. Praveen Seshadri. Building Notification Services with Microsoft SQLServer. Proceedings of SIGMOD 2003, 635-636.
3. Hansjörg Zeller. NonStop SQL/MX Publish/Subscribe: Continuous Data Streams in Transaction Processing. Proceedings of SIGMOD 2003, 636.
4. Marcos Kawazoe Aguilera, Robert E. Strom, Daniel C. Sturman, Mark Astley, and Tushar Deepak Chandra. Matching Events in a Content-Based Subscription System. Proceedings of PODC 1999, 53-61.
5. Mehmet Altinel and Michael J. Franklin. Efficient Filtering of XML Documents for Selective Dissemination of Information. Proceedings of VLDB 2000, 53-64.
6. Yanlei Diao, Peter M. Fischer, Michael J. Franklin, and Raymond To. YFilter: Efficient and Scalable Filtering of XML Documents. Proceedings of ICDE 2002, 341.
7. Chee Yong Chan, Pascal Felber, Minos N. Garofalakis, and Rajeev Rastogi. Efficient Filtering of XML Documents with XPath Expressions. Proceedings of ICDE 2002, 235-244.
8. Chee Yong Chan, Pascal Felber, Minos N. Garofalakis, and Rajeev Rastogi. Efficient filtering of XML documents with XPath Expressions. VLDB Journal. 2002, 11(4), 354-379.
9. Todd J. Green, Gerome Miklau, Makoto Onizuka, and Dan Suciu. Processing XML Streams with Deterministic Automata. Proceedings of ICDT 2003, 173-189.
10. Feng Peng and Sudarshan S. Chawathe. XPath Queries on Streaming Data. Proceedings of SIGMOD 2003, 431-442.
11. Ashish Kumar Gupta and Dan Suciu. Stream Processing of XPath Queries with Predicates. Proceedings of SIGMOD 2003, 419-430.
12. Feng Tian, Berthold Reinwald, Hamid Pirahesh, Tobias Mayr, and Jussi Myllymaki. Implementing A Scalable XML Publish/Subscribe System Using Relational Database Systems. Proceedings of SIGMOD 2004, 479-490.
13. Masatoshi Yoshikawa, Toshiyuki Amagasa, Takeyuki Shimura, and Shunsuke Uemura. XRel: A Path-based Approach to Storage and Retrieval of XML Documents Using Relational Databases. ACM Transactions on Internet Technology. 2001, 1(1), 110-141.
14. Toshiyuki Amagasa, Masatoshi Yoshikawa, and Shunsuke Uemura. QRS: A Robust Numbering Scheme for XML Documents. Proceedings of ICDE 2003, 705-707.
15. Igor Tatarinov, Stratis Viglas, Kevin S. Beyer, Jayavel Shanmugasundaram, Eugene J. Shekita, and Chun Zhang. Storing and Querying Ordered XML Using a Relational Database System. Proceedings of SIGMOD 2002, 204-215.
16. Edith Cohen, Haim Kaplan, and Tova Milo. Labeling Dynamic XML Trees. Proceedings of PODS 2002, 271-281.
17. Xiaodong Wu, Mong-Li Lee, and Wynne Hsu. A Prime Number Labeling Scheme for Dynamic Ordered XML Trees. Proceedings of ICDE 2004, 66-78.
18. Laurent Mignet, Denilson Barbosa, and Pierangelo Veltri. Web XML: A First Study. Proceedings of WWW 2003, 500-510.
19. http://tarantella.gsfc.nasa.gov/xml/dataset_dtd.txt
20. http://www.alphaworks.ibm.com/tech/xmlgenerator

Retrieving Arbitrary XML Fragments from Structured Peer-to-Peer Networks

Toshiyuki Amagasa[1,2], Chunhui Wu[1], and Hiroyuki Kitagawa[1,2]

[1] Graduate School of Systems and Information Engineering,
Department of Computer Science
[2] Center for Computational Sciences,
University of Tsukuba
1–1–1 Tennodai, Tsukuba 305–8573, Japan
{amagasa,kitagawa}@cs.tsukuba.ac.jp, chunhuiwu@kde.cs.tsukuba.ac.jp

Abstract. This paper proposes a novel method for the storage and retrieval of XML in structured P2P networks. Peer-to-Peer (P2P), a new paradigm in the field of distributed computing, is attracting much attention. In particular, DHT (Distributed Hash Table) offers effective and powerful search through peer collaboration. However, query retrieval of arbitrary XML fragments using DHTs is a challenging issue because of the impedance mismatch between hashing, which is the core mechanism of the DHTs, and the complex structure of XML data and the XML query model. To address this problem, we propose a novel scheme for managing XML data in DHT. Specifically, we (virtually) deploy two kinds of DHTs, called Contents-DHT (C-DHT) and Structure-DHT (S-DHT), to map textual information and structural information, respectively. Queries against stored XML are processed using the DHTs concurrently. We can, therefore, retrieve arbitrary XML fragments. Effectiveness of the proposed scheme is demonstrated with a series of experiments.

1 Introduction

With the rapid development of high-performance processors and the diffusion of high-speed networks, distributed computing environments, where performance, reliability, and availability are all achieved, are gaining much attention. In particular, tailored networks built on top of other (existing) networks, called *overlay networks*, are emerging as a practical way to construct application-oriented networks on user demand. Nodes (or peers) in an overlay are virtually (or logically) connected by links provided by the underlying network. Peer-to-peer (P2P) networks are also a kind of overlay network, as they are constructed on top of the Internet.

P2P networks can be categorized roughly into two types, hybrid and pure, according to the way peers belonging to the network are organized. A hybrid P2P network has a central server, which is responsible for managing information about peers and shared resources. In such systems, the central server could be both a bottle neck and single point of failure.

G. Dong et al. (Eds.): APWeb/WAIM 2007, LNCS 4505, pp. 317–328, 2007.

To cope with this problem, pure P2P systems are emerging. In contrast with hybrid P2P systems, peers in a pure P2P system act equally; hence, there is no central server or central router. Pure P2P systems can further be categorized as *unstructured* and *structured*. In the unstructured type, a query is forwarded to as many peers as possible. The network traffic caused by flooding grows as the number of peers becomes large, which is a major disadvantage of unstructured systems.

Structured P2P networks, on the other hand, try to overcome such problems using a Distributed Hash Table (DHT), which is a mechanism residing in data (or a peer) in the network. In a typical DHT, individual hosts or data are mapped to a key by a hash function, and then located at a point in a keyspace. The keyspace is divided into several subspaces, and each host is responsible for a specific portion of the network. In addition, a global protocol locates the peer responsible for a given key. In this way, a search is performed, without flooding, by forwarding the request in steps using the protocol.

For several years, demands have grown to share domain-specific data using P2P networks. In such networks, Extensible Markup Language (XML) [1] has usually been used as a de facto standard for data representation and exchange. For this reason, an efficient means for the storage and retrieval of XML data in structured P2P networks is important. However, managing XML data in a DHT is not a trivial task: XML data are modeled as trees, whereas DHTs assume complete matching between a given search key and data, given the nature of hash functions. That means flexible search of XML data using DHTs is a challenging task [2]. Several papers have addressed this problem [3,4].

This paper proposes a novel scheme for managing XML data using DHT. Key features follow:

1. We decompose XML data based on textual values, such as text and attribute nodes (and leaf nodes for empty elements) and store them in a DHT. We show that we can sufficiently reconstruct XML fragments from that information. The extracted data is stored in a DHT, called Content-DHT (C-DHT), for querying textual values. Notice that they are indexed by element (attribute) names, which make descendant queries (//) efficient.
2. To allow internal nodes that are not stored in a C-DHT to be queried, we additionally store structural information of the stored XML data in a Structure DHT (S-DHT). The structural information is based on a path-prefix tree, which is a structural summary of XML data. An XPath [5] query is thus efficiently processed by collaborations of C-DHTs and the S-DHT.

We finally confirm the effectiveness of the proposed scheme with a series of experiments. Experimental results show that our proposed scheme performs well compared to existing methods.

2 Preliminaries

2.1 Distributed Hash Table (DHT)

The Distributed Hash Table (DHT) is a class of decentralized distributed systems. In a DHT, peers and objects are mapped to sets of keys in a space, called "keyspace." The mapping is usually done by a hash function, such as SHA-1. The keyspace is partitioned into several subspaces (keyspace partition), according to the scheme the DHT follows. Each participating peer in the network is responsible for maintaining the data in the assigned partition. When a request is issued, the message is forwarded from peer to peer until it reaches the peer that maintains the desired object. There are several variations of DHT according to the way keyspace is organized and how the routing table is maintained. Well-known implementations are Chord [6], Pastry [7], Tapestry [8], and CAN [9].

This work adopts Chord [6] as the underlying DHT. In Chord, a circular form of keyspace ("identifier space") with 2^N keys is assumed. Here, N is called "scale factor." A peer (data) is mapped to the identifier circle by a hash function. Supposing a peer is of the key k, it maintains the following information: 1) neighboring peers' information, $predecessor(k)$ and $successor(k)$, 2) the peer is responsible for the data within the range of $[predecessor(k) + 1, k]$. and 3) a routing table ("finger table") with N entries, each of which is a pointer to the peer, which is responsible for the data within the range of $[k+2^i, k+2^{i+1}-1](i = 0, \dots, N-1)$. From the configuration, the average number of hops to reach to the desired peer is $O(log_2)N$.

2.2 XML, XPath, and Path-Prefix Tree

XML [1] is a metalanguage for marking up semistructured data. XML data includes a properly nested tag structure, making it possible to extract a tree structure from XML data using syntax analysis. An XML tree is therefore modeled as a tree of $T = (V, E, r)$ where V denotes the set of nodes consisting of root, element, text, and attribute, E denotes the set of edges, and $r \in V$ represents the root of the tree.

To locate an arbitrary node (or subtree) in XML data, XPath [5] is used. An XPath expression consists of a number of *location steps* separated by '/', and a *location step* comprising an *axis*, *node test*, and optional *predicates*. In this paper, we employ a subset of XPath called core XPath, which allows child (/), descendant axes (//), and predicates ([]).

When managing XML data, path expressions play an important role, and we attempt to introduce a data structure, *path-prefix tree*, with which we can compactly represent the set of possible path expressions contained in an XML data. To introduce the path-prefix tree, let us first define the absolute path expression of an XML node.

Definition 1 (Absolute path expression). *For a given XML tree T, the absolute path expression of a node v is denoted as $PE(v) = /t_1/t_2/\dots/t_n$, where $t_1, \dots, t_n$ are the node (element) names from the root to v, and n is the depth of v.*

The following defines the path-prefix tree.

Definition 2 (Path-prefix tree). *The path-prefix tree of a given set of XML data is a tree T_p containing all absolute path expressions of the XML data.*

Note that a path-prefix tree of XML data is essentially the same as the Strong DataGuide [10]. We therefore do not go into detail on how to construct it. There is a simple one-path algorithm to construct the Strong DataGuide and we can borrow the idea to construct the path-prefix tree.

3 The Proposed Scheme

As discussed in Section 1, there are two technical challenges in the problem of XML data management using a DHT: 1) How to map XML data to a DHT for subsequent query processing; and 2) How to process a given query and retrieve the result as a set of arbitrary XML fragments.

When storing XML data in a DHT, a straightforward approach is to use a mapping scheme for mapping XML data to relational tables, and to store the resulting relational table in a DHT. Having constructed a DHT, given an XML query, it is decomposed and distributed around the peers in the DHT. [4] takes a similar strategy. It seems such approaches do not scale easily, i.e., even for a simple path query without predicates, a structural join [11] for each path step is needed, which is costly to process.

Another possibility is to shred the XML data being stored, and store the pieces in DHT with their path expressions as the keys [4]. This approach is quite effective if the XML schema is fixed, and there are a limited number of query patterns, by which we can decide where to partition the XML data. However, the disadvantage is that it lacks flexibility to cope with queries that include descendant axes (//), because there is no way to obtain appropriate search keys.

To cope with this problem, we propose a novel scheme for storing and querying XML data using DHTs. We map every leaf node, including text nodes, attribute nodes, and empty elements, to a DHT. In other words, we do not map internal nodes, because necessary information is all stored in the textual nodes and, by contrast, internal (element) nodes are acting as containers. By omitting those nodes, we can significantly reduce the number of keys stored in a DHT. Beyond that, to provide an entry point for querying (omitted) internal nodes, we construct a path-prefix tree and store it in another DHT. A query is processed as collaborations of the two DHTs.

3.1 Content-DHT (C-DHT)

In the proposed scheme, we use two kinds of DHTs. One is Content-DHT (C-DHT) for managing the contents of XML data. As discussed above, given XML data, we basically extract all leaf nodes, including text nodes, attribute nodes, or empty elements, and put them into a DHT with the following information with its *element (or attribute) name* as the key:

$$(peerID, docID, pathexp, label, value)$$

where *peerID* is the key of the peer that keeps the original XML data, *docID* is the document ID (or filename), *pathexp* is the absolute path expression of the node, *label* is the node label that represents structural information of the node, and *value* is the value of the node.

For the node label, there are many candidate schemes for encoding structural information of XML trees [12]. Small examples are preorder and postorder pair [13], range labeling [14], and the Dewey order [15]. In this work, we attempt to use the Dewey order for several reasons. One is that we can represent detailed structural relationships, not only ancestor-descendant relationship, but also parent-child and order relationships among sibling nodes. This is important to ensure that we can reconstruct an XML fragment from only its leaf nodes, even if we do not have all nodes of an XML subtree. Another feature is that we can replace it with another insert-friendly labeling scheme like ORDPATHs [16] to cope with updates in XML data.

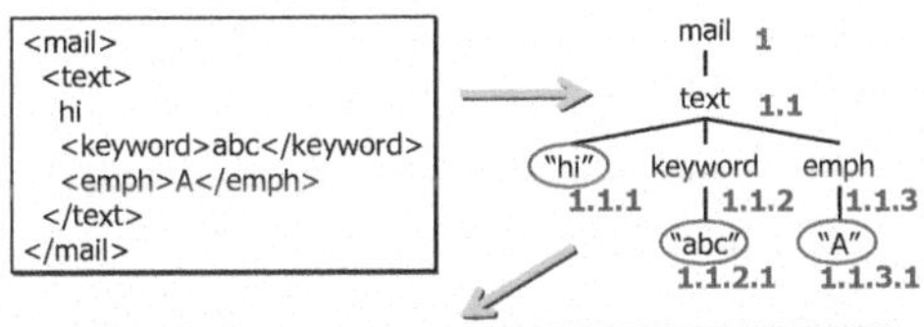

key	peerID	docID	pathexp	label	value
text	1	1	/mail/text	1.1.1	"hi"
keyword	1	1	/mail/text/keyword	1.1.2.1	"abc"
emph	1	1	/mail/text/emph	1.1.3.1	"A"

Fig. 1. Content-DHT (C-DHT)

Figure 1 depicts a storage example of C-DHT. An XML tree is constructed from XML data and node labels are assigned. Of the nodes, only three text nodes are extracted and put in a DHT. Notice that in the table below, we intend to show the entire list of indexes in the DHT, even if it is not distributed in the figure. In a working system, each row is distributed to peers according to the key value.

An important point here is that by omitting internal nodes, we can significantly reduce the number of entries to a DHT. We do not show the detailed statistics due to the page limitation, but we found that 40% to 50% of entries can be omitted for real XML data, such as DBLP Bibliography, Gene Ontology, and INEX. This gives a significant impact on both storage and search performance on the peers.

Reconstructing XML fragments from leaf nodes. As mentioned, we can reconstruct XML data from a list of leaf nodes even if we do not have a complete list of nodes that includes internal nodes. This becomes possible because we have the absolute path expression and the Dewey label for each leaf node.

Let us look at Figure 1. In the figure are three leaf nodes, whose absolute path expressions and Dewey labels are (/mail/text, 1.1.1), (/mail/text/keyword, 1.1.2.1), (/mail/text/emph, 1.1.3.1). All the internal nodes can be extracted by taking i-th prefixes of the path expressions and the Dewey labels, i.e., by taking the first prefix, we get (/mail, 1), and (/mail/text, 1.1) by the second prefix, and so on. In this way, we can reconstruct an XML fragment from a list of leaf nodes.

Query processing by C-DHT. Suppose p is an core XPath query of the form: $\mid e_1[p_1] \mid e_2[p_2] \mid \ldots \mid e_n[p_n]$ where "|" is a separator ("/" or "//"), e_i is an element (or attribute) name, and optional p_i is a predicate. For simplicity, let us begin with a simple path expression that does not contain any predicates. How to process queries including predicates will be discussed later. Query processing proceeds as follows:

1. First extract the element (or attribute) name at the bottom (e_n), and look up the C-DHT.
2. Then, for each candidate result with key e_n, check if the absolute path expression matches with the query, and filter out unqualified candidates. Specifically, rewrite the query using regular expression by replacing "//" with "*". As a consequence, ordinary string match functions can be used to perform the matching process.

For example, queries "/mail/text/emph", "//emph", and "/mail//emph" can be processed in the same way. Specifically, once candidate results are found, we translate the queries "//emph" and "/mail//emph" to regular expressions "/*/emph" and "/mail/*/emph" for subsequent string match with the absolute path expressions in the candidates. In addition, queries like "//abc/emph" will be omitted at the string match phase.

In this way, we can flexibly answer XPath queries, but we cannot query internal nodes because there is no entry point in a C-DHT for them. For this reason, we need an additional index called Structure-DHT (S-DHT).

3.2 Structure-DHT (S-DHT)

To provide a clue to querying internal nodes, we additionally maintain structural information about stored XML data. For that purpose, we use the path-prefix tree mentioned above. In practice, we can use it as a structural summary of a set of XML data because it captures all possible path expressions appearing in the XML data.

Given a path-prefix tree, we map every node to S-DHT with the element (or attribute) name as its key. The value is a tuple of the form:

$$(pathexp, child, count)$$

where *pathexp* represents the absolute path expression of the node in the path-prefix tree, *child* is a set of child nodes, and *count* shows the number of corresponding XML nodes in the stored XML data.

Figure 2 illustrates an example of S-DHT. Again, this table attempts to show the contents of a DHT, and each tuple is distributed to a peer with respect to the key value in a real system. From the figure, we see that the structure of the path-prefix tree is completely captured in the mapped table except for ordinal information among sibling nodes. Specifically, with the *child* attribute, we can determine the possible occurrences of child elements. Notice that "#TEXT" is a special symbol representing text node. If the node is an empty element, the child is "#EMPTY". Such information structure is similar to the content model of XML schema languages such as DTD. The *count* attribute maintains statistical information that can be used in query planning.

Query processing using S-DHT combined with C-DHT. Query processing is simple. For a given query $p = /e_1/\ldots/e_n$, we attempt to look for all leaf

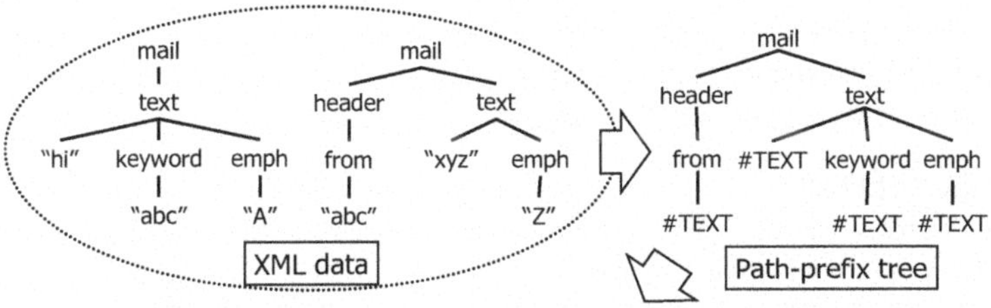

key	pathexp	child	count
mail	/mail	{header, text}	2
header	/mail/header	{from}	1
from	/mail/header/from	{#TEXT}	1
text	/mail/text	{#TEXT, keyword, emph}	2
keyword	/mail/text/keyword	{#TEXT}	1
emph	/mail/text/emph	{#TEXT}	2

Fig. 2. Structure-DHT (S-DHT)

nodes contained in subtrees that are rooted by p by referring to C-DHT and S-DHT concurrently. Specifically, we first look up C-DHT with search key e_n and find tuples that have path expressions conforming to p. We then put them in the result set. In the next step, we look up S-DHT with search key e_n looking for an internal node that matches the query p. If we find entries that match p, we rewrite the query by concatenating p and the child element's name, and continue until the complete list of leaf nodes is obtained.

Following is the algorithm *Retrieve* that processes a given core XPath query using C-DHT and S-DHT. Suppose we can refer to the a attribute of t tuple by $t.a$ in the following discussion, e.g., $t.pathexp$ denotes the path expression of t tuple.

Processing branching queries. We have discussed simple (non-branching) queries that do not contain predicates. We can deal with branching cases as an extension of the simple path expression case in that two or more simple path expressions are processed. Suppose we are given the following query $/mail/text[emph = "A"]$. By parsing the query, we can extract all simple path expressions contained in the query, i.e., $/mail/text$ and $/mail/text/emph$. We then call *Retrieve* for each that yields several lists of leaf nodes. For each list, we extract the root of each subtree by computing the common prefix of both the absolute path expression and the Dewey label. Finally, we perform structural join [11] on them to filter out unnecessary candidates.

3.3 Discussions

Load balancing. In the proposed scheme, the variation of keys in the DHTs is obviously bounded by the number of elements (attributes), which may cause an unbalanced load among peers. This means that even if there are 10,000 peers,

Algorithm 1. Retrieve(p)

Input: p {core XPath query $/e_1/\ldots/e_n$}
Output: R {set of resulting tuples}
  $R \leftarrow \emptyset$
  $T \leftarrow$ lookupCDHT(e_n)
  for all $t \in T$ **do**
    if $t.pathexp$ matches with p **then**
      $R \leftarrow R \cup c$
    end if
  end for
  $S \leftarrow$ lookupSDHT(e_n)
  for all $c \in s.child$ **do**
    if $c \notin \{\#\mathrm{TEXT}, \#\mathrm{EMPTY}\}$ **then**
      $R \leftarrow R\cup$ Retrieve(p/c)
    end if
  end for
  return R

approximately only 10 peers are under load if we have only 10 kinds of elements. From one perspective, this is an advantage because related information is clustered, which achieves efficient query processing. However, it may cause a problem if there are extremely large entries.

To cope with large entries, we might introduce parameter M, which is the maximum number of entries for each key (element or attribute name). If the number is too high, the peer tries to find another peer to share the load. When processing a query, the peer originally responsible for the key acts as the mediator to collect necessary information from the peers sharing the load, and to respond to the request.

4 Implementation

This section describes an implementation of the proposed scheme. Our prototype system is implemented using the Overlay Weaver [17], which is a toolkit for constructing P2P networks. It provides a wide range of functionalities, which are necessary for implementing P2P networks; it enables us to simulate a large-scale virtual P2P network as well as real P2P networks.

Since our scheme is based on the basic functionalities of DHTs, the following APIs are needed in the implementation:

put(key, value, ttl)] Put a *value* with *key*. *ttl* (time to live) specifies the expiration time of the value in the DHT.
get(key) Get the set of values with *key*.
remove(key, [value)] Remove all the values with *key* from the DHT. If optional *value* is specified, it only removes *value* with *key*.

The prototype system comprises two main modules, *publish* and *search*. The *publish* module, taking XML data as the input, parses the data to extract leaf

Table 1. Benchmark queries

No.	Query
1	/site/people/person/name
2	/site/regions/asia/item/description/keyword
3	/site/regions/europe/item[payment="cash"]
4	/site/people/person[5]/profile/age
5	//asia

nodes and also to construct the path-prefix tree. The extracted leaf nodes are put in the C-DHT. The constructed path-prefix tree is then traversed by putting each node to the S-DHT. If an entry with the same key and the same path expression is already registered, the *count* value is incremented by one. The *search* module basically does the same thing as described in Section 3.

5 Performance Study

Experimental setup. We have conducted a series of experiments to confirm effectiveness of the proposed scheme. The datasets contain two kinds of synthetic XML data generated by XMark [18] and XBench [19], and the sizes are 10MB and 50MB.

For benchmark queries, we used the five XPath queries given in Table 1. Queries 1 and 2 are simple path expressions, 3 and 4 contain predicates, and 5 contains descendant axes.

For the baseline, which is to be compared with the proposed scheme, we have also implemented XP2P [3] using Overlay Weaver, and compare the number of messages needed to process each query by changing the number of peers from 200 to 600.

All the experiments are made using a PC with 4-way Intel Xeon 3GHz processor and 6GB Memory running Linux kernel 2.6.9-24. The version of Java Runtime Environment is 1.5.0_07.

Experimental results. In Figures 3 a) to e), the horizontal and vertical axes show the number of peers and the number of messages, respectively. Viewing the figures, we observe that 1) the number of messages grows linearly as the number of peers grows; 2) our (proposed) scheme outperforms XP2P in that fewer messages are necessary; 3) it seems that the performance of our scheme is more stable, whereas XP2P's performance changes depending on queries, e.g., more messages are required if the query expression is long, because XP2P must navigate a query expression from the head to tail by looking up a DHT; and 4) even if a query contains predicates ([]) or descendant axes (//), our scheme can process it efficiently.

Figure 3 f) shows load distribution with 50MB datasets. As the figure reveals, from the nature of the scheme, the load distribution in our system is not balanced, whereas the load is balanced in XP2P. This behavior is considered beneficial in that related data are clustered, which results in better query

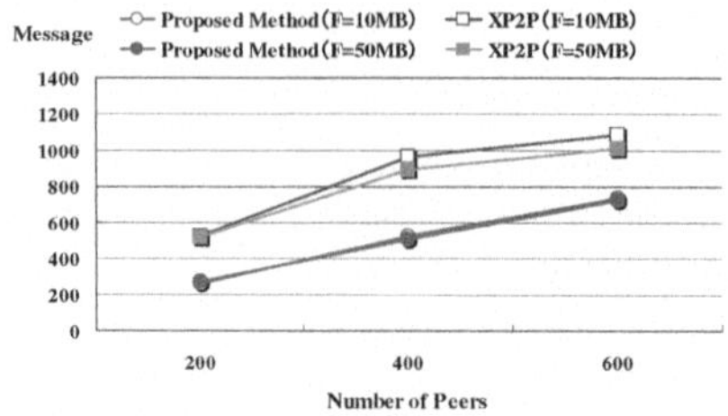

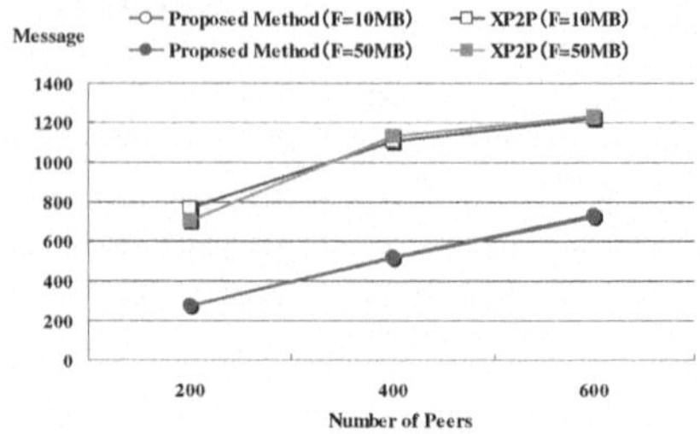

a) /site/people/person/name (left) and **b)** /site/regions/asia/item/description/ keyword (right)

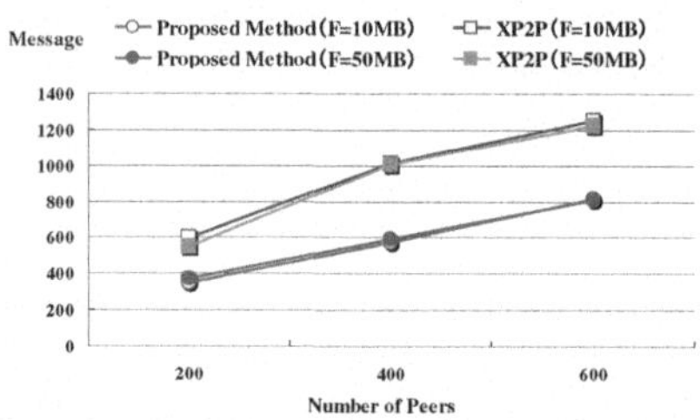

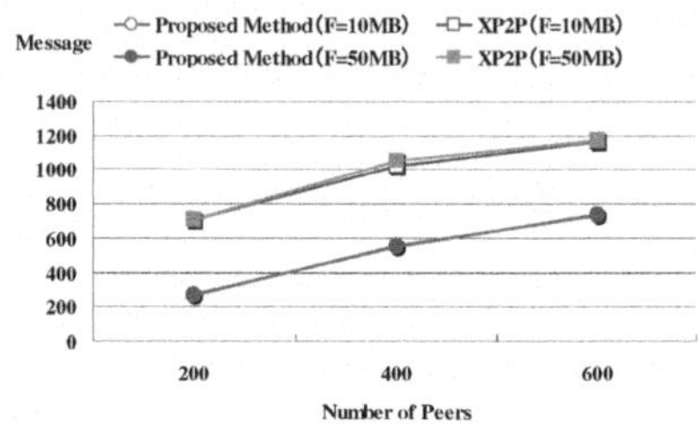

c) /site/regions/europe/item[payment='cash'] (left) and **d)** /site/people/person[5]/ profile/age (right)

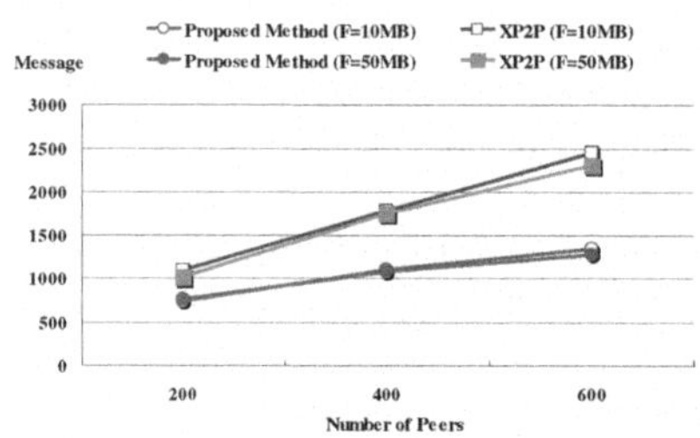

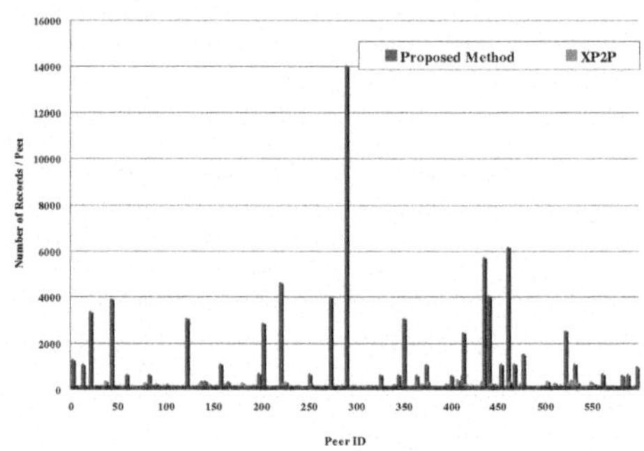

e) //asia (left) and**f)** Load distribution (# peers 600)

Fig. 3. Experimental results

performance. However, this also suggests the need to cope with extreme cases where most of the load is assigned to some (a limited number of) peers. We might introduce the technique discussed in Section 3.3.

6 Related Work

A number of papers have addressed the problem of XML data management in P2P networks [3,4,20]. XP2P [3] and XPeer [4] are based on DHT, while yet another XPeer [20] is based on hybrid P2P system. In XP2P [3], XML data is fragmented and stored in a DHT with its path expression. By contrast, XPeer [4] stores XML data using node-based decomposition. These other papers differ from our work in that they do not use DHT.

From a different perspective, many efforts attempt to deal with various kinds of data. [21] tries to deal with RDF (Resource Description Framework) using DHT [22]. RDF data is decomposed to a set of triples, which is the non-decomposable unit of RDF data, and then stored in DHT. [23,24] deal with relational data, trying to construct relational storage that can perform relational query processing in a P2P environment.

7 Conclusions

In this paper, we have proposed a novel scheme for storing and querying XML data using DHT. We introduced two kinds of DHTs: Content-DHT (C-DHT) and Structure-DHT (S-DHT) for maintaining both the content and structure of XML data. Using the DHTs, we decompose XML data based on textual values such as text and attribute nodes (and leaf nodes for empty elements) and store them in the C-DHT. Additionally, to maintain structural information of the XML data, we construct a path-prefix tree and map it to the S-DHT. Query processing is achieved using the DHTs concurrently. Experimental results show that the proposed scheme performs well, even for queries that contain predicates or descendant axes (//). In the future, we will extend the current work to cope with updates and more complex queries like XQuery.

Acknowledgements. This research is partly supported by the Grant-in-Aid for Scientific Research (#15300029 and #17700110) from Japan Society for the Promotion of Science (JSPS), Japan, and Grant-in Aid from Core Research for Evolutinal Science and Technology (CREST), Japan.

References

1. W3C: Extensible Markup Language (XML) 1.0 (Third Edition). `http://www.w3.org/TR/xml/` (April 2004) Recommendation.
2. Koloniari, G., Pitoura, E.: Peer-to-Peer Management of XML Data: Issues and Research Challenges. SIGMOD Record **34**(2) (2005) 6–17
3. Bonifati, A., Matrangolo, U., Cuzzocrea, A., Jain, M.: XPath Lookup Queries in P2P Networks. In: Proc. WIDM 2004. (2004) 48–55
4. Rao, W., Song, H., Ma, F.: Querying XML Data over DHT System Using XPeer. In: Proc. GCC 2004. (2004) 559–566
5. W3C: XML Path Language (XPath) Version 1.0. `http://www.w3.org/TR/xpath.html` (November 1999) Recommendation.
6. Stoica, I., Morris, R., Karger, D.R., Kaashoek, M.F., Balakrishnan, H.: Chord: A Scalable Peer-to-Peer Lookup Service for Internet Applications . In: Proc. ACM SIGCOMM 2001. (2001) 149–160

7. Rowstron, A., Druschel, P.: Pastry: Scalable, Decentralized Object Location, and Routing for Large-Scale Peer-to-Peer Systems. In: IFIP/ACM Int'l Conf. on Distributed Systems Platforms (Middleware). Volume 2218 of Lecture Notes in Computer Science. (2001) 329–350
8. Zhao, B.Y., Kubiatowicz, J.D., Joseph, A.D.: Tapestry: An Infrastructure for Fault-tolerant Wide-area Location and Routing. Technical Report UCB/CSD-01-1141, UC Berkeley (April 2001)
9. Ratnasamy, S., Francis, P., Handley, M., Karp, R., Shenker, S.: A Scalable Content-Addressable Network. In: Proc. ACM SIGCOMM 2001. (2001) 161–172
10. Goldman, R., Widom:, J.: DataGuides: Enabling Query Formulation and Optimization in Semistructured Databases. In: Proc. VLDB 1997. (1997) 436–445
11. Al-Khalifa, S., Jagadish, H.V., Patel, J.M., Wu, Y., Koudas, N., Srivastava, D.: Structural Joins: A Primitive for Efficient XML Query Pattern Matching. In: Proc. ICDE 2002. (2002) 141–152
12. Catania, B., Maddalena, A., Vakali, A.: XML Document Indexes: A Classification. IEEE Internet Computing **9**(5) (2005) 64–71
13. Dietz, P.F.: Maintaining order in a linked list. In: Proc. of the fourteenth annual ACM symposium on Theory of computing table of contents. (1982) 122–127
14. Yoshikawa, M., Amagasa, T., Shimura, T., Uemura, S.: XRel: a path-based approach to storage and retrieval of XML documents using relational databases. ACM Transactions on Internet Technology (TOIT) **1**(1) (2001) 110–141
15. Tatarinov, I., Viglas, S.D., Beyer, K., Shanmugasundaram, J., Shekita, E., Zhang, C.: Storing and Querying Ordered XML Using a Relational Database System. In: Proc. ACM SIGMOD 2002. (2002) 204–215
16. O'Neil, P.E., O'Neil, E.J., Pal, S., Cseri, I., Schaller, G., Westbury, N.: ORDPATHs: Insert-Friendly XML Node Labels. In: Proc. ACM SIGMOD 2004. (2004) 903–908
17. Shudo, K.: Overlay Weaver An Overlay Construction Toolkit. `http://overlayweaver.sourceforge.net/`
18. Project, X.A.X.B. `http://monetdb.cwi.nl/xml/`
19. of Benchmarks for XML DBMSs, X.A.F.: Xbench. `http://se.uwaterloo.ca/~ddbms/projects/xbench/`
20. Sartiani, C., Manghi, P., Ghelli, G., Conforti, G.: XPeer: A Self-Organizing XML P2P Database System. In: EDBT Workshops 2004. (2004) 456–465
21. Cai, M., Frank, M.R.: RDFPeers: a scalable distributed RDF repository based on a structured peer-to-peer network. In: Proc. WWW 2004. (2004) 650–657
22. W3C: Resource Description Framework (RDF): Concepts and Abstract Syntax. `http://www.w3.org/TR/2004/REC-rdf-concepts-20040210/` (feb 2004) Recommendation.
23. Huebsch, R., Hellerstein, J.M., Lanham, N., Loo, B.T., Shenker, S., Stoica, I.: Querying the Internet with PIER. In: Proc. VLDB 2003. (2003) 321–332
24. Ng, W.S., Ooi, B.C., Tan, K.L., Zhou, A.: PeerDB: A P2P-based System for Distributed Data Sharing. In: Proc. ICDE 2003. (2003) 633–644

Combining Smooth Graphs with Semi-supervised Learning

Liang Liu[1], Weijun Chen[2], and Jianmin Wang[3]

[1] School of Software, Tsinghua University,
Beijing, 100084, P.R. China
liang-liu05@mails.tsinghua.edu.cn

[2] School of Software, Tsinghua University,
Beijing, 100084, P.R. China
cwj@tsinghua.edu.cn

[3] School of Software, Tsinghua University,
Beijing, 100084, P.R. China
Jimwang@tsinghua.edu.cn

Abstract. The key points of the semi-supervised learning problem are the label smoothness and cluster assumptions. In graph-based semi-supervised learning, graph representations of the data are so important that different graph representations can affect the classification results heavily. We present a novel method to produce a graph called smooth Markov random walk graph which takes into account the two assumptions employed by semi-supervised learning. The new graph is achieved by modifying the eigenspectrum of the transition matrix of Markov random walk graph and is sufficiently smooth with respect to the intrinsic structure of labeled and unlabeled points. We believe the smoother graph will benefit semi-supervised learning. Experiments on artificial and real world dataset indicate that our method provides superior classification accuracy over several state-of-the-art methods.

1 Introduction

In many machine learning settings, there are only a few labeled examples but quite a number of unlabeled examples. Since labeling often requires expensive human labors, unlabeled examples are much easier to get compared with labeled ones. The goal of the learner is to predict the labels of the unlabeled data points. This learning method is often called a semi-supervised or transductive learning. A typical application of this learning method is web categorization. To train such a system to automatically classify web pages, one would typically rely on manually labeled web pages and the manual work highly increases the labeling cost. In contrast, the number of unlabeled web pages is large. To achieve good performance, the learning algorithm must take as much advantage of the unlabeled data as possible.

To make use of unlabeled data, two basic assumptions are employed directly or indirectly by semi-supervised learning. The first one is labels smoothness

G. Dong et al. (Eds.): APWeb/WAIM 2007, LNCS 4505, pp. 329–340, 2007.

assumption, which supposes that neighboring data points tend to obtain the same label. The other one is called "cluster assumption" which assumes that data points on the same structure (typically referred as a cluster) are likely to have the same label, or that the decision boundary should lie in regions of low density. These two assumptions are true in many real world settings and also true in human intuition.

Based on these two assumptions, some methods are successfully applied in semi-supervised learning. Seeger[1] and Zhu[2] reviewed the existing techniques of semi-supervised learning. Gaussian function[3] uses the average of the indicating function value f at neighboring points as the value of each unlabeled data point. In the consistency method[4], the indicating function values of unlabeled data points are smoothed by regularization method. A common aim of these methods is to make sure that neighboring points have approximately the same indicating function value. Corresponding to these methods, cluster kernels[6] are constructed such that induced distance is smaller for points in the same cluster and larger for points in different clusters. In [5], low density regions are identified to separate clusters. The main difference between various semi-supervised learning algorithms lies in their way they realize these assumptions.

Although these two assumptions seem different, in many semi-supervised learning settings they share the same meaning that is a good indicating function or decision function estimated by the semi-supervised learning algorithms should change slowly on the coherent structure aggregated by a quantity of data and greatly on the incoherent region. The assumptions also reflect the same data distribution requirement that the data of one cluster is usually distributed continuously in a relatively high density region, and low density regions usually separate different clusters[12].

However, real world data might not be distributed ideally and the assumptions might be damaged to some extent, especially in high dimension cases such as text data. In this case, the density of data varies largely, and low density regions may lie inside some clusters. Therefore, it is important to build a much smooth graph representation which highly reflects the nature of the data for graph-based semi-supervised learning just like Gaussian function method and the consistency method.

Most of these methods focus on the process of constructing the decision function itself to smooth the graph. We attempt to smooth the graph explicitly first, and then label the unlabeled data using any other graph-based semi-supervised learning methods. To overcome the difficulties of bad data distribution, we build a smooth graph using Markov random walk model. Our method can combine local and global distribution of labeled and unlabeled data to make the graphical representation fit labels smoothness and cluster assumption better than original representation.

The paper is organized as follows: Section 2 introduces Markov random walk model. Section 3 introduces some typical graph-based semi-supervised methods. In section 4, the paper explores the idea of how to build a novel graph representation which takes into account the labels smoothness and cluster assumption.

Section 5 presents the experimental results and finally we conclude this paper in Section 6.

2 Graph Based on Markov Random Walk

Many graph based methods which have done well in semi-supervised learning are a type of Markov random walk[11]. More formally, consider a point set $X = \{x_1, x_2, \ldots, x_l, x_{l+1}, \ldots, x_n\}$ and a label set $\{1, \ldots, c\}$ where the first l points have desired labels $\{y_1, y_2, \ldots, y_l\} \in \{1, \ldots, c\}$ and the remaining points are unlabeled. We use L and U to denote labeled and unlabeled data points respectively. Consider a graph $G(V, E)$ with nodes $\{v_1, v_2, \ldots, v_l\}$ corresponding to the labeled points with labels $\{y_1, y_2, \ldots, y_l\}$, and nodes $\{v_{l+1}, v_{l+2}, \ldots, v_n\}$ corresponding to the unlabeled points. An $n \times n$ symmetric weight matrix W on the edges of the graph is constructed and the weight of each undirected edge in the graph is defined as:

$$w_{ij} = exp(-\frac{||x_i - x_j||^2}{2\sigma^2}) \tag{1}$$

Where x_i is a vector representing the i-th data point. σ is a parameter for exponential weight. Thus, nearby points are assigned large edge weights. The one-step transition probabilities p_{ij} from data point i to j can be obtained directly from these weights:

$$p_{ij} = \frac{w_{ij}}{\sum_k w_{ik}} \tag{2}$$

We can get a transition matrix $P = [p_{ij}]$. Such a random walk graph can be viewed as a Markov chain. For classification, data points are mapped to nodes in the graph. The transition probability can be seen as the similarity between data point pairs. The labeled data points are assumed as the ending states for the Markov random walk. We can evaluate the probabilities that the Markov process starts from an unlabeled data point and ends up in a labeled point after t steps. If the probabilities are quite high, the unlabeled point belongs to the same category as the labeled point. The probabilities should stay low if both points belong to two different categories.

3 Label Propagation on the Graph

Such a Markov random walk graph representation is used in several semi-supervised learning methods [2][3][4][6][7]. These methods formulate the semi-supervised learning problem as a form of label propagation on the graph, where a node's label propagates to neighboring nodes according to their proximity. In the process, the labels on the labeled data are fixed, and thus labeled data act as sources that push out labels through unlabeled data.

3.1 The Gaussian Field and Harmonic Function

This method is present in [3]. The transition matrix P is calculated as above . Also define a $l \times c$ label matrix Y_L with $Y_{ij} = 1$ if x_i is labeled as $y_i = j$ and $Y_{ij} = 0$ otherwise. F is a $n \times c$ matrix and let $F = (F_L, F_U)^T$. The rows of F can be interpreted as the probability distribution over labels. The Gaussian Field and Harmonic Function is as follows:

1. $F^{(n)} = F^{(n-1)} \times P$
2. Clamp the labeled data $F_L = Y_L$.
3. Repeat from step 1 until F converges.

We are solely interested in F_U and split P into labeled and unlabeled sub-matrices.

$$P = \begin{bmatrix} P_{LL} & P_{LU} \\ P_{UL} & P_{UU} \end{bmatrix} \tag{3}$$

It can be shown that:

$$F_U = (I - P_{UU})^{-1} P_{UL} Y_L \tag{4}$$

F_U is the unique fixed point and the solution to iterative algorithm.

3.2 The Local and Global Consistency Method

The second label propagation algorithm is the local and global consistency method[4]. Define a $n \times c$ matrix Y with $Y_{ij} = 1$ if x_i is labeled as $y_i = j$ and $Y_{ij} = 0$ otherwise. Clearly, Y is consistent with the initial labels. Iterate the following equation until convergence.

$$F^{(n)} = \alpha S F^{(n-1)} + (1 - \alpha) Y \quad 0 < \alpha < 1 \tag{5}$$

Let F^* denote the limit of the sequence $F^{(n)}$. Suppose $F^{(0)} = Y$ and by the iteration equation, we have:

$$F^* = (I - \alpha P)^{-1} Y \tag{6}$$

The difference between the two typical label propagation methods is that in the local and global consistency method, each data point receives the information from its neighbors and also retains its initial information in each repetition. The parameter α specifies the relative amount of the information from its neighbors and its initial label information.

4 Smooth Markov Random Walk Graph

In label propagation we need a graph, represented by the transition matrix which reflects the structure of the graph. Data points usually propagate their labels to the ones nearby according to the possibilities on the edges, and the class boundaries will be pushed through high density regions and settled in low density gapes.

The smoothness here refers to the fact that the distribution of data changes slowly in one category and greatly between two separated categories. Based on Markov random walk graph, we propose the following framework which is consisted of two steps to construct a smooth graph under labels smoothness and cluster assumption in order to get a more accurate classification result.

4.1 Step 1: Graph Smoothing Step

This step aims to change the local structure of the graph and avoid low density regions inside some clusters in order to make data points get higher probabilities to the ones nearby and walk denser regions more likely than sparser ones.

As before, we can get a full connected graph G and its representation matrix W from labeled and unlabeled data points. Define D to be the diagonal matrix whose (i, i)-element is the sum of W's i-th row, and compute the transition matrix $P = D^{-1/2}WD^{-1/2}$. We propose the following graph as the substitute for the basic graph representation.

$$P^* = P^m \tag{7}$$

If we map the graph $G(V, E)$ into a Markov chain, the node set V is the state set I. It is easy to satisfy the following conditions: it is a finite-state Markov chain with no two disjoint closed sets and it is aperiodic. And we can have the following result [11]: there exists a probability distribution $\{\pi_j, j \in I\}$, such that for all $i, j \in I$

$$|p_{ij}^{(m)} - \pi_j| \leq \alpha\beta^{(m)} \tag{8}$$

Where $\alpha > 0$ and $0 < \beta < 1$. In particular, for all $i, j \in I$

$$\lim_{m->\infty} p_{ij}^{(m)} = \pi_j \tag{9}$$

In equation (8)(9), $p_{ij}^{(m)}$ is an element in matrix P^m. Therefore, we can prove that as m becomes larger, $p_{ij}^{(m)}$ gets closer to the fixed value π_j.

From previous section, we know that larger probabilities indicate larger weights in the graph. As the number of m gets larger, $p_{ij}^{(m)}$ between far points in the same cluster becomes larger, which means its corresponding weight gets larger. And in turn, it means the distance between the two points far away now becomes smaller. Furthermore, this kind of distance shortening is based on the density of the data. Distances between points in one cluster are shortened more but in different clusters are shortened less.

In addition, it should be noted that m here should not get too large. When $p \to \infty$, P^m will become an uniform distribution for each point, which provides little information about the learning processing.

4.2 Step 2: Graph Clustering Step

Graph clustering step aims to change the global structure of graph such that the distance between the points in different clusters is much larger than that between the points in the same cluster. In other word, data points in the same cluster are grouped together in the new representation. The detail of the step is as follows:

1. Find $r_1, r_2, \ldots, r_n$, the eigenvectors of P^*(chosen to be orthogonal to each other in the case of repeated eigenvalues)which correspond to eigenvalues $\lambda_1, \lambda_2, \ldots, \lambda_n$. Form the matrix $U = [r_1 r_2 \ldots r_n]$ by stacking the eigenvectors in columns and the matrix Λ by placing the corresponding eigenvalues on the diagonal and then $P^* = U \Lambda U^T$.
2. Let $\lambda_i^{'} = (\lambda_i)^{1/m}$ if $\lambda_i \geq \lambda_k$ and $(\lambda_i)^m$ otherwise, where λ_k is the k-th largest eigenvalue of P* and construct a new diagonal matrix $\Lambda^{'}$. This step tends to push those smaller eigenvalues to 0 and make larger eigenvalues getting bigger.
3. Compute $\tilde{P} = U \Lambda^{'} U^T$ and let $\tilde{P}_{ii} = 0$ which form a non-lazy random walk.

The smooth Markov walk random graph is represented by $\tilde{P}$. We then conduct a new transition matrix for semi-supervised classification.

The reason why this step may help implement the purpose of cluster assumption is described as follows. If there are k clusters in the dataset infinitely far apart from each other, it can be shown that the first k eigenvalues of the transition matrix will be 1 and the $(k+1)$-th eigenvalue will be strictly less than 1[10]. Also, there will be k-dimensional vectors $z_1, z_2, \ldots, z_k$ orthonormal to each other such that each data point is mapped to one of those k points depending on which cluster it belongs to. To illustrate this case, we consider an "ideal" case in which all data points are in k clusters, and the size of each cluster is $n_1, n_2, \ldots, n_k (X = X_1 \cup X_2 \cup \ldots \cup X_k, n = n_1 + n_2 + \ldots + n_k)$. All data points are separated far apart from each other and ordered by the cluster that they are in. Define $w_{ij} = 0$ if x_i and x_j belong to different clusters and $w_{ij} > 0 (i \neq j)$ if x_i and x_j in the same cluster. Note that W and P are block-diagonal:

$$W = \begin{bmatrix} W^{(11)} & 0 & 0 \\ 0 & \cdots & 0 \\ 0 & 0 & W^{(kk)} \end{bmatrix}; D = \begin{bmatrix} D^{(11)} & 0 & 0 \\ 0 & \cdots & 0 \\ 0 & 0 & D^{(kk)} \end{bmatrix}; P = \begin{bmatrix} P^{(11)} & 0 & 0 \\ 0 & \cdots & 0 \\ 0 & 0 & P^{(kk)} \end{bmatrix} \quad (10)$$

Where $P^{(ii)}$ is the matrix of "intra-cluster" transition possibilities and $P^{(ii)} = (D^{(ii)})^{-1/2} W^{(ii)} (D^{(ii)})^{-1/2}$. Since P is block diagonal, its eigenvalues and eigenvectors are the union of the eigenvalues and eigenvectors of its blocks. Thus, we can construct $\tilde{X}$ by stacking P's first k largest eigenvector in columns:

$$\tilde{X} = \begin{bmatrix} r_1^{(11)} & 0 & 0 \\ 0 & \cdots & 0 \\ 0 & 0 & r_1^{(kk)} \end{bmatrix} \quad (11)$$

Since 1 is a repeated eigenvalue of P, when we renormalize each row of $\tilde{X}$ to have unit length we could have picked k orthogonal vectors spanning the same subspace as $\tilde{X}$'s columns and defined them to be the first k eigenvectors. In other words, there are k mutually orthogonal points on the surface of the unit k-sphere around which P's rows will cluster.

In a general case, P's off-diagonal blocks are non-zero, but we can still make sure that similar to "ideal" case under certain assumptions. According to Matrix perturbation[8] and spectral graph[9][10] theory, the gap of the k-th and $(k+1)$-th eiganvlaue determines the stability of the eigenvectors of a matrix. The eigengap depends on how well each cluster is connected: the better they are connected, the larger the gap is. The subspace spanned by P's first k eigenvectors will only subject to a small change to P if and only if the difference between the k-th and $(k+1)$-th eigenvalues of P is large.

Let $\lambda_j^{(i)}$ be the j-th largest eigenvalue of $P^{(ii)}$ and $d^{(i)}$be the vector containing $D^{(ii)}$'s diagonal elements. There are four assumptions.

Assumption A1. There exists $\delta > 0$ so that, for all $i = 1, 2, \ldots, k$, $\lambda_2^{(i)} \leq 1 - \delta$.

Assumption A2. There is a fixed $\varepsilon_1 > 0$, so that for every $i_1, i_2 \in \{1, 2, \ldots, k\}$, $i_1 \neq i_2$we have that:

$$\sum_{j \in S_{i1}} \sum_{k \in S_{i2}} \frac{p_{jk}^2}{d_j d_k} \leq \varepsilon_1 \tag{12}$$

Assumption A3. For a fixed $\varepsilon_2 > 0$, for every $i \in \{1, 2, \ldots, k\}$, $j \in S_i$, we have:

$$\frac{\sum_{k:k \notin S_i} p_{jk}}{d_j} \leq \varepsilon_2 \left(\sum_{k,l \notin S_i} \frac{p_{kl}^2}{d_k d_j} \right)^{-1/2} \tag{13}$$

Assumption A4. There is a constant $C > 0$ so that for every $i \in \{1, 2, \ldots, n_i\}$, $j \in \{1, 2, \ldots, k\}$, we have:

$$d_j^{(i)} \geq \frac{\left(\sum_{k=1}^{n_i} d_k^{(i)}\right)}{C_{n_i}} \tag{14}$$

Informally, the assumptions $A1$, $A2$ and $A3$ require that all points must be connected to the points in the same cluster more than to the points in other clusters. And the last assumption $A4$ indicates that no points in a cluster are "too much less" connected than other points in the same cluster. In a word, each of these clusters in fact looks like a cluster and this is true in our case. Therefore, we can have the following theorem:

Theorem 1. *Under assumptions A1,A2,A3 and A4 hold. Set* $\varepsilon = \sqrt{k(k-1)\varepsilon_1 + k\varepsilon_2^2}$ *. If* $\delta > (2+\sqrt{2})\varepsilon$*, then there exist* k *orthogonal vectors* $r_1, r_2, \ldots, r_k$ $(r_i^T r_j = 1$ *if* $i = j$, 0 *otherwise) so that* P*'s rows satisfy:*

$$\frac{1}{n} \sum_{i=1}^{k} \sum_{j=1}^{n_i} ||p_j^{(i)} - r_i||^2 \leq 4C(4 + 2\sqrt{k})^2 \frac{\varepsilon^2}{(\delta - \sqrt{2}\varepsilon)^2} \tag{15}$$

Thus, the rows of P will generate tight clusters around k well-separated points on the surface of the k-sphere. Moreover, these clusters correspond exactly to the true clustering of the original data. Details for Matrix perturbation and spectral graph are given in [8][9][10].

5 Experiments

To take advantage of our smooth graph representation, we employ a graph based method, i.e. consistency method presented by [4] to propagate labels to unlabeled data points. We compare our method to standard consistency method [4] and Gaussian function [3] using artificial and real world dataset. The performance of the algorithm is measured by the accuracy rate on the unlabeled points.

5.1 Toy Problem

In this experiment we considered the toy problem which is the switch or two-moon data mentioned in many semi-supervised learning papers. The transition matrix is formed using equation (1). From Fig.1 we can see that the classification result of consistency method with smooth graph is completely the same as the ideal result and there is a slight difference of the classification results between the ideal case and Gaussian function. Furthermore, when $\sigma \in [0.2, 0.7]$, the accuracy of classification using Gaussian function is great than 98%. But, the consistency method with smooth graph can enlarge the range of parameter σ to $[0.2, 1.2]$ with k=2 and remain the accuracy at 100%.

The transition matrix of original graph and the smooth Markov walk random graph can be visualized in Fig.2.

There are 150 data points and two categories in the switch data set. The first 75 data points belongs to the same category. From Fig.2, we can see that the smooth Markov walk random graph present label smoothness and cluster assumptions better. The transition possibilities of two data points are extremely high if they are in the same cluster, on the contrary, the transition possibilities will be very low if the points are in the different clusters.

5.2 Image Recognition

In this experiment, we design a classification task using colored teapots images from different views. There are 135 examples for each category with a total of 270 examples. Each example is represented by a 23028-dimension vector. We use the following weight function on the edges:

$$w_{ij} = exp(-\frac{1}{0.03}(1 - \frac{x_i^T x_j}{|x_i||x_j|})) \tag{16}$$

From 1 to 5 labeled data points of each category are randomly selected to generate the labeled data set and the rest of the data points in the categories form the unlabeled data points. The values of k and m are all set to 2. The experimental

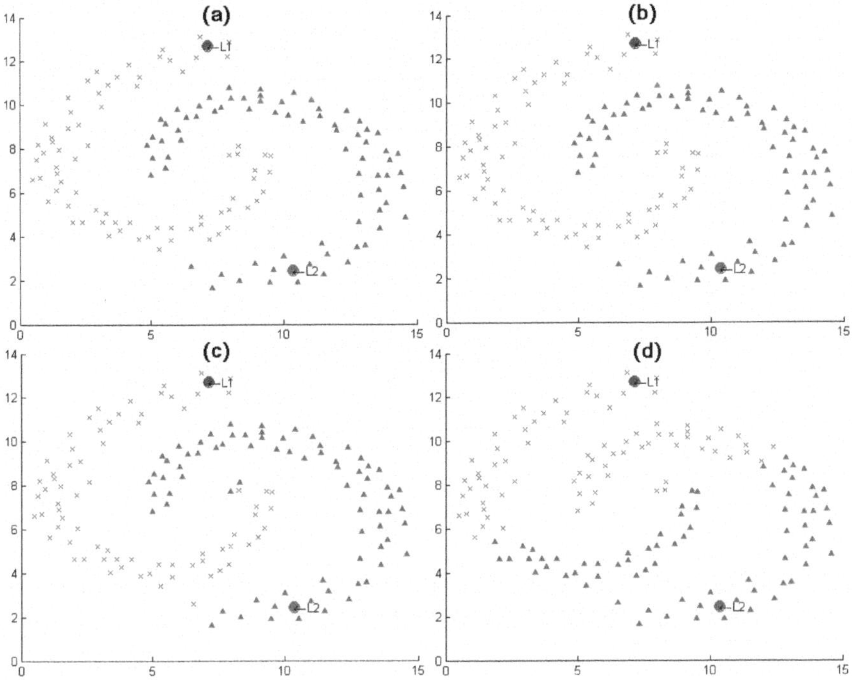

Fig. 1. Classification results of the pattern of two moons of ideal result, consistency method with smooth graph, harmonic function and standard consistency method shown from (a) to (d)

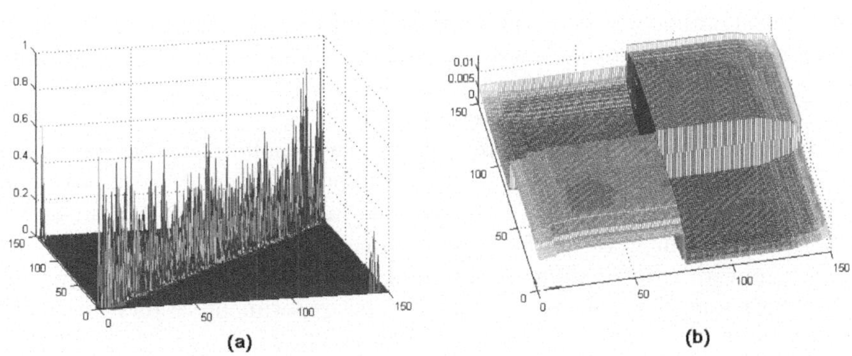

Fig. 2. (a) is the original graph, and (b) is the smooth Markov walk random graph generated based on (a)

results of the classification are shown in Figure 2. The results show that consistency method with smooth graph outperforms other methods on the teapots image dataset.

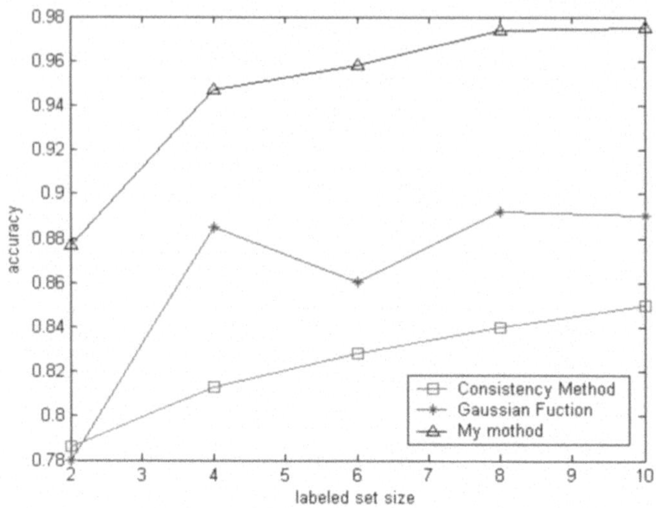

Fig. 3. Classification results of consistency method with smooth graph, harmonic function and standard consistency method on image dataset

5.3 Text Classification

In this experiment, we investigate the task of text classification using 20 Newsgroups dataset. The articles are processed by the Rainbow software package with below options: (1) passing all words through the porter stemmer before counting them; (2) excluding words in the stoplist of SMART system; (3) skipping any headers; (4) ignoring words that occur in 5 or fewer documents. No further preprocessing step is taken. After removing the empty documents, the documents are normalized into a TF.IDF representation.

We select the binary problems to compare the results: MS-Windows vs. Mac. For each category, we randomly label data points from 1 to 32 to generate the labeled data and the rest of the articles in the category remain to be the unlabeled data points. We performance ten runs for each dataset and count the average accurate rate of different methods. The parameter σ of equation (1) is 0.7. The value of k and m is 8 and 2. The experiment results are shown in Figure 3.

5.4 Result Evaluation

Difficulty to provide enough information about categories and rough data distribution are the challenges to any semi-supervised learning algorithms. The process of constructing the smooth graph is to reshape the data distribution and make clusters separated clearly. There are two steps for the construction. The first step tends to compress the distance between any two points in the

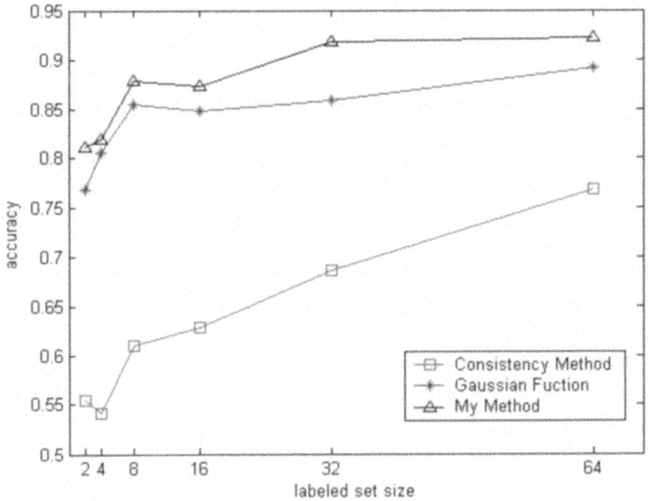

Fig. 4. Classification result on MS-Windows vs. MAC

same cluster, and the second step enlarges the gap between two clusters. It is quite similar with the notion of clustering. With the help of smooth graph, label propagation on the graph is simplified to label each cluster. Since the clustering accuracy has no relation to labeled data points, graph-based semi-supervised learning with smooth graph can still get high classification accuracy with quite few labeled data points. Our experiments show that smooth Markov walk random graph can improve the performance of graph-based semi-supervised methods.

6 Conclusion

This paper presents an approach to create graphs by graph smoothing and clustering step. The smoothed graph can capture the nature of data distribution and reflect smoothness labels and cluster assumptions more accurately. We combine the graph-based semi-supervised classification methods with our smooth graph to reduce the errors and biases caused by rough transition probabilities. Experimental results show that the graph-based semi-supervised classification methods with our smooth graph outperform the same methods with original graph representation.

Acknowledgments

The authors would like to thank Prof. Chunping Li for his invaluable advice on the paper.

References

1. Seeger, M.: Learning with labeled and unlabeled data. Technical report , Edinburgh University (2000).
2. Zhu, X.: Semi-Supervised Learning with Graphs. Doctoral Thesis. CMU-LTI-05-192 (2005).
3. Zhu, X., Lafferty, J., Ghahramani, Z.: Semi-Supervised Learning Using Gaussian Fields and Harmonic Function. In Proceedings of ICML-03 (2003).
4. Zhou, D. et al.: Learning with local and global consistency. In Advances in Neural Information Processing System 16 (2004).
5. Chapelle, O. and A. Zien: Semi-Supervised Classification by Low Density Separation. Proceedings of the Tenth International Workshop on Artificial Intelligence and Statistics (2005) 57-64.
6. Chapelle, O., Weston, J., Scholkopf, B.: Cluster Kernels for Semi-Supervised Learning. Advances in Neural Information Processing Systems 15, MIT Press (2003) 585-592.
7. Szummer, M., Jaakkola, T.: Partially labeled classification with Markov random walks. Neural Information Processing Systems (NIPS), Vol 14 (2001).
8. G.W.Stewart, J.G.Sun.:Matrix perturbation Theory. Academic Press, (1990).
9. F.Chung. Spectral Graph Theory. Number 92 in CBMS Regional Conference Series in Mathematics. American Mathematical Society, (1997).
10. A.Y.Ng, M.I.Jordan, Y.Weiss: on spectral clustering: Analysis and an algorithm. In Advances in Neural Information Proceeding Systems, volume 14, (2001).
11. Henk C. T.:Stochastic Models: An Algorithmic Approach, John Wiley & Sons (1994).
12. Xueyuan Zhou, Chunping Li: Combining Smooth Graphs with Semi-Supervised Classification. In The 10th Pacific-Asia Conference on Knowledge Discovery and Data Mining(2006) 400-409.

Extracting Trend of Time Series Based on Improved Empirical Mode Decomposition Method

Hui-ting Liu[1,2], Zhi-wei Ni[1], and Jian-yang Li[1,3]

[1] Institute of Computer Network System, Hefei University of Technology, 230009 Hefei, China
[2] School of Computer Science and Technology, Anhui University, 230039 Hefei, China
[3] Department of Computer Science, Longyan University, 364000 Longyan, China
wangph168@yahoo.com.cn

Abstract. Solving overshoot and undershoot problems existed in the spline interpolation of empirical mode decomposition (EMD), improving this method and extracting trend of time series with it are the main tasks of this paper. A method is devised by using simple means of successive extrema instead from the envelope average to form the mean envelope. In this way, only one spline interpolation is required rather than two during the course of sifting process of EMD. It is easier to implement, those problems can be alleviated and EMD method is improved. How to get the successive extrema of series and how to realize trend extraction are also expounded in the paper. Experimental results show that the improved EMD method is better at trend extraction than the original one.

Keywords: trend extraction, empirical mode decomposition, spline interpolation, overshoot and undershoot problems.

1 Introduction

Trend analysis is a useful approach to extract information from databases [1], for example, stock forecasting and weather prediction [2]. Literature [3] proposed the concept of "trend" of time series. The trend of a time series is a higher-level pattern of directions that an original sequence moves and indicates up, crossover, or down movements of the series.

There have been many methods to extract trend of series. Free hand and least square methods are the techniques commonly used, but the former depends on experience of users and the latter is difficult to use when original series are very irregular [4]. Empirical mode decomposition (EMD) is also an effective trend-extracting method [5], however, it has some weaknesses. The spline interpolation, which is the core of the sifting process of EMD, has both overshoot and undershoot problems [6]. Using higher-order spline can alleviate these problems [7], but it would be more time consuming.

G. Dong et al. (Eds.): APWeb/WAIM 2007, LNCS 4505, pp. 341–349, 2007.

This paper mainly discusses extracting trend of time series based on improved EMD method. As the higher-order spline is time consuming, an efficient method is devised by using means of successive extrema instead from the envelope mean to form the mean envelope. In this way, only one spline interpolation is required rather than two in each loop of the sifting process. Time complexity is reduced, overshoot and undershoot problems are alleviated, and EMD method is improved. Then time series can be decomposed by the improved EMD method and trends of them are obtained. To get means of successive extrema, one should find all maxima and minima of a series at first, including finding all the extrema that the series actually has, additional boundary and interior extrema. These added extrema can solve overshoot and undershoot problems to some extent.

2 Empirical Mode Decomposition

In 1998, Huang et al [6] presented the use of EMD to decompose any multi-component series into the trend of it and a set of nearly mono-component sequences that are referred as intrinsic mode functions (IMFs). It is the best algorithm to get trend of series by now [5].

The procedure to decompose time series is called sifting process. If the number of extrema and the number of zero crossings of a sequence $x(t)$ differ more than one, or the mean value of the upper envelope and the lower envelope is not zero at every point, the sequence will be decomposed into a few IMFs $c(t)$ and the trend $r(t)$ by the sifting process [8]. The sifting process can be expressed as follows [9].

1. Initialize: $r_0(t)=x(t)$, $i=1$
2. Extract the ith IMF:
   a) Initialize: $h_0(t)=r_{i-1}(t)$, $k=1$
   b) Extract the local maxima and minima of $h_{k-1}(t)$
   c) Interpolate the local maxima and local minima by cubic splines to form upper and lower envelopes of $h_{k-1}(t)$
   d) Calculate mean of the upper and lower envelopes of $h_{k-1}(t)$ to get the mean envelope $m_{k-1}(t)$
   e) Define: $h_k(t)=h_{k-1}(t)-m_{k-1}(t)$
   f) If IMF criteria is satisfied, then set $C_i(t)=h_k(t)$ else go to b) with $k=k+1$
3. Define $r_i(t)=r_{i-1}(t)-C_i(t)$
4. If $r_i(t)$ still has at least two extrema, then go to 2 with $i=i+1$; else the decomposition is completed and $r_i(t)$ is the trend of the series $x(t)$.

3 Extracting Trend of Series Based on Improved EMD Method

In this section, problems of the spline interpolation are pointed out at first, then how to solve the problems is expounded. At last, trends of time series are extracted successfully based on the improved EMD method.

3.1 Problems of the Spline Interpolation

In general, the goal of a spline interpolation is to create a function, which achieves the best possible approximation to a given data set. A popular and practically the most applied choice is the piecewise cubic spline polynomial.

A set of n+1 points $(x_0, y_0),\ldots,(x_n, y_n)$, where n≥3 [10] can be linked by a third order spline polynomial that is generally defined by three conditions:

1. $p(x)$ in each interval $[x_i, x_{i+1}]$ $i = 0,1,\ldots,n-1$, is given by a polynomial $p_i(x)$ of a third order degree:

$$p_i(x) = \alpha_i + \beta_i(x - x_i) + \gamma_i(x - x_i)^2 + \delta_i(x - x_i)^3 . \tag{1}$$

2. In each interval $[x_i, x_{i+1}]$ $i = 0,1,\ldots,n-1$, the polynomial $p_i(x)$ complies with the following two boundary conditions:

$$p_i(x_i) = y_i \text{ and } p_i(x_{i+1}) = y_{i+1} . \tag{2}$$

3. Two neighbored polynomials $p_{i-1}(x)$ and $p_i(x)$ fulfill the following two compatibility conditions:

$$p'_{i-1}(x_i) = p'_i(x_i) \text{ for } i = 1,2,\ldots,n-1 . \tag{3}$$

$$p''_{i-1}(x_i) = p''_i(x_i) \text{ for } i = 1,2,\ldots,n-1 . \tag{4}$$

Concerning the EMD, the spline interpolation to define the upper and the lower envelopes of a series rather seems to be an easy task. On the practical side, it is prone to have large swings [9] as shown in figure 1.

From figure 1, we can see serious problems occur at the ends of the lower envelope and in the middle of the upper envelope. These large swings can eventually corrupt the whole data series and, consequently, eliminate the natural embedded characteristic of the data. In this case, empirical mode decomposition gets totally disrupted. How to solve the problems, how to improve EMD method and how to extract trend of time series based on improved EMD seem to be complex tasks and they are all the issues needed to discuss in the following.

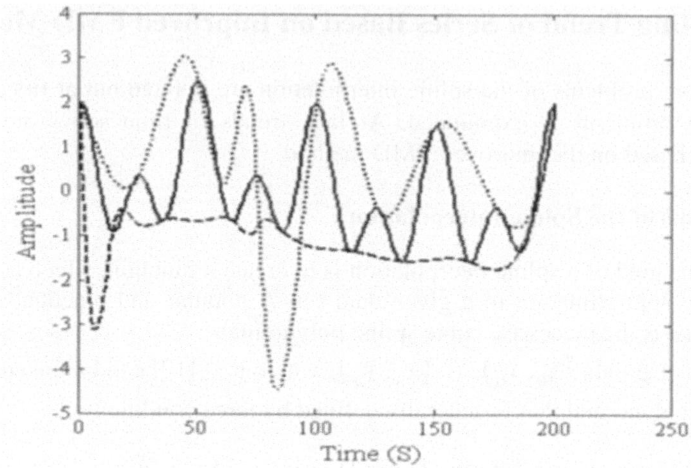

Fig. 1. Time series (*solid line*) and upper envelope (*dot line*), lower envelope (*dash line*) of it

3.2 A Solution of the Problems

Because overshoots and undershoots are phenomena inherent in the spline interpolation, the problems described in figure 1 are difficult to solve, decreasing times of spline interpolations in each loop of the sifting process and adding data points are better ideas to improve EMD. The previous method shown in section two of this paper utilizes average of the upper and lower envelopes to get the mean envelope. In this paper, we use a cubic spline to interpolate means of successive extrema to form the mean envelope. Times of the spline interpolation is reduced, meanwhile, additional boundary and interior data points are created in the course of finding successive extrema. So overshoot and undershoot problems can be alleviated and EMD method can be improved by the proposed method.

Now, we will expound the proposed method. Firstly, successive extrema must be found, because the proposed method gets mean envelopes by utilizing means of them. As we know, time series has no maximum if it has a minimum at any point, vice versa. So we should create an approximate maximum or minimum at that point the series has only one minimum or maximum. From figure 1, we can see there are two kinds of swings, at the ends of the series and in the middle of it. Therefore methods of obtaining additional extrema can be divided into two categories, the method to create additional boundary data points and the method to find additional interior knots.

The method to get additional boundary points has been given in literature [11] by Liu et al. If a series has a maximum at the left end, it will not have a minimum at the same time. In this case, we will connect a few minima that having existed near the left end of series with a cubic polynomial, and obtain approximation of the minima sequence at the left end using the polynomial. This procedure will repeat until four extrema of both ends are gained. The cubic spline, which interpolates all the extrema, will not swing widely at both ends of time series when it has fixed values to keep.

We will illustrate how to find additional interior extrema by the following figure.

Fig. 2. Time series (*solid line*) and maxima (‘*o*’), minima (‘+’) of them

In figure 2, solid line donates the time series, ‘o’ means maxima of it and ‘+’ means minima. In subgraph (a), abscissa of the first maximum is smaller than that of the first minimum. In subgraph (b), abscissa of the first maximum is bigger.

Suppose a series has n maxima $(\max_1, \max_2, \ldots, \max_n)$ and m minima $(\min_1, \min_2, \ldots, \min_m)$, their abscissas and vertical coordinates can be expressed as $(x_{\max_1}, x_{\max_2}, \ldots, x_{\max_n})$, $(y_{\max_1}, y_{\max_2}, \ldots, y_{\max_n})$, $(x_{\min_1}, x_{\min_2}, \ldots, x_{\min_m})$ and $(y_{\min_1}, y_{\min_2}, \ldots, y_{\min_m})$. To get the additional maximum at the point time series has an accurate minimum $\min_i$, two cases should be taken into account.

When abscissa of the first maximum is smaller than that of the first minimum, the new maximum $\max_i'$ can be got by formula (5).

$$y_{\max_i'} = y_{\max_i} + (y_{\max_{i+1}} - y_{\max_i})/(x_{\max_{i+1}} - x_{\max_i}) * (x_{\min_i} - x_{\max_i})$$

$$x_{\max_i'} = x_{\min_i} \quad i = 1,2,\ldots,m \text{ if } \min_m \text{ is not the end} \tag{5}$$

$$\text{or } i = 1,2,\ldots,m-1 \text{ if it is.}$$

If abscissa of the first maximum is larger than that of the first minimum, the new maximum $\max_i'$ will be gained by formula (6).

$$y_{\max_i'} = y_{\max_{i-1}} + (y_{\max_i} - y_{\max_{i-1}})/(x_{\max_i} - x_{\max_{i-1}}) * (x_{\min_i} - x_{\max_{i-1}})$$

$$x_{\max_i'} = x_{\min_i} \quad i = 2,3,\ldots,m \text{ if } \min_m \text{ is not the end} \tag{6}$$

$$\text{or } i = 2,3,\ldots,m-1 \text{ if it is.}$$

The similar algorithm can obtain additional minima of a series where it has accurate maxima. After successive extrema have been found, mean envelopes can be easily got by the spline interpolation, and step 2 of the sifting process can be modified as follows.

a) Initialize: $h_0(t) = r_{i-1}(t)$, $k = 1$

b) Extract the local maxima and minima of $h_{k-1}(t)$

c) Get additional boundary and interior extrema of $h_{k-1}(t)$

d) Calculate means of successive extrema and interpolate them by a cubic spline to form the mean envelope of $h_{k-1}(t)$, that is, $m_{k-1}(t)$

e) Define: $h_k(t) = h_{k-1}(t) - m_{k-1}(t)$

f) If IMF criteria is satisfied, then set $C_i(t) = h_k(t)$ else go to b) with $k = k+1$

In the new sifting process, additional extrema are extracted and times of the spline interpolation of each loop is decreased to only once, so overshoot and undershoot problems are greatly alleviated.

3.3 Extracting Trend of Time Series

Problems existed in the spline interpolation are alleviated, EMD method is improved and trend of time series can be extracted by now.

In improved EMD method, time series are decomposed into the trend of it and a few IMFs by the new sifting process. The process can be stopped by any of the following criteria: either when the component $c(t)$ or the residue $r(t)$ becomes less than the predetermined value, or when $r(t)$ becomes a monotonic function from which no more IMF can be extracted [6]. After sifting process, the last residue is the trend of original series $x(t)$, and it can be expressed as the following equation.

$$r(t) = x(t) - \sum_{i=1}^{n} c_i(t) . \tag{7}$$

The improved EMD method can extract trend of time series perfectly and it will be proved by experiments.

4 Performance Study

In this section, we present the superiority of the improved EMD method over the original one via comparing their time complexity and accuracy.

The paper will compare two methods by experiments. The data used in experiments are stock data from http://kumo.swcp.com/stocks. There are 467 kinds of stock data after pretreatment, and they are 253-days transaction data from 2004/8/26 to 2005/8/25. We use opening prices in the following experiments.

The first experiment utilizes the original EMD method to extract trends of sequences, that is, it gets mean envelopes from average of upper and lower envelopes. Results of this algorithm will be shown in table one and figure 3.

The second experiment uses the improved EMD method to extract trends of time series, that is to say, it gets mean envelopes from average of successive extrema. Results of this algorithm are in table one and figure 3, too.

Table 1. Execution time needed in the trend-extraction procedure

	Number of series=100	Number of series=200	Number of series=300	Number of series=467
Original EMD method	31.99 Sec.	51.76 Sec.	74.87 Sec.	124.97 Sec.
Improved EMD method	11.13 Sec.	21.96 Sec.	33.25 Sec.	52.37 Sec.

Table 1 shows execution time per trend extraction for different values of the number of sequences. Clearly, improved EMD method outperforms original one. As the number of sequences increases, the gain of improved EMD method increases, making it even more attractive for larger databases.

Fig. 3. Time series (*solid line*), trend extracted by original EMD method (*dash line*), and trend extracted by improved EMD method (*dot line*)

From figure 3, we can see trends extracted by the improved EMD method can indicate up, crossover, or down movements of sequences clearly and exactly. The trend extracted by the original EMD algorithm shown in subgraph (a) has serious overshoot and undershoot problems at both ends of time series. And the trend got by the previous EMD method shown in subgraph (b) differs a lot with the real trend at the left end of series. The actual trend of the series in subgraph (b) is relatively low at the beginning and becomes higher and higher although there is some decline in the middle of it, but the trend got by the original EMD method begins at a high level and descends in the middle of it. So trend got by the previous EMD method cannot give correct directions that an original data sequence moves, and the improved EMD method is better at trend extraction than it.

5 Conclusion

This paper discusses extracting trend of time series based on improved EMD method, that is, solving overshoot and undershoot problems existed in the spline interpolation of EMD and finishing trend extraction with it are the main tasks.

Phenomena of overshoot and undershoot are obvious, if we use the algorithm described in section two to get mean envelopes of time series. To decrease these phenomena, a novel method is proposed in this paper. We obtain mean envelopes by using simple means of successive extrema instead from the envelope mean, the times of the spline interpolation is decreased to only once a loop, and more data points are added during the course of finding successive extrema. Problems are alleviated and EMD method is improved. Then trend of times series can be extracted successfully. Experimental results show that improved EMD method outperforms the original one when they are used in trend extraction.

Acknowledgments. This paper is funded by natural science foundation of Anhui province (050460402), natural science foundation of Fujian province (A0640001), scientific research funds of education office, Anhui province (2006sk010), and planned projects for young teachers' research funds of Anhui province (2005jq1035).

References

1. Sylvie, C., Carlos, G.B., Catherine, C., et al: Trends extraction and analysis for complex system monitoring and decision support. Engineering Applications of Artificial Intelligence 18 (2005) 21–36
2. Min, Z., Yan, P.Z., Jia, X.C.: Hierarchical Algorithm to Match Similar Time Series Pattern. Journal of computer-aided design & Computer graphics 117 (2005) 1480–1485
3. Yong, J.P., Lee, J., Kim, S.: Trend Similarity and Prediction in Time-Series Databases. In: Proceedings of SPIE on Data Mining and Knowledge Discovery: Theory, Tools, and Technology II, Vol. 4057. SPIE, Washington (2000) 201–212
4. Fan, Y., Zhi, J.W., Yuan, S.L.: Improvement in Time-Series Trend Analysis. Computer technology and development 16 (2006) 82–84
5. Yong, J.D., Wei, W., et al: Boundary-Processing technique in EMD method and Hilbert transform. Chinese Science Bulletin 46 (2001) 954–961
6. Huang, N.E., Shen, Z., Long, S.R., et al: The empirical mode decomposition and the Hilbert spectrum for nonlinear and non-stationary time series analysis. Proceedings of the Royal Society A: Mathematical, Physical and Engineering Sciences 454 (1998) 903–995
7. Shi, X.Y., Jing, S.H., Zhao, T.W., et al: Study of empirical mode decomposition based on high-order spline interpolation. Journal of Zhejiang University (Engineering Science) 38 (2004) 267–270
8. Peng, Z.K., Tse, P.W., Chu, F.L.: An improved Hilbert–Huang transform and its application in vibration signal analysis. Journal of Sound and Vibration 286 (2005) 187–205

9. Marcus, D., Torsten, S.: Performance and limitations of the Hilbert-Huang transformation (HHT) with an application to irregular water waves. Ocean Engineering 31 (2004) 1783–1834
10. Weisstein, E.W.: Cubic splines. Eric Weisstein's World of Mathematics. Available from http://mathworld.wolfram.com/Isometry.html (2001)
11. Hui, T.L., Min, Z., Jia, X.C.: Dealing with the End Issue of EMD Based on Polynomial Fitting Algorithm. Computer Engineering and Applications 40 (2004) 84–86

Spectral Edit Distance Method for Image Clustering

Nian Wang[1], Jun Tang[1], Jiang Zhang[1], Yi-Zheng Fan[1,2], and Dong Liang[1]

[1] Key Laboratory of Intelligent Computing & Signal Processing, Anhui University, Education Ministry, Hefei, 230039, P.R. China
[2] Department of Mathematics, Anhui University, Hefei , 230039, P.R. China
{wn_xlb,tangjun,jiangzh,fanyz,dliang}@ahu.edu.cn

Abstract. The spectral graph theories have been widely used in the domain of image clustering where editing distances between graphs are critical. This paper presents a method for spectral edit distance between the graphs constructed on the images. Using the feature points of each image, we define a weighted adjacency matrix of the relational graph and obtain a covariance matrix based on the spectra of all the graphs. Then we project the vectorized spectrum of each graph to the eigenspace of the covariance matrix, and derive the distances between pairwise graphs. We also conduct some theoretical analyses to support our method. Experiments on both synthetic data and real-world images demonstrate the effectiveness of our approach.

Keywords: Edit distance, Graph, Spectrum, Clustering.

1 Introduction

Many image clustering problems in the field of data mining can be abstracted using relational graphs. Examples include the use of the graph constructed on the feature points of the images to represent object structure which is often applied to successive frames in a motion sequence or image databases. Broadly speaking, there are two aspects of image clustering. Some researchers work on dividing the content of a single image into some basic classes which is appropriate to the classification of remote sensing images. For this kind of image, we can assign every pixel to some class such as mountain, river, cloud, and so on [1]. On the other hand, some researchers focus on the classification of images sequences. They devote to classifying similarity image into given classes accurately [2]. And the stage of building feature vectors of images and constructing right classifiers is critical for the image clustering works. Most previous works use color, texture or frequency domain as feature vectors [3] and adopt neural networks or probabilistic method as classifier [4].

Recently there has been some research interest in applying central clustering techniques to clustering graphs. Rather than characterizing them in a statistical manner, a structural characterization is adopted. For instance, both Lozano and Escolano [5], and Bunke et al. [6] classify the data by using a super-graph. Each

G. Dong et al. (Eds.): APWeb/WAIM 2007, LNCS 4505, pp. 350–357, 2007.

sample can be obtained from the super-graph using edit operations. However, the way in which the super-graph is learned or estimated is not statistical in nature. Bagdanov and Worring [7] have overcome some of the computational difficulties associated with the above method by using continuous Gaussian distributions. Qiu and Hancock [8] have reported that the partitions delivered by the Fiedler vector can be used to simplify the graph clustering problem. They use the Fiedler vector to decompose graphs by partitioning them into structural units. And the partitions can simplify the problem of graph clustering in a hierarchical manner. Luo et al. [9] have shown how to construct a linear deformable model for graph structure by performing principal components analysis (PCA) on the vectorized adjacency matrix. They commence by using correspondence information to place the nodes of every one among a set of graphs in a standard reference order. With using the correspondence order, they convert the adjacency matrices to long-vectors and compute the long-vector covariance matrix. Through projecting the vectorized adjacency matrices onto the leading eigenvectors of the covariance matrix, the graphs can be embedded in a pattern-space.

These methods in all the literatures mentioned in this section concern about the eigenvectors of graph corresponding to the object structure. In a different manner, this paper provides a clustering method of images using spectra of graphs, i.e. the eigenvalues of graphs. Firstly, with using the feature points of images, we define a weighted adjacency matrix (representing intra-image point proximity) with a Euclidean distance between points in the same image. Secondly, the normalized feature vector formed from eigenvalues can be projected to eigenspace alternatively to compute the distance between pairwise images. Thirdly, with the distances available, we can cluster the images easily. Finally, the experiments on both synthetic data and real-world images demonstrate the feasibility of our method.

The outline of this paper is as follows. Section 2 details a spectral method for clustering and a theoretical analysis is given to support our approach. In Section 3, we conduct some experiments to validate out method. Finally, Section 4 summarizes some conclusions.

2 Spectral Edit Distance Between Images

2.1 Spectral Representation of Feature of Image

Given clustered image sequence $P_1, P_2, \cdots, P_k, \cdots P_m$, let V_k denote the feature point-set of image P_k. It represents the structural feature of image P_k. Graph G_k is denoted as $G_k = (V_k, E_K)$ where $E_K \subseteq V_k \times V_k$ is the edge-set. For each G_k, we compute the weighted adjacency matrix $\boldsymbol{A}_k$. $\boldsymbol{A}_k$ is a $|V_k| \times |V_k|$ matrix whose element with row indexed i and column indexed j is defined as:

$$A_k(i,j) = \left\| V_{k_i} - V_{k_j} \right\| \qquad i,j = 1,2,\cdots,|V_k| \tag{1}$$

With the matrix $\boldsymbol{A}_k$ available, we can calculate the eigenvalues λ_k^η by solving the equation $|\boldsymbol{A}_k - \lambda_k^\eta \boldsymbol{I}| = 0$ where $\eta = 1, 2, \cdots, |V_k|$ is the eigenvalue index. We order these eigenvalues in descending order, i.e. $|\lambda_k^1| \ge |\lambda_k^2| \ge \cdots \ge |\lambda_k^{|V_k|}|$. Furthermore, we can obtain the spectrum corresponding to the relational graph constructed on the image, i.e. $\boldsymbol{\lambda}_k = (\lambda_k^1, \lambda_k^2, \cdots, \lambda_k^{|V_k|})^T$, which is defined as the feature vector corresponding to its image. In other words, we can use spectrum to represent the object structure.

2.2 Spectral Edit Distance

With the feature vector $\boldsymbol{\lambda}_k = (\lambda_k^1, \lambda_k^2, \cdots, \lambda_k^{|V_k|})^T$ at hand, regularizing $\boldsymbol{\lambda}_k$, we can compute the mean feature vector and the covariance matrix. The feature vector is:

$$\bar{\boldsymbol{\lambda}} = \frac{1}{m} \sum_{k=1}^{m} \boldsymbol{\lambda}_k . \tag{2}$$

And the covariance matrix is

$$\Sigma = \frac{1}{m} \sum_{k=1}^{m} (\boldsymbol{\lambda}_k - \bar{\boldsymbol{\lambda}})(\boldsymbol{\lambda}_k - \bar{\boldsymbol{\lambda}})^T . \tag{3}$$

Σ can be decomposed as:

$$\Sigma = \boldsymbol{U \Delta U}^T \tag{4}$$

where $\boldsymbol{\Delta} = \mathrm{diag}\{\gamma_1, \cdots, \gamma_m\}$ ($\gamma_1 \ge \cdots \ge \gamma_{m-1} > \gamma_m$) is a diagonal matrix whose diagonal entries are eigenvalues of Σ, and $\boldsymbol{U} = (\boldsymbol{u}_1, \cdots \boldsymbol{u}_m)$ is an orthogonal matrix whose column $\boldsymbol{u}_t$ is an eigenvector of Σ associated with eigenvalue γ_t. Then we obtain an orthogonal vector set $\boldsymbol{U}^T$.

The graph spectrum can be projected onto referenced set $\boldsymbol{U}^T$ of the covariance matrix. So the centered projection of the feature vector for the graph indexed k is $\boldsymbol{Z}_k = \boldsymbol{U}^T \boldsymbol{\lambda}_k$. For each pair of graphs G_k and G_l, we compute the Euclidean with distance $D_{k,l} = \|\boldsymbol{Z}_k - \boldsymbol{Z}_l\| = \sqrt{(\boldsymbol{Z}_k - \boldsymbol{Z}_l)^T (\boldsymbol{Z}_k - \boldsymbol{Z}_l)}$.

2.3 Theoretical Analysis

Let I_1, J_1 represent two image planes containing n points. Applying $(m-1)$ similarity transformations $\{\psi_2, \psi_3, \cdots, \psi_m\}$ to image planes I_1, J_1, we obtain m images $I_1, I_2, \cdots, I_m$ and m images $J_1, J_2, \cdots, J_m$ as well as containing n points. Let $\psi_k = s_k \varphi_k$, where $s_k > 0$ is the magnification coefficient, and φ_k denotes an equiform transformation. Construct $2m$ graphs $G_1^I, G_2^I, \cdots, G_m^I, G_1^J, G_2^J, \cdots, G_m^J$ using points of these $2m$ images, and let $\boldsymbol{A}_1 = \boldsymbol{A}(G_1^I)$, $\boldsymbol{B}_1 = \boldsymbol{B}(G_1^J)$. Then

$$\boldsymbol{A}_k = \boldsymbol{A}(G_k^I) = s_k \boldsymbol{P}_k^T \boldsymbol{A}_1 \boldsymbol{P}_k, \boldsymbol{B}_k = \boldsymbol{B}(G_k^J) = s_k \boldsymbol{Q}_k^T \boldsymbol{B}_1 \boldsymbol{Q}_k . \tag{5}$$

For some permutation matrices $\boldsymbol{P}_k, \boldsymbol{Q}_k$ with order n, we can obtain

$$\lambda_k = s_k \lambda_1, \gamma_k = s_k \gamma_1. \tag{6}$$

With eigenvalues λ_k, γ_k available, we form these eigenvalues to a vector and normalize it, then

$$\lambda_k = \lambda_1 =: \lambda, \gamma_k = \gamma_1 =: \gamma \tag{7}$$

and the mean of this vector is

$$\beta = \frac{\lambda + \gamma}{2} \tag{8}$$

So we have the covariance matrix as follows:

$$\begin{aligned}
\Sigma &= \frac{1}{2m}\left(\sum_{k=1}^{m}(\lambda_k - \beta)(\lambda_k - \beta)^T + \sum_{k=1}^{m}(\gamma_k - \beta)(\gamma_k - \beta)^T\right) \\
&= \frac{1}{2m}\left(\sum_{k=1}^{m}(\lambda - \frac{\lambda+\gamma}{2})(\lambda - \frac{\lambda+\gamma}{2})^T + \sum_{k=1}^{m}(\gamma - \frac{\lambda+\gamma}{2})(\gamma - \frac{\lambda+\gamma}{2})^T\right) \\
&= \frac{1}{2m}\left(\frac{1}{4}\sum_{k=1}^{m}(\lambda-\gamma)(\lambda-\gamma)^T + \frac{1}{4}\sum_{k=1}^{m}(\gamma-\lambda)(\gamma-\lambda)^T\right) \\
&= \frac{1}{4}(\lambda-\gamma)(\lambda-\gamma)^T \\
&= \boldsymbol{U}\boldsymbol{\Lambda}\boldsymbol{U}^T,
\end{aligned} \tag{9}$$

where $\boldsymbol{U} = \{\frac{\lambda-\gamma}{\|\lambda-\gamma\|}, \boldsymbol{u}_2, \cdots, \boldsymbol{u}_n\}$ is an orthogonal matrix whose column vectors are normalized and orthogonal to each other, and $\boldsymbol{\Lambda} = \mathrm{diag}\{\frac{1}{4}\|\lambda-\gamma\|^2, 0, \cdots, 0\}$ is a diagonal matrix. Then

$$\boldsymbol{Z}_k^I = \boldsymbol{U}^T \lambda_k = \boldsymbol{U}^T \lambda = (\frac{(\lambda-\gamma)\lambda}{\|\lambda-\gamma\|}, \boldsymbol{u}_2^T\lambda, \cdots, \boldsymbol{u}_m^T\lambda)^T. \tag{10}$$

$$\boldsymbol{Z}_k^J = \boldsymbol{U}^T \gamma_k = \boldsymbol{U}^T \gamma = (\frac{(\lambda-\gamma)\gamma}{\|\lambda-\gamma\|}, \boldsymbol{u}_2^T\gamma, \cdots, \boldsymbol{u}_m^T\gamma)^T. \tag{11}$$

Note that $(\lambda-\gamma)^T \boldsymbol{u}_k = 0$ for each $k = 2, \cdots, n$ since $\boldsymbol{U}$ is an orthogonal matrix, and we have

$$\begin{aligned}
\|\boldsymbol{U}^T\lambda - \boldsymbol{U}^T\gamma\| &= \left\|\left(\frac{(\lambda-\gamma)^T(\lambda-\gamma)}{\|\lambda-\gamma\|^2}, \boldsymbol{u}_2^T(\lambda-\gamma), \cdots, \boldsymbol{u}_m^T(\lambda-\gamma)\right)\right\| \\
&= \|(1, 0, \cdots, 0)\| \\
&= 1,
\end{aligned} \tag{12}$$

Consequently we can obtain a distance matrix as follows:

$$D = \begin{array}{c} \\ I_1 \\ \vdots \\ I_m \\ J_1 \\ \vdots \\ J_m \end{array} \begin{array}{c} \begin{array}{cccccc} I_1 & \cdots & I_m & J_1 & \cdots & J_m \end{array} \\ \left[\begin{array}{ccc|ccc} 0 & \cdots & 0 & 1 & \cdots & 1 \\ \vdots & \ddots & \vdots & \vdots & \ddots & \vdots \\ 0 & \cdots & 0 & 1 & \cdots & 1 \\ \hline 1 & \cdots & 1 & 0 & \cdots & 0 \\ \vdots & \ddots & \vdots & \vdots & \ddots & \vdots \\ 1 & \cdots & 1 & 0 & \cdots & 0 \end{array}\right] \end{array} = \begin{bmatrix} \mathbf{0}_{m\times m} & \mathbf{1}_{m\times m} \\ \mathbf{1}_{m\times m} & \mathbf{0}_{m\times m} \end{bmatrix}. \tag{13}$$

By means of the above methods, we can easily cluster the $2m$ images $I_1, I_2, \cdots, I_m$ and $J_1, J_2, \cdots, J_m$ into two groups with the former m images in one group and the latter m images in the other.

If we apply affine transformations $\{\psi_2, \psi_3, \cdots, \psi_m\}$ approximate to equiform transformations to the images I_1 and J_1 respectively, similarly, we can obtain the following matrix:

$$D \to \begin{bmatrix} \mathbf{0}_{m\times m} & \mathbf{1}_{m\times m} \\ \mathbf{1}_{m\times m} & \mathbf{0}_{m\times m} \end{bmatrix} \tag{14}$$

and those $2m$ images can also be clustered via the method discussed earlier.

3 Experiments

In this section, two kinds of experiments are carried out in which synthetic data and real-world data are used to explain variations of graph structure associated with graph spectrum.

3.1 Synthetic Data Analysis

First of all, we investigate the relationship between the graph structure and the spectral distance of two graphs by using synthetic data. Two sets of synthetic data shown in Fig.1 are generated by performing a series of similar transformations on the letters 'W' and 'Z'. And we obtain 20 synthetic images respectively. Then we acquire the feature vectors formed from those eigenvalues by performing SVD on the weighted adjacency matrix and normalizing them. Based on the above discussion, we know that the images in a specific group have the same feature vectors. Therefore, two different feature vectors are obtained. Next we calculate the mean of them and the covariance matrix. And then we derive a distance matrix. Fig. 2 gives the distribution of the distance in two-dimension for these two groups of synthetic images. Moreover, another group of synthetic data is generated by performing a series of affine transformations on a house shown in Fig. 3, and the projection of the distance matrix in the two-dimensional space is shown in Fig. 4.

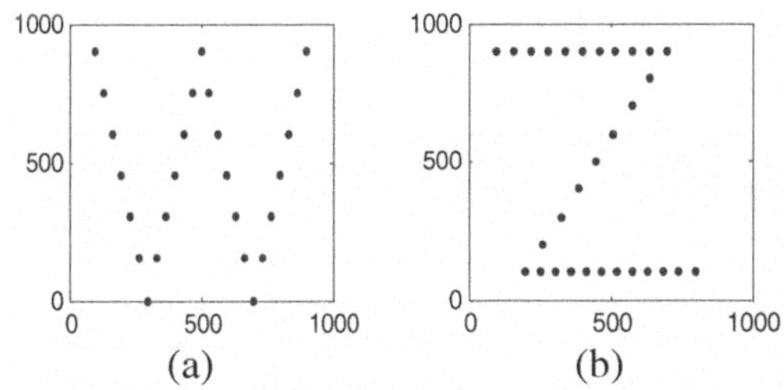

Fig. 1. Synthetic data of the letters 'W' and 'Z'

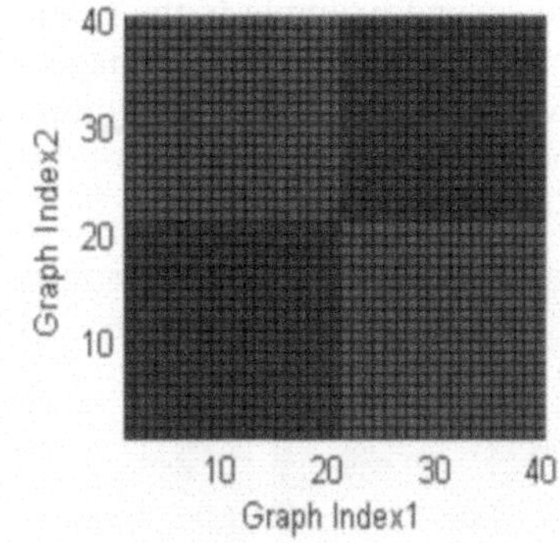

Fig. 2. Distribution of the distance matrix

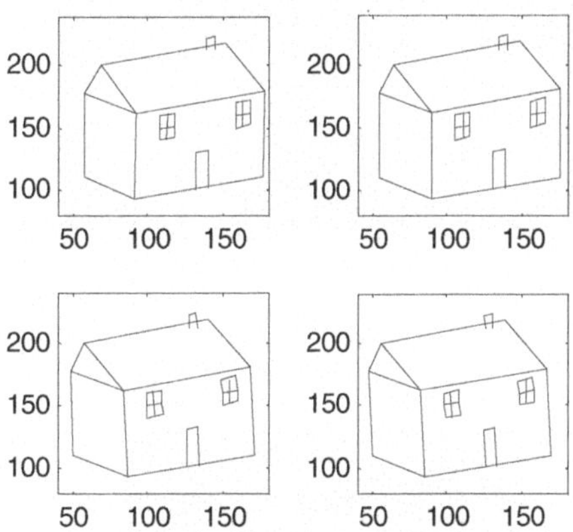

Fig. 3. Synthetic data sequence

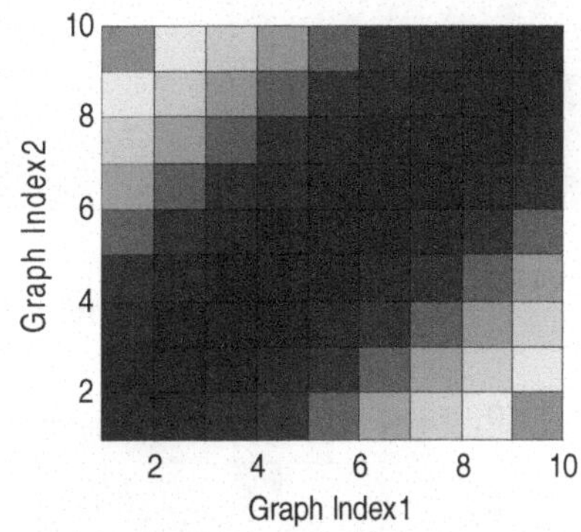

Fig. 4. Distribution of the distance matrix

From Fig. 4, we can see that if the structural change of two images does not become larger, their Euclidean distance is smaller, too. In this figure, the dark blue areas denote that the value of Euclidean distance is 0. When the color approaches dark blue, the value of Euclidean distance approaches 0; the color approaches light blue, the value of Euclidean distance becomes larger. When the color is not blue, it indicates that the structural variation of these two images is extremely large.

3.2 Real-World Image Experiments

Now we use the real-world data with unknown correspondences to study the relationship between the graph structure and spectral distance. The first example is the MOVI image sequence, which is shown in Fig. 5 (a). We select 10 images from the MOVI image sequence, and extract 30 feature points from each image respectively. We compute the covariance matrix and the distance matrix. Fig 6(a) demonstrates the distribution of the distance matrix in the two-dimensional space.

The second example is the CMU image sequence shown in Fig. 5 (b). 40 images are selected from the CMU image sequence and 30 feature points are detected from each image respectively. We compute the covariance matrix and the distance matrix. Fig. 6(b) shows the projection of distance matrix in two-dimension.

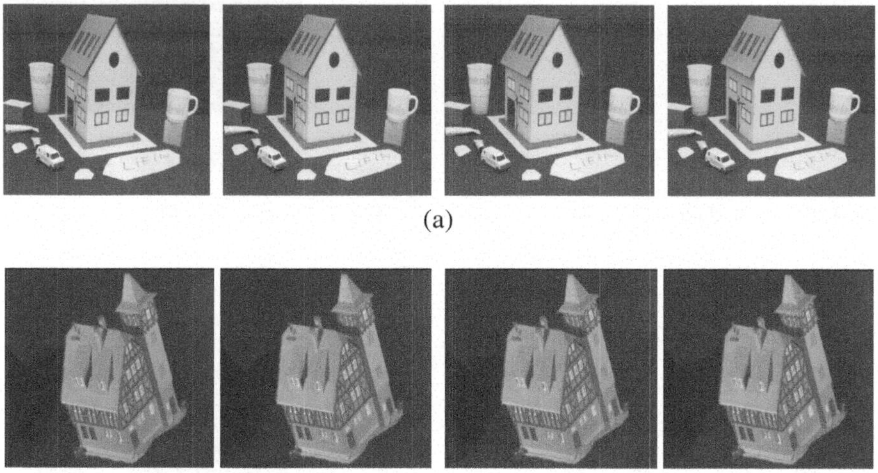

(a)

(b)

Fig. 5. Image sequence

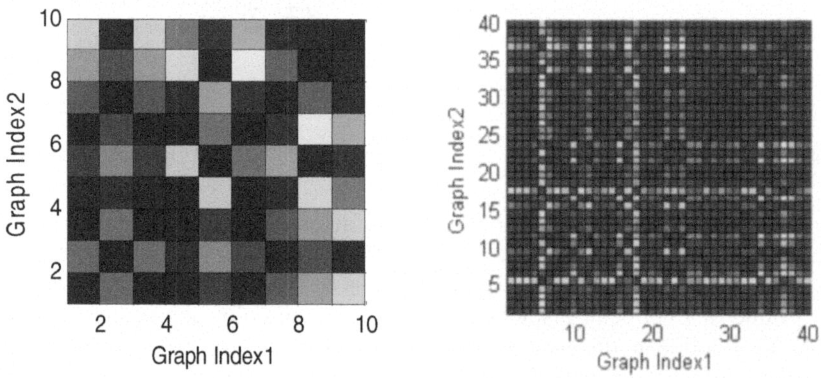

(a) (b)

Fig. 6. The distribution of the distance matrix

From the experiments of the real-world images, we can see that the Euclidean distance between two images increases when the viewing angle changes. Fig. 6 shows the tendency of the Euclidean distance. When two graph indices are nearer, the value of Euclidean distance is closer to 0, and the color of which approaches dark blue. Whereas, the color approaches light blue, the Euclidean distance is larger. That is to say, the structures of two graphs have significant variation. So we can learn that the graph structure is related to the spectral distance.

4 Conclusions and Future Work

This paper presents a method of the edit distance by using spectra of the relational graphs. Using the feature points of each image, we define a weighted adjacency matrix of the relational graph constructed on an image and obtain a covariance matrix based on the spectra of all graphs. We then project the vectorized spectra of each graph to the eigenspace of the covariance matrix and derive the distances between pairwise images. Experiments of synthetic data and real image demonstrate the feasibility of our method.

Our future plans involve studying in more detail the object structure resulting from our spectral features. We also intend to investigate whether the spectral attributes studied here can be used for the purpose of organizing image databases

Acknowledgments. This work is supported by National Science Foundation of China (Grant No. 10601001), Anhui Provincial Natural Science Foundation of China (Grant No. 050460102 070412065), Natural Science Foundation of Anhui Provincial Education department (Grant No. 2006KJ030B) and Innovative research team of 211 project in Anhui University.

References

1. D. Guillamet, J. Vitri, B. Schiele: Introducing a weighted non-negative matrix factorization for image classification. Pattern Recognition Letters(2003).Vol.24. 2447-2454.
2. S. Belongie, J. Malik, J. Puzicha: Shape matching and object recognition using shape contexts. IEEE Transactions on Pattern Analysis and Machine Intelligence (2002). Vol. 24. No.24. 509–522.
3. S. B. Park et al.: Content-based image classification using a neural network. Pattern Recognition Letters (2004). Vol.25. 287–300.
4. G. Giacinto et al.: Combination of neural and statistical algorithms for supervised classification of remote-sensing images. Pattern Recognition Letters (2000).Vol. 21.385–397.
5. M. A. Lozano, F. Escolano: ACM attributed graph clustering for learning classes of images. in: Graph Based Representations in Pattern Recognition, Lecture Notes in Computer Science (2003), Vol. 2726, 247–258.
6. H. Bunke et al.: Graph clustering using the weighted minimum common supergraph. in: Graph Based Representations in Pattern Recognition, Lecture Notes in Computer Science(2003), Vol. 2726, 235–246
7. A. D. Bagdanov, M. Worring: First order Gaussian graphs for efficient structure classification. Pattern Recognition (2003) .36.1311–1324
8. H. Qiu, E. R. Hancock: Graph matching and clustering using spectral partitions. Pattern Recognition.(2006).39.22-34.
9. B. Luo, R. C. Wilson, E. R. Hancock: A spectral approach to learning structural variations in graphs. Pattern Recognition (2006). 39. 1188–1198

Mining Invisible Tasks from Event Logs

Lijie Wen, Jianmin Wang, and Jiaguang Sun

School of Software, Tsinghua University, Beijing 100084, China
wenlj00@mails.tsinghua.edu.cn, {jimwang,sunjg}@tsinghua.edu.cn

Abstract. Most existing process mining algorithms have problems in dealing with invisible tasks. In this paper, a new process mining algorithm named $\alpha^{\#}$ is proposed, which extends the mining capacity of the classical α algorithm by supporting the detection of invisible tasks from event logs. Invisible tasks are first divided into four types according to their functional features, i.e., SIDE, SKIP, REDO and SWITCH. After that, the new ordering relation for detecting mendacious dependencies between tasks that reflects invisible tasks is introduced. Then the construction algorithms for invisible tasks of SIDE and SKIP/REDO/SWITCH types are proposed respectively. Finally, the $\alpha^{\#}$ algorithm constructs the mined process models incorporating invisible tasks in WF-net. A lot of experiments are done to evaluate the mining quality of the proposed $\alpha^{\#}$ algorithm and the results are promising.

1 Introduction

Although quite a lot of work has been done on process mining, there are still some challenging problems left [9,10,11], i.e., short loops, duplicated tasks, invisible tasks, non-free-choice constructs, time, noise, incompleteness, etc. The issue of short loops is solved in [2]. For discussion about duplicated tasks, readers are referred to [1,3]. [7,12] attempts to resolve most kinds of non-free-choice constructs. Time issue is partially considered in [8]. Noise and incompleteness are discussed in [5]. Here, we will investigate how to mine invisible tasks from event logs.

Invisible tasks are difficult to mine because they do not appear in any event trace. The following situations can lead to invisible tasks:

- There are meaningless tasks for routing purpose only in process models.
- There are real tasks that have been executed lost in some event traces.
- The enactment services of process models allow skipping or redoing current task and jumping to any previous task. But such execution logic is not expressed in the control logic of the process model.

The problems encountered when mining process models using the classical α algorithm from event logs containing invisible tasks will be investigated here. In Figure 1, N_1 to N_6 are the original WF-nets and N_1' to N_6' are the corresponding mined models derived from the complete event logs W_1 to W_6 respectively by using α algorithm. The black block transitions without labels represent invisible

G. Dong et al. (Eds.): APWeb/WAIM 2007, LNCS 4505, pp. 358–365, 2007.

tasks. All the original WF-nets are sound, but the mined models have various issues. Tasks B and C as well as B and D are parallel in N_1, while they are mutually exclusive in N_1'. Here N_1' is not a sound WF-net because a deadlock will always occur. Although N_2' is a sound WF-net, C cannot directly follow A and there is an implicit place in it. As a result, the behavior of N_2' is not equivalent with that of N_2. In N_3, B behaves like a length-one-loop task. However, C never directly follows A. There will not a place connecting A and C in N_3', which should be the only place associated with B. Here N_3' is not a WF-net at all. N_4', N_5' and N_6' are all WF-nets but not sound. N_4 is more general than N_2 and N_3 is a special case of N_5. The steps for constructing the mined model in α algorithm result in the above issues. In this paper, new mining algorithm will be proposed based on α algorithm, in which invisible tasks can be derived correctly.

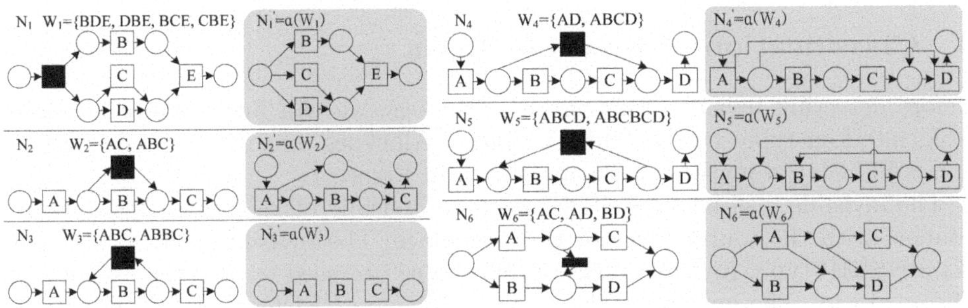

(a) case one (b) case two

Fig. 1. Problems encountered when mining process models using α algorithm

The remainder of the paper is organized as follows. Section 2 introduces related work on mining invisible tasks. Section 3 gives the classification of invisible tasks according to their functional features. The detection methods of invisible tasks are proposed in Section 4. The new mining algorithm $\alpha^{\#}$ is illustrated thoroughly in Section 5. Section 6 shows the evaluation results on the new algorithm. Section 7 concludes the paper and gives future work.

2 Related Work

Here only process mining algorithms based on Petri net are considered. For other mining algorithms, their emphases are focused on the efficient identification of relationships between each pair of input/output arcs of the same task.

A Synchro-net based mining algorithm is proposed in [4]. The authors state that short loops and invisible tasks can be handled with ease. However, neither the original model nor the mined model contains any invisible task.

[6] attempts to mine decisions from process logs, which emphasizes detecting data dependencies that affect the routings of cases. When interpreting the

semantics of the control flows in the mined decisions, the authors propose a descriptive method to identify decision branches starting from invisible tasks. This method cannot handle all kinds of invisible tasks. Even when there are other decision points with join nodes on one decision branch, the method fails.

By far the genetic mining algorithm is the only method that natively supports the detection of invisible tasks [7]. It uses the basic idea of the genetic algorithm and defines two genetic operators for process mining, i.e., crossover and mutation. It aims at supporting duplicated tasks, invisible tasks, non-free-choice constructs. However, this algorithm needs many user-defined parameters and it cannot always guarantee to return the most appropriate results.

In summary, there is still not any efficient solution that can handle invisible tasks well. This paper will focus on mining process models from event logs with invisible tasks based on the classical α algorithm proposed in [11].

3 Classification of Invisible Tasks

Before detecting invisible tasks from event logs, we will first classify invisible tasks into four types by their functionalities. All types (i.e., SIDE, SKIP, REDO and SWITCH) of invisible tasks are shown in Figure 1.

The invisible task in N_1 is SIDE type and invisible tasks of this type directly connect with the source place or the sink place. The invisible tasks in N_2 and N_4 are SKIP type. Invisible tasks of SKIP type are used to skip the executions of some tasks. The invisible tasks in N_3 and N_5 are REDO type. Invisible tasks of REDO type are used to repeat the executions of some tasks. The invisible task in N_6 is SWITCH type and invisible tasks of this type are used to switch the execution rights among multiple alternative branches.

4 Detection of Invisible Tasks

When there are invisible tasks in process models, the causal dependencies between tasks detected from event logs are not always correct any more. Such dependencies are called *mendacious dependencies*. The most important step of detecting invisible tasks from event logs is identifying all the mendacious dependencies out of the causal dependencies. The basic ordering relations between tasks derived from event logs are first listed below. For more detailed explanation about these basic ordering relations, readers are referred to [11,2].

Definition 1 (Basic ordering relations). *Let $N = (P, T, F)$ be a sound WF-net, W be an event log of N (i.e., $W \subseteq T^*$). Let $a, b \in T$, then:*

- $a >_W b$ *iff* $\exists \sigma = t_1 t_2 t_3 \cdots t_n \in W, i \in \{1, \ldots, n-1\} : t_i = a \wedge t_{i+1} = b$,
- $a \triangle_W b$ *iff* $\exists \sigma = t_1 t_2 t_3 \cdots t_n \in W, i \in \{1, \ldots, n-2\} : t_i = t_{i+2} = a \wedge t_{i+1} = b$,
- $a \diamond_W b$ *iff* $a \triangle_W b \wedge b \triangle_W a$,
- $a \rightarrow_W b$ *iff* $a >_W b \wedge (b \not>_W a \vee a \diamond_W b)$,
- $a \#_W b$ *iff* $a \not>_W b \wedge b \not>_W b$, *and*
- $a \parallel_W b$ *iff* $a >_W b \wedge b >_W b \wedge a \not\diamond_W b$.

The requirement for the completeness of an event log is the same as the one proposed in [2] (i.e., loop-complete). Now advanced ordering relations for mendacious dependencies can be derived from the basic ordering relations.

Definition 2 (Advanced ordering relations). *Let $N = (P, T, F)$ be a sound WF-net, W be a loop-complete event log of N (i.e., $W \subseteq T^*$). Let $a, b \in T$, then:* $a \rightsquigarrow_W b$ *iff* $a \rightarrow_W b \wedge \exists x, y \in T : a \rightarrow_W x \wedge y \rightarrow_W b \wedge y \not\succ_W x \wedge x \nparallel_W b$.

$\rightsquigarrow_W$ reflects the mendacious dependencies associated with invisible tasks of SKIP,REDO and SWITCH types and this kind of ordering relation can be used to construct invisible tasks. From the logs shown in Figure 1, $A \rightsquigarrow_W C$, $B \rightsquigarrow_W B$, $A \rightsquigarrow_W D$, $C \rightsquigarrow_W B$ and $A \rightsquigarrow_W D$ can be detected from W_2 to W_6 respectively. To illustrate the derivation of $\rightsquigarrow_W$, see Figure 2.

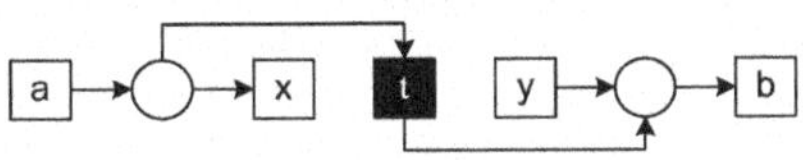

Fig. 2. Illustration for the derivation of $\rightsquigarrow_W$

In Figure 2, there is an invisible task t in the snippet of a WF-net and assume that t can be detected from the corresponding log. The correctness of the detection method corresponding to $\rightsquigarrow_W$ can be proved theoretically based on SWF-net. If y is equal to x, t is S-SKIP type. If y is reachable from x, t is L-SKIP type. If a is equal to b, t is S-REDO type. If a is reachable from b, t is L-REDO type. Otherwise, t is SWITCH type, i.e., a to x and y to b are two alternative paths. Detecting invisible tasks of SIDE type is relatively direct (see Section 5). However, invisible tasks of SIDE type should be detected first because they will affect the detection of invisible tasks of other types.

After detecting all mendacious dependencies between tasks, the corresponding causal dependencies should be eliminated, i.e., $a \rightsquigarrow_W b \Rightarrow a \nrightarrow_W b$. More precisely, the equation $a \rightsquigarrow_W b \Rightarrow a \not\succ_W b$ holds.

5 The Mining Algorithm $\alpha^\#$

For process models containing only causal relations between tasks, there is a one-to-one correspondence between invisible tasks and mendacious dependencies. However, this is not always the case because selective and parallel relations are so common in real-life processes. Constructing invisible tasks is not such a trivial task. See Figure 3 for detail explanation.

The process model N_9 is a sound SWF-net and one of its complete log is $W_9 = \{ACDDFGHI, BCEEFHGI, ADEDEGHI, AEDGHI, BEDHGI, BDEHGI\}$. t_1 corresponds to $D \rightsquigarrow_W D$, $D \rightsquigarrow_W E$, $E \rightsquigarrow_W D$ and $E \rightsquigarrow_W E$. Similar things happen to t_2 and t_3. Furthermore, there are situations where multiple invisible tasks correspond to one mendacious dependency.

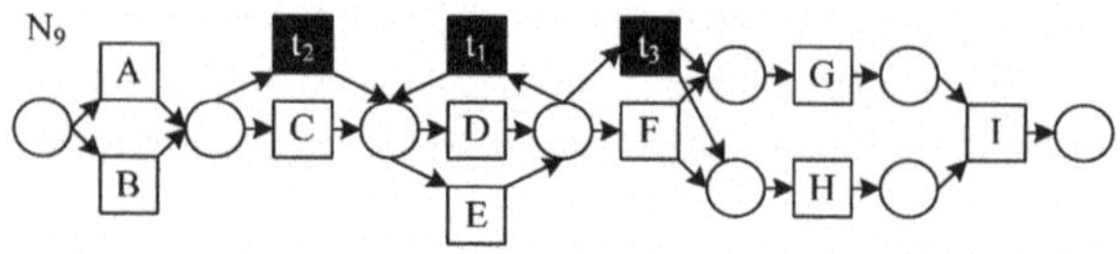

Fig. 3. The one-to-multi correspondence between invisible tasks and $\rightsquigarrow_W$

The construction algorithm for invisible tasks of SIDE type (i.e., $ConSideIT$) is omitted here. The algorithm for constructing invisible tasks of all types is given below, which is the core of the $\alpha^{\#}$ algorithm. The two functions $PreSet$ and $PostSet$ are used to construct the input and output places of a task.

Definition 3 (Construction algorithm ConIT). *Let W be a loop-complete event log over a task set T (i.e., $W \subseteq T^*$). ConIT(W) is defined as follows.*

1. $(T_W, T_I, T_O, D_S) = ConSideIT(W)$,
2. $D_M = \{(a,b) | a \in T_W \wedge b \in T_W \wedge a \rightsquigarrow_W b\}$,
3. $X_I = \{(P_{in}, P_{out}) | (\forall (A,X) \in P_{in}, (Y,B) \in P_{out} : (\forall a \in A, b \in B : (a,b) \in D_M \wedge (A,X) \in PostSet(a) \wedge (Y,B) \in PreSet(b)) \wedge (\forall x \in X, y \in Y : x \nparallel_W y)) \wedge (\forall (A_1,X_1),(A_2,X_2) \in P_{in} : \exists a_1 \in A_1, a_2 \in A_2 : a_1 \parallel_W a_2) \wedge (\forall (Y_1,B_1),(Y_2,B_2) \in P_{out} : \exists b_1 \in B_1, b_2 \in B_2 : b_1 \parallel_W b_2)\}$,
4. $Y_I = \{(P_{in}, P_{out}) \in X_I | \forall (P'_{in}, P'_{out}) \in X_I : P_{in} \subseteq P'_{in} \wedge P_{out} \subseteq P'_{out} \Rightarrow (P_{in}, P_{out}) = (P'_{in}, P'_{out})\}$,
5. $D_S = D_S \cup \{(t_{(P_{in},P_{out})}, t_{(P'_{in},P'_{out})}) | (P_{in}, P_{out}), (P'_{in}, P'_{out}) \in Y_I \wedge P_{out} \cap P'_{in} \neq \emptyset\} \cup \{(a, t_{(P_{in},P_{out})}) | (P_{in}, P_{out}) \in Y_I \wedge \exists (A,X) \in P_{in} : a \in A\} \cup \{(t_{(P_{in},P_{out})}, b) | (P_{in}, P_{out}) \in Y_I \wedge \exists (Y,B) \in P_{out} : b \in B\}$,
6. $D_P = \{(t_{(P_{in},P_{out})}, t_{(P'_{in},P'_{out})}) | (P_{in}, P_{out}), (P'_{in}, P'_{out}) \in Y_I \wedge \forall (A,X) \in P_{in}, (A',X') \in P'_{in} : \exists a \in A, a' \in A', x \in X, x' \in X' : a \parallel_W a' \vee x \parallel_W x'\} \cup \{(t, t_{(P_{in},P_{out})}) | t \in T_W \wedge (P_{in}, P_{out}) \in Y_I \wedge \forall (A,X) \in P_{in} : \exists a \in A, x \in X : a \parallel_W t \vee x \parallel_W t\} \cup \{(t_{(P_{in},P_{out})}, t) | t \in T_W \wedge (P_{in}, P_{out}) \in Y_I \wedge \forall (A,X) \in P_{in} : \exists a \in A, x \in X : a \parallel_W t \vee x \parallel_W t\}$,
7. $T_W = T_W \cup \{t_{(P_{in},P_{out})} | (P_{in}, P_{out}) \in Y_I\}$, *and*
8. $ConIT(W) = (T_W, T_I, T_O, D_S, D_P, D_M)$.

The algorithm $ConIT$ works as follows. Step 1 invokes the algorithm $ConSide$ IT to construct invisible tasks of SIDE type, fix first and last task sets and add necessary causal relations. All mendacious dependencies between tasks are detected in step 2. Steps 3 and 4 are used to construct invisible tasks of SKIP/REDO/SWITCH types (stored in Y_I) reflected by the mendacious dependencies. These two steps are the most important ones in the whole algorithm. In steps 5 and 6, new causal and parallel relations between invisible tasks as well as the ones between invisible tasks and visible tasks are added. Finally, the task set T_W are extended by new constructed invisible tasks in Step 7 and Step 8 returns the necessary results.

Based on the algorithms proposed above, the mining algorithm named $\alpha^{\#}$ can be defined as follows. It returns the mined model in WF-net.

Definition 4 (Mining algorithm $\alpha^{\#}$). *Let W be a loop-complete event log over a task set T (i.e., $W \subseteq T^*$). $\alpha^{\#}(W)$ is defined as follows.*

1. $(T_W, T_I, T_O, D_S, D_P, D_M) = ConIT(W)$,
2. $X_W = \{(A,B) | A \subseteq T_W \wedge B \subseteq T_W \wedge (\forall a \in A, b \in B : (a \rightarrow_W b \wedge (a,b) \notin D_M) \vee (a,b) \in D_S) \wedge (\forall a_1, a_2 \in A : (a_1 \#_W a_2 \wedge (a_1,a_2) \notin D_P) \vee (a_1 \rightarrow_W a_2 \wedge a_2 >_W a_2) \vee (a_2 \rightarrow_W a_1 \wedge a_1 >_W a_1)) \wedge (\forall b_1, b_2 \in B : (b_1 \#_W b_2 \wedge (b_1,b_2) \notin D_P) \vee (b_1 \rightarrow_W b_2 \wedge b_1 >_W b_1) \vee (b_2 \rightarrow_W b_1 \wedge b_2 >_W b_2))\}$,
3. $Y_W = \{(A,B) \in X_W | \forall (A',B') \in X_W : A \subseteq A' \wedge B \subseteq B' \Rightarrow (A,B) = (A',B')\}$,
4. $P_W = \{P_{(A,B)} | (A,B) \in Y_W\} \cup \{i_W, o_W\}$,
5. $F_W = \{(a, P_{(A,B)}) | (A,B) \in Y_W \wedge a \in A\} \cup \{(P_{(A,B)}, b) | (A,B) \in Y_W \wedge b \in B\} \cup \{(i_W, t) | t \in T_I\} \cup \{(t, o_W) | t \in T_O\}$, *and*
6. $N_W = (P_W, T_W, F_W)$.

The $\alpha^{\#}$ algorithm is relatively simple and easy to understand, which works as follows. Step 1 invokes the algorithm *ConIT* to construct all invisible tasks and fix the causal/parallel relations between tasks. All pairs of task sets related to possible places are constructed in Step 2. This step takes into account invisible tasks and length-one-loop tasks at the same time. Steps 3 to 6 are directly borrowed from [11], in which the places together with the connecting arcs are constructed and the mined process model in WF-net is returned.

6 Experimental Evaluation

The $\alpha^{\#}$ algorithm has been implemented as a ProM plug-in and can be downloaded from `www.processmining.org`.

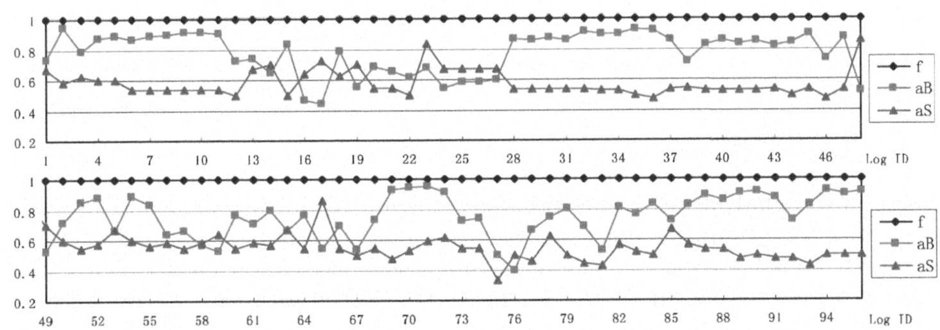

Fig. 4. Evaluation of $\alpha^{\#}$ algorithm using 96 artificial logs

The $\alpha^{\#}$ plug-in of ProM has been applied to several real-life logs and many smaller artificial logs. The evaluation criteria is the three conformance testing metrics between a given event log and a process model proposed in [7], i.e., f

(fitness), aB (behavioral appropriateness) and aS (structural appropriateness). For any successful mining, the value of f should be 1.0 and the values of aB and aS should be as big as possible.

There are totally 96 artificial examples in WF-nets evaluated. The corresponding complete logs are generated manually. The maximum number of tasks in one process model is less than 20 and the number of cases in one event log is less than 30. Although thirty process models are not SWF-nets, the evaluation results show that all the mined process models fit the corresponding logs. The conformance testing results are shown in Figure 4.

Ten real-life logs are obtained from Kinglong Company in Xiamen, Fujian province, China, which are all about processes for routing engineering documents. All the conformance testing results are shown in Table **??**. It is obvious that all the experiments are successful. The proportion for invisible tasks out of all tasks is 59/(59+88)=40.1%. The evaluation results show that so long as the event logs are complete, the $\alpha^{\#}$ algorithm can mine all necessary invisible tasks in all SWF-nets and most WF-nets successfully.

Table 1. Conformance testing results based on real-life logs: f-fitness, aB-behavioral appropriateness, aS-structural appropriateness, NoI-the number of invisible tasks, NoC-the number of cases, NoE-the number of events, NoT-the number of visible tasks

	$L1$	$L2$	$L3$	$L4$	$L5$	$L6$	$L7$	$L8$	$L9$	$L10$
f	1.0	1.0	1.0	1.0	1.0	1.0	1.0	1.0	1.0	1.0
aB	0.732	0.983	0.729	0.902	0.811	0.933	0.747	0.953	0.753	0.786
aS	0.429	0.5	0.462	0.5	0.45	0.478	0.333	0.529	0.45	0.342
NoI	6	3	6	3	5	4	11	2	5	14
NoC	8	6	5	11	40	42	30	42	297	50
NoE	43	52	59	84	324	469	221	288	2020	537
NoT	7	15	10	7	7	9	8	7	7	11

7 Conclusion

Based on the analysis of mining problems encountered using the classical α algorithm, a new mining algorithm based on Petri net named $\alpha^{\#}$ algorithm is proposed. Invisible tasks are classified into four types according to their functionalities for the first time, i.e., SIDE, SKIP, REDO and SWITCH. The universal detection method for invisible tasks of SKIP/REDO/SWITCH types is illustrated in detail and the correctness of the method can be proved theoretically. The construction algorithms for all types of invisible tasks and the process models in WF-nets are proposed and explained too. The $\alpha^{\#}$ algorithm has been implemented as a plug-in of ProM and evaluated using a lot of artificial logs and a few real-life logs. The evaluation results show that the algorithm is pragmatic.

Our future work will mainly focus on the following two aspects. Firstly, more real-life logs will be gathered for further evaluating the $\alpha^{\#}$ algorithm and the implemented plug-in. Secondly, theoretical analysis will be done to explore the exact mining capacity of the $\alpha^{\#}$ algorithm.

Acknowledgement

This work is supported by the 973 Project of China (No. 2002CB312006) and the Project of National Natural Science Foundation of China (No. 60373011).

References

1. J.E. Cook, Z.D. Du, C.B. Liu, and A.L. Wolf. Discovering models of behavior for concurrent workflows. *Computers in Industry*, 53(3):297–319, 2004.
2. A.K.A. de Medeiros, B.F. van Dongen, W.M.P. van der Aalst, and A.J.M.M. Weijters. Process Mining for Ubiquitous Mobile Systems: An Overview and a Concrete Algorithm. In L. Baresi, S. Dustdar, H. Gall, and M. Matera, editors, *Ubiquitous Mobile Information and Collaboration Systems*, pages 154–168, 2004.
3. J. Herbst and D. Karagiannis. Workflow Mining with InWoLvE. *Computers in Industry*, 53(3):245–264, 2004.
4. X.Q. Huang, L.F. Wang, W. Zhao, S.K. Zhang, and C.Y. Yuan. A workflow process mining algorithm based on synchro-net. *Journal of Computer Science and Technology*, 21(1):66–71, 2006.
5. L. Maruster, A.J.M.M. Weijters, W.M.P. van der Aalst, and A. van der Bosch. A Rule-Based Approach for Process Discovery: Dealing with Noise and Imbalance in Process Logs. *Data Mining and Knowledge Discovery*, 13(1):67–87, 2006.
6. A. Rozinat and W.M.P. van der Aalst. Decision Mining in Business Processes. BETA Working Paper Series, WP 164, Eindhoven University of Technology, 2006.
7. W.M.P. van der Aalst, A.K.A. de Medeiros, and A.J.M.M. Weijters. Genetic Process Mining. In G. Ciardo and P. Darondeau, editors, *Applications and Theory of Petri Nets*, volume 3536 of *LNCS*, pages 48–69. Springer-Verlag, Berlin, 2005.
8. W.M.P. van der Aalst and B.F. van Dongen. Discovering Workflow Performance Models from Timed Logs. In Y. Han, S. Tai, and D. Wikarski, editors, *International Conference on Engineering and Deployment of Cooperative Information Systems*, volume 2480 of *LNCS*, pages 45–63. Springer-Verlag, Berlin, 2002.
9. W.M.P. van der Aalst, B.F. van Dongen, J. Herbst, L. Maruster, G. Schimm, and A.J.M.M. Weijters. Workflow Mining: A Survey of Issues and Approaches. *Data and Knowledge Engineering*, 47(2):237–267, 2003.
10. W.M.P. van der Aalst and A.J.M.M. Weijters. Process Mining: A Research Agenda. *Computers in Industry*, 53(3):231–244, 2004.
11. W.M.P. van der Aalst, A.J.M.M. Weijters, and L. Maruster. Workflow Mining: Discovering Process Models from Event Logs. *IEEE Transactions on Knowledge and Data Engineering*, 16(9):1128–1142, 2004.
12. L.J. Wen, J.M. Wang, and J.G. Sun. Detecting implicit dependencies between tasks from event logs. In X. Zhou, X. Lin, and H. Lu et al., editors, *APWeb 2006*, volume 3841 of *LNCS*, pages 591–603. Springer-Verlag, Berlin, 2006.

The Selection of Tunable DBMS Resources Using the Incremental/Decremental Relationship*

Jeong Seok Oh[1], Hyun Woong Shin[2], and Sang Ho Lee[3]

[1] Korea Gas Safety Corporation, Shihung-Shi, Gyounggi-Do, Korea
dbstar@kgs.or.kr
[2] Samsung Electronics Co. LTD, Suwon-Shi, Gyounggi-Do, Korea
hws52.shin@samsung.com
[3] School of Computing, Soongsil University, Seoul, Korea
shlee@comp.ssu.ac.kr

Abstract. The DBMS performance might change by allocating resources and by performing a specific kind of workload. Database administrators should be able to identify relative resources that can change DBMS performance in order to effectively manage database systems. This paper aims to identify the relative resources that can affect the DBMS performance depending on the different kinds of workload. The relative resource is identified by the incremental or the decremental relationship between the performance indicator and the resource. The relationship is determined by the Pearson's correlation coefficient with the t-test. We identify the relative resources that have an impact on the DBMS performance under TPC-C and TPC-W benchmarks using our proposed method. As a result, the data buffer and the shared memory could affect the DBMS performance in TPC-C, and only the data buffer in TPC-W. In order to verify our works, we measure the maximum load that can be executed in the individual system for TPC-C and TPC-W.

1 Introduction

Database workload characteristics can be different depending on database applications. Since database applications become more complex and various, database administrators need to carefully consider peculiar workload characteristics [2, 4, 6, 9, 10, 11, 14]. Changing DBMS resources can differ in resource usages and the DBMS performance depending on the different kinds of workload. Resource usages should be taken by performance indicators in database systems for tracking the current DBMS state.

Several studies on database workloads have been reported [2, 6, 1]. These studies take no account of analyzing relative inter-relationships between performance indicators and resources, or use only a few resources in order to analyze workloads. In our previous studies, some performance indicators are shown to be affected by particular resource changes. For example, the data buffer ratio was increased by expanding the

* This work was supported by Korea Research Foundation Grant (KRF-2006-005-J03803).

G. Dong et al. (Eds.): APWeb/WAIM 2007, LNCS 4505, pp. 366–373, 2007.

data buffer, but it seemed to remain unchanged by expanding other resources, in particular workload type. We, however, need to consider a precise decision method for identifying relative resources that can have an impact on DBMS performance depending on the different kinds of workloads.

The goal of this paper is to identify the resources that can have an impact on the DBMS performance through analyzing relationships between performance indicators and resources in a workload type. To construct a standard workload environment, the TPC-C and TPC-W [12, 13] were used. The workload data, which are represented by fourteen performance indicators, were investigated in response to changes of four tuning parameters (data buffer, private memory, shared memory, and I/O process). The relationship is determined by the Pearson's correlation coefficient with the t-test. Furthermore, we identify certain resources that can affect the DBMS performance within the TPC-C and TPC-W workloads. To verify our hypothesis and methodology, we measure the maximum load that can be executed in the individual system for the TPC-C and TPC-W. This research can aid database tuning and automatic DBMS managements.

This paper is composed as follows. Section 2 explains collecting workload data using TPC-C and TPC-W benchmarks. Section 3 introduces the method that can analyze the relationship between performance indicators and resources. Section 4 describes the relationship results using TPC-C and TPC-W and shows experiments in order to support the proposed method. Section 5 presents the conclusions and plan for future work.

2 Collecting Workload Data

In the collection phase, we need to determine target resources, adjust the resource size, and select performance indicators. For the target resources, we selected four database resources: data buffer, private memory, shared memory, and I/O processes [1, 2, 3, 4, 6, 9, 10, 11]. The shared memory contains processing information and executable plans of frequently used queries. The data in shared memory is shared. The private memory generally keeps data for join, sort, and cursor operations. The data in the private memory is not shared.

The resource size is changed by the database tuning parameter. Resources (data buffer, shared memory, private memory, and I/O processes) can be adjusted by tuning four system parameters (db_cache_size, shared_pool_size, pga_aggregate_target, and dbwr_io_slave), respectively [3].

Table 1 shows how a parameter is changed during workload collection. The initial values and incremental values of parameters are set as default values of the DBMS. During workload collection, only one parameter is subject to change at a time, while leaving the others unchanged. For instance, the value of db_cache_size increases from 32MB to 480MB by 32MB intervals, while other parameters remain as initial values. The workload data is collected 114 times.

To analyze workloads of database systems, we use 14 performance indicators of a database system: the data buffer hit ratio; the shared memory hit ratio; the system catalog hit ratio; the latch contention ratio, the memory sort ratio; the memory parsing ratio; the data variance ratio; the data buffer reads; the non-data buffer reads; the data buffer writes; the non-data buffer writes; the disk writes with checkpoints; the disk writes without checkpoints; and the redo size [1, 2, 4, 8].

Table 1. How to change parameters

Parameter	initial	incremental	Final
db_cache_size	32MB	32MB	480MB
shared_pool_size	32MB	32MB	480MB
pga_aggregate_target	20MB	20MB	300MB
dbwr_io_slave	1	1	15

3 The Incremental and Decremental Relationship

This paper proposes a new method to effectively identify relative resources, which is able to define the incremental or decremental relationship between the resource and the performance indicator using the Pearson's correlation coefficient and the t-test. The Pearson's correlation coefficient measures the degree of which two variables are linearly related. The Pearson's correlation coefficient is used to test this relationship, and the equation is shown below [5, 7]. The equation is defined as the covariance of *X* with *Y* divided by the product of the standard deviation of *X* and the standard deviation of *Y*. $\overline{X}$ and $\overline{Y}$ indicates the mean of the variable and *n* represents the number of variable values. Pearson's correlation coefficient is defined as a value between +1 and -1. If the correlation coefficient is positive, it indicates that one variable increases accordingly when the other variable increases. If the correlation coefficient is negative, it indicates that another variable increases accordingly when the variable decreases.

$$COE(X,Y) = \frac{\sum_{i=1}^{n}(X_i - \overline{X})(Y_i - \overline{Y})}{\sqrt{\sum_{i=1}^{n}(X_i - \overline{X})\sum_{i=1}^{n}(Y_i - \overline{Y})}} \quad (1)$$

In our research, *X* and *Y* are the resource and the performance indicator, respectively. Therefore, if the coefficient is positive, the performance indicator seems to be increased by expanding the resource. If the coefficient is negative, the performance indicator seems to be decreased by expanding the resource. Since not all correlations show the incremental or decremental relationship, we apply significance level test using the t-test in order to define the real incremental or decremental relationship.

The t-test sets null hypothesis, alternative hypothesis, and significance level and finds the critical value that can distinguish rejection area and acceptance area using the t-distribution table. If null hypothesis is accepted when the t-test value exists in the acceptance area, the relationship between resource and performance indicator is meaningless. Otherwise, the relationship is meaningful. The t-test shows in Equation (2). *r* is the correlation coefficient between two variables, and *n* is the number of data.

$$t = \frac{r}{\sqrt{1-r^2}}\sqrt{n-2} \tag{2}$$

Example 1. Let us obtain the incremental or decremental relationship between values of performance indicator (P, Q) and values (in megabytes) of resource K. The assumptions for t-test are followings:

K={32, 64, 96, 128, 160, 192, 224, 256, 288, 320}
P={27.21, 27.49, 27.45, 27.22, 27.43, 26.5, 26.95, 27.1, 27.03}
Q={74.32, 76.79, 78.25, 80.63, 81.69, 81,95, 84.3, 84.61, 87.7, 89.41}
Null hypothesis(H0): correlation between resource and performance indicator do not exist (ρ =0)
Alternative hypothesis: correlation between resource and performance indicator exist (ρ =0)
Significance level (α): 0.01
Critical value : $t_{a/2}(n-2)$

[Identifying relationship between resource K and indicator P]

$$COE(K,P) = \frac{COV(K,P)}{STD(K) \times STD(P)} = -0.48634$$

$$t - test = \frac{-0.48634}{\sqrt{1-(-0.48634)^2}}\sqrt{10-2} = -1.5743$$

The correlation coefficient between k and P is about -0.48634. We expect decremental relationship from the coefficient. The t-test value about the correlation coefficient between k and p is about -1.5743. At the 0.01 significance level, the critical value calculates 3.355 using the t-distribution table. Since the acceptance area is $-3.355 \leq t \geq 3.355$, null hypothesis is accepted. In consequence, resource K and indicator P do not have the meaningful decremental relationship.

[Identifying the relationship between resource K and indicator Q]

$$COE(K,Q) = \frac{COV(K,P)}{STD(K) \times STD(P)} = 0.98948$$

$$t - test = \frac{0.98948}{\sqrt{1-(0.98948)^2}}\sqrt{10-2} = 19.34525$$

The correlation coefficient between K and P is about 0.98948. We expect incremental relationship from the coefficient. The t-test value about the correlation coefficient between K and Q is about 19.34525. At the 0.01 significance level, the critical value calculates 3.355 using the t-distribution table. Since the t-test value exists within the reject area, null hypothesis is rejected. Therefore, resource K and indicator P have the significant incremental relationship.

4 Experimental Result

Using TPC-C and TPC-W workloads, we calculate correlation coefficients between expanded resources and recorded performance indicators, and obtain the t-test values of correlation coefficients. At the 0.05 significance level, the relationship between the resource and the performance indicator can be accepted when the t-test value exists in the rejection area; otherwise the relationship cannot be accepted. Table 2 describes null and alternative hypothesis, rejection and acceptance area for significance level test.

Table 2. Assumptions for t-test

Null hypothesis(H0): correlation between resource and performance indicator do not exist (ρ =0)
Alternative hypothesis: correlation between resource and performance indicator exist (ρ =0)
Significance level (α): 0.05
Critical value: $\lvert 3.372 \rvert$
Rejection area of null hypothesis: $t < -3.372$ or $t > 3.372$
Acceptance area of null hypothesis: $-3.372 \leq t \leq 3.372$

Table 3 shows performance indicators that have the meaningful relationship between the 14 performance indicators and expansions of db_cache_size in TPC-C. There are five performance indicators that exist in the rejection area of null hypothesis. There exists the significant incremental relationship in the data buffer hit ratio and the disk writes with checkpoints. On the other hand, the significant decremental relationships are shown in the data buffer reads, the data buffer and the disk writes.

Table 3. Five indicators significantly affected by db_cache_size in TPC-C

Indicator	t-test value
Data buffer hit ratio	40.70532
Data buffer reads	-55.5472
Data buffer writes	-8.27179
Disk writes with checkpoints	6.392314
Disk writes without checkpoints	-8.62323

Table 4 shows performance indicators that have the meaningful relationship between the 14 performance indicators and expansions of shared_pool_size in TPC-C. There are seven performance indicators that exist in the rejection area of null hypothesis. All of the seven performance indicators show the significant decremental relationships. In expansions of other parameters, no performance indicator exists in the rejection area of null hypothesis.

Table 4. Seven indicators significantly affected by shared_pool_size in TPC-C

Indicator	t-test value
Data variance ratio	-6.29063
Data buffer hit ratio	-4.802897
Latch contention ratio	-3.43032
Data buffer writes	-11.0534
Non data buffer writes	-3.479
Disk writes without checkpoints	-15.3574
Redo size	-11.652

For TPC-W, a similar methodology is applied. Table 5 shows performance indicators that have the meaningful relationship between the 14 performance indicators and expansions of db_cache_size in TPC-W. There are two performance indicators that exist in the rejection area of null hypothesis. The significant incremental relationship is shown in the data buffer hit ratio is, and the significant decremental relationship is in the data buffer reads. For the remaining parameters, we apply a similar method, and it turns out that no performance indicator exists in the rejection area of null hypothesis.

Table 5. Two indicators significantly affected by db_cache_size in TPC-W

Indicator	t-test value
Data buffer hit ratio	4.510149
Data buffer reads	-4.42021

Because the incremental or decremental relationship can detect the change of the performance indicator to expanding the parameter, it is used for identifying resources that can affect the DBMS performance. The data buffer and the shared memory in the TPC-C can have an impact on the DBMS performance because the significant incremental or decremental relationship exists. Only the data buffer can affect the DBMS performance in the TPC-W.

In order to prove our claim, we measured the maximum number of warehouses and emulated browsers that can be executed in the individual system for the TPC-C and TPC-W. The measured results are shown in Figure 1.

In the case of TPC-C, expanding the data buffer can change the maximum number of warehouses (from 16 to 17) over the 12th phases. The expansion of this resource might reduce disk I/O and enhance DBMS performance. However, as update queries occur frequently in the TPC-C environment, the expansion of data buffer over 416MB can change the maximum number of warehouses in our experiments.

Expanding the shared memory can change the maximum number of warehouses (from 16 to 11) over the 2nd phases. The expansion of the resource can down the maximum number of warehouses. Oracle database management system is used for our test. The excessively large shared memory in Oracle reduces DBMS performance of updating queries, because the time is delayed about searching free list and allocating, reallocating or deallocating objects. Other resources had no influence on the number of allowable warehouses.

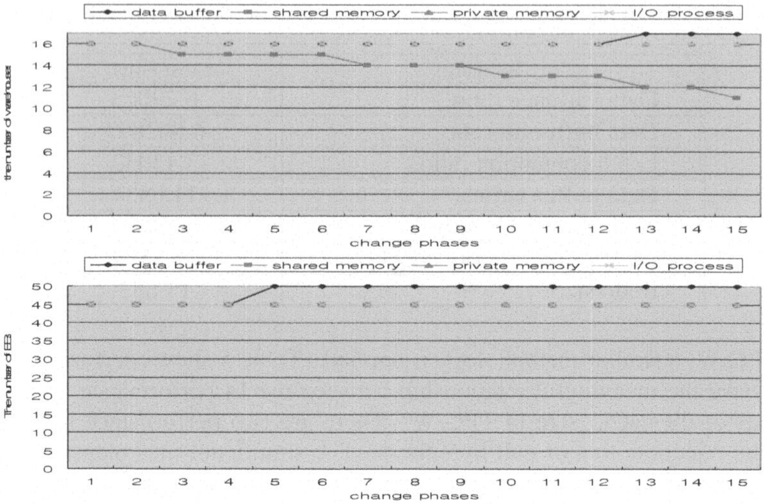

Fig. 1. The maximum allowable load

In the case of TPC-W, only the data buffer affects the number of allowable emulated browsers while others do not. That is, expanding the data buffer can change the maximum number of emulated browsers (from 45 to 50) over the 4th phases. Expending the shared memory has not an impact on DBMS performance in TPC-W environments, because the time is not delayed about searching free list and allocating, reallocating or deallocating objects. The private memory and the I/O processes had no influence on the number of allowable warehouses. N_k .

5 Conclusion

This paper identified the relative resources that can have an impact upon DBMS performance by analyzing the relationship between the resource and the performance indicator in TPC-C and TPC-W workloads, respectively. Pearson's correlation coefficient and the t-test with the .05 significance level were used to detect the lineal relationship.

The relationship is shown by expanding db_cache_size and shared_pool_size in the TPC-C, while it is shown by expanding db_cache_size in the TPC-W. Therefore, the data buffer and the shared memory are the resources that can be tuned to enhance the database performance in the TPC-C, while only the data buffer can be a resource in the TPC-W.

This study produced evidences which prove our previous information of effective database tuning as it identifies relative tunable resources depending on the different workload type. Furthermore, this study will pave the way toward automatic database management that reduces human intervention and burden.

References

[1] D. G. Benoit, "Automated Diagnosis and Control of DBMS resources", Ph. D Workshop on EDBT Conference, 2000.

[2] K. P. Brown, M. Mehta, M. J. Carey, and M. Livny, "Towards Automated Performance Tuning for Complex Workloads", Proceedings of 20th VLDB Conference, pp 72-84, Santiago, 1994.

[3] M. Cyran, "Oracle 9i: Database Performance Guide and Reference, Release 2(9.2)", Oracle Corporation, 2001.

[4] S. Elnaffar, P. Martin, and R. Horman, "Automatically Classifying Database Workloads", Proceedings of 11th CKIM Conference, pp 622-624, McLean, 2002.

[5] D. M. Lane, "Hyperstat Online: An Introductory Statistics Textbook and Online Tutorial for Help in Statistic", http://davidmlane.com/hyperstat/index.html

[6] P. Martin, H. Y. Li, M. Zheng, K. Romanufa, and W. Powley, "Managing Database Server Performance to Meet QoS Requirements in Electronic Commerce Systems", International Journal on Digital Libraries, Vol. 3, No. 4, pp 316-324, 2002.

[7] D. S. Moore, "Statistics Concepts and Controversies (the fifth edition)", W.H.Freeman and Company, 2001.

[8] T. Morals and D. Lorentz, "Oracle 9i: Database Reference, Release 2(9.2)", Oracle Corporation, 2001.

[9] J. S. Oh and S. H. Lee, "Resource Selection for Autonomic Database Tuning", Proceedings of IEEE International Workshop on Self-Managing Database Systems, pp 66-73, Tokyo, 2005.

[10] J. S. Oh and S. H. Lee, "Database Workload Analysis: empirical study", Journal of KISS, Vol. 11-D, No. 4, pp 747-754, 2004.

[11] V. Signhal and A. J. Smith, "Analysis of Locking Behavior in Three Real Database Systems", The VLDB Journal, Vol. 6, No. 1, pp 40-52, 1997.

[12] TPC Benchmark C Specification (Revision 5.0), 2001, http://www.tpc.org/tpcc/default.asp

[13] TPC Benchmark W (Web Commerce) Specification (version 1.8), 2002, http://www.tpc.org/tpcw/default.asp

[14] G. Weikum, A. C. Konig, A. Krasis, and M. Sinnewell, "Towards Self Tuning Memory Management of Data Servers", Bulletin of the Technical Committee on Data Engineering, IEEE Computer Society, Vol. 22, No. 2, pp 3-11, 1999.

Hyperclique Pattern Based Off-Topic Detection

Tianming Hu, Qingui Xu, Huaqiang Yuan, Jiali Hou, and Chao Qu

Department of Computer Science, DongGuan University of Technology
DongGuan, 523808, China
tmhu05@gmail.com

Abstract. This paper addresses the problem of detecting access to off-topic documents by exploiting user profiles. Existing methods usually store a few prototype off-topic documents as the profile and label their top nearest neighbors in the test set as suspects. This is based on the common assumption that nearby documents are from the same class. However, due to the inherent sparseness of high-dimensional space, a document and its nearest neighbors may not belong to the same class. To this end, we develop a hyperclique pattern based off-topic detection method for selecting which ones to label. Hyperclique patterns consider joint similarity among a set of objects instead of the traditional pairwise similarity. As a result, the objects from hypercliques are more reliable as seeds for classifying their neighbors. Indeed, our experimental results on real world document data favorably demonstrate the effectiveness of our technique over the existing methods in terms of detection precision.

1 Introduction

With the rapid development of online information retrieval systems, especially search engines on the Internet, more and more information is becoming readily accessible. Meanwhile, it also poses a challenge to appropriate management and protection against misuse and intrusion. Generally speaking, intrusion is performed by unauthorized people who are outside an organization. Misuse, on the other hand, refers to the situation that an authorized user tries to misuse the authorization to retrieve documents that is considered "off-topic" to his predefined area of interest and thus should not be viewed by him. Such misuse is the second most prevalent form of computer crime after viruses, according to the recent government studies [1].

The problem of off-topic detection is often addressed by exploiting the user profile. Depending on the particular applications, the profile defines his legitimate or illegitimate scope of interest. For instance, for the children browsing the Internet at home, they should be allowed to view any webpages except those inappropriate such as violent or porn. The profile in this case consists of those off-topic webpages. By comparing the test documents with the profile, the primary goal is to detect if there is any off-topic document in the test set. Existing methods usually employ a Top k Nearest Neighbors (TKNN) approach by regarding the off-topic documents in the profile as seeds and seeking their top k

G. Dong et al. (Eds.): APWeb/WAIM 2007, LNCS 4505, pp. 374–381, 2007.

nearest neighbors in the test set as suspects. This is based on the common assumption that nearby documents are from the same class. However, due to the inherent sparseness of high-dimensional space, a document and its nearest neighbors may not belong to the same class. To that end, in this paper, we develop a HYperclique Pattern based Off-topic Detection (HYPOD) approach for selecting which ones to label. Hyperclique patterns consider joint similarity among a set of objects instead of the traditional pairwise similarity. As a result, the objects from hypercliques are more reliable as prototypes for classifying their neighbors. Indeed, our experimental results on real world document data favorably demonstrate the effectiveness of our technique over the existing methods in terms of detection precision. Note that high precision is much more important than recall here, for a false access violation accusation unfairly subjects the user to scrutiny. The detection system is expected to sort out several suspects that are highly likely to be off-topic. It is the human who is responsible to monitor the prediction and make a final decision.

Overview. Section 2 reviews the related work. Section 3 introduces the TKNN approach, which is typical among the pairwise similarity based approaches. Then Section 4 presents our HYPOD approach that employs hyperclique patterns for prediction. Comparative results are reported in Section 5. Finally, we draw conclusions in Section 6.

2 Related Work

Previous work on detection can be divided into intrusion detection and off-topic detection. Some work on intrusion detection has been in the area of pattern matching [2] and text (programs) clustering [3]. Off-topic detection techniques generally fall in system based and content based approaches. System based approaches rely on system characteristics, such as file name, size and storage location, to detect a deviation from normal behavior [4]. Predetermining the mapping of documents to allowable users, however, is highly difficult in large and dynamic document collections. In contrast, content-based approaches check if the content being accessed matches a valid scope of interest, which is usually defined by a user profile of on-topic content. Along this line, information retrieval techniques have been extensively employed, e.g., clustering query results [5], relevance feedback [6] and fusion of warnings from individual methods [7].

Off-topic detection is also related to outlier detection, which has also been explored in the data mining community [8]. Compared to the totally unsupervised techniques for outlier detection, profile based methods for off-topic detection are semi-supervised in that the class labels of documents in the profile are available for the construction of the prediction model. Both classes of methods try to assign an off-topic (outlier) degree to every test object. In contrast, here our goal is to detect the presence of off-topic documents in the test set. Therefore, the predicted set of off-topic content should possess a high precision while recall is relatively less important.

Input

D: a test document dataset $\{d_1, ..., d_n\}$.

O: a set of seed off-topic documents $\{o_1, ..., o_m\}$ in the profile.

k: the number of documents to be predicted in D.

Output

P: a result set of k documents predicted as off-topic.

Steps

1. Compute similarity matrix $S(i, j) = sim(o_i, d_j)$, which stores similarity between seed documents and test documents.
2. **For** each test document d_j **Do**
3. $sim(O, d_j) = \max_i S(i, j)$
4. **End of for**
5. $P = \{d_j : ||\{d_i : sim(O, d_i) > sim(O, d_j)\}|| < k\}$

Fig. 1. Overview of the TKNN approach

3 The TKNN Approach

In this section, we briefly review the TKNN approach, which is commonly used as a principled method in outlier and off-topic detection [9,7]. It is outlined in Fig. 1 and explained in detail below.

For a system monitoring many users, the allocated space for each user profile is small. We assume that the profile is represented by $O = \{o_1, ..., o_m\}$, a set of m seed off-topic documents, which are either assigned or learned over a period of use, e.g., by random sampling, clustering the whole document set or query results. For each test document $d_i \in D$, TKNN first assigns it to the seed $o_i \in O$ with which it shares the largest similarity (line 1-4). Regarding this similarity also as the similarity to the whole profile, TKNN finally selects the top k documents d_i with highest similarity to O, which is shown in line 5 where $|| \cdot ||$ denotes set cardinality. TKNN is simple and easy to implement. It assigns different predictive power to different seeds and can avoid prediction errors when some seeds are noise. On the other hand, TKNN has several weaknesses. The similarity to the seed set is still based on pairwise similarity. Besides, since the final selection of top k nearest neighbors is on a global scale, it is possible that a seed from a sparse cluster never gets used in prediction.

4 Off-Topic Detection Based on Hyperclique Patterns

In this paper, we employ hyperclique patterns to find nearest neighbors to the seed set. This section first briefly describe the concept of hyperclique patterns, then presents our HYPOD approach for off-topic detection.

4.1 Hyperclique Patterns

Let $I = \{i_1, i_2, ..., i_n\}$ be a set of distinct items. Each transaction T in database D is a subset of I. We call $X \subseteq I$ an itemset. The support of X, denoted by $supp(X)$, is the fraction of transactions containing X. If $supp(X)$ is no less than a user-specified threshold, X is called a frequent itemset. The confidence of association rule $X_1 \rightarrow X_2$ is defined as $conf(X_1 \rightarrow X_2) = supp(X_1 \cup X_2)/supp(X_1)$. It estimates the likelihood that the presence of an itemset X_1 implies the presence of the other itemset X_2 in the same transaction.

Because frequent itemsets only consider support, they may include items with very different support values. To measure the overall affinity among items within an itemset, the h-confidence was proposed in [10]. Formally, the h-confidence of an itemset $P = \{i_1, ..., i_m\}$ is defined as $hconf(P) = min_k\{conf(\{i_k\} \rightarrow P-\{i_k\})\}$. Given a set of items I and a minimum h-confidence threshold h_c, an itemset $P \subseteq I$ is a hyperclique pattern if and only if $hconf(P) \geq h_c$. A hyperclique pattern P can be interpreted as that the presence of any item $i \in P$ in a transaction implies the presence of all other items $P - \{i\}$ in the same transaction with probability at least h_c. A hyperclique pattern is a maximal hyperclique pattern if no superset of this pattern is a hyperclique pattern. With *cosine* as similarity measure and with documents represented by binary vectors indicating which words occur, it is easy to show that the similarity between any two documents in a hyperclique is lower bounded by the hyperclique's h-confidence.

4.2 The HYPOD Approach

In the HYPOD approach, instead of selecting the top k nearest neighbors to the seed set on a global scale, the main idea is to label only the documents in the hypercliques that contain the seed documents. As shown in line 2 of Fig. 2, we first mine the maximal hyperclique patterns DP from the document set $D \cup O$ that contain at least one seed document. This is implemented on top of an efficient algorithm [11] for mining all maximal hyperclique patterns. With similarity between O and all test documents initialized to 0, we only update the similarity for the test documents that appear in DP. There are two cases that deserve special attention. First, as shown in line 5, if several seeds appear in the same pattern dp, then $sim_{dp}(O, d)$, the similarity between O and $d \in dp$ w.r.t. dp, is the maximal similarity value with those seeds. Second, as shown in line 6, if a test document appears in several patterns, the final $sim(O, d)$ is the maximal similarity value with those patterns. Finally, we select the top k test documents with largest similarity with O.

The HYPOD approach has several advantages. First, it only predicts documents strongly connected to the seeds, which is also shared by the TKNN approach. Second, unlike TKNN, it considers the similarity among groups of documents instead of just pairs of documents. Thus it is able to label test documents in the sparse clusters where TKNN might fail. Fig. 3 illustrates such an example, where o_1 and o_2 are seed documents, d_1-d_5 are test documents with d_1-d_4 truly off-topic but d_5 on-topic. Their pairwise similarities are indicated as the corresponding edge weights.

Input

D: a test document dataset $\{d_1, ..., d_n\}$.
O: a set of seed off-topic documents $\{o_1, ..., o_m\}$ in the profile.
k: the number of documents to be predicted in D.
α: a minimum support threshold for documents.
θ: a minimum h-confidence threshold for documents.

Variable

DP: the set of maximal document hyperclique patterns.

Output

P: a result set of k documents predicted as off-topic.

Steps

1. Initialize $sim(O, d) = 0$, for all $d \in D$.
2. DP = MaximalHypercliquePattern(α, θ, D, O)
3. **For** each document pattern $dp \in DP$ **Do**
4. Partition dp into seed set $O(dp)$ and test set $D(dp)$.
5. $sim_{dp}(O, d) = \max_{o \in O(dp)} sim(o, d)$, for all $d \in D(dp)$.
6. $sim(O, d) = \max\{sim(O, d), sim_{dp}(O, d)\}$, for all $d \in D(dp)$.
7. **End of for**
8. $P = \{d_j : \|\{d_i : sim(O, d_i) > sim(O, d_j)\}\| < k\}$

Fig. 2. Overview of the HYPOD approach

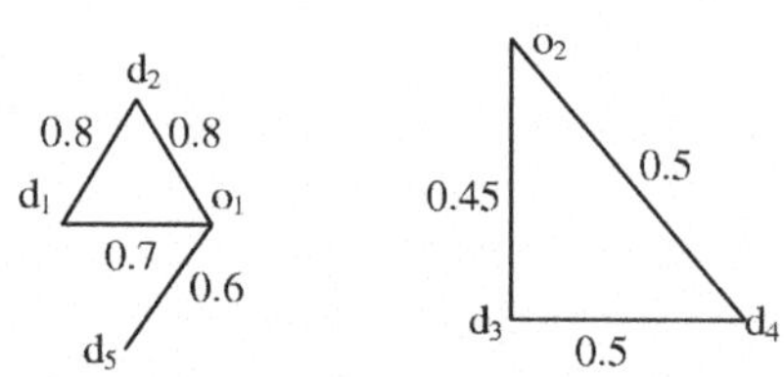

Fig. 3. An illustrative example

If we set $k = 3$, TKNN would only label documents from the dense cluster. To label d_3 and d_4 from the sparse cluster, we have to increase k. However, in terms of pairwise similarity alone, they can appear in the final prediction set only after d_5 appears in it. In contrast, by setting the proper threshold, HYPOD can find two patterns, $\{o_1, d_1, d_2\}$ and $\{o_2, d_3, d_4\}$ simultaneously and thus label all off-topic documents with o_1 and o_2 respectively.

5 Experimental Evaluation

In this section, we first introduce the real world document datasets used in our experiments and then illustrate the purity of hyperclique patterns. Finally we report comparative results on these datasets.

Table 1. Characteristics of datasets

data	RE0	RE1	K1	WAP	TR31	TR41
source	Reuters-21578		WebACE		TREC-6,7	
#doc	1504	1657	2340	1560	927	878
#word	2886	3758	4592	8460	4703	7454
#class	13	25	6	20	7	10
#OffClass	8	14	7	8	3	5
OffClass%	14	18	8	11	9	14
all(%)	80	76	76	69	92	90
hypercliques(%)	83	84	92	81	97	95

5.1 Hyperclique Patterns in the Document Datasets

In our experiments, we used six datasets them from different sources to ensure diversity. Similar to outlier detection [12] that treats rare class objects as outliers, we divide each dataset into on-topic and off-topic, where off-topic includes those classes whose size is less than half of average class size. Some characteristics of these datasets are shown in Table 1, where the 6th and 7th rows give the number of classes used as off-topic and their total fraction, respectively. For all datasets, we used a stoplist to remove common words, stemmed the remaining words using Porter's suffix-stripping algorithm and removed those words with extreme low document frequencies. The datasets are given in the transactional form and *cosine* is used as the pairwise similarity measure.

By considering group similarity instead of pairwise similarity, the discovered documents in the same hyperclique tend to belong to the same class. Now we demonstrate such purity. By regarding words as transactions and documents as items, Fig. 4(a) illustrates the average entropy of the document hypercliques from K1. We can see that as the minimum h-confidence threshold increases, the entropy of hyperclique patterns decreases dramatically, especially at low support values. This indicates that hyperclique patterns include objects from the same class above certain h-confidence levels. Besides, in high dimensional datasets like documents, two objects can often be nearest neighbors without belonging to the same class. The last 2nd row in Table 1 shows the percent of documents whose nearest neighbor is from the same class, which is clearly below 1 for all datasets. In contrast, taking advantage of the hyperclique purity, we can use only documents in the hypercliques to label their nearest neighbors. As indicated in the last row of Table 1, the corresponding fractions of correct prediction increase for all datasets, though to a varying degree. Of course this fraction usually depends on the specific thresholds used and Fig. 4(b) shows details for K1.

5.2 Comparative Results

Given the condition that the profile for each user is quite limited, it can only contain a very small amount of off-topic documents. To simulate this condition,

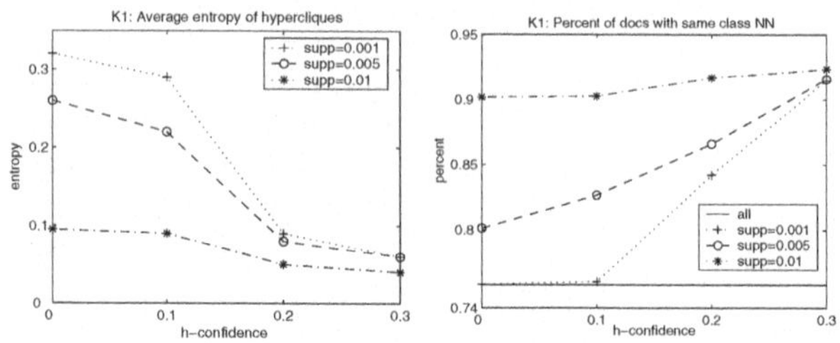

(a) (b)

Fig. 4. Analysis of dataset K1

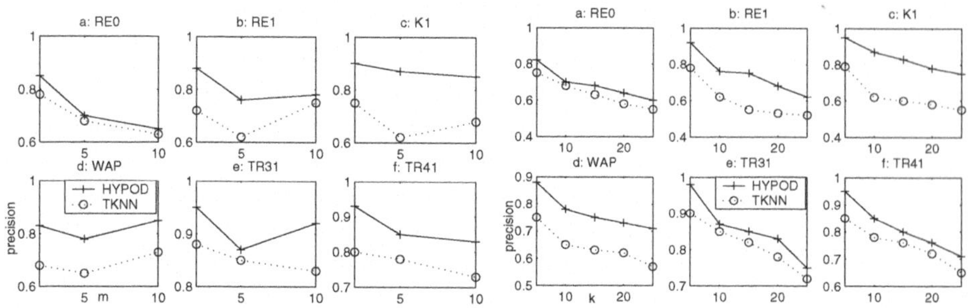

(a) (b)

Fig. 5. Comparison of prediction precision by varying the profile size m (a) or the prediction size k (b)

we randomly choose m off-topic documents as the profile, with $m = 2, 5, 10$. Treating the rest of documents as the test set, we apply both TPKNN and HYPOD to predict $k = 2m$ test documents. The average results of 10 runs are illustrated in Fig. 5(a). One can see that HYPOD yields better precision than TKNN. Fixing $m = 5$, we also evaluate them by varying $k = 5, 10, 15, 20, 25$. As shown in Fig. 5(b), the average precision of 10 runs generally decreases as k increases, since we need to use seeds to predict more neighbors farther away. Nevertheless, HYPOD still gives better results than TPKNN.

6 Conclusions

In this paper, we proposed a HYperclique Pattern based Off-topic Detection (HYPOD) approach to detecting access to off-topic documents by exploiting user profiles. Conventional methods like TKNN are usually based on pairwise similarity alone. However, in the high dimensional space like documents, two objects can often be nearest neighbors without belonging to the same class. On the other hand, hyperclique patterns consider group similarity and thus items in them are more reliable as seeds for predicting other items in the same pattern. Our experimental results on real world datasets favorably confirmed the advantage of HYPOD in terms of detection precision.

References

1. Dept. of Justice of United States: Press releases. http://www.usdoj.gov/criminal/cybercrime/ (2006)
2. Kumar, S., Spafford, E.H.: A pattern matching model for misuse intrusion detection. In: Proc. the 17th National Computer Security Conf. (1994) 11–21
3. Liao, Y., Vemuri, V.R.: Using text categorization techniques for intrusion detection. In: Proc. of the 11th USENIX Security Symp. (2002) 51–59
4. Chung, C.Y., Gertz, M., Levitt, K.: DEMIDS: A misuse detection system for database systems. In: Proc. the 3rd Int. IFIP TC-11 WG11.5 Working Conf. Integrity and Internal Control in Information Systems. (1999) 159–178
5. Goharian, N., Platt, A.: Detection using clustering query results. In: Proc. IEEE Int. Conf. Intelligence and Security Informatics. (2006) 671–673
6. Goharian, N., Ma, L.: On off-topic access detection in information systems. In: Proc. ACM CIKM. (2005) 353–354
7. Cathey, R., Ma, L., Goharian, N., Grossman, D.A.: Misuse detection for information retrieval systems. In: Proc. ACM CIKM. (2003) 183–190
8. Breunig, M.M., Kriegel, H.P., Ng, R.T., Sander, J.: LOF: Identifying density-based local outliers. In: Proc. ACM SIGMOD. (2000) 93–104
9. Ramaswamy, S., Rastogi, R., Shim, K.: Efficient algorithms for mining outliers from large data sets. In: Proc. ACM SIGMOD. (2000) 427–438
10. Xiong, H., Tan, P.N., Kumar, V.: Mining strong affinity association patterns in data sets with skewed support distribution. In: Proc. IEEE ICDM. (2003) 387–394
11. Huang, Y., Xiong, H., Wu, W., Zhang, Z.: A hybrid approach for mining maximal hyperclique patterns. In: Proc. IEEE ICTAI. (2004) 354–361
12. Aggarwal, C.C., Yu, P.S.: Outlier detection for high dimensional data. In: Proc. ACM SIGMOD. (2001) 37–46

An Energy Efficient Connected Coverage Protocol in Wireless Sensor Networks

Yingchi Mao[1], Zhuoming Xu[1], and Yi Liang[2]

[1] College of Computer and Information Engineering, Hohai University, 210098 Nanjing, China
{yingchimao,zmxu}@hhu.edu.cn
[2] Automation Research Institute, State Power Corporation of China, 210009 Nanjing, China
liangy@naritech.cn

Abstract. The connected coverage is one of the most important problems in Wireless Sensor Networks. However, most existing approaches to connected coverage require knowledge of accurate location information. This paper solves a challenging problem: without accurate location information, how to schedule sensor nodes to save energy and meet both constraints of sensing area coverage and network connectivity. Our solution is based on the theoretical analysis of the sensing area coverage property of minimal dominating set. We establish the relationship between point coverage and area coverage, and derive the upper and lower bound that point coverage is equivalent to area coverage in random geometric graphs. Based on the analytical results and the existing algorithms which construct the connected dominating set, an Energy Efficient Connected Coverage Protocol (EECCP) is proposed. Extensive simulation studies show that the proposed connected coverage protocol can effectively maintain both high quality sensing coverage and connectivity for a long time.

Keywords: Wireless Sensor Networks, Connected Coverage, Dominating Set.

1 Introduction

Wireless sensor networks (WSNs) consist of a large number of sensors. They can sense and collect information from all kinds of objects in the monitored area. Furthermore, they can process the gathered information and send it back to users. Therefore, they are being widely employed for military fields, national security, environmental monitoring, and disaster prevention and recovery [1]. However, due to their extremely small dimension, sensors have very limited energy supply. In addition, it is usually hard to recharge the battery after deployment, either because the deployment area is hostile, or because the number of sensor nodes is too large. Once deployed, a WSN is expected to keep working for several weeks or months. Therefore, energy efficiency becomes the essential requirement in WSNs. Sensor scheduling plays a critical role for energy efficiency in WSNS.

On the other hand, it is required that WSNs can provide the high quality of sensing area coverage and ensure the network connectivity. The connected coverage is one of the most important problems in WSNs. The existing algorithms in maintaining

G. Dong et al. (Eds.): APWeb/WAIM 2007, LNCS 4505, pp. 382–394, 2007.

connected coverage usually rely on the availability of accurate location information such as that obtained with the GPS systems and the directional antenna technology. However, the energy cost and system complexity involved in obtaining the accurate location information may compromise the effectiveness of proposed solution as a whole. Furthermore, it is still a very difficult issue to estimate sensors' locations, since GPS and other complicated hardware devices consume too much energy and the costs are too high for tiny sensors [2]. Therefore, we solve a challenging problem: without accurate location information, how to energy-efficiently schedule sensor nodes to meet both constraints of sensing area coverage and network connectivity.

Our work is based on the (connected) dominating set (CDS), which has extensively been investigated sparse structure in ad hoc wireless networks. Much work has been done to derive distributed algorithms [3],[4],[5],[6],[7] to construct minimal (connected) dominating set (MDS) with different objective functions. It is obvious that MDS provide coverage of points in an area (termed point coverage) when the sensing range (*Rs*) equals the transmission range (*Rt*). Although it is considered that point coverage is equivalent to area sensing area coverage in the high density of deployed sensors [8]. However, point coverage is in general not equivalent to area coverage. In this paper, we analyze the relationship between them, and derive the upper and lower bound that the point coverage of MDS is equivalent to area coverage in random geometric graphs. Based on the theoretical results, we propose an energy efficient connected coverage protocol (EECCP) that incorporates existing connected dominating set (CDS) selection algorithms to determine the set of active sensor nodes. In EECCP, every sensor nodes can adjust its own *Rt* based on local nodes density, and schedule its own status to construct CDS without accurate location information. The active nodes can maintain both high quality of *Rs* and network connectivity.

The remainder of the paper is organized as follows. Section 2 presents a review of related work in connected coverage. The next section introduces some necessary notations and preliminaries. Section 4 establishes the analytical results on relationship between the point coverage and sensing are coverage. A connected coverage protocol, EECCP is proposed and evaluated in Section 5 and Section 6, respectively. Finally, we conclude the paper.

2 Related Work

Recently, researches have considered connectivity and coverage in an integrated platform. In [9], the authors consider an unreliable sensor network and sensor nodes are deployed strictly in grids. The necessary and sufficient conditions for the area coverage and network connectivity with high probability are provided. In [10], it is proved that to ensure that the full sensing area coverage of a convex area also guarantee the connectivity of the active nodes, the communication range should be at least twice of the sensing range. Therefore, the connected coverage problem is simplified to maintain a complete coverage. Based on the analytical results, Zhang et al. proposed a distributed, localized density control algorithm named OGDC [10]. Wang et al. draw the same conclusion in [11], and presented a unified framework to study both coverage and connectivity problems. They used the SPAN [12] protocol to maintain coverage, and a separate CCP protocol to maintain coverage. Carle et al.

proposed SCR-CADS algorithm to maintain both sensing coverage and network connectivity when Rs equals to Rt [13]. Gupta et al defined the connected sensor cover, and demonstrated that the calculation of the smallest connected sensor cover is NP-hard and they proposed both centralized and distributed approximate algorithms to solve it and provide the performance bounds as well [14]. Zou et al considered the variable radii k1-connected k2-cover problem by adjusting the sensing range and transmission range [15]. The authors proposed a distributed and local Voronoi–based and relative neighborhood graph (RNG)-based algorithm to preserve k1 connectivity and k2 coverage. However, unlike our approach, the above algorithms all require each sensor node to be aware of its precise location in order to check its local coverage redundancy. In this paper, we address how to ensure both sensing area coverage and network connectivity without accurate location information.

3 Preliminary

A dominating set (DS) of an undirected graph G(V, E) is a subset V' of the vertex set V such that every vertex in V-V' is adjacent to a vertex in V'. A minimal dominating set (MDS) is a dominating set which ceases to be a dominating set if any vertex is removed from it. A dominating set V' is connected if for any two vertices u and v $\in$V', a path (u, v_1), (v_1, v_2), ..., (v_n, v) ($v_i \in$ V', $1 \le i \le n$) in E exists. Some notations are as follows.

(1) $B_i(r_i)$: let $B_i(r_i)$ be a disk centered at a point $z_i \in \mathbf{R}^2$ with radius r_i, B_i for short.
(2) $\pi_i(x)$: the power distance of a point $x \in \mathbf{R}^2$ from B_i, defined as $\pi_i(x)=\|x-z_i\|^2-r_i^2$.
(3) B_i^j: $B_i^j=\{x\in\mathbf{R}^2 | \pi_j(x)\le\pi_i(x)\le 0\}$, B_i^j is the portion of B_i on B_j's side of the bisector.
(4) P_i, P_i^j, P_j^{ik}: let P_i, P_i^j, P_j^{ik} be length of the circle arcs in the boundaries of B_i, B_i^j and $B_j^i \cap B_j^k$.
(5) A_i^j, A_j^{ik}: let A_i^j, A_j^{ik} be area of B_i^j and $B_j^i \cap B_j^k$, that is $A_i^j = Area(B_i^j)$ and A_j^{ik}=Area($B_j^i \cap B_j^k$).
(6) $\mathbf{r}=[r_1,r_2,\ldots,r_n]$, $\mathbf{z}=[z_1,z_2,\ldots,z_n]$: let $\mathbf{r}$ and $\mathbf{z}$ denote vectors of disk radii and positions of centers, respectively.
(7) $A\mathbf{r}$: $A\mathbf{r}$ is a function of the position of the disks as well as their radii, that is $A\mathbf{r}=f(\mathbf{r},\mathbf{z})$. $A\mathbf{r}$ denotes the numeric area of the union of n disks.

4 Problem Analysis

According to the definition of MDS, the nodes in the MDS set can control the nodes in the non-MDS set. When *Rt=Rs*, the MDS can provide coverage of points in the region M (termed point coverage). In addition, MDS$\subseteq$CDS, CDS not only meet the requirement of point coverage, but can ensure the network connectivity. However, point coverage is in general not equivalent to area coverage. In this section, we analyze the relationship between the point coverage and sensing area coverage, and derive the upper and lower bound that point coverage is equivalent to area coverage in random geometric graphs.

Theorem 1. *Given sensing range Rs and transmission range Rt, let Area(V) be the geometric area covered by all nodes in the network and Area(MDS) be the geometric area covered by the MDS. We have* $\frac{Rs^2}{(Rs+Rt)^2} \le \frac{Area(MDS)}{Area(V)} \le 1$.

The main idea is from the work on the area and perimeter derivatives of a union of disks [16]. Before delving to the proof, we first introduce some lemmas.

Lemma 1. Assume *A* be the geometric area covered by n disks deployed in a plane $\mathbf{R}^2$. The derivative of the numeric area *A***r** of a union of n disks with radius vector **r** is $DA\mathbf{r} = \frac{\partial f(\mathbf{r},\mathbf{z})}{\partial \mathbf{r}} = 2\pi\sum_{i=1}^{n} r_i\sigma_i$, where $\sigma_i = 1-(\sum_j P_i^j - \sum_{j,k} P_i^{jk})/2\pi r_i, i=1,2,...,n$.

In lemma 1, σ_i denotes the contribution of disk B_i's boundary to the perimeter of *A*. In particular, if a disk is fully contained inside *A*, the disk does not share boundary with *A*, and σ_i=0.The proof of lemma 1 refers to [16].

Lemma 2. *Let A(r) be the numeric area of the union of n disks with the same radius r on a plane. Its derivative with respect to r is given by* $\frac{dA(r)}{dr} = \sum_{i=1}^{n} 2\pi r\sigma_i$, *where* $\sigma_i = 1-(\sum_j P_i^j - \sum_{j,k} P_i^{jk})/2\pi r, (i=1,2,...,n)$.

Lemma 3. *For a differentiable function* $g: R^+ \to R^+$, *if* $\forall x \in R^+$, $g'(x) \le \frac{2g(x)}{x}$, *then* $\frac{g(x_1)}{g(x_2)} \le \frac{x_1^2}{x_2^2}$, $0 < x_2 \le x_1 \in R^+$.

Lemma 4. *Assume two disks B_i and B_j of radius r_i and r_j centered as z_i and z_j respectively. Let S_i (S_j) be the sector defined by the arc P_i-P_i^j (P_j-P_j^i) and the line segments that join the two end points of P_i-P_i^j (P_j-P_j^i) and its center z_i (z_j).Then Area($S_i \cap S_j$)=0.*

Due to the limit of the length of paper, we omit the proof of Lemma 2, 3 and 4.

Now we are in the position to prove Theorem 1.

Proof. (Theorem 1)

A) $\frac{Area(MDS)}{Area(V)} \le 1$: Based on the definition of MDS, MDS can control all of non-MDS. Thus, $|MDS| \le |V|$, $\bigcup_{i=1}^{i=m} B_i(Rs) \subseteq \bigcup_{i=1}^{i=N} B_i(Rs)$, where *m*=|*MDS*|, *N*=|*V*|. Therefore, $\frac{Area(MDS)}{Area(V)} \le 1$.

B) $\frac{Rs^2}{(Rs+Rt)^2} \le \frac{Area(MDS)}{Area(V)}$: Consider a dominator node *u*. Define D_u be the set of nodes that are dominated by *u* including *u* itself, $D_u=\{v\in V | d(u,v)\le Rt\}$. A point *p* is covered by nodes in D_u only if *p* is within (*Rs*+*Rt*) from node *u*. Therefore, the maximum possible geometric area covered by D_u is Area(D_u)=$\pi(Rs+Rt)^2$. The numeric area covered by node *u* is $Area(u)=\pi Rs^2$. Therefore, the ratio between the numeric areas covered by u and D_u is $\frac{Area(u)}{Area(D_u)} \geq \frac{Rs^2}{(Rs+Rt)^2}$. By definition of the DS, we have $\bigcup_{i=1}^{i=m} D_i = V$, where m=|MDS|. There are two cases.

(1) None of the coverage area of the nodes in MDS overlap. Summing up coverage area of all the nodes in MDS, we have $Area(MDS) = Area(\bigcup_{i=1}^{i=m} B_i(Rs)) = m \cdot \pi Rs^2$. On the other hand, the coverage area of some nodes in $D_i (i = 1,2,...,m)$ may overlap. We get $Area(V) \leq Area(\bigcup_{i=1}^{i=m} B_i(Rs+Rt)) = m\pi(Rs+Rt)^2$. Therefore $\frac{Area(MDS)}{Area(V)} \geq \frac{Rs^2}{(Rs+Rt)^2}$.

(2) The coverage area of some nodes in MDS overlap. The area function of $A(r)$ is $A(r) = \left|\bigcup_{i=1}^{i=n} B_i(r)\right|$. Let $A(Rs) = Area(MDS(Rs))$ and $A(Rs+Rt) = Area(MDS(Rs+Rt))$. While $Area(V) \leq Area(\bigcup_{i=1}^{i=m} B_i(Rs+Rt))$, we have $\frac{Area(MDS)}{Area(V)} \geq \frac{A(Rs)}{A(Rs+Rt)}$. Therefore, to prove $\frac{Area(MDS)}{Area(V)} \geq \frac{Rs^2}{(Rs+Rt)^2}$, it is sufficient to prove $\frac{A(Rs)}{A(Rs+Rt)} \geq \frac{Rs^2}{(Rs+Rt)^2}$, that is to say, we need to prove that $\forall 0 < r_2 \leq r_1 \in R^+, \frac{A(r_1)}{A(r_2)} \geq \frac{r_1^2}{r_2^2}$. By Lemma 3, it is sufficient to show that the derivative $A'(r) \leq \frac{2A(r)}{r}, \forall r > 0$.

From Lemma 2, we have

$$\frac{dA(r)}{dr} = \sum_{i=1}^{n} 2\pi r \sigma_i = 2\pi r \sum_{i=1}^{n} (1 - (\sum_j P_i^j - \sum_{j,k} P_i^{jk}) / 2\pi r) \tag{1}$$

The term $\sigma_i = 1 - (\sum_j P_i^j - \sum_{j,k} P_i^{jk}) / 2\pi r, (i = 1,2,...,n)$ gives the portion of the perimeter of B_i that is at the boundary of $A(r)$.

Let $S_{i,mi}$ be the set of sectors defined by the arc segment of P_i at the boundary of $A(r)$ and the line segments that join the end points of the arc segments and its center z_i, where m_i is the number of disks which intersect with disk B_i. Therefore,

$$\pi r^2 \sigma_i = Area(\bigcap_{l=1}^{m_i} S_{i,l}) = Area(S_{i,m_i}) \tag{2}$$

As the same way, $\pi r^2 \sigma_j$ =Area($S_{j,mj}$), where m_j is the number of disks which intersect with disk B_j. Since $S_{i,mi} \subseteq S_i$ and $S_{j,mj} \subseteq S_j$, From Lemma 4,$\forall$i,j, $Area(S_i \cap S_j)$=0. Therefore, $Area(S_{i,mi})+Area(S_{j,mj})=Area(S_{i,mi} \cup S_{j,mj})$.

Since $S_{i,mi} \subseteq S_i \subseteq B_i$, combining Equation (1) and (2), we get

$$A'(r) \times r/2 = (2\pi r \sum_{i=1}^{n} \sigma_i) \times r/2 = \sum_{i=1}^{n} \pi r^2 \sigma_i = \sum_{i=1}^{n} Area(S_{i,m_i}) = Area(\bigcup_{i=1}^{n} S_{i,m_i}) \leq Area(\bigcup_{i=1}^{n} S_i) \leq Area(\bigcup_{i=1}^{n} B_i) = A(r)$$

Thus, $\forall r > 0, A'(r) \leq \frac{2A(r)}{r}$. Therefore, $\frac{Area(MDS)}{Area(V)} \geq \frac{A(Rs)}{A(Rs+Rt)} \geq \frac{Rs^2}{(Rs+Rt)^2}$. □

By definition of MDS, nodes in MDS can dominate all of the nodes in the network. Obviously, if the nodes are densely deployed, MDS can coverage almost of the monitored region when *Rt*=*Rs*. That is to say, the denser the deployment of sensors is, the better sensing coverage the MDS can provide. On the other hand, from Theorem 1, we can see as long as *Rt* is significantly small, MDS can provide comparable sensing coverage as that of all sensor nodes. However, as *Rt* goes smaller, the number of nodes in MDS becomes larger, which results in the more energy consumption. Therefore, it is necessary to adjust the appropriate *Rt* for each sensor node to ensure MDS can provide the high quality of sensing coverage.

The upper and lower bound that point coverage is equivalent to sensing coverage in random geometric graphs will be discussed as follows.

Theorem 2. *Suppose sensor nodes follow Poisson point process of density* λ *in the plane* $\mathbf{R}^2$. *For* $\forall\varepsilon>0$ *and* $\forall p\in\mathbf{R}^2$, *then there* $\exists r$ *s.t.* $P(\exists u\in MDS \wedge d(u,p)\le r+Rt)\ge 1-\varepsilon$.

Proof. Consider an arbitrary point $p\in\mathbf{R}^2$. A sufficient condition that point *p* is within (*r*+*Rt*) distance from any one dominator node *v* is that there exists a node *u* within *r* distance from point *p*.

Since there is a node *u* within *r* distance from point *p*, i.e., $d(u,p)\le r$, there are two cases. Case one: If $u\in$ MDS, point *p* must lie within (*r*+*Rt*) distance from the node *u*, that is $d(u,p)\le r+Rt$. Case two: If $u\notin$ MDS, there must exist one dominator node *v* which can dominate node *u* by the definition of MDS. That is to say, node *u* lies within *Rt* distance from dominator *v*, $d(u,v)\le Rt$. Combining two cases, we get $d(v,p)\le d(u,v)+d(v,p)\le Rt+r$, namely, point *p* is within (*r*+*Rt*) distance from any one dominator node *v* (as illustrated in Fig 1). The sufficient condition holds.

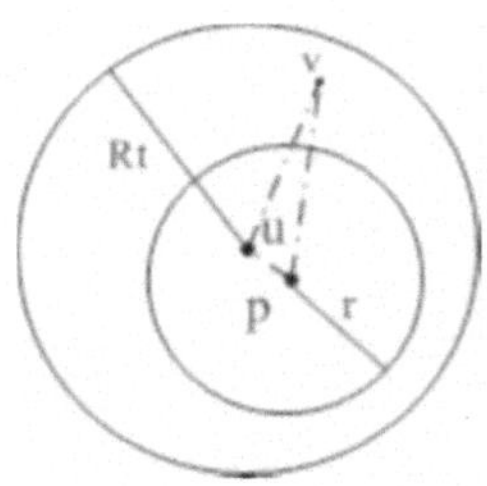

Fig. 1. Illustration of the sufficient condition

From the sufficient condition, we have $P(\exists v\in MDS$, s.t.$d(v,p)\le r+Rt)\ge P(\exists u\in V$, s.t. $d(u,p)<r)$. Since $\forall p\in\mathbf{R}^2$, there $\exists u\in V$, s.t. $d(u,p)\le Rs$, where *V* is the set of all sensor nodes. Thus, we have $P(\exists u\in V$, s.t. $d(u,p)<r)=1$ when $r\ge Rs$. On the other hand, since nodes follow a Poisson point process at density λ, the sensing coverage of such a sensor network is $f_a = 1-e^{-\pi\lambda Rs^2}$ [17], we have $P(\exists u\in V$, s.t. $d(u,p)<r)=\dfrac{1-e^{-\pi\lambda r^2}}{1-e^{-\pi\lambda Rs^2}}$ when $d(u,p)<r<Rs$. Therefore, the following equality holds,

$$P(\exists v\in MDS, \text{s.t.} d(v,p)\leq r+Rt)\geq P(\exists u\in V, \text{s.t. } d(u,p)<r)=\begin{cases}1, r\geq Rs\\ \dfrac{1-e^{-\pi\lambda r^2}}{1-e^{-\pi\lambda Rs^2}}, r<Rs\end{cases}.$$

Let $\dfrac{1-e^{-\pi\lambda r^2}}{1-e^{-\pi\lambda Rs^2}}\geq 1-\varepsilon$, we get $r\leq\sqrt{\dfrac{-\ln(\varepsilon+(1-\varepsilon)e^{-\pi\lambda Rs^2})}{\pi\lambda}}$. Therefore, $\forall\varepsilon>0$ *and* $\forall p\in \mathbf{R}^2$,

$\exists r=\min(Rs, \sqrt{\dfrac{-ln(\varepsilon+(1-\varepsilon)e^{-\pi\lambda Rs^2})}{\pi\lambda}})$, s.t. $P(\exists u\in MDS\wedge d(u,p)\leq r+Rt)\geq 1-\varepsilon$. □

From Theorem 2, we have the following result:

$$Rt = Rs - r = Rs - \min(Rs, \sqrt{\frac{-ln(\varepsilon+(1-\varepsilon)e^{-\pi\lambda Rs^2})}{\pi\lambda}}) \quad (3)$$

Theorem 3. *Given sensing range Rs, if all points in the monitored region are covered by sensor nodes, then there exists a point p not covered by MDS for Rt>2Rs.*

Proof. We use a proof by contradiction. Suppose every point in the region is covered by MDS for $Rt>2Rs$. Without loss of generality, consider an arbitrary dominator $v\in$ MDS. Let S_v be the set of nodes in MDS that are on the boundary or inside of disk B_v excluding node v, that is $S_v=\{u|d(u,v)\leq Rt\wedge u\in \text{MDS}\}-\{v\}$. Thus, $S_v\neq\emptyset$ for $Rt>2Rs$. This is because if $S_v=\emptyset$, the points with distance between Rs and $Rt/2$ from node v (the region is $B_v(Rt/2)$-$B_v(Rs)$, as illustrated in shadow part of Fig 2) are not covered by either node v (since $Rt/2>Rs$) or other nodes excluding node v in MDS (since these points are at least $Rt/2$ away from nodes in MDS-$\{v\}$). Hence these points are not covered by the MDS, which contradicts every point is covered by the MDS.

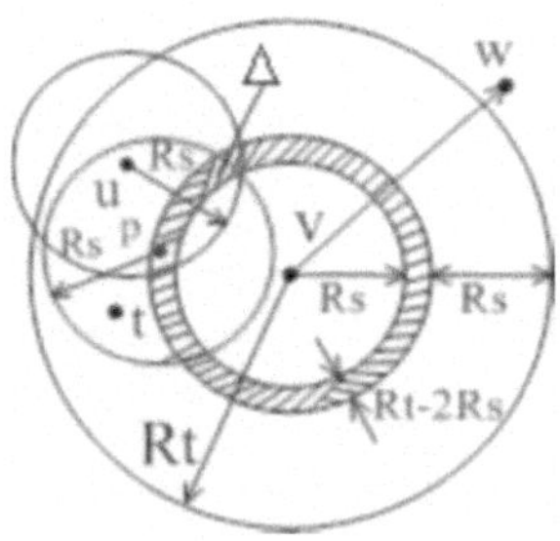

Fig. 2. Illustration of Theorem 3

Since $S_v\neq\emptyset$, S_v can dominate the node v. These are two cases about the relationship between S_v and D_v, Define $D_v=\{w|d(w,v)\leq Rt, w\in V\}$.

(1) *Case 1*: D_v-$\{v\}\subseteq S_v$, i.e., $(V\text{-}S_v\cup\{v\})\cap D_v=\emptyset$. If node v is removed from MDS, the nodes in S_v still dominate the nodes in D_v. In another word, the resulting set MDS-$\{v\}$ is still a dominating set. This contradicts with the definition of MDS.

(2) *Case 2*: D_v-$\{v\}\not\subset S_v$, i.e., there exists a node $u\in(V\text{-}S_v\cup\{v\})\cap D_v$. Define $\delta=Rt-2Rs$. Let △ be the area, where the sensing area of node u intersects with the two

concentric circles centered at node v with radius Rs and $Rs+\delta$, equivalently, $\triangle=B_u(Rs)\cap(B_v(Rs+\delta)-B_v(Rs))$ (shown in the red-marked part of Fig 2). Clearly, $\triangle\neq\emptyset$.

Since $d(v,p)>Rs$, for $\forall p\in\triangle$, point p is not covered by the dominator v. Second, for $\forall w\in$ MDS-S_v, since $d(w,v)>Rt$ and $Rs<d(v,p)<Rs+(Rt-2Rs)=Rt-Rs$, we have $d(w,p)>d(w,v)-d(v,p)>Rt-(Rt-Rs)=Rs$. Thus, the point p is not covered by node w. Based on the assumption that every point is covered by MDS, the point p must be covered by at least one node $\forall w\in S_v$, or equivalently, $\exists t\in S_v$ s.t. $d(t,p)\leq Rs$. Otherwise, point p is not covered by MDS.

Now, we have $d(u,t)\leq d(u,p)+d(p,t)\leq Rs+Rs<Rt$, in another word, dominator t can dominate node u. This implies $\forall u\in D_v$, $\exists t\in S_v$ s.t. $d(u,t)\leq Rt$. Therefore, if dominator v can be removed from MDS, the nodes in S_v still dominate the nodes in D_v. Equivalently, the resulting set MDS-$\{v\}$ is still a dominating set. This contradicts with the definition of MDS. By case 1 and case 2, Theorem 3 holds. □

5 EECCP Protocol

Theorem 2 and 3 give the up the upper and lower bound that transmission range of sensors respectively. By Theorem 2, adjusting the appropriate Rt, CDS can provide the same sensing coverage as that of all sensor nodes with probability at least 1-ε and ensure the connectivity. We propose an Energy Efficient Connected Coverage Protocol (EECCP) that provides good sensing coverage and ensures the network connectivity. EECCP protocol consists of two parts. First, nodes adjust the appropriate transmission range. Second, construct CDS using Rule K algorithm [7] to meet the requirement of connected coverage. It is proved that the resulting dominating set from the Rule K algorithm must be connected [7]. Therefore, EECCP protocol can provide the good sensing coverage and ensure the network connectivity. EECCP like other coverage algorithm also involves round-operation for energy-consumption balance. In every round, each sensor node adjusts the appropriate Rt and executes the Rule K algorithm to construct the CDS. The nodes in CDS become the working nodes to execute the monitoring task and transfer the sensory data.

5.1 Network Model

Assume N sensor nodes are deployed randomly with a uniform distribution on a square field M. The sensor density is high enough that any point can be monitored by at least one sensor, and the network is connected if all sensor nodes are in active mode. The sensing range of a sensor is modeled by a disk of radius Rs, and the radio transceiver can adjust its transmission range by changing the transmission power. All sensors have the same sensing range. Sensors are location-unaware.

5.2 Adjust the Transmission Range

By the Equation (3), we have $Rt=Rs-\min\left(Rs, \sqrt{\frac{-\ln(\varepsilon+(1-\varepsilon)e^{-\pi\lambda Rs^2})}{\pi\lambda}}\right)$, where λ is the density of sensors. With the execution of the WSN, sensors are prone to failure, the value of λ is changing. For a single node, it is very difficult to know the change of λ. Therefore, each sensor determines a local node density λ_{local} by counting the number of nodes in its k-hop neighborhood using the maximum transmission range $Rt_{\max}$. The local node density λ_{local} can be approximated as the network density λ and can reflect the variations of node density. λ_{local} can be computed as follows:

$$\lambda_{local} = \frac{|c|+1}{\pi (Rt_{\max})^2 k^2} \tag{4}$$

where $|c|$ is the number of k-hop neighbors.

The larger the value of k is, the more accurate the approximation of λ_{local} is. However, the increase of the number of messages will results in the more energy consumption. EECCP estimates the local density λ_{local} by counting the number of nodes in its 1-hop neighborhood. Let $\varepsilon_0 = \varepsilon + (1-\varepsilon)e^{-\pi\lambda Rs^2}$. Clearly, $\varepsilon_0 \in (0,1)$. Combing the Equation (3) and (4), we have

$$Rt = Rs - \min(Rs, \sqrt{\frac{-\ln \varepsilon_0}{\pi\lambda}}) = Rs - \min(Rs, Rt_{\max}\sqrt{\frac{-\ln \varepsilon_0}{|c|+1}}) \tag{5}$$

Therefore, each sensor can adjust the appropriate the transmission range by the number of nodes in its 1-hop neighbors and Equation (5).

6 Simulation Results

In this section, we evaluate EECCP's performance by conducting simulations in terms of evaluating the number of working nodes, coverage ratio, the network lifetime and control overhead. In the simulation, 400-800 sensor nodes are randomly deployed in a 100×100 (m^2) region, and each sensor has a sensing range of 10 meter and the maximum transmission range of 20 meter.

6.1 Effectiveness of EECCP

In the experiment, we vary ε_0 in the EECCP from 0.1 to 0.9 and investigate its impact on the sensing area coverage of the resulting working nodes (Fig. 3) and the number of working nodes (Fig. 4). Two scenarios are considered with sensor nodes 400 and 800 randomly distributed in the monitored area respectively.

From Fig. 3 and 4, we observe that the value of ε_0 has no much impact on the sensing coverage and the number working nodes for EECCP. In general, when ε_0=0.5, EECCP can provide the best sensing coverage and the least number of working nodes in the different node density. Therefore, we choose ε_0=0.5 when investigating the network lifetime and control overhead for EECCP. Compared to EECCP, OGDC and GAF is not relevant to the parameter ε_0=0.5. Thus, for OGDC and GAF, their curves of the number of working nodes and sensing coverage are parallel to X axis.

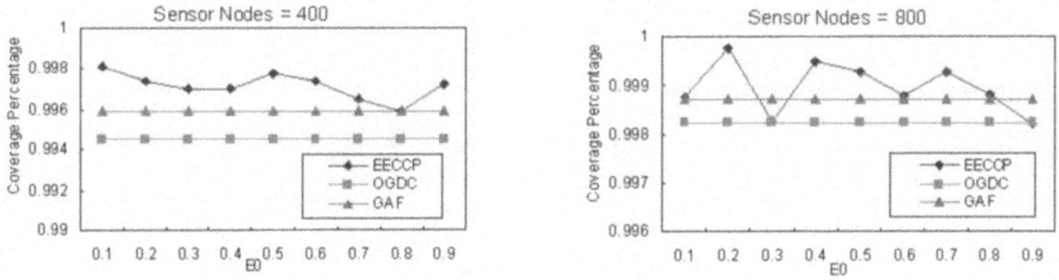

(a) sensor nodes = 400 (b) sensor nodes = 800

Fig. 3. coverage percentage vs.ε0 in the different density

On the other hand, in the most situations, EECCP can provide above 99.5% sensing coverage for the different node density. EECCP requires 2-hop neighboring information to provide the connected coverage without the location information of sensors; therefore the number of working nodes by EECCP is greater 13%~30% and 7%~26% than that of OGDC and GAF respectively. However, compared with the number of all deployed sensor nodes, the number of working nodes by EECCP is reasonable. In addition, the number of working nodes under EECCP does not greatly increases with the number of deployed nodes.

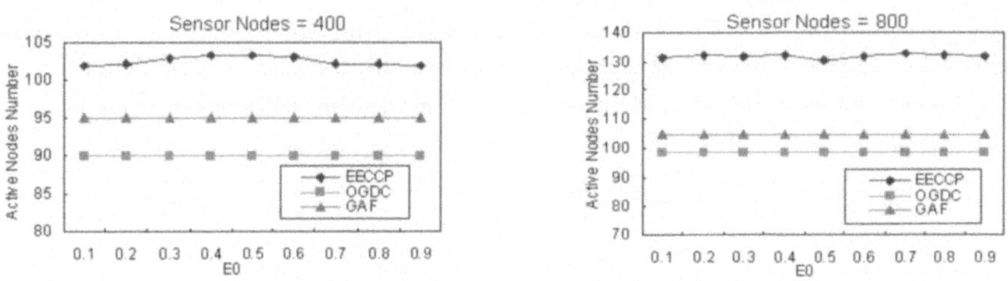

(a) sensor nodes = 400 (b) sensor nodes = 800

Fig. 4. the number of active nodes vs.ε0 in the different density

6.2 Network Lifetime

If the current sensing coverage is lower than the threshold or the network is disconnected, WSNs could not perform the monitoring task. Thus, we consider the lifetime of sensor network is over in such a situation. In our simulation, we choose the threshold of sensing coverage is 70%. In another word, if the sensing coverage is lower than 70 % or the network is disconnected, we think the network lifetime is over.

Fig. 5 shows the curves of lifetime for EECCP and OGDC with the different density of deployed sensors. From Fig. 5, for EECCP and OGDC, in the given network configuration, the network is disconnected before the sensing coverage drop to 70%. In addition, the coverage ratio drop more slowly in EECCP than that in OGDC, and the lifetime of EECCP is 35% longer than that of OGDC. This is due to

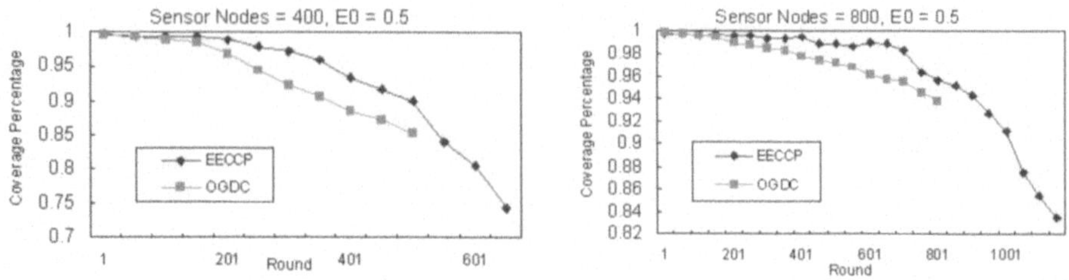

(a) sensor nodes = 400 (b) sensor nodes = 800

Fig. 5. network lifetime vs. time in the different density

the fact that every node must use the maximum transmission range to guarantee the network connectivity, while in EECCP, every node can adjust its own transmission range to make its decision, which consumes less energy than OGDC protocol.

6.3 Control Overhead

For EECCP, first, each node needs to count the number of 1-hop neighbors and compute the appropriate transmission. Thus, each node consumes energy to receive and send control messages. Second, each node needs to know the information of 2-hop neighbors during the construction of CDS. Thus, each node also consumes energy. In addition, for EECCP, the smaller sleeping time in each round should increase the number of control messages and consume more energy. Therefore, the round interval have an impact on the energy consumption of the network.

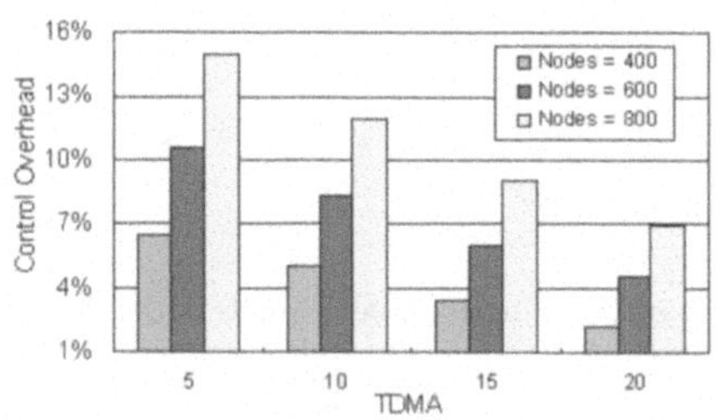

Fig. 6. control overhead vs. round interval

Fig. 6 shows the relationship among the deployed sensor nodes, the round interval and control overhead. The control overhead is calculated as the ratio of energy consumption in the sending and receiving the control messages to the total of energy consumption in the network. From figure 6, the control overhead is relatively sensitive the round interval, especially when the density of deployed nodes is large. For instance, for a network with 800 sensors, the control overhead is about 15% when the round interval is 5 TDMA frames, and about 7% when the round interval is 20 TDMA frames. Another observation is that a denser network tends to have larger control overhead. This is because a denser network has to exchange more messages to turn off more sensors for same coverage ratio. The increase of control overhead due to the increase of the network density, however, is not very large when the round interval is long. This demonstrates that although the network density has an impact on the control overhead, the overhead can be effectively controlled by setting a long round interval.

7 Discussion

In this paper, we studied the area coverage property of minimal dominating set, and established the relationship between point coverage and area coverage, and derive the upper and lower bound that point coverage is equivalent to area coverage in random geometric graphs. Based on the analytical results and Rule K algorithm, EECCP protocol is proposed. The experiment studies show that the proposed EECCP protocol can indeed provide comparable coverage using the similar number of working nodes as compared to other location-based solutions. EECCP protocol can effectively maintain both high quality sensing coverage and connectivity for a long time.

Acknowledgments. This research is partially supported by the NSF of China under Grant No. 60573132; the NSF of Jiangsu Province under Grant No. BK2006168, the National Basic Research Program of China (973) under Grant No. 2002CB312002.

References

1. I. F. Akyildiz, W Su, Y. Sankarasubramaniam, E. Cayirci. Wireless sensor networks: a survey. Computer Networks, 2002, 38(4): 393-422
2. I. Stojmenovic. Position based routing in ad hoc networks. IEEE Communications Magazine, 2002, 40(7):128-134.
3. P. J. Wan, K. M. Alzoubi, O. Frieder. Distributed construction of connected dominating set in wireless ad hoc networks. In: Proc. of IEEE INFOCOM 2002:1597-1604.
4. K. M. Alzoubi, P. J. Wan, O. Frieder. Message-optimal connected dominating sets in mobile ad hoc networks. In: Proc. of the ACM Int'l Symp. on MobiHoc, 2002:157-164
5. Y. Chen, A. Liestman. Approximating minimum size weakly-connected dominating sets for clustering mobile ad hoc networks. In: Proc. of the ACM Int'l Symp. on MobiHoc, 2002:165-172
6. J. Wu, F. Dai. Broadcasting in ad hoc networks based on self-pruning. In: Proc. of IEEE INFOCOM 2003. San Francisco, 2003:2240-2250
7. F. Dai, J. Wu. An extended localized algorithm for connected dominating set formation in ad hoc wireless networks. IEEE Transaction on Parallel and Distributed Systems, 2004, 15(10): 908-920
8. M. Cardei, J. Wu. Energy-efficient coverage problems in wireless ad-hoc sensor networks. Computer communication, 2006(29):413-420.
9. S. Shakkottai, R. Srikant, N. Shroff. Unreliable sensor grids: coverage, connectivity and diameter. In Proc. of IEEE INFOCOM 2003.
10. H. Zhang, J. C. Hou. Maintaining sensing coverage and connectivity in large sensor networks. Ad Hoc & Wireless Networks, 2005, 1(1). 89-124.

11. X. Wang, G. Xing, Y. Zhang, C. Lu, R. Pless, C. Gill. Integrated coverage and connectivity configuration in wireless sensor networks. In: Proceedings of the ACM Int'l Conf. on Embedded Networked Sensor Systems (SenSys), 2003:28-39.
12. B. Chen, K. Jamieson, H. Balakrishnan, R. Morris. SPAN: An energy-efficient coordination algorithm for topology maintenance in ad hoc wireless networks. In: Proceedings of the 7th ACM Int'l Conf. on Mobile Computing and Networking (MobiCom), 2001:85-96
13. J. Carle, A. Gallais, D. Simplot-Ryl. Preserving area coverage in wireless sensor networks by using surface coverage relay dominating sets. In: Pro. of the 10th IEEE Symp. on Computers and Communications (ISCC), 2005:347-352
14. H. Gupta, S. R. Das, Q. Gu. Connected sensor cover: Self-Organization of sensor networks for efficient query execution. In: Proc. of the ACM Int'l Symp. on MobiHoc, 2003:189~200.
15. Z. Zhou, S. R. Das, H. Gupta. Fault tolerant connected sensor cover with variable sensing and transmission ranges. In: Proc. of 2nd Annual IEEE Communications Society Conf. on Sensor and Ad Hoc Communications and Networks (SECON), 2005:594-604.
16. H. L. Cheng, H. Edelsbrunner. Area and perimeter derivatives of a union of disks. Computer Science in Perspective, Lecture Notes Computer Science. Springer, 2003, 2598:88–97.
17. B. Liu, D. Towsley. A study of the coverage of large-scale sensor networks. In: Proc. of the first IEEE Int'l Conf. on Mobile Ad hoc and Sensor Systems (MASS), 2004: 475-483.

A Clustered Routing Protocol with Distributed Intrusion Detection for Wireless Sensor Networks

Lan Yao, Na An, Fuxiang Gao, and Ge Yu

Northeastern University, Shenyang 110004, P.R. China
{yaolan,anna,gaofuxiang,yuge}@mail.neu.edu.cn

Abstract. Clustered routing protocols in wireless sensor networks (WSN) provide significant advantages in energy saving and data preprocessing. However, the communication greatly depends on cluster heads, which leads to severe security problems. In order to solve this problem, a clustered routing protocol with distributed intrusion detection for WSN (WSN_DID) is proposed in this paper. In WSN_DID all nodes are organized to identify intruding cluster heads in a distributed way by pre-distributed random keys. The experiments demonstrate that compared with classic cluster-based routing protocol in WSN, energy dissipation in WSN_DID is rather less and the toleration of intrusion is elevated.

1 Introduction

A Wireless Sensor Network(WSN) consists of considerable amount of sensor nodes with sensing, computing and communication abilities. The nodes cooperate to acquire environmental information in a multi-hop, self-organized way [1], and then send it to the terminal users. WSN has a wide application future, such as wearable sensors [2]. Some WSNs are deployed in unattended and often adversarial environments such as a battlefield. Some strategies have been proposed to solve the inherent insecurity of mobile wireless communication [3]. However, providing routing security to WSN is particularly challenging, due to its restricted energy amount, computing ability and storage capacity.

Security mechanisms in WSN are required to provide confidentiality, authentication, integrity, freshness and authentication for broadcast. Cryptography has already been intensively studied [4], in which the key distribution is a critical point. Though many key distribution schemes have been proposed, one solution that fits in with a particular WSN architecture may not be suitable for another due to the different communication patterns. In general, two types of WSN architectures are widely accepted: flat and hierarchical (cluster-based). The latter one usually aggregates the local messages before sending them to the upper nodes, which is quite suitable for applications such as data query process [5] [6].

In cluster-based WSN, nodes are typically organized into clusters. With cluster heads(CHs) playing the role of aggregating messages and relaying messages from non-CHs to a Base Station(BS). With rotating CHs (like LEACH [7]),

G. Dong et al. (Eds.): APWeb/WAIM 2007, LNCS 4505, pp. 395–406, 2007.

cluster-based WSNs have both advantages and disadvantages in security. The CHs, which have more priorities in the network communications than ordinary nodes, are more prominent to become targets for adversaries; while rotating from one node to another periodically makes it harder for an adversary to identify the routing elements and corrupt them. Therefore, it is challenging to provide such protocols with secure strategies when dynamic and periodic rearranging of the network's clustering is considered.

In this paper, we focus on presenting a secure routing protocol to provide intrusion detection for cluster-based WSNs without any participation of the BS. Our protocol is based on pre-distributed random keys, and with limited message exchanges it is able to recognize the Malicious Cluster Heads(MCH). We give a detailed description of our protocol first, followed by the security analysis, and we finally evaluate the protocol implementation. Our main idea is that nodes in WSN detect MCHs in a distributed way, which is totally independent on BS. Through setting various parameter values, we trade off between connectivity, energy consumption and security levels.

The rest of this paper is organized as follows. In section 2, we analyze clustered routing protocols and their vulnerabilities. Several security clustered routing protocols are also discussed here.In section 3, we present the detailed description and security analysis of our protocol. In section 4 we evaluate the performance of our protocol. We give our concluding remarks in section 5.

2 Security Analysis of Clustered Routing Protocols

Hierarchical routing protocols have advantages over flat routing protocols in security as mentioned above, for instance, the attacks often cause cluster invalidation. However, they are vulnerable to a number of attacks as well, including Sink-hole attack, Selective forwarding, Sybil attack [8], etc. Due to its clustered feature, the communications in the network fundamentally rely on the CHs. As a result, it will cause damaging in the worst case if the target of intruders is CHs. For example, the intruding node pretends to be a CH and sends advertisement message (*adv*) in a tremendous sending power. Then all nodes may join this cluster, which will lead to a global invalidation. Once the intruders manage to become CHs, Sink-hole and Selective forwarding at-tack can be carried out. Of course, the intruders or corrupted nodes can broadcast bogus data throughout the whole network. They can also perform eavesdropping.

LEACH (Low Energy Adaptive Clustering Hierarchy) [7], a classical clustered routing protocol, was first proposed to leverage the energy consumption among nodes in WSN. In this section, we introduce LEACH and analyze its vulnerabilities. Afterwards, two existent secure clustered routing protocols F-LEACH and SecLEACH are discussed.

2.1 LEACH and Its Security Vulnerabilities

LEACH works in rounds(a predetermined duration lasting from the cluster formation to the end of cluster communication). Each round is divided into two phases: a setup phase and a steady-state phase. The setup phase consists of three steps, namely CHs selection, cluster formation and scheduling. The details of each step are described below.

1) CHs selection: nodes probabilistically decide whether to become the CHs in current round or not. There're two ways for the nodes to make the decision: one depends on whether a node has already become a CH recently, with the energy even assumption; the other lies on nodes' rest energy level, without the above assumption. Those self-elected CHs then broadcast their advertisements(*adv*) to all nodes in the network, while non-CHs merely wait.

2) Cluster formation: each non-CH chooses a cluster to join in based on the received signal strength by sending a join require message (*join_req*). When the CH receives all its join requirements, it comes to the next step.

3) Scheduling: each CH broadcasts the TDMS schedule including a slot time for each cluster member. Once the clusters are set up, the steady-state phase starts. Cluster members send their sensing reports to CH strictly according to its slot time. The CHs receive all the data from their members, perform data aggregation and send the results back to the BS. The steady-state operation is often broken into frames.

Essentially, LEACH doesn't include any security mechanism. The communications between BS and CHs in LEACH are in single hop way, which prevent intruders from performing some kinds of relay attacks. However, a MCH may broadcast its *adv* message with a large power, which leads to non-CHs' confusion. Non-CHs will select the MCH as the only CH in this WSN. Under this condition, the whole network is seized.

2.2 F-LEACH and SecLEACH

Because of the hierarchical routing protocols' extraordinary performance on energy-saving and applications, people start to pay more attention to providing them with security mechanisms. Cryptographic protection for clustered networks has been studied intensively by researchers. A.C. Ferreira et al. [9] proposed F-LEACH. Using two keys for each node,a unique key shared with BS and the last key of the one-way key chain generated by BS. The secure mechanism provided by F-LEACH is that each self-elected CH broadcasts its *adv* message as well as a Message Authentication Code (MAC) which is calculated using the unique key shared with BS. BS collects all the *adv* messages, validates the MAC field, and then broadcasts the list of legitimate CHs using μTESLA [10]. Ordinary nodes then choose their CHs according to the list.

Leonardo B. Oliveira et al. proposed SecLEACH [11], which uses pre-distributed random key. In SecLEACH, each node have two kinds of keys: one is unique shared with the BS, the other is a ring of pre-distributed keys with ids drawn randomly from the key pool. Each ordinary node only chooses CH which shares at least one common key with it to join in. The algorithm is almost the same as LEACH, except that the messages sent from ordinary nodes are protected by MAC which is produced by the common key they share. The communications between BS and CHs are also protected by MAC calculated by the unique shared key.

F-LEACH achieves secure communication in clustered routing protocol based on BS's participation.It is tolerant to the most typical attack staged to hierarchical routing protocols in which the intruders pretend to be CHs. However, F-LEACH doesn't provide a complete and efficient solution for node-to-CH authentication. Furthermore, F-LEACH over relys on the BS, and makes the expansibility of WSN impossible.

SecLEACH implemented the secure node-to-node communications in clustered routing(LEACH-based) networks; however the challenging problem in secure hierarchical routing protocols remains unsolved. SecLEACH doesn't manage to authenticate the CHs; thus attacks on CHs cannot be defended at all.

3 WSN_DID

3.1 Random Key Pre-distribution in Nodes

In our solution, we propose to generate a pool of random keys with their indexes. We calculate a key index ring for each node using the node's id as seed, and then the key index ring as well as the real keys are assigned to each node prior to the network deployment. Thus several components need to be added to each node: a PseudoRandom Sequence number Generator(PRSG), which is used to calculate the key index ring; a PseudoRandom number Generator(PRG), which is for generating a nonce for validating CHs; a Neighbor Table (NT) to register the neighbors' information, and a Malicious cluster Head Table(MHT) to record the suspicious intruding CHs. Assume a N-node WSN. The pre-distributed random key strategy [12]used in our protocol is composed of the following steps:

1. Generate a pool of P random keys $V = \{v_p^1, ...v_p^p\}$ and its corresponding indexes $IDX = \{idx_p^1, ..., idx_p^p\}$;
2. Generate k indexes using PRSG with a node's id as the seed, draw the k corresponding keys from the key pool, and assign the k keys and their indexes to the node. Hence, each node keeps k key ring $V_\alpha = \{v_\alpha^1, ..., v_\alpha^k\}$, and corresponding k index ring $IDX_\alpha = \{idx_\alpha^1, ..., idx_\alpha^1\}$.

Every node can calculate any other's key ring using the node's id, and thus it can get the shared keys. This key discovery procedure requires no message exchange; therefore it will not cause extra traffic and downgrade the security level. A unique key shared between the BS and the node also needs to be pre-deployed used for the direct communication between BS and CHs.

3.2 Cluster Head Selection

Cluster head selection is the core of clustered routing protocol. There are three methods that have been widely applied to select CHs: a node itself decides whether to become a CH or not(such as LEACH); nodes cooperate to elect CHs(such as TEEN [13]); the BS decides the CHs(such as LEACH-C [7]). Several elements should be taken into consideration here, such as the remaining energy level, the location, the distance between nodes and the BS, etc. In order to determine how to select CHs, we need to consider these elements and do tradeoff analysis among all these elements [14]. For example, if only taking energy into account, we should choose the node with highest rest energy level as the CH. Despite all the tradeoffs, it is important for a secure cluster head selection algorithm to choose CH randomly, so that even when a malicious node manages to become a CH, it can't be a CH forever.

Based on the above discussion, our cluster head selection algorithm is proposed as follow. The basic idea of this algorithm is that each node generates a random number between 1 and 0. If this number is less than C(n), the node elects itself to be the CH. C(n) is defined as follow:

$$C(n) = \begin{cases} \frac{p}{1 - p[r \bmod (1/p)]}[\eta + (i \times p)(1 - \eta)], \ n \in G \\ 0, \ \text{otherwise} \end{cases} \quad (1)$$

where p is the percentage of CHs; r represents the current round number; η is the percentage of remaining energy and initial energy; i is the sum of rounds that a node is idle, and it is reset to 0 when the node becomes CH. G is the collection of nodes that haven't been CHs in the lately $1/p$ round.

This algorithm takes both randomicity and energy into consideration, and fundamentally guarantees that nodes that haven't been CHs recently are able to become CHs randomly.

3.3 Cluster Formation

Once a node elects itself as a CH, it will start to build their clusters. Steps are given in detail below. In step 1, a self-elected CH H broadcasts its id id_H to inform the whole network. When the non-CH nodes A_i receive the message (step 2.1), they compute the set of H's key index ring (using PRSG with H's id as the seed) and choose one or more(δ) CHs with which they share at least a key to validate, based on the concrete demands (which will be discussed later in section 3.3). If there're more than one shared keys between H and A_i, the key for current communication $key_{(r)}$ will be calculated by the operation XOR of all the shared keys. A_i also generate a *nonce* and send it to the CH along with the MAC produced using $key_{(r)}$(val message). Note that A_i can eavesdrop neighbors' *ids* in this step as well, and record them as their neighbors. In step 2.2, CH H receives nodes A_i's *val* messages. If the CH is a legitimate one, it can calculate the key index it shares with A_i, thus get $key_{(r)}$ (the same way as A_i do). Then the CH can decrypt the messages using $key_{(r)}$, get *nonce*, and send it back to A_i . Otherwise, the malicious CH has no way to calculate $key_{(r)}$, unless it has all the pre-distributed keys shared with nodes A_i. Node A_i wait for H's to reply in step 2.3. If the *nonce* in the reply is wrong or no reply is sent back, A_i take H as a malicious CH and send this message (*mch* message) to the neighbors (recorded in step 2.1) encrypted using the key shared with neighbors, and then wait. If the reply is correct, A_i just wait to get messages from others and transmit it to their neighbors. When A_i receive more than m *mch* messages from different nodes suspecting the same CH, this CH will be recorded as an unauthentic CH, and it will not be chosen as the CH in the following rounds. Now, nodes A_i choose an authentic CH and join in by sending a $join - req$ message. CHs wait till the ordinary nodes validate MCHs in a distributed way. After a CH receives all requirements(step 3), it sets up a TDMA(Time Division Multiple Ac-cess) schedule and transmits this schedule to the nodes in the cluster. Thus authentic clusters are formed without any participation of the BS.

Setup phrase

1. $H \Rightarrow G: \; id_H, adv$

2.1 $A_i \rightarrow H: \; id_{A_i}, id_H, val, mac_{key_{(r)}}(nonce)$

A_i: $addNT(A_i, sig)$

2.2 $H \rightarrow A_i: \; id_H, id_{A_i}, val - rpl, nonce$

2.3 $A_i \rightarrow N: \; id_{A_i}, id_{N_i}, mch, mac_{key_{(r)}}(< id_{A_i}, id_H > |anc)$

A_i: $addMHT(< id_{A_i}, id_H >)$

2.4 $A_i \rightarrow H: \; id_{A_i}, id_H, join - req, mac_{key_{(r)}}(id_{A_i}|id_H|join - req)$

3. $H \Rightarrow G: \; id_H, sched, (...., < id_{A_i}, t_{A_i} >, ...)$

Steady - stage phrase

4. $A_i \rightarrow H: \; id_{A_i}, id_H, d_{A_i}, mac_{key_{(r)}}(id_{A_i}|id_H|d_{A_i})$

5. $H \rightarrow BS: \; id_H, id_{BS}, mac_{key_{unq}}(F(..., d_{A_i}, ...))$

Symbols as previously defined, with the following additions

N : The set of node A_i's neighbor;

$val, val - rpl, mch$: Message type identifiers;

$addNT(A_i, sig)$: Add Node A_i's id to Neighbor Table;

key_{unq} : Unique shared key between BS and Node;

$addMHT(< id_{A_i}, id_H >)$: Add bogus CH's id and detecting node A_i's id to MCH Table;

3.4 In-Cluster Communication

After the cluster formation, members of clusters start to communicate according to their TDMA slot time. The messages sent from ordinary nodes to the CH are protected by same key calculated in step 2.1 ($key_{(r)}$), while those between CHs and the BS are encrypted using the unique key key_{unq}.

3.5 Security Analysis of WSN_DID

In this section, we analyze the security of WSN_DID, and discuss how to obtain different security levels when given different restriction of the networks. We bring forward the attacks considered in this paper: 1) exterior nodes intruding: when the exterior nodes launch attacks to the WSN, they'll pretend to be legitimate notes, especially CHs, and also eavesdrop them; 2) interior nodes corrupting: when the inner nodes are captured, they'll surrender all the information and keys, including PRSG, PRG and pre-distributed key ring, as well as the index ring. When more than one node are corrupted, they will collude. In this paper, we emphasize on when both of the two attacks mentioned above happen at the same time, for this is the worst attack of all. That is, corrupted nods collude with each other, giving the intruders all their keys and corresponding indexes, and the intruders pretend to be CHs. As for the corrupted nodes themselves work together to malign legitimate CHs, that's equal to the above situation. The messages sent to or from the corrupted nodes also can be tampered. Since several strategies have been proposed to solve this problem [15], we won't discuss it in our paper. Furthermore, because of the rotation of the CHs, even when the corrupted nodes manage to become CHs, they can't be the CHs forever. Notations employed throughout this paper are given below.

P : Size of the key pool; k : Size of the key pool for each node;
N : Network size, in nodes; c : Number of Cluster Head(CH);
s : Number of nodes in each Cluster(N/c);
m : Number of nodes cooperated to detect MCH;
w : Number of intruding nodes; l : Number of rounds a MCH won't be used;
m' : Number of corrupted nodes; δ : Number of rounds a MCH won't be used;

In the rest of this section, we focus on how the different parameters' values affect the security level when the attacks mentioned above are launched. In the setup phase, ordinary nodes receive all self-elected CHs' advertisement. When come to validation, nodes have several options. For the sake of energy saving, the non-CH nodes should choose the closest CH; while for the sake of security, the CH with whom they share the most keys should be chosen. When nodes are set to choose more than one CH in this step, such as δ ($\delta \leq c$), the two options above can be combined. The value of parameter δ affects node orphan rate (nodes not belong to any cluster are called orphans). The larger δ is, the less possible that an ordinary node becomes an orphan. Because even when part of the CHs are proved to be malicious, there're still legitimate ones left to join in. However, the traffic (energy consumption) increases with δ at the same time.

As for the messages exchanging among ordinary nodes, we show how the values of parameters m and l should be set to trade off the accuracy of intrusion detection and the energy consumption. $m(m \leq k^2N/P)$ is called tolerance of intrusion, that is the amount of intruders the WSN can tolerate. Generally, the amount of corrupted nodes the WSN can bear is less than m , that is $m' \leq m - 1$. The reason is that if m is larger than m', even when the m' nodes collude to send bogus mch messages, other nodes will not be cheated. m essentially represents the times at most one node has to transmit about the same MCH with different finders. Besides, taking m and δ into consideration together, we conclude that corrupted nodes are unable to incriminate more than δ legitimate nodes in one round. Because if a node receives more than δ different mch messages from the same node, they record the node, and stop transmitting its messages. Apparently, the increment of m also leads to more energy consumption. l determines the number of

Table 1. Analysis on immunity

Protocol type Attack type	F-LEACH	SecLEACH	WSN_DID
Bogus Routing Information	√	√	√
Selective Forwarding	√	√	×
Sink-Hole attack	√	×	√
Sybil attack	×	×	√
Worm-Hole	√	√	√
HELLO Flood	√	√	√
Eavesdropping	×	√	√

following rounds,in which a MCH will not be considered as a CH. The transmission is also well suited for setting a TTL according to clustered networks' feature.

In the worst situation where the CHs validated by A_i are all proved to be MCHs, A_i turn to sleep in this round.

Using node id as the seed to generate the key index ring for every node provides the same extra security benefits as stated in [12]. The data freshness can be guaranteed by adding a clock field to each message. The clock value is generated randomly at the begging of each round, and increased by 1 in the current round [11].

Analysis of Immunity. Table 1 compares the performances of F-LEACH, SecLEACH and WSN_DID in immunity. Since all of these protocols adapt single-hop communication between BS and CHs which is protected by the unique key shared only between them, none of the following attacks would cause serious damages to them. The following figure shows that although WSN_DID is unable to resist the selective forwarding attack while spreading *mch* messages, the damage caused by this vulnerability can be limited by setting a proper value of m . WSN_DID, with respect to useful sensor data, is insusceptible to Selective Forwarding attack.

Connectivity. Note that with pre-distributed random key as our basis, the communication between two arbitrary nodes is only possible if at least one common key is shared between them. Obviously, for a given P, a larger k will lead to higher connectivity in Equation(2). While k=P, the connectivity equals 1. With the increment of k, the security decreases. That is because once few nodes are captured, the adversary almost obtains all the pre-distributed keys. The probability that two arbitrary nodes share at least one common key is:

$$p_{con} = 1 - \overline{p}_{con} = 1 - \frac{C^k_{P-k}}{C^k_P} - 1 - \frac{[(P-k)!]^2}{P!(P-2k)!} \quad (2)$$

We can conclude from the above analysis that, the connectivity between two arbitrary nodes has no relationship with the size of the network (n). It depends on the size of the key pool (P) and the key ring (k) for each node. We assume $\alpha = k/P (0 < \alpha \leq 1)$, Fig 1(a) shows the value of p_{con} while $0.01 \leq \alpha \leq 0.1$,

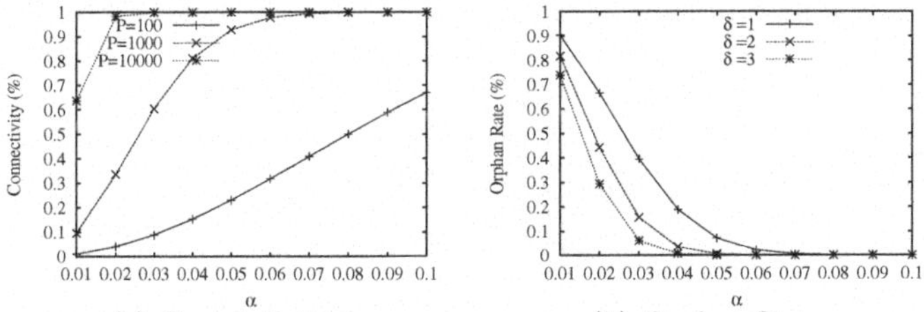

(a) Connectivity (b) Orphan Rate

Fig. 1. Connectivity and Orphan Rate

p=100, 1000, 10000 respectively. Each non-CH sends the mch message to δ CHs. With different values of δ, the orphan rate for each non-CH is as follow.

$$p_{con} = (\overline{p}_{con})^{\delta} \tag{3}$$

Fig 1(b) shows the probability of becoming an orphan for each non-CH while P=1000, δ=1, 2, 3 respectively. When $\alpha \geq 0.05$, the orphan rate is negligible. The security of our network is approximately regarded as $P_{\text{sec}} = 1 - kP$. That means while k/P is small, the network is in a high security level. Thus, when only take connectivity into account, we can get a well performed secure network through our pre-distributed random key strategy.

Energy Consumption. For the sake of simplicity, we assume the size of our network is N, which is randomly deployed at $M \times M$ range. A BS locates outside the square. The node transmitting energy consumption model is as [7]. The energy consumption for sending a data packet with the length l is:

$$E_{Tx}(l,d) = E_{Tx-elec}(l) + E_{Tx-amp}(l,d) = \begin{cases} lE_{elec} + l\varepsilon_{fs}d^2, d < d_0 \\ lE_{elec} + l\varepsilon_{mp}d^4, d \geq d_0 \end{cases} \tag{4}$$

The energy consumption for receiving a data packet with the length l is:

$$E_{Rx}(l) = E_{Rx-elec}(l) = lE_{elec} \tag{5}$$

The energy consumption for transmitting a data packet with the length l is :

$$E_{RT}(l,d) = E_{Rx}(l) + E_{Tx}(l,d) = 2lE_{elec} + l\varepsilon_{fs}d^2 \tag{6}$$

The energy consumption of a CH for one setup phase in WSN_DID is:

$$E_{CH} = E_{Tx}(l,d) + \delta(s-1)E_{RT}(l,d) + (s-1)E_{RT}(l,d) \tag{7}$$

The energy consumption of a non-CH is:

$$E_{non-CH} = cE_{Rx}(l) + \delta E_{RT}(l,d) + mwE_{RT}(l,d) + E_{Tx}(l,d) + E_{Rx}(l) \tag{8}$$

The average energy consumption for each cluster is:

$$E_{cluster} = E_{CH} + (s-1)E_{non-CH} \tag{9}$$

The total energy consumption of the whole network in one setup phase is:

$$\begin{aligned} E_{total} = cE_{cluster} &= l[c(E_{elec} + \varepsilon_{mp} d^4_{toAll}) + c(N-c)E_{elec} \\ &\quad +(\mu+2)(N-c)(2E_{elec} + \varepsilon_{fs} d^2_{toNode})] \\ &= l[c(N-2\mu-3)E_{elec} + c\varepsilon_{mp} d^4_{toAll} - c^2 E_{elec} \\ &\quad + \tfrac{N(\mu+2)\varepsilon_{fs}M^2}{2\pi c} - \tfrac{(\mu+2)\varepsilon_{fs}M^2}{2\pi} + 2(\mu+2)NE_{elec}] \end{aligned} \tag{10}$$

We assume $\mu = 2\delta + wm$, where $m-1$ is the number of corrupted nodes that WSN_DID can tolerate; w is the number of intruding nodes, meanwhile the number of legitimate CHs that $m-1$ corrupted nodes collude to frame, $w \leq \delta$ in this situation (each non-CH node is only allowed to verify δ CHs).

d^2_{toNode} is the distance from a node to BS or to neighbor.

$$E[d^2_{toNode}] = \frac{1}{2\pi}\frac{M^2}{c} \tag{11}$$

We get the value of E_{total} from (10) and (11). l here represents the length of all the command packets. In order to study the energy consumption increment, we compare our protocol with LEACH. In LEACH, the total energy consumption in once cluster setup phase is:

$$\begin{aligned} E_{Ltotal} &= l[c(E_{elec} + \varepsilon_{mp} d^4_{toAll}) + c(N-c)E_{elec} \\ &\quad +2(N-c)(2E_{elec} + \varepsilon_{fs} d^2_{toNode})] \\ &= l[c(N-3)E_{elec} + c\varepsilon_{mp} d^4_{toAll} - c^2 E_{elec} + \tfrac{N\varepsilon_{fs}M^2}{\pi c} - \tfrac{\varepsilon_{fs}M^2}{\pi} + 4NE_{elec}] \end{aligned} \tag{12}$$

Combining (10) and (12), we find that LEACH is a special case of WSN_DID with $\mu = 0$:

$$E_{total} = E_{Ltotal} + \mu(N-c)E_{RT}(l,d) \tag{13}$$

The total energy consumption of LEACH in one frame is:

$$E_{Ptotal} = l'[2NE_{elec} + NE_{DA} + c\varepsilon_{mp} d^4_{toBS} + N\varepsilon_{fs} d^2_{toNode}] \tag{14}$$

Where l' is the data packet length.

4 Experiment

We set different values of μ to study the tradeoff between energy consumption and intrusion tolerant of WSN_DID. Here we make the same assumption for our experiment as the original LEACH paper. We set BS at locations (50,175), N=100, M=100m, E_{elec}=50nj/bit, ε_{fs}=10pj/bit/m^2, ε_{mp}=0.0013pj/bit/m^4, E_{DA}=5nj/bit/signal, d_toAll=141m, 75m< d_{toBS} <185m, l=36byte, l'=500byte.

Figures show the energy consumption in once cluster setup with different number of CH (c), intruding nodes (w), CHs that each node verifies (δ), and corrupted nodes ($m'(m-1)$) , respectively.

Fig. 2.(a) above shows that when δ is fixed, the energy consumptions go up according to the number of intruding nodes. However, they are lower than energy

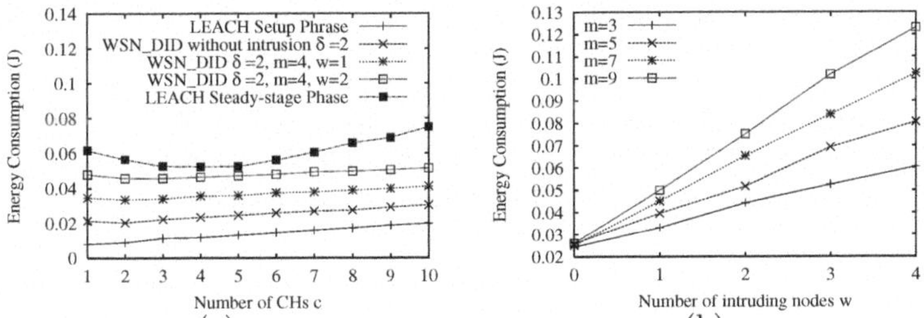

(a) (b)

Fig. 2. 100 Nodes

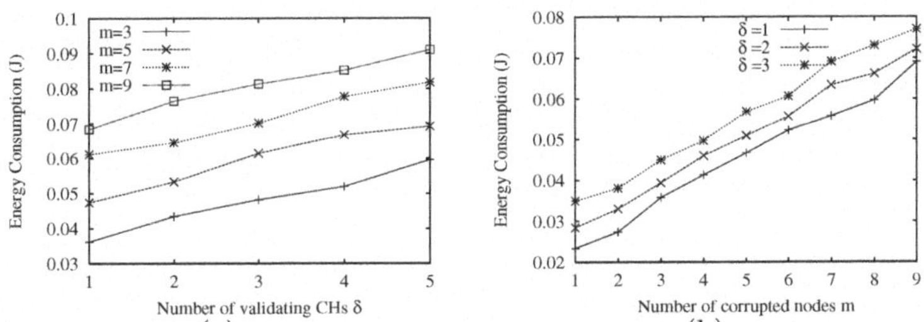

(a) (b)

Fig. 3. 100 Nodes, number of CHs is 5 and number of intruding CHs is 2

dissipation in LEACH communication step. we experiment the energy consumption based on w's variety when m is evaluated in Fig. 2.(b). Here we choose m and w some rather small values. For a WSN containing 100 nodes, four MCHs and 9 corrupted nodes(tolerated by the WSN) are serious intrusion for it. In WSN_DID, nodes can detect and tolerate more. However, the energy dissipation is unacceptable.

Fig. 3.(a) shows the relationship between δ and energy consumption. Compared with Fig. 2.(b), the slope is smaller. It indicates that when m is fixed, w has a greater effect than δ on energy consumption. Fig. 3.(b) shows the impact on energy consumption caused by the number of corrupted nodes(m). The results shows either w or m infects the energy consumption linearly. The increment of m, w or δ in WSNs leads to more energy consumption in linear rule. They are significant factors in energy consumption formulas.

5 Conclusion

In this paper, we present WSN_DID, a secure clustered routing protocol with intrusion detection initiated by general nodes in WSN. These nodes bootstrap the immunity to intruding CHs with the pre-distributed random keys. WSN_DID has different security levels depending on its emphasized parameters (such as δ,w and m) .Our experiments show the trade-off between network connectivity, energy consumption and security level under the different parameters. Though WSN_DID provides more security mechanism, its energy consumption is linear with the security level in a small slope and does not exceed the classic clustered routing protocols.

Acknowledgement. This work is partially supported by the National Natural Science Foundation of China under Grant Nos. 60503036 and 60473073, and Fok Ying Tong Education Foundation under Grant No. 104027.

References

1. Jianzhong Li, Jin-Bao Li, Sheng-Fei Shi: Concepts, Issues and Advance of Sensor Networks and Data Management of Sensor Networks. Journal of Software, Vol. 14, No.10 (2003) 1717-1727
2. N. Kern, B. Schiele, A. Schmidt: Wearable sensor, Multi-sensor Activity Context Detection for Wearable Computing. EUSAI , Veldhoven, The Netherlands, Nov. 3.-4. (2003) 220-232
3. Y. Villate, A. Illarramendi, E. Piroura: Keep Your Data Safe and Available While Roaming. Mobil Networks and Applications, Vol. 7, No.4 (2002) 315-328(14)
4. WenLiang Du, Jing Deng: A Pairwise Key Predistribution Scheme for Wireless Sensor Networks. ACM TISS, Vol. 8, No. 2 (2005) 228-258
5. A. Deshpande, C. Guestrin, S. R. Madden, J. M. Hellerstein, Wei Hong: Model-Driven Data Acquisition in Sensor Networks. Proceedings of the 30th VLDB Conference, Toronto, Canada (2004) 588-599
6. Alfredo Cuzzocrea, Filippo Furfaro, Giuseppe M. Mazzeo, Domenico Sacc : A Grid Framework for Approximate Aggregate Query Answering on Summarized Sesor Network Readings. Proc. GADA'04, Larnaca, Cyprus, 25 Oct 25-29 (2004)
7. Wendi B. Heinzelman, Anantha P. Chandrakasan: An Application-Specific Protocol Architecture for Wireless Microsensor Networks. IEEE Trans. On Wireless Comunication, 2002, 1(4), 660-670
8. Chris Karlof, David Wagner: Secure routing in wireless sensor networks: attacks and countermeasures. Ad Hoc Networks (2003) 293-315
9. A. C. Ferreira, M. A. Vilaca, L. B. Oliveira, E. Habib, H. C. Wong, and A. A. F. Loureiro: On the security of cluster-based communication protocols for wireless sensor networks. ICN'05, LNCS, Vol. 3420, Reunion Island (2005) 449-458
10. Adrian Perrig, Robert Szewczyk: SPINS: Security Protocols for Sensor Networks. Wireless Networks, Vol. 8 (2002) 521-534
11. Leonardo B. Oliveira Hao C. Wong, M. Bern: SecLEACH–A Random Key Distribution Solution for Securing Clustered Sensor Networks. Fifth IEEE Interna-tional Symposium on Network Computing and Applications (2006)
12. Roberto Di Pietro, Luigi V. Mancini: Random Key Assignment for Secure Wireless Sensor Networks. In: Proc. of the 1st ACM Workshop Security of Ad Hoc and Sensor Networks Fairfax.Virginia, ACM Press (2003)
13. Manjeshwar A, Grawal DP: TEEN: A protocol for enhanced efficiency in wireless sensor networks. In:Proc. of the 15th Paralleland Distributed Processing Symp. San Francisco: IEEE Computer Society (2001) 2009-2015
14. Handy MJ, Haase M, Timmermann D: Low energy adaptive clustering hierarchy with deterministic cluster-head selection. In: Proc. of the 4th IEEE Conf. on Mobile and Wireless Communications Networks. Stockholm: IEEE Communications Society, (2002) 368-372.
15. Radu Sion, Mikhail Atallah, Sunil Prabhakar: Resilient Rights Protection for Sensor Streams. Proceedings of the 30th VLDB Conference, Toronto, Canada (2004)

Continuous Approximate Window Queries in Wireless Sensor Networks*

Bin Wang, Xiaochun Yang, Guoren Wang, and Ge Yu

School of Information Science and Engineering,
Northeastern University, 110004 China
{binwang,yangxc,wanggr,yuge}@mail.neu.edu.cn

Abstract. Wireless sensor networks are demonstrating the valuable potential in a variety of settings, e.g. data analysis and statistics. How to minimize energy consumption and prolong network lifetime become one of the key challenges. In this paper, we propose a continuous approximate window query technique called CWQ to approximately collect data in a specified region in a wireless sensor network. CWQ clusters sensors in the network into several clusters, in which sensors have similar readings. We study how to discover a minimal path to concatenate clusters under non-failure as well as failure cases. Our experiments show that CWQ can provide significant communication (and hence energy) savings.

1 Introduction

Wireless sensor networks are utilized in a wide variety of scenarios, such as monitoring farm belt, large-scale habitat, or environmental disaster, tracing abnormities or events for military surveillance and reconnaissance tasks, scheduling real-time supply chain management and so on.

As we known, sensor nodes are often powered by limited batteries. Energy-efficient query processing techniques [1,2,3,4,5,6] and routing protocols [7] have been widely explored recently. In some applications, people may want to collect sensor data from a specified region [8,9], for example, "retrieving the average humidity from region A." Such query is called window query. For simplicity, we assume the region is shaped as a two-dimensional rectangle.

Often a user wants to continuously monitor environments in the specified region for a long time, take an example, "retrieving the average rainfall from region X every ten seconds, for three months." To answer such a continuous window query, it is not desired to forwarding all sensor readings in the region to the sink for each time step, since the energy consumption is huge. In this paper, we address the problem of providing an energy-efficient approximate window query approach. That is, given a sensor network with a fixed sink, a user specifies

* Supported by the 973 Program under Grant No. 2006CB303103, the Natural Science Foundation of China under Grant Nos.60503036, 60473074, 60573089, and the Fok Ying Tong Education Foundation of China under Grant No. 104027.

G. Dong et al. (Eds.): APWeb/WAIM 2007, LNCS 4505, pp. 407–418, 2007.

a window $R[(x_1, y_1), (x_2, y_2)]$ and a query "SELECT * FREQ f WITHIN $\pm\delta$ OVER $R[(x_1, y_1), (x_2, y_2)]$."

With the consideration of spatial correlation between sensors [8], we partition the query region into several clusters for each time step. Each cluster constructs a clique based on their similar sensor readings. This Clique-based Window Query (CWQ) approach does not require to collect or forward all sensor readings to the sink, instead, only part of sensors are chosen to report their readings. The key technical challenges of this work are listed as follows.

- A sensor may join or leave from a cluster when its reading exceeds the value interval of the cluster. Maintaining the cluster also consume energies. We show that we only use small number of messages to maintain the clusters.
- We provide techniques to find a minimal forwarding path to routing all clusters.

1.1 Related Work

Existing window-based queries processing techniques can be divided into infrastructure-based query execution [2,10,11] and infrastructure-free query execution [12]. Based on the size of the infrastructure, infrastructure-based query execution can be further classified into network spanning infrastructure (NSI) and window spanning infrastructure (WSI). In [2] (shown in Fig. 1(a)), an NSI is usually built when deploying the network and is maintained during the entire network lifetime. For window-query, NSI involves many irrelevant nodes outside the query window for query processing, which costs many extra resources. WSI approaches [10,11] (shown in Fig. 1(b)) adopt Geo-routing protocol (e.g., GPSR [13]) to transmit the query from the query originator node to query window. Once the query reaches the query window, an infrastructure within the query window is built along with the query propagation. Data collection execute along the same infrastructure (e.g., a tree or a cluster). However, constructing and maintaining the infrastructure also incurs noticeable overhead and time latency.

To overcome the problems of infrastructure-based query execution, an infrastructure-free window-based query processing technique for sensor networks is proposed in [12], called itinerary-based window query execution (IWQE), shown in Fig. 1(c). [12] combines query propagation and data collection into one single stage and a well-designed itinerary inside the query window is used to navigate the query processing. Differ from infrastructure-based techniques, IWQE does not maintain any infrastructure and hence improves energy efficiency and reduces the query response time.

The most closed work to ours is snapshot query proposed by Kotidis [14], which exploits local models and correlations, and elects a small set of representative nodes to answer the queries. There are two major difference between snapshot and CWQ, which are (i) snapshot need spend energy to choose representative node, whereas CWQ can randomly choose a sensor in the cluster as routing node to forward sensor readings, and (ii) each representative node using

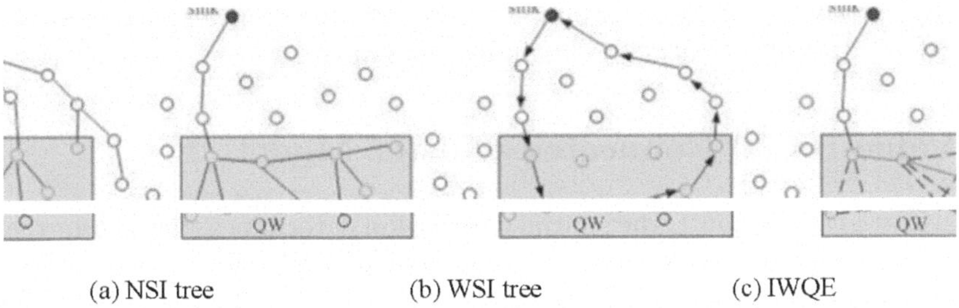

Fig. 1. Related work

snapshot needs choose a minimal path from each cluster to the sink to forward readings, whereas, CWQ tries to find a global minimal routing path in the query region. These two aspects makes CWQ more energy-efficient than snapshot.

2 Approximate Window Query Framework

We begin by formalizing the problem and proposing an approximate window query framework. To illustrate it, we use a network model as follows. Note, however, that our approach can be applied to more complicated environments.

Sensor Network. A sensor network consists of a set of fixed-location sensors, each of which has a unique ID. Two sensors are said to be neighbors if they are within the transmission range of each other. Each sensor knows its location as well as the location of its neighbors. Without loss of generality, we assume that the sink node is fixed, has unlimited energy, and initially broadcasts the locations of all sensors.

The primary consumption of energy in sensor nodes is for radio communication, whereas computation cost is relatively much lower than communication cost [15]. Given a window $R[(x_1, y_1), (x_2, y_2)]$, the sink initiates a window query "SELECT * FREQ f WITHIN $\pm\epsilon$ OVER $R[(x_1, y_1), (x_2, y_2)]$" to the network.

Problem Definition. Given a sensor network that continuously sensing values at each time step and a sink that requires an ϵ-loss approximation of the sensing data in a given region $R[(x_1, y_1), (x_2, y_2)]$ at all times, design a data collection protocol in the region R that costs network energy as less as possible.

In this paper, we propose an approximate continuous query processing method to fulfill the requirements of the above problem. Specifically, by utilizing the spatial correlation among sensor readings, sensors are first grouped into value-based connected clusters, called VCC, with a distributed clustering algorithm, which will be discussed in Section 3. Then, a single routing path will be chosen to connect all VCCs and forwards readings of the VCCs to the sink. A valid data route is proposed in Section 4 to guide the sensors inside each VCC to send their readings to the sink with a "minimal" route. We show our experimental results in Section 5. Finally, Section 6 concludes the paper.

3 Value-Based Connected Cluster Model

As illustrated in Section 1, the spatial correlation determines that sensor readings in a certain region are the same or quiet similar. We can separate the whole sensing region into several disjoint regions according to their sensor readings.

In order to simplify the presentation, we assume that there is a single measurement s_i collected on each sensor node. In practice there can be as many measurements as the number of sensing elements installed on a node. Our approach can also solve the problem.

Definition 1. *(Value-based Cluster) Given a region $R[(x_1, y_1), (x_2, y_2)]$ in the wireless sensor network, a set of sensor measurements $S(R) = \{s_1, ..., s_m\}$ belong to a value-based cluster C, iff for any two sensor measurements $s_i, s_j \in S(R)$, they are neighbors, and $|v_t(s_i) - v_t(s_j)| \leq \epsilon$, where $v_t(s)$ denotes sensing value of s at time step t.*

Definition 2. *(Disjoint cluster model) A cluster model is denoted as $C_v = C_1, ..., C_k$, where k is number of VCCs and for any two C_i and C_j, they are disjoint.*

For a region $R[(x_1, y_1), (x_2, y_2)]$, it is desired that sensor measurements in R can construct different VCCs and discovery a "good" data routing to forward their readings to the sink using as less energies as possible. In this section, we present our cluster construction approach, we then propose an approach to concatenate VCCs in Section 4.

3.1 Clusters Construction

Once a continuous query is submitted to the sink, the sink can choose a set of sensors in the specified region R to construct disjoint VCCs. Each chosen sensor actively invites their neighbors to join in its cluster. We call such chosen sensors *seeds*.

In order to let all sensors to involve in the clustering procedure simultaneously, seeds are desired to distribute uniformly in the network. That is, for any a sensor s in the network, either s is a seed, or s is a neighbor to a seed. The problem of choosing seeds is a typical problem of constructing a dominating set over the topology graph in the region R. We can use a greedy approach to choose seeds. That is, we (i) start with the network and an empty set S; (ii) pick a sensor with most neighbors not already in the set S and add it to S; (iii) repeat step (ii) until all sensors are in S or adjacent to the sensors in S. This greedy approach can guarantee $O(n)$ complexity, where n is the number of sensors in R.

Algorithm 1. Seeds construct disjoint VCCs
Input: a set of seeds S, time step t
1: Each seed s in S broadcasts $v_t(s)$ and ϵ to its neighbors;
2: Seed s gets messages from its neighbors $s'_1, \ldots, s'_h$;
3: **for** each response message in the response sensors S' **do**
4:    **if** $|v_t(s'_i) - v_t(s)| \leq \frac{1}{2}\epsilon$ **then**
5:       s constructs its VCC contains s'_i; $S' = S' - \{s'_i\}$;
6:    **end if**
7: **end for**
8: calculate $V_I(s) = [v_l, v_u]$, s.t. $v_l = min_{i=1}^{k}\{v_t(s), v_t(s'_i)\}$, $v_u = max_{i=1}^{k}\{v_t(s), v(s'_i)\}$;
9: ranking sensors in S into L using $dist(V_I(s), v_t(s'_i))$ in ascending order;
10: **while** $|v_u, v_l| \leq \epsilon$ **do**
11:    choose the first sensor s' in S';
12:    **if** $dist(V_I(s), v_t(s'_i)) \leq \epsilon$ **then**
13:       extend $V_I(s)$ using $dist(V_I(s), v_t(s'_i))$;
14:       $S' = S' - \{s'_i\}$;
15:    **end if**
16: **end while**
17: s notifies $V_I(s)$ and its member IDs to sensors in its VCC;
18: **if** s receives an invitation message from other seeds **then**
19:    s responses the invitation according to Algorithm 2;
20: **else if** no neighbors send messages **then**
21:    s builds a VCC contains itself;
22: **end if**

Algorithms 1 and 2 are used to construct disjoint VCCs. Algorithm 1 illustrates that a seed constructs a set of disjoint VCCs. After receiving the start message sent from the sink, each seed broadcasts its sensor reading and receives neighbors responses. Lines 3, 18, and 20 in Algorithm 1 show behaviors of a seed s according to its neighbors' responses. If s gets responses, then it clusters its neighbors who satisfy the value constraints into its group. Lines 4 to 6 in Algorithm 1 show a conservative way of adding member sensor measurements to the VCC of s. Lines 8 to 16 greedy add more member sensors in the VCC, where the value distance between $V_I(s) = [v_l, v_u]$ and $v_t(s')$ is defined in Equation 1. If s does not receive any response, then it either joins to another seed's group (Lines 18 to 19) or builds a VCC (Lines 20 to 21) that contains itself. Using Algorithm 1, a seed only need to send two messages to construct a VCC. Algorithm 2 describes that a non-seed behaviors. Line 2 in Algorithm 2 describes the case where a non-seed receives invitations from seeds. It chooses a seed who has the closest value with its reading. If there is no invitation, then the non-seed sensor changes it to a seed sensor and uses Algorithm 1 to construct its VCC. We can see a non-seed uses at most four messages to make it a member of a VCC.

$$dist(V_I(s), v_t(s')) = \begin{cases} v_u - v, & \text{if } v_t(s') < v_l \\ v - v_l, & \text{if } v_t(s') > v_u \\ 0. & \text{otherwise} \end{cases} \tag{1}$$

Algorithm 2. Non-seeds join in disjoint VCCs
Input: a set of non-seeds S', time step t
1: Each $s' \in S'$ listens messages from its neighbors;
2: **if** s' receives $v_t(s)$ from sensors S at time step t **then**
3: s' chooses a sensor $s \in S$ such that $|v_t(s') - v_t(s)|$ is minimize and less than ϵ;
4: s' sends $v_t(s')$ to s;
5: **if** s' receives message $V_I(s)$ from s **then**
6: s' stores $V_I(s)$;
7: **else**
8: s' resent $v(s')$ to s; //in case of forwarding failure
9: go to Line 1;
10: **end if**
11: **else if** there is no messages from its neighbors **then**
12: s' changes it to a seed;
13: **end if**

For example, Fig. 2(a) shows the sensor topology in a query region. Fig. 2(b) shows three constructed clusters using seeds s_1, s_7, and s_8.

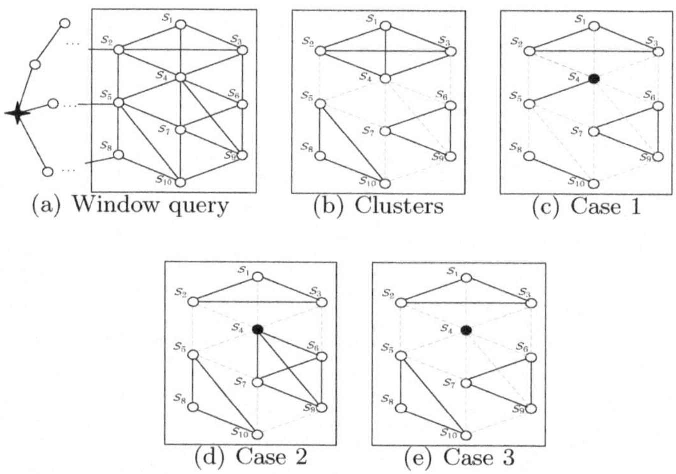

(a) Window query (b) Clusters (c) Case 1

(d) Case 2 (e) Case 3

Fig. 2. Clusters construction and maintenance

3.2 Clusters Maintenance

At time step t, if a sensor s in C detects an update, i.e. $v_t(s) \notin V_{I_{t-1}}(C)$, then s forwards $v_t(s)$ to the sink along its routing path (We show how to find a routing path in Section 4). The neighbors of s can listen such information using Gossip protocol [16]. Generally, neighbor sensors detects a similar value to $v_t(s)$. If the neighbors of s cannot detect such updates, then we assume that $v_t(s)$ is a noise (or outlier). If there is at least one neighbor s' detects update and $|v_t(s) - v_t(s')| \le \epsilon$, then s uses Algorithm 1 to build up a new VCC.

Figs. 2(c), (d), and (e) show the possible three cases when $dist(V_{I_{t_1}}(s), v_t(s)) > \epsilon$. In Fig. 2(c), s_4 and s_5 leave from their clusters, respectively and construct a new cluster; In Fig. 2(d), s_4 joins in cluster $\{s_6, s_7, s_9\}$, and Fig. 2(e) shows s_4 cannot find a cluster to join in, so s_4 might produce a noise and there is no need to forward its update to the sink.

4 Cluster-Based Data Routing

After we get a set of VCCs, a crucial problem that we need to address is how to find a "minimal" route to catenate all VCCs.

As discussed in [12], one path routing saves more energy than a tree routing. Ideally, sensors can independently find one *minimal route* to convey sensor readings of all VCCs to the sink.

Definition 3. *(Valid data route) Given a set of disjoint VCCs $C_1, ..., C_k$ and sink s, a route $s_1 \rightarrow ... \rightarrow s_m \rightarrow s$ is valid, if and only if for each VCC C_i, there exists at least one routing sensor $s_j \in C_i$ $(1 \leq j \leq m)$.*

Minimal valid data route. Given a set of disjoint VCCs $C_1, ..., C_k$ and a start sensor s in the specified region R, find a valid data route $s \rightarrow s_1 \rightarrow \ldots \rightarrow s_m \rightarrow s'$, where s' is a sensor outside of the region. The valid route is minimal if (i) s_1 is disconnected to s after removing s_j $(1 \leq j \leq m)$ from the route, and (ii) only updates are transmitted to s.

In this section, we propose a self-adaptive routing protocol to decrease numbers and sizes of forwarding messages, and balance the energy consumption of sensors in the specified region R. We show that using our proposed routing, sensors are smart to find a "minimal" valid route to forward messages.

4.1 Choosing Routing Sensors

In the region R, no sensor has global knowledge about other VCCs. In order to let sensors intelligently build up a routing tree, we extend the TAG tree technique.

Each sensor in the region R has a routing table to record a tuple <*level*, *routed-clusters*>, where *level* denotes the length of the route path in R, and *routed-clusters* denotes the routed clusters represented using seed in each cluster. For example, the seeds in Fig. 2 are s_1, s_4, s_7, and s_8. For a start sensor s, s forwards the query request q to the region R. It marks itself as the 0 level routing node and adds its cluster's seed to its routing table. Then s broadcasts its routing message to its neighbors. For each sensor in R, the routing strategy is listed as follows.

- For any two routing messages $< l_1, Cset_1 >$ and $< l_2, Cset_2 >$ that sensor s_i receives, if $l_1 < l_2$, then s_i chooses a sensor with $< l_1, Cset_1 >$ as its routing parent; if $l_1 = l_2$ and $Cset_1 \subset Cset_2$, then sensor s_i chooses a sensor with $< l_2, Cset_2 >$ as its routing parent.

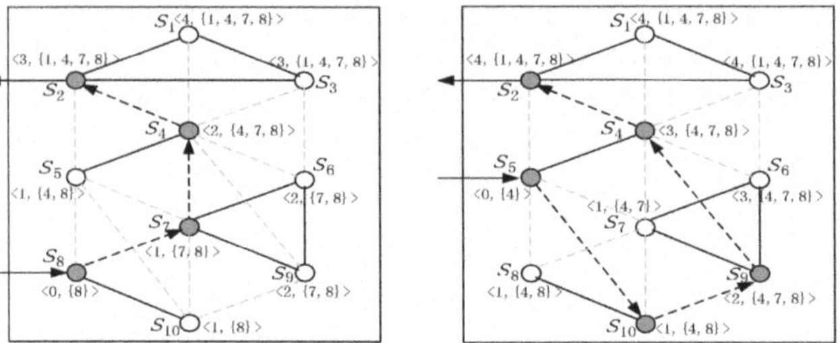

(a) Start from s_8 (b) Start from s_5

Fig. 3. Routing paths

- If non sensor in a cluster is routed, then sensors in the cluster broadcasts requests to its neighbors who do not belong to its sensor and chooses one response sensor as its routing parent.
- Each sensor increases its routing parent's level by 1 and add its cluster's seed into the its parent's routed-clusters.

Fig. 3 shows two routing paths start from different sensors. The routing path in Fig. 3(a) is $s_8 \rightarrow s_7 \rightarrow s_4 \rightarrow s_2$. The routing path in Fig. 3(b) is $s_5 \rightarrow s_{10} \rightarrow s_9 \rightarrow s_4 \rightarrow s_2$.

4.2 Continuous Query

In order to balance the energy consumption of sensors, CWQ builds up different routes periodically. During one period of time T, the route is expected to be the same so that only updates need forward. We first describe how to forward messages when a new route is built up. Then, we then discuss forwarding updates to the sink along the constructed route. Rule 1 describes how to forward messages along the route at the time a new route is built up.

Rule 1. *If there are more than one sensors in a VCC has been routed, then only the last routing sensor needs append message.*

For different period of routing time, sensors finds a different route to balance the energy consumption in the network. Note that for each period, we only build a single route, not multiple routes, thus, the energy consumption is approximately the same.

We next discuss how CWQ self-adaptively builds up a new valid route to balance the energy consumption in the network. Each sensor in the network keeps its routing history and is aware of its remaining energy. Also, CWQ utilizes Gossip protocol [16] to intelligently find the new valid route. For instance, at the first step of a new time period T, a new routing sensor s_a is chosen and selects sensor s_b as its downstream routing node. s_b checks its routing history and remaining energy. In case that s_b serves a routing sensor in recently time period or has not enough energy to forward messages, it will not forward messages further. After s_a sends message to s_b, it can listen messages sent from s_b although those messages are not sent to s_a according to Gossip protocol. If s_a cannot listen message from s_b, then s_a knows that s_b does not forward its message. So, s_a chooses another sensor s'_b. In this way, a sensor can self-determine its role of routing according to its own status. There is one exception that s_a failed to monitor the broadcast message from s_b, then s_a chooses another routing sensor and meanwhile s_b continues to forward messages. Therefore, there are at least two routes arrive the sink.

5 Experimental Results

We conducted experiments to evaluate the effect of our proposed CWQ technique by simulating light monitoring experience in our lab. In this section we report the experimental results.

All the experiments were implemented in C++, and run on an Intel Pentium IV 2.0GHz PC with 512MB RAM.

In the simulation, we generated the synthetic data we generated values that follow a random walk pattern. In order to simulate a network with large number of sensors, we developed a simulator to construct a network by expanding the deployment of sensors in our lab. We built up a 100×100 two-dimensional area, where N=100 sensor nodes are randomly placed in the area. We generated a density sensor network such that for each sensor, its 70% neighbors are connected with each other.

We consider two types of data sets and analyze the effect of them on CWQ. For each type of data set, we generated 5 groups of data to compute the average values.

Type 1: Updates of sensor readings in the region are similar.
Type 2: Small percentage of updates change significantly than the other sensor readings.

5.1 Comparison Results

We compared energy consumptions of the tree-based approach, cluster-based approach, and CWQ using two types of data sets. In the tree-based approach, the network topology is a tree in the query window. Similarly, in the cluster-based approach, a cluster head is chosen to answer the query.

We let ϵ be 1, a transmit range (maximal length of each sensor transmission) L be 30 in the 100×100 query region. For a query request q in the query region, we collected sensor readings continuously every 4 time steps.

Fig. 4 shows the energy consumption of the three approaches using two types of data sets. Figs.4(a) and (b) show that the performance of CWQ is better than the other two approaches. At the first time step, CWQ needs construct clusters and choose valid date route, which more energies than tree-based and cluster-based approaches. In the 40th time steps, the clusters and data routes get stable, therefore, CWQ does not require all sensors to forward their readings, which consumes less energy. The energy consumption of the tree-based approach is the highest,

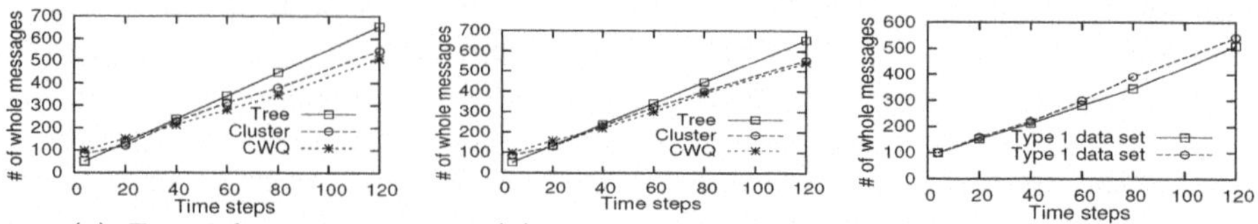

(a) *Type 1* data set (b) *Type 2* data set (c) *Type 1* and *2* data sets

Fig. 4. Comparison results

since it probes all sensors in the query region. Comparison of energy consumptions between the effects of two different types of data sets using CWQ is depicted in Fig. 4(c). As we can see, *Type 1* data set consumes less energy than *Type 2* data set does, since updates in the former data set are similar. The figure also shows that CWQ performs good when some sensor readings changes lot.

5.2 Effect of Transmission Range and Error Threshold

Consider the definition of clique, we know that the error threshold ϵ and transmission rang L are of great importance to partitioning clique. Fig. 5(a) shows the effect of transmission range L. Let the error threshold be 1. When increasing the transmission range, the density of sensors in the network increase, which results in high probability to cluster more sensors into one group. Fig. 5(b) shows the effect of varying ϵ when fixing the transmission range $L = 30$. It is not surprising to see that under the same sensor distribution, the number of clusters decreases when increasing ϵ.

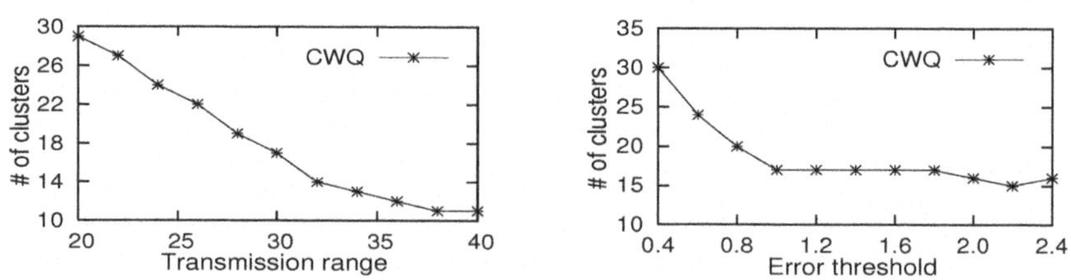

(a) (b)

Fig. 5. Effect of ϵ and L

5.3 Effect of Sensor Density

Fig. 6 shows that the energy consumption of the three approaches: tree-based, cluster-based approaches and CWQ. We let the transmission range $L = 30$ and $\epsilon = 1$. We calculated the average energy consumption (# of forwarding messages) of all sensors in the region. CWQ consumes much less energy than the tree-based approach. However, as the node density decreases, every cluster contains less

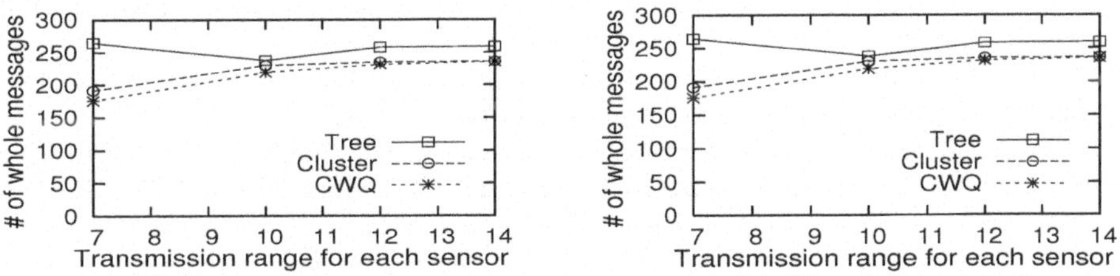

(a) *Type 1* data set

(b) *Type 2* data set

Fig. 6. Effect of sensor density

nodes. When the node density decreases to a small enough value, the network topology becomes a tree structure, which results in CWQ and cluster-based approach behaving like tree-based approach, therefore, the energy consumption become closer.

6 Conclusion

In this paper, we described the processing problem of window-based approximate continuous queries in wireless sensor network and proposed cluster-based approximate continuous window-based query processing technique (CWQ). Existing techniques for window-based do not support continuous queries with long lifetime. In CWQ, the network is partitioned into some clusters and a small number of sensors are intelligently chosen as routing nodes to forward sensing readings. The experimental results showed that CWQ performs better in terms of energy consumption under different factors.

References

1. Chu, D., Deshpande, A., Hellerstein, J. M., Hong, W.: Approximate Data Collection in Sensor Networks using Probabilistic Models. In: Liu, L., Reuter, A., Whang, K.-Y., Zhang, J. (eds.): Proceedings of the 22nd Int. Conf. on Data Engineering (2006) 48
2. Madden, S., Franklin, M. J., Hellerstein, J. M., Hong, W.: The Design of an Acquisitional Query Processor for Sensor Networks. In Proceedings of the ACM SIGMOD Int. Conf. on Management of Data (2003) 491–502
3. Madden, S., Franklin, M. J., Hellerstein, J. M., Hong, W.: TAG: a Tiny AGgregation Service for Ad-Hoc Sensor Networks. In Proceedings of the 5th Symposium on Operating System Design and Implementation (OSDI) (2002)
4. Manjhi, A., Nath, S., Gibbons, P. B.: Tributaries and Deltas: Efficient and Robust Aggregation in Sensor Network Streams. In: Özcan, F. (ed.): Proceedings of the ACM SIGMOD Int. Conf. on Management of Data (2005) 287–298
5. Yang, X., Li, L., Ng, Y.-K., Wang, B., Yu, G.: Associated Load Shedding Strategies for Computing Multi-Joins in Sensor Networks. In: Lee, M.L., Tan, K. L., Wuwongse, V. (eds.): Proceedings of the 11th Int. Conf. on Data Systems for Advanced Applications (DASFAA). Lecture Notes in Computer Science, Vol. 3882. Springer-Verlag, Berlin Heidelberg New York (2006) 50–64

6. Yao, Y., Gehrke. J.: Query Processing in Sensor Networks. In Proceedings of CIDR Conference (2003)
7. Al-Karaki J. N., Kamal A. E.: Routing Techniques in Wireless Sensor Networks: a Survey. IEEE Wireless Communications, 11:6–28, (2004)
8. Silberstein, A., Braynard, R., Yang, J.: Constraint Chaining: On Energy-Efficient Continuous Monitoring in Sensor Networks. In Proceedings of the ACM SIGMOD Int. Conf. on Management of Data (2006) 157–168
9. Tan, H., Körpeoğlu, İ.: Power Efficient Data Gathering and Aggregation in Wireless Sensor Networks. SIGMOD Record, 32(4):66–71 (2003)
10. Alexandru, C., Mario, A., Jorg, S.: A Framework for Spatio-Temporal Query Processing over Wireless Sensor Networks. In: Alexandros L, Samuel M, eds. Proceedings of the 1st Int. Workshop on Data Management for Sensor Networks in Conjunction with VLDB New York: ACM Press (2004) 104–110
11. Xu, Y., Lee, W.-C.: Window Query Processing in Highly Dynamic Geosensor Networks: Issues and Solutions. GeoSensor Networks (2004) 31–52
12. Xu, Y., Lee, W.-C., Xu, J., Mitchell, G.: Processing Window Queries in Wireless Sensor Networks. In Proceedings of the 22nd IEEE Int. Conf. on Data Engineering (ICDE), Atlanta, GA, (2006) 70
13. Karp, B., Kung, H.: GPSR: Greedy Perimeter Stateless Routing for Wireless Networks. In Proceedings of ACM MobiCom (2000) 243–254
14. Kotidis, Y.: Snapshot Queries: Towards Data-Centric Sensor Networks. In Proceedings of the 21st IEEE Int. Conf. on Data Engineering (ICDE) (2005) 131–142
15. Pottie, G., Kaiser, W.: Wireless Integrated Network Sensors. Communications of the ACM, 43(5):51–58 (2000)
16. Kempe, D., Kleinberg, J., Demers, A.: Spatial Gossip and Resource Location Protocols. In Proceedings on 33rd Annual ACM Symposium on Theory of Computing, Heraklion, Crete, Greece (2001) 163–172

A Survey of Job Scheduling in Grids

Congfeng Jiang, Cheng Wang, Xiaohu Liu, and Yinghui Zhao

Engineering Computing and Simulation Institute, Huazhong University of Science and Technology, Wuhan 430074, China
congfeng.jiang@smail.hust.edu.cn, wangch@mail.hust.edu.cn, xhliu@mail.hust.edu.cn, zhaoyinghuihust@gmail.com

Abstract. The problem of optimally scheduling tasks onto heterogeneous resources in grids, minimizing the makespan of these tasks, has proved to be NP-complete. There is no best scheduling algorithm for all grid computing systems. An alternative is to select an appropriate scheduling algorithm to use in a given grid environment because of the characteristics of the tasks, machines and network connectivity. In this paper a survey is presented on the problem and the different aspects of job scheduling in grids such as (a) fault-tolerance; (b) security; and (c) simulation of grid job scheduling strategies are discussed. This paper also presents a discussion on the future research topics and the challenges of job scheduling in grids.

Keywords: heterogeneous computing, task scheduling, fault-tolerance, security, simulation, load-balancing.

1 Introduction

Grid computing [1] is emerging as a popular way of providing high performance computing for many data intensive, scientific applications using heterogeneous computing resources. The process of mapping tasks onto grid system consists of assigning tasks to resources available in time and space, minimizing the makespan of the tasks. The available resources, system load and computing power fluctuate in grids because of the heterogeneity, dynamicity, and autonomy of it. The problem of mapping tasks onto heterogeneous resources, minimizing the makespan of these tasks, has proved to be NP-complete [2]. There are a lot of algorithms developed to solve this problem. However, there is no best scheduling algorithm for all grid environments. An alternative is to select a rather better scheduling algorithm to use in a given grid environment because of the heterogeneity among the tasks, machines and network connectivity.

Although many scheduling techniques for various computing systems exist [3, 4, 5, 6, 7], traditional scheduling systems are inappropriate for scheduling tasks onto grid resources. First, grid resources are geographically distributed and heterogeneous in nature. One of the central concepts of a grid is that of a virtual organization (VO) [1], which is a group of consumers and producers united in their secure use of distributed high-end computational resources. Second, these grid resources have decentralized ownership and different local scheduling policies

G. Dong et al. (Eds.): APWeb/WAIM 2007, LNCS 4505, pp. 419–427, 2007.

dependent on their VO. Third, the dynamic load and availability of the resources require mechanisms for discovering and characterizing their status continually [1].

Moreover, the failure rate increases when the scale of the grid system becomes larger. Consequently, the mapping of tasks onto grid sites must be adaptive, scalable and fault-tolerant because of the dynamicity of the grid environment. The mapping also must be load-balancing-effective to leverage the impact of imbalance caused by the heterogeneity among tasks and machines, and the arrival rate of the tasks.

Job executions are carried out across the domain boundaries in grids. Realistic platforms for grid systems are facing security threats from network attacks and system vulnerability. To enable more effective job scheduling, the job scheduling algorithm must be security-aware and risk-resilient. Therefore, special mechanisms for security and fault-tolerance are needed [7].

Experiments with real applications on real resources are often performed to evaluate the scheduling algorithms. However, modern computing platforms are increasingly distributed and often span multiple administrative domains. Therefore, resource availability fluctuations make it impossible to conduct repeatable experiments for relatively long running applications. Another problem is that the number of platform configurations that can be explored is limited [8]. As a result of these difficulties with real experimentations, most researchers have resorted to discrete-event simulation. Simulation has been used extensively as a way to evaluate and compare scheduling strategies because simulation experiments are configurable, repeatable, and generally fast [8].

In this paper, we discuss several aspects of job scheduling in grids. The rest of the paper is organized as follows: Section 2 presents the basic concepts of job scheduling in grids. In Section 3, some kinds of job scheduling algorithms and models are discussed. In Section 4, we describe and evaluate the fault-tolerant framework of task scheduling in grids. We present the existing study of security assurance for task scheduling in Section 5. In Section 6, we introduce some advances in simulation of grid job scheduling. Finally, we summarize the study of task scheduling in grids and make some remarks on further research in Section 7.

2 Basic Concepts

Let M denote hosts set, $M = \{m_j \mid j = 1,2,3,...,m\}$,T denote tasks set, $T=\{ t_i \mid i=1,2,3,...,n\}$, The expected execution time e_{ij} of task t_i on machine m_j is defined as the amount of time taken by m_j to execute t_i given m_j has no load when t_i is assigned. The expected completion time c_{ij} of task t_i on machine m_j is defined as the wall-clock time at which m_j completes t_i. Let the arrival time of the task t_i be a_i, and let the time t_i begins execution be b_i. From the above definitions, $c_{ij} = b_i + e_{ij}$.

The most common objective function of task scheduling problems is makespan. However, on a computational grid, the second optimal makespan may be much longer than the optimal makespan because the computing power of a grid varies over time [9].Consequently, if the performance measure is makespan, there is no approximation algorithm in general for scheduling onto a grid. In [9], a criterion of a schedule is proposed which is called Total Processor Cycle Consumption (TPCC), the total number of instructions the grid could compute until the completion time of the schedule.

The following assumptions are commonly made when evaluating the job scheduling algorithms in grids:

(1)A large application has been partitioned to some sub tasks, and the sub tasks are independent. This assumption is commonly made and it is prevalent in the job scheduling for grids (e.g., [3, 4, 6, 10]). Note that scheduling dependent jobs with precedence constraints, or DAG(Directed Acyclic Graph) topologies can be found in [11,12].

(2) The tasks have no deadlines or priorities associated with them.

(3) The real-time states of the resources are known. This can be achieved by using some network or grid services like NWS (Network Weather Service) [13] and MDS (Monitoring and Discovery System) [14] when scheduling.

(4) The execution time of the tasks is known. These estimates can be supplied before a task is submitted for execution, or at the time it is submitted. Thomas [15] proposed a job allocation scheme in distributed systems (TAG) using the Markovian process algebra PEPA where the scheme requires no prior knowledge of job size and had been shown to be more efficient than round robin and random allocation when the job size distribution is heavy tailed and the load is not high

(5) Communication delay between sender and receiver are not considered.

(6) Service strategy on a host is FCFS (First Comes First Served) [16], and the hosts execute tasks one at a time

(7) Sites are cooperative in the grid environments.

To evaluate the performance of various scheduling algorithms, the following metrics [6] are commonly used:

(1) *Makespan*: the total running time of all jobs;

(2) *Scheduling success rate*: the percentage of jobs successfully completed in the grid;

(3) *Grid utilization*: defined by the percentage of processing power allocated to user jobs out of total processing power available over all grid sites;

(4) *Average waiting time*: the average waiting time spent by a job in the grid.

3 Scheduling Algorithms

Heuristics are the main strategy to guide job scheduling in grids. Braun *et al.* [3] selected a collection of 11 heuristics and adapted, implemented, and analyzed them under one set of common assumptions. The 11 heuristics examined are Opportunistic Load Balancing, Minimum Execution Time, Minimum Completion Time, Min-min, Max-min, Duplex, Genetic Algorithm, Simulated Annealing, Genetic Simulated Annealing, Tabu, and A*. It is shown that for the cases studied, the relatively simple Min-min heuristic performs well in comparison to the other techniques. Braun *et al.* [3] also proposed two types of mapping heuristics, immediate mode and batch mode heuristics. The immediate mode dynamic heuristics consider task affinity for different machines and machine ready times. The batch mode dynamic heuristics consider these factors, as well as aging of tasks waiting to execute. The simulation results revealed that the choice of which dynamic mapping heuristic to use in a given heterogeneous environment depends on parameters such as [3] (a) the structure of the heterogeneity among tasks and machines, and (b) the arrival rate of the tasks. Thus, in a real grid system, there must be several scheduling algorithms to be selected in different environments respectively. For large-scale grid platforms, global coordination by a centralized scheduler may be unrealistic. Beaumont *et al.* [17] presented decentralized schedulers that use only local information at each participating resource.

Buyya [18] identified challenges in managing resources in a grid computing environment and proposed computational economy as a metaphor for effective management of resources and application scheduling. The literature also identified distributed resource management challenges and requirements of economy-based grid systems, and discussed various representative economy-based systems, both historical and emerging, for cooperative and competitive trading of resources such as CPU cycles, storage, and network bandwidth.

Due to the development of new applications and the increasing number of users with diverse needs, providing users with quality of service (QoS) guarantees while executing applications has become a crucial problem that needs to be addressed [5]. This problem is referred to as the QoS-based scheduling problem and proved to be NP-hard. Dogana *et al.* [5] investigated the problem of scheduling a set of independent tasks with multiple QoS needs, which may include timeliness, reliability, security, data accuracy, and priority, in grids. And a computationally efficient static scheduling algorithm (QSMTS_IP) which assumes time-invariant penalty functions is developed.

In order to satisfy the QoS requirements of tasks, the status of the grid systems must be monitored and the performance data should be recorded, such as resource utilization (CPU utilization, memory utilization, disk utilization, etc.), network connectivity. However, in a large grid, the collection of grid performance data, the matching of grid sites and QoS requirements will consume a large amount of computation and communication overhead. This is not acceptable for a low-end grid. Thus, efficient monitoring and discovering technologies must be developed. However, this exceeds the scope of scheduling.

In a real life situation, asking the grid users to fully specify their QoS requirements such as security demand is an unreasonable burden. For example, job user only need to specify a security level such as low, middle, or high when submitting a remote job rather than the numerical value of the security conditions. Thus, how to evaluate the qualitative analysis and quantitative analysis is a key factor that impacts the matching of user jobs and grid sites heavily.

4 Fault Tolerance

The jobs scheduled to the grid sites are subject to failure easier than in a centrally controlled or locally controlled environment. The failure rate grows higher when the scale of the grid becomes larger. And the whole application will fail due to some key tasks or sites failure. This is not acceptable for some small granularity, large scale and long running grid applications. Consequently, the mapping of tasks onto grid sites must be adaptive and fault-tolerant because of the dynamicity of it.

Hwang *et al.* [19] presented a failure detection service (FDS) and a flexible failure handling framework (Grid-WFS) as a fault tolerance mechanism on the grid. The FDS enables the detection of both task crashes and user-defined exceptions. A notification mechanism is proposed which is based on the interpretation of notification messages being delivered from the underlying grid resources. The paper also described how to achieve the flexibility by the use of workflow structure as a high-level recovery policy specification, which enables support for multiple failure recovery techniques, the separation of failure handling strategies from the application code, and user-defined exception handlings.

A resource monitoring center is always used to provide performance information. A central server collects performance information from all other sites periodically and monitors the status of jobs submitted or being run. Every remote host running a job will communicate the execution status of the job replications to the monitoring center periodically. However, the centralized monitoring can't scale well when the number of grid sites increases. One solution to this problem is to provide monitoring knowledge that correlates alarms and provides fewer, but more meaningful advisory information. Moreover, such a solution could reduce operator load further by taking automatic corrective action [20].

On failure there are a number of different possibilities: the entire queue may be lost, the job in service may be lost, or the entire queue may be retained [21]. In the case where the entire contents of the queue are lost on breakdown, the jobs will be lost from the system if they continue to be sent to a broken node. However, most existing models assume that the presence of breakdowns is immediately known by any router or scheduler directing jobs to a queue. Thus, communication latency must be tolerated in scheduling.

Job replications [6, 10] are commonly used to provide fault-tolerant scheduling in grids. However, existing job replication algorithms use a fixed-number replication. Abawajy [10] presented a Distributed Fault-Tolerant Scheduling (DFTS) to provide fault tolerance for job execution in a grid environment. Their algorithm uses fixed number replications of jobs at multiple sites to guarantee successful job executions. Fixed-number replications in scheduling strategies may utilize excessive hosts or resources. This makes the makespan and average waiting time of tasks rather longer. Thus an adaptive replication strategy is necessary in a real grid with dynamic security level.

5 Security Assurance

In a large-scale grid, job executions are usually carried out between many virtual organizations in business applications or scientific applications for faster execution or remote interaction. However, security is a main hurdle to make the job scheduling secure, reliable and fault-tolerant. If a host in the grid is under attack or malicious usage, its resources may not be accessible from remote sites. Thus, the jobs scheduled for that host may be delayed or failed because of the system infections or crashes. Unfortunately, most of the existing proposed scheduling algorithms had ignored the security problem while scheduling jobs onto geographically distributed grid sites, with a handful of exceptions. In a real life scenario, security threats always exist and the jobs are subject to failures or delays caused by infected hardware, software vulnerability, and distrusted security policy [6]. Consequently, the assumption that the grid environments are safe and the resources are 100% reliable is no longer applicable for job scheduling in real grids.

Arenas [22] presented an overview of the different concepts and technologies relevant to trust and security in grid systems and analyzed the relation between trust and security, described trust and security challenges in the grid, and introduced the existing mechanisms for managing trust and security.

Song *et al.* [6] developed three risk-resilient strategies, preemptive, replication, and delay-tolerant to provide security assurance. In addition to extending from known scheduling heuristics, they developed a new space-time genetic algorithm (STGA) based on faster searching and protected chromosome formation. Song *et al.* [23, 24] developed a security-binding scheme through site reputation assessment and trust integration across grid sites. They applied fuzzy theory to handle the fuzziness or uncertainties behind all trust attributes. The binding is achieved by periodic exchange of site security information and matchmaking to satisfy user job demands.

The USC GridSec [25] project develops distributed security infrastructure and self-defense capabilities to secure wide-area networked resource sites participating in a grid application. It proposes a grid security infrastructure including trust modeling, security-binding methodology, and defense architecture against intrusions, worms, and flooding attacks. Power *et al* [26] presented an approach to the facilitation of system-wide security that enables fine-grained access control within systems in which third party web services are deployed. In that paper, the OGSA-DAI grid services were secured via XACML-based access control policies.

Matching of security demand and trust level is commonly used to provide security assurance when scheduling jobs in a risky or failure-prone grid environment. However, it is difficult to describe or specify the security demand of a job. In other words, it is difficult to construct a reasonable, usable, and efficient trust model (or reputation model) for job scheduling. A too simple trust model is not reliable enough and a too complicated trust model will consume excessive computation and communication.

6 Simulation of Scheduling Strategies

Simulation has been used extensively as a way to evaluate and compare scheduling strategies because simulation experiments are configurable, repeatable, and generally fast [8]. Legrand [8] outlines that there are two main limitations to the simulation methodology used for scheduling research: (1) there is no simulation standard in the scheduling research community; and (2) traditional models and assumptions about computing platforms are no longer valid for modern platforms because the simplistic network models used in most scheduling literature do not hold for modern computing platforms. Consequently, there is a need for a simulation framework designed for conducting research on distributed application scheduling. And Legrand [8] outlined the objectives that a useful framework must have: (1) good usability; (2) possibility to run fast simulations;(3) possibility to build configurable, tunable, and extensible simulations; and (4) scalability and the sustaining of simulations with tens of thousands of resources and application tasks.

When simulation is conducted for a grid job scheduling, the corresponding modeling must be constructed, such as grid instance model, network model, hosts model and task model. Zanikolas *et al.*. [27] described a grid instance along the following parameters: (1) the number of grid sites; (2) the number of hosts that are shared via the grid; (3) the distribution of Internet connection capacity of grid sites; (4) the distribution of hosts among grid sites; (5) the mapping of the set of Internet connection capacities to that of grid sites; (6) the distribution of host types, such as PCs, Clusters, SCs, and other special-purpose instruments; and (7) the distribution and characteristics of resource types within hosts.

Networks are immensely complex due to diversity at all levels: end hosts with various implementations of a TCP/IP protocol stack, communication via diverse network devices over different mediums of various characteristics [8]. Moreover, the distribution of Internet connection capacity of grid sites and that of hosts among sites is highly skewed. In traditional grids, most grid sites are well-connected and have a considerable number of hosts; thus, a uniform distribution seems appropriate. Actually, in a real grid, there are a considerably smaller number of high performance sites.

Bolosky *et al.* [28] constructed an empirical model for downtime intervals of hosts in a corporate environment. The proposed model is a mixture of two uniform distributions for 14- and 64-hour downtime intervals, and a gamma distribution for the hosts that do not have a cyclical availability behavior.

In general, almost all algorithms assume that jobs arrive in the queue in a Poisson stream and are served in FIFO (First In First Out) order with service times negative-exponentially distributed. In a real grid, jobs may have QoS requirements, such as timeliness, priority, and security demands [6].Thus, these QoS requirements must be quantified based on some tasks model.

Through the above we can see that the modeling of hosts and jobs (or tasks) is not accurate or precise. And the proposed model can only be used for qualitative analysis, not for quantitative computation. For example, in a large dispersed grid, site A may need massive data on site B to compute and the computational results may be transferred to site C for visualization. The modeling of network connectivity is critical for efficient computation and remote interaction.

7 Conclusions and Future Work

In the past few years, grids have emerged as an important methodology to utilize dispersed resources over the Internet or local area network. Grids have shown global popularity and harnessed the aggregate computing power to hundreds of TFlops. Thus, scalability must be taken into account when scheduling jobs in a large-scale grid.

Job dependency is also a key factor that must be considered for job scheduling. In a large distributed environment, some jobs must be executed sequentially. For example, in a real lift scenario, job X must be executed before the execution of job Y. How to quantitate the QoS requirements of jobs is a main challenge for job scheduling. The combination of heuristics, security, and QoS specification may be a solution. Summarily, the performance of job scheduling could be improved if the modeling of jobs, network connectivity, and hosts (or grid sites), gets more accurate and precise.

Acknowledgments. The funding support of this work by Innovation Fund of Huazhong University of Science and Technology (HUST) under contract No. HF04012006271 is appreciated.

References

1. Foster, I., Kesselman, C., Tuecke, S.: The anatomy of the grid: Enabling scalable virtual organizations. International Journal of High Performance Computing Applications,**15**(3) (2001)200-222
2. Ibarra, O.H., Kim, C.E.: Heuristic algorithms for scheduling independent tasks on nonidentical processors. Journal of Association of Computing Machine, **24**(2) (1977)280-289
3. Braun, T.D., Siegel, H.J., Beck, N., Boloni, L.L., Maheswaran, M., Reuther, A.I., Robertson, J.P., Theys, M.D., Yao, B.: A Comparison of Eleven Static Heuristics for Mapping a Class of Independent Tasks onto Heterogeneous Distributed Computing Systems. Journal of Parallel and Distributed Computing, **61**(6) (2001)810-837
4. Kwok, Y.K., Maciejewski, A.A., Siegel, H.J., Ahmad, I., Ghafoor, A.: A semi-static approach to mapping dynamic iterative tasks onto heterogeneous computing system. Journal of Parallel and Distributed Computing, **66**(2006)77-98
5. Dogana, A., Özgüner, F.: Scheduling of a meta-task with QoS requirements in heterogeneous computing systems. Journal of Parallel and Distributed Computing, **66**(2) (2006)181–196
6. Song, S., Hwang, K., Kwok, Y.K.: Risk-Resilient Heuristics and Genetic Algorithms for Security-Assured Grid Job scheduling. IEEE Transactions on Computers, **55**(6) (2006)703-719
7. Hamscher, V., Schwiegelshohn, U., Streit, A., Yahyapour, R.: Evaluation of job-scheduling strategies for grid computing. In: Proceedings of 1st IEEE/ACM International Workshop on Grid Computing (GRID 2000). LNCS Vol.1971 (2000)191-202
8. Legrand, A., Quinson, M., Casanova, H., Fujiwara, K.: The SIMGRID project simulation and deployment of distributed applications. In: Proceedings of 15th IEEE International Symposium on High Performance Distributed Computing (HPDC06) (2006)385-386
9. Fujimoto,N., Hagihara, K.: Near-Optimal Dynamic Task Scheduling of Independent Coarse-Grained Tasks onto a Computational Grid, In: Proceedings of International Conference on Parallel Processing(ICPP2003)(2003)391-398
10. Abawajy, J.H.: Fault-Tolerant Dynamic Job Scheduling Policy. In: M. Hobbs, A. Goscinski, and W. Zhou. (Eds.): Proceedings of ICA3PP'05, LNCS Vol.3719 (2005)165–173
11. Kaya, K., Aykanat, C.: Iterative-improvement-based heuristics for adaptive scheduling of tasks sharing files on heterogeneous master–slave platforms. IEEE Transactions on Parallel and Distributed Systems, **17**(8) (2006)883–896
12. Shivle, S., Siegel, H.J., *et al.* Static allocation of resources to communicating subtasks in a heterogeneous ad hoc grid environment, Journal of Parallel and Distributed Computing,**66** (2006) 600–611

13. Wolski, R.: Forecasting network performance to support dynamic scheduling using the Network Weather Service. In: Proceedings of the 6th IEEE International Symposium on High Performance Distributed Computing (HPDC97) (1997)316-325
14. Schopf, J.M., D'Arcy, M., Miller, N., *et al.*: Monitoring and Discovery in a Web Services Framework: Functionality and Performance of the Globus Toolkit's MDS4, Available at http://www-unix.mcs.anl.gov/~schopf/Pubs/mds-sc05.pdf
15. Thomas, N.: Modeling job allocation where service duration is unknown. In: Proceedings of 20th IEEE International Parallel and Distributed Processing Symposium (IPDPS'06), 2006
16. Ernemann, C., Hamscher, V., Schwiegelshohn, U., Yahyapour, R., Streit, A.: On advantages of grid computing for parallel job scheduling. In: Proceedings of 2nd IEEE/ACM International Symposium on Cluster Computing and the Grid (CCGRID 2002) (2002)39-46
17. Beaumont, O., Carter, L., Ferrante, J., *et al.*: Centralized versus distributed schedulers for multiple bag-of-task applications. In: Proceedings of 20th IEEE International Parallel and Distributed Processing Symposium (IPDPS'06), 2006
18. Buyya R., Abramson D., Venugopal S.: The Grid Economy. Proceedings of the IEEE, **93**(3) (2005)698-714
19. S. Hwang, C. Kesselman.: A Flexible Framework for Fault Tolerance in the Grid, Journal of Grid Computing, **1**(3) (2003)251-272
20. Jones, P.L., Harrison, A.: The application of knowledge-based techniques to the monitoring of computers in a large heterogeneous distributed environment. Knowledge-Based Systems, **19**(7) (2006)565-575
21. Thomas, N., Bradley, J.T., Knottenbelt, W.J.: Stochastic analysis of scheduling strategies in a Grid-based resource model. IEE Proceedings: Software, **151**(5) (2004)232-239
22. Arenas, A.: State of the art survey on trust and security in Grid computing systems. Technical Report (RAL-TR-2006-008), 2006, CCLRC (ISSN 1358-6254)
23. Song, S., Hwang, K.: Trusted grid computing with security assurance and resource optimization. In: Proceedings of ISCA 17th International Conference on Parallel and Distributed Computing Systems (ISCA PDCS04) (2004)110-117
24. Song, S., Hwang, K., Kwok, Y.K.: Trusted grid computing with security binding and trust integration. Journal of Grid Computing, **3**(1) (2005)53-73
25. Hwang, K., Kwok, Y.K., Song, S., *et al.*: GridSec: Trusted grid computing with security binding and self-defense against network worms and DDoS attacks. In: Proceedings of 5th International Conference on Computational Science, LNCS Vol.3516 (2005)187-195
26. Power, D.J., Politou, E.A., Slaymaker, M.A., *et al.*: Securing web services for deployment in health grids. Future Generation Computer Systems, **22**(2006)547–570
27. Zanikolas, S., Sakellariou, R.: Application-Level Simulation Modeling of Large Grids. In: Proceedings of 5th International Symposium on Cluster Computing and Grid (CCGrid'05) (2005)582-589
28. W. J Bolosky, J.R. Douceur, D. Ely, *et al.*: Feasibility of a Serverless Distributed File System Deployed on an Existing Set of Desktop PCs. In: 2000 ACM SIGMETRICS, International Conference on Measurement and Modeling of Computer Systems, **28**(1) (2000)34–43

Relational Nested Optional Join for Efficient Semantic Web Query Processing

Artem Chebotko, Mustafa Atay, Shiyong Lu, and Farshad Fotouhi

Department of Computer Science
Wayne State University
5143 Cass Avenue, Detroit, Michigan 48202, USA
{artem,matay,shiyong,fotouhi}@cs.wayne.edu

Abstract. Increasing amount of RDF data on the Web drives the need for its efficient and effective management. In this light, numerous researchers have proposed to use RDBMSs to store and query RDF annotations using the SQL and SPARQL query languages. The first few attempts at SPARQL-to-SQL translation revealed non-trivial challenges related to correctness and efficiency of such translation in the presence of nested optional graph patterns. In this paper, we propose to extend relational databases with a novel relational operator, *nested optional join (NOJ)*, that is more efficient than left outer join in processing nested optional graph patterns. We design three efficient algorithms to implement the new operator in relational databases: (1) nested-loops NOJ algorithm, NL-NOJ, (2) sort-merge NOJ algorithm, SM-NOJ, and (3) simple hash NOJ algorithm, SH-NOJ. Based on a real life RDF dataset, we demonstrate the efficiency of our algorithms by comparing them with the corresponding left outer join implementations.

Keywords: Nested Optional Join, Relational Join, Relational Operator, RDBMS, SPARQL, RDF, Semantic Web, Query Processing.

1 Introduction

The Semantic Web [6] has recently gained tremendous momentum due to its great potential for providing a common framework that allows data to be shared and reused across application, enterprise, and community boundaries. Semantic data is represented in Resource Description Framework (RDF) [1], the standard language for annotating resources on the Web, and queried using the SPARQL [3] query language for RDF that has been recently proposed by the World Wide Web Consortium. RDF data is a collection of statements, called *triples*, of the form `<s,p,o>`, where `s` is a subject, `p` is a predicate and `o` is an object, and each triple states the relation between the subject and the object. Such collection of triples can be represented as a directed graph, in which nodes represent subjects and objects, and edges represent predicates connecting from subject nodes to object nodes. SPARQL allows the specification of triple and graph patterns to be matched over RDF graphs.

G. Dong et al. (Eds.): APWeb/WAIM 2007, LNCS 4505, pp. 428–439, 2007.

Increasing amount of RDF data on the Web drives the need for its efficient and effective management. In this light, numerous researchers [12,9,20,19,14,22,7,21,18] have proposed to use RDBMSs to store and query RDF data using the SQL and SPARQL query languages. One of the most challenging problems in such an approach is the translation of SPARQL queries into relational algebra and SQL. The first few attempts [9,12] at the SPARQL-to-SQL translation, although successful, revealed serious difficulties related to correctness and efficiency of such translation in the presence of nested optional graph patterns. The challenges of the SPARQL query processing in the presence of nested `OPTIONAL` patterns include:

- *Basic semantics of `OPTIONAL` patterns.* The evaluation of an `OPTIONAL` clause is not obligated to succeed, and in the case of failure, no value will be returned for those unbound variables in the `SELECT` clause.
- *Semantics of shared variables in `OPTIONAL` patterns.* In general, shared variables must be bound to the same values. Variables can be shared among subjects, predicates, objects, and across each other.
- *Semantics of nested `OPTIONAL` patterns.* Before a nested `OPTIONAL` clause is evaluated, all containing `OPTIONAL` clauses must have succeeded.

In existing SPARQL-to-SQL translation work [9,12,14,20], the handling of these three semantics in a relational database relies on the use of the left outer join (LOJ) defined in the relational algebra and SQL: (1) basic semantics of `OPTIONAL` patterns is captured by LOJ; (2) semantics of shared variables is treated with the conjunction of equalities of corresponding relational attributes in the LOJ condition; (3) semantics of nested `OPTIONAL` patterns is preserved by the `NOT NULL` check in the LOJ condition for one of the attributes/variables that correspond to the parent of a nested `OPTIONAL` clause. In the following, we present our running example to illustrate the translation of a SPARQL query with a nested `OPTIONAL` into a relational algebra expression, in which LOJ is used for implementing nested optional graph patterns; the example motivates the introduction of a new relational operator for a more efficient implementation.

Example 1. (Sample SPARQL query and its relational equivalent)
Consider the RDF graph presented in Figure 1(a). The graph describes academic relations among professors and graduate students in a university. The RDF schema defines two concepts/classes (*Professor* and *GradStudent*) and two relations/properties (*hasAdvisor* and *hasCoadvisor*). Each relation has the *GradStudent* class as a domain and the *Professor* class as a range. Additionally, two instances of *Professor*, two instances of *GradStudent* and relations among these instances are defined as shown in the figure.

We design a SPARQL query that returns (1) every graduate student in the RDF graph; (2) the student's advisor if this information is available; and (3) the student's coadvisor if this information is available and if the student's advisor has been successfully retrieved in the previous step. In other words, the query returns students and as many advisors as possible; there is no point to return a coadvisor if there is even no advisor for a student. The SPARQL representation of the query is as follows.

```
SELECT ?stu ?adv ?coadv
WHERE {
  ?stu rdf:type :GradStudent .     /* R1(stu) */
  OPTIONAL {
    ?stu :hasAdvisor ?adv .        /* R2(stu,adv) */
    OPTIONAL {
      ?stu :hasCoadvisor ?coadv ./* R3(stu,coadv) */
} } }
```

The query has three variables: `?stu` for the student, `?adv` for the advisor, and `?coadv` for the coadvisor. There are two `OPTIONAL` clauses, where the innermost one is the nested `OPTIONAL` clause.

Based on our translation strategy in [9], we translate the SPARQL query into a relational query as follows. Matching triples for the triple patterns `?stu rdf:type :GradStudent`, `?stu :hasAdvisor ?adv`, and `?stu :hasCoadvisor ?coadv` are retrieved into relations R_1, R_2, and R_3, respectively. Note that the triple patterns are annotated with the corresponding relations and relational schemas in the SPARQL query above. Then the equivalent relational algebra representation is

$$R_4 = \Pi_{R_1.stu, R_2.adv}(R_1 ⟕_{R_1.stu=R_2.stu} R_2),$$
$$R_{res} = \Pi_{R_4.stu, R_4.adv, R_3.coadv}(R_4 ⟕_{R_4.stu=R_3.stu \wedge R_4.adv \text{ IS NOT NULL}} R_3).$$

Each `OPTIONAL` clause corresponds to the left outer join, the shared variable `?stu` participates in the join conditions, and the nested `OPTIONAL` implements the `NOT NULL` check on the *adv* attribute to ensure that its parent clause has indeed succeeded. The graphical representation of the relational query is shown in Figure 1(b); the projection operators are not shown for ease of presentation.

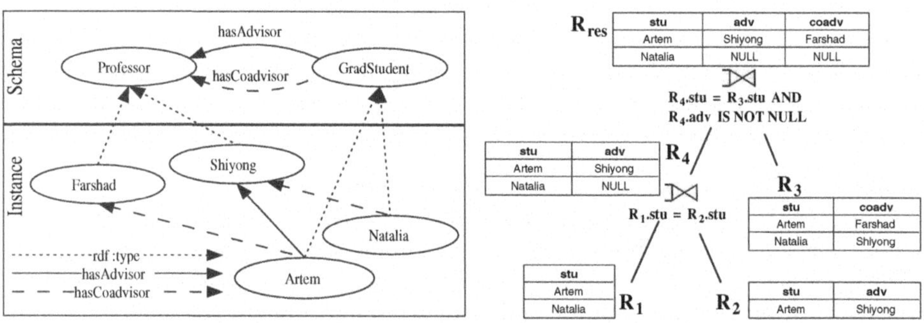

(a) Sample RDF graph (b) Relational query with LOJs

Fig. 1. Sample RDF graph and relational query over the graph

The running example motivates our research. The following is our insight to how the LOJ based query in Figure 1(b) wastes some computations: (1) Based on the result of the first LOJ and the semantics of the nested `OPTIONAL` pattern, we know that the `NULL` padded tuple (*Natalia*, `NULL`) will also be `NULL` padded in the second LOJ. After all, there is no need for this tuple to participate in the second LOJ condition. (2) On the other hand, we know that the successful match in the tuple (*Artem*, *Shiyong*) contains no NULLs. There is no need to apply the NOT NULL check to this tuple.

In this paper, we propose to extend relational databases with an innovative operator that mimics the nested optional pattern semantics of SPARQL to enable efficient processing of nested optional patterns in RDBMSs. The main contributions of our work include:

- We propose to extend relational databases with a novel relational operator, *nested optional join (NOJ)*, that is more efficient than left outer join in processing nested optional graph patterns. The computational advantage of NOJ over the currently used LOJ-based implementations comes from the two superior characteristics of NOJ: (1) NOJ allows the processing of the tuples that are guaranteed to be NULL padded very efficiently (in linear time) and (2) NOJ does not require the NOT NULL check to return correct results. In addition, (3) NOJ significantly simplifies the SPARQL-to-SQL translation.
- We design three efficient algorithms to implement the new operator in relational databases: (1) nested-loops NOJ algorithm, NL-NOJ, (2) sort-merge NOJ algorithm, SM-NOJ, and (3) simple hash NOJ algorithm, SH-NOJ.
- Based on a real life RDF dataset, we demonstrate the efficiency of our algorithms by comparing them with the corresponding left outer join implementations. The experimental results are very promising; NOJ is a favorable alternative to the LOJ-based evaluation of nested optional patterns in the Semantic Web query processing with RDBMSs.

Organization. The rest of the paper is organized as follows. In Section 2, we present our NOJ operator and highlight its advantages. We design three algorithms to implement NOJ in relational databases in Section 3 and report the results of our performance study in Section 4. In Section 5, we discuss related work. Finally, we provide our conclusions and future work in Section 6.

2 Nested Optional Join Operator

In this section, we present our nested optional join operator that is to be used to evaluate nested OPTIONAL patterns in relational databases and highlight its advantages over the left outer join.

The operands of NOJ are *twin relations* instead of conventional relations. The notion of twin relation is introduced as follows.

Definition 1 ***(Twin Relation).*** *A twin relation, denoted as* (R_b,R_o)*, is a pair of conventional relations with identical relational schemas and disjoint sets of tuples. The schema of a twin relation is denoted as* $\xi(R_b,R_o)$*.* R_b *with the schema* $\xi(R_b,R_o)$ *is called a* base relation *and* R_o *with the schema* $\xi(R_b,R_o)$ *is called an* optional relation*. A distinguished tuple* $n_{\xi(R_b,R_o)}$ *is defined as a tuple of* $\xi(R_b,R_o)$ *in which each attribute takes a NULL value.*

Intuitively, a base relation is used to store tuples that have a potential to satisfy a join condition of a nested optional join. An optional relation is used to store tuples that are guaranteed to fail a join condition of a nested optional join. We incorporate the twin relation into the relational algebra by introducing the following additional operators, $\uplus$ and $\ddagger$, such that

- $\uplus(R_b, R_o) = R_b \cup R_o$, and
- $\ddagger(R) = (R, \phi)$, where ϕ is an instance of empty relation with the same relational schema of R.

Note that $\ddagger$ is not a reversed operator of $\uplus$, because $\ddagger(\uplus(R_b, R_o)) \neq (R_b, R_o)$ in general.

We also extend the *projection* and *selection* operators to a twin relation as $\pi[(R_b, R_o)] = (\pi[R_b], \pi[R_o])$ and $\sigma[(R_b, R_o)] = (\sigma[R_b], \sigma[R_o])$, respectively. The definition of a complete algebra for a twin relation is not our focus in this paper; π and σ are sufficient for our running example and experimental study and, as we believe, for most SPARQL-to-SQL translations.

In the following, we define a novel relational operator, *nested optional join*, using the tuple calculus.

Definition 2 ***(Nested Optional Join).*** *A nested optional join of two twin relations, denoted as* $\exists\!\bowtie$, *yields a twin relation, such that* $(R_b, R_o) \exists\!\bowtie_{r(a)=s(b)} (S_b, S_o) = (Q_b, Q_o)$, *where* $Q_b = \{t | t = rs \wedge r \in R_b \wedge (s \in S_b \vee s \in S_o) \wedge r(a) = s(b)\}$ *and* $Q_o = \{t | t = rn \wedge (r \in R_o \vee (r \in R_b \wedge \neg\exists s[(s \in S_b \vee s \in S_o) \wedge r(a) = s(b)]))\}$, *where* $r(a) = s(b)$ *is a join predicate,* $r(a) \subseteq \xi(R_b, R_o)$ *and* $s(b) \subseteq \xi(S_b, S_o)$ *are join attributes,* $n = n_{\xi(S_b, S_o)}$.

In other words, the result base relation Q_b contains tuples t made up of two parts, r and s, where r must be a tuple in relation R_b and s must be a tuple in S_b or S_o. In each tuple t, the values of the join attributes $t(a)$, belonging to r, are identical in all respects to the values of join attributes $t(b)$, belonging to s. The result optional relation Q_o contains tuples t made up of two parts, r and n, where r must be a tuple in R_o with no other conditions enforced, or r must be a tuple in R_b and there must not exist a tuple s in S_b or S_o that can be combined with r based on the predicate $r(a) = s(b)$.

The graphical illustration of the NOJ operator is shown in Figure 2. Note how well it emphasizes one of the advantages of NOJ: the flow of tuples from R_o to Q_o bypasses the join condition and does not interact with tuples from any other relation. Obviously, the behavior of this flow can be implemented to have linear time performance in the worst case – the property that is, in general, not available in the LOJ implementations. The second important advantage of NOJ – no need for the `NOT NULL` check – is discussed in the following example that describes the translation of our sample SPARQL query into a relational algebra expression with our extensions.

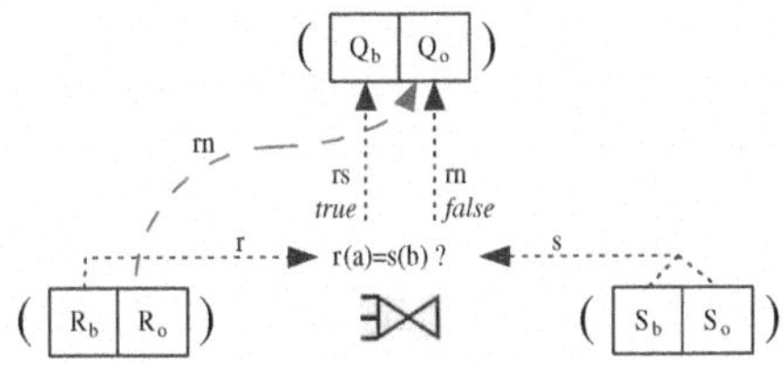

Fig. 2. Nested optional join

Example 2 (Evaluation of the sample SPARQL query using NOJs)
We use the same RDF graph as presented in Figure 1(a) and the SPARQL query described in Example 1. The translation strategy is similar to the one illustrated in Example 1 except that we use NOJ instead of LOJ. Matching triples for the triple patterns `?stu rdf:type :GradStudent`, `?stu :hasAdvisor ?adv`, and `?stu :hasCoadvisor ?coadv` are retrieved into relations R_1, R_2, and R_3, respectively. Then the equivalent relational algebra representation using NOJ is

$$(R_b^1, R_o^1) = \ddagger(R_1), (R_b^2, R_o^2) = \ddagger(R_2), (R_b^3, R_o^3) = \ddagger(R_3),$$
$$(R_b^4, R_o^4) = \Pi_{(R_b^1,R_o^1).stu,(R_b^2,R_o^2).adv}((R_b^1, R_o^1) \exists\!\!\bowtie_{(R_b^1,R_o^1).stu=(R_b^2,R_o^2).stu} (R_b^2, R_o^2)),$$
$$(R_b^{res}, R_o^{res}) = \Pi_{(R_b^4,R_o^4).stu,(R_b^4,R_o^4).adv,(R_b^3,R_o^3).coadv}((R_b^4, R_o^4) \exists\!\!\bowtie_{(R_b^4,R_o^4).stu=(R_b^3,R_o^3).stu} (R_b^3, R_o^3)),$$
$$R_{res} = \uplus(R_b^{res}, R_o^{res}).$$

The graphical representation of the relational query is shown in Figure 3; the conversion and projection operators are not shown for ease of presentation. Note that this query does not contain the `NOT NULL` check, because all the tuples that have not succeeded in the first join are padded with `NULL` values and stored into the optional relation R_o^4; the tuples of R_o^4 bypass the second join condition and are copied directly to R_o^{res} with additional `NULL`-padding.

Therefore, NOJ is superior to LOJ when we apply them to translate SPARQL nested `OPTIONAL` clauses to relational queries. The main advantages of NOJ are

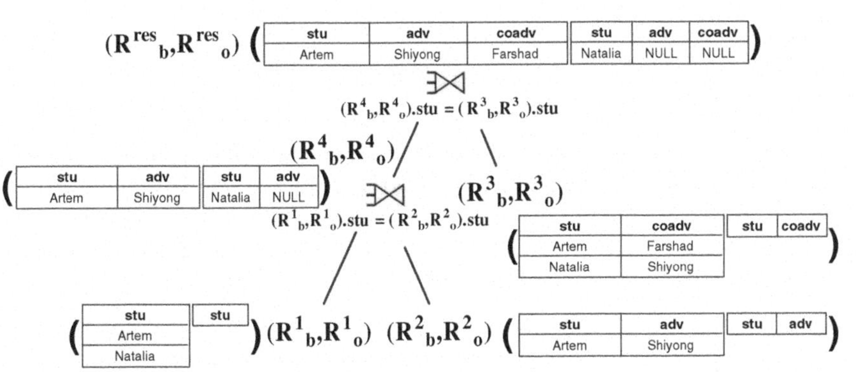

Fig. 3. Nested optional join based evaluation of the SPARQL query in Example 1

(1) NOJ allows the processing of the tuples that are guaranteed to be `NULL` padded very efficiently (in linear time), (2) NOJ does not require the `NOT NULL` check to return correct results, and (3) NOJ significantly simplifies the SPARQL-to-SQL translation, eliminating the need to choose a relational attribute for the `NOT NULL` check[1] and, in some cases (see [9]) when such an attribute can not be chosen from available ones, the need to introduce a new variable and even a new triple pattern into a SPARQL query. Our performance study showed that these advantages bring substantial speedup to the query evaluation.

3 Nested Optional Join Algorithms

Previously, we defined NOJ through the tuple calculus, but it is also possible to express the NOJ result relations using standard operators of the relational algebra: $Q_b = R_b \bowtie_{r(a)=s(b)} (S_b \cup S_o)$ and $Q_o = R_o \cup [(R_b \text{⟕}_{r(a)=s(b)} (S_b \cup S_o)) - (R_b \bowtie_{r(a)=s(b)} (S_b \cup S_o))]$. However, it should be evident that this direct translation will be inefficient if implemented. Therefore, in this section, we design our own algorithms to implement NOJ in a relational database. Our algorithms, NL-NOJ, SM-NOJ, and SH-NOJ, employ the classic methods used to implement relational joins: nested-loops, sort-merge, and hash-based join methods, respectively.

3.1 Nested-Loops Nested Optional Join Algorithm

The simplest algorithm to perform the NOJ operation is the nested-loops NOJ algorithm, denoted as NL-NOJ. The algorithm (see Figure 4) is self-descriptive, and thus we only clarify some important details. Note that for efficiency, the inner loop in line 07 should iterate over the tuples of a (twin) relation with higher cardinality. This remark is only valid when I/O operations are involved, and can be ignored for in-memory join processing. In the figure, we assume that (S_b, S_o) has more tuples than R_b. Also, note that the tuples of R_o are processed in linear time in lines 17-19.

The results of our complexity and applicability analysis are

- *NL-NOJ complexity*: $\Theta(|R_b| \times (|S_b| + |S_o|) + |R_o|)$.
- *NL-NOJ applicability*: NOJs with high selectivity factors (see our performance study for more details).

The comprehensive analysis of the performance and applicability of the nested-loops join method is presented in [17]. In the join processing literature, there is a number of optimizations on the nested-loops join method that are also applicable to NL-NOJ: e.g., the block nested-loops join method [15,13] and "rocking" the inner relation optimization [16] that reduce the number of I/O operations.

[1] Note that an attribute that serves as an indicator whether the parent `OPTIONAL` clause has succeeded should be carefully chosen as discussed in [9]. In a nutshell, such an attribute may not be bound in any clause that precedes the parent `OPTIONAL`, otherwise the `NOT NULL` check may succeed even if the parent `OPTIONAL` fails.

Algorithm: NL-NOJ
Input: twin relations (R_b, R_o) and (S_b, S_o)
Output: twin relation $(Q_b, Q_o) = (R_b, R_o) \sqsupset\!\!\bowtie_{r(a)=s(b)} (S_b, S_o)$
Begin
For each tuple $r \in R_b$ **do**
pad = *true*
For each tuple $s \in (S_b, S_o)$ **do**
If $r(a) = s(b)$ **then**
place tuple rs in relation Q_b
pad = *false*
End If
End For
If *pad* **then**
place tuple $rn_{\xi(S_b,S_o)}$ in relation Q_o
End If
End For
For each tuple $r \in R_o$ **do**
place tuple $rn_{\xi(S_b,S_o)}$ in relation Q_o
End For
Return (Q_b, Q_o)
End Algorithm

Fig. 4. Algorithm NL-NOJ

3.2 Sort-Merge and Simple Hash Nested Optional Join Algorithms

Due to the space limitation, we omit (see [8] for details) the description of the sort-merge NOJ algorithm, SM-NOJ, and the simple hash NOJ algorithm, SH-NOJ.

The results of our complexity and applicability analysis are

- *SM-NOJ complexity*: $\Omega(|R_b| \times log|R_b| + (|S_b| + |S_o|) \times log(|S_b| + |S_o|) + |R_o|)$, $O(|R_b| \times (|S_b| + |S_o|) + |R_o|)$.
- *SM-NOJ applicability*: SM-NOJ is the best choice when NL-NOJ or SH-NOJ is not selected as the best performer; NOJs with median selectivity factors.
- *SH-NOJ complexity*: $\Omega(|R_b| + |R_o| + |S_b| + |S_o|)$, $O(|R_b| \times (|S_b| + |S_o|) + |R_o|)$, depends on the efficiency of a hash function h.
- *SH-NOJ applicability*: NOJs with low selectivity factors.

The comprehensive analysis of the performance and applicability of these join methods is presented in [17].

4 Performance Study

This section reports the performance experiments conducted using the NOJ algorithms and an in-memory relational database. The performance of the NOJ algorithms is compared with the performance of the corresponding LOJ-based implementations. In addition, the behavior of the NOJ algorithms with respect to the NOJ selectivity factor is explored and reported in [8].

4.1 Experimental Setup

We implemented in-memory representations of a relation and a twin relation, such that each relation was represented by a double-linked list of tuples, where each tuple corresponds to an array of pointers to attribute data values, and each twin relation was represented by pointers to two conventional relations. The memory to store relations and their tuples was allocated dynamically in the heap.

Our algorithms NL-NOJ, SM-NOJ and SH-NOJ were implemented in C++ using MS Visual C++ 6.0. To compare the performance of queries evaluated with our algorithms and corresponding left outer join algorithms, we implemented nested-loops LOJ (NL-LOJ), sort-merge LOJ (SM-LOJ), and simple hash LOJ (SH-LOJ) algorithms (see, e.g., [17]).

The experiments were conducted on the PC with one 2.4 GHz Pentium IV CPU and 1024 MB of main memory operated by MS Windows XP Professional.

The timings reported below are the mean result from five or more trials with warm caches.

4.2 Dataset and Queries

We conducted the experiments using the OWL representation of WordNet [5] (version: 1.2; author: Claudia Ciorascu), a lexical database for the English language.

We chose nine SPARQL [3] queries to evaluate in our experiments based on the following criteria: (1) most queries should have nested `OPTIONAL` clauses; (2) the input, intermediate, and output (twin) relations involved in the query evaluation should fit into the main memory; and (3) some queries should have common patterns to reveal performance changes with increasing complexity of the queries.

An important characteristic of the test queries is that they only involve joins, whose selectivity factors are less than 0.0002 and, for most joins, are less than 0.00002. Join selectivity factor (JSF) is a factor to represent the ratio of the cardinality of a join result to the cross product of the cardinalities of the two join (twin) relations. The reason why we chose queries with only joins with low selectivity factors is that the result of a join should fit into the main memory.

4.3 Experiments

Figure 5 shows the results of four experiments that measure query evaluation time for the NOJ and LOJ algorithms. Note that the NOJ algorithms outperformed the LOJ-based implementations for all queries except for Q1 and Q2, because Q1 and Q2 contained no nested `OPTIONAL`s and thus could not benefit from NOJ. Corresponding NOJ and LOJ based implementations showed equal performance for Q1 and Q2.

The description of the algorithm performance for individual test queries is available in [8].

In summary, NOJ, $(R_b,R_o)\equiv\!\bowtie(S_b,S_o)$, has the performance advantage over the left outer join counterpart when used to evaluate nested `OPTIONAL`s, because (1) R_o is always processed in linear time by a NOJ algorithm and (2) NOJ does not require the `NOT NULL` check. Our experiments on the real life dataset showed that this advantage is significant.

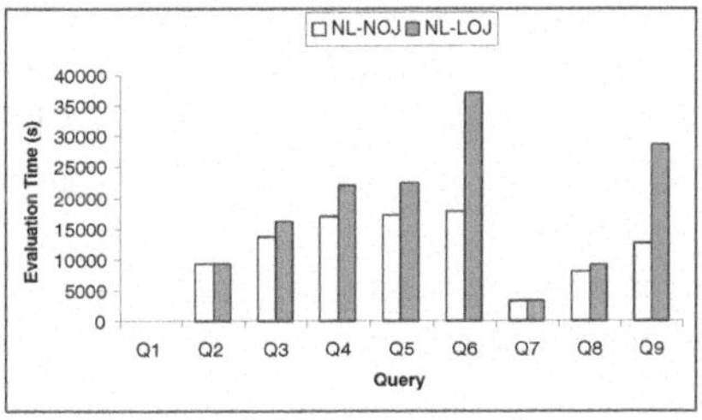

(a) Comparison of NL-NOJ and NL-LOJ

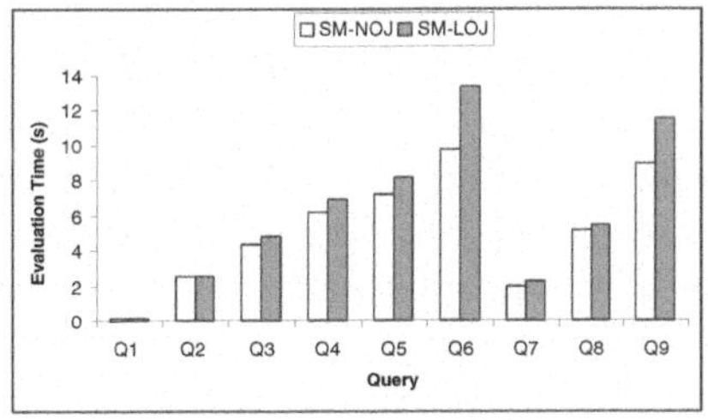

(b) Comparison of SM-NOJ and SM-LOJ

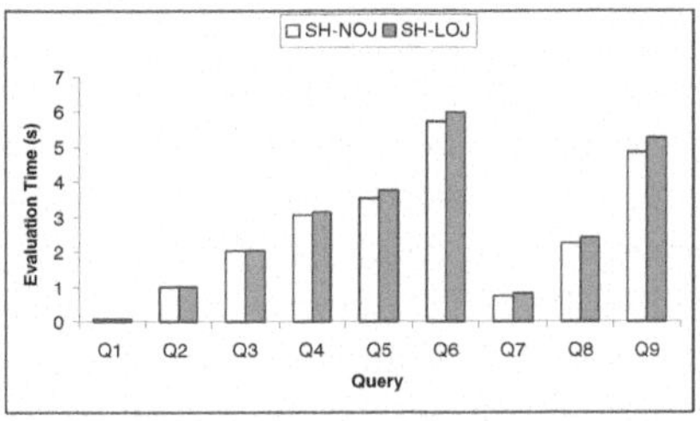

(c) Comparison of SH-NOJ and SH-LOJ

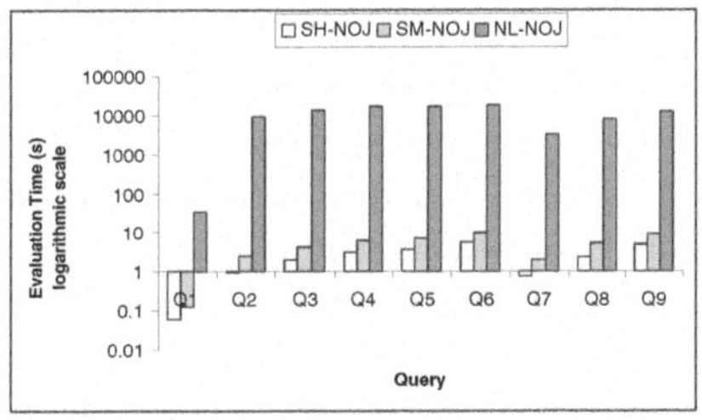

(d) Comparison of SH-NOJ, SM-NOJ and NL-NOJ

Fig. 5. Evaluation of the NOJ and LOJ algorithms using the WordNet test queries

5 Related Work

The join operation defined in the relational data model [10,11] is used to combine tuples from two or more relations based on a specified join condition. Several types of joins, such as theta-join, equi-join, natural join, semi-join, self-join, full outer join, left outer join, and right outer join, are studied in database courses [15,13] and implemented in RDBMSs. We introduce a new type of join, nested optional join, whose semantics mimics the semantics of optional graph patterns in SPARQL [3]. NOJ is defined on two twin relations, where each twin relation contains a base relation and an optional relation; therefore, NOJ can be viewed as a join of four conventional relations. The result of NOJ is also a twin relation, whose base relation stores tuples that have been concatenated and whose optional relation stores tuples that have been `NULL` padded. The above semantic and structural characteristics differentiate NOJ from any other join defined in the literature. We propose NOJ as a favorable alternative to the LOJ-based implementations for the nested optional graph pattern processing with relational databases. Note that NOJ is not a replacement of LOJ: their semantics are different, such as LOJ needs a special `NOT NULL` check to return similar results to the NOJ results and this check is not part of NOJ.

The join processing in relational databases has been an important research for over 30 years and the related literature is abundant [17]. To design algorithms for NOJ, we use three classical methods for implementing joins in RDBMSs:

nested-loops, sort-merge, and hash-based join methods [15,13]. These methods have numerous optimizations which are out of the scope of this paper.

In this paper, for the SPARQL-to-SQL translation, we used our SPARQL-toSQL algorithm presented in [9]; SPARQLtoSQL translates SPARQL queries with arbitrary complex optional patterns into SQL.

It is worthwhile to mention that SPARQL is not the only RDF query language that supports optional graph patterns. Other examples include SeRQL [4] and RDFQL [2], and NOJ is useful for these languages, too.

6 Conclusions and Future Work

We have proposed a novel relational operator, nested optional join, that enables efficient processing of Semantic Web queries with nested optional patterns. The computational advantage of NOJ over the currently used LOJ-based implementations comes from the two superior characteristics of NOJ: (1) NOJ allows the processing of the tuples that are guaranteed to be `NULL` padded very efficiently (in linear time) and (2) NOJ does not require the `NOT NULL` check to return correct results. In addition, (3) NOJ significantly simplifies the SPARQL-to-SQL translation. To facilitate the implementation of NOJ in relational databases, we have designed three efficient algorithms: (1) nested-loops NOJ algorithm, NL-NOJ, (2) sort-merge NOJ algorithm, SM-NOJ, and (3) simple hash NOJ algorithm, SH-NOJ. Based on the real life RDF dataset, we verified the efficiency of our algorithms by comparing them with the corresponding left outer join implementations. The experimental results are very promising; NOJ is a favorable alternative to the LOJ-based evaluation of nested optional patterns in the Semantic Web query processing with relational databases.

The future work includes the introduction of a parallel optional join operator for parallel `OPTIONAL`s in SPARQL and the definition of a relational algebra for SPARQL with these novel operators.

References

1. RDF Primer. W3C Recommendation, 10 February 2004. `http://www.w3.org/TR/rdf-primer/`.
2. RDFQL database command reference. `http://www.intellidimension.com/pages/rdfgateway/reference/db/default.rsp`.
3. SPARQL Query Language for RDF. W3C Working Draft, 4 October 2006. `http://www.w3.org/TR/2006/WD-rdf-sparql-query-20061004/`.
4. User guide for Sesame. Updated for Sesame release 1.2.3. `http://www.openrdf.org/doc/sesame/users/index.html`.
5. WordNet, a lexical database for the English language. `http://wordnet.princeton.edu/`.
6. T. Berners-Lee, J. Hendler, and O. Lassila. The Semantic Web. *Scientific American*, May 2001.
7. J. Broekstra, A. Kampman, and F. van Harmelen. Sesame: A generic architecture for storing and querying RDF and RDF Schema. In *ISWC*, 2002.

8. A. Chebotko, M. Atay, S. Lu, and F. Fotouhi. Extending relational databases with a nested optional join for efficient Semantic Web query processing. Technical Report TR-DB-112006-CALF. November 2006. http://www.cs.wayne.edu/~artem/main/research/TR-DB-112006-CALF.pdf.
9. A. Chebotko, S. Lu, H. M. Jamil, and F. Fotouhi. Semantics preserving SPARQL-to-SQL query translation for optional graph patterns. Technical Report TR-DB-052006-CLJF. May 2006. http://www.cs.wayne.edu/~artem/main/research/TR-DB-052006-CLJF.pdf.
10. E. F. Codd. A relational model of data for large shared data banks. *Commun. ACM*, 13(6):377–387, 1970.
11. E. F. Codd. Relational completeness of data base sublanguages. *In: R. Rustin (ed.): Database Systems: 65-98, Prentice Hall and IBM Research Report RJ 987, San Jose, California*, 1972.
12. R. Cyganiak. A relational algebra for SPARQL. Technical Report HPL-2005-170. 2005. http://www.hpl.hp.com/techreports/2005/HPL-2005-170.html.
13. R. Elmasri and S. B. Navathe. *Fundamentals of Database Systems*. Addison-Wesley, 2004.
14. S. Harris and N. Shadbolt. SPARQL query processing with conventional relational database systems. In *SSWS*, 2005.
15. M. Kifer, A. Bernstein, and P. M. Lewis. *Database Systems: An Application Oriented Approach*. Addison-Wesley, 2006.
16. W. Kim. A new way to compute the product and join of relations. In *SIGMOD*, pages 179–187, 1980.
17. P. Mishra and M. H. Eich. Join processing in relational databases. *ACM Computing Surveys*, 24(1):63–113, 1992.
18. Z. Pan and J. Heflin. DLDB: Extending relational databases to support Semantic Web queries. In *PSSS*, 2003.
19. E. Prud'hommeaux. *Notes on Adding SPARQL to MySQL*. http://www.w3.org/2005/05/22-SPARQL-MySQL/.
20. E. Prud'hommeaux. *Optimal RDF Access to Relational Databases*. http://www.w3.org/2004/04/30-RDF-RDB-access/.
21. R. Volz, D. Oberle, B. Motik, and S. Staab. KAON SERVER -A Semantic Web Management System. In *WWW, Alternate Tracks - Practice and Experience*, 2003.
22. K. Wilkinson, C. Sayers, H. Kuno, and D. Reynolds. Efficient RDF storage and retrieval in Jena2. In *SWDB*, 2003.

Efficient Processing of Relational Queries with Sum Constraints

Svetlozar Nestorov, Chuang Liu, and Ian Foster

Department of Computer Science, The University of Chicago,
1100 E 58th street, Chicago, IL 60637
{evtimov,chliu,foster}@cs.uchicago.edu

Abstract. In this paper, we consider relational queries involving one or more constraints over the sum of multiple attributes (sum constraint queries). We develop rewriting techniques to transform a sum constraint query in order to enable its efficient processing by conventional relational database engines. We also consider the problem of producing partial results for sum constraint queries. We propose a framework for ranking tuples in a relation according to their likelihood of contributing to a tuple in the result of a sum constraint query. Sorting tuples using this framework provides an alternative to traditional sorting based on single attribute value.

1 Introduction

As the amount of data stored and available electronically continues to grow rapidly, relational databases and their applications also proliferate. As a result, new types of relational queries emerge. In some instances, these new queries challenge traditional query processing and optimization, as illustrated by the following example.

Meal Example. *Consider a database storing nutritional information for single servings of different kinds of food in the following 4 relations: Meats, Vegetables, Fruits, and Beverages. All four relations have the same schema: name, cal, Vb6, Vc, fat, chol. Suppose that a meal consists of single servings of each of the four kinds of food. We are interested in finding meals that satisfy various nutritional requirements, such as restrictions on the number of calories, grams of saturated fat, and amount of Vitamin C. For example, the daily USDA recommendations for a 30-year old female, who is moderately active, are 1800-2200 calories, less than 18g of saturated fat and 300mg of cholesterol, and at least 4mg of Vitamin B6 and 76mg of Vitamin C. Assuming that a main meal carries about half of the daily nutritional values, we can find all such meals with the following SQL query:*

```
SELECT    M.name, V.name, B.name, F.name
FROM      Meats AS M, Vegetables AS V,
          Beverages AS B, Fruits AS F
WHERE     M.cal + V.cal + B.cal + F.cal > 900
AND       M.cal + V.cal + B.cal + F.cal < 1100
AND       M.Vb6 + V.Vb6 + B.Vb6 + F.Vb6 > 2
```

G. Dong et al. (Eds.): APWeb/WAIM 2007, LNCS 4505, pp. 440–451, 2007.

```
AND        M.Vc + V.Vc + B.Vc + F.Vc > 38
AND        M.fat + V.fat + B.fat + F.fat < 9
AND        M.chol + V.chol + B.chol + F.chol < 150
```

This type of query is commonly encountered in the context of document retrieval [1], multimedia data retrieval [2], geographic information system [3, 4], e-commerce [5], dynamic resource allocation on the grid [6], and supply chain management. For example, in the context of supply chain management, a product supplier may constrain the total cost according to a budget by carefully combining different service providers involved in producing, shipping and distributing products to construct a supply chain. Guha [7] shows that the decision-making process for supply chain management can be implemented as queries for a set of tuples with constraints over an aggregate of attribute values. In the context of Internet computing [8], applications need to find a set of computation resources with desired total memory size and CPU speed to run in order to get good performance and high efficiency. Assuming computation resources are stored in relations, Liu [9] modeled resource allocation as relational database queries seeking to identify tuples from relations with constraints over the sum of attribute values.

In this paper, we use the term *sum constraint* for a constraint over the sum of multiple attributes. A *sum constraint query* is a query containing multiple sum constraints in its query condition.

Conventional approaches implement this type of query by composing pair-wise joins. Because each condition refers to attributes from more than two relations, pair-wise join operators may fail to remove intermediate results based on these conditions, thus producing Cartesian products of relations that can lead to high memory and computation costs. Previous work improves the efficiency of query processing by either extending current database query processing engines (e.g. introducing new search algorithms [2] and new join operators [1, 9]), or by using sophisticated indexes [7]. In this paper we address the following question: Can we introduce efficient support for such query conditions into relational database systems without requiring significant modifications to the underlying database engine and building extra index structures? Note that our approach does not intend to replace previous methods. Instead, our method can be combined with previous methods to allow more efficient execution of sum constraint queries on relational database systems. In this paper, we make the following contributions:

We introduce query rewriting techniques for sum constraint queries. The techniques create new query conditions that can be used by join operators to remove intermediate results that do not lead to any results at the early stage of the execution. Our experimental results show that the rewriting achieves significant reductions in query response time.

We propose a framework for sorting tuples based on their likelihood of satisfying all sum constraints in a query. Using this framework, we consider the problem of producing partial results for a sum constraint query. We compare our method with traditional sorting algorithms that sort tuples based on only one attribute, and show that our method is more robust and efficient in handling sum constraint queries.

2 Related Work

The problem of processing sum constraint queries is closely related to the multi-join problem. A multi-join operation combines information from multiple relations to determine if an aggregate of tuples from those relations satisfies search conditions. Multi-joins are usually implemented by a set of pair-wise operations [10]. This method is efficient when the join condition consists of only equality or inequality comparisons of attributes from two join relations, as each individual join can eliminate tuples that do not satisfy its condition. Algorithms have been proposed for ordering pair-wise joins so as to obtain a minimal search space [11, 12]. (It is also possible to speed up queries by using multiple processors to parallelize the query processing [13, 14].) However, this pair-wise strategy does not work well for sum constraint queries. Since a sum constraint involves tuples from multiple relations, a pair-wise join operator cannot test the satisfiability of intermediate results based on a sum constraint until all attributes in the constraint are determined. Thus, a purely pair-wise strategy may generate many intermediate results that cannot lead to any solutions.

Guha et al. [7] and Agarwal et al. [15] address queries with sum constraints by building a sophisticated index. This index can be used to return results (approximate results in [7] or precise results in [15]) for sum constraint queries efficiently. Creating this index requires complex computations (e.g. solving a series of knapsack problems). Thus, these approaches may cause an extra load to maintain the index on dynamic data. Also, the focus of [7] and [15] is on finding tuples from the same relation with constraints involving two or three attributes. Our approach can be applied to sum constraint queries for tuples from many different relations.

Algorithms proposed by Fagin et al. [2] and Ilyas et al. [1] for top-k queries can be extended to implement queries with a constraint on the value of a monotone function, such as $A.attribute_1 + B.attribute_2 + C.attribute_3 > N$. They sort all join relations based on the value of attributes in the constraint, and then check combinations of tuples in the order of the value of the monotone function. However, this method can only use one constraint to guide the search process, which is not efficient when there are multiple constraints. In this paper, we introduce an approach using all constraints to guide the search process, and show that our method can be integrated with existing algorithms to further improve the search performance.

Searching for multiple tuples satisfying sum constraints can also be considered as a combinatorial search problem. Combinatorial search problem has been widely studied in areas such as artificial intelligence, operations research, and job scheduling etc. Constraint programming [16] and mathematical programming [17] have been developed to solve this problem. Liu et al. [9] integrated constraint-programming techniques with traditional database techniques to solve sum constraint queries by modifying existing nest-loop join operators. Our rewriting techniques are based on similar principles but can be implemented it without modifying existing database engines.

3 Query Rewriting

Most current database systems use a left-deep query plan with pair-wise joins to implement a sum constraint query. As an example, Fig. 1(a) shows the execution plan for the sum constraint query from the Meal example. The problem with this execution plan is that join operators cannot remove intermediate results based on a sum constraint until the values of all attributes involved in this sum constraint are decided. Therefore, all join operators except the last one (at the top level) will compute the Cartesian product of all involved relations.

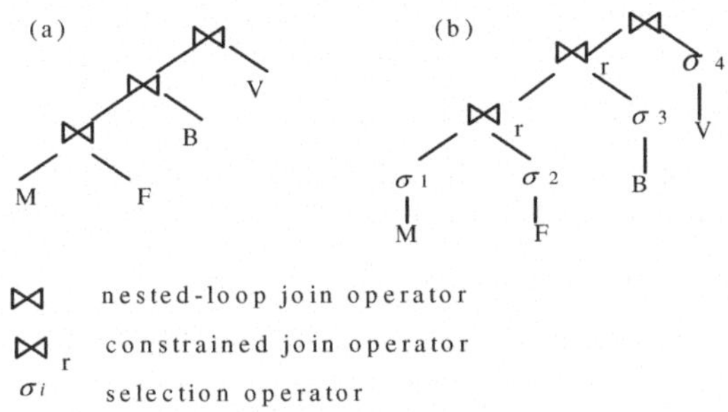

Fig. 1. (a) Conventional and (b) improved execution plans for a sum constraint query

To solve this problem, we want an execution plan as shown in Fig. 1(b). This new plan differs from the original in two respects. First, we add selection operators that filter tuples that cannot lead to a solution. Second, we introduce new query conditions that can be used by join operators to remove intermediate results.

3.1 Rewriting Techniques

The essence of our method is rewriting each sum constraint in the query condition as a set of simpler constraints that can be used to create the execution plan shown in Fig. 1(b). In other words, we need to provide constraints for selection operators to filter tuples, and provide constraints for join operators to reduce intermediate results.

3.1.1 Selection Operators

Selection operators use range constraints on one attribute to filter tuples from relations. We present a method to rewrite a sum constraint into a set of range constraints as follow.

A sum constraint has the following general form:

$$A_1 + A_2 + \ldots + A_n \; comp\_op \; c \qquad (1)$$

Here, c is a constant, A_i $(i = 1..n)$ are attributes that appear in this constraint, and *comp_op* represents a comparison operator that could be '>', '<', '≥', and '≤'. From (1) we can derive the following:

$$A_k \; comp\_op \; c - \sum_{i=1..n, i \neq k} A_i \quad for\ k = 1..n \qquad (2)$$

We can calculate the lower and upper bounds of the expression on the right side of *comp_op* in (2) as follows:

$$L_k = c - \sum_{i=1..n, i \neq k} \max(A_i) \quad \textit{for } k = 1..n \tag{3}$$

$$H_k = c - \sum_{i=1..n, i \neq k} \min(A_i) \quad \textit{for } k = 1..n \tag{4}$$

Here, $\max(A_i)$ and $\min(A_i)$ represent the maximum and minimum of attribute A_i in the database instance. Using L_k and H_k we can create a range constraints for attribute A_k depending on the *comp_op* in (1). If *comp_op* is '>' or '≥', we shall call such constraints *greater-than* constraints, and derive the following range constraint for A_k:

$$A_k \; comp\_op \; L_k \quad \textit{for } k = 1..n \tag{5}$$

If *comp_op* is '<' or '≤', we shall call such constraints *less-than* constraints, and derive the following range constraint for A_k:

$$A_k \; comp\_op \; H_k \quad \textit{for } k = 1..n \tag{6}$$

Any value not in the derived range cannot possibly satisfy the sum constraint shown in (1). Through (3) and (4), we obtain a range constraint for each attribute in the sum constraint. Selection operators, as shown in Fig. 1(b), can use these range constraints to filter out tuples that will not lead to any query result.

3.1.2 Join Operators

Even with these selection operators, the execution of a constraint query is still very expensive because it needs to calculate the Cartesian product of the involved relations after being processed by the selection operators. Another possible improvement is to use the join operators in order to remove intermediate results not leading to any query results. However, current join operators cannot remove intermediate results until all values of the attributes involved in a constraint are determined.

To this end, for each join operator, we rewrite a sum constraint into new constraints that only contain attributes appearing in intermediate results produced by this join operator. For the example in Fig. 1, we create, for the join operator between relation M and F, constraints containing attributes from only these two relations. Join operators can evaluate these constraints and remove intermediate results that do not satisfy these constraints.

For a join operator containing k attributes A_j (j=1…k), we create new constraints in two steps. First, we rewrite the sum constraint (1) by moving all attributes not in A_i (i = k+1...n) to the right side of the formula. The result is shown below:

$$\sum_{i=1..k} A_i \; comp\_op \; c - \sum_{i=k+1..n} A_i \quad \textit{for } k = 1..n\text{-}1 \tag{7}$$

The left side is the sum of the first k attributes. We can calculate the bounds (SH_k and SL_k as upper and low bound respectively) of the expression on the right side of this formula:

$$SL_k = c - \sum_{i=k+1..n} \max(A_i) \quad \textit{for } k = 1..n\text{-}1 \tag{8}$$

$$SH_k = c - \sum_{i=k+1..n} \min(A_i) \quad \textit{for } k = 1..n\text{-}1 \tag{9}$$

Then we can create constraints on the sum of the first k attributes as follows. If the original sum constraint was *greater-than*:

$$\sum_{i=1..k} A_i \; comp\_op \; SL_k \quad \textit{for } k = 1..n\text{-}1 \tag{10}$$

Or, if the original sum constraint was *less-than*:

$$\sum_{i=1..k} A_i \; comp\_op \; SH_k \; for\; k = 1..n\text{-}1 \tag{11}$$

New constraints contain only attribute A_j (j=1…k), and can be used by the targeted join operator to remove intermediate results. Using this rewriting technique, we rewrite the meal query, and show a part of the new query:

```
SELECT     M.name, V.name, B.name, F.name
FROM       Meats AS M, Vegetables AS V,
           Beverages AS B, Fruits AS F
WHERE      M.Cal < 1077 AND F.Cal < 1030
AND        B.Cal < 1010 AND V.Cal < 1013
AND        M.Cal + F.Cal < 1097
AND        M.Cal + F.Cal + B.Cal < 1097
AND        M.Cal + F.Cal + B.Cal + V.Cal < 1100
   M.Cal + F.Cal + B.Cal > 493                AND
   M.Cal + F.Cal + B.Cal + V.Cal > 900        AND ...
```

We only show the newly created constraints from the two sum constraints on the *Cal* attributes due to space limitation. Besides the original sum constraints, the new query contains range constraints on *Cal* attributes that can be used by selection operators to filter tuples, and sum constraints containing attributes from two and three relations, which can be used by the join operator between M and F, and the join operator between M, F, and B to remove intermediate results.

When rewriting a query, we assume that we already know the order of the join operators. To decide the join order, we can calculate the number of tuples in each relation after it is filtered by the selection operators, and decide the join order based on the size of filtered relations [10].

In summary, we rewrite a sum constraint into a set of simpler constraints that can be used by selection operators and join operators to improve the query performance. Although this rewriting technique adds more constraints to the query condition that may cause some extra computations to validate a result, this cost is trivial compared to the gains archived by reducing tuple reading operations and by reducing the number of the intermediate results. In Section 5 we discuss a benchmark we run in order to quantify the performance improvements. In contrast to previous method proposed in [1, 2, 7, 15], the rewriting techniques proposed in this paper do not require modifications of database query engines or building complex index structures, and can be easily deployed on current database systems.

4 Computing Partial Results

Many applications, instead of asking for all query results, are only interested in a given number K results. Besides the query rewriting techniques, we propose an approach to find any K results of a sum constraint query. We focus on finding any K results instead of top-K results since the notion of result ordering is not uniquely defined for our problem.

4.1 Tuple Quality

A conventional way to find K results is to sort tuples based on one constraint and consider 'good' tuples first. For Meal Example in order to find meals with enough vitamin C (represented as *Vc*), we sort relation Fruits, Meats, Vegetables, and Beverages based on *Vc*, and combine tuples with bigger *Vc* first.

However, a query may contain multiple 'conflicting' sum constraints. For example, a query may ask for meals with calories and vitamin C more than some required values. The constraint on vitamin C 'conflicts' with the constraint on calories because foods containing more vitamin C usually have less calories. Therefore, trying combinations of foods with more Vitamin C (or calories) does not necessarily help to improve the performance of the query for K results. As some of our experimental results show (Section 5.2) sorting tuples based on only one constraint can sometimes hurt performance.

For a query with multiple sum constraints, instead of ordering tuples based on just one constraint, we need to consider all constraints, and try combinations of tuples that are likely to satisfy all constraints first. We call these tuples *high quality tuples*. The quality of a tuple for a query is an aggregate of its quality for each constraint in the query.

For a *greater-than* constraint C requiring the sum of attributes A *to be* greater than a value d, we define a quality of tuple t as:

$$\text{quality}(t, C) = \min(t.A/d, 1) \tag{12}$$

This quality value is between 0 and 1, assuming only positive attribute values, and can be understood as the degree to which this tuple contributes to satisfying this constraint. If the attribute value is more than d, this tuple fully satisfies this constraint by itself. We represent its quality as 1. If the attribute value is less than d, this tuple satisfies t.A/d fraction of the requirement.

For a *less-than* constraint C requiring the sum of attributes A to be less than a value d, we define a quality of tuple t as:

$$\text{quality}(t, C) = \text{-MAX\_INT} \ \textit{if}\ t.A > d,\ \textit{otherwise}\ -t.A/d \tag{13}$$

Because the constraint requiring the sum of attribute values less than a value d, tuples with positive attribute values contributes negatively to the satisfaction of this constraint. If the attribute value is more than d, this tuple cannot satisfy this constraint assuming all attribute are positive. We represent its contribution as a minimal value. If the attribute value is less than d, this tuple uses up t.A/d percent of the range specified in the constraint.

We calculate the quality of a tuple for a query by aggregating its quality value to each constraint. Among all constraints, some are difficult to satisfy than others. Intuitively, tuples that contribute to the satisfaction of difficult constraints are more valuable than tuples that contribute to the satisfaction of easy constraints.

4.2 Constraint Difficulty

We evaluate the difficulty of a constraint based on attribute statistics. We use the Meal example to illustrate our method. Below, we show the median values of the attributes of all four relations.

	Cal	Vb6	Vc	Fat	Chol
Meats	221	0.37	1.16	4.95	105
Vegetables	63	0.18	26.4	0.23	0.69
Beverages	142	0.22	67.6	0.7	1.46
Fruits	86	0.07	31	0.07	0

We represent the difficulty of a constraint by H. For a *greater-than* constraint C, we define H as the ratio between the sum of median values of the attributes and d.

$$H(C) = d/\sum_{i=1..n} \text{median}(A_i) \qquad (14)$$

For a *less-than* constraint, we define H as:

$$H(C) = \sum_{i=1..n} \text{median}(A_i)/d \qquad (15)$$

The intuition behind the quantity H is that the bigger H is, the smaller the number of tuples that can satisfy this constraint.

The difficulty (H) values of the six constraints of the Meal query are shown below:

	Cal	Vb6	Vc	Fat	Chol	Cal
D	>900	>2	>28	<9	<150	<1100
Difficulty (H)	1.75	2.38	0.3	0.67	0.71	0.47

We calculate the quality of a tuple for a query by aggregating its quality values for each constraint using (16). Using the constraint difficulty as a factor when summing up the quality for each constraint, this formula gives more weight to tuples with a high quality for difficult constraints. We sort tuples in relations according to their quality for the query, and join higher quality tuples first. In this way, we check combinations of tuples that most likely satisfy all sum constraints in a query first.

$$\text{quality}(t) = \sum_{i=1..m} \text{quality}(t, Ci) * H(Ci) \qquad (16)$$

5 Experimental Results

In order to evaluate the rewriting technique and sorting framework we propose in this paper, we performed several experiments using real-world data.

We use a food database published by the USDA [18]. This database contains nutritional facts for more than 7000 types of food. From this database we created four relations: Fruits, Vegetables, Beverages, and Meats containing information about their eponymous types of food. The sizes of these relations are as follows: Fruits 273, Vegetables 717, Beverages 199, and Meats 1602.

We created this database and run all queries on a PostgreSQL 8.0.3 relational database system. The system is running on a Linux machine with 4 Intel Xeon CPU 3GHz processors, and 2GB of main memory.

5.1 Evaluating the Query Rewriting Technique

We evaluated the efficiency of our rewriting technique by comparing the response time of several sum constraint queries with and without rewriting. Another way to improve the query performance is to consider tuples in the order of their quality. In our experiments we studied how these two methods interact under different scenarios.

Table 1. Execution plans

Tuple ordering	Random	Ascending	Descending
No rewriting	Plan I	Plan III	Plan V
With rewriting	Plan II	Plan IV	Plan VI

We compared six different combinations of these two methods as shown in Table 3. For example, plan I does not sort tuples in the relations, and does not use rewriting. Plan VI, in contrast, sorts tuples in descending order of their attribute values, and uses rewriting. In this section, we focus on the traditional sorting methods: ordering tuples according to one attribute. We consider our new sorting algorithm in the next section.

We evaluated these 6 plans on several sum constraint queries. In this section, we report on two of them, with representative performance, shown in Fig. 2. QA contains one constraint requiring the sum of attributes *cal* to be greater than Min_cal. QB contains two constraints requiring the value of the sum of attributes *cal* to be in the range of Min_cal and Min_cal + 200.

```
SELECT M.name, V.name, B.name, F.name
FROM Meats AS M, Vegetables AS V, Beverages AS B, Fruits AS F
WHERE M.cal+V.cal+B.cal+F.cal > Min_cal
AND   M.cal+V.cal+B.cal+F.cal < Min_cal+200 (QB only)
LIMIT K
```

Fig. 2. Benchmark queries QA and QB. The only difference between the two queries is the additional *less-than* sum constraint in QB.

First, consider QA, containing only the constraint that specifies the minimal requirement of the total calories in a meal. In this case, combinations of foods with larger number of calories have a better chance of satisfying this constraint. We only consider four plans (plans I, II, V, and VI in Table 1) for this query because it is obviously not an appropriate choice to consider foods with lower number of calories first (plans III, IV).

We ran QA with *K=20*. We varied the required calories (represented by *Min_cal*) in order to change the selectivity of this query, and show the experimental results in Fig. 3(a).

We note that the rewriting technique improves the query performance by comparing plan II, which rewrites the query, and plan I, which does not use any optimizations. When the number of required calories (Min_cal) is relatively small (around 700), the two plans have similar response times because there are many query results. When total number of results decreases with the increase of Min_cal, plan I

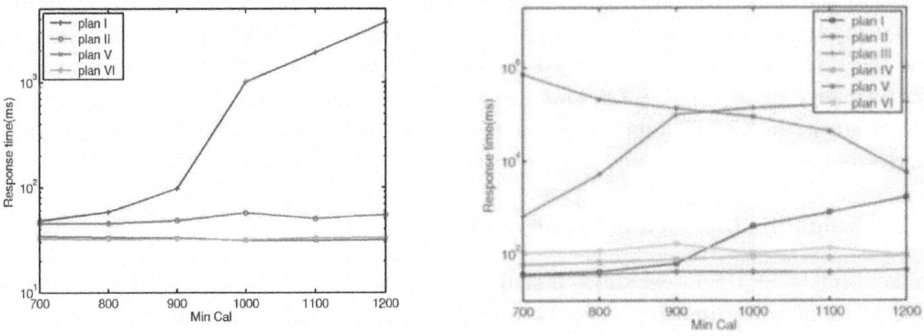

Fig. 3. (a) and (b) Performance results for plans from Table 1

needs to check more combinations of foods to find 20 meals, and shows increasingly high response time. In comparison, plan II consistently shows low response time.

We also consider which one of the two approaches (sorting and query rewriting) is more effective in improving query performance. Although both approaches outperform plan I, the sorting-based method (plan III) consistently outperforms the rewriting-based approach (plan II) for QA. Note that plan IV, which uses both query rewriting and sorting, does not show any significant performance gain or loss compared to plan III. Thus, the query rewriting technique does not further improve the performance of the sort-based method for QA, which contains only one sum constraint.

However, we found that rewriting does improve the performance of queries containing multiple conflicting sum constraints. We use QB as the benchmark for this type of queries. QB contains two 'conflicting' constraints that specify an upper bound and a lower bound on the sum of the attribute *cal*. These two constraints 'conflict' because the lower bound constraint favors tuples with higher calories, while the upper bound constraint favors tuples with lower calories. Since we cannot tell intuitively which sorting order is better for this query, we compare all six execution plans for this query, and show the results in Fig. 3(b).

We first show that sorting approaches do not always improve the query performance for queries containing 'conflicting' constraints. Plan III (Plan V), which ranks tuples based on the attribute *cal* and considers foods with higher (lower) calories first, shows long response time when Min_cal is small (large). Since QB asks for foods with total calories in a range, plan III (plan V) unnecessarily checks foods that violate the upper (lower) bound of the range at the beginning of the search process. Plan I, which does not use any optimization, performs better than plan III and IV. This shows that sorting actually hurts the query performance of QB.

In comparison, plan II, which uses the rewriting technique, shows a stable query performance and consistently outperforms sorting based methods.

From the experimental results of the two benchmark queries, we can see that our proposed rewriting techniques improve the query performance consistently for different types of sum constraint queries. Sorting-based plans, on the other hand, show only good performance for queries with one sum constraint (QA in our case).

Also, by comparing sorting only plans (plan III and Plan V) with plans using both rewriting and sorting (plan IV and plan VI), we can see that rewriting does not hurt the performance of sorting-based plans for QA, and remarkably improves the performance of sorting-based plans for QB. Therefore, we can conclude that rewriting techniques can be efficiently used with sorting-based approaches to improve query performance.

5.2 Evaluating the Sorting Framework

In sections 4 and 5.1, we show that traditional sorting methods have limitations for queries containing multiple 'conflicting' constraints. In this section, we evaluate the performance of our sorting framework that takes into account all constraints.

We use the Meal query as the benchmark query. This query asks for K combinations of single servings of meats, vegetables, fruits, and beverages with the required nutritional values. We compared the response time of 6 execution plans, shown in Fig. 4(a). Plan I through plan V sort tuples based on the values of one attribute as traditional sorting methods do. For example, plan I sorts tuples based on the attribute *cal* in descending order because the first constraint of the benchmark query requires the sum of the attribute *cal* to be greater than 900. Plan N, on the other hand, sorts tuples based on their quality which takes into account all constraints.

Plans	Description
I	descending order of cal
II	descending order of Vc
III	descending order of Vb6
IV	ascending order of fat
V	ascending order of chol
N	new sorting technique

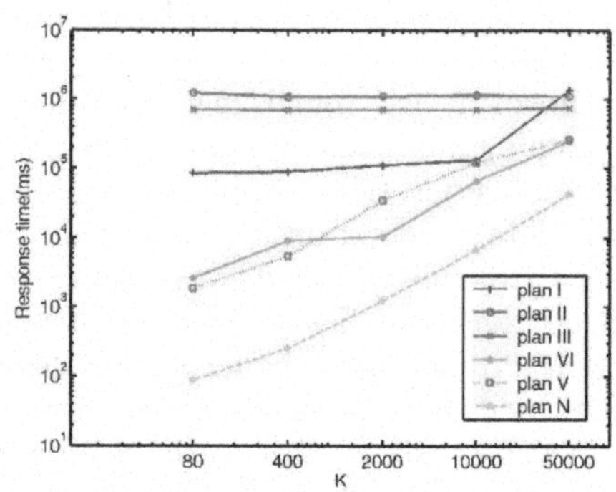

Fig. 4. (a) Query plans with sorting and (b) their performance

We rewrite this query and vary the number of required query results (K) in order to change the selectivity of the query. The experimental results are shown in Fig. 4(b). The response time is the average query response time over 100 runs.

Our approach (Plan N) outperforms all other plans by a large margin when K is small. It shows that the proposed sorting method, which takes into account all constraints, improves the performance of queries with multiple conflicting constraints. The experimental results also show that the performance differences between plans shrink with the increase of K. Intuitively, sorting-based approaches only help when users ask for a small number of query results. In the extreme case when a query asks for all results, sorting-based approaches do not help at all.

References

1. I. F. Ilyas, W. G. Aref, and A. K. Elmagarmid: Supporting Top-K Join Queries in Relational Databases. In VLDB, Berlin, Germany (2003)
2. R. Fagin, A. Lotem, and M. Naor: Optimal aggregation algorithms for middleware. In Journal of Computer and System Sciences, vol. 66, pp. 614-656 (2003)
3. J. Goldstein, R. Ramakrishnan, U. Shaft, and J. B. Yu: Processing queries by linear-constraints. In ACM Symp. Principles of Database Systems (1997)
4. F. Dumortier, M. Gyssens, and L. Vandeurzen: On the decidability of semi-linearity for semi-algebraic sets and its implications for spatial databases. In PODS (1997)
5. M. Stonebraker and J. M. Hellerstein: Content integration for e-business. In PODS, Santa Barbara, California, United States (2001)
6. C. Liu, L. Yang, I. Foster, and D. Angulo: Design and Evaluation of a Resource Selection Framework. In IEEE Intl. Symp. on High Performance Distributed Computing (2002)
7. S. Guha,.D. Gunopoulos, N. Koudas, D. Srivastava, and M. Vlachos: Efficient approximation of optimization queries under parametric aggregation constraints. In VLDB, Berlin, Germany (2003)
8. I. Foster and C. Kesselman: The Grid: Blueprint for A New Computing Infrastructure. San Francisco: Morgan Kaufmann Publishers (1999)
9. C. Liu, L. Yang, and I. Foster: Efficient relational joins with arithmetic constraints on multiple attributes. In Intl. Database Appl. & Engr. Symp., Montreal, Canada (2005)
10. H. Garcia-Molina, J. D. Ullman, and J. Widom: Database Systems: the Complete Book. Upper Saddle River, NJ: Prentice Hall (2002)
11. R. Avnur and J. M. Hellerstein: Eddies: continuously adaptive query processing. In SIGMOD, Dallas, TX, USA (2000)
12. P. G. Selinger, M. M. Astrahan, D. D. Chamberlin, R. A. Lorie, and T. G. Price: Access path selection in a relational database management system. In SIGMOD, Boston, USA (1979)
13. E. J. Shekita, K. L. Tan, and H. C. Young. Multi-join optimization for symmetric multiprocessors. In VLDB, Dublin, Ireland (1993)
14. A. N. Wilschut, J. Flokstra, and P. M. G. Apers: Parallel evaluation of multi-join queries. In SIGMOD, San Jose, USA (1995)
15. P. K. Agarwal, L. Arge, J. Erickson, P. G. Franciosa, and J. S. Vitter: Efficient searching with linear constraints. In PODS, Seatle, Washington, United States (1998)
16. K. Marriott and P. J. Stuckey: Programming with Constraints: An Introduction. Cambridge, Massachusetts: The MIT Press (1998)
17. L. A. Wolsey and G. L. Nemhauser: Integer and Combinatorial Optimization, first ed: Wiley-Interscience (1999)
18. USDA national nutrient database for standard reference: http://www.nal.usda.gov/fnic/

A Theoretical Framework of Natural Computing – M Good Lattice Points (GLP) Method*

Jia-xing Cheng[1], Ling Zhang[1], and Bo Zhang[2]

[1] Key Lab of IC&SP, Ministry of Education, Anhui University, Hefei, China
[2] Dept. of Computer Science & Technology, Tsinghua University, Beijing, China
cjx@ahu.edu.cn, zling@ahu.edu.cn, dcszb@mail.tsinghua.edu.cn

Abstract. This paper analyses several currently used computing methods inspired by the nature and concludes their common properties and their disadvantages. It then proposes a more abstract computing model inspired by the nature according to theoretical results on number theory. We also present a good lattice points method based on the number theory for problem solving, of which the discrepancy of the new method is minimized in the sense when the number of points are fixed. This method is dimensional independent and can be used to solve high dimensional problems. A typical algorithm is proposed to apply Genetic Algorithm and Immume Algorithm. Some comparable examples are given to show the advantages of our new method.

Keywords: genetic algorithm, ant colony algorithm, particle swarm optimizer algorithm, immunity algorithm, semi-supervising learning, discrepancy, good lattice point method.

1 Introduction

Currently different methods of natural computing are based on some natural phenomenon and the corresponding optimization procedures. In this paper, we propose a new computational method, which definitely improves the currently used methods, inspired by the nature. For example, Genetic algorithm (GA) [1], Ant colony algorithm (ACA) [2][3], Particle swarm optimizer algorithm (PSO) [5][6], Immunity algorithm (IA) [4], and for all natural computing methods mentioned above can be stated as follows [7][10]： Firstly choose a set K_1 from X, measure its property F_1; then choose a set K_2 according to some regulation (generally it is inspired from the regulations of the nature), measure its property F_2; repeat the above procedure until the termination conditions are satisfied. This framework is very similar to increment learning. However, it is not easy to comment on the advantages and the disadvantages of those methods based on natural computing because (1) human beings lack well understanding to natural phenomenon, (2) those methods are subjectively proposed, and (3) the methods and the algorithms only modify some parameters and then simulate on some given data, while the essence of the problem is still unknown or only known a little. In this paper, we will deal with natural computing problems at a more abstract level and give some

* This work is partly supported by a grant of Ministry of Education, P.R. China (20040357002).

G. Dong et al. (Eds.): APWeb/WAIM 2007, LNCS 4505, pp. 452–462, 2007.

objective criteria in judging different optimization methods. We propose a Good Lattice Points method (GAGLP) based on natural computing. Some experiment results are given to illustrate to advantages of the new method.

This paper is organized as follows. Section 2 comments on current methods of natural computing and points out that it should have criteria to judge advantages and disadvantages for comparing different methods. Section 3 proposes a theoretical model for natural computing and shows the advantage of the proposed new method. A constructive method with uniform distribution and non-uniform distribution of good lattice points is described in detail and an algorithm of the model is also given. Section 4 is the related comparison works with some typical works and section 5 is conclusion and suggestion for further research work.

2 Analysis on Some Computing Methods Inspired by the Nature

Let us look at some methods based on natural computing such as Genetic algorithm (GA), Ant colony algorithm (ACA), Particle swarm optimizer algorithm (PSO) and Immunity algorithm (IA). For Genetic algorithm (GA), An algorithm based on imitating natural evolution procedure is to: 1) Select a subset P1 from universe X; 2) Comput the values of $f(x)$ for all x in P_1; 3) Do: genetic operators, obtaining a new subset P_2; 4) Repeat above procedures until the termination conditions are satisfied. For Ant colony algorithm (ACA), an algorithm based on imitating natural evolution procedure is to: 1) For a given graph G, let P_1 be a node set in G, V1 the set of moving direction for each node in P_1; 2) Describe the distribution of P_1 after moving of V_1 by using hormone effected on all nodes; 3) Determine new moving direction set V_2 according to the distribution of P_2; 4) Repeat the above procedures until the termination conditions are satisfied. Again for Particle swarm optimizer algorithm (PSO):1) Select a subset P_1 from X, let （ X_1,V_1） be the position and velocity for each point in P_1, 2) Obtain X_2 after moving, measure the object function value for each point; 3) Determine the new velocity V_2 and obtain （X_2, $V2$）; 4) Repeat above procedures until the termination conditions are satisfied. Lastly for Immunity algorithm (IA): 1) For a given intruding subset (antigen) Y_1; 2) Construct the corresponding subset (antibody) X_1 according some regulations; 3) When a new intruding subset (antigen) Y_2 is given, construct X_2 according to the situation of （X_1, $Y1$） *and* Y_2; 4) Repeat above procedures until the termination conditions are satisfied.

From the above algorithms we can see that most current methods of natural computing are from regulations of the nature, and different methods are proposed according to different "subjective understandings" . The given methods tend to be "subjectively" proposed and no criterion is given to judge the advantages and disadvantages. Hence most research works are only based on experiment calculation with some examples.

It is necessary to identify criteria for comparing different methods so that the new natural computing method should have a solid theoretical basis. Then a new method based on the above criteria can be sought to optimize the given problem in higher efficiency. We have found that discrepancy on number theory is a very useful criterion to judge different methods of natural computing inspired by nature and can be used to propose a new method to improve current algorithms such as genetic algorithm.

3 Theoretical Model for Natural Computing

3.1 The Problem Statement

Given a domain X and an object function f on X, find x^*: such that $f(x*)=$ $\max f(x)$. We assume that for any $f(x)$ and its domain $u(x)$; it is proportional for the probability of x^* in $u(x)$ to the value of $f(x)$ and the measurement of $u(x)$. Intuitively, for finding the maximum, the probability at the neighbor of a larger object function is bigger than that of a smaller one [16].

For any domain X, an intuitive method [16] to choose n points Y from X that may contain the optimal point with the largest possibility is chosen in such a way that the n points are "most" uniformly distributed in X if no any knowledge obtained from X. Because the more uniformly distributed the n points are, the larger the probability to choose the optimal point. Now how to define the "most" uniform distribution ? What is the most uniform distribution ? The answer is the least "discrepancy" from the number theory [8].

3.2 Definition of Discrepancy and Good Lattice Point with Properties

Definition 1 [8]: Let G_t be unit cuboids with t dimension, γ be a subset of G_t, then

$$\phi(n) = \sup_n \left| \frac{N(\gamma)}{n} - |\gamma| \right|$$

is called discrepancy of γ with respect to G_t, where $N(\gamma)$ is the number of points $P_n(i)$ in γ, $|\gamma|$ denotes the measure of γ and $P_n(i)$ the set points with number n, $i = 1, 2, ...n$.

Definition 2 [11] : Let $r \in G_t$, a set of points $P_n(i) = \{(\{r_1 \times i\}, ..., \{r_t \times i\})\}$ ($i = 1, 2, ..., n$) is called a good lattice point set and r the good lattice point (GLP) if

$$\forall n, \varepsilon > 0, \textit{ the discrepancy of } P_n(i) \textit{ is}: \ \phi(n)$$

Where $\phi(n) = C(r,\varepsilon) n^{(-1+\varepsilon)}$, C(r, ε) only depends on r, ε (ε is an arbitrary small positive number).

Theorem 1[11]: Let $P_n(i)$ *(1≤i≤n)* is with discrepancy of $\phi(n)$, $f \in B_t$ (t-dimensional bounded variation functions), then

$$\left| \int_{x \in G_t} f(x)dx - \sum \frac{f(P_n(i))}{n} \right| \le V(f)\phi(n) \tag{1}$$

Where $V(f)$ is total value of bounded variable f.

Theorem 2[11]: If (1) holds for all $f \in B_t$, then $P_n(i)$ $(1 \le i \le n)$ are a point set with discrepancy not more than $\phi(n)$.

Theorem 3[11]: If $f(x)$ satisfies

$$|f| \le L, \left|\frac{\partial f}{\partial x_i}\right| \le L, \quad 1 \le i \le n, \ldots, \left|\frac{\partial^2 f}{\partial x_i \partial x_j}\right| \le L, \ldots, \left|\frac{\partial^t f}{\partial x_1 .. \partial x_t}\right| \le L$$

Then the discrepancy is at least of $O(n^{-1})$ for a summation of values of f on any given n points with any weighted approximates the integral of f on G_t.

Remark 1. Theorem 3 shows why good lattice points are defined because it is impossible to get a result with the discrepancy less than $O(n^{-1})$, while the discrepancy of GLP is $O(n^{-1+\varepsilon})$, so it is reasonable to define the GLP set, where ε is an arbitrary small positive number.

Remark 2. From theorems 1, 2 and 3, we can see that discrepancy only depends on the number of points n but not on dimension t of the space. That is why it will give a very useful algorithm for solving the high dimensional problem. That means our algorithm is dimensional independent to problem solving.

Remark 3. From theorems 1, 2 and 3, we can see that it will be the best method for the n points chosen in a GLP way, if the linear combination of function f approximates the integral f on G_t.

Theorem 4 Kiefer [9] : Let $x_1, x_2, \ldots, x_n$ be chosen in G_t with independent identical distribution, $P_n = \{x_1, x_2, \ldots, x_n\}$, the probability is 1 for discrepancy $\phi(n) = O(n^{-1/2}(\log\log n)^{1/2})$ of P_n.

Remark 4: Theorem 4 shows that if we know nothing about a problem and choose an n-point set, then the discrepancy will be of $O(n^{-1/2})$. But if we choose the set in a GLP way, then discrepancy is of $O(n^{-1+\varepsilon})$. For instance, if n=10,000, then discrepancy of the former is of $O(10^{-2})$, while the latter is of $O(10^{-4})$. The discrepancy is much smaller for point set chosen in a GLP way than for which chosen randomly. That is the theoretical reason why an algorithm with GLP method converges much faster!

3.3 Estimation for Distribution of Object Function

However, for a domain X if point sets are not uniformly distributed, we can define good lattice point sets (GLP sets) as follows.

Definition 3. Let probability density of space G_t be $p(x) = \prod_i p_i(x_i), 0 \le p_i(x_i) \le 1, p(0) = 0, p(1) = 1, 0 \le x_i \le 1$, and let a set with n

points in G_t be $P_n(i)=\{x_1^{(n)}(i),...,x_t^{(n)}(i)\}, i=1,...,n, 0\le x_k^{(n)}\le 1, k=1,...,t$, for an arbitrary point $r=(r_1,r_2,...,r_t)$ in G_t, let $N_n(r)=N_n(r_1,...,r_t)$ be the number of $P_n(i)$ that satisfies the following inequalities: $0\le x_j^{(n)}(i)<r_j, \quad j=1,...,t$. Let $\phi(n)=\underset{r}{Sup}\left|N_n(r)/n-|r|\right|, \quad where \quad |r|=p_1(r_1)p_2(r_2)\cdots p_t(r_t)$ then $P_n(i)$ is of discrepancy $\phi(n)$ for $(G_t, p(x))$.

Let probability density of G_t be: $p(x)=\prod_i p_i(x_i), x=(x\ ,...,x_t)$, where $p(x)$ is the measurement in cuboids $\coprod [0^{,} x^{!}]$. We now can define a good lattice point set for space G_t with measurement $p(x)$.

Definition 4: Given any space G_t and its measurement $p(x)$, for any point $r=(r_1,r_2,...,r_t)$ in G_t, let $N_n(r)=N_n(r_1,...,r_t)$ be the number of $P_n(i)$ that that satisfies the following inequalities $0\le x_j^{(n)}(i)<r_j, \ j=1,...,t.$ let $\phi(n)=\underset{r}{Sup}\left|N_n(r)/n-|r|\right|, \quad where \quad |r|=p_1(r_1)p_2(r_2)\cdots p_t(r_t)$.

If discrepancy $\phi(n)$ satisfies $\phi(n)=C(r,\varepsilon)n^{(-1+\varepsilon)}$, where $C(r,\varepsilon)$ is a constant that only depends on r,ε (εis an arbitrary small positive number), then $P_n(i)$ is a good lattice point set for $(G_t, p(x))$.

***Remark 5*:** If we regard $p_i(x_i)$ as measurement of the *i*-th edge of G_t which is non-uniformly distributed, then the physical meaning of discrepancy is the same as in the case of uniform distribution. By choosing $r_k=\{2\cos 2k\pi/p\}, 1\le k\le t$, where p is the smallest prime number satisfying $(p-3)/2\ge t$, then r is a GLP set, or equivalently by choosing $r_k=\{e^k\}, 1\le k\le t$, r is also a GLP set. Again $\{\ x\ \}$ denotes the fractional part of x and we have

Theorem 5: Let cuboids G_t be of density distribution $p(x)=\prod_i p_i(x_i)$, p be the smallest prime number which satisfies inequality $(p-3)/2\ge t$, then $P_n(i)=(p_1^{-1}\{2\cos\frac{2\pi}{p}i\},...,p_t^{-1}\{2\cos\frac{2\pi t}{p}i\}), 1\le i\le n$ is a good lattice points set with $(G_t, p(x))$.

Proof: Given γ, by definition we know that the number of points which satisfies $0\le p_k^{-1}\{2\cos\frac{2\pi k}{p}i\}<r_k, \quad k=1,...,t$ in $P_n(i)$ is $N_n(\gamma)$, we have $0\le\{2\cos(2k\pi/p)i\}<p_k(r_k), \quad k=1,...,t, \quad i=1,...,n.$

Also since $\{2\cos(2ke/p)i\}=\{\{2\cos(2k\pi/p)\}i\}$, By Property 1 , We can regard $0\le\{\{2\cos 2k\pi/p\}i\}<p_k(r_k),\quad k=1,...,t,i=1,...,n$ as the conditions that the length of each edge of cuboids G_t is $p_j(\gamma_j)$,since $(\{2\cos 2\pi/p\},...,\{2\cos 2k\pi/p\})$ are good lattice point, $\phi(n)=\underset{\gamma}{Sup}\left|N_n(r)/n-|\gamma|\right|=C(r,\varepsilon)n^{(-1+\varepsilon)}$, *where* $|\gamma|=p_1(\gamma_1)p_2(\gamma_2)\cdots p_t(\gamma_t)$

and hence $P_n(i)=(p_1^{-1}\{2\cos\frac{2\pi}{p}i\},...,p_t^{-1}\{2\cos\frac{2\pi t}{p}i\}),1\le i\le n$ is a GLP set for $(G_t,p(x))$. This completes the proof.

Definition 5. Given a point set $P_n(i)$ and calculate its object function values $\{f(x(i)).i=1,...,n\}$. Fixed j, choose $\{(x_j(i),f(x(i)).i=1,...,n\}$, rearrange components of x : $\{x_j^*(1),x_j^*(2),...,x_j^*(n)\}$ in a rising order. Let

$$a_j(k)=x_j^*(k),\quad b(k)=\frac{\sum_{i=1}^{k}f(x_j^*(i))}{\sum_{i=1}^{n}f(x_j^*(i))},k=1,...,n$$

And let $(a(0),b(0))=(0,0),(a(n+1),b(n+1))=(1,1)$.

Now Joining $(a(j-1),b(j-1))$ and $(a(j),b(j)),\ j=1,...,n+1$ and obtain a broken line denoted by $p_j(x_j),0\le x_j\le 1$. For each j, obtain $p_j(x_j),0\le x_j\le 1,j=1,...,t$, according to the above manner, let $p(x)=\prod_{j=1}^{t}p_j(x_j),x=(x_1,...x_t)$, then we call $p(x)$ empirical estimation distribution of G_t .

3.4 Construction of GLP Set

Property 1. Choose $r_k=\{2\cos(2k\pi/p)\},1\le k\le t,p$ is the smallest prime satisfying $(p-3)/2\ge t$, then r_k *k=1,2,…,t,* is a GLP .

Property 2. Choose $r_k=\{e^k\},1\le k\le t$ *t*, then r_k *k=1,2,…,t,* is also a GLP. Again *{x}* denotes the fractional part of *x*.

For example, let t =12, from $(p-3)/2\ge t$, we can know that p = 29, then $r_k=\{2\cos(2k\pi/29)\},1\le k\le t$, from Property 1, or $r_k=\{e^k\},1\le k\le t$ from Property 2.

3.5 A Model of New Computing Method Inspired by the Nature

Now let us consider the following problem. For a given set G_t, and an object function $f(x)$, $\forall x \in G_t$, find x^*, such that $f(x^*) = \max_{x \in G_t} f(x)$. If we know nothing for distribution of $f(x)$ on G_t or no any regulation of the distribution is provided, then it is hard for us to find optimal value for $f(x)$. So it is necessary for us to give some assumption for distribution of $f(x)$.

Assumption 1: For $\forall x \in G_t$, and a neighborhood of $u(x)$, then probability of $f(x^*)$ in $u(x)$ is proportional to that of $f(x)$ in $u(x)$.

Now we can estimate empirical estimation distribution of object function $f(x)$ according to object function values on some subset of G_t.

The model is as follows. "Problem statement: Given $G_t, \forall x \in G_t, f(x)$, find $x^*, f(x^*) = \max_{x \in G} f(x)$".

Algorithm: 1) Given n, find P_1 in G_t, where n is the number of points to be chosen;

2) Measure object function values $\{f(x), x \in P_1\}$ on P_1;

3) Find empirical distribution $p(x)$ of G_t for $\{f(x), x \in P_1\}$;

4) Find P_2 according to distribution of $p(x)$ in G_t, and return to step 2)

5) Stop until terminate conditions are satisfied. The maximum of the last step is x^*.

Remark 6. The most important step of the above algorithm is in step 4). How to find point set P_2 from P_1? Most current algorithms involving the computing methods did not consider discrepancy when choosing the next generation point set P_2. We applying discrepancy on number theory and hence it can guarantee point set P_2 to be chosen in a "best" way and the algorithm converges much faster. Another advantage of the above is that it can be used to some problem solving in which a group of parameters is to be optimized. In this case we regard these parameters as the point set P, and find P_2 from P_1 in a similar way. For example, we can apply this idea to Immune algorithm in which a number of parameters are to be optimized to problem solving. In the next section we will show some examples using the above algorithm.

4 Example for Testing Functions

We propose a good lattice point method and combine it with the genetic algorithm (GA) which is mainly changed on the cross-over operator to produce the next generation and to apply some optimization problems such as standard function optimization, TSP, SAT and knapsack problem. The method is as the following.

Let A_1, A_2 be two individuals produced by standard GA, revise crossover operator as follows. For simplicity, assume that only the first t components of A_1 and A_2 are different. Let $A_1=(a_1^1,...a_L^1)$, $A_2=(a_2^1,...a_2^1)$ and $J=\{i|\ a_i^1\neq a_i^2,\ 1\leq i\leq L\}$. Now we find the new sample by using GLP. Let $H=\{(x_1,..., x_L)\ |\ i\in J, x=*;\ i\notin J, x_i=a_i^1\}$. We regard the first t components as t-dimensional cuboids, which are also denoted as H first. Then we seek good lattice points on H, or group the t components into s groups in which each has d components and each group denotes a real number, and so the s groups will denote s real numbers, and hence space H is squeezed into an s dimensional space. It is easy to get s-dimensional cuboids by normornization. Now we find the good lattice points on the s-dimensional cuboids. For example, let $(1,1,0,0,1,0,1,0,1,1,0,1)$ be a point in a 12-dimensional cuboids, we can squeeze it into a point in a 3-dimensional cuboids with real numbers. Each real number denotes 4 components as $(1100,1010,1101) = (12,10,13)$, and then normalizes to get $(12/16,10/16,13/16) = (3/4,5/8,13/16)$, which is a point in 3- dimensional cuboids.

Next find good lattice point as follows. Construct GLP with points of n in t-dimensional cuboids t (let H be a binary cuboids). Where $r_k=\{2\cos 2\pi k/p\}$, $1\leq k\leq t$,

$$P_n(i) = \{\ (\{r_1\times i\},...,\{r_t\times i\}),\ i=1,2,...,n\}$$

$$where\ \ b_m^k = \begin{cases} a_m^1,\ \ m\notin J \\ \{r_{t_j}\times k\},\ m=t_j\in J,\ 1\leq k\leq n,\ 1\leq m\leq L,\ 1\leq j\leq s. \end{cases}$$

p is the least prime number satisfying $p\geq 2t+3$, or $r_k=\{\ e^k\ \}$, $1\leq k\leq t$. Where $\{a\}$ denotes the fractional part of a[11]. Let the k-th chromosome of the generations of n after cross-over operator be

$$b^k = (b_1^k,...,b_L^k)$$

In this way, we produce the generations of n after crossover operator, and then select the largest or some largest ones from them as the next generation after crossover operator.

The construction of GLP set in the situation of nonuniformly distribution is almost the same as in the case of uniformly distribution. Now we can show some experiment results in the following subsections.

The following examples are chosen from [18] which are somewhat difficult to find the maximum values since they are or of multi-maximum values, or have discontinued in their domain. We wish to find the maximum for:

1. $\max f_1(x) = \sum_{i=1}^{5} \mathrm{integer}(x_i)$ s.t. $-5.12\leq x_i\leq 5.12$

2. $\max f_2(x) =$

$$[1+(x_1+x_2+1)^2(19-14x_1+3x_1^2-14x_2+6x_1x_2+3x_2^2)].$$
$$[30+(2x_1-3x_2)^2(18-32x_1+12x_1^2+48x_2-36x_1x_2+27x_2^2)]$$
$$\text{s.t. } -2\leq x_i\leq 2,\ i=1,2.$$

3. max$f_3(x)$=

$$\frac{\pi}{3}\{10\sin^2(\pi y_1)+\sum_{i=1}^{2}(y_i-k_2)^2[1+10\sin^2(\pi y_{i+1})]+(y_3-1)^2\}$$

s.t. $-10 \le x_i \le 10$, where $y_i = 1+\frac{1}{4}(x_i+1)$, $i=1,2.$

4. max $f_4(x)$ =

$$\sin^2(3\pi x_1)+\sum_{i=1}^{4}(x_i-1)^2\left[1+\sin^2(3\pi x_{i+1})\right]$$
$$+(x_5-1)^2\left[1+\sin^2(2\pi x_5)\right]$$

s.t. $-5 \le x_i \le 5$, $i=1,2,...,5$

We find the maximum of functions by Good Lattice Point Genetic Algorithm (GAGLP) with fitness function f (x)   g (x)+C, such that f (x)>0. We found that the average times of iteration with GAGLP for finding the maximum are only half of Genetic Algorithm (GA). The results are shown in Table 1.

Table 1. Computation with GAGLP

function	Average times	Optimal	
f_1	5	x = (5.11, 5.03, 5.01, 5.11, 5.08)	f^* = 25
f_2	10	x = (-0.6667, 1.9995)	f^* = 509084.2342
f_3	8	x = (9.976196, 9.180981)	f^* = 171.3242
f_4	10	x = (-4.9993, -4.8946, -4.7894, -4.7998, -4.7636)	f^* = 91.2026

We also worked TSP on 144 cities in China with GAGLP with the result 30356.0km, while the currently best result is : 30385.1km[12][13][16]. Some postgraduate students[17] worked on knapsack problem with GAGLP and found that GAGLP is much better than GA on the accuracy of solutions, and the speed of finding the optimal without the precocity. A new method combined with a heuristic greedy algorithm and a hybrid genetic algorithm has been working on this problem and got the optimal solutions. We guess that might be the best solution to the problem.

5 Conclusions and Future Work

Some computing methods inspired by the nature are analyzed and a more abstract model is proposed in this paper. In addition, a new method is proposed to problem solving by using a good lattice point (GLP) set method from the number theory. The advantage of our new method is that (1) the discrepancy of GLP set is minimized of fixed number of *n* points to be chosen, (2) the discrepancy only depends on *n* (the number of points to be produced) but not on dimension t of the problem solving space. This is the reason why it gives a very useful algorithm for solving high dimensional problems. Currently high dimensional problems are usually hard to be solved and almost all methods are not given a satisfactory solution. We believe that our method gives theoretical criteria and can be used to judge the advantages and disadvantages of different methods. We also believe that this method can be used to some other computing methods inspired by the nature.

For future work we will use our new model to deal with some other computing methods inspired by nature and applied to bioinformatics such as protein folding, protein forecasting and some other NP-hard problems.

Acknowledgement

This work is partly supported by a grant of Ministry of Education, P. R. China (20040357002). The authors also thank our post-graduate students C.Y. Zhao, J.S. Cheng, W.L. Wu and W. Li for the program to calculate some numerical examples.

References

1. Holland, J.H.: Adaptation in Natural and Artificial Systems. The University of Michigan Press, (1975)
2. Dorigo, M., Maniezzo, V., Colony, A.: The Ant System: Optimization by a colony of cooperating agents. IEEE Trans. on Systems, Man and Cyber metrics, (1996), 26 (1): 1 - 13.
3. Dorigo, M., Gambardella, L M.: Ant Colony System: A Cooperative Learning Approach to the Traveling Salesman Problem. IEEE Trans. on Evolutionary Computation, (1997), 1(1): 53 - 66.
4. Castro, L. N., Timmis, J.: Artificial immune system: a new computational intelligence app roach. Springer-Verlag, (2002)
5. Kennedy, J., Eberhart, R. C.: Particle Swarm Optimization, IEEE Service Centered.IEEE International Conference on Neural Networks IV, Piscataway: IEEE Press, (1995): 1942-1948
6. Clerc, M., Kennedy, J.: The Particle Swarm—Explosion, Stability and Convergence in a Multidimensional Complex Space. IEEE Trans. on Evolutionary Computation, (2002), 6(1): 58-73
7. Rumelhart, D.E. and McClelland, J.L.: Parallel Distributed Processing Vol.I, Cambridge, MA: MIT Press, (1986)
8. Hua, LG. and Wang Y.: Applications to approximation analysis on number theory, Science Press, Beijing, (1978)
9. Kiefer, J.: On large deviations of the empiric d.f. Of vector chance variables and a law of the iterated logarithm, acific J.Math.11 (3), (1961) 649-660,
10. Vapnik, V.N.: The nature of statistical learning theory, Berlin: Springer-Verlag, (1995)
11. Zhang, L. and Zhang, B.: A new genetic algorithm with good lattice point method, Journal of computer, 24(9), (2001) 917-922
12. Zhao, C.Y. and Zhang, L.: Solving Chinese TSP based on good lattice point method, Journal of Nanjing University, Vol.36, (2000) Computer Issue

13. Zhao, C.Y. and Zhang, L.: Solving TSP with good lattice point method, Journal of Computer Engineering 37(3), (2001) :83-84
14. Cheng JS, and Zhang L.: Solving Job-Shop problem based on good lattice point method, Journal of Computer Engineering, (2002) :29(4) : 67-68
15. Li, W. and Huang, W. Q.: A mathematical and physical method for solving internation norm, China Science, 24(11), (1994): 1208-1217.
16. Zhang, H., Zhang, L.: The statistic genetic algorithms for combinatorial optimization problem. In Proceeding of IWCSE'97, Hefei, China, (1997), 267-269
17. Zhao C., Zhang L.: A solution to knapsack problem based on good-point set GA, In Proceeding of PAICMA'2000, Hefei, China, (2000): 256-258
18. Ma, SP. et al.: Artificial Intelligence, Tsinghua University Press, (2004)

Building Data Synopses Within a Known Maximum Error Bound

Chaoyi Pang, Qing Zhang, David Hansen, and Anthony Maeder

eHealth Research Centre, ICT CSIRO, Australia
{chaoyi.pang,qing.zhang,david.hansen,anthony.maeder}@csiro.au

Abstract. The constructions of Haar wavelet synopses for large data sets have proven to be useful tools for data approximation. Recently, research on constructing wavelet synopses with a guaranteed maximum error has gained attention. Two relevant problems have been proposed: One is the *size bounded* problem that requires the construction of a synopsis of a given size to minimize the maximum error. Another is the *error bounded* problem that requires a minimum sized synopsis be built to satisfy a given error bound. The optimum algorithms for these two problems take $O(N^2)$ time complexity. In this paper, we provide new algorithms for building error-bounded synopses. We first provide several property-based pruning techniques, which can greatly improve the performance of optimal error bounded synopses construction. We then demonstrate the efficiencies and effectiveness of our techniques through extensive experiments.

1 Introduction

Approximate Query Processing (AQP) has been extensively studied and used to deal with massive data sets in decision support and data exploration applications. As AQP usually relies on the pre-computed data synopses to compute the approximate results of queries over original data, research on improving the accuracy of approximate results inevitably focuses on finding good data synopses construction methods. Many techniques have been proposed on constructing data synopses [4,1]. Among them, the wavelet technique has been considered very promising as it was first adopted by Matias et al. to process range query approximation in relational database [11]. The basic idea of constructing a wavelet synopsis of a data vector, with size N, is to first transform the data vector into a representation with respect to a wavelet basis. Then it is approximated by retaining M coefficients as the wavelet synopsis and setting remains to 0 implicitly. The procedure of choosing M coefficients is called *coefficients thresholding*. A conventional approach is to find M coefficients to minimize the overall mean squared error [13]. This can be easily solved by applying the Parseval's theorem. However, the main drawback of this synopsis is that users have no method in which they can control the approximation error of individual elements in the data vector. This severely impedes further applications of the wavelet approximation. To alleviate this, researches have made efforts on constructing wavelet synopses with error guarantee [2]. Two dual approaches have been taken: one is to construct size bounded

G. Dong et al. (Eds.): APWeb/WAIM 2007, LNCS 4505, pp. 463–470, 2007.

synopses which would minimize the maximum approximation error of single data elements [3] whilst the other is to construct the smallest size of synopses such that the maximum approximation error does not exceed a given error bound [12]. The optimal synopses construction for both approaches has $O(N^2)$ time complexity. Although several methods have been proposed to improve the performance of constructing size bounded synopses [5,7,9], there are no investigations in the literature on improving the performance of constructing error bounded synopses. The approximate algorithm for size bounded synopses construction [9] can be easily extended to approximately solve the construction of the error bounded synopses but it may incur large approximation error in some situations as we will illustrate later on. Indeed, there exist some nice features that can greatly improve the performance of error bounded synopses's construction, which are not very obvious applicable on the size bounded synopses's construction. Motivated by this, in this paper we develop fast wavelet synopses construction method, which aims at minimizing synopses size under a given error bound.

Our contributions can be summarized as follows: We have obtained nice features based on the error tree structure used in Haar wavelet transformation. We propose pruning strategies that can greatly improve the performance of the original optimal algorithm. With these properties, we give a nontrivial low bound on the size of an optimal synopsis in linear time.

The rest of the paper is organized as follows. Section 2 defines the problem and enumerates related works. Section 3 investigates the error tree structure and proposes our pruning strategy to improve the original optimal algorithm. Section 4 reports our experiment results on applying our pruning strategy. Section 5 concludes this paper.

2 Background Information

In this section, we first introduce Haar wavelet transformation and coefficients thresholding. Then we present the two types of synopses and review related techniques. In Table 1 we summarize the math notation used throughout the paper.

Table 1. Notations

Symbol $i \in \{0..N-1\}$	**Description**
D, $[d_0, \ldots, d_{N-1}]$	Original data vector
W_D	Haar wavelet transformation on D
c_i	(coefficient) node
d_i, $\hat{d}_i$	(leaf, data) node and its reconstruction
$path(u)$	All ancestors of node u in the error tree
T, $T(c)$	Error tree and its subtree rooted at c
$T_L(c)/T_R(c)$	The subtree rooted at c's left/right child
Δ	A given error bound

2.1 Haar Wavelet Transformation and Thresholding

Approximate query processing using Haar wavelet is first introduced in [11] by Matias et al. The basic idea of Haar wavelet transformation is to recursively find the average and difference of two adjacent data of the data vector D. The final average value, i.e. the average of all data in D, together with those differences form a new vector W_D. The data elements of W_D are named wavelet coefficients. We use the following example to illustrate details. Given $D = [12, 6, 4, 2, 5, 1, 2, 0]$, we transform D to $W_D = [4, 2, 3, 1, 3, 1, 2, 1]$. Figure 1(i) shows details.

Each internal node c_i represents a wavelet coefficient whilst each leaf node d_i represents an original data item. l represents the corresponding resolution listed in Figure 1(i). Given a node u (internal or leaf), we define $path(u)$ as the set of nodes that lie on the path from the root to u (excluding u); $T(u)$ as the subtree rooted at u; $T_L(u)$ as the subtree rooted at the left child of u, if it exists; and $T_R(u)$ as the subtree rooted at the right child of u, if it exists. To reconstruct any leaf node d_i through the error tree T, we only need to compute the summation of nodes belong to $path(d_i)$. That is, $d_i = \sum_{c_j \in path(d_i)} \delta_{ij} c_j$, where $\delta_{ij} = +1$ if $d_i \in T_L(c_j)$ and $\delta_{ij} = -1$ if $d_i \in T_R(c_j)$. For example, d_2 can be reconstructed through the nodes of $path(d_2)$, i.e. $d_2 = 4 + 2 + (-3) + 1$. It is easy to see that reconstructing any original value of an error tree with N internal nodes, requires $O(\log N)$ time complexity.

The idea of wavelet synopses construction is to only keep a certain number of exact coefficients of W_D, while setting values of the remains as a constant number - zero is the normally implicit one. The goal is to find an optimal synopses that minimize the approximation error under certain metrics. Two commonly adopted error metrics ones are the mean squared error (L_2) and the maximum absolute error (L_∞). More formally, let $\hat{d}_i$ be the approximate value of d_i. Minimizing L_2 is to minimize $\sqrt{\frac{1}{N}\sum_i (\hat{d}_i - d_i)^2}$. Finding an optimal solution to minimize L_2 leads to a simple graceful algorithm due to the energy preserving property of wavelet transformations [13]. However this error metric is arguably not the best choice for approximate query processing [2]. One of the main drawbacks of this error metric is that users have no way of knowing the accuracy of any individual value approximation. Thus techniques on minimizing L_∞, i.e. $\max_i\{\left|\hat{d}_i - d_i\right|\}$, have been developed in recent years.

Resolution	Averages	Coefficients
3	[12, 6, 4, 2, 5, 1, 2, 0]	
2	[9, 3, 3, 1]	[3,1,2,1]
1	[6,2]	[3,1]
0	[4]	[2]

(i)

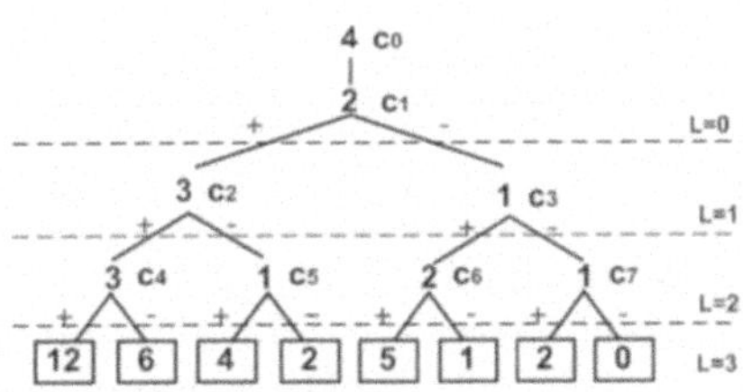

(ii)

Fig. 1. Haar Wavelet Decomposition and Error Tree

One approach is to to minimize L_∞ under a fixed number of coefficients to construct the *size bounded* synopses, also called B-bound. The first solution were proposed in [2]. This probabilistic method, however, has flaws due to its questionable expectation guarantees[8]. In [3], Garofalakis and Kumar propose a deterministic solution for constructing B-bound synopses. Given a data vector D with N elements, their algorithm takes $O(N^2 B \log B)$ time complexity and $O(N^2 B)$ space complexity. Guha improved the space requirements of this deterministic algorithm to $O(N)$ with a divide and conqueror idea in [5]. To make the construction of B-bound synopses applicable in a data stream environment, Karras and Mamoulis propose a greedy algorithm [9]. Meanwhile, Guha extended this problem from the original restricted version to unrestricted version, where the stored set B could be any set of real numbers without being limited to the wavelet coefficients [7,6]. The details are out of the scope of this paper.

The other approach aims at minimizing the number of necessary coefficients under an error bound (Δ) to construct the *error bounded* synopses. It is also called Δ-bound. Instead of fixing the wavelet synopsis size, it fixes the error tolerance through constructing a synopsis that satisfies $L_\infty < \Delta$. The goal is to find a synopsis with the smallest set of coefficients B among all possible solutions that would satisfy the Δ bound. This model is also very important and promising on providing approximate answers with good quality. Interestingly, it was only mentioned recently by Muthukrishnan and Guha in [12,5]. They proposed an optimal solution which however takes $O(N^2)$ time complexity.

In the next section, we will provide several important properties that can greatly improve the optimal error bounded synopses construction of [12,5].

3 Path Traverse Pruning for Synopses Construction

We start this section by first reviewing the existing optimal wavelet synopses construction algorithm, we then introduce our pruning techniques to accelerate the synopses construction for Δ-bounded approximation.

The optimal algorithm for Δ-bounded approximation has been proposed in [9]. Briefly, its idea can be described as follows. Assume there is a subtree $T(v)$ rooted at node v and set S of retained nodes on *path*(v). Let $B(T(v), \Delta, S)$ denote the least number of retained wavelet coefficients of $T(v)$ that satisfies Δ-bound for the approximation. The algorithm of [9] uses the following two equations to compute $B(T(v), \Delta, S)$: use Equation (1) if node v is to be retained, otherwise, use Equation (2).

$$B(T(v), \Delta, S) = B(T_L(v), \Delta, S \cup v) + B(T_R(v), \Delta, S \cup v) + 1 \tag{1}$$

$$B(T(v), \Delta, S) = B(T_L(v), \Delta, S) + B(T_R(v), \Delta, S) \tag{2}$$

Therefore, the final $B(T(v), \Delta, S)$ is the minimum of the above two possibilities. This method shares the same dynamic programming idea as the one published in [3], where an optimal algorithm of synopses construction for B-bound approximation was proposed.

However, due to special characteristics of the Δ-bounded problem, we can exploit typical Δ related features to improve the performance of the optimal

Algorithm 1. $minMax(c_r, v_k, v_d)$

Input:

c_r is the root node of a subtree; v_k is the summation of kept nodes on $path(c_r)$; v_d is the summation of discarded nodes on $path(c_r)$

Output:

The optimal set of kept coefficients in T_{c_r} that satisfies Δ bound

Description:

initialize OPT, the optimum results set
if c_r is leaf node **then**
  if $|v_k - c_r| < \Delta$ **then**
    $OPT.bucketNumber = 0$
  end if
else
  $OPT.bucketNumber = +\infty$; //indicate the retained set no valid
end if
if c_r is an inner node **then**
  $L(c_r)(R(c_r))$ = left (right) child of c_r
  $L1_{OPT}$, $L2_{OPT}$ ($R1_{OPT}$, $R2_{OPT}$), left (right) subtree's optimum result
  //pruning criteria, if not satisfied, c_r must be kept
  if $|c_r| + |v_d| < \Delta$ **then**
    $L1_{OPT} = minMax(L(c_r), v_k, v_d + c_r)$
    $R1_{OPT} = minMax(R(c_r), v_k, v_d - c_r)$
  end if
  b_1 = left + right //bucket numbers of not keeping c_r
  $L2_{OPT} = minMax(L(c_r), v_k + c_r, v_d)$
  $R2_{OPT} = minMax(R(c_r), v_k - c_r, v_d)$
  b_2 = left + right + 1 //bucket numbers of keeping c_r
  find $min(b_1, b_2)$ and combine subtree results to get OPT, accordingly.
  return OPT
end if

algorithm in some cases. For example, in Equation (1) and (2), set S can be constrained to satisfy certain conditions rather than being an arbitrary subset of nodes on *path*(v). In the following, we will describe some properties of Δ-bounded synopses which will be used in our algorithm.

Let M_∞ be an optimal Δ-bounded synopsis on error tree T and denote *diff*(d_i) as $\sum_{c_j \in path(d_i) - M_\infty} \delta_{ij} c_j$.

By definition, we know that for any viable solution that satisfies the Δ bound, the summation of deleted nodes along any path (from root to a leaf node) of the error tree is less than Δ. That is, $|diff(d_i)| < \Delta$ holds for $i = 0, 1, \ldots N-1$.

For coefficient $c_j \in path(d_i)$, we define *diff*(c_j) to be the summation of deleted ancestor nodes along $path(d_i)$, which is

$$diff(c_j) = \sum_{c_k \in path(c_j) - M_\infty} \delta_{ik} c_k.$$

Based on these formulae, we develop the following three properties:

Property 1. Let T be the error tree on $D = [d_0, d_1, \ldots, d_{N-1}]$ and M_∞ be an optimal Δ-bounded synopsis on T. Suppose $c_j \in path(d_i)$. Then

(i) $|diff(c_j)| < \Delta$;
(ii) $|diff(c_j)| + |c_j| < \Delta$ if $c_j \notin M_\infty$;
(iii) For any node $c_k \in T$, $c_k \in M_\infty$ if $|c_k| \geq \Delta$.

Proof. Suppose there are l leaf nodes in $T(c_j)$, ranging from d_h to d_{h+l-1}.
The proof of (i): It is easy to verify that discarding any inner node of $T(c_j)$ will not change the summation of the difference of the l leaf nodes [10]. That is:

$$\sum_{i=h}^{h+l-1} diff(d_i) = l \times diff(c_i)$$

Since $|diff(d_i)| < \Delta$, we have: $l \times |diff(c_i)| < l \times \Delta$. Thus $|diff(c_i)| < \Delta$ is proven.
The proof of (ii): From (i), we have

$$|diff(c_j) + c_j| < \Delta \text{ and } |diff(c_j) - c_j| < \Delta,$$

which implies (ii).
The proof of (iii): A contradiction will be derived from (ii) if $c_j \notin M_\infty$ and $c_j \geq \Delta$ are assumed. From (ii) can be derived straightforwardly from the above formulas.

We propose an optimal algorithm with $minMax(c_r, v_r, v_d)$ as the key procedure (Algorithm 1). $minMax(c_r, v_r, v_d)$ has three parameters: c_r is the currently considered node; v_r is the summation of the retained nodes that are on the path from the root node to node c_r (excluding c_r); and v_d is the summation of discarded nodes that are on the path from the root node to node c_r (excluding c_r). The function returns the set of coefficients that represents an optimal Δ-bounded synopsis in the subtree T_{c_r} under the two given values v_r and v_d.

Property 1(i) and (ii) can be used to check the Δ condition dynamically. Property 1(ii) is used at Steps 15-18 of $minMax(c_r, v_r, v_d)$ to prune unnecessary data expansion: node c_r can not be discarded if $|v_d| + |c_r| \geq \Delta$.

While the time complexity of our property-based algorithm is still of $O(N^2)$ in theory, the extensive experiments, as described in Section 4, indicate that this algorithm is more efficient than the existing algorithms in many situations and no worse in others. Refer to Section 4 for details.

Additionally, Property 1(iii) also gives a lower bound on M_∞ which can be used for a rough estimation on the size of M_∞. That is,

Corollary 1. *Let T be the error tree on $D = [d_0, d_1, \ldots, d_{N-1}]$ and M_∞ be an optimal Δ-bounded synopses on T. Then*

$$|M_\infty| \geq |\{c_i | (c_i \in T) \wedge (|c_i| \geq \Delta)\}| .$$

Clearly, $\{c_i | (c_i \in T) \wedge (|c_i| \geq \Delta)\}$ can be obtained in $O(|T|)$ time.

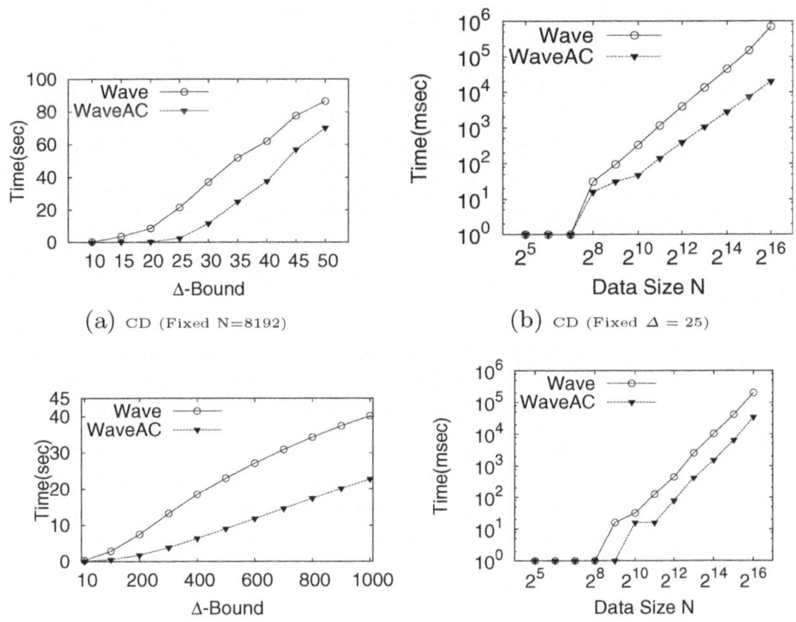

(a) CD (Fixed N=8192)
(b) CD (Fixed $\Delta = 25$)
(c) OD (Fixed N=8192)
(d) OD (Fixed $\Delta = 25$)

Fig. 2. The comparisons of WaveAC and WaveAC

4 Experimental Evaluation

In this section, we evaluate the effectiveness of our pruning techniques. We implement our algorithms through VC++ .NET. All the experiments were performed on a Pentium IV 3.6GHZ machine with 2 GB memory.

Two types of synthetic data sets are generated for our experiments on constructing Δ-bounded synopses: the coefficient-data set (CD) and the original-data set (OD). The CD data set contains data uniformly selected from $[10, 20]$ as a set of wavelet coefficients (W_D). It is actually an error tree and we can directly construct the Δ-bounded synopses from the CD. The OD data set contains data uniformly selected from $[0, 10000]$ as a vector data (D). It is the original data set and we need to apply the Haar wavelet transformation on it before we can construct the synopses.

We conducted experiments to evaluate the efficiency of the two algorithms in generating optimal Δ-bounded synopses, one with our pruning techniques (named as WaveAC) and one without it (named as Wave), i.e., the original algorithm mentioned at [12]. Their comparisons on computation time are depicted in Figure 2.

In Figure 2(1) and (2), the experimental results were on CD data. Figure 2(1) is on a fixed size CD data ($|D| = 8192$) under varied Δ ranging from 10 to 50. Figure 2(2) is on a fixed Δ ($\Delta = 25$) under varied data size D ranging from 32 to 65536 nodes. The experiments on OD data under the same scenario are given in Figure 2(3) and (4).

From the experiments, we have the following observations. On a fixed size data set, as indicated in Figure 2(1), the pruning technique (WaveAC) can improve the speed up to 25 times faster when Δ is between 10 and 20, which is the range of the values of the coefficients. The increase of speed will drop to 1.5 times as Δ increases. Figure 2(2) shows a comparison of varied data size for a fixed Δ. The improvement caused by the pruning techniques increased the speed to up to 35 times faster. These facts are further supported with the results of Figure 2(3) and (4).

5 Conclusion

In this paper, we have proposed new algorithms on the construction of Δ-bounded synopses to minimize maximum error. Our approach is based on the intrinsic properties of W_D upon a Δ error bound.

Our future work is to improve and extend this work in several ways: to apply the obtained properties in different ways to derive better results; to investigate more properties that can lead more efficient algorithms on construction of Δ-bounded synopses and to support streaming data processing and applications.

References

1. S. Chaudhuri, R. Motwani, and V. Narasayya, *Random sampling for histogram construction: How much is enough?*, ACM SIGMOD'98, pp. 436–447.
2. M. Garofalakis and P. B. Gibbons, *Wavelet synopses with error guarantees*, ACM SIGMOD'02, pp. 476–487.
3. M. Garofalakis and A. Kumar, *Deterministic wavelet thresholding for maximum-error metrics*, ACM PODS'04, pp. 166–176.
4. A. C. Gilbert, Y. Kotidis, S. Muthukrishnan, and M. Strauss, *Optimal and approximate computation of summary statistics for range aggregates*, ACM PODS'01, pp. 227–236.
5. S. Guha, *Space efficiency in synopsis construction algorithms*, VLDB'05, pp. 409–420.
6. S. Guha and B.Harb, *Approximation algorithms for wavelet transform coding of data streams*, SODA, 2006, pp. 698–707.
7. S. Guha and B. Harb, *Wavelet synopsis for data streams: minimizing non-euclidean error*, ACM SIGKDD, 2005, pp. 88–97.
8. S. Guha, K. Shim, and J. Woo, *Rehist: Relative error histogram construction algorithms.*, VLDB'04, pp. 300–311.
9. P. Karras and N. Mamoulis, *One-pass wavelet synopses for maximum-error metrics*, VLDB'05, pp. 421–432.
10. Y. Matias and D. Urieli, *Inner-product based wavelet synopses for range-sum queries.*, ESA, 2006, pp. 504–515.
11. Y. Matias, J. S. Vitter, and M. Wang, *Wavelet-based histograms for selectivity estimation*, ACM SIGMOD'98, pp. 448–459.
12. S. Muthukrishnan, *Subquadratic algorithms for workload-aware haar wavelet synopses.*, FSTTCS, 2005, pp. 285–296.
13. E. J. Stollnitz, T. D. Derose, and D. H. Salesin, *Wavelets for computer graphics: theory and applications*, Morgan Kaufmann Publishers Inc., 1996.

Exploiting the Structure of Update Fragments for Efficient XML Index Maintenance*

Katharina Grün and Michael Schrefl

Department of Business Informatics - Data & Knowledge Engineering
Johannes Kepler University Linz, Austria
{gruen,schrefl}@dke.uni-linz.ac.at

Abstract. XML databases provide index structures to accelerate queries on the content and structure of XML documents. As index structures must be consistent with the documents on which they are defined, updates on documents need to be propagated to affected index structures. This paper presents an index maintenance algorithm that is solely based on index definitions and update fragments instead of on the maintenance of auxiliary data structures. The use of index definitions assures that the algorithm supports arbitrary index structures defined on arbitrary document fragments. By exploiting the structure of update fragments, the algorithm directly extracts the nodes which are required for index maintenance from the fragments. Source queries are only necessary if the fragment does not contain all nodes required for indexing. The presented performance studies demonstrate the advantages of this approach over previous work that propagates each updated node individually.

1 Introduction

The increasing number and size of XML documents require efficient techniques for querying and updating the content and structure of XML data stored in databases. To accelerate query processing, databases use secondary index structures, which provide a search function to select nodes without scanning all data. Each entry in an index consists of a list of index keys, specifying the search condition, and the nodes to be returned by the search function. Incremental index maintenance refers to the process of determining which index structures need to be updated with which index entries when updating document fragments.

State-of-the-art indexing approaches (cf. [1]) mostly consider primary index structures that can be updated with specific maintenance algorithms. To adapt indices to the query workload, an XML database however requires various secondary index structures, e.g. b-trees and multidimensional indices. To reduce the index size, the database needs to support indices on arbitrary document fragments instead of on whole documents only. This implies that update fragments need not correspond to indexed fragments and thus do not necessarily contain all nodes required for index maintenance. To incrementally maintain index structures, either (i) each index structure maintains required nodes in an

* This work was supported by FIT-IT under grant 809262/9315-KA/HN.

G. Dong et al. (Eds.): APWeb/WAIM 2007, LNCS 4505, pp. 471–478, 2007.

auxiliary data structure, or (ii) a generic maintenance algorithm extracts index entries directly from update fragments and decouples the generation of index entries from the index structures as such. The approach presented in this paper follows approach (ii) as it is widely applicable and does not need to persist and maintain auxiliary data structures.

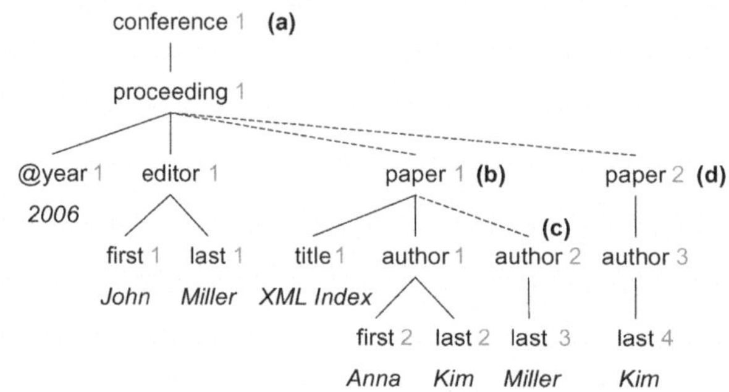

Fig. 1. Sample update fragments (a-d)

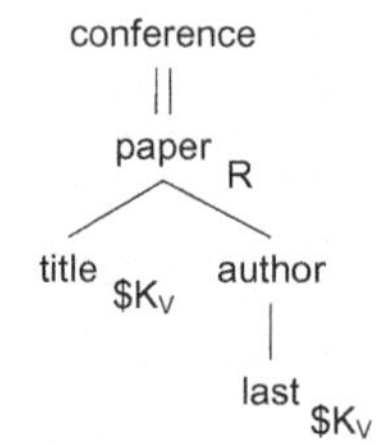

Fig. 2. Sample index

Example 1. As running example, an XML database storing conference proceedings is used. The sample document of Fig. 1 consists of four fragments (a-d), which are connected via dashed lines. The name and the number associated with each node uniquely identify the node. Figure 2 depicts a sample index definition, represented as a tree pattern. Symbol $\$K_V$ indicates to use the value of a node as index key. Symbol R specifies which nodes to return by the index. Single lines represent parent-child edges, double lines are used for ancestor-descendant edges. The index selects papers (return R) by specifying their title and their authors' last name (index keys K). Each index entry can be written as [(K) → R]. Assume that we successively insert fragments (a-c) into the database. The insertion of fragment (a) does not affect the index. The insertion of fragments (b) and (c) triggers the insertion of the index entries [('`XML Index`', '`Kim`') → `paper1`] and [('`XML Index`', '`Miller`') → `paper1`], respectively. Note that fragment (c) misses the title which is required to create the index entry.

This paper proposes a novel algorithm to extract index entries from update fragments based on index definitions. Index definitions uncouple the algorithm from specific index structures. They are represented as tree patterns to support indices on arbitrary document fragments selected by an arbitrary index definition language, e.g. with the XPath fragment {/, //, *, []}. The main idea of the algorithm is to find embeddings of index patterns in update fragments and then generate index entries from these embeddings. By exploiting the structure of the update fragment, the proposed algorithm minimizes the number of additional queries without relying on auxiliary data structures. It only needs to retrieve those nodes from the database that are required for indexing but are not contained in the update fragment. If all index entries can be inferred from the update fragment, the algorithm is completely self-maintainable. The update of index structures with index entries is not subject of this work, as there exist specific algorithms to update the index structures as such. The developed techniques are not restricted to index maintenance, but can be carried over to the maintenance of caches, views or related problems.

The paper is structured as follows. Section 2 reviews related work and Sect. 3 defines update fragments, index patterns and embeddings of patterns in documents. The maintenance algorithm is presented in Sect. 4 and analyzed in Sect. 5. Section 6 concludes the paper.

2 Related Work

Relational databases define indices on columns of relations and simply need to update an index when the values of a tuple change. Object-oriented databases use specific index structures for efficient navigation along aggregation hierarchies, i.e. nested index and multi-index [2]. When one of the objects changes, all objects on the indexed path are required for performing the update operation. For this purpose, the index structures maintain the relevant objects in an auxiliary data structure. Various native XML databases [3] integrate indices into query processing. They support simple structural and value indices but cannot index nodes along multiple axes, e.g. for multidimensional indexing.

To the best of our knowledge, only the approach of Hammerschmidt et al. [4] supports incremental index maintenance based on more complex XML index definitions. The algorithm decomposes each index definition into a set of linearized path expressions and matches them with each updated node. If the path of an updated node intersects with a path of an index definition, queries on the remaining paths of the index definition are executed to retrieve all nodes that constitute an index entry. As the algorithm processes each updated node individually, it may generate the same index entry several times, which may lead to inconsistencies, and executes a large number of source queries (cf. Example 2).

Example 2. Assume that the index of Fig. 2 is updated with fragment (d) of Fig. 1. The algorithm generates the index entry $[(\perp, \text{'Kim'}) \rightarrow$ `paper2`] twice, once for node `paper2` and once for node `last4`. The algorithm executes source queries to retrieve the remaining index keys and/or the return of the index entries. The update however only affects one entry and would not require any source query as all index entries can be inferred from the update fragment.

Maintaining XML views is closely related to maintaining XML indices since views also need to be updated when documents change. Existing approaches on XML view maintenance either only support updates containing all relevant nodes (e.g. [5]) or maintain relevant nodes in an auxiliary data structure (e.g. [6]).

Pattern matching (e.g. [7]) and tree inclusion (e.g. [8]) algorithms find embeddings of query patterns in XML documents. XML filtering techniques (e.g. [9]) evaluate queries on XML documents on-the-fly. This work applies ideas of these query processing techniques to index maintenance. The main differences are that the index maintenance algorithm needs to (i) find embeddings in update fragments instead of in whole documents, (ii) issue source queries if the fragment misses nodes which are required for indexing, (iii) associate the index keys with the nodes to be returned and generate index entries.

3 Embeddings of Index Patterns in Documents

The main idea of our approach is to find embeddings of index patterns in update fragments and then generate index entries from these embeddings. Each embedding consists of nodes in the document that structurally correspond to the index pattern and represents an index entry. Before describing embeddings in more detail, we define update fragments and index patterns.

A *document* D is an ordered tree of nodes N_D connected via directed edges F_D, $D = (N_D, F_D)$. Each node is identified by a unique node id, returned by function *nid*. Function *label* returns the name of a node. Leaf nodes may have associated a value. Function *root* returns the root node of a document. The *parent* of n is the node whose edge leads to n. The *children* of n are all nodes whose edges emanate from n. The *descendants* of n correspond to the transitive closure of the *children* of n, the *ancestors* function is its inverse. The *path* of n is the sequence of nodes and edges from the root to n. Function *labelpath* of n applies function *label* on each node of the *path* of n and returns the sequence of node names and edges that connect the nodes. A *schema* is a structural summary consisting of all distinct labelpaths of a document. Note that generating a schema does not require a schema file in form of a DTD or an XML Schema.

A document *fragment* D_F is a subtree of D consisting of a node $n \in N_D$, which is the root of D_F, plus all descendants of n in D. An *update fragment* consists of the nodes that are either inserted into or deleted from a document. Modifications can be executed as a deletion followed by an insertion. The update operation specifies the location of the update fragment, i.e. the path from the root of the document to the root of the update fragment. This implies that the labelpath of each node in the update fragment is known.

An *index* I consists of a set of index entries, $I = (E_I)$. Each index entry maps a list of index keys K to a list of nodes $R \subseteq N_D$, which are identified by their node ids, $E_I = [(K_1, ..., K_n) \rightarrow (nid(n_1), ..., nid(n_m))]$. An index key can be any property of a node that can be indexed (e.g. its value or type). While the index entries of an index contain all nodes belonging to certain index keys, each index entry of an update operation only consists of one node and its index keys. This node is then either inserted into or deleted from the index entry of the index having the same keys. The number of index keys of an index entry determines the dimensionality of the index. A multidimensional index may allow one or several (but not all) index keys to be null, represented as $\perp$. An *index structure* is the specific data structure used to represent an index, e.g. a b-tree.

An index *pattern* P represents an index definition in a language-independent way. It is an unordered tree of pattern nodes N_P connected via parent-child (/) or ancestor-descendant (//) edges F_P, $P = (N_P, F_P)$. Each pattern node has a name, which is either a concrete name or a wildcard (*), and may have associated index variables. Index variables $\$K$ determine which properties of a node to use as index keys. The order of index variables in the index definition determines the order of index keys in an index entry. Each index pattern consists of one distinguished pattern node specifying the return value R of an index entry. We assume that each index is defined on one document, but the approach can easily be extended to document collections. We define the same functions on a pattern as on a document. If any pattern node has more than one child, the patterns is called a *twig index pattern*, otherwise it is referred to as *path index pattern*.

Nodes of a document that structurally correspond to the pattern nodes of a pattern are referred to as embedding. Formally, an *embedding* of a pattern $P = (N_P, F_P)$ in a document $D = (N_D, F_D)$ is a mapping from N_P to N_D, $emb : N_P \rightarrow N_D$, such that $\forall x, y \in N_P$ (i) if $label(x) \neq * \rightarrow label(x) = label(emb(x))$, (ii) if $x/y \in F_P \rightarrow emb(x) = parent(emb(y))$, (iii) if $x//y \in F_P \rightarrow emb(x) \in ancestors(emb(y))$. The nodes of an embedding are called *matching nodes*. The pattern nodes that match a node are referred to as *matching pattern nodes*.

4 Maintenance Process

The proposed algorithm consists of three steps: (1) find embeddings of patterns in update fragments, (2) query required nodes that are missing in the fragment and (3) generate index entries from the embeddings. The index entries are then forwarded to the affected index structures, which update their data structure with proprietary algorithms. Insertions and deletions can be handled analogously as they only differ in the update operation executed on the index structure.

The algorithm uses an intermediate data structure to compactly store the nodes of an embedding up to the time of generating index entries. The data structure resembles the one used in pattern matching algorithms (e.g. [7]) and associates with each pattern node a stack. Each entry in a stack consists of a node of the document and a pointer to its ancestor node in the stack of the parent pattern node, encoding the structural relationships between the nodes.

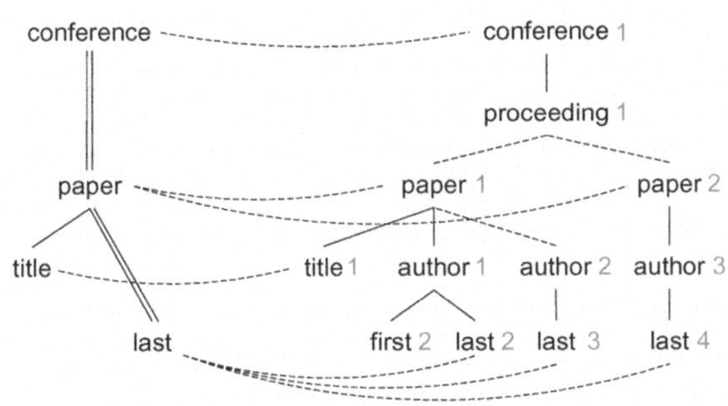

Fig. 3. Sample embeddings

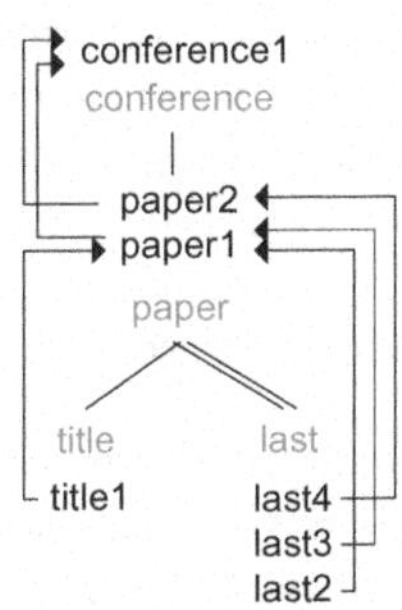

Fig. 4. Stack encoding

Example 3. Figure 3 visualizes embeddings of a pattern in a document and Fig. 4 represents these embeddings in the intermediate data structure. The data structure encodes the embeddings (`conference1`, `paper1`, `title1`, `last2`), (`conference1`, `paper1`, `title1`, `last3`) and (`conference1`, `paper2`, $\perp$, `last4`).

Step 1 - Find embeddings. To find embeddings of an index pattern in an update fragment, step 1 of the algorithm traverses the fragment once in document order. If a fragment is relevant, the algorithm matches its root with the pattern and puts it onto the stacks of matching pattern nodes. It then recursively processes the children of the root. Nodes of a fragment that is not relevant are not further processed. The details of step 1 are shown in Algorithm 1.

Procedure *relevant* determines whether a fragment is relevant for a pattern by comparing the labelpath of its root with the labelpaths of the pattern nodes. Procedure *match* finds matching pattern nodes for a node (cf. embedding in Sect. 3). Procedure *addToStack* puts the node onto the stack of each matching pattern node. Additionally, it creates a pointer from the node to the previously added node in the parent stack to connect the node with its ancestor.

Algorithm 1. Find embeddings of a pattern in a fragment
Input: update fragment $D_F = (N_D, F_D)$, index pattern $P = (N_P, F_P)$
Output: embeddings of P in D_F, encoded via stacks associated with N_P

procedure FINDEMBEDDINGS(D_F,P)
    if $relevant(P, D_F)$ **then** ▷ process fragment if it is relevant
        $N_{P_M} = match(root(D_F), P)$ ▷ determine matching pattern nodes N_{P_M}
        $addToStack(N_{P_M}, root(D_F))$ ▷ add root to stacks of N_{P_M}

        for all $child \in children(root(D_F))$ **do**
            $findEmbeddings(child, P)$ ▷ recursively process child fragments
        end for
    end if
end procedure

The first two procedures can benefit from schema information by associating with each pattern node the matching labelpaths of the document. It is then possible to pregenerate (i) a list of labelpaths that are relevant for a pattern, (ii) a map from labelpaths to matching pattern nodes. The algorithm then does not need to compare each node with each pattern node but can directly determine whether a fragment is relevant as well as matching pattern nodes for a node.

Example 4. When updating the index of Fig. 2 with the document of Fig. 1, the algorithm proceeds as follows (cf. Fig. 3 and 4). Node `conference1` matches pattern node `conference`. The fragment rooted at `proceeding1` is relevant, but its root does not match a pattern node. With a schema, the algorithm can exclude that an editor has a paper as descendant and need not process this fragment.

The fragment rooted at `paper1` is relevant and its root matches pattern node `paper`. The algorithm performs the same steps on the remaining nodes.

The algorithm can be extended in several ways (details are omitted for space reasons): (i) Each embedding requires a matching node for the pattern node specifying the return. Also other pattern nodes may be marked as required or may specify properties that matching nodes must fulfill. To improve space complexity, the algorithm should not keep partial embeddings for which required nodes cannot exist in the intermediate data structure. (ii) In case of a large number of index patterns, the algorithm should avoid matching the same sub patterns multiple times by pattern sharing.

Step 2 - Execute queries. To retrieve the nodes that are part of index entries but not contained in the update fragment, the algorithm recursively processes the pattern nodes in a postorder traversal to query missing ancestors. When adding an ancestor to a pattern node, the algorithm recursively traverses its child pattern nodes in a preorder traversal to query missing descendants.

Example 5. When updating the index of Fig. 2 with fragment (c) of Fig. 1, step 1 of the algorithm associates node `last3` with pattern node `last`. Starting from this pattern node, the query execution algorithm retrieves the corresponding paper (`paper1`) and its title (`title1`).

The algorithm can also be used to retrieve all embeddings from a document when creating or deleting an index on an existing document. The efficiency of this step can be improved with labeling schemes that allow for navigating along certain axes without accessing base data. The update of whole documents is always self-maintainable as no queries are executed. A schema-aware labeling scheme (e.g. [10]) also makes the update of path patterns self-maintainable as it allows for navigating to ancestors without source queries.

Step 3 - Generate index entries. To generate index entries, the algorithm extracts the embeddings from the intermediate data structure (cf. [7]). The index entries need to be inserted into the index structure in case of an insert operation, and deleted in case of a delete operation. Special attention needs to be paid when an update operation adds/removes keys to/from existing index entries in a multidimensional index. These keys always stem from the nodes that have not been queried and either replace null values when executing an insert or generate null values in case of a delete operation.

Example 6. The first embedding of Example 3 is converted into the index entry [('`XML index`', '`Kim`') $\rightarrow$ `paper1`]. Assume that we add a title to the second paper and update the sample index. The existing index entry [($\perp$, '`Kim`') $\rightarrow$ `paper2`] then needs to be modified, i.e. the null value needs to be replaced with the new title '`XML Index`'.

5 Performance Studies

We benchmarked our approach (a) of exploiting the structure of update fragments against the approach (b) of updating individual nodes (cf. Sect. 2, [4]). As sample dataset, we generated three fragments (cf. Fig. 1): (i) a proceeding with 100 papers, (ii) a paper and (iii) an author with a last name. When updating the sample index of Fig. 2 with fragments (i) and (ii), approach (b) generates each index entry twice and executes a large number of source queries. Matching whole fragments makes our approach 9 times faster, as it does not issue any source queries when updating the index with these fragments. Both approaches perform equally for fragment (iii), which only contains one relevant node.

6 Conclusion

We presented an algorithm to extract index entries from update fragments. Indices are represented as index patterns to handle arbitrary index structures defined on arbitrary document fragments. The algorithm (1) finds embeddings of index patterns in update fragments, (2) executes queries if nodes are required for indexing that are not part of the update fragment, (3) generates index entries from the embeddings and forwards them to affected index structures. By exploiting the structure of update fragments, the algorithm minimizes the number of source queries, resulting in an improved update performance.

References

1. Catania, B., Maddalena, A., Vakali, A.: XML Document Indexes: A Classification. IEEE Internet Computing **9**(5) (2005) 64–71
2. Bertino, E., Foscoli, P.: Index Organizations for Object-Oriented Database Systems. IEEE Transactions on Knowledge and Data Engineering **7**(2) (1995) 193–209
3. Chaudhri, A., Zicari, R., Rashid, A.: Xml Data Management: Native XML and XML-Enabled Database Systems. Addison-Wesley Longman Publishing (2003)
4. Hammerschmidt, B.C.: KeyX: Selective Key-Oriented Indexing in Native XML-Databases. PhD thesis
5. Liefke, H., Davidson, S.B.: View Maintenance for Hierarchical Semistructured Data. In: DaWaK, Springer (2000) 114–125
6. El-Sayed, M., Wang, L., Ding, L., Rundensteiner, E.: An Algebraic Approach for Incremental Maintenance of Materialized XQuery Views. In: WIDM, ACM (2002) 88–91
7. Yao, J., Zhang, M.: A Fast Tree Pattern Matching Algorithm for XML Query. In: Web Intelligence, IEEE Computer Society (2004) 235–241
8. Bille, P., Gørtz, I.L.: The Tree Inclusion Problem: In Optimal Space and Faster. In: ICALP, Springer (2005) 66–77
9. Diao, Y., Altinel, M., Franklin, M.J., Zhang, H., Fischer:, P.M.: Path Sharing and Predicate Evaluation for High-Performance XML Filtering. ACM Transactions on Database Systems (TODS) **28**(4) (2003) 467–516
10. Grün, K., Karlinger, M., Schrefl, M.: Schema-aware Labelling of XML Documents for Efficient Query and Update Processing in SemCrypt. Computer Systems Science and Engineering **21**(1) (2006) 65–82

Improvements of HITS Algorithms for Spam Links

Yasuhito Asano[1], Yu Tezuka[2], and Takao Nishizeki[2]

[1] Tokyo Denki University, Ishizaka, Hatoyama-Cho, Hiki-Gun, Saitama, Japan
asano@j.dendai.ac.jp
[2] Tohoku University, Aza-Aoba 6-6-05, Aoba-Ku, Sendai, Miyagi, Japan

Abstract. The HITS algorithm proposed by Kleinberg is one of the representative methods of scoring Web pages by using hyperlinks. In the days when the algorithm was proposed, most of the pages given high score by the algorithm were really related to a given topic, and hence the algorithm could be used to find related pages. However, the algorithm and the variants including BHITS proposed by Bharat and Henzinger cannot be used to find related pages any more on today's Web, due to an increase of spam links. In this paper, we first propose three methods to find "linkfarms," that is, sets of spam links forming a densely connected subgraph of a Web graph. We then present an algorithm, called a trust-score algorithm, to give high scores to pages which are not spam pages with a high probability. Combining the three methods and the trust-score algorithm with BHITS, we obtain several variants of the HITS algorithm. We ascertain by experiments that one of them, named TaN+BHITS using the trust-score algorithm and the method of finding linkfarm by employing name servers, is most suitable for finding related pages on today's Web. Our algorithms take time and memory no more than those required by the original HITS algorithm, and can be executed on a PC with a small amount of main memory.

1 Introduction

Search engines are widely used as a tool for obtaining information on a topic from the Web. Given keywords specifying a topic, search engines score Web pages containing the keywords by using several scoring algorithms, and output the pages in decending order of the score. For example, PageRank proposed by Brin and Page [3] has been used as a scoring algorithm by Google. Another scoring algorithm HITS proposed by Kleinberg [9] has the following three significant features (1)-(3). (1) The HITS algorithm gives high scores to pages related to the topic specified by given keywords even if the pages do not contain the keywords. (2) The HITS algorithm can be executed on a PC with a small amount of main memory, because it needs data of a quite small number of pages, compared with the PageRank algorithm and most of the scoring algorithms used by search engines. (3) The HITS algorithm can be executed on demand, because it needs data small enough to be collected through the network on demand. On the

G. Dong et al. (Eds.): APWeb/WAIM 2007, LNCS 4505, pp. 479–490, 2007.

other hand, the PageRank algorithm takes several weeks to collect data, and hence cannot be executed on demand.

The HITS algorithm worked well in the days when it was proposed. Several HITS-based algorithms have been proposed since [1,10,11,13]. However, the original HITS algorithm and the HITS-based algorithms no longer work well on today's Web due to an increase of spam links. Several methods of finding spam links have recently been developed [4,5,6], but they require too large data of pages to be executed on demand with a PC. For example, the methods proposed by Fetterly *et al.* [5,6] require the data of the contents of pages, which are much larger than the data of the links of the pages used by the HITS algorithm.

In this paper, we first propose three methods to find linkfarms by using network information; a *linkfarm* is a set of spam links which form a densely connected subgraph of a Web graph; a *Web graph* is a directed graph whose vertex set is a set of Web pages, and whose edge set is a set of links between pages. Our methods find more linkfarms than the method proposed by Wu and Davison [14]. We then propose a *trust-score* algorithm to give high scores to pages which are not spam pages with a high probability, by extending the ideas used by the TrustRank algorithm [7]. We then construct four scoring algorithms; the first one is obtained by combining our trust-score algorithm with BHITS algorithm proposed by Bharat and Henzinger [1]; the remaining three are obtained by combining each of our three methods of finding linkfarms with the trust-score algorithm and BHITS. We finally evaluate our algorithms and several HITS-based algorithms by experiments.

In order to evaluate various scoring algorithms, we use the "quality of top ten authorities" found by the algorithm for a given topic; the *top ten authorities* are pages of the top ten high score given by the algorithm, the *quality* of top ten authorities is measured by the number of pages related to the topic among the top ten authorities, and hence the quality of top ten authorities is at most ten. We examine the quality of top ten authorities by computational experiments using fourteen topics. For almost all the topics, our algorithms find top ten authorities of higher quality than those found by the existing algorithms. Particularly, one of our algorithms, called TaN+BHITS, employing the trust-score algorithm and a method of finding linkfarm by using name servers, finds top ten authorities of the best average quality 8.79, while the existing algorithms find top ten authorities of average quality at most 3.07 (see Table 1 in Section 4).

Our TaN+BHITS algorithm can be used to find pages related to a given topic on today's Web; most of the pages given high score by the algorithm are truly related to a given topic for almost all the topics used in our experiments. Our four algorithms including TaN+BHITS require no data of pages other than the data collected by the original HITS algorithm, and hence can be executed on demand for a given topic on a PC with a small amount of main memory.

2 Preliminaries

We first present the definition of a host name and a domain name in Section 2.1, then summarize the original HITS algorithm in Section 2.2, and finally outline the BHITS algorithm proposed by Bharat and Henzinger in Section 2.3.

2.1 Terms

Since a term "host name" is sometimes confused with a term "domain name," we use the following definitions throughout the paper.

The *host name* of a Web page p is the name of the host containing p. As a host name of p, we use a substring of p's URI between `http://` and the next slash symbol. For example, if a page p has URI `http://www.geocities.jp/ken/a.html`, then the host name of p is `www.geocities.jp`. Let $domlevel(p)$ be one plus the number of dot symbols in the host name of page p. Thus, $domlevel(p) = 3$ for the page p above.

Divide a host name by dot symbols into a number $domlevel(p)$ of substrings, then the i-th substring from the right is called the *i-th level domain*. For example, if the host name of a page p is `www.geocities.jp`, then `geocities` is the second level domain of p. We say that two pages p and q have the *same domain name* if either p and q have the same host name or $domlevel(p) = domlevel(q) \geq 3$ and the i-th level domain of p is equal to the i-th level domain of q for each i, $1 \leq i \leq domlevel(p) - 1$. For example, if page p has URI `http://news.infoseek.jp/` and page q has URI `http://music.infoseek.jp/`, then p and q have the same domain name, because $domlevel(p) = domlevel(q) = 3$ and p and q have the same first and second level domains, `jp` and `infoseek`, respectively. On the other hand, if page p has URI `http://ask.jp` and page q has URI `http://slashdot.jp`, then p and q do not have the same domain, because they do not have the same host name and $domlevel(p) = domlevel(q) = 2$.

2.2 Original HITS Algorithm

The HITS algorithm proposed by Kleinberg [9] finds *authorities* and *hubs* for a topic specified by given keywords. The algorithm regards a page linked from many pages as an *authority*, and regards a page having links to many authorities as a *hub*, as outlined as follows.

1. Let the *root set* R be a set of top x pages of the result of some search engine for given keywords, where x is an integer parameter. Kleinberg [9] usually set $x = 200$.
2. Construct a Web graph $G = (V, E)$, where the vertex set V is a union of R and the set of all pages that either link to or are linked from pages in R, and the edge set E consists of all the links between pages in V except links between pages having the same hostname.
3. For each vertex $v_i \in V$, $1 \leq i \leq |V|$, set an *authority score* a_i, to 1 and set a *hub score* h_i, to 1.
4. Repeat the following procedures (a)-(c) ℓ times, where ℓ is a given integer parameter.
   (a) For each $v_i \in V$, $a_i := \sum_{(v_j, v_i) \in E} h_j$;
   (b) For each $v_i \in V$, $h_i := \sum_{(v_i, v_j) \in E} a_j$; and
   (c) Normalize vectors $\mathbf{a} = (a_1, a_2, ..., a_{|V|})$ and $\mathbf{h} = (h_1, h_2, ..., h_{|V|})$ in the L_2-norm so that $\sum_{i=1}^{|V|} a_i = 1$ and $\sum_{i=1}^{|V|} h_i = 1$.
5. Output vectors $\mathbf{a}$ and $\mathbf{h}$.

Kleinberg usually set $\ell = 50$, because the vectors **a** and **h** almost converge after repeating (a)-(c) in Step 4 fifty times.

Throughout the paper, the "top ten authorities" found by a HITS-based algorithm mean the ten pages of the heighest authority score among the pages found by the algorithm, and we measure the "quality" of the top ten authorities by the number of pages related to a given topic among the top ten authorities. For example, if the top ten authorities found by a HITS-based algorithm contain seven pages related to a given topic, then the quality of the top ten authorities is seven. Similar measures have been used by several researchers on HITS-based algorithms [1,11].

The original HITS algorithm could find top ten authorities of practically good quality and could be used to find pages related to a given topic in the days when it was proposed. Subsequently, several researchers pointed out a *mutual reinforcing* problem and a *topic drift* problem of the original HITS, and proposed various HITS-based algorithms with effective solutions to these problems [1,2,8,10,11,12,13,15]. We will describe the mutual reinforcing problem in Section 2.3. The original HITS algorithm sometimes wrongly gives high scores to pages not related to a given topic. This is called a topic drift problem.

The number of pages on the Web has been exponentially increasing since the Web was born, and the so-called spam links have been increasing, too. On today's Web, the HITS algorithm and the variants cannot find top ten authorities of good quality any more due to the increase of spam links.

Some authors [4,5,6] define a spam link as follows: a link from a page p to a page q is a *spam link* if the content of q is irrelevant to the content of p and the link is created to force a scoring method to give a high score to the site containing the page p. It is difficult to find spam links according to this definition, and hence the authors try to find spam links approximately by using heuristics. For example, Fetterly *et al.* [5,6] proposed several heuristics to find spam pages. The heuristics require the content information of pages, although the original HITS uses only the link information of pages. Costa Carvalho *et al.* [4] proposed several heuristics to find spam links by using characteristic graph structures of *inter-host links*, each between a page in a host and a page in another host. All these heuristics require data of many pages other than the pages used by the original HITS. Thus, the existing methods of finding spam links require much larger data than those required by the original HITS, and hence they are not suitable for our objective of establishing a scoring algorithm which can be executed on demand with a PC.

2.3 BHITS Algorithm

The orignal HITS algorithm suffers from the following problem. If a malicious person creates a host containing a number of pages linking to the same page p in another site, then the authority score of the page p becomes unfairly high. One can easily create such a host. Thus, the top ten authorities found by the original HITS algorithm could be made unreliable by a malicious person. This problem is called a *mutually reinforcing problem.*

Slightly modifying the original HITS algorithm, Bharat and Henzinger [1] proposed the BHITS algorithm, which almost resolves the mutually reinforcing problem. We now introduce several notations to explain their modifications. For each vertex v_i in the base set V used in the original HITS, let $\Gamma^-(v_i)$ be the set of vertices $\{v_j \in V \mid (v_j, v_i) \in E\}$, and let $\Gamma^+(v_i)$ be the set of vertices $\{v_j \in V \mid (v_i, v_j) \in E\}$. Let $H^-(v_i)$ (or $H^+(v_i)$) be the set of all hosts containing pages corresponding to vertices in $\Gamma^-(v_i)$ (or $\Gamma^+(v_i)$, respectively). For each host $host_k \in H^-(v_i)$, $1 \le k \le |H^-(v_i)|$, let $r^-(host_k, v_i)$ be the number of links going from pages in $host_k$ to the page v_i. Similarly, for each host $host_k \in H^+(v_i)$, $1 \le k \le |H^+(v_i)|$, let $r^+(host_k, v_i)$ be the number of links going from page v_i to those in $host_k$. The BHITS algorithm replaces (a) and (b) in Step 4 of the original HITS algorithm by the following (a)′ and (b)′, respectively.

(a)′ For each $v_i \in V$, $a_i := \sum_{(v_j,v_i)\in E} \frac{h_j}{r^-(host(v_j),v_i)}$.
(b)′ For each $v_i \in V$, $h_i := \sum_{(v_i,v_j)\in E} \frac{a_j}{r^+(host(v_j),v_i)}$.

Thus, even if a number of pages in a host link to the same page p, the authority score of the page p computed by the BHITS algorithm does not become unfairly high, and hence the mutually reinforcing problem is almost resolved.

For most of the topics used in their experiments, the BHITS algorithm found top ten authorities of better quality than those obtained by the original HITS algorithm [1]. We will hence construct several variants of the BHITS algorithm instead of the original HITS algorithm.

3 Improvements

3.1 Removing Linkfarms

Wu and Davison [14] define a linkfarm as a set of spam links which form a densely connected subgraph on a Web graph, and say that a page belongs to a linkfarm if the page is an end of a spam link in the linkfarm. The HITS algorithm gives high authority scores and high hub scores to pages belonging to a densely connected subgraph, and hence pages belonging to a linkfarm would get high authority score and high hub score. Consequently, the top ten authorities obtained by the HITS algorithm would contain a number of pages belonging to the linkfarm.

Wu and Davison proposed an algorithm for finding linkfarms, and evaluated how effective their algorithm is for improving scoring algorithms. For the evaluation, they used a simple scoring algorithm which gives each page a score equal to the number of links entering to the page. For most of the topics used in their experiments, the simple scoring algorithm could obtain top ten authorities of good quality if it ignores all the links in the linkfarms found by their algorithm. Thus, their algorithm is effective for improving the simple scoring algorithm. However, they did not confirm whether their algorithm is effective or not for improving the HITS-based algorithms, which are more sophisticated than the simple scoring algorithm.

We evaluate how effective the algorithm of Wu and Davison is for improving the BHITS algorithm. Let **LF+BHITS** be the BHITS algorithm which ignores all the links in the linkfarms found by their algorithm. We will evaluate the quality of top ten authorities found by LF+BHITS in Section 4.

We discover that a typical linkfarm consists of pages sharing some kinds of network information, such as a host name, a domain name, an IP address, and a name server. (Both the IP address of the host containing a given page and the name of the name server used by the host can be easily obtained by several methods including `nslookup` UNIX command.) (A domain name is defined in Section 2.)

The original HITS algorithm ignores every link between two pages belonging to the same host, but it still suffers from a number of linkfarms. Authors of References [4,5,6] mentioned that a set of spam pages created by a marlicious person frequently shares some kinds of network information even if the pages do not have the same host name. Investigating a number of pages sharing the same host name, domain name, or IP address, we found that they usually share the same name server, too.

We propose an algorithm **N+BHITS**, which is the BHITS algorithm with the following two modifications:

(1) In Step 2 of the original HITS algorithm, remove every link between two pages sharing the same name server from the Web graph G; and
(2) Use a name server instead of a host name in Step 4 (a)$'$ and (b)$'$ of the BHITS algorithm.

Let Algorithm **I+BHITS** be the same as Algorithm N+BHITS except that it uses an IP address instead of a name server. Similarly, let Algorithm **D+BHITS** be the same as N+BHITS except that it uses a domain name instead of a name server. It is not new ideas to ignore links between two pages sharing the same IP address or domain name [15], but we compare effectiveness of these ideas with that of the method proposed by Wu and Davison. We will evaluate the quality of top ten authorities found by these algorithms in Section 4.

3.2 Trust-Score

Gyongyi *et al.* [7] proposed the TrustRank algorithm which approximately finds *reliable pages*, which are not created with malicious motives. The algorithm gives a *reliability score* to a page on the Web. The main idea used by the algorithm is to note that "a page linked from reliable pages would be reliable." This idea is similar to the main idea used by the PageRank [3] algorithm: "a page linked from important pages would be important."

The TrustRank algorithm requires a huge number of pages on the Web similarly as the PageRank algorithm, and it cannot be directly used to score the pages on a small Web graph used by the HITS algorithm. Thus, we cannot directly use the TrustRank algorithm for improving the HITS algorithm.

We propose an algorithm to give a *trust-score* to every page contained in the base set used by HITS. Employing the ideas of the HITS algorithm and the TrustRank algorithm, we say that a page is reliable if it has links to many reliable pages related to a given topic, and that it would be a good hub page for the topic. For most of the topics used in our preliminary experiments, we found that more than half of the pages in the root set are reliable and related to the given topic. Our trust-score algorithm thus regards a page u as a *reliable hub page* if the page u links to many pages in the root set, and regards a page v as a *reliable authority page* if the page v is linked from many reliable hub pages. More precisely, if there are two or more hosts, each containing a page which belongs to the root set and is linked from a page u, then the trust-score algorithm gives u a *trust hub score* $T_{hub}(u)$ equal to the number of these hosts; otherwise, the algorithm gives u a trust hub score $T_{hub}(u) = 0$. The algorithm also gives a page v a *trust authority score* $T_{auth}(v)$ equal to the sum of trust hub scores (divied by $|H^+(u)|$) of the pages u linking to v. The *trust-score* of v is its trust authority score normalized by the sum of all trust authority scores. The trust hub score of a page u is a measure to indicate how reliable u is as a hub page, and the trust-score of a page v is a measure to indicate how reliable v is as an authority page. Thus our trust-score algorithm is as follows.

Let R, V and E be the root set, the vertex set and the edge set, respectively, used in the HITS algorithm.

1. For each page $u \in V$, let $\{s_1, s_2, ..., s_k\}$ be the set of all pages that are linked from u and belong to the root set R, that is, $(u, s_i) \in E$ and $s_i \in R$, $1 \leq i \leq k$. Let $host(u)$ be the number of distinct host names of $s_1, s_2, ..., s_k$. If $host(u) \geq 2$, then set a trust hub score $T_{hub}(u) = host(u)$. Otherwise, set $T_{hub}(u) = 0$.
2. For each page $v \in V$, set a trust authority score $T_{auth}(v)$ to $\sum_{(u,v)\in E} \frac{T_{hub}(u)}{|H^+(u)|}$.
3. For each page $v \in V$, output $\frac{T_{auth}(v)}{\sum_{w\in V} T_{auth}(w)}$ as the trust-score of page v.

3.3 Our Four Scoring Algorithms

We propose the following four algorithms using the trust-score algorithm. Let **T+BHITS** be the algorithm which gives a page p a score $t(p)+a(p)$, where $t(p)$ is the trust-score of the page p, and $a(p)$ is the authority score of p computed by BHITS. Similarly, let **TaI+BHITS**, **TaD+BHITS**, and **TaN+BHITS** be the algorithms which give a page p a score $t(p) + a(p)$, where $a(p)$ is the authority score of p computed by I+BHITS, D+BHITS, and N+BHITS, respectively. Using two weight factors x and y, one can construct an algorithm which gives a page p a score $x \cdot t(p) + y \cdot a(p)$. We have made some preliminary experiments varying the weights x and y, and confirmed that, for most of the used topics, top ten authorities of the best quality are obtained when $x = y = 1$. We will evaluate the quality of top ten authorities found by the four algorithms in Section 4.

4 Experimental Results

Table 1 depicts the results of our experiments. For the experiments, we use the following eleven HITS-based algorithms. The first three are existing HITS-based

Table 1. The experimental results

	Existing Algorithms			Variants of BHITS				Ours			
Algorithm	**H**	**B**	**WB**	**LB**	**DB**	**IB**	**NB**	**T**	**TaD**	**TaI**	**TaN**
Topic 1	0	7	9	7	8	8	8	10	10	10	10
Topic 2	0	0	0	0	0	0	0	10	0	0	0
Topic 3	0	0	0	0	0	10	10	10	10	10	10
Topic 4	0	0	0	0	0	0	0	7	9	5	9
Topic 5	0	5	10	5	10	10	10	7	10	10	10
Topic 6	0	1	1	1	1	1	1	10	9	10	9
Topic 7	0	2	2	2	10	10	10	6	9	9	9
Topic 8	4	10	10	10	10	2	2	10	10	9	9
Topic 9	10	0	0	0	10	10	10	8	10	10	10
Topic 10	0	0	1	0	10	0	10	7	10	7	10
Topic 11	10	10	10	10	10	10	10	10	10	10	10
Topic 12	0	0	0	0	0	0	10	2	5	2	10
Topic 13	0	0	0	10	0	10	10	3	3	10	10
Topic 14	0	0	0	5	0	0	6	8	7	5	7
Average	**1.71**	**2.50**	**3.07**	**3.57**	**4.93**	**5.07**	**6.93**	**7.71**	**8.00**	**7.64**	**8.79**
Sufficient	**2**	**2**	**4**	**3**	**6**	**6**	**8**	**6**	**10**	**9**	**12**
Non-Root	**10**	**16**	**12**	**21**	**30**	**19**	**39**	**13**	**20**	**17**	**25**

H: HITS, B: BHITS, WB: WBHITS, LB: LF+BHITS, DB: D+BHITS, IB: I+BHITS, NB: N+BHITS, T: T+BHITS, TaD: TaD+BHITS, TaI: TaI+BHITS, TaN: TaN+BHITS.

algorithms: HITS, BHITS, and WBHITS proposed by Li *et al.* [11]; they are abbreviated as H, B and WB, respectively. The next four algorithms LF+BHITS, D+BHITS, I+BHITS and N+BHITS, abbreviated as LB, DB, IB and NB respectively, are presented in Section 3.1, and are variants of the BHITS algorithm using methods of finding linkfarms; LF+BHITS uses Wu and Davison's method, D+BHITS uses a domain name, I+BHITS uses an IP address, and N+BHITS uses a name server. The last four algorithms T+BHITS, TaD+BHITS, TaI+BHITS and TaN+BHITS, abbreviated as T, TaD, TaI and TaN respectively, are our algorithms using the trust-score in Section 3.2; T+BHITS combines the trust-score algorithm with BHITS, and similarly, the three algorithms TaD+BHITS, TaI+BHITS and TaN+BHITS combine the trust-score algorithm with D+BHITS, I+BHITS and N+BHITS, respectively.

Each of the given fourteen topics Topic 1, Topic 2, ..., Topic 14 is specified by one of the following fourteen keywords: "Canoe," "Cheese," "F1 (Formula One)," "Gardening," "Iriomote-Cat (one of the endangered species)," "Kabuki (Japanese traditional performance)," "Monochrome-Photograph," "Movie," "Olympic," "Pipe-Organ," "Railway," "Rock-Climbing," "Tennis," and "Wine." For each topic, the root set used in our experiments consists of top two hundred pages of the result of Google for the topic. The results for other topics are similar to those for the topics used in the experiments, and the results obtained by using other search engines are also similar to those obtained by using Google.

Each cell in the rows from "Topic 1" to "Topic 14" shows the quality of the top ten authorities found by each algorithm for the given topic; as we have already described, the quality of top ten authorities is defined as the number of pages which are related to the given topic and belong to the top ten authorities found by each algorithm. It is judged by manual inspection of human subjects whether a page is truly related to a given topic. For each algorithm, "Average" row shows the average quality of the top ten authorities obtained for the fourteen topics. For each algorithm, "Sufficient" row shows the number of topics for which the obtained top ten authorities has sufficient quality; we say that the quality of top ten authorities is *sufficient* if the top ten authorities contain only at most one page not related to the given topic. For example, HITS finds top ten authorities of sufficient quality for only two topics among the fourteen topics, and our TaN+BHITS finds top ten authorities of sufficient quality for twelve.

The original HITS algorithm and the BHITS algorithm find top ten authorities containing few pages related to the given topics; the top ten authorities found by HITS contain no related pages for eleven topics among the fourteen topics, and the top ten authorities found by BHITS contain no related pages for eight topics. It was reported that the WBHITS algorithm is one of the best HITS-based algorithms [11], but WBHITS finds top ten authorities containing no related pages for seven topics among the fourteen topics. These three algorithms cannot be used for finding related pages on today's Web.

We say that two hosts $host_1$ and $host_2$ are *mutually linked* if there are both a link from a page in $host_1$ to a page in $host_2$ and a link from a page in $host_2$ to a page in $host_1$, and say that a link from a page in a host to a page in another host is a *mutual inter-host link* if these hosts are mutually linked. For four topics Topic 3, 9, 10 and 12 we found a number of linkfarms, each containing few mutual inter-host links. These linkfarms cannot be found by LF+BHITS using Wu and Davison's method of finding linkfarms, because their method can find only linkfarms containing a number of mutual inter-host links [14]. On the other hand, each of the linkfarms consists of links between pages sharing some kind of network information, and hence at least one of our algorithms D+BHITSCI+BHITS, and N+BHITS finds such linkfarms. This is the reason why our proposed algorithms D+BHITS, I+BHITS, and N+BHITS find top ten authorities of better quality than those found by LF+BHITS. Thus, our proposed methods of finding linkfarms are more effective for improving the BHITS algorithm than Wu and Davison's method. For most of the fourteen topics, N+BHITS finds top ten authorities of better quality than those found by D+BHITS or I+BHITS, and hence the method of finding linkfarms by utilizing the name server is better than the other two methods for improving the BHITS algorithm.

For most of the fourteen topics, our four algorithms using the trust-score algorithm, T+BHITS, TaD+BHITS, TaI+BHITS and TaN+BHITS, find top ten authorities of much better quality than those found by the other algorithms, hence the trust-score algorithm is effective for improving the BHITS algorithm. Particularly, for most of the topics, TaN+BHITS obtains top ten authorities of better quality than those obtained by the other algorithms.

The average quality 8.79 of the top ten authorities found by TaN+BHITS is significantly better than the average quality 1.71, 2.50, and 3.07 of the top ten authorities found by HITS, BHITS, and WBHITS, respectively. TaN+BHITS also finds top ten authorities of sufficient quality for most of the topics, twelve topics among the fourteen, although the existing HITS-based algorithms, HITS, BHITS and WBHITS, find top ten authorities of sufficient quality for very few topics. For each of the following four topics Topic 6, 9, 10 and 12 the top ten results of Google contain some pages not related to the topic, but all the pages in the top ten authorities found by TaN+BHITS are related to the topic.

The HITS algorithm finds some of the related pages which cannot be found by Google, as we described in Section 1. We now evaluate how many such pages the eleven algorithms find. For each algorithm, the bottom row, "Non-Root," of Table 1 denotes the total number of pages satisfying the following three conditions (1)-(3): (1) they are related to the given topic; (2) they belong to the top ten authorities found by the algorithm; and (3) they do not belong to the root set. If an algorithm has a large value in the bottom row, then the algorithm finds a number of related pages which could not be found by Google, because the root set consists of the top two hundred pages obtained by Google.

One can observe the following three facts (a)-(c): (a) the algorithms using our methods of finding linkfarms, DB, IB, NB, TaD, TaI, and TaN, have larger value in the bottom row than the existing algorithms, H, B, and WB; (b) DB has larger value in the bottom row than TaD which is the algorithm obtained by combining DB with the trust-score. Similarly, IB (or NB) has larger value than TaI (or TaN) which is the algorithm combining IB (or NB, respectively) with the trust-score; and (c) TaN has larger value in the bottom row than the other algorithms using the trust-score.

The fact (a) implies that our methods of finding linkfarms are effective for finding related pages which cannot be found by Google. The fact (b) can be explained as follows: the trust-score algorithm frequently gives high scores to pages belonging to the root set, and hence all the algorithms using the trust-score frequently find top ten authorities containing a number of pages belonging to the root set; consequently these top ten authorities contain few pages which do not belong to the root set; hence TaD (or TaI, TaN) using the trust-score has smaller value in the bottom row than DB (or IB, NB, respectively) without the trust-score. As we described above, the algorithms with the trust-score find top ten authorities of better quality than those found by the algorithms without the trust-score, and thus the fact (b) implies a trade-off between the quality of top ten authorities and the value in the bottom row. The quality of top ten authorities is generally more important than the value in the bottom row, and hence the fact (c) implies that TaN+BHITS is the best choice for finding both top ten authorities of sufficient quality and a number of related pages which cannot be found by Google.

We now discuss a remaining problem of our HITS-based algorithms. For Topic 2, all the algorithms except T+BHITS find top ten authorities containing no related pages. On the Web graph used by T+BHITS for Topic 2, there are several links emanating from pages with large trust hub scores and entering to some related pages in the top ten authorities, and hence these related pages are given high trust scores. Most of these links join pages sharing some kind of network information other than a host name. These links are ignored by DB, IB, NB, TaD, TaI, and TaN, and hence the top ten authorities found by these algorithms do not contain such related pages. In other words, these algorithms wrongly regard the links as linkfarms and remove them. It is one of the remaining problems to distinguish such links from actual linkfarms.

For each of the fourteen topics, all the algorithms use the same vertex set. On the other hand, each of the algorithms uses an edge set slightly different from the edge sets used by other algorithms, because each algorithm ignores several links in its own manner. H, B, WB and T use the same edge set for each topic, because all of them ignores only links between pages having the same host name. Similarly, each of the following three pairs of algorithms (DB, TaD), (IB, TaI), and (NB, TaN) uses the same edge set, which excludes links between pages having the same domain name, IP address, and name server, respectively.

The amount of memory used by each of our proposed algorithms is almost equal to that used by the original HITS algorithm. Most part of the running time of the HITS-based algorithms is spent to collect data of pages through the Internet. Every algorithm uses data of the same set of pages, and hence the running times of our propsed algorithms are almost equal to that of the original HITS algorithms. In our experiments, all the algorithms outputs top ten authorities in a few seconds once the required data is collected through the Internet. The algorithms use about 10MB of memory for the Topic 8, which is larger than the amount of memory used for any of other topics. Our scoring algorithm TaN+BHITS can find top ten authorities of good quality on today's Web and can be executed on demand with a PC. The algorithm particularly finds top ten authorities of the best quality for most of the used topics.

5 Concluding Remarks

We have proposed several improved variants of the HITS algorithm, by proposing the trust-score algorithm and three methods of finding linkfarms. One of our algorithms, named TaN+BHITS, finds top ten authorities of much better quality than those found by the existing algorithms, and the top ten authorities found by TaN+BHITS contain a number of related pages which cannot be found by Google. We have hence succeeded in developing a HITS-based algorithm which can find top ten authorities of sufficiently good quality on today's Web. Our algorithms take almost the same amount of memory and running time as those taken by the original HITS algorithm, and hence our algorithms can be executed on demand with a PC having a small amount of main memory.

References

1. K. Bharat and M. R. Henzinger. "Improved algorithms for topic distillation in a hyperlinked environment," *Proc. 21st ACM SIGIR Conference*, pp. 104–111, 1998.
2. A. Borodin, G. O. Roberts, J. S. Rosenthal and P. Tsaparas, "Finding authorities and hubs from link structures on the World Wide Web," *Proc. 10th WWW Conference*, pp. 415–429, 2001.
3. S. Brin and L. Page, The anatomy of a large-scale hypertextual Web search engine. *Proc. 7th WWW Conference*, pp. 14–18, 1998.
4. A. Costa Carvalho, P. Chirita, E. Moura, P. Calado and W. Nejdl, "Site level noise removal for search engines," *Proc. 15th WWW Conference*, pp. 73–82, 2006.
5. D. Fetterly, M. Manasse and M. Najork. "Spam, damn spam, and statistics: Using statistical analysis to locate spam Web pages," *Proc. 7th International Workshop on the Web and Databases*, pp. 1–6, 2004.
6. D. Fetterly, M. Manasse, M. Najork and A. Ntoulas. "Detecting spam Web pages through content analysis," *Proc. 15th WWW Conference*, pp. 83–92, 2006.
7. Z. Gyongyi, H. Garcia-Molina and J. Pedersen. "Combating Web spam with TrustRank," *Proc. 30th VLDB Conference*, pp. 576–587, 2004.
8. G. Jeh and J. Widom, "SimRank: A measure of structual-context similarity," *Proc. 8th ACM SIGKDD Conference*, pp. 538–543, 2002.
9. J. Kleinberg, "Authoritative sources in a hyperlinked environment," *Proc. 9th ACM-SIAM Symposium on Discrete Algorithms*, pp. 668–677, 1998.
10. R. Lempel and S. Moran, "The stochastic approach for link-structure analysis (SALSA) and the tkc effect," *Proc. 9th WWW Conference*, pp. 387–401, 2000.
11. L. Li, Y. Shang and W. Zhang, "Improvement of HITS-based algorithms on Web documents," *Proc. 11th WWW Conference*, pp. 527–535, 2002.
12. M. Wang. "A significant improvement to Clever algorithm in hyperlinked environment". *Poster Proc. 11th WWW Conference*, 2002.
13. X. Wang, Z. Lu and A. Zhou. "Topic exploration and distillation for web search by a similarity-based analysis," *Proc. 3rd WAIM Conference*, pp. 316–327, 2002.
14. B. Wu and B. D. Davison, "Identifying link farm spam pages," *Proc. 14th WWW Conference*, pp. 820–829, 2005.
15. Y. Zhang, J. X. Yu and J. Hou. "Web Communities : Analysis and Construction," Springer, Berlin, 2006.

Efficient Keyword Search over Data-Centric XML Documents

Guoliang Li, Jianhua Feng, Na Ta, and Lizhu Zhou

Department of Computer Science and Technology, Tsinghua University, Beijing 100084, P.R. China
{liguoliang,fengjh,dcszlz}@tsinghua.edu.cn, dan04@mails.tsinghua.edu.cn

Abstract. We in this paper investigate keyword search over data-centric XML documents. We first present a novel method to divide an XML document into self-integrated subtrees, which are connected subtrees and can capture different structural information of the XML document. We then propose the meaningful self-integrated trees, which contain all the keywords and describe how the keywords are interrelated, to answer keyword search over XML documents. In addition, we introduce the B^+-tree index to accelerate the retrieval of those meaningful self-integrated trees. Moreover, to further enhance the performance of keyword search, we present Bloom Filter to improve the efficiency of generating those meaningful self-integrated trees. Finally, we conducted extensive experiments to evaluate the performance of our method, and the experimental results demonstrate that our method achieves high efficiency and outperforms the existing approaches significantly.

1 Introduction

Keyword search is a proven and widely accepted mechanism for querying in document systems and World Wide Web. Database research community has recently recognized the benefits of keyword search and has been introducing keyword search capability into relational databases [3,7,10,15,17,19,23] and XML databases [4,5,6,9,11,12,13,16,18,20,21,22,25,26,28].

Traditional query processing approaches on relational and XML databases are constrained by the query constructs imposed by the languages such as SQL and XQuery. Firstly, the query languages themselves are hard to comprehend for non-database users. For example, the XQuery is fairly complicated to grasp. Secondly, these query languages require the queries to be posed against the underlying, sometimes complex, database schemas. These traditional querying methods are powerful but unfriendly to the non-expert users. Keyword search is proposed as an alternative means for querying the databases, which is simple and yet familiar to most internet users as it only requires the input of some keywords. While keyword search has been proven to be effective for text documents (e.g. HTML documents), the problem of keyword search on the structured data (e.g. relational databases) and the semi-structured data (e.g. XML databases) is not straightforward and well studied.

G. Dong et al. (Eds.): APWeb/WAIM 2007, LNCS 4505, pp. 491–502, 2007.

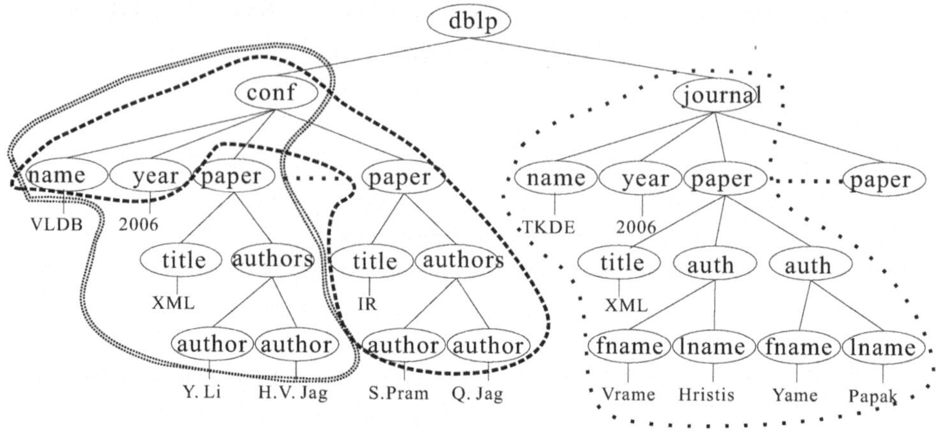

Fig. 1. An example XML document

Keyword search in text documents takes as the answers the documents that are more interrelated with the input keywords, while in relational databases it takes the correlative tuples in the database that contain all (or a part of) the keywords as the answers. However, it still remains an open problem that, for XML documents, what should be the answer for keyword search? The notion of Lowest Common Ancestor (LCA) has been introduced to answer keyword queries over XML documents [13]. And more recently, Meaningful LCA (MLCA), Smallest LCA(SLCA), Grouped Distance Minimum Connecting Tree(GDMCT) have been proposed to improve the efficiency and effectiveness of keyword search against LCA in [22,28,16], respectively. However, the answer of keyword search should not be limited to just the LCAs, as LCAs themselves cannot explain how the keywords are connected. Although the subtrees proposed in some existing methods [16,22], which are composed of LCAs and their relevant keywords, may be more meaningful as the answer of keyword search over XML documents, these subtrees are not meaningful enough to capture the overall structural information to answer keyword queries.

Intuitively, for keyword search over text documents, it is evident that the documents that contain all the keywords should be the answer. Furthermore, some other relevant and meaningful information is also contained in the answer. Similarly, for relational databases, the interrelated tuples connected by primary-foreign key relationships and containing all (or a part of) the keywords should be the answer. Moreover, some other relevant and meaningful data besides the elements that contain some keywords are also adhered to those tuples by the way. However, for XML documents, it is not straightforward to retrieve the subtrees connected by the *content nodes* that directly contain some keywords. Accordingly, it is much harder to retrieve the integrated subtrees (like the documents and interrelated tuples) connected by the content nodes and *complementary nodes* that do not contain any keyword but contain some relevant and meaningful data as the complementarity to answer keyword queries meaningfully, since it is rather difficult to determine which nodes are complementary nodes and can be adhered to the answer. For example, in Figure 1, if a user inputs keywords of {"XML","2006","VLDB"}, he expects to get the subtree circled by the double dotted lines as the answer, and the two authors should be adhered to the answer as they are complementary nodes and relevant to the keyword query.

Therefore, we in this paper investigate how to retrieve those subtrees, namely self-integrated trees, which contain some complementary nodes as the complementarity to capture the focuses of keyword queries, besides the content nodes. More importantly, each self-integrated tree can represent an integrated meaning to answer a keyword query. In addition, to accelerate the retrieval of those meaningful self-integrated trees, we introduce the B^+-Tree index and Bloom Filter to improve the efficiency. To the best of our knowledge, this is the first paper to generate the meaningful self-integrated trees with complementary nodes as the answer of keyword search over XML documents.

To summarize, we make the following contributions,

- We propose how to divide a data-centric XML document into self-integrated subtrees and take those meaningful self-integrated trees as the answer of keyword search over XML documents.
- We introduce the B^+-Tree index and Bloom Filter to accelerate the retrieval of those meaningful self-integrated trees, which can improve the efficiency of keyword search over XML documents significantly.
- We conducted a set of extensive experiments to evaluate our method, and the experimental results obtained demonstrate that our method achieves high efficiency and outperforms the existing approaches.

The rest of this paper is structured as follows. Section 2 proposes the meaningful self-integrated trees as the answer of keyword search over XML documents. We then introduce the B^+-Tree index and Bloom Filter to accelerate the retrieval of those meaningful self-integrated trees in Section 3. Section 4 provides the extensive experimental study. We review the related work in Section 5 and make a conclusion in Section 6.

2 Self-integrated Trees

We first briefly outline the XML data model in Section 2.1 and then introduce the concept of *Self-Integrated Trees* in Section 2.2. Finally we give an efficient algorithm to generate those meaningful self-integrated trees in Section 2.3.

2.1 Overview

An XML document can be modelled as a rooted, ordered, and labeled tree. We in this paper mainly consider the data-centric XML document, which has a concrete schema that is mainly used to describe how the structure information is organized. For instance, DBLP [1] is a typical data-centric XML document.

Most of existing studies take the subtrees, which contain all the keywords (e.g., the conjunctive predicate) or a part of the keywords (e.g., the disjunctive predicate), as the answer of a given keyword query. They mainly first compute the lowest common ancestors (LCAs) of the content nodes that directly contain at least one keyword and then take the subtrees rooted at LCAs as the answer. However, they involve huge computation and thus lead to inefficiency. In addition, the existing approaches cannot introduce some relevant nodes that do not contain any keyword into the answer. For example, in Figure 1, suppose, a user wants to retrieve the `author` of those papers, which are published in 2006 and entitled with a keyword "XML", but the user does not know how to input the keyword w.r.t. `author`, as `conf` and `journal` have different structures about the tag/label of `author`. Thus, all the existing studies cannot deal with this situation as it is hard to select suitable keywords as input.

Therefore, we in this paper mainly discuss how to generate those self-integrated and meaningful subtrees, which capture all the relevant data, as the answers of keyword queries, even with limited keywords, over XML documents.

2.2 Meaningful Self-integrated Trees

We introduce how to divide the XML document into different meaningful subtrees. Each subtree itself can represent an integrated meaning while different subtrees have little interrelationship. For example, in Figure 1, we can divide the XML document into three subtrees as circled by the dotted lines. We observe that the three subtrees can describe integrated meanings respectively. Accordingly, for any keyword query, if any of these three subtrees contains all the keywords, it should be an answer.

In addition, to meaningfully divide an XML document into some integrated subtrees, the division should capture the following properties,

1. Each subtree represents an integrated meaning and thus can capture all the relevant information to answer a keyword query.
2. Each subtree contains some other relevant nodes as the complementarity to meaningfully answer a keyword query, besides those content nodes.
3. All the subtrees can be generated and indexed in advance, thus the answer, i.e., the subtrees that contain all the keywords, can be gotten efficiently.

To achieve our goal, we give the concept of *Self-Integrated Trees* as follows.

Definition 1 ***(Self-Integrated Trees).*** *Given a data-centric XML document $\mathcal{D}$ and its schema (or DTD) $\mathcal{S}$, a Self-Integrated tree w.r.t. $\mathcal{D}$ and $\mathcal{S}$ is the subtree of $\mathcal{D}$, which can represent an integrated and meaningful structural information.*

A self-integrated tree is a subtree of an XML document, which can represent an integrated meaning. For example, in Figure 1, the three circled subtrees are self-integrated subtrees. Accordingly, we take the *meaningful self-integrated subtrees* that contain all the keywords as the answer of a keyword query.

We can generate all the self-integrated subtrees according to the semantics of the schema as follows. We first analyze and divide the schema into meaningful parts according to the semantics, and accordingly divide the XML document into

Algorithm 1: `Merge-Join` Algorithm

Input: A keyword query $\mathcal{K}=\{k_1,k_2,...,k_q\}$ and our inverted indices of a given XML document $\mathcal{D}$.

Output: `SITreeSet`, composed of all the meaningful self-integrated subtrees w.r.t. $\mathcal{K}$ and $\mathcal{D}$, each of which contains all the keywords.

begin
  `SIIDSet`$\leftarrow\phi$;
  `SITreeSet`$\leftarrow\phi$;
  $\mathcal{I}_i$=$IDList_i$=$\{n_{SI}|n_{SI}$ is the ID of a self-integrated tree that contains $k_i\}$;
  while *each* $\mathcal{I}_i$ *is not empty* **do**
    if $\mathcal{I}_1$*.first()*=$\mathcal{I}_2$*.first()*=...=$\mathcal{I}_q$*.first()* **then**
      `SIIDSet`$\leftarrow\{\mathcal{I}_1$.first()$\}$;
      for i=1 *to* q **do**
        $\mathcal{I}_i$.pop_front();
    for *any* $j\neq k$ *and* $1\leq j\leq q$ *and* $1\leq k\leq q$ **do**
      if $\mathcal{I}_j$*.first()*$<\mathcal{I}_k$*.first()* **then**
        $\mathcal{I}_j$.pop_front();
  retrieve `SITreeSet` according to `SIIDSet`;
  return `SITreeSet`;
end

Fig. 2. `Merge-Join` algorithm to generate all the meaningful self-integrated trees

some self-integrated subtrees based on the division of the schema. Moreover, all the self-integrated subtrees can be generated and indexed off-line.

In addition, for any term in the XML document, k_i, the existing studies first retrieve the inverted lists of the content nodes w.r.t. k_i, and then compute the LCA of each combination of the content nodes, finally return the subtrees rooted at LCAs as the answer. However, it is inefficient to compute all the LCAs as there are many combinations of the content nodes. Therefore, we in this paper propose a novel inverted index to accelerate the retrieval of the meaningful self-integrated subtrees. We assign each self-integrated subtree with a distinct identifier(ID), and then for any term, k_i, we record the IDList of all the IDs which contain k_i. In addition, we maintain the statistical information for k_i, i.e., the number of IDs that contain k_i, and this statistics will be used to accelerate the retrieval of the meaningful self-integrated subtrees in Section 3.2. Accordingly, given a keyword query $\mathcal{K}=\{k_1,k_2,...,k_q\}$, we first retrieve each IDList $\mathcal{I}_i$ w.r.t. $k_i(1\leq i\leq q)$, and then compute `SIIDSet`$=\bigcap_{i=1}^{q}\mathcal{I}_i$, which is exactly the set composed of the IDs that contain all the keywords. Finally, we can retrieve the meaningful self-integrated subtrees as the answer according to those IDs in `SIIDSet`.

2.3 Algorithm to Generate the Meaningful Self-integrated Trees

We in this section introduce an efficient algorithm, `Merge-Join`, to generate the meaningful self-integrated trees. Without loss of generality, we suppose all the

IDLists w.r.t. a keyword query are sorted by their IDs and thus we can merge-join these IDLists with one scan of each IDList to generate the intersection of these IDLists, which is composed of all the IDs that appear in every IDList. The detailed algorithm is illustrated in Figure 2.

`Merge-Join` first retrieves the IDLists, each of which is composed of the IDs that contain a keyword (line 4), and then while each IDList is not empty (line 5), if an ID w.r.t. a self-integrated subtree appears in every IDList (line 6), the self-integrated tree w.r.t. it must contain all the keywords and thus this ID is added into `SIIDSet` (line 7). In addition, the first element of each IDList is popped from the corresponding IDList (lines 8-9). On the contrary, if the first element in an IDList, e.g., $\mathcal{I}_j.\mathit{first}()$, is smaller than that of another IDList, e.g., $\mathcal{I}_k.\mathit{first}()$, this element ($\mathcal{I}_j.\mathit{first}()$) should be popped from the corresponding IDList (lines 10-12). To compute the intersection of those IDLists, `SIIDSet`, we repeat these steps until one IDList is empty. Finally, we retrieve the meaningful self-integrated trees, `SITreeSet`, according to `SIIDSet` (line 13).

3 Optimizations

3.1 B^+-Tree Indices

We in this section propose the B^+-tree index to accelerate the retrieval of the meaningful self-integrated trees. For a keyword query, the sizes of keyword lists generally are not the same, and even the size of one keyword list is far less than those of the other ones. For example, suppose a user wants to retrieve the paper entitled with a keyword "XML" and having an author of "Tom". The number of the self-integrated subtrees w.r.t. "Tom" generally be far smaller than those w.r.t. *title* or *author*. Accordingly, some self-integrated trees w.r.t. *author* and *title*, which cannot contain all the keywords, should not be in the answer, and therefore, we propose the B^+-tree index to skip those self-integrated trees. However, the B^+-tree index only improves the CPU cost while the bottleneck is the I/O cost to retrieve the self-integrated trees associated with the input keywords, and hence we present Bloom Filter to further improve the performance in the next section.

3.2 Bloom Filter

We in this section propose Bloom Filter to improve the efficiency of retrieving the meaningful self-integrated trees. As we all know, Bloom Filter is a probabilistic method to quickly test membership in a large set by using multiple hash functions and thus is efficient to test whether an element is in a set, therefore, we borrow it to efficiently retrieve the meaningful self-integrated trees.

A Bloom Filter employs a vector for representing a set of n elements (also called keys) so as to support membership queries. It was invented by Burton Bloom in 1970 [8] and was proposed for use in the web context by Marais and Bharat [24] as a mechanism for identifying which pages have associated comments stored within a common knowledge server.

The idea behind Bloom Filter is to allocate a vector v of m bits, initially all set to 0, and then choose k independent hash functions, h_1,h_2,...,h_k, each with range {1,2,...,m}. For each element e, the bits at positions $h_1(e)$,$h_2(e)$,...,$h_k(e)$ in v are set to 1 (A particular bit might be set to 1 multiple times.). Given a query for testing whether e' is in the set, we check the bits at positions $h_1(e')$, $h_2(e')$,...,$h_k(e')$. If any of them is 0, then certainly e' is not in the set. Otherwise we conjecture that e' is in the set although there is a certain probability that we are wrong. This is called a "false positive", or, for historical reasons, a "false drop". The parameters k and m should be chosen such that the probability of a false positive (and hence a false hit) is acceptable.

The salient feature of Bloom filters is that there is a clear tradeoff between m and the probability of a false positive. Observe that after inserting n keys into a set of size m, the probability that a particular bit is still 0 is exactly $(1-\frac{1}{m})^{kn}$. Hence the probability of a false positive in this situation is,

$$(1-(1-\frac{1}{m})^{kn})^k \approx (1-e^{-kn/m})^k.$$

The right hand side is minimized for $k = ln2 * m/n$, in which case it becomes $(1/2)^k$=$(0.6185)^{m/n}$. In fact k must be an integer and in practice we might chose a value less than the optimal to reduce computational overhead. In addition, we give two examples about this probability, for example, if m/n=20 and k=10, the probability is 8.89e-05. While if m/n=24 and k=8, the probability is 4.17e-05.

To achieve our goal to retrieve all the meaningful self-integrated subtrees according to bloom filter, we maintain a bloom filter for each self-integrated tree, and the bloom filter preserves the keywords in this self-integrated tree, that is, we insert all the keywords in this self-integrated tree into the set of the bloom filter. Accordingly, we can test whether a self-integrated tree contains all the keywords of a keyword query. However, there may be many candidate self-integrated trees but only few of them contain all the keywords. Therefore, we introduce an optimization to improve the retrieval of the meaningful self-integrated subtrees as follows. For each term, we record the number of the self-integrated trees that contain it and thus we can get the minimal IDList w.r.t. a keyword query, which has the minimal size. Subsequently we retrieve the bloom filters w.r.t. the self-integrated trees in the minimal IDList and then test whether each candidate self-integrated tree contains all the keywords, if so, this subtree should be an answer; otherwise it cannot be. Finally, we can generate all the meaningful self-integrated subtrees through filtering out false positives (if any).

Moreover, we give the complexity of Bloom Filter here. Given a keyword query, $\mathcal{K}$=$\{k_1,k_2,...,k_q\}$, $\mathcal{I}_i$ is the IDList w.r.t. k_i $(1 \le i \le q)$, suppose $|\mathcal{I}_1| \le |\mathcal{I}_2| \le ... \le |\mathcal{I}_q|$. The I/O cost of the method based on bloom filter is to retrieve IDs in the minimal IDList, e.g., $\mathcal{I}_1$, and the corresponding bloom filters, thus it is very efficient when one keyword list has smaller size, e.g., $|\mathcal{I}_1| \ll \sum_{i=2}^{q} |\mathcal{I}_i|$. Therefore, bloom filter improves the I/O cost significantly. In addition, the complexity of CPU cost is at most k*(q-1)*$|\mathcal{I}_1|$, as we only test whether each keyword in $\{k_2,...,k_q\}$ is contained in a candidate self-integrated tree through computing hash values of k hash functions. Therefore, if k*(q-1)*$|\mathcal{I}_1| < \sum_{i=1}^{q} |\mathcal{I}_i|$, bloom filter further improves the efficiency of retrieving the meaningful self-integrated trees, and we will experimentally demonstrate that bloom filter in practice outperforms B^+-tree, which in turns is better than `Merge-Join` in Section 4.

Example 1. Given a keyword query {"XML","2006","author"}, and suppose $|\mathcal{I}_{XML}|$=10, $|\mathcal{I}_{2006}|$=100,000 and $|\mathcal{I}_{author}|$=1000,000. The method based on bloom filter first retrieves the IDList of "XML", e.g., {1,12,16,66,80}, subsequently, it needs to test whether "2006" and "author" are in these five candidate self-integrated trees. Each bloom filter w.r.t. a self-integrated tree preserves an m-bits vector, and we compute the hash values of the given k (here, e.g., 3) hash functions on "XML", and if one of the hash values is 0, the keyword "XML" is not in the self-integrated tree; otherwise, the self-integrated tree may contain keyword "XML". Then, we test whether the candidate self-integrated trees, which contain both "XML" and "2006", contain keyword "author", if so, these self-integrated trees should contain all the keywords and may be the answer.

4 Experimental Study

We have designed and performed a comprehensive set of experiments to evaluate the performance of our approach. We used the real dataset DBLP [1] in our experiments, and the raw file was about 340MB. The experiments were conducted on an Intel(R) Pentium(R) 2.4GHz computer with 512MB of RAM running Windows XP Professional. The algorithms were implemented in Java and the parsing of the XML files was performed using the SAX API of the Xerces Java Parser. We compared our approach with the state-of-the-art approaches, GDMCT[16] and SLCA [28]. We implemented the Indexed Lookup Eager algorithm(IL) for SLCA since IL was the best one among the three proposed algorithms of SLCA.

Furthermore, to better understand the performance of our method on different keyword queries with various characteristics, we selected several keyword queries with distinct keyword numbers from 2 to 6, respectively. To further evaluate the performance of our algorithms for the keyword queries with different selectivities, we performed our experiments using various sets of keywords with different frequencies, namely low, medium and high, with respective keyword frequency range of 1 to 100, 101 to 10000, and above 10000. Accordingly, we varied the frequencies of keywords and computed the corresponding elapsed time to compare our methods with the existing approaches.

4.1 Evaluation of Our Three Methods

We first evaluated our proposed methods, `Merge-Join`, B^+-tree and Bloom Filter, on different queries with various characteristics. In Figure 3, each query contains a keyword of small frequency (10 in Figure 3(a),100 in Figure 3(b), 1000 in Figure 3(c)), and all other keywords of keyword queries have high frequency of 1000,000. For example, each query of Figure 3(b) in the #keywords=5 category contains five keywords where one of them has frequency of 100 and the other four have frequency of 1000,000.

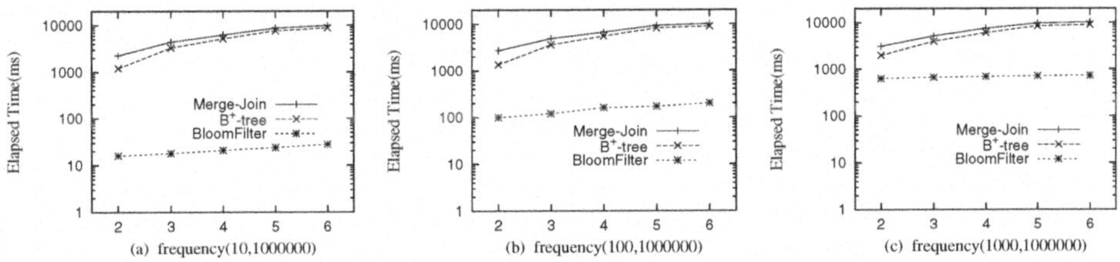

Fig. 3. Varying the number of keywords and keeping large frequencies constant

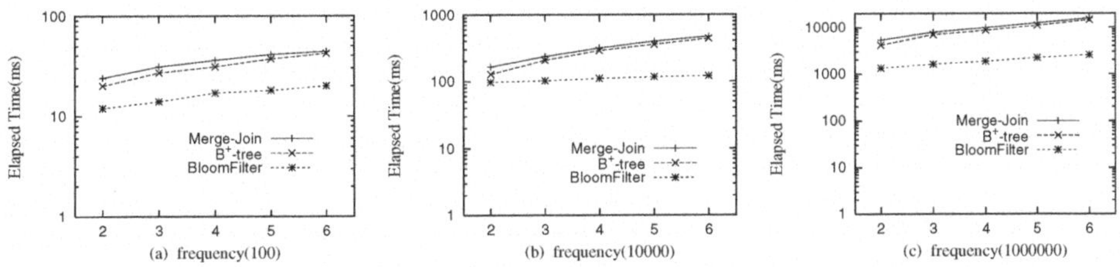

Fig. 4. Varying the number of keywords and all keyword lists having the same frequency

We can observe that, the performance of B^+-tree is more efficient than `Merge-Join` as B^+-tree can skip many irrelevant elements. Moreover, Bloom Filter only depends on the minimal IDList with fewest elements but is essentially independent of other IDLists with high frequencies, therefore, it outperforms B^+-tree and `Merge-Join` significantly, especially when one keyword list has low frequency while others have high frequencies.

Keywords in Figure 4 have the same frequency as shown in the caption. B^+-tree and Bloom Filter perform a little better than `Merge-Join` in most experiments since the former two algorithms perform best when the frequencies of keyword lists vary greatly. While all the keyword lists in Figure 4 have the same size, thus B^+-tree is a little better than `Merge-Join` as B^+-tree lookup is more likely efficient than the scan of `Merge-Join`. In addition, Bloom Filter improves the I/O cost and thus is still better than the other two methods.

4.2 Comparison with the Existing Approaches

We in this section compared B^+-tree and Bloom Filter with the existing approaches, and we still employed the keyword queries with different keyword frequencies and with the same keyword frequency as illustrated in Figure 5 and Figure 6 respectively. We can observe that both B^+-tree and Bloom filter are better than GDMCT and SLCA, and Bloom Filter outperforms the other approaches significantly when the keywords have distinct frequencies. In addition, although when the keywords have the same frequency or have little difference, our two methods still beat the other two existing approaches. More importantly, Bloom Filter outperforms all the other methods for various keyword queries, and this further reflects that Bloom Filter achieves high efficiency and performs best for keyword search over data-centric XML documents.

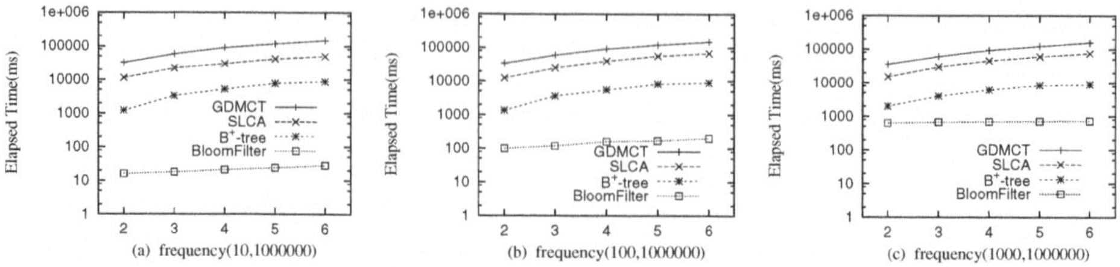

Fig. 5. Comparison:varying the number of keywords with large frequencies constant

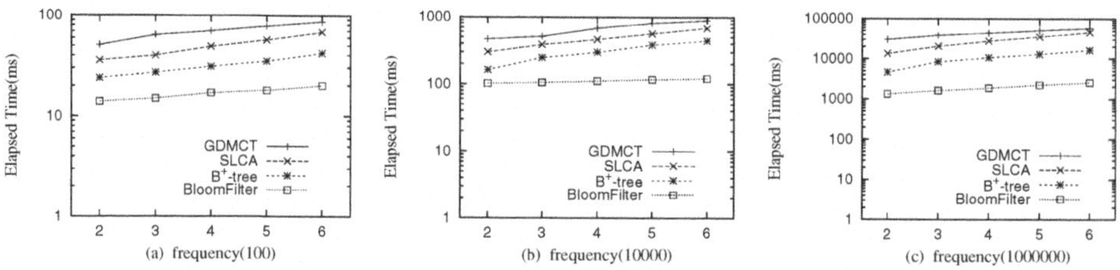

Fig. 6. Comparison:varying the number of keywords with the same frequency

5 Related Work

The first area of research related to our work is the computation of the LCA of two or more nodes, which has been studied in [14,27]. As an extension of LCA, MLCA, SLCA and GDMCT have recently been proposed to answer keyword queries over XML documents in [22], [28] and [16] respectively. On the other hand, XRANK [13] and XSEarch [12] are systems facilitating keyword search for XML documents, which return subtrees as answers for the keyword queries. However, XRANK does not return connected trees to explain how the keywords are connected to each other. Furthermore, only the most specific result is output. They also present a ranking method, which, given a tree $\mathcal{T}$ containing all the keywords, assigns a score to $\mathcal{T}$ using an adaptation of PageRank for XML documents. Their ranking techniques are orthogonal to the retrieval method, and thus can easily be incorporated in other works. XSEarch focuses on the semantics and the ranking of the results, and during execution it uses an all-pairs interconnection index (i.e., for any two nodes in an XML document, it has to record whether they are interrelated or not) to check the connectivity between the nodes, which is not efficient for large XML documents and thus is impracticable in practice. XKeyword [18] is a system that offers keyword proximity search over XML documents that conform to an XML schema, however, it needs to compute candidate networks and thus is constrained by the schemas.

In addition, DISCOVER [17], BANKS [7] and DBXplorer [3] are systems built on top of relational databases. DISCOVER and DBXplorer output trees of tuples connected through primary-foreign key relationships that contain all the keywords of the query, while BANKS identifies connected trees in a labelled graph by using an approximation to the Steiner tree problem, which is an

NP-hard problem. More recently, Liu et al. [23] proposed a novel ranking strategy to solve the effectiveness problem for relational databases, which employs phrase-based and concept-based models to improve search effectiveness. While Kimefeld et al. [19] demonstrated keyword proximity search in relational databases, which shows that the answer of keyword proximity search can be enumerated in ranked order with polynomial delay, under data complexity.

Furthermore, various XML full-text query languages have also been proposed, such as, [4,9,20,21], and a workshop INEX [2], INitiative for the Evaluation of XML Retrieval, aiming at evaluating XML retrieval effectiveness, has also been organized. More recently, two algebras for keyword search over XML documents have been proposed [5,26]. [5] presents the XFT algebra that accounts for element nesting in XML document structure to evaluate queries with complex full-text predicates, while [26] demonstrates several optimization techniques that guarantees better efficiency for keyword search over tree-structured documents.

6 Conclusion

We propose the self-integrated trees to answer keyword queries over data-centric XML documents. Each meaningful self-integrated tree contains all the keywords and can represent a compact and integrated meaning to answer keyword queries. More importantly, besides the content nodes which directly contain at least one keyword, the meaningful self-integrated tree still contains some other relevant nodes as the complementarity to explain how the keywords are connected and how the self-integrated tree answers the keyword query.

In addition, we propose the B^+-tree index to accelerate the retrieval of all the meaningful self-integrated trees. Moreover, we introduce Bloom Filter to further enhance the performance of retrieving those meaningful self-integrated trees, which can improve the I/O cost dramatically and thus achieves much higher efficiency. Finally, we conducted a set of extensive experiments and the experimental results demonstrate that our method achieves high efficiency and outperforms the existing approaches significantly.

Acknowledgments

This work is in part supported by the National Natural Science Foundation of China under Grant No.60573094, the National Grand Fundamental Research 973 Program of China under Grant No.2006CB303103, the National High Technology Development 863 Program of China under Grant No.2006AA01A101, Tsinghua Basic Research Foundation under Grant No. JCqn2005022, Basic Research Foundation of Tsinghua National Laboratory for Information Science and Technology (TNList), and Zhejiang Natural Science Foundation under Grant No. Y105230.

References

1. http://dblp.uni-trier.de/xml/.
2. http://inex.is.informatik.uni-duisburg.de/2006/index.html.

3. S. Agrawal, S. Chaudhuri, and G. Das. Dbxplorer: A system for keyword-based search over relational databases. In *ICDE*, pages 5–16, 2002.
4. S. Amer-Yahia, C. Botev, J. Dorre, and J. Shanmugasundaram. Xquery full-text extensions explained. In *IBM Systems Journal 45(2)*, pages 335–352, 2006.
5. S. Amer-Yahia, E. Curtmola, and A. Deutsch. Flexible and efficient xml search with complex full-text predicates. In *SIGMOD*, pages 575–586, 2006.
6. S. Amer-Yahia, N. Koudas, A. Marian, D. Srivastava, and D. Toman. Structure and content scoring for xml. In *VLDB*, pages 361–372, 2005.
7. G. Bhalotia, A. Hulgeri, C. Nakhe, S. Chakrabarti, and S. Sudarshan. Keyword searching and browsing in databases using banks. In *ICDE*, pages 431–440, 2002.
8. B. Bloom. Space/time trade-offs in hash coding with allowable errors. In *Communications of ACM, pages 13(7)*, pages 422–426, 1970.
9. C. Botev, S. Amer-Yahia, and J. Shanmugasundaram. Expressiveness and performance of full-text search languages. In *EDBT*, pages 349–367, 2006.
10. S. Chaudhuri, G. Das, V. Hristidis, and G. Weikum. Probabilistic ranking of database query results. In *VLDB*, pages 888–899, 2004.
11. S. Cohen, Y. Kanza, B. Kimelfeld, and Y. Sagiv. Interconnection semantics for keyword search in xml. In *CIKM*, pages 389–396, 2005.
12. S. Cohen, J. Mamou, Y. Kanza, and Y. Sagiv. Xsearch: A semantic search engine for xml. In *VLDB*, pages 45–56, 2003.
13. L. Guo, F. Shao, C. Botev, and J. Shanmugasundaram. Xrank: Ranked keyword search over xml documents. In *SIGMOD*, pages 16–27, 2003.
14. D. Harel and R. E. Tarjan. Fast algorithms for finding nearest common ancestors. In *SIAM J. Comput. 13(2)*, pages 338–355, 1984.
15. V. Hristidis, L. Gravano, and Y. Papakonstantinou. Efficient ir-style keyword search over relational databases. In *VLDB*, pages 850–861, 2003.
16. V. Hristidis, N. Koudas, Y. Papakonstantinou, and D. Srivastava. Keyword proximity search in xml trees. In *IEEE Trans. Knowl. Data Eng. 18(4)*, pages 525–539, 2006.
17. V. Hristidis and Y. Papakonstantinou. Discover: Keyword search in relational databases. In *VLDB*, pages 670–681, 2002.
18. V. Hristidis, Y. Papakonstantinou, and A. Balmin. Keyword proximity search on xml graphs. In *ICDE*, pages 367–378, 2003.
19. B. Kimelfeld and Y. Sagiv. Finding and approximating top-k answers in keyword proximity search. In *PODS*, pages 173–182, 2006.
20. Y. Li, H. Yang, and H. V. Jagadish. Nalix: an interactive natural language interface for querying xml. In *SIGMOD*, pages 900–902, 2005.
21. Y. Li, H. Yang, and H. V. Jagadish. Constructing a generic natural language interface for an xml database. In *EDBT*, pages 737–754, 2006.
22. Y. Li, C. Yu, and H. V. Jagadish. Schema-free xquery. In *VLDB*, pages 72–84, 2004.
23. F. Liu, C. Yu, W. Meng, and A. Chowdhury. Effective keyword search in relational databases. In *SIGMOD*, pages 563–574, 2006.
24. J. Marais and K. Bharat. Supporting cooperative and personal surfing with a desktop assistant. In *ACM UIST*, 1997.
25. A. Marian, S. Amer-Yahia, N. Koudas, and D. Srivastava. Adaptive processing of top-k queries in xml. In *ICDE*, pages 162–173, 2005.
26. S. Pradhan. An algebraic query model for effective and efficient retrieval of xml fragments. In *VLDB*, pages 295–306, 2006.
27. B. Schieber and U. Vishkin. On finding lowest common ancestors: Simplification and parallelization. In *SIAM J. Comput. 17(6)*, pages 1253–1262, 1988.
28. Y. Xu and Y. Papakonstantinou. Efficient keyword search for smallest lcas in xml databases. In *SIGMOD*, pages 527–538, 2005.

Promotional Ranking of Search Engine Results: Giving New Web Pages a Chance to Prove Their Values*

Yizhen Zhu[1], Mingda Wu[1], Yan Zhang[1,**], and Xiaoming Li[2]

[1] National Laboratory on Machine Perception, Peking Univ., Beijing, 100871, China
{zhuyz,wumd,zhy}@cis.pku.edu.cn

[2] Department of Computer Science and Technology, Peking Univ., Beijing, 100871, China
lxm@pku.edu.cn

Abstract. Recent studies show that the link-structure-based measures of Web page popularity prolong the procedure of new pages reaching their deserved ranking. In this paper we propose a promotional ranking scheme that offers an opportunity for newly-created pages to be recognized.We conduct a simulation to evaluate our method. Experimental results show that our method remarkably raises the probability for new pages to obtain user-awareness.

1 Introduction

As the Web grows dramatically, search engine has become an indispensable tool for accessing online information. According to [9], by April 2006, Google and Yahoo respectively receive over 150 million search requests per day. Due to the enormous number of matching outcomes, normally people are only interested in the few top retrieved URLs on the list. This kind of user's preference leads to a higher popularity for the currently-popular pages and a deeper entrenchment for the newly-built ones to span. Ideally, search engines are supposed to present the results ordered by their *quality*. Nevertheless, the quality of a web page is too subjective to be measured directly, which impels search engines to turn to other substitutes for quality. Popularity is one of the most widely used substitutes which could be directly measured by number of in-links and clicks, or by PageRank, etc.

Whereas, some researchers have found that the accumulation of popularity differs by the current popularity of the page [2] and search engines do inhibit the discovery of new pages. On the other hand, the real web environment is very dynamic with high rates of birth, death and replacement of both the web pages and hyperlinks between them [4]. However, the query results of search engines remain relatively stale under the rank schemes like PageRank, in-degree, or number of visits. The primitive motivation of our work is to solve this problem: to offer an opportunity for those newly-created Web pages to be noticed and visited.

Section 2 is devoted to related work.Section 3 explains the web evolution from the perspective of search engine. Section 4 describes the details of our method step by step. Section 5 demonstrates the details of our experiments.

* Supported by NSFC with Grant No.60673129 and 60573166, Natural Science Foundation of Beijing with Grant No.4073034 and Guangdong SCUT Key Lab Open Grant.

** Corresponding author.

G. Dong et al. (Eds.): APWeb/WAIM 2007, LNCS 4505, pp. 503–510, 2007.

2 Related Work

A. Ntoulas et al. found that existing pages were being removed from the Web and replaced by new ones at a very rapid rate[5,4]. In [1] researchers showed several relations between the macro structure of the web, page and site age, and quality of pages and sites. J. Cho et al. presented the impact of search engines on the page popularity by introducing two surfer models[2]. It is estimated that when search engine ranked pages based on their popularity, it took several orders of magnitude more time for a new page to become popular even if it was of high quality. In [3] J. Cho et al. further investigated the problem of page quality and they proposed a reasonable definition for page quality and derived a practical way of estimating the quality.

The biggest difference between our method and [6] is that there exists an inevitable trade-off problem in their scheme: exploration vs exploitation. We doubt if it is secure to substitute a popular result for a new page whose quality is unpredictable.

3 The Evolution of Web from the Perspective of Search Engine

The web is very dynamic with high rates of birth and death. According to [4], it is estimated that the new pages are created at the rate of 8% per week and only 40% of the pages will survive one year. As for the links, 25% new links are created per week. We conduct a survey by ourselves to investigate the turnover rate by retrieving a historical data set of URLs which is obtained two years ago. We find that only about $1/3$ of the URLs are still available, which coincides with the conclusions in [4]. However, due to the limitation of disk space, band width and cost, it's impossible for a search engine to frequently update its repository to make it consistent with the real web. According to [8], Google refreshes its repository and pagerank every 28 days and other search engines from a week to a month. According to the birth rate of web pages and the refresh cycle of search engines, we estimate that between each two crawling processes, the new pages consist of the search engine's repository with a fraction from 15% to 20%. Here we assume a page to be new if it is not crawled in the last repository update.

PageRank was first introduced as an index of the probability that a page is visited by a web user who randomly and repeatedly jumps into a new page and follows its link to the next page. Nowadays Web users are more relying on search engines to find useful information. Whats more, users tend to choose the results with a high rank. Under such a circumstance, the probability of a page being visited is much more connected to its current ranking. Many high-quality pages are ignored just because they are too new and low-ranked to be noticed. Recent studies [2,3]warn that search engines prolong the time for the new pages to reach the ranking they deserve.

Now with a dynamic web environment and a highly search-engine-driven browsing habit, it is urgent to develop a new ranking method to alleviate the problem and offer the new pages a chance to prove their values.

4 Our Promotion Method

We assume that search results are listed by page with a certain number of results in each page. Our idea is to display some relatively new results together with the mature ones in a friendly way, thus keep the standard ordering of the older pages and promote the new ones as well.The scheme in [6] promotes new pages at the cost of deserting several old pages. However, we think that insert promoted pages into a consistent ranking can be feasible and acceptable. Here are some questions we need to answer:

- How to evaluate the initial rank of the new pages?
- How many pages should be promoted each time? Which pages should be promoted?
- How to display the promoted pages?

Before we start our discussion, a concept proposed in [6] need to be explained. The concept of community is designed to model the relationship between user queries and search engine results. If P is a set of pages related to a topic T, and U is a set of users interested in topic T, we say that the users U and pages P corresponding to topic T constitute a Web community. Under the assumption that queries and topics have a one-to-one relationship, it is easy to notice each query returns the same set of pages from the corresponding community and the only thing that can vary is the sequence of the results. The concept of community is crucial if we want to study the effect of a certain ranking method via simulation because in real web query results are basically unstable. In our case, we divide the pages into different categories according to their topics. Thus, the users looking into a particular category and all the pages within that category make up a community.

4.1 Initial Ranking of the New Pages

The initial ranking of the new pages is very important in selecting the pages to be promoted for that it is the basic knowledge we have about these pages. An uneven distribution of the initial ranking will lead to the bias toward some few pages, an over-even distribution of the initial ranking will lead to the indistinguishability between new pages with high and low quality.

We assume that the original pagerank of these new pages have been calculated beforehand which is reasonable and feasible for most of the search engines. Their original pagerank is a vivid index of the popularity of the pages pointing to them or the site they belong to. Since they are new to the web environment, it is unlikely that pages from another site will spare outlinks to the newly built pages. Therefore, we can make the assumption that the distribution of the new pages' pagerank is similar to that of their parent pages'.

4.2 Selection of Candidates

In Section 3 we've discussed the proportion of new pages in a search engine's repository. We use the ratio of new pages to old ones to decide how many new pages should be promoted each time. We narrow down our attention from the search engine repository to a specific topic and the corresponding community of pages. Assuming that within a specific community, the ratio of new pages to all pages is r and the number of pages listed in a batch is N, we choose the quota of promotion denoted as N_p to hold the proportion of new pages in one batch. Ideally, we want $r = N_p/(N + N_p)$, thus we have that $N_p = r \cdot N/(1 - r)$.

Before explaining our selection scheme, we model our selection process as follows. All the new pages in the community constitute a set, known as the *candidate pool*. Let

P denote the candidate pool with a size of m. Supposed the quota of a promotion is n, then the selection will be conducted n times on P until we collect all the n pages to be recommended. If a new page is selected, it will be automatically wiped off from P. A straightforward selection scheme is the *randomized selection*. Each time we randomly pick out a new page from the candidate pool with equal probability.

With the diversity of new pages quality, however, it will be disturbing and confusing if we randomly choose the candidates without taking their quality into account. In order to balance between impartially treating all new pages and maintaining the general quality of results, we propose a second selection scheme, *the probabilistic selection*, inspired by the study in [2]. J. Cho and S. Roy present an algorithm of visit popularity which takes the impact of search engines into consideration. We transplant this algorithm into our second selection scheme in order to model the behavior of page visit. First we sequence all pages in P by a descending order of their initial ranking. Suppose r_i is the sequence number of page i in P and M is the size of candidate pool,the probability of i being selected is:

$$P(i) = c \cdot r_i^{-3/2}, \quad c = \left(\sum_{i=1}^{M} i^{-3/2}\right)^{-1}$$

4.3 Combination of Promotional Ranking

We propose two schemes to combine the pages selected from the candidate pool with the original results. No matter which scheme we adopt, the ranking of the original result remains intact. And the selected pages queue in the order of how they are picked out one after another from the candidate pool. The only difference is how we mix the two ranking. The first scheme is called implicit promotion. As we've discussed before, we are going to insert N_p promoted pages into the original N results in a batch. Firstly, we randomly choose N_p positions in a sequence of $N + N_p$, then we fill the N_p positions respectively with the N_p promoted pages in their order. We name it implicit simply because the promoted pages are mixed with the original results, so when a user clicks on a page, he is unaware if the page is already popular or promoted by our scheme.

The other one is *explicit* promotion. Under this scheme users are informed of the status of the results. We append the N_p promoted pages after the N original results in each batch and we clearly label the new pages so that we leave users the opportunity to decide whether or not to click on a promoted result.

5 Experiments

5.1 Experimental Setup

As we've discussed in Section 4, simulation is an inevitable process to evaluate our method. In [7,6] simulations prove to be efficient. In our simulation, firstly we build a website where visitors can retrieve pages from different categories sorted by different promotion modes. Then we learn users' awareness and fondness of the pages from log analysis. The combination of 2 selection schemes and 2 display schemes produces 4 promotion modes: *random-implicit-promotion*, *random-explicit-promotion*, *probabilistic-implicit-promotion* and *probabilistic-explicit-promotion*. Together with *no-promotion*

Table 1. Corresponding relationship between our simulation and the real web

Our website	**Search engine website**
Photograph work	Web page
Photo preview	Abstract of web page
Photo information(title, author, date)	Page information(title, site name, last update)
Category of photography	Community of web pages

mode, there are totally 5 modes to be evaluated. Nevertheless, the quality of the promoted pages is unstable in random-implicit-promotion, which may disappoint the users and even bring a negative effect; meanwhile *probabilistic-implicit-promotion* does not make good use of the previous efforts such as estimating new pages' quality. Due to the above reasons and the limited popularity of our temporary web site (more modes will lead to less users assigned to the group for a single mode, see below), we evaluate 3 modes: *no-promotion*, *random-implicit-promotion* and *probabilistic-explicit-promotion*.

No-promotion serves as a baseline to compare with the other two modes. *Random-implicit-promotion* can be viewed as a proximity of the "*random shuffling*" method [6], except for 12 results with 2 new ones per batch, which is made to make a fair-play with *probabilistic-explicit-promotion*, one of the methods we proposed.

We establish a website composed of $6,912$ Web pages, each containing a photograph work. All the photos were downloaded in Mar. 2006 from www.altphotos.com, a popular photography website, each with a smaller preview, an original rank and a brief introduction. The original photos were uploaded by the owners into corresponding categories and graded by visitors of the site. First, we chose six different categories, namely architecture, essay, people, photojournalism, vehicle, places. Then we downloaded 400 to 700 most popular photos and 600 newly-uploaded ones from each category.

To study the effect of our promotion method, visitors of our website, or *users* are randomly assigned to one of the three groups since their first visit. The user interface of each group differs in the display of results.We record the IP address and the group number of every user so that if a user visits our site from the same IP again, the corresponding interface will be presented. During a period of 47 days, we've attracted 455 visitors and recorded over $4,000$ actions of viewing and rating the pictures.Table 1 shows the corresponding relationship between our simulation and the real web environment.

Definition (*democell*): We call the preview plus the brief introduction of a photo a *democell*. A democell is corresponding to the web page with a full view of the photo. Democells are the basic elements we used to display the results to users.

After dividing the photos into two groups: already-popular and newly uploaded, we assign initial ranks to both groups. Already-popular pages' initial rank is their original rank while new pages' is 20% of their original rank. This step extends the gap between the new pages and the already popular ones. As we've discussed in Section 3 that the new pages constitute 15% to 20% of total amount in a community, and we've proposed the quota calculation in Section 4 that $N_p = r \cdot N/(1 - r)$. With $N = 10$, $15\% \le r \le 20\%$, we assign the quota of promotion to be 2.Under each category, one batch of the results presented contains 12 democells listed in 6 rows with 2 democells in each. We display 12 results per batch because normally search engines present 10 urls per page and we promote 2 pages each time.

- *no-promotion* All the democells, both popular and unpopular, are ranked according to their initial ranking.
- *random-implicit-promotion* We *randomly* select 10 photos from the candidate pool of 600 photos. They are implicitly inserted into the popular results in the first 5 batches, 2 for each batch. The locations are randomly determined from 2 to 12. We avoid the first position to preserve the most popular result.
- *probabilistic-explicit-promotion* The 10 new pages are selected with the probability of $P(i) = c \cdot r_i^{-3/2}$. They are displayed at the last row of the batch with a label of "newly discovered photos".

5.2 Results Analysis

After log information analysis, we decide to use user click as an index of user's awareness of a certain page. Let G_i denote user group i (only including those who have at least a single click on a certain photo), U_j denote user j, C_k denote photo category k and P_l denote photo l. We use Probability-Of-Hit (POH) to estimate the chance that a user from a certain group i visits a new page:

$$POH_i = \frac{1}{|G_i|} \sum_{U_j \in G_i} v_j, \quad v_j = \begin{cases} 1 & \text{if user } j \text{ has visited a new photo,} \\ 0 & \text{if not.} \end{cases}$$

The chance that a user from group i visits a new page from category k is:

$$POH_{i,k} = \frac{1}{|G_{i,k}|} \sum_{U_j \in G_{i,k}} v_{j,k}, \quad v_{j,k} = \begin{cases} 1 & \text{if user } j \text{ has visited a new photo of } C_k, \\ 0 & \text{if not.} \end{cases}$$

By calculating both POH_i and $POH_{i,k}$, we can learn how new pages are becoming more accessible with our promotional ranking. From Figure 1, we see that both *random-implicit* promotion and *probabilistic-explicit* promotion yield good effect in making new pages more accessible. Furthermore, we notice that *probabilistic-explicit* promotion outperforms *random-implicit* promotion. There might be two reasons to explain.Firstly, *probabilistic-explicit* promotion may stimulate the user to click on the labeled new page.Secondly, the *probabilistic* candidate selection scheme is more sensitive to the potential quality of new pages,thus the pages promoted probabilistically are inherently more attractive than those promoted randomly.We've also used other evaluation metrics, but only our key results are shown in this paper due to page limit.

5.3 Sorting New Pages

Results in Section 5.2 demonstrate the advantage of a probabilistic selection based on the potential quality of new pages. In the above simulation experiment, we utilize the original user-made-grade of the photos to produce their initial rank. In order to discover the hidden quality of new pages, we further conduct another experiment to evaluate the method of sorting new pages by estimating pagerank for new pages.

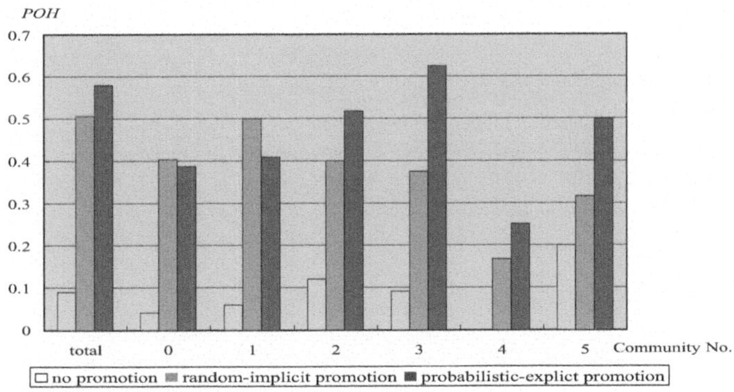

Fig. 1. POH in each category

Our estimation method originates from the idea that good pages (none-spamming ones) tend to link to pages with comparative quality. A new page is likely to be of high quality if its siblings (sharing the same parent with p) are of high quality. Therefore, we adopt ASP (the Average Siblings' PageRank) as an index to estimate the quality of new pages. Meanwhile, to avoid the situation that one or two parent pages with too many outlinks may bias the estimated value, we assign ACP (the Average of Children PageRank) to each pages, and calculate ASP via ACP.

$$ACP(q) = \frac{\sum_{q \to p} PR(p)}{outdegree(q)}, \quad ASP(p) = \frac{\sum_{q \to p} ACP(q)}{indegree(p)}$$

To evaluate this method, we take a snapshot of $1,631,483$ web pages (PS2) and compare it with another snapshot of the same set of pages (PS1) 22 months ago. First, we calculate the ordinary pageranks on both sets of pages – the pagerank output by PS2 is supposed to be a measure of inherent quality of the pages. Then we randomly pick 5 sets of "new pages", each set including $160,000$ pages that are of low pagerank and have only 1 or 2 inlinks each, and calculate ASP for each set of new pages and then sort them by ordinary pagerank and ASP respectively. Let $R_{PS2}(p)$ be the rank position of p in all new pages ordered by PageRank calculated on PS2, $R_{naive}(p)$ the rank position of p in all new pages ordered by PageRank calculated on PS1 and $R_{ASP}(p)$ the rank position of page p in all new pages ordered by ASP. Now we have the performance evaluation functions:

$$F_{naive}(N) = average(R_{naive}(p))/\text{number of newpages}, \quad \text{in which } p \text{ sat. } R_{PS2}(p) \leq N$$
$$F_{ASP}(N) = average(R_{ASP}(p))/\text{number of newpages}, \quad \text{in which } p \text{ sat. } R_{PS2}(p) \leq N$$

We run the calculation of $F_{naive}(N)$ and $F_{ASP}(N)$ on the five sets of pages. The results share a similar pattern. Figure 2 presents the five values on average. Obviously, ordinary PageRank hardly convey the potential of new pages while ASP does upgrade the rank position of pages with high quality. Whereas, due to the limit of our coarse selection on "new pages" in this experiment, the results may not be as good as we've expected. We hope to implement further approaches to evaluate the ASP algorithm.

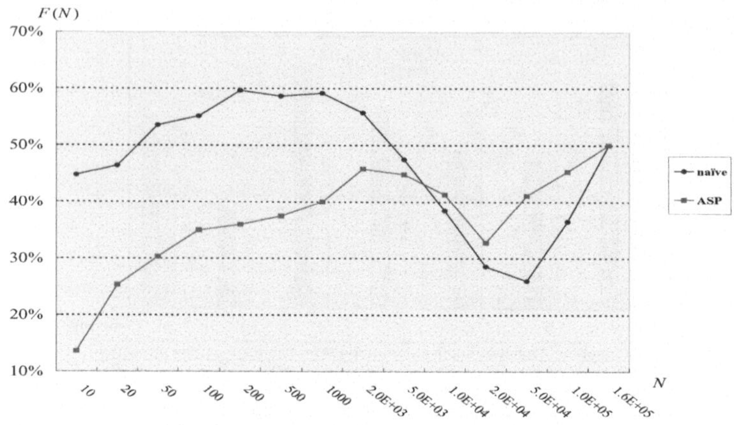

Fig. 2. Performance of ASP

6 Conclusion and Future Work

We propose a promotional ranking scheme in the paper in order to give the new pages a chance to prove their values. The experimental results show that our methods really improve the result quality and the probability for new pages to be noticed. Due to the infeasibility of testing on a real commercial search engine, we conduct a simulation for the evaluation. For that a search engine is more like a recommendation system retrieving fuzzy answers than a QA system that offers the precise answer, we believe that our simulation, though primitive, still has its merit in demonstrating our method.

In section 4.1, we make an assumption of "distribution of the new pages' PageRank is similar to that of their parent pages'" to help ease the sorting of new pages which may cause popular sites to become more popular. Our future work will be focused on finding less biased ways of ranking new pages. We are also looking forward to upgrading our simulation by using real web pages instead of photography works.

References

1. R. Baeza-Yates, F. Saint-Jean, and C. Castillo. Web structure, dynamics and page quality. In *Proc. String Processing and Information Retrieval*, 2002.
2. J. Cho and S. Roy. Impact of search engines on page popularity. In *WWW2004*, May 2004.
3. J. Cho, S. Roy, and R. E. Adams. Page quality: In search of an unbiased web ranking. In *SIGMOD 2005*, June 2005.
4. A. Ntoulas, J. Cho, H. K. Cho, H. Cho, and Y.-J. Cho. A study on the evolution of the web. In *the 2005 UKC Conference*, august 2005.
5. A. Ntoulas, J. Cho, and C. Olston. What's new on the web? the evolution of the web from a search engine perspective. In *WWW2004*, May 2004.
6. S. Pandey, S. Roy, C. Olston, J. Cho, and S. Chakrabarti. Shuffling a stacked deck: The case for partially randomized ranking of search engine results. In *VLDB2005*, August 2005.
7. F. Qiu and J. Cho. Automatic identification of user interest for personalized search. In *WWW2006*, May 2006.
8. C. Sherman. Meet the search engines. http://searchenginewatch.com/searchday/article.php/2157701.
9. D. Sullivan. Searches per day. http://searchenginewatch.com/reports/article.php/2156461.

Adaptive Scheduling Strategy for Data Stream Management System

Guangzhong Sun, Yipeng Zhou, Yu Huang, and Yinghua Zhou

Anhui Province-MOST Key Co-Lab of High Performance Computing and Its Applications, Department of Computer Science, University of Science and Technology of China, Hefei, 230027, P.R. China

Abstract. More and more applications involve processing continuous data streams, and the data stream management system (DSMS) is designed to deal with such data streams. Due to features of large volume and stochastic arrival, DSMS must process data stream efficiently in order to avoid system memory exhaustion and reduce the data access latency, to satisfy requirements of the application requirement. One of the key factors, which significantly impact the system performance significantly, is the scheduling strategy adopted by the DSMS. Chain scheduling is an operator-based scheduling strategy for DSMS, which has near-optimal in terms of run-time memory usage. FIFO strategy achieves optimal performance in terms of data access latency. Inspired by the two important scheduling strategies, Chain and FIFO, we propose two novel adaptive strategies for DSMS, ASCF and CSS, which efficiently deal with the varying input load in terms of both memory usage and data access latency. To give a fair comparison performance with other competing strategies, we design thorough simulation experiment and run different strategies under the same system environment. The outcomes of simulation experiment demonstrate the potential benefits and advantages of ASCF and CSC.

Keywords: Data stream management system, scheduling, adaptive strategy, chain scheduling, FIFO.

1 Introduction

Data stream is a sequence data tuples with large volume and the the data arrival is busty without any discipline. The data rate fluctuates over time. Recently, due to its distinguishing characteristics and growing demands from a wide range of applications, the continuous query processing system, named Data Stream Management System (DSMS), over data streams has received more and more attention. There are a large various representative applications of DSMS, such as sensor network, network traffic analysis, financial ticket[1], electrical train service[2], and so on. These applications have motivated a considerable body of research ranging from algorithms and architectures for stream processing to a full-fledged data stream processing such as Telegraph[3], STREAM[4], Aurora[5], Niagara[6], Tapestry[7], and Tribeca[8].

G. Dong et al. (Eds.): APWeb/WAIM 2007, LNCS 4505, pp. 511–521, 2007.

Due to the special features of data stream, the DSMS differs from a traditional static database in the following two ways. (i) the input data tuples are high volume with intermittent busts. (ii) the system should provide fast response time[2], [9], [10], [11], [12]. So it is not appropriate to apply DBMS to these applications. These unique characteristics compared with traditional database motivate us to develop newer techniques and methods in order to process input data efficiently and effectively. DSMS just is developed to solve these tricky problems, only by scanning these data one time or several times according to their arrival order. At the same time, under the memory constraint, the DSMS reduces the data latency with all its efforts. The scheduling strategy is a key factor that impacts the system performance evidently. In this paper, we mainly focus on the operator scheduling problem in DSMS.

The scheduling strategy determines which operator needs to be scheduled at each time unit in multiple, continuous query processing system. DSMS scheduling strategy has two main objects: memory consumed and data latency, first DSMS must reduce the memory requirement because the input volume is large, and it is impossible to store all these data in computer memory, second, applications request the DSMS must response the query within the time limitation of applications. In fact, the two metrics are tradeoff. Some scheduling strategies have been proposed in some literatures. However, most of them ignore the variety of input data. Our main contribution is designing two novel adaptive scheduling strategies by combining Chain and FIFO strategies, which adjust process strategy according to input load and system free resources. The outcomes of simulation experiment indicate our strategies indeed surpass other existed competing strategies.

The rest of the paper is organized as follows. In section two, we discuss two important strategies (Chain and FIFO) and another competing strategy such as Chain-Flush, and these strategies will be compared with our novel strategies in experiment section. In section three, we construct the model of DSMS. Some assumptions and notions are defined also. In section four, new strategies and some conclusive discussion are shown. In the next section, we design simulation experiments and compare the performance of our strategies with other existed strategies. Conclusion and more discussion is in the last section.

2 Related Work

In this section we will discuss several previous strategies. First we introduce Chain [11], which is nearly optimized in memory usage. Second we introduce FIFO, which is latency optimized strategy. Both Chain and FIFO strategies of theory merit are referred to by our novel strategies. However, due to some bad characteristics, nearly no real DSMS adopts FIFO or Chain alone. At last we will show another strategy Chain Flush of real system, which will be compared with our novel strategies.

Chain Strategy: At any time instant, consider all tuples that are currently in the system. Of these, schedule for a single time unit the tuple that lies on the segment with the steepest slope in its lower envelope simulation. If there are multiple such tuples, select the tuple which has the earliest arrival time[11].

The difference between the Chain Strategy and the strategy which consumes the least memory is only a constant [11]. The problem of the optimization of consumed memory is NP-complete [11]. Therefore, Chain Strategy is our best choice unless $P = NP$, if the system memory is limited. However, there is nearly no system adopting this strategy alone, due to its severe tuple latency. An example will be shown following to make these metrics clearly.

FIFO Strategy: Tuples are processed in the order that they arrive. A tuple is passed through both operators in two consecutive units of time, during which time no other tuple is processed.

It is not hard to justify that FIFO is the best strategy in data latency [9]. But, it ignores the different selectivities of different operators on one operator path and just process corresponding to their arrival order. Though the data latency is optimized, there is nearly no system is available to satisfy the high memory requirement.

Chain Flush Strategy: First, assume there is a latency constraint assumed for each data tuple. The Chain Flush Strategy proceeds exactly like Chain until deadline constraints become tight, forcing a change in strategy. For example, there are n tuples totally in the system, and the i_{th} tuple is on the verge of missing its deadline. At this point, no more processing on the later arriving tuples can be performed until the i_{th} tuple and all earlier tuples have been completely processed. For more details, please refer to literature [11]. Chain Flush Strategy is applied in real systems, designed by Stanford University. But Chain Flush ignores the variation of input load and system free resources. The performance comparison with novel strategies will be shown in section five through the obtained data of simulation experiment.

3 Model and Assumption

In this section, the model of DSMS is constructed and some important assumptions and definitions are introduced mainly after [11].

Data stream systems are characterized by the presence of multiple continuous queries [13],[14],[15],[16]. Tuple is the minimum unit of input streaming data. Query execution can be conceptualized as a data flow diagram as in [14], which is a directed acyclic graph(DAG) of nodes and edges, where the nodes are pipelined operators that process tuples and edges represent composition of operators. An edge from node A to node B indicates that the output of operator A is an input to node B. The edge (A, B) also represent an input queue that buffers the output of operator A before it is input to the operator B. Input data streams are represented as leaf nodes that have no input edges and query outputs are represent as root nodes that have no output edges[1],[3],[9],[10].

As mentioned earlier, we assume that all operators execute in a streaming or pipelined manner. Operators like select and project naturally fit in this category. Every tuple that enters the system must pass through a unique path of operators,

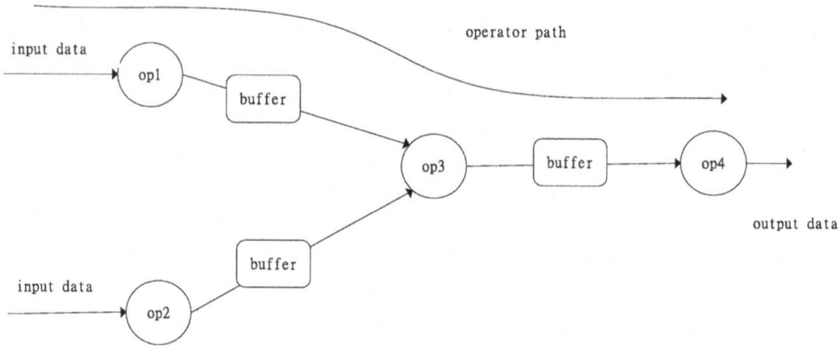

Fig. 1. DAG of DSMS

referred to as an operator path. And multiple continuous queries input can be divided into several independent single stream queries [11]. Therefore, we only consider the characteristics of a single operator path. There is the Figure 1 following to show the structure of DSMS and operator path clearly. There are two operator path: the first is op_1, op_3, op_4 and the second is op_2, op_3, op_4.

The operators that we consider act like filters, which operate on a tuple and produce s tuples, where s is the selectivity of the operator. The selectivity s is less than 1 for select and project operators, but it may be greater than 1 for a join operator. For the rest of the paper, we view the operator as a filter that produces s tuples on processing 1 tuple. Obviously, the selectivity assumption does not hold at the granularity of a single tuple but is merely a convenient abstraction to capture the average behavior of the operator. In terms of this model, the goal of operator scheduling is to minimize the total memory requirement and reduce the tuple latency at the same time.

Selectivity and per-tuple processing time are the most important parameters of operators. Obviously, it is important to consider the different sizes of a tuple as it progresses through its operator path. We assume that the selectivities and per-tuple processing times are known for each operator [14] and capture this using the notation of a progress chart, illustrated in Figure 2.

The horizontal axis of the progress chart represents time and the vertical axis represents tuple size. The $n+1$ operator points $(t_0, s_0), (t_1, s_1), ..., (t_n, s_n)$, where $(0 < t_0 < t_1 < t_2 ... < t_n)$ are positive integers, represent an operator path consisting of n operators, where the $i_{th}(1 \leq i \leq n)$ operator takes $t_i - t_{i-1}$ units of time to process a tuple of size s_{i-1} at the end of which it produces a tuple of size s_i. The selectivity of operator i is s_i/s_{i-1} [11]. From the progress chart, we can obtain the lower envelope simulation for the progress chart through making the convex curve, which just encloses the lower side of the progress chart. The lower envelope simulation cuts the progress chart into several segments. Segment will be important element in the following scheduling strategies. In figure 2, progress chart and lower envelope simulation are shown clearly; the lower envelope is represented by a dashed line and it cuts the path into three segments.

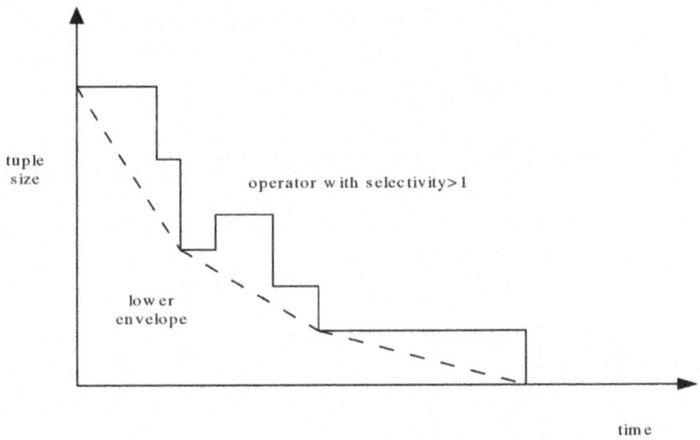

Fig. 2. Progress Chart and Lower Envelope [11]

4 New Scheduling Strategies

In this section, we first present an potential example, which gives us some inspiration. Then two novel scheduling strategies are introduced.

4.1 A Potential Example

As mentioned above, the Chain Strategy is the memory consumed nearly optimized strategy and the FIFO Strategy can obtain the minimum data tuple latency. But, both Chain and FIFO have their own advantages and disadvantages. We should consider Chain and FIFO strategies together to achieve our objection, which guarantees consumed memory not exceed the upper limit of DSMS and reduce streaming data latency at the same time. Following is an interesting example, which will give some inspiration on how to combine the two strategies.

Assume an operator path are composed by there operators $(denoted O_1, O_2, O_3)$ and the selectivity of O_1, O_2 is 0.3, 0.6667 separately. O_3 is the last operator that will not output any data. The time cost for each operator to process one unit data is $0.2s, 0.2s$ and $0.6s$,respectively. Suppose there are 11 unit data tuples input the system, the arrival times distribute as are shown in the following Table 1. We propose two strategies to deal with these tuples (suppose S_1, S_2), S_1 processes the first five tuples by Chain Strategy and the other six tuples by FIFO Strategy; inversely, S_2 use FIFO Strategy to deal with the first five tuples and the left by Chain Strategy. The following Table 2 and Table 3 show their memory consumed and data tuples latency.

Table 1. Arrival Time of Input Data

	1	2	3	4	5	6	7	8	9	10	11
arrive time	0	0.4	0.8	1.2	1.6	2.2	2.4	2.6	2.8	3.0	3.2

Table 2. Memory consumed

Time	0	0.4	0.8	1.2	1.6	2	2.2	2.4	2.6	2.8	3	3.2
S_1	1	1.2	1.4	1.6	1.8	1	2	3	3.8	4.8	5.8	5.6
S_2	1	1.2	2.2	2.3	3.2	3	3.3	3.6	3.9	4.2	4.5	4.8

Table 3. Data tuple latency

Tuple	1	2	3	4	5	6	7	8	9	10	11
S_1	2.6	3.2	6.2	6.8	7.4	8	8.6	9.2	9.8	10.4	11.0
S_2	1.0	2.0	4.0	4.8	5.6	6.4	7.2	8	9	10	11.0

From the above two tables, we can make the conclusion that the second strategy S_2 not only reduce the maximum memory requirement, also get much lower tuple latency than strategy S_1. An intuitive explanation is that: When the input load is light, strategy S_2 adopts FIFO to get a better tuple latency, and while the input load is heavy it uses Chain to reduce the consumed memory. This fact inspires us to proposal two novel strategies based on the input data distribution.

4.2 Novel Strategies

Before our new strategies are introduced, some useful notations are defined as follows:

- M is the upper limit of system memory, the consumed memory should never surpass it.
- m represents the currently consumed memory of DSMS.
- s_i represents the selectivity of the ith operator.
- t_i represents the time for the *ith* operator to process one tuple and there are n operators totally in the system.
- t, γ are the parameter of DSMS, whose values are to be determined
- f is an approximation of the data arrival rate.
- *size* is the memory consumed by one tuple.

Now, we introduce the first strategy, Adaptive Switch Chain and FIFO Strategy (ASCF). ASCF process input streaming data by FIFO Strategy to get a good latency if the input load is not tight and free source is abundant, else it will execute Chain Strategy to save memory. The following expression (1) is an approximation of the input load and system sources. If the inequation is satisfied, that means the indicates input load is not heavy and the system left free sources is abundant, thus FIFO Strategy is executed, or else Chain Strategy will be executed to save memory. The advantage of ASCF is that ASCF is a kind of adaptive strategy, and it will change policy along with fluctuation of the input data.

$$\frac{t}{f} \times size + m < \gamma \times M, \quad 0 < \gamma < 1 \tag{1}$$

Then we introduce the second scheduling strategy. It has been proved that if the slope of segments along an operator path is decreased [11], the last several segments of an operator path will cause severity data latency but have little contribution to reduce consumed memory. Therefore, a good data latency and consumed memory can be obtained through modifying the lower envelope and combining the last several segments into one. The more segments are combined, the lower data latency can be obtained but more system memory will be required. Obviously, that how many segments of an operator path should be combined depends on the system free recourse and input load. Before to do the determination, we define some variable: $K_i = \frac{\prod_{j=i}^{n} s_j}{T_i}$ and $k_i = \frac{s_i}{t_i}$. K_i is an approximation of the process velocity if segments $i, i+1, ...n$ are combined. k_i is the process velocity of the i_{th} segment. The second strategy, Chain with Segments Combined Strategy (CSC), is proposed base on the following two conditions, which approximately computes the system free resources and input data load.

$$\sum_{i=l}^{n} t_i \leq f < \sum_{j=l-1}^{n} t_j \tag{2}$$

$$t \times (t_i - K_i) < \gamma \times (M - m) \tag{3}$$

We get l and i by solving expressions (2) and (3). Denote w equals to $max(l, i)$. At last we determine the last w segments (e.g. operator $w, w+1, ..., n$) combined. Through the above expressions, the more free sources and lighter input load, the more segments will be combined. The performance will be evaluated in simulation experiment.

5 Experiments

To better evaluate our newer strategies, various simulation experiments have been designed to check the metrics and performance of these strategies under the same environment. All the input data are generated randomly. The DSMS separately executes the two new strategies, and other competing strategies under the same environment.

First step, the experiment environment is set up as follow:

- all the distributions of the input data of different strategies are uniform.
- Our designed system adopts the Poisson Stochastic, which has a broad application, as the input data distribution and the parameter of Poisson Stochastic is set to 10^{-3}
- The number of operator path, operator and the selectivity of our system are all in the scope of 5 to 20 and these numbers are generated randomly. The expected time for each operator path to deal with one unit data are assumed to about $1200us$.
- The total input data volume is more than 200 tuples, while the memory limit of DSMS is 20 unit tuples.

- The parameter of the strategies γ is set to 0.6 and 0.8 separately, but the parameter can equal to any float numerical less than 1. The effect of different parameters will be depicted in the following experiment outcomes.

To evaluate our novel strategies, test the performance and show an impartial comparison with other strategies, three kinds of experiments are proposed: First, compare ASCF, CSC with the important basic strategies Chain and FIFO; the second kind experiment will vary the parameter of each strategy but still under the same environment; At last, the comparison novel strategies with strategy of real systems will be given as a convictive proof that the proposed strategies possess practical worthy.

5.1 Compare with Basic Strategies

Due to the importance of Chain Strategy and FIFO Strategy, First, the memory consumed and data latency of novel strategies and basic strategies are compared and we lay out the outcome by the following charts, Figure 3 and Figure 4.

Just as we predict above, ASCF and CSC reduce the data latency as possible as they can and not break the memory constraint. Avoid breaking down the system memory, DSMS gets good data latency at the same time. It is more practical and valuable to use them in real DSMS.

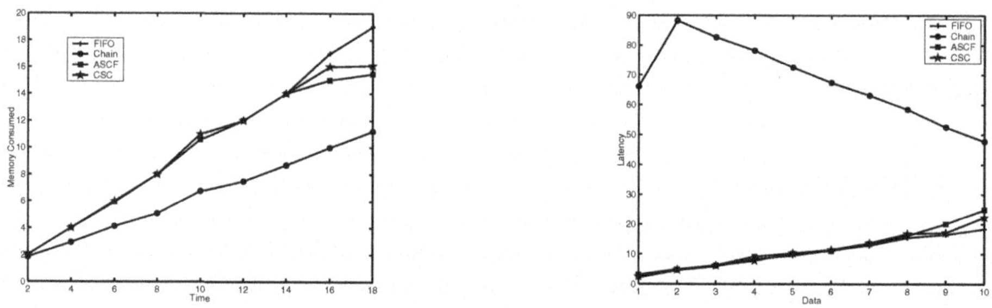

Fig. 3. Comparison Result with FIFO and Chain

5.2 The Influence of Important Parameter

Parameter γ is a key factor in our strategies that will affect the metrics significantly. From the following experiment, it is meaningful to make the conclusion that the smaller parameter γ is set, the less memory is consumed but the data latency increase at the same time. To better evaluate the affection caused by parameter, γ is set to 0.8 and 0.6 respectively. The performance comparison is depicted in the following graphs under different parameters with the same strategy and the same environment. It is easy to analysis the difference when the parameter changes. It is available and convenient for designers of DSMS to adjust the strategy on the base of different environment by varying parameters.

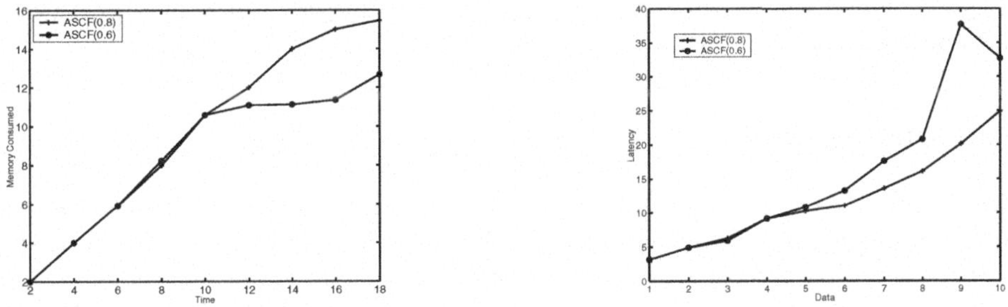

Fig. 4. Comparison with Different Parameters of ASCF

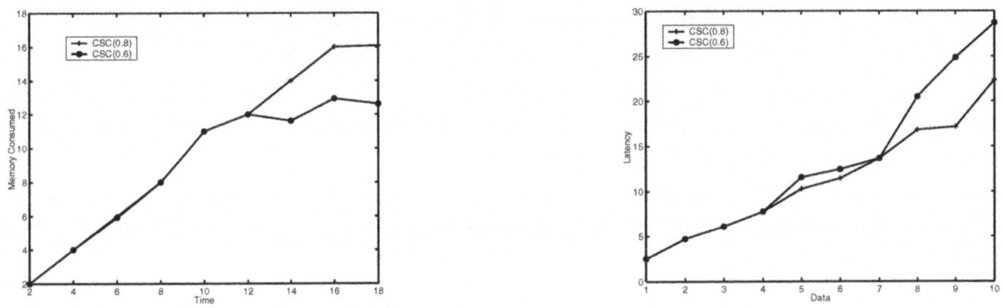

Fig. 5. Comparison with Different Parameters of CSC

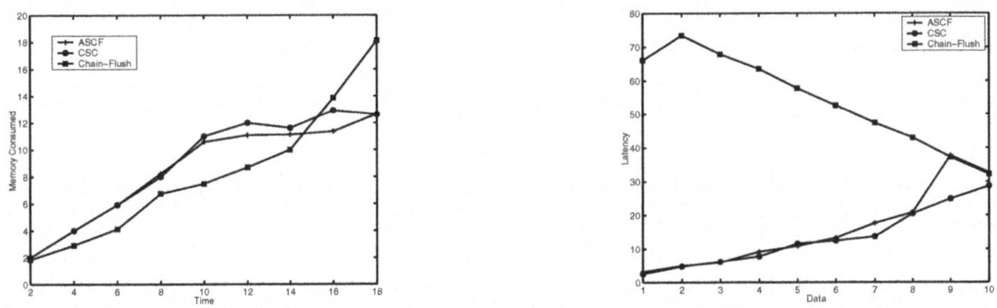

Fig. 6. Comparison with Chain-Flush

5.3 Compare with Chain-Flush

Chain-Flush is an excellent scheduling strategy for DSMS which has been introduced in section three. For more details please refer to literature [11]. Here we compare the performance of Chain-Flush with ASCF and CSC. According to the following result, ASCF and CSC use less maximum memory usage than Chain-Flush and get much lower data latency. Because at first when data load is not heavy, Chain-Flush adopt strategy to save memory but when the data load is heavy it make effort to reduce latency. Chain-Flush strategy is passive in this situation. This is a good evidence to indicate the merits and benefits of ASCF and CSC strategies.

6 Conclusion

Most of previous research work pays much attention to the architecture of DSMS, query language and so on and ignored the scheduling problems. However, different scheduling strategies can affect the system performance significantly. It is quite important and challenge for researcher to design scheduling strategies for DSMS. Most important two strategies are Chain and FIFO. Chain is nearly memory optimized, and FIFO is tuple latency optimized. Through combining Chain and FIFO, we have presented two new adaptive scheduling strategies, ASCF and CSC, which will adjust scheduling strategies according to the distribution of input data. The next step work includes doing simulation experiment to compare performance of different strategies with the same input data and computation ability. The comparison of novel strategies and other strategies shows the excellence and merit of our strategies.

Besides the memory consumed problem and data latency problem, there are still many problems worthy of attentions. Some problems are as below, which will be our future work.

Starvation problem: In many static priority systems, some low priority tuples may stay in system too long without processed. Because the priority will never change. Even if there is only one tuple with higher priority, the lower priority tuples will not be executed. ASCF and CSC successfully avoid this problem due to their adaptation and agility. If it is possible, they will assign free source to reduce the data latency to avoid starvation.

Scheduling overhead problem: If each new tuples enter the system, DSMS will reschedule the tuple, these scheduling strategies themselves will exhaust CPU resource. ASCF and CSC both are operator-based strategies not tuple-based and in most DSMS the number of operators is not very large, which can still be sustained according to the computation ability of modern CPU. Therefore the scheduling cost will not depress the performance of system severely.

Dynamic priority problem: ASCF and CSC belong to the kind of adaptive strategies and the priorities of operators are different according to the varying input data and system free source. Therefore ASCF and CSC possess more flexibility and can solve the starvation and long waiting problem.

Acknowledgements

This work is supported by the National Science Foundation of China under the grant No.60533020. We would like to thank the anonymous referees for their useful suggestions to improve the presentation of this paper.

References

[1] Lukasz Golab and M. Tamer Ozsu. Issues in Data Stream Management. SIGMOD RECORD, 32(2):5-14, 2003.

[2] Sven Schmidt, Thomas Legler, Daniel Schaller, Wolfgang Lehner. Real-time Scheduling for Data Stream Management System. Proceding of the 17th Euromicro Conference on Real-Time Systems,167-176, 2005.

[3] J. Hellerstein, M. Franklin, et al. Adaptive query processing: Technology in evolution. IEEE Data Engineering. Bulletin, 23(2):7-18, June 2000.

[4] R. Motwani, J.Widom, and et al. Query processing, approximation, and resource management in a data stream management system. In Proc. First Biennial Conf. on Innovative. Data Systems Research, Jan 2003.

[5] D. Carney, U. Cetintemel, M. Cherniack, C. Convey, S. Lee, G. Seidman, M. Stonebraker, N. TatbuL, and S. Zdonik. Monitoring streams - a new class of data management applications. In Proc. Of the 2002 Intl. Conf. On Very Large Data Bases, 2002.

[6] J. Chen, D. Dewitt, F. Tian, and Y. Wang. Niagaracq: A scalable continuous query system for internet databases. In Proc. of the 2000 ACM SIGMOD Intl. Conf. on Management of Data, 379-390, 2000.

[7] D. Terry, D. Goldberg, D. Nichols, and B. Oki. Continuous queries over append-only databases. In Proc. of the 1992 ACM SIGMOD Intl. Conf. on Management of Data, 321-330, June 1992.

[8] M. Sullivan. Tribeca: A stream database manager for network traffic analysis. In Proc. of the 1996 Intl. Conf. on Very Large Data Bases, 594, Sept 1996.

[9] Q. Jiang, S. Chakravarthy. Scheduling Strategies for a Data Stream Management System. BNCOD 2004. 16-30.

[10] Mohamed A.Sharaf, Panos K. Chrysanthis, Alexandros Labrinidis. Preemptive Rate-Based Operator Scheduling in a Data Stream Management System. The 3rd ACS/IEEE International Conference on Computer Systems and Applications, 2005.

[11] Brian Babcock, Shivnath Babu, Mayur Datar, Rajeev Motwani, Dilys Thomas. Operator Scheduling in Data Stream System.VLDB J. 13(4), 2004. 333-353.

[12] Brian Babcock, Shivnath Babu, Mayur Datar, Rajeev Motwani, Jennifer Widom. Models and Issues in Data Stream Systems.PODS 2002. 1-16

[13] D. Carney, U. Cetintemel, M. Cherniack, C.convey, S. Lee, G.Seidman, M. Stonebraker, N. Tatbul, and S. Zdonik. Monitoring streams-a new class of data management application. In Pro. 28th intl. Conf. on Very Large Data Bases, Augst 2002.

[14] S. Chandrasekaran and M. Franklin. Streaming queries over streaming data. In Pro. 28th Intl. Conf. on Very Large Data Bases, August 2002.

[15] S. Madden, M.Shan, J.N.Hellerstein, and V.Raman. Continuously adaptive continuous queries over streams. In Proc of the 2002 ACM SIGMOD Intl. Conf. on Management of Data, June 2002.

[16] R. Avnur and J.M. Hellerstein. Eddies: Continuously adaptive query processing. In Proc. Of the 2000 ACM SIGMOD Intl. Conf. on Management of Data, pages 261-272, May 2000.

A QoS-Guaranteeing Scheduling Algorithm for Continuous Queries over Streams*

Shanshan Wu, Yanfei Lv, Ge Yu, Yu Gu, and Xiaojing Li

Northeastern University, Shenyang, China 110004
wssmail@126.com

Abstract. The increasing demand on streaming data processing has motivated the study of providing Quality of Service (QoS) for data stream processing. Especially for mission-critical applications, deterministic QoS requirements are always desired in order for the results to be useful. However, the best-effort QoS providing of most existing data stream systems would bring about considerable uncertainty to the query results. This paper attempts to provide a new insight into the problem of providing QoS guarantee for continuous queries over streams. Based on the proposed QoS model of stream processing, a QoS guaranteeing scheduling algorithm is proposed. Experimental results are presented to characterize the efficiency and effectiveness of our approaches.

1 Introduction

New applications that must deal with vast number of input streams are becoming more common recently. For instance, data feeds from sensor applications, financial analysis applications that monitor streams of stock data, web server log records, environmental monitoring in chemical reactions and so on. Most of these applications have sophisticated Quality of Service (QoS) requirements that need to be met under unbounded, high-volume and time-varying data streams. The uncertainty of best-effort services probably return outdated answers or task failures, which may leads to low performance or even catastrophic results, and therefore is not adequate for many applications. To give an example of fire alarm application, the measures of temperature and smoke density are delivered from sensors and processed by DSMS for fire detection. Once the temperature and smoke density exceed certain threshold, there's probably a potential fire event, and DSMS has to take actions in alarming within given time limit. Even if the average response delay obeys the real-time constraint, those responses exceeding delay constraint would probably lead to the fire happening without alarm. However, most stream processing techniques provide best-effort services for data stream processing[1,2,3], and they can only provide statistics QoS, such as average response delay, overall throughput, *etc*. Therefore, deterministic QoS guarantee rather than best-effort performance providing is critical to these stream-based applications.

* Partially supported by National Science Foundation (60473073, 60503036), and Fok Ying Tung Education Foundation under No. 104027.

G. Dong et al. (Eds.): APWeb/WAIM 2007, LNCS 4505, pp. 522–533, 2007.

QoS for data stream processing can be defined in a variety of ways and includes a diverse set of service requirements such as performance, availability, reliability, security, etc. All these service requirements are important aspects of a comprehensive data stream processing QoS service offering. Taking the unbounded, large volume and real-time properties of data stream into consideration, we take a more performance-centric view of data stream processing and focus primarily on the issues in providing time-related and space-related performance guarantee. In this paper, we dedicate to the query scheduling method to guarantee the QoS over data stream processing based on the theory of *Network Calculus* [4,5]. Network Calculus provides a sound theoretical foundation for network scheduling, traffic control and analyzing delays and backlog in a network. It has played an important role in the development of algorithms that support QoS guaranteeing in packet networks. The main contributions we made in this paper are as follows:

- A QoS model for stream processing is proposed, including QoS-aware task definitions;
- A task scheduling strategy, named QED, is provided to guarantee the QoS requirements of each continuous query in a DSMS;
- Extensions of batch scheduling and window task preempting are suggested to improve the efficiency.

The rest of this paper is organized as follows. We begin Section 2 by presenting QoS model for the stream processing. Our QoS-guaranteeing scheduling algorithm is discussed in Section 3. Then the algorithm is deployment over our proposed architecture in Section 4. Thereafter, experimental results are presented in Section 5. Related work is described in Section 6 before ending with conclusions in Section 7.

2 QoS Modeling for Stream Processing

Firstly, a QoS-aware stream processing model is suggested. Afterwards, scheduling tasks in a DSMS are defined before introducing our scheduling algorithm. Without loss of generality, we assume that time referred in this paper is divided into slots numbered 0,1,2…

2.1 QoS-Aware Stream Processing Model

Take a stream processing system as a blackbox with several continuous queries registered inside, as illustrated in Fig.1. The blackbox deals with the incoming data and thus, produces permanent results for each query on the fly. For a continuous query, cumulative functions are used to describe the amount of data that the query has received and processed. Terminologically, let:

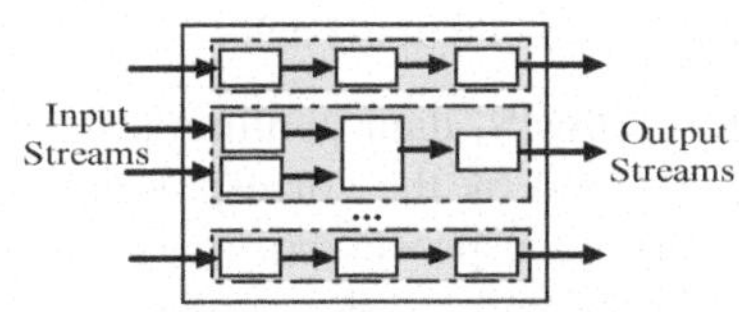

Fig. 1. Queries in DSMS

- $R(t)$ denote the number of tuples that have arrived in the input stream during the interval $[0,t]$. By convention, we take $R(0)=0$;
- $R^*(t)$ denotes the number of tuples that have been served (processed) by the query during $[0,t]$.

Note that $R^*(t)$ is measured in the number of tuples processed by the query rather than the real number of result tuples, due to the query selectivity. Thus, a stream processing system can be regarded as a lossless service node, in which the query engine acts as a service provider, and the registered queries are service consumers. In this lossless stream processing system, the number of tuples waiting in the input queue at time point t can be easily expressed by the vertical distance of $R(t)$ and $R^*(t)$, *i.e.* $R(t)-R^*(t)$. Similarly, the horizontal distance of $R(t)$ and $R^*(t)$ at any function value n, denoted by $(R^*)^{-1}(n)-(R)^{-1}(n)$, represents the delay suffered by the nth tuple. For deterministic QoS guarantee, a formal definition of the QoS includes: (1) the characteristics of the input given by the service consumer; and (2) the QoS requirement demanded by the service consumer. With the information above, the service provider is capable of determining the required resource to meet the QoS requirements. Here, the terms of arrival curve and service curve in Network Calculus are introduced to describe the characteristics of input streams and QoS constraints, respectively. Arrival curve defines a constraint on input burstiness as follows.

Definition 1 (Arrival Curve). Given a wide-sense increasing function α, we say that the stream is constrained by α if and only if $R(t)-(s)\le\alpha(t-s)$ for all $s\le t$. Function α is called an arrival curve for the stream, and the stream is also called α-smooth.

Definition 2 (Service Curve). Consider a system S and a continuous query to be processed by S with input R and output R^*. We say that S offers to the query a service curve β if and only if β is wide sense increasing, $\beta(0)=0$ and $R^*\ge R\otimes\beta$ [1].

Service curves were defined as a measure for partially characterizing the service provided to a query. Roughly speaking, a service curve is a function that lower bounds the amount of data that has been processed by a DSMS within some specific interval. For example, if a query requires that the response delay be bounded by d, the service curve for the query is defined as $\beta=\delta_d=\begin{cases}0, 0\le t\le d\\ \infty, \quad t>d\end{cases}$. It is obvious that for all t we have $R^*(t)\ge(R\otimes\delta_d)(t)=R(t-d)$, which means the response delay is no more than d.

According to Theorem 1.4.1 and 1.4.2 in [5], once the arrival curve and service curve of a continuous query are given, the input queue size and the response delay are bounded. Therefore, any QoS requirements related to time and space is deterministic.

2.2 QoS-Aware Tasks

In a DSMS, data continuously arrive in the input queue to be processed by the corresponding queries. According to how the queries are processed, two scenarios are considered to catalog the scheduling tasks for data stream processing:

Table 1. Notations

	Meaning
T_i^n	n-th task of Qi
a_i^n	the arrival time of T_i^n
d_i^n	the finish time of T_i^n
D_i^n	the deadline of T_i^n
r_i^n	n-th input tuple of Qi

- ***SPJ* query:** SPJ is a fairly common class of queries in data stream applications, which typically involve selections and projections over a single stream, and may involve joins with one or more static stored relations. For

[1] $\otimes$ is convolution operator, $(f\otimes g)(t)=\inf_{s:0\le s\le t}\{f(t-s)+g(s)\}$.

these queries, query operations could be independently applied to each single tuple directly, and the QoS requirements are always posed on single tuple processing. For example, asks for smoke density of the sensors whose temperatures exceed 50°C, and the results should be returned within 1 min. Therefore, we define the scheduling task for SPJ query on the granularity of single input tuple as follows.

Definition 3 (Basic Task). The processing of a tuple by a query is defined as a basic task.

- **Aggregation query:** A distinguishing feature of this class of query is that the query is unable to produce the first tuple of its output until it has seen its entire input. For data streams, such queries are typically used to get aggregations of sliding window, such as SUM, COUNT, MIN, MAX, and AVG. In such queries, each window is an entity with certain query semantics, and the performance requirements are generally posed on window processing. For example, every 5 mins return the average temperature of the sensors within 10 mins, and response the query within 1 min. The real-time constraint requires that the result of each window be output at most 1min after the window is built up. Therefore, each window is regarded as a data unit, defined as a window tasks in Definition 4.

Definition 4 (Window Task). The processing of a window by a continuous query is called a window task.

Without loss of generality, we assume that each stream is related to one query and each query has exactly one related stream. In the real DSMS, data sharing techniques could be applied to a stream participating multiple queries, and each query is related to a logic stream. And for those queries with multi-input, input streams are combined into one logic input stream. Thus, the task processing in a DSMS is independent of each other. By the way, some of the notations used in the paper are listed in Table 1.

3 QoS-Guaranteeing Scheduling

In this section, we first talk about the main idea of our QoS-Guaranteeing scheduling. Then the key techniques used in the scheduling strategy are developed one by one.

3.1 QED Overview

The scheduling of continuous queries may follow different strategies, depending on the overall QoS expected. Since all the QoS requirements are abstracted in the form of service curves, the absolute goal of our scheduling is to find a schedulable and efficient strategy to guarantee the respective service curve for each continuous query.

Motivated by the service curve allocation method in Network Calculus, named SCED (Service Curve Earliest Deadline)[6], we assign a QoS-aware deadline to each task, and tasks are scheduled in EDF (Earliest Deadline First) strategy. Therefore, the scheduling algorithm is named as QED (QoS Earliest Deadline). If the deadlines are met, the service curves required by the queries are guaranteed, so are the QoS requirements demanded by each query. The deadline satisfaction is guaranteed by schedulability verification, which is carried out with the knowledge of queries' QoS requirements and the system capability.

Considering the respective properties of basic task and window task, optimizations are carried out to improve the scheduling efficiency. As for basic tasks, due to the high volume characteristics of data streams, scheduling in *a-tuple-a-time* fashion would lead to heavy overhead of scheduling and context switching. Therefore, a dynamical batch scheduling is proposed. Each time a query is triggered, scheduler picks up tasks in its task queue as much as possible into running, on the premise that the schedulability is not violated. As for window tasks, if the window task under running is costly, new arrived tasks with more urgent deadline have to wait a relative longer time. In order to guarantee QoS, each query is allocated a larger service curve, which would lead to much resource reservation. Therefore, a preemptive adaptation is suggested for window task scheduling, with QoS satisfaction not violated.

3.2 Deadline Allocation

In network calculus, SCED policy defines output deadline to each packet according to the service curve that the flow requests, as given in Definition 3. Theorem 5 in [7] tells that if the flows are schedulable in a service node with EDF strategy, the delay of each packet is bounded. All the packets are served before or at their deadlines plus l_{max}/r, where l_{max} is the maximum packet size, and r is the capacity of the server. This is because packets are transmitted non-preemptively in EDF strategy.

Definition 3 (SCED). If service curve β_i is allocated to flow i, the deadline for n^{th} data of flow i is defined by:

$$D_i^n = \left(R_i^*\right)^{-1}(n) = \min\left\{t \in N : R_i^n(t) \geq n\right\} \tag{1}$$

where $R_i^n(t) = \inf_{s \in [0, a_i^n]} \left[R_i(s) + \beta_i(t-s)\right]$ and a_i^n is the arrival time of the data.

In the circumstance of data stream processing, since the processing cost of each query differs from one another, the service received by each query should be measured in a uniform comparable measurement rather than the number of tuples that have been served by each query. Therefore, the computation resource is standardized as follows. Assume that the query engine has the ability to perform C basic operations per time slot, and it takes at most C_i basic operations for query Q_i to process a tuple. Let $\Delta_i = C_i/C$, which represents per tuple processing cost bound for query Q_i. Assume the costs can be known in advance by permanently monitoring the stream processing during the warm-up stage. Supposing there are N continuous queries in a DSMS sharing computation resources, query Q_i requires service curve β_i be guaranteed, for $i=1,\ldots,N$. Theorem 1 tells how the deadlines are allocated to tasks in a DSMS.

Theorem 1 (Deadline Allocation in DSMS). In a DSMS, query Q_i requires β_i as a guaranteed service curve, and per tuple processing cost bound for query Q_i is Δ_i. Deadlines are defined to tasks in the system with SCED policy, with service curve allocated to query Q_i as:

$$\beta_i^* = \begin{cases} 0, & t = 0 \\ \beta_i(t + \Delta_{\max}), & t > 0 \end{cases} \tag{2}$$

where $\Delta_{max} = \max\{\Delta_i\}$.

Proof: Suppose that the task T_i^n arrives at a_i^n and finishs at d_i^n. If the deadlines D_i^n are defined with service curve β_i^*, the tasks depart no later than $D_i^n+\Delta_{max}$. That is, $d_i^n-\Delta_{max}\leq D_i^n$. Observe that from Equation (2), if a time t is less than or equal to D_i^n,

$$\inf_{s\in[0,a_i^n]}\left[R_i(s)+\beta_i^*(t-s)\right]\leq R_i^*(d_i^n)$$

Thus,

$$\inf_{s\in[0,a_i^n]}\left[R_i(s)+\beta_i^*(d_i^n-\Delta_{\max}-s)\right]\leq R_i^*(d_i^n)$$

Due to the expression of β_i^*, we have:

$$R_i^*(d_i^n)\geq \inf_{s\in[0,a_i^n]}\left[R_i(s)+\beta_i(d_i^n-s)\right].$$

According to the definition of service curve, β_i is guaranteed to query Q_i.

$R_i^n(t)$ in SCED policy gives the number of tasks of query Q_i that should be output by t. Suppose tasks T_i^n could be output at a later time point t' with service curve β_i guaranteed, then $R_i^n(t')\leq n$. With the definition of D_i^n, we can know that $R_i^n(D_i^n)\geq n$, so that $R_i^*(t')\leq R_i^n(D_i^n)$, which contradicts the definition of service curve. Therefore, D_i^n is proved to be the latest time bound that the task should finish, in order to guarantee service curve β_i. On the other hand, conclusion in [8] proved that EDF is known to be optimal for any independent process scheduling algorithms upon uniprocessors. As a result, if the tasks are scheduled in EDF strategy with deadlines allocated by SCED, they are scheduled in an optimal discipline to guarantee the required service curves. In other words, if the EDF strategy with deadline allocated by SCED could not schedule the query tasks in a system, there is no other scheduling strategy that could make all the required service curves guaranteed. And this is why the SCED in network calculus is introduced in our QoS-guaranteeing scheduling.

3.3 Schedulability Verification

Since the capability of a certain query engine is limited, when a set of queries run in a single DSMS, they compete for computation resources with one another. So how to find the condition under which the shared query engine can simultaneously serve all the queries with their QoS guaranteed is a necessary problem to discuss. In Theorem 2, we provide a condition under which if all the tasks never miss their deadlines, the desired service curve is guaranteed to each query.

Theorem 2 (Schedulability Condition in DSMS). In a DSMS, query Q_i requires βi as a guaranteed service curve, and per tuple processing cost bound for query Q_i is Δ_i. If the input stream to query Q_i is α_i-smooth, then service curve β_i is guaranteed to query Q_i, $i=1,...,N$, if Equation (3) is satisfied:

$$\sum_{i=1}^{N}(\alpha_i\otimes\beta_i^*)\times\Delta_i\leq t,\quad \forall t\geq 0 \tag{3}$$

Proof: Only if the overall required computation capability is always not greater than the capability that the DSMS could provide, the service curves could be guaranteed to each query. According to Theorem 1, in order to guarantee service curve β_i to query Q_i, the service curve assigned in SCED should be β_i^*. Therefore, the convolution of arrival curve α_i and service curve β_i^* are used to get the required computation capability to guarantee the service curve in the worst case. Therefore, according to Proposition 2.3.3. in [5], the deadlines allocated by SCED policy are satisfied if the following equation is satisfied:

$$\sum_{i=1}^{N}(\alpha_i \otimes \beta_i^*)\times C_i \le Ct, \quad \forall t \ge 0$$

Thus, we get the schedulability condition for a DSMS as given in Equation (3).

3.4 Batch Size Decision

To introduce batch scheduling for QoS-guaranteeing stream processing, the following aspects should be taken into consideration:

1. The introduce of batch should not disturb the property of QoS guaranteeing;
2. The involved computation cost due to batch scheduling could not be too heavy to affect the QoS guaranteeing;
3. The batch size should be as large as possible.

To obey the above rules, the main idea of our batch size decision is as follows. On the premise that the schedulability condition is satisfied, allocate a larger service curve to the query to be scheduled; and the batch size is decided dynamically according to the improved service curve and the backlog at that time.

Aiming at improving query Q_i's service curve β_i to β_i^{max}, the service curve could be improved to the extreme condition in which the system is just schedulable. Therefore, the improved service curve could be obtained from Equation (4):

$$\sum_{j\ne i, i=1}^{N}((\alpha_j \otimes \beta_j^*)(t)\times \Delta_j) + (\alpha_i \otimes \beta_i^{\max})(t)\times \Delta_i = t, \forall t \ge 0 \tag{4}$$

Batch size decision is performed each time query engine is switched to a query. Suppose query Q_i gets the query engine at time point t, and the task to be scheduled is T_i^n. At that time, the task with the earliest deadline in other task queues is T_j^m, that is to say, the next query to be processed is Q_i, where $i{\ne}j$. Then, Equation (5) tells the overall number of tasks that query Q_i could process before scheduling T_j^m, with β_i^{max} as the guaranteed service curve for query Q_i.

$$BatchSize(i,t) = (\alpha_i \otimes \beta_i^{\max})(D_j^m - t) \tag{5}$$

During the processing of query Q_i, if $T_k^l(k{\ne}i{\ne}j)$ arrives with higher priority than T_j^m, *i.e.*, $D_k^l{<}D_j^m$. In order to comply to the EDF strategy, *Batchsize* should be updated to the tasks that should be output before scheduling T_k^l.

Carrying the idea farther, the work of batch scheduling could be divided into offline part and online part.

Offline Part. As for the decision of batch size suggested above, it is not necessary to get the expression of the improved service curve for each query. In Equation (5), only the convolution of α_i and β_i^{max} is required. Therefore, if each $\alpha_i \otimes \beta_i^*$ is derived offline, $\alpha_i \otimes \beta_i^{max}$ could be available in advance by using Equation (4). Though convolution is not a trivial work, it could be performed offline before system running. Therefore, the computation incurred by batch scheduling does not aggravate the runtime cost.

Online Part. During system running at time *t*, *BatchSize* is calculated by get the function value of $\alpha_i \otimes \beta_i^*$ at $D_j^m - t$. Only if a more urgent tuple arrives, *BatchSize* are modified. Therefore, the computation complexity of online part is quite trivial.

3.5 Preemptive Adaptation for Window Tasks

The scheduling algorithm proposed above works in a non-preemptive fashion. During the processing of a task, if another task with a higher priority arrives, it has to wait till the current task finishes. Since window tasks always involve processing several tuples, the cost of a window task is relative large. The task with higher priority has to wait for a relative long time if a window task is being scheduled, which may lead to lower real time property of task processing. On the other hand, aiming at guaranteeing the QoS requirements, the worse case is considered when verify schedulability. For each query, the offset of service curve β_i^* would be relative large compared to the original service curve β_i, even if only one heavy-cost windowed query exists. This would lead to much resource reservation in the system. So, we try to break down the window tasks and suggest a preemptive scheduling extension as follows.

When a window task is being scheduled, if a new task with higher priority arrives, the window task is interrupted, and the new arrived task is scheduled. The intermediate results of the window task are preserved, and the rest of the window task is replaced into the task queue as a new window task with the same deadline as before. Since the task's deadline is not changed, the window task seems to be split into several sub-tasks with the same deadline. Therefore, the schedulability verification condition is not broken, and so it is schedulable in a preemptive fashion.

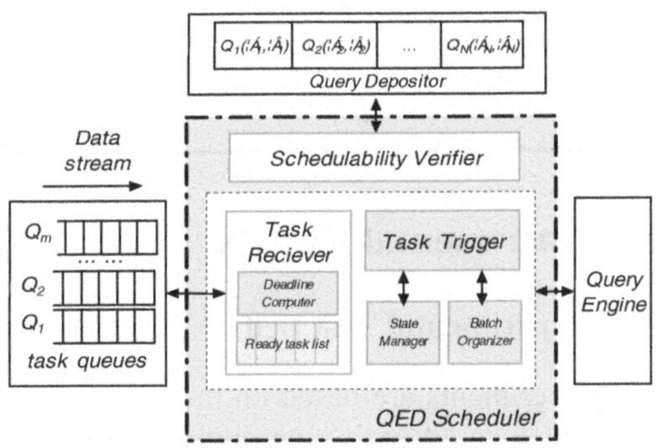

Fig. 2. Architecture of QED Scheduler

4 Algorithm Deployment

As shown in Fig.2, the architecture of QED scheduler includes four components: *Schedulability Verifier*, *Task Receiver*, *Task Trigger* and *Batch Organizer*.

After the queries together with their respective QoS requirements are registered to the system, *Query Depositor* models the QoS requirements with service curves and arrival curves. Then *Schedulability Verifier* checks whether all the queries could be admitted to the system, with their QoS requirements satisfied. Once the schedulability condition is obeyed, The *QED Scheduler* begins to trigger tasks into processing.

In order to work in EDF fashion, *Priority Order List* maintains the priority of each query in increasing order. The priority of a query is assigned as the deadline of the task in the most front of its task queues. Each time *Task Trigger* chooses the query with the highest priority to execute, *Batch Organizer* tells *Task Trigger* how many tasks the query could process for the basic tasks. As for window tasks, if a window tasks is preempted, the intermediate results are kept in *State Manager*. The scheduling procedures of each component in QED scheduler are listed in following Table 2.

Table 2. QED scheduling Algorithm Deployment

Task Trigger:

while(TRUE){
    Get the query with the highest priority from Priority Order List, denoted by Q_i;
    Call **Batch Organizer**;
    while (queue size ≤ BatchSize)
        Remove a task from the most front of the queue and Process the task;
    if (task queue i is empty)
        Set priority of Q_i to ∞ and reorder Priority Order List;
}

Task Receiver:

Triggered Condition: T_i^n arrives in the most front of task queue i;

Compute Deadline D_i^n;
Assign D_i^n as the priority of query Q_i;
Order Query Priority List;
Case current running task type of
    basic task:
        If (P_i is the second highest priority) {
            Call **Batch Organizer**;
            Update current BatchSize;
        }
    window task:
        If(P_i is the highest priority){
            Save intermediate result of current task and Put the rest of work back to task queue;
            Schedule T_i^n;
        }

Batch Organizer:

Offline:

(1) $U_i(t)=(\alpha_i \otimes \beta_i)\times\Delta_i,\ i=1,\dots,N;$

(2) $(\alpha_i \otimes \beta_i^{max})(t) = \frac{\Delta_i}{1}\left(t - \sum_{j=1, j\neq i}^{N} U_j\right), i=1,\dots,N$

Online:

(1) $\tau = P_j - t;$

(2) $BatchSize(i,t) = (\alpha_i \otimes \beta_i^{max})(\tau)$

(3) *Return BatchSize;*

5 Experimental Results

5.1 Experimental Test Bed

The experiments are based on the following Example 1. For ease of experimentation, we implemented each continuous query as a black box whose per-tuple processing cost is randomly generated obeying normal distributions. To illustrate the significance of extensions of batch and preempting, the two extensions are added to the algorithm gradually. We call the algorithm only with batch scheduling for basic tasks as B-QED, and the version with both extensions are named P-QED. Moreover, we implement FIFO and SPT (Shortest Processing Time First) algorithms as well to make comparisons.

The software environment for the following experiments is Windows XP using Java, and the hardware environment is PC with 1.73GHz PentiumM processor and 512M of memory. The machine is isolated from other tasks except the three continuous queries.

Example 1: Suppose the user registers three continuous queries to a DSMS. The performance requirements and input characteristics of these queries are described as follows:

(1) Q_1 is an aggregation query. It asks for throughput at least 30 windows per second, and the response delay at the beginning of backlogged period is no more than 7ms. As for the input stream, average tuple arrival interval is 2ms. However, tuples may arrive 4ms earlier or 6ms later than the expected interval. And the minimum interval constraint is 1ms.

(2) Q2 requires that any tuple delay be bounded by 6ms. Peak rate of input stream is limited by 350 tuples per second.

(3) *Q3 is an aggregation query too. It asks for throughput at least 20 windows per second, and the response delay at the beginning of backlogged period is no more than 8ms. Input stream is also peak rate limited, with peak rate limit 200 tuples/sec.*

According to the characteristics, we give α, β and other information listed in Table 3. The schedulability verification is discussed for B-QED and P-QED as follow.

- B-QED:

$$\sum_{i=1}^{3}(\Delta_{\max i} * (\alpha_i \otimes \beta_i^*)) = \begin{cases} 0 & t = 0 \\ 0.55t + 0.6 & t \le 2 \\ 0.95t - 0.2 & 2 < t \end{cases}$$

When $t \le 1$, $\sum_{i=1}^{3}(\Delta_{\max i} * (\alpha_i \otimes \beta_i^*)) = 0.55t + 0.6 > t$, So B-QED is unschedulable in this example. But from long point of view, $\sum_{i=1}^{3}(\Delta_{\max i} * (\alpha_i \otimes \beta_i^*)) = 0.95t - 0.2 < t$, that is, the system resource is much sufficient than that needed. It is not strange since the unschedulablity is caused by non-preemption.

Table 3. Query Information

	Q_1	Q_2	Q_3
Δ_{max}	1	1	2
Δ^w_{max}	6	--	10
α	$\begin{cases} 0 & t=0 \\ t+1 & 0<t\le 10 \\ 0.5t+6 & 10<t \end{cases}$	$0.35t$	$0.25t$
β	$\begin{cases} 0 & t\le 7 \\ 0.2(t-7) & 7<t \end{cases}$	$\begin{cases} 0 & t\le 6 \\ +\infty & 6<t \end{cases}$	$\begin{cases} 0 & t\le 8 \\ 0.2(t-8) & 8<t \end{cases}$
β^*	$\begin{cases} 0 & t=0 \\ 0.2(t+3) & 0<t \end{cases}$	$0.35t$	$\begin{cases} 0 & t\le 2 \\ 0.2(t-2) & 2<t \end{cases}$
β^{P*}	$\begin{cases} 0 & t\le 5 \\ 0.2(t-5) & 5<t \end{cases}$	$\begin{cases} 0 & t\le 4 \\ +\infty & 4<t \end{cases}$	$\begin{cases} 0 & t\le 7 \\ 0.2(t-7) & 7<t \end{cases}$

- P-QED:

$$\sum_{i=1}^{3}(\Delta_{\max i} * (\alpha_i \otimes \beta_i^{P*})) = \begin{cases} 0 & t \le 4 \\ 0.35t - 1.4 & 4 < t \le 5 \\ 0.55t - 2.4 & 5 < t \le 7 \\ 0.95t - 5.2 & 7 < t \end{cases}$$

It holds that: $\sum_{i=1}^{3}(\Delta_{\max i} * (\alpha_i \otimes \beta_i^{P*})) < t \quad \forall t > 0$, therefore using P-QED algorithm, all the three Query's QoS can be satisfied.

5.2 Experimental Results

- **Schedulable Situation**

We decrease the average interval of tuple arrival for query Q_1, that is, increase the average speed of Q_1 input stream. From Section 5.1 we can know system is always schedulable in this process. As it is illustrated in Fig.3, P-QED can totally guarantees QoS of all the queries while other can't, however B-QED is much better than FIFO and SPT. This result is consistent with our above analysis. Take a deep insight into the case, when the load of Q_1 becomes greater with the decrease of T, chances of scheduling are more for Q_1 in FIFO strategy, therefore tuples of Q_2 and Q_3 tend to miss their QoS requirements. As for SPT, the processing time of Q_2 is always the least, so it is always scheduled with high priorities.

- **Unschedulable Situation**

Now let's see the algorithms' performance in unschedulable situation illustrated in Fig.4. We increase the peak rate of Q_2. The system can be verified schedulable if the rate limit is less than 0.45, but as it is shown in experiments, while the peak rate is less than 0.55, the P-QED algorithm can still provide QoS guaranteeing service. This phenomenon suggests P-QED has resource reservation. When the peak rate exceeds 0.55, our strategy provides a rather fair performance. Notice that SPT can guarantee Q_2's QoS to some extent even when the peak rate is very high. So in unschedulable situation, our algorithm is quite fair but not excellent for QoS guaranteeing.

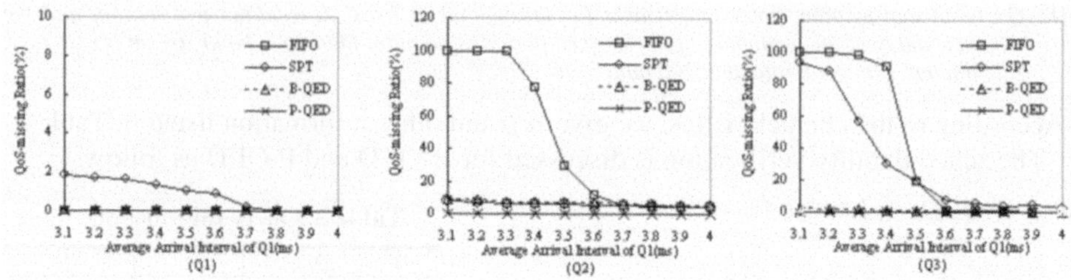

Fig. 3. QoS-missing Ratio in Schedulable Situation

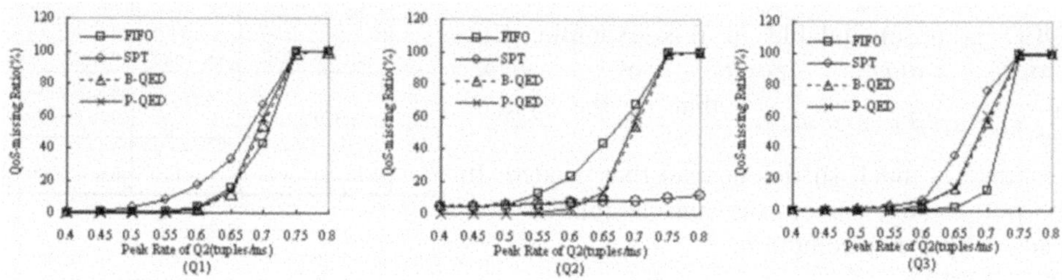

Fig. 4. QoS-missing Ratio in Unschedulable Situation

6 Related Work

Scheduling problem over data streams has motivated a considerable research recently [1,2,3]. [2] proposes a chain strategy with a goal of minimizing the total internal queue size. [9] provides scheduling strategies to minimize tuple latency as well as total memory requirement. The Aurora project employs a two-level scheduling approach [1]: the first level handles the scheduling of superboxes that are a set of operators, and the second level decides how to schedule a box within a superbox. [10] designs a preemptive rate-based scheduling policy that handles the asynchronous nature of tuple arrival and the heterogeneity in the query plan. Eddies [3] tuple-at-a-time scheduling does away with fixed query plans and provides extreme adaptability. However, all above scheduling strategies try their best to gain better processing efficiency, and the issue of QoS guarantee has been given little consideration. To the best of our knowledge, the scheduling method [11] used in Qstream system is the only one pays attention to time-related QoS guarantee. Since it is implemented on a RTOS, its deterministic behavior regarding the processing times relies on the real-time operating system environment. It cannot provide a general QoS guaranteeing for data stream processing.

7 Conclusions and Future Work

This paper has investigated scheduling algorithms for stream data processing in order to satisfy QoS requirements of continuous queries in a multi-query environment. A QoS model containing input and performance requirement for stream processing is proposed based on the theory of network calculus, and an algorithm according to EDF strategy is proposed to guarantee QoS requirements. Then we optimize the algorithm for SPJ query and aggregation query respectively in order to improve the stream processing efficiency. At last we obtain a preemptive QoS-guaranteeing batch processing strategy. Experimental results confirm that our claims are effective and efficient. Our current strategy considers a query as a black box, next we can extend our algorithms to address operator-level scheduling problem. In addition, some probability model could be imported in order to reduce system resource reservation.

References

1. D. Carney, U. Centintemel, *et al.* Operator scheduling in a data stream manager. In Proc. of the 29th VLDB Conf., Berlin, Germany, Sept.2003, 838-849.
2. B. Babcock, S. Babu, M. Datar, R. Motwani. Chain: Operator Scheduling for Memory Minimization in Data Stream Systems. In Proc. of the ACM SIGMOD 2003, San Diego, California, USA, June 2003, 253–264.
3. R. Avnur, J. M. Hellerstein. Eddies: Continuously Adaptive Query Processing. In Proc. of the ACM SIGMOD 2000, Dallas, TX, May 2000, 261–272.
4. R. L. Cruz. Quality of service guarantees in virtual circuit switched networks. IEEE J. Select. Aareas Commun., vol. 13, 1048–1056, Aug. 1995.
5. L. Boudec, J. Yves, P. Thiran. Network calculus: Springer Lecture Notes in Computer Science, Vol.2050, 2001.
6. H. Sariowan, R. L. Cruz, G. C. Polyzos. SCED: A generalized scheduling policy for guaranteeing quality-of-service. IEEE/ACM Trans. On Networking.1999, 7(5): 669-684.
7. K. Pyun, J. Song, H. K. Lee. Two service curve allocation schemes for achieving both high network utilization and a constant deadline computation in a service curve algorithm. Technical report CS/TR-2003-190 of KAIST Department of Computer Science, April 2003.
8. C. L. Liu and J. W. Layland. Scheduling algorithms for multi-programming in a hard real-time environment. Journal of the Association for Computing Machinery, January 1973,20(1): 40-61.
9. J. Qingchun, C. Sharma. Scheduling Strategies for Processing ContinuousQueries over Streams. In Proc. of 21st British National Conference on Databases (BNCOD 2004), Edinburgh, UK, July 2004, 16-30.
10. A. M. Sharaf, K. P. Chrysanthis, A. Labrinidis. Preemptive Rate-based Operator Scheduling in a Data Stream Management System. In Proc. of the Third ACS/IEEE International Conference on Computer Systems and Applications (AICCSA), Cairo, Egypt, January 2005, 46-54.
11. S. Schmidt, T. Legler, D. Schaller, W. Lehner. Real-Time Scheduling for Data Stream Management Systems. In Proc. of the 17th Euromicro Conference on Real-Time Systems (ECRTS'05), Palma de Mallorca, Spain, July 2005, 167-176.

A Simple But Effective Event-Driven Model for Data Stream Queries*

Yu Gu, Ge Yu, Shanshan Wu, Xiaojing Li, Yanfei Lv, and Dejun Yue

Northeastern University, Shenyang, China 110004
gy_neu@163.com

Abstract. With expanding of query requirements, event-relative semantic has appeared for data stream applications. This paper builds a simple but effective event-driven data stream model *EQM* based on event semantics and features over data stream. Furthermore, some improved approaches over event detection and queries as well as relative efficiency evaluation are discussed. Experiments show that our approaches gain better performance than available data stream model and processing approaches as far as event-relative problem is concerned.

1 Introdution

As a new kind of data form, data stream has been paid much attention. Query processing over data stream is different from that in traditional database. In data stream applications, data are active, and queries are positive. Generally, queries are registered beforehand, and on-line monitoring results are output continually with the arrival of each tuple coming from the data stream sources. Such kind of query is usually called continuous query and the trigger fashion is called tuple-driven. Besides, the fashion that results are output over a period is called time-driven. Presently, some representative DSMSs[1][2][3][4] have been designed and realized in this manner.

However, with expanding of application requirements over data stream, some event-relative semantic has been proposed. Therefore, the available trigger fashions can't describe the application requirements well and can't process data from event relative applications efficiently. For example, the event that when there is a car accident on a segment, we want to detect the speed change of those cars on the same segment is hard to describe by using simple continuous queries over data stream. From a new perspective into trigger fashions, this paper proposes a more general model and classification methods to solve event-relative problems over data streams.

The rest of the paper is organized as follows. Section 2 reviews related work. Section 3 presents event semantics and an event-driven model *EQM* over data stream. Section 4 mainly proposes two relative optimized algorithms to solve some typical query processing problems. Section 5 shows our model and algorithm are efficient by examples and experimental analysis. The last section concludes the paper.

* Partially supported by National Science Foundation (60473073, 60503036), and Fok Ying Tung Education Foundation under No. 104027.

G. Dong et al. (Eds.): APWeb/WAIM 2007, LNCS 4505, pp. 534–541, 2007.

2 Related Work

At present, most of the existing data stream models are extensions of relation algebra and based on characteristic of data stream, some operators like sliding window are included[1][2]. In these models, query trigger manner mainly includes tuple-driven and time-driven, and languages are simple extensions of *SQL* in order to support stream model. Also, the query processing is similar to that in traditional database in which a query plan is a net of operators connected by queues and correct results are output by communication and synchronization methods. There are also some particular methods in the processing of increment maintenance, such as direct maintenance methods and negative tuple method[5][6]. *STREAM*[1]proposed by Standford is a DBMS which can describe models and query processing mentioned above well and especially *CQL*[7] is a representative powerful query language over data stream, but it can't support query processing over complicated events well.

Traditional active database models[8][9]mostly adopt ECA fashion which has studied event models and semantics in depth. However, ECA mainly focuses on static data set and pays less attention to dynamic and real-time characters over data stream and real-time detection into events, so it can't address query requests over data stream. Paper[10] focuses on how to correlate a series of events over data stream which is irrelative with our devotion about event-driven model.

3 Event-Driven Models over Data Stream

In this section, we introduce the definitions of basic events, semantic composite events and time sequence composite events separately, which can help express various event semantics. Also, the event-driven model named as *EQM*(Event Query Model) is proposed.

3.1 Basic Events Definition over Data Stream

In our model, basic events include *Time Event* and *Content Event*.

Definition 1: *TE*(*Time Event*). Suppose that the time domain T of data stream system S is modeled as $T=\{t_i|i=0,1,2,3\ldots\}$, t_i denotes the ith time point, and t_{now} denotes the current time point. A *TE* is defined to occur at time point t_i when $t_{now} = t_i$. *TE* occurring at t_i is denoted as $TE(t_i)$.

Definition 2: *CE*(*Content Event*). Suppose that d_{ij} denotes the jth-arrival tuple over the ith data stream and *Comp* denotes the comparison between two tuples or a tuple and a constant value. According to different semantics and complexity degree, we divide *CE* into three categories further.

- Arrival Event(CE_{arr}). If $d_{ij}.t=t_{now}$, a CE_{arr} occurs.
- *Transversal Comparison Event*(CE_{trans}).If $Comp(d_{ij}, m) \wedge (d_{ij}.t=t_{now})$, a CE_{trans} occurs. m denotes a threshold or tuple value over other data streams. For example, when temperature exceeds 50 degree, a CE_{trans} occurs.

- *Consecutive Comparison Event*(CE_{cons}).If $Comp(d_{ij}, d_{ik}) \wedge (d_{ij}.t=t_{now})$, a CE_{cons} occurs. d_{ik} denotes some tuple arriving before d_{ij} over the same data stream *i*. For example, when temperature increases by 30 degree, a *RE* occurs. This situation is quite hard to describe by using relational algebra.

Due to the time feature owned by stream data, *TE* and *CE* have relations between each other. A *CE* occurring at *t* is denoted as *CE*(*t*) which can be expressed by *TE*(*t*) and necessary event information. A *TE*(*t*) can be considered as an tuple-arrival event whose attributes are set null except the timestamp attribute.

3.2 Semantic Composite Events

In order to express semantics better, in our model, stream events are further classified into instant event, lasting event, and periodic event according to event correlation over data stream and semantic granularity.

- *IE* (*Instant Event*). It's composed of single *TE*(*t*) or *CE*(*t*). It's the smallest event granularity over data stream, and an isolated *IE* expresses one independent event occurs.
- *LE*(*Lasting Event*). It expresses a time or semantic range [$IE(t_1)$, $IE(t_2)$] composed of two semantic-related instant events $IE(t_1)$ and $IE(t_2)$, where $t_1<t_2$. $IE(t_1)$ is a *Beginning Event* denoted as *B*, and $IE(t_2)$ is an *Ending Event* denoted as *E*.
- *PE* (*Periodic Event*). A group of *IEs* that occur periodically and express the same semantics together institute a periodic event, denoted as *PE*(*B*(*t*),*E*(*t*),*I*), where *I* denotes the time interval between consecutive *IEs*. Clock *PE* is the most ordinary periodic event over data stream which can also be denoted as $PE(TE(t_i))$ and in this paper, we only consider it.

3.3 Time Sequence Composite Events

Based on the definitions of basic events and semantic-composite events, we can make composites further according to time sequence relation between events in order to express more complicated semantics. @ denotes time sequence-composite operation.

- *IE@IE*. It expresses that two *IEs* occurring at the same time. If $t_1=t_2$, two instant events $IE(t_1)$ and $IE(t_2)$ institute a time sequence-composite event $IE(t_1)@IE(t_2)$.
- *IE@LE*. It expresses that an instant event occurs during a lasting event. If $t_1<t_3<t_2$, a lasting event $LE(IE(t_1),IE(t_2))$ and an instant event $IE(t_3)$ institute a time sequence-composite event $LE(IE(t_1),IE(t_2))@IE(t_3)$.
- *LE@LE*. It expresses that one lasting event occurs during another lasting event. If $t_1<t_3$ and $t_4<t_2$, two lasting events $LE(IE(t_1),IE(t_2))$ and $LE(IE(t_3),IE(t_4))$ make up of a time sequence-composite event $LE(IE(t_1),IE(t_2))@LE(IE(t_3),IE(t_4))$.
- *PE@LE*. It expresses that clock periodic event occurs during a lasting event. If $t_1<t_3$ and $t_4<t_2$, a lasting event $LE(IE(t_1),IE(t_2))$ and a periodical event $PE(B(t_3),E(t_4),I)$ constitute $LE(IE(t_1),IE(t_2))$ @ $PE(B(t_3),E(t_4),I)$.

3.4 Event-Driven Models over Data Stream

However, some applications over data streams are composed of complicated events that can not be addressed by simple query fashions mentioned above. Based on the event definitions above, we propose a novel model *EQM*. *EQM* first sorts the data streams into *event-detecting stream* and *event-driven stream*, events are detected over the *event-detecting stream* and event-driven queries are processed over *event-driven stream* separately. Furthermore, driven models over *event-driven stream* are classified in *EQM* as follows.

- *AEC* (Arrival Event Continuous) *Model* and *TEC* (Time Event Continuous) *Model.* They are equivalent to tuple-driven model and time-driven model respectively to solve ordinary stream applications. In this sense, we can infer *EQM* can describe various application requirements in a more general manner. In a time-driven window query, *Hop* can be used to express the frequency of window sliding.
- *IES* (Instant Event Snapshot) *Model.* It is defined as that when an *IE* occurs, the queries related with the values at the current time point or during the current time interval are triggered. For example, when somebody enters a room, the current room temperature or the average temperature over previous 5 seconds is queried.
- *LES* (Lasting Event Snapshot) *Model.* It is defined as that during the interval determined by a *LE*, a one-time query is run. For example, during the interval that somebody enters and leaves a room, the average room temperature is queried.
- *LEC* (Lasting Event Continuous) *Model.* It is defined as that during the interval determined by a *LE*, a continuous query is run. Also Sliding Window and *Hop* can be added to this kind of query. For example, during the interval that a soldier stops moving and don't begin to move again, for every 10 seconds, heartbeat times over the recent 30 seconds are queried.

Based on *EQM*, we also propose a query language named as *EQL* with the function of event detection, event definition and event driven query. Due to space limitation, concrete *EQL* definition and illustration will not be given further in this paper.

4 Processing Algorithms for *EQM*

For *EQM*, *AEC* and *TEC* can be realized using available continuous query approaches. For *IES*, *LES* and *LEC*, improved processing approaches have to be proposed. We adopt event interruption manner to trigger evaluation over query stream rather than join operation, which will improve average processing time. Due to space limitation, we only discuss *LES* which is quite typical in stream event relative application. *IES* and *LEC* query problems can be analyzed in a similar way.

For *LES*, data sharing is very common and a key problem to be considered. All the *LEs* having shared data with $LE_i(B_i,E_i)$ can be defined in equation (1). For the data sharing relation $\hat{R}$, if $LE_j \in R_{share}(LE_i)$, $(LE_i)\hat{R}(LE_j)$ holds.

$$R_{share}(LE_i) = \{LE_j \mid B_i.t < B_j.t \Rightarrow E_i.t > E_j.t\} \tag{1}$$

For a *LE* series *P*, we define the sharing ratio of *P* denoted as $\delta(P)$ in eqution (2).

$$\delta(P) = Count\left(\bigcup_{R_{share}(LE_i) \neq \varnothing} LE_i \right) \Big/ Count(P), \ where \ LE_i \in P \tag{2}$$

Furthermore, we can construct a graph using the definition above as $G = \left(P, \hat{R}\right)$. G can be divided into *N* connected graph $G' = \left(P', \hat{R}\right)$. Each P' is defined as a *SLER* (*Shared LE Range*). B_i and E_i divide each *SLER* into different intervals *I*. Our approaches named as *SLES* (*shared LES*) and *SLES** only evaluate data based on *I* in a shared and incremental manner in each *SLER*. Fig.1 illustrates the data structure and procedure of *SLES* and concrete algorithm is described in table 1. Its main data structure is *LEL* (*LE Linked list*) which is a linked list whose node reserves *LE* aggregation value up to now.

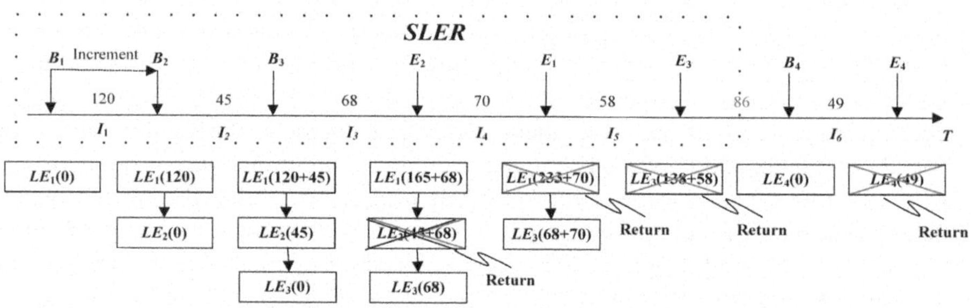

Fig. 1. Illustration of *SLES* Algorithm

Table 1. *SLES* Algorithm

```
Input: detected LE series and processed data
Output: results for each LES
1 if( event.type="begin"){
2 LEL.add(LEi);
3  For each LEi in LEL
4  Increment Calculate(LEi,Ij);
5 }//endif 1
6 else if(event.type="end"){
7  For each LEi in LEL
8    Increment Calculate(LEi,Ij);
9   Output corresponding LEi results;
10 LEL.remove(LEi);
11 }//endif 6
12 if(LEL is not empty){
13 j++;
14 calcucate Ij incrementally;
15}//endif12
```

Table 2. *SLES** Algotithm

```
Input: detected TLE series and processed data
Output: results for each LES
1. if( event.type="begin" and counter!=0){
2.    MIL.add(MIk);
3.    Result=Result+Ij;
4.    Couter++;k++;
5. }//endif 1
6. else if(event.type="end"){
7.    Result=Result+Ij;
8.   Output Result; Couter--;
9.   if(Counter!=0){
10.       Result=Result-MIL.head->value;
11.       MIL.remove(head);
12.   }//endif9
13.   else MIL.removeAll();
14. }//endif6
15. if(Counter!=0){
16.   j++;
17.   calcucate MIk and Ij incrementally;
18. }//endif15
```

For All $LE_i(B_i,E_i)$, if E_i is a *TE* type, which often expresses some time point after m points of B_i's occurrence. We name this kind of *LES* as *TLES*. For example, detect the average temperature in a room during a period of 20ms after a person comes in. In this case, all *LE*s hold equation(3), which means all *LE*s conform to the FIFO fashion. Based on the feature, improved method *SLES** Algorithm is given in Table 2. Its main data structure is *MIL*(Merged Interval Linked list)whose node only reserves some merged interval values.

$$B_i.t < B_j.t \Leftrightarrow E_i.t < E_j.t, \quad \forall LE_i, LE_j \tag{3}$$

No available algorithm over data stream can solve the applications based on common *LES*. But in *CQL*, *DWM* (*delayed window method*) can be used to solve only *TLES*, which depends on *Istream* Operation to delay the occurrence of each B_i and use *Join* to trigger the query. Main structures for *DWM*, *SLES* and *SLES** are *sliding window*, *LEL* and *MIL*.

5 Experiments Results

We begin with a description of application example and our simulation framework in section 5.1. Experiments results and analysis are presented in section 5.2.

5.1 Experiment Setup

In this section, we first give a specific application example from Stream Query Repository[11], which is used to monitor transients' living habits. We choose one of the queries based on events as our example. The query is described as follows: when some infrared sensor reports the event it detects a bird, detecting sensors whose distance from the infrared sensor is not more than 10 meters begin to collect data about temperature and light intensity and the sampling rate is set to be 2 times/s. After 30 seconds, it's requested to output the average values about temperature and light intensity over this period. Table 3 shows the descriptions for the application in *EQL*. Corresponding *CQL* based on *DWM* can be directly found in Query Repository[11].

Table 3. Transient monitoring appliction in *EQL*

Define IEvent E Select * From events Where type='BIRD'	Since E To E+30 Select location,avg(light),avg(temp) From sensors S Where dist(E.location,S.location)<10 Sample period 2s;

Table 4. Original Parameters List

#sensor	30
#event	200
$\delta(P)$	0.5
λ	0.23
Time delay（D）	30s
Sampling rate	2/s

Compared to *CQL*, *EQL* is simple and pellucid. Further more, we will compare the efficiency of *DWM* and our approaches by experiments. Our Experimental environment is that PentiumIV2.4GHz, 512MDDRAM and Windows XP and experiments are realized in Java. Temperature, light intensity and the event that a bird arrives are simulated. The inter-arrival time interval of the events is considered under exponential distribution.

Obviously, $\delta(P)$ is very difficult to simulate directly. But because we can prove Equation (4) based on the definition of $\delta(P)$, we can adjust λ according to Equation(4) to simulate $\delta(P)$. Original experiments Parameters are listed in Table 4.

$$\delta(P) = F(Y) = \int_0^D \lambda e^{-\lambda x} dx = 1 - e^{-\lambda D} \Rightarrow \lambda = \ln\left(1/1 - \delta(P)\right) \Big/ D \tag{4}$$

5.2 Experiment Results

Our experiments compare *DWM*, *SLES* and *SLES** in the aspects of response delay and memory cost by adjusting original experimental parameters.

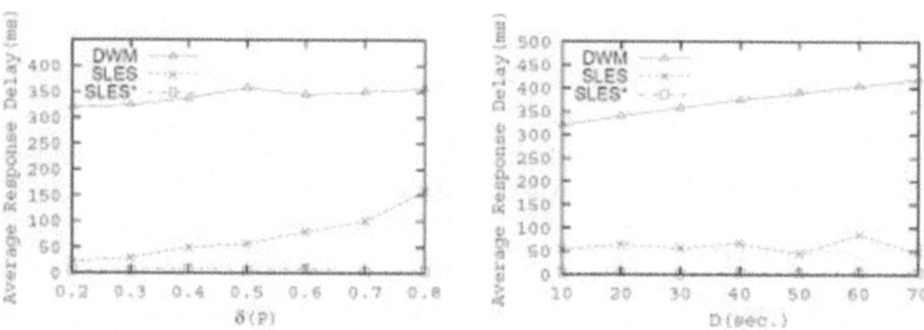

(a) $\delta(P)$ effects on response delay (b) *D* effects on response delay

Fig. 2. Comparison of response time between different methods

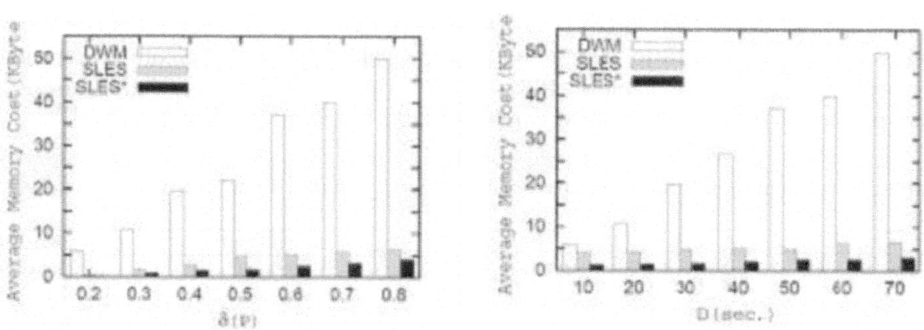

(a) $\delta(P)$ effects on memory cost (b) *D* effects on memory cost

Fig. 3. Comparison of response memory cost different methods

Experiment 1-2 test the response delay of the three approaches by adjusting *D* and $\delta(P)$. From the corresponding experiment results shown in Fig. 2, we can find *SLES* and *SLES** have obvious advantage in response delay beyond *DWM* and *SLES** can respond nearly without any delay.

Experiment 3-4 test the memory cost of the three approaches by adjusting D and $\delta(P)$. From the corresponding experiment results shown in Fig. 3, we can find *SLES* and $SLES^*$ have obvious advantage in memory cost beyond *DWM* and $SLES^*$ is even better.

As experiments show, because *SLES* and $SLES^*$ make improvements in trigger fashions, incremental maintenance and data sharing, they can offer better performances in the response delay and memory utilization. Also, *SLES* can be used for other *LES* queries.

6 Conclusions and Future Work

This paper introduces event concept into queries over data stream and proposes a query model *EQM* over data stream based on event-driven fashion. Relative algorithms especially about *LES* are discussed. The experiments to simulate a real-world application have proved our algorithms' efficiency. We plan to make further studies in improved algorithms about *LEC* and the real-time response algorithm. Finally, a *DSMS* to support complicated event application is in the plan.

References

1. Arasu A, Babcock B, Babu S, Datar M, Ito K, Nishizawa I, Rosenstein J, Widom J. STREAM: the standford stream data manager. In Proceedings of ACM SIGMOD, 2003: 665-668
2. Abadi D J, Carney D, Cetintemel U, et al. Aurora: a new model and architecture for data stream management. The VLDB Journal, 2003, 12(2): 120-139
3. Chandrasekaran S and Franklin M J. PSoup: a system for streaming queries over streaming data.VLDB Journal, 2003,12(2):140-156
4. Chen J, De Witt D J, Tian F and Wang Y. NiagaraCQ: a scalable continuous query system for internet databases. In Proceedings of ACM SIGMOD, 2000: 379-390
5. Hammad M, Aref W, Franklin M, Mokbel M, Elmagarmid A. Efficient execution of sliding window queries over data streams. Technical Report CSD TR 03-035, Purdue University, 2003
6. Golab L, Tamer Özsu M. Update-pattern-aware modeling and processing of continuous queries. In Proceedings of SIGMOD, 2005: 658-669
7. Arasu A, Babu S, andWidom J. The CQL continuous query language: Semantic foundations and query execution. The VLDB Journal, 2006, 15(2):121-142
8. Gehani N H, Jagadish H V, and Shmueli O. Composite Event Specification in Active Databases: Model & Implementation. In proceedings of VLDB, 1992:327-338
9. Gehani N H, Jagadish H V, and Shmueli O. Event Specification in an Object-Oriented Database. In Proceedings of SIGMOD, 1992:81-90
10. Wu E, Diao Y, Rizvi S. High-performance complex event processing over streams. In Proceedings of ACM SIGMOD, 2006: 26-29
11. A stream query repository. http://www.db.standford.edu/stream/sqr.

Efficient Difference NN Queries for Moving Objects*

Bin Wang, Xiaochun Yang, Guoren Wang, and Ge Yu

School of Information Science and Engineering, Northeastern University, 110004 China
{binwang,yangxc,wanggr,yuge}@mail.neu.edu.cn

Abstract. Group Nearest Neighbor query is a relatively prevalent application in spatial databases and overlay network. Unlike the traditional KNN queries, GNN queries maintain several query points and allow aggregate operations among them. Our paper proposes a novel approach for dealing with difference operation of GNN queries on multiple query points. Difference nearest neighbor (DNN) plays an important role on statistical analysis and engineer computation. Seldom existing approaches consider DNN queries. In our paper, we use the properties of hyperbola to efficiently solve DNN queries. A hyperbola divides the query space into several subspaces. Such properties can help us to prune the search spaces. However, the computation cost using hyperbola is not desirable since it is difficult to estimate spaces using curves. Therefore, we adopt asymptotes of hyperbola to simplify the hyperbola-based pruning strategy to reduce the computation cost and the search space. Our experimental results show that the proposed approaches can efficiently solve DNN queries.

1 Introduction

Group Nearest neighbor (GNN) queries [1,2,3,4] are utilized in many applications, such as geographic information systems (GIS) [5], CAD/CAM, multi-media [6], knowledge discovery [7], data mining [8], and etc. As an extension of the traditional KNN queries, GNN queries maintain several query points and allow aggregate operations over the query points. Such distinguishes make GNN queries more complicated than traditional KNN queries.

GNN queries mainly focus on *aggregate* operations, which includes the most widely used functions like *maximum*, *minimum*, *summation,* and etc. However, these functions cannot satisfy all GNN applications. For example, several old friends living in different addresses want to find a place to meet together. They list a set of candidate meeting places and hope to choose a "good" place such that they can use similar time to arrive at the meeting place and do not want any of them to wait too long time. Such problem has not been efficiently solved so far. In this paper, we present a novel approach to process the problem. We define the above query problem as a *difference nearest neighbors* (denoted DNN for short) query. The general description DNN are defined as follows.

---

* Supported by the 973 Program under Grant No. 2006CB303103, the Natural Science Foundation of China under Grant Nos.60503036, 60473074, 60573089, and the Fok Ying Tong Education Foundation of China under Grant No. 104027.

G. Dong et al. (Eds.): APWeb/WAIM 2007, LNCS 4505, pp. 542–553, 2007.

DNN query. Given two query points q_1 and q_2, a set of objects O and a threshold a, find top-k objects in O, such that $|dist(q_1, o_i) - dist(q_2, o_i)| > a$ (or $<a$), where $dist(q, o_i)$ is the distance between query q and object o_i ($o_i \in O$).

The rest of the paper is organized as follows. Section 2 reviews related work. Section 3 proposes a pruning approach based on hyperbola for DNN queries. In Section 4, we use asymptote to further improve hyperbola-based approach. Section 5 experimentally evaluates our proposed approaches. Finally, Section 6 concludes the paper.

2 Related Work

In the procedure of monitoring moving objects, NN queries have been intensively researched in recent years [9,10,11,12]. The target of NN queries is to find the nearest objects to a query q to under a specific query condition. The existing algorithms mainly adopt GRID [4,13,14,12] or its variations (e.g. CPM-CNN [14]) as index structure.

In YPK-CNN algorithm [15], objects are assumed to fit in main memory and are indexed with a regular grid of cells with size $\delta \times \delta$. YPK-CNN does not process updates as moving objects arrive, but directly applies their updates on the grid. Each *continuous* NN (CNN) query installed in the system is re-evaluated in every T time unit. When a query q is evaluated for the first time, a two-step NN search technique retrieves its result. The initial step keeps visiting the cells in a square R until k objects are found. Fig. 1(a) shows an example of a single NN query where the first candidate NN is p_1 and the distance between p_1 and q is d. p_1 may not be an answer to the NN query, since there may exist another object (e.g., p_2) outside R, such that the distance between p_2 and q is less than d. To avoid miss such objects, the second step searches the cells in square SR (shaded area in Fig. 1(a)) that is centered at c_q, the cell who contains q, with side length $2d+\delta$. In this way, the actual NN answers to q can be got. In Fig. 1(a), YPK-CNN processes the six points from p_1 to p_6 and returns p_2 as the actual NN answer. When re-evaluating an existing query q, YPK-CNN makes use of its previous result in order to restrict the search space. If a query point moves, the approach needs to compute the new NN answers.

SEA-CNN [13] focuses on monitoring the NN changes exclusively without including a module for the first-time evaluation of an arriving query q (i.e., it assumes that the initial result is available). p_2 shown in Fig. 1(b) is the 1-NN object. Objects are stored in secondary memory, indexed using a regular grid. The *answer region* of a query q is defined as a circle, where q is the center point and the distance between q and the kth NN object (*best_dist*) is the radius. When object moves, SEA-CNN considers the moving status to determine a circular search region SR (shaded area in Fig. 1(b)) around q, and computes the newest NN objects to q. If the query q moves to a new location q', then SEA-CNN sets $r = best\_dist + dist(q, q')$, and computes the newest NN objects of q by processing all the objects in the circle centered at q' with radius r.

Similar to the previous CNN query algorithms [15,13,16,17], CPM-CNN algorithm [14] uses the GRID structure to index objects. But CPM divides the search space helically centered at c_q as shown in Fig. 1(c). In particular, CPM initializes an empty heap H and inserts (i) the cell c_q with key *mindist* $(c_q, q) = 0$, and (ii) the level zero rectangles for each direction DIR, with key *mindist* (DIR_0, q). Then, it starts de-heaping entries iteratively and adjusts its answer set *best_NN* accordingly. The algorithm terminates when the next entry in the heap H (corresponding either to a cell or a rectangle) has a key that is greater than or equal to the current shorted distance to the query q. The approach can guarantee that no results are in the outer level of the stripped region. When object moves (e.g. some objects in NN list moving to outside), it first searches the stripped region to get new NN results.

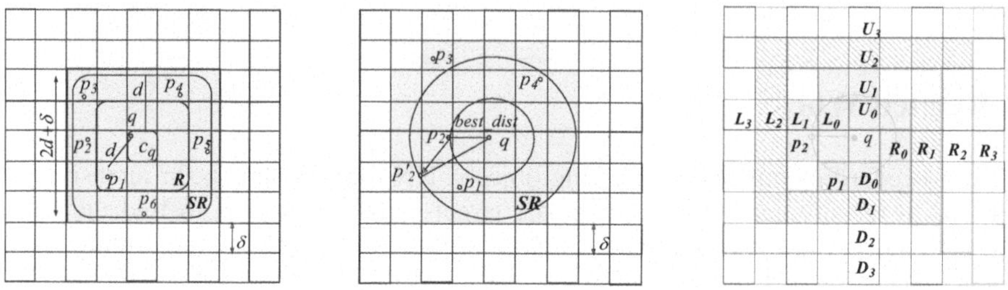

(*a*) NN query in YPK-CNN (*b*) NN query in SEA-CNN and p_2 issues an update (*c*) The space partitioning in CPM-CNN and its NN query

Fig. 1. An example of a CNN query using different index structures

Aiming at the DNN aggregate query, those CNN query approaches cannot efficiently answer the query. In this paper, we make use of the curvilinear geometry property to prune the query space, and reduce the vast computation.

3 Hyperbola-Based Difference Query

In order to simplify the discussion, we assume data objects and queries are in 2D space. Objects change their locations frequently as unpredictable manners. In this section, we present a basic hyperbola-based algorithm using grid index to answer DNN queries. Without loss of generality, we use two query points and we store objects in disk.

Table 1. Frequently used symbols and functions

Symbol	Description
P	The set of moving objects
N	Number of objects in P
G	The grid that indexes P
Δ	Cell side length
Q	The query point
c_q	The cell containing q
N	The number of queries installed in the system
$dist(p,q)$	Euclidean distance from object p to query point q
$R(c_q,l)$	Rectangle centered at c_q,$2l+\delta$ as the length of cell
$PL(i,j)$	The objects list in the cell(i,j)
$D(q_1,q_2)$	The k nearest neighbors of q_1 according with the specific *difference* operation
$Rcrit$	The rectangle search region containing at least k nearest neighbors

Similar to existing approaches (e.g., YPK-CNN, SEACNN), we use a grid index, since a more complicated data structure (e.g., main memory R-tree [18]) is expensive to maintain dynamically. We update the top-k list in terms of the given interval τ periodically, i.e., in the specific time interval, we could ensure the correctness of the result of the *difference* query, once interval τ is out of date, the algorithm re-computes the current top-k. Notations in this paper are shown Table 1.

3.1 Overview of the Hyperbola-Based Approach

Property of Hyperbola. Fig. 2 shows a hyperbola. Let C be the hyperbola, q_1 and q_2 be its foci, and p be any point of C. The unsigned difference between the distances dist(p, q_1) and dist(p, q_2) is constant and equal to $2a$ (a is shown in Fig. 2).

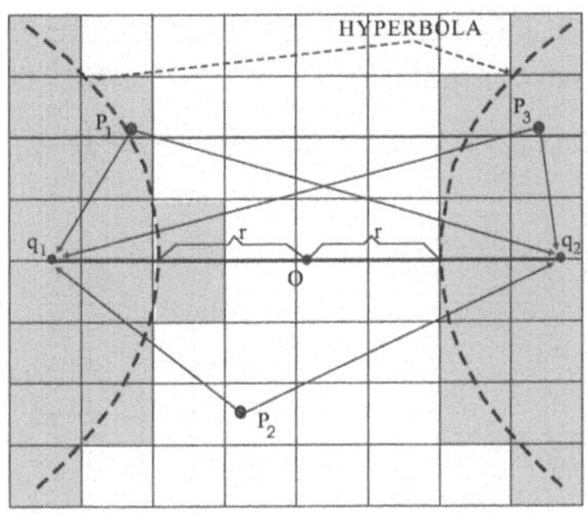

Fig. 2. Hyperbola-based pruning strategy

We make use of the property of hyperbola to solve DNN. We set the two query points as two foci of the hyperbola, set the distance between these two foci as the focus c of the hyperbola, set the given value in the aggregate operation as the constant a in the hyperbola. In this way one branch of the hyperbola could be contained (In terms of the specific definition of the *difference* aggregate operation in our algorithm, we could only consider one branch of the hyperbola). The cells comprised by this branch of hyperbola will contain vast suitable query objects with high possibility in terms of the property of the hyperbola. Therefore, we could directly delete the search region outside and non-intersect with the hyperbola. Then we only search the remaining space and could spare a lot of computation workload and CPU response time.

For example, in Fig. 2, there are the dataset $P = \{p_1, p_2, p_3\}$, and two query points are q_1 and q_2. We can partition the whole search space into 9×9 cells. Thus, the objects in P and two query points can map the corresponding cells as shown in Fig. 2. Then, we can get a hyperbola formula using the threshold a defined by users. According to the hyperbola property, we compute the results among the objects and query points using Equation 1.

$$\begin{aligned} \partial 1 &= |dist(p_1, q_1) - dist(p_1, q_2)|, \partial 2 = |dist(p_2, q_1) - dist(p_2, q_2)|, \\ \partial 3 &= |dist(p_3, q_1) - dist(p_3, q_2)|. \end{aligned} \tag{1}$$

Then, we should get the relations: $\partial 1 = a, \partial 1 > \partial 2, \partial 1 < \partial 3$. From the results, we easily prune the cells in the blank region, and only need consider those objects in the cells of dark region as candidate objects. Hence, we make used of the hyperbola to reduce the search space and improve the performance of DNN queries.

For clearly using the hyperbola property, in Section 3.2, we introduce grid-based index structure and how to efficiently maintain the index. We give a hyperbola-based algorithm to introduce the whole query processing in detail.

3.2 Cell-Based Index Construct and Update

As shown in Fig. 3(*a*), our search space range is $[0,1]^2$. the side of every cell is $\delta \times \delta$. In Fig. 3(*b*), we propose $P(t)$ as the objects position list at time t. Each object has a specific identifier $p(t) \in P(t)$. Each cell is denoted $cell(i, j)$, where i is the row number of the cell and j is the line number of the cell. The cell $cell(i, j)$ has a list $PL(i, j)$ that stores identifiers of all objects in the cell. That is, $PL(i, j) = \{p(t) \in P(t) : \langle p(t)_x, p(t)_y \rangle \in [i\delta, (i+1)\delta) \times [j\delta, (j+1) \times \delta)\}$.

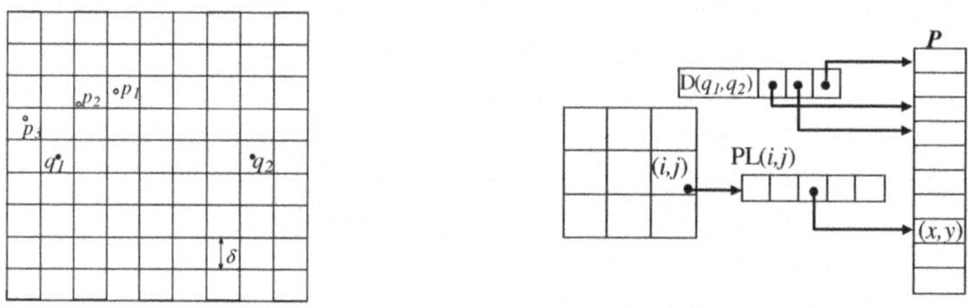

(a) Search space (b) Index structure of the search objects

Fig. 3. Search space and the index structure of the search objects

The process of object index's construction includes two steps: (i) scanning all of the query objects $p(t) \in P(t)$, (ii) inserting their *id* into the corresponding list of $cell(\lfloor p(t)_x / \delta \rfloor, \lfloor p(t)_y / \delta \rfloor)$, i.e., $PL(\lfloor p(t)_x / \delta \rfloor, \lfloor p(t)_y / \delta \rfloor)$. We introduce a list $D(q_1, q_2)$ to express the results of *DNN* queries. The algorithm of object index construction is shown in Algorithm 1.

Algorithm 1. Object index construction

Input: list of all the query objects $P(t)$;search space GRID; $PL(i,j)$ of $cell(i,j)$;
Algorithm description:
1: **for** each $p(t) \in P(t)$ {
2: $x := p(t)_x$; $y := p(t)_y$;
3: $i := \lfloor x / \delta \rfloor$; $j := \lfloor y / \delta \rfloor$;
4: insert $p(t)$ into PL(i,j);
5: }

When the system surpasses the interval, the objects may be moved to another position. We must consider the update of the object index. The previous objects in $P(t)$ maybe change to the new $P(t')$. That is, there are some query objects leaving or incoming into the cells. This must incur the update of object index. In the case of new

$p(t')$ does not remain in the same cell of $p(t)$, we delete the $p(t)$ from the original cell's $PL(i, j)$, and insert it into the new cell's $PL(i, j)$. In the case that the new position still remains in the previous cell, no update is arisen. The update algorithm of the object index is shown in Algorithm 2.

Algorithm 2. Object index update

Input: objects list $P(t)$ and $P(t')$ at time t and t';search space GRID;
objects list $PL(i,j)$ of every $cell(i,j)$;
Algorithm Description:
for each $p(t') \in P(t')$ {
  $x'=p(t)_x$; $y'=p(t)_y$;
  $i'=\lfloor x/\delta \rfloor$; $j'=\lfloor y/\delta \rfloor$;
  if $c(i,j)$<>$c(i',j')$ {
    insert $p(t')$ into PL(i',j');
    delete $p(t)$ from PL(i,j);
  }
}

3.3 Hyperbola-Based DNN Algorithm

We first set q_1's search region *Rcrit*, which contains at least k objects. Then use the hyperbola to prune the unsuitable objects. The remaining objects are sorted according to their distance difference and stored into list $D(q_1,q_2)$ ascending. In a special case, we found it is possible that the region *Rcrit* contains less than k objects, which fulfill the results of the *DNN*. As the number of objects in *Rcrit* is at least k originally, But after the *DNN* query pruning, the number of remaining objects in this region is lower than k

Algorithm 3. Hyperbola-based *Difference* Nearest Neighbor query(DNN)

Input: $PL(i,j)$ in $c(i,j)$; the coordinates of query points q_1 and q_2;
k as the number of query results; the constant a in DNN;
Output: top-k of $D(q_1,q_2)$;
Algorithm Description:
l_0:=0; R_0:=$R(c_q, l_0)$;
while($|R_0 \cap P(t)| < k$){
  l_0:=l_0+1; R_0:=$R(c_q, l_0)$;
}
$l$crit:=$||q_1\text{-far}_{q1}(R_0 \cap P(t))||$ Rcrit:=$R(c_q, \lceil \text{lcrit}/\ \delta \rceil)$;
while($|D(q_1,q_2)| < k$){
 for each cell $c \in R$crit {
  for each object $p(t) \in PL(c)$ {
    if ($dist(q_1, p)$ - $dist(q_2, p) > a$)
    insert $p(t)$ into the answer list of D(q_1,q_2) based on distance comparison;
  }
 }
 if ($|D(q_1,q_2)|<k$) {
  l:= l+1; $Rcrit$:=$R(c_q, \lceil \text{lcrit}/\ \delta \rceil)$; // l is the updated objects in $D(q_1, q_2)$.
 }
}

at all. In order to guarantee the correctness of the algorithm, we have to expand the search region *Rcrit* and repeat the pruning strategy. Absolutely, this enlarging process increases the complexity of our algorithm. However, our major concern is a large and wide environment with a great deal of mobile objects. Therefore, every cell contains a certain number of objects, and the region *Rcrit* contains objects farther than *k*.

Furthermore, we have to maintain the objects' index structure using Algorithm 2 and make use of the objects information and the original DNN results, find the object moving farthest, calculate the newest distance between this object and the query point q_1 and q_2 and update this new result as the search region *Rcrit*. Finally, sort the new DNN in the *Rcrit*. Algorithm 3 is the implementation given in detail.

4 Asymptote-Based DNN Algorithm

Based on the basic algorithm presented above, we could get the search region *Rcrit*. Then we make use of the particularity of the aggregate algorithm and the relevant geometry knowledge to simplify our query algorithm.

Though the definition and property of hyperbola could prune the search space, this property itself is related to quadratic operation with quite high computation cost. Therefore, we introduce the asymptote of hyperbola to prune the search space. This can both reduce the search space and reduce the computation workload. We use the obtained distance between two foci as the focus *c* of the hyperbola, and set the given value as the constant *a* of the hyperbola. We only concern the particular condition: two foci have the same vertical coordinate, the center of the two foci is set to be the origin. Therefore, we get the hyperbola equation: $\frac{x^2}{a^2}-\frac{y^2}{b^2}=1$. Its asymptote equation is $y=\pm\frac{b}{a}x$, where $b=\sqrt{c^2-a^2}$ and the vertical asymptote is $x=\pm\frac{a}{2}$.

As shown in Fig. 4, the cells in the starboard of the hyperbola asymptote (including the top-right and down-right section) can be directly pruned, the cells in the right of the vertical asymptote can be deleted too. We could only consider the objects in the cells residing in the left of asymptote or intersected with the asymptotes (the shaded region as shown in Fig. 4). Then find the DNN query results from the cells remained after the pruning strategy. The cells intersected with the asymptotes shouldn't be omitted being tested whether they could satisfy the DNN. Compared with the non-pruning strategy algorithm, our adapted algorithm is more excellent both in computation workload and the query response time.

Especially, without loss of generality, suppose there are lots of objects in the left of the asymptote as shown in Fig. 4. Furthermore, we could expand the special conditions to more common conditions, and adapt the above hyperbola into the common equation, which is shown in Equation 2.

$$\frac{[\cos\theta(x-x_c)+\sin\theta(y-y_c)]^2}{a^2}-\frac{[-\sin\theta(x-x_c)+\cos\theta(y-y_c)]^2}{b^2}=1, \textit{where}$$
$$\theta=\arctan\left(\frac{y_2-y_1}{x_2-x_1}\right), x_c=\frac{x_1+x_2}{2}, y_c=\frac{y_1+y_2}{2}, c=\frac{\sqrt{(y_2-y_1)^2+(x_2-x_1)^2}}{2}, b=\sqrt{c^2-a^2}. \quad (2)$$

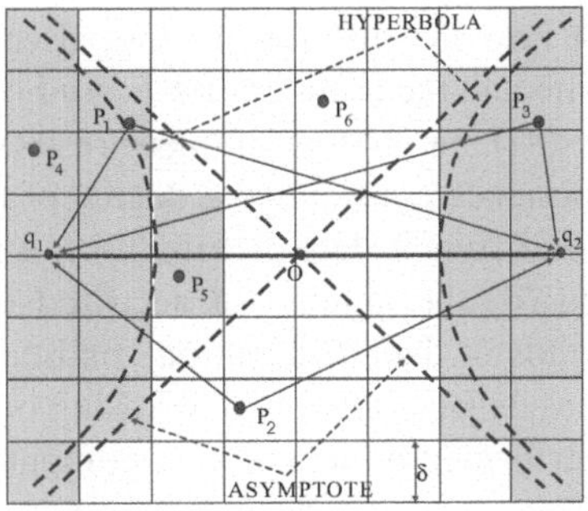

Fig. 4. Search region based on the hyperbolic property pruning strategy

The asymptote equation could be deduced through getting the derivative from the original hyperbola, we will not dwell on it at this paper. The Hyperbola-based pruned algorithm is shown in Algorithm 4.

Algorithm 4. Asymptote-based *Difference* Nearest Neighbor query (ADNN)

Input: $PL(i,j)$ in $c(i,j)$;the coordinates of query points q_1 and q_2;
    k as the number of query results; the constant a in DNN;
Output: top-k of $D(q_1,q_2)$;
Algorithm Description:
l_0:=0; R_0:=$R(c_q, l_0)$;
while($|R_0 \cap P(t)| < k$){
    l_0:=l_0+1; R_0:=$R(c_q, l_0)$;
}
$l$crit:=$\|q_1$-far$_{q1}(R_0 \cap P(t))\|$; Rcrit:=$R(c_q, \lceil \text{lcrit}/\ \delta \rceil)$;
Construct the asymptotes lines based on d:=$dist(q_1,q_2)$ and a;
while($|D(q_1,q_2)| < k$){
  for each cell $c \in R$crit {
    if (($c \cap$lines!=NULL) or (c is in the left of lines)){
      for each object $p(t)$ $PL(c)$&& (c is in the left of lines)
        Insert $p(t)$ into the answer list of $D(q_1,q_2)$ based on distance comparison;
      for each object $p(t) \in$ PL(c)&& ($c \cap$lines!=NULL)
        if($dist(q_1,p)$ - $dist(q_2,p) > a$)
          Insert $p(t)$ into the answer list of $D(q_1,q_2)$ based on distance comparison;
    }
  }
  if ($|D(q_1,q_2)|$<k) {
    l:= l+1; Rcrit:=$R(c_q, \lceil \text{lcrit}/\ \delta \rceil)$; // l is the updated objects in $D(q_1, q_2)$.
  }
}

---

5 Experimental Results

In this section, we make use of Algorithms 1 and 2 to maintain the index structure of our object data. Then we run Algorithms 3 and 4 based on the same datasets from

Algorithms 1 and 2. We test how the different factors affect the performance in the aspects of search response time and the number of accessed objects.

We mainly test two factors: (i) the number of query results, (ii) the size of the cells in the search region. The experience considers the degree of the two parameters influenced by the above factors in different data distributions. These two parameters are: (i) search response time, and (ii) the number of accessed objects. We evaluate our performance based on two kinds of data distribution conditions: uniform distribution; and zipf distribution. The datasets were all generated using our program randomly.

All the experiments are performed on a notebook computer with Intel Pentium IV CPU 2.0G, 256MB RAM. The operation system is Windows XP, and all the algorithms are implemented in Microsoft Visual C++ 6.0 and compiled by gcc 3.2.

5.1 Performance Influenced by *k*

First we consider the effect from the number of query results. Fig. 5 shows the difference of search response time influenced by the value *k* in the uniform and zipf distribution conditions. In Fig. 5, we find that the search response time in both of the algorithms present the trend of ascending no matter what distribution of the datasets. But our adapted algorithm has introduced the pruning strategy based on the property of hyperbola, the performance of the experiment is much super. Especially in the zipf distribution of datasets, the advantage of our adapted algorithm is excellent evidently. That is, the pruning strategy based on the hyperbola has deleted a great deal of data. So Algorithm 4 (ADNN) is much better than Algorithm 3 (DNN) in the aspect of search response time.

We could conclude the search response time is insensitive to *k*. Although we introduce the gradual enlarging strategy of the search region, the network simulated in our experiment contains vast data, when the size of the search region enlarge a little, it still maybe contain many objects into the search region, then our algorithm will take all these data into memory and sort ascending in order to maintain DNN. Apparently the value k has little effect on the response time.

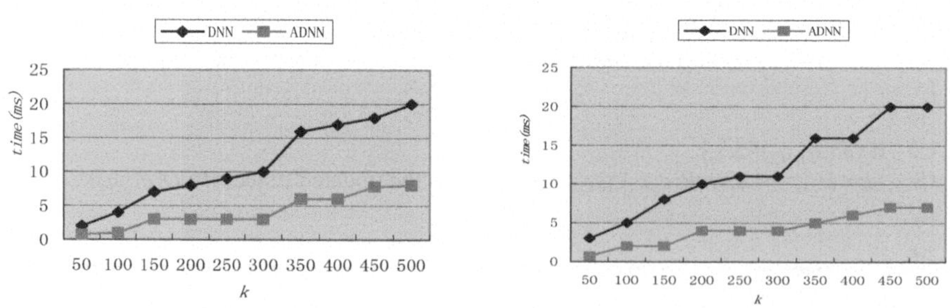

(a) Uniform distribution

(b) Zipf distribution

Fig. 5. Search response time influenced by *k*

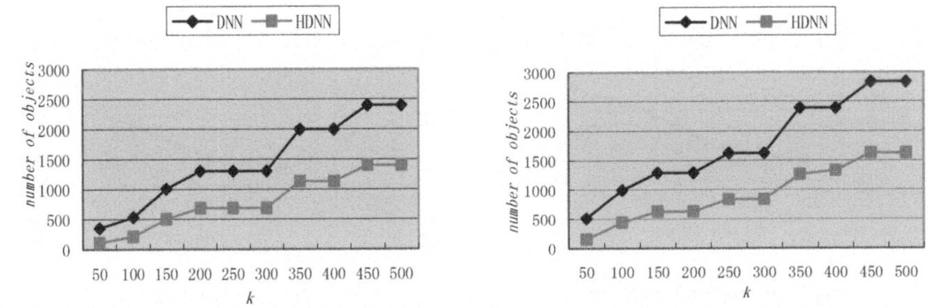

(a) Uniform distribution (b) Zipf distribution

Fig. 6. The number of accessed objects influenced by *k*

Fig. 6 shows the number of accessed objects influenced by *k* in different data distribution. As the increasing of *k*, we must access more objects in order to find much results. So along with the change of *k*, the number of accessed objects and the search response time have the same trend. The number of accessed objects is also steady. As we have introduced the pruning strategy into Algorithm 4 (ADNN), it accesses much less objects than Algorithm 3 (DNN).

5.2 Performance Influenced by Cell's Size

We have assumed the size of search region is $[0,1)^2$, and set the number of query results as 400, the query target objects maintain stable at the query time. In specific time interval, the total number of the objects in the network is steady. Then cells with different size contain different number of objects. The bigger cells have lots of objects, the index structure of objects need less attention for maintaining their indices' efficiency. But in the pruning period based on the property of the hyperbolic asymptote, our algorithm must contain much data, which could be deleted in the environment with small size cells, the unnecessary waste in the time and space become unavoidable. Correspondingly the maintaining of objects' index structure in the environment with small cells is complex a little, but when come to the pruning session, the pruning function is much stronger, and the search region could be much accurate.

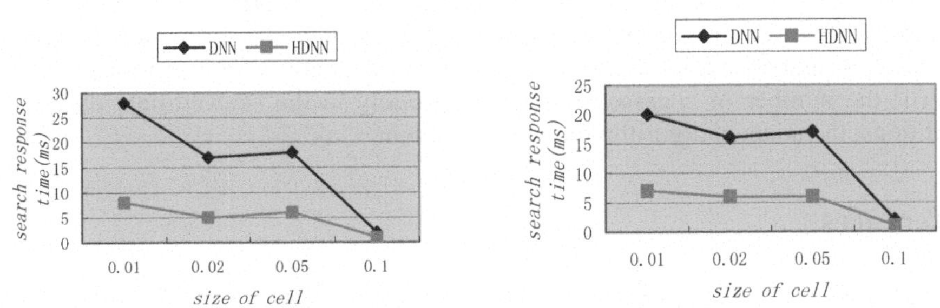

(a) Uniform distribution (b) Zipf distribution

Fig. 7. Search response time influenced by cell's size

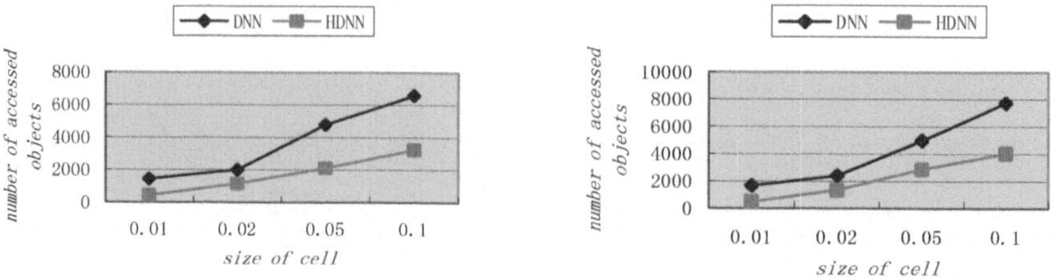

(a) Uniform distribution (b) Zipf distribution

Fig. 8. The number of accessed objects influenced by cell's size

Fig. 7 shows the condition of the search response time influenced by cell's size in different data distributions. Search response time have a trend of decrease along with the enlarging of cells. Form our analysis, every time we access the objects in a cell, the algorithm will incur a I/O operation because the objects in every cells are indexed in the corresponding file in disk. The smaller cell contain less objects, every time we access such cells, we could only obtain a little objects. So the network with smaller cells needs much more I/O operations than lager cells network. Since the cost of I/O operation is relative high than the computation in memory, the response time decreases along with the enlargement of cells in the network. Since Algorithm 4 (ADNN) adopts the pruning strategy and deletes lot of cells impossible to maintain query results. The high cost I/O operations could be decreased a lot. Algorithm 4 (ADNN) induces less I/O disk operations than Algorithm 3 (DNN), so the superiority of Algorithm 4 is much obvious.

Apparently, our adapted Algorithm 4 (ADNN) is much more excellent than algorithm 3 (DNN) at both the search response time and the number of accessed objects. It proves that the pruning strategy introduced in our algorithm is very efficient in practice.

6 Conclusions

We bring forward a basic aggregate query algorithm, and adapt the basic algorithm into a algorithm with the pruning strategy based on the property of hyperbolic asymptote. From the evaluation of our experiments, the adapted algorithm with the pruning strategy has improved a lot in the aspects of two parameters: (i) search response time and (ii) the number of accessed objects. Especially in the skewed data distribution conditions, the adapted algorithm is much excellent.

References

1. Hjaltason, G., Samet, H.: Distance Browsing in Spatial Database. ACM Trans. on Database Systems, 24(2): 265-318 (1999)
2. Tao, Y., Zhang, J., Papadias, D., Mamoulis, N.: An Efficient Cost Model for Optimization of Nearest Neighbor Search in Low and Medium Dimensional Spaces. IEEE Trans. Knowl. Data Eng. 16(10): 1169-1184 (2004)

3. Ferhatosmanoglu, H., Stanoi, I., Agrawal, D., Abbadi, A. E.: Constrained Nearest Neighbor Queries. In Proceedings of Symposium on Spatial and Temporal Databases (SSTD) (2001) 257-278
4. Papadias, D., Shen, Q., Tao, Y., Mouratidis, K.: Group Nearest Neighbor Queries. In Proceedings of the 20th IEEE Int. Conf. on Data Engineering (ICDE) (2004) 301-312
5. Aggrawal, C., Yu, P.: Outlier detection for high dimensional data. In Proceedings of the ACM SIGMOD Int. Conf. on Management of Data (2001) 37-46
6. Faloutsos, C., Ranganathan, M., Manolopoulos, Y.: Fast Subsequence Matching in Time-Series Databases. In Proceedings of the ACM SIGMOD Int. Conf. on Management of Data (1994) 419-429
7. Ester, M., Kriegel, H.-P., Sander, J.: Knowledge Discovery in Spatial Databases. Invited paper at German Conf. on Artificial Intelligence (1999)
8. Jain, A., Murthy, M., Flynn, P.: Data Clustering: A review. ACM Comp. Surveys, 31(3):64-323 (1999)
9. Roussopoulos, N., Kelley, S., Vincent, F.: Nearest Neighbor Queries. In Proceedings of the ACM SIGMOD Int. Conf. on Management of Data (1995) 71-79
10. Arya, S., Mount, D. M., Netanyahu, N. S., Silverman, R., Wu, A.Y.: An Optimal Algorithm for Approximate Nearest Neighbor Searching Fixed Dimensions. Journal of ACM, 45(6): 891-923 (1998)
11. Nakano, K., Olariu, S.: An Optimal Algorithm for the Angle-Restricted All Nearest Neighbor Problem on the Reconfigurable Mesh, with Applications. IEEE Trans. on Parallel and Distributed Systems, 8(9): 983-990 (1997)
12. Benetis, R., Jensen, C. S., Karciauskas, G., Saltenis, S.: Nearest and reverse nearest neighbor queries for moving objects. VLDB J. 15(3): 229-249 (2006)
13. Xiong, X., Mokbel, M., Aref, W.: SEA-CNN: Scalable Processing of Continuous K-Nearest Neighbor Queries in Spatio-temporal Databases. In Proceedings of the 21st IEEE Int. Conf. on Data Engineering (ICDE) (2005) 643-654
14. Kyriakos, M., Marios H., Dimitris P.: Conceptual Partitioning: An Efficient Method for Continuous Nearest Neighbor Monitoring. In Proceedings of the ACM SIGMOD Int. Conf. on Management of Data (2005) 634-645
15. Yu, X., Pu, K., Koudas, N.: Monitoring K-Nearest Neighbor Queries Over Moving Objects. In Proceedings of the 21st IEEE Int. Conf. on Data Engineering (ICDE) (2005) 631-642
16. Papadopoulos, A., Manolopoulos, Y.: Performance of Nearest Neighbor Queries in R-trees. In Proceedings of Int. Conf. on Database Theory (ICDT) (1997) 394-408
17. Beckmann, N., Kriegel, H. P., Schneider, R., Seeger, B.: The R*-tree: An Efficient and Robust Access Method for Points and Rectangles. In Proceedings of the ACM SIGMOD Int. Conf. on Management of Data (1990) 322-331
18. Guttman, A.: R-tree: a Dynamic Index Structure for Spatial Searching. In Proceedings of the ACM SIGMOD Int. Conf. on Management of Data (1984) 47- 57

APCAS: An Approximate Approach to Adaptively Segment Time Series Stream

Li Junkui and Wang Yuanzhen

College of Computer Science and Technology,
Huazhong University of Science and Technology,
Wuhan, Hubei, 430074, China
{jklfuture,wangyz2005}@163.com

Abstract. We study the problem of segmenting time series stream. Existing segmenting methods for time series mainly focus on the static data, and may be infeasible under the circumstance of time series stream. We propose an approximate method of APCAS(*Adaptive Piecewise Constant Approximate Segmentation*) to adaptively segment time series stream, which works in linear time. Extensive experiments, both on synthetic and real datasets, show that our approach is efficient and effective.

Keywords: Data Mining, Time Series Stream, Online Segmenting.

1 Introduction

Time series is a sequence of real values, each of which represents a value measured at a point in time. With the growing popularity of time series data in science, engineering and business, there is an increasing demand to apply data mining approaches to gain a more clear insight about the data. Segmentation is one of the important tasks in mining time series[7] and is useful in its own right as a tool for delimiting similar neighboring sequential data[3] and change detection, etc. The problem of segmentation arises in many data mining applications, including bioinformatics and context-aware systems. However, on one hand, existing methods mainly focused on the static data(i.e. the data in database)[4][6], and little work was done on the streaming time series; on the other hand, for most of the time, time series data in real life are typically streaming data, such as the stock prices, currency exchange rates, electrocardiograms, etc, it is desirable to segment time series stream in many applications.

As noted in [1], time series streams arrive online, and are usually in an unknown order and unbounded mode, significant challenges on segmenting time series stream lie on the one-pass real processing, and the process should finish in linear time with limited memory consumption. However, as we shall see, most of the existing methods on static time series data try their best to segment the data into optimal subsequences, and may be infeasible in segmenting the high speed time series streams. For most purposes, especially in the context of time series stream, an exact optimal segmenting is not required[5]. We, therefore, use an approximate method instead. The original method, works by applying an adaptively segmenting tool to minimize the reconstruction error.

G. Dong et al. (Eds.): APWeb/WAIM 2007, LNCS 4505, pp. 554–565, 2007.

In this work, we introduce an approach to segment time series stream. Our contributions can be simply summarized as follows.

1. We show most of the methods on segmenting static time series can be infeasible in segmenting streaming time series, and new approaches should be devised.
2. We propose an approximate technique, called *Adaptive Piecewise Constant Approximate Segmentation*(APCAS), to segment time series stream. The method works in linear time, and can be applied into time series stream segmentation.
3. We demonstrate the effectiveness of the segmentation method by extensive empirical experiments on both generated and real datasets. The experiments validate the utility of the approach we propose.

The rest of the paper is organized as follows. Section 2 reviews the related work and provides a background for our work. In Section 3 we present the ways to produce APCA segmenting on static time series. We propose the method APCAS to segment time series stream in Section 4. In Section 5 we give experimental results and make discussions about the results. Finally we offer conclusions in Section 6.

2 Related Work and Backgrounds

2.1 Review of Segmentation Algorithms

Segmenting static time series data has been extensively studied in the past decades. However, there is no unified definition on time series segmentation[6]. Generally, time series segmentation is to divide long time series sequences into several non-overlapping short subsequences, and data lie into the short sequence have certain properties, such that the data in the same subsequence are more similar to each other, the reconstruction error is minimal to a given threshold etc. Keogh et al.[6] gave a general survey of different segmenting time series methods.

There are three major approaches to segment time series[6], namely, *sliding window*, *top-down* and *bottom-up segmentation*, respectively.

- The *sliding window* method extends iteratively the segment in the current window to include more data points.
- The *top-down* method breaks the series down iteratively into even-smaller segments.
- The *bottom-up* method merges neighboring segments iteratively into even-larger ones.

These approaches have different features that may act as strengths or weaknesses in the static time series segmentation, and some algorithms[4][6], appear to be more complex, are a combination of one or two of these approaches. As we are now in the context of time series stream segmentation, the data in the whole stream is usually unavailable at one time, and we pay more our attention to whether the method can be finished in linear time. With such considerations, the *sliding window* method is more appropriate in segmenting time series stream, as it has relative lower computational complexity and can be solved in acceptable time.

In terms of the result of segmentation, all the segmentation methods need a stopping criterion to terminate iteration. Given a time series T, existing time series segmenting methods, try to produce the best segmentations[6] as,

- exact K segments.
- maximum reconstruction error of each subsequence does not exceed some threshold, max_error, specified by user.
- total reconstruction error of the whole segmentation does not exceed some threshold, $total\_error$, specified by user.

As observed by Keogh et al.[6], the performance of the algorithms is tightly related with the parameters(i.e., the K, max_error, $total\_error$ etc.), different parameters may produce different segmentations. However, as time series stream is generally endless with time, and the overall image of time series stream is usually unavailable, thus it is undesirable to require the user to provide parameters about the whole time series stream(such as the K, $total\_error$). In other words, the parameters specified by user should reflect the local properties, rather than the global properties, of time series stream.

Another part in time series segmentation is with regard to the choice of which approximate line to fit the data. Some researches utilized the line regression, or line interpolation method. The difference between the two methods, is that in the line regression, the result line joins the end points of the subsequence, while that is not a must in the line interpolation. We choose APCA to segment time series stream. We observe that, in the relative smooth streaming data, the methods of line regression and line interpolation may produce many redundant segmentations, as illustrated in the example of line interpolation in Fig. 1. However, as we will demonstrate later, the exact APCA on static time series is not suitable to stream segmentation directly, we give an approximate computation method to work in linear time, and apply it to the time series stream segmentation.

2.2 Backgrounds for APCA

In this section, we review some background material for APCA. Since we will constantly reference to the APCA approximation for the rest of the paper, we first describe the technique. For more detailed description, interested readers are recommended to [5]. The notations used in the paper are listed in Table 1.

Given a time series T, the APCA approximation of T is a sequence of $2M$ values in the following format:

$$T' = \{< tv_1, tr_1 >, < tv_2, tr_2 >, \ldots, < tv_M, tr_M >\}, tv_0 = 0 \quad (1)$$

Table 1. Symbolic Notation Used in the Paper

Notation	Description
t_i	value of the ith data point.
T	a time series sequence of length n, $T = t_1, t_2, \ldots, t_n$.
$T[i,j]$	subsequence of T from i to $j(i \leq j)$, $T[i,j] = t_i, t_{i+1}, \ldots, t_j$.
T'	the APCA approximation of T.
c_avg	current mean data value in the segment.
c_avg'	new mean data value in the segment when a data point arrives.
c_sum	current sum of data value in the segment.
c_sum'	new sum of data value when a data point arrives.
c_error	current sum of deviation of data value to c_avg in the segment.
c_error'	new sum of deviation of data value to c_avg' when a data point arrives.
δ	threshold of cumulate errors in the segment, specified by user.

where M is the number of variable-length segments, $tv_i (1 \leq i \leq M)$ is the mean value of data points in the ith segment,

$$tv_i = \frac{\sum_{j=tr_{i-1}+1}^{tr_i} t_j}{tr_i - tr_{i-1}} \tag{2}$$

and $tr_i (1 \leq i \leq M)$ is the right end point in the ith segment. Fig.2 shows an example of APCA segmentation.

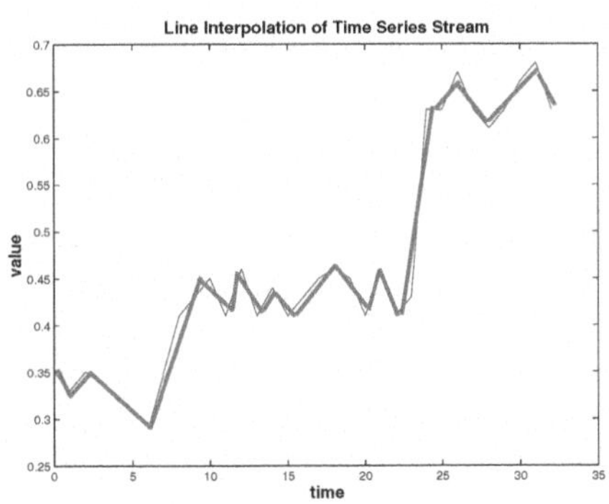

Fig. 1. Example of line interpolation of time series stream. The method produces many redundant segments on the minor fluctuation of values.

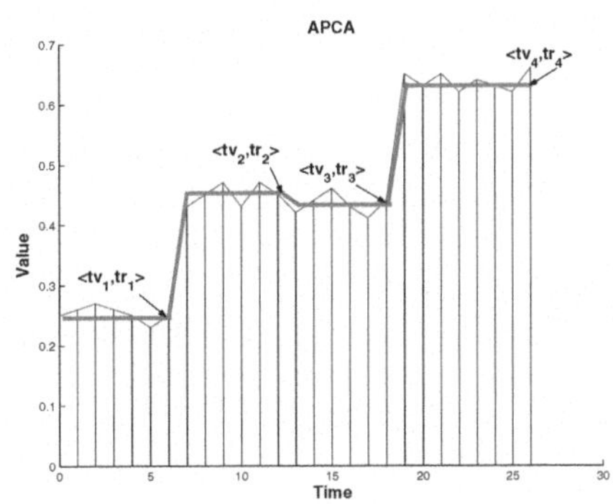

Fig. 2. Illustration of APCA, $M = 4$

3 Ways to Produce APCA Segmentation on Static Time Series

In this section, we discuss the ways to produce the APCA segmentation.

Table 2. The straight way to produce APCA segmentation

Algorithm: A straight way to produce APCA segmentation
Input: T: Time Series of length n. δ: max error in one subsequence. **Output:** T': APCA segmentation of T.

$left \leftarrow 1$; $right \leftarrow 2$;
while $right \le n$ **do** //whole segmentation
    $c\_avg \leftarrow$ calculate average of $T[left, right]$;
    $c\_error \leftarrow 0$;
    for $i = left$ **to** $right$ **do**
        $c\_error \leftarrow c\_error + \lvert t_i - c\_avg \rvert$; // calculate sum of errors in the segment
    end;
    if $c\_error > \delta$ **then** // exceed the threshold
        segment at position $right$; // $< c\_avg, right >$
        $left \leftarrow right + 1$; // new segment
    end;
    $right \leftarrow right + 1$;// continue with next point in the subsequence
end;

For a time series T, the most straight way to produce its APCA segmentation is to slide a window on T, and calculate the error *c_error* point by point. If *c_error* overcomes the threshold δ, then segmentation occurs. The idea is illustrated in Table 2.

To get the N optimal APCA segments from a time series of length n, requires a $O(Nn^2)$ time complexity with dynamic programming[2].

Keogh et al.[5] pointed out that exact optimal approximation is usually not required, and they proposed a method to produce an approximation in $O(nlogn)$, which uses a wavelet transformation and converts the problem into a wavelet compression problem.

These methods, even with the approximate APCA, all require more than one-pass computation of the data in the sliding window and suffer from the non-linear computational complexity, thus are not fit in the context of stream.

4 Proposed Methods to Produce APCAS on Time Series Stream

In this section, we give the approximate method APCAS to segment time series stream. Before that, we first give two realistic consumptions about time series stream:

- The time series stream is in large volume and in high speed;
- The memory of handling system is relative low, the size of sliding window is small, and can be enough to handle a small number of subsequences while segmenting.

Let us look back to the algorithm to produce exact APCA segmentation on static time series in Table 2. When a new data point enters into sliding window, the c_avg and c_error will be updated accordingly. However, the computation of c_error' is a loop through the data points in the segment, and calculating the sum of deviation value to the c_avg'(i.e.$\sum_{i=1}^{k} |t_i - c\_avg'|$) of each data point in the segment. We observed that, when a new data point arrives smoothly, the new value of c_avg changes slowly.

Proposition 1. *Suppose there is a segment in sliding window with length k, and the next data point with value of t_{k+1} arrives. If $t_{k+1} \approx c\_avg$, then $c\_error' \approx c\_error + |t_{k+1} - c\_avg'|$.*

Proof. For simplicity, we denote the segment in sliding window as $T_s = t_1, t_2, \ldots, t_k$. From the computation of APCA in Table 2, we have

$$c\_avg = \frac{\sum_{i=1}^{k} t_i}{k}, \tag{3}$$

$$c\_avg' = \frac{\sum_{i=1}^{k+1} t_i}{k+1} = \frac{k}{k+1} c\_avg + \frac{t_{k+1}}{k+1}, \tag{4}$$

$$c\_error = \sum_{i=1}^{k} |t_i - c\_avg|, \tag{5}$$

$$c\_error' = \sum_{i=1}^{k+1} |t_i - c\_avg'| = \sum_{i=1}^{k} |t_i - c\_avg'| + |t_{k+1} - c\_avg'|, \tag{6}$$

if $t_{k+1} \approx c\_avg$, then from (4), we get

$$c\_avg' \approx c\_avg, \tag{7}$$

thus from (6) and (7), we have

$$c\_error' \approx \sum_{i=1}^{k} |t_i - c\_avg| + |t_{k+1} - c\_avg'| = c\_error + |t_{k+1} - c\_avg'|. \tag{8}$$

The idea behind such approximation is illustrated in Fig.3. As $c\_avg \approx c\_avg'$, we calculate the c_error' with equation (8), and the step can be finished in constant time.

After the replacement, the algorithm to produce APCA in Table 2 then, changes to the one shown in Table 3.

One thing, however, should be noted in the approximate algorithm that, there is an inherent loop in calculation of the new average value of $T[left, right]$(i.e. c_avg'), which prevents the algorithm from finishing in linear time. To solve the problem, we set a variable c_sum to temporarily store the sum of values of data points in the segment. Since

$$c\_sum = \sum_{i=1}^{k} t_i, c\_sum' = \sum_{i=1}^{k+1} t_i = c\_sum + t_{k+1}, \tag{9}$$

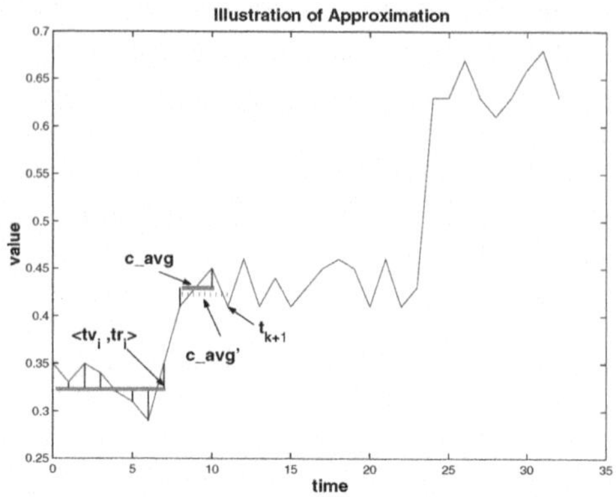

Fig. 3. Illustration of APCA Appproximation, $c\_avg' \approx c\_avg$

Table 3. The approximate algorithm to produce APCA segmentation

Algorithm: An approximate algorithm to produce APCA segmentation
Input:
T: Time Series of length n.
δ: max error in one subsequence.
Output:
T': APCA segmentation of T.
1. $c\_error \leftarrow 0$; $left \leftarrow 1$; $right \leftarrow 2$;
2. **while** $right \leq n$ **do** //whole segmentation
3. $c\_avg \leftarrow$ calculate average of $T[left, right]$;
4. $c\_error \leftarrow c\_error + |t_{k+1} - c\_avg|$; // calculate sum of errors in the segment
5. **if** $c\_error > \delta$ **then** // exceed the threshold
6. segment at position $right$; // $< c\_avg, right >$
7. $left \leftarrow right + 1$; // new segment
8. $right \leftarrow left$; $c\_error \leftarrow 0$;
9. **end**;
10. $right \leftarrow right + 1$;// continue with next point in the subsequence
11. **end**;

$$c\_avg = \frac{c\_sum}{k}, c\_avg' = \frac{c\_sum'}{k+1}, \tag{10}$$

and the number of data points in the current segment can be calculated as $right - left + 1$, thus we don't need to recalculate the c_avg' by looping forwardly through the data points in the segment when a new data point arrives, we just update the value of c_sum' with equation (9), and this step needs $O(1)$.

As data in time series stream arrives with time, hence differs from the segmentation on static time series, the segmentation on time series stream should also continue with data, and should not make any hypothesis of the data volume(such as length of n). In the approximate algorithm of APCAS shown in

Table 4. The approximate algorithm to produce APCAS on time series stream

Algorithm: An approximate algorithm to produce APCAS on time series stream
Input:
T: Time Series stream.
δ: max error in one subsequence.
Output:
T': APCAS segmentation of T.
1. $c\_error \leftarrow 0$; $left \leftarrow 1$; $right \leftarrow 2$; $c\_sum \leftarrow t_{left}$;
2.**while not** stop_by_user **do** //continue processing till stopped by user
3. $c\_sum \leftarrow c\_sum + t_{right}$;
4. $c\_avg \leftarrow \frac{c\_sum}{right-left+1}$;
5. $c\_error \leftarrow c\_error + |t_{k+1} - c\_avg|$; // calculate sum of errors in the segment
6. **if** $c\_error > \delta$ **then** // exceed the threshold
7. segment at position $right$; // mark in stream: $< c\_avg, right >$
8. $left \leftarrow right + 1$; $right \leftarrow left$; // new segment
9. $c\_error \leftarrow 0$; $c\_sum \leftarrow t_{left}$;
10. **end**;
11. $right \leftarrow right + 1$;// continue with next point in the subsequence
12.**end**;

Table 4, we set a boolean flag **stop_by_user** to let the user decide whether or not to stop the segmentation process.

The algorithm in Table 4 is a one-pass computation, and can be applied into time series stream segmentation. To put the point more clearly, we design a Finite State Automata(FSA) to model the process of APCAS segmentation, which is shown in Fig. 4. In FSA, the initial state is *EnterNewSegment*, which indicates that a new segment starts, when the first data point enters into the sliding window, the state changes to *WithinSegment*, and with more data points arrives, FSA calculates the c_error' continuously. If $c\_error' < \delta$, the data point is not the segment boundary, and the state does not change; and if $c\_error' \geq \delta$, the data point is recognized and marked by FSA as the right boundary of the segment, the state changes back to *EnterNewSegment* to start a new segment segmentation.

There still remains a problem to be settled, as readers may have noted, the hypothesis of Proposition 1 is that the stream runs smoothly, $t_{k+1} \approx c\_avg$, but what if $t_{k+1} \gg c\_avg$? Now we take time to explain here, which we have omitted previously. As seen from equation (8), the c_avg' is not replaced with c_avg in

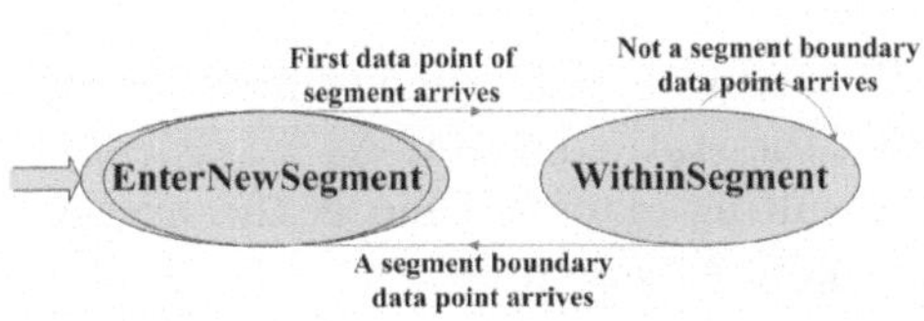

Fig. 4. The Finite State Automata to model APCAS

the part of $|t_{k+1} - c\_avg'|$. As in calculating the c_avg' when the data point with value of t_{k+1} arrives, the data points $t_1, t_2, \ldots, t_k$ in the sliding window are assumed to be in the same relative smooth segment, the $c\_avg \approx c\_avg'$, thus $\sum_{i=1}^{k} |t_i - c\_avg| \approx \sum_{i=1}^{k} |t_i - c\_avg'|$. However, if $t_{k+1} \gg c\_avg$, the c_error' increases quickly and the FSA may recognize the new data point t_{k+1} as the right segment boundary, our experiments validate such conclusion.

5 Experiment Research

In this section, we empirically demonstrate the utility of APCAS in time series stream segmentation with a comprehensive set of experiments. For these experiments, we used a PC with a Pentinum 3-866 processor, 256MB RAM and 40GB disk space. The source code for the experiments is written in C++ Language(with g++ 3.2.3 as the compiler). Note that in order to allow reproducibility, all the source code and datasets are freely available, interested readers may contact with the authors by email.

For completeness, we implemented all the methods proposed in this work. We have taken great care to create high quality implementations of all techniques. All approaches are optimized as much as possible.

5.1 Experiment Metrics

In our experiments, we evaluated the efficiency of different techniques using three metrics. We measured the *memory usage* and the *segment precision* as the main factors affecting overall performance of the segmentation. In addition, we measured the *elapsed time* as the performance metric directly perceived by the user.

- **Memory usage.** This is memory consumption during the segmentation. As the handle system is relatively memory constraint in the light of large data volume of time series stream, memory usage is one of the key factors that we must take great care of.
- **Segment precision.** As APCAS is an approximate method to APCA, we compare the segment precision to the exact APCA segmentation with different size of data specified. We use the following equation to denote the segment precision:

$$precision\_of\_segmentation = \frac{total\_segmentation\_error}{total\_number\_of\_segments} \quad (11)$$

- **Elapsed time.** We used wall-clock time to measure the elapsed time during the segmentation on the static dataset. As we repeated each experiments in several time, the resulting elapsed time is the average of the experiments with the same parameters configured.

5.2 Evaluation on Artificial Data

The data sets used in this section were created using a random time series generator that produces n time series, one of the example and its APCA segmentation are shown in Fig. 5, the approximate APCAS segmentation of the dataset is shown in Fig. 6.

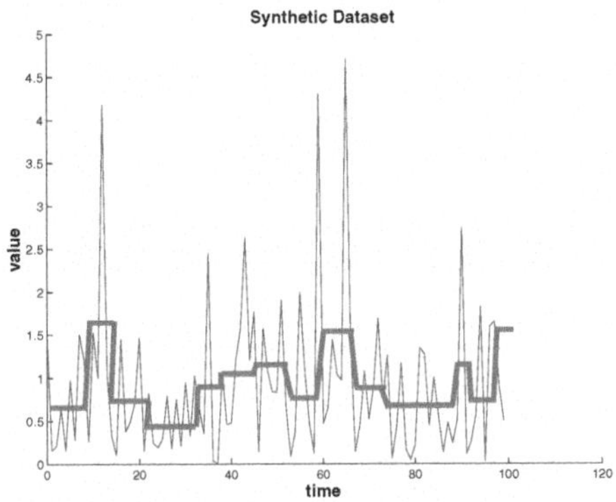

Fig. 5. Example of exact APCA segmentation on synthetic dataset: $\delta = 13.0$

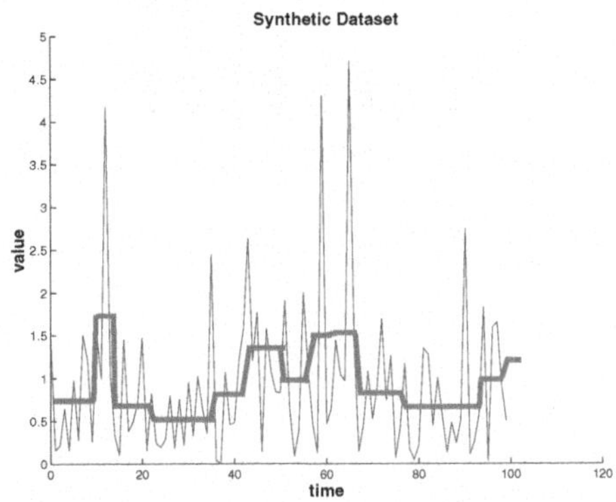

Fig. 6. Example of APCAS segmentation on synthetic dataset: $\delta = 13.0$

The memory usage during segmentation is shown in Fig. 7, the comparison of segmentation precision is shown in Fig. 8 and the elapsed time is shown in Fig. 9. As the results indicate, APCAS outperforms the exact APCA method in the memory consumption and elapsed time dramatically. However, compared with the exact computation of APCA, APCAS suffers from the precision loss. Nonetheless, the precision loss is acceptable in real application(the max loss is roughly within 8%).

5.3 Evaluation on Real Data

Our real data are the measure data from a signal processing device. A small sketch of the data, together with the results of exact APCA and approximate APCAS are shown in Fig. 10 and Fig. 11, respectively. One observation can be made based on the results, the right location of each segment in APCAS is very near to that in APCA(i.e. $36 - 35$, $60 - 60$, $92 - 91$, $103 - 101$, and $126 - 126$), which indicates that, as expected, the FSA recognizes the data points with a large fluctuation and performs the segmentation on the data points.

The memory usage on segmenting real dataset is shown in Fig. 12. The same trend is observed from the results. With time elapses, the advantage of APCAS over memory increases. Fig. 13 compares the segmentation precision of the exact APCA and APCAS methods on real data segmentation. The precision loss is lower than 9% and is acceptable in stream.

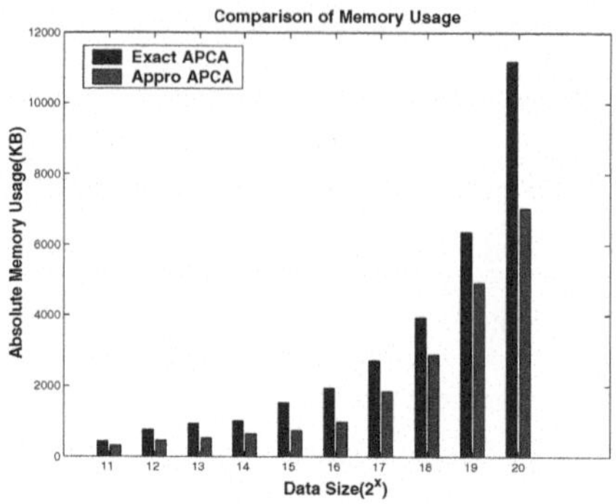

Fig. 7. Comparison of memory usage on synthetic dataset: $\delta = 13.0$

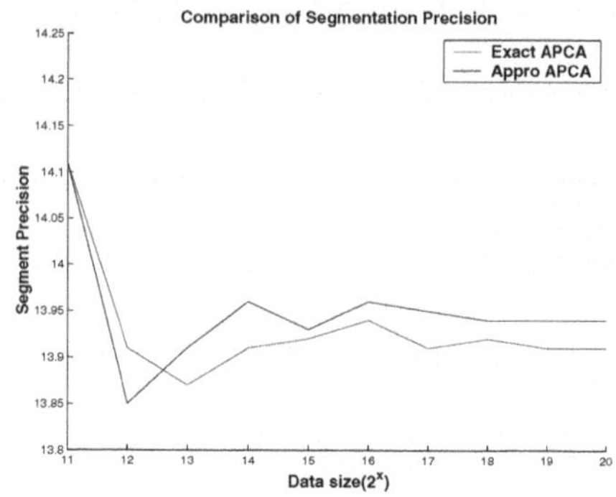

Fig. 8. Comparison of precision of segmentation on synthetic dataset: $\delta = 13.0$

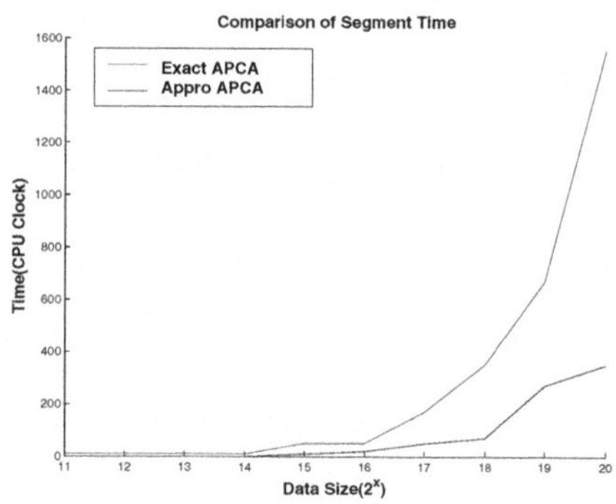

Fig. 9. Comparison of elapsed time of segmentation on synthetic dataset: $\delta = 13.0$

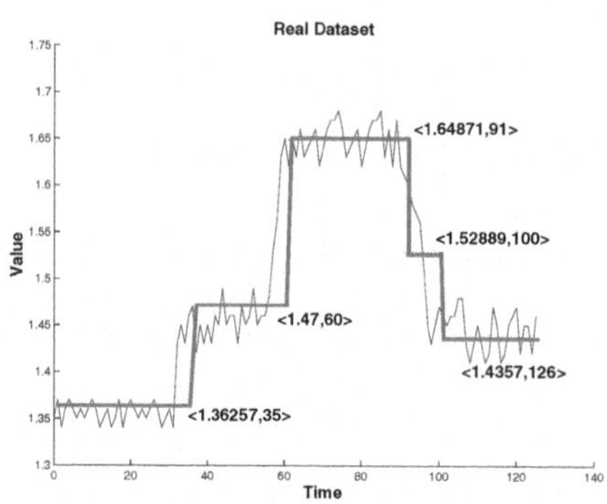

Fig. 10. Example of exact APCA on real dataset: $\delta = 0.5$

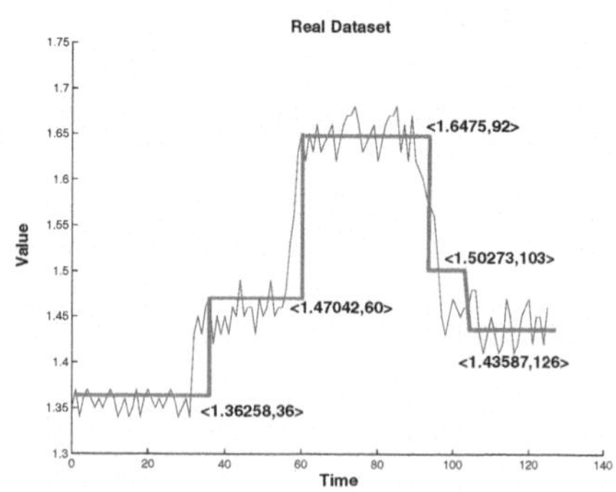

Fig. 11. Example of APCAS on real dataset: $\delta = 0.5$

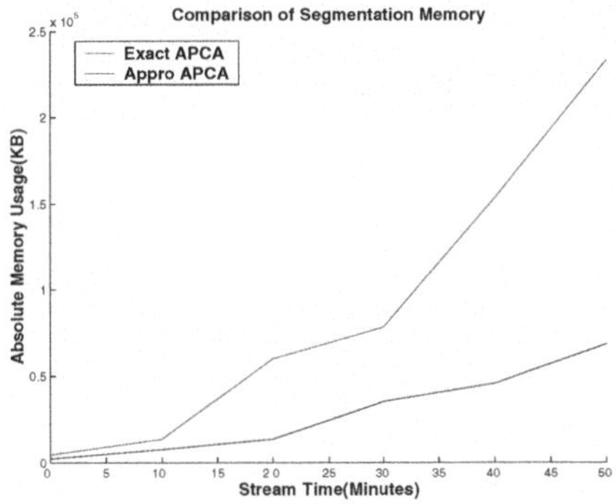

Fig. 12. Comparison of memory usage on real dataset: $\delta = 0.5$

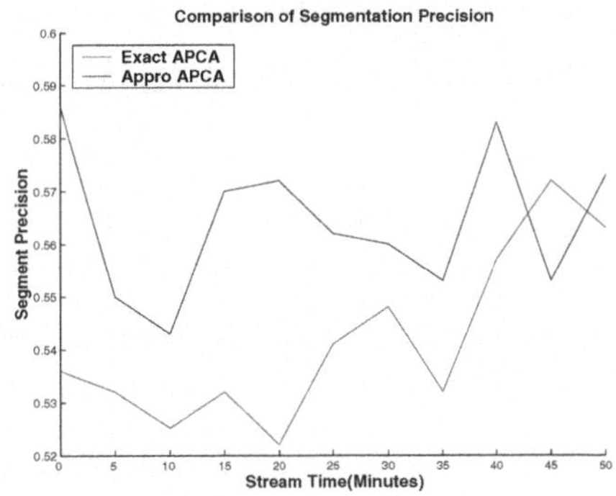

Fig. 13. Comparison of precision of segmentation on real dataset: $\delta = 0.5$

6 Conclusion

In this paper we propose an approximate method, namely, APCAS, to segment time series stream.We devised a Finite State Automata to model the process. The method works in linear time, has its wide applications in streaming data. Experiments show, the APCAS is a fast method and can be easily run in the memory constraint applications. The loss of precision of the method, compared with the exact APCA, is in acceptable scope. For future research, we plan to apply the method APCAS to change detection and signal processing in time series stream to further examine the applicability of this approach.

References

1. Brian B., Shivnath B., Mayur D., et al. Models and Issues in Data Stream Systems. In *Proc. of 21st ACM Symposium on Principles of Database Systems*, pages 1-16, New York: ACM Process, 2002.
2. Faloutsos C., Jagadish H., Mendelzon A., et al. A signaure technique for similarity-based queries. In *SEQUENCES 97*, Pousitano-Salerno, Italy, 1997.
3. Ingrid S., Mathias P., Bram G. Toward Automated Segmentation of the Pathological Lung in CT. In *IEEE Transactions on Medical Imaging*, 24(8), pages 1025-1038, 2005.
4. Li Aiguo, Qin Zheng. On-Line Segmentation of Time-Series Data. In *Journal of Software*, 15(11), pages 1671-1679, 2004.
5. Keogh E., Chakrabarti K., Pazzani M. Locally adaptive dimensionality reduction for indexing large time series databases. In *Proc. of ACM SIGMOD Conf. on Management of Data*, pages 151-162, 2001.
6. Keogh E., Selina C., David H., et al. An Online Algorithm for Segmenting Time Series. In *Intl' Conf. of Data Mining*, pages 289-296, USA, 2001.
7. Keogh E., Kasetty S. On the need for time series data mining benchmarks: a survey and empirical demonostration. In *the 8th ACM SIGKDD Int'l Conf. on Knowledge Discovery and Data Mining*, pages 102-111, Edmonton, Canada, 2002.

Continuous k-Nearest Neighbor Search Under Mobile Environment

Jun Feng[1], Linyan Wu[1], Yuelong Zhu[1], Naoto Mukai[2], and Toyohide Watanabe[2]

[1] Hohai University, Nanjing, Jiangsu 210098 China
{fengjun,ylzhu}@hhu.edu.cn
[2] Nagoya University, Nagoya, Aichi 464-8603 Japan
{watanabe,naoto}@is.nagoya-u.ac.jp

Abstract. Continuous K nearest neighbor (CkNN) queries under mobile environment are defined as the k nearest neighbors search for a query object at a serial query time, and all the query object and the target objects are moving. In this paper, we propose method to solve this problem, focusing on finding out the relations between the continuous queries, and proposing decision method for relative moving trend among moving objects. Our experiments show that our approach outperforms the original straightforward method, specially when the query interval is small.

1 Introduction

With the improvements of geographic positioning technology and the popularity of wireless communication, it is easy to trace and record the position of moving objects. New personal services are proposed and realized, many of which serve the user with desired functionality by considering the user's geo-location. This kind of service is also called as location-based service (LBS). For example, in Intelligent Transportation Systems (ITS), to find out the 3-nearest vehicles around an autonomous vehicle, continuously; or in the war field, to keep contact with the 5-nearest partners all the time and so on. To provide such kinds of services, the research on nearest neighbor search (NN-search), especially continuous kNN-search (CkNN) for mobile objects is arousing more and more attentions.

CkNN search under mobile environment can be defined as: given a query object q and a set of target objects $\{p_1, p_2, ..., p_n\}$ $(n \geq k)$, q and p_i $(1 \leq i \leq n)$ are moving in 2D (or 3D) space. Suppose query object q searches its k nearest neighbors every Δt time, this kind of continuous query is called CkNN under mobile environment. Because all the objects are moving, it is more complex than continuous k nearest static objects search, i.e., search the 3 nearest gas stations for a moving car. In that situation, there are methods based on voronoi division [1,2,3] or based on the properties of the query object's moving path [4,5,6,7,8]. All these methods cannot be applied to mobile environment CkNN search directly.

Up to now, there are not so much research for solving CkNN under mobile environment. Li et al. [9] discussed this problem with the condition of that

G. Dong et al. (Eds.): APWeb/WAIM 2007, LNCS 4505, pp. 566–573, 2007.

all the moving functions of the mobile objects are known before starting the continuous kNN search, therefore, they can only focus on the number k objects at all the time. Actually, there are more examples for moving functions are unknown in the real world. Jensen et al. [10] proposed method for ordered k-nearest neighbor queries for query and target objects that are moving in road networks. They employ a client-server architecture that partitions the NN search. First, a preliminary search for a Nearest Neighbor Candidate set (NNC set) is performed on the server. Then, the maintenance of the active query result is done on the client, which recomputes distances between data points in the NNC set and the query point, sorts the distances, and periodically refreshes the NNC set to avoid significant imprecision. The problem of this method is that the update interval is difficult to decide: a too short interval brings intensive computations, while a too long interval brings distrustful results.

In this paper, we propose CkNN search method for mobile objects. The focus of our research is to find out the relations between the continuous queries, in other words, we wish the current query result could help to increase the efficiency of the next query. The main contributions of our paper are as follows:

- proposes a method for generating candidate set for continuous queries, and assures the efficiency of the searching;
- proposes a quadrant representation method for deciding relative moving trend among objects and the candidate set under mobile environment;
- proposes an algorithm for CkNN search and carries experiments on three kinds of data sets.

The remainder of this paper is organized as follows. Section 2 describes our query method, including the generation method of candidate set, the decision method of relative moving trend. Section 3 analyzes our method based on experiments and Section 4 makes a conclusion on our work.

2 Method of CkNN Search

In this section, we propose a method for CkNN search under mobile environment, where the query object and the target objects are moving in 2D space, and the moving functions of them need not to be known before starting the query. All the query object and the target objects are moving all the time, therefore, at every query time, the query object may be at a new position and the target objects around it may be "away" or "close" to it. A straightforward method for CkNN search is to find out kNN for the query object at every query time from all the target objects. The computing cost at every time is decided by the efficiency of the access method for the target objects. For example, by using TPR*-tree [11] to index the target objects, and employing a priority queue for keeping the NN sequence of the tree nodes and the objects, the CkNN query under mobile environment is changed into a series of kNN search for static objects [12]. This method is effective but not efficient, because if the query interval is not too long, the continuous queries may share the same or part of the results. Here, we just try to find out the relations between continuous queries, in order to decrease the re-computations and limit the query into a relative small target object set. In the sequal, a generation method of kNN query candidate set and the decision method of the objects' distribution are given.

2.1 Selection of Candidates

To find out a proper candidate set based on up to now CkNN query results is an effective way for achieving an efficiency search. Song et al. [5] proposed methods for kNN search for a mobile object. However, their method is limited to all the data objects are static. When both the query and the data objects are moving, the decision should take the moving parameters into consideration. We give two theorems to cope with this situation.

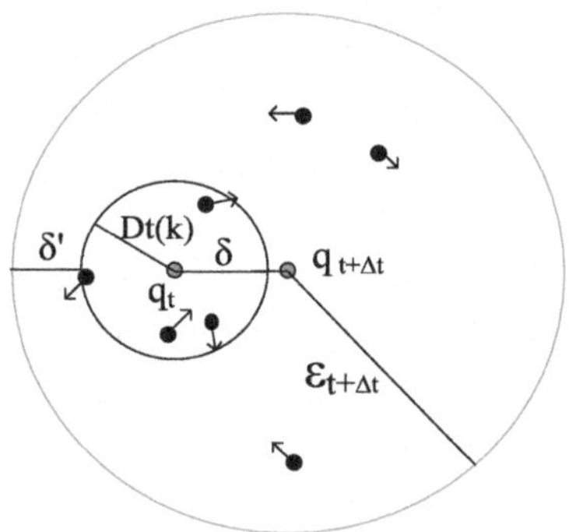

Fig. 1. Search region at time $t + \Delta t$ under mobil environment

Theorem 1. Suppose at time t, the kNN query result set of query point at location q_t are $\{p_1, p_2, ..., p_k\}$ and $D_t(k)$ is the maximum distance of these sites to q_t. At time $t + \Delta t$, time interval is Δt, the query point moves to $q_{t+\Delta t}$. We claim that $\varepsilon_{t+\Delta t} = D_t(k) + \delta + \delta'$ is legal (there are at least k sites are within the distance $\varepsilon_{t+\Delta t}$ from $q_{t+\Delta t}$), where δ is the distance between q_t and $q_{t+\Delta t}$, and $\delta' = v_{rs} \times \Delta t$, v_{rs} is the maximum speed of p_i $(1 \leq i \leq k)$ in the result set [5].

Proof. By definition of kNN we have $dist(q_t, p_i) \leq D_t(k), (1 \leq i \leq k)$. If the next position of p_i is p_i', then $dist(q_t, p_i') \leq dist(q_t, p_i) + v_{pi} \times \Delta t, (1 \leq i \leq k)$, and as v_{rs} is the maximum speed: $v_{pi} \times \Delta t \leq v_{rs} \times \Delta t, (1 \leq i \leq k)$. By definition, we have $\delta' = v_{rs} \times \Delta t$, then: $dist(q_t, p_i') \leq dist(q_t, p_i) + \delta'$. According to triangular inequality, we have: $dist(q_{t+\Delta t}, p_i') \leq dist(q_t, p_i') + \delta, (1 \leq i \leq k)$, Then $dist(q_{t+\Delta t}, p_i') \leq D_t(k) + \delta + \delta', (1 \leq i \leq k)$, which means that at time $t + \Delta t$, there are at least k objects whose distances to $q_{t+\Delta t}$ are not longer than $D_t(k) + \delta + \delta'$. By definition, we know that $\varepsilon_{t+\Delta t} = D_t(k) + \delta + \delta'$ is legal. □

Though the kNNs for time t are still inside the candidate set for time $t + \Delta t$, it does not mean that they "are" just the kNNs for time $t+\Delta t$. It only assures that kNNs for time $t + \Delta t$ can be found inside this candidate set, and the sequence of them still needs to be computed.

Theorem 2. Suppose at time t, the speed of query point at location q_t is v_q, the k nearest neighbors are $\{p_1, p_2, ..., p_k\}$, $D_t(k)$ is the maximum distance of these objects to q_t, and v_{rs} is the maximum speed of these objects, the query interval is Δt. p is the object with the minimum distance of the objects outside that set, and $D_t(p)$ is the distance between p and q_t. Then the update time of the candidate set is:

$$update\_time = \frac{D_t(p) - D_t(k) - (v_q + v_{rs}) \times \Delta t}{v_p + v_q + v_{rs}}. \tag{1}$$

Proof. If the expand speed of the candidate region v_{region}:

$$v_{region} = -\frac{\varepsilon_{t+\Delta t} - D_t(k)}{\Delta t} \tag{2}$$

The expand speed of candidate set is calculated as the centripetal speed related to the query point. Its size is the same as that of kNNs', the "minus" sign means the direction of the expand speed is in the direction of departing the center q.

Based on the definition of p, p is the nearest object to the candidate set at time t. The candidate set and p are moving all the time. If at time $t + \Delta t$, p is just appearing on the boundary of candidate set, then the distance it traveled is $(D_t(p) - \varepsilon_{t+\Delta t})$, its relative speed to the boundary of candidate set is $(v_p - v_{region})$. Considering that when p enters the candidate set, the candidate set needs to be updated, and the update time is:

$$update\_time = \frac{D_t(p) - \varepsilon_{t+\Delta t}}{v_p - v_{region}}. \tag{3}$$

Based on Theorem 1, the definition of δ and δ', we have $v_{region} = -(v_q + v_{rs})$, and

$$update\_time = \frac{D_t(p) - D_t(k) - \delta - \delta'}{v_p + v_q + v_{rs}}. \tag{4}$$

Then:

$$update\_time = \frac{D_t(p) - D_t(k) - (v_q + v_{rs}) \times \Delta t}{v_p + v_q + v_{rs}}. \tag{5}$$

After the first time of query, we can generate candidate set based on Theorem 1, calculate the update time of the candidate set based on Theorem 2. In the following search, after checking whether the candidate set need to be updated or not, the kNN search can be done inside the candidate set or an updated candidate set. The efficiency of the continuous kNN search can be assured.

2.2 Relative Moving Trend

The circle region limited by the candidate set is called search region. The relative moving trend between the object and the search region decides the object inside the region will leave off or the object outside the region will come in. Though the center of the search region is the position of the query object (q), at the next query time, the relative moving trend between the data object and the search region is different with that between the data object and q. This is because the size of search region and its moving speed are changing all the time, for example, one object p is closing to q and p is in the search region, when there are more data objects come into the region with higher speed, then p will be extruded from the region; on the other hand, when there are more objects leave off the region, p will enter the region.

In this subsection, we propose a quadrant method for representing the relative moving trend in the query. First give method for representing the relative position between two objects, then discuss the relative moving trend between them. There are three kinds of trend: closing, leaving and relative static. To simplify the discussion, we consider "static" as one kind of "closing" in the follows.

Given a query object $q(x, y)$ and a data object $p(x, y)$ in 2D space, the relation between them can be represented by using the quadrant, depicted by Fig. 2: q is the reference point, the parallel lines of x-axis and y-axis split the space into quadrant, and use 1 or -1 to decode the positive or negative of the location related to q.

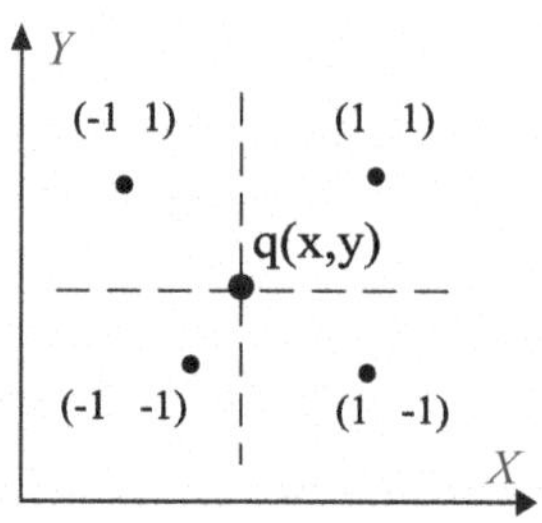

Fig. 2. Quadrant representation of relation between objects

The relation p and q can be represented by quadrant method, and there are:

- If $p_x > q_x$, then $q_x \cap p_x = 1$;
- If $p_x < q_x$, then $q_x \cap p_x = -1$;
- If $p_y > q_y$, then $q_y \cap p_y = 1$;
- If $p_y < q_y$, then $q_y \cap p_y = -1$.

The relative moving trend is decided by the relative position and the relative speed between objects. We use a NOT(Exclusive-OR) expression for deciding the moving trend: $(v_{qx} - v_{px}) \bigodot (q_x \cap p_x)$, the values of this expression is given in Table 1. Value 1 means in x-axis two objects are closing; while -1 means they are away. Therefore, there are three kinds of relations between two objects: when two object are closing in both x-axis and y-axis, then they are closing; when they are away in both axes, they are away; otherwise, the situation of them cannot be decided.

Table 1. Values of $(v_{qx} - v_{px}) \bigodot (q_x \cap p_x)$

$q_x \cap p_x$	$v_{qx} - v_{px}$	$(v_{qx} - v_{px}) \bigodot (q_x \cap p_x)$
1	1	1
1	-1	-1
-1	1	-1
-1	-1	1

kNNs are selected from the candidate set at every query time, and the candidate set is maintained by testing the relative moving trend between the object inside the candidate set and the search region. For example, if an object leaves q with a speed faster than that of the search region, then this object will be deleted from the candidate set. By using this method, the number of objects inside the candidate set will be decreased, and the kNN search can be more efficient.

3 Experiment and Analysis

We conducted several experiments to compare our method with the original straightforward method which finds out kNN for the query object at every query time from all the target objects. There are three data sets, A, B and C, used for the experiments: 1) in data set A, 10,000 objects are in a uniform distribution on a 100×100 grid, which are moving with the same speed; 2) in data set B, 10,000 objects are in a uniform distribution on a 100×100 grid, which are moving with random speed; 3) in data set C, 10,000 objects are in random distribution, which are moving with random speed. The experiments were performed on a PC with 1.70GHz CPU, 256MB of RAM, and Visual C++ 6.0 as the programming language. We record the executing time (seconds) of the CkNN queries for different values of query interval Δt, query number num_Query and different values of k. We present the average results of 10 runs of continuous k nearest neighbor queries, and compared them with those in the original method (in Fig. 3, "Original" refers to the straightforward method, while "New" refers to our method).

- Effect of Δt. We analyze the relation of the query interval and the algorithm efficiency in the condition of query times num_Query is 20 and the neighbor number k is 10. Fig. 3 gives the results. The original algorithm keeps a stable high cost all the time, while our algorithm is more efficient when the query interval becomes smaller, because when Δt is short, there are fewer objects come in or leave off the candidate set.

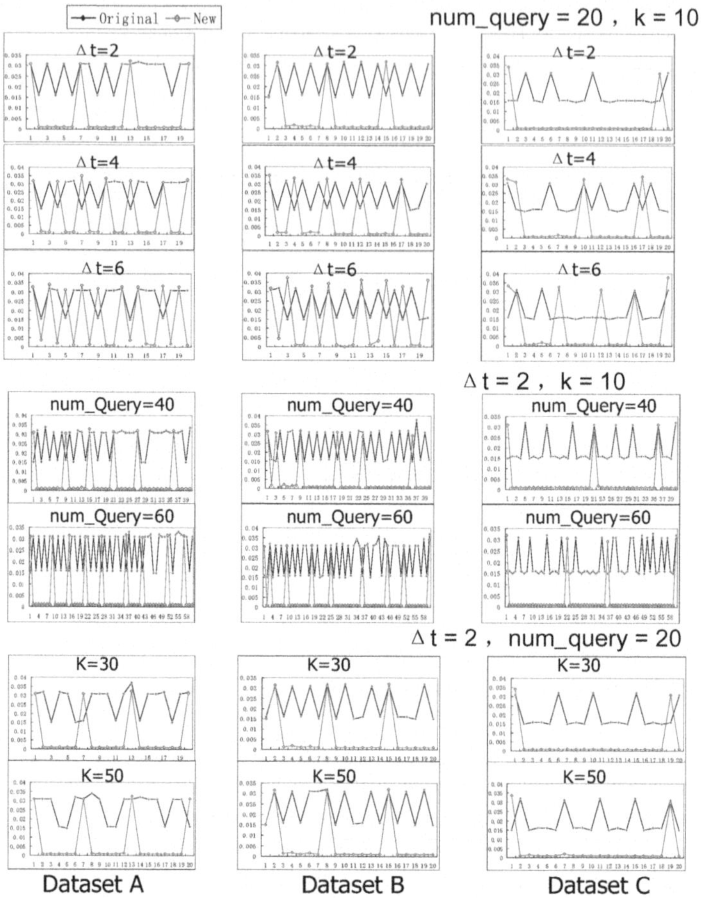

Fig. 3. Experiments on Δt, *num_Query* and k

- Effect of *num_Query*. Given that the query interval Δt is 2 and the neighbor number k is 10. For all the three data sets, the two methods behave regularly. This is because in our method, the current query is only based on the just before query result, and makes a contribution to the just after query. In the original method, kNNs are searched by expanding the priority queue every time. The query number has no effect to the query cost.
- Effect of k. Here the query interval Δt is 2 and the query number *num_Query* is 20. There is no obvious change in all the k situations.

4 Conclusion

In this paper, we proposed a search method for CkNN under mobile environment. In our method, the CkNN search can be executed in a relative small search region decided by the candidate set. We proposed method for deciding the update time of the candidate set dynamically. Especially, we proposed the relative moving trend decision method for the maintenance of the candidate set, and assured the candidate set is legal and minimum. Experiments showed that our method outperforms the original straightforward method in 2D mobile environment. In our future work, we will discuss the CkNN under mobile environment for moving-constraint situations: i.e., for the cars moving on road networks.

Acknowledgements

This research is supported by NSFC 60673141 (Research on the Index Structure of Spatial Network-based Moving Objects).

References

1. M. R. Kolahdouzan and C. Shahabi. Voronoi-based k nearest neighbor search for spatial network databases. *Proceedings of the 30th Very Large DataBase*, pages 840–851, 2004.
2. S. Bespamyatnikh and J. Snoeyink. Queries with segments in voronoi diagrams. *SODA*, 1999.
3. M. R. Kolahdouzan and C. Shahabi. Continuous k nearest neighbor queries in spatial network databases. *Proceedings of the Second Workshop on Spatio-Temporal Database Management*, pages 33–40, 2004.
4. Y. F. Tao, D. Papadias, and Q. M. Shen. Continuous nearest neighbor search. *Proc. of VLDB'02*, pages 287–298, 2002.
5. Z. X. Song and N. Roussopoulos. K-nearest neighbor search for moving query point. *Proc. of SSTD'01*, pages 79–96, 2001.
6. J. Feng and T. Watanabe. A fast method for continuous nearest target objects query on road network. *Proc. of VSMM'02*, pages 182–191, 2002.
7. J. Feng, N. Mukai, and T. Watanabe. Stepwise optimization method for k-cnn search for location-based service. *Proc. of SOFSEM 2005, LNCS*, 3381:363–366, 2005.
8. J. Feng, N. Mukai, and T. Watanabe. Search on transportation network for location-based service. *Proc. of IEA/AIE 2005, LNAI*, 3533:657–666, 2005.
9. Yifan Li, Jiong Yang, and Jiawei Han. Continuous k-nearest neighbor search for moving objects. In *SSDBM*, pages 123–126, 2004.
10. Christian S. Jensen, Jan Kolvr, Torben Bach Pedersen, and Igor Timko. Nearest neighbor queries in road networks. In *GIS '03: Proceedings of the 11th ACM international symposium on Advances in geographic information systems*, pages 1–8, New York, NY, USA, 2003. ACM Press.
11. Y. F. Tao, D. Papadias, and J. M. Sun. The tpr*-tree: An optimized spatio-temporal access method for predictive queries. *Proc. of VLDB'03*, pages 790–801, 2003.
12. G.R. Hjaltson and H. Samet. Distance browsing in spatial databases. *ACM Transactions on Database Systems*, 24(2):265–318, 1999.

Record Extraction Based on User Feedback and Document Selection

Jianwei Zhang[1], Yoshiharu Ishikawa[2], and Hiroyuki Kitagawa[1,3]

[1] Department of Computer Science, Graduate School of Systems and Information Engineering, University of Tsukuba
1-1-1 Tennohdai, Tsukuba, Ibaraki, 305-8573, Japan
zjw@kde.cs.tsukuba.ac.jp

[2] Information Technology Center, Nagoya University
Furo-cho, Chikusa-ku, Nagoya, Aichi, 464-8601, Japan
ishikawa@itc.nagoya-u.ac.jp

[3] Center for Computational Sciences, University of Tsukuba
1-1-1 Tennohdai, Tsukuba, Ibaraki, 305-8573, Japan
kitagawa@cs.tsukuba.ac.jp

Abstract. In recent years, the research of record extraction from large document data is becoming popular. However there still exist some problems in record extraction. 1) when large document data is used for the target of information extraction, the process usually becomes very expensive. 2) it is also likely that extracted records may not pertain to the user's interest on the aspect of the topic. To address these problems, in this paper we propose a method to efficiently extract those records whose topics agree with the user's interest. To improve the efficiency of the information extraction system, our method identifies documents from which useful records are probably extracted. We make use of user feedback on extraction results to find topic-related documents and records. Our experiments show that our system achieves high extraction accuracy across different extraction targets.

1 Introduction

With the recent progress of information delivery services, electronic text data is increasing rapidly. Useful information often exists in these text documents. However, computers can not easily process the information because it is usually hidden in unstructured texts. Therefore *information extraction* is becoming an important technique to find useful information from a large amount of text documents. Especially a lot of researches analyze the document structures and contexts to construct relational tables. Among many approaches, the *bootstrapping* extraction methods [1,2] have attracted a lot of research interests. These approaches expand the target relation by exploiting the duality between patterns and relations starting from only a small sample. The extracted information, which we call *records*, can be used as a relational table for answering SQL queries or being integrated with other databases.

G. Dong et al. (Eds.): APWeb/WAIM 2007, LNCS 4505, pp. 574–585, 2007.

Two problems exist in the previous approaches of information extraction. In general, an information extraction system needs to preprocess the documents (e.g. attaching named entity tagger to recognize person names, organization names and location descriptions etc.) and scan the documents. First, when the text document set is very large, processing all the documents is quite expensive. Second, records whose topics are not desirable for a user may also be extracted by only using pattern matching. For example, for a user who wants to acquire the information of IT companies and their locations, he is not satisfied with other topic-unrelated pairs (e.g. automobile companies and locations).

To solve these two problems, we propose a method to efficiently extract information suitable for the user's intention. In general, only a part of the documents in a large data set is useful for the extraction task. We manage to specify the documents that are likely to contain desirable records as the target documents for extraction. The efficiency is improved by processing not all the documents, but just a subset of them. From the selected documents related to the required topic, topic-related records are more likely to be extracted.

The rest of this paper is structured as follows. Section 2 reviews the related work. In Section 3, the overview of the proposed system is first presented, and then the procedure for extracting records and several document selection methods are described. Section 4 shows the experimental results and their evaluations. Finally, we conclude this paper and discuss the future work in Section 5.

2 Related Work

There have been many researches on information extraction from unstructured and semi-structured documents such as Web and news archives. *Lixto* [3] is a visual web information extraction system that allows a user to specify the extraction patterns. [4] applies a machine learning method to learn extraction rules, given a set of manually labeled pages. As opposed to these approaches dealing with one web page or some similarly structured web pages, *bootstrapping methods* [1,2,5] are proposed to extract information from documents whose structures are very different in a scalable manner. *DIPRE* (Dual Iterative Pattern Relation Extraction) [1] exploits the duality between patterns and relations from web pages. For example, beginning with a small seed set consisting of several (author, title) pairs, DIPRE generates patterns which are used to find new books. This technique is proved that it works well because relation pairs tend to appear in similar contexts in the Web environment. [5] proposes a method to estimate the coverage of extracted records to reduce iterations of extraction loops, and a technique to estimate the error rate of extracted information to improve the extraction quality. *Snowball* [2] considers the problem of extracting relation pairs from plain-text documents. This method improves the DIPRE method by using novel pattern representation including named entity tags, and precise evaluation measure of patterns and records so that more reliable results can be extracted. In this paper, we extend the basic framework of DIPRE and Snowball methods for the record extraction.

QProber [6] uses a small number of query probes to automatically classify hidden web databases. Chakrabarti et al. propose a topic-focused web crawling method through relevance feedback [7]. The focused crawler in [8] based on a hypertext classifier classifies crawled pages with categories in a topic taxonomy. We take the hint from these researches to prefer selecting useful documents as extraction targets, just as the focused crawler fetches relevant web pages and discards irrelevant ones. Agichtein et al. present a method [9] to retrieve documents and from them extract information from an efficiency viewpoint, but they do not consider whether the extraction results satisfy the user's interest with respect to the topic. In the best of our knowledge, our system is the first to provide topic-related information extraction facility using an interactive approach.

3 Record Extraction Incorporating Document Selection

3.1 System Overview

In this section, we describe the proposed system architecture (Fig. 1).

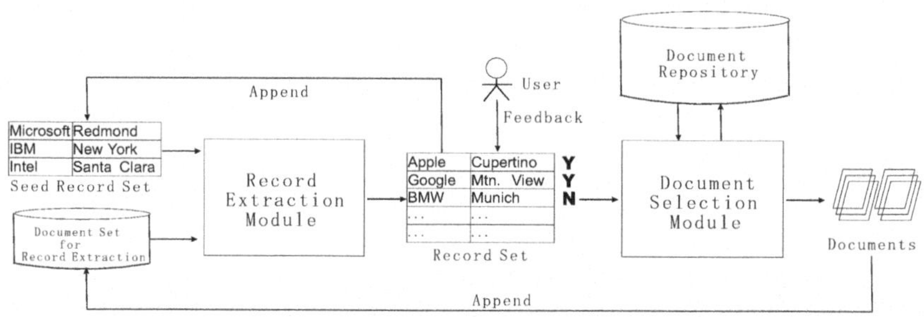

Fig. 1. System Components

In this paper, we also focus on the problem of extracting a relation of companies and their locations defined in Snowball[2]. Different from its scenario, we consider a user usually prefers the extraction results on a specific topic (e.g., "IT" companies and their locations), to all the extractable records. Extracted pairs on other topics are unwantedly troublesome. We present a method to solve this problem. Suppose that several samples (Seed Record Set in Fig. 1) are given as initial knowledge and they also reflect the topic that a user is interested in.

Document repository is large collections of text documents such as newspaper archives. *Document Set for Record Extraction* (DSRE) is a subset of document repository, consisting of documents where records on the related topic may exist. We index the documents using a full-text search system so that given appropriate queries the corresponding documents can be returned. In our proposed system, not all the documents in the document repository are scanned at one time by the extraction system. Considering that records related to the specific topic are more likely to be extracted from the topic-related documents, those documents are preceded as the extraction target. The efficiency is improved by only processing the documents worthy of analysis. Initially the DSRE set can be retrieved by using the sample records as the query. This DSRE set is continuously extended with the newly selected documents, i.e. the output of *Document Selection Module* to be explained later.

Record Extraction Module is to extract records from the DSRE set. We extend the bootstrapping framework of the DIPRE and Snowball systems. Beginning with the seed records given from a user, the program finds the occurrences of those records. Then the occurrences are analyzed to generate patterns. Using the patterns, the program searches the documents to match new records. This process is repeated until some termination criterion is met. In this way, a number of records can be obtained with a minimal sample from the user. We describe the details in section 3.2.

Next we consider the *User Feedback* process occurs after the records are extracted. Because the number of extracted records is generally large, it is not feasible that the user judges all of them. Therefore it is necessary to lighten the user's work. For the user, records with no or little noise are apt to judge. Thus we sort the extracted records in terms of a reliability measure so that the reliable ones are brought to the top place. What the user has to do is to check only the top ranked ones.

There are five kinds of feedback on the extracted records.

1. Desirable Record: The records have no noise, correct corresponding relationship and related topic.
2. Unrelated Topic: Although the extracted records are the right pairs of valid companies and locations, their topics are not what the user is concerned about. For example, for the user who is interested in the IT information, the (*BMW*, *Munich*) pair is not satisfactory and marked as "Unrelated Topic".
3. Incorrect Tag Recognition: Company names and location descriptions in the extracted records may not be valid due to wrong entity assignment from a named entity tagger. In the experiments described later, we observed some noisy pairs such as (*Com Corp.*, *Santa Clara*), (*Cupertino*, *Calif.*) were also extracted. The named entity tagger should take the blame for the misidentified companies and locations. In the above examples, 3*Com Corp.* was intercepted as *Com Corp.* and *Cupertino* that is really a city name was mistaken for a company name.
4. Wrong Relation: This means although both company and location are valid ones, the location is not the place where the company is located in.
5. Unknown: The user can not judge whether the extracted record is a desired one.

Based on the experiments, we observed that the third and fourth cases rarely appeared in the top ranked records after we sorted them. Therefore, we do not discuss those two cases too much, and pay more attention to the first and second kinds of feedback to select documents for extraction.

Document Selection Module receives the user feedback and selects documents useful for extraction from the document repository. The selected documents are appended to the DSRE set for subsequent record extraction. In section 3.3, we describe four methods of document selection in detail.

3.2 Record Extraction

For record extraction, our approach is based on the bootstrapping approach (Algorithm 1). In this algorithm, the document repository may be considered as static and relatively small. We extend the process of record extraction incorporating the document selection process in next section.

Algorithm 1. Record Extraction Based on Bootstrapping Approach

$Seed$: $Seed\ Record\ Set$
Doc : $Document\ Repository$
$Doc\_tag = attach\_tag(Doc)$
repeat
  $Occ = find\_occurrences(Doc\_tag, Seed)$
  $Pat = generate\_patterns(Occ)$
  $Rec = extract\_records(Doc\_tag, Pat)$
  $Top\_Rec = sort\_records(Rec)$
  $Seed = Seed + Top\_Rec$
until *termination criterion is met*
return Rec

The process flow is as follows.

1. **Providing Seed Records and Attaching Named Entity Tags:** A seed record set (e.g., the example in Fig. 1) is first given by a user. This set should declare the target relation the user wants to obtain and reflect the topic he cares for (e.g., he is interested in IT companies and their locations). As a preprocessing, we can use a named entity tagger to recognize person, organization, location, etc. occurring in the documents.
2. **Finding Occurrences:** Then we find *occurrences* of the records in the seed record set from the document repository. Occurrences are the contexts surrounding the attributes of the records. They are defined as the following style for our example case:

$$(company,\ location,\ o\_prefix,\ tag1,\ o\_middle,\ tag2,\ o\_suffix),$$

   where *company* is a company name and *location* represents its location. For the seed set in Fig. 1, *company* and *location* correspond to $Microsoft$ and *Redmond*, respectively. o_prefix is the context preceding the attribute (*company* or *location*) appearing first, and o_suffix is the context following the last attribute. o_middle is the string between two attributes. $tag1$ and $tag2$ are the named entity tags. For the task in this paper, we pay attention to the ORGANIZATION and LOCATION tags. Other kinds of tags, such as PERSON, are not considered.

3. **Generating Patterns:** Next patterns are generated by analyzing the occurrences. Patterns are defined as the following style:

$$(p\_prefix,\ tag1,\ p\_middle,\ tag2,\ p\_suffix).$$

First the occurrences are partitioned into groups. The occurrences in each group have the same $tag1$, o_middle and $tag2$. If the number of occurrences in a group is less than a threshold, the group is deleted. Patterns are generated for the remaining groups. The $tag1$, p_middle and $tag2$ of a pattern is same with those of the occurrences in the group. For each group, the longest common suffix of all the $o_prefix$es becomes the p_prefix of the pattern, and the longest common prefix of all the $o_suffix$es is the p_suffix of the pattern.
4. **Extracting Records:** Using the generated patterns in the previous step, the document set is scanned again to find new records matching the patterns.
5. **Sorting Records:** For selecting new records to be appended to the seed set and picking up records to receive feedback from the user, we sort the extracted records. Generally the probability that a record has noise is small if it is extracted by multiple patterns. Furthermore the more documents an extracted record appears in, the more reliable it is. Thus we give the order to the extracted records according to their numbers of patterns and numbers of documents where they occur. That is to say, records are first sorted in the descending order of the numbers of patterns extracting them, and in the case that the numbers of patterns are equivalent, the numbers of documents containing them are then considered.

   The top ranked records are used as the new seeds and then a new loop is begun. This repeated procedure terminates until a given condition is met (e.g., convergence, which means no more records can be extracted). In this way, a large amount of records can be obtained starting from a small sample set.

3.3 Document Selection

The previous extraction technique is supposed to examine all the documents in the document repository. When the repository is very large, the extraction process is time consuming. It is also unavoidable to extract undesired records from unrelated documents. Therefore we consider choosing documents for desirable extraction. The process of record extraction combined with document selection is shown in Algorithm 2. The main difference from Algorithm 1 described in Section 3.2 is that the document selection (Step 5-7) is done before the record extraction. At each iteration of the repeated procedure, new documents are chosen and only these documents are passed through a named entity tagger. The same documents are processed by the named entity tagger only once. Receiving feedback from a user (Step 12) is performed only for the third and fourth methods described later, not for the first and second ones. The termination criterion may be that a certain number of selected documents is reached, and based on them the extraction process converges.

Algorithm 2. Record Extraction Incorporating Document Selection

```
Seed : Seed Record Set
Doc : Document Repository
Doc_tag = φ : Tagged Documents Set
repeat
    D = select_documents(Doc)
    D_tag = attach_tag(D)
    Doc_tag = Doc_tag + D_tag
    Occ = find_occurrences(Doc_tag, Seed)
    Pat = generate_patterns(Occ)
    Rec = extract_records(Doc_tag, Pat)
    Top_Rec = sort_records(Rec)
    Top_Rec = review_feedback(Top_Rec)
    {This step may be disregarded for different document selection methods}
    Seed = Seed + Top_Rec
until termination criterion is met
return Rec
```

We assume that the document repository is indexed. Given a query, corresponding documents can be retrieved. For comparison, we present four methods of document selection. In the rest of this section, we discuss how they work.

1. **Baseline:** This simplest method is to choose documents randomly as the target of extraction from the document repository.
2. **Records without Feedback:** This method simply employs the words appearing in the top ranked records as the query. The query is composed of the disjunction of the attribute values of the records. For the extraction example in Fig. 2, the query is "(*Apple* AND *Cupertino*) OR (*Google* AND *Mtn.* AND *View*) OR (*BMW* AND *Munich*) OR (*NEC* AND *Tokyo*)".

Company	Location
Apple	Cupertino
Google	Mtn. View
BMW	Munich
NEC	Tokyo
⋮	⋮

Fig. 2. Extraction Example

Company	Location	Feedback
Apple	Cupertino	Yes
Google	Mtn. View	Yes
BMW	Munich	No
NEC	Tokyo	Yes
⋮	⋮	⋮

Fig. 3. User Feedback

3. **Records with Feedback:** Our experimental observation is that most of the top ranked records used as the query in the previous method are noiseless ones. However the records on different topics sometimes come to the top place. For example, the (BMW, Munich) pair is unsatisfactory for a user who wants to obtain the IT information. If the records on inappropriate topic are popular, many topic-unrelated documents may be retrieved and in turn undesired records are extracted from these documents. In the third

method, we consider eliminating the records on different topics with the help of the user. A user gives his feedback about whether a record is topic-related or not. Only the records judged as good ones by the user are used as the query. For the exaction example in Fig. 3, the query is "($Apple$ AND $Cupertino$) OR ($Google$ AND $Mtn.$ AND $View$) OR (NEC AND $Tokyo$)", where (BMW, $Munich$) is not contained.

4. **Learning:** In the fourth method, the top ranked records also receive the feedback from the user. At the next step unlike the above methods in which the words of records are directly used as the query, we manage to identify feature words that represent the concerned topic as the query. Based on the feedback results, we first select a training document set consisting of relevant documents and irrelevant ones. Then the training document set is used to generate an ordered list of words appearing in the relevant documents. The top ranked words tend to represent the topic that a user is interested in. The disjunction of top k words constitutes the query.
   (a) Selecting relevant and irrelevant documents
   First the documents from which more than one record are extracted are detected as *Relevant Document Candidates* (RDC). Then we assign scores to the documents in the RDC set using the following formula:

$$score(d) = \frac{r + pu}{r + w + u} * \log(r + pu + 1) \tag{1}$$

   where d represents a document in RDC, r is the number of records on right topic, w is the number of records on wrong topic, u is the number of records whose topics a user did not (or could not) decide, and p is the probability that an unjudged record agrees with the concerned topic. For different tasks, p may be given a different value empirically. The top n documents with the highest scores are used as the *relevant documents.* In this way, the relevant documents tend to be the ones from which many desirable records are extracted and the ratios of them to all the extracted records are high. We also randomly select the documents that do not overlap the RDC set as the *irrelevant documents.*
   (b) Learning feature words
   Then each word t in the relevant documents is assigned the Okapi [10] weight:

$$score(t) = \frac{(r_t + 0.5)/(R - r_t + 0.5)}{(n_t - r_t + 0.5)/(N - n_t - R + r_t + 0.5)} \tag{2}$$

   where r_t is the number of relevant documents containing t, n_t is the number of documents containing t, R is the number of relevant documents, and N is the number of documents in both the relevant document set and the irrelevant document set. Intuitively the word t that appears in many relevant documents and rarely appears in the irrelevant documents, can get a high score. We observed that most of the top ranked words were the names of popular companies and their locations, or the words representing the concerned topic in our experiments.

4 Experiments

4.1 Experimental Setting

For our experiments, we use the document repository of Wall Street Journal from 1986 to 1992 consisting of 173,039 documents. For named entity recognition, we use the named entity tagger released by University of Illinois [11]. It can recognize PERSON, LOCATION, ORGANIZATION and MISC. entities in English. For the information retrieval system, we use a full-text search engine *Namazu* [12], that is popular in Japan. Namazu supports a Boolean retrieval model with tf-idf ranking, which simply adds up the tf-idf values of the words appearing in the query for each document as the score of the document.

4.2 Experimental Results

In this section, we introduce two extraction targets on different topic. We experimentally compare the extraction results performed on the documents selected by the four different document selection methods.

Extraction of IT companies and locations: For this target, we assume a user provides five examples as Table 1 shows. We select 5000 documents from the document repository respectively using the four document selection methods. The extraction results are shown in Table 2. From the documents selected by the *baseline* method, few patterns are generated and consequently not so many records are extracted. This is because the probability that randomly selected documents may contain occurrences and patterns of records is relatively small. In contrast, about 10 patterns are generated and about 3000 records are extracted from other three 5000 documents set chosen by other document selection methods.

We sort the extracted records and manually evaluate the top ranked 50 ones. Table 3 shows the evaluation results. The numbers in the first row are those of records without noise among the checked 50 records. The second row represents the numbers of topic-related records among the records without noise. Because the test collection does not exist, the first author investigates introductions of extracted companies and decides whether extracted records are topic-related or not by his subjective judgement. As we can see, less than one half of records has the concerned topic for the *Baseline* and *Records without Feedback* methods, while the ratios of the records on the desirable topic extracted from the documents selected by the *Record with Feedback* and *Learning* methods are much higher. Notice that the document selection methods of *Records with Feedback* and *Learning* require the user's feedback. The top ranked records in the extraction results may also include the ones that have received feedback from the user. After eliminating the records that have been judged midway, the evaluation results are reported in the brackets. For example, after sorting the records extracted from the 5000 documents selected by the *Records with Feedback* method, we pick up the top 50 records that the user did not give his feedback and evaluate them. Among the 50 records, 43 ones have no noise and the 43 records include 33 IT pairs.

Table 1. Example Records for IT Target

IT Company	Location
Xerox	Stamford
Intel	Santa Clara
Apple	Cupertino
Compaq	Houston
Sun	Mountain View

Table 2. The Numbers of Patterns and Records for IT Target

	Baseline	Rec without Fdbk	Rec with Fdbk	Learning
#patterns	2	9	13	10
#records	130	2800	3815	2909

Table 3. Extraction Quality for IT Target

	Baseline	Rec without Fdbk	Rec with Fdbk	Learning
#records without noise	48	47	50 (43)	50 (49)
#records on IT topic	9	20	44 (33)	41 (37)

In summary, *Records without Feedback*, *Records with Feedback* and *Learning* tend to find the documents where more patterns and records can be generated than *Baseline*. With feedback incorporated, *Records with Feedback* and *Learning* help select useful documents and feed them to the extraction system so that more topic-related records are recognized than *Baseline* and *Records without Feedback*.

Extraction of Biotechnology companies and locations: We also do experiments for another topic of biotechnology to further test the generality of different document selection methods' effects. Because the biotechnology topic is not as popular as the IT topic, we limit the number of documents as the extraction target to 1000. For this target, the seed record set (Table 4) also consists of only five pairs of biotechnology companies and locations. Extraction results and qualities are shown in Table 5 and Table 6. As we can see, they have the trends similar to the case of the IT topic.

Discussion: As we can see, the numbers of patterns and records (Table 2 and Table 5) extracted from the documents chosen by the *Records with Feedback* and *Learning* methods are close, and their qualities (Table 3 and Table 6) nearly draw. However, for obtaining the same number of documents, the user's labors that two methods require are quite different. In the IT experiments, for the document selection method of *Records with Feedback*, the user looks through the sorted records from the top ones at each iteration, and totally gives 124 feedback, which includes 75 desirable records and 49 ones with noise or on unrelated topic. The *Records with Feedback* method uses the records as the query. Therefore

Table 4. Example Records for Bio Target

Bio Company	Location
Amgen	Thousand Oaks
Genentech	South San Francisco
Biogen	Cambridge
Chiron	Emeryville
Gilead Sciences	Foster City

Table 5. The Numbers of Patterns and Records for Bio Target

	Baseline	Rec without Fdbk	Rec with Fdbk	Learning
#patterns	2	8	10	10
#records	6	1103	743	714

Table 6. Extraction Quality for Bio Target

	Baseline	Rec without Fdbk	Rec with Fdbk	Learning
#records without noise	6	29	29 (23)	30 (30)
#records on bio topic	2	15	28 (10)	24 (17)

for obtaining a relatively large number of documents, enough judged records are indispensable. In contrast, the *Learning* method uses the feature words representing the concerned topic as the query so that selecting a specific number of documents does not directly depend on the feedback number (i.e., the number of records judged as desirable). For the *Learning* method, the user feedback is only used to choose the relevant documents to construct the training data. The learning result, a ranked word list, is used to generate the query for retrieving documents. The requirement for a larger number of documents can be solved by expanding the query with the disjunction of more feature words, not by checking more records. Actually the feedback number for the *Learning* method is 21, which is even smaller than 124 of *Records with Feedback*, but causes the rivalrous extraction results.

5 Conclusions and Future Work

In this paper, we proposed a record extraction method incorporated with document selection to efficiently acquire topic-related records. We showed the significant improvement of extraction qualities by using feedback to select documents as extraction target.

The current experiments are restricted in the extraction of (company, location) pairs. In our future work, we will make more attempts to extract other relations. It is also an interesting work to put the effect of feedback on pattern and record evaluation. Moreover not only the feedback from a user but also the integration of extraction results with other existent databases is also considerable. Our ongoing research will address these questions.

Acknowledgement

This research is partly supported by the Grant-in-Aid for Scientific Research (16500048) from Japan Society for Promotion of Science (JSPS), the Grant-in-Aid for Scientific Research on Priority Areas (18049005) from the Ministry of Education, Culture, Sports, Science and Technology (MEXT), and the Grant-in-Aid for Core Research for Evolutional Science and Technology (CREST) from Japan Science and Technology Agency (JST). In addition, this work is supported by the grants from Kayamori Foundation of Information Science Advancement, Secom Science and Technology Foundation, and Hoso Bunka Foundation.

References

1. S. Brin, Extracting Patterns and Relations from the World Wide Web. *Proc. WebDB*, 1998.
2. E. Agichtein and L. Gravano, Snowball: Extracting Relations from Large Plain-Text Collections. *Proc. ACM SIGMOD*, 2001.
3. R. Baumgartner, S. Flesca and G. Gottlob, Visual Web Information Extraction with Lixto. *Proc. VLDB*, pp. 119–128, 2001.
4. N. Kushmerick, Wrapper Induction: Efficiency And Expressiveness. *Artificial Intelligence*, Vol. 118, No. 1-2, pp. 15–68, 2000.
5. R. Y. Zhang, L. V.S. Lakshmanan and R. H. Zamar, Extracting Relational Data from HTML Repositories. *SIGKDD Explorations*, 2004.
6. L. Gravano, P. Ipeirotis and M. Sahami, QProber: A System for Automatic Classification of Hidden-web Databases. *ACM Trans. Inf. Syst.*, Vol. 21, No. 1, pp. 1–41, 2003.
7. S. Chakrabarti, K. Punera and M. Subramanyam, Accelerated Focused Crawling through Online Relevance Feedback. *Proc. WWW*, pp. 148–159, 2002.
8. S. Chakrabarti, M. van den Berg, and B. Dom, Focused Crawling: A New Approach to Topic-specific Web Resource Discovery. *Computer Networks*, Vol. 31, No. 11-16, pp. 1623-1640, 1999.
9. E. Agichtein and L. Gravano, Querying Text Databases for Efficient Information Extraction. *Proc. ICDE*, pp. 113–124, 2003.
10. S. E. Robertson, Overview of the Okapi projects. *Journal of the American Society for Information Science*, Vol. 53, No. 1, pp. 3–7, 1997.
11. Named Entity Tagger: `http://l2r.cs.uiuc.edu/~cogcomp/asoftware.php?skey=NE`
12. Namazu: `http://www.namazu.org/index.html.en`

Density Analysis of Winnowing on Non-uniform Distributions

Xiaoming Yu[1,2], Yue Liu[1], and Hongbo Xu[1]

[1] Institute of Computing Technology, Chinese Academy of Sciences, Beijing, P.R. China, 100080
{yuxiaoming,yliu,hbxu}@software.ict.ac.cn
[2] Graduate School, Chinese Academy of Sciences, Beijing, P.R. China, 100039

Abstract. The increasing copies of digital documents make detecting duplicates an important problem. Among the techniques proposed so far, Winnowing fingerprinting algorithm [5] is one of the most efficient. However, the previous density analysis leave the performance of Winnowing unwarranted in real systems, because the assumption of uniformly distributed *k-grams* is far from true in practice. In this paper, an improved density analysis method is introduced. Compared with the previous, our method needs only identically distributed *k-grams* to get the prediction. This means our theoretical result can be safely used on highly non-uniformly distributed data which are common in real systems. Extensive experiments are performed on both artificial data and real data. The experiment results agree with the theoretical predictions well.

1 Introduction

For kinds of reasons, digital documents are copied completely or partially. Web sites are mirrored. Some students plagiarize their homework or papers from the Web. The increasing copies make detecting duplicates among a set of digital documents an important problem.

Full copies of documents can be easily detected by comparing document checksums. As to partial copy detection, it is not trivial. Many techniques have been proposed to address the problem, such as [1,2,3,4]. Most of them use the following idea: Store small sketch of document, i.e. fingerprints of document, so that by comparing the fingerprints between two documents, copies can be identified. In these methods, it is vital for fingerprinting algorithms to select representative fingerprints, because they have to find copies without knowing, in advance, which documents and which parts of document are involved. It is also important for fingerprinting algorithms to have high performance, i.e. to select as small fingerprints as possible, if recall that there are usually a great deal of documents to be compared with in a real system.

Among fingerprinting algorithms proposed so far, Winnowing [5] is one of the most efficient. Winnowing is a fingerprinting algorithm based on *k-grams* and selects some hashes of *k-gram* as fingerprints of document. Compared with

G. Dong et al. (Eds.): APWeb/WAIM 2007, LNCS 4505, pp. 586–593, 2007.

other fingerprinting algorithms, it provides guarantee to detect copies longer than a user predefined length. This makes fingerprints selected by Winnowing representative, and thus makes the detection results more reliable.

In order to measure performance of Winnowing, *density*, which is defined by (1), is used:

$$Density = E\left[\frac{Number\ of\ Hashes\ Selected}{Number\ of\ k\text{-}grams}\right] \quad (1)$$

where E[·] denotes expectation over distribution of *k-grams*. The previous density analysis of Winnowing needs uniformly distributed *k-grams* [5]. But the assumption of uniformity seems too strong for real data. Instead, more and more researchers report highly non-uniform distributions [5,6,7,8,9]. The gap leaves behavior of Winnowing unwarranted in real systems and separates theory from reality. From the view of practice, it would be valuable to predict behavior of Winnowing on highly non-uniformly distributed data effectively. Using the prediction, copy detection systems can be optimized. So the questions whether we can theoretically predict the density of Winnowing on highly non-uniformly distributed data and the above prediction, if any, can work on real data, may be asked. We will answer these questions in this paper. Our contributions are the following:

- We carry out a careful theoretical study that expands the applicability of density analysis of Winnowing from uniform to identical distribution. The method can be safely used on highly non-uniformly distributed data.
- We verify the theoretical results with extensive experiments using both artificial data and real data. The experimental results agree well with the theoretical predictions.

The rest of the paper is organized as follows. The next section describes background and related work. In section 3, we provide our theoretical density analysis method of Winnowing. In Section 4, experimental results which verify the theoretical findings are given. Section 5 is concludes.

2 Background and Related Work

Winnowing is a fingerprinting algorithm based on *k-grams*. A *k-gram* is a contiguous character sequence of length k. Distinct *k-grams* may overlap, and there are almost as many *k-grams* as characters in document. Fig. 1(a-c) gives an example.

In Winnowing, there are two user specified parameters, k and $t\,(k \leqslant t)$, to control the process of fingerprinting. Parameter k is called *noise threshold.* Any duplicate shorter than k will not be detected. Parameter t is called *guarantee threshold.* Any duplicate longer than t is guaranteed to be found. Winnowing avoid matching duplicates shorter than *noise threshold* by considering hashes of *k-gram.* Many hash functions can be used here, such as MD5 [10]. But the most popular is Rabin's algorithm [11], because of it's computational efficiency. Fig. 1(d) shows hashes of *7-grams* in Fig. 1(c), calculated by a hypothetical hash function. To provide the detection guarantee, Winnowing uses the idea

`Mickey Mouse And Donacdduck`
(a) Some Text

`mickeymouseanddonacdduck`
(b) Canonical Form of the Text

`mickeym ickeymo ckeymou keymous eymouse ymousea mousean ouseand useandd`
`seanddo eanddon anddona nddonac ddonacd donacdd onacddu nacdduc acdduck`
(c) 7-Grams Derived from the Text

`17 42 98 50 43 77 58 24 54 90 25 64 10 37 80 22 14 60`
(d) Hypothetical Hashes of the 7-Grams

`(17 42 98 50)(42 98 50 43)(98 50 43 77)(50 43 77 58)(43 77 58 24)`
`(77 58 24 54)(58 24 54 90)(24 54 90 25)(54 90 25 64)(90 25 64 10)`
`(25 64 10 37)(64 10 37 80)(10 37 80 22)(37 80 22 14)(80 22 14 60)`
(e) Windows of Size 4

`17 42 43 24 25 10 14`
(f) Fingerprints Selected by Winnowing

Fig. 1. Fingerprinting Some Text Using Winnowing

of *local algorithms*, which select fingerprints depending only on the contents of a local window. The *local* property guarantees that the same fingerprints are selected no matter where window appears. Specifically, in Winnowing, a sliding window of size w is used, where w=$t-k+1$, and window of size w is defined to be w contiguous hashes. As the window starting from the beginning of document slides hash by hash, Winnowing selects the minimum hash in each window (If there are more than one minimum hashes, select the rightmost). This means that, for any contiguous t characters in document, at least one fingerprint is selected, and duplicates containing them can be detected. Note that the same hash may be selected from adjacent windows, only distinct hashes are stored as fingerprints. This process is illustrated in Fig. 1(e-f).

Assume independent and uniformly distributed input *k-grams*. The density of Winnowing can be associated with w, the size of windows. In this case, the density of Winnowing is $\frac{2}{w+1}$, given that the possibility of multiple minimum hashes appear in a small window can be ignored [5].

Uniform distribution simplifies analysis, but the assumption seems so strong that real data collections can seldom satisfy. Instead, many researchers report highly non-uniform distributions of real data. Zipf observed that frequency of occurrence of words in English documents, as a function of the rank when the rank is determined by the above frequency of occurrence, is a power-law function with the exponent close to -1 [7]. According to [8,9], other language and language unit can also be charactered by Zipf's Law. Furthermore, power-law relationship between frequency and rank is observed in data collections consisting of Web pages [5], contents of packets traveling through networks [6], and so on. It seems that, in real data collections, power law phenomenon, which means the distribution is far from uniform, is pervasive. These observations leaves the performance of Winnowing in practice unwarranted. Designers of copy detection system are not sure how Winnowing will behave in their systems.

3 Density Analysis of Winnowing

In this section, our theoretical density analysis of Winnowing will be given. Now, lets' introduce Lemma 1 as start.

Lemma 1. *Assume that* $h_1, h_2, \cdots, h_n$ *are independent and identically distributed random variables, whose sample space are* $\mathbb{Z}$*, a finite subset of integer. Then* $P_{min}(h_1) = P_{min}(h_2) = \cdots = P_{min}(h_n)$*, where* $P_{min}(h_i)$ *denotes the probability that* h_i *takes the smallest value within sample point in joint sample space* $\mathbb{Z}^n$*.*

Proof. Define relation $\mathfrak{R}$ on $\mathbb{Z}^n$: For sample points $\hat{h} = (\hat{h_1}, \hat{h_2}, \cdots, \hat{h_n})$ and $\check{h} = (\check{h_1}, \check{h_2}, \cdots, \check{h_n})$, $\hat{h}\ \mathfrak{R}\ \check{h}$ iff increasingly order integers in $\hat{h}$ and $\check{h}$ individually and get the same integer sequence. It is obvious that $\mathfrak{R}$ is an equivalence relation. So $\mathbb{Z}^n$ can be divided into equivalence classes $E_1, E_2, \cdots, E_Q$ by $\mathfrak{R}$, where Q is the number of equivalence classes.

Let $S_{m,j}$ be a set of sample points in $\mathbb{Z}^n$, $\bar{h} \in S_{m,j}$ iff $\bar{h} \in E_m$ and h_j takes the smallest value within $\bar{h}$. Given integer $m \in [1, Q]$, we say $|S_{m,i}| = |S_{m,j}|$ for any $1 \leqslant i \leqslant n$ and $1 \leqslant j \leqslant n$, where $|S_{m,i}|$ and $|S_{m,j}|$ denote cardinality of $S_{m,i}$ and $S_{m,j}$ individually. To prove this, assume $\hat{h} = (\hat{h_1}, \cdots, \hat{h_i}, \cdots, \hat{h_j}, \cdots, \hat{h_n}) \in S_{m,i}$. By definition of $S_{m,i}$, $\hat{h_i}$ must be the smallest integer within $\hat{h}$. Exchange $\hat{h_i}$ and $\hat{h_j}$, we get a new sample point $\hat{h}' = (\hat{h_1}, \cdots, \hat{h_j}, \cdots, \hat{h_i}, \cdots, \hat{h_n})$. It is obvious that $\hat{h}' \in S_{m,j}$. Given two distinct sample points $\hat{h} \in S_{m,i}$ and $\check{h} \in S_{m,i}$, after the above exchange, we get two new sample points $\hat{h}'$ and $\check{h}'$. $\hat{h}'$ and $\check{h}'$ must be distinct, otherwise $\hat{h}$ and $\check{h}$ are the same. Thus, $|S_{m,i}| \leqslant |S_{m,j}|$. Because the selection of i and j is arbitrary, we get $|S_{m,i}| \geqslant |S_{m,j}|$ in the same way. Therefore, $|S_{m,i}| = |S_{m,j}|$, and furthermore $|S_{m,1}| = |S_{m,2}| = \cdots = |S_{m,n}|$.

Consider the definition of E_m and the independent and identically distributed random variables. In any given E_m, the probability that each sample point occurs is the same. Recall $|S_{m,1}| = |S_{m,2}| = \cdots = |S_{m,n}|$, we get $P(S_{m,1}) = P(S_{m,2}) = \cdots = P(S_{m,n})$ for any E_m. Consider $P_{min}(h_i) = \sum_{m=1}^{Q} P(S_{m,i})$. It follows that $P_{min}(h_1) = P_{min}(h_2) = \cdots = P_{min}(h_n)$. □

Theorem 1. *Given input hashes* $h_1, h_2, \cdots, h_n, \cdots$*, if the hashes are independently and identically distributed, the density of Winnowing is* $\frac{2}{w+1}$*, provided that the probability that there are more than one smallest hash among contiguous* $w + 1$ *input hashes is small enough to be ignored.*

Proof. Consider the function FS that maps the position of window, which is defined to be the position of leftmost hash in it, to the position of fingerprint the window selected. We say function FS is monotonic increasing, namely if $i < j$, then $FS(i) \leqslant FS(j)$. To prove this, consider two cases. If windows W_i and W_j do not overlap, $FS(i)$ is less than the position of any hash in W_j. So $FS(i) \leqslant FS(j)$. If W_i and W_j overlap, denote $W_i = (h_i, h_{i+1}, \cdots, h_j, \cdots, h_{i+w-1})$ and $W_j = (h_j, h_{j+1}, \cdots, h_{i+w-1}, \cdots, h_{j+w-1})$, where $i < j \leqslant i + w - 1$. Then the maximum value of $FS(i)$ is q, where q is the position of minimum hash among $h_j, \cdots, h_{i+w-1}$; On the contrary, the minimum value of $FS(j)$ is also q. This means $FS(i) \leqslant FS(j)$. Thus the function FS is monotonic increasing.

Consider an indicator random variable X_i that is one iff W_i selects a fingerprint which is not selected by any previous window. Consider two contiguous windows W_{i-1} and W_i. The two windows overlap except the leftmost hash h_{i-1} and the rightmost hash h_{i+w-1}. Consider the position p containing the smallest hash in $h_{i-1}, h_i, \cdots, h_{i+w-1}$. There are three cases:

a). If $p = i - 1$, then W_{i-1} selects it and W_i must select another position q, where $q > p$. Because of the monotonicity of function FS, W_i must be the first window that selects q. Thus, $X_i = 1$.

b). If $p = i + w - 1$, then W_i selects it. Because W_i is the first window containing p, it must also be the first window that selects it. Again, $X_i = 1$.

c). If $i - 1 < p < i + w - 1$, both W_{i-1} and W_i select it. So W_i cannot be the first window that selects p. Therefore, $X_i = 0$.

The first two cases happen with probability $P_{\min}(h_{i-1})$ and $P_{\min}(h_{i+w-1})$, in interval $h_{i-1}, h_i, \cdots, h_{i+w-1}$, individually. In this interval, according to Lemma 1, we have $P_{\min}(h_{i-1}) = P_{\min}(h_i) = \cdots = P_{\min}(h_{i+w-1})$. We also have $P_{\min}(h_{i-1}) + P_{\min}(h_i) + \cdots + P_{\min}(h_{i+w-1}) = 1$, because the probability that there are more than one smallest hash in contiguous $w + 1$ hashes can be ignored. The above equations imply $P_{\min}(h_{i-1}) = P_{\min}(h_{i+w-1}) = \frac{1}{w+1}$. So the expected value of X_i is $\frac{2}{w+1}$. Recall that the sum of the expected values is the expected value of the sum, even if the random variables are not independent. Therefore, we get the density. □

4 Experiments

4.1 Experiments with Artificial Data

In this set of experiments, we use artificial data which are randomly generated hashes. The purpose of the experiments is to verify our method on non-uniformly distributed data in ideal cases.

In the experiments, normal, exponential and uniform distributions are used. All hashes involved are generated by pseudo-random number generators independently. Every experiment here is carried out in the following way: Generate hash sequence of length 10^8; Perform Winnowing fingerprinting with commonly used window sizes; Repeat 100 times and get the average. The experiment results are shown in Table 1. The first column of the table gives sizes of window used by Winnowing. The second to the fourth columns show relative deviations of observed densities from theoretical predictions on uniform, normal and exponential distributions individually. Relative deviations are calculated by:

$$Deviation = \left| \frac{Observation - Prediction}{Prediction} \right| \tag{2}$$

From the table, we can see the deviations are quite small in all cases. Note that, this is true not only on uniform distribution, but also on non-uniform distributions. This result fits our theoretical predictions perfectly, and which in

Table 1. Relative Deviations on Randomly Generated Data

Window Size	Uniform	Normal	Exponential
50	1.91699^{-5}	5.23020^{-5}	3.12212^{-5}
100	2.25486^{-5}	2.50780^{-5}	4.99442^{-5}
150	1.02097^{-4}	5.76124^{-5}	5.70462^{-5}
200	3.23767^{-5}	9.26162^{-6}	6.80743^{-5}
250	1.40780^{-4}	5.82259^{-5}	1.52545^{-5}
300	5.73415^{-5}	5.95974^{-5}	5.01325^{-5}
350	4.38731^{-5}	7.86397^{-5}	9.55797^{-5}
400	3.23351^{-5}	4.49667^{-5}	8.58287^{-5}
450	3.56767^{-5}	1.42965^{-4}	8.60762^{-5}
500	9.30560^{-6}	2.47866^{-5}	3.23277^{-5}

turn implies that performance of Winnowing on non-uniform distributions can be predicted by our method.

According to our theorem, no significant performance difference should be observed between uniformly and non-uniformly distributed hashes. In order to verify this conclusion, paired *t-tests* are conducted. We calculate *t-statistics* between densities of 100 runs on uniformly distributed hashes and densities of 100 runs on non-uniformly distributed hashes, etc. normally and exponentially distributed hashes. If set significant level to be 0.05, the hypotheses that no significant differences occur can be accepted safely. (We don't report the calculated *t-statistics* just for space constraint).

4.2 Experiments with Real Data

TDT-5(Topic Detection and Tracking) [13] is chosen as test collection in this set of experiments (We also conduct experiments on Reuters-21578, Request for Comments (RFC) documents and TDT-4, but don't report the results for space constraint). TDT-5 is used for evaluation of topic detection and tracking. The sources of it include radio and television broadcasts as well as newswire services. TDT-5 is a multilingual collection. We only use all English documents of it, about 600MB in size, for only a simple preprocessor of documents is employed in our experiments.

Before the experiments, we are interested in how *k-grams* in TDT-5 are distributed. We elaborately compute frequency of every *k-gram* occurring in TDT-5 for distinct k, i.e. 16, 32, 48 and 64; Then sort *k-grams* according to frequency in monotonically decreasing order; Take the position of a *k-gram* in the ordered list as it's rank. We plot the frequency-rank relations in log-log scale, as what is shown in Fig. 2 (If more than one *k-grams* have the same frequency, only the first *k-gram* and the last *k-gram* of that frequency in the ordered list are plotted, and the two points are linked by a dashed). From the figure, highly non-uniformly distributed *k-grams* are observed. The distributions of *k-grams* vary greatly with different values of k. For all values of k, power law relation between frequency

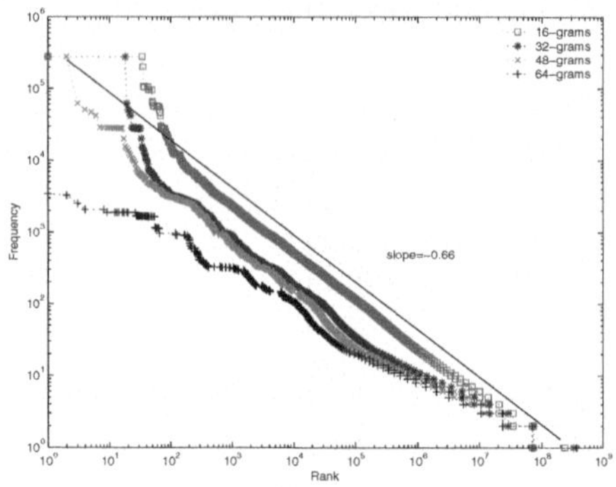

Fig. 2. Log-log Plot of Frequency-Rank for All *k-Grams* on TDT-5

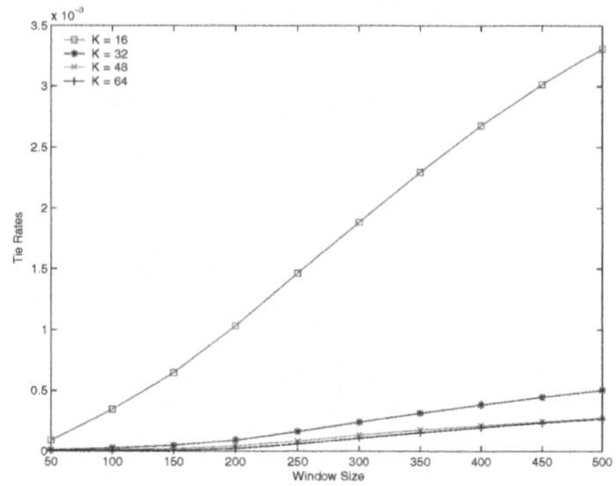

Fig. 3. Tie Window Rates of TDT-5

and rank is exhibited. At the same time, one can find that smaller k makes *k-gram* of the same rank higher frequency. This is because small k tends to make *k-grams* repeated frequently.

As in previous section, we calculate relative density deviations for different k and different window size. The results are listed in Table 2. From the tables, we find the experimental results agree with our theoretical predictions well. Compared with artificial data, relatively large deviations are found. One reason may be the assumption of independent and identical distribution of all *k-grams*. In real documents, *k-grams* are not completely independently and identically distributed. For example, there are idioms of language, and there are some relationships and differences among different paragraphs of document.

Another factor that may influence the densities is the tie hashes. This is because in our theorem the probability that tie minimum hashes occur should be small. We count the number of windows, size of $w + 1$, within which tie minimum hashes are observed. The results are illustrated in Fig. 3. In the figure, we plot rate of window that tie occur as function of window size. From the

Table 2. Relative Density Deviations of TDT-5

Window Size	$K = 16$	$K = 32$	$K = 48$	$K = 64$
50	1.18707^{-3}	1.18934^{-3}	1.73217^{-4}	2.76092^{-4}
100	8.10199^{-5}	1.51451^{-3}	4.25587^{-4}	7.50228^{-4}
150	2.65735^{-4}	1.05881^{-3}	6.29951^{-4}	8.57446^{-4}
200	5.56679^{-4}	1.23981^{-3}	7.90873^{-4}	7.40191^{-4}
250	3.02702^{-4}	1.29101^{-3}	5.84161^{-4}	8.08966^{-4}
300	2.09958^{-5}	1.02671^{-3}	6.71869^{-4}	1.03020^{-3}
350	1.40222^{-4}	1.49041^{-3}	6.30496^{-4}	1.49034^{-3}
400	3.72315^{-4}	1.84705^{-3}	8.83788^{-4}	1.39691^{-3}
450	4.75044^{-4}	1.92255^{-3}	1.42308^{-3}	1.16798^{-3}
500	1.01456^{-3}	1.87038^{-3}	1.99081^{-3}	1.18230^{-3}

figure, We find something interesting. First, the rates of tie window increase with increasing window sizes. This may be caused by the reason that the same tie hashes appear in more windows if window size is large. Another interesting thing is that the rates are sensitive to the value of k. When k increases, the rates are decreased in general. If $K > 32$, the influence of k is relatively small. The reason may be that larger k makes the probability of tie smaller. These findings suggest something valuable: From the point of tie hashes, small size of window with large k trends to make the theoretical prediction of density precise.

5 Conclusion

In this paper, an improved density analysis method of Winnowing is proposed. The new method removes the strong assumption of uniformly distributed *k-grams* from the previous. Instead, only identical distribution is required. Using the weaker assumption, our method can be safely used on highly non-uniformly distributed data which seems common in practice. Finally, we report experiments on artificial data and real data. The experiment results verify our theoretical findings.

References

1. Sergey Brin, James Davis, Hector Garcia-Molina. Copy Detection Mechanisms for Digital Documents. In Proceedings of ACM SIGMOD 1995, pages 398-409.
2. Udi Manber. Finding Similar Files in a Large File System. In Proceedings of Winter USENIX Conference 1994, pages 1-10.
3. George Forman, Kave Eshghi, Stephane Chiocchetti. Finding similar files in large document repositories. In Proceedings of ACM SIGKDD 2005, pages 394-400.
4. Nevin Heintze. Scalable Document Fingerprinting. In Proceedings of the Second USENIX Workshop on Electronic Commerce, 1996, pages 191-200.
5. Saul Schleimer, Daniel S. Wilkerson, Alex Aiken. Winnowing : Local Algorithms for Document Fingerprinting. In Proceedings of ACM SIGMOD 2003, pages 76-85.
6. Sumeet Singh, Cristian Estan, George Varghese, Stefan Savage. The EarlyBird System for Real-time Detection of Unknown Worms. Technical Report CS2003-0761, University of California, San Diego, 2003.
7. George K. Zipf. The Psychobiology of Language. Houghton Mifflin, Boston, 1935.
8. Guan Yi, Wang Xiaolong, Zhang Kai. The Frequency-Rank Relation of Language Units in Modern Chinese Computational Language Model. Journal of Chinese Information Processing, 1999, 13(2): 8-15.
9. G.A. Miller, E.B. Newman. Tests of Statistical Explanation of the Rank-Frequency Relation for Words in Writen English. The American Journal of Psychology, 1958, 71(1): 209-218.
10. RFC1321. The MD5 Message-Digest Algorithm.
11. M.O. Rabin. Fingerprinting by Random Polynomials. Technical Report TR-15-81, Harvard Aiken Computation Laboratory, 1981.
12. William H. Press, Saul A. Teukolsky, William T. Vetterling, and Brian P. Flannery. Numerical Recipies in C : the Art of Scientific Computing. Cambridge University Press, 2nd Edition, 1992.
13. TDT. http://projects.ldc.upenn.edu/TDT/

Error-Based Collaborative Filtering Algorithm for Top-N Recommendation

Heung-Nam Kim[1], Ae-Ttie Ji[1], Hyun-Jun Kim[2], and Geun-Sik Jo[3]

[1] Intelligent E-Commerce Systems Lab., Inha University, Incheon, Korea
{nami,aerry13}@eslab.inha.ac.kr
[2] Samsung Electronics, Corporate Technology Operations, R&D IT Infra Group
dannis@eslab.inha.ac.kr
[3] School of Computer Science & Engineering, Inha University, Incheon, Korea
gsjo@inha.ac.kr

Abstract. Collaborative Filtering recommender system, one of the most representative systems for personalized recommendations in E-commerce, is a system assisting users in easily finding useful information. However, traditional collaborative filtering systems are typically unable to make good quality recommendations in the situation where users have presented few opinions; this is known as the cold start problem. In addition, the existing systems suffer some weaknesses with regard to quality evaluation: the sparsity of the data and scalability problem. To address these issues, we present a novel approach to provide enhanced recommendation quality supporting incremental updating of a model through the use of explicit user feedback. A model-based approach is employed to overcome the sparsity and scalability problems. The proposed approach first identifies errors of prior predictions and subsequently constructs a model, namely the *user-item error matrix*, for recommendations. An experimental evaluation on *MovieLens* datasets shows that the proposed method offers significant advantages both in terms of improving the recommendation quality and in dealing with cold start users.

1 Introduction

With the explosive growth of the Internet, recommender systems have been proposed as a solution for the problem of information overload. Recommender systems assist users in finding the information most relevant to their preferences [11]. One of the most successful technologies among recommender systems is Collaborative Filtering (CF). Numerous commercial systems (e.g. amazon.com, half.com, cdnow.com) apply this technology to serve recommendations to their customers. The traditional task in CF is to predict the utility of a certain item for the target user (often called an active user) from the user's previous preferences or the opinions of other similar users, and thereby make appropriate recommendations [2]. However, despite its success and popularity, traditional CF suffers from several problems, including sparsity, scalability, cold start, shilling problem. A number of studies have attempted to address the problems related to collaborative filtering [2, 5, 6, 7, 10, 14].

G. Dong et al. (Eds.): APWeb/WAIM 2007, LNCS 4505, pp. 594–605, 2007.

One notable challenge in a CF is the cold start problem, which can be divided into *cold start items* and *cold start users* [13]. The *cold start user*, the focus of the present research, describes a new user that joins a CF-based recommender system and has not yet expressed ratings (i.e., the user has no purchase history). With this situation, the system cannot generate predictions or provide recommendations for the user. In addition, the system is generally unable to make high quality recommendations when users have presented few opinions [14]. In this paper, we present a novel approach to provide enhanced recommendation quality derived from explicit ratings. The main objective of this research is to develop an effective approach that provides high-quality predictions and recommendations even when users have little rating information. A model-based approach is employed to overcome the sparsity and scalability problems [2, 12]. The proposed approach first determines similarities between the items and subsequently identifies errors of past predictions, which are used in the process of online predictions and recommendations. This paper also presents a method of updating the model incrementally, collectively called *a user-item error matrix*, from explicit user feedback.

The subsequent sections of this paper are organized as follows: The next section contains a brief overview of some related studies. In section 3, the approach for CF, an error-based CF algorithm, is described. An experimental evaluation is presented in section 4. Finally, conclusions and future works are presented.

2 Related Work

This section briefly explains previous studies related to CF-based recommender systems. The various approaches developed in this area can be divided into two classes: Memory-based CF (also known as *nearest-neighbor* or *user-based* CF) and Model-based CF [1]. Following the proposal of GroupLens [3], the first system to generate automated recommendations, user-based approaches have seen the widest use in recommendation systems. User-based CF uses a similarity measurement between neighbors and the target users to learn and predict the target user's preferences for new items or unrated products. Though user-based CF algorithms tend to produce more accurate recommendations, they have some serious problems stemming from the complexity of computing each recommendation as the number of users and items grow. In order to improve scalability and real-time performance in large applications, a variety of model-based recommendation techniques have been developed [2, 3, 12]. In particular, a new class of item-based CF, which is one of model-based CF approaches, has been proposed. This model-based approach, which is the focus of the present work, provides item recommendations by first developing a model of user ratings. In comparison to user-based approaches, item-based CF is typically faster in terms of recommendation time, although the method may require expensive learning or a model building process [4]. Instead of computing similarities between users, item-based CF reviews a set of items rated by the target user and selects k most similar items, based on the similarities between the items. Sarwar et al. [2] evaluated various methods to compute similarity and approaches to limit the set of item-to-item similarities that must be considered. Deshpande et al. [5] proposed item-based top-N recommendation algorithms that are similar to previous item-based schemes. They separated the algorithms into two distinct parts for building a model of item-to-item similarities and deriving top-N recommendations using this pre-computed model. While item-based CF algorithms are effective, they still have some weaknesses with regard to cold start users and ratings of malicious users. Hence, a number of recent research efforts have focused on the use of trust concepts during the recommendation process [7, 14]. In addition, distributed recommender systems have been proposed to deal with the existing weaknesses [10, 12].

3 Error-Based Collaborative Filtering Algorithm

Fig. 1 provides a brief overview of the proposed approach. The proposed method is divided into two phases, an offline phase and an online phase. The offline phase is a model building phase to support fast online predictions and the online phase is either a prediction or recommendation phase.

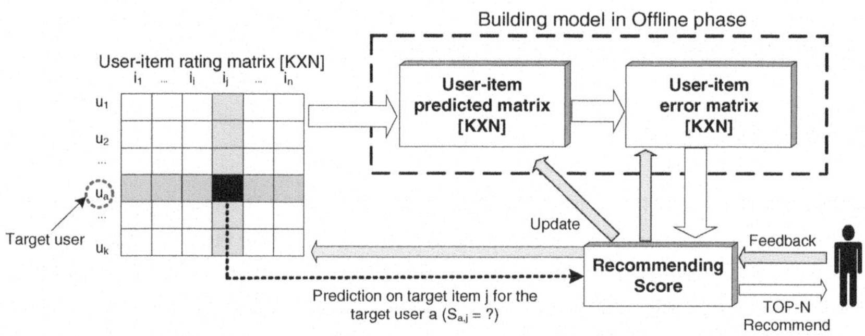

Fig. 1. Overview of the proposed method for generating predictions and updating models

3.1 Building an Error Matrix

Before describing the algorithms, some definitions of the matrices are introduced.

Definition 1 (User-item actual rating matrix, A). If there is a list of k users $U=\{u_1,u_2,\ldots,u_k\}$, a list of n items $I=\{i_1,i_2,\ldots i_n\}$ mapping between user-item pairs, and explicit ratings, $k \times n$ user-item data can be represented as a rating matrix. This matrix is called a *User-item actual rating matrix*, *A*. The matrix rows represent users, the columns represent items, and $A_{u,j}$ represents the rating of a user u on item j. Some of the entries are not filled, as there are items not rated by some users.

Definition 2 (User-item predicted rating matrix, P). This is a matrix of users and items that have predicted values for users on items. From matrix A, the system can predict $P_{u,j}$ for a given target item j that **has already been rated** by target user u. This matrix is called a *User-item predicted rating matrix*, *P*. The matrix rows also represent users, the columns represent items.

Definition 3 (User-item error matrix, E). From the given set of actual and predicted rating pairs $<A_{u,j}, P_{u,j}>$ for all the data in matrices *A* and *P*, a *User-item error matrix*, *E*, can be filled as errors, which can be computed by subtracting the predicted ratings for users on items from the actual rating for users on items. Each entry $E_{u,j}$ in *E* represents the prediction error of the *u*th user on the *j*th item. Some of the entries are not filled, as there are items not rated by some users.

For constructing matrix *E*, firstly the user's rating should be predicted for an item that has already been rated. To this end, an item-based prediction measure is used as presented in equation (1) [7]. The prior prediction for the target user *u* on item *i*, $P_{u,i}$, is obtained as follows:

$$P_{u,i} = \overline{A_i} + \frac{\sum_{j \in MSI(u)} (A_{u,j} - \overline{A_j}) \cdot sim(i,j)}{\sum_{j \in MSI(u)} | sim(i,j) |} \tag{1}$$

where *MSI(u)* is the set of *k* most similar items to the target item *i* among items rated by the user *u* and $A_{u,j}$ is the rating of user *u* on item *j*. In addition, $\overline{A_i}$ and $\overline{A_j}$ refer to the average rating of item *i* and *j*. *sim(i, j)* represents the similarity between items *i* and *j*, which can be calculated using diverse similarity algorithms such as cosine-based similarity, correlation-based similarity, and adjusted cosine similarity [2]. However, we also consider the number of users' ratings for items in generating item-to-item similarities, namely *the inverse item frequency*.

When *the inverse item frequency* is applied to the cosine similarity technique, the similarity between two items, *i* and *j* is measured by equation (2).

$$Sim(i,j) = \frac{\sum_{u \in User} (A_{u,i} \cdot f_u)(A_{u,j} \cdot f_u)}{\sqrt{\sum_{u \in User} (A_{u,i} \cdot f_u)^2} \sqrt{\sum_{u \in User} (A_{u,j} \cdot f_u)^2}} \tag{2}$$

where *User* is a set of users who both rated *i* and *j*, $A_{u,i}$ is the rating of user *u* on item *i*, and $A_{u,j}$ is the rating of user *u* on item *j*. The inverse item frequency of user *u*, f_u is defined as $log(n/n_u)$, where n_u is the number of items rated by user *u* and *n* is the total number of items in the database. If user *u* rated all items, then the value of f_u is 0. Likewise, the inverse user frequency, the main concept of the inverse item frequency dictates that users rating numerous items present less contribution with regard to similarity than users rating a smaller number of items [7].

Once the predictions for users on items are represented on a user-item predicted rating matrix, the error of each prediction can be computed for constructing a user-item error matrix. Given the set of actual and predicted rating pairs $<A_{u,j}, P_{u,j}>$ for all data in the user-item matrices, the prediction error $E_{u,j}$ is calculated as:

$$E_{u,j} = A_{u,j} - P_{u,j}$$

Fig. 2 illustrates the process of a *user-item error matrix* construction.

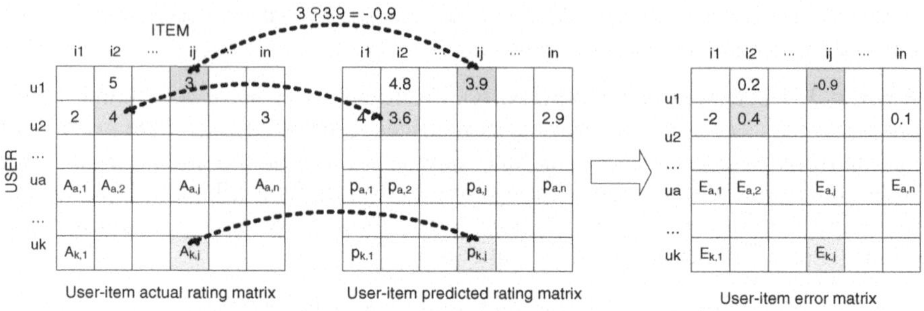

Fig. 2. Process of a *user-item error matrix* construction. The prediction error can be calculated by subtracting the predicted rating from the actual rating.

3.2 Prediction Computation Using Prior Prediction Errors

As noted previously, the proposed CF approach constructs an error model (*an Error Matrix, E*), which can be accomplished offline, prior to online prediction or recommendation. Since most tasks can be conducted in the offline phase, this approach can result in fast online performance. In addition, model-based approach assists in solving the sparsity and scalability problems [2, 5]. The proposed method also provides another advantage, the ability to overcome the cold start users.

The most important task in CF is to generate a prediction, this is, attempting to speculate upon the rating that a user would provide for an item [2]. In order to compute the recommending score[1] of the target user *u* for the target item *i* that **has not yet been rated** by user *u*, the sum of the prior prediction errors of user *u* are used as defined in equation (3).

$$S_{u,i} = \overline{A_i} + \frac{\sum_{j \in LEI(u)} E_{u,j}}{|LEI(u)|} \tag{3}$$

where *LEI(u)* is a set of Lowest Error Items that a user *u* rated and $E_{u,j}$ is the prior prediction error of user *u* on item *j*. In addition, $\overline{A_i}$ refers to the average rating of the items *i*.

From the simple example given in Table 1 and Table 2, suppose the system is trying to compute how much *Alice* will prefer *Titanic*. In the case of $\varepsilon = 0.4$, a set of Lowest Error Items on *Alice*, LEI(Alice) = {Seven, Toy Story, Matrix}. Therefore, we can calculate the score of *Alice* for *Titanic* as follows: $S_{Alice,\ Titanic} = 3 + (-0.3-0.4+0.1)/3 = 2.8$. Likewise, the score of *John* for *Toy Story*, $S_{John,\ Toy\ Story} = 2.5 + (-0.1 + 0.2 - 0.4 - 0.3)/4 = 2.35$.

Once all items that the target user *u* has not yet rated are predicted, the items are sorted in the order of descending the recommending score. Thereafter, a set of *N* items that have obtained higher recommending scores are identified for user *u* (*top*-N recommendation).

[1] In collaborative filtering research literature, "predicted rating" is more commonly used. This term, however, is somewhat confusing as the term "prior predicted rating" in Definition 2. Hence, we use the term "recommending score."

Table 1. An example of a user-item actual rating matrix, A

	Seven	Star Wars	Titanic	Toy Story	AI	Matrix
Alice	3	5	?	1	?	5
Bob	4	1	5	4	3	?
John	2	2	1	?	4	4
Item Avg.	3	2.67	3	2.5	3.5	4.5

Table 2. An example of a user-item error matrix, E

	Seven	Star Wars	Titanic	Toy Story	AI	Matrix
Alice	-0.3	0.8		-0.4		0.1
Bob	0.1	-2	0	2	-0.02	
John	-0.1	0.2	-0.4		2	-0.3

Definition 4 (Lowest Error Item). Given a user-item error matrix E and a set of items I_u that have been rated by target user u, item j is deemed the lowest error item of user u if and only if the absolute error of the uth user on item $j \in I_u$, $|E_{u,j}|$, is within a predetermined error value ε ($|E_{u,j}| \leq \varepsilon$).

Definition 5 (Top-N recommendation). Let I be a set of all items, I_u an item list that user u has already expressed his opinions about, and IP_u an item list that user u has not yet rated, $IP_u = I - I_u$ and $I_u \cap IP_u = \varnothing$. Given two items i and j, $i \in IP_u$ $j \in IP_u$, item i will be of more interest to user u than item j if and only if the recommending score $S_{u,i}$ of item i is higher than that of item j, $S_{u,i} > S_{u,j}$. Top-N recommendation is a ordered set of N items, $TopN_u$, that will be of interest to user u, $|TopN_u| \leq$ N and $TopN_u \subseteq IP_u$.

3.3 Incremental Updates of the Error Matrix

In comparison to user-based approaches, model-based approaches, by using pre-computed models in the offline phase, are typically faster in terms of recommendation time. In addition, these approaches help alleviate the sparsity and scalability problems. However, model-based CF tends to require expensive learning time for building a model [4]. Moreover, once the model is built, it is difficult to reflect user feedback despite its significance in the recommender system.

In order to alleviate the weak points of model-based CF, the proposed approach is designed such that the model is updated effectively and users' new opinions are reflected incrementally, even when users present explicit feedback. As illustrated in Fig. 3, the target user a can provide explicit feedback about the target item j, which the system recommended before according it a recommending score. Subsequently, the model, an error matrix E, can easily update the error value, which is computed by subtracting the score from the rating of feedback rating. Therefore, the proposed method can update information instantaneously for new predictions as well as make enhanced recommendations regarding user preferences.

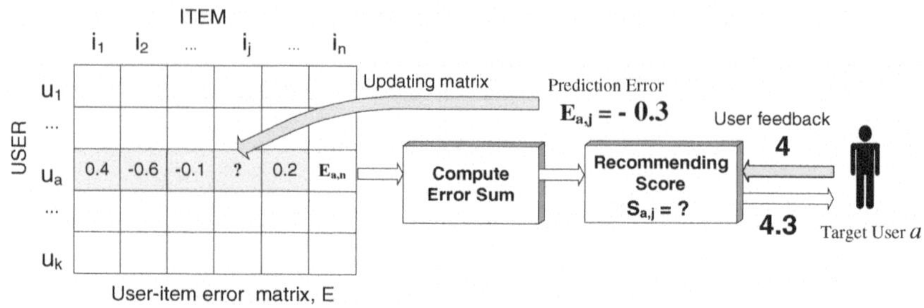

Fig. 3. Updating the error matrix incrementally by using user feedback

4 Experimental Evaluation

In this section, experimental results of the proposed method are presented. The performance evaluation is divided into two dimensions. The quality of the prediction based on prior errors is first evaluated, and then the quality of the *top*-N recommendations is evaluated. All experiments were carried out on a Pentium IV 3.0GHz with 2GB RAM, running a MS-Window 2003 server. In addition, the recommendation system for the web was implemented using MySQL 5.0 and PHP 5.1 on an Apache 2.0 environment.

4.1 Data Set and Evaluation Metrics

The experimental data is taken from *MovieLens*, a web-based research recommendation system (www.movielens.org). The data set contains 100,000 ratings of 1682 movies rated by 943 users (943 rows and 1682 columns of a *user-item matrix A*).

Evaluation of Prediction Quality. For evaluating the quality of the prediction measurement, the total ratings were divided into two groups: 80% of the data (80,000 ratings) was used as *a training set* and 20% of the data (20,000 ratings) was used as *a test set*. Prior to evaluating the accuracy of the proposed method, a *user-item error matrix E* should first be constructed. Therefore, the training data set was further subdivided into training and testing portions, and a matrix *E* was generated using a 5-fold cross validation scheme. In order to measure the accuracy of the predictions, *mean absolute error* (MAE), which is widely used for statistical accuracy measurements in diverse algorithms [1, 2, 7, 9], was adopted. The mean absolute error for user *u* is defined as:

$$MAUE(u) = \frac{\sum_{i \in I_u} |A_{u,i} - S_{u,i}|}{|Test_u|}$$

where $Test_u$ is a item list of user *u* in the test data, and $<A_{u,i}, S_{u,i}>$ is the actual rating and the recommending score (predicted rating in benchmark algorithms) pairs of user *u* in the test data. Finally, the MAE of all users in the test set is computed as:

$$MAE = \frac{\sum_{u=1}^{k} MAUE(u)}{k}$$

Evaluation of TOP-N Recommendation. To evaluate the performance of *top*-N recommendations, we divide the data set into *a test set* (9,430 ratings) with exactly 10 ratings per user in the test set and *a training set* (90,570 ratings) with the remaining ratings. In addition, the training data set is also further subdivided into training and testing portions to build a *user-item error matrix E*. In the context of *top*-N recommendations, *recall*, a measure of how often a list of recommendations contains an item that the user has actually rated, was used for the evaluation metric [5, 6, 11, 12]. The *hit-ratio* for user *u* is defined as:

$$hit - ratio(u) = \frac{|Test_u \cap TopN_u|}{|Test_u|}$$

where $Test_u$ is an item list of user *u* in the test data and $TopN_u$ is a *top*-N recommended item list for user *u*. Finally, the overall *recall* of the *top*-N recommendation is computed as:

$$recall = \frac{\sum_{u=1}^{k} hit - ratio(u)}{k} \times 100$$

Benchmark Algorithms. In order to compare the performance of the proposed method, a user-based CF algorithm, wherein the similarity is computed by the well known Pearson correlation coefficient (denoted as *UserCF*) [3], and the item-based CF approach of [2], which employs cosine-based similarity (denoted as *ItemCF*), were implemented. The prediction quality and the *top*-N recommendation of an error-based strategy (*ErrorCF+x*, where *x* is the similarity method used in building an error matrix *E*) were evaluated in comparison with benchmark algorithms.

4.2 Experimental Results

As noted previously, *a user-item error matrix*, *E*, which is closely connected with the similarity algorithm, should first be built for an error-based CF algorithm. Thereby, four error matrices are constructed using different item-item similarity algorithms such as cosine-based similarity (denoted as *ErrorCF+Cos*), correlation-based similarity (denoted as *ErrorCF+Corr*) as described in [2], cosine-based similarity with inverse item frequency (denoted as *ErrorCF+CosIIF*), and correlation-based similarity with inverse item frequency (denoted as *ErrorCF+CorrIIF*) as described in Section 3.1.

4.2.1 Experiments with the Prior Prediction Errors

As stated in Section 3.2, the prediction quality in an error-based CF is affected by the set of Lowest Error Items (cf. Definition 4). Fig. 4 presents the variation of MAE obtained by changing the ε value used for selecting the Lowest Error Item. It is observed from the graph that the ε value affects the prediction quality and the four methods demonstrate similar types of charts. The approaches of *the inverse item frequency* applied to the similarity (*ErrorCF+CosIIF* and *ErrorCF+CorrIIF*) elevate the prediction quality as the ε value increases from 0.2 to 1.4; beyond this point, the quality deteriorates. Likewise, *ErrorCF+Cos* and *ErrorCF+Corr* improved until a ε value of 1.2 and 1.6, respectively. When we compare the best MAE of original similarity methods with those of the similarity methods applying the inverse item frequency, it is found that *ErrorCF+CosIIF* (MAE of 0.8058) and *ErrorCF+CorrIIF* (MAE of 0.8043) perform better than *ErrorCF+Cos* (MAE of 0.8064) and *ErrorCF+Corr* (MAE of 0.8056), respectively.

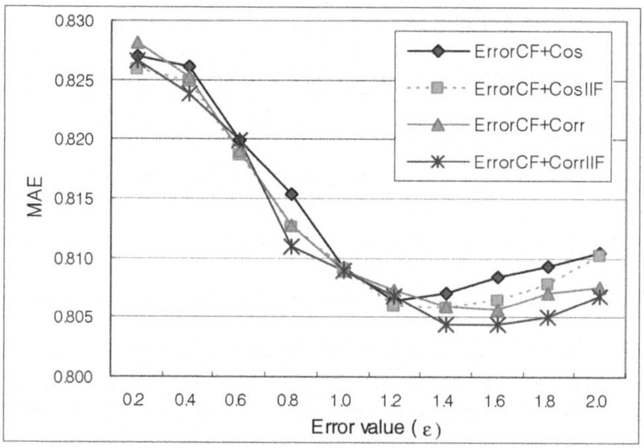

Fig. 4. MAE according to variation of the error value (ε) used in generating the prediction

4.2.2 Experiments with k Neighbor Size

The following experiments investigate the effect of the neighbor size on the performance of collaborative filtering, especially in relation to the cold start problem. The size of the neighborhood has a significant influence on the recommendation performance [7, 11]. Therefore, different numbers of user/item neighbors *k* were used for the prediction generation: 2, 4, 6, 8, 10, and 30. In the case of *ErrorCF+x*, the parameter *k* denotes the number of Lowest Error Items (*k* lowest error items) whereas it is the size of the nearest users for *UserCF* (*k* nearest neighbors) and the size of the most similar items for *ItemCF* (*k* most similar items). Table 3 summarizes the results of the experiment while Fig. 5 depicts the improved MAE of error-based CF over item-based (left graph) and user-based CF (right graph) algorithms.

The result demonstrates that, for a small neighborhood size, the proposed algorithm provides more accurate predictions than the traditional user-based and item-based algorithms. For example, when the neighborhood size is 2, *ErrorCF+CosIIF* yields an MAE of 0.827, which is the best prediction quality within the *ErrorCF+x*, whereas *ItemCF* and *UserCF* give an MAE of 1.042 and 0.909, respectively. In other words, the proposed method achieves 26% and 9% improvement compared to *ItemCF* and *UserCF*, respectively. However, the classic user-based scheme provides better quality as the neighbor size increases. Although the prediction quality of the *ErrorCF+x* is slightly worse than that of *UserCF* at a large neighborhood size, notably the proposed methods provides better quality than the traditional user-based and item-based algorithms in the event of cold start users.

Table 3. Comparison of MAE as neighbor size (*k*) grows

Algorithms	2	4	6	8	10	30
ErrorCF+Cos	0.82727	0.82855	0.82904	0.82874	0.82866	0.82375
ErrorCF+CosIIF	0.82712	0.82785	0.82831	0.82761	0.82712	0.82193
ErrorCF+Corr	0.83022	0.82949	0.82988	0.82957	0.82782	0.81958
ErrorCF+CorrIIF	0.83064	0.83041	0.82944	0.82919	0.82693	0.81864
ItemCF	1.04274	0.95708	0.91633	0.89165	0.87373	0.82743
UserCF	0.90953	0.83059	0.80064	0.78507	0.77674	0.75605

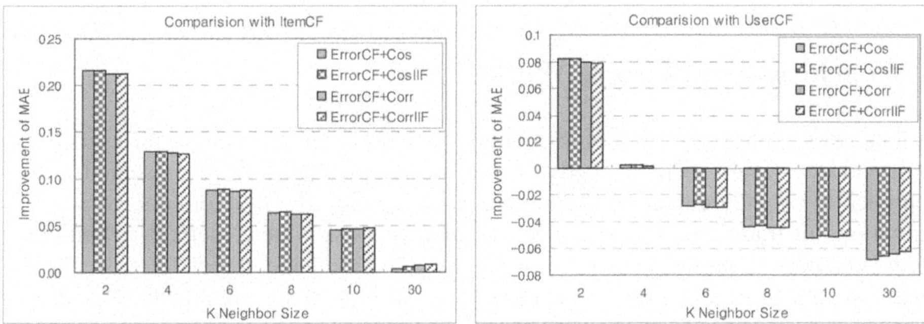

Fig. 5. Improvement of error-based CF over item-based CF (left) and user-based CF (right)

4.2.3 Top-N Recommendation Evaluation

For evaluating the *top*-N recommendation, the number of recommended items (the value of N) were increased, and we calculated the recall achieved by *UserCF*, *ErrorCF+CosIIF*, and *ItemCF* using a neighborhood size of 30. Fig. 6 presents the results of the experiment. The recall was measured by generating a list of the top-N recommended items for each user and then testing whether an item in the test set is provided on the *top*-N list recommended for each user.

In general, with the growth of recommended items N, recall tends increase. As we can see from Fig 6, *ErrorCF+CosIIF* shows considerably improved performance on all occasions compared to *UserCF*. In addition, comparing the results achieved by *ErrorCF+CosIIF* and *ItemCF*, the recommendation quality of the former is superior to that of *ItemCF* as top-N is increased, although the performance is diminished slightly in the case of *top*-10. For example, the proposed method achieves 34% improvement in the case of *top*-50 (N=50), and 45% improvement in the case of *top*-100 (N=100) compared to the item-based scheme. We conclude from this experiment that the proposed prediction strategy for top-N recommendation is effective in terms of improving the performance.

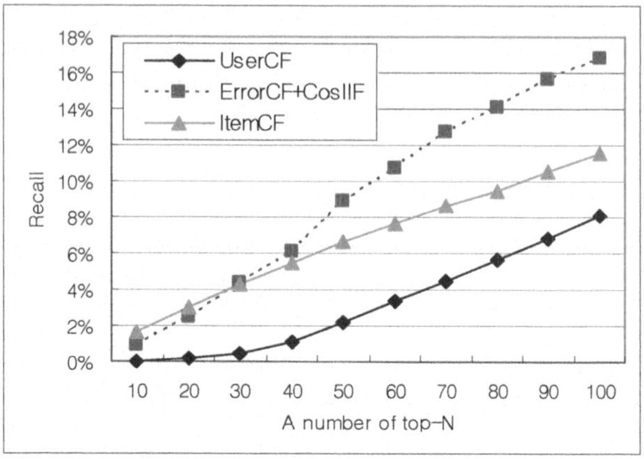

Fig. 6. Recall as the value of N (number of recommended items) increases

5 Conclusion and Future Work

Collaborative Filtering for Recommendations is a powerful technology for users to find information relevant to their needs. In the present work, we have presented a novel approach to provide enhanced recommendation quality and to overcome some of the limitations in traditional CF systems. We also propose a new method of building a model, namely *a user-item error matrix*, for CF-based recommender systems. The major advantage of the proposed approach is that it supports incremental updating of the model by using explicit user feedback. The experimental results demonstrate that the proposed method offers significant advantages both in terms of improving the recommendation quality and in dealing with cold start users as compared to traditional CF algorithms. However, the proposed method gives slightly worse prediction than user-based CF in the case of a sufficient neighborhood.

A research area that is attracted attention at present is a recommender system based on collaborative tagging. We are currently extending our algorithm to allow for personalized recommendation using semantic tagging information. Therefore, we plan to further study the impact of using a folksonomy or collective intelligence for a recommender system.

References

1. Breese, J.S., Heckerman, D., Kadie, C.: Empirical Analysis of Predictive Algorithms for Collaborative Filtering. In Proc. of the 14th Conf. on Uncertainty in Artificial Intelligence (1998) 43–52
2. Sarwar, B., Karypis, G., Konstan, J., Reidl, J.: Item-based Collaborative Filtering Recommendation Algorithms. In Proc. of the 10th Int. Conf. on World Wide Web (2001)
3. Resnick, P., Iacovou, N., Suchak, M., Bergstrom, P. Riedl, J.: GroupLens: an open architecture for collaborative filtering of netnews. In Proc. of the ACM Conf. on Computer supported Cooperative Work (1994) 175–186

4. Lemire, D., Maclachlan, A.: Slope One Predictors for Online Rating-Based Collaborative Filtering. In Proc. of SIAM Data Mining (2005)
5. Deshpande, M., Karypis, G.: Item-based top-N recommendation algorithms. ACM Transactions on Information Systems, Vol. 22 (2004) 143–177
6. Ziegler, C. N, Mcnee, S. M., Konstan, J. A., Lausen, G.: Improving Recommendation Lists Through Topic Diversification. In Proc. of 14th Int. Conf. on World Wide Web (2005)
7. Kim, H. N, Ji, A. T., Jo, G. S.: Enhanced Prediction Algorithm for Item-based Collaborative Filtering Recommendation. Lecture Notes in Computer Science, Vol. 4082. Springer-Verlag, Berlin Heidelberg (2006) 41–50
8. Wang, J., de Vries, A. P., Reinders, M. J. T.: Unifying User-based and Item-based Collaborative Filtering Approaches by Similarity Fusion. In Proc. of the 29th Int. ACM SIGIR Conf. on Research and Development in Information Retrieval (2006) 501–508
9. Mobasher, B., Jin, X., Zhou, Y.: Semantically Enhanced Collaborative Filtering On the Web. Lecture Notes in Computer Science, Vol. 3209. Springer-Verlag, Berlin Heidelberg (2004) 57–76
10. Kim, H. J., Jung, J. J., Jo, G. S.: Conceptual Framework for Recommendation System based on Distributed User Ratings. In Proc. of the 2nd Int. Workshop on Grid and Cooperative Computing (2003)
11. Sarwar, B., Karypis, G., Konstan, J., Riedl, J.: Analysis of recommendation algorithms for E-commerce. In Proc. of ACM'00 Conf. on Electronic Commerce (2000) 158–167
12. Miller, B. N., Konstan, J. A., Riedl, J.: PocketLens: Toward a personal recommender system. ACM Transactions on Information Systems, Vol. 22 (2004) 437–476
13. Schein, A. I., Popescul, A., Ungar, L. H.: Methods and Metrics for Cold-Start Recommendations. In Proc. of the 25th Int. ACM Conf. on Research and Development in Information Retrieval (2002)
14. Massa, P., Bhattacharjee, B.: Using Trust in Recommender Systems: An Experimental Analysis. Lecture Notes in Computer Science, Vol. 2995. Springer-Verlag, Berlin Heidelberg (2006) 221–235

A PLSA-Based Approach for Building User Profile and Implementing Personalized Recommendation*

Dongling Chen[1,2], Daling Wang[1], Ge Yu[1], and Fang Yu[1]

[1] School of Information Science and Engineering, Northeastern University
Shenyang 110004, P.R. China
[2] School of Information, Shenyang University, Shenyang 110044, P.R. China

Abstract. This paper proposes a method based on Probability Latent Semantic Analysis (PLSA) to analyze web pages that are of interest to the user and the user query co-occurrence relationship, and utilize the latent factors between the two co-occurrence data for building user profile. To make the weight of web pages that user isn't interested decay rapidly, a Fibonacci function is designed as the decay factor for representing the user's interests more exactly. The personalized recommendation is implemented according to the score of web pages. The experimental results showed that our approach was more effective than the other typical approaches to construct user profile.

1 Introduction

Traditional information retrieval approaches are usually based on term matching. For example, the vector space model (VSM) [1], probability-based method such as Baeza-Yates approach [2], time series technique-based method such as hidden Markov model (HMM) [3], language-based method [1,4] and so on. However, those approaches are suffered from the problem of word usage diversity, namely, words mismatch. In general, the query and its relevant documents are using quite different sets of words, which will make the retrieval performance degrade severely.

For solving the problem, the concept matching method is proposed. In contrast to term matching methods, it retrieves text documents semantically relevant to the query but not necessarily "look like" or "sound like" the user query. The Latent Semantic Analysis (LSA) model and the Probability Latent Semantic Analysis (PLSA) are two typical examples [5, 6, 7]. But for LSA, it only operates on "bag of terms". In contrast, PLSA relies on the likelihood function of multinomial sampling and aims at an explicit maximization of the predictive power of the model [5]. Based on this point, in this paper we propose a PLSA-based method to analyze web pages that are of interest to the user and the user query co-occurrence relationship, and utilize the latent factors between the two co-occurrence data for building user profile.

Nowadays, the most general method for implementing personalization through constructing user profile is utilizing user's web usage information. There are two ways utilizing web usage information: clustering or classifying user web usage data.

* This work is supported by National Natural Science Foundation of China (No. 60573090, 60673139).

G. Dong et al. (Eds.): APWeb/WAIM 2007, LNCS 4505, pp. 606–613, 2007.

For example, Micro Speretta [8] and Fang Liu [9] utilized user's query history, Trajkova [10] and Qiu [20] utilized user's browsing history to create user profile respectively. Perkowitz [11] utilized the classify techniques, Mobasher et al. [12, 13] utilized the clustering techniques to mine web usage data and build user profile for web personalization.

These methods have been proven to be efficient, but they have to face up the problem that a user had multiple crossed interests. Moreover, they can not reveal the underlying characteristics of user usage information.

Addressing those problems aforementioned, we proposed the method adopting the algorithm based on Probability Latent Semantic Analysis (PLSA) to cluster user's web pages and construct user profile for implementing personalized recommendation.

The rest of this paper is organized as follows. Section 2 gives a short overview of PLSA. Section 3 describes the user profile constructing, learning and updating algorithm, and how to implement the personalized recommendation. The experiments and evaluation are given in Section 4. Finally, the paper is concluded in Section 5.

2 The PLSA Model

The Probabilistic Latent Semantic Analysis (PLSA) model is a statistical latent variable analysis model. It has shown excellent results in several IR related tasks [14, 15,16,17]. The core of PLSA is an aspect model [5]. It is assumed that there exist a set of hidden factors underlying the co-occurrence among two sets of object. The relationship between the hidden factors and the two sets of object can be estimated by the Expectation-Maximization (EM) algorithm. PLSA represents the joint probability of a document d and a word w based on a latent class variable z:

$$P(w,d) = P(d)P(w \mid d) = \sum_z P(z)P(w \mid z)P(d \mid z) \tag{1}$$

The probability that a word w in a particular document d is explained by the latent class corresponding to z is estimated during the E-step as:

$$P(z \mid d,w) = \frac{P(z)P(d \mid z)P(w \mid z)}{\sum_{z'} P(z')P(d \mid z')P(w \mid z')} \tag{2}$$

And the M-step consists of:

$$P(w \mid z) = \frac{\sum_d f(d,w)P(z \mid d,w)}{\sum_{d,w'} f(d,w')P(z \mid d,w')} \tag{3}$$

$$P(d \mid z) = \frac{\sum_w f(d,w)P(z \mid d,w)}{\sum_{d',w} f(d',w)P(z \mid d',w)} \tag{4}$$

$$P(z) = \frac{\sum_{d,w} f(d,w)P(z \mid d,w)}{\sum_{d,w} f(d,w)} \tag{5}$$

In order to avoid overtraining, Hoffman [5] proposed the method using tempered Expectation Maximization (TEM) for fitting the algorithm. Thus the E-step is modified in Eq.(4) by introducing a control parameter β, as follows:

$$P_\beta(z \mid d,w) = \frac{P(z)[P(d \mid z)P(w \mid z)]^\beta}{\sum_{z'} P(z')[P(d \mid z')P(w \mid z')]^\beta} \tag{6}$$

According to all, the conditional distributions $P(\cdot|d)$ can be approximated by a multinomial and represented as a convex combination of the class conditional $P(z|\cdot)$. Moreover, in a geometrical, the weights $P(z|d)$ are the coordinates of a document in that sub-simplex, which is identified as a probabilistic latent semantic space [7].

3 User Profile Descriptions and Study

In our approach, we desire to build a topic level user interest profile based on the latent topics, which consists of four hierarchies: the first level denotes different topics, the second level denotes the different web pages under their corresponding topic attributes, the third level is the weights of each topic respectively, and the fourth level is the sequentially visited times. Utilizing the visited times sequentially of each topic, we construct a user interest decay factor which is represented with Fibonacci function. The Fibonacci function can be represented as:

$$Fibonacci[i] = \begin{cases} 0 & i = 0 \\ 1 & i = 1,2 \\ Fib[i-1] + Fib[i-2] & i \geq 2 \end{cases}$$

For example, the $topic_i$ has not been visited sequentially for five times, its weight will be decreased by $Fib[5]$, namely, $weight=weight-Fib[5]$.

In initial user profile, we assign each topic an initial weight value 100, and settle each topic not visited times as 0. The structure of user profile is shown as table 1.

Table 1. Initial User Profile Sketch

$Topic_1$	$Topic_2$	…	$Topic_k$
$P_{11}, P_{12}, \ldots, P_{1k}$	$P_{21}, P_{22}, \ldots, P_{2m}$	…	$P_{k1}, P_{k2}, \ldots, P_{kt}$
100	100	100	100
0	0	0	0

3.1 Building User Profile Based on PLSA

Unlike the traditional clustering algorithm, the Probability Latent Semantic Analysis is a generative clustering technique that explicitly models document hidden topics, which are underlying factors. The algorithm of building user profile is described as Algorithm 1.

Algorithm 1. Building User Profile

Input: 1) The set D of pages d interesting to user and the keyword list extracted from D; 2) A predefined threshold μ;

Output: A set of topic categories (there are different number pages in each topic and the different topics have their uniform weight).

Method:

1) Preprocess the keyword list including filtering the stop words and stemming;

2) For each web page that are interesting to user;

3) Compute probability distribution $P(topic_k|D)$:

$$P(topic_k \mid D) = \sum_{w_i} P(w_i)P(topic_k \mid D, w_i) = \frac{\sum_{w_i} f(w_i, D)P(topic_k \mid w_i, D)}{|D|} \quad (7)$$

In order to avoid overfitting, we adopt TEM [5] algorithm for fitting the model. The computing process includes E-step and M-step, and the E-step is as follows:

$$= \sum_{w_i} f(w_i, D) \frac{P(topic_k)[P(D \mid topic_k)P(w_i \mid topic_k)]^\beta}{\mid D \mid \sum_{topic_{k'}} P(topic_{k'})[P(D \mid topic_{k'})P(q_i \mid topic_{k'})]^\beta} \quad (8)$$

Where M-step consists of:

$$P(D \mid topic_k) = \frac{\sum_{w_i} f(D, w_i)[P(topic_k \mid D, w_i)]^\beta}{\sum_{D', w_i} f(D', w_{i'})[P(topic_k \mid D', w_{i'})]^\beta} \quad (9)$$

$$P(w_i \mid topic_k) = \frac{\sum_{D} f(D, w_i)[P(topic_k \mid D, w_i)]^\beta}{\sum_{D, w_{i'}} f(D, w_{i'})[P(topic_k \mid D, w_{i'})]^\beta} \quad (10)$$

Iterating between the E-step and M-step until a local optimal limit is got, we can get the final probabilities of $P(topic_k|D)$;

4) Classify the D according to probability distribution $P(topic_k|D)$;

5) If $P(topic_k|D)>\mu$ then put the pages D into the $topic_k$ categories;

6) Else compute next web page;

7) Give a uniform positive initial weight to each category, and the pages in this categories have the corresponding weight;

8) Return the user profile.

Noted that among the topics in user profile, there may be the overlapping, that means that a page maybe belongs to several categories (topics).

According to Algorithm1, we can construct the initial user profile as Table 1.

3.2 Learning and Updating User Profile

We think the probability $P(topic_k|Q)$ denotes the topic of query, $P(topic_k|Q)$ is the probability distribution of query Q over the latent topics. In order to decrease quickly the weight of pages not visited recently by users, we design Fibonacci function as interest decay factor. In detail, on the foundation of initial weight in user profile, each user issue a query, according to the topic probability distribution of query, we add 1 to the corresponding visited topic weight, represented as: *weight=weight*+1 is visited, the corresponding weight will be added. The user profile learning and updating algorithm is described as Algorithm2.

Algorithm 2. Learning and Updating User Profile

Input: 1) The initial User Profile; 2) Query Q, namely, a set of keyword; 3) A predefined threshold θ;

Output: Updated User Profile.

Method:

1) For each Q, preprocess the Q, including a set of keyword $\{q_1, q_2, \ldots, q_n\}$;

2) Judge the user's latent topic based on iteratively using EM algorithm, the E-step just as Eq.(7) and Eq.(8), only replace the D, d with Q, q respectively. Where $P(q_i|topic_k)$ denotes the probability with respect to the query term q_i occurring in specific latent topics $topic_k$, and the M-step just as Eq.(9) and Eq.(10), only replace the D, d with Q, q respectively. Iterating between the E-step and M-step, the probability value $P(topic_k|Q)$ is obtained;

3) Classify the Q according to probability distribution $P(topic_k|Q)$;

4) If $P(topic_k|Q)>\theta$, then $weight\ topic_k=weight\ topic_k+k$ (k is a constant), and in user profile, select all the $topic_k$, which probability distribution over latent topic $P(topic_k|Q)\leq\theta$, let theirs not visited times add 1, and *weight=weight-Fib*[not visited times];

5) If some topic's weight value less than zero, then delete all the web pages in those topics;

6) Return updated User Profile.

3.3 Implementing Personalized Recommendation

In order to incorporate an individual user's interests to personalized web search, we recommend more relevant web pages to individuals according to their user profile. The detailed algorithm is described as Algorithm 3.

Algorithm 3. Scoring Page P and Implementing Personalized Recommendation

---

Input: 1) The User Profile; 2) A predefined threshold ψ;

Output: The top N pages recommended to individual user.

Method:

1) If $P(topic_k|Q)>\psi$, then $topic_k$ is matching topic, maybe which are one topic or multi topics;

2) For every matching topic, which includes several web pages and these web pages have the same weight with corresponding the topic, record these web pages as $PageSet=\{p_1, p_2, \ldots, p_n\}$, noted that the pages in *PageSet* maybe belong to different topics;

3) Calculate the score of each page in *PageSet*. For example, page p_1 belongs to

$topic_1$ and $topic_2$, representing as $p_1\left\langle \begin{matrix} topic_1 \\ topic_2 \end{matrix} \right.$, then $score\ p_1 = \sum_{i=1,2} weight\ topic_i$;

4) Sort the pages on their score in *PageSet* decreasingly, then recommend top *N* web pages to user.

4 Experiments and Evaluations

We design the experiments in two parts: the first part is to evaluate whether user profile constructing method based on PLSA can improve the performance of personalized recommendation, the second part is to evaluate the precision of classification algorithm. In each part, we also compared the results with baselines.

The data set is based on a random sample of users visiting CTI web server (http://www.cs.depaul.edu). CTI denotes a department of Computer Science, Telecommunications and Information Systems of DEPAUL University. It contains a two week period during April of 2002 web log files. The original data contains a total of 20950 sessions and 5446 users. This dataset is referred to as the "CTI data".

4.1 The Performance Evaluation

In order to filtering invalid the sessions, we preprocess the dataset and get a dataset including 181 users, whose valid sessions are more than 10. For each user's sessions, we construct the user profiles through identifying latent topics in each web page they have visited. In order to evaluate the effectiveness of user profile, we partition the each user's sessions into four parts and adopt 4-fold cross-validation for experiments.

Take user id=2885 as the example, we constructing his user profile. Firstly, compute $P(topic_k|D)$ for each web page in his sessions. Here, for convenience, we only list user id= 2885 parts profile as Table 2.

Table 2. Partial Profile Example with User id=2885

people	facultyinfo.asp?id=268; evalgrid.asp	87	5
course	syllabilist.asp; default.asp; printsyllabus.asp	98	2
news	default.asp; news.asp	80	0
program	courses.asp?section=courses	89	1
...	...	...	...

We download the ready-made software package of LSA [18] and the software SVDLIBC [19] library to perform the SVD transformation. To implement LSA method, we extract keywords from web pages and execute Singular Value Decomposition (SVD). In order to simplify the compare process, we only truncate latent space for 10 ranks, i.e., for each web page, we only find its most important topics. For PLSA, we only select the first 10 biggest weight topics.

From the view of evaluating user profile performance, the effectiveness of our method is evaluated by the precision of the recommendation. For evaluating it, we use a metric called Recommendation Ratio in the context of top-*N* recommendations. For each user session in the training set, we generate a top-*N* recommendation set, and then compare the recommendations with the pages in the test sessions, and a match is considered a Recommendation Success. We define the Recommendation Ratio as the total number of recommendation web pages to divide the numbers of success recommendation. The results are depicted in Fig.1.

4.2 Precisions of Cluster Document Evaluation

In order to make a comparative experiment to evaluate the accuracy of cluster the web page based on PLSA, we select the algorithm in [12], which creates an aggregate profiles. To utilizing this method, we first cluster the user sessions into a set of clusters. For example, we depict as $\{Cluster_1, Cluster_2, \ldots, Cluster_n\}$, then get the centroid of each cluster to build aggregate profile.

We adopt the evaluation criteria presented in [12], which is named Weighted Average Visit Percentage (WAVP). This evaluation method is assessing each user profile individually according to the likelihood that a user who visits web page in the session will visit the rest of the pages in that session during the same session. Fig.2 shows the comparative result.

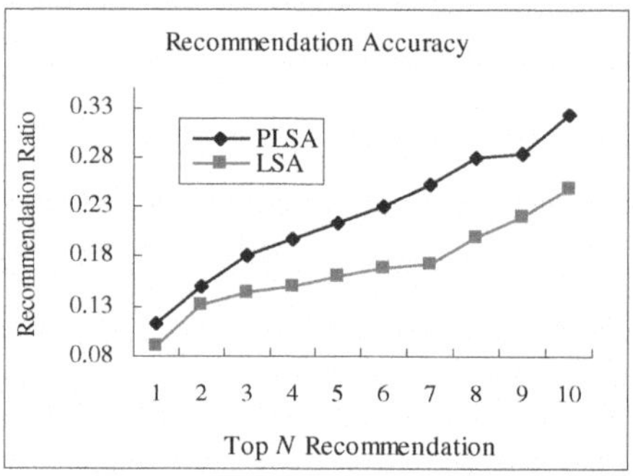

Fig. 1. Recommendation Comparison

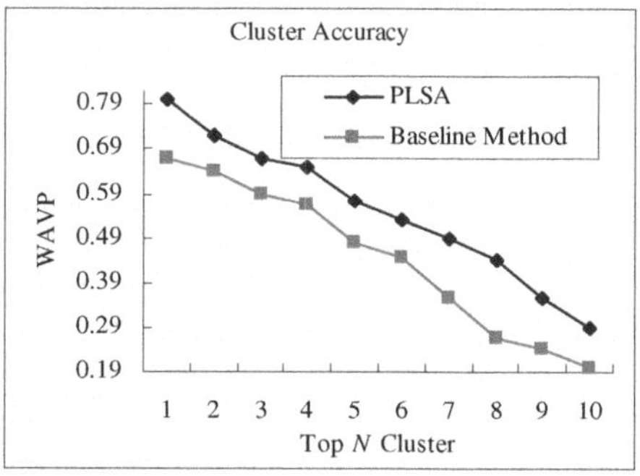

Fig. 2. Cluster Comparison

As shown in Fig. 1, 2, two experimental results are consistent. In Fig. 1, PLSA-based method behaves better than the baseline in recommendation accuracy and the average improvement of performance is 7.3% when top 10 recommendations. In Fig. 2, the experimental result inspects that the PLSA-based method rises about 17.6% than the baseline when cluster number is 8.

5 Conclusion and Future Work

In this paper, we presented an approach based on Probability Latent Semantic Analysis to construct user profile for personalized recommendation. In the process of constructing user profile, we utilized Bayesian probability equation to calculate the latent factor probability distribution under the co-occurrence data. We built user profiles in terms of those underlying factors. In order to explicitly display those latent factors, we made a hypothesis that latent factor space was latent topic class. Experimental results proved that our approach was more effective than the other typical approaches to construct user profile.

In the future, we will combine user's long-term interests with short-term interests for updating user profile more effectively. Furthermore, because of the experiments we conducted are all in objective datasets, which means that we can only objectively evaluate the whole performance, which decide we can't judge how well our method in according with actual user interests. So in future work, we will make a subjective experiment on real world dataset, which will be collected by ourselves. At the same time, under PLSA framework, we will combine other classify methods or cluster methods to explore personalized recommendation.

References

1. B. Chen. Research Summary http://berlin.csie.ntnu.edu.tw/Berlin_Research/Resarch_Berlin Chen2007.pdf
2. A. Ricardo and A. Berthier. Modern Information Retrieval. ACM Press. 1999.
3. D. Miller, T. Leek and R. Schwartz.. A Hidden Markov Model Information Retrieval System. SIGIR1999: 214-221
4. J. Ponte, W. Croft. A Language Modeling approach to information retrieval. SIGIR1998: 275-281
5. T. Hofmann. Probabilistic Latent Semantic Analysis. UAI1999: 289-296
6. Y. Hsieh, Y. Huang, C. Wang. and L. Lee. Improved spoken document retrieval with dynamic key term lexicon and probabilistic latent semantic analysis (PLSA). ICASSP2006, 961-964
7. Y. Zhang, G. Xu, and X. Zhou. A Latent Usage Approach for Clustering Web Transaction and Building User Profile. ADMA2005: 31-42
8. S. Micro and G. Susan. Personalized Search Based on User Search Histories. Web Intelligence 2005: 622-628
9. F. Liu, C. Yu, and W. Meng. Personalized Web Search for Improving Retrieval Effectiveness. IEEE Trans. Knowl. Data Eng. 2004, 16(1): 28-40
10. J. Trajkova, G. Susan. Improving Ontology-based User Profiles. http://www.ittc.ku.edu/ keyconcept/publications/ RIAO 2004.pdf
11. M. Perkowitz. Adaptive Web Sites: Cluster Mining and Conceptual Clustering for Index Page Synthesis, PhD thesis, University of Washington, Computer Science and Engineering.
12. B. Mobasher, H. Dai, T. Luo, and M. Nakagawa. Discovery and Evaluation of Aggregate Usage Profiles for Web Personalization. Data Mining and Knowledge Discovery, 2002, 6(1): 61-82
13. B. Mobasher. Web Usage Mining and Personalization. Practical Handbook of Internet Computing, M.P. Singh, Editor. 2004, CRC Press.
14. T. Hofmann. Latent Semantic Models for Collaborative Filtering. ACM Transactions on Information Systems, 2004. 22(1): 89-115.
15. T. Hofmann. Unsupervised Learning by Probabilistic Latent Semantic Analysis. Machine Learning Journal, 2001. 42(1): 177-196.
16. D. Cohn and H. Chang. Learning to Probabilistically Identify Authoritative Documents. ICML2000: 167-174
17. D. Cohn and T. Hofmann. The missing link - A probabilistic model of document content and hypertext connectivity. NIPS 2000: 430-436
18. http://lsi.research.telcordia.com
19. http://tedlab.mit.edu/~dr/SVDLIBC
20. F. Qiu and J. Cho. Automatic identification of user interest for personalized search. WWW2006: 727-736

CoXML: A Cooperative XML Query Answering System

Shaorong Liu[1] and Wesley W. Chu[2]

[1] IBM Silicon Valley Lab, San Jose, CA, 95141, USA
shaorongliu@gmail.com
[2] UCLA Computer Science Department, Los Angeles, CA, 90095, USA
wwc@cs.ucla.edu

Abstract. The heterogeneity nature of XML data creates the need for approximate query answering. In this paper, we present an XML system that cooperates with users to provide user-specific approximate query answering. The key features of the system include: 1) a query language that allows users to specify approximate conditions and relaxation controls; 2) a relaxation index structure, XTAH, that enables the system to provide user-desired relaxations as specified in the queries; and 3) a ranking model that incorporates both content and structure similarities in evaluating the relevancy of approximate answers. We evaluate our system with the INEX 05 test collections. The results reveal the expressiveness of the language, show XTAH's capability in providing user-desired relaxation control and demonstrate the effectiveness of the ranking model.

1 Introduction

The growing use of XML in scientific data repositories, digital libraries and Web applications has increased the need for flexible and effective XML search methods. There are two types of queries for searching XML data: content-only (CO) queries and content-and-structure (CAS) queries. CAS queries are more expressive and thus yield more accurate searches than CO queries. XML structures, however, are usually heterogeneous due to the flexible nature of its data model. It is often difficult and unrealistic for users to completely grasp the structural properties of data and specify exact query structure constraints. Thus, XML approximate query answering is desired, which can be achieved by relaxing query conditions.

Query relaxation is often user-specific. For a given query, different users may have different specifications about which conditions to relax and how to relax them. Most existing approaches on XML query relaxation (e.g., [1]) do not provide control during relaxation, which may yield undesired approximate answers. To provide user-specific approximate query answering, it is essential for an XML system to have a relaxation language that allows users to specify their relaxation control requirements and to have the capability to control the query relaxation process.

Furthermore, query relaxation returns a set of approximate answers. These answers should be ranked based on their relevancy to both the structure and content conditions of the posed query. Many existing ranking models (e.g., [2], [3]) only measure the content similarities between queries and answers, and thus are inadequate for ranking approximate answers that use structure relaxations. Recently, [4] proposed a family of structure

G. Dong et al. (Eds.): APWeb/WAIM 2007, LNCS 4505, pp. 614–621, 2007.

scoring functions based on the occurrence frequencies of query structures among data without considering data semantics. Clearly, using the rich semantics provided in XML data in design scoring functions can improve ranking accuracy.

To remedy these shortcomings, we propose a new paradigm for XML approximate query answering that places users and their demands in the center of design approach. Based on this paradigm, we develop a cooperative XML system that provides user-specific approximate query answering. More specifically:

- First, we develop a relaxation language that allows users to specify approximate conditions and control requirements in queries (e.g., preferred or unacceptable relaxations, non-relaxable conditions and relaxation orders). (Section 3)
- Second, we introduce a relaxation index structure that clusters twigs (as introduced in Sec 2.1) into multi-level groups based on relaxation types and distances. By such clustering, the index structure enables a systematic control of the relaxation processing based on users' specifications in queries. (Section 4)
- Third, we propose a semantic-based tree editing distance to evaluate XML structure similarities based on not only the number of relaxations but also relaxation semantics. We also develop a model that combines both structure and content similarities in evaluating the overall relevancy [5].
- Finally, our experimental studies using the INEX 05 benchmark test collection[1] demonstrate the effectiveness of our proposed methodology.

2 XML Query Relaxation Background

2.1 Query Model

A fundamental construct in most existing XML query languages is the tree-pattern query or *twig*, which selects elements and/or attributes with tree-like structures. Thus, we use twig as the basic query model. Fig. 1(a) illustrates a sample twig, which searches for articles with a title on "*data mining,*" a year in *2000* and a body section about "*frequent itemset.*" Each twig node is associated with an unique *id*, shown in italic beside the node. The IDs are not needed when all the node labels are distinct. The text under a node is the content constraint on the node.

For a twig T, we use $T.V$ and $T.E$ to represent its nodes and edges respectively. For a twig node v ($v \in T.V$), we use $v.label$ to denote the node label. We use $e_{u,v}$ to denote the edge from nodes u to v, either parent-to-child ("/") or ancestor-to-descendant ("//").

2.2 Query Relaxation

In the XML model, there are two types of query relaxations: value relaxation and structure relaxation. Value relaxation, successfully used in relational models (e.g., [6]), is orthogonal to structure relaxation. In this paper, we focus on structure relaxation. Many structure relaxation types have been proposed ([7], [8], [1]). We use the following three structure relaxation types, similar to the ones in [1], which capture most of the relaxation types proposed in previous work.

---

[1] http://inex.is.informatik.uni-duisburg.de/

- **Node Relabel.** A node can be relabeled to similar or equivalent labels according to domain knowledge. For example, the twig in Fig. 1(a) can be relaxed to that in Fig. 1(b) by relabeling node *section* to *paragraph*.
- **Edge Generalization.** A parent-to-child edge can be generalized to an ancestor-to-descendant edge. For example, the twig in Fig. 1(a) can be relaxed to that in Fig. 1(c) by relaxing *body/section* to *body//section*.
- **Node Deletion.** A node v may be deleted to relax the structure constraint. If v is an internal node, then the children of v will be connected to the parent of v with ancestor-to-descendant edges. For instance, the twig in Fig. 1(a) can be relaxed to that in Fig. 1(d) by deleting the internal node *body*. We assume that the root node of a twig cannot be deleted since it represents the search context.

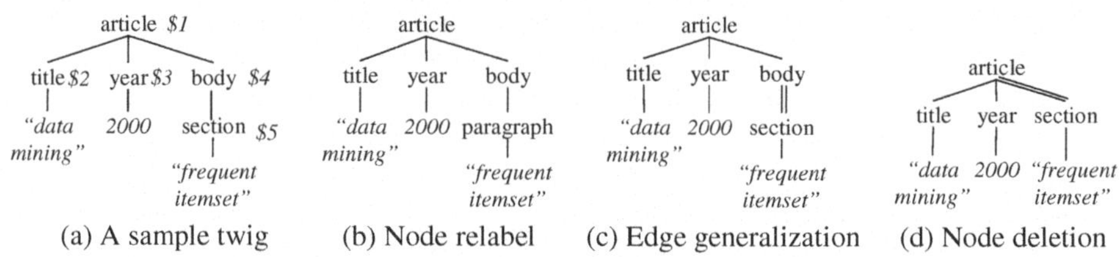

Fig. 1. A sample twig and its relaxed twigs

Given a twig T, a *relaxed twig* can be generated by applying one or more relaxation operations to T. Let m be the number of relaxation operations applicable to T, then there are at most $\binom{m}{1} + ... + \binom{m}{m} = 2^m$ relaxation combinations, i.e., 2^m relaxed twigs.

3 XML Query Relaxation Language

A number of XML approximate search languages have been proposed. Most extend standard query languages with constructs for approximate text search (e.g., XIRQL [3], TeXQuery [9]). XXL [10] provides users with constructs for users to specify both approximate structure and content conditions, which however, does not allow users to control the relaxation process. Users may often want to specify their preferred or rejected relaxations, non-relaxable query conditions, or to control the relaxation orders among multiple relaxable conditions.

To remedy this shortcoming, we propose an XML relaxation language that allows users to both specify approximate conditions and to control the relaxation process. A relaxation-enabled query $\mathcal{Q}$ is a tuple $(\mathcal{T}, \mathcal{R}, \mathcal{C}, \mathcal{S})$, where:

- $\mathcal{T}$ is a twig as described as Section 2.1;
- $\mathcal{R}$ is a set of relaxation constructs specifying which conditions in $\mathcal{T}$ may be approximated when needed;
- $\mathcal{C}$ is a boolean combination of controls stating how the query shall be relaxed;
- $\mathcal{S}$ is a stop condition indicating when to terminate the relaxation process.

The execution semantics for a relaxation-enabled query are: we first search for answers exactly matching the twig; we then test the stop condition to check whether relaxation is needed. If not, we repeatedly relax the twig based on the relaxation constructs and control until either the stop condition is met or the twig cannot be further relaxed.

Given a relaxation-enabled query $\mathcal{Q}$, we use $\mathcal{Q}.\mathcal{T}$, $\mathcal{Q}.\mathcal{R}$, $\mathcal{Q}.\mathcal{C}$ and $\mathcal{Q}.\mathcal{S}$ to represent its twig, relaxation constructs, control and stop condition respectively. Note that a twig is required to specify a query, while relaxation constructs, control and stop condition are optional. When only a twig is present, we iteratively relax the query based on similarity metrics until the query cannot be further relaxed.

A relaxation construct for a query $\mathcal{Q}$ is either a specific or a generic relaxation operation in any of the following forms:

- $rel(u, -)$, where $u \in \mathcal{Q}.\mathcal{T}.V$, specifies that node u may be relabeled when needed;
- $del(u)$, where $u \in \mathcal{Q}.\mathcal{T}.V$, specifies that node u may be deleted if necessary;
- $gen(e_{u,v})$, where $e_{u,v} \in \mathcal{Q}.\mathcal{T}.E$, specifies that edge $e_{u,v}$ may be generalized.

The relaxation control for a query $\mathcal{Q}$ is a conjunction of any of the following forms:

- Non-relaxable condition $!r$, where $r \in \{rel(u, -), del(u), gen(e_{u,v}) \mid u, v \in \mathcal{Q}.\mathcal{T}.V, e_{u,v} \in \mathcal{Q}.\mathcal{T}.E\}$, specifies that node u cannot be relabeled or deleted, or edge $e_{u,v}$ cannot be generalized;
- $Prefer(u, l_1, ..., l_n)$, where $u \in \mathcal{Q}.\mathcal{T}.V$ and l_i is a label ($1 \leq i \leq n$), specifies that node u is preferred to be relabeled to the labels in the order of (l_1, ..., l_n);
- $Reject(u, l_1, ..., l_n)$, where $u \in \mathcal{Q}.\mathcal{T}.V$, specifies a set of unacceptable labels for node u;
- $RelaxOrder(r_1, ..., r_n)$, where $r_i \in \mathcal{Q}.\mathcal{R}$ ($1 \leq i \leq n$), specifies the relaxation orders for the constructs in R to be (r_1, ..., r_n);
- $UseRType(rt_1, ..., rt_k)$, where $rt_i \in \{$*node_relabel, node_delete, edge_generalize*$\}$ ($1 \leq i \leq k \leq 3$), specifies the set of relaxation types allowed to be used. By default, all three relaxation types may be used.

A stop condition $\mathcal{S}$ is either:

- $AtLeast(n)$, where n is a positive integer, specifies the minimum number of answers to be returned; or
- $d(\mathcal{Q}.\mathcal{T}, T') \leq \tau$, where T' stands for a relaxed twig and τ a distance threshold, specifies that the relaxation should be terminated when the distance between the original twig and a relaxed twig exceeds the threshold.

We now present an example of using our relaxation language to express INEX 05 topic 267 (Fig. 2(a)). The topic consists of three parts: *castitle* (i.e., the query formulated in an XPath-like syntax), *description* and *narrative*. The *narrative* part contains the detailed description of a user's information needs and is used for judging result relevancy. The topic author considers an article's title, i.e., *atl*, non-relaxable and regards titles about "digital libraries" under the bibliography part, i.e., *bb*, irrelevant. Based on this narrative, we formulate this topic using our relaxation language as in Fig. 2(b).

We have developed a GUI interface for users to specify relaxations. Users may first input the twig using an XPath-like syntax. Based on the input twig, the interface automatically generates a set of relaxation candidates. Users can then specify relaxation constructs and controls by selecting relaxations from the candidate set.

<inex_topic topic_id="267" query_type="CAS" ct_no="113" >
<castitle>//article//fm//atl[about(., "digital libraries")]</castitle>
<description>Articles containing "digital libraries" in their title.</description>
<narrative>I'm interested in articles discussing Digital Libraries as their main subject. Therefore I require that the title of any relevant article mentions "digital library" explicitly. Documents that mention digital libraries only under the bibliography are not relevant, as well as documents that do not have the phrase "digital library" in their title.</narrative></inex_topic>

(a)

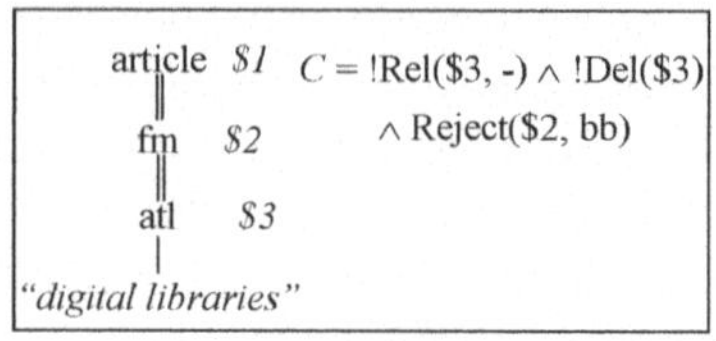

(b)

Fig. 2. Topic 267 in INEX 05 (a) & specifying the topic with our relaxation language (b)

4 XML Relaxation Index

4.1 XML Type Abstraction Hierarchy - XTAH

Several approaches for relaxing XML or graph queries have been proposed ([7], [4], [11], [1], [12]). Most focus on efficient algorithms for deriving top-k approximate answers without relaxation control. To remedy this condition, we propose an XML relaxation index structure, XML Type Abstraction Hierarchy (XTAH), that clusters relaxed twigs into multi-level groups based on relaxation types and distances. Each group consists of twigs using similar types of relaxations. Thus, XTAH enables systematic relaxation control based on users' specifications. For example, *Reject* can be implemented by pruning groups of twigs with unacceptable relaxations. *RelaxOrder* can be implemented by selecting the relaxed twigs from groups based on the specified order.

An XTAH for a twig structure T, denoted as XT_T, is a hierarchical cluster that represents relaxed twigs of T at different levels of relaxations based on the types of operations used by the twigs and the distances between them. More specifically, an XTAH is a multi-level labeled cluster with two types of nodes: internal and leaf nodes. A leaf node is a relaxed twig of T. An internal node represents a cluster of relaxed twigs that use similar operations and are closer to each other by a given distance metric. The label of an internal node is the common relaxation operations (or types) used by the twigs in the cluster. The higher level an internal node in the XTAH, the more general the label of the node, the less relaxed the twigs in the internal node.

Fig. 3 shows an XTAH for the sample twig in Fig. 1(a).[2] For ease of reference, we associate each node in the XTAH with an unique ID, where the IDs of internal nodes are prefixed with *I* and the IDs of leaf nodes are prefixed with *T'*.

Given a relaxation operation r, let I_r be an internal node with a label $\{r\}$. That is, I_r represents a cluster of relaxed twigs whose common relaxation operation is r. Due to the tree-like organization of clusters, each relaxed twig belongs to only one cluster, while the twig may use multiple relaxation operations. Thus, it may be the case that not all the relaxed twigs that use the relaxation operation r are within the group I_r. For example, the relaxed twig T_2', which uses two operations $gen(e_{\$1,\$2})$ and $gen(e_{\$4,\$5})$, is not included in the internal node that represents$\{gen(e_{\$4,\$5})\}$, I_7. This is because T_2' may belong to either group I_4 or group I_7 but is closer to the twigs in group I_4.

To support efficient searching or pruning of relaxed twigs in an XTAH that use an operation r, we add a virtual link from internal node I_r to internal node I_k where I_k is

[2] Due to space limitations, we only show part of the XTAH here.

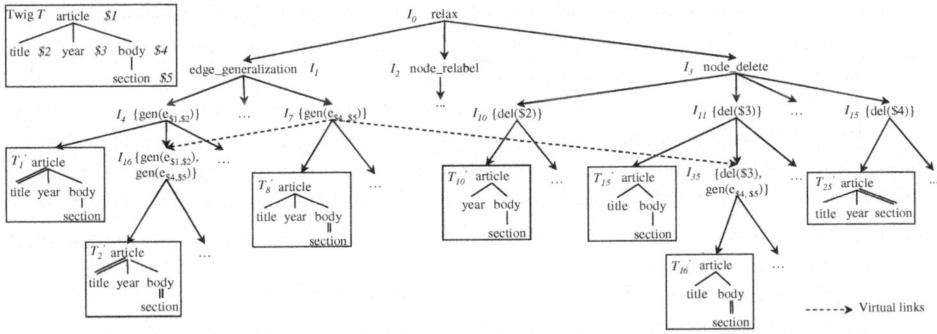

Fig. 3. An example of XML relaxation index structure for the twig T

not a descendant of I_r but all the twigs within I_k use operation r. By doing so, relaxed twigs that use operation r are either within group I_r or within the groups connected to I_r by virtual links. For example, internal node I_7 is connected to internal nodes I_{16} and I_{35} via virtual links. Thus, all the relaxed twigs using the operation $gen(e_{\$4,\$5})$ are within the groups I_7, I_{16} and I_{35}.

XTAH provides several significant advantages: 1) we can efficiently relax a query based on relaxation constructs by fetching relaxed twigs from internal nodes whose labels satisfy the constructs; 2) we can relax a query at different granularities by traversing up and down an XTAH; and 3) we can control and schedule query relaxation based on users' relaxation control requirements. For example, relaxation control such as non-relaxable conditions, *Reject* or *UseRType* can be implemented by pruning XTAH internal nodes corresponding to unacceptable operations or types.

Due to space limitations, the algorithm for building XTAH is not presented here. Interested readers should refer to [5] for details.

4.2 XTAH-Guided Query Relaxation Process

Fig. 4 presents the control flow of a relaxation process based on XTAH and relaxation specifications in a query. The *Relaxation Control* module prunes irrelevant XTAH groups corresponding to unacceptable relaxation operations or types and schedules relaxation operations such as *Prefer* and *RelaxOrder*, as specified in the query. More specifically, the process first searches for exactly matched answers. If there are enough

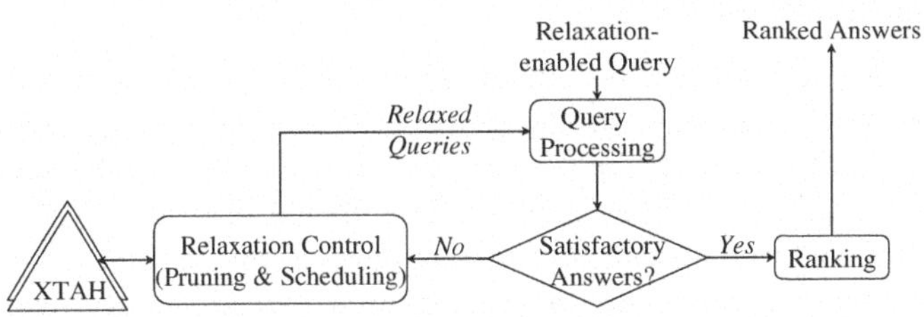

Fig. 4. Query relaxation control flow

number of answers available, there is no need for relaxation and the answers are returned. Otherwise, based on the relaxation control, the algorithm prunes XTAH internal nodes that correspond to unacceptable operations such as non-relaxable twig nodes (or edges), unacceptable node relabels and rejected relaxation types. This step can be efficiently carried out by using internal node labels and virtual links. After pruning disqualified internal groups, based on relaxation constructs and control such as *RelaxOrder* and *Prefer*, the *Relaxation Control* module schedules and searches for the relaxed query that best satisfies users' specifications from the XTAH. This step terminates when either the stop condition is met or all the constructs have been processed. If further relaxation is needed, the process then iteratively searches for the relaxed query that is closest to the original query by distance, which may use relaxation operations in addition to those specified in the query. This process terminates when either the stop condition holds or the query cannot be further relaxed. Finally, the process outputs approximate answers.

5 Experimental Evaluations

5.1 Experiment Setup

We have implemented the CoXML system in Java, which consists of a relaxation language parser, an XTAH builder, a relaxation controller and a ranking module. The ranking model evaluates the relevancy of an answer $\mathcal{A}$ to a query $\mathcal{Q}$, denoted as $sim(\mathcal{A}, \mathcal{Q})$, based on two factors: the structure distance between $\mathcal{A}$ and $\mathcal{Q}$, $struct\_dist(\mathcal{A}, \mathcal{Q})$, and the content similarity between $\mathcal{A}$ and $\mathcal{Q}$, denoted as $cont\_sim(\mathcal{A}, \mathcal{Q})$, as shown in (1).

$$sim(\mathcal{A}, \mathcal{Q}) = \alpha^{struct\_dist(\mathcal{A}, \mathcal{Q})} * cont\_sim(\mathcal{A}, \mathcal{Q}) \quad (1)$$

where α is a constant between 0 and 1; $cont\_sim(\mathcal{A}, \mathcal{Q})$ is an extended vector space model for evaluating XML content similarity [2]; and $struct\_dist(\mathcal{A}, \mathcal{Q})$ is a tree editing distance metric that evaluates relaxation cost based on its semantics [5].

We use INEX 05 document collection, content-and-structure queries and relevance assessment ("gold standard") to study the effectiveness of approximate answers returned by our system. We use the INEX 05 evaluation metric to evaluate experimental results: normalized extended cumulative gain (nxCG). For a given rank i, the value of nxCG@i reflects the relative gain an user accumulated up to that rank, compared to the gain the user could have obtained if the system would have produced the optimum best ranking. For any rank i, the ideal nxCG@i performance is 1.

5.2 Experimental Results

The first experiment evaluates the effectiveness of our semantic-based tree editing distance for evaluating structure similarity. We used the 22 single-branch content-and-structure queries in INEX 05 for the experiment. Table 1 presents the nxCG@10 evaluation results (averaged over the 22 queries) with the semantics-based tree editing distance as compared to that with uniform-cost tree editing distance. The results validate that differentiating the operation cost improves relaxation performance.

The second experiment tests the effectiveness of relaxation control by comparing the results with relaxation control against the results without relaxation control for topic

Table 1. The nxCG@10 evaluations of the first experiment results with semantic vs. uniform tree editing distance

Cost Model \ α	0.1	0.3	0.5	0.7	0.9
Uniform	0.2584	0.2616	0.2828	0.2894	0.2916
Semantic	0.3319	0.3190	0.3196	0.3068	0.2957

Table 2. The evaluations of the second experiment results with vs. w/o relaxation controls ($\alpha = 0.1$)

Control? \ Metric	nxCG@10	nxCG@25
Yes	1.0	0.8986
No	0.1013	0.2365

267 in INEX 05 (Fig. 2). The evaluation result in Table 2 demonstrates that relaxation specifications enable the system to control the relaxation process and thus yield results with greater relevancy.

6 Conclusion

In this paper, we have developed an XML system that cooperates with users to provide user-specific approximate query answering. More specifically, we first introduce a relaxation language that allows users to specify approximate conditions and relaxation control requirements in a posed query. We then propose a relaxation index structure, XTAH, that clusters relaxed twigs into multi-level groups based on relaxation types and their inter-distances. XTAH enables the system to provide user-desired relaxation control as specified in the query. Our experimental studies with INEX 05 test collection reveal the expressiveness of the relaxation language and the effectiveness of using XTAH for providing user-desired relaxation control.

References

1. S. Amer-Yahia, S. Cho, and D. Srivastava. XML Tree Pattern Relaxation. In *EDBT*, 2002.
2. S. Liu, Q. Zou, and W.W. Chu. Configurable Indexing and Ranking for XML Information Retrieval. In *SIGIR*, 2004.
3. N. Fuhr and K. Großjohann. XIRQL: A Query Language for Information Retrieval in XML Documents. In *SIGIR*, 2001.
4. S. Amer-Yahia, N. Koudas, A. Marian, D. Srivastava, and D. Toman. Structure and Content Scoring for XML. In *VLDB*, 2005.
5. W. W. Chu and S. Liu. CoXML: A Cooperative XML Query Answering System. In *The Encyclopedia of Computer Science and Engineering, Edit by B. Wah*. John Wiley & Sons, Inc, 2007.
6. W.W. Chu, H. Yang, K. Chiang, M. Minock, G. Chow, and C. Larson. CoBase: A Scalable and Extensible Cooperative Information System. *J. Intell. Inform. Syst.*, 6(11), 1996.
7. Y. Kanza and Y. Sagiv. Flexible Queries Over Semistructured Data. In *PODS*, 2001.
8. T. Schlieder. Schema-Driven Evaluations of Approximate Tree Pattern Queries. In *EDBT*, 2002.
9. S. Amer-Yahia, C. Botev, and J. Shanmugasundaram. TeXQuery: A Full-Text Search Extension to XQuery. In *WWW*, 2004.
10. A. Theobald and G. Weikum. Adding Relevance to XML. In *WebDB*, 2000.
11. A. Marian, S. Amer-Yahia, N. Koudas, and D. Srivastava. Adaptive Processing of Top-k Queries in XML. In *ICDE*, 2005.
12. I. Manolescu, D. Florescu, and D. Kossmann. Answering XML Queries on Heterogeneous Data Sources. In *VLDB*, 2001.

Concept-Based Query Transformation Based on Semantic Centrality in Semantic Peer-to-Peer Environment

Jason J. Jung[1,2], Antoine Zimmerman[2], and Jérôme Euzenat[2]

[1] Inha University, Korea
j2jung@intelligent.pe.kr
[2] INRIA Rhône-Alpes, France
{Antoine.Zimmerman,Jerome.Euzenat}@inrialpes.fr

Abstract. Query transformation is a serious hurdle on semantic peer-to-peer environment. The problem is that the transformed queries might lose some information from the original one, as continuously traveling p2p networks. We mainly consider two factors; i) number of transformations and ii) quality of ontology alignment. In this paper, we propose semantic centrality (SC) measurement meaning the power of semantic bridging on semantic p2p environment. Thereby, we want to build semantically cohesive user subgroups, and find out the best peers for query transformation, i.e., minimizing information loss. We have shown an example for retrieving image resources annotated on p2p environment by using query transformation based on SC.

1 Introduction

Information retrieval process on the p2p networks has been performed by propagating a certain message containing a certain queries to neighbor peers and their neighbors. We assume that the queries for interactions between peers (from source peer to destination peer) are simply represented as a set of concepts derived from the ontology of source peer. For high accessibility, the queries can be transformed into the concepts of destination peer ontology. The concepts in the original query can be replaced to the correspondent concepts resulting from ontology alignment between peer ontologies. More importantly, we propose a novel measurement of semantic centrality (SC), which expresses the power of controlling *semantic* information on semantic p2p network, and show that it is applied to search for the most proper peers for concept-based query transformation.

Thereby, in this study, we introduce a three-layered structure[1] made of superposed networks that are assumed to be strongly linked:

Social layer relating peers (or people) on the basis of common interest;
Ontology layer relating ontologies on the basis of explicit import relationships or implicit similarity;

[1] Please refer to [1] for more description in detail.

G. Dong et al. (Eds.): APWeb/WAIM 2007, LNCS 4505, pp. 622–629, 2007.

Concept layer relating concepts on the basis of explicit ontological relationships or implicit similarity.

We may call this stack of interlinked networks a semantic social space.

Generally, the networks will be characterized here as a set of objects (or nodes) and a set of relations. A network $\langle N, E^1, \ldots E^n \rangle$ is made of a set N of nodes and n sets of object pairs $E^i \subseteq N \times N$ the set of relations between these nodes. These networks can express the relationships between people or many other sort of items. As usual, a path p between node e and e' is a sequence of edges $\langle e_0, e_1 \rangle, \langle e_1, e_2 \rangle, \ldots, \langle e_{k-1}, e_k \rangle$ in which $e_0 = e$ and $e_k = e'$. The length of a path is its number of edges (here k) and the shortest path distance $spd(e, e')$ between two nodes e and e' is the length of the shortest path between them. By convention, $spd(e, e) = 0$.

Definition 1 (Distance network). *A distance network $\langle N, E^1, \ldots, E^n \rangle$ is made of a set N of nodes and n sets of distance functions $E^i : N \times N \longrightarrow [0\ 1]$ defining the distance between nodes (so satisfying symmetry, positiveness, minimality, and triangular inequality).*

It is clear that any network is a weighted network which attributes either 0 or 1 as a weight. The definition of social network analysis can be adapted to distance networks if each time the cardinality of a set of edges if used, it is replaced by the sum of its distances. The distance of a path is obtained by summing the distances of its edges.

In the three-layered model we design to propagate the relational information (e.g., the distance or similarity) not only within a layer but also between layers. We have provided the principles for extracting similarity between concepts in different ontologies and propagating this similarity to a distance and an alignment relation between ontologies. We compute semantic affinities between peers, so that the semantic subgroups can be discovered. By using topological features of the discovered subgroups, two centrality measurements (e.g., local and global centralities) can be obtained. Finally, these centralities are applied to determine the best path on which the queries can travel in p2p network.

2 Inferring Relationships

The numerous relationships that can be found by construction of the concept layer, new relationships can be inferred between the entities. One particularly interesting relationship is similarity: in order to find relationship between concepts from different ontologies, identifying the entities denoting the same concept is a very important feature. As a matter of fact, most of the matching algorithms use some similarity measure or distance in order to match entities.

A distance between two ontologies can be established by finding a maximal matching maximising similarity between the elements of this ontology and computing a global measure which can be further normalised:

Definition 2 (Ontology distance). *Given a set of ontologies N_O, a set of entities N_C provided with a distance function $E_C^{dist} : N_C \times N_C \longrightarrow [0\ 1]$ and a relation $Defines$:*

$N_O \times N_C$, *the distance function* $E_O^{dist} : N_O \times N_O \longrightarrow [0\ 1]$ *is defined as:*

$$E_O^{dist}(o,o') = \frac{max(\sum_{\langle c,c'\rangle \in Pairing(Defines(o),Defines(o'))} E_C^{dist}(c,c'))}{max(|Defines(o)|, |Defines(o')|}$$

The resulting measure is minimal ($\forall o \in N_O, E_O^{dist}(o,o) = 0$), but is not guarantee to be a distance unless we apply a closure with the triangular inequality.

This is the measure that is used in the OLA algorithm for deciding which alignment is available between two ontologies [2]. However, other distances can be used such as the well known single, average and multiple linkage distances.

This ontology distance introduces a new relation on the ontology layer which provides a good idea of the distances between ontologies. It is, in turn, a clue of the difficulty to find an alignment between ontologies. It can be used for choosing to match the closest ontologies with regard to this distance. This can help a newcomer in a community to choose the best contact point: the one with whom ease of understanding will be maximised.

It can however happen that people have similar but different ontologies. In order for them to exchange their annotations, they need to know the alignments existing within the ontology network. As the result of applying alignment algorithms, the similarity or distance on the network is the basis for many matching algorithms [2]. Manually extracted alignments can also be added to this relation.

As a result, from concept similarity these algorithms will define a new relation E^{align} at the ontology level.

Definition 3 (Alignment relation). *Given a set of ontologies* N_O*, a set of entities* N_C *provided with a relation* $E_C^{dist} : N_C \times N_C$*, and a matching algorithm* $Match$ *based on* E_C^{dist}*, the alignment relation* $E^{align} \subseteq N_O \times N_O$ *is defined as:*

$$\langle o,o'\rangle \in E^{align} \textit{ iff } Match(o,o') \neq \emptyset$$

If one has a measure of the difficulty to use an alignment or of its quality, this network can also be turned into a distance network on which all these measures can be performed. Of course, when an alignment exists between all the ontologies used by two peers, there is at least some chance that they can talk to each others. This can be further used in the social network.

This new relation in the ontology layer allows a new agents to choose the ontology that it will align with first. Indeed, the ontologies with maximal hub centrality and closeness for the alignment network are those for which the benefit to align to will be the highest because they are aligned with more ontologies. In the peer-to-peer sharing application, choosing such an ontology will bring the maximum answers to queries.

This is the occasion to note the difference between the relations in the same network: in the ontology network, the hub ontologies for the import relation are rather complete ontologies that cover many aspects of the domains, while hub ontologies for the E^{align} relation are those which will offer access to more answers.

Once these measures on ontologies are obtained, this distance can be further used on the social layer. As we proposed it is possible to think that people using the same ontologies should be close to each other. The affinity between people can be measured from the similarity between the ontology they use.

Definition 4 (Affinity). *Given a set of people N_S, a set of ontologies N_O provided with a distance $E_O^{dist}: N_O \times N_O \longrightarrow [0\ 1]$ and a relation $Uses: N_S \times N_O$, the affinity is the similarity measure defined as*

$$E^{aff}(p, p') = 1 - \frac{max\left(\sum_{\langle o,o'\rangle \in Pairing(Use(p),Use(p'))} 1 - E_O^{dist}(o, o')\right)}{max(|Use(p)|, |Use(p')|)} \quad (1)$$

Since this measure is normalised, it can be again converted to a distance measure through complementation to 1.

Introducing the distance corresponding to affinity in the social network allows to compute the affinity relationships between people with regard to their knowledge structure. Bottom-up inference from $\mathcal{C}$ allows to find out the semantic relationships between users based on this space.

3 Transformation Path Selection

Affinity measurements between people (in Equ. 1) can play a role of the strength of social tie on a semantic social network. Then, we can apply various social network analysis methods to discover meaningful patterns from the social layer $\mathcal{S}$. In this study, by using cohesive subgroups (communities) identification [3], the linkages on the p2p network should be re-organized to discriminate which peers are more proper to support interoperability among peers.

Basically, the interactions between peers are based on exchanging messages, including either a certain query or answer sets. To make queries understandable on heterogeneous peers, the queries have to be transformed with referring to the corresponding peer ontologies. The peer sending queries should select some other neighbor peers to ask query transformation with their own peer ontologies.

Definition 5 (Query). *A query q can be embedded into a message $\langle p_{src}, p_{dest}, q\rangle$ sent from peer p_{src} to p_{dest}. The ontologies of two peers are denoted as $o_{src} = Use(p_{src})$ and o_{dest}. The query grammar is simply given by $q ::= c|\neg q|q \wedge q|q \vee q$ where $c \in Define(o)$.*

In this study, we are interested in queries consisting of a set of concepts from the peer ontologies, so that the queries can be transformed by *concept replacement* strategy based on correspondences discovered by ontology alignment.

Definition 6 (Correspondence). *A set of correspondences discovered ontology alignment process between two ontologies o_i and o_j is given by*

$$\{\langle c_i, c_j, rel\rangle | E^{align}(o_i, o_j), c_i \in Define(o_i), c_j \in Define(o_j)\} \quad (2)$$

where rel indicates a relation between two classes (e.g., equivalence, subclass, superclass, and so on).

For example, if there exist correspondences $\{\langle c_\alpha^1, c_\beta^3, =\rangle, \langle c_\alpha^2, c_\beta^4, =\rangle\}$ between peer ontologies o_α and o_β, a peer query "$q_\alpha = c_1 \vee c_2$" from α can be transformed to "$q_\beta = c_3 \vee c_4$" for β.

However, we have to deal with the problems;

- what if the correspondences are not enough to transform the queries sent?
- which peers can efficiently help this transformation process?

Thereby, main scheme of our approach is to find out the best transformation path, minimizing information loss from ontology alignment process. In order to reduce information loss caused by ontology mismatching during transforming queries, we can intuitively consider two heuristic criterion; i) minimizing the number of transformations (or length of transformation path), and ii) maximizing the semantic similarities (or correspondences) with neighbors. Instead of meeting these two objectives, we focus on searching for the most *powerful* peer, most likely to help them communicate with each other.

3.1 Measuring Semantic Centrality

When sending a query on semantic p2p network, we need to find out which peer (more exactly, peer ontology) is most useful to transform the query for interoperability between source and destination peer. Thereby, SC of each peer is measured by peer ontology alignment. By mapping peer ontologies, consensual ontology can be built and applied to semantic community identification.

Based on the strengths of social ties E^{aff} between pairs of peers, we can apply a non-parametric approach, e.g., nearest neighborhood method [4]. As extending [3], this task is to maximize "semantic" modularity function $Q^{\diamond}$ on social layer $\mathcal{S}$. With the number of communities k predefined, we find out that the given peer set in a social layer $\mathcal{S}$ can be partitioned into a set of communities (or subgroup) $\mathcal{G} = \{g_1, \ldots, g_k\}$. The users can be involved in more than one community. It means that a certain peer p in g_i can also be taken as one of members of g_j, because the semantics in his ontology is relatively close to both consensus semantics of g_i and g_j. Thus, the modularity function $Q^{\diamond}$ is formulated by $Q^{\diamond}(\mathcal{S}) = \sum_{i=1}^{k} \frac{\sum_{p_a \in g_i, p_b \in g_i} E^{aff}(p_a, p_b)}{|g_i|}$. The only pairs of peers where $E^{aff}(p_a, p_b) \geq \tau_{aff}$ should be considered. Thus, $\mathcal{G}(\mathcal{S})$ can be discovered when $Q^{\diamond}(\mathcal{S})$ is maximized. For computing this, in this paper, we applied an iterative k-nearest neighborhood methods. As changing k, consequently, the social layer is hierarchically re-organized.

Generally, centrality measures of a user are computed by using several features on the social network, and applied to determine the structural power. So far, in order to extract the structural information from a given social network, various measurements such as centrality [5], pair closeness [6], and authoritative [7] have been studied to realize the social relationships among a set of users. Especially, the centrality can be a way of representing the geometrical power of controlling *information flow* among participants on p2p network.

We define two kinds of semantic centralities, with respect to the scope and the topologies of communities;

- *Local* semantic centrality $C_L^{\diamond}$, meaning the power of semantic bridging between the members within the same community, and
- *Global* semantic centrality $C_G^{\diamond}$, implying the power of bridging for a certain target community.

Local SC of peer $p \in g_i$ is easily measured by $C_L^{\diamond}(p, g_i) = \frac{\sum_{p,p' \in g_i, p \neq p'} E^{aff}(p,p')}{|g_i|}$, because we are concerning only $E^{aff}(p_a, p_b) \geq \tau_{aff}$ and regarding them as most potential transformation paths. This is similar to the closeness centrality.

On the other hand, global SC $C_G^{\diamond}$ of peer $p \in g_i$ toward a certain target community g_X is based on three factors; $i)$ the number of available transformation paths (s.t. $E^{aff} \geq \tau_{aff}$), $ii)$ the strength of each path E^{aff}, and $iii)$ the local SC of the peer in target community. Thus, we formulate it as three different ways;

$$C_G^{\diamond}(p, g_X) = \frac{\sum_{p' \in g_X} E^{aff}(p, p') \times C_L^{\diamond}(p', g_X)}{|g_X|} \quad (3)$$

$$= \frac{\left[\max_{p' \in g_X} E^{aff}(p, p')\right] \times C_L^{\diamond}(p', g_X)}{|g_X|} \quad (4)$$

$$= \frac{\max_{p' \in g_X} \left[E^{aff}(p, p') \times C_L^{\diamond}(p', g_X)\right]}{|g_X|} \quad (5)$$

While Equ. 3 can take into account all possible paths to taget community by measuring the average centrality, Equ. 4 and Equ. 5 are focusing on only the maximum affinity path. We empirically evaluated these three different heuristic functions in Sect. 4.

3.2 Query Transformation Strategy

We establish query transformation strategy in accordance with the semantic position of peers in social layer $\mathcal{S}$. Query transformation between heterogeneous peers should be conducted by referring to the following strategies;

- If the peers p and p' are located in a same semantic community, a set of transformation paths $TP_L(p, p')$ between them can be evaluated (or ranked) by $\frac{\sum_{p'' \in TP_L} C_L^{\diamond}(p'')}{\exp(1+|TP_L|)}$ where p'' is on the transformation path TP_L. It means the best transformation path has to be chosen, as the length of the path is shorter and local semantic centralities of the peers on the path are higher.
- If the peers $p_i \in g_i, p_j \in g_j$ are in different semantic communities, a set of transformation paths $TP_G(p_i, p_j)$ between them can be evaluated (or ranked) by $TP_{L_i}(p_i, p'_i) + C_G^{\diamond}(p'_i, g_j) + TP_{L_j}(p'_j, p_j)$, and this can be expanded as $\frac{\sum_{p''_i \in TP_{L_i}} C_L^{\diamond}(p''_i)}{\exp(1+|TP_{L_i}|)} + C_G^{\diamond}(p'_i, g_j) + \frac{\sum_{p''_j \in TP_{L_j}} C_L^{\diamond}(p''_j)}{\exp(1+|TP_{L_j}|)}$. A global transformation path is decomposed into two local transformation path and a transformation path with best global centrality. Exceptionally, when there is no path between communities, the social layer should be re-organized as decreasing the number of communities k.

Thereby, the best transformation path have to be selected by comparing all candidate ones.

4 Experimental Results

In order to evaluate the proposed approach, we invited seven students and asked them to annotate a given set of images by referring to any other standard ontologies (e.g.,

SUMO, WordNet and ODP). While annotating the images, we could collect peer ontologies for building semantic social space.

4.1 From Peer Ontologies to Social Ties

Here, we want to show the experimental results of building social semantic space by ontology alignment. They are compared with simple co-occurrence patterns between the annotated images by Mika's social centrality C_M [8], which is formulated by $C_M(U_i) = \frac{\sum \frac{\cap_{k=1,k\neq i}^{|U|}(\mathcal{R}_{U_k},\mathcal{R}_{U_i})}{\mathcal{R}_{U_i}}}{|U|-1}$ where $|U|$ is the total number of peers (or people) on social network. The results are shown in Table 1. We found out that the number

Table 1. Experimental results of a) closeness centrality by co-occurrence patterns, and b) semantic affinity E^{aff} and centrality in semantic social network

(a/b)	AS	AZ	FAK	JE	JJ	JP	SL	C_M	$C_L^\diamond$
AS	-	0.98/0.65	0.62/0.33	0.94/0.73	1.00/0.26	0.60/0.32	0.23/0.62	0.73	0.49
AZ	0.98	-	0.62/0.49	**0.94/0.825**	0.98/0.31	0.62/0.3	0.26/0.52	0.73	0.52
FAK	0.78	0.78	-	0.70/0.57	0.78/0.28	0.54/0.22	0.30/0.32	0.65	0.37
JE	0.90	**0.90**	0.53	-	0.90/0.46	0.57/0.49	0.16/0.75	0.66	**0.64**
JJ	1.00	0.98	0.62	0.94	-	0.60/0.72	0.23/0.39	0.73	0.40
JP	0.93	0.97	0.67	0.93	0.93	-	0.13/0.51	**0.76**	0.43
SL	0.44	0.48	0.44	0.32	0.44	0.16	-	0.38	0.52

of annotated resources are barely related to the social centrality. SL annotated the least number of resources, so that his centrality also lowest among people. But, even though JE's annotations were the largest one, JP has shown the most powerful centrality.

4.2 Heterogeneous Query Processing

From the organized three groups $g_A = \{JE, AZ\}$, $g_B = \{JJ, JP\}$, and $g_C = \{AS, FAK, SL\}$ (the number of communities $k = 3$), we compared the image results retrieved by ten concept-based queries generated by every peers, according to the transformation strategies. In Table 2, we show "Precision" performance, because we are emphasizing the information loss effected from query transformation. We found out that Equ. 3 has outperform the others by about 19 % and 11%.

Table 2. Precision performance on query transformation strategies; *stp* means the simple shortest path on social layer

	g_A				g_B				g_C			
	stp	Equ. 3	Equ. 4	Equ. 5	*stp*	Equ. 3	Equ. 4	Equ. 5	*stp*	Equ. 3	Equ. 4	Equ. 5
g_A	0.72	0.75 by Local SC			0.36	0.71	0.57	0.64	0.36	0.74	0.59	0.67
g_B	0.317	0.67	0.54	0.6	0.64	0.69 by Local SC			0.34	0.78	0.62	0.7
g_C	0.425	0.63	0.51	0.57	0.41	0.68	0.54	0.64	0.685	0.67 by Local SC		

5 Discussion and Concluding Remark

Semantic overlay network Various applications (Edutella, Bibster, and Oyster) for sharing resources on p2p network have been released. Most similarly, semantic overlay network [9] concerns query processing for information sharing on p2p network, but it is based on simple keyword matching to estimate the relationships between nodes.

As another important issue, we want to carefully discuss information loss by semantic transformation. While equivalent correspondences (e.g., $\langle c, c', = \rangle$) are acceptable, subsumption correspondences make the transformed queries more specific, and the resources retrieved from peers may (possibly) show higher precision and lower recall results.

As a conclusion, in this paper, we claim a new centrality measurement for providing query-based interactions on p2p network. Especially, we found out very efficient transformation path selection mechanism (e.g., Equ. 3). Moreover, by peer ontology alignment, consensus ontology has been built and applied to identify some semantic communities. We believe that it will play a role of generating semantic geometry to quantify social roles on p2p network.

References

1. Jung, J.J., Euzenat, J.: From personal ontologies to semantic social space. In: Poster of the 4th European Semantic Web Conference (ESWC 2006). (2006)
2. Euzenat, J., Valtchev, P.: Similarity-based ontology alignment in OWL-Lite. In de Mántaras, R.L., Saitta, L., eds.: Proceedings of the 16th European Conference on Artificial Intelligence (ECAI'2004), Valencia, Spain, August 22-27, 2004, IOS Press (2004) 333–337
3. Newman, M.E.J.: Fast algorithm for detecting community structure in networks. Physical Review E **69** (2004) 066133
4. Gowda, K.C., Krishna, G.: Agglomerative clustering using the concept of mutual nearest neighbourhood. Pattern Recognition **10**(2) (1978) 105–112
5. Haga, P., Harary, F.: Eccentricity and centrality in networks. Social Networks **17**(1) (1995) 57–63
6. Girvan, M., Newman, M.E.J.: Community structure in social and biological networks. Proceedings of the National Academy of Sciences **99** (2002) 7821–7826
7. Kleinberg, J.M.: Authoritative sources in a hyperlinked environment. Journal of ACM **46**(5) (1999) 604–632
8. Mika, P.: Ontologies are us: A unified model of social networks and semantics. In Gil, Y., Motta, E., Benjamins, V.R., Musen, M.A., eds.: Proceedings of the 4th International Semantic Web Conference (ISWC 2005), November 6-10, 2005. Volume 3729 of Lecture Notes in Computer Science., Springer (2005) 522–536
9. Crespo, A., Garcia-Molina, H.: Semantic overlay networks for p2p systems. In Moro, G., Bergamaschi, S., Aberer, K., eds.: Proceedings of the 3rd International Workshop Agents and Peer-to-Peer Computing (AP2PC 2004), July 19, 2004. Volume 3601 of Lecture Notes in Computer Science., Springer (2005) 1–13

Mining Infrequently-Accessed File Correlations in Distributed File System

Lihua Yu, Gang Chen, and Jinxiang Dong

College Of Computer Science, Zhejiang University,
Hangzhou, Zhejiang, P.R. China, 310027
peterylh@163.com, cg@zju.edu.cn, djx@zju.edu.cn

Abstract. File correlation mining, as a technique to enhance file system performance, can usually be exploited for many purposes such as to improve the effectiveness of cache, to optimize file layout, as well as to enable disk file prefetching. While most research works on file correlations focus on traditional stand-alone file systems, this paper investigates the problem of mining file correlations in a distributed environment. We present a parallel data mining algorithm called PFC-Miner (Parallel File Correlation Miner), which is based on Locality Sensitive Hashing. PFC-Miner can efficiently discover correlations between infrequently-accessed files which are more valuable for web applications. Experimental results show that PFC-Miner can efficiently discover file correlations in distributed file systems without compromising the accuracy, and that the proposed approach has good scalability.

Keywords: File Correlation, Parallel Data Mining, Distributed File System.

1 Introduction

Nowadays, due to the dramatic growth of the number of Internet users, the response time has become the central performance barrier for many websites. Analysis of statistics of web sites has shown that the file system as one of the fundamental infrastructures of web applications, often becomes the bottleneck. With the development of high bandwidth devices, such as RAID and SAN, throughput of file systems has seen remarkable improvements. However, the access latency still remains a problem, which is not likely to improve significantly due to the physical limitations of storage devices and network transfer latencies [3].

Several techniques, such as caching, prefetching, disk layout optimization and so on, can be used to reduce access latency. Being automatic or not, these techniques all heavily depend on file access patterns. The automatic methods usually deploy techniques such as caching, prefetching, or layout optimization based on predictions of file access patterns. In comparison, non-automatic methods are based on access pattern hints given by applications. Simple access patterns, such as temporal locality, spatial locality and sequential locality, have been heavily exploited in commodity storage systems, file systems and database systems. However, file access patterns in real system may be more complex, making it difficult to be captured accurately.

G. Dong et al. (Eds.): APWeb/WAIM 2007, LNCS 4505, pp. 630–641, 2007.

Moreover, access latency may be exacerbated, when the predicted access patterns mismatch access patterns of application.

File correlations are one of the common access patterns in file systems. Obviously, files belong to same application are usually correlated by application semantics. For example, C source code file is correlated to its corresponding header files, and pictures in a web page are correlated to other pictures in the same page. Because correlated files are usually accessed together, correlations can be harnessed to improve storage cache hit ratio, optimize file layout, prefetch more accurately and reduce file access latency. Correlations are also helpful in file hoarding in the mobile environments, which lead to better support for disconnected operation [4][19].

There are already some techniques in the literature, which aims at mining file correlations in stand-alone file systems, such as *semantic distances* [4][7], *dependency graph* [3], *stable successor* (or Noah) [5] and *recent popularity* [6], etc. However, to the best of our knowledge, these methods are not designed for searching correlated files in distributed file systems. Finding file correlations in distributed storage systems is more challenging: Firstly, the huge number of files desires fast and scalable algorithm. Secondly, the existences of both inter and intra server correlated files require efficient communication between servers. Finally, the distributed environments make it difficult to maintain global data structures required by most previous approaches.

Another major disadvantage of previous methods is that they focus on high support correlated files. (Support is defined in Section 3, higher support indicates higher access frequency). However low support correlated files are more valuable for web applications, especially for web 2.0 applications. For example, a typical Bloger access his Blog once a day, which leads to low file access frequency. Consequently, the correlation supports of typical user's files are low. Thereby, user feeling of Blog System can be improved dramatically by prefetching correlated files of a typical user. In contrast, files with high support probably exist in system cache, making prefetching useless. In Order to create better user feeling, we emphasize on higher correlation instead of higher correlation support.

This paper proposes PFC-Miner (Parallel File Correlations Miner) algorithm to discover file correlations efficiently in distributed file systems using file access logs. The algorithm is based on M-LSH (Min Locality Sensitive Hash) mining algorithm [2], which is efficient in finding interesting association rules without support threshold. The algorithm is parallel and designed for execution in shared nothing distributed clusters. Instead of maintaining global data structures, the algorithm exchanges candidates between file servers using signature join with low communication overhead. The algorithm is both fast and scalable, making it a practical method for discovering file correlations in large file systems. Additionally, this algorithm can find file correlations with very low support which are more valuable for web applications.

The main contributions of this paper are as following:

1. Analysis and definition of file correlations in distributed file systems.
2. A parallel correlation mining algorithm to discover file correlations without support threshold.
3. Evaluation of the scalability and accuracy of the algorithm.

The remainder of the paper is organized as follows. In Section 2, we introduce file correlation and related techniques. In Section 3, we present our correlation mining algorithm. The experiment results are given in Section 4. In Section 5, we discuss related works. Section 6 concludes the paper.

2 Preliminary

2.1 File Correlations and Block Correlation

File correlations are ubiquitous in file systems. A set of files are correlated if they are "linked" or "glued" together semantically. For example, an HTML file is correlated to images embedded in it. At the same time, the images are correlated to each other, because they are glued together by this web page. Other examples include images in same photo albums, C source files, etc.

Correlated files tend to be accessed together during their life time, because correlations reflect application level semantics. So correlations can be exploited to improve storage system performance by directing prefetching, optimizing file layout in disk and increasing cache hit ratio.

The concept of file correlations is similar to block correlations [8][9]. They both explore the semantic correlations between objects to enhance storage system performance. Obviously, the first difference between file correlations and block correlations is level of abstraction. The former interests in files, while the latter focuses on blocks. As a result, file correlations are more useful in applications with simple block access patterns, such as web applications. Neighboring blocks prefetching found in commodity file systems is sufficient in these cases, because files are sequentially accessed in most web applications. Conversely, block correlations are more useful in applications with complex block access patterns, such as databases. The second major difference is scalability. The number of blocks in storage system is much larger than the number of files. As a result, it is more difficult to discover block correlations in distributed environments.

2.2 Obtaining File Correlation

There are two categories of file correlations obtaining approach, namely, informed and automatic approaches. In Informed approaches, applications inform the system about the set of correlated files. Dynamic Set [11] is an informed approach proposed to reduce file system latency. In this scheme, application puts correlated files into dynamic sets which are created on demand to enhance performance.

In automatic approach, system automatically reveals file correlations of application. Griffioen and Appleton presented a file prefetching scheme relying on dependency graph [3]. The graph tracks file accesses observed within a certain window after the current access. For each file access, the probability of its different followers observed within the window is used to make prefetching decisions. Their simulations show that different combinations of window and threshold values will largely affect the performance. Kuenning and Popek propose Semantic Distance method, which use temporal, sequence and lifetime semantic distance to measure correlation [4]. Yeh et al.

investigated Program-Based model that identifies the relationships between files through identification of the programs accessing them [14].

The approaches mentioned above work well in traditional file systems, however they may not be suitable in distributed file systems for web applications. There are three reasons. First, it is difficult and error prone to maintain global data structures in distributed system. Second problem is scalability. Space and time overhead of graph-based and Semantic Distance methods is high when the number of files is huge [9]. Third, they are not efficient in finding correlated files with low support. Finally, it is difficult to obtain process information of application for Program-Based approach, because the application code may be running on other servers.

2.3 Interesting File Correlations

Researchers have found that the object access model roughly follows Zipf-like distributions [14], which state that access frequency of the ith% popular object p_i is estimated to be:

$$p_i = k/i, where\ k = 1/\sum_i 1/i^{\alpha}, 0 < \alpha \leq 1 \tag{1}$$

As shown in formula 1, the access frequencies of majority files are small. However, the file correlations between these files are more interesting to us. Firstly, the probability that frequent accessed file is in cache is high. Secondly, these kinds of files represent the majority of users' access behavior, for typical users accessing web server once or several times a day. Therefore we interest in Files with high correlations but with very low support.

3 PFC-Miner

In this section, we describe distributed file system model and the PFC-Mine algorithm. The summary of notations used is listed in table 1.

Table 1. Notations summary

Notation	Description
CoS^*	Correlation support threshold
f_i	The ith file
k	The number of random numbers generated per log record
r	The row count of M's submatrix
l	Sub-matrix count of M
M	The k-Min-Hash matrix
OSN_i	The ith Object based Storage Node
H_x	Hash signature on x^{th} OSN
T_x	Inverted signature table on x^{th} OSN
C_{xy}	Correlation candidates between x^{th} and y^{th} OSN

3.1 File System Model

This section describes distributed file system model used in this paper. As shown in figure 1, it is a prevalent distributed object store based file system [12][13]. The system runs on commodity hardware, uses Object based Storage Node (OSN) for storage and metadata servers (MDS) for storing metadata. Unlike traditional files system, it has flat file naming method, that is, files are accessed using unique file ID instead of file names. To enhance performance, file data transfers directly between applications and OSNs. So naturally, file access logs are distributed on each OSN.

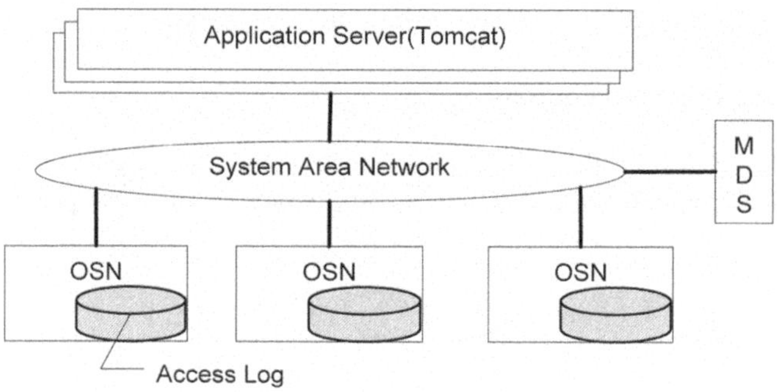

Fig. 1. Distributed Object Store based File System Model

3.2 Correlation Measurements

Access log consists of log entries, and a log entry in this paper is a tuple as <*ts*, *fileid*>, where *fileid* is the ID of the accessed file, *ts* is the timestamp of the access. Log entries are preprocessed and grouped into log records, and all log entries in the same record are regarded as accessed together. Because files stored on different OSNs may be accessed together, a log record is naturally partitioned among all OSNs. Furthermore, each log record is assigned an ascending record ID starting from 1.

Correlated files tend to be accessed together, so naturally, we regard files accessed together frequently as correlated files. Intuitively, we define correlation among a pair of files as: count of concurrent access / total access count of both file. Suppose there are two files f_i and f_j, the set of records containing f_i is R_i, the set of records containing f_j is R_j. Correlation of the two files $Co(f_i, f_j)$ is defined as:

$$Co(f_i, f_j) = \frac{|R_i \cap R_j|}{|R_i \cup R_j|} . \tag{2}$$

The value of $Co(f_i, f_j)$ is between 0 and 1, larger value indicating higher correlation between f_i and f_j. To make correlation more meaningful, we define correlation support, $CoS(f_i, f_j)$, as minimum access count of f_i and f_j:

$$CoS(f_i, f_j) = \min\{|R_i|, |R_j|\} \tag{3}$$

Correlation make no sense unless $CoS(i, j)$ is larger than a relative small threshold. For example, suppose f_i is accessed only once, i.e. $|R_i| = 1$. Obviously the correlation of f_i is not meaningful, because there is not enough information about that file.

3.3 Preprocessing

Association rule mining algorithms including M-LSH are designed to discover rules in databases rather than streams of access logs. Thereby, just like other log mining algorithms, preprocess is necessary to break logs into short access records.

Figure 2 shows four cutting methods to preprocess access logs into access records. These methods can be explained in two dimensions. In one dimension, break by count splits log into records with fixed number of log entries. And break by time method tries to group log entries whose timestamps within a same time interval together into a record. In other dimension, non-overlapped method divides the log into non-overlapping records with same size. In contrast, consecutive records overlap in overlapping method.

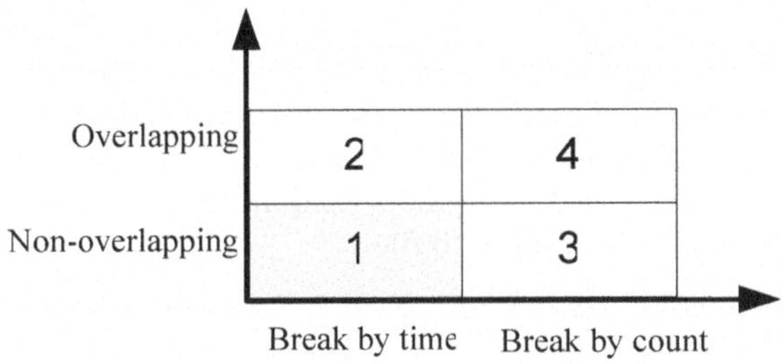

Fig. 2. Preprocessing Methods

We use break by time and non-overlapping methods in this paper. As mentioned above, a single access record may be distributed on many OSNs, because files on different OSNs may be accessed together. Assuming that OSNs have loosely synchronized clocks (which can be easily synchronized using SNTP protocol), each OSN can preprocess its own log without interacting with others in break by time method. Further, we choose non-overlapping to improve efficiency. With relative large time interval, the amount of lost information due to non-overlapping cutting is relative small.

3.4 Core Algorithm

The algorithm for identifying correlated files consists of three stages: compute Min-Random values; compute hash signatures and create inverted signature index; perform signature join and generate intra-server results. In the first stage, each OSN scans its local log records, prunes files with too low access count and generates a small set of Min-Random values for each file. The set of Min-Random values is expected to be small and will fit into memory. Then in the second stage, we generate even smaller hash signatures from Min-Random values, create inverted signature index which maps a hash signature to a set of file IDs. Finally, in the last stage, we join different files on different OSNs using hash signatures, generate candidates using inverted signature index, calculate candidates' correlations and output correlated files.

Although PFC-Miner is based on M-LSH [2], there are three major differences between them. First, it is a parallel algorithm working in distributed environments. Second, PFC-Miner uses pre-computed inverted signature index and signature join to reduce the number of candidates. Third, to improve performance, we compute correlations using Min-Random values without another scan of access log to prune false positives. As shown in experiments, the additional number of false positives is small, especially when correlation support is low.

Algorithm 1. Hash signature computing

```
Input: k-Min-Random values matrix M
Output: inverted signature table T, hash signature H
for each f_i in M {
   For j from 0 to l-1 {
      h_i= hash(M_{j*l,1}, M_{j*l+1,i},..., M_{j*l+r-1,i})
     H = H    { h_i }
     Insert into (h_i ,f_i) into T
   }
}
```

The first phrase algorithm is similar to M-LSH, we choose k random functions, $h_1,\ldots, h_k$. Then each OSN scans in parallel through local access records and generates k random values for each record using the k random functions. At the same time, we calculate k-Min-Random values for each file.

Definition 1. *Min-Random value, $h_x(f_i)$, is the minimum random value of the records containing file f_i, which is generated by hash function h_x. k-Min Random values of f_i is defined as all k Min-Random values, i.e. $h_1(f_i)$, $h_2(f_i),\ldots,h_k(f_i)$.*

All k-Min-Random values of $f_1, f_2, \ldots, f_m$ make up a k-Min-Random matrix M, with k rows and m columns, where $M_{ij} = h_i(f_j)$. M can be viewed as a compact representation of access log. The correlation of two files f_i and f_j is captured by their column similarity $S_M(f_i, f_j)$.

Definition 2. *$S_M(f_i, f_j)$ is the fraction of Min-Random values that are identical for f_i and f_j.* As proven in M-LSH, $S_M(f_i, f_j)$ is good estimation of $Co(f_i,f_j)$, when k is relatively large.

$$S_M(f_i, f_j) = \frac{|\{l \mid 1 \le l \le k \cap M_{li} = M_{lj} \mid}{k} \,. \tag{4}$$

In second stage, we apply Locality Sensitive Hashing (LSH) technique to k-Min-Random values matrix M. As shown in Algorithm 1, the algorithm first splits M horizontally into l sub-matrices of r rows. Recall that M has k rows, so $k = l \times r$. Then for each the l sub-matrices, we repeat following. A hash signature value *sig* is generated for each file f_i using as hashing key the concatenation of all r values, then (sig, f_i) is added to inverted signature table T. Consequently, the algorithm generates total l hash signature values for each file f_i, and these l values are defined as the hash signature of f_i. At the same time, hash signature, H, is computed as the set of distinct hash values of all files. Because correlated files tend to have same hash signature values, files with at least one same hash signature values are considered as candidates in the third stage.

As shown in Algorithm 2, the third stage algorithm performs hash signature join between OSNs. Each OSN first broadcasts local hash signature, H_x, to all other OSNs. On receiving hash signature H_y from OSN y, local inverted signature table T_x is searched using hash signature values in H_y, inter-server correlation candidates C_{xy} which is the set of files with at least one hash signature values in H_y is generated, and is sent back to OSN y. When candidate set C_{yx} is received from OSN y, correlated file pairs can now be generated using C_{yx} and T_x. Note that, it isn't necessary to broadcast H_x to all other OSNs, because file correlation is symmetric. When the number of OSNs is odd, we need only broadcast H_x to half of the other OSNs. When there are even OSNs (more than two) in system, we add a fake node, messages broadcasted to whom are not really sent. These two cases are shown in figure 3.

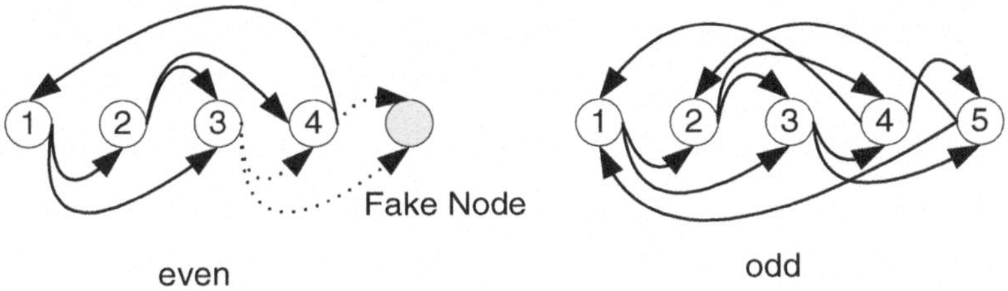

Fig. 3. Broadcast relationships

Signature join can significantly reduce search space for correlated files, with adjustable false negatives and false positives. As proven in M-LSH, both the false positive and false negative ratio are small by choose relatively large r and l.

Algorithm 2. hash signature join

Input: k-Min-Random values matrix M
Ouput: correlated files
perform in parallel at each OSN:
  broad cast local signature H_x to all other selected OSN in system
  for each received H_y from other OSN y {
  for each h_i in H_y {
    find the set of file $F_x(h_i)$ in T_x with signature values h_i
      $C_{xy} = C_{xy} \cup \{ <h_i, F_x(h_i)> \}$
    }
    send C_{xy} to server y
  }
  for each received C_{yx} from OSN y {
    for each $<h_i, F_y(h_i)>$ in C_{yx} {
      find $F_x(h_i)$ in T_x with hash values h_i
      Calculate correlated pair of files from $F_x(h_i)$ and $F_y(h_i)$
    }
  }

4 Experiments

We evaluated real world benefits, scalability and accuracy of PFC-Miner. In all tests, we use the same hardware and software environments: Xeon 2.0G CPU, 2 GB memories, 1 GB/s link, Debian Linux operation system.

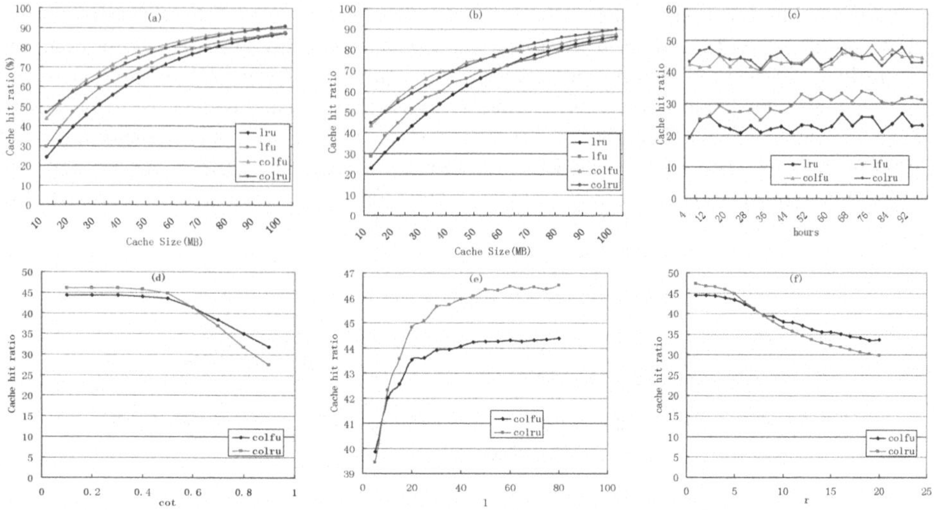

Fig. 4. (a) Cache hit ratio of first 3 day trace. (b) Cache hit ratio of another 4 day trace. (c) Correlation stability. (d), (e), (f) the choice of mining parameters' impact on cache hit ratio.

4.1 Correlation Guided Prefetch

To show the benefits of file correlation, we implemented a storage cache simulator with both lru (Least Recently Used) and lfu (Least Frequently Used) cache replacement policy, and evaluated the improvements on cache hit ratio by using correlation guided prefetch. In this test, we use trace driven simulation with publicly available web access logs: ClarkNet, which is a week of web log of ClarkNet WWW server and is available at http://ita.ee.lbl.gov/html/contrib/ClarkNet-HTTP.html. We divide web log into two parts, and use only the first three days of web logs to mine file correlations. Then, we trace both the two part of log guided by mined correlations. As shown in figure 4(a) and figure 4(b), colru (lru with prefetch) and colfu (lfu with prefetch) performance significantly better than lru and lfu, especially when cache size is relatively small.

Figure 4(c) shows the distribution of cache hit ratio over time when cache size equals 10M. The results of colru and colfu remain stable for next 4 day, which is always better than lru and lfu. This implies that file correlation is effective for a relatively long period of time. Therefore, there is no need to mine file correlation online continuously.

All three parameter *r*, *l* and *cot* might affect the benefits of mining. Figure (d) shows the effects of correlation threshold, i.e. *cot*. As shown in the figure, cache hit ratio drops when *cot* > 0.5, because higher *cot* leading to fewer correlated files. We can also see that the line if flat when *cot* < 0.5, which implies that low correlations are not as useful as high correlations. Figure (e) shows the impacts of sub-matrix count, i.e. *l*. Cache hit ratio grows quickly when *l* < 20, and becomes flat afterward. This can be quite expected, as correlation accuracy increase quickly with *l* and approaches real value after some threshold. Figure (f) describes that *r* has negative effect on cache hit ratio, because the mining algorithm becomes more coarse-grained when *r* grows.

4.2 Performance and Scalability

Our performance and scalability study uses synthetic data which is generated using Zipf distribution. In the scalability test, we use a data set consists of 50k different files and more than 1M accesses. We choose the data set size so that a single server is capable to compute file correlation in main memory, and hence its performance can be directly compared to parallel version of algorithm. The access log is evenly distributed across all servers to balance workload among servers. Execution time and speed up with one to eight servers are shown in figure 5(a). As the charts indicate, the algorithm is efficient and has good scalability. Nearly linear speed up is achieved when server count is between 2 and 6. Speed up is relatively small when server count is small and large. When server count is small, the large size of hash signature leads to relatively large communication time. When server count is large, local computation cost is small, leading to relatively large communication cost. Figure 5(b) shows the performance by varying data set size under 8 servers. This result indicates that execution time grows linearly with data set size.

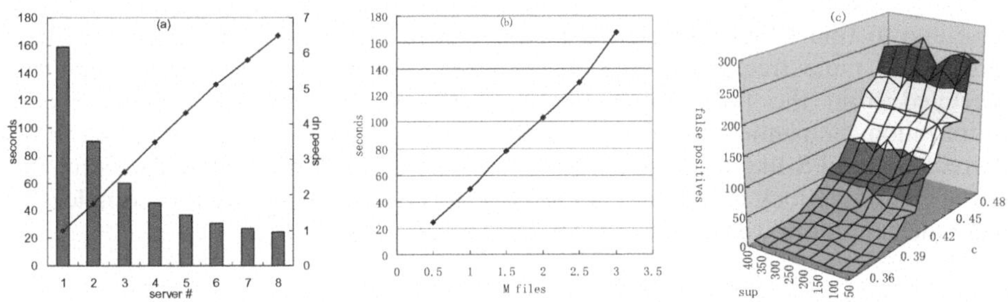

Fig. 5. (a) Elapsed time and speed up by varying the number of mining server. (b) Execution time grows linearly with data set size. (c) Mining accuracy.

4.3 Correlation Accuracy

As shown in Section 3, our algorithm estimates correlation using Min-Random values, which introduces additional false positives compared to M-LSH algorithm. Therefore we evaluate correlation accuracy by counting the number of extra false positives. The corresponding data sets consists of 1000 files, 500 correlated file pairs with correlation *c* and *sup* accesses per file. To count false positive number, we run PFC-Miner with $r = 5$, l=20, *cot*=0.5 which is larger than *c*. All correlated pairs found will be regarded as false positives, because there are no file pairs in the data set with correlation larger than the threshold. False positive numbers with different parameter *c*, *sup* are drawn in figure 5(c). As shown in the figure, false positives count drops quickly when data set's correlation *c* decreases, but remains stable when *sup* changes. For example, false positive count drops below 20 when $c = 0.4$ and equals 0 where $c < 0.36$. In a word, the mine results contain few false positives with large $cot - c$, and modest number of false positives when *cot-l* is small, which is accurate enough for to be used in real world systems.

5 Related Works

File correlation has been studied extensively in file systems. SEER [4] cluster correlated file to support better disconnected operation. They record every file a semantic distance between several closest related files. Then, correlation among files is computed using the number of shared neighbors. Griffioen and Appleton [3] uses weighed probability graph to represent relationships between files. Ahmed Amer et al [16] investigate a group based file cached management approach. The main idea is grouping a file together with its immediate and transitive successors. There are also many works [17][18] recording file relations and access patterns using trees with each node representing the sequence of consecutive file accessed from root to the node. Access tree [15] use access tree to capture relationships and dependencies between files of a user process. Several access trees are maintained for each program representing its access patterns. Then an access tree is matched using current access activity of the program, which is then used to direct file prefetching. Most of these approaches work well in traditional file systems, but may not practical in distributed file systems with huge number of files.

In the spectrum of block correlation, C-Miner [8][9] proposes an effective algorithm to find Block Correlation in file system. They use a modified version of CloSpan to mine frequent closed block access sequences. Then association rules generated from the found sequences are used to guide block prefetch, layout optimization. Semantically-Smart Disk System [10] can effectively discover Block Correlation in FFS-like file systems based on knowledge of file system data structures and operations. Block correlation is useful when block access pattern is complex, but is not meaningful in web applications with sequential block access.

6 Conclusion

The research of file correlations is important for improving file system performance. However, existing methods for extracting correlations are challenged by the increasing volume of data available nowadays. We have proposed PFC-Miner algorithm to inferring file correlations with low support threshold in distributed file systems. The algorithm computes file correlations efficiently using Min-Random Values, exchanges candidates between file servers using signature join with low communication overhead. Experiments demonstrate that the algorithm is fast and scalable with reasonable accuracy, making it practical to be used in distributed file system with large volume of files. This paper pays mostly attention to performance and accuracy up to the present, future research will focus on how to apply PFC-Miner algorithm to real world distributed file systems.

References

1. Agrawal, R., Imielinski T., Swami, A.: Mining association rules between sets of items in large databases. Proc. 1993 ACM-SIGMOD Int. Conf. Management of Data. Washington, D.C. (1993) 207-216
2. Li, J., Zhang, X.: Efficient mining of high confidence association rules without support thresholds. Proc. 3rd European Conf. Principles and Practice of Knowledge Discovery in Databases. Prague (1999)

3. Griffioen, J., Appleton, R.. Reducing file system latency using a predictive approach. In USENIX Summer Technical Conference (1994) 197–207
4. Kuenning, G., Popek, G. Automated hoarding for mobile computers. In Proceedings of the 16th ACM Symposium on Operating Systems Principles (1997)
5. Amer, A., Long, D. D. E.. Noah: Low-cost file access prediction through pairs. In Proc. 20th International Performance, Computing, and Communications Conference (2001) 27–33
6. Amer, A., Long, D. D. E., Pâris, J.-F., Burns, R. C.. File access prediction with adjustable accuracy. In Proc.21st International Performance of Computers and Communication Conference (2002) 131-140
7. Kuenning, G. H., Popek, G. J.. Automated hoarding for mobile computers. In Proceedings of the 15th Symposium on Operating Systems Principles. St. Malo, France (1997) 264–275
8. Zhenmin Li, Zhifeng Chen, Sudarshan M. Srinivasan, Yuanyuan Zhou. C-Miner: Mining Block Correlations in Storage Systems. In Proceedings of the 3rd USENIX Conference on File and Storage Technologies. San Francisco, CA (2004) 173 – 186
9. Zhenmin Li, Zhifeng Chen, Yuanyuan Zhou. Mining Block Correlations in Storage Systems. In ACM Transactions on Storage (TOS). New York, NY, USA (2005) 213 – 245
10. Sivathanu, M., Prabhakaran, V., Popovici, F., Denehy, T. E., Arpaci-Dusseau, A. C., Arpaci-Dusseau, R. H.. Semantically-Smart Disk Systems. In Proceedings of the Second USENIX Conference on File and Storage Technologies. San Francisco, CA (2003) 73 – 88
11. David Cappers Steere, Mahadev Satyanarayanan. Using dynamic sets to reduce the aggregate latency of data access. (1997)
12. Cluster File Systems. Lustre: A Scalable, High-Performance File System. http://www.lustre.org/docs/whitepaper.pdf
13. Garth A. Gibson, David F. Nagle , Khalil Amiri , Jeff Butler , Fay W. Chang , Howard Gobioff, Charles Hardin , Erik Riedel , David Rochberg , Jim Zelenka. A cost-effective, high-bandwidth storage architecture. In Proceedings of the eighth international conference on Architectural support for programming languages and operating systems. San Jose, California, United States (1998) 92-103
14. Breslau, L., Cao, P., Fan, L., Philips, G., Shenker, S.. Web Caching and Zipf-like Distributions: Evidence and Implications. In Proc. IEEE Infocom. New York, NY (1999) 126-134
15. Lei, H., Duchamp, D.. An analytical approach to file prefetching. In USENIX Annual Technical Conference. Anaheim, CA (1997).
16. Amer, A., Long, D. D. E., Pâris, J.-F., Burns, R. C.. Group-Based Management of Distributed File Caches. In Proc. International Conference on Distributed Computing Systems. Vienna, Austria (2002)
17. Vellanki, V., Chervenak, A.. A cost-benefit scheme for high performance predictive prefetching. In Proceedings of SC99: High Performance Networking and Computing. Portland (1999)
18. Kroeger, T. M., Long, D. D. E.. Predicting file-system actions from prior events. In USENIX Annual Technical Conference. (1996) 319–328
19. Trifonova, A. and Ronchetti, M. Hoarding content for mobile learning. Int. J. Mobile Communications, Vol. 4, No. 4. (2006) 459–476.

Learning-Based Trust Model for Optimization of Selecting Web Services

Janarbek Matai[1,2] and Dong Soo Han[1,*]

[1] Information and Communications University,
P.O.Box 77, Yusong, Daejeon 305-600, Korea
[2] Electronics and Telecommunications Research Institute
161 Gajeong-dong, Yuseong-Gu, Daejeon Korea
{janarbek,dshan}@icu.ac.kr

Abstract. As the deployment of Web services increases in complex business application integration, it is becoming inevitable that several Web services may have the same or similar functionalities each holds different Quality of Services(QoS). In this context, it is becoming a challenge for Web services developers to select most pertinent services for their businesses. In this paper, we suggest a novel method so called Trust Model for selecting provider Web services among several similar providers but with different QoS. Our Trust Model is a function of historically gathered and learned QoS values, references or feedbacks from other services and honesty degree of this service. Client Web services satisfaction degree function was devised for checking the satisfaction degree when using our Trust Model. The proposed model is validated with initial result by comparing two cases: when using our Trust Model for selecting Web services and when not using our Trust Model.

1 Introduction

Web services standards and technologies are expected to contribute in reducing the cost and complexity of application integration within an enterprise and across enterprise boundaries. As the deployment of Web services increases in complex business application integration, it becomes inevitable that several Web services may have the same or similar functionalities each holds different Quality of Services(QoS) [2],[7],[3],[4],[8],[5]. Due to the increasing number of Web services with the same or similar functionalities, it is getting difficult for web services consumers to find most pertinent services for their businesses. When selecting services in either dynamic or static environment, there are some issues Web service clients have to consider such as input-output dependency, semantics of services and QoS requirements of services. Among these issues the last one, QoS requirements of clients, is essential for clients to select their providers wisely [10], [11]. Since the nature of web services is very dynamic, it is hard to predict providers' QoS without testing. It leads Web services clients to face a question

* Corresponding author.

G. Dong et al. (Eds.): APWeb/WAIM 2007, LNCS 4505, pp. 642–649, 2007.

of how to select provider web service among the same or similar providers which can able to provide required services that satisfying QoS requirements of client. Moreover, clients, which request the same services may have different perspective on each required QoS of provider Web services. For instance, for the time-critical applications response time of provider is critical for clients regardless of cost of service. On the other hand, in some cases, response time is not so important but cost of service is essential. In other words, clients requiring the same type of service may request different requirements for QoS factos such as response time, availability, cost,...etc. Many researches on selecting services based on QoS metrics have been performed either in static or in dynamic environments[2],[7],[6]. However, most of the works are still in their immature state. Three approaches such as QoS broker approach, UDDIe and QoS-model are main trends so for. But none of the approaches are able to handle problem of selecting services in dynamic environment where dynamic nature of web services are fully exposed. Proposed approaches for selection of services up to today are lack of following points: Using reliable historical data, incorporating reliable feedbacks from other services, a technique for motivating client Web services to give feedbacks to their providers and user-preference-awareness feature. Furthermore, there is no universally accepted method for specifying QoS of Web services fully taking into account of dynamic nature of Web services. It incurs additional coding and cost for specifying QoS of Web services with current suggested ways. Currently, QoS can be specified by WSDL file of Web services with additional coding of providers. Specifying QoS in WSDL file has several disadvantages: Firstly, in dynamic selection of provider Web services, it requires additional WSDL parser for extracting QoS. Secondly, providers needed to update the QoS of their services frequently in order to provide reliable QoS for their consumers.

In this paper, we suggest a mechanism for specifying QoS value of Web services. The mechanism is based on a model called Trust Model and implemented as QoS Broker Web services. Proposed Trust Model is based on learning experiences of past actions of individual Web services and reliable feedbacks from other services to this certain provider Web service. Feedbacks from other clients are checked in order to gather only honest and useful information for later use. In this way, we can collect up to date information about QoS values of provider Web services in reliable manner. In proposed model, each client Web service supposed to give feedback to its providers by our QoS Broker Web services. If client gives really honest and useful feedback its reputation increases otherwise decreases. This technique can be a way to motivate client Web services to give reliable feedbacks to their providers. The usage of QoS Broker Web service is up to clients either to use or not to use. Since our model is independent from other applications it will be easy to use our model with existing approaches as suggested [6]. This paper is organized as follows:

The next section addresses three trends of researches related to proposed approach. Section 3 describes proposed Trust Model in detail. Section 4 illustrates validation method, case study and result of validation. Finally, Section 5 and Section 6 concludes our work and proposes some discussions.

2 Related Work

QoS Broker based approaches: Yutu et al.,[2] introduced open, fair and dynamic QoS computation model by means of central QoS registery. Their Broker architecture is human-oriented which means every time a consumer Web services should give feedback to provider web services which leads to not collecting reliable feedback data. Gao et al.,[7] used ANN to predict dynamic performance of web services using historically collected data. However, they collected data for evaluating performance of providers which makes the data unreliable. This paper can be good solution only performance evaluation, but still there are holes like customers don't given a chance to contribute their providers performance.

UDDIe based approaches: Ali et al.,[12] introduced UDDIe, extension of UDDI, which has ability to express QoS information by means of blue pages in UDDI. This information allows other Web services to discover Web services based on QoS values. Although, it solves limitation of UDDI their work is missing two points: updating QoS values in blue page and guaranteeing the reliability of QoS values in blue page.

QoS-Model based approaches: Mou et al.,[14] defined extensible QoS model in which authors suggested multifaceted, fuzzy, dynamic and configurable QoS metrics for Web services. They tried to incorporate above characteristics of QoS of Web services with the description language. However, their model requires extra parser for dynamic Web service selection or dynamic web service coordination for extracting QoS metrics. Le et al.,[15] presented Quality of Service based service selection and ranking method with reputation management. Although they present the idea of discovering dishonest provider which is similar to our idea in come sense, their method is obscure in presenting detailed framework and identifying honest clients from dishonest ones.

3 Learning-Based Trust Model

Our Trust Model, *T(x)*, is a combination of three sub function such as *QTM()*, *MTM()* and *DH()*. *T(x,y)* ={*QTM(x,y)*, *MTM(x,y)*. ±*DH(x)*}

- QTM(*S1*,*S2*) is the function for evaluating *S1* Web services abilities in order to identify how *S1* satisfy QoS requirements of requester service *S2* based on historically gathered data
- MTM(*S1*,*S2*) is the function which evaluates how *S1* is suitable for the requirements of service *S2* according to other clients of *S1*.
- DH(*S1*) is the function for evaluating reputation of service S1. Basically, DH(*S1*) calculates the number of honest feedbacks given by S1 to its providers. By incorporating DH(*S1*) into our Trust Model, web services are motivated to give honest feedbacks.

In general, our Trust Model is defined by following equation 1.

$$T(s1, s2) = w_1^{s2} * QTM^n(s1, s2) + w_2^{s2} * MTM^n(s1, s2) + w_3^{s2} * DH^n(s1) \quad (1)$$

When using *T()*, users can give weights for requested QoS values from providers. In this way, our framework can provide *user-preference-aware* feature.

3.1 QoS Trust Matrix: QTM

Construction of QTM matrix is composed of three steps: In first step, Web services that have similar or same functionalities are grouped. In second step, the quality of service factors for each group are identified. The last step is to construct QTM matrix for each group of Web services. let's make following two assumptions: First, there are three groups of web services such as *A, B* and *C*. Second, Web service group *A* has two QoS factors, *B* group has four QoS factors, *C* group has three QoS factors. If we construct QTM Matrix for Web services group *B* it looks like in figure 1*(a)*. Values of QTM matrix is collected periodically by automatic QoS Broker from provider Web services. Thus, we can make QoS repository for QoS of Web services which is collected historically. For instance, let's assume *S1* is provider Web service and *S2* is requester Web service. Then *QTM* function value is calculated as in equation 2.

$QTM(S_1, S_2) =$

$$w_{rt}^{s2} * \frac{RT_r^n - RT_p^n}{RT_p^n} + w_c^{s2} * \frac{C_r^n - C_p^n}{C_p^n} + w_a^{s2} * \frac{A_p^n - A_r^n}{A_p^n} + w_r^{s2} * \frac{R_p^n - R_r^n}{RT_p^n} \tag{2}$$

Where $w_{rt}^{s2}, w_c^{s2}, w_a^{s2}, w_r^{s2}$ are respectively weight values for required response time, cost, availability and reliability of requester service $S2$. Meanwhile, $RT_r^n C_r^n$, A_r^n, R_r^n are normalized response time, cost, availability and reliability of of requester service *S2* and $RT_p^n C_p^n, A_p^n, R_p^n$ are normalized response time, cost, availability and reliability of service *S1* in QTM matrix.

3.2 Mutual Trust Matrix: MTM

Each Web service has MTM matrix which is constructed by feedbacks from its client Web services. MTM construction has two steps: In first step, all Web services connected to that specific Web services is identified. In second step, based on client Web services feedback and QoS factors of that Web service MTM matrix is constructed. In general, we collect feedbacks for each provider Web service from its clients. Collected feedbacks are stored in feedback repository or in MTM database. After a some time, our Feedback Manager module which we implemented in QoS Broker, updates MTM matrix with only honest feedbacks. For example, let's assume that we have identified *Sb1* has four client Web services such as *Sa1, Sa2, Sc1* and *Sc2*..

Then, all feedbacks given to *Sb1* are checked by Feedback Manager module and only honest feedbacks are used to construct MTM matrix or update MTM matrix of *Sb1*. The MTM matrix for above example is shown in figure 1*(b)*. Following equation evaluates data in MTM matrix for provider Web service *S1* for the sake of requester service *S2*.

QoS Factors	Sb_1	Sb_2	Sb_3	Sb_4
F_1 (RT)	0.5	0.3	0.7	0.1
F_2 (Cost)	10	21	20	15
F_3(A vailabilit)	78%	89%	98%	79%
F_4(Reliability)	77%	100%	99%	97%

(a. QTM Example)

QoS Factors	Sa_1	Sa_2	Sc_1	Sc_2
F_1 (Latency)	0.5	0.3	0.7	0.1
F_2 (Cost)	10	21	20	15
F_3 (Availability)	78%	89%	98%	79%
F_4 (Reliability)	77%	100%	99%	97%

(b. MTM Example)

Fig. 1. QTM matrix for *WSG B* and MTM matrix for Web services *B*

$$MTM(S_1, S_2) =$$

$$= w_{rt}^{s2} * \frac{RT_r^n - RT_{re}^n}{RT_{re}^n} + w_c^{s2} * \frac{C_r^n - C_{re}^n}{C_{re}^n} + w_a^{s2} * \frac{A_{re}^n - A_r^n}{A_{re}^n} + w_r^{s2} * \frac{R_{re}^n - R_r^n}{RT_{re}^n} \quad (3)$$

3.3 DH: Degree of Honesty

Degree of Honesty*DH()* is used for evaluating client Web services feedback when constructing MTM Matrix or updating MTM Matrix. DH() function in this work was proposed to identify reputation of Web services based on their honesty degree. Identifying reputation and truthfulness in distributed environment is vital research on its own [9], [13]. To simplify our research, we used a normal distribution with historically gathered data for identifying truthfulness of feedbacks. *DH()* can be *(+)* or *(-)* which means *DH()* can be a function encouraging client Web services to give only honest feedbacks. The *DH()* function for service *S1* is given in equation 4.

$$DH(S1) = \sum \frac{N_{RT}^h}{N_f} + \sum \frac{N_A^h}{N_f} + \sum \frac{N_R^h}{N_f} + \sum \frac{N_C^h}{N_f} \quad (4)$$

where N_f is the number of feedbacks given by service *S1* to other services, and $N_{rt}^h, N_a^h, N_r^h, N_c^h$ are number of honest feedbacks given by *S1* to response time, availability, reliability and cost of other services.

3.4 QoS Broker Architecture

The proposed architecture in figure 2 has three different phases. In phase one, constructing QTM matrices is the vital task. The first task in phase 1 is to get all available Web services information from UDDI and making WSGs based on their QoS metrics values and functional similarities. The second task in phase 1 is to construct QTM matrices for each group. The third task in phase 1 is to gather QoS values for available Web services by automated Web services monitoring tool. In phase two, there are also three tasks: First, Web services Manager module identifies all interactions between any two Web services. Secondly, Mutual Trust Evaluation module gathers feedbacks from clients for each provider Web services and constructs MTM matrix for each provider Web services. Finally, Trust check module identifies truthfulness of feedback gathered by Mutual Trust Evaluation

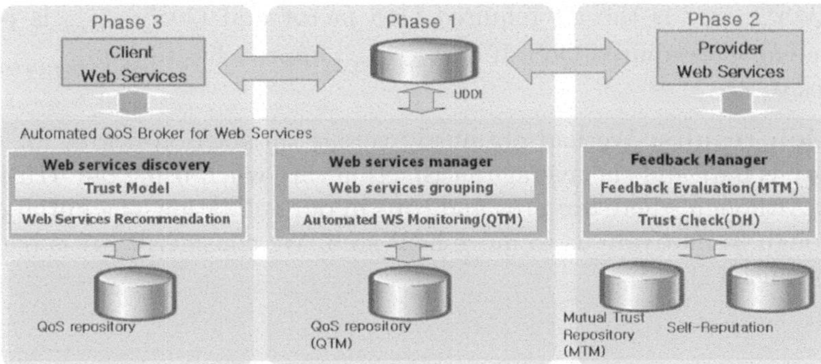

Fig. 2. QoS-broker Architecture

Module. According to feedbacks from clients the MTM matrix updated by only honest feedbacks. At the same time, the clients giving honest feedbacks are rewarded by increasing their value of *DH()* function. In phase three, there are two tasks: In first task, client Web service's required QoS is evaluated and most pertinent provider Web services is returned based on Trust Model. Secondly, any requests from clients *(give feedback or to calculate T())* should be handled in phase three. All the functions in our architecture are implemented as Web services. In this way, our approach can be easily used by Web services developer.

4 Validation

In order to show applicability of our method in real life case we have taken a four groups of Web services such as Tour, Ticketing, Banking and Delivering for a case study. Each group may or may not have ten or more than ten member Web services each holding different response time, availability, reliability and cost. Following sequences of tasks performed for each Web services selected from each Web services group(WSG)[1].

$$Client \rightarrow Tour \rightarrow Ticket; Tour \rightarrow Bank \rightarrow Ticket \rightarrow Delivery \rightarrow Tour.$$

Validation method: Validation was done by comparing two cases: when using our Trust model and when not using our Trust model. Conventionally, provider Web services is selected among candidate provider Web services based on promised QoS of providers. In our Trust Model, provider Web services is selected based on value calculated by *T(x)* function. We have applied both conventional way and our Trust Model for the case study and identified satisfaction of client Web services by following satisfaction degree equation.

Satisfaction Degree(SD(n)) =

$$n - \sum_{i=1}^{n} \frac{|QoS^i_{required} - QoS^i_{provided}|}{QoS^i_{provided}} \quad (5)$$

[1] WSG=Web Services Group.

where $QoS^i_{required}$ is the ith required QoS factor and $QoS^i_{provided}$ is provided QoS value for i^{th} required QoS. If $QoS^i_{provided}$ approaches to $QoS^i_{required}$ $SD(n)$ converges to n.

Validation results: We had obtained average satisfaction degree for conventional way is 0.61 and the average satisfaction rate when using our Trust Model is 0.701. Although, there are pros and cons in our validation, the initial result is good enough to motivate us to move next step. We consider 0.701 is reasonable result for initial step.

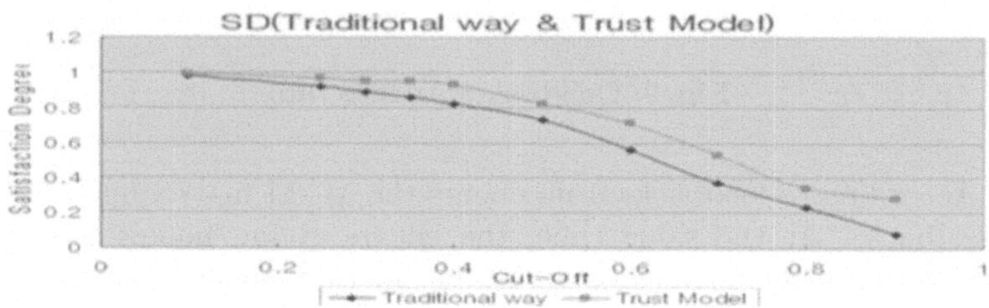

Fig. 3. *Graph 1.* Satisfaction rate. Conventional way vs. Trust Model

5 Discussion

In this work, we considered dynamic nature of Web services and tried to suggest a novel method for selecting Web services which we consider more trustable than other methods. However, still our work is in its infancy and there are several issues should be discussed in this research.

First issue is the dynamic nature of Web services. Web services may be newly be published or disappeared from UDDI or entirely from the net. Some researchers suggested to use Web services monitoring and testing tool for handling this issue. However, if the number of Web services increases to be monitored it decreases the reliability of the tool. Our work suggests to use historically gathered data by Web services monitoring tool, feedback and *DH()*. In this way, we can gather more reliable data and more up to date information.

Secondly, validation was done in local machine with self implemented Web services and artificially generated QoS values. Our result was obtained by taking average result of 400 experiments under different cases *(increasing/decreasing overhead of machine)*.

Thirdly, we used only response time for the validation. Although response time is critical factor, there is also need to consider other QoS factors for our validation. However, our initial validation motivates us to do next validation with multiple QoS factors. Finally, honest feedbacks are identified by historically gathered data which is not wise way. However, for initial research our way to identify honest feedbacks supports our approach.

6 Conclusion

In this paper, we suggested a mechanism for specifying QoS(Quality of Services) for Web services based on so-called Trust Model. Trust Model is a function *T(x)* where it consists of three sub functions such as *QTM(), MTM()* and *DH()*. *QTM()*, which is a function for evaluating history of certain Web services by using data gathered by Web services monitoring tool. *MTM()* is a function for evaluating other services references to certain provider Web services. *DH()* is a function for encouraging and forcing client Web services to give honest feedbacks to their providers. Automatic, learning, dynamic QoS Broker Web services framework is suggested in detail and implemented with some essential functions. Our suggested model is implemented as a Web service and validated with initial result which is showing and supporting our approach. Also, a method for identifying client Web services satisfaction degree in Web services environment is suggested. Finally, proposed approach was validated using client Web services satisfaction degree function with initial feasible result.

References

1. Adel Serhani, Rachida Dissouli, Hafid, Houari Sahraoui, A QoS Broker based architecture for efficient web services selection, IEEE, Vol.1 (2005), 113- 120.
2. Yutu Liu, Anne .H, Liangzhao Zeng, QoS computation and Policing in Dynamic web services selection, WWW2004, NY, May, 2004. USA, May 17–20, 2004. pp.
3. Tao You and Kwei-Jay Lin, The Design of QoS Broker Algorithms for QoS-Capable Web Services, 2004 IEEE, IC on e-Commerce and e-Service, Pages: 17 - 24
4. Tao You and Kwei-Jay Lin, Service Selection Algorithms for Web Services with End-to-end QoS Constraints, IS and E-Business Management, 3.2(2005): 103-126
5. Dong-Soo Han, Sung Joon, Park, Exception based Dynamic Service Coordination Framework for Web Services, Florida, Lighthouse Point, 2006.
6. Dong-Soo Han, Sungdoke Lee, Inyoung Ko, A Feedback Based Framework for Semi-Automatic Composition of Web Services, LNCS Vol. 3841,2006
7. Zhengdong Gao; Gengfeng Wu, Combining QoS-based service selection with performance prediction, e-Business Engineering, ICEBE 2005. Page(s):611 - 614
8. L. Zeng, B. Boualem, Ngu, A.H.H. Dumas, M. Kalagnanam, J. Chang, H, QoS-Aware Middleware for Web Services Composition, IEEE, SWE, 30(5): 311-327, 2004
9. Li Xiong Ling Liu , PeerTrust: Supporting Reputation-Based Trust for Peer-to-Peer Electronic Communities, IEEE Transactions on KDE,Vol. 16, No. 7, 2004
10. Marco Comerio, Flavio De Paoli, Simone Grega, "Quality Composition in Web-Service Design," icdcsw, p. 72, 26th IEEE (ICDCSW'06), 2006.
11. M. Tian, A. Gramm, H. Ritter, J. Schiller: Efficient Selection and Monitoring of QoS-aware Web services with the WS-QoS Framework.(WI'04), Sep.2004, Beijing.
12. Ali ShaikhAli, Omer F. Rana, Rashid Al-Ali, David W. Walker, "UDDIe: An Extended Registry for Web Services," saint-w, p. 85, (SAINT'03 Workshops), 2003.
13. Li Xiong, Ling Liu, A Reputation-Based Trust Model For Peer-To-Peer Ecommerce Communities, IEEE Conference on E-Commerce (CEC'03)", 2003
14. Mou Yu-jie Cao Jian Zhang Shen-sheng Zhang Jian-hong , Interactive Web service choice-making based on extended QoS model, 5th ICCIT, pp 1130 - 1134 , 2005
15. Le, Hauswirth, Manfred, Aberer, Karl, QoS-based Service Selection and Ranking with Trust and Reputation Management, ODBASE 2005, LNCS 3760, pp. 466-483.

SeCED-FS: A New Approach for the Classification and Discovery of Significant Regions in Medical Images

Hui Li[1], Hanhu Wang[1], Mei Chen[1], Teng Wang[1], and Xuejian Wang[2]

[1] Department of Computer Science & Technology, Guizhou University,
550025 Guiyang, P.R. China
{LiHui_gzu,HanhuWang}@tom.com
[2] Department of Radiology, Affiliated Hospital of Guiyang Medical College,
550025 Guiyang, P.R. China

Abstract. A novel diagnosis method named SeCED-FS is proposed in this paper. The method combines the clusterer ensemble and feature selection technique to improve the diagnosis performance. At first, selective clusterer ensemble with feature selection technique is utilized to perform the classification of medical images in the two-level architecture. Then, the Regions of Interest in positively identified image are outlined by using an ensemble of Fuzzy C-Means algorithm. Case studies on real data experiments show that, the SeCED-FS holds the improved generalization ability and achieved a satisfactory result not only in the accuracy of classification but also correctly labeling the significant regions.

Keywords: medical image mining, computer aided diagnosis, selective ensemble, Clustering, feature selection, ROI.

1 Introduction

In the last decade, the digital medical modality and PACS (Picture Archiving and Communication System) have been pervasive in medical care, and they were raised the possibility of discovering potentially new and useful knowledge through mining in medical images. Due to the particular characteristics of data mining technique that could be used as a stand-alone tool to gain an insight into the relationship and patterns hidden in the data, many researchers have devoted to this field to develop novel medical imaged based computer aided diagnosis method [3-7, 9, 11, 12].

Generally speaking, there are two critical challenges in medical image based computer aided diagnosis: classify the medical images and label the Regions of Interest (ROI) in it. It is not easy for the following reasons: firstly, classification of medical images involved many complex techniques, such as image feature extraction, subset feature selection, classification, and much effort should be put into combine these techniques to perform the classification so as to get an acceptable result; secondly, it is difficult to discover the significant regions in medical images through segmentation or other methods, especially when performing this task on those images which in the early stage that will critical to patient's treatment and survival. As we

G. Dong et al. (Eds.): APWeb/WAIM 2007, LNCS 4505, pp. 650–657, 2007.

can see, due to its interdisciplinarity and complexity, in this field, there remain many problems not being solved very well.

In this paper, a novel automatic pathological diagnosis procedure named SeCED-FS is proposed. It focuses on combine the clusterer ensemble and feature selection technique together and applied in the diagnosis process. At first, it utilizes the selective clusterer ensemble with feature selection technique to perform the classification of medical images in the two-level architecture. Then, the ROI in positively identified images are outlined by using an ensemble of Fuzzy C-Means algorithm [14]. Furthermore, in each ensemble level of SeCED-FS, the k-Medoids algorithm [2] is utilized to select a subset of clusterers and voting to the final result for the purpose of improving the diagnosis performance. Case studies on real data experiments show that, benefit from adopting the SeCED-FS method as the core mechanism of MiCADS, this diagnosis system proved to be effective in classifying and recognizing the significant regions on medical images.

The rest of this paper is organized as follows. In section 2, a brief introduction of MiCADS is presented. Then, in section 3, methodology of SeCED-FS is proposed. In section 4, the experimental results will illustrate in detail. Finally, the conclusions are drawn in section 5.

2 MiCADS

Utilizing computer to aid the diagnosis is an important and challenging task. In this paper, an automatic pathological diagnosis system named MiCADS (Medical Image based Computer Aided Diagnosis System) which has implemented SeCED-FS as the core mechanism, is developed. It mainly includes 5 modules: DICOM [16] gateway, image pre-precossing, image feature extraction, lesions diagnosis and ROI determination. The architecture of MiCADS is depicted in Figure 1.

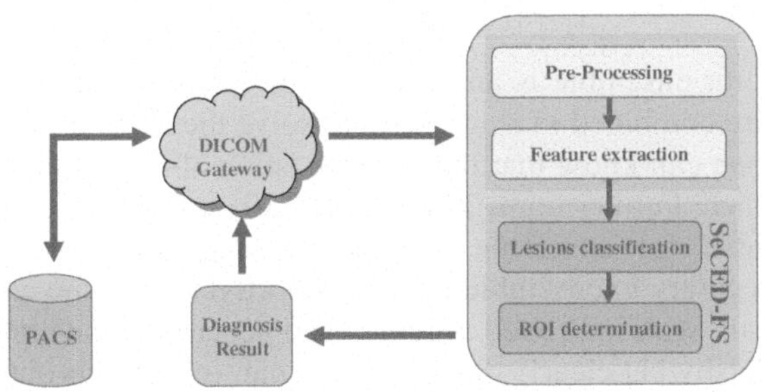

Fig. 1. Architecture of MiCADS

The DICOM gateway is used to get the medical images produced by digital medical modality. All the imputed images will be performing noise filtering, smoothing and enhancement in the image pre-processing module. After that, in the feature extraction module of MiCADS, texture characteristics are extracted from the images based on the techniques of gray level co-occurrence matrix and gray-grads level co-occurrence matrix by introducing the factor of direction. Then, the lesions classification procedure is employed through using those features as the input. Finally, the ROI in the positively identified images will be labeled.

3 SeCED-FS

In this section, the diagnosis method named SeCED-FS (Selective Clusterer Ensemble based Diagnosis combined with Feature Selection) is proposed, which utilizes an ensemble based two-level architecture to classify the lesions for the purpose of decreasing the false negative error rate as much as possible, and an one-level ensemble which uses the Fuzzy C-Means algorithm as the base learner is designed for ROI identification.

3.1 Motivation

The motivation of SeCED-FS is based on the two critical problems mentioned above: improve the generality ability and decrease the false negative error rate. In order to achieve these targets, the following techniques: feature selection and selective clusterer ensemble, are utilized.

Feature selection is a technique that used to identify and remove the irrelevant and redundant attributes. It will make the learning algorithm faster and more effective. In our work, we use the prediction risk criteria [13] for the feature selection. It is a strategy that evaluates one feature through estimating prediction error of the data sets when the values of all examples of this feature are replaced by their mean value. It could be illustrated as formula 1, where *Err* is the training error, and $Err(\bar{x}_i)$ is the test error on the training set with the mean value of *i*th feature. At last, the feature holding the smallest value of S_i will be deleted, due to this feature only causes the smallest error and is the least important one.

$$S_i = Err(\bar{x}_i) - Err \tag{1}$$

Selective clusterer ensemble is an unsupervised learning paradigm where selected several clusterer are combined in some ways to solve the same problem. From Hansen and Salamon [8], we can know that the generalization ability of an ensemble is better than a single learner. Furthermore, Zhou *et al.* [10] proved that, not all the trained learners are suitable for ensemble, so called "many could be better than all". Therefore, it is necessary to ensemble learners selectively in some circumstances so as to gain a better result with improved accuracy and generalization ability.

Krough & Vedelsby [1] clearly demonstrates that, in order to obtain a high quality ensemble, the bias of individual learner should be reduced and the diversity between learners should be increased. Therefore, one of the keys to improve the accuracy and generalization ability is how to define the inter-clusterer similarity (*ICS*) and utilize it to construct an ensemble. In our work, ICS is defined as formula 2.

Suppose the size of data set is *N*, let $n_i(a,b)$ denote the number of common objects in the *i*th cluster between the clusterer C_a and C_b. It should be noted that, since the clustering is an unsupervised learning process, the class label of *i*th cluster in C_a may differ from C_b, in order to ensure the *i*th cluster in C_a and C_b are hold the same class label while we calculate the inter-clusterer similarity, the clusters should be aligned and matched based on class labels in advance.

$$ICS(a,b) = \frac{\sum_{i=1}^{k} n_i(a,b)}{N} \tag{2}$$

Suppose there are n candidate clusterers for ensemble. In order to improve the generalization ability of SeCED-FS, firstly, we utilize formula 2 to form a $n \times n$ matrix *M* and consider the element *M(a,b)* as the distance between C_a and C_b, the meaning of *M(a,b)* is the same with *ICS(a,b)*, then, we employ the *k*-Medoids algorithm to divide the *n* candidate clusterers into *k* groups and select the medoid of each group to join the clusterer ensemble. In the first level ensemble, the full voting strategy, a strong strategy that considers the label of normal to be the final output only when all the individuals hold the same result, is used to combine the multiple predictions, and in the second level, the plurality voting method is employed. Plurality voting is a strategy that judges a prediction to be the final result if this prediction ranks first according to the number of votes. Based on the knowledge that the objects within same cluster are similar and are dissimilar to the objects in other cluster, the component clusterers selected by *k*-Medoids algorithm are with high diversity and the ensemble combined by them would have improved generalization ability.

In order to overcome the problem of recognizing inconspicuous focus in the early stage medical image, an ensemble of segmentation procedure is also constructed in the same way. The ROI identification procedure is as following: Firstly, using enough trained Fuzzy C-Means models segment separately on the target image. Secondly, labeling the properties of segmented regions by using the rules mentioned in [15]. Then, the segmented regions are labeled as background, brain, skull, hemorrhage or calcification. Finally, make a comparison on the (overlapped) regions that have been labeled certain property by these models, if multiple models are given the same label, then the region is considered to hold this property with a high confidence. It is evident that this method provides an intuitional way to depict the ROI.

3.2 Diagnosis Process

Based on the techniques and ideas illustrated above, we devise the SeCED-FS method as follows.

Hypothese: F is the number of selected subset features, L denotes the number of trained learners, and S indicates the size of selected ensemble;

Stage A: Classification of lesions
Step 1: Using prediction risk criteria to select an optimal subset features (size is F) for the clusterer construction;
Step 2: Generate a training set on the selected features through sampled with replacement, and then train a k-Means clusterer on it;
Step 3: Repeat step 2 until the L model are setup;

Step 4: Utilize k-Medoids algorithm to select S clusterers which are with high diversity for the result combination based on the ICS;
Step 5: Using full voting strategy to yield the classification result in the first ensemble level;
Step 6: Using plurality voting strategy to produce the final result in the second ensemble level;

Stage B: Identification of ROI
Step 7: Generate a training set through sample with replacement, and then train a Fuzzy C-Means clusterer on it;
Step 8: Repeat step 7 until the L model are setup;
Step 9: Utilize k-Medoids algorithm to select S clusterer for the result combination based on the ICS;
Step10: Identify ROI based on the segmented (overlapped) region's property through utilize Fuzzy C-Means ensemble.

In order to illustrate the idea of SeCED-FS more clearly, let's see the example that using SeCED-FS to aid in the brain lesions diagnosis based on CT images, its flowchart is shown as Figure 2. It should be mentioned that, while classifying the brain lesions, although both of the ensemble levels could produce the label of normal as the diagnosis result, owing to the adoption of full voting strategy, the first-level has a higher confidence on this output than the second-level ensemble, it decreased the false negative error rate indeed, the detailed experimental result is given in section 4.

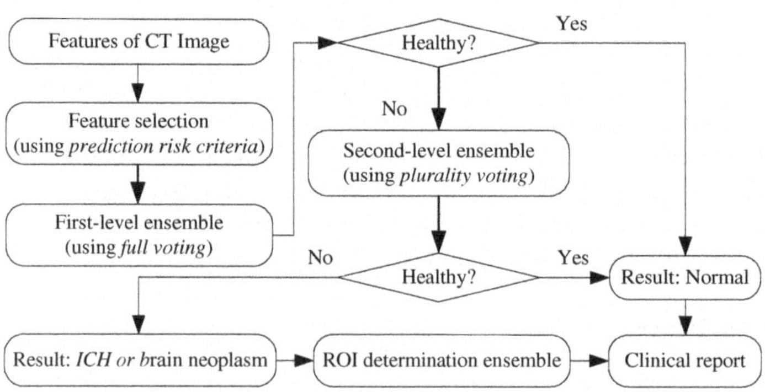

Fig. 2. Flowchart of utilize SeCED-FS in the diagnosis of brain lesions

4 Experiments

Three case studies, on brain lesions, hepatitis, and diabetes, respectively, have been analyzed. The data set of brain lesions is from the Affiliated Hospital of Guiyang Medical College, including 517 series of head CT images, among them, 163 series of images belong to brain neoplasm (about 31.5%), 217 series of images are ICH (about 42%) and the rest 137 series of image are normal (about 26.5%). The hepatitis and diabetes data set are getting from UCI Machine Learning Repository [17], there are 155 instances (with 20 attributes) in the hepatitis dataset which belonging to 2 classes, the data set of diabetes contains 768 instances with 9 features, and it is also categorized into 2 classes.

4.1 Experimental Results on Lesions Classification

In this section, the effectiveness of classifying lesions through SeCED-FS is evaluated. The experimental results are compared among SeCED-FS, CED (Clusterer Ensemble based Diagnosis), SeCED (Selective Clusterer Ensemble based Diagnosis), k-Means, Naïve Bayesian, Bayesian Networks and C4.5. The detailed experimental data is given below.

Table 1. Experimental result of 4 classical algorithms

%	brain lesions	hepatitis	diabetes
C4.5	43.57/12.6	16.13/11.61	26.17/14.06
k-Means	45.3/13.12	26.45/5.16	33.2/17.58
Naïve Bayesian	39.72/10.85	15.48/6.45	23.7/13.54
Bayesian Networks	42.34/11.53	16.77/6.45	25.65/13.67

In the experiments, results are validated by 10-fold cross-validation. The detailed results of 4 classical learning algorithms are listed in table 1. These number separated by the slash represent the overall error rate and false negative error rate in corresponding algorithm.

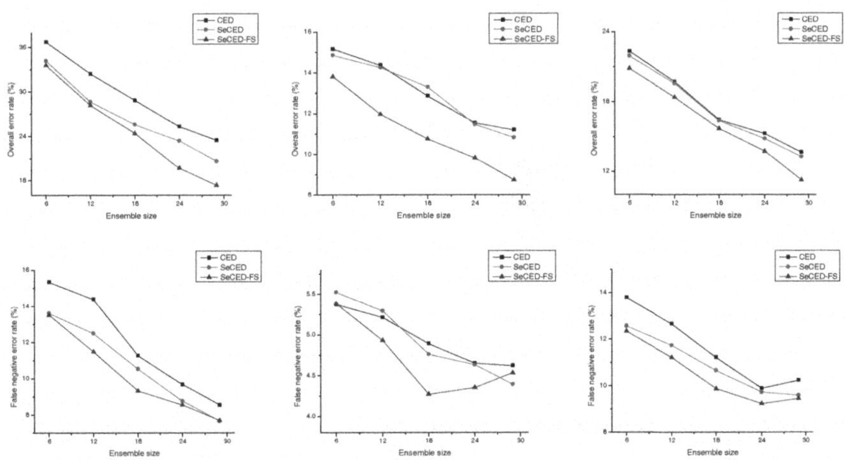

(a) brain lesions (b) hepatitis (c) diabetes

Fig. 3. The overall error rate and false negative error rate of CED, SeCED and SeCED-FS

From table 1 we can know that, the k-Means only achieved a depressed result in lesions classification. However, while we employ it as the base learner of CED, SeCED and SeCED-FS, these 3 methods hold a certain improved result through using the ensemble learning technique. In the estimation of SeCED and SeCED-FS, 32, 15 and 7 attributes of the 3 data set is selected respectively, 150 clusterers were trained in total, and several of it will be selected through k-Medoids algorithm to join the two-level based ensemble architecture. The classification results on 3 data sets by these 3 methods are shown as Figure 3. It is interesting that while the ensemble size is large enough, the false negative error rate is almost equally in these 3 algorithms. Based on the comparison between SeCED-FS and other frequently used classical algorithms on the overall error rate and false negative error rate, SeCED-FS could be considered as a preferable approach in medical image based lesions diagnosis.

4.2 Experimental Results on Discovery of ROI

In this work, the ROI is outlined by an ensemble of Fuzzy C-Means model based on the method illustrated in section 3.1. Our medical image data is from brain lesions data set and The Whole Brain Atlas [18], 82 pairs of images was selected for the experimental purpose. Among them, 58 images were produced by CT scanner, and 24 images were produced by MRI analyzer. We define 3 levels to measure the effect of discovering ROI: primary correct, partial correct and wrong. Primary correct means that the primary significant regions labeled in the atlas or illustrated in clinical report, has been found; the term partial correct denotes that only part of ROI was labeled by the ensemble; the term wrong indicates the result that no correct target area was identified. The detailed results are tabulated in table 2. These number separated by the slash denotes the result perform by a single Fuzzy C-Means learner and its ensemble.

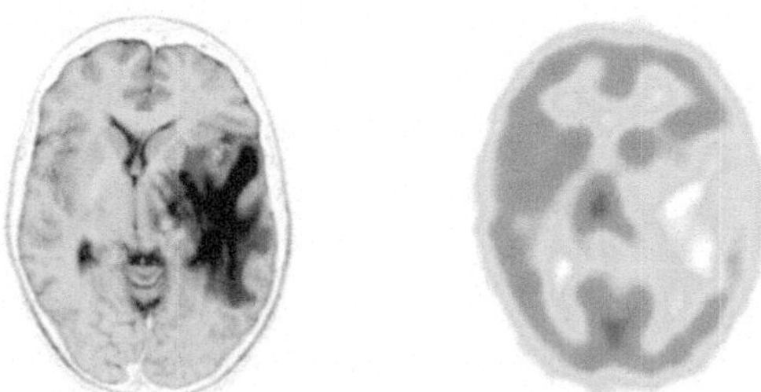

Fig. 4. An example of the origin image and its atlas

Table 2. ROI determination by using a single Fuzzy C-Means and its ensemble

	CT Images	MRI Images
Wrong	18/11	11/8
Partly Correct	23/21	7/9
Primary correct	17/26	6/7

Analyzing the experimental data list in table 2, it is obvious that the generalization ability was boosted for the adoption of ensemble learning technique.

5 Conclusion

In this paper, a medical image based automatic pathological diagnosis procedure named SeCED-FS is proposed and verified on the real cases. Medical image classification and discovery of ROI is just the first step to survey in this field, there are remain lots of works on discover patterns and associations need to be investigated.

Acknowledgments. The authors also wish to thank their collaborators in this work: Cai Peng, Jiang Honghui, Ma Dan, Jiang Hua and Shan Jingsong.

References

1. Krogh A, Vedelsby J. Neural network ensembles, cross validation, and active learning. Advances in Neural Information Processing Systems, 7, (1995) 231-238
2. L. Kaufman and P.J. Rousseeuw. Finding Groups in Data: An Introduction to Cluster Analysis. New York: John Wiley & Sons, (1990)
3. Guo-Zheng Li, Tian-Yu Liu, Victor S. Cheng. Classification of Brain Glioma by Using SVMs Bagging with Feature Selection, LNBI 3916, (2006) 124-130
4. Megalooikonomou, V., Davatzikos, C. and Herskovits, E., Mining lesion-deficit associations in a brain image database, SIGKDD'99, 347-351
5. Cocosco C A, Zijden.bos A P, Evans A C. A fully automatic and robust brain MRI tissue classification method. Medical Image Analysis, 7, Vol. 4, (2003) 513-527
6. Megalooikonomou, V., Ford, J., Shen, L., Makedon, F., Saykin, F.: Data mining in brain imaging, Statistical Methods in Medical Research, 9, (2000) 359-394
7. Zhou, Z.-H., Jiang, Y.: Medical diagnosis with C4.5 rule preceded by artificial neural network ensemble. IEEE TITB 7, (2003), 37-42
8. Hansen L. K., & Salamon P. Neural network ensembles. IEEE TPAMI, 12, (1990) 993-1001
9. Zhou, Z.-H., Jiang, Y., Yang Y.-B., Chen S.-F.: Lung cancer cell identification based on artificial neural network ensembles. Artificial Intelligence in Medicine, 24, (2002) 25-36
10. Zhou, Z.-H., Wu, J., and Tang, W., Ensembling neural networks: Many could be better than all. Artificial Intelligence, 137, Vol. 1-2, (2002) 239-263
11. Mitsuru Kakimoto, Chie Morita, Hiroshi Tsukimoto. Data mining from functional brain images. MDM/KDD 2000, 91-97
12. Rong Chen, Edward Herskovits. A Bayesian network classifier with inverse tree structure for voxelwise magnetic resonance image analysis. SIGKDD'05, 4-12
13. Moody, J., Utans, J.: Principled architecture selection for neural networks: Application to corporate bond rating prediction. Advances in Neural Information Processing Systems. Vol. 4, (1992) 683-690
14. I. Gath, A. Geva. Unsupervised optimal Fuzzy clustering. IEEE TPAMI, 11, (1989) 773-781
15. D. Cosic, S. Loncaric, Rule-based labeling of CT head image, LNAI 1211, (1999) 453-456
16. DICOM standard. http://medical.nema.org/dicom/2006
17. UCI Machine Learning Repository. http://www.ics.uci.edu/~mlearn/MLRepository.html
18. The Whole Brain Atlas. http://www.med.harvard.edu/AANLIB/home.html

Context-Aware Search Inside e-Learning Materials Using Textbook Ontologies

Nimit Pattanasri, Adam Jatowt, and Katsumi Tanaka

Department of Social Informatics, Kyoto University
{nimit,adam,tanaka}@dl.kuis.kyoto-u.ac.jp

Abstract. One of the main problems of delivering online course materials to learners lies in a deficiency of search capability. We propose a method for querying inside lecture materials exploiting a textbook ontology, which is automatically constructed from a back-of-the-book index and table of contents. The use of a textbook ontology is two-fold: (i) to help users formulate query more efficiently, and (ii) to discriminate query results that are relevant to user information needs from those that merely contain query terms but are not informative.

Keywords: e-Learning, context-aware search, PowerPoint, lecture video, table of contents, back-of-the-book index, OWL, XDD, reasoning.

1 Introduction

e-Learning is a learner-centered, self-paced learning environment where materials are delivered to learners in most cases through the Web. For example, MIT's OpenCourseware project offers freely online learning materials including lecture notes and teaching videos. Nevertheless, very few online courses provide search capability inside those materials [12]. As it has been recently evident in Web2.0, social evolution will advance e-Learning to the stage that not only learning materials are created by courseware authors themselves but by learners as well [1]. Therefore, the need for search capability is increasingly important.

Still, most fundamental learning materials take the form of lecture notes, which are an important source for learners to understand concepts being taught by instructors. Lecture notes are usually a sort of summaries of a textbook which is used for their construction. Information in lecture notes contains the most important concepts that students should understand and remember after completing a course. Typically, an academic course consists of a series of lecture notes, each of which, in most cases, is a sequence of lecture slides such as PowerPoint. Manually searching for learning materials of interest from a whole course can be, however, tedious and time consuming. Therefore, there is a need for lecture retrieval systems (e.g. [3,5,9,6]).

This paper exploits a textbook index and table of contents to bridge a semantic gap between user queries and information contained in lecture materials. An advantage of using textbook indexes is that an indexer, who provides the index for a textbook, usually provides the conceptual structure of technical terms in

G. Dong et al. (Eds.): APWeb/WAIM 2007, LNCS 4505, pp. 658–669, 2007.

an index. This is particularly useful, in this paper, for analyzing the context of user queries and lecture materials.

One of the main problems of searching inside learning materials lies in a deficiency of information retrieval systems that cannot *systematically* interpret and identify the context of both user queries and fragments of learning materials. For example, almost every lecture slide in a database course will be obtained for a query "I want to know about database". In this case, query context such as "query language" plays two important roles: to help users formulate queries efficiently and to prevent them from information overload. Consider another query "Give me a definition of database". A system may return no result for such a query if no slide contains both terms "database" and "definition". Context (i.e. "definition", in this case) can play another essential role in a lecture slide as its implicit meaning. These requirements thus call for a system that can automatically extract the context of lecture materials.

The contribution of this paper is three-fold. This paper is, to the best of our knowledge, the first to exploit textbook metadata to solve the problem of context interpretation of user queries and context identification in lecture materials. Second, exploiting a textbook's indexer viewpoint, our approach can automatically offer query contexts to help users efficiently refine searches. Query results can be categorized by their contexts so that users can investigate each of them easily in order to confirm an answer. In addition, incorporating textbook index with table of contents, we can discriminate query results (i.e. lecture fragments) that are relevant from those that merely contain query terms but are not informative. Third, a unified framework for lecture slide and video fragment retrieval is proposed to provide a single access point for both types of learning materials.

The rest of the paper is organized as follows. Section 2 explains the basic idea for identifying context in both user queries and lecture materials. Section 3 proposes a method for querying lecture materials based on inference rules using textbook ontologies. Section 4 reports some preliminary experimental results. Section 5 discusses the related work. Section 6 concludes the paper.

2 Basic Concepts

In this paper, TOC and textbook index (in short, index) refer to table of contents and back-of-the-book index of a particular textbook, respectively. All index and TOC examples used in this paper are taken from [4]. Terminologies of book index elements are borrowed from [7].

2.1 Textbook Ontology

Figure 1 shows a part of index, TOC, and its corresponding representation in OWL[1]. An index entry is composed of a main heading with its subheadings. Main heading is a main access point to a particular topic being found in a book while its subheadings are more specific aspects (or contexts) of such a topic. Maximally

[1] http://www.w3.org/TR/owl-features/

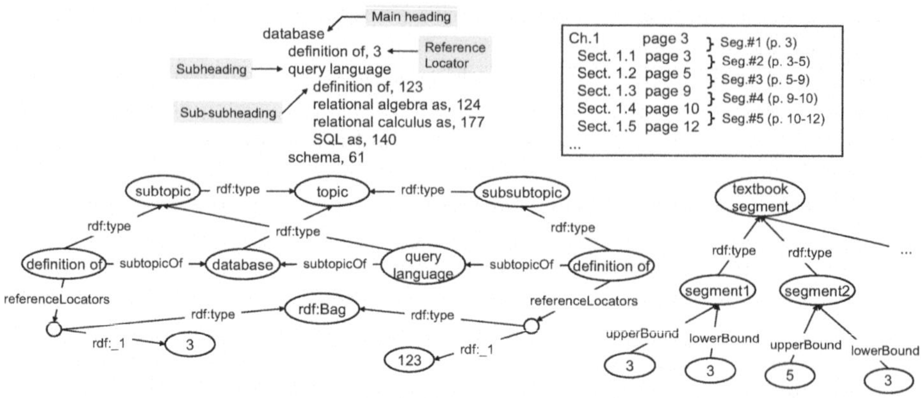

Fig. 1. A part of textbook index, TOC, and its representation in OWL

three-level index headings are allowed in this paper: main heading, subheading, and sub-subheading.[2] Each index heading may contain several reference locators, each of which is usually a page number pointing to the location of information to which the heading refers. In this paper, TOC is used to partition a textbook into non-overlapped subsegments, called *textbook segments*, determined by book page intervals. Identification of the *subtopicOf* relationship in a textbook ontology is straightforward. An index subheading is a subtopic of its index main heading, and an index sub-subheading is a subtopic of its index subheading.

2.2 Context of User Queries and Lecture Slides

In order to help users formulate queries efficiently for lecture materials, we exploit existing book index structure; a subheading can be considered a context of the main heading. When a query is formulated without context or subtopics (e.g. "database"), we can offer available query contexts so that a user can easily narrow down the search (e.g. "definition of", "query language", or "schema"). Not only do these contexts help users refine the search but also prevent them from information overload. Note also that, in this paper, a user query is a combination of a topic and its subtopics. A topic is likely to be found in an index main heading while its subtopics can be located in its subheadings of a particular index entry.

We exploit a TOC together with an associated index to identify context of lecture slides. Specifically, page numbers in a textbook index and TOC can be associated for a particular topic so that we know which textbook segments contain information about the topic. Imagine further that if such a topic appears in a particular lecture slide and there also exists a given mapping from such a slide to a textbook segment (i.e. a book page interval in TOC) to which the topic is mapped, we may conclude that the lecture slide contains information

[2] A book index containing index entries whose indexing level exceeds three is considered hard to use and not helpful [10].

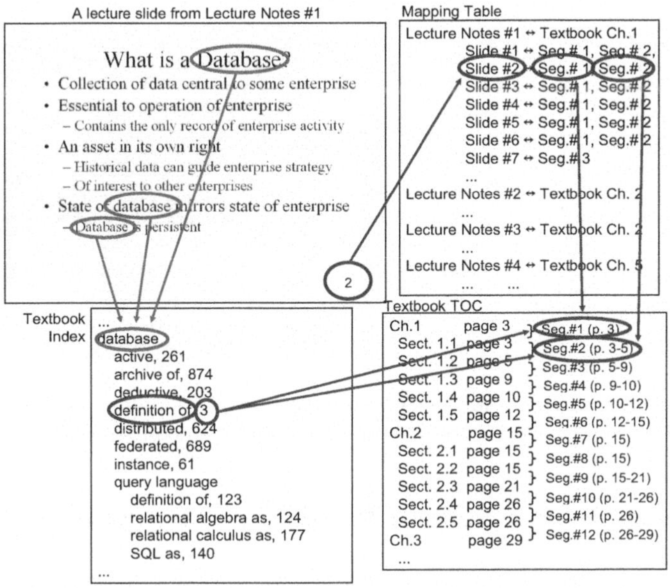

Fig. 2. The basic idea for identifying context of lecture slides

about the topic. Consider an example in Fig. 2. Since a "database" term is found in the lecture slide which can be mapped to textbook segments, Seg.#1 or Seg.#2, in Ch. 1, it is likely that the context of the lecture slide is "definition of" "database". On the contrary, a "database" term that appears in other slides elsewhere but cannot be mapped to Seg.#1 or Seg.#2 in Ch. 1 is *unlikely* to contain information about "definition of" "database". Furthermore, if a lecture slide, which can be mapped to other chapters such as Ch. 2, also contains a term "database" but none of the reference locators in the index heading of "database" and its subheadings point to Ch. 2, it is likely that such a slide uses a "database" term to explain some other terms or concepts (i.e. such a slide does not provide useful information about "database" itself). Therefore, using information from TOC and index, we can systematically distinguish information that is informative from one that is just passing mention.

Note also that a topic (i.e. technical terms such as "database") tends to be found exactly in lecture slides while its contexts or subtopics (i.e. general terms such as "definition") do not necessarily appear. We also assume that lecture slides partially follow the structure of a particular textbook. In this respect, mapping from lecture notes to textbook chapters is easy and often provided by courseware authors.[3] Therefore, lecture-notes-to-textbook-chapters mapping needs not always be detected automatically. Automatic mapping from a lecture slide to textbook segments inside a particular chapter requires more effort and is deferred to Sect. 3.5.

[3] See an example at http://csis.pace.edu/∼scharff/cs623/ref/cs623indexref.html

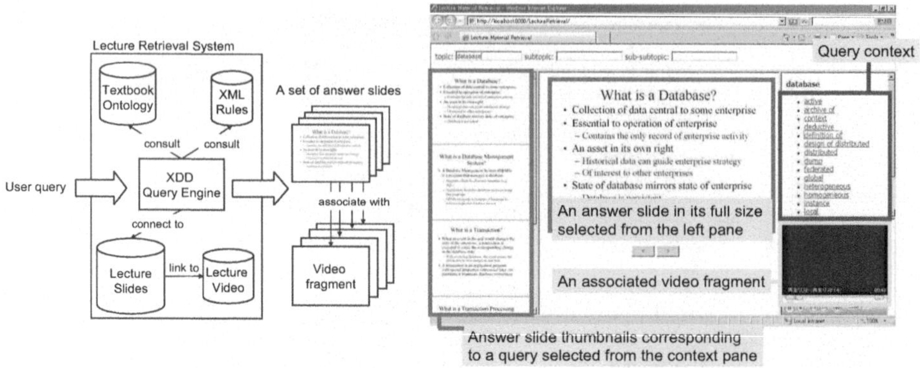

Fig. 3. A unified framework for lecture slide and video fragment retrieval (on the left) and screen capture of a prototype system for a database course (on the right)

3 Context-Aware Search Inside Lecture Materials

The left side of Fig. 3 shows a unified framework for lecture slide and video fragment retrieval where the input is a user query which is a combination of a topic and its subtopics, and the output is a set of answer slides classified by their context. Each answer slide is associated with a corresponding lecture video fragment, if it exists. The XDD query engine is responsible for identifying answer slides for queries by consulting rules and a textbook ontology.

3.1 Data Preparation

A course lecture material consists of a series of lecture notes. Each lecture note is a sequence of PowerPoint slides that can be exported to the OpenDocument XML format [8] through OpenOffice Presentation[4]. For textbook ontology construction, there are two ways for extracting the TOC and index for a particular textbook: (i) to scan them from the hardcopy, or (ii) to obtain a search result by submitting the textbook title as a query to Google Book Search[5] and then later by accessing the TOC and index from the result. Next, OCR software is used to extract text characters so that the scanned documents (i.e. TOC and index) can be further used for ontology construction as illustrated in Fig. 1. Once the whole process is complete, XML data including PowerPoint slides and a textbook ontology is ready to be used.

3.2 XML Declarative Description

Although XML can provide a schema for representing a textbook ontology through OWL and describing PowerPoint slides through OpenDocument, it cannot explicitly express relationships, rules and constraints among lecture slides, a

[4] http://www.openoffice.org/
[5] http://books.google.com/

textbook ontology, and user queries. *XML Declarative Description (XDD)* [11] is a language for modeling XML databases capable of expressing explicit and implicit information through XML expressions (variable-embedded XML elements), and relationships, constraints, and rules through XML clauses. XML clause is of the form: H $\leftarrow$ $B_1,...,B_n$. where n $\geq$ 0, and H and B_i are XML expressions or constraints. When n=0, a clause is called unit clause, otherwise non-unit clause. XML documents (or elements) such as PowerPoint slides and a textbook ontology can immediately become ground XML unit clauses. Given a database, XDB, consisting of XML unit and non-unit clauses and a query, Q, formulated in terms of a non-unit clause, the answers, A, can be obtained by transforming Q repetitively using XML clauses in XDB.

3.3 Query Model

An XML database consists of PowerPoint documents (XML unit clauses) representing course lecture notes, an OWL document (XML unit clauses) describing TOC and indexes of a particular textbook associated with the course, and a set of inference rules (XML non-unit clauses) for identifying answer slides together with their context for user queries.

Definition 1 (Query context). *Let T, ST, SST be terms. A query context of T is a tuple, $\mathcal{C}$=(T, ST, SST), where ST is a subtopic of T (determined by the subtopicOf relationship in a textbook ontology), and SST is a subtopic of ST, and a subtopic cannot exist without a direct parent topic. 'null' is used to specify the absence of each of ST, and SST.*

R1, R2, and R3 identify query context by considering the *subtopicOf* relationships in a textbook ontology. Consider a textbook ontology in Fig. 1. By R3, one of the query contexts of "database" can be identified as shown below.

```
<queryContext>
  <T>database</T>
  <ST>query language</ST>
  <SST>definition of</SST>
</queryContext>
```

```
R1: queryContext(T,ST,SST) ← mainTopic(T,L), [ST = null], [SST = null].
R2: queryContext(T,ST,SST) ← subTopic(T,ST,L), [SST = null].
R3: queryContext(T,ST,SST) ← subsubTopic(T,ST,SST,L).
where inside a bracket is a constraint, and the semantics of each predicate is as follows.
  queryContext(T,ST,SST) ... C=(T,ST,SST) is a query context of T.
  mainTopic(T,L) ... T is a topic and one of its reference locators is a book page number, L.
                     There is no reference locator if L is 'null'.
  subTopic(T,ST,L) ... ST is a subtopic of T, and one of the reference locators of ST is
                       a book page number, L.
  subsubTopic(T,ST,SST,L) ... SST is a subtopic of ST and ST is a subtopic of T, and
                              one of the reference locators of SST is a book page number, L.
```

Definition 2 (Slide topic). *Let T, ST, and SST be terms, SL a lecture slide, SEG a textbook segment of a particular textbook chapter, CH. A slide topic of SL is a tuple, $\mathcal{T}$=(T,ST,SST), where all of the three conditions are satisfied: (i) a term, T, ST, or SST, appears in SL, (ii) $\mathcal{T}$ is indexed in SEG, and (iii) there exists a mapping from SL to SEG in the mapping table.*

To put it simply, a slide topic is a combination of a topic and its subtopics. A lecture slide may contain several slide topics (i.e. information about the topics), which can be identified by R4. The underlying assumption is that, in a particular slide, contextual terms of a topic may be omitted. For example, a topic (e.g. database) tends to appear in a lecture slide while its subtopic (e.g. definition) may not necessarily appear. Such a subtopic can be identified by R4. As another case, a subtopic (e.g. query language) may appear in a lecture slide while its topic (e.g. database) and sub-subtopic (e.g. definition) disappear. R4 is also responsible for identifying an implicit topic and sub-subtopic of such a subtopic. R5, R6, and R7 use page numbers associated between a book index and TOC to determine mapping between (a combination of) terms and a textbook segment.

```
R4: slideTopic(SL, T, ST, SST)
   ← slide(SL,CH),[T,ST,or SST ∈ SL], // there exists SL such that T,ST,or SST appears in SL
      TtoSeg(T,ST,SST,CH,SEG),       // SEG of CH contains information of 𝒯=(T, ST, SST)
      SLtoSeg(SL,CH,SEG).            // SL can be mapped to SEG
R5: TtoSeg(T,ST,SST,CH,SEG)
   ← mainTopic(T,L), [L ≠ null],
      toc(CH,SEG,PB,PE), [PB ≤ L ≤ PE].  // T is indexed in SEG of CH
R6: TtoSeg(T,ST,SST,CH,SEG)
   ← subTopic(T,ST,L), [L ≠ null],
      toc(CH,SEG,PB,PE), [PB ≤ L ≤ PE].  // As a subtopic of T, ST is indexed in SEG of CH
R7: TtoSeg(T,ST,SST,CH,SEG)
   ← subsubTopic(T,ST,SST,L), [L ≠ null],
      toc(CH,SEG,PB,PE), [PB ≤ L ≤ PE].  // As a subtopic of ST, SST is indexed in SEG of CH
where the semantics of each predicate is as follows.
slideTopic(SL,T,ST,SST) ... A lecture slide, SL, contains information about 𝒯=(T, ST, SST).
slide(SL,CH) ... There exists SL in the database and SL can be mapped
                 to a textbook chapter, CH (our pre-assumption).
SLtoSeg(SL,CH,SEG) ... SL can be mapped to a textbook segment, SEG, of a textbook chapter,CH,
                       by using the mapping table (see, e.g., in Fig. 2).
TtoSeg(T,ST,SST,CH,SEG) ... there exists a mapping from 𝒯=(T, ST, SST) to SEG of CH.
toc(CH,SEG,PB,PE)... SEG is a textbook segment of CH with the page boundary [PB,PE] where PB
                     is the beginning page and PE is the ending page.
```

Consider an example in Fig. 4. By R4 and R6, we obtain that the *description of relational algebra* is a slide topic (i.e. $\mathcal{T}$=("relational algebra", "description of", *null*)). Also by R4 and R7, the *definition of database query language* is identified as a slide topic (i.e. $\mathcal{T}$=("database", "query langauge", "definition of")). Note that the slide topics can be identified although a term, "definition" does not appear in the slide. On the contrary, other slides that contain a term, for example, "query language" are considered irrelevant to the context of "definition of query language" if they cannot be mapped to Seg.#58, Seg.#59, or Seg.#60.

A query submitted by a user can be formulated as a non-unit XML clause below. Let T be "database", ST "query language", SST "definition of". The lecture slide in Fig. 4 will be obtained as an answer. If, for example, SST is left unspecified, lecture slides with the slide topics where SST is "definition of", "relational algebra as", "relational calculus as", or "SQL as" will be obtained (i.e. all possible subtopics of "query language" in Fig. 1)).

```
Q: answer(SL) ← queryContext(T,ST,SST), slideTopic(SL,T,ST,SST).
```

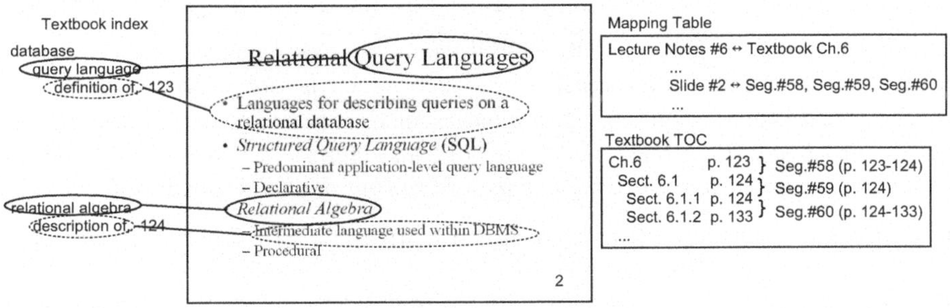

Fig. 4. Slide topics of a lecture slide

3.4 Prototype

The right side of Fig. 3 shows a prototype system for lecture material retrieval in a database course. Users can input a combination of a topic and its context as a query. The output is a set of answer slides, automatically classified by query contexts. Note that if a query context is specified but is not exactly matched with any index subheading, no results will be returned (because the evaluation of the queryContext predicate is failed). In this case, the system will automatically reformulate the query by leaving the ST (or SST) argument unspecified in the query clause so that all possible results in any context can be obtained, if they exist. In this way, users are able to compare themselves their specified context (subtopics) with those suggested by the system. Since lecture slides are assumed to be associated with corresponding video fragments (i.e. spoken words), our system can also be considered as lecture video segment retrieval that uses the content of PowerPoint slides as underlying metadata.

3.5 Building the Mapping Table

Recall from Sect. 2.2 that it is not difficult to manually establish mappings from a lecture note (i.e. a sequence of lecture slides) to a textbook chapter. Thus, the problem is reduced to that of mapping from a slide to textbook segments, given a chapter mapping. In this paper, two main heuristics are used to determine automatic mapping from lecture slides to textbook segments. One important clue is slide titles whose terms can indicate main topics (or subtopics) being discussed in the slides. Associated through page numbers in the book index and TOC, such terms can finally be mapped to particular textbook segments. Another helpful hint is information about mapping from neighboring lecture slides. For example, if we know that several surrounding slides of a slide of interest can be mapped to a particular textbook segment, it is likely that such a slide should also be mapped to the same segment, regardless of the content in the slide.

It is also tempting to exploit the keyword matching between slide titles and titles of TOC in a textbook. However, terms used in titles of TOC may not accurately reflect content, and thus restrict complete topic indication while slide titles contain more specific terms representing topics or subtopics. Thus, a book index is more appropriate for the matching.[6]

Definition 3 (Main slide topic). *Let T be a term and SL a lecture slide. T is a main slide topic of SL iff (i) T appears in the slide title of SL, and (ii) T is exactly matched with a topic, subtopic, or sub-subtopic in the textbook ontology (i.e. main heading, subheading, or sub-subheading of the textbook index).*

Definition 4 (Ad-hoc slide-to-page mapping). *Let SL be a lecture slide, P a book page number, O a textbook ontology. Ad-hoc mapping from a lecture slide to a book page number is represented by a tuple, $\mathcal{M}$=(SL, P), where one of the five conditions is satisfied:*

1. *$P = L_t$ if there exists a reference locator, L_t, of a topic, T, in O such that T is a main slide topic of SL.*
2. *$P = L_{st}$ or $P = L_{sst}$ if a topic, T, is a main slide topic of SL but there exists no reference locator of T in O. Instead, there exists a reference locator, L_{st}, of a subtopic of T, or a reference locator, L_{sst}, of a sub-subtopic of T.*
3. *$P = L_{st}$ if there exists a reference locator, L_{st}, of a subtopic, ST, in O such that ST is a main slide topic of SL.*
4. *$P = L_{sst}$ if a subtopic, ST, is a main slide topic of SL, but there exists no reference locator of ST in O. Instead, there exists a reference locator, L_{sst}, of a subtopic of ST.*
5. *$P = L_{sst}$ if there exists a reference locator, L_{sst}, of a sub-subtopic, SST, in O such that SST is a main slide topic of SL.*

Terms appearing in the slide title can, in most cases, indicate main topics being described in a particular slide. Such terms can be mapped to book pages through the ontology. If a term appearing in the slide title is found in the ontology as, for example, a topic but without reference locators being specified, its subtopics (and/or sub-subtopics if they exist) are possibly (implicit) slide topics of that slide. This is the reason why all reference locators appearing under a particular topic (which does not have reference locators) are taken into consideration in this case (consider conditions 2 and 4 in Definition 4).

Consider Fig. 5 and let SL_i be a lecture slide number i of the lecture note Ch. 6, and P a book page number. $\mathcal{M}=(SL_2, P_{SL_2})$ is obtained as the mapping result for SL_2 where $P_{SL_2} \in \{123, 124, 140\}$. For SL_3, there is no mapping (i.e. $\mathcal{M}=(SL_3, \text{null})$) since SL_3 has no main slide topic. Nevertheless, given the assumption that the lecture note follows the structure of a textbook, it is possible to estimate the mapping for SL_3 by considering the mapping of its surrounding lecture slides (i.e. SL_2 and SL_4 in this case). Specifically, given that $\mathcal{M}=(SL_3, P_{SL_3})$ is the mapping for SL_3, the minimum page number and the maximum page number from $\mathcal{M}=(SL_2, P_{SL_2})$ and $\mathcal{M}=(SL_4, P_{SL_4})$ are chosen to be the lower bound and the upper bound of P_{SL_3}, respectively. In this case, we obtain $\mathcal{M}=(SL_3, P_{SL_3})$ where $P_{SL_3} \in [123, 137]$ as the mapping result. Note also that SL_2 and SL_4 are the nearest preceding lecture slide and the nearest succeeding lecture slide of SL_3, respectively, which contains a mapping.

[6] Exploiting TOC titles as an additional heuristic is beyond the scope of this paper.

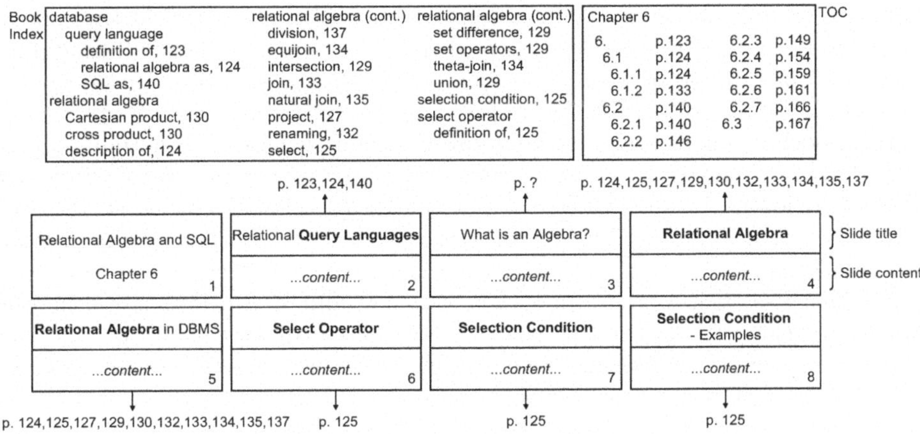

Fig. 5. Main slide topics as an important clue to determine the ad-hoc mapping

Definition 5 (Mapping conflict). *Let SL_i and SL_j be lecture slides that can be mapped to the same textbook chapter where $i \neq j$, and P a book page number. Given a mapping $\mathcal{M}=(SL_i, P_{SL_i})$, a mapping conflict between SL_i and SL_j exists for P_{SL_i} being selected iff one of the two conditions is satisfied:*

1. *SL_i precedes SL_j, and there exists no P_{SL_j} for all $\mathcal{M}=(SL_j, P_{SL_j})$ such that $P_{SL_j} \geq P_{SL_i}$, or*
2. *SL_i follows SL_j, and there exists no P_{SL_j} for all $\mathcal{M}=(SL_j, P_{SL_j})$ such that $P_{SL_j} \leq P_{SL_i}$.*

As mentioned earlier, another helpful heuristic to determine the mapping from a lecture slide to textbook segments is the mapping of surrounding slides. Consider Fig. 5 and suppose we choose $\mathcal{M}=(\text{SL}_2, 140)$ as a mapping for SL_2. This mapping contradicts with our initial assumption in that the presentation flow must follow the structure of a textbook; none of the mappings of SL_3, SL_4, SL_5, SL_6, SL_7, SL_8 yields book page numbers that are greater than or equal to 140. In this way, it is possible to *approximately* calculate the total number of conflicts for each mapping with all other slides (e.g. the total numbers of conflicts for $\mathcal{M}=(\text{SL}_2, 140)$ being selected is 6). Finally, we can retain only those mappings with minimum number of conflicts decided by predefined threshold. For example, if the threshold is 0, only $\mathcal{M}=(\text{SL}_2, 123)$ and $\mathcal{M}=(\text{SL}_2, 124)$ are retained for SL_2. Then, we obtain that SL_2 can be mapped to textbook segments, Seg.#58 (p. 123-124), Seg.#59 (p. 124), and Seg.#60 (p. 124-133) (see also Fig. 4). Note that we allow one-to-many mapping from a slide to textbook segments.

4 Empirical Evaluation

We conducted a preliminary experiment comparing our system with Lucene[7] where PowerPoint slides (i.e. each slide constitutes one document) are indexed.

[7] A full-featured text search engine freely available at http://lucene.apache.org

Test data are taken from a database course web site[8] containing 7 lecture notes, 275 slides in total. While our system returns the answer slide out of 5 candidate slides without ranking provided for a query, *definition of database*, Lucene yields the answer at the 12th rank. As expected, the top-ranked slide from Lucene contains both terms *database* and *definition* although the slide is actually about *SQL*. This is because some terms such as *database* appear in several slides but do not provide useful information about the terms themselves (i.e. such terms are used instead for describing some other terms). As another example, for a query such as *OLAP*, Lucene can return the correct answer at the top rank. This is because such a term appears in very few slides. An implication here is that a traditional IR approach may not be able to distinguish terms that are informative from ones that are just passing mention. Nevertheless, in order to prove this hypothesis, extensive study and necessary experiments must be done.

5 Related Work

[6] develops a system for context-based searching in lecture slides. A number of lecture notes are used as a training set, incorporated with a machine learning technique, to discover patterns of lecture presentation flow. The probabilistic state diagram for lectures (where each state represents a context) is used to recognize the context of lecture slides. Limited types of context can be detected such as "intro", "definition" and "example" while our approach deals with broader context based on subheadings in a textbook index. Unlike our approach, [6] does not distinguish between informative and passing mention terms, which may lead to several irrelevant results being obtained. [9] performs mapping from handwritten lectures to textbook chapters. Handwritten annotations are automatically recognized and used as keywords to compare similarity with TOC of an associated textbook using a traditional IR approach. [3] mainly focuses on enhancing user interfaces for information browsing and filtering in lecture videos and the synchronized lecture slides. Search capability is also implemented using *tf-idf* weighting with additional heuristics such as increasing the matching score of a term appearing in a slide if its duration (according to that of the lecture video segment associated) is longer than that of another slide containing the same term. A similar idea to our approach that uses textbook metadata can be found in [2]. TOC and textbook index are exploited to generate a topic vector profile in order to help classify messages being discussed in online discussion boards. A topic vector is simply a set of index terms that can be associated (through page numbers) with a particular textbook section. The authors do not, however, introduce this idea to lecture materials nor do they address the problem of context interpretation and identification.

6 Conclusions and Future Work

Although the original purpose of a textbook index is to allow quick and easy access to information in a textbook, we incorporate it with table of contents

[8] http://csis.pace.edu/~scharff/cs623/ref/cs623indexref.html

in the way that integrates context from user queries and lecture slides, thus enabling effective search capability inside lecture materials. Since a textbook index is unable to cover all possible query terms issued by users as well as new terms introduced in only lecture slides but not in the textbook, generalization of our approach is necessary. Omitted also in this paper is a ranking algorithm for answer slides, which forms part of our future work.

Acknowledgement. This research was supported by MEXT Grant-in-Aid for Scientific Research on Priority Areas: "Cyber Infrastructure for the Information-explosion Era", Planning Research: "Contents Fusion and Seamless Search for Information Explosion" (Project Leader: Katsumi Tanaka, A01-00-02, Grant#: 18049041) and by MEXT Grant-in-Aid for Young Scientists B "Information Retrieval and Mining in Web Archives" (Grant#: 18700111).

References

1. Downes, S. E-learning 2.0. eLearn, 2005(10), 1, ACM Press. (2005)
2. Feng, D., Kim, J., Shaw, E., Hovy, E. Towards Modeling Threaded Discussions Using Induced Ontology Knowledge. In Proceedings of AAAI'06. (2006)
3. Hurst, W., Muller, R., Mayer, C. Multimedia Information Retrieval from Recorded Presentations. In Proceedings of ACM SIGIR'00. (2000)
4. Lewis, P., M., Bernstein, A., Kifer, M. Databases and Transaction Processing: an Application-Oriented Approach. Addison Wesley. (2002)
5. Mertens, R., *et al.* Hypermedia Navigation Concepts for Lecture Recordings. World Conf. on E-Learning in Corp., Gov., Healthcare & Higher Education. (2004)
6. Mittal, A., *et al.* Content Classification and Context-Based Retrieval System for E-learning. Educational Technology & Society, 9(1), 349–358. (2006)
7. Mulvany, N., C. Indexing Books (Chicago Guides to Writing, Editing, and Publishing), 2nd edition. (2005)
8. Open Document Format for Office Applications (OpenDocument), v. 1.0, ISO/IEC 26300. http://www.oasis-open.org/committees/download.php/12572/OpenDocument-v1.0-os.pdf
9. Tang, L., Kender, J., R. Educational Video Understanding: Mapping Handwritten Text to Textbook Chapters. In Proc. of Int. Conf. on Document Analysis and Recognition. (2005)
10. Walsh, N., Muellner, L. DocBook: The Definitive Guide. O'Reilly & Assoc. (2006)
11. Wuwongse, V., Akama, K., Anutariya, C., Nantajeewarawat, E. A Data Model for XML Databases. Journal of Intelligent Information Systems, 20(1), 63–80. (2003)
12. Zhang, D., Zhao, J., L., Zhou, L., Nunamaker, J., F. Can e-Learning Replace Classroom Learning? Commun. ACM, 47(5), 75–79. (2004)

Activate Interaction Relationships Between Students Acceptance Behavior and E-Learning

Fong-Ling Fu[1], Hung-Gi Chou[2], and Sheng-Chin Yu[3]

[1,2] Department of Information Management, National Cheng-Chi University, Taipei 11605, Taiwan
flfu@.nccu.edu.tw
[3] Department of Business Administration, Tung-Nan Technology Institute, Taipei 22202, Taiwan
jeanyu@mis.nccu.edu.tw

Abstract. The purpose and characteristics of e-learning are different from other Web applications. This paper uses an extended Technology Acceptance Model (TAM) to explain the motivation, attitude and acceptance behind participants of e-learning. The factors that determine Web quality include the external variables: system functionality, interface design, pedagogic and contents, as well as community.. Perceived enjoyment was also added to this model as a factor. . The empirical results indicated that the extended TAM explains the acceptability of on-line learning systems with high reliability, validity, and model fitness. All "beliefs" (user perceptions) – perceived usefulness, ease of use, and enjoyment – are good predictors of attitude and acceptance. Pedagogic, and content as well as community are important external factors that predict e-learning acceptance.

Keywords: On-line learning, TAM, Web quality, Perceived enjoyment.

1 Introduction

E-learning has a distinctly different purpose from other Web applications [13], which is particularly in self-learning through fluent Web material and collaborative learning with a virtual community. Websites are capable of providing a richer degree of knowledge and multimedia content. They also employ new pedagogic strategies and different students' assessment of the learning [19]. Therefore, the attitudes and perceptions which students hold toward their learning experiences become increasingly important [22].

The Technology Acceptance Model (TAM) explored the intentional usage of new information technology [5] while the Extended TAM employed the same purpose but provided a more complete and clear explanation of the determinants of users' online usage [12]. Besides perceived usefulness and ease of use, on-line multimedia learning material may create the feelings of being challenge and having fun within students, which are intrinsic motivations to obtaining education [18]. Therefore, the factor of perceived enjoyment played a key role in the Extended TAM due to the importance of leisureliness in online usage and students' ability to cultivate their learning experience [7].

G. Dong et al. (Eds.): APWeb/WAIM 2007, LNCS 4505, pp. 670–677, 2007.

Another concern in the study is fact that the extended TAMs have used some different critical quality variables on Internet applications as their external variables [2, 12, 17], even though a consistent and complete summary on external variables in TAMs would make the model more comprehensible. The comprehensive measurements – Website Quality [2], which has long progressed from SERQUAL [17], and information system quality measurement [12] – may be suitable for external variables of the Extended TAM. On the other hand, although empirical studies found little support for the importance of perceived enjoyment in Website access, this lack of identifiable relationships could be attributed to the characteristics of Website quality which indirectly influence the effectiveness of students' learning [14, 19]. Therefore, in this study we not only validate the role played by perceived enjoyment through the examination of e-learning – but also develop a complete model, particularly from students' point of view, in order to explain more proper the interaction relationships among the factors for e-learning system acceptance.

2 Extended TAM and Web Quality in E-Learning

2.1 The Students' Attitude Toward the E-Learning System Acceptance

The TAM was a very simple but effective model to predict and explain user acceptance of information technology. It suggested that external variables would influence one's beliefs, attitudes and intentions regarding an information system. In the model, two beliefs – "perceived usefulness" and "perceived ease to use" – were found to be positively related to the usage. Perceived usefulness was defined as "the degree of one's job performance that would be improved by a specific system." Perceived ease to use was defined as "the degree of lack of difficulty in using a specific system" [5]. Besides extrinsic motivation, such as perceived usefulness, the intrinsic motivation of perceived enjoyment is defined as "all enjoyment generated from participating in the computer-based activity itself, independent of any other predictable result of the activity" [18, 21]. Since browsing behavior on Websites was being controlled more by the intentions of users,"user enjoyment" was placed in the spotlight in current empirical studies [10].

2.2 Web-Site Quality for On-Line Course

External variables in TAM played important roles in the process of understanding the relationship between internal beliefs, attitudes and intentions [5]. To evaluate the effectiveness of or students' satisfaction with information technology products in particular learning contexts, researchers suggested that three other factors – the student's background, pedagogic, and content presentation – should be considered together [22]. In online learning, the system functions that enhance flexibility and interaction as well as learning materials, indeed also influence perceived usefulness [17]. "Perceived visual attractiveness" is positively related to perceived usefulness, perceived ease to use and perceived enjoyment [7].

Ample research states that Website quality influences users' perception of effectiveness [11]. The factors that determine the quality of e-commerce Websites include information content, content reliability, Website attractiveness, navigation speed, security, and customer service [11]. Another body of research on Website quality did not originate from studies on the quality of information systems, but from PZB measurement (SERQUAL). SERQUAL contains five dimensions: physic, service reliability, responsiveness, assurance, and sympathy [17]. Service quality has been used as the third factor besides system quality and information quality in the model of effectiveness on information systems [9].

3 Research Method

3.1 Framework and Hypotheses of Interaction Activated TAM on E-Learning

The research framework was adjusted from previous TAM and added an extra belief, perceived enjoyment to accommodate the application of e-learning [7, 14], as shown in Figure 1. The adjusted measurement of Website quality consisted of system functionality, interface design, pedagogic and content design, and community as external variables [1, 2, 19]. Therefore it was hypothesized that:

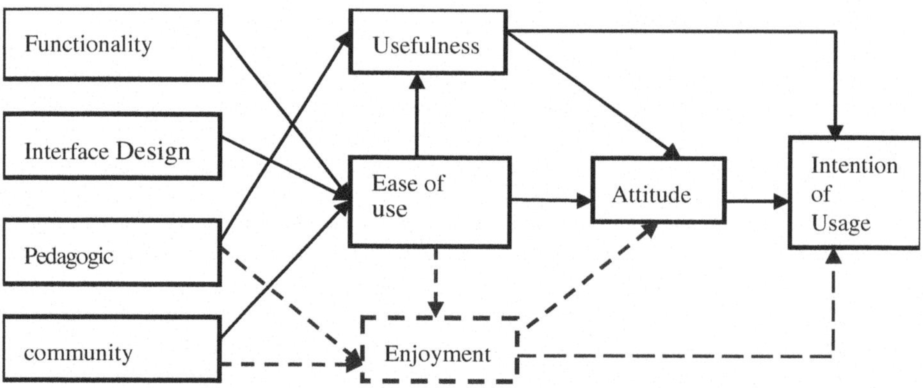

Fig. 1. Research Framework

H1: The system functionality will positively influence perceived ease to use.
H2: Interface design will positively influence perceived ease to use.
H3: Pedagogic and content will positively influence perceived usefulness.
H4: Pedagogic and content will positively influence perceived enjoyment.
H5: Community will positively influence perceived ease of use.
H6: Community will positively influence perceived enjoyment.
H7: Perceived ease of use will positively influence perceived usefulness.
H8: Perceived ease of use will positively influence attitude.
H9: Perceived usefulness will positively influence attitude.
H10: Perceived usefulness will positively influence intention of usage.
H11: Attitude will positively influence intention of usage.
H12: Perceived ease of use will positively influence perceived enjoyment.
H13: Perceived enjoyment will positively influence attitude.
H14: Perceived enjoyment will positively influence intention of usage.

3.2 Design of Questionnaire

The operation definition of the extended TAM was an adjusted version of the models of Lin & Lu [12] and Davis [6]. The external variables of TAM in the study that measured Website quality were based on Swan's [19] suggestions and included system functionality [2, 12, 19], interface design [4, 19], pedagogic and content design [19], and community [2]. All items were measured using a seven-point Likert scale.

Table 1. Reliability and validity of the questionnaire

Construction	Content of Item	Factor loading	Extrac -tion of σ	α
System functionality	The Website is reliable	0.82	0.69	0.87
	Satisfied with the waiting time to connect	0.74		
	Links on the Website are working and correct	0.77		
Interface Design	The appearance of the Website is attractive	0.87	0.77	0.91
	The method of use is consistent	0.84		
	The interface design of the Website is consistent	0.90		
Pedagogic and content	The content is rich in quantity and quality	0.94	0.72	0.88
	The content is neither too easy nor too difficult	0.94		
	The content is clear and easy to read	0.85		
Community	Easy to get support from staff and classmate	0.88	0.67	0.86
	The facilities support peer interaction	0.87		
	The Website served as a learning community to me	0.88		
Perceived Ease to Use	It is easy to browse the Website	0.82	0.62	0.82
	It is easy to access to the Website	0.74		
	It is easy to search the materials in the Website	0.77		
Perceived Usefulness	The Website is useful	0.87	0.76	0.90
	The Website helps me learn more effectively	0.84		
	Using the Website improved my performance	0.90		
Perceived Enjoyment	Using the Website is an enjoyable experience	0.91	0.87	0.95
	Using the Website is a happy experience	0.96		
	Using the Website is an interesting experience	0.93		
Attitude	I like to use the Website	0.94	0.83	0.94
	I feel comfortable to learn with the Website	0.94		
	I have positive attitude toward using the Website	0.85		

The reliability and validity of the measurement were tested through confirmation factor analyses using the Structural Equation Model. The confirmatory factor analyses were used for validity tests. According to the LAMBCA value calculated by the software LISREL, all items reported factor loadings greater than 0.7, indicating high validity (Table 1). The values of total extraction of variance were greater than 0.6 for each dimension (Table 1) – higher than the acceptable of level 0.5. This also indicated that the measurement was valid. Coefficients of internal consistency (Cronbach α) were greater than 0.8 (Table 1) for each of the dimensions, further indicating that these measurements are reliable.

4 Data Analyses

4.1 Sample Analyses

An online survey was conducted in a Taiwanese university. The subjects taking the survey were volunteer students who had taken at least one online course. The total valid sample was 451, 39% of which majored in business and 23.9% of which majored in social science. The majority were not new Internet users, as 72.5% reported having more than 3 years of online surfing experience.

4.2 Evaluation of the Model

Results from all three fit indexes of the Structural Equation Model were good, indicating that the extended TAM model is applicable to e-learning. The indexes are listed as follows:

1. Absolute Fit Measures: Absolute fitness could be measured through coefficient of RMSEA. The acceptable value is said to be either smaller than 0.06 [8] or smaller then 0.08 [16]. The value of RMSEA in the study was 0.06. The absolute fitness could also be indicated by the value of SRMR, which should be smaller than 0.08 [8]. The value of RMSR in this study was 0.058.
2. Incremental Fit Measures: The common measurement is the value of CFI, which should be equal to or greater than 0.95 [3]. Other indicators are NFI or NNFI. Their values are always between 0 and 1. At the same time, the model cannot be considered as meeting the standards unless the value of NFI or NNFI is greater than 0.9 [8]. In this study, the value of CFI was 0.98, greater 0.95. Both the value of NNFI and NFI were 0.98, greater than 0.9.
3. Parsimonious Fit Measures: The number of estimates that fulfill a specific level of appropriateness for the model. The model was considered good if the value of PGFI was greater than 0.5 [15]. The value of PGFI in this study was 0.7.

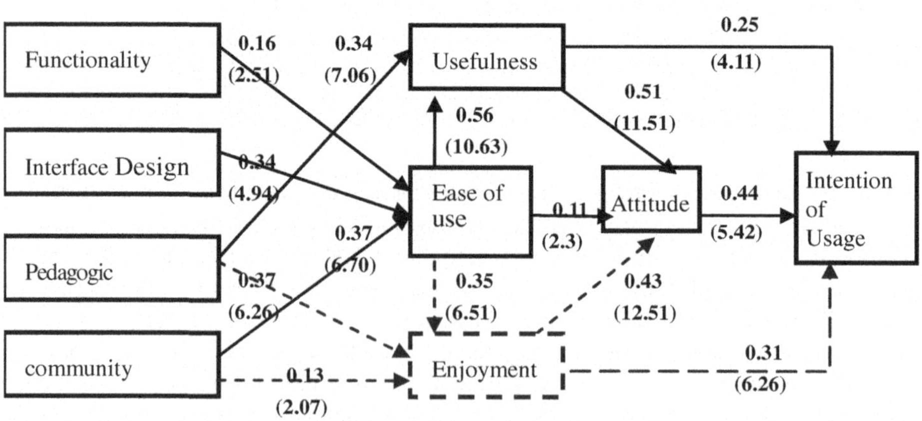

Fig. 2. Path Analyses

4.3 Results of the Model

In Fig. 2, coefficients of the paths are demonstrated by the arrows, with t-values listed between the brackets. All t-values in Fig. 2 were greater than the value 1.96 at the significant level of 0.05, indicating that all null hypotheses were rejected. For example, for the Hypothesis 1, **the c**oefficient of path =0.16 while t-value=2.51, implying that students who consider the e-learning system as more quickly accessible and more reliable will perceive the system as easier to use. The other hypotheses can also be proven according to their coefficients of path and t-values.

5 Conclusions

5.1 Perceived Usefulness, Ease to Use, and Enjoyment on TAM in E-Learning

It is concluded that the extended TAM developed by the study is valid for understanding students' acceptance of e-learning. The variable "perceived enjoyment" should be included in the TAM of e-learning. Both the reliability and validity of the questionnaire were acceptable. All hypotheses were supported. The fit tests revealed that the model could explain the attitude toward and the intention of students' usage of an e-learning course. By using Squared Multiple Correlations for Structural Equations, the model could explain 84% variance of attitude and 83% variance of intention of usage. Perceived usefulness is more important in predicting the acceptance of e-learning than ease to use and enjoyment (Table 2).

Table 2. Squared Multiple Correlations for Structural Equations

Perceived Ease to Use	Perceived Usefulness	Perceived Enjoyment	Attitude	Intention of Usage
0.53	0.62	0.54	0.84	0.83

5.2 Importance of Web Qualities

This research illustrated the key factors of Web quality from students' point of views-the importance of online content and community. The pedagogic and content online yielded a higher influence on attitude and intention of usage than all other external variables. The coefficients of regression were 0.33 and 0.34 respectively (Table 3). The second most influential external variable in regards to attitude and intention of usage was "community," with a regression coefficient of 0.26 and 0.24. Online interactions among class members, establish a convenient communication channel for sharing learning experiences.

Table 3. Regression Matrix ETA on KSI

	Functionality	Interface Design	Pedagogic and content	Community
Perceived ease to use	0.16	0.34	--	0.37
Perceived usefulness	0.09	0.19	0.34	0.21
Perceived enjoyment	0.06	0.12	0.37	0.26
Attitude	0.09	0.18	0.33	0.26
Intention of Usage	0.08	0.16	0.34	0.24

5.3 Contribution and Further Study

Previous studies, whether they concentrated on extended TAM or Web quality, focused on the application of online transactions or services. The extended model of e-learning developed in this study provides a model which is simple, easy to understand, and able to effectively predict students' attitudes and levels of acceptance. It not only combines the factor of perceived enjoyment, but also synthesizes and analyzes the insights of the studies of Website quality on e-learning. The empirical results of the study proved the importance of the Website quality on the attitude and acceptance of e-learning.

The attitude and acceptance of Websites may not be the same in other Web-technology applications due to their different purposes. Even though the main purpose of learning is not to have fun, applying the variable of "perceived enjoyment" into TAM is acceptable because the behavior of online learning is more voluntary and controllable from the user's perspective. Further studies are needed to investigate how the variable of "perceived enjoyment" as well as the factors on Web quality affect other applications, such as online gaming where the main motive behind Internet usage is for recreational purposes.

References

1. Bagozzi, R.P., Yi, Y.: On the Evaluation of Structural Equation Models", Journal of Academy of Marketing Science, Vol.16, No.1, (1988) 74-94.
2. Barnes, S. J., Vidgen, R.T.: Measuring Web Site Quality Improvements: a Case Study of the Forum on Strategic Management Knowledge Exchange. Industrial Management & Data Systems, (2003) 297-309.
3. Bentler, P.M.: EQS Structural Equations Program Manual. Encino, CA: Multivariate Software, (1995)
4. Cox, J., Dale, B.G.:, Key Quality Factors in Web Site Design and Use: an Examination. International Journal of Quality & Reliability Management, 19, (2002) 862-888
5. Davis, F.D., Bagozzi, R.P., Warshaw, P.R.:X: User Acceptance of Computer Technology : A Comparison of Two Theoretical Models. Management Science, Vol.35, No.8, (2002)
6. Davis, F.D.: Perceived Usefulness, Perceived Ease of Use, and User Acceptance of Information Technology. MIS Quarterly, Sep, Vol.13, No.3, (1989) 319-339
7. Heijden, H.: Factors Influencing the Usage of Websites: the Case of a Generic Portal in The Netherlands. Information & Management, 40, (2003) 541-549
8. Hu, L., Bentler, P. M.: Cutoff Criteria for Fit Indexes in Covariance Structural Equation Modeling, Vol.6, No.1,(1999) 1-55
9. Jiang, J. J., Klein, G. C., Christopher L.: Measuring Information System Service Quality: SERVQUAL from the Other Side, MIS Quarterly, Vol.26, No.2, (2002)145-166
10. Johnson, R.A., Hignite, M.A.: Applying the Technology Acceptance Model to the WWW. Academy of Information and Management Sciences Journal, Vol.3, No.2, (2000) 130-142
11. Kim, Sung-Eon, Shaw, T., Schneider, H.: Web Site Design Benchmarking within Industry Groups. Internet Research, Vol.13, No.1, (2003) 17-26
12. Lin, Chuan-Chuan, Lu, H.: Towards an Understanding of the Behavioural Intention to Use a Web Site. International Journal of Information Management, Vol.20, (2000) 197-208

13. Meuter, M. L., Ostrom, A. L., Roundtree, R. I. , Binter M. J.: Self-service Technologies: Understanding Customer Satisfaction with Technology-based Service Encounters. Journal of Marketing, Vol. 64, No. 3, (2000) 50-64.
14. Moon, Ji-Won, Kim, Young-Gul,: Extending the TAM for a World-Wide-Web context. Information & Management, Vol.38, (2001) 217-230
15. Mulaik, S.A., James, L.R., Van Altine, J., Bennett, N., Lind, S., Stilwell, C.D.: Evaluation of Goodness-of-fit Indices for Structural Equation Models. Psychological Bulletin, Vol.105, (1989) 430-445
16. McDonald, R.P., Ho, M.R.: Principles and Practice in Reporting Structural Equation Analysis. Psychological Methods, Vol.7, (2002) 64-82
17. Parasuraman, A., Zeitham, V.A., Berry, L.L.: Service Quality: A Multiple-item Scale for Measuring Consumer Perceptions of Service Quality. Journal of Retailing, Vol. 64, No. 1. (1988)
18. Prensky, M.: Digital game-based learning, USA7 McGraw-Hill. (2001)
19. Swan, K.: Relationships between Interactions and Learning in Online Environments. Available: http://www.sloan-c.org/publications/books/interactions.pdf (2004)
20. Landry, B. J., Griffeth, R., Hartman, S.: Measuring Student Perceptions of Blackboard Using the Technology Acceptance Model. Decision Science Journal of Innovative Education, Vol.4, No.1, January, (2006)
21. Kiili, K.: Digital games-based learning: Towards an experiential gaming model. The Internet and Higher Education, Vol.8, Issue 3, 3rd Quarter, (2005)13-24.
22. Chao, T., Saj, T., Tessier, F.: Establishing a Quality Review for Online Courses. EDUCAUSE Quarterly, No.3. (2006).

Semantic-Based Grouping of Search Engine Results Using WordNet

Reza Hemayati[1], Weiyi Meng[1], and Clement Yu[2]

[1] Department of Computer Science
State University of New York at Binghamton
Binghamton, NY 13902, USA
{rtaghiz1,meng}@binghamton.edu
[2] Department of Computer Science
University of Illinois at Chicago
Chicago, IL 60607, USA
yu@cs.uic.edu

Abstract. Terms used in search queries often have multiple meanings. Consequently, search results corresponding to different meanings may be retrieved, making identifying relevant results inconvenient and time-consuming. In this paper, we propose a new solution to address this issue. Our method groups the search results based on the different meanings of the query. It utilizes the semantic dictionary WordNet to determine the basic meanings or senses of each query term and similar senses are merged to improve grouping quality. Our grouping algorithm employs a combination of categorization and clustering techniques. Our experimental results indicate that our method can achieve high grouping accuracy.

Keywords: Categorization, Clustering, WordNet, Search Engine.

1 Introduction

Most Web users use search engines to find the information they want from the Web. One common complaint about the current search engines is that they return too many useless results for users' queries. Both the search engines and the users contribute to this problem. On the one hand, current search engines make little effort to understand users' intentions and they retrieve documents that match query words literally and syntactically. On the other hand, Internet users tend to submit very short queries (average length is about 2.3 terms and 30% have a single term [8]). One way to tackle this problem is to group the search results for a query into multiple categories such that all results in the same category corresponds to the same meaning of the query.

In this paper we propose a new technique to group the *search result records* (**SRR**s) returned from any search engine. Our focus will be on SRRs retrieved by *single term queries*. For queries with multiple terms, the specific meaning of each term is easier to determine because other terms in the same query can provide the context [9]. Our technique differs from existing techniques in the following aspects. First, we use a semantic electronic dictionary WordNet [5, 13] to provide the basic

G. Dong et al. (Eds.): APWeb/WAIM 2007, LNCS 4505, pp. 678–686, 2007.

meanings of each query term. Second, we apply a merging algorithm to merge synsets that have very close meanings into a super-synset. Third, we employ a two-step process to categorize SRRs into super-synsets. Fourth, our method also deals with SRRs that do not correspond to any WordNet-provided synsets of the query terms by clustering them. For example, when a word is used as a name, like "Apple" and "Jaguar", it does not have its traditional meanings. The rest of the paper is organized as follows. In section 2, we review some related work. In section 3, we describe our grouping algorithm. Preliminary evaluation results are reported in section 4. The future work and conclusion are presented in section 5.

2 Related Work

The general problem of document clustering and categorization has been studied extensively [7] and they will not be reviewed in this paper. Instead, we focus on related works that deal with the clustering and categorization of the search result records (SRRs) returned from search or metasearch engines.

Techniques for clustering web documents and SRRs have been reported in many papers and systems such as [14, 17, 18]. However, these techniques perform clustering based on the syntactical similarity but not semantic similarity. In contrast, our method employs both categorization and clustering techniques and it also utilizes similarities that are computed using both syntactical and semantic information.

Techniques for clustering and categorizing web documents using WordNet or other ontologies have also been extensively studied (e.g., [3, 4, 6, 12]) and some of them (e.g., [3, 4]) also tried to categorize SRRs based on the meanings of the query term. However, our approach differs from these techniques significantly. First, we use more features of WordNet such as hypernym, hyponym, synonym and domain. Second, we employ a sense-merging algorithm to merge similar senses before grouping. Third, our SRR grouping algorithm employs both categorization and clustering in a unique way. Fourth, our method also copes with SRRs that do not match any sense of the query term in WordNet. In other words, we utilize senses provided by WordNet but are not limited by them. Different techniques for clustering WordNet word senses are presented in [1] but they do not actually perform sense merging. These techniques can potentially be used for merging WordNet senses, e.g., merging the senses that are in the same cluster. However, our method is specific to the senses for the same query term, not for general sense merging. Consequently, our technique can be more efficiently applied to grouping SRRs. In addition, our merging algorithm is also different from the existing ones.

3 Grouping SRRs Using WordNet

Our method groups the SRRs not only based on their syntactically similar words but also semantically similar words. It is possible that two SRRs talk about very similar topics but they have less similar words while two other SRRs have more common words but they are less similar in reality. Since we compare the meanings and semantic relations between words, our method is more likely to yield better-grouped SRRs compared to current methods.

3.1 Method Overview

Our SRR grouping system for a user query *Q* consists of the following steps (Fig.1):

1. Send *Q* to a search engine/metasearch engine and process the returned SRRs.
2. While the query is being evaluated and the SRRs are being processed, send *Q* to WordNet to obtain its synonym sets (synsets) as well as terms that have certain semantic relationships with each synset.
3. Merge similar synsets into *super-synsets.*
4. Categorize SRRs by assigning each SRR to the most similar super-synset if the similarity is greater than a threshold T1. Temporary categories are obtained based on the current assignments and the remaining SRRs form another temporary category.
5. Further categorize the remaining SRRs by assigning each such SRR to the most similar temporary category.
6. Cluster the remaining SRRs.

We will explain each step in our algorithm in detail in the following subsections.

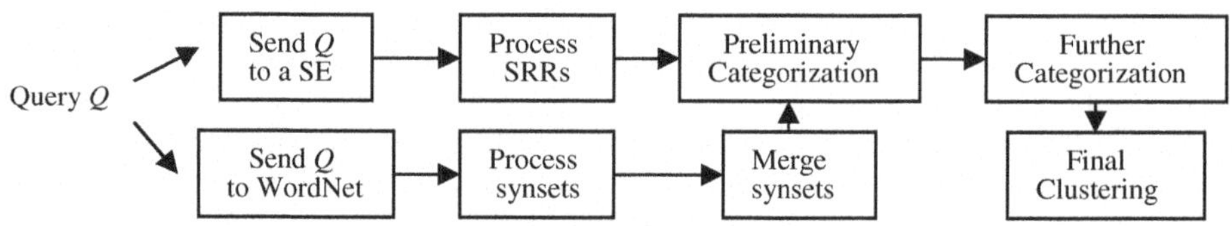

Fig. 1. Overview of our system

3.2 Submitting Query and Processing Results

For each user query, the top *k* (*k* = 50) distinct results (duplicates are removed) are retrieved and are used as input to our SRR grouping algorithm. Each result (SRR) usually consists of three different items: title, URL and snippet. Only the title and snippet of each SRR will be utilized to perform the grouping in our current approach.

For each SRR, we first remove the stop words and stem each remaining word. Next, the SRR is converted as a vector of terms. For each term, its term frequency (*tf*) in the SRR is recorded. The words in the title are considered to be more important than words in the snippet (we currently double the *tf* of each term in the title).

3.3 Sending Query to WordNet

Sending a user's query to WordNet means that certain information about the query term is obtained from WordNet and processed. This step is done in parallel to sending the query to the search engine and processing the returned SRRs. The fact that relationships between synsets are explicit is the motivation behind using WordNet in our approach. Synsets are linked using various types of relationship links. In our current approach, the following types of synsets are utilized for a given synset *S*:

- Hypernyms: Synsets that are more general in meaning than *S*
- Hyponyms: Synsets that are more specific in meaning than *S*

- Domains: Synsets that represent the domain of *S*
- Synonyms: Keywords that have the same meaning with the user query

This step (i.e., sending the query to WordNet) involves two procedures:

1. Get the senses/meanings for the query term. For each sense (synset), the synonyms, direct hypernyms/hyponyms, and the words in the definition and examples of the sense are all included in the representation of the synset. After removing stop words and stemming, each synset is represented as a vector of terms with weights that are computed from the term frequency of each term.
2. Merge similar senses if it's applicable. This step will be explained in section 3.4.

3.4 Sense Merging

It is often the case that some of the senses in WordNet fundamentally refer to the same concept or very similar concepts. The example below illustrates one such a case.

Example 1. Consider the following two synsets for query term "web"
Sense 1: web: (an intricate network suggesting something that was formed by weaving or interweaving; "the trees cast a delicate web of shadows over the lawn")

Sense 2: web, entanglement: (an intricate trap that entangles or ensnares its victim)

These two senses are very similar because both talk about physical webs with a subtle difference that the former emphasizes how the web is formed and the latter emphasizes how the web is used. The following is another sense of "web":

Sense 3: World Wide Web, WWW, web

It is easy to see that this sense is very different from the first two senses. The presence of synsets with similar meanings poses challenges to the SRR grouping algorithm as well as to the users who consume the grouped results. We propose to tackle this problem by merging the similar senses. Our sense-merging algorithm consists of five merging rules, each of which gives one condition under which two senses S1 and S2 can be merged. The five rules are given below:

Rule 1. If S1 and S2 have the same direct hypernym synset or one is a direct hypernym of the other, then merge S1 and S2.

Rule 2. If S1 and S2 have the same direct hyponym synset or one is a direct hyponym of the other, then merge S1 and S2.

Rule 3. If S1 and S2 have the same coordinate terms (i.e., there exist a synset S3 such that S1 and S3 share a direct hypernyn, and S2 and S3 also share a direct hypernym), then merge S1 and S2.

Rule 4. If S1 and S2 have common synonyms, then merge S1 and S2.

Rule 5. If S1 and S2 have the same direct domain synset or one is the domain of the other, then merge S1 and S2.

Intuitively, each condition in the above rules indicates that S1 and S2 are semantically similar.

3.5 Computing the Similarity Between SRRs and Super-Synsets

In this paper, we use a revised *Okapi function* to compute the similarity between SRRs and super-synsets. We made the changes to the original *Okapi function* [11] to fit our situation where two sets of documents are compared, one is the SRR set R* and the other is the super-synset set S*, whereas in the traditional information retrieval context, one document (the query) is compared with a set of documents. Our revised *Okapi function* for computing the similarity between an SRR *R* and a super-synset *S* is:

$$sim(R,S) = \sum_{T \in R \cap S} \frac{w_1 + w_2}{2} * w(T,R) * w(T,S) \qquad (1)$$

with $w_i = \log \dfrac{N_i - n_i + 0.5}{n_i + 0.5}, i = 1,2$, $w(T,R) = \dfrac{(k+1) * tf(R)}{K(R) + tf(R)}$, and

$$K(R) = k * ((1-b) + b * \frac{dl(R)}{avgdl(R^*)})$$

where N_1 and N_2 are the numbers of SRRs in R* and super-synsets in S*, respectively, for the current query; n_1 and n_2 are the numbers of SRRs in R* and super-synsets in S* that contain term *T*, respectively; w_1 and w_2 reflect the importance of term *T* with respect to the SRRs in R* and the super-synsets in S*, respectively (they are similar to the *idf* weight in information retrieval); *w*(*T*, *R*) computes the importance of term *T* in *R*; *tf*(*R*) is the term frequency of *T* in *R*; *dl*(*R*) is the length of *R*, and *avgdl*(*R**) is the average length of all the SRRs in *R**; $k = 1.2$ and $b = 0.75$ are two constants. Finally, *w*(*T*, *S*) is similar to *w*(*T*, *R*) except *R* is replaced *S* and *R** is replaced by *S**.

3.6 SRR Grouping Algorithm

Our SRR grouping algorithm (Algorithm **CCC**) consists of the following three steps:

1. *Preliminary* ***C****ategorization*. Categorize SRRs based on their similarities with super-synsets. Specifically, for each SRR *R*, find the super-synset *S* that is most similar to *R* among all super-synsets. If sim(*R*, *S*) is greater than a threshold T1, then categorize *R* to *S*. At the end of this step, super-synsets with no SRRs categorized to them are removed. For each remaining super-synset *S*, its term vector and all the SRRs that are categorized into it are merged (i.e., the term sets are unioned and for each term, its term frequencies in *S* and the SRRs are added). Let the merged result be called the *expanded synset* from *S*.
2. *Further* ***C****ategorization*. Categorize the remaining SRRs from step 1 (i.e., those SRRs whose similarity with every super-synset is less than or equal to T1). Let RR be the set of remaining SRRs. Let *R* be an SRR in RR and *C* = RR – {*R*}. Find the expanded synset *S* that is most similar to *R* among all expanded synsets. If sim(*R*, *S*) > sim(*R*, *C*), then add *R* to the category corresponding to *S* and remove *R* from RR; otherwise, keep *R* in RR. When computing sim(*R*, *C*), the SRRs in *C* are temporarily merged.
3. *Final* ***C****lustering*. If there are still uncategorized SRRs left, we cluster them using a two-step clustering algorithm similar to the one in [10]. In the first step, use the

first SRR to form a cluster by itself and for each subsequent SRR *R*, place it in the most similar current cluster if the similarity is higher than a threshold T2; else create a new cluster based on *R*. This step is order sensitive and may leave some SRRs in less fitting clusters. In the second step, for each SRR *R*, we compute its similarity with the centroid of each cluster and move it to the cluster whose centroid is the most similar to *R* if *R* was not in this cluster. This is repeated until no *R* can be moved. *Cosine* similarity function is used in SRR clustering.

In our current implementation, the two thresholds T1 and T2 are determined using a training set. When training T1, we try to find the value that achieves the maximum possible *recall* under the condition that nearly every categorized SRR is assigned to the correct synset (i.e., close to 100% *precision*). In step 1, we try to be more conservative since we will have another chance to categorize the remaining SRRs in step 2. Consequently, after the SRRs categorized into a synset are merged at the end of step 1, each category is as accurately represented as possible. In step 2, the cluster *C* is considered because we want each remaining SRR to have a fair chance to be categorized or stay uncategorized, as it is possible that some SRRs do not match any senses from the WordNet. Step 3 is needed because many English words have non-standard uses in practice (such as used as a name of a company) that do not match any senses the WordNet has about these words and a query term may be a non-standard English word (such as "allinone").

4 Evaluation

We implemented our algorithm using Java. We use JWNL to connect to WordNet 2.0. Two datasets are used in this paper and each dataset contains 10 single-term queries and the 500 SRRs (50 unique SRRs per query) from search engine Yahoo. The 10 queries for the first dataset DS1 are (notebook, jaguar, mouse, metabolism, piracy, suicide, magnetism, web, people, salmon), and the 10 queries for the second dataset DS2 are (apple, dish, trademark, map, music, car, game, tie, poker, mold). DS1 is used for training to obtain the thresholds (T1 = 4 and T2 = 0.1 are obtained).

4.1 Alternative Solutions

As mentioned before for some terms, there are categories that are not covered in WordNet. For example, for query "jaguar", there are two categories not in WordNet, the first is the brand name for car and the second is unknown (some company names).

For our evaluation, we also compare the SRR grouping algorithm described in section 3.6 with two other intuitively reasonable solutions. Basically, each of the two alternative solutions replaces the last two steps (*Further categorization* and *Final clustering*) while the first step (*Preliminary categorization*) remains the same. In WordNet, a *frequency of use* is associated with each sense of a word and this value indicates how widely this sense of the word is used relative to other senses of the word. Our first alternative solution is based on the frequency of use. Note that during sense merging, the frequency of use of a super-synset is computed as the sum of the frequencies of use of all the individual synsets it contains.

- *Largest frequency of use* (**LF**): Assign all remaining SRRs (after the *Preliminary categorization* step) to the super-synset that has the largest frequency of use.

The rationale for this method is that the super-synset with the largest frequency of use represents the most common sense of the term among those covered by WordNet.

Our second alternative solution is based on the intuition that if, after the preliminary categorization step, a category has the most SRRs, then this category is probably the most popular category for the retrieved SRRs.

- *Largest category* (**LC**): Assign the remaining SRRs to the category that has the largest number of SRRs after the preliminary categorization step.

4.2 Performance Measures

We evaluate the sense-merging algorithm as well as the three SRR grouping algorithms (CCC, LF and LC). For all algorithms, we use the *recall*, *precision* and *F1 measure* (which combines recall and precision) as the performance measures. For the merging algorithm, we define precision = $|A \cap B|/|B|$ and recall = $|A \cap B|/|A|$, where A is the set of merges that should be performed as judged by a human expert and B is the set of merges our merging algorithm performed. All the 20 queries in both datasets are used. Note that our merging algorithm does not need any training. For the SRR grouping algorithms, the recall and precision are defined below [10]:

- *Precision p*: For a given category, the precision is the ratio of the number of SRRs that are correctly grouped over the number of SRRs that are assigned to the group. For example, if among the 5 SRRs assigned to a group, only 4 are correct, then the precision for this group is 4/5 or 80%. The overall precision for all groups is the average of the precisions for all groups weighted by the size of each group. Specifically, the formula for computing the overall precision is

$$p = \sum_{i=1}^{n} p_i * \frac{N_i}{N}$$

  where p_i is the precision of the i-th group, N_i is the number of SRRs in the i-th group, N is the total number of SRRs (= 50) and n is the total number of groups.
- *Recall r*: For a given group, recall is the ratio of the number of SRRs that are correctly grouped over the number of SRRs that should have been grouped. For example, if a group should have 5 SRRs but an algorithm puts only 3 of them into this group, then the recall for this group is 3/5 or 60%. The overall recall for all clusters is the average of the recalls for all groups weighted by the size of each group. The formula for computing the overall recall is:

$$r = \sum_{i=1}^{n} r_i * \frac{N_i}{N}$$

  where r_i is the recall of the i-th group, and N_i, N and n are the same as in the definition of precision. When evaluating the SRR grouping algorithms, the final precision and recall are averaged over all queries.

Once precision p and recall r are computed, the *F1 measure* can be computed by $2*p*r / (p + r)$. The *F1 measure* is high only when both precision and recall are high.

4.3 Experimental Results

Our sense merging algorithm has a precision of 100, recall of 66 and F1-measure of 80. The results show that all merged senses are correct, but our algorithm still couldn't find all possible merges. Tables 1 and 3 show the results for the three SRR grouping algorithms based on DS1 and DS2 when merged senses are used. It can be seen that Algorithm CCC performs significantly better than Algorithms LF and LC. This is mainly due to the fact that the former can group the SRRs beyond the synsets in WordNet while the latter two methods force the SRRs that do not match any WordNet senses into incorrect categories. Another observation is that the results for the testing dataset DS2 are only slightly lower than the results for the training set DS1, indicating that the trained thresholds are reasonably robust. Tables 2 and 4 show the results when un-merged senses are used. It can be seen that sense merging helped the performance improve by approximately 5 percentage points. One of the reasons that causes incorrect grouping is the lack of common terms between some SRRs and the correct synset representations. We plan to investigate this problem in the future.

Table 1. With merged senses for DS1

Algorithm	Precision	Recall	F1
CCC	93%	91%	92%
LF	75%	77%	76%
LC	78%	80%	79%

Table 2. Without merged senses for DS1

Algorithm	Precision	Recall	F1
CCC	89%	86%	87%
LF	68%	70%	70%
LC	73%	74%	73%

Table 3. With merged senses for DS2

Algorithm	Precision	Recall	F1
CCC	90%	89%	89%
LF	74%	77%	75%
LC	77%	79%	78%

Table 4. Without merged senses for DS2

Algorithm	Precision	Recall	F1
CCC	85%	83%	84%
LF	65%	67%	66%
LC	69%	70%	69%

5 Conclusions and Future Work

In this paper, we investigated the problem of how to group the search result records from search engines (or metasearch engines) for single-term queries. Single-term queries are often ambiguous because many English words have multiple meanings. By grouping the search results based on the different meanings of the query term, it makes it easier for users to identify relevant results from the retrieved results. We proposed a novel three-step grouping algorithm that combines both categorization and clustering techniques. We also proposed an algorithm to merge similar senses returned from WordNet. Our preliminary experimental results indicated that our SRR grouping algorithm is effective, achieving an accuracy of about 90%. We also showed that this algorithm is significantly better than two other possible solutions and our sense-merging algorithm can improve grouping accuracy by about 5%.

We plan to continue this research in the following directions. First, we plan to conduct more experiments using a significantly larger dataset. Second, we will try to improve our sense-merging algorithm and SRR grouping algorithm as there are still rooms for improvement. Third, while WordNet is very useful, it is far from perfect in providing all the senses for many words. We plan to see if other online semantic dictionaries, such as Wikipedia, can also be utilized. Finally, we also plan to develop good SRR grouping solutions for multi-term queries.

Acknowledgment. This work is supported in part by the following NSF grants: IIS-0414981, IIS-0414939 and CNS-0454298.

References

1. E. Agirre, E. Alfonseca, O. Lopez. Approximating Hierarchy-based Similarity for WordNet Nominal Synsets Using Topic Signatures. Global WordNet Conference, 2004.
2. G. Attardi, A. Cisternino, F. Formica, M. Simi, A. Tommasi. PiQASso 2002. TREC11 2002.
3. E. W. De Luca and A. Nürnberger, O. von-Guericke. Ontology-Based Semantic Online Classification of Documents: Supporting Users in Searching the Web. University of Magdeburg, Universitätsplatz 2, 39106 Magdeburg, Germany, AMR, 2004.
4. T. de Simone and D. Kazakov. Using WordNet Similarity and Antonymy Relations to Aid Document Retrieval. Recent Advances in Natural Language Processing (RANLP), 2005.
5. C. Fellbaum (edited). WordNet: An Electronic Lexical Database (Language, Speech & Communication). The MIT Press, 1998.
6. A. Hotho, S. Staab, G. Stumme. WordNet Improves Text Document Clustering. ACM SIGIR Semantic Web Workshop, 2003.
7. A.K. Jain, M.N. Murty. Data Clustering: A Review. ACM Computing Surveys, 1999.
8. B. Jansen, B. Spink, J. Bateman, T. Saraceric. Real Life Information Retrieval: A Study of User Queries of the Web. ACM SIGIR Forum, 32(1), pp.5-17, 1998.
9. S. Liu, C. Yu, and W. Meng. Word Sense Disambiguation in Queries. CIKM, 2005.
10. Q. Peng, W. Meng, H. He, and C. Yu. WISE-Cluster: Clustering E-Commerce Search Engines Automatically. WIDM, 2004.
11. S. Robertson, S. Walker, M. Beaulieu. Okapi at Trec-7: Automatic Ad Hoc, Filtering, Vlc, and Interactive Track. 7th Text REtrieval Conference, 1999, pp.253-264.
12. M.H. Song, S.Y. Lim, D.J. Kang, S.J. Lee. Ontology-Based Automatic Classification of Web Documents. ICIC (2) 2006: 690-700.
13. WordNet; http://wordnet.princeton.edu/
14. Vivisimo, http://www.vivisimo.com
15. Y. Yang, S. Slattery and R. Ghani. A Study of Approaches to Hypertext Categorization, Journal of Intelligent Information Systems, 18(2), March 2002.
16. Y. Yang. A Study of Thresholding Strategies for Text Categorization. ACM SIGIR, 2001.
17. O. Zamir, O. Etzioni, Grouper: A Dynamic Clustering Interface to Web Search Results. World Wide Web Conference, 1999.
18. H. Zeng, Q. He, Z. Chen, and W. Ma. Learning To Cluster Web Search Results. ACM SIGIR, 2004.

Static Verification of Access Control Model for AXML Documents*

Il-Gon Kim

Korea Information Security Agency,
78, Garak-Dong, Songpa-Gu, Seoul, Korea 138-803
igkim@kisa.or.kr

Abstract. Reasoning about the access control model for AXML documents is a non-trivial topic because of its own challenging issues: the hierarchical nature of XML with embedded service call and query transformation. In this paper, we present a methodology to specify an access control model (GUP^{ster}) for AXML (Active XML) documents by translating a query, schema, and access control policy in CSP language. Then, we show how to verify access control policies of AXML documents, by illustrating the running example, with the FDR model checker. Finally, the examples demonstrate that our automated static verification is efficient to analyze security problems, not only whether the policies give legitimate users enough permissions to read data, but also whether the policies prevent unauthorized users from reading sensitive data.

1 Introduction

The management and access control of distributed sensitive data such as patient records are increasingly becoming challenging issues; however existing approaches [4,6] focus on the centralized regulation of access to data. For this reason, Abiteboul et al.[1] provided a novel solution to unify the enforcement of access control mechanism based on GUP^{ster}(*Generic User Profile+Napster*)[10] and the distributed data integration by AXML (Active XML) documents[3] in a P2P architecture.

Verification of the access control model for AXML is non-trivial due to some challenging characteristics in itself (embedded service call, query language, and query filtering). The primary motivation of our approach is to transform the semantics of the access control model for AXML to CSP (Communicating Sequential Processes)[9] model and verify correctness of query filtering for secure data access and the enforcement of combined rules using FDR (Failure-Divergence Refinements)[5] model checker. This approach ensures that evaluation of query q over AXML document D returns only information in D that the peer is allowed to see.

* This work was supported by the INRIA projects ARC-ASAX.

G. Dong et al. (Eds.): APWeb/WAIM 2007, LNCS 4505, pp. 687–696, 2007.

The main contributions of this paper are the following:

We present a static verification method of the access control model for AXML documents. To conduct this, first, we present how to describe an equivalent CSP model against an XSquirrel path expression, in order to express a hierarchical nature of a schema with embedded calls, query, and access control policy. Second, we construct CSP processes for a query expression, an access control policy, and a schema specific to AXML documents, by assigning the concurrency semantics of CSP language. Third, we show how to model a rewritten query created by the filtering service combined with the regulation of access control rules. Finally, we verify access control policies against queries over AXML documents using the FDR model checking tool, by illustrating how to elicit a property model and a system model.

The remainder of this paper is organized as follows. Section 2 describes some related works. In Section 3, we illustrate the scenario of our running example. In Section 4, we define the semantics of XSquirrel expression, access control model in GUP^{ster}, and query transformation. In Section 5, we show how to construct the CSP models for schema, query, policy, and query transformation. In Section 6, we illustrate how to perform static verification of the access control policies for AXML documents. Finally, we give our conclusions in Section 7.

2 Related Works

Access control for generic XML documents is a non-trivial issue, so fine-grained access control for them has been studied by many researchers.

Sahuguet et al.[10] developed the access control model named GUP^{ster}, which could support for 1) integration of highly distributed user profile XML data in a convergence network, and 2) access control written in the XSquirrel query language which expresses users permissions and data sources (e.g., the location of resources mapping to a query).

There has been several works to verify access control policies in the research area of databases and Web services[2].

Bryans[2] showed how to reason about access control policies by taking into account a standard access control language, XACML. Furthermore, the author presented how to use CSP semantics to describe the semantics of XACML and showed how the two access control polices can be compared with the FDR model checking.

We consider the query transformation and access control semantics of GUP^{ster} for modelling and analyzing access control policies using CSP and FDR. In this regard, our research is different from related works.

3 Illustrative Example

We give a brief overview of major concepts in an AXML document, schema, and GUP^{ster} access control mechanism through a running example in Fig. 1. Armed with this example, we will also explain the concept of *query filtering service* to protect sensitive AXML data by filtering a query according to a schema-level access control policy in GUP^{ster}.

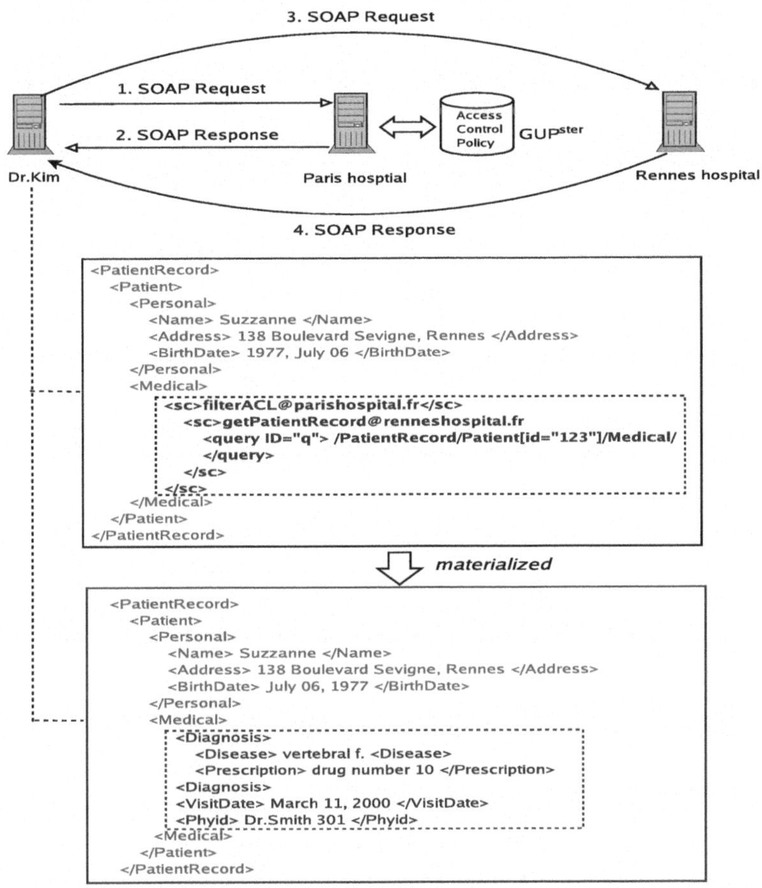

Fig. 1. Embedded service calls and a filtering service in the AXML document for a patient record

In this example, we assume that the peer Dr. Kim wants to see the patient record for a patient named "Suzzane" before diagnosing her. Dr. Kim is already registered in Paris hospital so that both of them are in a trust relationship. Dr. Kim already knows that the patient ID is "123" and a relevant patient record for the patient is stored in Rennes hospital after consultation with the patient.

Step 1: Dr. Kim logs on the portal site of the Paris hospital and sends a request for the query q by invoking the call "*getPatientRecord@renneshospital.fr*". It is nested by another call *filterACL@parishospital* which is a filtering service to enforce a list of access control rules associated with Dr. Kim and filter the query.
Step 2: After *filterACL@parishospital* call has been invoked, Paris hospital performs the filtering service by enforcing the relevant access control rules, so the query q gets rewritten into q'.

Step 3: The filtered query is returned to Dr. Kim after signing it with a private key of Paris hospital.
Step 4: After Dr. Kim sends the delegated filtered query to Rennes hospital by invoking a service call, the evaluation of the filtered query *q'* is executed and answered data to *q'* will be returned to Dr. Kim's AXML document.

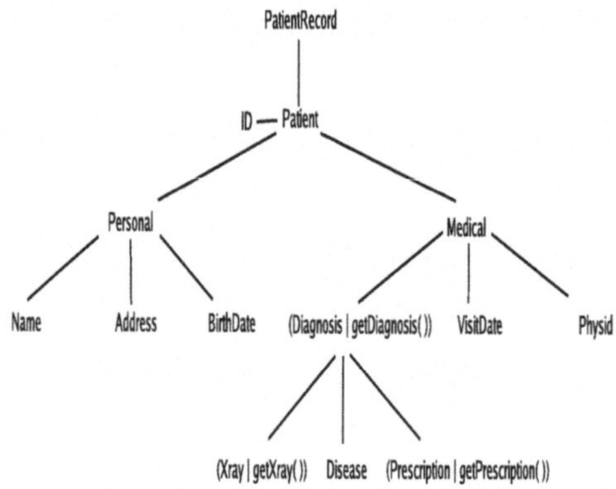

Fig. 2. The schema for a patient record AXML document

4 Language and Semantics for the Access Control Model

4.1 XSquirrel Expression

XSquirrel[10] is a simple XML path expression language defined in GUP^{ster} in order to express: 1) the mappings between global document's schema and remote sources, 2) access control rules which users are allowed to access specified objects, and 3) queries over documents. This language uses a similar syntax to XPath, however it can express the view of more than one path in AXML documents by allowing a more flexible expression such as the ♯ (union) operator. For the rest of this paper, we use the simplified XSquirrel expression and semantics.

Definition 1. *Given an absolute XSquirrel expression q consisting of path expression* $l_1/l_2/(l_{31} \sharp l_{32})$*, on an AXML document D, it is separated by the path character '/' as in XPath expression, or '♯ (union operator)'. In addition, the notation q[D] stands for an answer set of q, so it denotes the the sub-document which is presented after evaluation of semantics of the path expression. It is composed of all descendant leaf nodes of the requested nodes and their ancestors up to the root of the initial document. The notation Eval(q[D]) returns the mapping nodes against query evaluation of q.*

For example, given an absolute XSquirrel query q :

```
/PatientRecord/Patient/Medical/(VisitDate ♯ Physid)
```

over the document D (in Fig. 1), *q[D]* is the union of each XPath expression, as given below:

```
/PatientRecord/Patient/Medical/VisitDate ∪
/PatientRecord/Patient/Medical/Physid
```

Then, *q[D]* covers two branches for patient records and still retains the original document structure unlike XPath .

```
<PatientRecord>
    <Patient>
        <Medical>
            <VisitDate> March 11, 2000 </VisitDate>
            <Physid> Dr.Smith 301 </Physid>
        <Medical>
    </Patient>
</PatientRecord>
```

However, *Eval(q[D])* returns only two nodes `VisitDate` and `Physid`, and their atomic values:

```
<VisitDate> March 11, 2000 </VisitDate>
<Physid> Dr.Smith 301 </Physid>
```

In the remainder of this paper, we use the notations q and *q[D]* indiscriminately. The path expression may allow a conditional expression (also called *predicates*). For example, `/PatientRecord/Patient[ID="123"]` returns the sub-document specific to the patient ID="123".

4.2 Query Transformation

A *query transformation* (*query filtering*) in the combined framework of AXML documents and GUP^{ster} is a technique which 1) protects privacy-conscious resource from unauthorized requester by enforcing the access control rules, and 2) leads a peer to access a third-peer holding a requested resource by giving a filtered query.

Recall the motivating example in Fig. 1. When Dr. Kim sends a request query q for medical resources to Paris hospital, q is rewritten into q' through the *filterACL* service call enforcing the access control rules on a schema s in Paris hospital. A document of filtered query q', *q'[D]*, is the authorized view that a peer is permitted to read a requested resource. Such an authorized view of a requester on a document depends on the access permissions defined by an access control policy. Then, Paris hospital returns a filtered query to Dr. Kim and he can receive the restricted parts of the requested resource from Rennes hospital, within the context that the permissions are allowed (see Fig. 1).

A simplified version of the schema for patient records is shown in Fig. 2. We assume that the data mapping and access control rules are defined in XSquirrel

Mapping Rule	Sub-document	Resource Location
1	/PatientRecord/Patient[ID="1234"]/	www.renneshospital.com

(a) Data mappings between the sub-document for patient records and remote source location

Rule	Requester	Action	Resource
1	Kim	read	/PatientRecord/Patient/Medical/VisitDate
2	Kim	read	/PatientRecord/Patient/Medical/(VisitDate # Physid # Diagnosis/(Disease # Prescription))
3	Lee	read	/PatientRecord/Patient/Medical/VisitDate

(b) Access control rules

Query q from Dr. Kim : /PatientRecord/Patient/Medical/(VisitDate # Physid # Diagnosis)

⇩ *query rewriting*

Rule1 = /PatientRecord/Patient/Medical/VisitDate
Rule2 = /PatientRecord/Medical/(VisitDate # Physid # Diagnosis(Disease # Prescription))

q' = q o (Rule1 U Rule2)

Filtered query q' : /PatientRecord/Patient/Medical/(VisitDate # Physid # Diagnosis/(Disease # Prescription))

(c) Query filtering

Fig. 3. Data integration, access control rules, and query filtering

language as given in Fig. 3. The flow of generating a filtered query *q'* from a query *q*, against access control rules, can be sequentialized as below:

1) GUP^{ster} finds all of the relevant access control rules, $\bigcup_i ACR_i$, against a requester's query *q*.
2) It rewrites the query *q* into *q'* by applying the relevant access control rules. The overall process for creating a filtered query *q'* from *q* against ACR is naively expressed as:
   $q' ::= q \circ \bigcup_i ACR_i$
   where $\circ$ is an intersection operator, *q'* returns a subdocument of both *q* and $\bigcup_i ACR_i$. Note that query rewriting is executed statically by accessing a virtual document (schema), and not accessing the document itself.
3) Then, it performs data integration by mapping the sub-document of a query to a remote resource which actually stores related data.

As an example, suppose that *q* from Dr. Kim is:

```
/PatientRecord/Patient/Medical/(VisitDate ♯ Physid ♯ Diagnosis)
```

Next, all the relevant rules $Rule_1$ and $Rule_2$ for Dr. Kim are unified based on *grant overwrites*[1]. Then, the combined rules are composed with the query *q* and

[1] This algorithm grants a request to a node if a grant access rule to the same node is defined.

the filtered query *q'* is obtained with the intersection of *q* and the combined rules (see Fig. 3(c)):

q'=`/PatientRecord/Patient/Medical/(VisitDate`$\sharp$`Physid`$\sharp$`Diagnosis/`
(Disease $\sharp$ Prescription))

5 CSP Models

In this section, we show how to describe CSP models in a tree structure for a schema, query, access control policy and filtered query, based on the automata theory in [8]. We also assume that the conditional expression in XSquirrel is always satisfied when constructing CSP models. For example, the expression `/PatientRecord/Patient[ID="123"]` for CSP model will be simplified as `/PatientRecord/Patient`.

5.1 Modelling Schema

Thus, the motivating example shown in Fig. 2 can be modelled in CSP language:

$$
\begin{aligned}
&S = patientrecord \rightarrow PR\\
&PR = patient \rightarrow P\\
&P = personal \rightarrow PERS \,\Box\, medical \rightarrow MD\\
&PERS = name \rightarrow N \,\Box\, address \rightarrow ADDR \,\Box\, birth \rightarrow Birth\\
&MD = visit \rightarrow VISIT \,\Box\, physid \rightarrow PHYSID \,\Box\, (diagnosis \rightarrow DIAG \,\Box\\
&\qquad getdiagnosis \rightarrow GD)\\
&GD = (xray \rightarrow XRAY \,\Box\, getxray \rightarrow GX) \,\Box\, (disease \rightarrow DIS \,\Box\\
&\qquad (prescription \rightarrow PRES \,\Box\, getprescription \rightarrow GPRES)
\end{aligned}
$$

For the sake of readability, we will omit the description of sub-processes in the following CSP models. Instead, we will use the parenthesis () to represent a tree structure in CSP description. In addition, the termination process ($N_S, ADDRS_S, BIRTH_S, VISIT_S,\ PHYSID_S, DIAG_S$) will be replaced with *STOP* or *SKIP*.

5.2 Modelling Query

Given a query expression *q*, it can be modelled as a CSP process by describing the sub-document after interpreting a path expression, in XSquirrel language. For example, let us assume that there is the query *q* as shown in Fig. 3(c):

q = `/PatientRecord/Patient/Medical/(VisitDate`$\sharp$`Physid`$\sharp$`Diagnosis)`

The answer set of *q*, *q[D]*, can be modelled in CSP language as the following:

$$
\begin{aligned}
&Q = patientrecord \rightarrow patient \rightarrow medical \rightarrow (visit \rightarrow SKIP \,\Box\, physid \rightarrow SKIP\\
&\Box\, diagnosis \rightarrow (xray \rightarrow SKIP \,\Box\, disease \rightarrow SKIP \,\Box\, prescription \rightarrow SKIP))
\end{aligned}
$$

5.3 Modelling Access Control Policy

An access control policy (ACP) is the composition of access control rules (ACR). Thus, the tuple of <*Requester, Resource, Action*> shown in section 6.2 can be modelled using a compound channel in CSP. Here, we define the action *read* as the channel name, declare the type of the channel to be *Requester*, and model ACP as a *Resource* in a tree data structure to which can be accessed after performing *read* event:

$$PEER = 1..N$$
$$channel\ read : PEER$$
$$PEER_1 = read \rightarrow ACP$$

where $PEER_{1..n} \in Requester$, and a peer can take a form of events as $read.p_1$ on the *read* channel.

As mentioned earlier, ACP is the view of a document restricted by all of relevant access control rules $\bigcup_i ACR_i$. Therefore, we construct a CSP model for ACP by using interleaving parallel operator $|||$ to reflect on the *grant-overwrites* semantics in ACP. For example, let us assume that there is the peer *Dr. Kim* who sends a query request for some resources of patients record documents. Then, the ACP model for *Dr. Kim* can be modelled in CSP language as below:

$$ACP = ACR_1\ |||\ ACR_2$$
$$ACR_1 = patientrecord \rightarrow patient \rightarrow medical \rightarrow visit \rightarrow STOP$$
$$ACR_2 = patientrecord \rightarrow patient \rightarrow medical \rightarrow (visit \rightarrow STOP \,\square\, physid \rightarrow STOP \,\square\, diagnosis \rightarrow (disease \rightarrow STOP \,\square\, prescription \rightarrow STOP))$$

where ACR_1 and ACR_2 are the CSP processes for $Rule_1$ and $Rule_2$, respectively, in Fig. 3(b).

5.4 Modelling Query Transformation

Given a query expression *q*, a rewritten query *q'* against ACRs is created using the intersection operator $\circ$ and union operator $\bigcup$, respectively:

$$q' ::= q \circ \bigcup_i ACR_i$$

In Section 5.3, we have already shown that the union operator $\bigcup$ is well suited with an interleaving parallel operator $|||$ in CSP expression. Here, we insist that the semantics of $\circ$ operator is well described with a synchronized parallel operator $|[\, A \,]|$ where A is a set of events in a process model.

For a better understanding, let us consider an example of showing how the CSP model *Q'* for a rewritten query *q'* denoted in Section 4.3, can be modelled using with the synchronized parallel operator:

$$\mathrm{Q'} = \mathrm{Q}\ |[\, A \,]|\ (\mathrm{S}\ |[\, A \,]|\ (\mathrm{ACR}_1\ |||\ \mathrm{ACR}_2))$$

where A is a set of events such that $A = \alpha(Q) \cup \alpha(S) \cup \alpha(ACR_1) \cup \alpha(ACR_2)$. Note that the parallel composition between *S*, and the union of ACR_1 and ACR_2 represents a schema-level access control model for AXML documents.

6 Verification

6.1 Analysis of Access Control Policy

In this subsection, we show how to specify *Spec* model and *System* model to analyze desired properties such as:

- A query is evaluated against an access control policy.
- A policy should be defined in order to prevent the leakage of sensitive data from unauthorized peer.
- A policy should be defined in order to allow a peer to access authorized data.

The requester's query q against an access control policy is *always-granted* if a set of trace events of the query process Q is a subset of those of both the schema process S and the policy process ACP:

$$assert\ S \ |[\,A\,]|\ ACP \sqsubseteq_T Q$$

where A is the event set between S and ACP such that $A = \alpha(S) \cup \alpha(ACP)$, and *assert* is the reserved word for *equivalence checking* in FDR. Note that the parallel process $S \ |[\,A\,]|\ ACP$ represents a schema-level access control model for AXML documents. For example, let us assume that the query q from *Dr. Kim* is :

```
/Patientrecord/Patient/Medical/VisitDate
```

Then, we can confirm that $S \ |[\,A\,]|\ ACP \sqsubseteq_T Q$ is satisfied using FDR tool. However, if the query q is:

```
/Patientrecord/Patient/Medical/
```

Then the CSP model Q for its answer set $q[D]$, means the following process semantically, according to the definition in Section 4.1:

$$Q = patientrecord \rightarrow patient \rightarrow medical \rightarrow (visit \rightarrow STOP \ \Box \ physid \rightarrow STOP \ \Box \ diagnosis \rightarrow (xray \rightarrow STOP \ \Box \ disease \rightarrow STOP \ \Box \ prescription \rightarrow STOP))$$

In this case, the FDR model checker shows the counterexample of: $\langle patientrecord, patient, medical, diagnosis, xray\rangle$ in the model Q. This trace means that the access to the node `Xray` is not permitted against the query q. As a result, the access control policy rewrites q into q' to protect privacy-conscious `Xray` data by enforcing all the relevant rules, and then the filtered query q' is:

`/PatientRecord/Patient/Medical/(Diagnosis/(Disease` ♯ `Prescription)` ♯ VisitDate ♯ Physid))

7 Conclusion

We have presented a verification method for analyzing an access control model for restricting query access to AXML documents by modelling query, schema, and access control policy as tree data structure in CSP language. We have also shown how to translate the declarative semantics of an access control policy of GUP^{ster} and the rewritten query created by a filtering service to CSP models.

Thereby, given CSP models for a query, access control policy, and schema for AXML document, our static verification can not only determine whether the requested query is permitted by the schema-level access control policy or not, but also show a hierarchical path if access to data is allowed or not.

References

1. S. Abiteboul, B. Alxe, O. Benjelloun, B. Cautis, I. Fundulaki, T. Milo, and A. Sahuguet. "*An Electronic Patient Record "On Steroids": Distributed, Peer-to-Peer, Secure and Privacy-conscious*", *Proceedings of VLDB Conference*, 2004.
2. J. Bryans. "*Reasoning about XACML policies using CSP*", *Proceedings of SWS Workshop*, pp.28-35, 2005.
3. Active XML (AXML) Home Page , *http://activexml.net*, 2004.
4. E. Damiani. S. De Capitani di Vimercati, S. Paraboschi, and P. Samarati. "*A Fine-Grained Access Control System for XML Documents*", *TISSEC*, 5(2):169-202, 2002.
5. Formal Systems Ltd. FDR2 User Manual, Aug. 1999.
6. A. Gabillion and E. Bruno. "*Regulating Access to XML Documents*", *Proceedings of Working Conference on Database and Application Security*, 2001.
7. S. Godik, and T. Moses. *eXtensible Access Control Markup Language(XACML) version 1.0*, Technical Report, OASIS, 2003.
8. M. Murata, A. Tozawa, and M. Kudo. "*XML Access Control Using Static Analysis*", *Proceedings of CCS Conference*, pp.73-84, 2002.
9. A.W. Roscoe. *The Theory and Practice of Concurrency*, Prentice Hall, 1997.
10. A. Sahuguet, R. Hull, D.F. Lieuwen, and M. Xiong. "*Enter Once, Share Everywhere: User Profile Management in Converged Networks, Proceedings of CIDR Conference*", 2003.

SAM: An Efficient Algorithm for F&B-Index Construction*

Xianmin Liu, Jianzhong Li, and Hongzhi Wang

Harbin Institute of Technology, Heilongjiang, China
{liuxianmin,lijzh,wangzh}@hit.edu.cn

Abstract. Using index to process structural queries on XML data is a natural way. F&B-Index has been proven to be the smallest index which covers all branching path queries. One disadvantage which prevents the wide usage of F&B-Index is that its construction requires lots of time and very large main memory. However, few works focus on this problem. In this paper, we propose an effective and efficient F&B-Index construction algorithm, SAM, for DAG-structured XML data. By maintaining only a small part of index, SAM can save required space of construction. Avoiding complex computation of the selection of nodes to process, SAM takes less time cost than existing algorithms. Theoretical analysis and experimental results show that SAM is correct, effective and efficient.

Keywords: XML, F&B-Index, SAM, construction.

1 Introduction

Increasing popularity of XML in recent years has generated much interest in query processing over graph-structured data. To summarize the structure of XML data and support query processing efficiently, some structural indexes have been proposed [1,2,3,4]. F&B-Index is the smallest structural index covering all branching path queries [6].

For wide usage of F&B-Index, efficient algorithm for building F&B-Index is in great demand. In [4], based on graph model, PT algorithm which is first proposed in [7] is extended to build F&B-Index. Even though PT algorithm can construct F&B index efficiently when the size of XML document is small, it has two problems:

- **Space Problem.** PT algorithm have to maintain both the whole XML data nodes and F&B-Index in memory. When XML data becomes very large, main memory may not be enough to support this algorithm. And try to extend PT algorithm using disk is unfeasible, because PT need to frequently visit data nodes by some order we couldn't know in advance.
- **Time Problem.** PT algorithm selects nodes to partition by way of searching among all data nodes. When XML data becomes very large, the node will be searched in very large space and this procedure will take huge time cost.

* Research supported by the key Program National Natural Science Foundation of China (NSFC) under Grant No. 60533110, NSFC under Grant No. 60473075, National Grand Fundamental research 973 Program of China under Grant No. 2006CB303000 and Program for New Century Excellent Talents in University (NCET) under Grant No. NCET-05-0333.

G. Dong et al. (Eds.): APWeb/WAIM 2007, LNCS 4505, pp. 697–708, 2007.

Since space is the key problem of PT, to find an algorithm which can build F&B-Index part by part is a natural idea. However, in graph model, a node can have relation with all other nodes in worst case and relations can be transmitted along circles in graph. These two observations make it impossible to find an algorithm efficiently in graph model.

Surprisingly, although some researches model XML data as graphs, most of XML data can be represented as directed acyclic graphs (DAGs) especially in applications. For example, the gene ontology data available at [9] can be modeled as DAGs with nodes representing gene terms and edges denoting their is-a and part-of relationships. Such a DAG model makes it possible to provide an efficient F&B-Index construction algorithm for most XML data in practice.

In this paper, we focus on how to construct F&B-Index efficiently over XML DAG model. Very few works have been done to solve this problem. Based on DAG and stream model, we propose a novel and efficient F&B-Index construction algorithm called Scan-And-Merge algorithm (SAM for brief) which can save much space and time cost. Given a DAG G, SAM loads data nodes from streams and build F-partition and B-partition of G part by part. Then these two partitions are merged to form the index nodes of F&B-Index. Finally, F&B-Index is formed by adding edges between those index nodes. With maintaining partial index instead of the whole graph and index in main memory, SAM can save much space cost. Searching nodes within a few nodes, SAM can save much time cost. Our major contributions include:

- **Novel Algorithm.** Based on XML DAG stream model, we propose an efficient F&B index construction algorithm by merging F-partition and B-partition.
- **Show Correctness.** With analysis on correctness and complexity of time and space, SAM is proved to be correct and very efficient on both time and space.
- **Performance Study.** With extensive experiments, we show SAM algorithm outperforms PT algorithm and has good scalability.

Organization. The rest of this paper is organized as follows. We present background in section 2 and describe SAM algorithm in detail in section 3. The theoretical analysis is shown in section 4. We report the experimental results in section 5. Related work and conclusion are presented in section 6 and 7.

2 Background

2.1 XML DAG Stream Model

To design efficient F&B-Index construction algorithm, we try to combine XML DAG model with XML stream model. In XML DAG stream model, each node v has an attribute $depth(v)$ which denotes length of the longest path from *root* to v, and each stream S_i is associated with graph nodes whose depth is i; Since XML data can be modeled as an directed acyclic graph, $depth(v)$ can be easily determined during the procedure of topology sort [12]. The operations of stream S_i are: $top(S_i)$, $advance(S_i)$ and $isempty(S_i)$. $Top(S_i)$ returns the first node of S_i; $advance(S_i)$ moves S_i forward; $isempty(S_i)$ return the boolean value which shows whether S_i is empty.

```
<root>
 <m aref= "a2" >
  <a id= "a1" bref= "b3" >
   <b id= "b1" />
   <b id= "b2" > <c id= "c1" /> <c id= "c2" />
   </b>
  </a>
 </m>
 <n>
  <a id= "a2" bref= "b1" bref= "b2" >
   <b id= "b3" > <c id= "c3" /> </b>
  </a>
 </n>
</root>
```

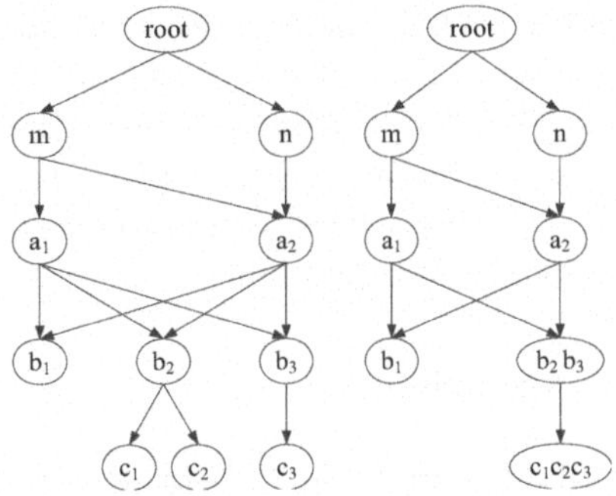

(a) (b) (c)

Fig. 1. (a) Example XML document (b) XML data DAG and (c) XML F&B-Index

Example 1. The DAG stream model S of graph in Figure 1(b) can be described as follows: $S = \{S_0, S_1, S_2, S_3, S_4\}$, $S_0 = \{root\}$, $S_1 = \{m, n\}$, $S_2 = \{a_1, a_2\}$, $S_3 = \{b_1, b_2, b_3\}$, $S_4 = \{c_1, c_2, c_3\}$.

2.2 F&B-Index

Structural index. A structure index for a DAG graph takes the form of another DAG graph (V_I, G_I), which can be built by following steps: (1) according to some equivalence relation, partition all data nodes into some equivalence classes, (2) form an index node for every class, and (3) add an edge between two index nodes *i* and *j* if there is an edge between some data node in *i* and *j*. Obviously, the relation used to partition data nodes determines the kind of the structure index. And how to partition nodes into classes determines the cost of this procedure.

F&B-bisimulation. F&B-bisimulation is a binary relation which can be used to construct structure index. It can be defined as follows:

Definition 1. *Given a XML DAG, for graph nodes* n_1 *and* n_2*, we say* n_1 *and* n_2 *satisfy F&B-bisimulation, that is* $n_1 \approx_{FB} n_2$*, if:*

1. ***[label condition]*** *Both of* n_1 *and* n_2 *are root, or* $label(n_1) = label(n_2)$.
2. ***[outgoing edge condition]*** *For every edge* (n_1, n_1')*, there exits an edge* (n_2, n_2') *such that* $n_1' \approx_{FB} n_2'$*, and vice versa.*
3. ***[incoming edge condition]*** *For every edge* (n_1', n_1)*, there exits an edge* (n_2', n_2) *such that* $n_1' \approx_{FB} n_2'$*, and vice versa.*

Obviously, F&B-bisimulation is reflexive, symmetric and transitive, so it is an equivalence relation. Moreover, if n_1 and n_2 satisfy label condition and incoming edge condition, we say n_1 and n_2 satisfy **B-bisimulation** that is $n_1 \approx_B n_2$; if n_1 and n_2 satisfy label condition and outgoing edge condition, we say n_1 and n_2 satisfy **F-bisimulation** that is $n_1 \approx_F n_2$.

Theorem 1. *Node* n_1 *and* n_2 *satisfy F&B-bisimulation if and only if they satisfy both F-bisimulation and B-bisimulation.* □

F&B-Index. While building structure index, if we use F&B-bisimulation to partition nodes, the set of equivalence classes we got is F&B-partition, and the structure index we got is *F&B-Index*. Moreover, using F-bisimulation or B-bisimulation to partition graph nodes, *F-Index* or *B-Index* which is also one kind of structure index can be constructed. And during these two procedures, the two sets of equivalence classes we got are F-partition and B-partition.

2.3 PT Algorithm

PT algorithm is proposed by Paige and Tarjan [7] to solve the relational coarsest partition problem, and this algorithm can be extended to solve the problem of building F&B-Index. For a given graph $G = (V_G, E_G)$, this algorithm can be implemented in $O(|E_G| \log |V_G|)$ time cost and $O(|E_G| + |V_G|)$ space cost. Because of searching node to partition in large space, the coefficient of $|E_G| \log |V_G|$ is usually very large.

3 SAM Algorithm

In this section, we describe SAM algorithm in detail. First, we make a outline of SAM algorithm, then we introduce SAM step by step.

3.1 Outline of SAM Algorithm and Notations

Since in XML DAG model data nodes come from streams with different depth, we can load nodes according to their depth. Such a model makes it possible to process data nodes part by part. In this paper, to construct F&B-Index efficiently based on XML DAG stream model, SAM (Scan-And-Merge) has been proposed. It includes three steps. First, SAM scans all nodes through XML stream to build F-partition and B-partition. Then, these two partitions are merged to compute F&B-partition. At last, SAM forms an index node for each set in F&B-partition and add edges between index nodes.

To make the description simple and clear, we define four kinds of nodes used during the execution of SAM:

- **DNode:** a data node from stream or a graph node in DAG;
- **FNode:** an index node in F-Index or an equivalence class in F-partition;
- **BNode:** an index node in B-Index or an equivalence class in B-partition;
- **FBNode:** an index node in F&-Index or an equivalence class in F&B-partition.

Example 2. Consider DAG in Figure 1(b). Its XML DAG stream is described in Figure 2. We can see that each node with depth i is put into stream S_i *and nodes in the same stream have the same depth. Nodes are scanned from stream* S_0 *to* S_4 *and B-partition described in Figure 2 is built. In B-partition, all nodes satisfying B-bisimulation are put into the same set. Nodes are scanned from stream* S_4 *to* S_0 *and F-partition described in Figure 2 is built. In F-partition, all nodes satisfying F-bisimulation are put into the same set. Data structures in Figure 2 supports these two partitions' construction.*

3.2 Scan and Partition

In this subsection, the first step of SAM is presented. We first make an overview, then introduce some data structures we used and describe two algorithms for building B-partition and F-partition at last.The goal of this step is to build F-partition and B-partition efficiently while scanning nodes in stream. The stream nodes will be scanned

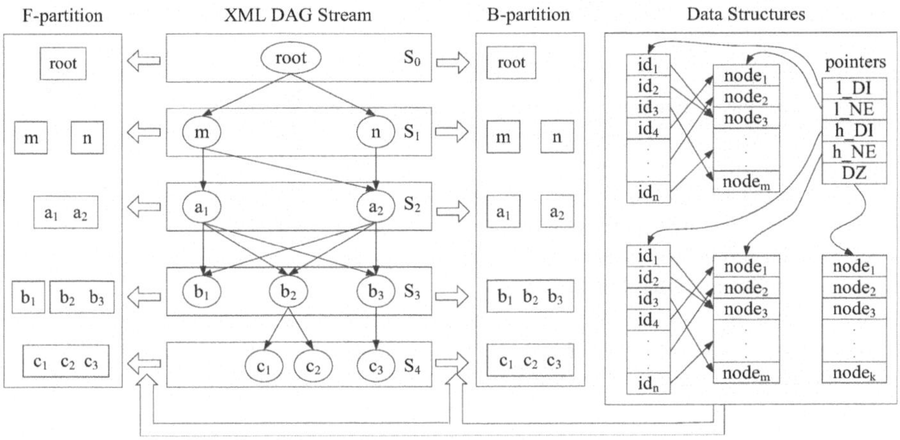

Fig. 2. Scan and build F-partition and B-partition

twice in this step. First, SAM scans streams from top to bottom and build B-partition. Then, it scans in the opposite order and build F-partition. During building partitions, many streams with different depth will be scanned, but at one time only two streams are maintained. For example, during building B-partition, if SAM is scanning stream S_i whose depth is i, there are only S_i and S_{i-1} in memory. Each stream S_j $(0 \leq j < i)$ has been deleted and each stream S_j $(j > i)$ has not been loaded into memory.

Data Structures. During this step, three kinds of tables are used to support building partition. These tables can support both algorithms for building B-partition and F-partition. During following description, we use index node to represent a BNode or FNode.

- **D**atanode to **I**ndexnode **T**able: DI-Table for short, it is used to map one data node to its index node. The index of DI-Table is id of data node, and the content is id of the index node corresponding to some data node.
- **N**ode **E**xample **T**able: NE-Table for short, it is used to store one of the data nodes which belongs to some index node. The index of NE-Table is id of an index node, and the content is one of its data nodes.
- **D**egree **Z**ero **T**able: DZ-Table for short, it is used to store nodes with no parents or children. The index of DZ-Table is a label, and the content is one of those data nodes with the same label.

Algorithm 1. ScanBuildBpartition(S)

```
1: Initialize l_DI, l_NE, h_DI, h_NE and DZ with NULL
2: for (i = 0 to n)
3:     output and delete the partitions of high stream
4:     move the low stream to high and initialize DZ with NULL
5:     While (¬ isempty(S_i))
6:         ¬ isempty(S_i)q=top(S_i), MergeNode(q)
7:         advance(S_i)
 Function MergeNode(DNode&q)
1: if (parent(q) == ∅)
2:     if (DZ[label(q)] ==NULL)
3:         generate a new partition p for q and insert p to l_NE and DZ
4:     else insert q into DZ[label(q)]'s partition
5: else temp=FindNodeToMerge(q)
6:     if (temp ==NULL)
7:         generate a new partition p for q and insert p to l_NE and DZ
8:     else insert q into temp's partition
 Function FindNodeToMerge(DNode&q)
1: parentid = id(q_parent), where q_parent has the minimum number of children
2: for(each q_parent's child q' , where label(q') = label(q) and |parent(q)| = |parent(q')|)
3:     ifparent(q) = parent(q') return q'
4: return NULL
```

As Figure 2 shows, five tables are used to build partitions, and their pointers are stored in a 5-array **pointers**. For more, *l_DI* and *l_NE* point to the DI-Table and NE-Table of the low stream whose depth is bigger; *h_DI* and *h_NE* point to the DI-Table and NE-Table of the high stream whose depth is smaller; *DZ* is the pointer of DZ-Table.

ScanBuildBpartition Algorithm. The algorithm for building B-partition is shown in Algorithm 1. The key idea of this step is to scan DNodes through XML DAG stream and decide whether to build a new BNode or to insert this DNode into some BNode according to bisimulation relations. Line 2, in Algorithm 1, controls the order of visiting stream nodes. Lines 3 deletes nodes of high stream, because they won't be used any more. Lines 6 gets a node q from low stream S_i and calls Function to merge q with nodes in the same set as q in B-partition. Function MergeNode finishes such a task according to rules of B-bisimulation. In Function MergeNode, lines 3 and 7 build a new BNode, and lines 4 and 8 insert q into some BNode. Function FindNodeToMerge searches only in children of one parent of q not children of all parents of q, and find node to merge q with. To save space, Algorithm 1 only maintains two of all streams which is called high and low stream; to save time, Algorithm 1 only makes searching in children of q's one parent. The detail in complexity will be analyzed in section 4.2.

ScanBuildFpartition Algorithm. ScanBuildFpartition Algorithm can be easily got by modifying ScanBuildBpartition Algorithm: reverse the order of scanning streams and swap operations of *parent* and *children* in Algorithm. By limits of space, we do not introduce such a similar algorithm in detail.

Algorithm 2. MergePartition

```
Input: F-partition A, B-partition B
Output: F&B-partition C
k = 0 and for each smallest id i of set j in A or B, set AN[i] or BN[i] as j.
While(k < AN.size())
    if (AN[k] ≥ 0)
        Output A[AN[k]] ∩ B[BN[k]] to C, delete these nodes from A[AN[k]] and B[BN[k]]
        if (A[AN[k]] = NULL) AN[k] = −1
        else AN[smallest id of A[AN[k]]]=AN[k]
            if (k ≠ smallest if of A[AN[k]]) AN[k] = −1
            Do the same operations to BN and B
    else k++
return C
```

3.3 Merge and Build F&B-partition

By scanning step described in 3.2, we can get two partitions of data nodes, F-partition and B-partition. In this subsection, we introduce how to get the F&B-partition of XML data with F-partition and B-partition. The key idea of this procedure comes from Theorem 1: if node a and b belong to the same set in both F-partition and B-partition, they belong to the same set in F&B-partition; otherwise they belong to different sets.

This algorithm of merging is sketched in Algorithm 2. Algorithm 2 accepts F-partition A and B-partition B as inputs and outputs F&B-partition C. During execution, the set with the current smallest id is selected to process iteratively. AN and BN are used to maintain the smallest id of A and B, and will be updated while a set of C has been computed. In each iteration, the sets with the smallest id in A and B are selected and joined to form a set q of C. Then, q's element is deleted from A and B, and AN and BN is updated. This procedure is executed until all nodes are processed.

Example 3. Consider the given F-partition and B-partition in Figure 3. When Merge Partition algorithm finishes, F&B-partition C is got. We can see, nodes a_1 and a_2 don't belong to the same set in F&B-partition because they don't belong to the same set in B-partition; node c_1, c_2 and c_3 belong to the same set in F&B-partition because they belong to the same set in both F-partition and B-partition.

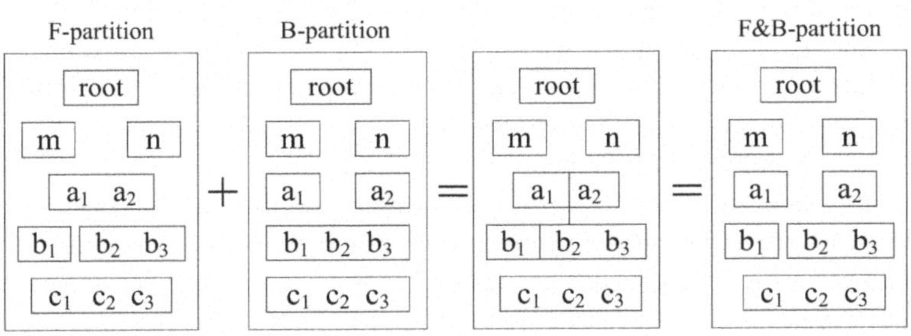

Fig. 3. Merge F-partition and B-partition to build F&B-partition

3.4 Form F&B-Index

In this step, we first form an index node for each set of F&B-partition, then add edges between index nodes. The rule of adding edges is that: given index node a and b, if there is a data node a' in a and b' in b such that an edge (a', b') exists in data graph, we add an edge (a, b) between a and b. Although this step can be implemented very efficiently in naive way, there are also some techniques to save space. When adding edges, if we maintain two streams as previous steps and scan the edges between them, this step can be easy to implement with $O(max|S_i|)$ space complexity and $O(|E|)$ time complexity. By limits of space we won't describe it in detail.

4 Analysis of SAM

In this section, first we prove the correctness of SAM. Then we will analyze the space and time complexity of SAM.

4.1 Correctness of SAM

In this part, first we propose lemma 1 which is used by other lemmas and theorems. Then, we prove lemma 2 and theorem 2 to show that SAM can build B-partition correctly. Third, we prove lemma 3 and theorem 3 to show that SAM can build F-partition correctly. Finally, we prove theorem 4 to show the correctness of SAM.

Lemma 1. Given a DAG-structured XML data G, if node m and n satisfy one of following three formulae: $m \approx_B n$, $m \approx_F n$, $m \approx_{FB} n$, we know $depth(m) = depth(n)$. □

Lemma 2. Given a DAG-structured XML data G, based on DAG stream model, node m and n can be put into the same set by Algorithm 1, if and only if $m \approx_B n$.

Proof. To prove this lemma, we only need to show correctness of two propositions:

⇒: If node m and n are put into the same set in Algorithm 1, line 5 or line 12 must be executed to merge m and n.

- **(a)** If line 5 are executed, we can know: (1)both m and n have no parents, so they satisfy incoming edge condition; (2)since m and n have the same index of DE, m and n have the same label, and they satisfy label condition.
- **(b)** If line 12 are executed , we can know: (1) $parent(m) = parent(n)$, so they satisfy incoming edge condition; (2) by implementation of FindNodeToMerge, we know m and n have the parent, and they satisfy label condition.

Therefore, m and n satisfy B-bisimulation, that is $m \approx_B n$.

⇐: If $m \approx_B n$, there must be $depth(m) = depth(n)$ according to Lemma 1. Suppose m is visited before n, we can prove this part by mathematical induction:

- **(a)** For m and n whose depth are 0, it is easy to know that $parent(n) = \emptyset$, and DE[$label(n)$] must have one node m because $label(m) = label(n), depth(m) = depth(n) and both of m and n have no parent$. So line 5 in MergeNode must be executed and m and n must be put into the same set.

(b) If two nodes $m \approx_B n$, suppose Algorithm 1 can put them into a same set while $depth(m) = depth(n) <= i$, we try to prove this proposition is correct for $depth(m) = depth(n) = i + 1$. By induced condition, we know Algorithm 1 could find m' which belongs to the same set as m. Obviously, Algorithm 1 can put n into m', that is put m and n into a same set.

Finally, we can know node m and n can be put into the same set by Algorithm 1, if and only if $m \approx_B n$. □

By lemma 2, we can easily get the following theorem.

Theorem 2. *Given a DAG-structured XML data G, based on DAG stream model, Algorithm 1 ScanBuildBpartition can correctly return B-partition of G.* □

Similarly, we can get the following theorem.

Theorem 3. *Given a DAG-structured XML data G, based on DAG stream model, Algorithm ScanBuildFpartition can correctly return F-partition of G.* □

Theorem 4. *Given a DAG-structured XML data G, based on DAG stream model, Algorithm SAM can return the correct F&B-Index.*

Proof. **[Sketch]** Given a DAG-structured XML data G, there are three steps to process G in SAM. By theorem 2 and 3, we can know step 1 can build F-partition and B-partition of G correctly; by theorem 1, it is easy to see that step 2 can build correct F&B-partition based on step 1; step 3 is a standard step of build F&B-Index, so its correctness is obvious. Finally, we know that given a DAG-structured XML data G, based on DAG stream model, Algorithm SAM can return the correct F&B-Index. □

4.2 Complexity of SAM

To analysis complexity of Algorithm SAM, we describe an XML document as DAG $G = \{V_G, E_G\}$ where V_G is the node set of G, and E_G is the edge set of G. Moreover,

- G's F&B-Index is G_{FB}, whose node set is V_{FB} and edge set is E_{FB}.
- G's B-Index is G_B, whose node set is V_B and edge set is E_B.
- G's F-Index is G_F, whose node set is V_F and edge set is E_F.

For more precise analysis, we define V_i (i.e., S_i), $V_{FB,i}$, $V_{F,i}$ and $V_{B,i}$ as those corresponding nodes with depth i. *Label_num* denotes the total number of distinct labels. We identify the max number of one index node's children as f_{maxc} and identify the max number of one index node's parents as f_{maxp}. Both f_{maxc} and f_{maxp} are small integers for most XML data. So we have the following theorem.

Theorem 5. *Given an XML DAG G, Algorithm 1 has the worst-case space complexity of* $|V_i| + |V_{i-1}| + label\_num + min(|V_{B,i-1}| \times f_{maxc}, |V_{B,i}| \times f_{maxp})$ *which can be bounded by* $O(max(f_{maxc}, f_{maxp}) \times max|V_i|) = O(max|V_i|)$ *and the worst-case time complexity of* $O(f_{maxc} \times f_{maxp}|V|)$ *which can be simplified to be* $O(|V|)$. *For more, Algorithm ScanAndBuildFartition has the same space and time complexity as Algorithm 1.* □

Theorem 6. *Given an XML DAG G, Algorithm 2 has the worst-case space complexity of $O(max(|V_i|))$ and time complexity of $O(|V|)$.* □

Theorem 7. *Given an XML DAG G, in SAM Algorithm, the procedure of forming F&B-Index has space complexity of $O(max|V_i|)$ and time complexity of $O(|E|)$.* □

Theorem 8. *Given an DAG G, to build F&B-Index, the worst-case space complexity of SAM is $O(max|V_i|)$, and the worst-case time complexity of SAM is $O(|V| + |E|)$.* □

Comparing with PT Algorithm. The space complexity of PT is $O(|V| + |E|)$ which is larger than $O(max|V_i|)$, and the time complexity of PT is $O(|E| \log |V|)$ which is larger than $O(|V| + |E|)$ in graph model.

5 Experiments

5.1 Experiment Setting

We implemented SAM in C++ and all experiments were run on a 1.7Ghz Pentium IV processor with 256MB of main memory. We two typical data sets from real world to test SAM. One data set is DAG-structured gene ontology data [9] and the other one is XML data generated from XMark benchmark [10] by deleting some edges.

5.2 Comparison

In this section we compare the efficiency of SAM and PT Algorithm. We use XMark10M, XMark50M, XMark100M and gene ontology (30M) as data set and run SAM and PT separately on data set. Gene ontology and XMark50M are both treated as normal (not small or large) real world data. Because the size of XMark can be changed, we use XMark10M as the small data and XMark100M as the large data.

Space Cost. We use the number of nodes and edges that must be maintained during execution of algorithm to measure the space cost of two algorithms. Figure 4 shows the space cost of SAM and PT. Note that when PT is running on XMark100M, by memory limit, it is busying on swapping buffers between memory and disk, and can't finish building F&B-Index, so we just record the number as what it was until we killed PT algorithm. This appearance just shows that PT is unpractical for large XML data and SAM can be used to build F&B-Index for very large XML data. It's easy to see that SAM takes much less space cost than PT, SAM is more efficient in space. Based on our proof of complexity, SAM has less space complexity than PT, so it is natural.

Time Cost. Figure 5 shows the execution time of SAM and PT. Note that the time is in log scale. In this testing experiment, after waiting very long time, PT still couldn't finish building F&B-Index, so we just record a max number which is much smaller than it took. This appearance shows that PT is unpractical for large XML data, and SAM can be used to build F&B-Index for very large XML data. We can see that for all data sets SAM takes much less execution time than PT, SAM is more efficient in time. Based on our proof of complexity, SAM has less time complexity than PT, so it is natural.

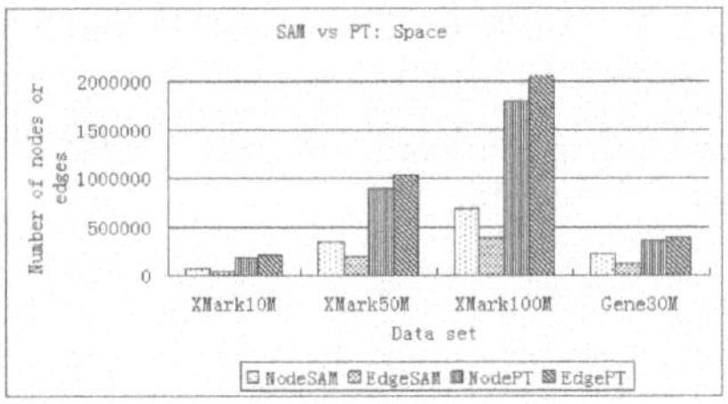

Fig. 4. Compare SAM with PT on Space

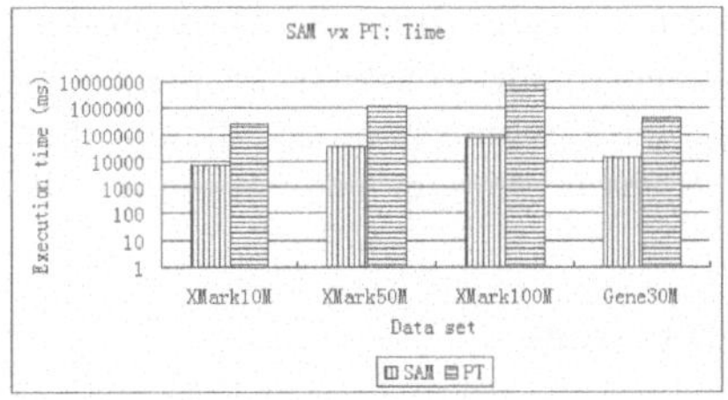

Fig. 5. Compare SAM with PT on Time

5.3 Scalability

In this section, to study the scalability of SAM, we vary the size of XMark data set from 1M to 100M, run experiments and record execution time and the numbers of nodes and

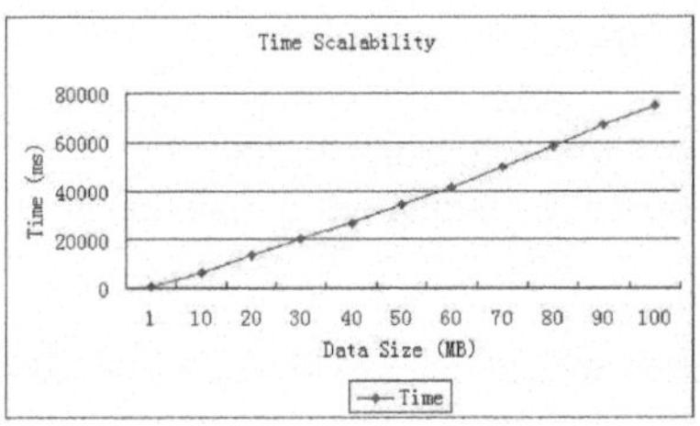

Fig. 6. Time Scalability

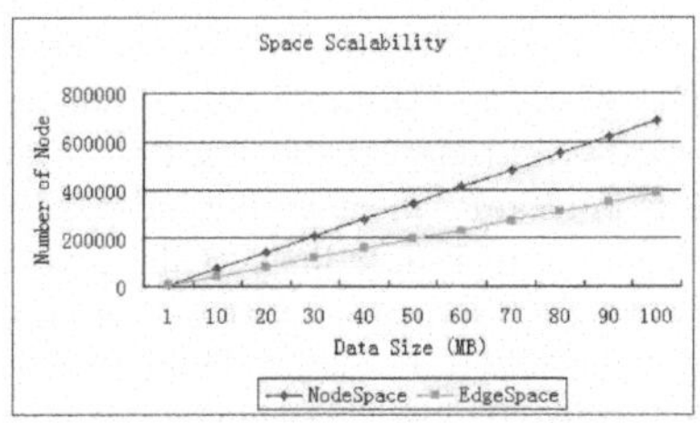

Fig. 7. Space Scalability

edges maintained in memory. Figure 6 and 7 show the result. We can observe that all three parameters increase linearly with the increase of the size of data set. SAM can scale very well for such large XML data as that.

6 Related Work

Structure index for XML have been widely used for indexing, query processing and selectivity estimation. DataGuides [14] was one of first used structure index in query processing. Simulation and bisimulation [13,16] are two notions in graph theory which are used to build relations on vertices. The idea of simulation were first applied in processing semistructured data in [15]. Later, 1-Index [1] and A(k)-Index [2] were proposed to support query processing. F&B-Index based on bisimulation were first proposed in [4].

7 Conclusion

Efficient F&B-Index construction is an important and interesting problem. In this paper, based on XML DAG and stream model, we propose a novel and efficient F&B-Index

construction algorithm, SAM (Scan-And-Merge). Theoretical analysis shows that SAM is correct and has $O(max|V_i|)$ space complexity and $O(|V| + |E|)$ time complexity in the worst case, which is more efficient that previous methods. Experimental results show that SAM is efficient and can scale well for large XML data.

References

1. T. Milo and D. Suciu. Index structures for path expressions. In Proceedings of 7th International Conference on Database Theory (ICDT 1999), pages 277-295, Jerusalem, Israel, 1999.
2. Wei Wang. PBiTree Coding and Efficient Processing of Containment Joins. The 19th International Conference on Data Engineering (ICDE 2003), pages 391-402, Bangalore, India, 2003.
3. Qun Chen, Andrew Lim, Kian Win Ong. D(K)-Index: An Adaptive Structural Summary for Graph-Structured Data. In Proceedings of the 22nd ACM International Conference on Management of Data (SIGMOD 2003), pages 134-144, San Diego, California, USA, 2003.
4. R. Kaushik. Covering Indexes for Branching Path Queries. In Proceedings of the ACM SIGMOD Conference, pages 133-144, Madison, USA, 2002.
5. Wei Wang, and Hongzhi Wang. Efficient Processing of XML Path Queries Using the Disk-based F&B Index. In the 31st Proc. of VLDB, pages 145-156, Norway, 2005.
6. Prakash Ramanan. Covering Indexes for XML Queries: Bisimulation−Simulation= Negation. In Proceedings of the 29th International Conference on Very Large Data Bases (VLDB 2003), pages 165-176, Berlin, German, 2003.
7. R. Paige and R. E. Tarjan. Three Partition refinement algorithms. SIAM Journal on Computing, 16(6): 973-989, December 1987.
8. Xianmin Liu, Jianzhong Li, and Hongzhi Wang. SAJ: An F&B-Index Construction Algorithm with Optimized Space Cost. In Proc. of NDBC, pages 413-417, Guangzhou, China, 2006.
9. Gene Ontology. http://www.geneontology.org.
10. XMark. The xml-benchmark project. http://www.xml-benchmark.org, Apr. 2001.
11. N. Bruno, N. Koudas, and D. Srivastava. Holistic twig joins: optimal xml pattern matching. In SIGMOD, pages 310-321, San Jose, CA, 2002.
12. T. H. Cormen and et. al. Introduction to Algorithms (ISBN 0-262-530-910).MIT Press, 1994.
13. D. Park. Concurrency and Automata on Infinite Sequences. Proc. 5th GI-Conf. LNCS Vol. 104. Springer-Verlag, NY, 1981.
14. R. Goldman and J. Widom. DataGuides: Enabling query formulation and optimization in semistructured databases. In Proc.Of the 23rd VLDB Conf, pages 436-445, Greece, 1997.
15. M. F. Fernandez. Optimizing regular path expressions using graph schemas. In Proc.of the 14th Int.Conf.on Data Engineering (ICDE 1998), pages 14-23, Florida, USA, 1998.
16. Milner. A Calculus for Communicating Processes. LNCS, Vol.92, Springer-Verlag, 1980.

BUXMiner: An Efficient Bottom-Up Approach to Mining XML Query Patterns

Yijun Bei, Gang Chen, and Jinxiang Dong

College of Computer Science, Zhejiang University, Hangzhou, P.R. China 310027
alphabyj@yahoo.com.cn, cg@zju.edu.cn, djx@zju.edu.cn

Abstract. Discovery of frequent XML query patterns in the history log of XML queries can be used to expedite XML query processing, as the answers to these queries can be cached and reused when the future queries "hit" such frequent patterns. In this paper, we propose an efficient bottom-up mining approach to finding frequent query patterns in XML queries. We merge all queries into a summarizing structure named global tree guide (GTG). We refine GTG by pruning infrequent nodes and clustering adjacent nodes in the queries to obtain a Compressed GTG (known as CGTG). We employ a bottom-up traversal scheme based on CGTG to generate frequent query patterns for each node till the root of CGTG. Experiments show that our proposed method is efficient and outperforms the previous mining algorithms of XML queries, such as XQPMinerTID and FastXMiner.

Keywords: XML Query Patterns, Mining, Bottom-Up.

1 Introduction

With the proliferation of XML as a standard for data representation and data exchanging, efficient querying techniques of XML data becomes an important topic for the database community. Many studies have focused on the indexing of XML data to expedite the querying process. Besides index strategy, caching has also played an import part in improving performance of XML query processing, especial for repeated or similar queries [1, 2, 3, 4]. Users can obtain answer right away if the query results have been computed and cached. To cache useful queries, one of the most effective approaches is to discover frequent query patterns from the user queries, in respect that the frequent query patterns contain a wealth of information of user queries. Basically, the frequent query pattern mining problem is considered as finding a set of rooted subtrees that occurs frequently over a set of queries which can be modeled as trees.

In this paper, we present an efficient algorithm called BUXMiner to discover frequent query patterns using a bottom-up enumerating method. We introduce a novel data structure called compressed global tree guide (CGTG) to accelerate candidate generation and infrequent tree pruning. We note that previous algorithms such as XQPMiner, XQPMinerTID and FastXMiner [3, 5] employ a rightmost branch expansion enumeration approach to generate candidates from top to bottom. In contrast, we

G. Dong et al. (Eds.): APWeb/WAIM 2007, LNCS 4505, pp. 709–720, 2007.

perform an efficient bottom-up candidate generation process. We remove all infrequent nodes from the global tree guide before candidate enumeration and generate candidates within each prefix equivalence class. In this way, our approach will eliminate unnecessary candidates. Once more, previous methods have no guarantee on the number of dataset scans required. In our approach we no longer need dataset scan after the constructing of CGTG since the supports of query pattern trees can be computed through CGTG. Like mining algorithms XQPMiner, XQPMinerTID and FastXMiner, we do not consider XML queries that contain sibling repetitions either. Experiments results on public datasets show that our method outperforms the previous algorithms. Meanwhile, we will show that our approach has good scalability.

The rest of the paper is organized as follows. In section 2 we discuss previous work related to query pattern mining approaches. In section 3, we describe some concepts used in our mining approach. We propose the bottom-up algorithm BUXMiner in section 4. Section 5 gives the results of our experiments and we make a conclusion in section 6.

2 Related Work

Many efficient tree mining approaches have been developed [6, 7, 8, 9, 10, 11] to find tree-like patterns. Basically, there are two main steps for generating frequent trees in a database. First of all, a systematic way should be conceived for generating non-redundant candidate trees. Secondly, an efficient way is needed to compute the support of each candidate tree and determine whether a tree is frequent. Anyhow, they adopt a straight-forward generate-and-test strategy. Various algorithms have been brought forward to solve various forms of tree-structure such as rooted ordered tree, rooted unordered tree, free tree, etc. Asai et al. present a rooted ordered and rooted unordered tree mining approach in [6] and [7] respectively. Zaki gives ordered and unordered embedded tree mining algorithms in [8, 9]. Chi et al. bring forward algorithms for mining rooted unordered and free trees in [10, 11]. However, these mining approaches mainly treat with general trees. They don't take schema information into account when dealing with special trees like XML query pattern trees.

XQPMiner and XQPMinerTID presented in [5] are two approaches to frequent query pattern discovering exploiting schema information to guide the enumeration of candidates. Both of them employ rightmost branch expansion enumeration approach to generate candidates from top to down. XQPMinerTID outperforms XQPMiner due to less dataset scans. XQPMiner will scan dataset for each candidate, while XQPMinerTID only scans dataset when expansion happens on the leaf node. Mining algorithm FastXMiner in [3] also generates candidate trees with the help of schema information. It needs dataset scans only when the candidate tree is a single branch tree. The support of non-single branch tree can be computed based on its joined rooted subtrees. In this way, FastXMiner needs less dataset scans than algorithms XQPMiner and XQPMinerTID. Frequent query access patterns can be made use of for caching to accelerate retrieval of query results. Liang et al. in [3] employ FastXMiner to discover frequent XML query patterns and demonstrate how the frequent query patterns can be used to improve caching performance. Ling et al. in [4] take into account the temporal features of user queries to discover association rules. They cluster XML queries according to their semantics first and then mine association rules between clusters for caching.

3 Preliminaries

3.1 Query Pattern Tree

Query Pattern Tree. A query pattern tree is a tree QPT = <R, N, E> where:

- R is the root node.
- N is the node set. Each node n has a label whose value is in {"*", "//"} ∪ labelSet where the labelSet is the label set of all elements, attributes.
- E is the edge set. For each edge e = (n_1, n_2), node n_1 is the parent of n_2.

Rooted Query Pattern Subtree. Given two query pattern trees *T* and *S*, we say that *S* is a rooted query pattern subtree of *T* iff there exists a one-to-one mapping φ: $V_S \rightarrow V_T$, such that satisfies the following conditions:

- φ preserves the root of trees, i.e., $R(S) = R(T)$.
- φ preserves the labels, i.e., $L(v) = L(\varphi(v))$, $\forall v \in V_S$.
- φ preserves the parent relation, i.e., $(u,v) \in E_S$ iff $(\varphi(u), \varphi(v)) \in E_T$.

Frequent Rooted Query Pattern Tree. Let *D* denote all the query pattern trees of the issued queries and d_T be an indicator variable with $d_T(S) = 1$ if query pattern tree *S* is a rooted query pattern subtree of *T* and $d_T(S) = 0$ if tree *S* is not a rooted subtree of *T*. The support of query pattern tree *S* in *D* can be defined as $\sigma(S) = \sum_{T\in D} d_T(S) / \sum_{T\in D}$, i.e., the percent of the number of trees in *D* that contain tree *S*. A rooted query pattern tree is frequent if its support is more than or equal to a user-specified minimum support.

3.2 Global Tree Guide

We associate each QPT a unique ID, denoted as QPT.ID, which will be used for construction of global tree guide and mining process.

Global Tree Guide. By merging all the query pattern tress, we construct a global tree guide (GTG), with each node recording ID list of query pattern trees. Each ID indicates that a query pattern tree with such ID contains the path from root to current node. In Figure 1 we show a GTG constructed using 10 query pattern trees. The QPT list for node Java means five queries contain the path order/items/book/title/java. To deal with special label like wildcard "*" and descendent path "//", we combine the special label and the following label to produce a new label. For example, a QPT order/items//Java in the GTG will be considered as a single path tree with nodes order, items and //_Java.

We denote a subtree rooted at arbitrary node in the GTG as OT, a subtree rooted at the root node of the GTG as RT, and a single path staring at the root node as SRT.

Frequent Node. The support of node in GTG is the number of QPTs that contain the path from root to current node, namely the size of QPT list, to the number of all

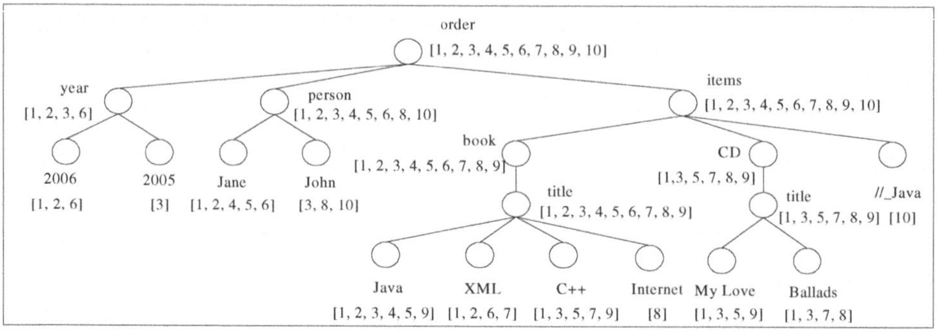

Fig. 1. Global Tree Guide

QPTs. For example, the support of node Java is 6 / 10 = 0.6. A node in GTG is frequent if its support is no less than the minimum support.

Lemma 1: The support of the node is no less than the support of its descendent node.

Proof: A descendent node can be reached only through its ancestor in a QPT. If a QPT contains the root path tree from root to the descendent node, then it must be contain the path from root to the ancestor node. Therefore, the support of ancestor is more than or equal to the descendent.

Lemma 2: If a node is infrequent in the GTG, then an RT including it will not be a frequent tree.

Proof: Since the support of an RT will be no more than the support of a node in the RT, then an RT will be infrequent if an included node is not frequent.

Lemma 3: If a node is frequent in the GTG, a SRT including it as the leaf node must be a frequent tree.

Proof: As the support of a SRT equals to the support of the leaf node, a SRT will be frequent if the node is frequent.

Assume the minimum support is 0.2. In Figure 1 the node Internet is infrequent, and an RT order/items/book/title/Internet is also an infrequent one. The node XML is a frequent node, and the SRT order/items/book/title/XML is also frequent.

If a node is infrequent, then all its descendent nodes are infrequent as well due to the less support of the descendent nodes. As a result, we can prune all the infrequent nodes in the GTG using a top to down traversal. We search the GTG starting at the root level by level and prune infrequent nodes along with all its descendent nodes once we find an infrequent node. For example, in Figure 1 node Internet can be pruned since it is an infrequent node.

Furthermore, to reduce memory space, we can compress the node and its child node into a single node with the following satisfaction: 1) the parent node has only one child; 2) the parent node and the child node have the same ID list of QPTs. For instance, in Figure 1 we can compress node book and child node title into a single node book/title.

Compressed Global Tree Guide. Employing the infrequent nodes pruning scheme and nodes compressing scheme, we compress the GTG into a compressed global tree guide (CGTG). Figure 2 presents a CGTG transformed from GTG in Figure 1.

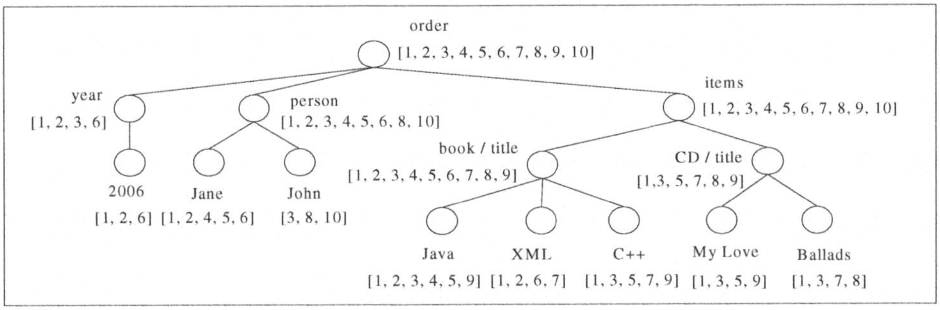

Fig. 2. Compressed Global Tree Guide

4 Mining Query Pattern Trees

BUXMiner performs a bottom-up process for generating frequent RTs in the CGTG. To generate frequent RTs rooted at node n in the CGTG, we will have to generate all frequent RTs rooted at all children of n firstly, and then merge discovered RTs rooted at child nodes. Figure 3 shows the high level structure of BUXMiner. We first scan all query pattern trees and construct a GTG. We then generate a CGTG by way of pruning and compressing nodes of GTG. We use the root node of the CGTG as an input to the recursive generating algorithm to obtain all frequent RTs.

```
Algorithm BUXMiner (D, minsupp)
Input: A set of query pattern trees
       Specified minimum support
Output: A set of Frequent query pattern trees FRTS
   GTG = ConstructGTG(D);
   CGTG = CompressGTG(GTG, minsupp);
   root = root node of CGTG;
   FRTS=GenerateFrequentRT(root,minsupp);
   return FRTS
```

Fig. 3. Query Patten Tree Mining Algorithm

4.1 Generating Frequent Query Pattern Trees

Query Pattern Tree Encoding. In stead of the standard data structure, such as the adjacency-matrix, the adjacency-list representation, we adopt a string encoding scheme to represent trees, which is first introduced by Luccio [12]. This encoding scheme is more space-efficient and is simpler to be manipulated [8]. To generate the string encoding, a depth-first preorder search is performed on the tree from the root.

When a node is encountered, the node label is appended to the last of the string. Whenever a backtracking occurs from a child to its parent a distinguished label (-1 is used here) not existing in tree node labels needs to be appended to the string. For example, the tree OT_1 in Figure 4 can be encoded as a string “items, book, title, Java, -1, XML, -1, -1, -1, CD, title, -1, -1, -1”.

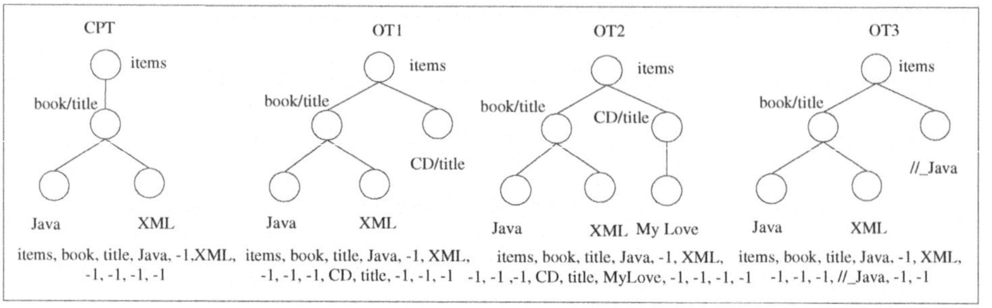

Fig. 4. Prefix Equivalence Class

Prefix Equivalence Class. We say that OTs in the CGTG are in the same prefix equivalence class, if and only if they share a maximal common prefix tree. Thus using previous tree encoding scheme any two members of an equivalence class has a same prefix string which represents prefix tree. If the prefix tree is represented as a string “*Labels*,-1”, then trees in the equivalence class must have the prefix “*Labels*”. For example, in Figure 4 trees OT_1, OT_2, OT_3 are in a same equivalence class since they share a common prefix tree CPT.

In Figure 5 we show the algorithm for generating frequent rooted trees in the CGTG, which employs a bottom-up approach. Frequent trees rooted at a given node are generated using the following steps:

Adding Root Node. Since we perform the searching process in a CGTG which has pruned all infrequent nodes, the tree with only root node is a frequent RT. We construct a new tree represented with string “root.label, -1” and copy the ID list of QPT from the root node (Lines 1-2).

Obtaining Children FRTs. To obtain frequent rooted trees at a specified node, we firstly generate all frequent rooted trees at the children and then employ a merging strategy on the frequent trees rooted at child nodes. FRT is represented as a frequent rooted tree. We denote $FRTS_i$ as the set of frequent trees rooted at the ith child. Trees in $FRTS_i$ constitute an equivalence class since they have a common prefix tree which is the ith child node itself.

- If the root node has only one child, then for each FRT_{child} in the $FRTS_{child}$ we add the root node to the FRT_{child} and generate a new FRT which is “root.label, FRT, -1” (Lines 8-12).
- If the root node has more than one child, firstly we generate all frequent rooted trees by add the root node to FRT_{child} (Lines 14-18). Then we merge frequent rooted trees from different equivalence class to construct new candidates and determine whether they are frequent (Lines 20-29). Suppose

the root node has *n* children and there exist *n* sets of $FRTS_{child}$ from $FRTS_1$ to $FRTS_n$. We pick up a FRT from $FRTS_i$ and join it with all FRTs from $FRTS_{i+1}$ to $FRTS_n$ (Lines 26-29). For each FRT and an equivalence class $FRTS_{child}$, we generate a new equivalence class whose common prefix tree is a combined tree of FRT and prefix tree of $FRTS_{child}$. After joining the FRT from $FRTS_i$ with all frequent trees from $FRTS_{i+1}$ to $FRTS_n$, we generate $(n\text{-}i)$ equivalence class and regard them as a new group to perform a next merging process.

Algorithm GenerateFrequentRT (root, minsupp)
Input: Root node of new generated frequent tree root
Specified minimum support
Output: A set of Frequent query pattern trees FRTS rooted at the root node

```
1   FRT_root.label = ((root.lable, -1)); FRT_root.IDList = root.IDList;
2   FRTS = {FRT_root};  // add the tree having only root node
3   CandSet = Φ;
4   if (root have at least one child) then
5       for (each child of root) do // FRTS belonging to same child construct a new class
6           FRTS_child = GenerateFrequentRT(child,minsupp);
7           CandSet = CandSet ∪ {FRTS_child};
8       if (|CandSet| == 1) then // only one set FRTS of child,
9           for (each FRT_child in CanSet[0]) do   // using FRT_child to generate new FRT
10              NewFRT.label = (root.label, FRT_child.label, -1);
11              NewFRT.IDList = FRT_child.IDList;
12              FRTS = FRTS ∪ {NewFRT};
13      else if (|CandSet| >= 2) then // more than one set FRTS of children
14          for each ChildSet in CandSet do
15              for each FRT_child in ChildSet do
16                  NewFRT.label = (root.label,FRT_child.label,-1);
17                  NewFRT.IDList = FRT_child.IDList;
18                  FRTS = FRTS ∪ {NewFRT};
19          MergeList = {CandSet}; //merge trees in candset to generate new FRT
20          while (|MergeList| > 0) do
21              CandSet = MergeList[0];
22              MergeList = MergeList - CandSet;
23              while (|CandSet| > 0) do
24                  ChildSet = CandSet[0];
25                  CandSet = CandSet - ChildSet;
26                  for (each FRT_child in ChildSet) {
27                      NewCandSet= MergeTree(FRT_child, CandSet, minsupp, FRTS);
28                      if (|NewCandSet| >= 2) then
29                          MergetList = MergeList ∪ {NewCandSet};
30  return FRTS;
```

Fig. 5. Frequent Query Patten Tree Generating Algorithm

4.2 Merging Query Pattern Trees

We show the tree merging process in Figure 6. We merge a prefix tree with all suffix trees from a set of equivalence classes. To avoid unnecessary computing, a pruning process is performed before support computing of candidate trees. We prune the new candidate *k*-size tree if it has infrequent (*k*-1)-size rooted subtrees. We then compute the support of candidate tree using the ID list of QPTs. Since all IDs in the list are ordered in an ascending order, the ID list of new tree can be computed quickly by way of merge join. If the new merged tree is a frequent one, we add it into both the *FRTS* set and the new equivalence class whose prefix tree is a combined tree of *prefixTree* and the common prefix tree of *EQClassSet*. The *FRTS* set contains all discovered frequent trees. And the *CandSet* contains all new generated equivalence classes which will be used in a next merging process.

```
Algorithm MergeTree (prefixTree, suffixSet, minsupp, FRTS)
Input: Prefix Tree of the new generated tree
       Suffix Tree Set of the new
       Specified minimum support
       Frequent Rooted Tree Set
Output: Sets of Frequent query pattern trees CandSet
  CandSet = Φ;
  for (each EQClassSet in suffixSet) do
      NewEQClassSet = Φ;
      for (each suffixTree in EQClassSet) do
          NewTree = ConstructTree(prefixTree,suffixTree);
          if (!isPruned(NewTree) && |NewTree.IDList| >= minsupp * |D|) then
              FRTS = FRTS ∪ {NewTree};
              NewClassSet = NewEQClassSet ∪ {NewTree};
      if (|NewClassSet| > 0) then
          CandSet = CandSet ∪ {NewEQClassSet};
  return CanSet
```

Fig. 6. Tree Merging Algorithm

Tree Merging. Given a prefix tree T_1, a suffix tree T_2, and a common prefix tree CT of T_1 and T_2, we merge the two trees T_1 and T_2 and produce a new tree with prefix T_1. We denote the merging process as $T = T_1 \cup_{CT} T_2$. The ID list of the created tree is the join result of two ID lists of the merged trees. Suppose the CT is represented as string "*CT_Labels*, -1". Then we can denote T_1 as "*CT_Labels*, *T_1_Follow_Labels*, -1" and T2 as "*CT_Labels*, *T_2_Follow_Labels*, -1". The constructed tree is represented as "*CT_Labels*, *T_1_Follow_Lables*, *T_2_Follow_Labels*, -1". In Figure 7 we shows a tree merging process, where the *CT_Lables* is the string "order, year, 2006, -1, -1", *T_1_Follow_Labels* is "person, Jane, -1, -1", and *T_2_Follow_Labels* is "items, book, title, Java, -1, XML, -1, -1, -1, -1".

Automatic Ordering. If trees in an equivalence class set EQ are ordered according to node order in CGTG, then trees in a new equivalence class set NEQ constructed by means of merging a prefix tree with all trees in EQ are still ordered in node order. Because the tree merging process do not change the node label order, and only append a new prefix tree to all suffix trees. When generating frequent trees we insert new frequent trees into the new equivalence class according to the original order, which

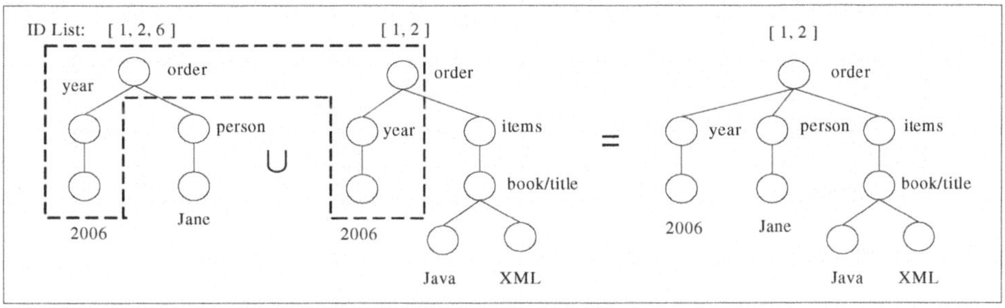

Fig. 7. Tree Merging

result in an automatic ordering of the trees in each equivalence class. Once more, trees also generated after their rooted subtrees based on our candidate generating process.

Candidate Pruning. Before computing the support of a k-size candidate tree, we carry out a pruning test to make sure that all its rooted subtrees are frequent. For the sake of saving time, we only check whether its (k-1)-subtrees are frequent. According automatic ordering property of our candidate generating method, we can make sure all (k-1)-subtrees have been enumerated. To efficiently perform the pruning step, during creation of frequent RTs, we add each frequent tree into a hash table. The key of each entry in hash table is the string representation of the frequent RT. Thus it takes O(1) time to check for each rooted subtree.

Space Reducing. The main consumption of space is the ID list of QPT for each frequent rooted tree. If a parent node has been computed, then all frequent trees rooted at the child nodes can be removed. In this way, the space consumption of our algorithm is the whole CGTG plus ID list for FRTs rooted at the current node and its children.

5 Experiments

In this section, we present experimental results of our BUXMiner algorithm compared to previous algorithms XQPMinerTID and FastXMiner. We implemented both our mining algorithm and previous algorithms in Java language and carried out all experiments on an Intel Xeon 2.8GHz computer with 3GB of RAM running operating system RedHat Linux 9.0. When performing experiments, all QPTs are loaded into memory. Thus there is no disk access when scanning datasets.

5.1 Workload

We use the DBLP.DTD [13] and XMARK.DTD [14] as the schemas to generate QPTs. DTDs are converted into DTD trees by introducing some wildcard "*" and descendent path "//" to make the query pattern trees more complex. We add 5 "*" and 5 "//" into DBLP.DTD, and 6 "*" and 6 "//" into XMARK.DTD. Table 1 shows the characteristics of 100,000 query pattern trees generated using DBLP and XMARK respectively.

Table 1. Characteristics of Datasets

QPT / Dataset	Average Nodes	Max Nodes	Average Depth	Max Depth	Max Fanout
DBLP	9.471	10	3.654	6	9
XMARK	10.152	11	4.964	9	10

5.2 Performance Comparison

We evaluate the performance of BUXMiner and compare it with XQPMinerTID and FastXMiner on both DBLP and XMARK with various dataset sizes and supports. Since we do not take semantic containment with wildcard and descendent path into consideration, we implement the Contains algorithm in XQPMinerTID and FastXMiner in a simple way, i.e., we just perform an ordinal subtree inclusion process.

In Figure 8 we show the response time for three algorithms with varying number of QPTs from 50,000 to 250,000. The minimum support is 1% for each dataset. From figure we can see that BUXMiner outperform XQPMinerTID and FastXMiner by 20% ~ 40%. On the one hand, we prune infrequent nodes before candidate generation and make less enumeration of candidates. On the other hand, we do not need dataset scans to compute supports of candidates. Although XQPMinerTID and FastXMiner employ various schemes to reduce dataset scans, they still need dataset scans in some particular situations like leaf node expansion. In despite of memory dataset scan in our experiments, it still spends much time. When the dataset is larger, the improvement is more obvious as the scan of dataset needs more time. For different characteristics of datasets, XQPMinerTID and FastXMiner show different performance. XQPMinerTID outperforms FastXMiner on the DBLP dataset. Nevertheless, FastXMiner performs more efficiently than XQPMinerTID on the XMARK dataset. In addition to its efficiency, BUXMiner scales linearly as the number of QPTs increases.

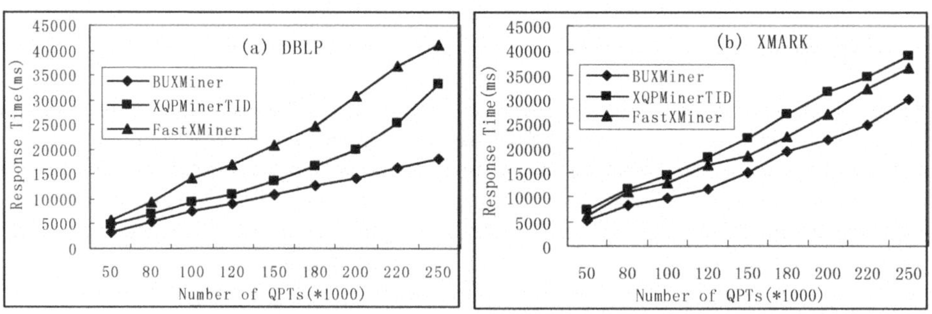

Fig. 8. Response Time with Varying Number of QPTs

In Figure 9, we show the performance of three algorithms with various supports from 0.1% to 1.5% on dataset with 100,000 QPTs. As previous results, BUXMiner still outperforms the other two algorithms on various supports. When the support is low, the improvement is more obvious. For a low support, more frequent rooted query patterns are discovered. Thus XQPMinerTID and FastXMiner will spend more time on scans of dataset. As the previous experiments, XQPMinerTID and FastXMiner present different performance on different datasets.

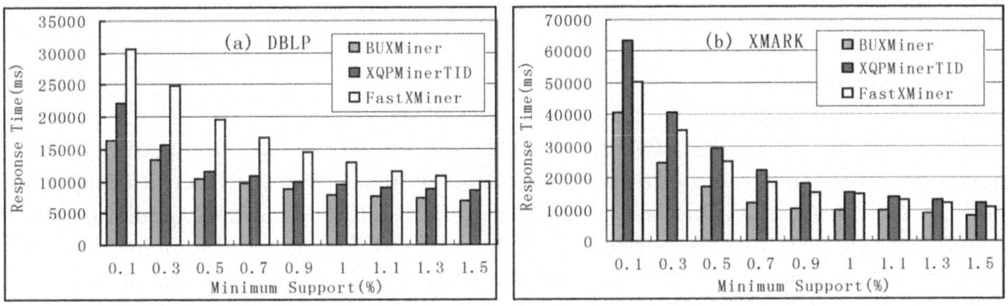

Fig. 9. Response Time with Varying Minimum Supports

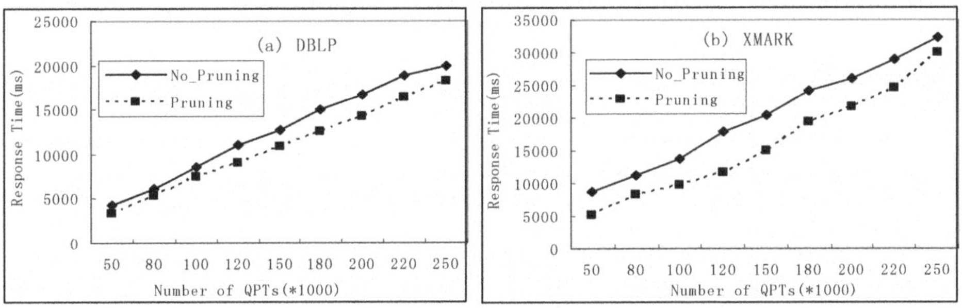

Fig. 10. Response Time with Pruning and without Pruning for Various datasets

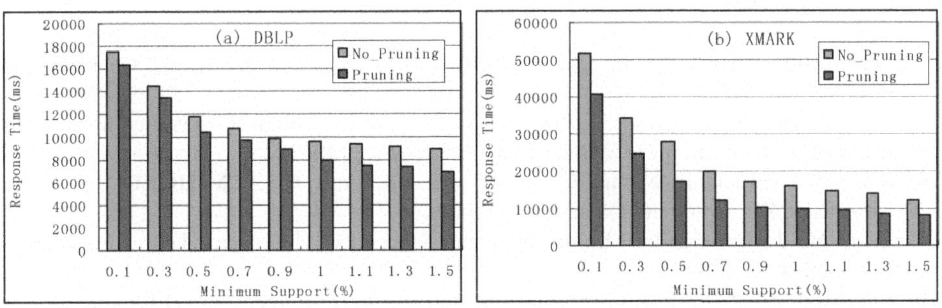

Fig. 11. Response Time with Pruning and without Pruning for Various Supports

We show the effect of pruning scheme for BUXMiner in Figure 10 and Figure 11. We adopt both the infrequent nodes pruning and (k-1)-size pruning schemes when mining frequent trees for Pruning experiments. On the contrary, we adopt neither of the above pruning schemes for No_Pruning experiments. In Figure 10 we show the effect of pruning scheme for various datasets from 50,000 to 250,000. Our pruning schemes improve the mining time about 10% ~30% with various size and supports. Figure 11 presents the effect of pruning scheme for various supports. Pruning scheme has greater impact on mining with higher supporpt. This is because we will prune more infrequent nodes in the CGTG with a higher support. Less nodes in the CGTG result in less enumeration of candidates.

6 Conclusion

In this paper, we present an efficient algorithm called BUXMiner to discover frequent query patterns using a bottom-up enumerating method. We introduce a compressed global tree guide (CGTG) as a schema to accelerate candidate generation and infrequent tree pruning. We perform an efficient candidate generation using a bottom-up approach. We remove all infrequent nodes from the compressed global tree guide before candidate generation and generate candidates within each prefix equivalence class. Our approach no longer needs dataset scan since the supports of query pattern trees can be computed through CGTG. Experiments show that our proposed methods outperform the mining algorithm XQPMinerTID, FastXMiner.

References

1. Chen, L., Rundensteiner, E.A., Wang S.: Xcache-a semantic caching system for xml queries. In Demo in ACM SIGMOD (2002).
2. Hristidis, V., Petropoulos, M.: Semantic caching of xml databases. In Proc. Of the 5th WebDB (2002).
3. Yang, L. H., Lee, M.L., Hsu W.: Efficient mining of xml query patterns for caching. In Proc. of 29th VLDB (2003).
4. Chen, L., Bhowmick, S.S., Chia, L.T.: Mining Positive and Negative Association Rules from XML Query Patterns for Caching. In DASFAA (2005) 736-747.
5. Yang, L. H., Lee, M.L., Hsu W., Acharya S.: Mining Frequent Query Patterns from XML Queries. In DASFAA (2003) 355-362.
6. Asai, T., Abe, K., Kawasoe, S., Arimura, H., Satamoto, H., Arikawa, S.: Efficient Substructure Discovery from Large Semi-structured Data, 2nd SIAM Int'l Conference on Data Mining (2002).
7. Asai, T., Arimura, H., Uno, T., Nakano, S.: Discovering Frequent Substructures in Large Unordered Trees, 6th Int'l Conf. on Discovery Science (2003).
8. Zaki, M. J.: Efficiently Mining Frequent Trees in a Forest, 8th ACM SIGKDD Int'l Conf. Knowledge Discovery and Data Mining (2002).
9. Zaki, M. J.: Efficiently Mining Frequent Embedded Unordered Trees, Fundamenta Informaticae (2005).
10. Chi, Y., Yang, Y., Muntz, R. R.: Indexing and Mining Free Trees, 3rd IEEE International Conference on Data Mining (2003).
11. Chi, Y., Yang, Y., Muntz, R. R.: HybridTreeMiner: An Efficient Algorihtm for Mining Frequent Rooted Trees and Free Trees Using Canonical Forms, 16th International Conference on Scientific and Statistical Database Management (2004).
12. Luccio, F., Enriquez, A. M., Rieumont, P. O., Pagli, L.: Exact Rooted Subtree Matching in Sublinear Time, Technical Report TR-01-14 (2001).
13. http://www.informatik.uni-trier.de/~ley/db/
14. http://monetdb.cwi.nl/xml/

A Web Service Architecture for Bidirectional XML Updating

Yasushi Hayashi, Dongxi Liu, Kento Emoto, Kazutaka Matsuda, Zhenjiang Hu, and Masato Takeichi

Graduate School of Information Science and Technology
University of Tokyo
{hayashi,emoto,kztk}@ipl.t.u-tokyo.ac.jp,
{liu,hu,takeichi}@mist.i.u-tokyo.ac.jp

Abstract. A Web service architecture is described for bidirectional XML updating. The updating mechanism exploits the power of bidirectional transformation so that users can update remote XML data by editing a view on the local machine that is generated by a transformation of the XML data. This architecture consists of three tiers: data viewer clients, a bidirectional transformation engine, and content servers accessible through the Internet. Due to the use of standard Web service technologies, the data viewer clients and content servers can be easily replaced with ones chosen by the user. Users can use this architecture to implement their own applications that exploit the power of bidirectional transformation without the burden of installing and maintaining a bidirectional language package.

1 Introduction

XML is the de facto standard format for data exchange. The role of XML as a format for data repositories on the Internet is becoming more important because the amount of resources stored in XML format is rapidly increasing. This trend arose for several reasons. First, many kinds of application software now support the export of data in XML format. Second, native XML databases are becoming widely used, and many relational database systems provide facilities for managing and generating XML data. Finally, various kinds of data that were not expressed in XML format are now being expressed in XML format. As many XML resources are available through the Internet and are used by various kinds of applications, an environment that provides an efficient and easy way to update XML data is critically important for efficient use of XML resources.

It has been observed that XML data are rarely retrieved for their own sake; rather, the data is typically transformed into another format for processing by a Web application. For example, a Web server might run an XSL [18] processor to generate HTML data from XML data for display in a Web browser. When users want to edit data being viewed in a transformed format, it would be more convenient and efficient if they could do so by editing the transformed data directly and then having the modifications reflected back into the source XML

G. Dong et al. (Eds.): APWeb/WAIM 2007, LNCS 4505, pp. 721–732, 2007.

data, rather than editing the original XML data. One technique for doing this is called "view updating" [6,1]. The source XML data is first transformed to another format, a "view," in which the user can more easily and intuitively understand its meaning. The modifications are made on the view directly and then are reflected back into the source data on the basis of some predefined updating policy.

Updating remote source data on the Internet through various views means that the updating mechanism should be adaptable to both the database system that stores the source data and the application that processes the views. This requires the updating system to be a modular and extendable component that works efficiently with other components in a distributed computing system. Web Services, a key Web technology, addresses this problem using standard technologies such as the simple object access protocol (SOAP) [15] and the Web services description language (WSDL) [16].

We have developed a Web service architecture with a communication protocol for bidirectional XML updating. It has high modularity and extendibility, so it can be used with various kinds of applications. This was achieved by combining two technologies mentioned above, that is, view updating and SOAP. A previously reported bidirectional transformation language, Bi-X [13,12] is used for the view updating. It was developed by extending the expressiveness of previous bidirectional languages [8,9] to make it usable in an architecture for general-purpose XML processing. It is used to obtain both the original source view and the updated source from the modified view. Our main contributions of this paper can be summarized as follows.

- We made the view updating technique based on our bidirectional language be available as a Web service.
- We proposed a novel SOAP-based XML updating server, which has highly modularity and extendibility for general use of XML updating.
- We designed a three tier architecture with a communication protocol to achieve bidirectional updating service.

The remainder of this paper is organized as follows. Section 2 gives a detailed explanation of the Bi-X service architecture. Section 3 describes the Bi-X bidirectional language. Section 4 explains the protocol used to achieve the Bi-X service in the three-tier architecture. Section 5 describes the implementation of the Bi-X server. Section 6 describes our application examples. Section 7 summarizes related work. Finally, Section 8 concludes with a brief summary and a look at future work.

2 Bi-X Service Architecture

2.1 Three Tiers

We start by explaining the structure of the Web service architecture behind the view updating process. Our Web service architecture consists of three tiers

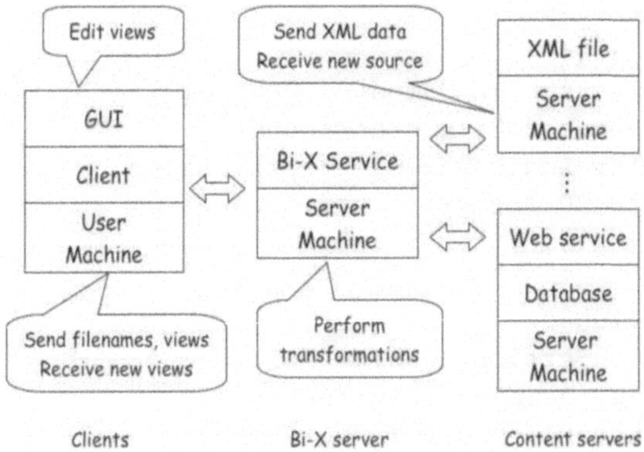

Fig. 1. Three-tier Architecture

(as shown in Figure 1): clients, a Bi-X server, and the content servers that provide XML data. The heart of this architecture is the Bi-X server, which has a bidirectional transformation engine based on an implementation of the Bi-X bidirectional transformation language. It receives a request from a client and either applies a forward transformation to the specified source data fetched from a content server in order to generate a view or applies a backward transformation to produce updated source data.

A client can be any XML data viewer. It is typically a Web application that will display transformed data in a formatted view that supports editing. It receives information from users, such as parameters specifying Bi-X code or specifying XML data to be fetched and edited, and then sends an appropriate request message to the Bi-X server.

A content server provides XML data and a Bi-X code for transformation. Also it must be able to accept modified files and update the corresponding XML data. For example, if the file the server sends to the Bi-X server is simply an XML file on a machine on the Internet, when the modified file comes back, the existing file is simply replaced with the modified one. If the file is obtained from a Web service by querying some XML database, the user must guarantee that the modified data in the file can be put back into the XML database, for instance, by preparing some special query for updating the database.

2.2 A Simple Example

A simple example will illustrate how the view updating with our service architecture works. Suppose the following XML data (bibliography information) can be accessed through the Internet by entering its uniform resource indicator (URI).

```
<bib>
    <book year="2005">
        <title>The Art of Computer Programming</title>
        <author>Donald E. Knuth</author>
        <publisher>Addison-Wesley</publisher>
        <price>19.99</price>
    </book>
     . . .
    <book year="1989">
        <title>Connection Machine</title>
        <author>W. Danny Hillis</author>
        <publisher>MIT Press</publisher>
        <price>25</price>
    </book>
</bib>
```

Also suppose we want to update only titles and authors. We access the source XML data by sending the appropriate URI to the Bi-X server, and we instruct the Bi-X server to transform the source XML data into an XHTML view showing only the titles and authors as an ordered list on the XHTML editor on the local machine by calling the *init* service provided by the server. The Bi-X code for this will be given in Section 3. The result is the following XHTML document.

```
<html>
  <body>
    <ol>
      <li>
        <div>The Art of Computer Programming</div>
        <div>Donald E. Knuth</div>
      </li>
       ...
      <li>
        <div>Connection Machine</div>
        <div>W. Danny Hillis</div>
      </li>
    </ol>
  </body>
</html>
```

From this view, we can modify the titles and/or authors as well as insert and/or delete items. We then have any modifications reflected back into the source data by invoking the *Update* service provided by the server. For example, suppose we change the title of the first book from "The Art of Computer Programming" to "The Art of Computer Programming, Volume 1" and insert a new item that includes the title and author of another book whose title is "The Art of Computer Programming, Volume 2" after the first item. After the update service has run, the corresponding title in the source data will be the new title,

X ::= BX | XC | CM
BX ::= `<xid>`[] | `<xconst>`[S] | `<xchild>`[]
XC ::= `<xseq>`[$X_1, ..., X_n$] | `<xchcont>`[$X_1, ..., X_n$]
| `<xmap>`[X] | `<xif>`[P, X_1, X_2]
CM ::= `<xstore>`[Var] | `<xload>`[Var]
| `<xfree>`[Var]
P ::= `<xwithtag>`[str] | X

Fig. 2. Syntax of Underlying Language

and the new book information will appear after that of the first one in the source data.

3 Bi-X: A Bidirectional Transformation Language

The current widely used XML transformation languages, such as XSLT [18] and XQuery [17], perform transformation in only one direction. In our XML transformation language, the code for a forward transformation is similar to that in XSLT and XQuery, and the code for the backward transformation is automatically derived. The derived code takes the modified target view and the original source as inputs and generates the updated source as output, propagating the modifications made in the view. Here we briefly summarize the basics of the language and discuss the properties of bidirectional transformations. A more detailed description of the Bi-X language can be found elsewhere [13].

3.1 Bi-X Syntax and Basic Transformations

The syntax for a fragment of the language underlying Bi-X is shown in Figure 2. Each language construct is an XML element with the end tags omitted. The contents are enclosed by brackets to save space. For example, <a>[b] represents `<a>b</a>`.

The basic transformations BX perform particular transformations on source data. **<xid>**[] transforms the source data into the same target data. **<xconst>**[S] transforms any source data into constant target data S. **<xchild>**[] accepts an element as source data and returns its contents.

Transformation combinators, XC, are used to build more complex transformations by combining simpler transformations. **<xseq>**[$X_1, ..., X_n$] is a composed transformation that applies its argument transformations, $X_i (1 \leq i \leq n)$, in sequence, and the target data of the transformation, X_i, is used as the source data for the next transformation, X_{i+1}. **<xchcont>**[$X_1, ..., X_n$] accepts an element as source data and returns it with its contents replaced by the result of applying transformations $X_i (1 \leq i \leq n)$ to empty values. **<xmap>**[X] transforms the sequence source data by applying X to each item in the sequence. **<xif>**[P, X_1, X_2] applies X_1 to the source data if the predicate P holds over this source data, otherwise X_2 is applied.

```
<xseq>[
 <xstore>[$src],
 <xconst>[<html>[]],
 <xchcont>[
   <xseq>[
     <xconst>[<body>[]],
       <xchcont>[
       <xseq>[
         <xconst>[<ol>[]],
         <xchcont>[
           <xseq>[
             <xload>[$src],
             <xchild>[],
             <xmap>[
               <xseq>[
                 <xstore>[$var],
                 <xconst>[<li>[]],
                 <xchcont>[code-a code-b],
                 <xfree>[$var],
               ]]]]]]]],
 <xfree>[$src]
]
```

Fig. 3. Bi-X Program Example

```
<xseq>[
  <xconst>[<div>[]],
  <xchcont>[
    <xseq>[
      <xload>[$var],
      <xchild>[],
      <xmap>[
        <xif>[
          <xwithtag>[title],
          <xchild>[],
          <xconst>[],
        ]
      ]
    ]
  ]
]
```

Fig. 4. *code-a*

The transformations, *CM*, are used to manage or use the transformation context. They provide a variable binding mechanism for the Bi-X language. <**xstore**>[*Var*] binds the source data to the variable *Var*, which is valid until it is released by <**xfree**>[*Var*]. <**xload**>[*Var*] accesses the bound value of a valid variable. The predicate <**xwithtag**>[*str*] holds if the source data is an element with tag *str*, and any transformation can be used as a predicate for <**xif**>$[P, X_1, X_2]$.

Using the Bi-X syntax, the Bi-X code needed to perform the transformation for the example given in section 2 is shown in Figure 3 and Figure 4. The code is divided into two parts for readability. *code-a* in Figure 3 represents the code given in Figure 4, which extracts the titles from the source data. *code-b* in Figure 3 represents the code for extracting the author and is not shown to save a space.

As can be seen from these figures, the code for Bi-X tends to be longer than that for the one-way XML transformation languages. An XQuery interpreter has been developed in order to reduce the coding effort [12]. Since the expressive power of Bi-X is almost the same as that of XQuery, a user can write an XQuery code for the forward transformation and automatically get the equivalent Bi-X code for the bidirectional transformation.

3.2 Bidirectional Property of Bi-X

In this section, the view updating property of the Bi-X language is illustrated informally to help users better understand the results of backward transformation.

That is, given an updated view, what should the updated source document look like after backward transformation? To shorten the presentation, we show only the modifications needed to update the XML text contents and tags. The more complex updatings, such as insertion and deletion, are described elsewhere [12].

During a session of forward and backward transformation, there are two pairs of documents: the original source document and the source document after updating, and the original view and the updated view. Each pair of documents has the same structure since we are interested in only modifications here. The property of Bi-X is defined on the differences between the original and updated documents. The differences are represented as a multiset of pairs, and each pair consists of two different strings, which are either element tags or text contents. A pair represents a modification; that is, the first component is changed to the second one.

To represent modifications more precisely, tags and text contents in source documents are assigned unique identifiers, while tags and text contents in `xconst` are associated with a specific identifier, say `c`. Identifiers are kept unchanged while transforming source documents and modifying views. A modification is called a bad modification if it contains strings with the `c` identifier. This means data from the transformation code cannot be modified. Two string components in a good modification must have the same identifiers, and no two good modifications in one document can have the same identifier. Two modifications are said to be equal if they make the same changes to strings with the same identifiers.

We write `diff`(*od*,*md*) for the differences between the original document, *od*, and its modified version, *md*. For two documents with the same structure, their differences can be easily obtained by traversing the document structure and comparing each tag and text content. The view updating property of Bi-X is as follows. Suppose *sd* is a source document, X a Bi-X transformation, *td* a target document of *sd* transformed by X, and td' is obtained from *td* with only good modifications. After backward transformation of td' using X, the following condition holds:

$$\texttt{diff}(sd, sd') = \texttt{diff}(td, td'),$$

where sd' is the updated *sd* generated by the backward transformation. Intuitively, this property means that, after a backward transformation, the modifications on the views are reflected back to the corresponding tags or text contents in the source documents.

4 Communication Protocol

The communication protocol in the data updating process, comprises two phases: *init* and *update*. They are performed by the init and update services, respectively, provided by the Bi-X service. Between the two phases, the user edits the view on the client. The steps in each phase are illustrated in Figure 5 and described below.

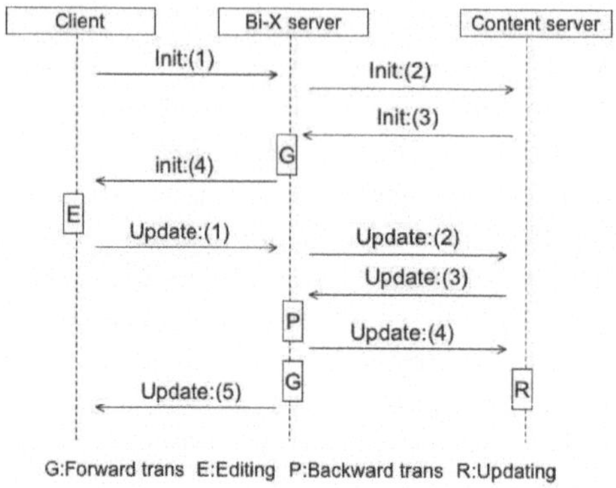

Fig. 5. Communication Patterns

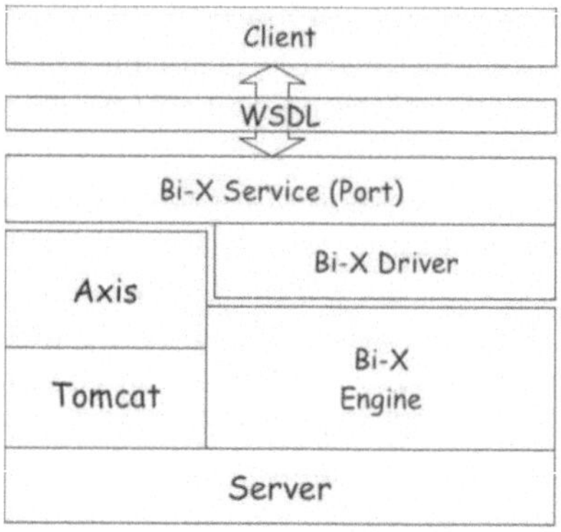

Fig. 6. Configuration of Implemented Bi-X Service

Init Phase

Init(1): Client sends init message to Bi-X server with two arguments: URI1 for source data to be transformed and URI2 for Bi-X code.

Init(2): Bi-X server requests files specified by URI1 and URI2 using HTTP Get method.

Init(3): Machines specified in URI1 and URI2 process HTTP Get method and return specified files.

Init(4): Bi-X server performs forward transformation and sends view to client.

Updating Phase

Update(1): After data is edited, client sends update message to Bi-X server with three arguments: URI1 for source data, URI2 for Bi-X code, and changed view.

Update(2): Bi-X server requests source data to be updated and code specified in URI1 and URI2 using HTTP Get method.

Update(3): Machines specified in URI1 and URI2 process HTTP Get method and return specified files.

Update(4): Bi-X server performs backward transformation to obtain updated source data and sends updated source data back to URI1 using HTTP POST method.

Update(5): Bi-X server performs forward transformation using updated source data and sends new view to client.

5 Bi-X Service Implementation

We implemented our Bi-X service in Java, using standard Web service technologies such as SOAP [15], the representational state transfer (REST) model [7], and WSDL [16]. The configuration is shown in Figure 6. Its application to a practical case is described in Section 6.

Axis and Tomcat. The Axis2 platform [2] is used to implement SOAP and the REST model. It runs on the Tomcat server engine [3]. Because the Bi-X service uses these standard technologies, its installation requires only that a Bi-X service archive file be registered to the containers. Users can thus easily install Bi-X service on their own machines.

Bi-X Driver. The Bi-X driver wraps Bi-X engine, which is a Java implementation of the Bi-X bidirectional transformation language. The driver is also written in Java. It provides the engine with the network communications used to transfer XML documents from and to the content servers. These communications use HTTP GET messages to retrieve XML documents (i.e., source and code) from content servers and HTTP POST messages to place modified XML documents (i.e., new source) on content servers.

Bi-X Service Port and WSDL. The Bi-X service port and WSDL enable users to use such methods as *init* and *update* through the Internet. The types of these methods and the data structures of their arguments are provided by WSDL. Users can easily construct SOAP clients for these methods by putting the WSDL to an automatic program generator such as WSDL2Java of Axis. Moreover, users can use REST interfaces for these methods thanks to the power of Axis2. If they do, they need only a method to access the target URLs to use the Bi-X service.

6 Application Examples

In our architecture, the client and the content servers simply need to satisfy the requirements given in Section 2. Here, we give an example of the client and content server, by which we have tested several use cases. We also show the usability of our system using one test case that uses CiteSeer [5] database.

6.1 Client and Content Server Example

A Bi-X service client that calls methods provided by the server can be easily prepared using standard Web service technologies. All necessary information for this can be obtained from the WSDL description of the Bi-X service. For example, a client program can be created by using the WSDL2java tool included in Axis. It generates client stub code for SOAP communication from the WSDL description. The client simply uses the code to invoke a Web service as if it was a regular Java object in the same address space.

As the interface for our client, we use Justsystem xfy [10], which is an "integrated XML application development environment" developed by Justsystem Corporation. An advantage of using xfy in our testing is its ability to handle various kinds of XML vocabularies in an optimized and sophisticated manner. For example, texts in XHTML vocabulary are directly editable in the xfy browser. We incorporated our client program with xfy so that it would work as an xfy plug-in. We create request messages on the xfy interface and sent them to the Bi-X server. The results from the server are displayed in the xfy browser. In the current update implementation, the entire document of the changed view

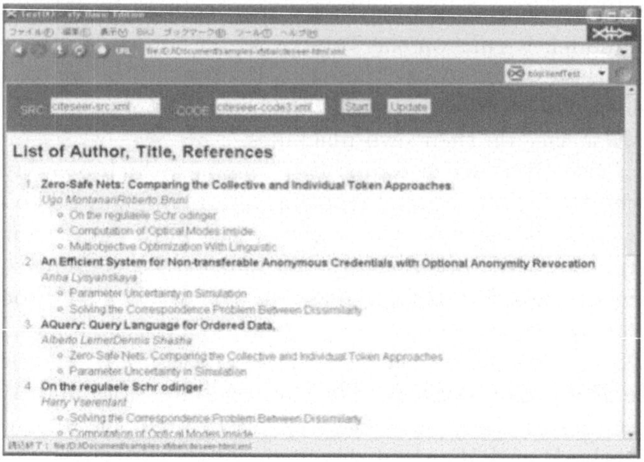

Fig. 7. CiteSeer View on xfy

is sent to the Bi-X server, and its well-formedness is checked in the client. The validity over a schema is checked in the Bi-X server when the URI of the schema definition file is given.

There are two basic requirements for a content server: be able to provide XML files and be able to accept modified files. For example, we can use the eXist XML DB [14] to provide source data. In this case, when receiving a request for source data, the content server extracts the source data from the DB with XQuery and sends them to the *transformation engine*. When the updated source data is returned, it updates the DB accordingly by executing *updating queries* prepared by the user. The XQuery in eXist extends the standard XQuery with some update statements that can be used to create *updating queries*.

6.2 CiteSeer

CiteSeer is a scientific literature digital library and search engine that focuses primarily on the literature in computer and information science. It crawls the Web and harvests academic and scientific documents. It uses autonomous citation indexing to permit querying by citation. The CiteSeer Web site has pages for correcting the information for a given document (title, abstract, summary, author(s), etc.). Any user can submit a correction through a form-based Web interface by editing the contents and submitting them. This kind of application is thus well suited to our view updating system.

To test the view updating, we saved part of the original XML data taken from the CiteSeer library and perform view updating using the Bi-X server. Figure 7 shows a snapshot of the view in the xfy browser. We provide the URIs of the source XML file and the Bi-X code needed to transform it. We then press the *Start* button to invoke the init service. The XHTML view is generated by the Bi-X code and displayed in the xfy browser. The view contains the document information (title, author, and titles of cited documents) in list format. We edit the information directly in the XHTML view provided by the xfy browser. The modifications are then reflected back to the source by pressing the *Update* button, which invokes the update service. Thus, users can create a view that includes only the contents of interest in a suitable format by creating an appropriate Bi-X code, edit the contents in the view, and update the source XML data.

7 Related Work

The Bi-X language has a bidirectional transformation style similar to that of the Harmony [8] and XEditor [9], which are both domain-specific. Harmony was designed for synchronizing tree-structured data, while XEditor is mainly used for editing tree-structured data. Bi-X extends their capabilities, so it can be used for general purpose XML processing. The differences between Bi-X and these languages are discussed in detail elsewhere [13].

In the relational database area, there has been some work on bidirectional mapping between a database and XML documents. In the approach of Braganholo et al. [4], the underlying relational database tables are updated directly rather than through views. In that of Knudsen et al. [11], the updates to query tree are transformed into SQL updates, and then the traditional view updating techniques are used to update relational databases. Obviously, these approaches are not suitable for updating native XML repositories.

Many XML updating systems that use a database are closely connected to the database system, so they are not easy to re-implement to work with a different system. The Bi-X server is a generic tool for XML updating, so it can be easily connected to content servers and web applications and can be reused.

8 Conclusion

In our Web service architecture for bidirectional XML updating, users can update remote source data by editing a target view on the local machine. This view is generated by some transformation of the source data. The user can create a view that includes only contents of interest in a suitable format by creating an appropriate Bi-X code, edit the contents in the view, and update the source XML data accordingly without coding a backward transformation. Due to the use of standard Web service technologies, the data viewer client and content servers can be easily replaced with ones chosen by users to implement their own applications.

There are a number of directions for future research to make the service architecture more practical and usable. Although we considered discrete updates in this work, concurrency control that would enable many updates to be made at the same time would make it more practical. A control policy needs to be defined for allowing access to the service.

Acknowledgments

We are grateful to Justsystem Corporation for providing us with helpful technical information about xfy. This research is supported by the *Comprehensive Development of e-Society Foundation Software* program of the Ministry of Education, Culture, Sports, Science and Technology, Japan.

References

1. Abiteboul, S.: On views and XML. In Proceedings of the 18th ACM SIGACT-SIGMOD-SIGART Symposium on Principles of Database Systems. (1999) 1–9
2. Apache Software Foundation: Apache Axis2/Java. http://ws.apache.org/axis2/.
3. Apache Software Foundation: Apache Tomcat. http://tomcat.apache.org/.
4. Braganholo, V., Davidson, S., Heuser, C.: From XML view updates to relational view updates: old solutions to a new problem. In Proceedings of International Conference on Very Large Databases. (2004) 276–287
5. College of Information Sciences and Technology, The Pennsylvania State University: CiteSeer. http://citeseer.ist.psu.edu/.
6. Dayal, U., Bernstein, P. A.: On the correct translation of update operations on relational views. ACM Trans. Database Syst. **7** (1982) 381–416
7. Fielding, R. T.: Architectural styles and the design of network-based software architectures. PhD thesis, University of California. (2000)
8. Foster, J. N., Greenwald, M. B., Moore, J. T., Pierce, B. C., Schmitt, A.: Combinators for bi-directional tree transformations: a linguistic approach to the view update problem. In Proceedings of the 32nd ACM SIGPLANSIGACT Symposium on Principles of Programming Languages. (2005) 233–246
9. Hu, Z., Mu, S-C., Takeichi, M.: A programmable editor for developing structured documents based on bidirectional transformations. In Proceedings of the 2004 ACM SIGPLAN Symposium on Partial Evaluation and Semantics-based Program Manipulation. (2004)
10. Justsystem Corporation: xfy technology. http://www.xfytec.com.
11. Knudsen, S. U., Pedersen, T. B., Thomsen. C, Torp, K.: RelaXML: bidirectional transfer between relational and XML data. Proceedings of the 9th International Database Engineering and Applications Symposium. (2005) 151–162
12. Liu, D., Hu, Z., Takeichi, M.: Bidirectional interpretation of XQuery. In Proceedings of the ACM SIGPLAN 2007 Workshop on Partial Evaluation and Program Manipulation. (2007)
13. Liu, D., Hu, Z., Takeichi, M., Kakehi, K., Wang, H.: A Java library for bidirectional XML transformation. JSSST Computer Software (to appear)
14. Meier, F.: eXist: Open Source Native XML Database. http://www.existdb.org/.
15. W3C: Simple Object Access Protocol (SOAP) 1.1. http://www.w3.org/TR/soap. (2000)
16. W3C: Web Services Description Language (WSDL) 1.1. http://www.w3.org/TR/wsdl. (2001)
17. W3C Draft: XML Query (XQuery). http://www.w3.org/XML/Query. (2005)
18. W3C Draft: XSL Transformations (XSLT) Version 2.0. http://www.w3.org/TR/xslt20/. (2005)

(α, k)-anonymity Based Privacy Preservation by Lossy Join

Raymond Chi-Wing Wong[1], Yubao Liu[2], Jian Yin[2], Zhilan Huang[2], Ada Wai-Chee Fu[1], and Jian Pei[3]

[1] Department of Computer Science and Engineering, the Chinese University of Hong Kong, Hong Kong
{cwwong,adafu}@cse.cuhk.edu.hk
[2] Department of Computer Science, Zhongshan University, China
{liuyubao,issjyin}@mail.sysu.edu.cn, santahzl@gmail.com
[3] School of Computing Science, Simon Fraser University, Canada
jpei@cs.sfu.ca

Abstract. Privacy-preserving data publication for data mining is to protect sensitive information of individuals in published data while the distortion to the data is minimized. Recently, it is shown that (α, k)-anonymity is a feasible technique when we are given some sensitive attribute(s) and quasi-identifier attributes. In previous work, generalization of the given data table has been used for the anonymization. In this paper, we show that we can project the data onto two tables for publishing in such a way that the privacy protection for (α, k)-anonymity can be achieved with less distortion. In the two tables, one table contains the undisturbed non-sensitive values and the other table contains the undisturbed sensitive values. Privacy preservation is guaranteed by the lossy join property of the two tables. We show by experiments that the results are better than previous approaches.

1 Introduction

Privacy-preserving data mining is about preserving the individual privacy and retaining as much as possible the information in a dataset to be released for mining. The perturbation approach [2] and the k-anonymity model [14,13,4,1] are two major techniques for this goal.

The k-anonymity model assumes a **quasi-identifier** (QID), which is a set of attributes that may serve as an identifier in the data set. In the simplest case, it is assumed that the dataset is a table and that each tuple corresponds to an individual. For example, in Table 1, attributes Job, Birth and Postcode form a quasi-identifier, where attribute Illness is a sensitive attribute. The privacy may be violated if some quasi-identifier values are unique in the released table. The assumption is that an attacker can have the knowledge of another table where the quasi-identifier values are linked with the identities of individuals. Therefore, a join of the released table with this background table will disclose the sensitive data of individuals. A real example is found in the voter registration records in

G. Dong et al. (Eds.): APWeb/WAIM 2007, LNCS 4505, pp. 733–744, 2007.

Table 1. Raw medical data set

Job	Birth	Postcode	Illness
clerk	1975	4350	HIV
manger	1955	4350	flu
clerk	1955	5432	flu
factory worker	1955	5432	fever
factory worker	1975	4350	flu
technical supporter	1940	4350	fever

Table 2. A $(0.5, 2)$-anonymous table of Table 1 by full-domain generalization

Job	Birth	Postcode	Illness
*	*	4350	HIV
*	*	4350	flu
*	*	5432	flu
*	*	5432	fever
*	*	4350	flu
*	*	4350	fever

Table 3. An (0.5,2)-anonymous data set of Table 1 by local recoding

Job	Birth	Postcode	Illness
white-collar	*	4350	HIV
white-collar	*	4350	flu
*	1955	5432	flu
*	1955	5432	fever
blue-collar	*	4350	flu
blue-collar	*	4350	fever

the United States, where the attributes of name, gender, zip code and date of birth are recorded. It is found that a high percentage of the population can be uniquely identified by the gender, date of birth and the zip code [12].

Let Q be the quasi-identifier (QID). An **equivalence class** set, called a QID-EC, for the same QID value of a table with respect to Q is a collection of all tuples in the table containing identical values of Q. For instance, Table 2 contains two QID-EC's. The first QID-EC contains the first two and the last two tuples because these tuples contain identical values of Q. Similarly, the second QID-EC contains the third and the fourth record.

A data set D is k-*anonymous* with respect to Q if the size of every QID-EC with respect to Q is k or more. As a result, it is less likely that any tuple in the released table can be linked to an individual and thus personal privacy is preserved. For example, each QID-EC in Table 2 has a size equal to or greater than 2. If $k = 2$, the data set in Table 2 is said to be k-anonymous.

We assume that each attribute follows a generalization hierarchy. In this hierarchy, a value in a lower level has a more specific meaning compared with a value in a higher level. For instance, Figure 1 shows a generalization hierarchy of attribute job.

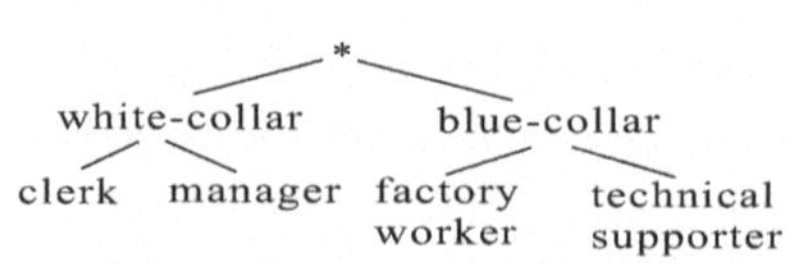

Fig. 1. Generalization hierarchy of attribute job

In order to achieve k-anonymity, we *generalize* some values in some attributes in the quasi-identifier by replacing the values in a lower level by the values in a higher level according to the generalization hierarchy. Table 2 is a generalization of Table 1.

2 (α, k)-anonymity

The k-anonymity model is proposed in order to prevent the re-identification of individuals in the released data set. However, it does not consider the *inference* relationship from the quasi-identifier to some sensitive attribute. We assume for simplicity that there is only one sensitive attribute and that some values of this attribute are sensitive values. Suppose all tuples in a QID-EC contain the same sensitive value in the released data set, even though the size of the QID-EC is greater than or equal to k, all tuples in this QID-EC are linked to this sensitive value in the released data set. Therefore, each individual that has the corresponding QID value will be linked to the sensitive value. Let us call such an attack an inference attack.

In order to overcome this attack, [9,17] proposed some privacy models and methods. [9] and [17] proposed an l-diversity model and an (α, k)-anonymity model, respectively, where α is a real number $\in [0, 1]$ and k and l are positive integers. As discussed in [17], it is difficult for users to set the parameters in the l-diversity model. In this paper, we focus on the (α, k)-anonymity model, which generates publishable data that is free from the inference attack. In addition to k-anonymity, this model requires that the value of the frequency (in fraction) of any sensitive value in any QID-EC is no more than α after anonymization.

There are two possible schemes of generalizations: *global recoding* and *local recoding*. With **global recoding** [13,8,3,11,7,16,4] all values of an attribute come from the same domain level in the hierarchy. In other words, all values come from the values in the same level in the generalization hierarchy. For example, all values in attribute job are in the lowest level (i.e. clerk, manager, factory worker and technical supporter), or all are in the top level (i.e. *). For example, a global recoding of Table 1 is Table 2. One advantage is that an anonymous view has uniform domains. But, it may lose more information, compared with local recoding (which will be discussed next), because it suffers from *over-generalization.*

Under the scheme of **local recoding** [14,13,1,10,6,5,19], values may be generalized to different levels in the domain. For example, Table 3 is a (0.5, 2)-anonymous table by local recoding. In fact, one can say that local recoding is a more general model and global recoding is a special case of local recoding. Note that, in the example, known values are replaced by unknown values (*). This is called *suppression*, which is one special case of generalization, which is in turn one of the ways of recoding.

It is easy to check that generalizing data to form QID-EC's in a released table is one possible way to achieve (α, k)-anonymity. However, it is not the only possible way, and we shall describe another method in the next section.

Table 4. Temp table

Job	Birth	Postcode	Illness	ClassID
clerk	1975	4350	HIV	1
manager	1955	4350	flu	1
clerk	1955	5432	flu	2
factory worker	1955	5432	fever	2
factory worker	1975	4350	flu	3
technical supporter	1940	4350	fever	3

Table 5. NSS Table

Job	Birth	Postcode	ClassID
clerk	1975	4350	1
manager	1955	4350	1
clerk	1955	5432	2
factory worker	1955	5432	2
factory worker	1975	4350	3
technical supporter	1940	4350	3

Table 6. SS Table

ClassID	Illness
1	HIV
1	flu
2	flu
2	fever
3	flu
3	fever

3 The Lossy Join Approach

In recent work, it has been found that lossy join of multiple tables is useful in privacy-preserving data publishing [18,15]. The idea is that if two tables with a join attribute are published, the join of the two tables can be lossy and this lossy join helps to conceal the private information.

In this paper, we make use of the idea of lossy join to derive a new mechanism for achieving a similar privacy preservation target as (α, k)-anonymization. Let us take a look at an example in Table 1. A (0.5, 2)-anonymization is given in Table 3. From this table, we can generate a table Temp as shown in Table 4. For each equivalence class E in the anonymized table, we assign a unique identifier (ID) to E and also to all tuples in E. Then, we attach the correspondence ID to each tuple in the original raw table and form a new table Temp. From the Temp table, we can generate two separate tables, Tables 5 and 6. The two tables share the attribute of ClassID. If we join these two tables by the ClassID, it is easy to see that the join is lossy and it is not possible to derive the table Temp after the join. The result of joining the two tables is given in Table 7. From the lossy join, each individual is linked to at least 2 values in the sensitive attribute. Therefore, the required privacy of individual can be guaranteed. Also, in the joined table, for each individual, there are at least 2 individuals that are linked to the same bag B of sensitive values, such that in terms of the sensitive values, they are not distinguishable. For example, the first record in the raw table (QID = (clerk, 1975, 4350)) is linked to bag {HIV,flu}. We find that the second individual (QID = (manager, 1955, 4350)) is also linked to the same bag B of sensitive values. This is the goal of k-anonymity for the protection of sensitive values.

Table 7. Join result table

Job	Birth	Postcode	Illness	ClassID
clerk	1975	4350	HIV	1
manager	1955	4350	HIV	1
clerk	1975	4350	flu	1
manager	1955	4350	flu	1
clerk	1955	5432	flu	2
factory worker	1955	5432	flu	2
clerk	1955	5432	fever	2
factory worker	1955	5432	fever	2
factory worker	1975	4350	flu	3
technical supporter	1940	4350	flu	3
factory worker	1975	4350	fever	3
technical supporter	1940	4350	fever	3

3.1 Contribution

[17] proposed to generate *one generalized* table which satisfies (α, k)-anonymity. Since the table is generalized, the data in the table is distorted. In this paper, we generalize the definition of (α, k)-anonymity to allow for the generation of *two* tables instead of *one generalized* table. In this way, the privacy protection for (α, k)-anonymity can be achieved with less distortion. In the two tables, one table contains the *undisturbed* non-sensitive values and the other table contains the *undisturbed* sensitive values. The privacy preservation is by the *lossy join* property of the two tables. We show that the results are better than previous approaches [17,18] in the experiments.

The rest of the paper is organized as follows. In Section 4, we re-visit (α, k)-anonymity and propose a genearlization model of (α, k)-anonymity. In Section 5, we describe how the lossy join can be adapted to the generalized (α, k)-anonymity model. We propose an algorithm which generates two tables satisfying (α, k)-anonymity in Section 6. A systematic performance study is reported in Section 7. The paper is concluded in Section 8.

4 Generalized (α, k)-anonymity

Let us re-examine the objectives of (α, k)-anonymity. With k-anonymity, we want to make sure that when an individual is mapped to some sensitive values, at least $k-1$ other individuals are also mapped to the same sensitive values. Let B be a bag of these sensitive values. For example, consider an individual with QID=(clerk, 1975, 4350) in Table 1. With 2-anonymity, since s/he is mapped to the first and the second tuple in Table 3, s/he is mapped to a bag $B = \{HIV, flu\}$. There is another individual with QID=(manager, 1955, 4350) in Table 1 that also is mapped to the same bag $B = \{HIV, flu\}$ in Table 3. (α, k)-anonymity further ensures that no sensitive value is sufficiently dominating in B so that an individual cannot be linked to any sensitive value in B with a high confidence. For instance, with $\alpha = 0.5$, since B contains HIV and flu, the frequency (in fraction) of each value in B is at most 0.5. Based on this observation, we generalize the definition of (α, k)-anonymity as follows:

Definition 1 (Generalized (α, k)-anonymity). *Consider a dataset D in which a set of attributes form the QID. We assume that the adversary only has the knowledge of an external table where the QID's are linked to individuals. A released data set D' generated from D satisfies generalized k-anonymity if, whenever an individual is linked to a bag B of sensitive values, at least $k - 1$ other individuals are also linked to B. In addition, if the frequency (in fraction) of any sensitive value in B is no more than α, then the released data satisfies generalized (α, k)-anonymity.*

5 Generalized (α, k)-anonymity by Lossy Join

Suppose we form an anonymized table in which some QID values are generalized. In the anonymized table, each set of tuples with the same QID values forms a QID-EC. However, instead of publishing one single table A with the generalized values, there is the possibility of separating the sensitive attribute from the non-sensitive attributes and generate two tables by projecting these two sets of attributes. Tuples in the two tables are linked if they belong to the same QID-EC in A. Hence we can publish two tables: (1) one table, called *non-sensitive table (NSS table)*, containing all the non-sensitive attributes together with QID equivalence class (QID-EC in A) IDs, and (2) the other table, called *sensitive table (SS table)*, containing the QID-EC ID and the sensitive attributes. The released tables are annotated with the remark that each tuple in each of the two published table corresponds to one record in the original single table. This is to ensure that a user will not mistakenly join the two tables and assume that the join result corresponds to the original table.

The schema of the non-sensitive table (NSS table) is shown as follows, where Class ID corresponds to QID-EC ID.

Original QID attributes	Class ID

The schema of the sensitive table (SS table) is shown as follows.

Class ID	Sensitive attribute

Let us consider the example in Table 1 again. We propose that Table 5 (NSS) and Table 6 (SS) can be published as the anonymized data.

Theorem 1. *The resulting published tables NSS and SS satisfy generalized (α, k)-anonymity.*

Proof: Given the QID information of individuals in a table T_I (which we assume that an attacker may possess) and the anonymized Table T_A (e.g. Table 3), we can "join" the two tables by matching each QID in T_A to its anonymized equivalence class and obtain a table T_{IA}. Since T_A satisfies (α, k)-anonymity, when the QID of an individual is linked to a bag B of values in the sensitive attribute, at least $k-1$ other QID's of other individuals are also linked to B. In addition, the frequency (in fraction) of any sensitive value in B is no more than α.

Now, suppose the adversary is given tables NSS and SS. Equipped with only table T_I, an adversary must join the tables NSS and SS on their common attribute in an attempt to link the QID's to the sensitive values. Let the join result be table T'_A, such as Table 7. Consider any QID-EC with class ID X. Let B_X be the bag of sensitive values that X is linked to in T_A and suppose there are a tuples in T_A belonging to X. In Table T'_A, there will be a^2 tuples generated for X. In Table T'_A, B_X becomes B'_X and each entry in B_X is duplicated a times in B'_X. In the a^2 tuples in T'_A, each original QID value in the given table T will now be linked to the bag B'_X. Besides, a individuals are involved in X, and each is linked to B'_X. The frequency of each sensitive value in B'_X is the same as that in B_X in T_{IA}. Hence, the tables NSS and SS release no more information as the table T_A in terms of the linkage of an individual to a bag B of sensitive values and in terms of the percentage of each sensitive value in B.

This shows that the privacy protection provided by the single anonymized table T_A is no stronger than that provided by the NSS and SS tables in terms of (α, k)-anonymity. Since T_A satisfies (α, k)-anonymity, tables NSS and SS also satisfy (α, k)-anonymity. ■

The example shown in Tables 3 to 7 demonstrates the ideas in the proof above. If we publish Tables 5 and 6, we can achieve similar privacy preservation objectives as if we publish Table 3 only.

6 Algorithm

Our method includes the following steps.

1. Construct an (α, k)-anonymous table $T*$ from the given raw table (which will be described in Algorithm 1), and assign each equivalence class in the resulting table a class ID.
2. Add a column for the class ID of the equivalence class in the original raw table, such that, for each tuple, the class ID is the ID of the equivalence class that the tuple belongs in $T*$. Call this new table the Temp table. Hence the Temp table contains the raw table plus one extra column.
3. Project the Temp table on the QID attributes and the Class ID column. The resulting table is the NSS table.
4. Project the Temp table on the sensitive attributes and the Class ID column. This results in the SS table.

The top-down approach has been found to be highly effective in k-anonymization [4]. In this approach, the table is first totally anonymized to the unknown values, and then attributes are specialized one at a time until we hit a point where the resulting table violates (α, k)-anonymity. We shall adopt

Algorithm 1. Top-Down Approach for Single Attribute

fully generalize all tuples such that all tuples are equal
let P be a set containing all these generalized tuples
$S \leftarrow \{P\}$; $O \leftarrow \emptyset$
repeat
  $S' \leftarrow \emptyset$
  for all $P \in S$ **do**
    specialize all tuples in P one level down in the generalization hierarchy such that a number of specialized child nodes are formed
    unspecialize the nodes which do not satisfy (α, k)-anonymity by moving the tuples back to the parent node
    if the parent P does not satisfy (α, k)-anonymity **then**
      unspecialize some tuples in the remaining child nodes so that the parent P satisfies (α, k)-anonymity
    for all non-empty branches B of P, **do** $S' \leftarrow S' \cup \{B\}$
    $S \leftarrow S'$
    if P is non-empty **then** $O \leftarrow O \cup \{P\}$
until $S = \emptyset$
return O

the top-down approach in [17] to tackle the first step of (α, k)-anonymization in the above. The idea of the algorithm is to first generalize all tuples *completely* so that, initially, all tuples are generalized to one equivalence class. Then, some values in the dataset are *specialized* in iterations. During the specialization, we must maintain (α, k)-anonymity. The process continues until we cannot specialize the tuples anymore without violating (α, k)-anonymity. The pseudo-code of the top-down approach is shown in Algorithm 1.

7 Experimental Results

The system platform we used is: Windows XP OS, Microsoft SQL Server 2000, Intel Celeron CPU 2.66GHz, 256MB Memory, 80G Hard disk. We implemented our proposed algorithm, the (α, k)-anonymity based privacy preservation by lossy join, in C/C++ language. Let us denote it by $Alpha(Lossy)$. We compared the proposed lossy-join algorithm with two algorithms in the literature. One is the original algorithm of (α, k)-anonymity [17] which generalizes the QID and forms one generalized table only. Let us denote the algorithm by *Alpha*. The other algorithm is the anatomy algorithm which makes use of the lossy join for the anonymization [18]. Let us denote the algorithm by *Anatomy*. Anatomy also generates two tables with a similar strategy of separating the sensitive data and the QID data. However, the goal of Anatomy is to create QID-EC's which satisfy the l-diversity requirement, without precaution in creating QID-EC's that also minimizes the effective distortion to the QID values. In other words, Anatomy does not consider the minimization of the variations in the QID values in each QID-EC when two tables are released. $Alpha(Lossy)$ takes care of this issue by the top-down anonymization algorithm and therefore results in less data distortion.

Table 8. Description of Adult Data Set

	Attribute	Distinct Values	Generalizations	Height
1	Age	74	5-, 10-, 20-year ranges	4
2	Work Class	7	Taxonomy Tree	3
3	Education	16	Taxonomy Tree	4
4	Martial Status	7	Taxonomy Tree	3
5	Race	5	Taxonomy Tree	2
6	Sex	2	Suppression	1
7	Native Country	41	Taxonomy Tree	3
8	Salary Class	2	Suppression	1
9	Occupation	14	Taxonomy Tree	2

The source code of this algorithm can be obtained from the author's website http://www.cs.cityu.edu.hk/~taoyf/paper/vldb06.html. In our experiments, we make some modifications on the ST files generated by the original anatomy algorithm such that ST table can be loaded into the Microsoft SQL Server 2000.

Similar to [4,8,17], we adopted the adult data set for the experiment, which can be downloaded in the UCIrvine Machine Learning Repository (http://www.ics.uci.edu/~mlearn/MLRepository.html). We eliminated the records with unknown values in this data set. The resulting data set contains 45,222 tuples. Nine of the attributes were chosen in our experiments, as shown in Table 8. By default, we set $k = 2$ and $\alpha = 0.33$. In Table 8, we set the first eight attributes and the last attribute as the quasi-identifer and the sensitive attribute, respectively.

We compare the algorithms in terms of effectiveness for aggregate queries. Similar to [18], the effectiveness of aggregate query is defined to be its average relative error in answering a query of the following form.

```
SELECT COUNT(*)
FROM Unknown-Microdata
WHERE pred(A_1^qi) AND ... AND pred(A_qd^qi) AND pred(A^s)
```

In the above query, `Unknown-Microdata` is an original data set or an anonymized data set. qd denotes the number of QID attributes to be queried and A^s denotes the sensitive attribute. For any attribute A, the predicate $pred(A)$ has the form

$$(A = x_1 \text{ OR } A = x_2 \text{ OR } \dots \text{ OR } A = x_b)$$

where x_i is a random value in the domain of A, for $1 \leq i \leq b$. The value of b depends on the *expected query selectivity* s

$$b = \lceil |A| \cdot s^{1/(qd+1)} \rceil$$

where $|A|$ is the domain size of A. If the value of s is set higher, the selection conditions in $pred(A)$ will be more.

We compare the anonymized tables generated by different algorithms in terms of *average relative error*, which is defined as follows. We perform the aggregate query with the original data set, called *Original*. That is,

SELECT COUNT(*)
FROM Original
WHERE $pred(A_1^{qi})$ AND ... AND $pred(A_{qd}^{qi})$ AND $pred(A^s)$

Let us call the count obtained above *act*. We execute the aggregate query with the anonymized data set as follows. As algorithm *Alpha(Lossy)* and algorithm *Anatomy* generates two tables, namely NSS and SS, we perform the query as follows.

SELECT COUNT(*)
FROM SS
WHERE SS.ClassID in (SELECT NSS.ClassID FROM NSS
WHERE $pred(A_1^{qi})$ AND ... AND $pred(A_{qd}^{qi})$) AND $pred(A^s)$)

Let us call the count obtained above *est*. As algorithm *Alpha* generates one anonymized table, we perform the first query by replacing `Unknown-Microdata` with the anonymized or generalized data. Then, we define the relative error to be $|act - est|/act$, where *act* is its actual count derived from the original, and *est* the estimated count computed from the anonymized table. In our experiments, we compare all algorithms by varying the following factors: (1) the number of QID-attributes d; (2) query dimensionality qd; (3) selectivity s and (4) dataset cardinality n.

For each setting, we performed 1000 queries on the anonymized tables and then reported the average query accuracy. By default, we set $qd = 4, s = 0.05$ and $n = 45222$. As we adopt the first eight attributes in Table 8 as the quasi-identifier, the default value of d is 8.

We study the effect of the number of QI-attributes as shown in Figure 2. The average relative error remains unchanged. Also, algorithm *Alpha(Lossy)* gives a lower average relative error compared with algorithm *Anatomy* and algorithm *Alpha*. This is because algorithm *Alpha(Lossy)* considers the minimization step of the distortion for the anonymization but algorithm *Anatomy* does not. Also, algorithm *Alpha(Lossy)* does not generalize the table but algorithm *Alpha* generalize the table, which makes the average relative error higher. On average, algorithm *Anatomy* gives lower average relative error compared with algorithm *Alpha*. The reason is similar. Algorithm *Alpha* generalizes the table, which distort the data much, but algorithm *Anatomy* does not.

We also studied the effect of query dimensionality qd as shown in Figure 3. Similarly, even though the average relative error of algorithm *Alpha(Lossy)* is smaller than that of algorithm *Anatomy* and algorithm *Alpha*, qd had little effect on the average relative error.

We also varied the selectivity s as shown in Figure 4 and found that the average relative error of all algorithms decreases when s increases. This is because, when s is larger, each attribute in the aggregate query involves more value matches. That means the actual count is larger. Note that the actual count is the denominator of the average relative error. Besides, if the generalized values in the anonymized

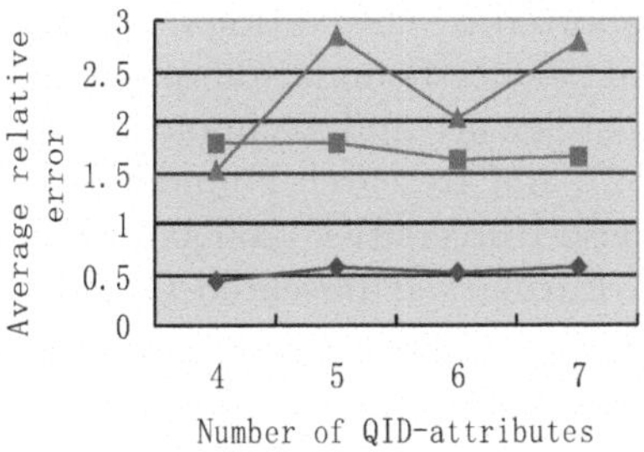

Fig. 2. Query accuracy vs. the number of QID-attributes d

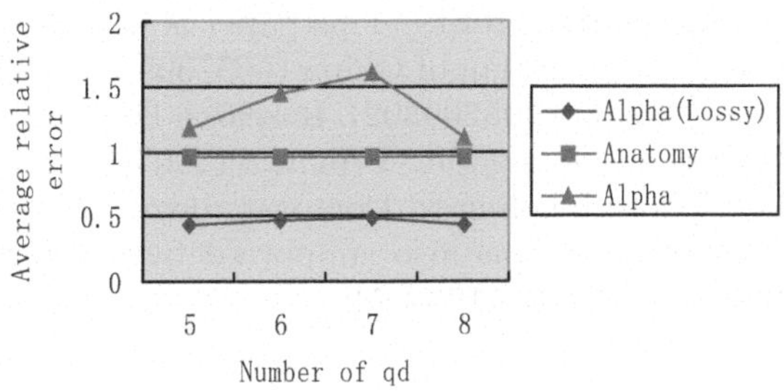

Fig. 3. Query accuracy vs. query dimensionality qd

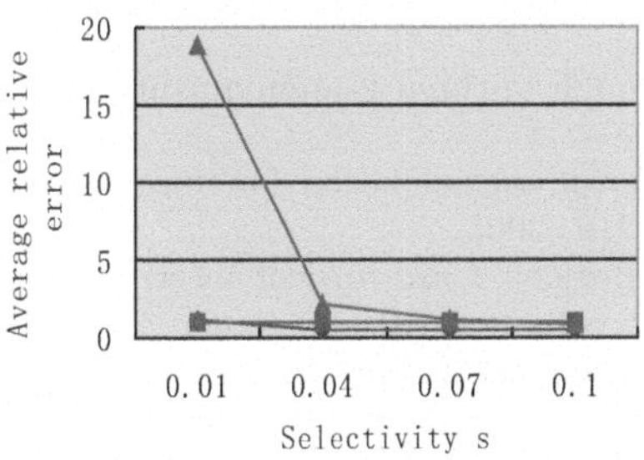

Fig. 4. Query accuracy vs. selectivity s

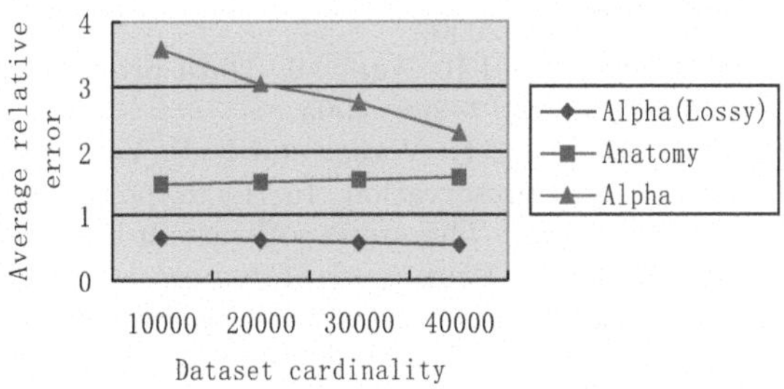

Fig. 5. Query accuracy vs. dataset cardinality

table match more aggregate values in the query, the estimated count will be more accurate. Thus, the overall average relative error decreases when s increases.

Figure 5 shows the average relative error against the data set cardinality n. We found that the average relative error of all algorithms decreases slightly when n increases. This is because, when n is larger, there is more chance that a tuple can be matched with an existing tuple in the data without much generalization. Similarly, algorithm *Alpha(Lossy)* gives a lower average relative error compared with algorithm *Anatomy* and algorithm *Alpha*.

8 Conclusion

In this paper, we proposed an (α, k)-anonymity based privacy preservation mechanism that reduce information loss by the use of lossy join. Instead of one generalized table, we generate two tables with a sharing attribute called ClassID, which corresponds to a unique identifier of an "equivalence class". One table contains the detailed information of the quasi-identifier and ClassID, and the other table contains ClassID and the sensitive attribute. By avoiding the generalization of the quasi-identifier in the first table, we achieve less information loss. We conducted some experiments and verified the improvement on information loss.

Acknowledgements: This paper is in part supported by the National Natural Science Foundation of China (60573097), Natural Science Foundation of Guangdong Province (05200302), Research Foundation of Science and Technology Plan Project in Guangdong Province (2005B10101032), and Research Foundation of Disciplines Leading to Doctorate degree of Chinese Universities (20050558017). This research was also supported by the RGC Earmarked Research Grant of HKSAR CUHK 4120/05E.

References

1. G. Aggarwal, T. Feder, K. Kenthapadi, R. Motwani, R. Panigrahy, D. Thomas, and A. Zhu. Anonymizing tables. In *ICDT*, pages 246–258, 2005.
2. R. Agrawal and R. Srikant. Privacy-preserving data mining. In *SIGMOD*, pages 439–450, May 2000.
3. R. Bayardo and R. Agrawal. Data privacy through optimal k-anonymization. In *ICDE*, pages 217–228, 2005.
4. B. C. M. Fung, K. Wang, and P. S. Yu. Top-down specialization for information and privacy preservation. In *ICDE*, pages 205–216, 2005.
5. A. Hundepool. The argus software in the casc-project: Casc project international workshop. In *Privacy in Statistical Databases*, volume 3050 of *Lecture Notes in Computer Science*, pages 323–335, Barcelona, Spain, 2004. Springer.
6. A. Hundepool and L. Willenborg. μ-and τ- argus: software for statistical disclosure control. In *Third international seminar on statsitcal confidentiality*, Bled, 1996.
7. V. S. Iyengar. Transforming data to satisfy privacy constraints. In *SIGKDD*, pages 279–288, 2002.
8. K. LeFevre, D. J. DeWitt, and R. Ramakrishnan. Incognito: Efficient full-domain k-anonymity. In *SIGMOD*, pages 49–60, 2005.
9. A. Machanavajjhala, J. Gehrke, and D. Kifer. l-diversity: privacy beyond k-anonymity. In *ICDE*, 2006.
10. A. Meyerson and R. Williams. On the complexity of optimal k-anonymity. In *PODS*, pages 223–228, 2004.
11. P. Samarati. Protecting respondents' identities in microdata release. *IEEE Transactions on Knowledge and Data Engineering*, 13(6):1010–1027, 2001.
12. L. Sweeney. Uniqueness of simple demographics in the u.s. population. *Technical Report, Carnegie Mellon University*, 2000.
13. L. Sweeney. Achieving k-anonymity privacy protection using generalization and suppression. *International journal on uncertainty, Fuzziness and knowldege based systems*, 10(5):571 – 588, 2002.
14. L. Sweeney. k-anonymity: a model for protecting privacy. *International journal on uncertainty, Fuzziness and knowldege based systems*, 10(5):557 – 570, 2002.
15. K. Wang and B. Fung. Anonymizing sequential releases. In *SIGKDD*, 2006.
16. K. Wang, P. S. Yu, and S. Chakraborty. Bottom-up generalization: A data mining solution to privacy protection. In *ICDM*, pages 249–256, 2004.
17. R. Wong, J. Li, A. Fu, and K. Wang. (alpha, k)-anonymity: An enhanced k-anonymity model for privacy-preserving data publishing. In *SIGKDD*, 2006.
18. X. Xiao and Y. Tao. Anatomy: Simple and effective privacy preservation. In *VLDB*, 2006.
19. J. Xu, W. Wang, J. Pei, X. Wang, B. Shi, and A. W.-C. Fu. Utility-based anonymization using local recoding. In *SIGKDD*, 2006.

Achieving k-Anonymity Via a Density-Based Clustering Method

Hua Zhu and Xiaojun Ye

School of Software, Tsinghua University,
Beijing, 100084, P. R. China
wuya1982513@vip.sina.com, yexj@tsinghua.edu.cn

Abstract. The key idea of our k-anonymity is to cluster the personal data based on the density which is measured by the k-Nearest-Neighbor (KNN) distance. We add a constraint that each cluster contains at least k records which is not the same as the traditional clustering methods, and provide an algorithm to come up with such a clustering. We also develop more appropriate metrics to measure the distance and information loss, which is suitable in both numeric and categorical attributes. Experiment results show that our algorithm causes significantly less information loss than previous proposed clustering algorithms.

1 Introduction

Society is experiencing exponential growth in the number and variety of data collections containing person-specific information as computer technology, network connectivity and disk storage space become increasingly affordable[9]. Many data holders publish their microdata for different purposes. However, they have difficulties in releasing information which does not compromise privacy. The difficulty is that data quality and data privacy conflict with each other.

Recently, a new approach of protecting data privacy called k-anonymity[8] has gained popularity. In a k-anonymized dataset, quasi-identifier attributes that leak information are suppressed or generalized so that, each record is indistinguishable from at least (k-1) other records with respect to the quasi-identifier. Since the k-anonymity is simple and practical, so a number of algorithms have been proposed[5][6].

The objective of this paper is to develop a new approach to achieve k-anonymity, where quasi-identifier attribute values are clustered and then published with these clusters. We view the k-anonymity problem as a clustering issue, and we add a constraint that each cluster contains at least k records, so that it satisfies k-anonymity requirements. The key idea is to cluster records based on density which is measured by the k-Nearest-Neighbor distance. We develop an algorithm to come up with such a clustering. To measure the information loss, we give some data quality metrics which are suitable in both numeric and categorical attributes.

G. Dong et al. (Eds.): APWeb/WAIM 2007, LNCS 4505, pp. 745–752, 2007.

2 Basic Concepts

The process of k-anonymization is to delete all the direct identifiers firstly, then generalize/suppress the quasi-identifiers by which most individuals may identified[8], and finally release the modified dataset which satisfies the k-anonymity constraint. For example, Table 1(left) is a raw mircodata of hospital and Table 1(right) is a 2-anonymity view of (left).

Table 1. Table of health data. Left: a raw table. Right: a 2-anonymity view.

Zip	Gender	Age	Disease
43520	Male	22	Cancer
43522	Male	25	Flu
43518	Male	23	Cancer
43533	Female	21	Obesity
43567	Female	30	Coryza
43562	Female	27	Flu

Zip	Gender	Age	Disease
4352*	Male	[21,25]	Cancer
4352*	Male	[21,25]	Flu
435**	Person	[21,25]	Cancer
435**	Person	[21,25]	Obesity
4356*	Female	[26,30]	Coryza
4356*	Female	[26,30]	Flu

Definition 1 (Quasi-Identifier). *A quasi-identifier is a minimal set Q of attributes in table T that can be joined with external information to re-identify individual records(with sufficiently high probability)[8].*

Definition 2 (Equivalence Class). *An equivalence class of a table with respect to the quasi-identifier is the set of all records in the table containing identical values for the quasi-identifier attributes.*

For example in Table 1, the attribute set {*Zip, Gender, Age*} is the quasi-identifier. Record 1 and 2 form an equivalence class in Table1(b), with respect to quasi-identifier {*ZIP, Gender, Age*} and their corresponding item values are identical.

Definition 3 (k-Anonymity). *Table T is said to satisfy k-anonymity if and only if each set of values in Q appears at least k times in T[8].*

For example, Table 1(b) is a 2-anonymity view of Table 1(a) since the minimum size of all equivalence classes is great than 2. So it can ensure that even though an intruder knows a particular individual is in the k-anonymous table T, he can not infer which record in T corresponds to the individual with a probability greater than $1/k$.

Clustering techniques used in k-anonymity issue do not require the number of clusters; instead, they need to satisfy a constraint that each cluster contains at least k records[1][3]. We define k-anonymity clustering issue as follow:

Definition 4 (k-Anonymity Clustering Issue). *The k-anonymity clustering issue is to cluster n points into a set of clusters with an information loss metric, such that each cluster contains at least k ($k \leq n$) data points and that the sum of information loss for all clusters is minimized.*

3 Distance and Information Loss Metrics

The distance metrics measure the dissimilarities among data points and minimizing the information loss for published microdata is the objective of anonymization issue. Distance metrics should handle records that consist of both numeric and categorical attributes. The earlier works[5][6] described generalizations for a categorical attribute by a taxonomy tree. Consider some sample in Table 2 and a taxonomy tree of attribute *workclass* in Fig.1. The leaf nodes depict all the distinct value of attribute *workclass*. These leaf nodes can be generalized at next level into *self-employed*, *government*, and *unemployed*. The level of a leaf node is 0 and the level of a root node is h_w, based on the notion tree height, [3] gives a distance definition between two categorical values.

Table 2. some sample patient records of a hospital

Age	Workclass	Disease
37	Self-emp-inc	Cancer
22	Self-emp-not-inc	Flu
31	Federal government	Cancer
21	State government	Obesity
54	Local government	Coryza
43	Private	Flu
25	Without pay	Flu
18	Never worked	Cancer

The priority of generalization should be considered such that the generalization near to the root should give greater information loss compared with the generalization far from the root[7]. Thus we reformulate the level weight scheme based on [3]. We define the weight distance between two categorical values as follow:

Definition 5 (Weight Distance Between Two Categorical Values). *Let C be a categorical attribute, and let h_w be the height of weight taxonomy tree of C. $w_{i,i+1}(0 \le i < h_w)$ is the weight from level i to level i+1. The weight distance between two values $v_i, v_j \in C$ is defined as:*

$$distCW(v_1, v_2) = \frac{\sum_{i=0}^{l_{12}-1} w_{i,i+1}}{\sum_{j=0}^{h_w-1} w_{j,j+1}} \tag{1}$$

where l_{12} is the level of the closet common ancestor of v_1 and v_2.

For example, the weight distance in Fig.1 between of *Federal* and *Local* is $1/(1+2) = 0.33$, while the distance between *Inc* and *Without pay* is $(1+2)/(1+2) = 1$.

Generalizing a numeric attribute (such as age in Table 2) is done by discretizing values into a set of disjoint intervals. How to choose possible end points

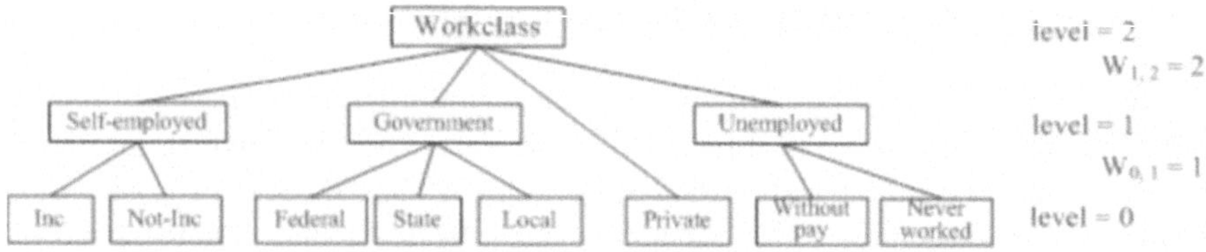

Fig. 1. A Taxonomy Tree of Attribute *workclass*

determines the granularity of the intervals. Intuitively, the difference between two numeric values indeed represents their distances on the k-anonymity clustering problem. We define the distance between two numeric values as follow:

Definition 6 (Distance between Two Numeric Values). *Let N be a finite numeric attribute domain. The distances between two numeric values v_1,v_2 is defined as:[3]*

$$distN(v_1, v_2) = \frac{|v_1 - v_2|}{|N_i|} \tag{2}$$

where $|N_i|$ is the size of numeric attribute $|N_i|$.

For example, we consider the *Age* attribute in Table 2. The distance between the first two records in the *Age* attribute is $|37 - 22|/|54 - 18| = 0.42$.

Definition 7 (Distance between Two Records). *Let $C_1, C_2, \ldots, C_m, N_1, N_2, \ldots, N_n$ be the quasi-identifier attributes in table T. $C_i (i = 1 \ldots m)$ is the categorical attribute and $N_j (j = 1 \ldots n)$ is the numeric attribute. The distance between two records is defined as:*

$$distance(r_1, r_2) = \sum_{i=1}^{m} distCW(r_1[C_i], r_2[C_i]) + \sum_{j=1}^{n} distN(r_1[N_j], r_2[N_j]) \tag{3}$$

For example, the distance between the first two records from Table 2 is $1/3 + 0.25 = 0.58$.

Based on the above distance definition between records, information loss for the anonymous table can be defined as follow:

Definition 8 (Information Loss). *Let $C_1, C_2, \ldots, C_m, N_1, N_2, \ldots, N_n$ be the quasi-identifier attributes. Let c be a cluster. We define information loss as follow:*

$$ilC_i = \sum_{i=0}^{level(v_{all})-1} w_{i,i+1} \tag{4}$$

$$ilN_j = \frac{|v_{max} - v_{min}|}{|N_j|} \tag{5}$$

$$IL(c) = |c|(\sum_{i=1}^{m} ilC_i + \sum_{j=1}^{n} ilN_j) \tag{6}$$

where ilC_i is the information loss for categorical attribute C_i and ilN_j is the information loss for numeric attribute N_j. v_{all} is the value of the closest common ancestor of all values in attribute C_i. v_{max} is the maximal value in N_i and v_{min} is the minimal value in N_i. $|N_i|$ represents the size in N_i. $IL(c)$ is the information loss of cluster c.

Thus, the total information loss of all clusters for the released microdata is:

$$TotalIL(R) = \sum_{c \in R} IL(c) \tag{7}$$

where R is a set of clusters.

4 k-Anonymity Clustering Algorithm

The choice of cluster center points can be based on the distribution density of data points. We pick a record whose density is the maximal and make it as the center of a cluster center *c*. Then we choose *k-1* records to c that make the information loss minimal. We note that there are two important issues in the algorithm:

1. The effect of clustering. We introduce a density metric called k-nearest-neighbor distance which is defined as follow:

Definition 9 (k-Nearest-Neighbor Distance). *Let R be a set of records and r be a record in R. Let $distK(i)(0 < i \leq k)$ be the minimal k values in all $distance(r, r_j)(0 < j \leq |R|)$. Then we define k-nearest-neighbor distance of r as follow:*

$$distKNN(r) = \frac{\sum_{i=1}^{k} distK(i)}{k} \tag{8}$$

where $|R|$ represents the size of R.

Definition 10 (Density). *Let $distKNN(r)$ be the k-nearest-neighbor distance of record r, we define the density of r as follow:*

$$dens(r) = \frac{1}{distKNN(r)} \tag{9}$$

The larger the density of r is, the smaller the distances between r and other records around it are. The record with larger density will be made as a cluster center with high probability because the cluster has a smaller information loss.

2. The process of clustering. How to choose the next cluster center is another important issue when one iteration has finished, because we consider that the next cluster center is a record which has the maximal density in remainder records. And the next cluster center is not in the k-nearest-neighbor records of this center, thus we define a principle as follow:

Definition 11 (Principle of Choosing the Next Cluster Center). *Let R be a set of records, r_c be a center of cluster c and r_cNext be the next cluster center. The $r_cNext \in \{R\text{-}c\}$ chosen must satisfy the follow two requirements at the same time:*

$$distance(r_c, r_cNext) > (distKNN(r_c) + distKNN(r_cNext)) \tag{10}$$

$$dens(r_cNext) = max\{dens(r_i), r_i \in \{R - c\}\} \tag{11}$$

So we propose an algorithm called density-based k-anonymity clustering (DBKC). We provide the pseudo code of the algorithm as follow:

Density-Based K-Anonymity Clustering (DBKC)

```
1:  compute density of each record in R and sort all records in a
decrease order according to density;
2:  choose the first records r(with the maximal density) in R and
make it as a cluster e1's center;
3:  while the size of R > k, do
4:      delete r from R;
5:      find k-1 best records in R and add them to cluster e1 and
delete them from R;
6:      find the next cluster center r in R and make it as a new
cluster e1's center;
7:  end while;
8:  while the size of R > 0, do
9:      insert each remainder record into best cluster;
10: end while;
```

In line1-2, we compute the density of each record and sort them. The density of each record is computed with Definition 10. Sorting algorithm chosen here is quick-sort()[4] because its time complexity is smallest.

Line3-7 we form one cluster whose size is k in each iteration . For one cluster center, we find k-1 best records to add them to cluster in line 5. The best record here is a record r_i in R such that $IL(e_1 \cup r_i)$ is minimal. Line 6 finds the next cluster center according to Definition 11.

After all iterations in line3-7, there are fewer than k records in R and these remainder records will be handled in line 8-10. We insert each remainder r_j into the best cluster in line 9. The best cluster here is a cluster e_1 from the set of clusters formed in line3-7 such that $IL(e_1 \cup r_j)$ is minimal.

For the sake of space, we do not provide the source codes of DBKC algorithm. We analysis the time complexity based on the source codes. Computing the density of all records in R needs $O((k + \log k + 1)n^2 \approx O(n^2)$ (when $k \ll n$); Sorting all records with quick-sort() needs $O(n \log n)$. In line 3-7, the execution times $ET = (n-1) + (n-2) + \ldots + k \approx n(n-1)/2$, thus ET is in $O(n^2)$. Line 8-9 need fewer than k passes. As a result of analysis above, the time complexity of density-based k-clustering algorithm is $O(n^2)$, when $k \ll n$.

5 Experimental Results

For experiments, we adopted the Adult dataset from the UC Irvine Machine Learning Repository[2]. Before the experiments, the Adult dataset was prepared similarly to[1][6]. Eight attributes were chosen as the quasi-identifier and two of them were treated as numeric attributes while the other were treated as categorical attributes.

We evaluate the algorithm in terms of two measurements: information loss and execution time, and compare DBKC algorithm with the k-means which was added only one constraint that each cluster contains at least k records. Fig.2 reports the results of these algorithms and shows that the total information loss in DBKC algorithm is 2.82 times lower than that in k-means algorithm for all k values on average. This result can be explained with the following reasons. First, the choice of the cluster center points in DBKC algorithm is based on the density, while the k-means algorithm used in our experiments chooses center points randomly. Secondly, DBKC algorithm chooses the closest point to the cluster in order to make information loss lowest, while k-means algorithm chooses the point to a cluster in order to make the distance between center point and it the shortest.

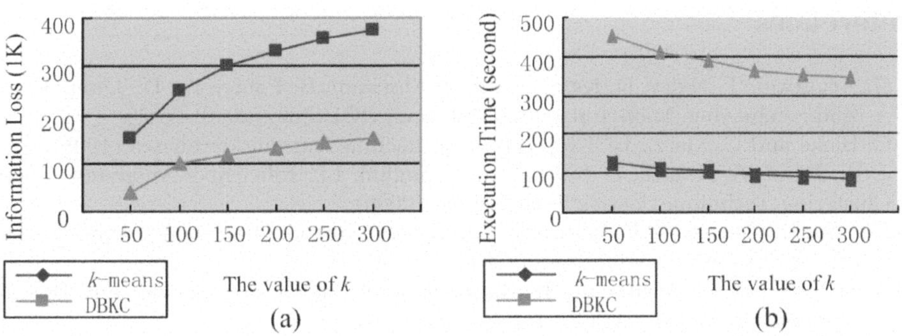

Fig. 2. Experiments result. (a): Information Loss Metric. (b): Execution Time.

As shown in Fig.2, the execution time of both algorithms decreases with the value of k. Although the execution time of the DBKC algorithm is larger than that of k-means algorithm, the time complexity of DBKC algorithm is $O(n^2)$(as discussed in Section 4)and that of k-means algorithm is also $O(n^2)$. The execution time of DBKC algorithm is acceptable in most cases considering its better performance on information loss, but it is not fully optimized and this is our future work.

The experiment result shows that the DBKC algorithm is acceptable on information loss and execution time. It is feasible to solve k-anonymity on using clustering methods based on density.

6 Conclusion

In this paper, we study the k-anonymity as a clustering problem and propose an algorithm based on density. We define the distance and information loss metrics, especially we discuss the advantage of weight distance in categorical attributes. We experimentally show that our algorithm causes significantly less information loss than traditional k-means clustering algorithm and we analyze the difference between two algorithms.

Our future work includes the following. Although the experiment result shows that DBKC algorithm has a better compromise between data quality and data privacy conflict. We believe that we can improve DBKC algorithm on time complexity. The key idea of DBKC algorithm is based on the density and we use k-nearest-neighbor distance to measure it, there may be a better density metric emergence in the future work. Because k-anonymity ensures relatively weak privacy protection, DBKC method should consider new privacy requirements such as l-diversity, personalized privacy preservation etc. in future.

Acknowledgement. This work was supported by NSFC 60673140 and NORPC 2004C B719400.

References

1. G. Aggarwal, T. Feder, K. Kenthapadi, R. Motwani, R. Panigrahy, D. Thomas, and A. Zhu.: Achieving Anonymity via Clustering. PODS'06, (2006) 26-28.
2. C. Blake and C. Merz.: UCI repository of machine learning databases (1998).
3. J.-W. Byun, A. Kamra, E. bertino, and Ninghui. Li.: Efficient k-Anonymity Using Clustering Technique. Cerias Tech Report (2006).
4. Thomas H. Cormen, C.E. Leiserson, RL. Rivest.: Introduction to algorithm, Second Edition, published by MIT press (2001).
5. K. L. Fevre, D. J. Dewitt, and R. Ramakrishnan.: Incognito: Efficient Full-Domain k-Aonymity. In SIGMOD 2005 June (2005) 14-16.
6. B.C.M. Fung, K. Wang, and P.S. Yu.: Top-down Specialization for Information and Privacy Preservation. In the twenty-first International Conference on Data Engineering (ICDE) (2005).
7. Jiuyong. Li, Raymond. Chi-Wing. W., Ada F., and Jian P.: Achieving k-Anonymity by Clustering in Attribute Hierarchical Structures. DaWaK 2006, LNCS 4081, (2006) 405-416.
8. L. Sweeney.: Achieving k-Anonymity Privacy Protection Using Generalization and Suppression. International Journal of Uncertainty, Fuzziness and Knowledge-Based System Vol. 10, No. 5 (2002) 571-588
9. K.Wang, P.S. Yu, and S. Chakraborty.: Bottom-up Generalization: A Data Mining Solution to Privacy Protection. In ICDM04: The fourth IEEE International Conference on Data Mining, (2004) 249-256.

k-Anonymization Without Q-S Associations

Weijia Yang[1] and Shangteng Huang[2]

[1] Shanghai Jiao Tong University, Shanghai 200030, China
weijia.yang@yahoo.com.cn
[2] Shanghai Jiao Tong University, Shanghai 200030, China
huang-st@cs.sjtu.edu.cn

Abstract. Privacy concerns on sensitive data are becoming indispensable in data publishing and knowledge discovering. The k-anonymization provides a way to protect the sensitivity without fabricating the data records. However, the anonymity can be breached by leveraging the associations between quasi-identifiers and sensitive attributes.

In this paper, we model the possible privacy breaches as Q-S associations using association and dissociation rules. We enhance the common k-anonymization methods by evaluating the Q-S associations. Moreover, we develop a greedy algorithm for rule hiding in order to remove all the Q-S associations in every anonymity-group. Our method can not only protect data from the privacy breaches but also minimize the data loss. We also make a comparison between our method and one of the common k-anonymization strategies.

1 Introduction

The researches on privacy preserving data mining starting from the work of [1] have been popular these years. Randomization is widely used upon the original datasets to hide sensitive values. In such a way, most of the data records have been "faked", tuples with real data can not be easily retrieved. The k-anonymization proposed in [2] provides an alternative way to preserve the sensitivity, it uses generalization to hide sensitive values while keeping the realness of data.

Most of the k-anonymization researches[2,3,4,5,6,7,8] focus on how to detach the individuals from their corresponding data records. In doing so, the individuals are hidden in groups sizing more than k. However, the frequent values in a group can break the defense setup by k-anonymization, which is first addressed in [9]. Furthermore, we find that once matching with the users' priori knowledge, the frequent patterns can lead to more serious sensitivity leakage.

For example: we derive a 5-anonymity dataset in Figure 1(b) from the original data in Figure 1(a). Statistically, users can only distinguish the right record of an individual with confidence less than 20%. But, without any priori knowledge, if Tom knows that Jennifer belongs to the first generalization group(all-female) in Figure 1(b), then he will get with 80% confidence that Jennifer's salary $\leq 50K$.

Moreover,when having some priori knowledge: Tom knows Jennifer works in a private company, and this time he ensures with 100% confidence about Jennifer's

G. Dong et al. (Eds.): APWeb/WAIM 2007, LNCS 4505, pp. 753–764, 2007.

	Quasi-identifier			Sensitive attributes		
id	age	sex	zipcode	work-class	marital-status	salary-class
T1	70	M	27000	Federal-gov	Divorced	≤50K
T2	33	F	18000	Federal-gov	Never-married	≤50K
T3	53	M	22000	State-gov	Married-civ-spouse	≤50K
T4	31	F	13000	Private	Divorced	≤50K
T5	35	F	14000	Local-gov	Widowed	>50K
T6	36	F	15000	Self-emp-not-inc	Married-civ-spouse	≤50K
T7	43	M	58000	Private	Married-civ-spouse	≤50K
T8	48	M	36000	Private	Married-civ-spouse	>50K
T9	39	F	23000	Private	Divorced	≤50K
T10	52	M	26000	Federal-gov	Widowed	>50K

(a)

	Quasi-identifier			Sensitive attributes		
id	age	sex	zipcode	work-class	marital-status	salary-class
T6	[31-39]	F	[13000-23000]	Self-emp-not-inc	Married-civ-spouse	≤50K
T2	[31-39]	F	[13000-23000]	Federal-gov	Never-married	≤50K
T9	[31-39]	F	[13000-23000]	Private	Divorced	≤50K
T4	[31-39]	F	[13000-23000]	Private	Divorced	≤50K
T5	[31-39]	F	[13000-23000]	Local-gov	Widowed	>50K
T1	[43-70]	M	[22000-58000]	Federal-gov	Divorced	≤50K
T7	[43-70]	M	[22000-58000]	Private	Married-civ-spouse	≤50K
T8	[43-70]	M	[22000-58000]	Private	Married-civ-spouse	>50K
T3	[43-70]	M	[22000-58000]	State-gov	Married-civ-spouse	≤50K
T10	[43-70]	M	[22000-58000]	Federal-gov	Widowed	>50K

(b)

Fig. 1. (a)Census data. (b)Generalized census data.

income and even her marital status. Because the rule "Private→Divorced,≤50K (100%)" exists in the group. The type of users' priori knowledge may either be negative or positive. Similarly, when Tom knows Michael is not married, he will find Michael works in federal government.

In this paper, we model those frequent values and patterns within groups using association and dissociation rules. We also try to lower them during the common anonymization process and hide them using our algorithm with minimal data loss.

This paper is organized as follows. In Section 2, we provide the works related to our topics. Some basic definitions for k-anonymization are presented in Section 3. In Section 4, we model the problem and give out our enhanced anonymization process. Section 5 is our hiding algorithm, and the experimental results are presented in Section 6. Finally, we summarize the conclusions of our study in Section 7.

2 Related Works

k-anonymization proposed in [2] has been a popular direction in protecting the sensitive information. Quite a few systems have been developed for this purpose: the μ-argus[5], the datafly[2], and the incognito[4] system,...

In the research of [3], the problem of optimal k-anonymization was proved to be NP-hard. And, various of strategies have been developed to approach this goal, such as the bottom-up generalization[8], the top-down anonymization[7], and the cell based approach[10], ...

Recently, the work of [9] first considered the problem of current k- anonymization methods: the associations between the quasi-identifier and the sensitive attributes will break the anonymity. It proposed the concept of "l-diversity" to measure such associations, and embedded the measurement into k-anonymization algorithm. However, their method is more applicable to handle tables with one sensitive attribute than more sensitive columns which is the practical condition. Tables with highly frequent attribute values are also beyond its competence.

Research [11] focused on implementing personalized anonymity requirements by generalizing both the quasi-identifier and sensitive values, by doing so, it also dissociated the associations mentioned above but only in one sensitive attribute condition.

The direction of association rule hiding has been proposed in [12,13]. And [13] is a summary of the authors' previous methods: SWA, IGA and DSA. Most of the researches developed heuristic methods to reduce the confidence or support of the sensitive rules by adding and removing rows.

3 Preliminary

Firstly, we inherit several basic definitions for k-anonymization from previous works mentioned in Section 2.

Definition 1. *(Generalization) Given a domain D consists of disjoint partitions $\{P_i\}(i = 1 \ldots n)$, and $\cup P_i = D$. On given value v, we call the generalization process as returning the only partition P_i containing v.*

Definition 2. *(Quasi-Identifier) Given a table $T(A_1, A_2, \ldots, A_n)$. If $\exists$ external table S, for $\forall$ record $t_i \in T$, by searching values of $t_i(A_j, \ldots, A_m)$ in S, t_i can be uniquely located, then we call the set of attributes $\{A_j, \ldots, A_m\}$ a quasi-identifier. ($i, j, m \leq n$, A_j is not the identifier attribute)*

Definition 3. *(k-Anonymity) Given a table $T(A_1, A_2, \ldots, A_n)$ and its quasi-identifier QI. If $\forall$ subset $C \subseteq QI$, $\forall$ record $t_i \in T$, $\exists$ at least $k-1$ other records that have the same values with t_i on attribute set C, then we call the Table T satisfies k-anonymity.*

Definition 4. *(Anonymity-Group) Given a table T and its quasi-identifier QI. We call an anonymity-group as the set of all records from T with the same values on QI.*

4 Enhanced Anonymization Process

4.1 Problem Modeling

From the example in Section 1, we notice that the users' all-positive priori knowledge has the same function as the antecedent of association rules while other inferable sensitive values as the consequent part. Similarly, the knowledge containing negative part can be represented by dissociation rules. Both types of rules obtained from sensitive data within an anonymity-group actually form the "inference paths" with respect to its quasi-identifier. These paths improve the attackers' ability to infer the sensitive values with high confidence in far excess of $1/k$.

We regard the anonymity breaking as two types: without priori knowledge and with priori knowledge. The first type can be represented as the frequent 1-itemsets within an anonymity-group. This is also the case discussed in [9], which checks the measurements of diversity and tries to make all values in each sensitive attribute evenly distributed in every anonymity-group. But it may not be feasible in some datasets with highly frequent itemsets.

We represent the second type of anonymity-breaking as association and dissociation rules with high confidence in the anonymity-group. Just as in the previous example, we have "Married-civ-spouse→Private(67%)" and "¬Married-civ-spouse → Federal-gov(100%)" in the second anonymity-group. Currently, we only deal with the dissociation rules in the form of $\neg A \rightarrow B$, more complex forms will be considered in our future work.

In doing so, we are trying to solve the problem of anonymity breaking in a different way. Our main idea is to lower the confidence of those association and dissociation rules, also the support of the frequent 1-itemsets. By developing our own rule hiding strategy, we will achieve this while generalizing the minimum number of sensitive data cells. Therefore, the inference probability can be controlled in the preset threshold, and datasets with all kinds of distributions can also be handled.

We combine the two types of anonymity breaking into our formal definition of the "quasi-identifier"-"sensitive attribute"(Q-S for short) association:

Definition 5. *(Q-S Associations) Given an anonymity-group AG, sensitive attribute set S and confidence threshold θ. Denote the 1-itemset as m, association rule as r, and dissociation rule as dr. We call Q-S associations for AG as* $\{m, r, dr | support(m) > \theta, confidence(r) > \theta, confidence(dr) > \theta \;\&\; m, r, dr \in AG(S)\}$.

We will carry out our anonymization process in two main steps:

1. Enhance the common k-anonymization process by evaluating and lowering the Q-S associations in all anonymity-groups.
2. After the anonymization, hide the Q-S associations in each k-anonymity group by sensitive value generalization.

In the first step, we evaluate the change of the Q-S associations brought by the candidate generalizations in each iteration. Combining with the measurements of anonymity and data loss, we use all of them to choose the best generalization in each iteration.

As for the rule discovering, there are works about it. We discover the rules in the way similar with the work of [14]. We treat the anonymity-groups as "partitions"[14], looking for rules in every group, and then form the "global" rules based on the local ones.

4.2 Data Structure

Each anonymity-group setups the structure "tree of inverted file". This structure together with the attached records id(outlined with dotted boundary in Figure 2) is indispensable in the "Q-S association hiding" step.

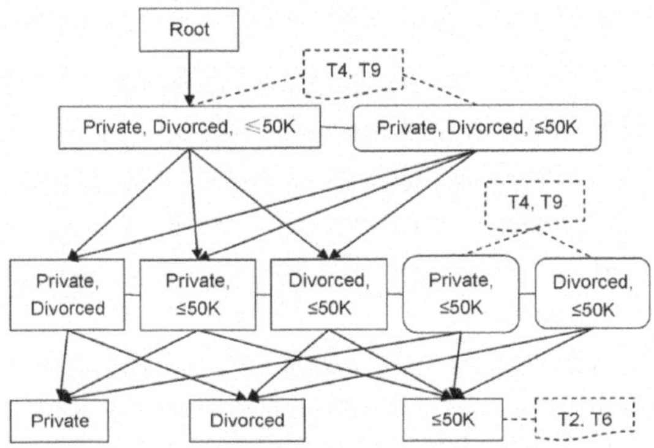

Fig. 2. Example tree of inverted file

In Figure 2, we give out an example tree structure for the first group in Figure 1(b)(we set support threshold to 25% and confidence threshold to 60%). The tree starts from the longest itemsets, we denote h as the height of the tree. Every node represents an itemset(rectangle for itemset containing association rules, and rounded rectangle for itemset associating dissociation rules), the lth layer consists of itemsets with length $h - l + 1$, and the leaf layer consists of frequent 1-itemsets. For association rules, the nodes in the subtree are the sub frequent itemsets of the root. Each node also contains the corresponding rules with their confidence(not included in Figure 2). Rather than having every itemset associate all id of the supporting records, we only store them in the root nodes of subtrees, and none of whose parents are supported by the records. As in Figure 2, record $T4, T9$ is not stored in any rectangle nodes under layer 2. Rules in child nodes can look up in their parent for all the supporting rows.

As for the dissociation rule $dr : \neg A \rightarrow B$, since A, B are also frequent itemsets[15], node of dr will link itemsets A and B as child nodes in the tree structure and records supporting the infrequent itemset $\{A, B\}$ will be attached.

4.3 Anonymization Metric

Let $\{r_1, r_2, \ldots, r_m\}$, $\{s_1, s_2, \ldots, s_n\}$ represent the Q-S associations of two anonymity-groups AG_1, AG_2 going to be merged. Suppose rule t length k(i.e. consist of k attributes) is one of their common Q-S associations. Let $conf(t)$ represent the confidence of t, $antec(t)$ be the antecedent itemset of t, $suppNum(t)$

be the number of records supporting t. It's fast in calculating the new confidence of t in the merged group without retrieving the dataset before really merge AG_1, AG_2:

$$new\_conf(t) = \frac{suppNum_{AG_1}(t) + suppNum_{AG_2}(t)}{suppNum_{AG_1}(antec(t)) + suppNum_{AG_2}(antec(t))} \tag{1}$$

If t does not exist in AG_2, we look for t's antecedent $antec(t)$ and other rules sharing the same itemset of t to calculate its new confidence. Furthermore, when AG_2 does not have a rule with the same k-itemset of t, we search AG_2 for the antecedent itemset of t, and have:

$$new\_conf \in [\frac{suppNum_{AG_1}(t) \cdot \theta}{suppNum_{AG_1}(antec(t)) + supNum_{AG_2}(antec(t))}, \frac{suppNum_{AG_1}(t) + suppNum_{AG_2}(antec(t)) \cdot \theta}{suppNum_{AG_1}(antec(t)) + suppNum_{AG_2}(antec(t))}) . \tag{2}$$

We use "Contribution" to quantify the effect of lowering each Q-S association in the candidate generalization.

Definition 6. *(Contribution) Given a table T, the confidence threshold θ, and a candidate generalization G. We denote all anonymity-groups involved in G as $\{AG_i\}$. For a single Q-S association t, we denote its number of records still to be generalized after applying G as $n\_after = suppNum(antec(t)) \cdot (new\_conf(t) - \theta)$, and that before G as $n\_before = \sum suppNum_{AG_i}(antec(t)) \cdot (conf_{AG_i}(t) - \theta)$. We have $1 - \frac{n\_after}{n\_before}$ as G's contribution to reduce t.*

When evaluating the candidate generalization, we denote the "average Q-S contribution" as the average of the contributions for all Q-S associations involved.

We will have contribution intervals when the specific Q-S association can not be found in all the anonymity-groups involved. We hold these intervals until overlaps exist in the interval comparing. Then, the data records contained in the corresponding groups will be retrieved in order to calculate the definite value of those contributions. The action of data retrieving will less happen with the growing of the minimum anonymity[8](i.e. the minimum size of the anonymity-groups), as every anonymity-group also maintains the global rules in its tree structure.

Therefore, for each candidate generalization G, we calculate $A(G)$ as the anonymity increase that G will make (i.e. the increase of the minimum size of the anonymity-groups); $DL(G)$ as the data loss after applying G, which can be quantified by entropy increase[8] or decrease of the distinct values in the taxonomy trees[6]; $Con(G)$ for average Q-S contribution. Thus, we evaluate G as:

$$\frac{A(G) \cdot Con(G)}{DL(G)} \tag{3}$$

We will choose the generalization with the largest value of Equation 3. More methods of evaluating $A(G), DL(G)$ can also be referenced from the works mentioned in Section 2.

Since the quasi-identifier uniquely identifies the individuals through external databases, the size of the initial anonymity-groups will be small. Thus, the time to start evaluating the Q-S associations has its effect on the balance among computing cost, memory taking, anonymity, and data loss. We choose to track the minimum anonymity[8] during the anonymization, if it reaches the preset $c \cdot k (c \in \mathbb{R})$, we bring in the Q-S association evaluation. The first round of Q-S association evaluation will have the most computational cost, as the tree of inverted file will be setup there. Afterwards, the evaluation will be quite fast, because most of the computation will be done without touching the original dataset. We will also prove this in the experiment part.

5 Q-S Association Hiding Algorithm

After the anonymization process, the anonymity-groups will have Q-S associations with relatively low confidence. Then we will try to generalize the sensitive values to totally hide the Q-S associations below the threshold.

As mentioned in Section 2, quite a few works about association rule hiding have been presented. However, most of them aim at removing the set of sensitive rules while preserving the remaining rules and introducing fewer new rules, i.e. to achieve less side effects and less artifactual patterns[13].

Although we also aim at hiding rules in the anonymity-groups, we have different goals and requirements:

1. Hide both association and dissociation rules.
2. Hide all rules exceeding the confidence threshold.
3. Minimize data loss during the sensitive value generalization.
4. Other than adding or deleting rows in the former studies[12,13], we use generalization.

Currently no work meets all the requirements above, while the problem handles by IGA [13] is the closest to ours. We will compare with it in the experiment.

5.1 Hiding Metrics

Since we use generalization to hide Q-S associations, the interest measure of rule t will be evaluated in minimum confidence, i.e. $\frac{\min suppNum(t)}{\max suppNum(antec(t))}$. For example, in Figure 1(b), suppose we generalize the marital-status of record $T9$ to "Any", the maximum confidence of the rule "Private, $\leq 50K$ →Divorced" decreases from 100% to 50%.

In our method, we try to hide all Q-S associations. And in each time, we reduce the confidence of only one Q-S association by choosing a generalization which generalize one attribute of it. We greedily choose the attributes to generalize in order to reduce the largest amount of other Q-S associations.

Lemma 1. *Given an anonymity-Group: AG, its tree of inverted file: $T(AG)$, the sensitive attributes: $\{S_1, S_2, \ldots, S_m\}$, and the confidence threshold: θ. Let*

NS represent the node of an arbitrary sensitive itemset in $T(AG)$, SR be the set of rules in both NS and the nodes $\subset NS$'s subtree . Then, $\forall$generalization $G_i \in SR, \exists$generalization $G_{ns} \in NS$, When generalize a fixed number of records, the contribution of G_{ns} to reduce the Q-S associations is no less than that of G_i.

Proof. Firstly, we derive the expressions of contribution in different cases. Suppose an association rule r in $NS : A \rightarrow B(A \cup B \subset NS \;\&\; A \cap B = \emptyset)$.

If the candidate generalization G for r is to generalize any attribute in A, and d records that support r will be affected. Then, the maximum possible number of records supporting A will not change, while the definite number of records that support $A \cup B$ will decrease by d.

We apply the concept of contribution here. And the generalization G contributes to the confidence reduction of r as:

$$contribution_G(r) = \frac{conf(r) - \frac{suppNum(r)-d}{suppNum(r)/conf(r)}}{conf(r) - \theta} \tag{4}$$

It is similar when we generalize any attribute in B.

As for the dissociation rule $dr : \neg A \rightarrow B$ which has itemsets A and B as its child nodes. Also, we generalize the itemset A, the maximum possible number of records that support $\neg A$ will increase, and the definite number of records supporting dr remains the same as before. The contribution will be:

$$contribution_G(dr) = \frac{conf(r) - \frac{suppNum(r)}{suppNum(r)/conf(r)+d}}{conf(r) - \theta} \tag{5}$$

When B is to be generalized, we shall only avoid those records attached in dr as they are supporting $A \cup B$. Generalizing records supporting dr produces the similar contribution as Equation 4.

Child nodes of NS will be affected by the candidate generalization, which leads to the confidence reduction in parts of association and dissociation rules within SR. We sum up all these contributions as the measurement to quantify G's effect of reducing Q-S associations:

$$wholeContribution_G(r) = \sum_{i=1}^{|SR|} contribution_G(r_i \in SR) \tag{6}$$

Therefore, if G' is a candidate generalization for rule r' in a child node of NS, which generalize the value in attribute S_j, let G take the same action for r, then we have $wholeContribution_G(r) \geq wholeContribution_{G'}(r')$. □

5.2 Hiding Algorithm

Based on Lemma 1, we develop our Q-S association hiding algorithm as below.

Algorithm 1: Q-S association hiding algorithm

Data: anonymity group: AG, inverted file tree for G: $T(AG)$, confidence threshold: θ

Result: anonymity-group without Q-S associations: AG'

```
begin
    foreach record ∈ AG do
        Store the record id in nodes of T using maximum matching;
    foreach level l ∈ T(AG) & l > 1 (top-down) do
        foreach node ∈ level l of T(AG) do
            Set s = node ∪ {node t|t ∈ subtree of node};
            mr ← {rule r|conf(r) = min conf(r_i), r_i ∈ node};
            H_mr ←candidate generations for generalizing each Attribute ∈ mr;
            wholeContribution(mr) ← zero vector with length(mr) dimensions;
            foreach rule rr ∈ s do
                wholeContribution(mr) ← contribution_{H_mr}(rr);
                if rr is the antecedent of dissociation dr then
                    wholeContribution(mr) ← contribution_{H_mr}(dr);
            attr ← max dimension ∈ wholeContribution(mr);
            foreach record row to be generalized do
                if attr in row is not generalized then
                    Generalize row by attribute attr;
                else if attr is generalized to D & row(attr) ∉ D then
                    Generalize attr into a higher position containing row(attr)
                    in the hierarchy.
                Recompute the confidence of other rules row supports;
    Generalize the remaining frequent 1-itemsets;
end
```

The Q-S association hiding algorithm proceeds as follows. Firstly, as in Section 4.2, we attach every record to the tree nodes. Then, start from the longest rule, we generate the candidate generalizations as generalizing each of its attribute. We test the candidates in the subtree to generate the vector of the "wholeContribution", each of whose dimension corresponds to one candidate. Afterwards, we select the generalization with the highest contribution sum to apply. Also, we notice that a data record may be stored in more than one itemsets which do not contain each other. This will lead to repeated generalization of the same column value in a record. Actually, we will check the status of the attribute and decide whether to generalize it to a higher domain or skip the record, and recompute the confidence of every missing rule (i.e. the rule outside the subtree but supported by the generalizing row).

Our algorithm choose the generalizing attribute through contribution comparing. Although the contribution calculation is limited to the subtree in the current study, it covers most of the generalization effect especially when handling long itemsets which rapidly reduce all the Q-S associations they contain.

Moreover, limiting the range of contribution calculation leads to a small memory requirement for our inverted file tree. Otherwise, we have to associate every row with all the Q-S associations it contains. Suppose the number of Q-S associations is n, since every rule is generalized after a traverse of its subtree, the time complexity of the algorithm is $O(n \log n)$.

It is also shown in Algorithm 1 that we deal with dissociation rules. When the antecedent node(also is the child node) of the dissociation rule is affected by a generalization, we also evaluate the effect and compute the contribution to that dissociation rule(as in Lemma 1) by the generalization.

6 Experiment Result

In our experiment, we use the "Adult Database" obtained from [16], which has 14 attributes, 48842 instances. The records with missing attribute values "?" are removed. We show in Table 1 the attributes we adopt, the number of the leaf nodes in their hierarchy trees, and the height of the trees. We make different combinations of quasi-identifier and sensitive columns to get the average experiment result.

Table 1. The Attributes Adopted

	Quasi-identifier/Sensitivity							
Attribute	Education	Occupation	Race	Sex	Work-class	Marital-status	Relationship	Native-country
Leaf Num.	16	14	5	2	8	7	6	41
Height	4	4	3	2	3	3	3	4

There are 2 steps in our implementation, we test them respectively. Due to the space limit, we can't list all the results of our experiments here.

For rule hiding, we make a comparison between our algorithm and the implementation of IGA[13] strategy using generalization. To be fair, we only choose to hide the association rules from the datasets. The support and confidence threshold are set to 20% and 50% respectively, and the hierarchy trees are constructed as height 2. Every time, we choose a different number of attributes, and compute the ratio of the cells generalized in our algorithm to that in IGA. We find in Figure 3(a) that under our requirement, our Q-S association hiding algorithm has smaller data loss. This is mainly attributed to that the item with the highest contribution reduce the most amount of Q-S associations.

Then, a comparison is made between the common k-anonymization and our enhanced version. The support and confidence threshold are 10%, 50% and $k = 250$. We implement the strategy in [8] as the common version. And we bring in the Q-S association evaluation to simulate our method at different values of minimum anonymity[8]: 25, 50, 100.... In Figure 3(b), we compare the "information loss", "performance" and "hiding efficiency" of both methods by calculating "entropy loss in anonymization", "execution time after building the inverted file

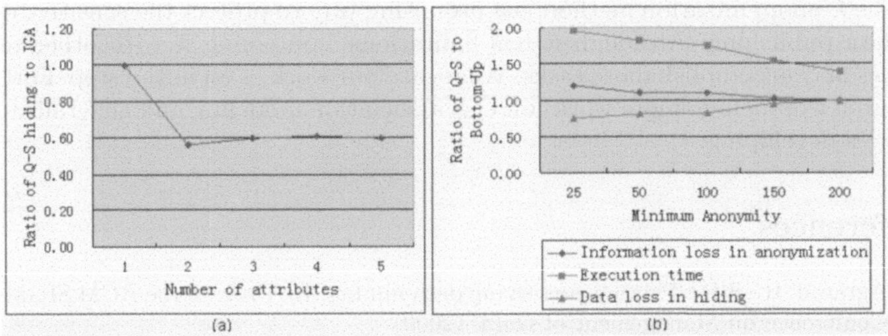

Fig. 3. Methods comparison (a)Comparison between Q-S association hiding and IGA. (b)Comparison of k-anonymization between our method and "bottom-up" strategy.

tree" and "data loss in hiding step", then we calculate their ratios of our method to the common k-anonymization.

As shown in Figure 3(b), our method approaches the optimal result of the "bottom-up" strategy when the minimum anonymity becomes larger. Currently, the Q-S associations, information loss and anonymity have the same weight in choosing the candidate generalizations. Therefore, when we start to evaluate the Q-S association with a small minimum anonymity, the anonymization will deviate from the optimal result in an early time. We can make different weights for these 3 metrics to relieve this phenomenon, which will be one of our future research directions. The inflexion on the curve of "information loss" shows the greedy characteristic of "bottom-up", which sometimes prevent it from getting the global optimal result. We also find in the series of "execution time comparison" that when the Q-S associations is evaluated early in the process, the performance after the tree construction decreases due to the increasing dataset accesses. Series of "data loss in hiding" shows that smaller number of cells have to be hidden when we bring in the Q-S association evaluation early in the anonymization. In order to have a balance among the performance, optimal k-anonymity result and cells to hide, we find it be better to start evaluating the Q-S associations when the minimum anonymity reaches either 50 or 100.

7 Conclusion

In this paper, we have introduced an enhanced k-anonymization method which detaches the links between quasi-identifiers and sensitive attributes. We've defined such links using frequent 1-itemsets, association and dissociation rules with high confidence within an anonymity-group. We've not only evaluated them in the k-anonymization process, but also removed them using our Q-S association hiding algorithm. In our research, k-anonymization is combined with rule hiding which is also a direction in privacy preserving data mining. By applying our greedy algorithm, we prevent the anonymity breaking from those "inference paths" with minimum data loss.

The k-anonymization method is a promising way to protect the sensitive data in data publishing. Although it has limitations, combining it with other techniques may accomplish more tasks. We regard our work as an initial step. Further research will include more works on Q-S association modeling and generalization metrics developing.

References

1. Agrawal, R., S.R.: Privacy-preserving data mining. In: Proc. of the ACM SIGMOD Conference on Management of Data. (2000)
2. Sweeney, L.: Achieving k-anonymity privacy protection using generalization and suppression. International Journal of Uncertainty **10**(5) (2002) 571–588
3. Meyerson, A., W.R.: On the complexity of optimal k-anonymity. In: Proc. of the 23th ACM SIGMOD-SIGACT-SIGART symposium on Principles of database systems. (2004)
4. LeFevre, K., D.D.R.R.: Incognito: Efficient fulldomain k-anonymity. In: Proc. of the 2005 ACM SIGMOD international conference on Management of data. (2005)
5. Hundpool, A., W.L.: Mu-argus and tau-argus: Software for statistical disclosure control. In: Proc. of the 3rd International Seminar on Statistical Confidentiality. (1996)
6. lyengar, V.: Transforming data to satisfy privacy constraints. In: Proc. of the 8th ACM SIGKDD international conference on Knowledge discovery and data mining. (2002)
7. Bayardo, R.J., A.R.: Data privacy through optimal k-anonymization. In: Proc. of the 21th International Conference on Data Engineering. (2005)
8. Wang, K., Y.P.C.S.: Bottom-up generalization a data mining solution to privacy protection. In: Proc. of the 4th IEEE International Conference on Data Mining. (2004)
9. Machanavajjhala, A., G.J.K.D.V.M.: l-diversity privacy beyond k-anonymity. In: Proc. of the 22th International Conference on Data Engineering. (2006)
10. Nergiz, M.E., C.C.: Thoughts on k-anonymization. In: Proc. of the 22th International Conference on Data Engineering Workshops. (2006) .
11. Xiao, X., T.Y.: Personalized privacy preservation. In: Proc. of the 2006 ACM SIGMOD international conference on Management of data. (2006)
12. Verykios, V.S., E.A.B.E.S.Y.D.E.: Association rule hiding. IEEE Transactions on Knowledge and Data Engineering **16**(4) (2004) 434–447
13. Oliveira, S.R.M., Z.O.: A unified framework for protecting sensitive association rules in business collaboration. International Journal of Business Intelligence and Data **1**(3) (2006) 247–287
14. Savasere, A., O.E.N.S.: An efficient algorithm for mining association rules in large databases. In: Proc. of the 21th International Conference on Very Large Data Bases. (1995)
15. Wu, X., Z.C.Z.S.: Mining both positive and negative association rules. In: Proc. of the 19th International Conference on Machine Learning. (2002)
16. Hettich, S., B.S.: The uci kdd archive. Univeristy of California, Irvine, Department of Information and Computer Science (1999)

Protecting and Recovering Database Systems Continuously

Yanlong Wang, Zhanhuai Li, and Juan Xu

School of Computer Science, Northwestern Polytechnical University,
No.127 West Youyi Road, Xi'an, Shaanxi, China 710072
{wangyl,xuj}@mail.nwpu.edu.cn, lizhh@nwpu.edu.cn

Abstract. Data protection is widely deployed in database systems, but the current technologies (e.g. backup, snapshot, mirroring and replication) can not restore database systems to any point in time. This means that data is less well protected than it ought to be. Continuous data protection (CDP) is a new way to protect and recover data, which changes the data protection focus from backup to recovery. We (1) present a taxonomy of the current CDP technologies and a strict definition of CDP, (2) describe a model of continuous data protection and recovery (CDP-R) that is implemented based on CDP technology, and (3) report a simple evaluation of CDP-R. We are confident that CDP-R continuously protect and recover database systems in the face of data loss, corruption and disaster, and that the key techniques of CDP-R are helpful to build a continuous data protections system, which can improve the reliability and availability of database systems and guarantee the business continuity.

1 Introduction

With the widespread use of computers, database systems are vital in human life and data stored in database systems is becoming companies' most valuable asset. Although we are careful to defend against all kinds of disasters, they still occur frequently. For example, hardware breaks, software has defects, viruses propagate, buildings catch fire, power fails and people make mistakes [1]. Data corruption and data loss by those disasters have become more dominant, accounting for up to 80% [2] of data loss. Recent high-profile data loss has raised awareness of the need to plan for recovery or continuity. In particular, it is a challenge that a large number of database systems must be continuously available and businesses also must be prepared to provide continued service in the event of disasters.

Many data protection solutions including fault-tolerance and disaster-tolerance techniques have been employed to increase database systems availability and to reduce the damage caused by data loss, corruption and disaster [3]. Backup [4] is the most popular solution which stores vital data on tape or disk. Basic backup include three modes: full backup, incremental backup and differential backup, all of which can be implemented offline and online. In addition, there are several solutions such as redundant disk arrays (RAID) [5], mirroring [6], snapshot [7] and replication [8].

G. Dong et al. (Eds.): APWeb/WAIM 2007, LNCS 4505, pp. 765–776, 2007.

However, conventional backup technologies have many drawbacks. First, offline backup (cold backup) requires that the application should be periodically (daily or weekly) down or completely offline, and although online backup (hot backup) allows to backup with database still running, it has to bare some performance penalty. Second, backup is time-consuming and takes a long time to recover data. Third, database systems can only be restored back to a pre-determined previous point, and data between backups are vulnerable to data loss. Recent research [1] has shown that data loss or data unavailability can cost up to millions of dollars per hour in many businesses. Other solutions have most of the same drawbacks as backup. Therefore, the traditional time-consuming techniques are no longer adequate for today's information age.

In order to remove backup window and resolve recovery point objective (RPO) and recovery time objective (RTO) issues, researchers present continuous data protection (CDP) [9]. CDP represents a major breakthrough in data protection and dramatically changes the data protection focus from backup to recovery. With CDP continuously capturing and protecting all data changes to the important data of database systems, this provides rapid recovery to any desired point in the past when disaster strikes and access to data at any point in time (APIT) [10] after recovery. CDP offers more flexible RPO and faster RTO than traditional data protection solutions which were designed to create, manage and store single-point-in-time (SPIT) [11] copies of data, thereby reducing data loss and eliminating costly downtime.

CDP has appeared only for a short while, so it isn't well understood. With our survey of the approaches used in practice, we found most of current CDP technologies aren't the real CDP and are only near-CDP. Our first contribution, then, is a taxonomy of current CDP technologies and a strict definition of CDP. Our second contribution is the design of a CDP model for database systems, referred to as continuous data protection and recovery model (CDP-R). It is built on block-level and provides continuous protecting and recovering of database systems' data. The final contribution is an evaluation of our CDP-R model, simply comparing it with other backup technologies.

2 CDP

2.1 Taxonomy

CDP is becoming a hot topic and there have been research efforts in some large IT companies, research institutions and emerging companies. There are several assessment criteria for CDP designs and we summarize the basic axes as data protection scheme, design level, storage repository and recovery mechanism.

Data protection scheme. Current CDP systems implement a continuous or near-continuous data protection scheme for retrieving even the most recently saved data:

1. *CDP systems*: save every change to data as it is made and let administrators or users recover files and other data such as email from any point in time, such as Peabody [12], TRAP-Array [13], CPS [14] and TimeData [15].

2. *Near-CDP systems*: do not have the detail of CDP and take snapshots of data at specified points in time and only allow customers to retrieve data from those times, not from seconds or even hours ago, such as Backup Exec 10d, DPM, Tivoli CDP for Files and LiveServe [16].

CDP syOstems can recover the primary to any point in time and near-CDP only can provide some scheduled point-in-time recovery. So we do not consider near-CDP in this paper.

Design level. CDP systems have been implemented at the block-, file- or application-level against disasters:

1. *Block-level CDP systems:* operate above the physical storage or logical volume management layer. As data blocks are written to the primary storage, copies of the writes are captured and stored to an independent location. Peabody [12] exposes virtual disks to recover any previous state of their sectors and shares backend storage to reduce the total amount of storage needed. TRAP-Array [13] designs a CDP prototype of the new RAID architecture and stores the time-stamped Exclusive-ORs of successive writes to provided timely recovery to any point in time. CPS [14] adopts time-addressable-storage (TAS) and adds time as a dimension of data storage.
2. *File-level CDP systems*: operate just above the file system. They capture and store file-system data and metadata events (such as file creation, modification, or deletion). For example, TimeData [15] keep the protected instances of files in their natural form and recover files to any point in time at file-level.
3. *Application-level CDP systems*: operate directly within the specific application that is being protected. Such solutions offer deep integration and are typically either built-in to the application itself or make use of special application APIs, which grant continuous access to the application's internal state as changes occur.

File- and application-level CDP systems provides CDP only for some fixed file systems or applications. Block-level CDP systems can take the advantage of supporting many different applications with the same general underlying approach. It can achieve high performance and help to build a multi-platform CDP engine to protect a variety of database systems. The recovery granularity of block is the most ideal and potential data loss is minimal. We discuss CDP at the block-level in this paper, although file- and application-level CDP could readily be implemented.

Storage repository. Storage repository provides the ability to store and manage the CDP data over time. CDP systems employ a distinct and dedicated node or the host itself as the storage repository:

1. *Distinct storage repository*: is architected in an independent location where all data changes are stored. The distinct node is available on the LAN, WAN or SAN. This kind of repository is employed by most of CDP systems.
2. *Self-storage repository*: is established on the protected host itself where changed data is written directly onto the independent CDP storage region, such as Peabody [12] and TRAP-Array [13].

We use the distinct storage repository to keep CDP data in the following text.

Recovery mechanism. Recovery mechanism determines the recovery procedure and can be implemented in two modes:

1. *Independent recovery*: is achieved only according to the storage repository where data includes the initial data set and the changed data set of the primary. Independent recovery makes it possible to reduce the cost of CDP recovery.
2. *Dependent recovery*: is achieved with the storage repository and an initial replica which increases the complexity of CDP recovery.

We use the independent recovery mechanism in the following text.

2.2 Definition

According to CDP systems and researchers, the SNIA (Storage Networking Industry Association) CDPSIG Continuous Data Protection Special Interest Group defines CDP as "*a methodology that continuously captures or tracks data modifications and stores changes independent of the primary data, enabling recovery points from any point in the past.*" [9]

While various CDP systems confuse us, the above definition is too simple to guide us to design a veritable CDP system. To describe CDP in details and rigorously, we define CDP in two aspects (i.e. protection and recovery) theoretically as follows:

Definition 1. $\tau^P(t)$ is the data image/view of the primary P at time $t(t \geq t_0)$ where $\tau^P(t_0)$ is the initial data image/view at the beginning time t_0. If $|\tau^P(t)|$ is the data set of the Primary P at time $t(t \geq t_0)$, $\delta^P(t) = |\tau^P(t)| - |\tau^P(t-\Delta t)|$ is the data set of all the changes of the primary P at time t where $\Delta t \to 0$ and $\delta^P(t_1, t_2) = \{\delta^P(t), t_1 \leq t \leq t_2\}$ is the sum of the changed data sets of the Primary P from time t_1 to time t_2. When $\delta^P(t_0, t)$ is stored to a distinct site, the backup B, the procedure is called as continuous data protection (CDP) from t_0 to t.

Definition 2. $\lambda^B(t)$ is the data set of the backup B corresponding to the data image/view of the primary P at time $t(t \geq t_0)$ where $\lambda^B(t_0)$ is the initial data set at the beginning time t_0. If the backup B receives the delta $\delta^P(t)$ from the primary P at time t, $\lambda^B(t) = \lambda^B(t-\Delta t) + \delta^P(t)$ where $\Delta t \to 0$ and inductively $\lambda^B(t) = \delta^P(t_0, t)$. If $|\lambda^B(t)|$ is the data set of the backup B after coalescing all the blocks with the same address at time $t(t \geq t_0)$, when $|\lambda^B(t)|$ is restored to the primary P and overwrites the blocks of the primary P according to the address of each block, the procedure is called as continuous data recovery (CDR) at time t.

3 CDP-R Model

In order to protect and recover database systems continuously, we set out to design a model of continuous protection and recovery (CDP-R). The goal of CDP-R model is to keep the copy of each block-level change of the database system in a distinct storage repository and make the data of the database system available despite both hardware failures and software failures, which could exploit a continuous protection and recovery of database systems.

CDP-R model is composed of *client*, *primary* and *backup* as shown in ***Fig. 1***:

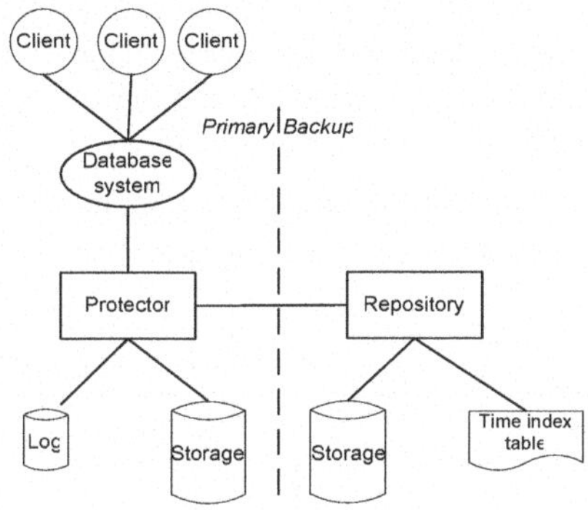

Fig. 1. An overview of CDP-R model

Client gives an intelligent management platform for users to operate database systems and configure CDP-R model. *Primary* includes database system, protector, storage and log, *Backup* includes repository, storage and time-index-table. The protector and the repository are main components of CDP-R model to protect and recover database systems as shown in ***Fig. 2***. The protector continuously captures every change of the primary and sends to the backup. The repository receives data from the primary and stores data over time in storage.

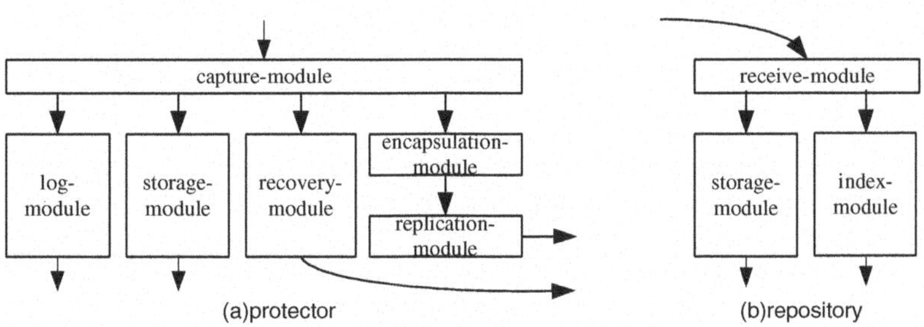

Fig. 2. Modules of protector and repository

3.2 Workflow

Normally, an operation at the primary causes a write record to be written synchronously in the primary log, and the block-level data can then be written to the primary storage. Simultaneously, CDP-R model implements three-step work:

1. *Capture*: After the capture-module gets every block-level change of database system, the encapsulation-module wraps the data block $data_i$ in a package with a timestamp t_i and other description information $disc_i$ (including storage address, size, etc.) and then forms a backup record $< t_i, disc_i, data_i >$.
2. *Backup*: The replication-module replicates every backup record to the backup synchronously or asynchronously. After the receive-module takes the backup record, the storage-module inserts an item into time-index-table and stores the record in the storage.
3. *Retrieve*: When database system needs recovering in case of data loss, corruption and disaster, clients can lookup the time-index-table and select a past point. Then we retrieve the primary from the appointed data of the backup and recreate the exact data state as it existed at any point in time.

The whole procedure of capture-backup-retrieve is implemented in the background automatically. ***Fig. 3*** shows the state transitions of a data block in CDP-R model.

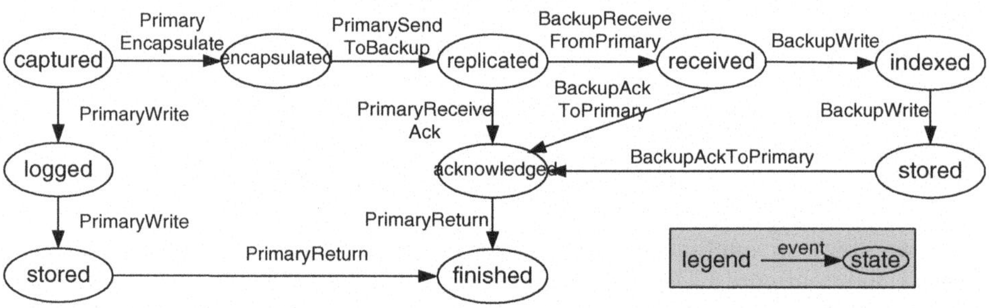

Fig. 3. Data block states in CDP-R model. The left part is the state transitions of the data block at the primary and the right part is the state transitions at the backup.

4 Key Technologies

CDP-R model is implemented by three key technologies (referred to as 3R) as follows:

1. *Replication*: To meet users' need and fit the networks situation, the primary must adopt an appropriate replication protocol and dynamically transmit the backup record to the backup within all types of TCP/IP networks (LAN, WAN, etc.). We implement two replication protocols, i.e. synchronous and asynchronous, and keep the data consistency between the primary and the backup.
2. *Repository*: To store and conveniently lookup every backup record, the backup must manage all backup records with an effective structure and an index dictionary. We architect a delta-chain to store backup records over time, and build a time-index-table to locate every record.
3. *Recovery*: To deal with a disaster, the primary must recover from the backup. We create an any-point-in-time incremental or full version of the backup, and use it to retrieve the primary rapidly.

To capture and encapsulate the changes of database system at the primary continuously, CDP-R model can adopt Loadable Kernel Modules (LKM) on Linux or Windows Driver Model (WDM) on Windows. We won't discuss this technology in more details here.

4.1 Replication

Replication mode. The replication protocol plays an important role in CDP-R model. It automatically transmits every backup record to the backup. It includes two modes of synchronous and asynchronous. We deal with a block-level change of database system by nine steps and implement the replication protocol in synchronous and asynchronous modes as shown in *Fig. 4*.

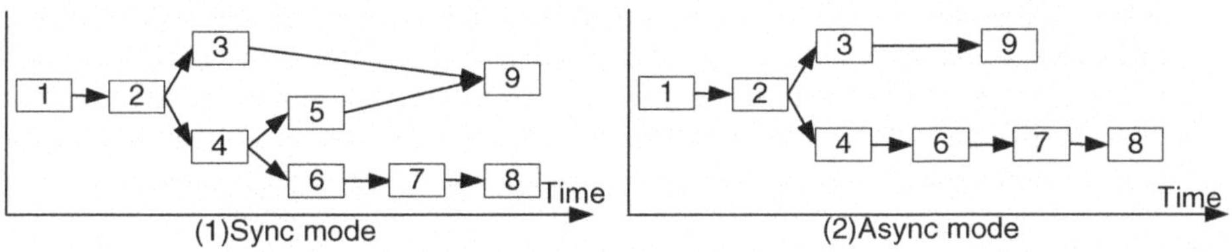

Fig. 4. Replication protocol of CDP-R model. 1-protector captures a block-level change of database system; 2-protector writes the change to the log; 3-protector writes the change to the storage; 4-protector encapsulates the change and sends the backup record to the repository; 5-repository returns the receiving acknowledgement; 6-repository writes the change to the time-index-table; 7-repository writes the change to the storage; 8-repository returns the completing acknowledgement; 9-protector returns success to database system.

We recast the traditional protocol into a new replication protocol and adopt some methods to increase the reliability and efficiency of the protocol. For example, we write the log/time-index-table before writing the storage. We also execute several steps in parallel and solve the block-level changes by pipelining.

Each replication mode deals with the block-level changes differently. Synchronous mode ensures that a backup record has been posted to the backup before the request of database system completes at the application-level. Database system performing an application may experience the response time degradation caused by each backup record incurring the cost of a network round-trip, but the backup is up to date. If a disaster occurs at the primary, data can be recovered from any surviving backup with minimal loss. Asynchronous mode completes an update when it has been recorded in the log and storage at the primary. The response time is shorter at the cost of the backup being potentially out of date. If a disaster strikes, it is likely that the most recent writes have not reached the backup. Therefore, the decision to use synchronous or asynchronous mode depends on users' requirements, the available network bandwidth, network latency and the number of backup servers.

Data consistency. Data is consistent if database system using it can be successfully restarted to a known, usable state. That is, data at the backup correctly reflects the data changes at the primary at some point in the past. CDP-R model maintains data consistency by two means:

1. *Send-queue and receive-queue*: The backup records are queued temporarily in a circular queue to be sent to the backup. When there is a surge in the block-level change rate, this queue may grow and will be continuously drained. After the backup records reach the backup, there the other circular queue keeps them temporarily and drains as fast as they are written to storage. Both queues try to keep the backup as consistent as the primary and achieve write-order-fidelity.
2. *Atomic replication and atomic write*: While data consistency in synchronous mode is not impacted by network failures, in asynchronous mode it tends to be impacted. Asynchronous mode makes it possible to fail to receive completing acknowledgements of some backup records when network problems occur, but in fact those backup records have been written to storage at the backup. When network is ok and we send those records again, the backup may be inconsistent with the primary. Thus, the primary sends them with atomic replication and the backup stores them with atomic write, which avoids the risk of inconsistency.

4.2 Repository

Time index table. When we want to recover the primary, we need select a past time t and then collect all backup records at that time. According to time t, time index table (See *Fig. 5*) is useful to build an index dictionary and find the target backup records stored in the storage. It just maps t to *address* in the storage.

In order to generate a unique fingerprint for every time, CDP-R model uses the Sha1 hash function[17] to build a large hash table as the time index table. Sha1 is a popular, efficient hash algorithm for many security systems and its output is a 160-bit hash value. Assuming the granularity of time is microsecond and random hash values has a uniform distribution, a collection of n different times and a hash function that generates 160 bits, the probability p that there will be one or more collisions is bounded by the number of pairs of times multiplied by the probability that a given pair will collide, i.e. $p \le \frac{n(n-1)}{2} * \frac{1}{2^{160}}$. If we keep backup records for one year that is enough for protecting common database systems, $n = 365 * 24 * 60 * 60 * 10^6 \approx 10^{14}$ and then p is less than 10^{-20}. Obviously, Sha1 is suitable for CDP-R model and the collision scenario can be ignored.

Although it is ideal that every backup record has a unique timestamp, in fact a series of backup records may have the same timestamp. For example, there may be some backup blocks with the same microsecond timestamp in current computer systems. Therefore, time index table locates the first of a series of backup records with the same timestamp. After receiving a backup record, repository extracts the timestamp from it and hashes the timestamp with Sha1. Then check whether the item has been in time index table. If yes, repository locates the address in the storage and scans the storage forwards to find a free space for the backup record; otherwise, repository fills a new address into the item and then stores the backup record in the storage according to the new address. Therefore, when getting a time t, we can collect a series of backup records with the same time.

DeltaChain. We present DeltaChain to manage the storage at the backup. DeltaChain is like a link list composed of a large number of segments, and a segment has a series of backup records with the same time (See ***Fig. 5.***). All of the backup records are stored continuously, referred to as continuous storage over time, which is not like Peabody [12] or TRAP-Array [13]. Continuous storage increases the speed of locating the address and reduces storage fragments in Peabody which stores every version of block continuously.

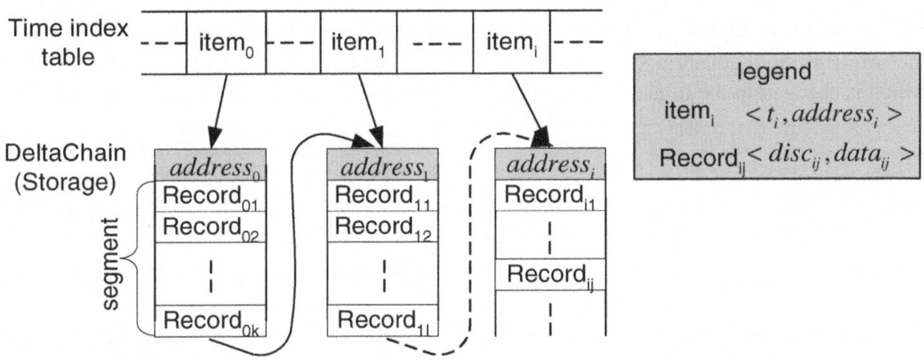

Fig. 5. Repository of CDP-R model. Every segment is stored continuously. In every segment, records have been coalesced if they have identical description.

Segment$_0$ is ready to store all the backup records from the primary at time t_0. If DeltaChain is fully initialized by $\tau^P(t_0)$ and k is equal to the number of all the data blocks of the primary, all the backup records in *Segment$_0$* is corresponding to all the data blocks of the primary at time t_0. If DeltaChain is partially initialized by $\tau^P(t_0)$, when data of a backup record (e.g. $< t_i, disc_{ij}, data_{ij} >$) is replicated to the repository for the first time, a backup record $< t_0, disc_{0r}, data_{0r} >$ also has to be replicated to the repository where $disc_{0r} = disc_{ij}$ and $data_{0r}$ is the data at the same address before being overwritten. Then the repository stores it as the $r-th$ backup record to *Segment$_0$* before storing $< t_i, disc_{ij}, data_{ij} >$. That is, only the data that will be overwritten has to be replicated and stored into *Segment$_0$*.

The other segments have the same function, and are used to store ordinary backup records. All the backup records in a segment have the same timestamp. A segment increases with receiving a new backup record.

4.3 Recovery

The primary faces several threat categories: data loss, data corruption and Data inaccessibility [1]. To limit the scope of this study, we focus on data loss events for the primary and map data corruption and inaccessibility threats into data loss.

After a failure, we can adopt one of the following continuous-data-recovery algorithms to restore the primary from the backup to any point in time and make it usable again. When we decide to restore the primary to a past time t_i, we find the newest version of each data block in the segments from time t_0 to time t_i, and send it to the primary to overwrite the data block according to the storage address of the description. The pseudo-code of recovery algorithms are shown in ***Table 1***.

Table 1. Recovery Algorithms. Full-recovery is used to recover the primary continuously and fully when Segment_0 keeps all data of the primary at time t_0. Fast-recovery is used to recover the primary continuously and fast when Segment_0 only keeps data of the primary at time t_0 which is changed in the future.

FULL_RECOVERY(t0)	FAST_RECOVERY(t0)
`S0:=GetSegment(t0);`	`S0:=GetSegment(t0);`
`B:=S0;`	`B:=S0;`
`S:=S0;`	`S:=S0;`
`repeat`	`repeat`
`  S:=GetNextSegment(S);`	`  S:=GetNextSegment(S);`
`  B:=Coalesce(B,S);`	`  B:=Coalesce(B,S);`
`until S==GetSegment(ti);`	`until S==GetSegment(ti);`
`P:=Recover(B,NULL);`	`P:=Recover(B,P);`
`return SUCCESS;`	`return SUCCESS;`

In ***Table 1***, the symbol S_i denotes the segment with time t_i, and S is a temporary variable to keep the segment S_i. The symbol B denotes the backup records that will be sent back to the primary and P denotes all the data blocks at the primary. *GetSegment(t)*, *GetNextSegment(S)*, *Coalesce(B,S)* and *Recover(B,P)* are APIs supplied by CDP-R model. *GetSegment(t)* is used to get the segment according to time t and *GetNextSegment(S)* is used to get the next segment of the current segment *S*. *Coalesce(B,S)* and *Recover(B,P)* are very important as shown in ***Fig. 6***:

1. *Coalesce(B,S)* : is used to coalesce the backup records of *B* and *S* with the same description and reserve the backup record of *S* as the newer version;
2. *Recover(B,P)*: is used to recover *P* from *B*. Data of each backup record of *B* is overwritten to one of *P* with the same storage address.

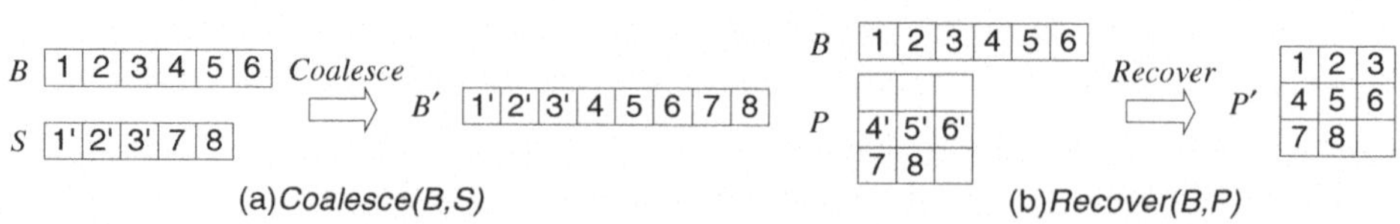

Fig. 6. Recovery APIs of CDP-R model

5 Evaluation

According to the above introduction, CDP is an innovative data protection technique and different from traditional data protection technologies, such as backup, mirroring, snapshot and replication (See ***Table 2***).

Table 2. Data Protection Technologie

Technology	Backup	Mirroring	Snapshot	Replication	CDP
Backup window	large	small	small	small	small
Recovery Point Objective (RPO)	large	small	medium	small	small
Recovery Time Objective (RTO)	large	medium	medium	medium	small
Recovery point	specified point in time	recent point in time	specified point in time	recent point in time	any point in time

CDP-R model supplies a new approach to protect and recover database by adopting the technology of CDP, and can be implemented on any platform, such Linux, Windows and Unix.

Here we just give an example to evaluate CDP-R model base on Logical Volume Manager on Linux:

If a database system (e.g. Oracle) build on CDP-R model at 8:00:00 a.m. and the time granularity of CDP-R is second, when the database system is in a disaster 14:00:00 p.m., we can restore the database system to a past time point between 8:00:00 and 13:59:59.

In CDP-R model, by coalescing the backup records in every segment of the repository, the storage space is reduced by up to 20%. By coalescing the backup records before restoring to the primary, the transmission bandwidth is reduced by up to 42%. In additional, fast recovery is 1~1.5 times faster than full recovery.

6 Conclusion and Future Work

Database systems are very important and require 24x7 availability. CDP transforms the backup/restore process to deliver high availability level of database system and keep business continuity. CDP is more comprehensive and cost-effective than any other solution, such as backup, snapshot, mirroring and replication.

CDP-R model adopts the CDP technology to solve the data restoration time-gap problem and to make true business continuity a realistic objective. It is presented based on the taxonomy and definition of CDP technology. CDP-R model synthesizes the technologies of block-level replication, repository and recovery to offer the ultimate solution. Therefore, CDP-R can provide days, weeks or months (even years) of protection with microsecond/second/minute/hour granularity. CDP-R model also can provide the business resiliency and the ability to rapidly restore to any point in time on the timeline. In addition, the attribute of CDP-R model built on block-level can achieve the high performance and satisfy all kinds of database systems.

CDP-R model complies with the needs of database system protection, but there still exits some future work. For example, we need to optimize the structure of

DeltaChain and the recovery algorithms. Furthermore, we are developing a prototype system based on CDP-R model and hope to explore many of these avenues.

Acknowledgments. This work is supported by the National Natural Science Foundation of China (60573096).

References

1. Kimberly Keeton, Cipriano A. Santos, Dirk Beyer, Jeffrey S. Chase, John Wilkes: Designing for Disasters. In: Proc of the 3rd USENIX Conf on File and Storage Technologies (FAST'04). (2004) 59–72
2. David Patterson, Aaron Brown and et al. Recovery oriented computing (ROC): Motivation, Definition, techniques, and Case Studies. Computer Science Technical Report, U.C. Berkeley (2002)
3. Manhoi Choy, Hong Va Leong, and Man Hon Wong. Disaster Recovery Techniques for Database systems. COMMUNICATIONS OF THE ACM. (2002) 272–280
4. A. L. Chervenak, V. Vellanki, and Z. Kurmas. Protecting file systems: A survey of backup techniques. In: Proc of Joint NASA and IEEE Mass Storage Conference. (1998)
5. D. A. Patterson, G. Gibson, and R. H. Katz. A case for redundant arrays of inexpensive disks (RAID). In: Proc of the ACM SIGMOD International Conference on Management of Data. (1988) 109–116
6. Minwen Ji, Alistair Veitch and John Wilkes. Seneca: Remote Mirroring Done Write. In: Proc of the 2nd USENIX Conf on File and Storage Technologies (FAST'03). (2003)
7. G. Duzy. Match snaps to apps. Storage, Special Issue on Managing the information that drives the enterprise. (2005) 46–52
8. H. M. Zou and P Jahanian. A real-time primary-backup replication service. IEEE Trans on Parallel and Distributed Systems. (1999) 533–548
9. Brian J. Olson and et al. CDP Buyers Guide: An overview of today's Continuous Data Protection (CDP) solutions. SNIA DMF CDP SIG. (2005) http://www.snia.org/
10. B. O'Neill. Any-point-in-time backups. Storage, Special Issue on Managing the Information that Drives the Enterprise. (2005)
11. Alain Azagury, Michael E. Factor and Julian Satran. Point-in-Time Copy: Yesterday, Today and Tomorrow. In: Proc of the 10th Goddard Conference on Mass Storage Systems and Technologies. (2002) 259–270
12. C. B. Morrey III and D. Grunwald. Peabody: The time traveling disk. In: Proceedings of IEEE Mass Storage Conference, San Diego, CA (2003)
13. Qing Yang, Weijun Xiao, and Jin Ren. TRAP-Array: A Disk Array Architecture Providing Timely Recovery to Any Point-in-time. In: Proc of the 33rd Annual International Symposium on Computer Architecture (ISCA'06), Boston, USA (2006)
14. Michael Rowan. Continuous Data Protection: A Technical Overview. Revivio, Inc. (2005) http://www.revivio.com/documents/CDP%20Technical%20Overview.pdf
15. Protecting Transaction Data: What Every IT Pro Should Know. TimeSpring Software Corp. (2004) http://www.timespring.com/ Protecting%20Transaction%20Data.pdf
16. Deni Connor. Continuous data protection finds Supporters. In: Network World. (2005) http://www.networkworld.com/news/2005/091605-continuous-data-protection.html
17. National Institute of Standards and Technology, FIPS 180-1. Secure Hash Standard. US Department of Commerce (1995)

Towards Web Services Composition Based on the Mining and Reasoning of Their Causal Relationships*

Kun Yue, Weiyi Liu, and Weihua Li

Department of Computer Science and Engineering, School of Information Science and Engineering, Yunnan University, 650091 Kunming, P.R. China
yue.kun@gmail.com

Abstract. In this paper, a probabilistic graphical modeling approach for Web services is proposed, and the Web services Bayesian network (WSBN) is constructed by mining the historical invocations among them. Further, the semantic guidance to Web services composition is generated based on the Markov blanket and causality reasoning in the WSBN. Preliminary experiments and performance analysis show that our approach is effective and feasible.

Keywords: Web Services, composition, Bayesian network, Markov blanket.

1 Introduction

To implement automatic Web services composition, an underlying model, the corresponding reasoning approach, and the measure for service associations are indispensable [1, 2, 3, 4]. Thus, the guidance of services composition can be obtained, and then the composition can be carried out automatically. Different approaches are proposed to address this problem, among which most are given at a syntactic level of services themselves, or annotated with ontologies, or based on keyword retrieval [12, 13, 14, 15, 16]. Actually, many services have nothing to do with the actual provision although they have the matched syntactic or keyword description [4]. This requires that the composition be done at the semantic level, and the reasoning among given services is necessary too. Therefore, towards automatic Web services composition, we should first develop a model to represent the implied semantic relationships among given services, and then the composition guidance can be derived.

Intuitively, by mining distributed historical service invocations, we can discover the knowledge or behavior rules, and learn the implied model of given services. In real paradigms, statistic computation is one of the frequently adopted approaches, and the Bayesian network (BN) [5] is such an effective model that can be used to represent the causal relationships implied among Web services. It is known that BNs are the graphical representation of probabilistic relationships between variables. They are widely used in nondeterministic knowledge representation and reasoning under conditions of uncertainty [5, 6, 7]. Modeling of Web services based on BN not only

* This work is supported by the Natural Science Foundation of Yunnan Province (No. 2005F0009Q), the Cultivating Scheme for Backbone Teachers in Yunnan University, and the Chun-Hui Project of the Educational Department of China (No. Z2005-2-65003).

G. Dong et al. (Eds.): APWeb/WAIM 2007, LNCS 4505, pp. 777–784, 2007.

can describe the causal dependencies with a graph structure, but also can give the quantitative measure of these dependencies.

In this paper, we focus on discovering causal relationships for elementary services, described as operations in WSDL documents. An approach to the probabilistic graphical modeling of Web services is proposed, and the method for constructing the Web services Bayesian network, denoted WSBN, is presented.

The Markov blanket (MB) of a variable *X* consists of *X*'s parents, *X*'s children, and parents of *X*'s children in a BN. Actually, MB describes the direct causes, direct effects, and the direct effects of direct causes of a variable [9, 10, 11]. In this paper, we develop a composition guidance of elementary services making use of the idea of MBs and corresponding reasoning mechanisms in the WSBN [5, 9, 11].

With preliminary experiments and performance analysis, the effectiveness and feasibility of the proposed method are verified.

The remainder of the paper is organized as follows: Section 2 introduces related work. Section 3 gives the method to constructing the WSBN. Section 4 presents the algorithm to developing the semantic guidance of services composition. Section 5 shows the experimental results. Section 6 concludes and discusses the future work.

2 Related Work

Similarity search for Web services is discussed in [4]. Firstly, the approaches to modeling Web services based on predefined rules and expert knowledge is discussed in [12, 13, 14, 15]. A lot of research work is oriented to the specific applications on Web services architectures [16, 17]. Secondly, the approaches to modeling Web services based on messages, events, activities and procedures are discussed in [2, 18].

However, both of these two classes of approaches are established on the pre-defined domain knowledge, which does not always make sense and is difficult to be updated and refined incrementally.

BNs have been used in many different intelligent applications [5, 6, 7]. Cheng et al. proposed the method for learning the BN from data based on information theory [8]. The concept of Markov blanket and its discovery are discussed in [5, 9, 10, 11].

Recently, there has been some research work of BN-based applications on Web services. In the semantic Web, BNs can be constructed from ontology by expanding OWL with probabilities [19]. The BN representing given domain knowledge is used to evaluate cost factors verses benefit factors of services [20]. In addition, Web services metadata are obtained based on the naïve Bayesian classifier [21]. According to our knowledge, the dynamic characteristic and inherent causal dependencies are almost not considered in these BN-based applications on Web services.

3 Modeling Elementary Services Based on the Bayesian Network

Following, we first give the definition of elementary services.

Definition 1. Let $ES=\{S_1, S_2, \ldots, S_n\}$ be the set of ordered elementary services in a given domain, in which S_i ($1 \le i \le n$) is a separate elementary service represented as an *operation* in the corresponding WSDL document.

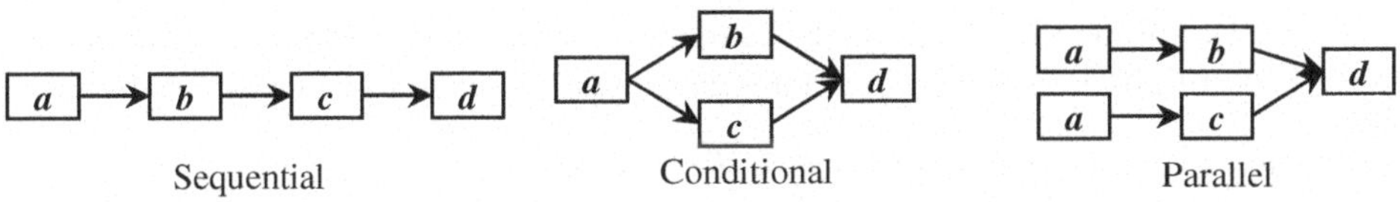

Fig. 1. Three basic types in Web services compositions

Fig. 1 shows the invocations of these three types with respect to elementary services *a*, *b*, *c* and *d*.

Now we give the following definition to universally describe service invocations.

Definition 2. Let $P=(id, p_s, c_s, \tau_b, \tau_e)$ represent direct invocations between two elementary services in composition procedures, and let T be a temporal domain of timestamps, in which *id* identifies a service composition procedure; p_s and c_s are the *parent* and *child* services in a service invocation respectively, $p_s \in ES$ and $c_s \in ES$; τ_b and τ_e are the *begin* and *end* times of the invocation from p_s to c_s respectively, and $\tau_b, \tau_e \in T$.

For any two instances p_1 and p_2 of P, if $p_1.id=p_2.id$, and $p_1.c_s=p_2.p_s$, then $p_1.\tau_e=p_2.\tau_b$. For example, $(1, a, b, \tau_1, \tau_2)$, $(1, b, c, \tau_2, \tau_4)$ and $(1, c, d, \tau_4, \tau_5)$ are instances of P containing direct invocations from the same procedure.

In this paper, for given elementary services, we will construct the semantic model from their historical invocations based on the BN.

Definition 3. A Bayesian network is a directed acyclic graph in which the following properties hold [5]: A set of random variables makes up the nodes of the network. A set of directed links connects pairs of nodes. An arrow from node *X* to node *Y* means that *X* has a direct influence on *Y*. Each node has a conditional probability table (CPT) that quantifies the effects that the parents have on the node. The parents of node *X* are all those that have arrows pointing to *X*.

A BN represents the joint probability distribution in products by the chain rule:

$$P(x_1,\cdots,x_n)=\prod_{i=1}^{n} P(x_i \mid Parents(x_i)) .$$

Based on the general definition of BNs, we will construct the elementary Web services Bayesian network (WSBN), $G=(ES, BE)$, to describe their implied causal relationships, in which *ES* represents the node set including given elementary services, and *BE* is the corresponding set of directed edges.

3.1 Fixpoint Deduction of Elementary Services Associations

The fixpoint of an initial data set can derive the fixed structure by a monotonic and iterative process of computation, and thus some indirect service associations can be deduced [22]. We adopt the basic idea of the fixpoint to obtain all the service associations completely by the deduction on the instances of *P*.

Definition 4. Let $\mathcal{L}=(id, p_s, c_s, \tau_b, \tau_e)$ represent all associations (direct and indirect) between any two elementary services, where $id, p_s, c_s, \tau_b, \tau_e$ are defined as those of *P* in Definition 2.

From Definition 3, $P \subseteq \mathcal{L}$ holds ultimately since only direct associations are described in P. In order to obtain $\mathcal{L}$ taking as input P, a recursive function is defined.

Definition 5. Let the function f from $(\mathcal{L}, P)$ to $\mathcal{L}$ be

$$\mathcal{L} = f(\mathcal{L}, P) = \pi_{1, 2, 8, 4, 10}(P \bowtie_{1=1 \wedge 3=2 \wedge 5=4} \mathcal{L}) \cup P, \tag{3-1}$$

where $P=(id, p_s, c_s, \tau_b, \tau_e)$, π and $\bowtie$ represent the *projection* and *join* operations respectively, similar to those in the relational algebra.

Initially, $\mathcal{L}$ is empty, i.e., $\mathcal{L}=\Phi$. We note P is given as a constant one, so equation 3-1 can be simplified to

$$\mathcal{L} = f(\mathcal{L}). \tag{3-2}$$

Clearly, f gives the recursive rule for defining the fixpoint computation [23]. The computation of f is iterative based on the result of the previous result, and f is monotonic. The instances of $\mathcal{L}$ are composed of two parts: the direct associations in P, and the indirect ones derived using equation 3-1. By using the above method, we can get the unique fixpoint given P, argued by Theorem 1.

Theorem 1. $\mathcal{L}$ that satisfies equation 3-2 is the least fixpoint of f.

□

For space limitation, the proof is ignored. For the monotonicity of f, we have $f^{\uparrow i}(\Phi) \subseteq f^{\uparrow i+1}(\Phi)$. Thus, let I_i be the instances of $\mathcal{L}$ after the i-th iteration, such that $I_i \subseteq I_{i+1}$ and suppose $I_{i+1} = I_i \cup \delta_{i+1}$, where δ_{i+1} is the incremental part. For any iteration in this process, the obtained instances of $\mathcal{L}$ must be included in the results of the next iteration. As well, we have $\delta_{i+1} = \pi_{1, 2, 8, 4, 10}(P \bowtie_{1=1 \wedge 3=2 \wedge 5=4} \delta_i) \cup P$. This idea is given in Algorithm 1. For the invocations of the first composition procedure given following Definition 2, $(1, a, c, \tau_1, \tau_4)$, $(1, b, d, \tau_2, \tau_5)$ will be obtained after the first iteration, and $(1, a, d, \tau_1, \tau_5)$ will be obtained after the second iteration.

3.2 Constructing the Elementary Web Services Bayesian Network

Based on the existing theory and approach, the WSBN will be constructed considering the specialty of Web services. It is well known that the most challenging and time-consuming operation is the tests of conditional independencies (CI tests). In this paper, we adopt the conditional mutual information to test whether X is independent of Y given Z, computed by the following equation:

$$I(X,Z,Y) = \sum_{\substack{x \in X \\ y \in Y \\ z \in Z}} P(x, y, z) \log\left(\frac{P(x, y \mid z)}{P(x \mid z) P(y \mid z)}\right). \tag{3-3}$$

If $I(X, Z, Y) \leq \varepsilon$, then X is conditionally independent of Y given Z, where ε is a given threshold. However, we note that $P(x, y, z)$, $P(x|z)$ and $P(y|z)$ in equation 3-3 cannot be computed directly from the sample data preprocessed by the deduction of fixpoint function. Thus, we first give a transformation for the sample data by augmenting the traces of service invocations. Let $M_{IST}=(m(i, j))_{|\mathcal{L}| \times n}$ $(1 \leq i \leq |\mathcal{L}|,\ 1 \leq j \leq n)$ be the spanning matrix of traces of invoked elementary services, in which $m(i, j)=1$ if S_j is in the trace of the i-th row of $\mathcal{L}$, and $m(i, j)=0$ otherwise. Fig. 2 gives an example of M_{IST}.

$$M_{IST} = \begin{bmatrix} a & b & c & d \\ 1 & 1 & 0 & 0 \\ 0 & 1 & 1 & 0 \\ 0 & 0 & 1 & 1 \\ 1 & 1 & 1 & 0 \\ 0 & 1 & 1 & 1 \\ 1 & 1 & 1 & 1 \end{bmatrix}$$

Fig. 2. A spanning matrix

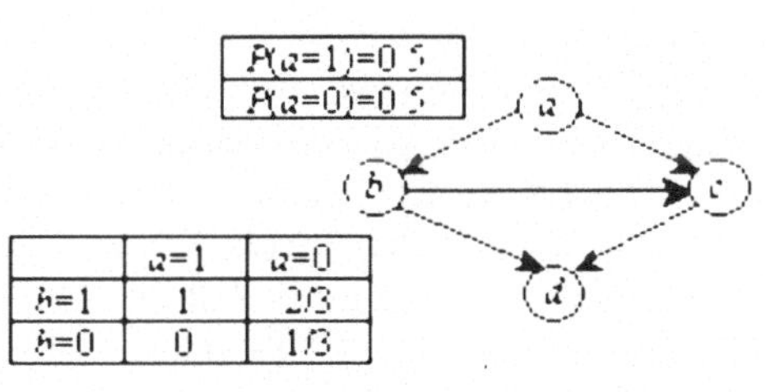

Fig. 3. The constructed WSBN

According to the general method to constructing a BN [5], the WSBN constructed from the M_{IST} in Fig. 2 is shown in Fig. 3, where the CPTs of *c* and *d* are ignored.

4 Generating Services Composition Guidance Based on the WSBN

Let us consider the WSBN of elementary services {*a*, *b*, *c*, *d*, *e*, *f*, *g*}, shown in Fig. 4 (the CPTs are ignored here). If *c* is one of the beginning services of a composition procedure, we can observe that *e* is likely to be concerned, since *e* is the child associated with *c* directly. As well, *d* is also likely to be concerned in the composition procedure, since it is another parent of *d*.

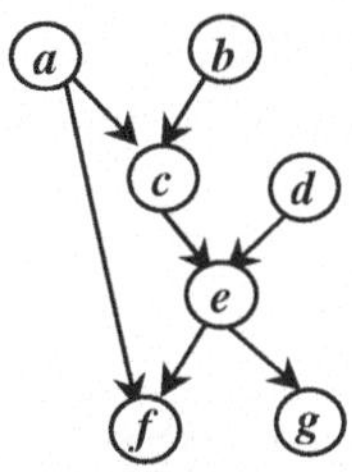

Fig. 4. An WSBN structure

We want to obtain the composition guidance that is universally suitable for the basic three types, composed of current node's children and the other parent nodes of these children step by step. Fortunately, the Markov blanket in the WSBN guarantees that the above two kinds of nodes are causally associated with the given node from the viewpoint of service invocation, while not associated with other nodes for conditional independence.

Definition 6. A Markov blanket (MB) *S* of an element $\alpha \in U$ (*U* is the set of elements in the BN) is any subset of elements for which $I(\alpha, S, U-S-\alpha)$ and $\alpha \notin S$.

The union of the following three types of neighbors is sufficient for forming a Markov blanket of node α: the direct parents of α, the direct successors of α, and all direct parents of α's direct successors [5]. The elements in the Markov blanket of an elementary service *S* ($S \in ES$) are causally associated with *S*. The invocation guidance is desired to demonstrate the immediate and subsequent services for each step. Additionally, the causal relationships among given services cannot be reversible when it comes to service invocations. Thus, we consider the associated services of *S* by the MB except its ancestors in the WSBN. For the WSBN in Fig. 4, *c* is directly associated with *e* and *d*, since *e* is *c*'s child and *d* is *e*'s another parent.

Definition 7. Let $Y_S=\{Y_1, Y_2, \ldots, Y_m\}$ be the children of *S* ($S \in ES$), and let F_j be the set of parents of Y_j ($1 \le j \le m$). Let $SN(S)=Y_S \cup F_1 \cup \ldots \cup F_m$ be the set of service neighbors of *S*. That is, $SN(S)=MB(S)-Parent(S)$, and each element in $SN(S)$ is called a service neighbor of *S*.

For example, $SN(c)=\{e, d\}$. Moreover, we always want to give the most probable or most associated services in each step instead of all possible ones. Although $SN(S)$

gives the associated services of S, how associated between them is not interpreted. We note that there exist following facts for the nodes in $SN(S)$:

(1) For each Y_j in Y_S, we consider the probability that Y_S may be invoked when S is invoked, $P(Y_S=1|S=1)$, which can be obtained directly from the CPT in the WSBN.

(2) If Y_j is likely to be invoked, we consider each f in F_j and the probability $P(f=1|Y_j=1)$. It is the posterior probability that can be computed by the Bayes formula based on the corresponding CPTs in the WSBN: $P(f=1|Y_j=1)=\left(P(Y_j=1|f=1)P(f=1)\right)/P(Y_j=1)$, in which $P(Y_j=1|f=1)$ and the marginal probabilities $P(f=1)$ and $P(Y_j=1)$ can be easily computed from the CPTs.

Definition 8. A service neighbor sn in $SN(S)$ is active if

(1) $sn\in Y_S$ and $P(Y_S=1|S=1)>ta_1$,

(2) $sn\in F_j$ and $P(Y_j=1|S=1)>ta_1$ and $P(sn=1|Y_j=1)>ta_2$,

where ta_1 and ta_2 are two given threshold values.

Definition 9. Given a WSBN $G=(ES, BE)$, let $SCG=(GB, GS, GE)$ be the services composition guidance, as a subgraph of G, in which

(1) GB is the set of beginning elementary services, and $GB\subseteq ES$;

(2) GS is the set of elementary services in SCG, and $GS\subseteq ES$. For each elementary service S in $GS-GB$, there is an elementary service S' ($S\neq S'$) in GS, such that S is an active service neighbor of S';

(3) GE is the set of directed edges in SCG.

Algorithm 1 gives the recursive method for constructing the WSBN.

Algorithm 1. GenerateSCG (*G*, *GB*): Generate *SCG* from the WSBN *G*

Initially, $GS=GB$ and $GE=\Phi$
1.for each S in GB do // starting from the elements in GB
2. for each y_s in $SN(S)$ do //consider the elements in $MB(S)-Parent(S)$
3. if $y_s\in Y_S$ and y_s is active then //if S' child is active
4. $GS\leftarrow GS\cup\{y_s\}$, $GE\leftarrow GE\cup\{(S, y_s)\}$
5. for each f_{ys} in F_{ys} do // consider the other parents of y_s
6. if f_{ys} is active then
7. $GS\leftarrow GS\cup\{f_{ys}\}$, $GE\leftarrow GE\cup\{(f_{ys}, y_s)\}$, GenerateSCG($G$, $\{f_{ys}\}$)
8. GenerateSCG(G, $\{y_s\}$)
9. output SCG

By Algorithm 1, services composition guidance can be generated. $SN(S)$ can be obtained in $O(n^2)$ time. Thus, Algorithm 2 can be done in $O(n^5)$ time for the worst case. Actually, less than $O(n^5)$ time will be necessary since the directed edges in WSBN are much less than those of the completely connected graph on ES.

5 Experimental Results

In this section, we mainly show the performance of constructing the WSBN. It is simulated on the machine with a 1.4GHZ P4 processor, 512M of main memory, running Windows 2000 server. The codes were written in JAVA, and JDBC-ODBC is used to communicate with DB2 (UDB 7.0). The elementary Web services and their invocations were generated by our program based on the real *City-Travel* services given by e-commerce Inc. [23], and revised considering the instances in [16].

Given 6 elementary services, the performance of generating M_{IST} and constructing the WSBN are shown in Fig. 5 and Fig. 6 respectively. Clearly, the time of generating M_{IST} is sensitively decreased with the increase of services composition procedures.

Meanwhile, by 50 services composition procedures and with the increase of the elementary services, the performance of preprocessing when generating M_{IST}, and the construction of WSBN are shown in Fig. 7 and Fig. 8 respectively. We note that by a certain number of composition procedures, the performance of generating M_{IST} only slightly decreases with the increase of elementary services, while the performance of constructing the WSBN on the generated M_{IST} is largely decreased.

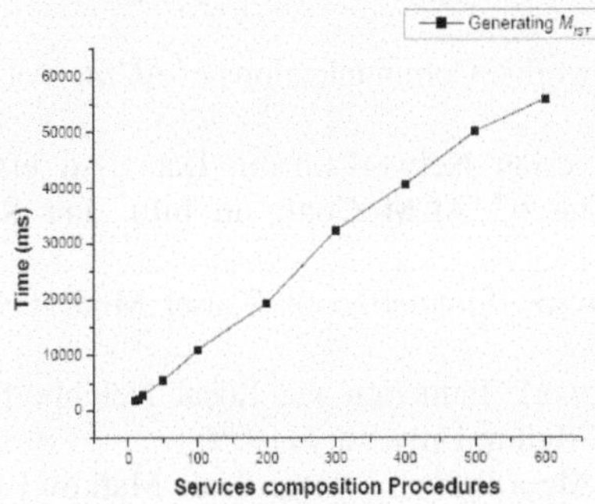

Fig. 5. Generating M_{IST} on 6 services

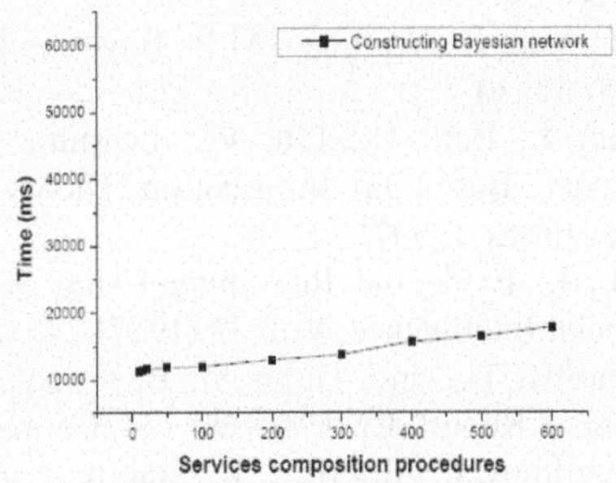

Fig. 6. Constructing WSBN on 6 services

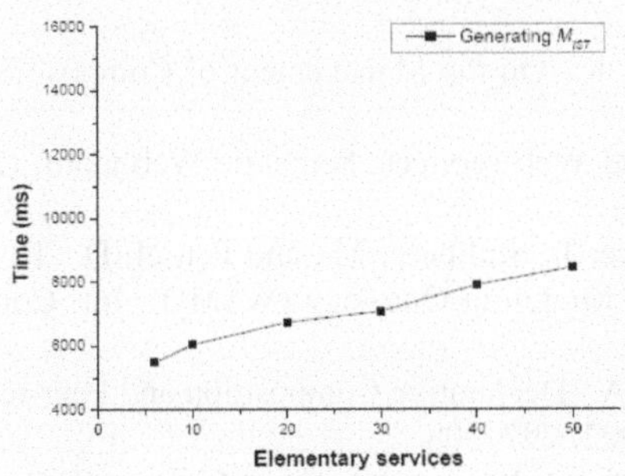

Fig. 7. Generating M_{IST} on increased elementary services

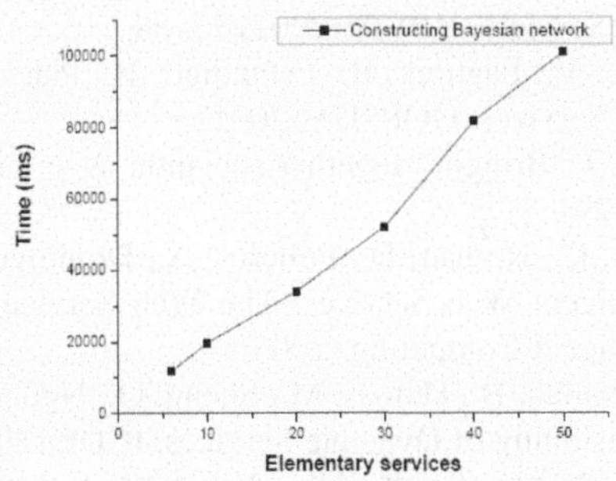

Fig. 8. Constructing WSBN on increased elementary services

Generally, the performance of our proposed method depends on the number of given elementary services and the size of historical services composition procedures. The experimental results show that our proposed approach is effective and feasible.

6 Conclusions and Future Work

In this paper, we propose an approach to the probabilistic graphical modeling of Web services based on the Bayesian network, and propose the services composition guidance based on the Markov Blankets in the WSBN. The proposed approach can be used to Web services clustering, intelligent services management, etc. As well, the behavior modeling of Web services to describe the inherent hierarchical, temporal and logical dependencies can be done based on the WSBN. These research issues are exactly our future work.

References

1. Yue, K., Wang, X., Zhou, A.: The Underlying Techniques for Web Services: A Survey. J. Software. Vol. 15. 3 (2004) 428–442
2. Dustdar, S. and Schreiner, W.: A Survey on Web Services Composition. Int. J. Web and Grid Services, Vol. 1. 1 (2005) 1–30
3. Hull, R., Su, J.: Tools for Design of Composite Web Services. SIGMOD (2004) 958–961
4. Dong, X., Halcvy, A., Madhavan, J., Ncmcs, E., Zhang J.: Similarity Search for Web Services. VLDB (2004)
5. Pearl, J.: Probabilistic Reasoning In Intelligent Systems: Networks of Plausible Inference. San Mateo. CA: Morgan Kaufmann Publishers, INC. (1988)
6. Pearl, J.: Propagation And Structuring In Belief Networks, Artificial Intelligence. Vol. 29. 3 (1986) 241–288
7. Heckerman, D., Wellman, M.P.: Bayesian Networks. Communications of ACM. Vol. 38. 3 (1995) 27–30
8. Cheng, J., Bell, D., Liu, W.: Learning Bayesian Networks from Data: An efficient Approach Based on Information Theory. The 6th ACM Conf. on Info. and Knowl. Management. (1997)
9. Pearl., J.: Evidential Reasoning Using Stochastic Simulation of Causal Models. Int. J. Artificial Intelligence. Vol. 32 (1987) 245–257
10. Margaritis, D., and Thrun S.: Bayesian Network Induction via Local Neighborhoods. Technical Report, CMU-CS-99-134, Carnegie Mellon University (1999)
11. Tsamardinos, I., Aliferis, C.F., Statnikov, A.: Algorithms for Large Scale Markov Blanket Discovery. 16th Int. FLAIRS Conf. (2003)
12. Narayanan, S., Mcilraith, S. A.: Simulation, Verification and Automated Composition of Web Services. WWW (2002) 77–88
13. Tosic, V., Pagurek, B., Esfandiari, B., Patel, K.: On the Management of Compositions of Web Services. OOPSLA (2001)
14. Peer, J.: Bringing together semantic Web and Web services. Semantic Web Conf. (2002) 279–291
15. Feier, C., Roman, D., Polleres, A., Domingue, J., Stollberg, M., and Fensel, D.: Towards Intelligent Web Services: The Web Service Modeling Ontology (WSMO). Int. Conf. on Intelligent Computing (2005)
16. Benetallah, B., Dumas, M., Sheng, Q., Ngu, A.: Declarative Composition and Peer-to-Peer Provisioning of Dynamic Services. ICDE (2002) 297–308
17. Amer-Yahia, S., Kotidis, Y.: A Web-Services Architecture for Efficient XML Data Exchange. ICDE (2004) 523–534
18. Bultan, T., Fu, X., Hull, R., Su, J.: Conversation Specification: A New Approach to Design and Analysis of E-Service Composition. WWW (2003)
19. Helsper, E. M., Van der Gaag, L. C.: Building Bayesian Network Through Ontologies. 15th European Conf. on Artificial Intelligence (2003)
20. Zhang, G., Bai, C., Lu, J., Zhang, C.: Bayesian Network Based Cost Benefit Factor Inference in E-Services. ICTITA (2004)
21. Heβ, A., Kushmerick, N.: Automatically attaching semantic metadata to Web services. IIWEB (2003)
22. Van Emden, M., Kowalski, R.: The Semantics of predicate logic as a programming language. JACM. Vol. 23. 4 (1976) 733–742
23. Web Services: Design, Travel, Shopping. http://www.ec-t.com

A Dynamically Adjustable Rule Engine for Agile Business Computing Environments

Yonghwan Lee[1], Junaid Ahsenali Chaudhry[2], Dugki Min[1], Sunyoung Han[1], and Seungkyu Park[2]

[1] School of Computer Science and Engineering, Konkuk University, Hwayang-dong, Kwangjin-gu, Seoul, 133-701, Korea
{yhlee,dkmin,syhan}@konkuk.ac.kr
[2] Graduate School of Information and Communication, Ajou University, Woncheon-dong, Paldal-gu, Suwon, 443-749, Korea
{junaid,sparky}@ajou.ac.kr

Abstract. Most agile applications have to deal with dynamic change of processes of automatic business policies, procedures, and logics. As a solution for the dynamic change of processes, rule-based software development is used. With the increase in complexity in modern day business system, the business rules have become harder to express hence require additional especially designed scripting languages. The high cost of modifying or updating those rules is our motivation in this paper. We propose a compilation-based dynamically adjustable rule engine that is used for rich rule expression and performance enhancement. Because of immense complications among and within business rules, we use Java language to create/modify rule instead of scripting languages. It gives us the facility of standardized syntax also. It separates the condition from action during run time which makes rule notification easier and quicker. According to experimental results, the proposed dynamically adjustable rule engine shows promising results when compared with contemporary script-based solutions.

1 Introduction

The revolution in computer systems and torrent of applications is led by growth in enabling technologies. The systems are increasing annually for 20 years by roughly a factor of 2 (disk capacity), 1.6 (Moore's Law), and 1.3 (personal networking; modem to Digital Subscriber Line (DSL)), respectively. The cost attached to manage the complex system of today is a lot more then the actual cost of the system. Among those applications (i.e. mission-critical applications, automatic process of business policies, procedures, and business logics) time is decisive. Better representation, organization and management of business processes in agile computing have helped optimize and fine tune the processes with the help of computer systems. Moreover, as the software industry has developed rapidly in various forms and smaller life cycles of software systems, companies need to produce highly competitive applications with many features like user adaptation, customization, software reusability, timeliness, low maintenance and fault free service etc. The component oriented software

G. Dong et al. (Eds.): APWeb/WAIM 2007, LNCS 4505, pp. 785–796, 2007.

engineering has stepped up and component-based software systems are growing in popularity. When software is divided into many dynamically connected components, the cost of immediate adjustment to new business processes or rearrangement of existing processes climbs high. So it is essential to develop software components that possess features of extensibility and flexibility, adapting to diverse requirements required upon each component's development and maintenance. Many researchers proposed a variety of adaptation methods for software components, emphasizing on extensibility and adaptability. However the application of those solutions in a real time application decreases performance that is the motivation of our work. In order to answer this weak point, the techniques of rule-based component development are proposed. For extensibility and adaptability of components, the techniques separate business variability [1] from the component's internal code by keeping separate rules. Upon the occurrence of requirement changes, a new requirement can be satisfied with changes in rules without changes in components. However, this technology usually needs some additional script language to describe rule expression, which has the limitation in expressing complex business rules. Also, this script-based rule handling is not suitable to the system that requires high performance.

In this paper, we propose the compilation-based rule engine for performance enhancement and improving rule expression to cope with dynamic system requiring runtime adjustments. Unlike interpretation-based rule engines proposed as contemporary solutions, our rule engine does not require any additional script language for expressing rules resulting into better performance in terms of time compilation and overall performance. Moreover, the solution we propose is able to use the current existing libraries for condition/action codes of rules in legacy systems, such as string, number, and logical expression etc, so that it may not only express complex condition or action statements but also easily integrate the existing systems developed in Java language. In agile business computing environments, computing systems have become highly capricious and complex. Our rule-based automatically changeable mechanism is an appropriate solution for bringing the benefits of automatic computing, trustworthy management, consistency, and easy maintenance in rule-based systems.

The remainder of this paper is organized as follows: In section 2 we present a scenario and functional features for better understanding. In section 3 we present the architecture of the rule engine proposed in this paper. We describe performance and compare the features of JSR-94 and the rule engine proposed in this paper in section 4. We discuss the related work in section 5 and lastly we conclude this paper along with the future work in section 6.

2 Solution of the Dynamically Adjustable Rule Engine

In order to apply a changing rule to a dynamically adjustable rule engine, it is an integral proposition that the rule engine should be adaptable to coupe up with regular updates and changes. The main procedure of our dynamically adjustable rule engine is that a rule writer composes a condition and an action part of a rule expression in Java language. The condition code and action code of a rule expression converts into condition and action object with *hook method* respectively and put them into an object pool. After finding a specific rule, our rule engine takes the condition and action objects specified by the rule's configuration from the object pool for rule execution. Processing a sample scenario is introduced in the following subsections.

2.1 A Sample Scenario of the Dynamically Adjustable Rules

Figure 1 shows the application example of customer's credit rule. Suppose that there is a rule of the customer's credit in import and export business domain.

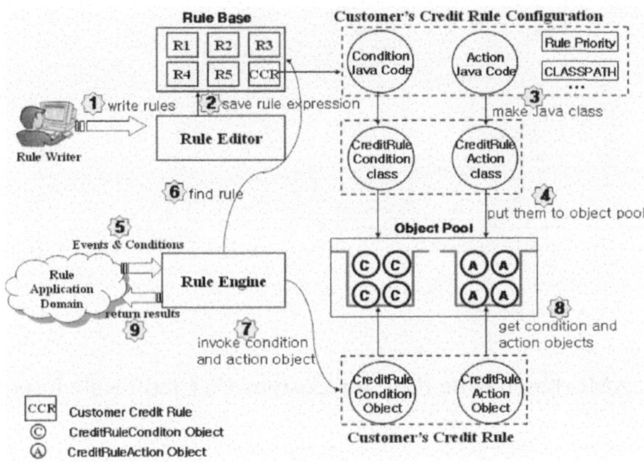

Fig. 1. Application Example of Customer's Credit Rule

Let us consider a simple credit rule: "*If a customer's credit limit is greater than the invoice amount and the status of the invoice is 'unpaid', the credit limit decreases by taking off the invoice amount and the status of the invoice becomes 'paid'.*"

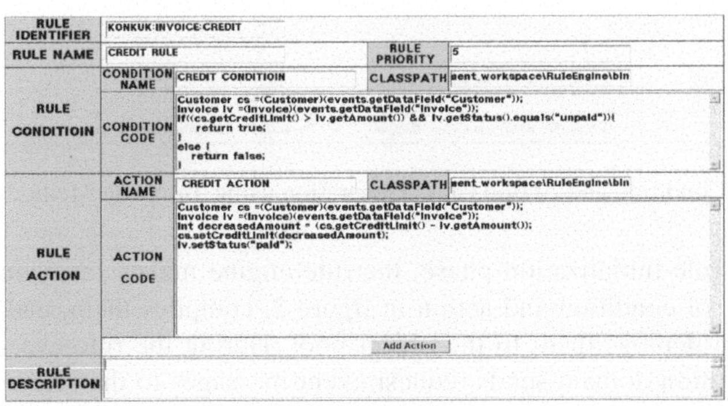

Fig. 2. Rule Expression for the Customer's Credit Rule with a Rule Editor

In this scenario, the process of applying the dynamically adjustable rules can be divided into 3 phases: 1) the rule expression phase, 2) the rule initialization phase, 3) and the rule execution phase. During the rule expression phase, a rule writer writes condition and action parts of the customer's credit rule using a rule editor as in figure 2. After writing the rule, the rule writer saves the customer's credit rule-related information to a rule base in form of an XML file. Figure 3 shows an example of XML-based rule base for the customer's credit rule.

```
<Configuration>
    <JavaClassPATH>E:\2006\development_workspace\RuleEngine\bin </JavaClassPATH>
    <JavaCompilerPATH>E:\2006\development_workspace\RuleEngine\bin </JavaCompilerPATH>

    <RuleDefinition>
        <ruleIdentifier>Konkuk:Invoice:Credit </ruleIdentifier>
        <ruleName>Credit Rule </ruleName>
        <rulePriority>5</rulePriority>
        <eventTokenID>Token1</eventTokenID>
        <conditionName>Credit Condition</conditionName>
        <ActionList>
            <ActionName>Credit Action</ActionName>
        </ActionList>
    </RuleDefinition>

    <ConditionDefinition>
        <ConditionName>Credit Condition</ConditionName>
        <CLASSPATH>.;E:\2006\development_workspace\RuleEngine\bin </CLASSPATH>
        <Code>
            Customer cs =(Customer)events.getDataField("Customer");
            Invoice iv =(Invoice)events.getDataField("Invoice");
            if(cs.getCreditLimit() > iv.getAmount() && iv.getStatus().equals("unpaid")) {
                return true;
            }
            else {
              return false;
            }
        </Code>
    </ConditionDefinition>

    <ActionDefinition>
        <ActionName>Credit Condition</ActionName>
        <CLASSPATH>.;E:\2006\development_workspace\RuleEngine\bin </CLASSPATH>
        <Code>
            Customer cs =(Customer)events.getDataField("Customer");
            Invoice iv =(Invoice)events.getDataField("Invoice");
            int decreasedAmount = (cs.getCreditLimit() - iv.getAmount());
            cs.setCreditLimit(decreasedAmount);
            iv.setStatus("paid");
        </Code>
    </ActionDefinition>

    <TokenDefinition>
        <TokenID>Token1</TokenID>
        <EventTyp>KONKUK:INVOICE:CREDIT</EventTyp>
    </TokenDefinition>
</Configuration>
```

Fig. 3. XML-based Rule Base for Customer's Credit Rule Expression

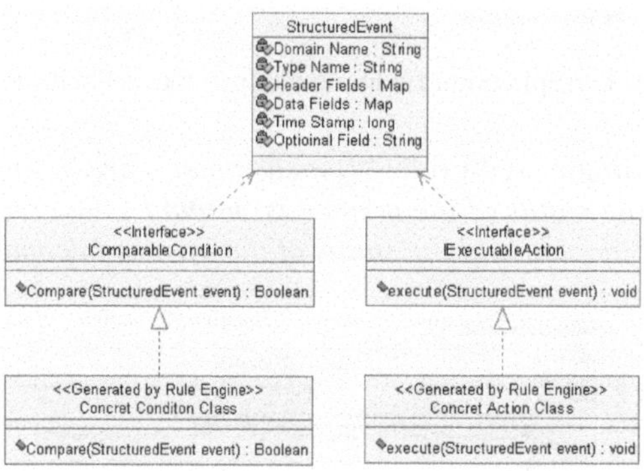

Fig. 4. Condition and Action Class Generation using Template Method Pattern

During the rule initialization phase, the rule engine makes Java source files from the Java codes of condition and action in figure 2, compiles them, makes *instance* of the classes and deploys them to the object pool. During the rule execution phase, if the rule application domain sends request event messages to the rule engine, the rule engine extracts the event identifier from the request event message. The rule engine finds the rule from a rule base by matching the event identifier. The rule engine takes condition and action objects from the object pool and invokes the *hook method* of the condition and action objects. In figure 2, the rule identifier is the unique name for finding the specified rule and the rule priority specifies the order of executing rules. It is also possible to use the existing libraries specified in *CLASSPATH*. If necessary, a rule writer can write multiple action codes for a rule.

2.2 Code Generation and Operation in the Rule Engine

In order to generate condition and action classes, the rule engine uses a template method pattern. Figure 4 shows the class diagram for applying the temple method pattern to our rule engine. The name of *hook method* for condition and action classes are "*Compare*" and "*Execute*", respectively. Figure 5 shows condition or action codes generated automatically through the template method pattern. The condition and action objects are made from the *CreditRuleCondition* and the *CreditRuleAction* class and put into an object pool to be used for executing the rule. When a rule application sends request events for rule execution to the rule engine, the rule engine extracts the event identifier from the request event message. The event identifier is the string of *"domain name: task identifier: rule name"*. The rule engine finds the rule from a rule base by matching the event identifier. The rule matched has rule configuration, such as rule identifier, rule name, condition or action class name, and rule priority. The rule engine takes condition and action objects from the object pool and invokes the hook method of the condition and action objects.

```
import dmics.ens.rule.*;
import dmics.ens.notification.*;

public class CreditRuleCondition implements IComparableCondition {
      public boolean compare(StructuredEvent events) {

            Customer cs =(Customer)(events.getDataField("Customer"));
            Invoice iv =(Invoice)(events.getDataField("Invoice"));
            if((cs.getCreditLimit() > iv.getAmount()) && iv.getStatus().equals("unpaid")) {
                  return true;
            }
            else {
                  return false;
            }
      }
}
```

An Example of the Condition Code

```
import dmics.ens.rule.*;
import dmics.ens.notification.*;

public class CreditRuleAction implements IExecutableAction {
      public void execute(StructuredEvent events)   {
            Customer cs =(Customer)(events.getDataField("Customer"));
            Invoice iv =(Invoice)(events.getDataField("Invoice"));
            int decreasedAmount = (cs.getCreditLimit() - iv.getAmount());
            cs.setCreditLimit(decreasedAmount);
            iv.setStatus("paid");
      }
}
```

An Example of the Action Code

Fig. 5. Condition and Action Code Generation for Customer's Credit Rule

3 Software Architecture of the Rule Engine

In the previous section, we studied a sample scenario with processing flow. This section introduces the architecture of the dynamically adjustable rule engine, which is operated based on compilation. Also we present flow of the initialization process in the rule engine and execution process of rules. In figure 6, we show the software architecture of the proposed rule engine. The rule engine is mainly comprised of three parts: the *Admin Console, Rule Repository,* and *Core Modules*. The *Admin Console* is the toolkit for the expressing and managing of rules. The *Rule Repository* saves the xml-based rule information expressed by the toolkit. The *Core Modules* are in charge of finding, paring, and executing rules. There are a number of modules in the *Core Modules*. The responsibility of the *Rule Engine* is to receive request message from a client and to execute rules. To find an appropriate rule, it sends the request message to the *Rule Parser*. The *Rule Parser* extracts the event identifier from the request message, compares it with the event identifier of a parsing table, and finds the rule. The event identifier is the string of "domain name: task identifier: rule name". After finding the rule, the *Rule Engine* knows the names of condition and action objects from the configuration of a rule and has the references of them from the *ObjectPool Manager*.

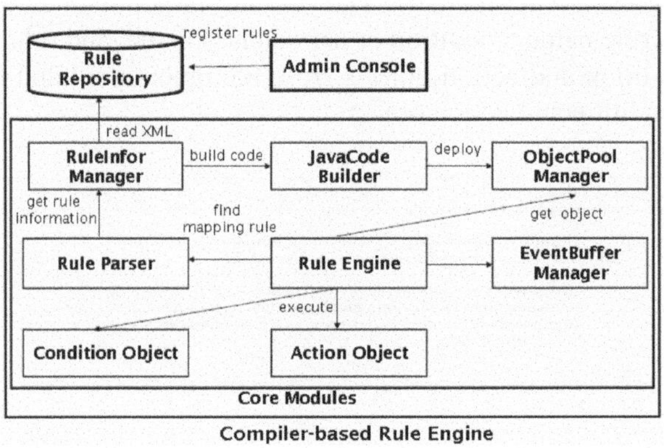

Fig. 6. Software Architecture of the proposed Rule Engine

The *Rule Parser* is responsible to find rules. The *ObjectPool Manager* manages the condition and action objects specified in rule expression. The *RuleInfor Manager* performs CRUD (Create, Read, Update, and Delete) action on the *Rule Repository*. The *JavaCode Builder* makes Java source files, compiles them, makes instances of the classes, and deploys them to the object pool. The *Condition* and *Action Objects* are the objects made from condition and action codes of rule expression.

The *Rule Engine* is required to initialize before executing rules. In figure 7, we show the collaboration diagram to show the flow of the process for rule engine initialization. The *Rule Engine* sends an initialization request to the *RuleInfor Manager*. The *RuleInfor Manager* reads rule information from the *Rule Repository* and save it to a buffer. Recursively, the *RuleInfor Manager* extracts condition and action codes of rules, makes object instances, and deploys to the object pool through the *ObjectPool Manager*. After the *Rule Engine* initializes the condition and action parts of rule, it calls the *Rule Parser* for building a parsing table. The *Rule Parser* gets a pair of rule identifiers and names from the *RuleInfor Manager*, and builds the parsing table with them for finding appropriate rules.

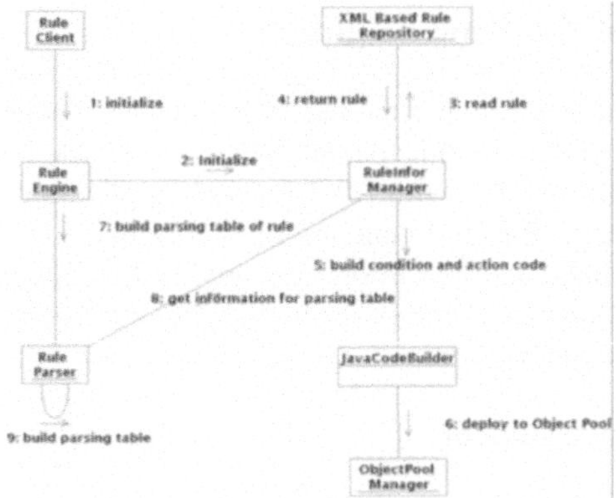

Fig. 7. Process Flow for Rule Initialization

Figure 8 presents the collaboration diagram to show the flow for rule execution. A client sends request messages to the *Rule Engine*. The *Rule Engine* saves it to a buffer through the *EventBuffer Manager* and then gets the request message with highest priority from the *EventBuffer Manager*.

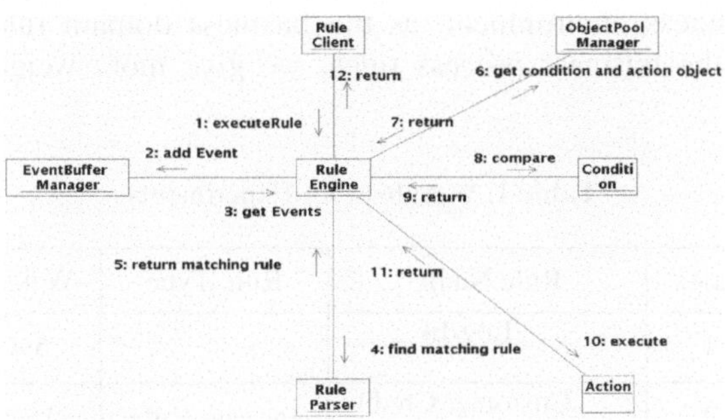

Fig. 8. Process Flow for Rule Execution

The *Rule Engine* calls the *Rule Parser* for finding the rule matched with the rule identifier. The *Rule Parser* searches the parsing table to find appropriate rules. After finding the rule, the *Rule Engine* calls the *ObjectPool Manager* to get the condition and action objects specified in the founded rule and then calls the "Compare" hook method of the condition object. If the result of invocation of the condition object is true, the *Rule Engine* calls the "Execute" hook method of the action object. If a rule has many action objects, the *Rule Engine* calls them according to the order of the action object specified in rule expression. The rule engine also supports the forward-chaining rule execution. It allows the action of one rule to cause the condition of other rules.

4 Performance of the Rule Engine

In this section, we show the experimental performance results of the compilation-based rule engine proposed in this paper. We use the Microsoft 2003 server for operation system, WebLogic 6.1 with SP 7 for web application server, and Oracle 9i for relational database. As for load generation, WebBench 5.0 tool is employed. TPS (Transactions per Seconds) and execution time are used for the metric of performance measurement. For performance comparison in J2EE environment, we use a servlet object as a client of the rule engine.

4.1 Experimental Environment

Before showing the performance results, we introduce the workloads that were used in the experiments. Generally, business rules are classified into business process rules and business domain rules. Business domain rules define the characteristics of variability and the variability methods which analyzes these characteristics for an object. Business process rules define the occupation type, sequence, and processing condition, which is necessary to process an operation. In the business process rule, the variability regulations for process flows are defined as the business process rules. Table 1 shows the workload configuration for experiments. Among the five rules, two rules are the business process rules and the other two rules are the business domain rules. In an e-business environment, as the business domain rules are more frequently used than the business process rules, we give more weight to the business domain rules.

Table 1. Workload for Experiments

Index	Rule Name	Rule Type	Weight
1	Log-In	-	5%
2	Customer Credit	Process Rule	15%
3	Customer Age	Domain Rule	30%
4	Interest Calculation	Process Rule	15%
5	Role Checking	Domain Rule	35%

The "*Customer Age*" rule measures a customer's age is according to problem request. The "*Interest Calculation*" rule calculates interest according to the interest rates. The "*Role Checking*" rule specifies the assertion of the "An authorized user can access certain resources." The rule engine takes role information from the profiles of the customer and decides whether the requesting jobs are accepted or not.

4.2 Performance Comparison

The performance of the proposed rule engine is compared with Java Rule Engine API (JSR-94) in figure 9. The proposed rule engine achieved 395 transactions per second

(TPS) in maximum workload. While JSR-94 achieved 150 TPS in maximum. The proposed rule engine in this paper processes 245 more transactions per second than JSR-94. We believe that the rule engine proposed achieved 2.5 times better performance than JSR 94 because of its special emphasis on features like ease in extensibility and highly levels of adjustability for rules that are used in a system. In order to compare performances of sub-modules of the rule engine, Figure 10 shows load analysis of two rule engines. Since the proposed rule engine operates on compilation-based rule processing, performance in the module of generating objects may take a long execution time, but there is not a big different in performance. Moreover, the proposed rule engine achieves better performance results in parsing and executing rules. It is because it divides the condition and action class into separate parts which gives ease at run time when rules are called in an object pool. . Moreover one does not have to define separate condition statement for multiple actions the proposed rule engines provides the facility of defining more than one executions for one condition which can help in fault tolerance in a hybrid environment.

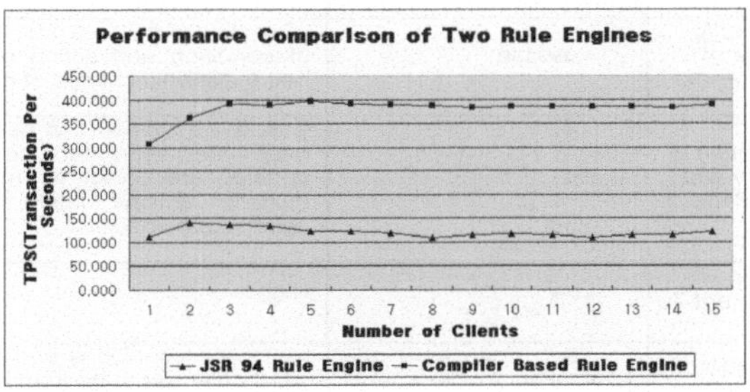

Fig. 9. Performance Comparisons with JSR-94

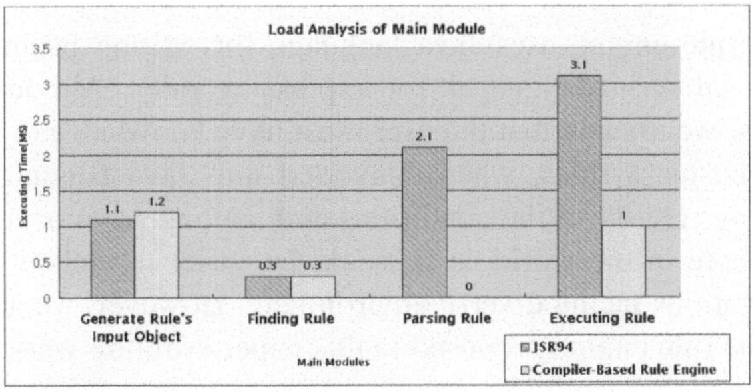

Fig. 10. Comparison of Load of Two Rule Engines

4.3 Feature Comparison

In table 2, we compare the features of the two rule engines. In contrast to JSR-94, the proposed rule engine expresses each business rule by a business task unit. If there are one or more rules in a task, each rule is categorized in a unique rule name.

Table 2. Feature Comparison between the Two Rule Engines

	JSR-94 Rule Engine	The proposed Rule Engine
Performance (Max TPS)	150 TPS	395 TPS (2.5 times better performance)
Rule Expression	A rule expression is confined by the JESS script rule language	- need to learn Java language - can express complex business rules using the Java language
Reusability of existing Libraries	Impossible	- possible by using the CLASSPATH in rule expression
Integration of existing system using rule engine	- needs additional rule expression for integrating existing systems	- easier to integrate with existing systems in Java language
Easy to Learn	- An application domain expert is easier to write rules. - needs to learn additional script-based rule language	- Any Java coder can be easier to write rules - Learning additional rule language is not required
Dynamic Change of Business Rules	Possible	- Possible (An object pool mechanism of condition and action objects can make dynamic change of rules.)
Separation of Condition and Action Parts	No	Yes, the condition and action part of rules are separated so that the updates are easier to manage and multiple actions could be taken against one condition.
Ease of embedment	Low	High
The condition/action dependability	Yes, causes rule evaluation to block until a condition becomes true or an event is raised	No, since conditions and events are 'physically' separate from each other, it gives the proposed engine an edge on time constraint.

The proposed rule engine uses Java language for writing business rules without using any additional script languages for expressing rules. Although it might seem odd to assume that we assume that the user must have knowledge of java language we foresee that the business rules, when converted into Java language, eliminates the fuzziness and brings clarity to the conditions and actions. Moreover syntax of java is the same everywhere in the world so it would be easier to embed the proposed rule engine into applications facing diverse environment. However, we aim to build a GUI based front end the rule engine proposed in this paper as future work.

Whenever executing each business rule in the proposed rule engine, the step for matching rule conditions is not required. In other words, after finding the required business rule from a rule base, the proposed rule engine executes it without parsing the rule and matching the rule conditions due to Java-based rule expression. The proposed rule engine converts the condition and action codes of a rule into condition and action objects, respectively and puts it into an object pool for improving performance and dynamic changeability. Thus, it can execute the newly changing business rule without restarting itself.

5 Related Works

The Business Rules Group [2] defines a business rule as "*a statement that defines and constraints some aspects of business*". It is intended to assert business structure or to control or influence the behavior of the business. The Object Management Group (OMG) is working on Business Rules Semantics [3]. Nevertheless, several classifications of different rule types have emerged [2, 4, 5]. In [4], business rules are classified into four different types, such as integrity rules, derivation rules, reaction rules, and demonic assignments. A well-known algorithm for matching rule conditions is RETE [6]. For business rule expression, rule markup language is needed. Currently, BRML (Business Rule Markup Language) [7], Rule Markup Language (RuleML) [8], and Semantic Web Rule Language (SWRL) [9] are proposed as rule markup languages. The IBM took initiative of developing Business Rule Markup Language (BRML) for the Electronic Commerce Project [7]. The BRML is an XML encoding which represents a broad subset of KIF. The Simple Rule Markup Language (SRML) [10] is a generic rule language consisting of a subset of language constructs common to the popular forward-chaining rule engines. Another rule markup approach is the Semantic Web Rule Language (SWRL), a member submission to the W3C. It is a combination of OWL DL an OWL Lite sublanguages of the OWL Web Ontology language [9]. The SWRL includes an abstract syntax for Horn-like rules in both of its sublanguages. Most recently, the Java Community Process finished the final version of their Java Rule Engine API. The JSR-094 (Java Specification Request) was developed in November 2000 to define a runtime API for different rule engines for the Java platform. The API prescribes a set of fundamental rule engine operations based on the assumption that clients need to be able to execute a basic multiple-step rule engine cycle (parsing the rules, adding objects to an engine, firing rules, and getting the results) [11]. It does not describe the content representation of the rules. The Java Rule API is already supported (at least partially) by a number of rule engine vendors (cf. Drools [12], ILOG [13] or JESS [14]) to support interoperability.

6 Concluding Remarks

As business applications become complex and changeable, rule-based mechanism is needed for automatic adaptive computing as well as trustworthy and easy maintenance. For this purpose, we propose a compilation-based rule engine that can easily express business rules in Java codes. It does not need additional script language for expressing rules. It can create and execute condition and action objects at run time. Moreover, it can use existing libraries for condition or action codes of rules (i.e., String, Number, and Logical Expression) so that it can not only express complex condition or action statements but also easily integrate the existing systems developed in Java. So the compilation-based rule engine, proposed in this paper, shows better performance than JSR-94, a generally used interpretation-based rule engine. According to our experiments, the proposed rule engine processes 245 more transactions per second than JSR-94. We intend to test the performance of the rule engine proposed in this research with different weights and in different conditions. This will not only gives us a better idea about the working capacity of the outcome of this research, it will give clear application of are for this rule engine too. Moreover we intend to develop a GUI that could assist the users who have limited knowledge of java in operating with this rule engine.

References

1. Lars Geyer and Martin Becker, "On the influence of Varaibilities on the Application-Engineering Process of a Product Famliy",Proceedings of SPLC2, 2002.
2. The Business Rules Group. Defining Business Rules – What Are They Really? http://www.businessrulesgroup.org/first paper/br01c0.htm, July 2000.
3. B. von Halle. Business Rules Applied. Wiley, 1 edition, 2001.
4. K. Taveter and G.Wagner. Agent-Oriented Enterprise Modeling Based on Business Rules? In Proceedings of 20th Int. Conf. on Conceptual Modeling (ER2001), LNCS,Yokohama, Japan, November 2001. Springer-Verlag.
5. S. Russell and P. Norvig. Artificial Intelligence –A Modern Approach. Prentice Hall, second edition, 2003.
6. C. Forgy. RETE: a fast algorithm for the many pattern/many object pattern atch problem. Artificial Intelligence, 19(1):17–37, 1982.
7. IBM T.J. Watson Research. Business Rules for Electronic Commerce Project. http://www.research. ibm.com/rules/home.html, 1999.
8. RuleML Initiative. Website. http://www.ruleml. org.
9. W3C. OWL Web Ontology Language Overview. http: //www.w3.org/TR/owl-features/. W3C Recommendation 10 February 2004.
10. ILOG. Simple Rule Markup Language (SRML). http://xml.coverpages.org/srml..html, 2001.
11. Java Community Process. JSR 94 - Java Rule Engine API. http://jcp.org/
12. aboutJava/communityprocess/final/jsr094/index. html, August 2004.
13. Drools. Java Rule Engine. http://www.drools.org.
14. ILOG. Website. http://www.ilog.com.
15. JESS. Java Rule Engine. http://herzberg.ca.sandia.gov/jess.

A Formal Design of Web Community Interactivity

Chima Adiele

University of Lethbridge
Lethbridge, Alberta, Canada
chima.adiele@uleth.ca

Abstract. Web Communities (WCs) are emerging as business enablers in the electronic marketplace. As the size of the community becomes increasingly large, there is a tendency for some members of the community to use resources provided by the community without necessarily making any contribution. It is, therefore, necessary to determine members' contributions towards sustaining the community. In this paper, we present a formal framework to dynamically measure the interactivity of members, and indeed the interactivity level of the community. This formal foundation is necessary to eliminate *ad hoc* approaches that characterize existing solutions, and provide a sound foundation for this new research area. We design an efficient interactivity algorithm, and also implement a prototype of the system.

Keywords: Formal specification, Web communities, and interactivity lifecycle.

1 Introduction

A Web community (WC) is a Web-enabled communication and social interaction between a group of people that have common interests. Rheingold [1] envisions a WC as a social phenomenon that has no business dimension. Recent advances in information and communication technologies, however, have given impetus to WCs as business enablers in the digital marketplace. Many organizations leverage virtual communities to attract new and retain old customers by identifying the needs and beliefs of their customer base, and hence, create value through intention-based customer relationships [2,3].

The main thrust of this paper is to provide a formal framework to measure interactivity of members in a WC, and also determine the community's interactivity level. Interactivity relates to the level of participation of a member in a given community, and the usefulness of such contributions to the needs of the community. To achieve the envisioned objectives, we leverage algebraic signatures to formally specify components of the interactivity model to provide a sound foundation. The use of formal and theoretical foundations is particularly important for this new research area to guarantee correctness and completeness of the system. We design an interactivity model that uses a common term vocabulary (CTV) to automatically filter irrelivant messages from the community. Automatically filtering irrelivant messages eliminates the manual process that is time consuming, labour intensive, and error prone. In addition, we provide an efficient interactivity algorithm and implement a prototype of the system.

G. Dong et al. (Eds.): APWeb/WAIM 2007, LNCS 4505, pp. 797–804, 2007.

The remaining part of this paper is structured as follows. In Section 2, we provide background information on our specification, and also discuss related work. Section 3 examines the dynamics of interactivity, and hence, presents a formal framework for a WC interactivity. We design an interactivity algorithm in Section 4, while Section 5 concludes the paper and provides insight into future work.

2 Background

The specification in this paper uses set notations ($\cap, \cup, \subseteq, \supseteq, \notin, \in, \mathrm{N}, \mathbb{P}$) to describe structural components, and predicate logic to describe pre- and post-conditions for any requirements. Pre- and post-conditions are stated as predicates. A simple predicate usually has one or more arguments, and is of the form P(x), where x is an argument that is used in the predicate P. The universal quantifier ($\forall$) and existential ($\exists$) quantifiers are in common use. Every declaration must satisfy a given constraint. In general, a quantified statement can be written in one of two forms:

1. <quantifier><declaration(s)> • <predicate>;
2. <quantifier><declaration(s)> | <constraint> •
   <predicate>

The symbols "|" and "•", which are part of the syntax mean "satisfying" and "such that", respectively. To create compound predicates, statements can be nested and combined together using one or more logical connectives, such as: and ($\wedge$), or ($\vee$), not ($\neg$), conditional ($\Longrightarrow$), and bi-conditional ($\Longleftrightarrow$). The formal specification of a requirement in this paper follows the general format of a quantified statement.

There are some previous research efforts that are tangentially related to our work. Lave and Wenger [4], and Menegon and D'Andrea [5] observe that members of a community develop shared practice by interacting around problems, solutions, and insights, and building a common store of knowledge. Blanchard and Markus [6] argue that "the success of community support platforms depends on the active participation of a significant percentage of the community members". Community participation is necessary for sustained interactivity. Some research [7,6] have examined the effect of size and under-contribution for online communities. These research work suggest ways of using concepts in social psychology to motivate contributions. In this paper, we provide a formal framework to dynamically measure the interactivity of members, and indeed the interactivity level of the community.

3 Formal Framework for a WC Interactivity

To discuss the formal framework of a WC interactivity model, we first examine its dynamics. We use the interactivity lifecycle in Figure 1 to discuss the dynamics of a WC interactivity. It is a multi-user, Web-based system designed to provide a WC where members can interact and exchange ideas. The system has several servers in a server farm to manage and display the different types of media (text, images, audio,and video). Video frames need to be transmitted quickly and in synchrony but at relatively low resolution to support video conferencing. Video contents may be compressed in a store, so the video server may handle video compression and decompression into different formats. There is also an audio server that facilitates teleconferencing. Both the audio and video servers are used to manage activities in the subset of conferencing activities. The other activities (such as post messages, read messages, reply messages, etc.) in the WC fall under message activities. There are different data servers used to manage messages and display member's interactivity records. These data servers provide support for extensive queries and scripting facilities to enable members interact.

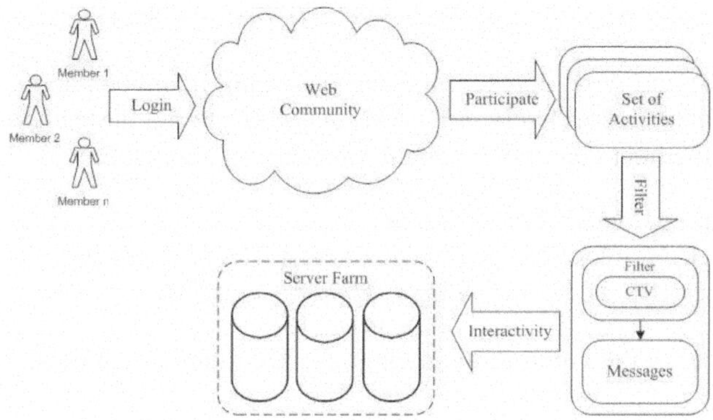

Fig. 1. WC interactivity Diagram

To address the issue of posting irrelivant messages that have nothing to do with the subject of discussion, some communities moderate messages posted. Manually moderating messages in large communities can be time consuming, labour intensive, and error prone. Therefore, there is a need to automate the process of filtering messages that are posted in a given community. We leverage a (CTV) to automatically filter messages before they are posted. A CTV is an ontology that contains primitive terms in a given domain and does not prescribe any structure for its designers [8]. When a member writes a message, that message has to pass through a filter mechanism. The filter mechanism, which uses the CTV, is an accepting device that either accepts a message and it is posted, or rejects otherwise [9].

3.1 Formal Foundation

Members loyalty to the community varies according to their level of participation in the community. Adiele and Ehikioya [8] identified three categories of membership, namely executive, senior and ordinary members with corresponding degrees of participation. Butler [7] identified similar categories of membership, namely leaders, active and silent users. Accordingly, we classify members into three groups, namely: Leading Members (LM)- these are members that make substantial contributions to the community by posting, responding and reading messages on a regular basis; Active Members (AM)- these are members that make some contributions to the community that are far less than the contributions of LM; and Non-active Members (NM)- these are members that make minimal or no contributions at all to the community.

We model Members participation as a function of class of membership. Accordingly, the following inequalities hold:

$$LM_{num} \leq AM_{num} \leq NM_{num}(\text{"num" is the number of members}) \quad (1)$$

$$LM_{cont} \geq AM_{cont} \geq NM_{cont}(\text{"cont" is the contributions of members}) \quad (2)$$

Let MEMBER be the basic type for members of a WC. Let *Mem* be a non-empty power set of members (i.e. *Mem*: $\mathbb{P}_1$ MEMBER). There are three classes of membership divided according to members' participation levels over a specified time window [7].

Let *LM*, *AM*, and *NM* represent the set of leading members, active members and non-active members respectively. *LM*, *AM*, and *NM* are the three classes of membership and every member can only belong to one class at a given time.

$$\begin{aligned} &\forall m_i : \text{MEMBER}|m_i \in Mem\bullet \\ &\quad \exists LM, AM, NM : \text{MEMBER}|LM, AM, NM \subset Mem\bullet \\ &\qquad \textstyle\bigcup_{j=1}^{n}(LM, AM, NM) = Mem \wedge \bigcap_{j=1}^{n}(LM, AM, NM) = \emptyset \end{aligned} \quad (3)$$

Every member in the community is unique. We capture this uniqueness formally as follows:

$$\begin{aligned} &\forall m_i, m_j : \text{MEMBER}|m_i, m_j \in Mem\bullet \\ &\quad m_i = m_j \Longrightarrow i = j \end{aligned} \quad (4)$$

Activity: In a WC, a member performs certain actions, which we call activities to contribute to the community. Different sets of activities have differnt parameters of measurements. For example, we count the number of messages that a member may have posted, read or replied to determine the member's contributions. While we measure the time a member spends on video conferencing or teleconferencing to determine the member's contributions. We refer to the former as message activities and the latter as conferencing activities. Let *MA* represent the set of message activities and *CA* represent the set of conferencing activities.Thus,

$$(MA \cup CA) = A; and \ \ (MA \cap CA) = \emptyset \quad (5)$$

Let ACTIVITY be the basic type for activities that members can participate (a formal definition of *Participate* is given in (8)) and *A*, a power set of activities, such that *A*: $\mathbb{P}_1$ ACTIVITY.

Definition 1: An activity, a_i, is an action that a member, m_j, undertakes in a WC to contribute to the community.

In every WC, an activity, $a_i \in A$ has a measure of importance. That importance is captured by the weight w_i. The weight of an activity is assigned relative to the importance of the activity in a given community. Let *W* be the set of weights for a corresponding set of activities *A*. Let VALUE be the basic type of values. The product of a_i and w_j

represents the value of the activity in a given community. We define a function *Value* that returns the value of each activity.

$$\begin{aligned}&Value : \text{ACTIVITY} \times \text{WEIGHT} \rightarrow \text{VALUE}\\&\quad \forall a_i : \text{ACTIVITY}|(a_i \in A) \bullet \exists w_j : \text{WEIGHT}|w_j \in W\bullet\\&\qquad Value(a_i, w_j) = (a_i * w_j)\end{aligned} \tag{6}$$

We define a function *Participate* that returns the activity a member participates in.

$$\begin{aligned}&Participate : \text{MEMBER} \rightarrow \text{ACTIVITY}\\&\quad \forall m_j : \text{MEMBER}|m_j \in Mem \bullet \exists a_i : \text{ACTIVITY}|a_i \in A\bullet\\&\qquad Participate(m_j) = a_i\end{aligned} \tag{7}$$

A member can only participate in one activity at a given time instance. Let t be a time instance of type TIME, we capture this constraint formally:

$$\begin{aligned}&\forall t : \text{TIME} \bullet (\exists m_j : \text{MEMBER} \wedge \exists_1 a_i : \text{ACTIVITY})\bullet\\&\quad Participate(m_j) = a_i\end{aligned} \tag{8}$$

To participate in a WC, a member has to log in to the system. We define status of members to facilitate the *Login* operation. $Status = \{ON, OFF\}$. Formally,

$$\begin{aligned}&Login : \text{MEMBER}\\&Login(m_j) = TRUE \Leftrightarrow \forall m_j : \text{MEMBER}|m_j \in Mem\bullet\\&\quad Status = ON\end{aligned} \tag{9}$$

A member who logs into the system can also log out at will. The definition of *Logout* follows.

$$\begin{aligned}&Logout : \text{MEMBER}\\&Logout(m_j) = TRUE \Leftrightarrow \forall m_j : \text{MEMBER}|m_j \in Mem\bullet\\&\quad Status = OFF\end{aligned} \tag{10}$$

To simplify our exposition and facilitate understandability, we discuss a subset of the activities. For example, start posts (sP) for a message that begins a thread; reply post (rP) for a message that responds to another message, thus building the thread; Reads R for messages read by a member. Let MESSAGE be the basic type for messages and MA be a set of messages for messages posted / replied to in a WC. MA is a subset of activities, such that $MA = \{sP, rP\}$, where $MA \subset A$. Let tM be the total number of messages, such that $sP + rP \leq tM$.

We specify a generic CTV that provides enterprise-wide definition of terms (called context labels) to automate the process of filtering messages. The CTV is organized hierarchically using linguistic relations to show how terms relate to one another. To capture these linguistic relations, we let CONTEXT-LABEL be the basic type for context labels. Let $LRI = \{synon, hyper, hypon, meron\}$ be the set of linguistic relationship identifiers, where $synon, hyper, hyponandmeron$ are synonym, hypernym, hyponym, and meronymy, respectively. To define the CTV, we first, define a context label, cl, as a primitive term (word) that has a unique meaning in the real world. A formal definition of linguistic relation follows. Let $\mathfrak{R}$ be a linguistic relation, then:

$$\mathfrak{R} : CL \times CL \longrightarrow LRI \tag{11}$$

Definition 2: A CTV is a pair $(CL, \mathfrak{R})$, where CL is a set of context labels and $\mathfrak{R}$ is a linguistic relation which shows that given $cl_i, cl_j \in CL$, then the relationship between cl_i and cl_j is one of $\{synon, hyper, hypon, meron\}$ (i.e., $\mathfrak{R}(cl_i, cl_j) \in LRI$).

Definition 3: A *filter mechanism*, FM is an accepting device, which uses the CTV to parse words in a message, and if the message meets a given acceptance standard the message is accepted, otherwise it is rejected.

To represent this partial function formally, we let DATABASE be the basic type of database. Only messages parsed by the filter mechanism are posted.

$$FM : MESSAGE \times CTV \longmapsto DATABASE \tag{12}$$

We define a function $Update$ that updates the database. To enable us define the function $Update$, we give the signature of $Write$, a function that writes into the database.

$$Write : \text{DATABASE} \rightarrow \text{DATABASE} \tag{13}$$

$$\begin{aligned}
&Update : \text{MEMBER} \times \text{ACTIVITY} \rightarrow \text{DATABASE}\\
&\forall m_i : \text{MEMBER} | m_i \in Mem \bullet\\
&\quad \exists a_i : \text{ACTIVITY} | a_i \in A \wedge tM : \text{MESSAGE} \bullet Update(m_i, a_i) \Longrightarrow\\
&\quad \forall a_i : \text{ACTIVITY} | (a_i = sP) \Longrightarrow Write(sP + 1) \vee\\
&\qquad \forall a_i : \text{ACTIVITY} | (a_i = rP \wedge (tM = sP \cup rP \wedge sP \cap rP = \emptyset)) \Longrightarrow\\
&\qquad Write(rP + 1) \vee \forall a_i : \text{ACTIVITY} | (a_i = R \wedge R < tM) \Longrightarrow Write(R + 1)
\end{aligned} \tag{14}$$

Interactivity: Let WC be a Web community, there exists a set of members Mem and a set of activities A, such that a member, $m_i \in Mem$ participates in activities, $a_i \in A$.

Definition 4: The interactivity of a member, m_j of a WC for a given time window, W (written, I_{WI}) is the sum of the values, v_k of the activities that m_j participates in over the width of W. Formally,

$$\begin{aligned}
&Interactivity \triangleq I_{WI} : \text{VALUE} \bullet \forall m_j : \text{MEMBER} | m_j \in Mem \bullet\\
&\quad (\exists a_i : \text{ACTIVITY} | (a_i \in A \wedge S : \text{TIME}) \bullet\\
&\qquad Participate(m_j) = a_i \wedge \exists w_k : \text{WEIGHT} | w_k \in W) \bullet\\
&\qquad I_{WI} = \Sigma_S(Value(a_i, w_k))
\end{aligned} \tag{15}$$

Definition 4 represents the interactivity of a member in a WC. We extend this definition to obtain the interactivity of a community. The interactivity of a community I_{WC} is the sum of the individual interactivity I_{WI} over the size of the community CS. Formally,

$$\begin{aligned}
&Interactivity \triangleq I_{WC} : \text{VALUE} \bullet\\
&I_{WC} = \Sigma_{CS}(I_{WI})
\end{aligned} \tag{16}$$

4 Overview of the System

In this section, we present an interactivity algorithm that describes how to capture the interactivity of members in a WC. We also describe a prototype of the system.

Algorithm: Measure-Interactivity
Input: (Unique member's ID and members's activities)
Output: (Member's interactivity level)

```
while login(Mid)
   Participate in activity
    if a_i ∈ MA and a_i = R then
       search(messages); read(messages);
       computeInteractivity(Mid);
    else if a_i is any of (sP, rP, Res) then
       filter(messages); updateDb( );
    else if a_i ∈ CM then
       T_1 = startTime(conferencing);
       T_2 = stopTime(conferencing);
       T = T_2 - T_1;
       computeInteractivity(Mid);
      updateDB(messages);
  end(while)
end.
```

We implemented a prototype of the WC on a client-server architecture using apache server 1.3.34 (Unix) as our Web server and Javascript as our main development language for the application server. Apache HTTP Server is a stable, efficient and portable open source HTTP web server for Unix-like systems. JavaScript permits easy vertical migration in future, and allows platform independence. We used CSS to specify the presentation of elements on the Web page independent of the document structure. At the backend, we used MySQL version 4.1.0 as the database and the application uses SQL query language to manipulate the database. Our prototype uses PHP to connect the client to database server and to run queries in the database from the client side.

Figure 2(a) is a screen shot of a discussion group showing messages that members posted. When a member posts a message, the filter mechanism uses the CTV to parse the message. Figure 2(b) shows how a member can search for posted messages.

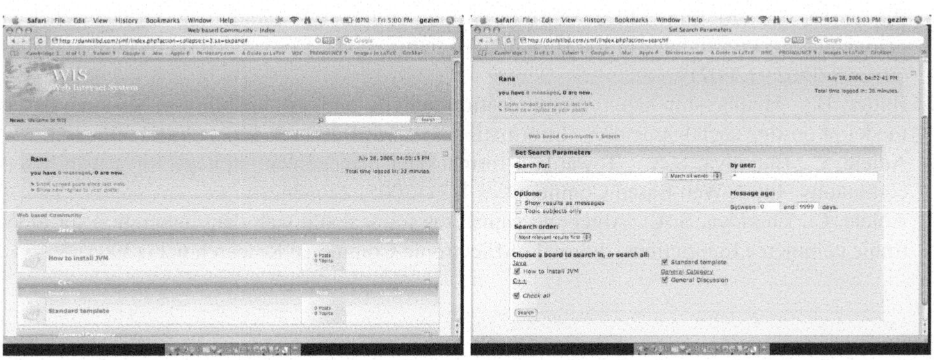

(a) (b)

Fig. 2. (a) Messages Posted in a Discussion Group; (b) Members Search for Messages Posted

Messages are indexed in the database according to subjects and titles. The system has an efficient search mechanism to enable members search for messages and respond to them.

5 Conclusions

In this paper, we formally specified components of an interactivity model to measure the contributions of members of a WC. The use of formal and theoretical foundations is particularly important for this new research area which, in the recent past, has been characterized mostly by *ad-hoc* solutions. We also designed an interactivity algorithm and provided a prototype of the Web community.

The model we presented dynamically measures individual member's interactivity, and indeed, the interactivity level of the community. These measurements will enable us to understand the dynamics of the community and also facilitate the classification of members into different groups according to their levels of participation. This classification provides a framework to address individual member's needs and reward deserving members.

References

1. Rheingold, H.: The Virtual Community: Homesteading on the Electronic Frontier. Revised edition edn. MIT Press (2000)
2. Boczkowski, P.J.: Mutula shaping of users and technology in a national virtual community. Journal of Communications **49**(2) (1999) 86–109
3. Romm, C., Pliskin, N., Clarke, R.: Virtual communities: Towards integrative three phase model. International Journal of Information Management **17**(4) (1997) 261–271
4. Lave, J., Wenger, E.: Situated Learning. Legitimate Peripheral Participation. Cambridge University Press (1991)
5. Menegon, F., D'Andrea, V.: Social processes and technology in an online community of practices. In: Proceedings of the International Conference on Web-based Communities (WBC2004). (2004) 115–122
6. Blanchard, A.L., Markus, M.L.: Sense of virtual community: Maintaining the experience of belonging. In: Proceedings of the Hawaii 35th International Conference on System Sciences (HICSS-3502). (2002)
7. Butler, B.: Membership size, communication activity and sustainability: a resource-based model of on-line social structures. Information Systems Research **12**(4) (2001) 346–362
8. Adiele, C., Ehikioya, S.A.: Towards a formal data management strategy for a web-based community. Int. J. Web Based Communities **1**(2) (2005) 226–242
9. Adiele, C., Ehikioya, S.A.: Algebraic signatures for scalable web data integration for electronic commerce transactions. Journal of Electronic Commerce Research **6**(1) (2005) 56–74

Towards a Type-2 Fuzzy Description Logic for Semantic Search Engine*

Ruixuan Li, Xiaolin Sun, Zhengding Lu, Kunmei Wen, and Yuhua Li

College of Computer Science and Technology,
Huazhong University of Science and Technology,
Wuhan 430074, Hubei, China
rxli@hust.edu.cn, arthursun@smail.hust.edu.cn,
{zdlu,kmwen,yhli3}@hust.edu.cn

Abstract. Classical description logics are limited in dealing with the crisp concepts and relationships, which makes it difficult to represent and process imprecise information in real applications. In this paper we present a type-2 fuzzy version of $\mathcal{ALC}$ and describe its syntax, semantics and reasoning algorithms, as well as the implementation of the logic with type-2 fuzzy OWL. Comparing with type-1 fuzzy $\mathcal{ALC}$, system based on type-2 fuzzy $\mathcal{ALC}$ can define imprecise knowledge more exactly by using membership degree interval. To evaluate the ability of type-2 fuzzy $\mathcal{ALC}$ for handling vague information, we apply it to semantic search engine for building the fuzzy ontology and carry out the experiments through comparing with other search schemes. The experimental results show that the type-2 fuzzy $\mathcal{ALC}$ based system can increase the number of relevant hits and improve the precision of semantic search engine.

Keywords: Semantic search engine, Description logic, Type-2 fuzzy $\mathcal{ALC}$, Fuzzy ontology.

1 Introduction

As the fundament of the semantic web [1,2], ontology is playing a very important role in many applications such as semantic search [3]. Being one of the logic supports of ontology, Description logics (DLs) [4] represent the knowledge of an application domain by defining the relevant concepts of the domain (terminologies) and using these concepts to specify properties of objects and individuals which belong to the domain (the world description). As one in the family of knowledge representation (KR) formalisms, the powerful ability of describing knowledge makes DLs express the information more easily in different application domains [5]. Being established by W3C in 2004, OWL (Web Ontology Language) [2,6] becomes the standard knowledge representation markup language for the semantic web.

* This work is supported by National Natural Science Foundation of China under Grant 60403027, Natural Science Foundation of Hubei Province under Grant 2005ABA258, Open Foundation of State Key Laboratory of Software Engineering under Grant SKLSE05-07, and a grant from Huawei Technologies Co., Ltd.

G. Dong et al. (Eds.): APWeb/WAIM 2007, LNCS 4505, pp. 805–812, 2007.

Being expected to be applied in semantic web, semantic search extended the search engine with ontology. Using general ontologies, the most current semantic search engines handle the information retrieval in semantic web based on classic DLs. University of Maryland proposed SHOE [7,8], which can find the semantic annotations from web pages. Tap [9,10,11] developed by Stanford University and IBM applies the technology of semantic web into Google, which augments the results in order to increase the quality of the retrieval. Swoogle [12,13,14] is designed for the information retrieval in structured documents such as RDF (Resource Description Framework), OWL and so on. At present, more and more semantic search systems are designed based on ontology that is supported by classic DLs.

But the classical DLs can only define the crisp concepts and properties, and the certain reasoning of classic DLs means that the answer of inference only returns "True" or "False", which cannot solve the fuzzy problem of ontology system in real world. Therefore, the fuzzy DLs are designed to expand the classic DLs to make it more applicable to ontology system. At present, most fuzzy logic systems (FLSs) are based on type-1 fuzzy sets, which were proposed by Zadeh in 1965 [15]. However, it was quite late when the fuzzy sets were applied to DLs and ontology System. Without reasoning algorithm, Meghini proposed a preliminary fuzzy DL as a tool for modeling multimedia document retrieval [16]. Straccia presented the formalized Fuzzy $\mathcal{ALC}$ ($\mathcal{FALC}$) [17] in 2001, which is a type-1 fuzzy extension of $\mathcal{ALC}$. Before long, Straccia extended the $\mathcal{SHOIN(D)}$, the corresponding DL of the standard ontology description language OWL DL, to a fuzzy version [18,19].

However, there are some limits in Type-1 fuzzy sets. For example the imprecision cannot be described by a crisp value clearly, which will result the loss of fuzzy information. To address the problem mentioned above, we propose a type-2 fuzzy $\mathcal{ALC}$ and try to apply it into semantic search engine. The contributions of the paper are as follows. First, we present the syntax and semantics of a type-2 fuzzy extension of $\mathcal{ALC}$, which can represent and reason fuzzy information with OWL, a formalized ontology description language. Besides the format of the axioms defined in Type-2 fuzzy $\mathcal{ALC}$, the reasoning algorithm is also proposed for semantic search. Finally, we design and realize the system of semantic search engine based on type-2 fuzzy $\mathcal{ALC}$ and carry out the experiments to evaluate the performance of the proposed search scheme.

The rest of the paper is organized as follows. Section 2 gives the condition of relative research and basic concepts of DL, typical $\mathcal{ALC}$ and type-1 fuzzy $\mathcal{ALC}$. Section 3 presents the format of the type-2 fuzzy $\mathcal{ALC}$ and the method of reasoning in type-2 fuzzy DL. Approaches for applying the type-2 fuzzy DL to deal with the description in fuzzy ontology for semantic search engine with OWL is addressed in section 4, followed by conclusions and future research of the paper.

2 Basic Concepts

$\mathcal{ALC}$ concepts and roles are built as follows. Use letter $\mathcal{A}$ for the set of atomic concepts, C for the set of complex concept defined by descriptions and $\mathcal{R}$ for the set of roles. Starting with: (1) A, B∈ $\mathcal{A}$ (2) C, D∈ $\mathcal{C}$ and (3) R∈ $\mathcal{R}$. The concept terms in TBox can be defined with the format as following inductively: C ⊑ f(A, B, R, ⊓, ⊔, ∀, ∃, ⊥, ⊤) (partial definition) and C ≡ f(A, B, R, ⊓, ⊔, ∀, ∃, ⊥, ⊤) (full definition). ⊥ and ⊤ are two special atomic concepts named "bottom concept" and "universe concept". The syntax and semantics of $\mathcal{ALC}$ constructors have been represented in [4].

For the reason we mentioned above, classic DL such as $\mathcal{ALC}$ cannot deal with the imprecise description. To solve this problem in DLs, Straccia presented $\mathcal{FALC}$, which is an extension of $\mathcal{ALC}$ with fuzzy features, to support fuzzy concept representation. Because Straccia used a certain number to describe the fuzzy concepts and individuals in $\mathcal{FALC}$, we call the $\mathcal{FALC}$ type-1 $\mathcal{FALC}$ [7].

3 Type-2 Fuzzy $\mathcal{ALC}$

3.1 Imprecise Axioms in Type-2 Fuzzy $\mathcal{ALC}$

Different from the type-1 fuzzy sets, type-2 fuzzy sets use an interval to show the membership. Each grade of the membership is an uncertain number in interval [0,1]. We denote the membership in type-2 fuzzy sets with $\overline{\mu_A}$ instead of μ_A in type-1, which is defined as following:

$$\overline{\mu_A}(x) = [\mu_A^L(x), \mu_A^U(x)] \tag{1}$$

In (1) we present: $\mu_A^L(x), \mu_A^U(x) : U \rightarrow [0,1]$, and $\forall x \in U, \mu_A^L(x) \le \mu_A^U(x)$. We call $\mu_A^L(x)$ and $\mu_A^U(x)$ the primary membership and secondary membership, and x is an instance in the fuzzy sets U. It is obvious that type-2 fuzzy sets can be reduced to type-1 fuzzy sets when the primary membership equals the secondary one. So a type-1 fuzzy set is embedded in a type-2 fuzzy set.

There are two fuzzy parts in type-2 fuzzy $\mathcal{ALC}$ presented in our paper, which are the imprecise terminological axiom (TBox) and fuzzy individual membership (ABox). To built a DL system, the first thing should be done in creating TBox is to define necessary atomic concepts and roles with some symbols. It is certainly that the base symbols exist in the DL system, but the name symbols are not. In other words, the atomic concepts defined by different axioms may be imprecise, which means that the axiom may not come into existence in type-2 fuzzy $\mathcal{ALC}$ TBox.

For example, given two base symbols named: Animal and FlyingObject, we can define the atomic concept Bird in TBox with the axiom (2).

$$\text{Bird}_{[0.9,0.95]} \equiv \text{Animal} \sqcap \text{FlyingObject} \tag{2}$$

(2) means that the probability value of that bird can be described with the conjunction of the Animal and FlyingObject is between 0.90 and 0.95.

Because of the certainty of the base symbols, the probability of atomic concepts Animal and FlyingObject are both 1, in the interval [1,1]. Instead of Animal $_{[1,1]}$ we define the certain atomic concept Animal without [1,1] concisely.

Type-2 fuzzy $\mathcal{ALC}$ can represent the vagueness in atomic concept with two properties, *fuzzy:LowerDegree* and *fuzzy:UpperDegree* to describe $\mu_A^L(x)$ and $\mu_A^U(x)$.

Because it can be considered true that every atomic concept (role) is independent, we can calculate the value of *fuzzy:LowerDegree* and *fuzzy:UpperDegree* of fuzzy concept if we do not know it beforehand. For example, we want to define an atomic concept Meat-eatingBird with base symbol Meat-eatingObject with axiom (3):

$$\text{Meat-eatingBird} \equiv \text{Bird}_{[0.9,0.95]} \sqcap \text{Meat-eatingObject} \quad (3)$$

when we apply the triangular norms $T(a,b) = ab/[1+(1-a)(1-b)]$, $S(a,b) = (a+b)/(1+ab)$, we can get the value of *fuzzy:LowerDegree* (*fuzzy:UpperDegree*) of Meat-eatingBird with the follow equation: μ^L (Meat-eatingBird)=T (μ^L (Bird), μ^L (Meat-eatingObject)), as mentioned above, we know that μ^L (Bird)=0.9 and μ^L (Meat-eatingObject)=1. So μ^L (Meat-eatingBird)= $(0.9\times1)/[1+(1-1)(1-0.9)] = 0.9$. So the membership of atomic concept Meat-eatingBird is in scope [0.9,0.95]. We call it transitivity of type-2 fuzzy $\mathcal{ALC}$.

In addition to the fuzzy TBox, the uncertainty still exists in ABox in fuzzy DLs. The assertion $\text{Bird}_{[0.9,0.95]}(\text{penguin})_{[0.65,0.9]}$ means the degree that the penguin can be considered as an instant of $\text{Bird}_{[0.9,0.95]}$ is in [0.65,0.90] in a given DL. Being similar with $\mathcal{FALC}$, the ABox assertions $C^I(d)=[a, b]$, in which $0\leq a\leq b\leq 1$. Take atomic Bird $_{[0.9,0.95]}$ for example, The Bird(penguin) being satisfied in ABox has two pre-conditions: (1) concept Bird should be satisfied in TBox; (2) penguin belongs to bird in ABox. So we can conclude that μ^L (Bird(penguin))= μ^L (Bird)$\times$ μ^L (penguin $\in$ Bird) = T(0.65,0.90)=0.565 (so do μ^U (Bird(penguin))). So the ABox can be denoted by a set of equations with form as: $C_{[a,b]}$ (a)=[c, d] Where C= f(A, B, R, $\sqcap$, $\sqcup$, $\forall$, $\exists$, $\bot$, $\top$). For example: $\text{Bird}_{[0.9,0.95]}$(penguin)=[0.65,0.95], or $\text{Bird}_{[0.9,0.95]}(\text{penguin})_{[0.65,0.98]}$.

3.3 The Syntax and Semantics of Type-2 Fuzzy $\mathcal{ALC}$

We define $\mathcal{A}$, $\mathcal{C}$ and $\mathcal{R}$ as the set of atomic concepts, complex concepts, and roles. $C\sqcap D$, $C\sqcup D$, $\neg C$, $\forall R.C$ and $\exists R.C$ are fuzzy concept. The fuzzy interpretation in type-2 fuzzy $\mathcal{ALC}$ is a pair $I = (\Delta^I, \cdot^I)$, and $\cdot^I$ is an interpretation function that map fuzzy concept and role into a membership degree interval: $C^I = \Delta^I \rightarrow [a, b]$ and $R^I = \Delta^I \times \Delta^I \rightarrow [a, b]$, and a, b must satisfy $0\leq a\leq b\leq 1$. The syntax and semantic of type-2 fuzzy $\mathcal{ALC}$ is shown in Table 1. Different from $\mathcal{FALC}$ the type-2 fuzzy $\mathcal{ALC}$, Δ^I is not a set of numbers in scope [0,1] but a set of pairs, which have the form like [a, b]. And it must satisfy the inequation: $0\leq a\leq b\leq 1$.

Table 1. The syntax and semantics of type-2 fuzzy $\mathcal{ALC}$ constructors

Constructor	Syntax	Semantics
Top (Universe)	$\top$	Δ^I
Bottom (Nothing)	$\bot$	Φ
Atomic Concept	$A_{[a,b]}$	$A^I_{[a,b]} \subseteq \Delta^I$
Atomic Role	$R_{[a,b]}$	$R^I_{[a,b]} \subseteq \Delta^I \times \Delta^I$
Conjunction	$C_{[a,b]} \sqcap D_{[c,d]}$	$(C \sqcap D)^I_{[T(a,c),T(b,d)]}$
Disjunction	$C_{[a,b]} \sqcup D_{[c,d]}$	$(C \sqcup D)^I_{[S(a,c),S(b,d)]}$
Negation	$\neg C_{[a,b]}$	$C^I_{[1-b,1-a]}$
Value restriction	$\forall R_{[a,b]}.C_{[c,d]}$	$\forall y.S(R_{[1-b,1-a]}(x,y), C_{[c,d]}(y))$
Full existential quantification	$\exists R_{[a,b]}.C_{[c,d]}$	$\exists y.T(R_{[a,b]}(x,y), C_{[c,d]}(y))$

3.4 Reasoning in Type-2 Fuzzy $\mathcal{ALC}$

Tableau algorithms use negation to reduce subsumption to (un)satisfiability of concept descriptions instead of testing subsumption of concept descriptions directly: $C \sqsubseteq D$ iff $\neg C \sqcap D = \bot$. The fuzzy tableau begin with an ABox $A_0=\{C_{[a,b]}(x)_{[c,d]}\}$ to check the (un)satisfiability of concept $C_{[a,b]}$. Since $\mathcal{ALC}$ has not number restrictions, here are 5 rules presented:

$\cap$ *-rule*: if A contains $C_{[a,b]}(x)_{[c,d]}$, and $C_{[e,f]}(x)_{[g,h]}$: if $[a, b] \cap [e, f] \neq \Phi$ and $[c, d] \cap [g, h] \neq \Phi$ algorithm should extend A to $A' = A - \{C_{[a,b]}(x)_{[c,d]}, C_{[e,f]}(x)_{[g,h]}\} \sqcup \{C_{[SO(a,e), TO(b,f)]}(x)_{[SO(c,g), TO(d,h)]}\}$, else $A' = A - \{C_{[a,b]}(x)_{[c,d]}, C_{[e,f]}(x)_{[g,h]}\}$

$\sqcap$ *-rule*: if A contains $(C'_{[e,f]} \sqcap C''_{[g,h]})_{[a,b]}(x)_{[c,d]} = (C' \sqcap C'')_{[T(T(e,f),a), T(T(g,h),b)]}(x)_{[c,d]}$, but not contains both $C'_{[e,f]}(x)_{[c,d]}$ and $C''_{[g,h]}(x)_{[c,d]}$, algorithm should extend A to $A' = A \sqcup \{C'_{[e,f]}(x)_{[c,d]}, C''_{[g,h]}(x)_{[c,d]}\}$.

$\sqcup$ *-rule*: if A contains $(C'_{[e,f]} \sqcup C''_{[g,h]})_{[a,b]}(x)_{[c,d]} = (C' \sqcup C'')_{[S(S(e,f),a)\ S(S(g,h),b)]}(x)_{[c,d]}$, but neither $C'_{[e,f]}(x)_{[c,d]}$ nor $C''_{[g,h]}(x)_{[c,d]}$, the algorithm should extend A to $A' = A \sqcup \{C'_{[e,f]}(x)_{[c,d]}\}$ or $A'' = A \sqcup \{C''_{[g,h]}(x)_{[c,d]}\}$.

$\exists$ *-rule*: if A contains $(\exists R_{[e,f]}.C_{[g,h]})(x)_{[c,d]}$, but no individuals z such that $R_{[e,f]}(x, z)_{[c,d]}$ and $C_{[g,h]}(z)_{[c,d]}$, the algorithm should extend A to $A' = A \sqcup \{R_{[e,f]}(x, y)_{[c,d]}, C_{[g,h]}(y)_{[c,d]}\}$ where y is an individual not occurring in A before.

$\forall$*-rule*: if A contains $(\forall R_{[e,f]}.C_{[g,h]})(x)_{[c,d]}$, and $R_{[e,f]}(x, y)_{[c,d]}$, but not $C_{[g,h]}(y)_{[c,d]}$, the algorithm should extend A to $A' = A \sqcup \{C_{[g,h]}(y)_{[c,d]}\}$.

Given two limit values: T_L and T_U, the way to decide whether the ABox in type-2 fuzzy $\mathcal{ALC}$ is unsatisfiable is different from typical tableau. In that

$\mu^{L(U)}(C) \le T_L \Leftrightarrow C_{[0,0]}$, $\mu^{L(U)}(C) \ge T_U \Leftrightarrow C_{[1,1]}$. So the process of tableau will stop when anyone of following conditions is established:

(1) Any obvious clash ($\perp$(x), ($C \sqcap \neg C$)(x), etc.) is found in process of algorithm.
(2) All rules ($\sqcap$-*rule*, etc.) have been executed.
(3) Any fuzzy clash ($C_{[0,0]}$ (x)=[*c, d*],$C_{[a,b]}$ (x)=[*c, d*], $C_{[c,d]}$ (x)=[*a, b*]with $a \le b \le T_L$, $C_{[a,b]}$ (x) and $C_{[c,d]}$ (x) with the intervals [*a, b*] and [*c, d*] do not overlap) happened in process of algorithm.

4 The Semantic Search Engine Based on Type-2 Fuzzy Ontology

4.1 Architecture of Type-2 Fuzzy Semantic Search Engine

The natural languages in daily communication often have imprecise information. We call the queries including fuzzy concepts fuzzy queries. To handle these fuzzy queries, semantic search engines based on ontology must extend their knowledge bases on fuzzy ontologies such as the fuzzy semantic search engine proposed in this paper. Fig. 1 shows the architecture of type-2 fuzzy semantic search engine.

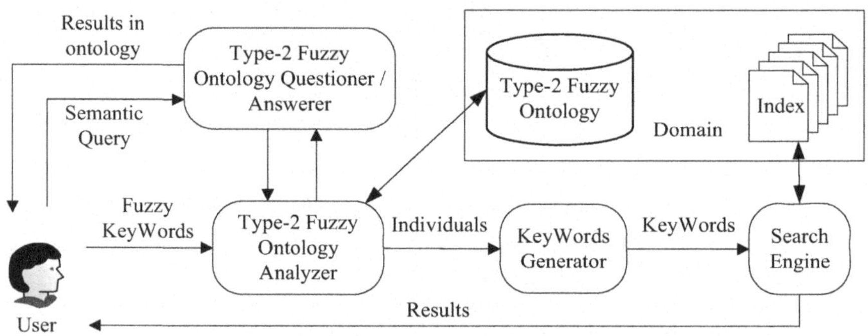

Fig. 1. Architecture of Type-2 fuzzy semantic search engine

In this framework, users can propose their query in two ways: they can ask the type-2 fuzzy ontology analyzer with keywords or fuzzy keywords. On the other hand, users can also search the ontology by issuing the semantic query to type-2 fuzzy ontology questioner (answerer) with keywords or other interfaces. Thus, users can communicate with ontology directly with the recalls formed by individuals or classes to make the queries precisely, which are sent to type-2 fuzzy ontology analyzer later. Then type-2 fuzzy ontology analyzer can generate individuals that satisfy to query and send these answers to keywords generator to combine proper keywords. At last, the traditional search engine will find the results from index with these keywords and return the hits to users.

4.2 Experiments and Analysis

Based on the framework introduced above, we have implemented the type-2 fuzzy search engine. Supported by the fuzzy ontology reasoner, the semantic search engine based on type-2 fuzzy $\mathcal{ALC}$ can improve the relativity of the responses to query. The experiment is carried out in the scope of all resources available in Huazhong University of Science and Technology, including almost 7000 web pages indexed from different departments and 2400 documents. The type-2 fuzzy ontology analyzer, answer, keywords generator and the search engine are all implemented with java. The ontology has built with protégé.

We chose a group of keywords to retrieve information from indexes, and then picked out the relevant hits (hits those are relevant to the retrieval) from result set and counted the average of them. Fig. 2 shows that the semantic search engine based on ontology (including classic and fuzzy ontology) can expand the relevant hits greatly when there is no imprecise information in keywords. The reason is that ontology generates more keywords with its individuals. However, the number of relevant hits of search engine based on classic ontology decreases rapidly when we add more fuzzy keywords such as "very", "young" into the keywords group. Compared to classic ontology, semantic search engine based on type-2 fuzzy ontology can accommodate itself to fuzzy keywords much better. We carry out an experiment on the precision (the fraction of the retrieved documents which is relevant) of the semantic search engine for that reason. Fig. 3 represents that the precision of semantic search engine based on classic ontology increases slower than the one based on type-2 fuzzy ontology when the number of nodes increases in otology. That means the precision of search engine will be improved if type-2 fuzzy $\mathcal{ALC}$ is applied.

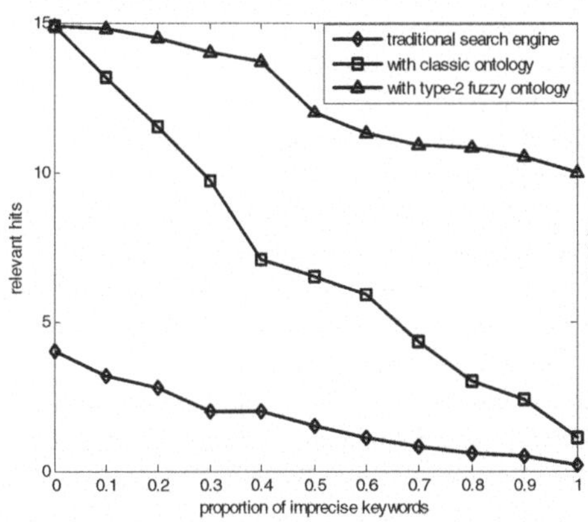

Fig. 2. Relevant hits -imprecision graph

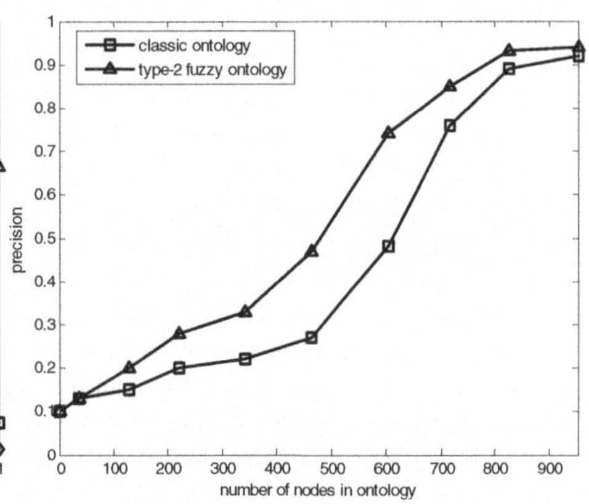

Fig. 3. Precision-nodes graph

5 Conclusions and Future Work

As the fundament of type-2 fuzzy DLs, type-2 $\mathcal{ALC}$ is introduced of its syntax, semantics, reasoning algorithm and application in this paper. Comparing with the type-1 fuzzy

$\mathcal{ALC}$, the type-2 fuzzy $\mathcal{ALC}$ can deal with the imprecise knowledge much better. Besides semantic search, there are many applications based on DLs need to handle fuzzy information such as trust management. Our approach can be applied in those domains to enrich its representation meaning and reasoning abilities. Future work includes the research of type-2 fuzzy $\mathcal{ALCN}$, $\mathcal{SHOIN(D)}$ and the reasoning algorithms.

References

1. Berners-Lee, T., Hendler, J., Lassila, O.: The Semantic Web. The Scientific American 284(5) (2001) 34-43
2. Horrocks, I., Patel-Schneider, P.F., van Harmelen, F.: From SHIQ and RDF to OWL: The Making of a Web Ontology Language. Journal of Web Semantics 1(1) (2003) 7-26
3. Guha, R., McCool, R., Miller, E.: Semantic Search. In: Proceedings of the 12th International World Wide Web Conference (WWW 2003). Budapest, Hungary (2003) 700-709
4. Baader, F., Calvanese, D., McGuinness, D.L., Nardi, D., Patel-Schneider, P.F.: The Description Logic Handbook: Theory, Implementation, and Applications. Cambridge University Press (2003) 47-100
5. Calvanese, D., Lenzerini, M., Nardi, D.: Unifying Class-Based Representation Formalisms. Journal of Artificial Intelligence Research 11(2) (1999) 199-240
6. Bechhofer, S., van Harmelen, F., Hendler, J., Horrocks, I., McGuinness, D.L., Patel-Schneider, P.F. (ed.): L.A.S.: OWL Web Ontology Language Reference (2004)
7. Heflin, J.D.: Towards the Semantic Web: Knowledge Representation in a Dynamic Distributed Environment. PhD Thesis, University of Maryland (2001)
8. Heflin, J., Hendler, J.: Searching the Web with Shoe. In: AAAI-2000 Workshop on AI for Web Search. Austin, Texas, USA (2000)
9. Guha, R., McCool, R.: TAP: A Semantic Web Test-bed. Journal of Web Semantics 1(1) (2003) 32-42
10. Guha, R., McCool, R.: The Tap Knowledge Base. http://tap.stanford.edu/
11. Guha, R., McCool, R.: Tap: Towards a Web of Data. http://tap.stanford.edu/
12. Ding, L., Finin, T., Joshi, A., et al.: Swoogle: A Search and Metadata Engine for the Semantic Web. In: CIKM'04. Washington DC, USA (2004)
13. Finin, T., Mayfield, J., Joshi, A., et al.: Information Retrieval and the Semantic Web. In: Proceedings of the 38th Hawaii International Conference on System Sciences (2005)
14. Mayfield, J. Finin, T.: Information Retrieval on the Semantic Web: Integrating Inference and Retrieval. In: 2004 SIGIR Workshop on the Semantic Web. Toronto (2004)
15. Zadeh, L. A.: Fuzzy Sets. Information and Control 8(3) (1965) 338-353
16. Meghini, C., Sebastiani, F., Straccia, U.: Reasoning about the Form and Content for Multimedia Objects. In: Proceedings of AAAI 1997 Spring Symposium on Intelligent Integration and Use of Text, Image, Video and Audio. California (1997) 89-94
17. Straccia, U.: Reasoning within Fuzzy Description Logics. Journal of Artificial Intelligence Research 14 (2001) 137-166
18. Straccia, U.: Transforming Fuzzy Description Logics into Classical Description Logics. In: Proceedings of the 9th European Conference on Logics in Artificial Intelligence. Lisbon, (2004) 385-399
19. Straccia, U.: Towards a Fuzzy Description Logic for the Semantic Web. In: Proceedings of the 1st Fuzzy Logic and the Semantic Web Workshop. Marseille (2005) 3-18

A Type-Based Analysis for Verifying Web Application*

Woosung Jung[1], Eunjoo Lee[2,**], Kapsu Kim[3], and Chisu Wu[1]

[1] School of Computer Science and Engineering, Seoul National University, Korea
{wsjung,wuchisu} @selab.snu.ac.kr
[2] Department of Computer Engineering, Kyungpook National University, Korea
ejlee@knu.ac.kr
[3] Department of Computer Education, Seoul National University of Education, Korea
kskim@snue.ac.kr

Abstract. Web applications have become standard for several areas, however, they tend to be poorly structured and do not have strongly-typed support. In this paper, we present a web application model and a process to extract the model using static and dynamic analysis. We show recurring problems regarding type and structure in web applications and formally describe algorithms to verify those problems. Finally, we show the potentials of our approach via tool support.

Keywords: Web application model, analysis, verification.

1 Introduction

It becomes more and more important to verify and validate web applications, because web applications have become standard for business and public areas [1]. Since web applications do not have strongly-typed support, type checking problem for web applications has arisen.

Several studies have been conducted on the verification of web applications using type information [2] [3] [4] [5], however, they concentrate on testing web applications and overlook such kinds of errors that occur frequently in using 'form's and resources.

In this paper, we present some practical recurring problems, such as frame structure, form-parameters' types, form-parameters' names, and resources' type. We convert them into type problems and try to solve them formally. At first, we define a model for web applications, and then, we formalize the algorithms for verifying the raised problems by using the model. A tool has been implemented to apply our approach.

The remainder of this paper is organized as follows: Section 2 defines a model. In section 3, we present four problems that are checked and define the verification

* This work was supported by the Brain Korea 21 Project and by the Korea Science and Engineering Foundation(KOSEF) grant funded by the Korea government(MOST) (No. R01-2006-000-11150-0).

** Corresponding author.

G. Dong et al. (Eds.): APWeb/WAIM 2007, LNCS 4505, pp. 813–820, 2007.

algorithm formally. Section 4 describes the results regarding type checking problems by using the tool we implemented. Finally, in section 5, conclusion and suggestions for future work are presented.

2 Web Application Model

We illustrate the web model in the ER diagram (Fig. 1). UML notations are adopted in many studies, however, we choose ER model because it enables the seamless modeling and verification using stored procedures in SQL. Also, it is more appropriate to apply the *fixed point* theory that we utilize.

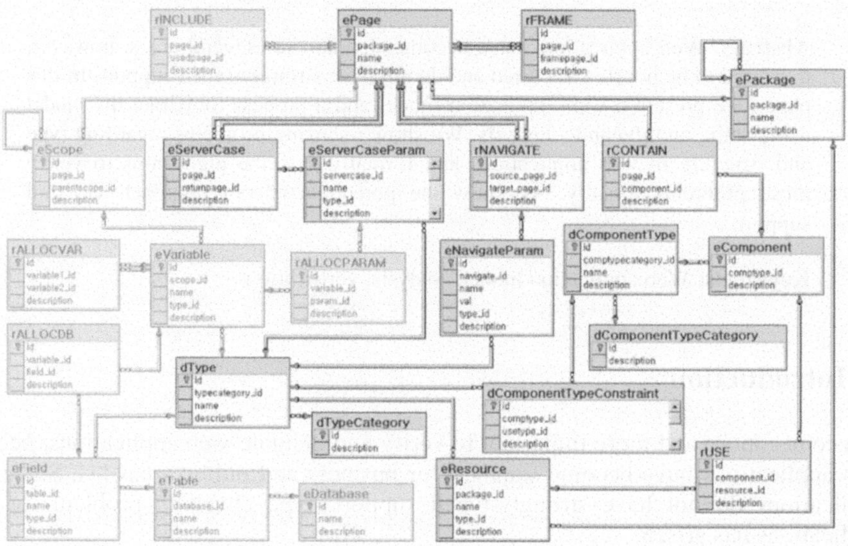

Fig. 1. Web application model in the ER-diagram

The entities in the DB schema are classified as follows:

- Static page structure: rINCLUDE, ePage, rFRAME, ePackage, eComponent, rCONTAIN, rUSE, and eResource
- Page behavior: eServerCase, eServerCaseParam, rNAVIGATE, and eNavigateParam
- Database: eField, eTable, and eDatabase
- Server-side allocation: eScope, rALLOCVAR, eVariable, rALLOCPARAM, and rALLOCDB
- Predefined environment: dComponentType, dComponentTypeCategory, dCom-ponent-TypeConstraint, dTypeCategory, and dType

3 Checking Algorithms with the Model

In this section, we introduce four frequent errors that happen in web applications and show checking algorithms.

3.1 Frame-Type Checking

If a user can navigate on a frame from a web page *p* to its upper page, the frame page may repeat throughout the entire web page. This is mostly because of the wrong 'target' in a frame tag or errors in the navigational structure. We call this kind of error as "frame-type error". We define the domain for frame-type checking in Fig. 2.

W: WebApplication
$P(W) = [[P]]W = \{p \mid \forall p \in Page, p \text{ is a page of } W\}$
$s \in Page, S \in 2^{page}$
$frameowner: 2^{page} \rightarrow 2^{page}$, $[[frameowner]] = \lambda S.\{p \mid p \in S, frameset(p) \neq \phi\}$
$frameset: Page \rightarrow 2^{page}$, $[[frameset]] = \lambda s.\{p \mid p \in Page, p \text{ is a frame page of } s\}$
$NavigationTargets: 2^{page} \rightarrow 2^{page}$,
$[[NavigationTargets]] =$
$\lambda s.\{p \mid p \in Page, p \text{ is a reachable page from } p' \in S \text{ with 1 navigation}\}$

Fig. 2. Domain definition for checking frame-type

We mark to all pages that have frames as '*visited*', to assure that a frame in a page cannot be navigable to its upper pages, including itself. This test is conducted on all web pages. When it is impossible for a page to navigate its upper pages, we can say that the frame-type of the web application is sound. Figure 3 shows the algorithm for frame-type checking.

```
for each p∈frameowner P(W) do
   P(W).visited=false
   T=NavigationTargets(frameset(p))
   if T = φ then <<Exit>>
            else if p∈T then <<FrameTypeError>>
                        else frameset(p).visited=true
   T'=NavigationTargets(T)
   if T' = φ then <<Exit>>
            else if p∈T' then <<FrameTypeError>>
                        else T.visited=true
   ...
end of for
```

Fig. 3. An algorithm for checking frame-type

We can describe semantics in frame-type checking by using part of the algorithm, as in Fig. 4.

[[*FrameTypeCheck*]] <*p*, *S*> =
  if $S = \phi$ then <<*Exit*>>
  else if $p \in$ [[*NavigationTargets*]]*S* then <<*FrameTypeError*>>
  else let *S.visited=true* $\in$ [[*FrameTypeCheck*]]<*p*, [[*NavigationTargets*]]*S*> *end*

Fig. 4. Semantics for checking frame-type

We regard semantics as an equation of "X=F(X)", then the algorithm can be described by using *fixed point* (Fig. 5).

[[*FrameTypeCheck*]]= *fixF* $\in$ *Page* x $2^{page} \rightarrow$ *Page* x 2^{page}
=*fix*(λX. (λS. (λp. if $S = \phi$ then << *Exit*>>
  else if $p \in$[[*NavigationTargets*]]*S* then <<*FrameTypeError*>>
  else let *S.visited=true* $\in X$<*p*, [[*NavigationTargets*]]*S*> *end*)))

Fig. 5. An Algorithm for checking frame-type using fixed point

To check the soundness of a frame-type, [[FrameTypeCheck]] is executed on all pages that have frames. The initial value is <*p*, *frameset*(*p*)>. That is, the algorithm is summarized as follows:

$\forall p \in$[[*frameowner*]]([[*p*]]*W*).[[*FrameTypeCheck*]]<*p*, [[*frameset*]]*p*>

3.2 Resource-type Checking

Resource-type checking tests the mismatch of resource-types. Each component in a web application has type constraints in the resources that it uses. For example, only an image type can be used in <IMG> tag. If 'AVI' resource is used in <IMG>, it generates a resource-type error. Resource-type errors are not revealed because web applications are not compiled. Also, it is difficult to find resource-type error in large web applications, however, this type of checking is not supported in existing web applications. We define the domain for resource-type checking (Fig. 6).

$c \in Comp$, $r \in Res$
component: $Page \rightarrow 2^{Comp}$,
  [[*component*]]= $\lambda s.\{c | c \in Comp$, *c is a component of page s*$\}$
resource: $Comp \rightarrow 2^{Res}$,
  [[*resource*]]= $\lambda c.\{r | r \in Res$, *r is a resource used by component c*$\}$

Fig. 6. Domain definition for checking resource-type

We define the function that checks the resource-type for a component in the web (Fig. 7).

[[*ResourceTypeCheck*]]*c*=
  if([[*resource*]]*c*).*type* $\notin$ ([[*constraint*]]*c*).*type* then <<*ResourceTypeError*>>

Fig. 7. A function for checking resource-type

Figure 8 shows the algorithm for resource-type checking.

```
for each p∈P(W) do
        for each c ∈component(p) do
          [[ResourceTypeCheck]]c
        end of for
end of for
```

Fig. 8. An algorithm for checking resource-type

The algorithm is summarized as follows:

$$\forall p \in [[P]]W,\ \forall c \in [[component]]p.[[ResourceTypeCheck]]c$$

If resource-type errors do not occur during checking, we can say that web application W is sound with regard to resource-type.

3.3 Form-Parameter Name Checking

When parameters are submitted by 'GET' or 'POST' in client side, the server pages may try to use some parameters that are not submitted from a client or that have different names from the submitter. For example, a form variable '*name*' in a web page is submitted and used as '*nama*' in other web page. This happens frequently in practical situations, however, it is difficult to find this kind of error in the web. We can uncover parameter-name mismatch error by static analysis based on the form. We define the domain for form-parameter name checking (Fig. 9)

$t \in Case,\ n \in NavigationCase$

$case : Page \rightarrow 2^{Case}$, $[[case]] = \lambda p.\{t | t \in Case$, *t is a case possible to happen in page p*$\}$

$Navigation: Page \rightarrow 2^{NavigationCase}$,

$[[Navigation]] =$

$\lambda p.\{n | n \in NavigationCase$, *n is a navigation case that can happen in a page p*$\}$

$NavigationParam: NavigationCase \rightarrow 2^{Param}$,

$[[NavigationParam]] = \lambda n.\{m | m \in Param$, *m is a submitted parameter in navigation case n*$\}$

$CaseParam: Case \rightarrow 2^{Param}$,

$[[CaseParam]] = \lambda t.\{m | m \in Param$, *expected parameters in case t*$\}$

Fig. 9. Domain definition for checking form-parameter names

We define a function that checks the resource-type for a component in the web (Fig. 10).

$[[FormNameCheck]]\langle t, n\rangle$

$=$ if $CaseParam(t).name \notin NavigationParam(n).name$ then $\ll FormsNameError \gg$

Fig. 10. A function for checking form-parameter names

Figure 11 shows the algorithm for resource-type checking. If form name errors do not happen for all web pages, we can say that the web application W is sound in form-parameter names.

```
for each p∈P(W) do
    for each n∈Navigation(p) do
        for each t∈case(n.TargetPage) do
            [[FormNameCheck]]<t, n>
        end of for
    end of for
end of for
```

Fig. 11. An algorithm for checking form-parameter names

The algorithm is summarized as follows:

$$\forall p \in [[P]]W,\ \forall n \in [[Navigation]]p,\ \forall t \in [[Case]](n.TargetPage).[[FormNameCheck]]\langle t, n \rangle$$

3.4 Form-Parameter Type Checking

In addition to parameter name, parameter type can be considered in form-parameter checking. Figure 12 describes a way to check the type mismatches between parameter m1 by a server and m2 by a client.

```
[[FormTypeCheck]]<m1, m2>=
        if m1.name=m2.name and m1.type≠m2.type then <<FormTypeError>>
```

Fig. 12. A function for checking form-parameter types

Figure 13 shows the algorithm for form-parameter type checking.

```
for each p∈P(W) do
    for each n∈Navigation(p) do
        for each t∈case(n.TargetPage) do
            for each m1∈CaseParam(t), m2∈NavigationParam(n) do
                [[FormTypeCheck]] <m1, m2>
            end of for
        end of for
    end of for
end of for
```

Fig. 13. An algorithm for checking form-parameter types

4 Implementation

We implemented a tool to support the static analysis of the web application model. We applied the tool to a sample web application. Figure 14 is a screenshot that illustrates the results of the verification. This tool supports four kinds of error checking as stated in section 3. Furthermore, the tool indicates information of errors including locations, reasons, and hints for debugging. In particular, frame-type error shows not only the page containing frames, but the navigational paths that may trigger the frame-type errors. The right side of the top shows the test results, which contain the number of errors and the validity of each type. The body of Fig. 14 shows the details of the result.

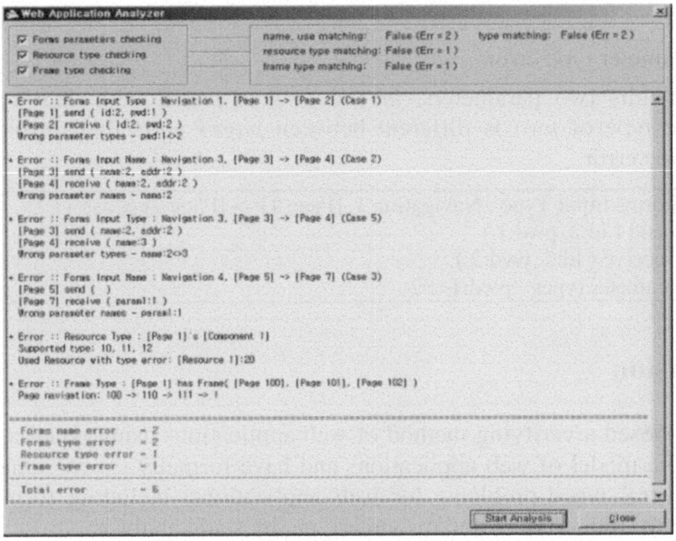

Fig. 14. The result of the analysis

We will explain part of the result in the following:

- Frame-type error

The following is part of Fig. 17, which indicates the frame-type error. This shows that *page1* has *page100*, *101* and *102* as its frame and there is a navigation path from *page100* to *page1* via *110* and *111*. This result in frame-type error; *Page1* is nested, which is undesirable.

```
* Error :: Frame-type : [Page 1] has Frame( [Page 100], [Page 101], [Page 102] )
  Page navigation: 100 → 110 → 111 → 1
```

- Resource-type error

The following results indicate that *component1* in *page1* can use *type10*, *11* and *12*, however, *component1* uses resource with *type 20*.

```
* Error :: Resource-type : [Page 1]'s [Component 1]
  Supported type: 10, 11, 12
  Used Resource with type error: [Resource 1]:20
```

- Form-parameter name error

Navigation3 in the first line of this example indicates that there is navigation from *page3* to *page4* and that the navigation id is three. *Page3* submits two form-parameters *name* and *addr* to *page4*. The attached number '2' (*name*:2, *addr*:2) is their type, but *page4* receives them as '*nama*' and '*addr*', which results in a wrong parameter name in '*nama*'.

```
* Error :: Forms Input Name : Navigation 3, [Page 3] -> [Page 4] (Case 2)
  [Page 3] send ( name:2, addr:2 )
  [Page 4] receive ( nama:2, addr:2 )
  Wrong parameter names - nama:2
```

- Form-parameter type error

Page1 submits two parameters, *id* and *pwd* to *page2* with their types. In this example, the type of *pwd* is different between *page1* and *page2*, which results in parameter-type error.

```
* Error :: Forms Input Type : Navigation 1, [Page 1] -> [Page 2] (Case 1)
  [Page 1] send ( id:2, pwd:1 )
  [Page 2] receive ( id:2, pwd:2 )
  Wrong parameter types - pwd:1<>2
```

5 Conclusion

We have proposed a verifying method of web applications using typed-approach. We have defined a model of web applications and have formally shown some algorithms to verify several typed-problems for web applications, including form-parameters, frame structure (frame-type), and resource-type, by static analysis.

It is expected that the proposed model can be used as a reference to obtain a web application structure with type information. The algorithms that have been presented formally can provide a type-verification method for problems that occur frequently in the field. Also, the verification cost can be decreased, because the checking processes are executed via tool support.

In future work, we will identify and verify other verification problems found in web applications using the model. Finally, we will extend our work for defining the framework to support model-driven development of web applications.

References

1. Tonella, P. and Ricca, F.: A 2-Layer Model for the White-Box Testing of Web Applications. Proc of the 6th IEEE International Workshop on Web Site Evolution, (2004).
2. Harmelen, F., Meer, J., Webmaster: Knowledge-based Verification of Web Pages. Proc of the 12th International Conference on Industrial and Engineering Applications of Artificial Intelligence and Expert Systems, (1999).
3. Despeyroux, T., Trousse, B.: Semantic Verification of Web Sites Using Natural Semantics. Proc of the 6th Conference on Content-Based Multimedia Information Access, (2000).
4. Despeyroux, T.: Practical Semantic Analysis of Web Sites and Documents, Proc of the 13th Conrefence on World Wide Web, (2004).
5. Draheim, D., Weber, G.: Strongly Typed Server Pages, Proc. of Next Generation Information Technologies and Systems, (2002).
6. http://www.antlr.org/
7. http://tidy.sourceforge.net/

Homomorphism Resolving of XPath Trees Based on Automata*

Ming Fu and Yu Zhang[1,2]

[1] Department of Computer Science & Technology,
University of Science & Technology of China, Hefei, 230027, China
[2] Laboratory of Computer Science, Chinese Academy of Sciences, Beijing, 100080, China
brightfu@ustc.edu, yuzhang@ustc.edu.cn

Abstract. As a query language for navigating XML trees and selecting a set of element nodes, XPath is ubiquitous in XML applications. One important issue of XPath queries is containment checking, which is known as a co-NP complete. The homomorphism relationship between two XPath trees, which is a PTIME problem, is a sufficient but not necessary condition for the containment relationship. We propose a new tree structure to depict XPath based on the level of the tree node and adopt a method of sharing the prefixes of multi-trees to construct incrementally the most effective automata, named XTHC (XPath Trees Homomorphism Checker). XTHC takes an XPath tree and produces the result of checking homomorphism relationship between an arbitrary tree in multi-trees and the input tree, thereinto the input tree is transformed into events which force the automata to run. Moreover, we consider and narrow the discrepancy between homomorphism relationship and containment relationship as possible as we can.

Keywords: XPath tree, containment, homomorphism, automata.

1 Introduction

XML has become the standard of exchanging a wide variety of data on Web and elsewhere. XML is essentially a directed labeled tree. XPath[1] is a simple and popular query language to navigate XML trees and extract information from them.

XPath expression p is said to contain another XPath expression q, denoted by $q \subseteq p$, if and only if for any XML document D, if the resulting set of p returned by querying on D contains the resulting set of q. Containment checking becomes one of the most important issues in XPath queries. Query containment is crucial in many contexts, such as query optimization and reformulation, information integration, integrity checking, etc. However, [2] shows that containment in fragment $XP^{\{[\,],*,//\}}$ is co-NP complete. The authors proposed a complete algorithm for containment, whose complexity is EXPTIME. The authors also proposed a sound but incomplete PTIME

* This work is supported by the National Natural Science Foundation of China under Grant No. 60673126, and the Foundation of Laboratory of Computer Science, Chinese Academy of Science under Grant No. SYSKF0502.

G. Dong et al. (Eds.): APWeb/WAIM 2007, LNCS 4505, pp. 821–828, 2007.

algorithm based on homomorphism. This algorithm may return false negatives because the homomorphism relationship between two XPath trees is a sufficient but not necessary condition for the containment relationship. In many practical situations containment can be replaced by homomorphism.

The homomorphism algorithms proposed in [2][3] are mainly focused on how to resolve the containment problem between two XPath expressions. In [3] the authors proposed hidden-conditioned homomorphism to further narrow the discrepancy between homomorphism and containment based on [2]. However, the homomorphism relationship was considered in these works only between two XPath trees. In practice we may need to verify the homomorphism relationship between an arbitrary tree in a set of XPath trees and the input XPath tree, such as filtering redundant queries in a large query set. It is inefficient to check one by one using the homomorphism algorithm, because the same prefix and branch in multi-trees will cause redundant computing. Although a method handling this was discussed in [4], it will return false negatives for some XPath expressions which have containment relationship, such as XPath expressions $p = /a//*/b$, $q = /a/*//b$ etc.

In this paper, we propose an efficient method to check homomorphism from multi-trees to a single XPath tree based on automata.We also narrow the discrepancy between homomorphism and containment as possible as we can. Our major contributions are: 1)We propose the fixed tree and alterable tree to describe the XPath tree, and define homomorphism based on them. 2)We define XTHC machine, a kind of indexed incremental automata with prefix-sharing of multi-trees, and our method can give the optimal automata. 3)We propose an algorithm to check homomorphism from multi-trees to a single tree based on XTHC machine. 4)The experiment results demonstrate both the practicability and efficiency of our techniques.

The rest of this paper is organized as follows. Section 2 gives some basic notations and definitions. Section 3 is the major part of our work, that is, how to construct XTHC machine and how to use XTHC to resolve the homomorphism problem. The last two sections present the experimental evaluation and conclusions, respectively.

2 Preliminaries

Each XPath expression has a corresponding XPath tree. The XPath tree given in [2] uses each node test in the XPath expression as a node in the tree, and classifies its edges into child-edge and descendant-edge according to the type of axes in the XPath expression. This description is straightforward and easy to understand, however, difficult to expand. If there is any backward axis (parent-axis or ancestor-axis) in the XPath expression, this method is no longer applicable to describing the tree structure.

We now give a different description of XPath tree, in which the level information between the adjacent two node tests is abstracted from the type of the axis between the node tests, and recorded at the corresponding node in the XPath tree. Our work is limited to $XP^{\{[\,],*,//\}}$ expression only.

Definition 1: For a given $XP^{\{[\,],*,//\}}$ expression q, we construct an XPath tree T. The root of T is independent of q. Every node test n in q can be described by a non-root node v. The relationship between v and its parental node v' is denoted by $L(v)=[a, b]$, where *a* and *b* are the minimum and maximum numbers of levels between *v* and *v'* respectively. The relationship between nodes in tree *T* is given as:

1) If *n* is a root node test, i.e. */n* or *//n*, there exists an edge in *T* between the node *v* in *T* that corresponds to *n*, and the root *r*, *edge*(*r*, *v*), where *r* is the parental node of *v*. When */n*, *L*(*v*)=[1, 1]; and *L*(*v*)=[1, ∞] when *//n*.

2) If *n* is not a root node test, there is an adjacent node test *n'* in *q* that satisfies *n'/n*, *n'*[*n*], *n'//n* or *n'*[*.//n*], therefore, there exists an edge in *T* between *v* and *v'* (corresponding to *n* and *n'* respectively), where *v'* is the parental node of *v*. When *n'/n* or *n'*[*n*], *L*(*v*)=[1, 1]; and *L*(*v*)=[1, ∞] when *n'//n* or *n'*[*.//n*].

Definition 2: Given an XPath tree *T*, let *NODES*(*T*) be the set of nodes in *T*, *EDGES*(*T*) be the set of edges in *T*, *ROOT*(*T*) be the root node of *T*. If there exists *v*∈ *NODES*(*T*), and the outdegree of *v* is greater than 1, or the outdegree or the indegree of *v* is 0, node *v* is then called *key node* of the XPath tree *T*. ∀ *edge*(*x*,*y*)∈ *EDGES*(*T*), where *x*,*y*∈ *NODES*(*T*), and *edge*(*x*,*y*) implies *x* is the parental node of *y*. If *nid* is the unique idtentifier of node *y* and *ln* is the label of node *y*, we then denote node *y* by $nid^{[a,b]}$, where [*a*,*b*] equals to *L*(*y*).

Informally, key nodes in an XPath tree are branching nodes (nodes with outdegree greater than 1), leaves, and root.

There are often some wildcard location steps without predicate used in an XPath expression, which are represented as non-branching nodes '*', such as the expression */a/*//*/b*. We can remove those wildcard nodes in the XPath tree for simplification, but have to revise the *L*(*v*) value of some related non-wildcard node *v* which is the descendent node of the removed wildcard node. Fig. 1(a) illustrates the two XPath trees of the expression */a/*//*/b* before and after removing non-branching wildcard nodes, where *L*(*b*) is revised. In the following context, all XPath trees are those trees from which the non-branching wildcard nodes are removed.

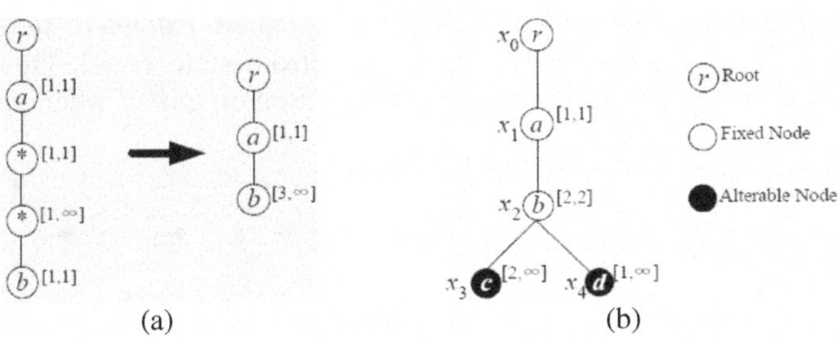

Fig. 1. (a)XPath Tree */a/*//*/b* ; (b)XPath Tree */a/*/b*[*.//*/c*]*//d*

Definition 3: Given an XPath tree *T*, let *CNODES*(*T*) be the set of alterable nodes, *FNODES*(*T*) be the set of fixed nodes, *NODES*(*T*)={*ROOT*(*T*)} ∪ *CNODES*(*T*) ∪ *FNODES*(*T*). ∀ *n*∈ *NODES*(*T*) and *n* ≠ *ROOT*(*T*), *L*(*n*) = [*a*,*b*]. If *a*=*b*, then *n*∈ *FNODES*(*T*); if *b*=∞, then *n*∈ *CNODES*(*T*). When *CNODES*(*T*) is not empty, the XPath tree *T* is an *alterable tree*, otherwise it is a *fixed tree*.

As an example, the XPath tree of the XPath expression */a/*/b*[*.//*/c*]*//d* is shown in Fig. 1(b). The set of level relationship between node x_2 and its parental node is $L(x_2)$=[2,2]. From definition 3, node x_2 is a fixed node. The set of level relationship between node x_3 and its parental node is $L(x_3)=[2,\infty]$, and node x_3 is an alterable node, so the corresponding XPath tree is an alterable tree.

Definition 4: Function h: $NODES(p) \rightarrow NODES(q)$ describes the homomorphism relationship from XPath tree p to XPath tree q:

1) $h(ROOT(p)) = ROOT(q)$;

2) For each $x \in NODES(p)$, $LABEL(x)$='*' or $LABEL(x) = LABEL(h(x))$;

3) For each $edge(x,y) \in EDGES(p)$, where $x,y \in NODES(p)$, $L(x,y) \supseteq L(h(x),h(y))$;

Fig. 2 shows the homomorphism mapping h from XPath tree p to XPath tree q based on XPath expressions /a/*//b and /a[c]//*/*//b.

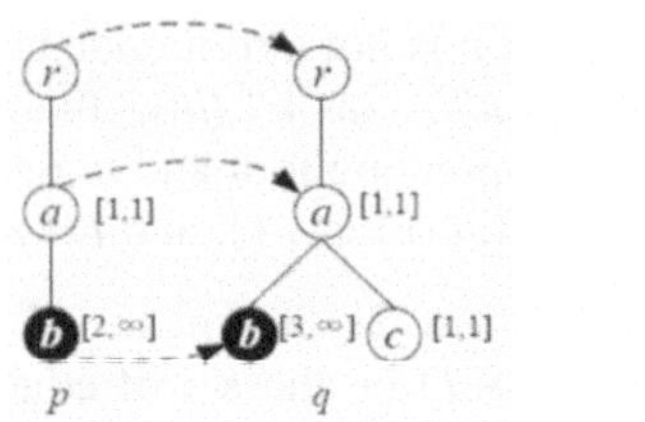

Fig. 2. Homomorphism mapping h:$p \rightarrow q$

3 Homomorphism Resolution Based on XTHC Machine

3.1 Construction of Basic XTHC Machine

We will incrementally construct NFA with prefix-sharing on the set of XPath trees $P=\{p_1,p_2...p_n\}$. Each node $nid^{[a,b]}$ in the XPath tree will be mapped to an automata fragment in NFA, and such a fragment has a unique start state and a unique end state. There are two cases while constructing the fragment from the node $nid^{[a,b]}$:

1. When $a=b$, $nid^{[a,b]}$ is a fixed node, the constructed automata fragment is shown in Fig.3(a). The states s-1 and s+a-1 are the start and end states of the fragment, respectively. Since a represents the minimum number of levels between node $nid^{[a,b]}$ and its parental node, starting from state s-1, we can construct in turn a-1 states along the arcs labeled '*', which are called *extended states*; we then construct state s+a-1 along the arc labeled *ln* from state s+a-2. Obviously there exist extended states in the automata fragment based on $nid^{[a,b]}$ when a>1.

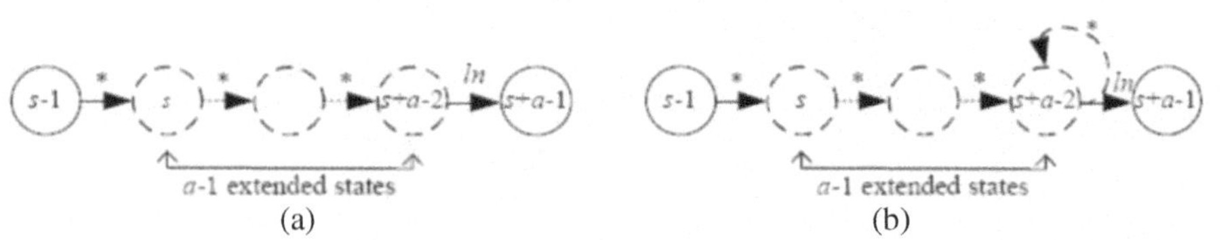

Fig. 3. (a) The automata fragment corresponding to the fixed node $nid^{[a,a]}$; (b) The automata fragment corresponding to the alterable node $nid^{[a,\infty]}$

2. When $b=\infty$, $nid^{[a,b]}$ is an alterable node, many kinds of automata fragment can be constructed, one example is shown in Fig.3(b). Similarly to that in case 1, we first construct a-1 extended states and the end state s+a-1, starting from state s-1. Since $b=\infty$, it is necessary to add self-looping arc, labeled by '*', in any one or more states from state s-1 or the following a-1 extended states. The chain consisting of the start state and the extended states, is denoted by *extended state-chain*. Fig.3(b) only shows one self-looping arc at last state of the extended state-chain. Obviously,

an automata fragment corresponding to an alterable node $nid^{[a,b]}$ (a>1) in an XPath tree p is optimal, if and only if there is only one state in the fragment that has a self-looping arc, and this state must be the last state along the extended state-chain.

Definition 5: Suppose the NFA constructed from set P of XPath trees is A, called the XHTC machine. We can create the following two index tables for each state s in A:

1) LP(s): list of leaf nodes. $\forall p \in P$, for each leaf node n_l in p, if s is the last state constructed from n_l, then $n_l \in$ LP(s). Only when s is a leaf state, LP(s) is non-empty.

2) LB(s): list of branching nodes. $\forall p \in P$, for each branching node n_b in p, if s is the last state constructed from n_b, then $n_b \in$ LB(s). Only when s is a branching state, LB(s) is non-empty.

Fig. 4(b) is the XTHC machine constructed from XPath trees p_1, p_2, and p_3 which are shown in Fig.4(a), $p_i.x$ represents node x in XPath tree p_i, a state is denoted by a circle. An arc implies state transition, where dashed lines represents transition of descendant-axis type, and solid lines represents transition of child-axis type. A label on an arc is a node test. State $S1$ has an arc to itself since it has a transition of descendant-axis type.

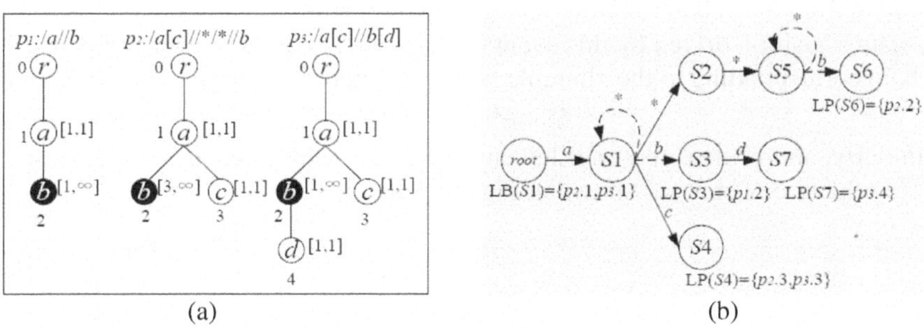

Fig. 4. (a) The XPath tree set P ; (b) The XTHC machine constructed from XPath tree set P

Definition 6: A basic non-deterministic XTHC machine A is defined as:

$$A = (Q^s, \Sigma, \delta, q^s_0, F, B, S_s)$$

where

- Q^s is the set of NFA states;
- Σ is the set of input symbols;
- q^s_0 is the *initial*(or *start*) NFA state of A, i.e. the root state;
- δ is the set of state transition functions, it contains at least the NFA state transition function, i.e. $t_{forward}$: $Q^s \times \Sigma \rightarrow 2^{Q^s}$;
- $F \subseteq Q^s$ is the set of *final* states, it is also the set of *leaf* states;
- $B \subseteq Q^s$ is the set of *branching* states;
- $\forall q^s \in Q^s$, we call q^s an NFA state of A, LP(q^s) and LB(q^s) are two index tables of q^s (see definition 5); S_s is the stack for state transition, the stack frame of S_s is a subset of Q^s.

3.2 Running an XTHC Machine

In order to resolve the homomorphous relationship using an XTHC machine, a depth-first traverse on the input XPath tree is required to generate SAX events. These events will be used as input to the XTHC machine for the XTHC machine running.

Four types of events will be generated at depth-first traverse on the input XPath tree *p*: startXPathTree, startElement, endElement and endXPathTree. Time of these events being generated is:

1) send startXPathTree event when entering root of *p*;
2) send startElement event when entering non-root node of *p*;
3) send endElement event when tracing back to non-root node of *p*;
4) send endXPathTree event when tracing back to root of *p*.

Since *a* and *b* are not always 1 in a node $nid^{[a,b]}$ of an XPath tree, more than one events are sent at entering or tracing back to node $nid^{[a,b]}$:

1) the startElement event sequence sent when *a*=*b* is shown in Fig.5(a);
2) the startElement event sequence sent when $b=\infty$ is shown in Fig.5(b).

In particular, there are some restrictions applied on a startElement('//') event:

1) it occurs only when node $nid^{[a,b]}$ is an alterable node;
2) state transition driven by this event occurs only at state *s* in the extended state-chain corresponding to the alterable node, and there is a unique state transition:

$$t_{forward}(s, \text{'//'}) \rightarrow s.$$

Similarly, more than one endElement event are sent when tracing back to node $nid^{[a,b]}$ in the tree, which are shown in Fig. 5(c) and 5(d).

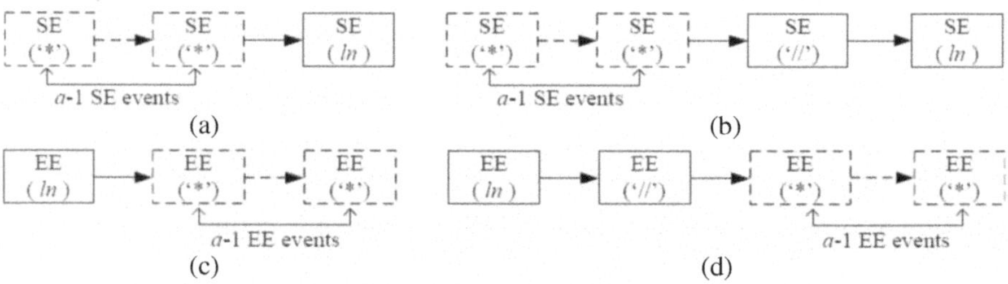

Fig. 5. (a) The startElement("SE" for short) event sequence of the fixed node $nid^{[a,a]}$; (b) The startElement event sequence of the alterable node $nid^{[a,\infty]}$; (c) The endElement("EE" for short) event sequence of the fixed node $nid^{[a,a]}$; (d) The endElement event sequence of the alterable node $nid^{[a,\infty]}$

Fig. 6 shows rules of processing SAX events in an XTHC machine. The homomorphism relationship between tree p_i in a set of XPath tree $P=\{p_1,p_2,\ldots,p_n\}$ and an input tree *q* can be resolved by running the XTHC machine. When the XTHC machine is running, $\forall p \in P$, homomorphism information between each node *v* in *p* and nodes in the input tree *q* is recorded. Let $v \in p$, *a* be the label of node *u* in the input XPath tree *q*. We define the following three operations to mark, deliver and reset information about the mapping in the XPath tree *p*:

1) mark(v, u): when the XTHC machine is running at a leaf state $q^s(q^s \in F)$, $\forall v \in LP(q^s)$, mark on v the information about the mapping from the leaf node v to the node u in the input XPath tree q;
2) deliver(v): when the machine traces back to a key state $q^s(q^s \in F \cup B)$, $\forall v \in LB(q^s) \cup LP(q^s)$, if information about the mapping was marked on node v, deliver the mapping information of v to the nearest ancestor key node to v in the XPath tree;
3) reset(v): when the machine traces back to a key state $q^s(q^s \in F \cup B)$, $\forall v \in LB(q^s) \cup LP(q^s)$, reset the mapping information on node v.

```
startXPathTree()
  push(Ss, {qs0}); other initialization
startElement(a)
  qsset ={ }; // current NFA state set
  u = getCurrentInputNode( );
  for each qs in peek(Ss)
    merge tforward(qs, a) into qsset
  push(Ss, qsset);
  for each qs in qsset
   if (qs∈ F)
   for each v in LP(qs)
    mark(v, u);

endElement(a)
  qsset = pop(Ss);
  for each qs in qsset
   if (qs∈ B or qs∈ F){
    for each v in LB(qs) or LP(qs)
     if exsit mapping of v{
       deliver(v);
       reset(v);
     }
   }
endXPathTree( )
  pop(Ss);
```

Fig. 6. The processing rules of SAX events in XTHC

The time complexity of the algorithm resolving homomorphism from one XPath tree p to another XPath tree q is $O(|p||q|^2)$[2]. Therefore, the time complexity from each tree p in a set of XPath tree $P=\{p_1,p_2,\ldots,p_n\}$ to q is $O(n|p||q|^2)$ without using prefix-sharing automata. However, if prefix-sharing automata is used, the time complexity is $O(m|q|^2)$, where m is the number of states in NFA. When XPath trees in P have common branches and prefixes, $n|p|$ is much greater than m, therefore, it is much more efficient to resolve homomorphism from multi-XPath trees to one single XPath tree using prefix-sharing automata.

4 Experiments

An algorithm resolving homomorphism based on the XTHC machine (XHO) was implemented using Java. The experimental platform is on Windows XP operation system, Pentium 4 CPU, with frequency of 1.6GHz and memory of 512MB. We compared several algorithms: the homomorphism algorithm (HO)[2], the complete algorithm in a cononical model (CM), branch homomorphism algorithm(BHO)[4], and the proposed XHO algorithm. We checked the scope of each algorithm at resolving containment of XPath expressions (see table 1, where T/F represents p containing/not containing q), and the time complexity of these algorithms(see Fig. 7).

This experiment shows XHO is as capable as existing homomorphism algorithms. Furthermore, XHO supports containment calculation from multi-XPath expressions to one single XPath expression. Although BHO also supports such calculation, it may give incorrect results in some cases as shown in Table 1. BHO gives a result that is rather different from the correct result CM gives. Compared to BHO, XHO gives smaller discrepancy between containment and homomorphism.

Table 1. Some pairs of XPath trees for experiments and containment results

No	*p*	*q*	HO	BHO	XHO	CM
no.1	*/a//*[.//c]//d*	*/a//b[c]//d*	T	T	T	T
no.2	*/a/*/*/c*	*/a/b[c]/e/c*	T	T	T	T
no.3	*/a//b[*//c]/b/*c	*/a//b[*//c]/b[b/c]//c*	T	T	T	T
no.4	*/a//*/b*	*/a/*//b*	T	F	T	T
no.5	*/a/*[.//b]//c*	*/a//*/b/c*	F	F	T	T
no.6	*/a[a//b[c/*//d]/b/c/d]*	*/a[a//b[c/*//d]/b[c//d]/b/c/d]*	F	F	F	T
no.7	*/a/*/*/*/c*	*/a//*/b//b/c*	F	F	F	F
no.8	*/a//b[c]/d*	*/a/b[.//c]//d*	F	F	F	F

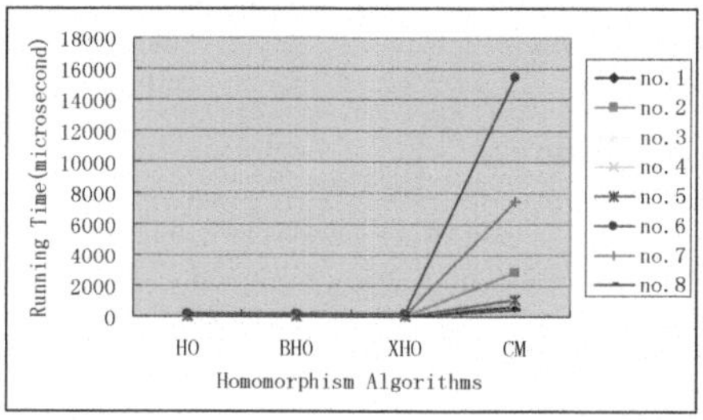

Fig. 7. The experimental results for some homomorphism algorithms

5 Conclusion

This paper considers an algorithm to resolve containment between multi-XPath expressions and one single XPath expression through homomorphism. While high efficiency is kept at calculating multi-containment relationships, we also consider discrepancy between containment and homomorphism. The algorithm works correctly on calculating containment of a special type of XPath expressions. Experiments showed that our algorithm is more complete than conventional homomorphism algorithms. Future research can be done on how to resolve homomorphism between one XPath tree and multi-XPath trees simultaneously.

References

[1] World Wide Web Consortium, XML Path Language (XPath) Version 1.0, http://www.w3.org/TR/xpath, W3C Recommendation, November 1999.

[2] G. Miklau and D. Suciu. Containment and equivalence for a fragment of XPath. Journal of the ACM, 51(1):2-45, 2004.

[3] Yuguo Liao, Jianhua Feng, Yong Zhang and Lizhu Zhou. Hidden conditioned homomorphism for XPath fragment containment. In DASFAA 2006, LNCS 3882, 454-467, 2006.

[4] Sanghyun Yoo, Jin Hyun Son and Myoung Ho Kim. Maintaining homomorphism information of XPath patterns. IASTED-DBA2005, 2005, 192-197.

An Efficient Overlay Multicast Routing Algorithm for Real-Time Multimedia Applications

Shan Jin, Yanyan Zhuang, Linfeng Liu, and Jiagao Wu

Key Laboratory of Computer Network and Information Integration (Southeast University), Ministry of Education, Nanjing 210096, China
{kingsoftseu, zhuangyanyan, liulf, jgwu}@seu.edu.cn

Abstract. Multicast services can be deployed either on network layer or application layer. Implementations of application-level multicast often provide more sophisticated features, and can provide multicast services where IP multicast services are not available. In this paper, we consider the degree and delay constrained routing problem in overlay multicast for real-time multimedia applications, and an efficient Distributed Tree Algorithm (DTA) is proposed. With DTA, end hosts can make trade-off between minimizing end-to-end delay and reducing local resource consumption by adjusting the heuristic parameters, and then self-organize into a scalable and robust multicast tree dynamically. By adopting distributed and tree-first schemes, a newcomer can adapt to different situations flexibly. The algorithm terminates when the newcomer reaches a leaf node, or joins the tree successfully. Simulation results show that the multicast tree has low node rejection rate by choosing appropriate values of the heuristic parameters.

Keywords: overlay multicast, routing algorithm, heuristic parameter.

1 Introduction

Multicast is a basic communication service for many new network applications, like the real-time multimedia transmission. When it comes to practical issues, however, full deployment of IP multicast [1] has long been postponed in the Internet for both technical and economic reasons [2]. Researchers wondered whether the network layer is appropriate for implementations of multicast functionality; therefore, overlay multicast [3] is proposed as an alternative to IP multicast. Overlay multicast deploys multicast services on hosts instead of core routers. The advantage of doing so is that the multicast services are easier to deploy, since there is no need to change the existing IP network infrastructure. From the architectural point of view, the overlay multicast systems can be classified as host-based architecture (like ALMI [4] and HMTP [5]), and proxy-based architecture (like Overcast [6] and Scattercast [7]). Both architectures face the same-natured problems when talking about overlay multicast routing. The overlay multicast routing problem in this paper is studied based on the host architecture, taking the common features of both architectures into consideration.

Since overlay multicast routing performance is usually not as efficient as that of network layer multicast, it is crucial to study degree and delay constrained overlay

G. Dong et al. (Eds.): APWeb/WAIM 2007, LNCS 4505, pp. 829–836, 2007.

multicast routing algorithms for real-time multimedia applications. In centralized algorithms [4, 8], a server, which is supposed to know the path latency between any nodes in an overlay network, constructs multicast tree according to an objective function. However, these algorithms do not consider the dynamic feature of multicast members, and ignore the problems of algorithm complexity and the single-point failure. In contrast, distributed algorithms that use local routing optimization bear great extensibility and dynamic flexibility. These algorithms can be classified into mesh-first [7, 9] and tree-first [5, 6] strategies. Studies show that none of the protocols above have considered the strict delay constraint in real-time multimedia applications, and how the multicast routing performance is affected by the dynamic end hosts also lacks sufficient study [10].

We introduce a novel distributed overlay multicast routing algorithm named Distributed Tree Algorithm (DTA). The algorithm adopts tree-first strategy in order to enhance multicast routing performance effectively and save system maintenance cost. By adjusting appropriate heuristic parameters, DTA can improve multicast routing performance and reduce node rejection rate considerably.

2 Problem Formulations and Design Objectives

The overlay multicast network is a logical network built on top of the Internet unicast infrastructure. It can be depicted as a complete directed graph, $G = (V, E)$, where V is the set of vertices and $E = V \times V$ is the set of edges. Each vertex in V represents a host. The directed edge from node i to node j in G represents a logical channel corresponding to a unicast path from host i to host j in the physical topology. The data delivery path will be a directed spanning tree T of G rooted at the source host, with the edges directed away from the root.

Definition 1 $d_{max}(v) \in N$: The out-degree constraint of host v in the overlay tree.

Definition 2 $l(u, v) \in R^+$: The unicast latency from host u to host v.

Definition 3 $delay(r, v) \in R^+$: The overlay latency from root r to host v. It is the sum of all the unicast latencies along the path from r to v in the spanning tree T.

We consider two optimization objectives: one seeks to minimize the maximum overlay latency in a multicast tree to reduce the session latency, taking the degree constraint at individual nodes into consideration; the other optimizes the bandwidth usage at each host to reduce the likelihood of bottleneck nodes and constructs a tree satisfying the constraint of the overlay maximum latency. $d_{used}(v)$ denotes the degree already used by node v, $d_{res}(v) = d_{max}(v) - d_{used}(v)$ denotes the residual degree of v, S denotes the set of all hosts in the tree, and L denotes the upper bound of the session latency. Then the two objectives are formulated as follows:

Problem 1 Minimum Maximum-Latency Degree-Bounded Directed Spanning Tree Problem (MMLDB): Given a complete directed graph $G = (V, E)$, a degree constraint $d_{max}(v)$ for each vertex $v \in V$, and a latency $l(u, v)$ for each edge $e(u, v) \in E$. Find a directed spanning tree T of G rooted at host r that minimizes the maximum $delay(r, v)$, and the degree constraint is satisfied at each node that $d_{used}(v) \leq d_{max}(v)$.

Problem 2 Residual-Balanced Degree and Latency-Bounded Directed Spanning Tree Problem (RBDLB): Given a complete directed graph $G = (V, E)$, a degree bound $d_{max}(v)$ for each vertex $v \in V$, and a latency $l(u, v)$ for each edge $e(u, v) \in E$. Find a directed spanning tree T of G rooted at host r that maximizes the minimum $d_{res}(v)$, satisfying both the degree constraint at each node that $d_{used}(v) \leq d_{max}(v)$ and the latency constraint of the session that $max_{v \in S}\, delay(r, v) \leq L$.

Both MMLDB and RBDLB problems are NP-complete [8]. Our design of DTA can make trade-off between minimizing end-to-end delay and reducing local resource consumption. Resultingly, both of the desired objectives are met.

3 Design of DTA

Each node only needs to maintain a local status set, $\{d_{max}(v), d_{used}(v), delay(r, v), Children(v), parent(v), l(parent(v), v)\}$. $Children(v)$ denotes the set of v's children and $parent(v)$ denotes v's parent, $l(parent(v), v)$ is the unicast latency from v's parent to v itself, which can be acquired by an end-to-end measuring tool.

3.1 Creating a Multicast Group

Each multicast group has a Rendezvous Point (RP) from which new members can learn about membership of the group so as to bootstrap themselves. The construction of a multicast group is as follows: 1) The host that sends out data acts as the creator, as well as the root, of the tree T once a multicast session commences. It sends to RP a CREATEREQUEST message. 2) When receiving the CREATEREQUEST message, RP adds the QoS parameters to its group list, then sends out a CREATEACK message to the corresponding requesting host.

3.2 Joining a Multicast Group

A newcomer v sends to RP a QUERYREQUEST message, containing the multicast group ID. On receiving the request message, RP checks its root list for the specific item, say r, of that group, then sends QUERYACK message containing r's IP address and the corresponding QoS parameters to v. Then v sets r as its tentative parent p_t and asks r for the list of r's children. Next, v queries r and its children for their latencies and bandwidth information to constitute its potential parents set $PP(p_t)$ defined in *Definition 4* (see below). From all nodes in $PP(p_t)$, v picks a local optimal parent according to function (1): Local Optimal Parent Selection (LOPS) Function. If the local optimal parent is not the tentative parent p_t, v replaces the old p_t with this parent, and repeats this process until a local optimal parent, u for instance, perseveres in its role as the tentative parent. Then v makes u its parent by sending JOINREQUEST message to u. On the contrary, if there is no potential parent of v, i.e., $PP(p_t)$ is empty, v selects a local optimal grandparent from p_t's children and sets this grandparent as a new tentative parent according to function (2): Local Optimal Grandparent Selection (LOGS) Function, then repeats the joining process.

Definition 4 $PP(p_t)$: Newcomer v's potential parents set. $PP(p_t) = \{n \,/\, d_{used}(n) < d_{max}(n) \wedge delay(r, n) + l(n, v) \leq L, n \in \{ p_t \} \cup Children(p_t)\}$.

Considering two different situations in which *PP(p_t)* is empty or not, DTA deals with them with either LOGS-Function or LOPS-Function mentioned above. The two functions are given as follows:

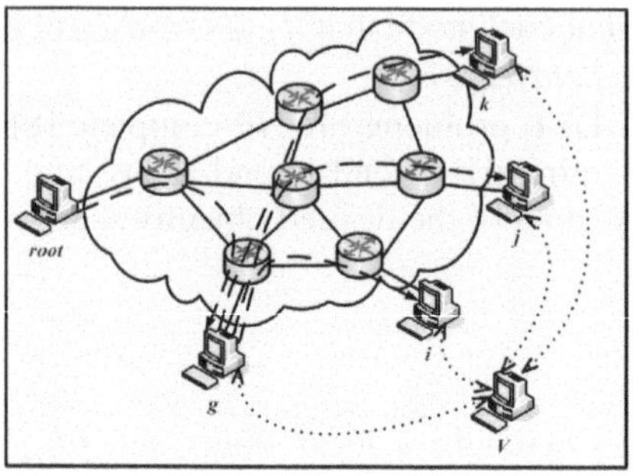

Fig. 1. An Illustration of LOPS-Function and LOGS-Function

Local Optimal Parent Selection (LOPS) Function:

$$Pfunc(p_t') = \min_{m \in PP(p_t)} Pfunc(m) \ . \tag{1}$$

Pfunc(m) reflects the efficiency in selecting a node from *PP(p_t)* as candidate for the newcomer's parent. It can be expressed as follows:

$$Pfunc(m) = \rho \cdot \frac{d_{used}(m)}{d_{\max}(m)} + (1-\rho) \cdot \frac{l(m,v)}{\max\limits_{n \in PP(p_t)} l(n,v)}, \rho \in [0,1] \ .$$

As shown in Fig. 1, v is a newcomer and g is its current tentative parent p_t. *PP(p_t)* = *{g} ∪ Children(g)* = *{g, i, j, k}*. v is now enquiring degree and latencies of all members in *PP(p_t)* to calculate the values of the corresponding *Pfunc()*. $\frac{l(m,v)}{\max\limits_{n \in PP(p_t)} l(n,v)}$ reflects how close node m is to node v. A smaller value denotes a shorter distance from a node in *PP(p_t)* to v. $\frac{d_{used}(m)}{d_{\max}(m)}$ reflects how many end system resources a node in *PP(p_t)* has used by now, and a smaller value denotes a smaller percentage of the resources that have been used. Weight ρ *is* a heuristic factor. We can trade off between minimizing end-to-end delay and reducing local resource consumption by adjusting the value of ρ between [0, 1].

Local Optimal Grandparent Selection (LOGS) Function:

$$Gfunc(p_t') = \max_{q \in Children(p_t)} Gfunc(q) \ . \tag{2}$$

Gfunc(q) is kind of forecasting of the joining action, and it can be expressed as follows:

$$Gfunc(q) = \frac{\max\limits_{m \in Children(p_t)} l(p_t, m)}{l(p_t, q)} \cdot \frac{d_{used}(q)}{\max\limits_{n \in Children(p_t)} d_{\max}(n)} \cdot \theta^{t(q)}, \theta \in (0,1), t(q) = 0,1,2,3,\cdots \ .$$

In Fig. 1, if g is v's current tentative parent p_t. A bigger value of $\frac{\max_{m\in Children(g)} l(g,m)}{l(g,q)} \cdot \frac{d_{used}(q)}{\max_{n\in Children(g)} d_{\max}(n)}$ denotes relatively small latency from node q to its parent g, and q itself has a relatively larger number of children. As a result, the tree's radius can be decreased and the node rejection rate will fall. $\theta^{t(q)}$ is a balancing factor with its value between (0,1). $t(q)$ records the number of times that node q ($q \in Children(p_t)$) has been selected as a local optimal grandparent. A smaller value of θ indicates a more likelihood of a newcomer's selecting a different node as its local optimal grandparent than the last time. Multiplying θ for $t(q)$ times, thus getting $\theta^{t(q)}$, is to prevent one single node from being selected as the local optimal grandparent all the time, which deteriorates the overall performance of the multicast tree. If all of g, i, j and k do not meet the degree and latency constraints, i.e., the set $PP(g) = \Phi$, v will use LOGS-Function to evaluate i, j, k in $Children(g)$ in order to decide which one will be the new tentative parent.

To summarize, a newcomer tries to find a "good" parent by searching a certain part of the tree. It stops when it joins the tree successfully or reaches a leaf node. The detailed algorithm is shown as follows:

Joining Algorithm

```
v finds root r by querying RP, let p_t = r;
while PP(p_t) == Φ
  Gfunc(p_t') = 0;
  foreach q ∈ Children(p_t)
    if Gfunc(q) > Gfunc(p_t')
      Gfunc(p_t') = Gfunc(q);
      p_t' = q;
  if Gfunc(p_t') == 0
    v returns JOINFAIL message to RP;
  p_t = p_t';
while true
  Pfunc(p_t') = +∞;
  foreach m ∈ PP(p_t)
    if Pfunc(m) < Pfunc(p_t')
      Pfunc(p_t') = Pfunc(m);
      p_t' = m;
  if p_t' == p_t
    v establishs a unicast tunnel to p_t;
    v returns JOINSUCCEED message to RP;
  p_t = p_t';
```

3.3 States Maintaining and Leaving a Multicast Group

Status in DTA is refreshed by periodic message exchanges between neighbors. Every child sends REFRESH message to its parent, and the parent replies this message by sending KEEPALIVE message back. Each member calculates the round-trip time (rtt) by these two messages. If a member cannot reach its parent any more or the rtt no longer meets the latency constraint, then the joining algorithm is triggered.

When a member leaves a group, it sends LEAVEREQUEST message to its parent and children, from whom it receives the LEAVEACK messages. Its parent simply deletes the leaving member from its children list. But the children of this leaving member must find new parents. A child looks for a new parent with the help of joining algorithm. If the root is leaving, it notifies RP and its children by sending CANCELGROUP message to them. RP then deletes the group information of this root from its group list. Other members in the tree pass the message on to their neighbors then all of them leave the group.

4 Performance Evaluation

4.1 Performance Metrics and Simulation Setup

We have done some simulations to evaluate the performance of DTA, concerning the node rejection rate defined as follows:

Definition 5 Node Rejection Rate R_r: $R_r = n / N$, in which n denotes the amount of nodes rejected by DTA and N denotes total amount of nodes.

Our simulations are based on a network that consists of 1000 routers. The network has a random flat topology generated by using the Waxman model [11]. The communication delay which is designated between [1ms, 50ms] between neighbor routers, is directly proportional to their geometric distance. Some additional nodes are generated as regular hosts and are randomly attached to these routers. The node degree is uniformly distributed between 4 and 8. Each node experiences 100 rounds of simulation and the average value is recorded as experimental result.

4.2 Simulation Results and Analyses

Fig. 2 and Fig. 3 show the node rejection rate versus the session delay constraint. There are 50 regular hosts which want to join the multicast group one by one in Fig. 2 and 200 in Fig. 3. We set the value of θ to 0.2, 0.5, and 0.8 respectively and adjust ρ's value among 0.0, 0.3, 0.7 and 1.0 with each different value of θ. Different curve in a chart denotes a different value combination of ρ and θ in the form of (ρ, θ). From all these charts, we can see that the rejection rate decreases as the session delay constraint increases. Furthermore, different combinations of (ρ, θ) also have an impact on system performance. Firstly, DTA approximates to minimizing local resource consumption strategy (RBDLB) when ρ is closer to 1, whereas it approximates to minimizing end-to-end delay strategy (MMLDB) when ρ approaches 0. The node rejection rate can not be decreased remarkably if only one of the two strategies is considered, i.e., only ρ's value equals to 0 or 1. Therefore, an appropriate value must be set to trade off between the two strategies. Secondly, if θ has a larger value, a newcomer is more likely to select some specific members as its local optimal grandparent, which could bring about overload in local area and the rejection rate will increase as a result; if θ is smaller, a newcomer is likely to select its local optimal grandparent among all its potential grandparents with a relatively equal probability, but some of the preferable ones may be ignored and the rejection rate could also increase as a result. From the six charts we can see that, DTA always performances best when the combination of (ρ, θ) is set to (0.3, 0.5). This result illustrates that our optimization strategy in DTA is much closer to end-to-end delay optimization strategy (ρ = *0.0*). By comparing Fig. 2 and Fig. 3, it is clear that the optimization objectives are better achieved when the number of multicast group members is larger. Therefore, DTA is more suitable for large-scale overlay multicast applications.

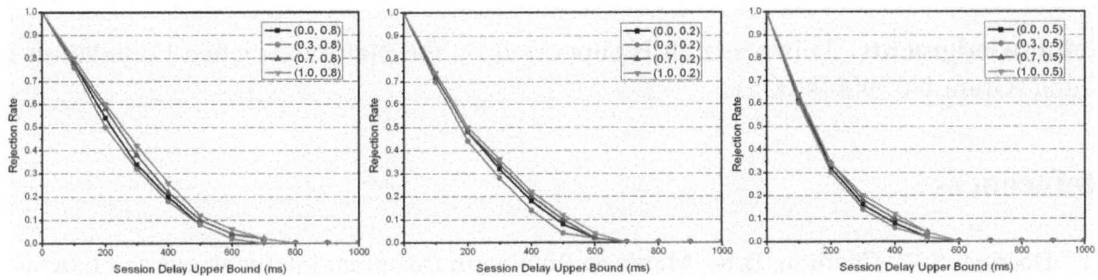

Fig. 2. Node rejection rate of DTA vs. Session delay upper bound. Group size = 50.

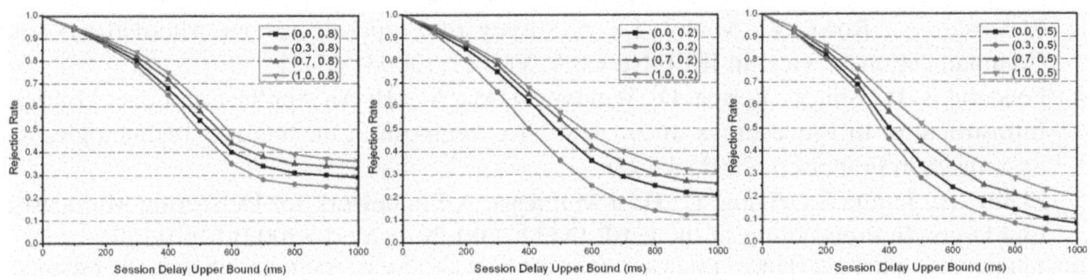

Fig. 3. Node rejection rate of DTA vs. Session delay upper bound. Group size = 200.

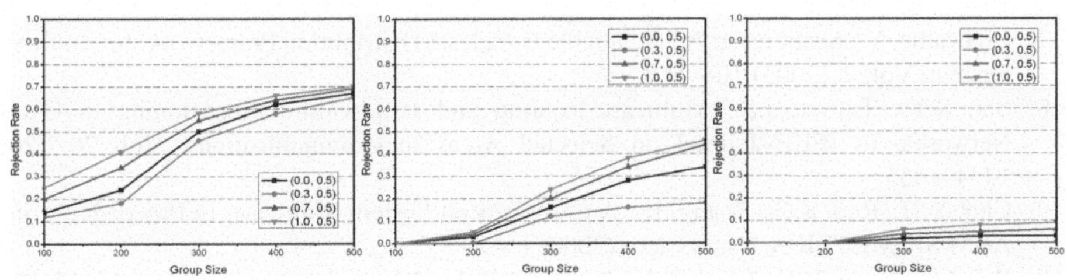

Fig. 4. Node rejection rate of DTA vs. Group size. Session delay upper bound = 600ms, 1400ms, 2000ms.

Fig. 4 shows the node rejection rate versus the multicast group size when θ is set to 0.5. We can see that the combination of (0.3, 0.5) also brings about the best performance. If the session delay constraint is set to too low (the chart on the left) or too high (the chart on the right) a value, then the change in the value of ρ will have less impact on the performance. But when we set session delay constraint to 1400ms (the chart in the middle), a better choice of (ρ, θ) will have a notable effect.

5 Conclusion

We study tree-first overlay multicast routing algorithm and propose an efficient distributed routing algorithm named DTA. Our algorithm seeks to make trade-off between minimizing end-to-end delay and minimizing local resource consumption. Simulations show that the performance of DTA is quite a satisfaction with node degree and end-to-end delay constraints by properly selecting (ρ, θ). Work for algorithm improvement and discussion on the best value of (ρ, θ) is left for the future.

Acknowledgments. This research is supported by the Natural Science Foundation of China (Grant No. 90604003).

References

1. Deering, S.E., Cheriton, D.R.: Multicast Routing in Datagram Internetworks and Extended LANs. In ACM Transactions on Computer Systems, Vol. 8. (1990) 85–110
2. Diot, C., Levine, B.N., Lyles, B., Kassem, H., Balensiefen, D.: Deployment Issues for the IP Multicast Service and Architecture. In IEEE Network, Vol. 14. (2000) 78–88
3. El-Sayed, A., Roca, V., Mathy, L.: A Survey of Proposals for an Alternative Group Communication Service. In IEEE Network, Vol. 17. (2003) 46–51
4. Pendarakis, D., Shi, S., Verma, D., Waldvogel, M.: ALMI: An Application Level Multicast Infrastructure. In Proceedings of 3rd USENIX Symposium on Internet Technologies and Systems, San Francisco (2001) 49–60
5. Zhang, B., Jamin, S., Zhang, L.: Host Multicast: A Framework for Delivering Multicast to End Users. In Proceedings of the IEEE INFOCOM, New York (2002) 1366–1375
6. Jannotti, J., Gifford, D.K., Johnson, K.L., Kaashoek, M.F., O'Toole, J.W., Jr.: Overcast: Reliable Multicasting with an Overlay Network. In Proceedings of the 4th USENIX Symposium on Operating Systems Design and Implementantion, San Diego (2000) 197–212
7. Chawathe, Y.: Scattercast: An Adaptable Broadcast Distribution Framework. In Multimedia System, Vol. 9. (2003) 104–118
8. Shi, S.Y., Turner, J.S.: Multicast Routing and Bandwidth Dimensioning in Overlay Networks. In IEEE Journal on Selected Areas in Communications, Vol. 20. (2002) 1444–1455
9. Chu, Y.H., Rao, S.G., Zhang, H.: A Case for End System Multicast. In Proceedings of the ACM SIGMETRICS, Santa Clara (2000) 1 – 12
10. Wu, J.G., Yang, Y.Y., Chen, Y.X., Ye, X.G.: Delay Constraint Supported Overlay Multicast Routing Protocol. In Journal on Communications, Vol. 26. (2005) 13–20
11. Zegura, E.W., Calvert, K.L., Bhattacharjee, S.: How to Model an Internetwork. In Proceedings of the IEEE INFOCOM, San Francisco (1996) 594–602

Novel NonGaussianity Measure Based BSS Algorithm for Dependent Signals

Fasong Wang[1], Hongwei Li[1], and Rui Li[2]

[1] School of Mathematics and Physics, China University of Geosciences
Wuhan 430074, P.R. China
fasongwang@126.com, hwli@cug.edu.cn
[2] School of Sciences, Henan University of Technology
Zhengzhou 450052, P.R. China
slxlr@haut.edu.cn

Abstract. The purpose of this paper is to develop novel Blind Source Separation (BSS) algorithms from linear mixtures of them, which enable to separate dependent source signals. Most of the proposed algorithms for solving BSS problem rely on independence or at least uncorrelation assumption of the source signals. Here, we show that maximization of the nonGaussianity(NG) measure can separate the statistically dependent source signals and the novel NG measure is given by the Hall Euclidean distance. The proposed separation algorithm can result in the famous FastICA algorithm. Simulation results show that the proposed separation algorithm is able to separate the dependent signals and yield ideal performance.

1 Introduction

Blind source separation(BSS) is typically based on the assumption that the observed signals are linear superpositions of underlying hidden source signals. When the source signals are mutual independent, the BSS can be solved by using the so called independent component analysis(ICA) method which has been attracted considerable attention in the signal processing and neural network fields and several efficient algorithms have been proposed (see for overview, e.g., [1-2]).

Despite the success of using standard ICA in many applications, the basic assumptions of ICA may not hold for some real-world situations, especially in biomedical signal processing and image processing, therefore the standard ICA cannot give the expected results. In fact, by definition, the standard ICA algorithms are not able to estimate statistically dependent original sources. Some authors [3] have proposed different approaches which take advantage of the nonstationarity of such sources in order to achieve better performance than the classical methods, but they still require their independence or uncorrelation.

Some extended data models have also been developed to relax the independence assumption in the standard ICA model, such as multidimensional ICA [4], independent subspace analysis [5] and subband decomposition ICA (SDICA) model [6].

G. Dong et al. (Eds.): APWeb/WAIM 2007, LNCS 4505, pp. 837–844, 2007.

As mentioned in [7], in the dependent sources situations, we can not resort minimization the mutual information(MI), but on the other hand we can maximization of NG to get the dependent sources. In this paper, based on the generalization of the central limit theorem(CLT) to special dependent variables, we will try to track the generalize ICA model-dependent component analysis problem by maximization NG measure. The NG quantity measure of arbitrary standardized probability density is defined by the L^2 norm in the L^2 space of the difference between the given density and the standard normal density.

This paper is organized as follows: Section 2 introduces briefly the dependent BSS model and NG measure; Then in section 3, we describe the novel NG measure using Hall distance in detail; In section 4, we use the NG measure to get the proposed separation algorithm and show that it is equivalent to the FastICA algorithm; Simulations illustrating the good performance of the proposed method are given in section 5; Finally, section 6 concludes the paper.

2 Dependent BSS Model and NG Measure

2.1 Dependent BSS Model

For our purposes, the problem of BSS can be formulated as: $\mathbf{x}(t) = \mathbf{A}\mathbf{s}(t) + \mathbf{n}(t)$, where $\mathbf{s}(t) = [s_1(t), s_2(t), \cdots, s_n(t)]^T$ is the unknown n-dimensional source vector. Matrix $\mathbf{A} \in R^{m \times n}$ is an unknown full column rank mixing matrix and $m \geq n$. The observed mixtures $\mathbf{x}(t) = [x_1(t), x_2(t), \cdots, x_m(t)]^T$ are called as sensor outputs and $\mathbf{n}(t) = [n_1(t), n_2(t), \cdots, n_m(t)]^T$ is a vector of additive noise that is assumed to be zero in this paper.

The task of BSS is to estimate the mixing matrix $\mathbf{A}$ or its pseudo inverse separating (unmixing) matrix $\mathbf{W} = \mathbf{A}^+$ in order to estimate the original source signals $\mathbf{s}(t)$, given only a finite number of observation data. There are two indeterminacies cannot be resolved in BSS without some *a priori* knowledge: scaling and permutation ambiguities. Thus, $(\hat{\mathbf{A}}; \hat{\mathbf{s}})$ and $(\mathbf{A}; \mathbf{s})$ are said to be related by a waveform-preserving relation.

A key factor in BSS is the assumption about the statistical properties of sources such as statistical independence. That is the reason why BSS is often confused with ICA. In this paper, we exploit some weaker conditions for separation of sources assuming that they have statistically dependent properties.

Throughout this paper the following assumptions are made unless stated otherwise: 1) The mixing matrix $\mathbf{A}$ is of full column rank; 2) Source signals are statistically dependent signals with zero-mean; 3) Additive noises $\{\mathbf{n}(t)\} = 0$.

So, the BSS model of this paper is simplified as

$$\mathbf{x}(t) = \mathbf{A}\mathbf{s}(t). \tag{1}$$

2.2 NG Measure

In ICA applications, NG measures are used based on the following fundamental idea: the outputs of a linear mixing process that preserves variances, have higher entropies than the inputs [7]. This general statement can be precisely expressed in mathematical terms as CLT which tell us that the linear mixture of N independent signals with finite variances will became asymptotically Gaussian (or *more nearly* Gaussian).

Since CLT is not valid for any set of dependent variables, we must be aware that we may not always recover the original sources using maximum NG criteria. [7] gives a very special condition on sources, for which the linear combinations of dependent signals are not more Gaussian than the components and therefore the maximum NG criteria fails, but fortunately this is not the case in most of real world scenarios.

The NG quantity measure of arbitrary standardized PDF is defined by the L^2 norm in the L^2 space of the difference between the given density and the normal density. This can be interpreted as the square-distance, with respect to some measure, between the two functions in the space of square integrable functions.

Let $\mathbf{x}$ be a random variable with PDF $f(x)$, We attempt to compute f's departure from Gaussianity by comparing it with its normal Gaussian counterpart: $g(x) = \frac{1}{\sqrt{2\pi}} exp(-\frac{x^2}{2})$. If one regards f and g as elements of the function space of PDF, the deviation of f from normality may be evaluated by an L^2 metric defined with some positive measure of the real line, $\mu(x)$:

$$D = \int_{-\infty}^{\infty} (f(x) - g(x))^2 w(x) dx, \tag{2}$$

where the $w(x)$ is given by $w(x) = d\mu(x)/dx$. This definition corresponds to the integrated square-difference between functions f and g, measured with the weight function $w(x)$. Although we leave $w(x)$ unspecified at this point, we assume that we choose w such that the integral converges for most reasonable densities.

We expand the function $f(x)$ in the integral (2) in terms of Hermite polynomials, a set of orthogonal functions on the entire real line with respect to an appropriate Gaussian weight. Following the notation in [8], two distinct families of Hermite polynomials, for $n = 0, 1, 2, \cdots$, are generated by the derivatives of the Gaussian PDF,

$$He_n(x) = (-1)^n e^{\frac{1}{2}x^2} \frac{d^n}{dx^n} e^{-\frac{1}{2}x^2}, \quad H_n(x) = (-1)^n e^{x^2} \frac{d^n}{dx^n} e^{-x^2}, \tag{3}$$

and $H_n(x) = \sqrt{2^n} He_n(x/\sqrt{2})$.

Following standard practice, we refer to the first set as Chebyshev-Hermite, and the second as Hermite polynomials. The first few polynomials are: $H_0(x) = 1, H_1(x) = 2x, H_2(x) = 4x^2 - 2x, H_3(x) = 8x^3 - 12x, H_4(x) = 16x^4 - 48x^2 + 12$.

Chebyshev-Hermite and Hermite polynomials satisfy an orthogonality relationship,

$$\int_{-\infty}^{\infty} He_n(x) He_m(x) g(x) dx = \delta_{nm} n!, \tag{4}$$

$$\int_{-\infty}^{\infty} H_n(x) H_m(x) g^2(x) dx = \delta_{nm} 2^{n-1} n! / \sqrt{\pi}. \tag{5}$$

with respect to the weight functions $g(x)$ for Chebyshev-Hermite polynomials $He_n(x)$, and $g^2(x)$ for Hermite polynomials $H_n(x)$. We will give a non-Gaussianity indices based on the squared functional distance [9]. The index is defined by a different form of orthogonal series expansion for arbitrary density $f(x)$, written in terms of either Chebyshev-Hermite or Hermite polynomials.

3 Hall Eudilean Distance Based Novel NG Measure

From the point of view of the L^2 metric space, perhaps the most natural weight is the uniform function $w(x) = 1$, which treats every point on the entire real line democratically. Hall [9] proposed such an index based on the L^2 Euclidean distance, $L^2(1)$, from the standard normal, called Hall distance.

$$D_H^2 = \int_{-\infty}^{\infty} (g(x) - f(x))^2 dx. \tag{6}$$

If f is a square integrable function (g certainly is, since g^2 is proportional to a Gaussian with variance $1/\sqrt{2}$), this integral is convergent. In such a case, we may expand f in terms of Hermite polynomials as follows:

$$f(x) = g(x) \sum_{n=0}^{\infty} \frac{b_n}{\sqrt{\kappa_n}} H_n(x), \tag{7}$$

where $b_n = \frac{1}{\sqrt{\kappa_n}} \int_{-\infty}^{\infty} f(x) H_n(x) g(x) dx$, and $\kappa_n = 2^{n-1} n! / \sqrt{\pi}$ is the normalization constant. This form of Hermite expansion is sometimes called the Gauss-Hermite series. Unlike the Gram-Charlier series, the polynomials used here are the Hermite polynomials (not Chebyshev-Hermite) and the Gaussian weight appears in both the decomposition and the reconstruction formulae. The Gauss-Hermite coefficients can also be considered as the expectation values,

$$b_n = \mathbf{E}\left[\frac{1}{\sqrt{\kappa_n}} H_n(X) g(X)\right] \approx \frac{1}{T\sqrt{\kappa_n}} \sum_{t=1}^{T} H_n(x_t) g(x_t), \tag{8}$$

and thus can be estimated from the samples x_t. In particular, one expects that these coefficients are robust against outliers, as large values of $|x_t|$ are attenuated by the tails of the Gaussian.

If we substitute the series representation (7) into the L^2 metric formula (6), and use the orthogonality conditions (4), we see that the Hall distance is

$$D_H^2 = (b_0 - \sqrt{\kappa_0})^2 + \sum_{n=1}^{\infty} (b_n)^2. \tag{9}$$

Again, the L^2 distance is expressed as the sum of squared Hermite coefficients, with a zeroth order correction because the origin is taken to be the standard normal. In general, we do not know a *priori* the first few terms of the sum as we did in the Gram-Charlier case, because the coefficients b_n are no longer directly linked to moments. However, this is only a minor computational disadvantage considering the benefit of the robustness gained by this formulation.

4 Proposed Algorithm of the Dependent Sources

4.1 Preprocessing

In order to apply the maximum NG method to dependent source separation, the research must restrict the separating matrix $\mathbf{W}$ which make the separated signals y_i have unit variance. A simple way to do this procedure is to apply first a spatial whitening filter to the mixtures $\mathbf{x}$, and then, to parameterize the new separation matrix as the one composed by unitnorm rows. We implement this spatial filter using Karhunen-Loeve transformation (KLT) [10] reaching to a new set of spatially uncorrelated data, $\mathbf{z} = \mathbf{V}\mathbf{\Lambda}^{-1/2}\mathbf{V}^T\mathbf{x}$, where $\mathbf{V}$ is a matrix of eigenvectors of the covariance matrix $\mathbf{R_{xx}} = \mathbf{E}[\mathbf{x}\mathbf{x}^T]$ and $\mathbf{\Lambda}$ is a diagonal matrix containing the eigenvalues of $\mathbf{R_{xx}}$ which are assumed to be non-zero.

Now, if we define $\mathbf{y} = \mathbf{U}\mathbf{z}$, the new separation matrix $\mathbf{U}$, must have the property of having unitnorm rows, which follows from the assumption of unit-variances of variables y_i ($\mathbf{R_{yy}} = \mathbf{E}[\mathbf{y}\mathbf{y}^T] = \mathbf{U}\mathbf{U}^T$). The "real" (original) separation matrix $\mathbf{W}$ can be calculated using $\mathbf{y} = \mathbf{U}\mathbf{z}$ and (10) as follows:

$$\mathbf{W} = \mathbf{U}\mathbf{V}\mathbf{\Lambda}^{-1/2}\mathbf{V}^T. \tag{10}$$

Note that source estimates may be permuted or sign changed versions of sources (scale ambiguity disappears since it is assumed that the sources have unit-variance).

4.2 The Main Algorithm

As mentioned in [7], in the dependent sources situations, we can not resort minimization the MI, but on the other hand we can maximization of NG to get the dependent sources.

So we view BSS algorithms as de-Gaussianization methods which based on other definitions of L^2 measurement, such as the Hall distances (6). For reasons stated above, we choose to use the Euclidean metric $L^2(1)$ to define a non-Gaussianity index. Note that each component x_i is a standardized random variable, $\mathbf{E}[\mathbf{x}(t)] = 0$, and $\mathbf{E}[\mathbf{x}(t)\mathbf{x}^T(t)] = \mathbf{I}$. A natural extension of the L^2 measurement is then given by the sum of $L^2(1)$ NG indices of x_i across all n dimensions,

$$D_H^2(\mathbf{x}) = \sum_{i=1}^{n} D_H^2(x_i), \tag{11}$$

where $D_H^2(x_i) = (b_0(x_i) - \sqrt{\kappa_0})^2 + \sum_{k=1}^{\infty} b_k^2(x_i)$. In particular, if we truncate the sum by taking only the 0-th order terms for each x_i, we can show

$$D_H^2(\mathbf{x}) \approx \sum_{i=1}^{n}(b_0(x_i) - \sqrt{\kappa_0})^2 \approx \frac{1}{\kappa_0}\sum_{i=1}^{n}(\mathbf{E}[g(x_i)] - \mathbf{E}[g(z)])^2. \tag{12}$$

Here, x_i is the standardized random variable with an unknown density f_k, z is a standard Gaussian random variable and g is the standard Gaussian PDF.

This truncated form of the multidimensional $L^2(1)$ distance is equivalent to an ICA contrast due to Hyvärinen, and the fixed-point iteration algorithm called FastICA was introduced in [2]. Then the main procedure of the basic form of the one unit FastICA algorithm can be concluded as follows:

step1. Choose an initial (e.g. random) weight vector $\mathbf{u}$.
step2. Let $\mathbf{u}^+ = \mathbf{E}\{\mathbf{z}g(\mathbf{u}^T\mathbf{z})\} - \mathbf{E}\{g'(\mathbf{u}^T\mathbf{z})\}\mathbf{u}$.
step3. Let $\mathbf{u} = \mathbf{u}^+/\|\mathbf{u}^+\|$.
step4. If not converged, go back to *step2*.

The one-unit algorithm estimates just one of the components. To estimate several components, we need to run the one-unit FastICA algorithm using several units (e.g. neurons) with weight vectors $\mathbf{u}_1; \cdots; \mathbf{u}_n$. To prevent different vectors from converging to the same maxima we must decorrelate the outputs $\mathbf{u}_1^T\mathbf{z}; \cdots; \mathbf{u}_n^T\mathbf{z}$ after every iteration. A simple way of achieving decorrelation is a deflation scheme based on a Gram-Schmidt-like decorrelation. This means that we estimate the components one by one. When we have estimated p components, or p vectors $\mathbf{u}_1; \cdots; \mathbf{u}_p$, we run the one-unit fixed point algorithm for $\mathbf{u}_{p+1}$, and after every iteration step subtract from $\mathbf{u}_{p+1}$ the "projections" $\mathbf{u}_{p+1}^T\mathbf{u}_j\mathbf{u}_j, j = 1, \cdots, p$ of the previously estimated p vectors, and then renormalize $\mathbf{u}_{p+1}$:

step1. Let $\mathbf{u}_{p+1} = \mathbf{u}_{p+1} - \sum_{j=1}^{p} \mathbf{u}_{p+1}^T\mathbf{u}_j\mathbf{u}_j$.
step2. Let $\mathbf{u}_{p+1} = \mathbf{u}_{p+1}/\sqrt{\mathbf{u}_{p+1}^T\mathbf{u}_{p+1}}$.

5 Simulation Results

In order to confirm the validity of the proposed Hall distance based BSS algorithm, simulations using Matlab were given below with four source signals which have different waveforms. The input signals were generated by mixing the four simulated sources with a 4×4 random mixing matrix in which the elements were distributed uniformly. The sources and mixtures are displayed in Figs. 1(a) and (b), respectively. The source signals correlation values are shown in Table 1.

Table 1. The Correlation Values Between Source Signals

	source 1	source 2	source 3	source 4
source 1	1	0.6027	0.3369	0.4113
source 2	0.6027	1	0.4375	0.4074
source 3	0.3369	0.4375	1	0.5376
source 4	0.4113	0.4074	0.5376	1

So the sources are not the i.i.d signals, the proposed NG measurement based BSS algorithm can separate the desired signals properly.

Next, for comparison we execute the mixed signals with different BSS algorithms: JADE Algorithm [11], SOBI algorithm [1],TDSEP algorithm [12] and AMUSE algorithm [1]. At the same convergent conditions, the proposed algorithm which we call it as NG-FastICA was compared along the criteria statistical whose performance was measured using a performance index called cross-talking error index E defined as [1]

$$E = \sum_{i=1}^{N}\Big(\sum_{j=1}^{N}\frac{|p_{ij}|}{max_k|p_{ik}|}-1\Big)+\sum_{j=1}^{N}\Big(\sum_{i=1}^{N}\frac{|p_{ij}|}{max_k|p_{kj}|}-1\Big),$$

where $\mathbf{P}(p_{ij})$ is the entries of $\mathbf{P} = \mathbf{WA}$ is the performance matrix. The separation results of the four different sources are shown in Table 2 for various BSS algorithms(averaged over 100 Monte Carlo simulation).

Table 2. The results of the separation are shown for various BSS algorithms

Algorithm	JADE	SOBI	TDSEP	AMUSE	NG-FastICA
E	0.4118	0.7844	0.4052	0.6685	0.3028

The waveforms of source signals, mixed signals and the separated signals are shown in Fig. 1(c)(the first 512 observations are given).

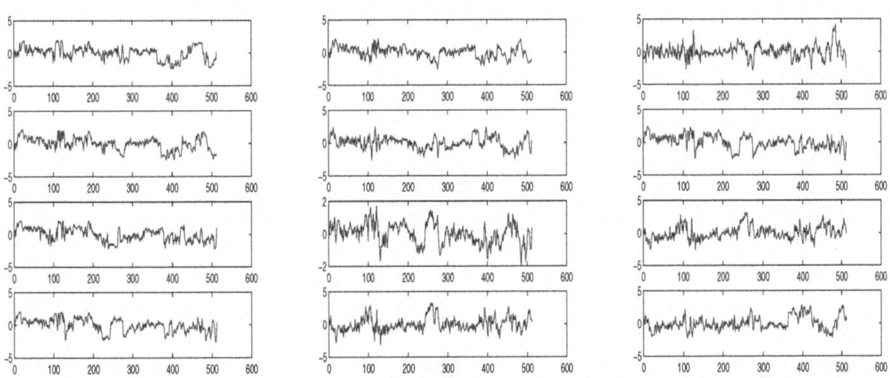

Fig. 1. The source signals, observed signals and experiment results showing the separation of correlated sources using the proposed NG-FastICA Algorithm

6 Conclusion

In this paper, we developed a novel Blind Source Separation (BSS) algorithms from linear mixtures of them, which enable to separate dependent source signals.

Most of the proposed algorithms for solving BSS problem rely on independence or at least uncorrelation assumption of source signals that is the independent component analysis algorithm. Here, we show that maximization of the nonGaussianity(NG) measure can separate the statistically dependent source signals and the novel NG measure is given by the Hall Euclidean distance. The proposed separation algorithm can result in the famous FastICA algorithm. Simulation results show that the proposed separation algorithm is able to separate the dependent signals and yield ideal performance.

Acknowledgment

This work is partially supported by National Natural Science Foundation of China(Grant No.60672049) and the Science Foundation of Henan University of Technology under Grant No.06XJC032.

References

1. Cichocki, A., Amari, S.: Adaptive Blind Siganl and Adaptive Blind Signal and Image Processing. John Wiley&Sons, New York (2002)
2. Hyvarinen, A., Karhunen, J., Oja, E.: Independent component analysis. John Wiley&Sons, New York (2001)
3. Hyvarinen, A.: Blind source separation by nonstationarity of variance: a cumulant-based approach. IEEE Trans. Neural Networks 12(6) (2001) 1471-1474.
4. Cardoso, J.F.: Multidimensional independent component analysis. In Proc. IEEE Int. Conf. on Acoustics, Speech and Signal Processing(ICASSP'98), Seattle, WA. (1998) 1941-1944.
5. Hyvarinen, A., Hoyer, P.O.: Emergence of phases and shift invariant features by decomposition of natural images into independent feature subspaces. Neural Computation 12(5) (2000) 1705-1720.
6. Zhang, K., Chan L.W.: An adaptive method for subband decomposition ICA. Neural Computation 18(1) (2006) 191-223
7. Caiafa,C.F., Proto, A.N.: Separation of statistically dependent sources using an L^2-distance non-Gaussianity measure. Signal Processing 86(11) 3404-3420
8. Yokoo,T., Knighty, B.W., Sirovich, L.: L^2 De-gaussianization and independent component analysis. In Proc. 4th Int. Sym. on ICA and BSS(ICA2003), Japan. (2003) 757-762.
9. Hall, P.: Polynomial Projection Pursuit. Annals of Statistics 17 (1989) 589-605.
10. Duda, R.O., Hart, P.E., Stork, D.G.: Pattern Classification. second ed., John Wiley&Sons, New York (2000).
11. Cardoso, J.F.: High-order contrasts for independent component analysis. Neural Computation 11(1)(1999) 157-192
12. Ziehe, A., Muller, K.R.: TDSEP-an efficient algorithm for blind separation using time structure. In Proc. ICANN'98 (1998) 675-680

HiBO: Mining Web's Favorites

Sofia Stamou, Lefteris Kozanidis, Paraskevi Tzekou, Nikos Zotos, and Dimitris Cristodoulakis

Computer Engineering and Informatics Department, Patras University, 26500 Patras, Greece
{stamou,kozanid,tzekou,zotosn,dxri}@ceid.upatras.gr

Abstract. HiBO is a bookmark management system that incorporates a number of Web mining techniques and offers new ways to search, browse, organize and share Web data. One of the most challenging features that HiBO incorporates is the automated hierarchical structuring of bookmarks that are shared across users. One way to go about organizing shared files is to use one of the existing collaborative filtering techniques, identify the common patterns in the user preferences and organize bookmarked files accordingly. However, collaborative filtering suffers from some intrinsic limitations, the most critical of which is the complexity of the collaborative filleting algorithms that inevitably leads to the latency in updating the user profiles. In this paper, we address the dynamic maintenance of personalized views to shared files from a bookmark management system perspective and we study ways of assisting Web users share their information space with the community. To evaluate the contribution of HiBO, we applied our Web mining techniques to manage a large pool of bookmarked pages that are shared across community members. Results demonstrate that HiBO has a significant potential in assisting users organize and manage their shared data across web-based social networks.

Keywords: Hierarchical Structures, Web Data Management, Bookmarks, System Architecture, Personalization.

1 Introduction

Millions of people today access the plentiful Web content to locate information that is of interest to them. However, as the Web grows larger there is an increasing need in helping users to keep track of the interesting Web pages that they have visited so that they can get back to them later. One way to address this need is by maintaining personalized local URL repositories, widely known as bookmarks [15]. Bookmarks, also called *favorites* in the Internet Explorer, enable users to store the location (address) of a Web page so that they can revisit it in the future without the need of remembering the page's exact address. People use bookmarks for various reasons [1]: some bookmark URLs for fast access, others bookmark URLs with long names that they find hard to remember, yet others bookmark their favorite Web pages in order to share them with a community of users with similar interests.

G. Dong et al. (Eds.): APWeb/WAIM 2007, LNCS 4505, pp. 845–856, 2007.

As the number of the pages that are available on the Web keeps growing, so does the number of the pages stored in personal Web repositories. Moreover, although users visit frequently their bookmarked URLs, they rarely delete them; which practically results into users keeping stale links in their personal Web repositories. As a consequence, people tend to maintain large, and possibly overwhelming, bookmark collections [16]. However, keeping a flat list of bookmark URLs is insufficient for tracking down previously visited pages, especially if we are dealing with a long list of favorites. As the size of the personal repositories increases, the need for organizing and managing bookmarks becomes prevalent. To assist users organize their bookmark URLs in a meaningful and useful manner, there exist quite a few bookmark management systems offering a variety of functionalities to their users. These functionalities enable users to store their bookmarks into folders and subfolders named for the sites they are found in or named for the information they contain, as well as to organize the folders in a tree-like structure. Moreover, commercial bookmark management tools, e.g. *BlinkPro* [2], *Bookmark Tracker* [3], *Check and Get* [4], *iKeepBookmarks* [5], provide users with a broad range of advanced features like detection of duplicate bookmarks and/or dead links, importing, exporting and synchronizing bookmarks across different Web browsers (Mozilla, Internet Explorer, Opera, Netscape), updating bookmarks and so forth.

In this paper, we present HiBO; an intelligent system that automatically organizes bookmarks into a hierarchical structure. HiBO is a powerful bookmark management system that exploits a multitude of Web mining techniques and offers a wide range of advanced services. Most importantly, HiBO is a non-commercial research project for managing the proliferating data in peoples' personal Web repositories without any user effort. The main difference between HiBO and the other available bookmark management systems (cf. [11], [14], [15]) is that HiBO uses a built-in subject hierarchy for automatically organizing bookmarks within both the users' local and shared Web repositories. The only input that our approach requires is a hierarchy of topics that one would like to use and a list of bookmark URLs that one would like to organize into these topics. Through the exploitation of the hierarchy, HiBO delivers personalized views to the shared files and eventually it assists Web users share their information space with the community.

The remainder of the paper is organized as follows: we begin our discussion with the description of HiBO's architecture. In Section 3, we give a detailed description of the functionalities and services that our bookmark management system offers. Experimental results are presented in Section 4. We finally review related work and conclude the paper in Section 6.

2 Overview of HiBO Architecture

HiBO evolved in the framework of a large research project that aimed at the automatic construction of Web directories through the use of subject hierarchies. The subject hierarchy that HiBO uses contains a total of 475 topics organized into 14 top level topics, borrowed from the top categories of the Open Directory Project (ODP) [6]. At a high level, the way in which HiBO organizes bookmarks proceeds as follows: firstly HiBO downloads all the Web pages that have been bookmarked by a user and process them one by one in order to identify the important terms inside every page. Important terms of a page are linked together formulating a lexical chain. Then, our system uses the subject hierarchy and the lexical chains to compute a suitable topic to assign to every page. Finally, HiBO sorts the Web pages organized into topics in terms of their relevance to the underlying topics. More specifically, given a URL (bookmark) HiBO performs a sequence of tasks as follows: (i) download the URL and parse the HTML page, (ii) segment the textual content of the page into shingles and extract the page's thematic words using the lexical chaining technique [8], (iii) map thematic words to the hierarchy's concepts and traverse the hierarchy's matching nodes upwards until reaching to one or more topic nodes, (iv) compute a relevance score of the page to each of the matching topics, (v) index the URL in the topic of the greatest relevance score. Figure 1 illustrates HiBO's architecture.

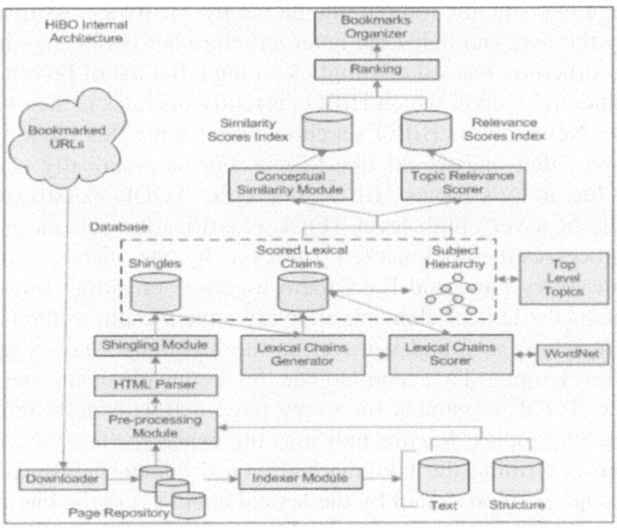

Fig. 1. Overview of HiBO architecture and functionality

In particular, after downloading and segmenting a Web page into shingles, HiBO generates a lexical chain for the page as follows: it selects a set of candidate terms from the page and for each candidate term it finds an appropriate chain relying on the type of links that are used in WordNet [7] for connecting the candidate term to the other terms that are already stored in existing lexical chains. If this is found, HiBO inserts the term in the chain and updates the latter accordingly. Lexical chains are then scored in terms of their elements' depth and similarity in WordNet, and their elements are mapped to the hierarchy's nodes. For each of the hierarchy's matching nodes, HiBO follows their hypernymy links until reaching a top level topic in which to categorize the Web page. Finally, HiBO sorts the Web pages categorized in each topic in terms of both the pages' conceptual similarity to one another and their relevance to the underlying topic. In estimating the pages' conceptual similarity, HiBO compares the elements in a page's lexical chain to the elements in the lexical chains of the other pages in the same topic, based on the assumption that the more elements the chains of two pages have in common, the more correlated the pages are to each other. On the other hand, in computing the pages' relevance to the hierarchy's topics, HiBO relies on the pages' lexical chains scores and the fraction of the chains' elements that match a given topic in the hierarchy. Based on this general and open architecture, HiBO explores a variety of Web mining techniques and provides users with a number of advanced functionalities that are presented below.

3 HiBO Functionalities

Organizing Bookmarks: Besides the conventional way to organize bookmarks into a hierarchy of user-defined folders and subfolders, HiBO also incorporates a built-in subject hierarchy and a classification module, which automatically assigns every bookmarked page to a suitable topic in the hierarchy. HiBO's classification module is set into forth by the user and helps the latter structure her bookmarks in a meaningful yet manageable structure, instead of simply keeping a flat list of favorite URLs.

The subject hierarchy upon which HiBO currently operates is the one introduced in the work of [19]. Nevertheless HiBO's architecture is quite flexible to incorporate any hierarchy of topics that one would like to use. For automatically classifying bookmarks into the hierarchy's topics HiBO adopts the TODE classification technique, reported in [20]. At a very high level TODE classification scheme proceeds as follows: First, it processes the bookmarked pages one by one, identifies the most important terms inside every page and links them together, creating "lexical chains" [8]. Thereafter, it maps the lexical elements in every page's chain to the hierarchy's concepts and if a matching is found it traverses the hierarchy's nodes upwards until it reaches a top level topic. To accommodate for chain elements matching multiple hierarchy topics, TODE computes for every page a Relatedness Score (*RScore*) to each of the matching topics. *RScore* indicates the expressiveness of each of the hierarchy's topics in describing the bookmarked pages' contents. Formally, the relatedness score of a page p_i (represented by the lexical chain C_i) to the hierarchy's topic T_k is determined by the fraction of words in the page's chain that are descendants (i.e. specializations) of T_k. Formally, the *RScore* of a page to each of the hierarchy's matching topics is given by:

$$\mathtt{RScore}_K(\mathtt{pi}) = \frac{|\text{thematic words in } p_i \text{ matching } K|}{|\text{thematic words in } p_i|} . \tag{1}$$

In the end, HiBO employs the topical category for which a bookmark has the highest relatedness score of all its *RScores* to describe that page's thematic content. By enabling bookmarks' automatic organization into a built-in hierarchical navigable structure, HiBO assists the user, who may be overwhelmed by the amount of her favorite pages organize and manage them instantly. Hierarchically organized bookmarks are stored locally on the user's site for future reference.

Moreover, HiBO supports personalized bookmarks' organization by enabling the user define the set of topics in which bookmarks would be organized. These topics can be either a subset of the hierarchy's topics or any other topic that the user decides. In case the user edits a new topic category in HiBO, she also needs to indicate a topic in HiBO's built-in hierarchy with which the newly inserted topic correlates. Through the HiBO interface, the user can view the topics available in HiBO as well as the number of bookmarks in each topic. The user can navigate through the hierarchical tree to locate bookmarks related to specific topics.

In the case of shared bookmarks across a user community, HiBO supports personalized bookmark management by providing different views across users or user groups. Personalized views, allow the user decide on the classification scheme in which her shared bookmarks will be displayed. For instance, a user might choose to view the bookmarks she shares with a Web community organized in her self-selected categories or alternatively organized in the system's built-in subject hierarchy. Optionally, a user might decide to view her shared bookmarks organized in the categories defined by another member of the community, who she trusts. To enable personalized views on shared bookmarks, HiBO's classification module re-assigns user favorites to the categories preferred by the user (self, community or system defined) following the categorization process described above. Additionally, HiBO enables bookmark organization by their file types.

Searching Bookmarks: HiBO incorporates a powerful search mechanism that allows users to explore bookmark collections. The queries that HiBO supports are of the following types: topic-specific search, site/domain search, temporal search and keyword search. Similarly to querying a search engine for finding information on the Web, querying HiBO for locating information within one's Web favorites enables users to issue queries and retrieve bookmark URLs that are relevant to the respective queries. Upon keyword-based search, the user submits a natural language query and the system's search mechanism looks for bookmarked pages that contain any of the user-typed keywords, simply by employing traditional IR string-matching techniques. Additionally, HiBO incorporates a query refinement module introduced in the work of [12] and provides information seekers with alternative query formulations. Alternative query wordings are determined based on the semantic similarity that they exhibit to the user selected keywords in WordNet hierarchy. Refined queries are visualized in a graphical representation, as illustrated in Figure 2 and allow the user pick any of the system suggested terms either for reformulating a query that returns few or no relevant pages, or for crystallizing an under-specified information need.

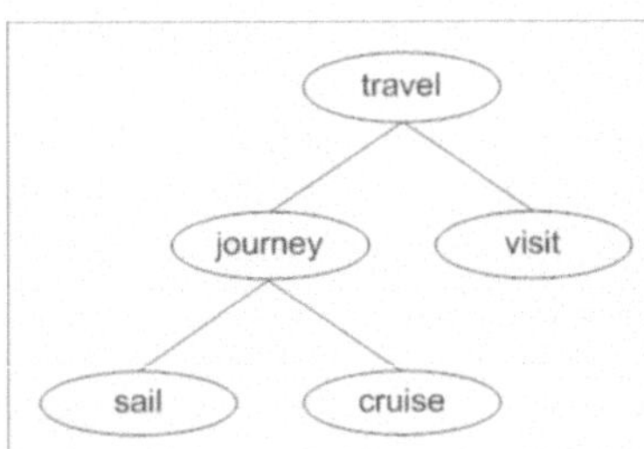

Fig. 2. A refined query graph example

Moreover, HiBO supports topic-specific searches by allowing users select the topical category (e.g. folder) out of which they wish to retrieve search results. Topic-specific searches greatly resemble the process of querying particular categories in

Web Directories, in the sense that the user firstly selects among the topics offered in the HiBO hierarchy the one that is of interest to her and thereafter she issues and executes the query against the index of the selected topic. Search results can be ranked according to the query-bookmark similarity values combined with any of the measures described in the following paragraph. If the user selects multiple ranking measures, then results are ranked by the product of their values. Conversely, if the user does not pick a particular ranking measure, results are ranked by the semantic similarity between the query keywords (either organic, i.e. user typed, or refined, i.e., system suggested) to the terms appearing in the bookmark pages that match the respective query.

Ranking Bookmarks: HiBO provides several options for sorting the bookmarks listed in each of the hierarchy's topics as well as for sorting bookmarks that are retrieved in response to a user query. For ranking bookmark URLs that are retrieved in response to some query *q*, HiBO relies on the semantic similarity between the query itself and the bookmark pages that contain any of the query terms. To measure the semantic similarity between the terms in a query and the terms in the pages that match the given query, we use the similarity measure presented in [18], which is established on the hypothesis that the more information two concepts share in common, the more similar they are. The information shared by two concepts is indicated by the information content of their most specific common subsumer. Formally the semantic similarity between words, w_1 and w_2, linked in WordNet via a relation r is given by:

$$sim_r(w_1, w_2) = -\log P\left(mscs(w_1, w_2)\right). \tag{2}$$

The measure of the most specific common subsumer (*mscs*) depends on: (i) the length of the shortest path from the root to the most specific common subsumer of w_1 and w_2 and (ii) the density of concepts on this path. Based on the semantic similarity values between the query terms and the terms in a page, we compute the average Query-Page similarity (QPsim) as:

$$sQPsim\left(q(t), P(t)\right) = \frac{\sum_{p=1}^{P(t)} sim\left(q(t), P(t)\right)}{|P(t)|}. \tag{3}$$

where q (t) denotes the terms in a query and P (t) denotes the terms in P that have some degree of similarity to the query terms. The greater the similarity value between the terms in a bookmark page and the terms in a query, the higher the ranking that the page will be given for that query.

On the other side of the spectrum, for ordering bookmarks in the hierarchy's topics, the default ranking that HiBO uses is the DirectoryRank (DR) metric [13], which determines the bookmarks' importance to particular topics as a combination of two factors: the bookmarks' relevance to their assigned topics and the semantic correlation that the bookmarks in the same topic exhibit to each other. In the DR scheme, a page's importance with respect to some topic is perceived as the amount of information that the page communicates about the topic. More precisely, to compute DR with respect to some topic *T*, we first compute the degree of the pages' relatedness to topic

T. Formally, the relatedness score of a page *p* (represented by a set of thematic terms[1]) to a hierarchy's topic *T* is defined as the fraction of the page's thematic words that are specializations of the concept describing *T* in the HiBO hierarchy, as given by Equation (1). The semantic correlation between pages p_1 and p_2 is determined by the degree of overlap between their thematic words, i.e. the common thematic words in p_1 and *p2* as given by:

$$\text{Sim}(p_1, p_2) = \frac{2 \bullet |\text{common words}|}{|\text{words in } p_1| + |\text{words in } p_2|} . \tag{4}$$

DR defines the importance of a page in a topic to be the sum of its topic relatedness score and its overall correlation to the fraction of pages with which it correlates in the given topic. Formally, consider that page p_i is indexed in topic T_k with some $RScore_k(i)$ and let $p_1, p_2, \ldots, p_n$ be pages in T_k with which p_i semantically correlates with scores of Sim (p_1, p_i), Sim (p_2, p_i), …, Sim (p_n, p_i), respectively. Then the DR of p_i is given by:

$$DR_{T_k}(p_i) = \text{RScore}_k(i) + \frac{\left[\text{Sim}(p_1, p_i) + \text{Sim}(p_2, p_i) + \ldots + \text{Sim}(p_n, p_i)\right]}{n} . \tag{5}$$

where *n* corresponds to the total number of pages in topic T_k with which p_i semantically correlates. Moreover, HiBO offers personalized bookmark sorting options such as the ordering of pages by their bookmark date or by their last update, as well as the ordering of bookmarks in terms of their popularity, where popularity is determined by the frequency with which a user or group of users sharing files, (re)visit bookmarks.

Sharing Bookmarks: Besides offering bookmark management services to individuals; HiBO constitutes a social bookmark network, as it allows community members share their Web favorites. In this perspective, HiBO operates as a bookmark recommendation system since it not only gathers and distributes individually collected URLs but it also organizes and processes them in a multi-faceted way. In particular, HiBO despite offering personalized views to shared bookmarks (cf. Organizing Bookmarks paragraph) it enables users annotate their preferred Web data, share their annotations with other members of the network and comment on others' annotations. To assist Web users exploit the knowledge accumulated in the bookmarks of others, HiBO goes beyond traditional collaborative filtering techniques and applies a multitude of Web mining techniques that exploit the hierarchical structure of the shared bookmarks. Such Web mining techniques range from the automatic classification of bookmark pages into a shared topical hierarchy, to the structuring of shared files according to their links and content similarity. Shared bookmarks' dynamic categorization is achieved through the utilization of the TODE categorization scheme, whereas bookmarks' structuring is supported by the different ranking algorithms that HiBO incorporates. Additionally, HiBO provides recommendation services to its users as it examines common patterns in the bookmarks of different community members and suggests interesting sites to users who might not have realized that they share common interests with others. HiBO communicates its recommendations in the form of

[1] The thematic terms in a page *p* are the lexical elements that formulate the lexical chain of *p*.

highlighted URLs that are associated to one's favorites, which are either stored in the system's hierarchy or retrieved in response to some query.

Keeping Bookmarks Fresh: Based on the observation that users rarely refresh their personal Web repositories, we equipped HiBO with a powerful update mechanism, which aims at maintaining the bookmarks index fresh. By fresh we mean that the index does not contain obsolete links among one's bookmarks, as well as that it reflects the current content of bookmarked pages. The update mechanism that HiBO uses performs a dual task: on the one hand it records the users' clickthrough data on their bookmarks and on the other it submits periodic requests to a built-in crawler for re-downloading the content of the bookmarked URLs. In case the system identifies bookmarks that have not been accessed for a long time, it posts a request to the user asking if she still wants to keep those bookmarks in her collection and/or if she still wants to share those bookmarks with other community members. Upon the user's negative answer, the system deletes those rarely visited URLs from the bookmark index and updates the latter accordingly, i.e. it re-orders pages etc. Similarly, if the system detects invalid, broken or obsolete URLs within a user's personal repository, it issues a notification to the user, who decides what to do with those links (either delete them, expunge them from her shared files, or keep them anyway). Furthermore, if the system detects a significant change in the current content of pages that had been bookmarked by a user some time ago, it issues an alert to the latter that her bookmarked URLs do not reflect the current content of their respective pages. It is then up to the user to decide whether she wants to keep the old or the new content of a bookmarked page. For content change detection, HiBO relies on the semantic similarity module discussed above, and uses a number of heuristics for deciding whether a page has significantly changed and therefore the user needs to be notified. HiBO's update mechanism although operates on a single user's site, it indirectly impacts the rest community members in the sense that upon changes in one's personal Web repository, these will be reflected on her shared files. Note that the update mechanism that HiBO embodies is optional to the user who might decide not to activate it and therefore not to be disturbed by the issued update alerts and notifications.

4 Experimental Setup

To evaluate HiBO's effectiveness in managing and organizing Web favorites, we launched a fully functional version of our bookmark management system and we contacted 25 postgraduate students from our school asking them to donate their bookmarks. Donating bookmarks pre-requisites that users register to the system by providing a valid e-mail address and they receive a personal code, which is used in all their transactions with the system. Upon code's receipt users obtain full rights on their personal bookmarks and they can also indicate the HiBO community with which they wish to share their preferred URLs. In the experiments reported here, all our 25 users formulated a single Web community sharing bookmarks. When users donate bookmarks, we use their agents to determine which browser and platform they are using in order to parse the files accordingly. We also use an SQL database server at the back-end of the system, where we store all the information handled by HiBO, i.e. users and user groups, URLs, bookmarks' structure at the user site, the subject hierarchy, time stamps, clickthrough data, queries, etc. In our experiments, we used a total set of 3,299 bookmarks donated by our subjects and we evaluated HiBO's performance in automatically categorizing bookmarks in the system's hierarchy, by comparing its classification accuracy to the accuracy of a Bayesian classifier and a Support Vector Machine (SVM) classifier. We also investigated the effectives of HiBO's ranking mechanisms in offering personalized rankings. Table 1 summarizes some statistics on our experimental dataset.

Table 1. Statistics on the experimental dataset

# of bookmark URLs	3,299
# of users	25
# of topics considered	86
# of queries	48
Avg. # of bookmarks per user	131.96
Avg. # of shared bookmarks per user	58
Avg. # of topics per user	21
Avg. # of shared topics	9.4
Avg. # of queries per user	7.5
Avg. # of visited pages per query	5.8
Avg. # of useful pages per query	3.5
Avg. # of terms per refined query	3.8

To evaluate HiBO's efficiency in categorizing bookmarks to the hierarchy's topics, we picked a random set of 1,350 pages from our experimental data that span 18 topics in the Open Directory that are also among our hierarchy's topics and we applied our categorization scheme. Obtained results were compared to the results delivered by both the SVM and the Bayesian classifier that we trained with the 90% of the same dataset. Classification results are reported in Table 2, where we can see clearly that HiBO's classifier significantly outperforms both Bayesian and SVM classification with a notable performance; reaching to a 90.70% overall classification accuracy.

In Table 3, we illustrate the different ranking measures of HiBO, using the results of both browsing and searching for spam. For comparison, we also present the pages that Google considers "important" to the query spam. Although, Google uses a number of non-disclosed factors for computing the importance of a page, with PageRank [17] being at the core, we assume that a combination of content and link analysis is employed. Obtained results demonstrate the differences between the two HiBO rankings examined. In particular, the rankings delivered by DR sort bookmark pages in terms of their content importance to the underlying topic, i.e. Spam. As we can see from the reported data, our DR ranking values highly pages of practical interest compared to the pages retrieved from Google, which are general sites that mainly provide definitions of spam. On the other hand, the similarity ranking orders the bookmarked pages that are retrieved in response to the query spam in terms of their content semantic closeness to the semantics of the query. As such the results retrieved by HiBO contain pages that even if they are not categorized in the topic Spam, their contents exhibit substantial semantic similarity to the issued query. Recall that our experiments were conducted towards a set of bookmarks that are shared across our subjects and as such reported results are influenced by our users' interests. This is exemplified by the appearance of Spam Filter for Outlook, Block Referrer Spam and Spam Fixer in the top ten results of DR and Similarity rankings respectively; sites that are naturally favored by computer science students as they contain information that is of practical use to them.

Table 2. Average classification accuracy between HiBO and Bayesian classifiers

Topics	HiBO classifier	Bayesian classifier	SVM classifier
Dance	97.05%	69.46%	71.58%
Music	94.37%	74.38%	78.49%
Artists	86.45%	83.59%	82.64%
Photography	81.68%	55.28%	69.03%
Architecture	79.77%	69.89%	72.11%
Art History	93.33%	78.47%	68.58%
Comics	95.45%	29.46%	45.24%
Costumes	89.06%	72.43%	69.77%
Design	90.79%	69.29%	55.08%
Literature	89.70%	59.26%	49.91%
Movies	94.59%	71.04%	68.97%
Performing Arts	87.34%	68.08%	65.06%
Collecting	92.87%	67.17%	53.88%
Writing	91.84%	69.56%	60.42%
Graphics	92.68%	79.80%	71.53%
Drawing	91.34%	59.55%	58.16%
Plastic Arts	90.86%	64.36%	62.07%
Mythology	93.58%	68.22%	64.93%
	90.70%	**67.18%**	**64.85%**

Table 3. Ordering bookmarks for `spam`

HiBO DR	HiBO Similarity	Google
Block Referrer Spam	Witchvox Article – That Pesky and Obnoxious Spam	www.spam.com
Referrer Log Spamming	Outlook Express Tutorial: Filter- how to stop spam	Fight Spam on the Internet
Spam Assassin	Message Cleaner – Stop viruses and spam emails now	Spam-Wikipedia
Stop Spam with Sneakmail 2.0	The Spammeister guide to spam	E-mail Spam-Wikipedia
Anti-Spam	Spamhuntress – Spam Cleaning for Blogs	FTC-Spam-Home Page
A Plan for Spam	Discuss Sam Forums-Learn how to eliminate and prevent spam	Coalition Against Unsolicited Commercial Email
Death to Spam	SpamFixer	SpamAssassin
Spam Filter for Outlook	Spam Email Discussion List	Spam Cop
The Spam Weblog	Emailabuse.org	What is Spam- Webopedia
Damn Spam	Spamcop.net	Spam Laws

5 Related Work

Bookmarks are essentially pointers to URLs that one would like to store in a personal Web repository for future reference and/or fast access. Today there exist many commercial bookmark management tools[2], providing users with a variety of functionalities in an attempt to assist them organize the list of their Web favorites [2] [3] [4] [5]. With the recent advent of social bookmarking, bookmarks[3] "*have become a means for users sharing similar interests to locate new websites that they might not have otherwise heard of; or to store their bookmarks in such a way that they are not tied to one specific computer*". In this light, there currently exist several Web sites that collect, share and process bookmarks. These include *Simpy*, *Furl*, *Del.icio.us*, *Spurl*, *Backflip*, *CiteULike* and *Connotea* and are reviewed by Hammond et al. [9]. Such social networks of bookmarks are being perceived as recommendation systems in the sense that they process shared files and, based on a combinational analysis of the files themselves and their contributors in the network, they suggest to other network members interesting sites submitted by a different community member. From a research point of view, there have been several studies on how shared bookmarks can be efficiently organized to serve communities. The work of [21] falls in this area and introduces *GiveALink*, an application that explores semantic similarity as a means to process collected data and determine similarity relations among all its users. Likewise, [10] suggest a novel distributed collaborative bookmark system that they call *CoWing* and which aims at helping people organize their shared bookmark files. To that end, the authors introduce the utilization of a bookmark agent, which *learns* the user strategy in classifying bookmarks and based on that knowledge it fetches new bookmarks that match the local user information need. In light of the above, we perceive our work on HiBO to be complementary to existing approaches. However, one aspect that differentiates our system from available bookmark management systems in that HiBO provides a built-in subject hierarchy that enables the automatic classification of bookmark URLs on the side of either an individual user or group of users. Through the subject hierarchy, HiBO ensures the dynamic maintenance of personalized views to shared files and as such it assists Web users share their information space with the community.

6 Concluding Remarks

In this paper we presented HiBO, a bookmark management system that automatically manages orders, retrieves and mines the data that is either stored in Web users' personal Web repositories or shared across community members. An obvious advantage of our system when compared to existing bookmark management tools is that HiBO uses a built-in subject hierarchy for dynamically grouping bookmarks thematically without any user effort. Another advantage of HiBO is the ordering of bookmarks into the hierarchy's topics in terms of their content importance to the underlying topics. Currently, we are working on privacy issues so as to motivate Web users donate their Web favorites to HiBO and therefore launch a powerful bookmark mining system to the community.

---

[2] For a complete list of available bookmark management systems we refer the reader to http://dmoz.org/Computers/Internet/On_the_Web/Web_Applications/Bookmark_Managers/

[3] http://en.wikipedia.org/wiki/Bookmark_%28computers%29

References

1. Abrams, D., Baecker, R. and Chignell, M. Information Archiving with Bookmarks: Personal Web Space Construction and Organization. In Proceedings of the Human Computer Interaction Conference, 1998, pp. 41-48.
2. BlinkPro: Powerful Bookmark Manager http://www.bookmarksplus.com/
3. Bookmark Tracker http://www.bookmarktracker.com/
4. Check and Get http://activeurls.com/en/
5. iKeepBookmarks http://www.ikeepbookmarks.com/
6. Open Directory Project: http://dmoz.org
7. WordNet 2.0: http://www.cogsci.princet on.edu/~wn/.
8. Barzilay, R. and Elhadad, M. Lexical chains for text summarization. In Advances in Automatic Text Summarization. MIT Press, 1999.
9. Hammond, T., Hannay, T., Lund, B. and Scott, J. Social Bookmarking Tools (I): A General Review. D-Lib Magazine, 11(4): doi:10.1045/april2005—hammond, 2005.
10. Kanawati, R., Malek, M., Klusch, M. and Zambonelli F. CoWing: A Collaborative Bookmark Management. In Lecture Notes in Computer Science, ISSN 0302-9743, 2001.
11. Karousos, N., Panaretou, I., Pandis, I. and Tzagarakis, M. Babylon Bookmarks: A Taxonomic Approach to the Management of WWW Bookmarks. In Proceedings of the Metainformatics Symposium 2002, 42-48.
12. Kozanidis, L., Tzekou, P., Zotos, N., Stamou, S., and Christodoulakis, D. Ontology-Based Adaptive Query Refinement. To appear in Proceedings of the 3rd International Conference on Web Information Systems and Technologies, 2007.
13. Krikos, V., Stamou, S., Ntoulas, A., Kokosis, P. and Christodoulakis, D. DirectoryRank: Ordering Pages in Web Directories. In Proceedings of the 7th ACM International Workshop on Web Information and Data Management (WIDM), Bremen, Germany, 2005.
14. Li, W.S., Vu, Q., Chang, E., Agrawal, D., Hirata, K., Mukherjea, S., Wu, Y.L., Bufi, C., Chang, C.K., Hara, Y., Ito, R., Kimura, Y., Shimazu, K. and Saito, Y. PowerBookmarks: A System for Personalizable Web Information Organization, Sharing and Management. In Proceedings of the ACM SIGMOD Conference, 1999, pp. 565-567.
15. Maarek, Y., and Shaul, I. Automatically Organizing Bookmarks per Contents. In Proceedings of the 5th Intl. World Wide Web Conference, 1996.
16. McKenzie, B. and Cockburn, A. An Empirical Analysis of Web Page Revisitation. In Proceedings of the 34th Hawaii Intl. Conference on System Sciences, 2001.
17. Page, L., Brin, S., Motwani, R. and Winograd, T. The PageRank citation ranking: Bringing order to the web. Available at: http://dbpubs.stanford.edu:8090/pub/1999-66, 1998.
18. Resnik, Ph. Using Information Content to Evaluate Semantic Similarity in a Taxonomy. In Proceedings of the 14th Intl. Joint Conference on Artificial Intelligence, 2005, pp. 448-453.
19. Stamou, S. and Christodoulakis, D. Integrating Domain Knowledge into a Generic Ontology. In Proceedings of the 2nd Meaning Workshop. Italy, 2005.
20. Stamou, S., Ntoulas, A., Krikos, V., Kokosis, P., and Christodoulakis, D. Classifying Web Data in Directory Structures. In Proceedings of the 8th Asia-Pacific Web Conference (APWeb), Harbin, China, 2006, pp. 238-249.
21. Stoilova, L., Holloway, T., Markines, B., Maguitman, A. and Mencezer, F. GiveALink: Mining a Semantic Network of Bookmarks for Web Search and Recommendation. In Proceedings of the LinkKDD Conference, Chicago, IL, USA, 2005.

Frequent Variable Sets Based Clustering for Artificial Neural Networks Particle Classification

Xin Jin and Rongfang Bie*

College of Information Science and Technology,
Beijing Normal University, Beijing 100875, P.R. China
xinjin4@yahoo.com, rfbie@bnu.edu.cn

Abstract. Particle classification is one of the major analyses in high-energy particle physics experiments. We design a classification framework combining classification and clustering for particle physics experiments data. The system involves classification by a set of Artificial Neural Networks (ANN); each using distinct subsets of samples selected from the general set. We use frequent variable sets based clustering for partitioning the train samples into several natural subsets, then standard back-propagation ANNs are trained on them. The final decision for each test case is a two-step process. First, the nearest cluster is found for the case, and then the decision is based on the ANN classifier trained on the specific cluster. Comparisons with other classification and clustering methods show that our method is promising.

1 Introduction

Classification (i.e. supervised learning) is a fundamental task in data mining. A classifier, built from the labeled train samples described by a set of features/attributes, is a function that chooses a class label (from a group of predefined labels) for test samples. Major classification algorithms include Artificial Neural Network (ANN) [2, 3, 11], Nearest Neighbor [17, 13], Naïve Bayes [1, 20], etc. Clustering (i.e. unsupervised learning) is another fundamental task in data mining [18]. Cluster analysis partition unlabeled samples into a number of groups using a measure of distance, so that the samples in one group are similar while samples belonging to different groups are not similar [15, 16, 19]. Many clustering algorithms have been proposed, among which *k*-means is one of the most popular [27].

Particle classification is an important analysis in particle physics experiments. Traditional method separates distinct particle events by application of a series of cuts, which act on projections of high-dimensional event parameter space onto orthogonal axes [11]. This procedure often fails to yield the optimum separation of distinct event classes. In this paper, we investigate the use of data mining technology for particle classification. We describe a clustering method FVC especially designed for particle analysis, and then present a classification framework combining ANNs and FVC to improve the high-energy particle classification performance.

* Corresponding author.

G. Dong et al. (Eds.): APWeb/WAIM 2007, LNCS 4505, pp. 857–867, 2007.

The remainder of this paper is organized as follows: We first describe an ANN classifier in Section 2. Section 3 describes the clustering method FVC. Section 4 describes the classification system combining ANNs and clustering. Section 5 describes methods for comparison. Section 6 presents the dataset, four evaluation measures and the experiment results. Conclusions are presented in Section 7.

2 Artificial Neural Networks

Artificial Neural Networks (ANN) is a network of perceptrons, which computes an output from multiple inputs by forming a linear combination according to its input weights and then putting the output through some activation function [4, 5]. Among many proposed ANN models, MLP, the multilayer feedforward network with a back-propagation-learning mechanism, is the most widely used [6]. MLP consists of an input layer of source nodes, one or more hidden layers of computation nodes, and an output layer of nodes. Data propagates through the network layer-by-layer. Fig. 1 shows the data flow of a MLP with two hidden layer.

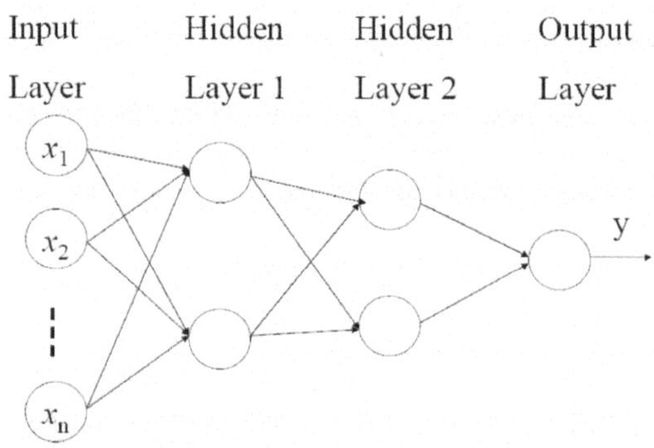

Fig. 1. Data-flow graph of a two hidden layer MLP

Define X as a vector of inputs and Y as a vector of outputs. Y, which may also be a 1-dimension vector, is typically obtained by:

$$Y = W_2 f_a(W_1 \cdot X + B_1) + B_2 \quad (1)$$

W_1 denotes the weight vector of the first layer and B_l the bias vector of the input layer. W_2 and B_2 are for the output layer. f_a denotes the activation function.

The classification problem of the MLP can be defined as follows: Given a training set of features-class/input-output pairs (x_i, c_i), MLP learns a model, the classifier, for the dependency between them by adapting the weights and biases to their optimal values for the given training set. Squared reconstruction error is commonly used as the criterion to be optimized.

MLP consists of iteration of two steps: (1) Forward - the predicted class corresponding to the given input are evaluated. (2) Backward - partial derivatives of the cost function with respect to the different parameters are propagated back through the network. The process stops when the weights and biases converge.

3 Frequent Variable Sets Based Clustering

In this section we describe a partitional clustering method Frequent Variable Sets based Clustering (FVC) to deal with the special characteristics of the high-energy particle data. It's based on frequent itemset mining and is based on the work of Fung B. et al [15], who developed a hierarchical document clustering using frequent itemsets. Frequent itemsets is a basis concept in association rule mining [8, 14]. Many different algorithms have been developed for that task, including the well known Apriori [10] and FP-growth [9].

Frequent item-based high-energy particle clustering is based on partitioning the particles according to their variables detected. Since we are dealing with particles other than transactions, we will use the notion of *variable* sets instead of *item* sets. A variable is any attribute describing a particle within physics experiments (high-energy particles collision, for example), and a particle can have some variables detected and others undetected due to inevitable changes in experimental environments or other reasons. Therefore, even the same kind of particles may have different set of detecting variables. Thus we assume that if we can cluster particles into different groups where each group has its own specific experimental environment (please note that the particles within each group will be in different classes because the group forming process is not based on the classes of the particles, particles of different classes may be under the same experimental situation thus have the same set of detected and undetected variables), then the classification model built from the particles in the same group will be a better distinguisher than the model built from the whole set of particles. Traditional clustering method, *k*-means for example, just group distance similar points thus is not suitable to find variable-oriented groups.

Instead of clustering the original high-dimensional space (for the data we used in this paper, the original space is 78-dimensional), FVC considers only the low-dimensional frequent variable sets as cluster candidates. We can say a frequent variable set (or variableset) is actually not a cluster (candidate) but only the description of a cluster (candidate), or the representational centroid of the cluster. The corresponding cluster itself consists of the set of particles containing all variables of the frequent variable set.

3.1 Binarizing

Original particle data have numeric attributes/variables; in order to find frequent variable sets we need to first convert them to binary attributes (1 for detected variable and 0 for undetected variable). For a particle, if one variable has a value other than 0, then we believe that the variable is detected for the particle (or we can say that it occurs in the particle) and the value will be converted to 1. If a variable has value 0 for the particle, then we believe that it is undetected and is set to 0 (unchanged). Some variables are very peculiar in that for one such kind of variables the value of it is very close to 0 (0.0001 for example) for one particle and 0 for another particle, then for the latter particle is hard to know whether the 0-value means detected or not. We simply assume that the variable is also detected for the particle and set the variable's value to 1. Table 1 shows an example data compose of four particles with five attributes/variables. Table 2 shows the converted data and its transaction representation.

Table 1. Original data. V1,..., V5 are five variables, P1,..., P4 are four particles.

ID	V1	V2	V3	V4	V5
P1	1.3546	0	2.5553	0.0001	0
P2	1.7865	2.3322	0	0	0
P3	0	0	0.0001	2.5343	2.3444
P4	0	0	0	4.7865	2.2211

Table 2. Binarized data and its transanction representation

ID	V1	V2	V3	V4	V5
P1	1	0	1	1	0
P2	1	1	1	1	0
P3	0	0	1	1	1
P4	0	0	1	1	1

ID	
P1	V1, V3, V4
P2	V1,V2,V3, V4
P3	V3, V4, V5
P4	V3, V4, V5

3.2 Representational Frequent Variableset

Assuming some variables occur in some particles, others occur in all particles. Let $P = \{P_1,\dots, P_n\}$ be a set of particles and $A = \{V1, V2,\dots\}$ be all the variables occurring in the particles of P. Each particle P_i can be represented by the set of variables occurring in it. For any set of variables (or call it variableset) S, define $C(S)$ as the set of particles containing all variables in S. For one particle, if just only a subset of S occurs in it, then the particle will not be in $C(S)$.

Define F_i as a representational frequent variableset, which is such kind of variableset that all variables in it appear together in more than a minimum and less than a maximum fraction of the whole particle set P. A minimum support (minsupp, in a percentage of all particles) and a maximum support (maxsupp, in a percentage of all particles) can be specified for this purpose. Define $F=\{F_1,\dots,F_m\}$ to be the set of all representational frequent variablesets in P with respect to minsupp and maxsupp, the variables in each F_i in at least minsupp and at most maxsupp percentage of the $|P|$ particles:

$$F = \{F_j \subseteq A \mid (\text{maxsupp}\times|P|) \geq |C(F_j)| \geq (\text{minsupp}\times|P|)\} \quad (2)$$

where $|P|$ is the number of particles.

A representational *frequent variable* is a variable that belongs to representational frequent variableset. A representational *frequent* k-variable*set* is a representational frequent variableset containing k variables.

The definition of representational frequent is different to the traditional definition of frequent in association rule mining where only minsupp is used. We introduce maxsupp in order to avoid too frequent variable sets because these variables occur in so many particles that they are not suitable for representing different kinds of particles (i.e., not representational).

In order to find representational frequent variablesets we first use a standard frequent itemset mining algorithm, such as Apriori or FP-growth, to find all frequent variablesets and then remove those whose support is beyond maxsupp and those who have any item/variable whose support is beyond maxsupp. For example, we define minsupp to be 10% and maxsupp 35%, suppose that variable $V1$'s support is 90%,

*V*2' s support is 30% and variableset {*V*1, *V*2} has a support of 30%, then frequent 1-variableset {*V*2} is representational frequent, but frequent 2-variableset {*V*1, *V*2} will not be representational frequent since {*V*1} is not representational frequent.

The method described above is simple but not optimized, we also provide an optimized way of mining representational frequent variablesets: to modify Apriori by adding a maxsupp threshold when finding frequent itemsets/variablesets. At the steps of finding candidate frequent k-variablesets C_k from frequent (k-1)-itemsets L_{k-1}, we remove those frequent (k-1)-variablesets whose support is beyond maxsupp. This method can reduce the size of C_k and can directly obtain representational frequent variablesets.

3.3 Obtaining Clusters

For each representational frequent variableset, we construct an initial cluster to contain all the particles that contain this variableset. One property of initial clusters is that all particles in a cluster contain all the items in the representational frequent variableset that defines the cluster, that is to say, these variables are mandatory for the cluster. We use this defining representational frequent variableset as the representational centroid to identify the cluster.

Initial clusters are not hard/disjoint because one particle may contain several representational frequent variablesets. We will need to merge the overlapping of clusters. The following are two steps for merging.

(I) Merging fully overlapped (or redundant) clusters. If two initial clusters are fully overlapping, that is, they have different representational centroids but the same set of particles; we will merge them and choose the largest representational centroid as resulting centroid. For example, if two representational frequent variablesets *V*1 and *V*2 are highly correlated (i.e. they always come together), then the three clusters, constructed by {*V*1}, {*V*2} and {*V*1, *V*2} respectively, will be merged and the resulting centroid is {*V*1, *V*2}.

(II) Merging partially overlapped clusters. If two initial clusters are partial overlapping, we assign particles in the overlapping area to the largest representational centroid. For example, if a particle belongs to two initial clusters, {*V*1, *V*2, *V*5} and {*V*8, *V*14}, we will assign the particle to {*V*1, *V*2, *V*5}.

The overall FVC clustering algorithm proceeds as follows.

1. Data binarizing.
2. Mining all representational frequent variablesets as the initial representational centroids and construct initial clusters.
3. Assign all points/particles to their representational centroids.
4. Merge overlapped clusters to disjoint clusters.

4 Classification Combining ANNs and FVC

We design a classification system combining ANNs and FVC. The system, we call Clustering-ANNs, involves classification by a set of ANNs, each using distinct subsets of samples selected from the general set using clustering algorithm FVC. More specifically, we use FVC for partitioning the train samples into several subsets, then train a standard back-propagation ANN for each subset. The final decision for a test case is a two-step process. First, the nearest cluster is found for the case, and then the decision is based on the ANN classifier trained on the specific cluster. The reason for using FVC before ANN is that FVC can partition particles into several groups according to their different experimental situation. Particles of different classes may be under the same experimental situation thus have the same set of detected and undetected variables. So each group will have particles within different classes.

5 Methods for Comparison

In this section we describe several classification methods for comparison.

5.1 Probability Learning

Naïve Bayes is a successful probability learning method which has been used in many applications [24, 25, 26]. For the task of Naïve Bayes based particle classification, we assume the particle data is generated by a parametric mixture model. Naïve Bayes estimates the parameters from labeled training samples since the true parameters of the mixture model are not known. Given a set of training particles $L = \{p_1, \ldots, p_N$, N is the number of training samples}, Naïve Bayes use maximum likelihood to estimate the class prior parameters as the fraction of training particles is c_i.

We describe the particle classification problem can be generally described as follows. By assuming one particle only belongs to one class (1 or 0 in our case), for a given particle p we search for a class c_i that maximizes the posterior probability by applying Bayes rule. The method assumes that the features of a particle are independent with each other. Fig. 2 shows the Naïve Bayes classifier for the 2-class and m-feature particle data.

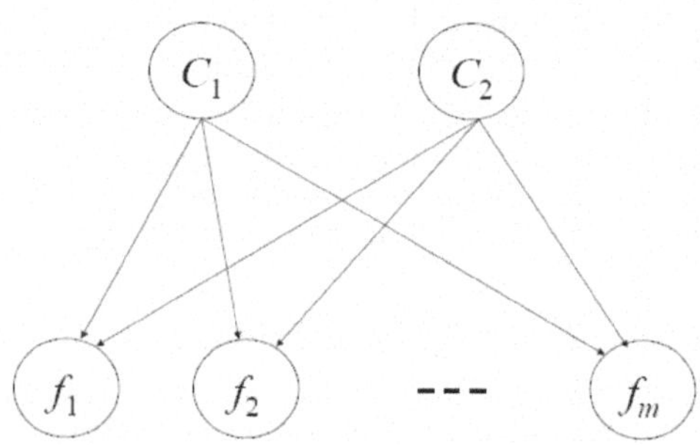

Fig. 2. Naïve Bayes classifier for the 2-class and m-feature particle data

5.2 Memory Learning

Memory based learning is a non-parametric inductive learning paradigm that stores training instances in a memory structure on which predictions of new instances are based [22]. It assumes that reasoning is based on direct reuse of kept experiences rather than on the knowledge, such as models, abstracted from experience. The similarity between the new instance and a sample in the memory is computed using a distance metric. We use the nearest neighbor (NN) classifier, a memory based learning method, that uses Euclidian distance metric [23] in the experiment.

For application in particle physics data, NN it treats all particles as points in the m-dimensional space (where m is the number of variables) and given an unseen particle, the algorithm classifies it by the nearest training particle.

5.3 Hard Partitional Clustering

Hard partitional clustering techniques create a one-level/unnested, partitioning of the data points. Defining k as the desired number of clusters, partitional approaches can find all k clusters at once. There are many such kind of techniques, among which the k-means algorithm is mostly widely used [21]. One of the basic ideas of k-means is that a center point can represent a cluster. Particularly, we use centroid, which is the mean (or median) point of a group of points.

The basic k-means clustering technique is summarized below.

1. Select k points as the initial centroids.
2. Assign all points to the closest centroid.
3. Re-compute the centroid of each cluster.
4. Repeat steps 2 and 3 until the centroids don't change or change little.

6 Experiments

6.1 Datasets

The high-energy particle physics dataset we used are publicly available on KDD website [7]. There are 50000 binary-labeled particles, 78 attributes for each particle. Since attributes 20, 21, 22, 29, 44, 45, 46 and 55 have many missing values, which may degrade the classification performance, we simply ignore these attributes. These particles fall into two classes: positive (1) and negative (0).

6.2 Evaluation Methods

We use four performance measures [12] for the particle classification problem:

Accuracy (ACC, to maximize): the number of cases predicted correctly, divided by the total number of cases.

Area Under the ROC Curve (AUC, to maximize): ROC is a plot of true positive rate vs. false positive rate as the prediction threshold sweeps through all the possible values. AUC is the area under this curve. AUC can measure how many times one would have to swap samples with their neighbors to repair the sort. AUC = 1 indicates perfect prediction, where all positive samples sorted above all negative samples. AUC = 0.5 indicates random prediction, where there is no relationship between the predicted values and actual values. AUC below 0.5 indicates there is a relationship between predicted values and actual values.

SLAC Q-Score (SLQ, to maximize): Researchers at the Stanford Linear Accellerator (SLAC) devised SLQ, a domain-specific performance metric, to measure the performance of predictions made for particle physics problems. SLQ breaks predictions interval into a series of bins. For the experiments we are using 100 equally sized bins within the 0 to 1 interval.

Cross-Entropy (CXE, to minimize): CXE measures how close predicted values are to actual values. It assumes the predicted values are probabilities on the interval of 0 to 1 that indicate the probability that the sample is with a certain class.

$$\text{CXE} = \text{Sum}\ ((A)^*\log(P) + (1\text{-}A)^*\log(1\text{-}P)) \tag{3}$$

where A is the actual class (in our case, 0 or 1) and P is the predicted probability that the sample is with the class. Mean CXE (the sum of the CXE for each sample divided by the total number of samples) is used to make CXE independent of data set size.

6.3 Results

6.3.1 Illustration with Random Subset of Data

We firstly provide an intuitional comparison between FCV and k-means. Fig.3 shows the results of FVC clustering on 100 randomly selected particles. Each column in the figure corresponds to a variable, and the rows represent particles, there are 65 columns and 100 rows. White in a grid means that the variable is detected for a particle, while black represents the variable is not detected for a particle. The number of clusters is decided automatically by FVC according to the nature of the data. In experiments, we found that the original particles (as shown in Fig. 3) will be partitioned into three natural groups as shown in Fig.4. We can see that FVC found natural groups of the dataset.

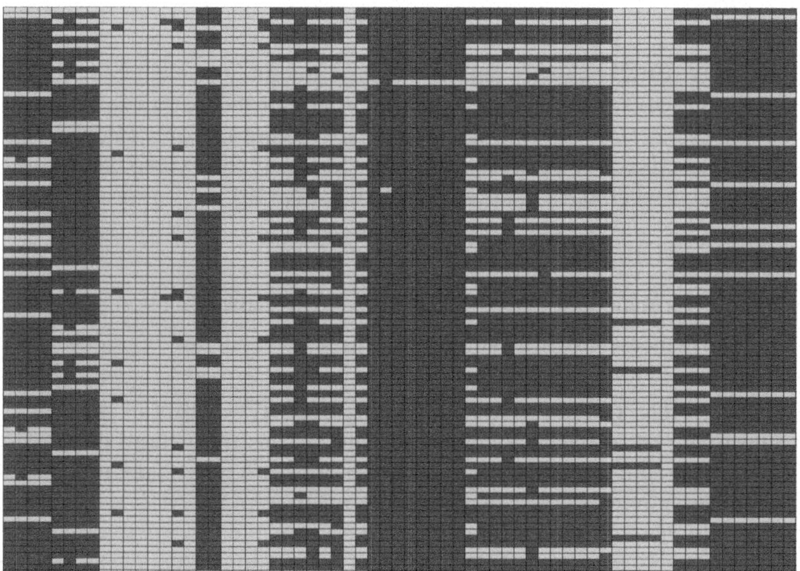

Fig. 3. The original 100 particles. The X-axis denotes the variables. The Y-axis denotes the particles.

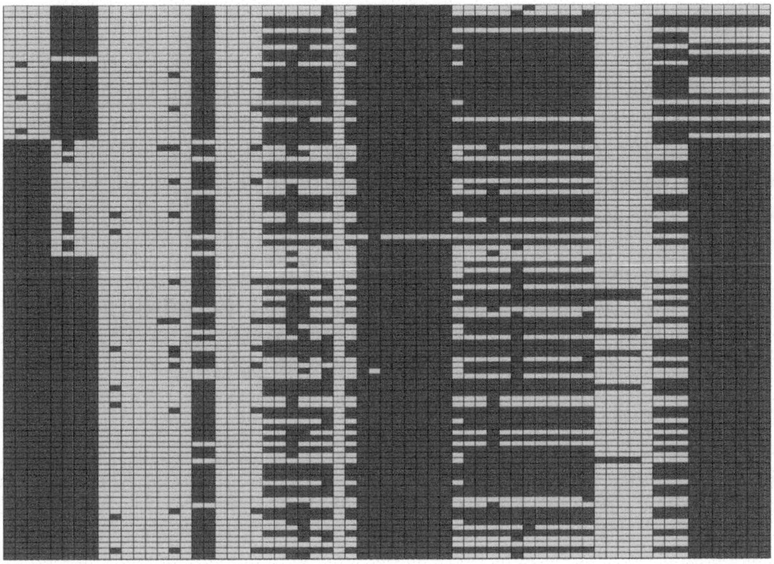

Fig. 4. FVC Clustering results on the 100 particles. The X-axis denotes the variables. The Y-axis denotes the particles.

6.3.2 Results on the Whole Dataset

Full experiments are done on the whole dataset which has 50000 samples. We use 10-fold cross-validation for estimating classification performance, so the four measures, ACC, AUC, SLQ and CXE, are averaged on the 10 runs. Table 3 shows the results.

The results show that ANN is better than Nearest Neighbor and Naive Bayes for particle classification. By combining clustering and ANNs, the proposed scheme Clustering-ANNs can get even better performance than ANN. Kmeans-ANNs is slightly better than ANNs for ACC and SLQ. By using clustering algorithm FVC which is especially designed for particles, we can get the best performance for all four measures.

The reason that FVC-ANNs is better than using a single ANN is that FVC can cluster particle data into different groups according to different experimental characteristics showed in the high-energy physics experiments. Different groups found by FCV have different set of variables. So more appropriate ANN can be trained for each group, this is better than just use a uniform ANN for all particles.

Table 3. Classification performance results of traditional classifers and Clustering-ANNs (Kmeans-ANNs and FVC-ANNs), results in bold type are the best performce

Methods	ACC	AUC	SLQ	CXE
Nearest Neighbor	0.653	0.730	0.253	1.033.
Naive Bayes	0.684	0.747	0.194	0.988
ANN	0.701	0.788	0.270	0.801
Kmeans-ANNs	0.703	0.788	0.272	0.800
FVC-ANNs	**0.719**	**0.801**	**0.293**	**0.787**

7 Conclusion

In this paper we describe a particle-oriented clustering method Frequent Variable Set based Clustering (FVC), and a framework Clustering-ANNs for the high-energy particle physics classification problem. The system involves classification by a set of artificial neural networks (ANNs), each using distinct subsets of samples selected from the general set using clustering algorithm. We use FVC clustering to partition the train samples into several subsets, then standard back-propagation ANNs are trained on them. Comparisons with other popular classification methods, Nearest Neighbor and Naive Bayes, show that ANN is the best for particle physics classification, and the proposed method FVC-ANNs can get even better performance.

Acknowledgments

The authors gratefully acknowledge the support of the National Science Foundation of China (Grant No. 60273015 and No. 10001006).

Reference

1. Jason D. Rennie, et al.: Tackling the Poor Assumptions of Naive Bayes Text Classifiers. Twentieth International Conference on Machine Learning. August 22 (2003)
2. Christopher Bishop: Neural Networks for Pattern Recognition. Oxford University Press, Oxford (1995)
3. Ken-ichi Funahashi: On the Approximate Realization of Continuous Mappings by Neural Networks. Neural Networks, 2(3):183-192 (1989)
4. Simon Haykin: Neural Networks - A Comprehensive Foundation, 2nd ed. Prentice-Hall, Englewood Cliffs (1998)
5. Sepp Hochreiter and Jürgen Schmidhuber: Feature Extraction Through LOCOCODE. Neural Computation, 11(3):679-714 (1999)
6. Kurt Hornik, Maxwell Stinchcombe, and Halbert White: Multilayer Feedforward Networks are Universal Approximators. Neural Networks, 2(5):359-366 (1989)
7. KDD Cup 2004, http://kodiak.cs.cornell.edu/kddcup/index.html (2004)
8. Hipp J., Guntzer U., Nakhaeizadeh G.: Algorithms for Association Rule Mining – a General Survey and Comparison, ACM SIGKDD Explorations, Vol.2, pp. 58-64. (2000)
9. J. Han, J. Pei, and Y. Yin: Mining Frequent Patterns without Candidate Generation. In Proc of ACM SIGMOD'00. (2000)
10. Agrawal, R., Srikant R.: Fast Algorithms for Mining Association Rules in Large Databases, Proc. VLDB 94, Santiago de Chile, Chile, pp. 487-499 (1994)
11. Marcel Kunze: Application of Artificial Neural Networks in the Analysis of Multi-Particle Data. In the Proceedings of the CORINNEII Conference (1994)
12. KDD Cup 2004 – Description of Performance Metrics: http://kodiak.cs.cornell.edu/kddcup/metrics.html (2006)
13. A. Statnikov, C. F. Aliferis, I. Tsamardinos, D. P. Hardin, S. Levy: A Comprehensive Evaluation of Multicategory Classification Methods for Microarray Gene Expression Cancer Diagnosis. Bioinformatics (2004)

14. J. Hipp, U. Guntzer, and G. Nakhaeizadeh: Algorithms for Association Rule Mining - a General Survey and Comparison. ACM SIGKDD Explorations, 2(1):58–64, July (2000)
15. Fung B., Wang K., Ester M.: Large Hierarchical Document Clustering Using Frequent Itemsets, Proc. SIAM International Conference on Data Mining 2003 (SDM '2003), San Francisco, CA. May (2003)
16. Florian Beil, Martin Ester, Xiaowei Xu: Frequent Term-based Text Clustering. KDD: 436-442 (2002)
17. Aha D., and D. Kibler: Instance-based Learning Algorithms, Machine Learning, Vol.6, 37-66 (1991)
18. I.Witten and E.Frank: Data Mining –Practical Machine Learning Tools and Techniques with Java Implementation. Morgan Kaufmann (2000)
19. R. C. Dubes and A. K. Jain: Algorithms for Clustering Data. Prentice Hall College Div, Englewood Cliffs, NJ, March (1998)
20. Karl-Michael Schneider: A Comparison of Event Models for Naive Bayes Anti-Spam E-Mail Filtering. In Proceedings of the 10th Conference of the European Chapter of the Association for Computational Linguistics, Budapest, Hungary, 307-314, April (2003)
21. Xin Jin, Anbang Xu, Rongfang Bie, Ping Guo: Kernel Independent Component Analysis for Gene Expression Data Clustering. ICA 2006: 454-461 (2006)
22. Aha, D., and D. Kibler: Instance-based Learning Algorithms, Machine Learning, Vol.6, 37-66 (1991)
23. Piotr Indyk: Nearest Neighbors in High-dimensional Spaces. In Jacob E. Goodman and Joseph O'Rourke, editors, Handbook of Discrete and Computational Geometry, chapter 39. CRC Press, 2rd edition, (2004)
24. George H. John and Pat Langley: Estimating Continuous Distributions in Bayesian Classifiers. In Proceedings of the Eleventh Conference on Uncertainty in Artificial Intelligence. Page 338-345. Morgan Kaufmann, San Mateo (1995)
25. Xiaoyong Chai, Lin Deng, Qiang Yang, Charles X. Ling: Test-Cost Sensitive Naive Bayes Classification. ICDM 2004: 51-58 (2004)
26. Peter A. Flach and Nicolas Lachiche: Naive Bayesian Classification of Structured Data. Machine Learning, Volume 57(3): 233--269, December (2004)
27. H. Wang, et al.: Clustering by Pattern Similarity in Large Data sets. SIGMOD, 394-405 (2002)

Attributes Reduction Based on GA-CFS Method

Zhiwei Ni, Fenggang Li, Shanling Yang, Xiao Liu, Weili Zhang, and Qin Luo

School of Management, Hefei University of Technology, Hefei 230009, China

Abstract. The selection and evaluation task of attributes is of great importance for knowledge-based systems. It is also a critical factor affecting systems' performance. By using the genetic operator as the searching approach and correlation-based heuristic strategy as the evaluating mechanism, this paper presents a GA-CFS method to select the optimal subset of attributes from a given case library. Based on the above, the classification performance is evaluated by employing the combination method of C4.5 algorithm with k-fold cross validation. The comparative experimental results indicate that the proposed method is capable of identifying the most related subset for classification and prediction with reducing the representation space of the attributes dramatically whilst hardly decreasing the classification precision.

Keywords: Attributes reduction, correlation-based feature selection (CFS), Genetic algorithm (GA), k-fold cross validation.

1 Introduction

In the research field of machine learning and data mining, significant attention has been paid to attributes reduction and evaluation. As an important task for knowledge-based systems, its key problem is how to identify the most related subset to the given target, and at the same time clear away irrelevant redundant attributes. Through the successful performance of this task, the reduction of data dimensions and the assumption space shall be achieved which would enable the algorithm to have quicker execution speed and higher efficiency. Attributes reduction and evaluation are also a NP-hard problem. Hence how to select a valid searching method is a critical aspect we should investigate. Genetic algorithm differentiates itself from other searching methods in its particular genetic operator. It can be well implemented to solve the problem of attributes searching. Another important factor in system design is how to measure the weight of attributes for classification and prediction. Correlation-based heuristic method can evaluate the degree of association among attributes and measure the contribution of attributes (subsets) to classification. It can serve as the evaluation criterion for attributes reduction.

This paper proposes a GA-CFS method by combining genetic algorithm with correlation-based evaluation. The proposed method solves not only the problem of searching efficiency caused by "combinatorial explosion" of attributes combination, but also the problem of correlation measure among attributes. Some researchers have implemented attributes reduction [1-3] using genetic mechanism without combining it

G. Dong et al. (Eds.): APWeb/WAIM 2007, LNCS 4505, pp. 868–875, 2007.

with correlation-based feature reduction. The problem of how to find the (approximate) optimal subset of attributes for a given case library is what the authors are devoted to in this paper.

The remainder of this paper is organized as follows. The next section describes the searching strategies as well as the evaluating strategies of attributes reduction and its formalization. Then section 3 describes the genetic algorithm in brief. Section 4 focuses on the process of attributes reduction using GA-CFS based on the genetic algorithm and correlation-based evaluation. Section 5 verifies the performance of GA-CFS with the method of combining C4.5 algorithm with K-fold cross validation, using the data from UCI database repository of the University of California. Finally, section 6 concludes this paper and points out future work.

2 Attributes Reduction

Attributes reduction is to select a subset from the attributes space which influences predicting or classifying results significantly. Its goal is to find those attributes or subsets that have the most classifiable ability. In general, attributes reduction includes two parts: (1) searching strategy in the attributes space; (2) evaluation strategy for the selected attributes subset. They are both the indispensable segments of the process.

2.1 Searching Strategy and Evaluation Strategy of Attributes Reduction

Attributes reduction is a combination optimization problem. It has high complexity and requires an efficient searching algorithm. Each searching state can be mapped as a subset in the searching space for a searching problem. An n-dimensional data set has a 2^n potential state space, so it is very important to select the originating point for searching and the searching strategy. Usually, we use heuristic searching strategies instead of exhaustive searching strategies to obtain the approximate optimal subset. Searching strategies of attributes include: best first [4], forward reduction, stochastic searching, exhaustive searching, genetic algorithm [1, 2], ordering method, etc.

From the viewpoint of an evaluation function, attributes evaluation figures out the opinion source of every potential attribute and then selects some attributes, which have the highest scores as the optimal subset. The evaluation function directly influences the final subset reduction. Based on different evaluation functions, different subsets will be formed. Generally, attributes evaluation methods include: information gain method [5], gain ratio method [6], correlation-based evaluation [7], principal component analysis, chi-mean square evaluation, etc. Attributes evaluation is the reduction problem of the evaluation function in a genetic algorithm.

2.2 Formalization of Attributes Reduction

Considering the attributes set to be attribute vectors, reduction is a process of selecting the subset of which the cardinal number is M from the attributes set of which the cardinal number is N ($M \leq N$).

Let F_N be the original attributes set, F_M be the selected subset. Thus, with respect to the optimized subset, the conditional probability $P(C_i / F_M = f_M)$ of each decision class C_i should be as equivalent to that of F_M as possible. It can be defined as:

$$C_i : P(C_i / F_M = f_M) \cong P(C_i / F_N = f_N) \tag{1}$$

Where, f_M represents the special attribute vector of attributes set F_M while f_N represents that of attributes set F_N. The process of attributes reduction is the process of searching for optimal or approximate optimal F_M.

3 Genetic Algorithm

Genetic algorithm is a searching approach [8] that is based on natural reduction and natural genetic mechanism. Following the strategy of "survival of the fittest" in the nature, the algorithm uses random genetic operators to generate several new solutions, eliminates the poorer, and keeps the better and promising ones. The information of the fittest solutions is constantly utilized to search for the new unknown area of the searching space. Since its effective use of historic information that makes every search moving forward according to the best direction, genetic algorithm is similar to simulated annealing algorithm and tabu searching. As a result, genetic algorithm is not only a random searching approach, but also a directing random searching approach.

Genetic algorithm can be formally defined as an 8-tuple:

$$GA = (P(0), N, l, s, g, p, f, t)$$

Where, $P(0) = (y_1(0), y_2(0), ..., y_N(0)) \in I^N$ denotes the initial population;

N is a positive integer, which denotes the number of individuals in a population;

l is also a positive integer, which denotes the length of the symbol string (chromosome);

$I = \Sigma^l$ represents the collectivity of the symbol strings of which the length is l. Σ is an alphabet. If binary coding is used, then $\Sigma = \{0,1\}$;

$s : I^N \to I^N$ represents reduction strategy;

g denotes genetic operators, which usually include duplicate operator $O_r : I \to I$, crossover operator $O_c : I \times I \to I \times I$ and mutation operator $O_m : I \to I$;

$f : I \to R^+$ is a fitness function;

$t : I^N \to \{0,1\}$ is a termination law.

Genetic algorithm presented by Holland initially adapted binary coding, that means $\Sigma = \{0,1\}$. But generally speaking, it can be expanded into any data structure. According to the needs of practical problems, Σ can be 0-1 bit string, as well as integer vector, Lisp expressions or neural networks. In this paper, we use binary-coded string to denote the attribute vector. The code '0' denotes that the represented attribute is not appeared in the search, while '1' denotes the opposite. The settings of genetic operators are given in section 4.2.

4 Attributes Reduction Based on GA-CFS Method

4.1 Reduction of Evaluation Method

CFS evaluation method based on correlation-based attributes reduction is a heuristic algorithm [7]. It can evaluate the 'merit' of the subset of attributes. Its main consideration is the class prediction ability of single attribute and their correlations. The heuristic algorithm is based on the hypothesis below: attributes that belong to quality subset F_M are highly correlated to class C_i while the attributes themselves are irrelevant to each other.

The irrelevant attributes in the subset are hardly related to the classification, so they can be ignored. The redundant attributes can also be eliminated for they are certain to have a correspondence to a high-correlated attribute. The acceptance degree of an attribute is due to its ability to predict the classification in the case library space while other attributes can not. The evaluation function CFS of the subset is defined as follows:

$$M_s = \frac{k\overline{r_{cf}}}{\sqrt{k + k(k-1)\overline{r_{ff}}}} \tag{2}$$

where M_s is the heuristic 'merit' when the subset includes k attributes; $\overline{r_{cf}}$ is the attribute-classification correlative average value (f∈S); and $\overline{r_{ff}}$ is the attribute-attribute correlation average value.

For successive value data, the relativity between attributes can be calculated as follows:

$$r_{XY} = \frac{\sum xy}{n\sigma_x\sigma_y} \tag{3}$$

where σ_x and σ_y denote quadratic mean deviation of the attributes of successive value data.

If one of the two attributes is successive and the other is discrete, the relativity can be calculated as follows:

$$r_{XY} = \sum_{i=1}^{k} p(X = x_i) r_{X_{bi}Y} \tag{4}$$

For X_{bi}, if $X = x_i$, then X_{bi} = 1, else X_{bi} = 0.

If both attributes are discrete, the relativity can be calculated as follows:

$$r_{XY} = \sum_{i=1}^{k} \sum_{j=1}^{l} p(X = x_i, Y = y_j) r_{X_{bi}Y_{bj}} \tag{5}$$

According to the above formulations, correlation of the attributes can be calculated no matter it is discrete or successive. Then it will be selected as the attributes reduction criterion in the next step of the genetic searching until the final criterion of the algorithm is met.

4.2 Settings of Genetic Operator

In order to obtain the reduction of attributes with genetic algorithm, the following operations need to be set:

1. *Initialization of the population.* Select N random initial points to form a population. The amount N of the individuals in a population is the population size. Each chromosome of the population is coded by binary string. Chromosomes denote the optimized parameters. Each initial individual denotes the initial solution.
2. *Reduction.* Select appropriate individuals according to the selective strategy of roulette wheel. The reduction should embody the principle of 'survival of the fittest'. On the basis of the fitness value of each individual, the best individual can be selected as the next generation population for repropagation.
3. *Crossover.* With the crossover probability p_c, new individuals can be generated. Thus searching can be effective in the solution space, meanwhile decrease the destruction to the effective scheme. Crossover is a mechanism of the information exchange between two chromosomes.
4. *Mutation.* According to the given mutation probability p_m, we can select some individuals randomly from the population while make the mutation calculation to the selected individuals in correspondence with certain strategy. The mutation calculation is an important factor to enlarge the population diversities. It enhances the ability for genetic algorithm to find the optimal solutions.

4.3 Evaluation Method of Attribution Reduction

In order to evaluate the performance of the subset of attributes F_M which is selected by the GA-CFS method that combines GA with correlation-based attributes reduction, this paper uses the method which combines C4.5 algorithm [6] with k-fold cross validation to verify the classification performance of F_M. Meanwhile, we compare the classification performance with that of the original subset F_N.

C4.5 algorithm is the improvement of ID3 [5]. It can deal with the following problems: the attributes of successive value, the deficiency and deterioration of attribute value, pruning of decision tree and the creation of rules, etc. Its core concept is to adapt the information-entropy-based sorting strategy of attributes.

K-fold cross validation is also called rotation estimation. It divides the whole set of case library (S) into k non-overlap and equal subsets (S_1 , S_2 , 。。。 , S_k) randomly. The classification model is trained and tested for k times. For each time ($t \in \{1,2,\cdots,k\}$), let ($S - S_t$) be the training subset. The cross validation precision is obtained from the average value by calculating the testing precision for k times separately:

$$CVA = \sum_{i=1}^{k} A_i \tag{6}$$

where, CVA denotes the precision of cross validation, k denotes the amount of the subset that has been used, and A_i is the precision rate of every subset. In the experiment described next in this paper, K =10 [9].

5 Experimental Results and Analysis

In order to evaluate the validation of GA-based attributes reduction, we use GA-CFS to compare the attributes sets selected by the means of combining genetic algorithm with correlation-based heuristic before attributes reduction.

Observing the variance of the number of attributes, the variance of precision and related performance values of subset, we can review the performances of the algorithm proposed in this paper. Our GA-CFS approach is implemented in Java and experiments were conducted on a Pentium(R) 4 CPU 2.80GHz with 256MB RAM running under Windows 2000.

In the experiment, we select 4 data sets from UCI ML database repository from the University of California. The detailed information is given in Table 1.

Table 1. Data set used in the experiment

Data set	Num. Of cases	Num. of attributes	Attributes deficiency (%)	Num. of classes
Anneal	798	38	73.2	5
Arrhythmia	452	279	0.32	13
Breast_cancer	286	9	0.3	2
Sick	3,772	30	5.4	2

Parametric setting of genetic algorithm is as follows: population scale N=20, crossover probability Pc=0.66, mutation probability Pm=0.033, the largest number of iteration is 20.

Use C4.5 algorithm to compute the classification precision before and after attributes reduction. Use k-fold cross validation to verify the computation of classification precision. Obtain computing results by averaging after executing 10 times. The experimental results are given in Table 2 and Table 3.

Table 2. Comparison before and after attributes selection

Data set	Num. of attributes before selection	Num. of attributes after selection	Reduction of attributes (%)	Correlation value of subset
Anneal	38	11	71.05	0.480,12
Arrhythmia	279	98	64.87	0.071,47
Breast_cancer	9	5	44.44	0.096,72
Sick	30	4	86.67	0.234,91

Table 3. Comparison of the classification accuracy before and after attributes selection

Data set	Accuracy before selection (%)	Accuracy after selection (%)	Decrease of accuracy (%)
Anneal	98.57	97.97	0.61
Arrhythmia	65.65	66.04	−0.59
Breast_cancer	74.28	73.08	1.62
Sick	98.72	97.39	1.35

The experimental results indicate that using GA-CFS to select subset, concerning the reduction of attributes, can reduce attributes in the 4 data sets by 44.44% at least, and by 86.67% at most as show in Table 2. So the reduction of dimensions is considerable. From the variance of classification accuracy of the 4 data sets after attributes reduction as shown in Table 3, we can see that the accuracy of anneal data set reduces less than 1%, breast_cancer and sick dataset reduce about 1% and arrhythmia data set even increases.

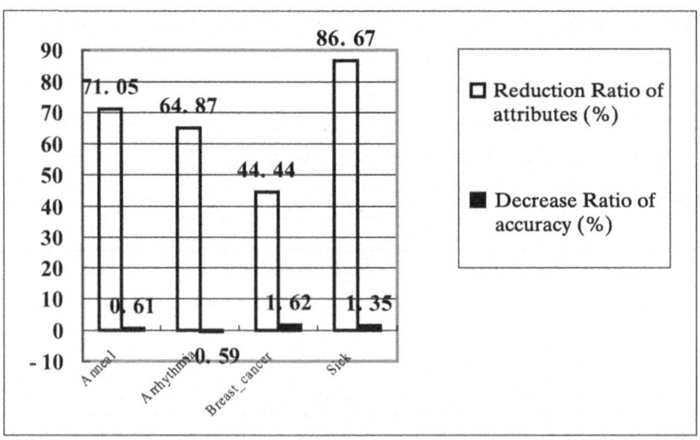

Fig. 1. Comparison of ratio between reduction of attributes and decrease of accuracy before and after attributes reduction

By analyzing the data set above we can conclude, by comparing with the original attributes, using the proposed attributes reduction method to optimize subset selection, the attributes are reduced by 70% on average, while precision decreases about 1% only, just as shown in Fig. 1

Hence, the proposed GA-CFS algorithm has achieved much better outcomes. It reduces the number of attributes dramatically whereas hardly decreases the classification precision.

6 Conclusions and Future Work

Attributes reduction and evaluation is an important task for knowledge-based systems. They can identify the most related attributes to the problems of the system, clear away irrelevant attributes, reduce the representation space of case library, decrease the complexity of systems, and improve the performance of systems.

We have proposed a GA-CFS method to guide the evolution of systems until it finds an approximate optimal subset. We have implemented the original searching approach using the genetic operator that introduces a correlation-based subset evaluation method as the evaluation function. By using C4.5 algorithm combined with k-fold cross validation to evaluate its performance, we have concluded that the GA-CFS method can identify the most related subset to classify and predict with reducing the representation space of the attributes dramatically whilst hardly decreasing classification precision.

In the future, we would like to do some benchmark work on attributes reduction. It relates to these theory and techniques, such as Rough Set (RS), Prime Component Analysis (PCA), entropy-based attributes reduction, etc. We believe that it would benefit the use of the various attributes reduction methods.

References

1. Yuan, C. A., Tang, C. J., Zuo, J., et al. Attribute reduction function mining algorithm based on gene expression programming, 5th International Conference on Machine Learning and Cybernetics, AUG 13-16, Vols. 1-7, (2006) 1007-1012
2. Hsu, W. H.:Genetic wrappers for feature reduction in decision tree induction and variable ordering in Bayesian network structure learning, Information Sciences, vol. 163, (2004) 103–122
3. Zhao, Y., Liu, W. Y.:GA-based feature reduction method, Computer engineering and application, vol. 15, (2004) 52-54
4. Kohavi, R., John, G. H.:Wrappers for feature subset reduction, Artificial Intelligence, (1997) 273-324
5. Quinlan, J. R.:Induction of decision trees, Machine Learning, vol. 1, No. 1, (1986) 81-106,
6. Quinlan, J. R.:C4.5: Programs for machine learning, Morgan Kaufmann, San Mateo. CA, (1993)
7. Hall, M. A.:Correlation-based feature reduction for discrete and numeric class machine learning, Proc. of the 17th International Conference on Machine Learning(2000)
8. Zhou, M.,Sun, S. D.: GA principle and application, National defense industry press, Beijing (1999)
9. Kohavi, R.:A study of cross-validation and bootstrap for accuracy estimation and model reduction. In: Wermter, S., Riloff, E., and Scheler, G., (eds.): The Fourteenth International Joint Conference on Artificial Intelligence (IJCAI), Morgan Kaufman, San Francisco, CA, (1995) 1137—1145

Towards High Performance and High Availability Clusters of Archived Stream*

Kai Du, Huaimin Wang, Shuqiang Yang, and Bo Deng

School of Computer Science, National University of Defense Technology
Changsha 410073, China
keyes.du@gmail.com, whm_w@163.com, sqyang9999@126.com,
dengbomail@gmail.com

Abstract. Some burgeoning web applications, such as web search engines, need to track, store and analyze massive real-time users' access logs with high availability of 24*7. The traditional high availability approaches towards general-purpose transaction applications are always not efficient enough to store these high-rate insertion-only archived streams. This paper presents an integrated approach to store these archived streams in a database cluster and recover it quickly. This approach is based on our simplified replication protocol and high performance data loading and query strategy. The experiments show that our approach can reach efficient data loading and query and get shorter recovery time than the traditional database cluster recovery methods.

1 Introduction

Some burgeoning applications have appeared which needs the high availability and extra high performance of data insertion operations. The records of web behavior, such as the records of personal search behavior in search engines, online stock transactions or call details, are the classical archived streams [11]. For instance, Google can improve the users' search experiences based on Personalized Search [3]. This information should be written into a large database in a real-time mode and queried repeatedly when the user uses the search engine again. All of these archived streams applications have the following common characteristics:

- A round-the-clock Internet company needs a high availability of 24*7. However high availability is a great challenge for a large-scale Internet company like Google since a large number of equipments are needed.
- High-rate data streams need a high performance and near real-time record insertion method. Google processes about 4200 requests every second [4] and needs a high performance insertion program to record all the users' behavior.
- The recorded data can be viewed as historical data because it will not be updated any more but only be queried repeatedly after being stored.

* Supported by the National Grand Fundamental Research 973 Program of China under Grant No.2005CB321804, and the National Science Fund for Distinguished Young Scholar of China under Grant No.60625203.

G. Dong et al. (Eds.): APWeb/WAIM 2007, LNCS 4505, pp. 876–883, 2007.

We call these applications as *log-intensive* web applications. [11] is the first one which optimizes querying on the live and archived streams, but doesn't study the insertion performance and system's availability. [14] studies the availability of an updatable data warehouse filled with less-update data. It bases on the general-purpose 2PC which is not efficient enough for the high-rate archived streams.

The first contribution of this paper is to optimize the insertion operations by writing no online-log and archived-log in databases and committing data in bulk. The second is providing a simple consistency protocol based on the no-update feature of the data. The third is designing an efficient recovery method for high-rate insertions.

The remains of this paper are: Section 2 is the problem statement and related work. Section 3 is transaction processing and consistency protocol. Section 4 introduces the recovery approach; Section 5 is the experiments; Section 6 is the conclusion.

2 Problem Statement and Related Work

Let's consider the classical *log-intensive* applications: when the users are accessing the web sites, all the users' behavior may be stored and a group of record items are generated at all times. These record items must be real-time stored and queried by subsequent web accesses. A high available and efficient system, such as a database cluster, needs to be built for these applications. A database cluster is m database servers, each having its own processors and disks, and running a "black-box" DBMS [1]. The "Read One Write All Available" policy [2] is always adopted. It means when a read request is received, it is dispatched to *any one* node in the available nodes.

In [8], bulk loading is adopted to optimize the insertion performance; however it doesn't focus on availability. The primary/secondary replica protocol [9] in commercial databases [10, 12] ships updates logs from the primary to the secondary. This way decreases the insertion performance for the IO access in *log-intensive* applications. The 2PC [2] keeps all replicas up-to-date, but has poor performance for its force-writes logs and poor recovery performance based on complex ARIES [7, 14].

In order to avoid force-writes, ClustRa [13] uses a neighbor logging technique, in which a node logs records to main memory both locally and on a remote neighbor; HARBOR[14] avoids logs by revising the 2PC protocol, but the revised 2PC is too complex to the insertion-intensive and no-update applications. [15, 16] is not based on 2PC and propose a simple protocol, but it needs to maintain an undo/redo log.

The object of this paper is to design an efficient integrated approach to solve the problem of high availability and high performance for these *log-intensive* applications. The basic idea is to insert the data in bulk without online log in databases and set a consistency fence for every table in the data processing phase.

3 Transaction Processing

All recovery approaches are based on the transaction processing. This section will introduce the details about insertion and query processing.

3.1 System Framework: Transaction Types and Unique External Timestamp

As is discussed in Section 1, in the *log-intensive* workloads, all the transactions can be classified as two types: *insertion* and *query* transactions since there are no update transactions. The *insertion* means inserting high-rate data into databases. The *query* means querying the massive non-update history data.

The following are adopted to reach our objects: 1) Buffer and insert the data into a database in bulk. The experiments show bulk insertions always outperform standard single insertions by an order of magnitude. 2) Write no online logs in databases for insertions. 3) Insert multiple objects in parallel. Eliminating the dependency of the insertions on different objects could be reached by simply canceling the foreign key constraints. 4) Recovery methods based on no-log must be developed.

According to 1), a coordinator is added upon a database cluster to buffer and insert data in bulk in Fig. 1. For every table, an *insertion* thread is always running since the coordinator processes the same data more easily than any underlying database. Thus only one thread for one table is enough. For a query request, a *query* thread dynamically starts and ends with that request. The *insertion* threads refresh the meta-info *TF* and *ANS* (introduced in Section 3.2) and the *query* threads read this on time.

Another mechanism, the unique external timestamp, is designed to implement the consistency protocol. Since a record data item usually has a time field *log_time*, we can construct a unique id for every record by adding a field *log_number* which can differentiate the different records with the same *log_time*. Thus every record has a virtual unique identifier *log_id* through binding *log_time* and *log_number*. The allied timestamp is also used in [14]. However it is generated in the database core when the insertion is committed which will destroy the autonomy of the underlying databases.

3.2 Insertion and Query Processing

The data insertion processing is illustrated in Fig. 2. The data is buffered into the input buffer B-in, and when B-in is full, it will be changed into output buffer B-out (① ② in Fig. 2). Then the data in *B-out* will be written into multiple database replicas simultaneously(③ in Fig. 2). After the insertion thread receives all the messages of replicas(④ in Fig. 2), it refreshes the *Time Fence (TF)* and *Available Node Set (ANS)*(⑤ in Fig. 2). Only if the insertion thread meets a database replica failure, it will write *B-out* into local log files (⑥ in Fig. 2). Before the failed replica is recovered, a group of *insertion log* files will be maintained.

The *Time Fence (TF)* is the *log_id* of the latest record inserted into the database. Every table has a *TF*. It is used to synchronize the query threads and insertion threads.

From the above analysis, it's obvious that no logs are generated on the coordinator node and database nodes as [14]. Since the volume of the log is at least larger than the data in a database, this method reduces at least 50% IO of the normal fashion. It is more efficient than [15] which stores logs both on middleware and database nodes.

The process of queries includes two steps. Step one is rewriting the SQL. In order to synchronize the result sets of every database replicas, an extra condition of *log_id* is added according to the *TF* of every table. The revising rule is as Table 1. Thus all query threads have a uniform logical view about the data in the replicas even though the same data may be not inserted synchronously by an insertion thread. Step two is dispatching the revised SQL to an available replica in the *ANS*. This can be done in terms of some load balance policy like current requests number.

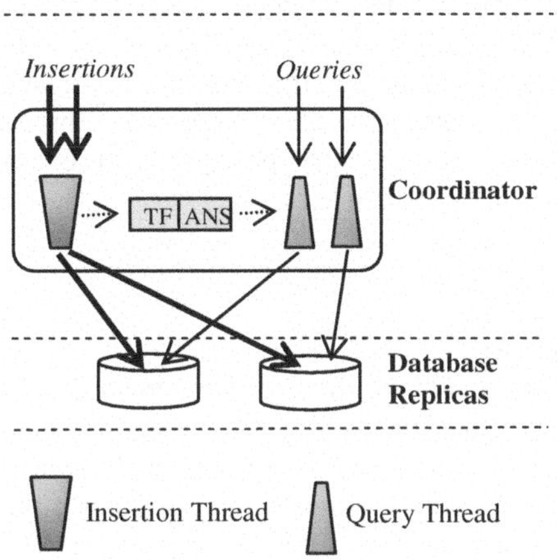

Fig. 1. System Framework

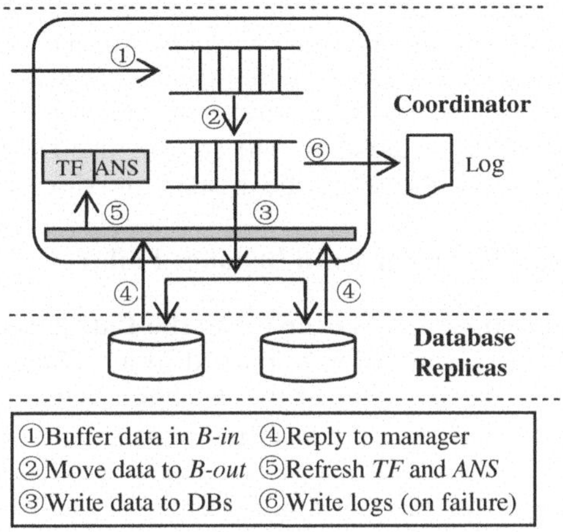

Fig. 2. Insertion Processing

Table 1. Rewriting Query Rule

Original	Rewritten
SELECT tuples FROM table_a WHERE original_predicates;	SELECT tuples FROM table_a WHERE original_predicates *AND* ***log_time < TF[table_a].log_time*** *AND* ***log_number <TF[table_a].log_number;***

3.3 Replication Protocol

The replication protocol is to keep copies (replicas) consistent despite updates [5]. The 2PC or its variation [14] can be used to synchronize the data, but it is too expensive for its communication overhead. Recently some efficient eager replication protocols[6] can partly solve the problem of throughput and scalability but not the latency. All these general-purpose protocols seem too complex for the simple transaction semantics of *log-intensive* workloads and inefficient for SQL log and complex locks.

In the *log-intensive* workload, the atomicity and consistency of an insertion transaction is guaranteed by a table's *TF*. When a *table_a*'s insertion thread receives the replies of every replica, it must wait until it attains an exclusive (write) lock of *table_a*'s *TF*. After that it can refresh the *table_a*'s *TF* and the *ANS*. Before a query thread revises a query SQL, it must wait until it attains a share (read) lock of *table_a*'s *TF*. Thus the committed data will not be seen until all replicas have committed it. This simply guarantees insertion transactions' atomicity because no query will see the data before the *TF* is changed.

4 Recovery Approach

The recovery approach is based on the insertion data log files (generated in ⑥ in Fig. 2). We design a recovery algorithm on a granularity of tables. This algorithm is constituted of a recovery manager thread rm_thread and many recovery threads recovery_thread(node_id, table_id). The rm_thread always runs on the background and monitor which failure database needs to be recovered. If it finds some one, it will create one recovery_thread for every table on that database. After a recovery_thread recovers a table, it will inform the rm_thread. The recovery procedure of every recovery_thread can be divided into three phases in Section 4.1.

4.1 Recovery from Instance Failure

(1) Phase 1: Recover From the Latest Save Point

When an insertion is pushed to a replica, the data will be directly written in pieces into the data files of the database. When the database meets an instance failure, one part of data of the insertion request is stored in the database while other in the memory is lost. In order to save the stored data and avoid duplicating it, we should get the *log_id* of the latest stored data. We call this *log_id* as "*the latest save point(LSP)*". The LSP can be got in this standard SQL clause:

SELECT MAX(log_time), MAX(log_number) INTO *LSP*.log_time, *LSP*.log_number FROM table_a;

Just as mentioned in Section 3.2, we can leverage the oldest insertion log file of the logs group. The pseudo code is just like:

LOAD DIRECT FILE= the oldest file of *table_id*
WHERE log_time ≤ *LSP*.log_time
AND log_number < *LSP*.log_number;

Thus all the data left in the oldest insertion log file is loaded into the recovering database. Then the other insertion log files can be directly loaded into the database.

From the above procedure, we can find that both the recovery of multiple tables in one database and the recovery of multiple failure databases can be done in parallel.

(2) Phase 2: Catch Up with Data Logs

This phase is a subsequent step of phase 1 and simpler. The pseudo code is:

LOAD DIRECT FILE=other files of table_id;

In this phase, we can optimize the recovery by merging several small files into big files. This can improve the recovery performance due to decreasing the access times to the recovering database. The size of every big file is determined by the load of network, disk, cpu of two sides. The effect of merging will be shown in Section 5.

(3)Phase 3: Catch Up with Current Insertion

After loading all the log files of *table_id*, the recovery_thread will inform the rm_thread and the insertion_thread(*table_id*). The insertion_thread will push the current insertion to the database of *node_id*. After the insertion_thread has completed this insertion, it will refresh the TF of *table_id* and added the recovered database into the ANS of *table_id*. From that time on, the insertion and query transaction can send to the table of *table_id* of the recovered database.

If all recovering databases have recovered on this table of *table_id*, the insertion_thread(*table_id*) will no longer write log files.

4.2 Recovery from Media Failure

When a database meets a media failure, such as some data files can't be read or written, the recovering procedure can be implemented like the following two steps: 1) Recover the data files based on *current* and *historical* partitions. This will not be introduced in details. 2) Recover the instance like Section 4.1.

5 Experiments

In these experiments, a database cluster of three nodes and a coordinator node is built. All the four nodes have two CPUs of Xeon 2G, 4G RAM, two 70G SCSI disks and are installed on Redhat AS 3.0. The three database nodes are installed with Oracle 10.1.0.4. And all the codes are written in GNU C++. The experimental data comes from the access records of some commercial search engine. A record item's size is about 329 bytes.

5.1 Runtime and Recovery Performance

Fig. 3, 4 are about runtime performance. Fig. 5, 6 are about recovery performance.

In Fig. 3, we can find two conclusions: 1) The optimized loading's performance is 50-100 fold of the standard INSERT SQL's and 2PC used in [14]. 2) As the results show, when a database node writes online and archived log, online log, no log, the time ratio in average is about 1.43:1.14:1. 3) The insertion time is proportional to the size of the data. In Fig. 4, the bulk size is 80MB and the time is the average processing time of multi-users. Three scenes are simulated: writing online log on databases and the coordinator (it happens when a database node fails), writing online log on databases and writing no log. The ratio is 1.28:1.11:1.

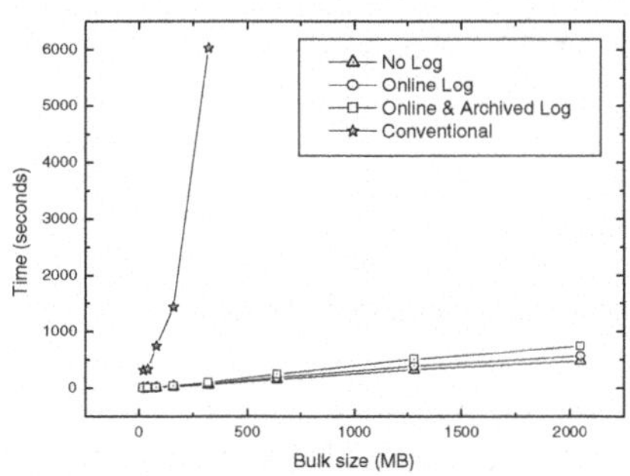

Fig. 3. Insertion performance and bulk size

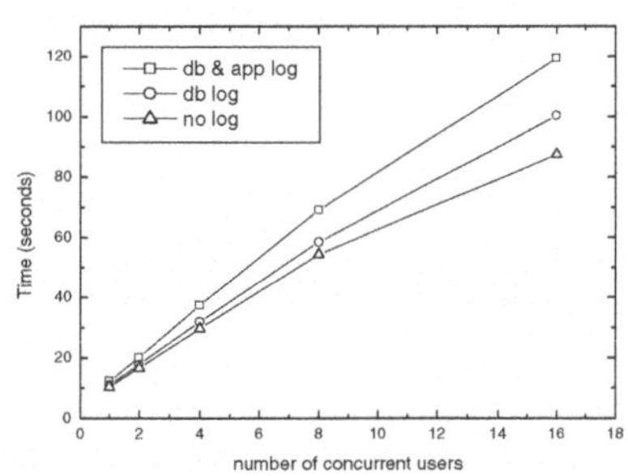

Fig. 4. Insertion performance and # of users

In Fig. 5, we compare the classical ARIES recovery method and ours. The results show that when the recovered data size is less than 4.5MB, the ARIES is better, but after that point our method gets a better performance. When the recovered data size is small, the startup cost of our method is greater than ARIES and later the complexity of ARIES leads to its long recovery time. In Fig. 6, the time of three recovery phases is shown. The startup time in phase 1 and the catching up time in phase 3 is a constant time, while the insertion time in phase 2 is proportional to the data to be recovered.

5.2 Performance During Failure and Recovery

The transaction processing performance during the databases' failure and recovery is another problem needed to be discussed. In Fig. 7, the x-axis is the time, the left y-axis is the insertion performance whose criterion is MB/s, and the right y-axis is the query performance whose criterion is the completed transactions per second.

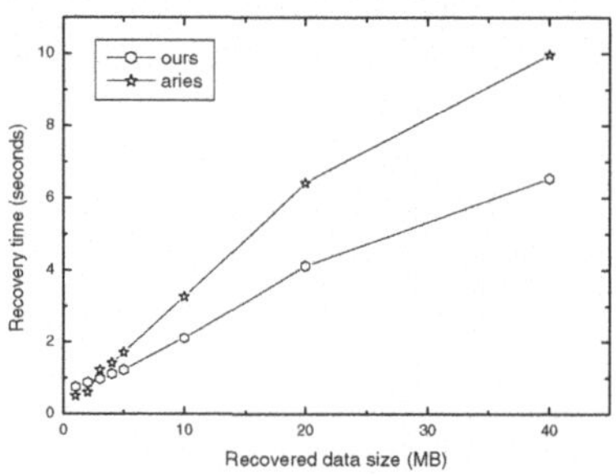

Fig. 5. Recovery performance and recovered size

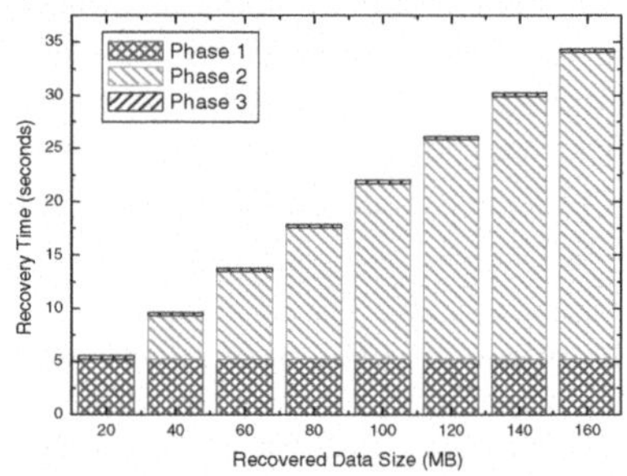

Fig. 6. Decomposition of recovery time

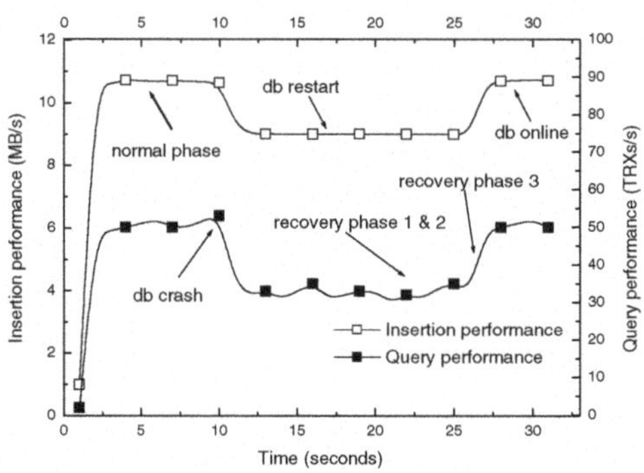

Fig. 7. Transaction processing performance during failure and recovery

Before the 10th second, the system runs in the normal state. At the 10th second, one of the three databases fails, the coordinator detects this and the DBA restarts the database at the 15th second. During this period, the insertion performance decreases a little about 13% because the log files need to be stored on the coordinator's disk, and the query performance decreases about 31% because 1/3 of the three nodes can't process the query requests. From 15th second to 25th second, the recovery phase 1 and 2 complete, and the performance is just as the 15th second because the recovery will not decrease the online performance. From 26th to 27th second, the phase 3 completes, and the performance return to the normal level.

From Fig. 7, we can find that there is no sharp performance degradation because other transactions will not be interrupted when one database fails.

6 Conclusion

In this paper we have studied the problem of how to store and recover high-rate archived streams in a database cluster. According to the *log-intensive* applications, we present an optimized data insertion method based on reducing the disk IO access cost and a simple and efficient consistency protocol. The experiments results show that our approach can reach efficient data loading and query and get shorter recovery time than the traditional database cluster recovery methods.

References

1. S. Ganarski, H. Naacke, E. Pacitti, P. Valduriez: Parallel Processing with Autonomous Databases in a Cluster System, CoopIS, 2002.
2. J. Gray, A. Reuter. Transaction Processing: Concepts and Techniques. Morgan Kaufman, 1992.
3. Google personalized search. http://www.google.com/psearch
4. http://news.com.com/Google,+eBay+Strategic+bedfellows/2100-1024_3-6110304.html
5. J. Gray and P. Helland and P. O'Neil and D. Shasha: The Danger of Replication and a Solution. ACM SIGMOD, 1996.
6. Matthias Wiesmann, Fernando Pedone, Andre Schiper, Bettina Kemme,Gustavo Alonso. Transaction Replication Techniques: a Three Parameter Classification. SRDS 2000.
7. C. Mohan, D. Haderle, B. Lindsay, H. Pirahesh, and P. Schwarz. ARIES: a transaction recovery method supporting fine-ranularity locking and partial rollbacks using write-ahead logging. ACM TODS, 17(1):94–162, 1992.
8. Y. Dora Cai, Ruth Aydt, Robert J. Brunner. "Optimized Data Loading for a Multi-Terabyte Sky Survey Repository". Super Computing 2005.
9. B. Liskov, S. Ghemawat, R. Gruber, P. Johnson, and L. Shrira. Replication in the harp file system. In SOSP, pages 226–238. ACM Press, 1991.
10. Microsoft Corp. Log shipping. http://www.microsoft.com/technet/prodtechnol/sql/2000/reskit/part4/c1361.mspx.
11. Sirish Chandrasekaran, Michael Franklin. Remembrance of Streams Past:Overload-Sensitive Management of Archived Streams. VLDB 2004.
12. Oracle Inc. Oracle database 10g Oracle Data Guard. http://www.oracle.com/technology/deploy/availability/htdocs/DataGuardOverview.html.
13. S.-O. Hvasshovd,Torbjrnsen, S. E. Bratsberg,and P. Holager. The clustra telecom database: High availability, high throughput, and real-time response. VLDB 1995.
14. Edmond Lau, Samuel Madden. An Integrated Approach to Recovery and High Availability in an Updatable, Distributed Data Warehouse. VLDB 2006.
15. R. Jim'enez-Peris, M. Patino-Martinez, and G. Alonso. An algorithm for non-intrusive, parallel recovery of replicated data and its correctness. In SRDS, 2002.
16. B. Kemme. Database Replication for Clusters of Workstations. PhD dissertation, Swiss Federal Institute of Technology, Zurich, Germany, 2000.

Continuously Matching Episode Rules for Predicting Future Events over Event Streams

Chung-Wen Cho[1], Ying Zheng[2], and Arbee L.P. Chen[3]

[1] Department of Computer Science, National Tsing Hua University, Taiwan, R.O.C.
[2] Department of Computer Science, Fudan University, China
[3] Department of Computer Science, National Chengchi University, Taiwan, R.O.C.
alpchen@cs.nccu.edu.tw

Abstract. Predicting future events has great importance in many applications. Generally, rules with *predicate events* and *consequent events* are mined out, and then current events are matched with the predicate ones to predict the occurrence of consequent events. Many previous works focus on the rule mining problem; however, little emphasis has been attached to the problem of predicate events matching. As events often arrive in a stream, how to design an efficient and effective event predictor becomes challenging. In this paper, we give a clear definition of this problem and propose our own method. We develop an event filter and incrementally maintain parts of the matching results. By running a series of experiments, we show that our method is efficient and effective in the stream environment.

Keywords: Continuous query, episode, event stream, prediction.

1 Introduction

In many applications, *events* such as specific TCP connections in an intrusion detection system [10] are recorded for the predicting of future events. Generally speaking, there are two steps for the event prediction problem. The first step is to derive event associations represented as rules from the past events. The second one is to use the discovered rules to predict future events when given a recent record of events. We now explain these two steps by an example, and show the motivation of our work.

Fig. 1 shows an example of the discovered rule in the form $\alpha \Rightarrow \beta$ where α is called the *predicate* and β the *consequent*. α and β are both represented by a directed acyclic graph, where each vertex represents an event, and each edge from vertex v to vertex u indicates that the event corresponding to vertex v should occur before that corresponding to vertex u. To be specific, according to the predicate α in Fig. 1, event a should precede events b and c, and event b should precede event d. Additionally, there are two time bounds associated with the rule and the predicate, respectively. For example, in Fig. 1, if all the events occur within the time bound of 7 time units in accord with the specified temporal orders in the predicate, we can predict that all the events in the rule will, with a certain probability, appear within 11 timestamps

G. Dong et al. (Eds.): APWeb/WAIM 2007, LNCS 4505, pp. 884–891, 2007.

according to their temporal orders indicated in the rule. The first step of the event prediction problem mines out such kind of rules with two time bounds (called *episode rule*) from the past events.

In the second step, whether all the events in the predicate have appeared according to their specified partial orders within the time bound (denoted as *rule matching* problem) is determined to predict the happening of the events in the corresponding consequents. For example, suppose the events coming from timestamp 1 to 9 are as depicted in Fig. 2, and we are to match the episode rule in Fig. 1. Notice that events a, b, c, and d occur within the time interval [3, 8), which satisfies the temporal constraint in the predicate. Thus, we should give the alarm that the consequent, or event f, may come within the time interval [8, 14) with a certain probability. We denote the occurrences of the events corresponding to the matching of the predicate as *predicate episode occurrence*.

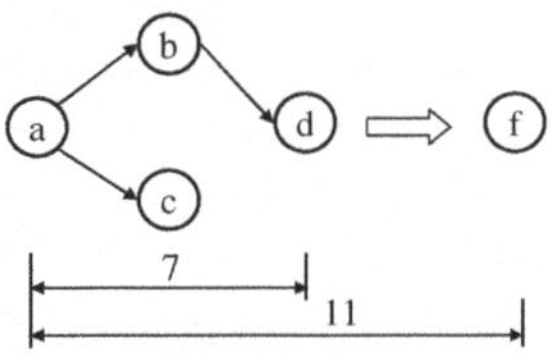

Fig. 1. An episode rule

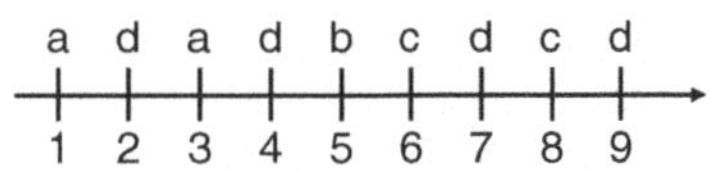

Fig. 2. A stream of events

The discovery of episode rules has been widely discussed over the past few years [6], [7]. However, little attention has been given to solve the important phase of episode rule matching. Since events arrive in streaming in all the applications mentioned above, how to efficiently match a number of episode rules in this environment becomes an important and difficult task. In this paper, we aim at this problem of continuously matching episode rules over the stream of events.

The main challenges of this problem are stated below. 1) Many predicate episode occurrences of a rule can exist simultaneously by sharing the same occurrences of events over the stream. However, only the occurrences that give non-repetitive predictions of the occurrences of the consequents are what we concern about. Take the example of the episode rule in Fig. 1 and the stream of events in Fig. 2. It can be seen that from the two predicate episode occurrences, {(a,1), (b,5), (c,6), (d,7)} and {(a,3), (b,5), (c,6), (d,7)} (we use (e,t) to denote the event e with an occurring time t), we predict [8,12) and [8, 14) as the occurring time interval of event f, respectively. Since the predicted interval [8,12) is included in [8,14) and becomes trivially redundant, the occurrence {(a,1), (b,5), (c,6), (d,7)} can be ignored. 2) The structure of the episodes can be complex. High precision should be emphasized to effectively deal with all the possible combinations of events within the specified time bounds to match the episodes. 3) There are a large number of predicate episodes in different rules to be matched simultaneously. Moreover, events usually come in bursts. There is only a limited time to make matches for all the rules. A prompt episode detector is hence required.

Our problem is related to three research topics. 1) Mining graph patterns from event or graph data sets [5], [6], [11]. The goals of these papers are essentially different from ours since we aim at continuous queries and they target on mining process. 2) Efficient graph indexing for pattern searching [3], [4]. Nevertheless, all these methods are applied to a static graph database searching, which is very different from our work of continuous retrieval in the streaming environment. 3) Graph filtering in the stream environment [8], [9] and the query of the temporal relations over DBMSs [2]. These works are similar to ours. However, we are to retrieve episodes within the specified time bounds and the repetitive reports of predicted time intervals should be avoided. So we cannot directly apply these algorithms to address our problem.

In this paper, we give a clear definition to the rule matching problem for event prediction, where the concepts of *minimal episode occurrence*, *latest episode occurrence* and *rejected event occurrence* were introduced to address the first challenge mentioned above. With the constraints in our problem definition, the retrieval of only user-required episode occurrences is assured. We then propose the method *ToFel* to solve this problem. ToFel makes use of the topological characteristic of the predicate episode, and develops its own pruning criteria. More specifically, ToFel finds the predicate episode occurrences by incrementally maintaining parts of the user-required episode occurrences, and thus avoid the backward scan of the stream. It constructs one event filter for each predicate episode to be matched. The filters continuously monitor the newly arrived events and only keep those which are likely to be parts of the predicate episode occurrences. By running a series of experiments with respect to different scales and distributions of the query set and the stream, we show that ToFel is efficient and effective in the stream environment.

The remainder of the paper is organized as follows. Section 2 gives a detailed description of the problem statement. Section 3 presents our rule matching algorithm. The experimental results are discussed in Section 4. We give the conclusion and future directions of our work in Section 5.

2 Problem Statement

The episode is a widely used representation for the associations of events. In this section, we first give the definitions related to the episode, and then present the basic concepts concerning the rule matching problem.

Episode: An *episode* is a directed acyclic graph g, where each vertex corresponds to an event, and each directed edge (u,v) indicates that the event corresponding to u must precede that corresponding to v. We call this precedence a temporal and transitive order $\prec$ between vertex u and vertex v. Denote *V(g)* as a vertex set, *E(g)* as an edge set, and $\mathcal{E}(v)$ as the event corresponding to vertex v. The *sink* of the graph g is defined as the vertex with out-edge degree being equal to 0. For convenience, we focus on episodes consisting of vertices corresponding to different events. However, our techniques can be extended to episodes containing two or more vertices corresponding to an identical event.

Episode occurrence: An *event stream* can be represented as $\hat{S} = <(a_1,t_1), (a_2,t_2), \ldots (a_n,t_n) \ldots>$, where (a_i,t_i) represents that event a_i occurs at time t_i , i = 1, 2, …, n, …, and $t_1<t_2< \ldots <t_n$ …. Given an *event sequence* $S = <(a'_1,t'_1), (a'_2,t'_2) \ldots (a'_m,t'_m)>$, where $t'_1<t'_2< \ldots <t'_m$, we define the *time interval* or *interval* of S as $[t'_1,t'_m+1)$, and

the *start time* and *end time* of S as t'_1 and t'_m+1, respectively. Given an episode α with its time bound ω_α, the *episode occurrence* or *occurrence* of α over $\hat{S}$ is an event sequence S with time interval $[t_s,t_e)$ satisfying that: 1) There exist m integers $i_1, i_2,\ldots, i_m$, such that $1\le i_1<i_2<\ldots<i_m$ and $\forall 1\le j\le m$: $a'_j=a_{i_j}$ and $t'_j=t_{ij}$; 2) Events corresponding to vertices in α have one-to-one mappings to events in S; 3) The occurrence orders of events in S are consistent with the $\prec$ constraints in α; and 4) $t_e - t_s \le \omega_\alpha$. We denote the *mapping occurrence* of the vertex v in α as the occurrence of the event corresponding to v in S.

Episode Rule: An *episode rule* R is a 5-tuple (α, β, ω_α, $\omega_{\alpha\beta}$, conf). Here, α and β are episodes representing the predicate and consequent of R, respectively. ω_α and $\omega_{\alpha\beta}$ ($\omega_\alpha<\omega_{\alpha\beta}$) correspond to the time bounds of α and $\alpha\beta$ ($\alpha\beta$ is an episode satisfying that each vertex in $\alpha \prec$ any vertex in β), respectively. The interpretation of R is that if α has an occurrence O with interval $[t_s,t_e)$, β will occur during interval $[t_e,t_s+\omega_{\alpha\beta})$ with probability conf. We denote $[t_e,t_s+\omega_{\alpha\beta})$ as the *predicted interval* of the occurrence O.

Given a set of episode rules, our problem is to continuously retrieve the episode occurrence of the predicate α with the time bound ω_α over the event stream, and give non-repetitive information of the predicted intervals for the consequent β. We now introduce the concepts of mi-latest occurrence and rejected event occurrence, and give a clear definition of the rule matching problem.

Definition 1. *Minimal occurrence.* A minimal occurrence O of a predicate episode α is an occurrence with predicted interval $[t_{s1},t_{e1})$ satisfying that there does not exist any other occurrence of α with predicted interval $[t_{s2},t_{e2})$, s.t. $t_{s1}\le t_{s2}<t_{e2}\le t_{e1}$, and $t_{e2}-t_{s2}<t_{e1}-t_{s1}$.

We can focus only on the minimal occurrences and report their predicted intervals without loss of any desired information of the prediction. Meanwhile, there may exist several different occurrences with the same time interval, which give the same predicted interval for the consequent. In the following, we introduce the concept of latest occurrences as a representative of the occurrences with the same interval, and match only the latest occurrences so as to avoid the repetitive reports of the same predicted interval.

Definition 2. *Latest occurrence of an event.* Let $\{(e,t_1), (e,t_2), \ldots (e,t_k)\}$ be the set of occurrences of event e in time interval $[t_s,t_e)$ on the event stream $\hat{S}$, where $t_s\le t_1<t_2<\ldots<t_k<t_e$. We call (e,t_k) the *latest occurrence of e* in $[t_s,t_e)$ on $\hat{S}$.

Definition 3. *Latest occurrence of an episode.* Given a predicate episode α with vertices $v_1, v_2, \ldots, v_n$ and its occurrence $O_r = <(\varepsilon(v_1),t_1), (\varepsilon(v_2),t_2), \ldots (\varepsilon(v_n),t_n)>$ in time interval $[t_s,t_e)$ on the event stream $\hat{S}$. O_r is called the *latest occurrence* of α in $[t_s,t_e)$ if both of the following conditions hold: 1) Let $v_{j1}, v_{j2}, \ldots v_{jx}$ be the sink vertices of α. $(\varepsilon(v_y),t_y)$ is the latest occurrence of $\varepsilon(v_y)$ in time interval $[t_s,t_e)$, $y=j_1, \ldots, j_x$; 2) for a non-sink vertex v_k of α, let $v_{k1}, v_{k2}, \ldots v_{km}$ be the children of v_k, where $1\le k<n$, and $t_s\le t_k<t_{k1}<t_{k2}<\ldots<t_{km}<t_e$. $(\varepsilon(v_k),t_k)$ is the latest occurrence of $\varepsilon(v_k)$ in time interval $[t_s, t_{k1})$.

Property 1. Let O be a latest occurrence of a predicate episode α with time interval $[t_s,t_e)$. If there exist minimal occurrences of α with the end time equal to t_e, O is one of the minimal occurrences of α.

For example, consider Fig. 1 and Fig. 2, where the latest occurrences of events a, b, c, and d in the interval [1,8) are (a,3), (b,5), (c,6), and (d,7), respectively. The latest occurrence of the predicate episode in the interval [1,8) is the occurrence O = <(a,3), (b,5), (c,6), (d,7)>, which is also a minimal occurrence.

Definition 4. *Mi-latest occurrence.* The *mi-latest occurrence* of a predicate episode α is defined as the occurrence which is both a minimal occurrence of α and a latest occurrence of α.

We define the rule matching problem by the concept of mi-latest occurrence. The rule matching problem is to give predicted intervals of only the mi-latest occurrences to the user.

Definition 5. *The rejected event occurrence.* Given a predicate episode α with vertices v_1, v_2, ... v_n and its mi-latest occurrence O = $<(\varepsilon(v_1),t_1), (\varepsilon(v_2),t_2), \ldots (\varepsilon(v_n),t_n)>$ on the event stream $\hat{S}$. The *rejected event occurrence* deduced from O is defined recursively as following: 1) $(\varepsilon(v_1),t_1)$ is a rejected event occurrence; 2) let v_{i1}, v_{i2}, ... v_{im} be the children of v_i, $1< i$, $i_1, i_2, \ldots, i_m \le n$. If $(\varepsilon(v_i),t_i)$ is a rejected event occurrence, $(\varepsilon(v_{ij}),t_{ij})$ is a rejected event, $\forall j=i_1, i_2, \ldots, i_m$, if there is no occurrence of $\varepsilon(v_i)$ in interval (t_i ,t_{ij}).

The essential of the rejected event occurrence deduced from the mi-latest occurrence O is that it can not be part of any other mi-latest occurrence that will appear later than O.

Lemma 1. Given a latest occurrence O = $<(e_1,t_1), (e_2,t_2), \ldots (e_n,t_n)>$ of a predicate episode α. If (e_i,t_i) is not a rejected event occurrence, $\forall 1 \le i \le n$, O is a minimal occurrence of α (For the detailed proof of the lemmas in this paper, please refer to our technical report [1]).

To conclude this section, only the latest occurrences with no rejected event occurrences are the mi-latest occurrences we are looking for. This is the basic of our approach, whose correctness can be guaranteed if such occurrences are always targeted during the rule matching process.

3 The Proposed Approach: ToFel

In the following, we present ToFel for the match of a given episode rule R = (α, β, ω_α, $\omega_{\alpha\beta}$, conf). ToFel builds a queue of event occurrences that are likely to be parts of mi-latest occurrences for α and maintains the queue at each timestamp. We first discuss which event occurrences should be kept in the queues and on which condition we should remove the stored occurrences from the queues in the process of continuously monitoring of the stream.

For each vertex of α, we implement a queue to store its corresponding event occurrences. Let Q_v be the queue for vertex $v \in V(\alpha)$. Intuitively, for any event $\varepsilon(v)$

arrives at time t, we should keep this event occurrence for contributing a mi-latest occurrence of α with another coming event occurrence.

As time passes and more and more events arrive, we maintain the queues and only keep those useful occurrences. Since the queues is to store only the occurrences likely to contribute to the results, the occurrences whose occurring time t' satisfies that t' + $\omega_\alpha \geq$ t (the current time) should be removed. In this condition, the maintenance of the queue is invoked. We call this kind of invocation of queue maintenance *time-out invocation*. Besides, as suggested in Definition 5 and Lemma 1, once we find out a mi-latest occurrence, we should adjust the queue by removing the rejected event occurrences. This condition is called *rejected-event invocation.* Both invocation forms are important for the correctness of our answer as well as the space saving.

Definition 6. *The nearest parent occurrence.* Given any two event occurrences (ε(v),t) and (ε(u),t'), where u, v$\in$V(α). If v is a parent of u and t<t', we call (ε(v),t) a *parent occurrence* of (ε(u),t'), and (ε(u),t') a *child occurrence* of (ε(v),t). Moreover, if (ε(v),t) is the latest occurrence of ε(v) in interval [t_s,t'), we call (ε(v),t) a *nearest parent occurrence* of (ε(u),t').

To easily maintain the queues, we establish a link called *nearest link*, or *nlink* to link from child occurrence to its nearest parent ones, and maintain a counter for each parent occurrence indicating the number of nlinks between it and its child occurrences.

In the following, we point out the principles for the maintenance of the queues and the further removal of other useless occurrences. A first, if there is no nlink between (ε(v),t') and any of v's child occurrences when (ε(v),t) arrives, it indicates that no child occurrence of (ε(v),t') occurs between time t' and t. Therefore, when a child occurrence of (ε(v),t') comes, its occurring time must be larger than t, and it will select (ε(v),t) as its nearest occurrence rather than (ε(v),t'). Therefore, the occurrence (ε(v),t') can not be linked by any of the later child occurrences, and can be removed without loss of any desired result. Moreover, if an occurrence is removed by above principle, the nlinks between it and its nearest parent occurrences are broken. The counts of those parent occurrences should be decreased by one accordingly. This may further cause the zero count of some of its parent occurrences. We should then adopt about principle again, and remove the parent occurrences with zero counts, except for the tail occurrences. Hence, the above procedure may be propagated from a child vertex to several of its ancestors. We call this queue maintenance principle the *parent-occurrence maintenance.* Moreover, when an occurrence (ε(v),t) is removed, the event occurrences that can be reached from (ε(v),t) by nlinks should be removed too. This maintenance principle is called *child-occurrence maintenance.*

To sum up, once an event ε(v) comes at time t, v$\in$V(α), if v is not a sink of α, we normally push (ε(v),t) into Q_v, and perform the parent-occurrence maintenance. Otherwise, if v is a sink of α, we detect whether the mi-latest occurrence of α exists. This includes three processes: 1) The time-out invocation calls the queue maintenance, 2) push (ε(v),t) into Q_v, and perform the parent-occurrence maintenance, and 3) check whether there exists a mi-latest occurrence according to Lemma 2 presented later. If a mi-latest occurrence exists, the rejected-event invocation will call the queue maintenance procedure. In the following, we show how to retrieve the mi-latest occurrence from the queues of event occurrences.

Property 2. Let v be a sink vertex of episode α. There is at most an occurrence $(\varepsilon(v),t)$ kept in Q_v, and $(\varepsilon(v),t)$ is the latest occurrence of $\varepsilon(v)$ so far.

Lemma 2. Let v_1, v_2, … v_n be the vertices of episode α and v_i, v_{i+1}, … v_n be the sinks of α. If there is a mapping occurrence of v_j kept in Q_{vj}, $\forall i \le j \le n$, there must exist a mi-latest occurrence O of α with interval $[t_1,t_n+1)$, and O must be $<(\varepsilon(v_1),t_1), (\varepsilon(v_2),t_2), \ldots (\varepsilon(v_n),t_n)>$, where $t_n+1-t_1 \le \omega_o$, and $(\varepsilon(v_k),t_k)$ is the 1st element in $Q_{vk}, \forall 1 \le k \le n$.

The correctness of ToFel can be proved as follows. Whenever a new mi-latest occurrence exists, its last element must correspond to a sink of α. Therefore, when each sink occurrence comes, we check if there exists a mi-latest occurrence by Lemma 2. And, we can prove that the time complexity of ToFel is O(n). [1]

4 Experimental Results

In this section, we evaluate the performances of DirectMatch [1] and ToFel by a series of experiments on synthetic data. The data is generated by the synthetic data generator [1]. We set various parameters to evaluate our method in the running time as well as the scalability on the structure of the episode and the size of the dataset. For the parameter settings, please refer to [1].

Fig. 3 shows the average execution time at each timestamp with respect to the number of episode rules to be matched. Though the time increases with the query number, ToFel always outperforms DirectMatch. And the increasing ratio in running time of ToFel is less than that of DirectMatch significantly. This can be explained that when matching an episode, ToFel only concerns the events likely to form the mi-latest occurrences, while DirectMatch usually repeatedly retrieves the kept events many of which are not even relevant to the episode. We also show the performance with respect to the number of vertices in the episode as shown in Fig. 4. And the result shows the slow increase in CPU time as well as the smaller time requirement of ToFel compared with DirectMatch. Finally, we compare the two approaches in their scalability with respect to the size of the event stream. As shown in Fig. 5, both approaches have the constant average running time at each timestamp no matter how the size of the stream changes.

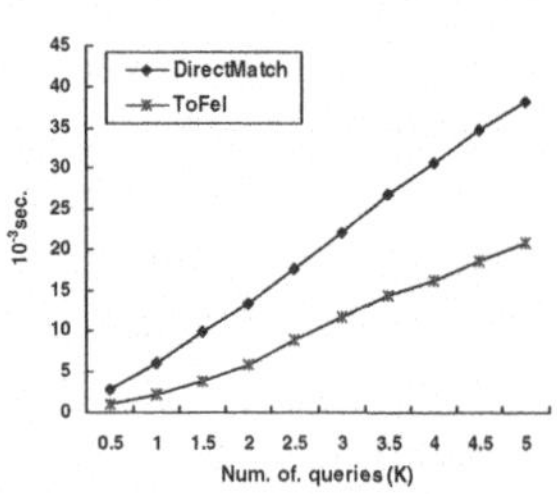

Fig. 3. Running time for different query numbers

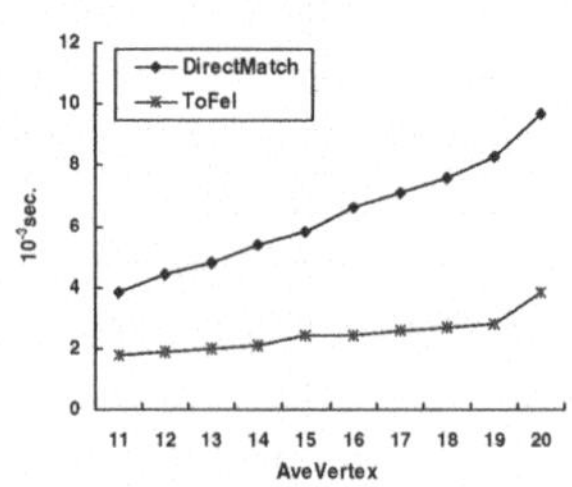

Fig. 4. Running time for different AveVertex values

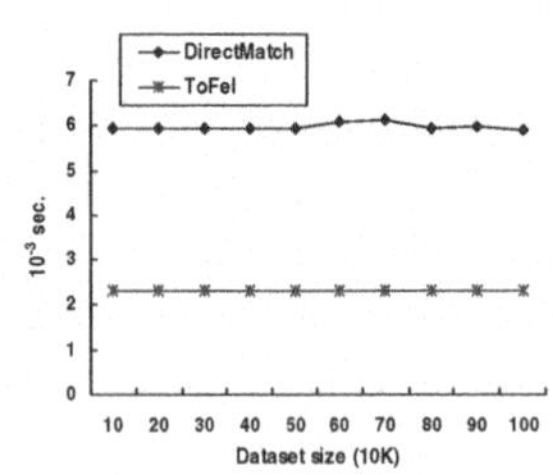

Fig. 5. Running time for different dataset sizes

5 Conclusion and Future Work

In this paper, we propose a novel, deterministic and efficient approach to continuously match the episode rules over event streams for the predicting of future events. We introduce the concepts of mi-latest occurrence and rejected event occurrences such that no repetitive predicted intervals are reported. Besides, we build and continuously maintain the queue of events which are likely to contribute to the desired occurrences in an efficient time once a new event arrives. This leads to a prompt reaction towards the desired report of episode occurrences that may burst at one timestamp. Moreover, a series of experiments demonstrate the high performance of our approach in real processing time as well as the stability with respect to the number of queries, the number of vertices, and the size of event streams. For the future work, we will focus on utilizing the common substructures among the predicate episodes so as to more efficiently process a batch of them simultaneously.

Acknowledgments. This work was partially supported by the NSC Program for Advanced Technologies and Applications for Next Generation Information Networks (II) under the grant number NSC 95-2752-E-007-004-PAE, and the NSC under the contract number 95-2627-E-004-002-.

References

1. Cho, C.W., Y. Zheng, and A.L.P. Chen. Continuously Matching Episode Rules for Predicting Future Events over Event Streams. Tech. Report CS-1006-05, Department of Computer Science, National Tsing Hua University, October 2006.
2. Chomicki J. History-less Checking of Dynamic Integrity Constraints. In Proceedings of the 8th International Conference on Data Engineering, 1992, 557-564.
3. Giugno, R. and D. Shasha. GraphGrep: A Fast and Universal Method for Querying Graphs. In 16th International Conference on Pattern Recognition, 2002, 112-115.
4. He, H. and A.K. Singh. Closure-Tree: An Index Structure for Graph Queries. In Proceedings of the 22nd International Conference on Data Engineering, 2006, p. 38.
5. Hsieh, C.E., Y.H. Wu and A.L.P. Chen. Discovering Frequent Tree Patterns over Data Streams. In Proceedings of the 6th SIAM International Conference on Data Mining, 2006.
6. Mannila, H., H. Toivonen, and A.I. Verkamo. Discovering Frequent Episodes in Sequences. In Proceedings of the 1st International Conference on Knowledge Discovery and Data Mining, 1995, 210-215.
7. Mannila, H., H. Toivonen, and A.I. Verkamo. Discovery of Frequent Episodes in Event Sequences. Data Mining and Knowledge Discovery, 1(3), 1997, 259-289.
8. Olteanu, D., T. Kiesling, and F. Bry. An Evaluation of Regular Path Expressions with Qualifiers against XML Streams. In Proceedings of the 19th International Conference on Data Engineering, 2003, 702-704.
9. Peng, F. and S.S. Chawathe. XPath Queries on Streaming Data. In Proceedings of the ACM SIGMOD International Conference on Management of Data, 2003, 431-442.
10. Qin, M. and K. Hwang. Frequent Episode Rules for Internet Anomaly Detection. IEEE International Symposium on Network Computing and Applications, 2004, 161-168.
11. Yan, X. and J. Han. gSpan: Graph-Based Substructure Pattern Mining. In Proceedings of the 2002 IEEE International Conference on Data Mining, 2002, 721-724.

Author Index

Lecture Notes in Computer Science

For information about Vols. 1–4419

please contact your bookseller or Springer

Vol. 4538: F. Escolano, M. Vento (Eds.), Graph-Based Representations in Pattern Recognition. XII, 416 pages. 2007.

Vol. 4534: I. Tomkos, F. Neri, J. Solé Pareta, X. Masip Bruin, S. Sánchez Lopez (Eds.), Optical Network Design and Modeling. XI, 460 pages. 2007.

Vol. 4531: J. Indulska, K. Raymond (Eds.), Distributed Applications and Interoperable Systems. XI, 337 pages. 2007.

Vol. 4530: D.H. Akehurst, R. Vogel, R.F. Paige (Eds.), Model Driven Architecture- Foundations and Applications. X, 219 pages. 2007.

Vol. 4526: M. Malek, M. Reitenspieß, A. van Moorsel (Eds.), Service Availability. X, 155 pages. 2007.

Vol. 4525: C. Demetrescu (Ed.), Experimental and Efficient Algorithms. XIII, 448 pages. 2007.

Vol. 4523: Y.-H. Lee, H.-N. Kim, J. Kim, Y. Park, L.T. Yang, S.W. Kim (Eds.), Embedded Software and Systems. XIX, 829 pages. 2007.

Vol. 4521: J. Katz, M. Yung (Eds.), Applied Cryptography and Network Security. XIII, 498 pages. 2007.

Vol. 4519: E. Franconi, M. Kifer, W. May (Eds.), The Semantic Web: Research and Applications. XVIII, 830 pages. 2007.

Vol. 4517: F. Boavida, E. Monteiro, S. Mascolo, Y. Koucheryavy (Eds.), Wired/Wireless Internet Communications. XIV, 382 pages. 2007.

Vol. 4515: M. Naor (Ed.), Advances in Cryptology - EUROCRYPT 2007. XIII, 591 pages. 2007.

Vol. 4514: S.N. Artemov, A. Nerode (Eds.), Logical Foundations of Computer Science. XI, 513 pages. 2007.

Vol. 4510: P. Van Hentenryck, L. Wolsey (Eds.), Integration of AI and OR Techniques in Constraint Programming for Combinatorial Optimization Problems. X, 391 pages. 2007.

Vol. 4509: Z. Kobti, D. Wu (Eds.), Advances in Artificial Intelligence. XII, 552 pages. 2007. (Sublibrary LNAI).

Vol. 4508: M.-Y. Kao, X.-Y. Li (Eds.), Algorithmic Applications in Management. VIII, 428 pages. 2007.

Vol. 4506: D. Zeng, I. Gotham, K. Komatsu, C. Lynch, M. Thurmond, D. Madigan, B. Lober, J. Kvach, H. Chen (Eds.), Intelligence and Security Informatics: Biosurveillance. XI, 234 pages. 2007.

Vol. 4505: G. Dong, X. Lin, W. Wang, Y. Yang, J.X. Yu (Eds.), Advances in Data and Web Management. XXII, 896 pages. 2007.

Vol. 4504: J. Huang, R. Kowalczyk, Z. Maamar, D. Martin, I. Müller, S. Stoutenburg, K.P. Sycara (Eds.), Service-Oriented Computing: Agents, Semantics, and Engineering. X, 175 pages. 2007.

Vol. 4501: J. Marques-Silva, K.A. Sakallah (Eds.), Theory and Applications of Satisfiability Testing – SAT 2007. XI, 384 pages. 2007.

Vol. 4500: N. Streitz, A. Kameas, I. Mavrommati (Eds.), The Disappearing Computer. XVIII, 304 pages. 2007.

Vol. 4496: N.T. Nguyen, A. Grzech, R.J. Howlett, L.C. Jain (Eds.), Agent and Multi-Agent Systems: Technologies and Applications. XXI, 1046 pages. 2007. (Sublibrary LNAI).

Vol. 4493: D. Liu, S. Fei, Z. Hou, H. Zhang, C. Sun (Eds.), Advances in Neural Networks – ISNN 2007, Part III. XXVI, 1215 pages. 2007.

Vol. 4492: D. Liu, S. Fei, Z. Hou, H. Zhang, C. Sun (Eds.), Advances in Neural Networks – ISNN 2007, Part II. XXVII, 1321 pages. 2007.

Vol. 4491: D. Liu, S. Fei, Z.-G. Hou, H. Zhang, C. Sun (Eds.), Advances in Neural Networks – ISNN 2007, Part I. LIV, 1365 pages. 2007.

Vol. 4490: Y. Shi, G.D. van Albada, J. Dongarra, P.M.A. Sloot (Eds.), Computational Science – ICCS 2007, Part IV. XXXVII, 1211 pages. 2007.

Vol. 4489: Y. Shi, G.D. van Albada, J. Dongarra, P.M.A. Sloot (Eds.), Computational Science – ICCS 2007, Part III. XXXVII, 1257 pages. 2007.

Vol. 4488: Y. Shi, G.D. van Albada, J. Dongarra, P.M.A. Sloot (Eds.), Computational Science – ICCS 2007, Part II. XXXV, 1251 pages. 2007.

Vol. 4487: Y. Shi, G.D. van Albada, J. Dongarra, P.M.A. Sloot (Eds.), Computational Science – ICCS 2007, Part I. LXXXI, 1275 pages. 2007.

Vol. 4486: M. Bernardo, J. Hillston (Eds.), Formal Methods for Performance Evaluation. VII, 469 pages. 2007.

Vol. 4485: F. Sgallari, A. Murli, N. Paragios (Eds.), Scale Space and Variational Methods in Computer Vision. XV, 931 pages. 2007.

Vol. 4484: J.-Y. Cai, S.B. Cooper, H. Zhu (Eds.), Theory and Applications of Models of Computation. XIII, 772 pages. 2007.

Vol. 4483: C. Baral, G. Brewka, J. Schlipf (Eds.), Logic Programming and Nonmonotonic Reasoning. IX, 327 pages. 2007. (Sublibrary LNAI).

Vol. 4482: A. An, J. Stefanowski, S. Ramanna, C.J. Butz, W. Pedrycz, G. Wang (Eds.), Rough Sets, Fuzzy Sets, Data Mining and Granular Computing. XIV, 585 pages. 2007. (Sublibrary LNAI).

Vol. 4481: J. Yao, P. Lingras, W.-Z. Wu, M. Szczuka, N.J. Cercone, D. Ślęzak (Eds.), Rough Sets and Knowledge Technology. XIV, 576 pages. 2007. (Sublibrary LNAI).

Vol. 4480: A. LaMarca, M. Langheinrich, K.N. Truong (Eds.), Pervasive Computing. XIII, 369 pages. 2007.

Vol. 4479: I.F. Akyildiz, R. Sivakumar, E. Ekici, J.C.d. Oliveira, J. McNair (Eds.), NETWORKING 2007. Ad Hoc and Sensor Networks, Wireless Networks, Next Generation Internet. XXVII, 1252 pages. 2007.

Vol. 4478: J. Martí, J.M. Benedí, A.M. Mendonça, J. Serrat (Eds.), Pattern Recognition and Image Analysis, Part II. XXVII, 657 pages. 2007.

Vol. 4477: J. Martí, J.M. Benedí, A.M. Mendonça, J. Serrat (Eds.), Pattern Recognition and Image Analysis, Part I. XXVII, 625 pages. 2007.

Vol. 4476: V. Gorodetsky, C. Zhang, V.A. Skormin, L. Cao (Eds.), Autonomous Intelligent Systems: Multi-Agents and Data Mining. XIII, 323 pages. 2007. (Sublibrary LNAI).

Vol. 4475: P. Crescenzi, G. Prencipe, G. Pucci (Eds.), Fun with Algorithms. X, 273 pages. 2007.

Vol. 4472: M. Haindl, J. Kittler, F. Roli (Eds.), Multiple Classifier Systems. XI, 524 pages. 2007.

Vol. 4471: P. Cesar, K. Chorianopoulos, J.F. Jensen (Eds.), Interactive TV: a Shared Experience. XIII, 236 pages. 2007.

Vol. 4470: Q. Wang, D. Pfahl, D.M. Raffo (Eds.), Software Process Dynamics and Agility. XI, 346 pages. 2007.

Vol. 4468: M.M. Bonsangue, E.B. Johnsen (Eds.), Formal Methods for Open Object-Based Distributed Systems. X, 317 pages. 2007.

Vol. 4467: A.L. Murphy, J. Vitek (Eds.), Coordination Models and Languages. X, 235 pages. 2007.

Vol. 4465: T. Chahed, B. Tuffin (Eds.), Network Control and Optimization. XIII, 305 pages. 2007.

Vol. 4464: E. Dawson, D.S. Wong (Eds.), Information Security Practice and Experience. XIII, 361 pages. 2007.

Vol. 4463: I. Măndoiu, A. Zelikovsky (Eds.), Bioinformatics Research and Applications. XV, 653 pages. 2007. (Sublibrary LNBI).

Vol. 4462: D. Sauveron, K. Markantonakis, A. Bilas, J.-J. Quisquater (Eds.), Information Security Theory and Practices. XII, 255 pages. 2007.

Vol. 4459: C. Cérin, K.-C. Li (Eds.), Advances in Grid and Pervasive Computing. XVI, 759 pages. 2007.

Vol. 4453: T. Speed, H. Huang (Eds.), Research in Computational Molecular Biology. XVI, 550 pages. 2007. (Sublibrary LNBI).

Vol. 4452: M. Fasli, O. Shehory (Eds.), Agent-Mediated Electronic Commerce. VIII, 249 pages. 2007. (Sublibrary LNAI).

Vol. 4451: T.S. Huang, A. Nijholt, M. Pantic, A. Pentland (Eds.), Artifical Intelligence for Human Computing. XVI, 359 pages. 2007. (Sublibrary LNAI).

Vol. 4450: T. Okamoto, X. Wang (Eds.), Public Key Cryptography – PKC 2007. XIII, 491 pages. 2007.

Vol. 4448: M. Giacobini et al. (Ed.), Applications of Evolutionary Computing. XXIII, 755 pages. 2007.

Vol. 4447: E. Marchiori, J.H. Moore, J.C. Rajapakse (Eds.), Evolutionary Computation,Machine Learning and Data Mining in Bioinformatics. XI, 302 pages. 2007.

Vol. 4446: C. Cotta, J. van Hemert (Eds.), Evolutionary Computation in Combinatorial Optimization. XII, 241 pages. 2007.

Vol. 4445: M. Ebner, M. O'Neill, A. Ekárt, L. Vanneschi, A.I. Esparcia-Alcázar (Eds.), Genetic Programming. XI, 382 pages. 2007.

Vol. 4444: T. Reps, M. Sagiv, J. Bauer (Eds.), Program Analysis and Compilation, Theory and Practice. X, 361 pages. 2007.

Vol. 4443: R. Kotagiri, P.R. Krishna, M. Mohania, E. Nantajeewarawat (Eds.), Advances in Databases: Concepts, Systems and Applications. XXI, 1126 pages. 2007.

Vol. 4440: B. Liblit, Cooperative Bug Isolation. XV, 101 pages. 2007.

Vol. 4439: W. Abramowicz (Ed.), Business Information Systems. XV, 654 pages. 2007.

Vol. 4438: L. Maicher, A. Sigel, L.M. Garshol (Eds.), Leveraging the Semantics of Topic Maps. X, 257 pages. 2007. (Sublibrary LNAI).

Vol. 4433: E. Şahin, W.M. Spears, A.F.T. Winfield (Eds.), Swarm Robotics. XII, 221 pages. 2007.

Vol. 4432: B. Beliczynski, A. Dzielinski, M. Iwanowski, B. Ribeiro (Eds.), Adaptive and Natural Computing Algorithms, Part II. XXVI, 761 pages. 2007.

Vol. 4431: B. Beliczynski, A. Dzielinski, M. Iwanowski, B. Ribeiro (Eds.), Adaptive and Natural Computing Algorithms, Part I. XXV, 851 pages. 2007.

Vol. 4430: C.C. Yang, D. Zeng, M. Chau, K. Chang, Q. Yang, X. Cheng, J. Wang, F.-Y. Wang, H. Chen (Eds.), Intelligence and Security Informatics. XII, 330 pages. 2007.

Vol. 4429: R. Lu, J.H. Siekmann, C. Ullrich (Eds.), Cognitive Systems. X, 161 pages. 2007. (Sublibrary LNAI).

Vol. 4427: S. Uhlig, K. Papagiannaki, O. Bonaventure (Eds.), Passive and Active Network Measurement. XI, 274 pages. 2007.

Vol. 4426: Z.-H. Zhou, H. Li, Q. Yang (Eds.), Advances in Knowledge Discovery and Data Mining. XXV, 1161 pages. 2007. (Sublibrary LNAI).

Vol. 4425: G. Amati, C. Carpineto, G. Romano (Eds.), Advances in Information Retrieval. XIX, 759 pages. 2007.

Vol. 4424: O. Grumberg, M. Huth (Eds.), Tools and Algorithms for the Construction and Analysis of Systems. XX, 738 pages. 2007.

Vol. 4423: H. Seidl (Ed.), Foundations of Software Science and Computational Structures. XVI, 379 pages. 2007.

Vol. 4422: M.B. Dwyer, A. Lopes (Eds.), Fundamental Approaches to Software Engineering. XV, 440 pages. 2007.

Vol. 4421: R. De Nicola (Ed.), Programming Languages and Systems. XVII, 538 pages. 2007.

Vol. 4420: S. Krishnamurthi, M. Odersky (Eds.), Compiler Construction. XIV, 233 pages. 2007.

The manufacturer's authorised representative in the EU is Springer Nature Customer Service Centre GmbH, Europaplatz 3, 69115 Heidelberg, Germany. If you have any concerns regarding our products, please contact ProductSafety@springernature.com

Printed and bound by CPI Group (UK) Ltd, Croydon, CR0 4YY
25/09/2026
02222831-0015